THE BOOK ®

Diesel Engine Systems & Data Book

Steve Rendle and Martynn Randall

The Haynes Manual providing descriptions and explanations of most modern diesel engine systems and their components. Maintenance, diagnostic and overhaul data for virtually all diesel-engined cars and light commercial vehicles available in the UK during the last ten years.

(3548-480)

© Haynes Publishing 2000

A book in the **Haynes Professional TechBook Series**

ISBN 1 85960 548 6

Printed by **J H Haynes & Co Ltd, Sparkford, Nr Yeovil, Somerset BA22 7JJ, England**

Haynes Publishing
Sparkford, Nr Yeovil, Somerset BA22 7JJ, England

Haynes North America, Inc
861 Lawrence Drive, Newbury Park, California 91320, USA

Editions Haynes S.A.
Tour Aurore - La Défense 2, 18 Place des Reflets, 92975 PARIS LA DEFENSE Cedex, France

Haynes Publishing Nordiska AB
Box 1504, 751 45 UPPSALA, Sweden

0•2 Contents

This manual provides descriptions and explanations of most modern diesel engine systems and their components. Maintenance, diagnostic and overhaul data is provided for virtually all diesel-engined cars and light commercial vehicles available in the UK during the last ten years. The book is intended for the small independent garage, the mobile tune-up specialist, and the enthusiastic DIY mechanic – that is to say, for people who are familiar with the repair and maintenance of vehicles, but may not have extensive specific experience of diesel engines and injection systems.

The manual is divided into ten main Chapters:

Chapter 1 provides a basic introduction to the diesel engine and its associated systems.

Chapter 2 looks in detail at the various diesel injection systems found on modern vehicles.

Chapter 3 describes the engine electrical systems, and provides maintenance and test procedures for the electrical system components.

Chapter 4 covers ancillary equipment such as turbochargers and vacuum pumps.

Chapter 5 gives details of the various emission control systems likely to encountered on modern diesel engines.

Chapter 6 looks at the tools and equipment needed for maintenance, diagnosis and repair.

Chapter 7 provides information on the various checks and adjustments which can be made on many systems.

Chapter 8 describes engine maintenance and repair operations which are unique to, or particularly important on, diesel engines.

Chapter 9 provides fault diagnosis charts and notes.

Chapter 10 gives tune-up, maintenance and repair data in tabular form. At the beginning of Chapter 10 a detailed explanation of the data presentation is provided – this should be read before using that Chapter.

Acknowledgements

Thanks are due to the various vehicle and equipment manufacturers and importers for providing technical literature, data and illustrations for this book. These include Lucas CAV Limited and Robert Bosch Limited for the use of their illustrations. Thanks are also due to all those people at Sparkford who helped in the production of this book, particularly Matthew Minter and Julian McGeoch who provided the basis from which this book was developed.

This book is not a direct reproduction of the vehicle manufacturers' data, and its publication should not be taken as implying any technical approval by the vehicle manufacturers or importers.

We take great pride in the accuracy of information given in this manual, but vehicle manufacturers make alterations and design changes during the production run of a particular vehicle of which they do not inform us. No liability can be accepted by the authors or publishers for loss, damage or injury caused by any errors in, or omissions from, the information given.

General

Although this book does not provide maintenance schedules and detailed maintenance procedures (although general procedures are provided in Chapters 7 and 8), the following points should be taken into account when working through the vehicle manufacturer's maintenance schedule.

Maintenance intervals

When both time and mileage intervals are specified by the manufacturer, the time interval should be followed if the specified mileage is not covered within the time stated. This is necessary because some fluids and systems deteriorate with time as well as with use. In particular, water trap draining and fuel filter changes should not be neglected on low-mileage vehicles. More water may accumulate in the fuel system of a vehicle which stands idle for long periods than in one which is in constant use. If water or dirt get past the fuel filter and into the injection system, serious damage may result. A clean filter is also less likely to suffer from waxing in cold weather.

Oil change intervals tend to be shorter for diesel engines than for comparable petrol engines, because more contamination and fuel dilution of the oil occurs in the diesel. Sulphur compounds in diesel fuel are particularly detrimental to the oil; if fuel with a higher than normal sulphur content has to be used, oil change intervals must be reduced. (The current EEC standard allows a maximum diesel content in diesel fuel for road vehicles of 0.2%, but in other countries it may be higher. A high sulphur content may also be found in fuels such as marine diesel.)

Adverse operating conditions

Vehicles used under adverse conditions may required more frequent maintenance. 'Adverse conditions' include the following:

Mainly short journeys
Full-time towing or taxi work
Operating in extremely hot or cold climates
Driving on unmade roads or in dusty conditions
Use of inferior fuel

Timing belt renewal

When a toothed belt is used to drive the camshaft and/or injection pump, periodic renewal is normally specified. *If a camshaft drivebelt breaks or slips in service, extensive engine damage will almost certainly result from the ensuing piston/valve contact.* Observe the specified intervals for inspection and renewal, even if the belt appears to be in good condition. Renew a belt which is obviously frayed, or which has been contaminated with oil or fuel, without question. Renew idler or tensioner rollers at the same time if they show shake or roughness when spun, and the sprockets if they are damaged.

Cooling system maintenance

Unless otherwise specified, the coolant antifreeze concentration should be checked at the beginning of each winter, and made good if necessary. Coolant should generally be renewed every two years, in order to maintain its corrosion-inhibiting qualities; note however, that some manufacturers claim to have a 'sealed-for-life' cooling system, often filled with their own brand of coolant (this coolant may not be compatible with other brands).

After draining the old coolant, take the opportunity to flush the system if necessary, and renew any hoses which are not in good condition.

Recommended lubricants and fluids

The following are general recommendations only. Observe the vehicle manufacturer's specifications when they differ from those given here.

Engine oil

The properties necessary in an oil for diesel engines are not identical to those needed for petrol engines. This is due to the higher mechanical loads imposed by compression ignition, and to the different effects of unburnt fuel and combustion products on the oil. When a turbocharger is fitted, the oil must also be able to cope with extremely high temperatures and rotational speeds.

For temperate climates, most manufacturers specify the use of multigrade engine oil to API CE, CCMC PD2/D4, ACEA B3-96, or equivalent (or higher) ratings.

API (American Petroleum Institute) ratings show the performance of the oil for both petrol and diesel applications. Petrol ratings begin with the letter 'S' for spark ignition, and 'C' for compression ignition. The second letter denotes the rating, with 'A' being the lowest. The higher the second letter in the alphabet, the better the rating.

CCMC (Constructors' Committee of the Common Market) ratings fall into three categories; 'G' for gasoline (petrol), 'D' for commercial diesel, and 'PD' for passenger diesel. Each rating is followed by a number. The higher the number, the better the rating.

ACEA (*Association des Constructeurs Europæen d'Automobiles*) ratings also fall into three categories 'A' for petrol engines, 'B' for 'light-duty' diesel engines, and 'E' for 'heavy-duty' diesel engines. ACEA standards replaced CCMC standards in Europe from January 1st 1996, and hence the ratings include '96', eg, 'B1-96' (this year code is likely to be updated in the future). Each rating letter is followed by a number ('1', '2' or '3' at the time of writing), and the higher the number, the better the rating.

Coolant

Modern engines often expose the coolant to several different metals - for instance iron, aluminium and copper - which in the presence of plain water will interact and rust or corrode rapidly. For this reason, it is essential that the coolant contains a corrosion inhibitor, even when freezing conditions are not expected. When hard water is used in the cooling system, a scale inhibitor is also required. The corrosion and scale inhibitors lose their effectiveness after a while, so coolant must be renewed periodically - typically every two years.

Antifreeze with a methanol content is particularly to be avoided. Methanol does lower the freezing point, but is highly poisonous and inflammable; it also tends to evaporate in use, so reducing the level of protection.

Some vehicle manufacturers (notably VW/Audi) use their own brand of antifreeze in vehicles when new. Certain of these antifreeze products will not mix with other brands, so the vehicle manufacturer's recommendations should always be followed when renewing antifreeze or topping up.

Working on your car can be dangerous. This page shows just some of the potential risks and hazards, with the aim of creating a safety-conscious attitude.

General hazards

Scalding

• Don't remove the radiator or expansion tank cap while the engine is hot.
• Engine oil, automatic transmission fluid or power steering fluid may also be dangerously hot if the engine has recently been running.

Burning

• Beware of burns from the exhaust system and from any part of the engine. Brake discs and drums can also be extremely hot immediately after use.

Crushing

• When working under or near a raised vehicle, always supplement the jack with axle stands, or use drive-on ramps. *Never venture under a car which is only supported by a jack.*
• Take care if loosening or tightening high-torque nuts when the vehicle is on stands. Initial loosening and final tightening should be done with the wheels on the ground.

Fire

• Fuel is highly flammable; fuel vapour is explosive.
• Don't let fuel spill onto a hot engine.
• Do not smoke or allow naked lights (including pilot lights) anywhere near a vehicle being worked on. Also beware of creating sparks (electrically or by use of tools).
• Fuel vapour is heavier than air, so don't work on the fuel system with the vehicle over an inspection pit.
• Another cause of fire is an electrical overload or short-circuit. Take care when repairing or modifying the vehicle wiring.
• Keep a fire extinguisher handy, of a type suitable for use on fuel and electrical fires.

Electric shock

• Ignition HT voltage can be dangerous, especially to people with heart problems or a pacemaker. Don't work on or near the ignition system with the engine running or the ignition switched on.

• Mains voltage is also dangerous. Make sure that any mains-operated equipment is correctly earthed. Mains power points should be protected by a residual current device (RCD) circuit breaker.

Fume or gas intoxication

• Exhaust fumes are poisonous; they often contain carbon monoxide, which is rapidly fatal if inhaled. Never run the engine in a confined space such as a garage with the doors shut.
• Fuel vapour is also poisonous, as are the vapours from some cleaning solvents and paint thinners.

Poisonous or irritant substances

• Avoid skin contact with battery acid and with any fuel, fluid or lubricant, especially antifreeze, brake hydraulic fluid and Diesel fuel. Don't syphon them by mouth. If such a substance is swallowed or gets into the eyes, seek medical advice.
• Prolonged contact with used engine oil can cause skin cancer. Wear gloves or use a barrier cream if necessary. Change out of oil-soaked clothes and do not keep oily rags in your pocket.
• Air conditioning refrigerant forms a poisonous gas if exposed to a naked flame (including a cigarette). It can also cause skin burns on contact.

Asbestos

• Asbestos dust can cause cancer if inhaled or swallowed. Asbestos may be found in gaskets and in brake and clutch linings. When dealing with such components it is safest to assume that they contain asbestos.

Special hazards

Hydrofluoric acid

• This extremely corrosive acid is formed when certain types of synthetic rubber, found in some O-rings, oil seals, fuel hoses etc, are exposed to temperatures above 400°C. The rubber changes into a charred or sticky substance containing the acid. *Once formed, the acid remains dangerous for years. If it gets onto the skin, it may be necessary to amputate the limb concerned.*
• When dealing with a vehicle which has suffered a fire, or with components salvaged from such a vehicle, wear protective gloves and discard them after use.

The battery

• Batteries contain sulphuric acid, which attacks clothing, eyes and skin. Take care when topping-up or carrying the battery.
• The hydrogen gas given off by the battery is highly explosive. Never cause a spark or allow a naked light nearby. Be careful when connecting and disconnecting battery chargers or jump leads.

Air bags

• Air bags can cause injury if they go off accidentally. Take care when removing the steering wheel and/or facia. Special storage instructions may apply.

Diesel injection equipment

• Diesel injection pumps supply fuel at very high pressure. Take care when working on the fuel injectors and fuel pipes.

⚠️ *Warning: Never expose the hands, face or any other part of the body to injector spray; the fuel can penetrate the skin with potentially fatal results.*

Remember...

DO

• Do use eye protection when using power tools, and when working under the vehicle.

• Do wear gloves or use barrier cream to protect your hands when necessary.

• Do get someone to check periodically that all is well when working alone on the vehicle.

• Do keep loose clothing and long hair well out of the way of moving mechanical parts.

• Do remove rings, wristwatch etc, before working on the vehicle – especially the electrical system.

• Do ensure that any lifting or jacking equipment has a safe working load rating adequate for the job.

DON'T

• Don't attempt to lift a heavy component which may be beyond your capability – get assistance.

• Don't rush to finish a job, or take unverified short cuts.

• Don't use ill-fitting tools which may slip and cause injury.

• Don't leave tools or parts lying around where someone can trip over them. Mop up oil and fuel spills at once.

• Don't allow children or pets to play in or near a vehicle being worked on.

Whenever servicing, repair or overhaul work is carried out on the car or its components, observe the following procedures and instructions. This will assist in carrying out the operation efficiently and to a professional standard of workmanship.

Joint mating faces and gaskets

When separating components at their mating faces, never insert screwdrivers or similar implements into the joint between the faces in order to prise them apart. This can cause severe damage which results in oil leaks, coolant leaks, etc upon reassembly. Separation is usually achieved by tapping along the joint with a soft-faced hammer in order to break the seal. However, note that this method may not be suitable where dowels are used for component location.

Where a gasket is used between the mating faces of two components, a new one must be fitted on reassembly; fit it dry unless otherwise stated in the repair procedure. Make sure that the mating faces are clean and dry, with all traces of old gasket removed. When cleaning a joint face, use a tool which is unlikely to score or damage the face, and remove any burrs or nicks with an oilstone or fine file.

Make sure that tapped holes are cleaned with a pipe cleaner, and keep them free of jointing compound, if this is being used, unless specifically instructed otherwise.

Ensure that all orifices, channels or pipes are clear, and blow through them, preferably using compressed air.

Oil seals

Oil seals can be removed by levering them out with a wide flat-bladed screwdriver or similar implement. Alternatively, a number of self-tapping screws may be screwed into the seal, and these used as a purchase for pliers or some similar device in order to pull the seal free.

Whenever an oil seal is removed from its working location, either individually or as part of an assembly, it should be renewed.

The very fine sealing lip of the seal is easily damaged, and will not seal if the surface it contacts is not completely clean and free from scratches, nicks or grooves. If the original sealing surface of the component cannot be restored, and the manufacturer has not made provision for slight relocation of the seal relative to the sealing surface, the component should be renewed.

Protect the lips of the seal from any surface which may damage them in the course of fitting. Use tape or a conical sleeve where possible. Lubricate the seal lips with oil before fitting and, on dual-lipped seals, fill the space between the lips with grease.

Unless otherwise stated, oil seals must be fitted with their sealing lips toward the lubricant to be sealed.

Use a tubular drift or block of wood of the appropriate size to install the seal and, if the seal housing is shouldered, drive the seal down to the shoulder. If the seal housing is unshouldered, the seal should be fitted with its face flush with the housing top face (unless otherwise instructed).

Screw threads and fastenings

Seized nuts, bolts and screws are quite a common occurrence where corrosion has set in, and the use of penetrating oil or releasing fluid will often overcome this problem if the offending item is soaked for a while before attempting to release it. The use of an impact driver may also provide a means of releasing such stubborn fastening devices, when used in conjunction with the appropriate screwdriver bit or socket. If none of these methods works, it may be necessary to resort to the careful application of heat, or the use of a hacksaw or nut splitter device.

Studs are usually removed by locking two nuts together on the threaded part, and then using a spanner on the lower nut to unscrew the stud. Studs or bolts which have broken off below the surface of the component in which they are mounted can sometimes be removed using a stud extractor. Always ensure that a blind tapped hole is completely free from oil, grease, water or other fluid before installing the bolt or stud. Failure to do this could cause the housing to crack due to the hydraulic action of the bolt or stud as it is screwed in.

When tightening a castellated nut to accept a split pin, tighten the nut to the specified torque, where applicable, and then tighten further to the next split pin hole. Never slacken the nut to align the split pin hole, unless stated in the repair procedure.

When checking or retightening a nut or bolt to a specified torque setting, slacken the nut or bolt by a quarter of a turn, and then retighten to the specified setting. However, this should not be attempted where angular tightening has been used.

For some screw fastenings, notably cylinder head bolts or nuts, torque wrench settings are no longer specified for the latter stages of tightening, "angle-tightening" being called up instead. Typically, a fairly low torque wrench setting will be applied to the bolts/nuts in the correct sequence, followed by one or more stages of tightening through specified angles.

Locknuts, locktabs and washers

Any fastening which will rotate against a component or housing during tightening should always have a washer between it and the relevant component or housing.

Spring or split washers should always be renewed when they are used to lock a critical component such as a big-end bearing retaining bolt or nut. Locktabs which are folded over to retain a nut or bolt should always be renewed.

Self-locking nuts can be re-used in non-critical areas, providing resistance can be felt when the locking portion passes over the bolt or stud thread. However, it should be noted that self-locking stiffnuts tend to lose their effectiveness after long periods of use, and should then be renewed as a matter of course.

Split pins must always be replaced with new ones of the correct size for the hole.

When thread-locking compound is found on the threads of a fastener which is to be re-used, it should be cleaned off with a wire brush and solvent, and fresh compound applied on reassembly.

Special tools

Some repair procedures in this manual entail the use of special tools such as a press, two or three-legged pullers, spring compressors, etc. Wherever possible, suitable readily-available alternatives to the manufacturer's special tools are described, and are shown in use. In some instances, where no alternative is possible, it has been necessary to resort to the use of a manufacturer's tool, and this has been done for reasons of safety as well as the efficient completion of the repair operation. Unless you are highly-skilled and have a thorough understanding of the procedures described, never attempt to bypass the use of any special tool when the procedure described specifies its use. Not only is there a very great risk of personal injury, but expensive damage could be caused to the components involved.

Environmental considerations

When disposing of used engine oil, brake fluid, antifreeze, etc, give due consideration to any detrimental environmental effects. Do not, for instance, pour any of the above liquids down drains into the general sewage system, or onto the ground to soak away. Many local council refuse tips provide a facility for waste oil disposal, as do some garages. If none of these facilities are available, consult your local Environmental Health Department, or the National Rivers Authority, for further advice.

With the universal tightening-up of legislation regarding the emission of environmentally-harmful substances from motor vehicles, most vehicles have tamperproof devices fitted to the main adjustment points of the fuel system. These devices are primarily designed to prevent unqualified persons from adjusting the fuel/air mixture, with the chance of a consequent increase in toxic emissions. If such devices are found during servicing or overhaul, they should, wherever possible, be renewed or refitted in accordance with the manufacturer's requirements or current legislation.

OIL CARE

FOLLOW THE CODE

OIL BANK LINE
0800 66 33 66

Note: It is antisocial and illegal to dump oil down the drain. To find the location of your local oil recycling bank, call this number free.

Length (distance)

Inches (in)	x 25.4	= Millimetres (mm)	x 0.0394	= Inches (in)
Feet (ft)	x 0.305	= Metres (m)	x 3.281	= Feet (ft)
Miles	x 1.609	= Kilometres (km)	x 0.621	= Miles

Volume (capacity)

Cubic inches (cu in; in³)	x 16.387	= Cubic centimetres (cc; cm³)	x 0.061	= Cubic inches (cu in; in³)
Imperial pints (Imp pt)	x 0.568	= Litres (l)	x 1.76	= Imperial pints (Imp pt)
Imperial quarts (Imp qt)	x 1.137	= Litres (l)	x 0.88	= Imperial quarts (Imp qt)
Imperial quarts (Imp qt)	x 1.201	= US quarts (US qt)	x 0.833	= Imperial quarts (Imp qt)
US quarts (US qt)	x 0.946	= Litres (l)	x 1.057	= US quarts (US qt)
Imperial gallons (Imp gal)	x 4.546	= Litres (l)	x 0.22	= Imperial gallons (Imp gal)
Imperial gallons (Imp gal)	x 1.201	= US gallons (US gal)	x 0.833	= Imperial gallons (Imp gal)
US gallons (US gal)	x 3.785	= Litres (l)	x 0.264	= US gallons (US gal)

Mass (weight)

Ounces (oz)	x 28.35	= Grams (g)	x 0.035	= Ounces (oz)
Pounds (lb)	x 0.454	= Kilograms (kg)	x 2.205	= Pounds (lb)

Force

Ounces-force (ozf; oz)	x 0.278	= Newtons (N)	x 3.6	= Ounces-force (ozf; oz)
Pounds-force (lbf; lb)	x 4.448	= Newtons (N)	x 0.225	= Pounds-force (lbf; lb)
Newtons (N)	x 0.1	= Kilograms-force (kgf; kg)	x 9.81	= Newtons (N)

Pressure

Pounds-force per square inch (psi; lbf/in²; lb/in²)	x 0.070	= Kilograms-force per square centimetre (kgf/cm²; kg/cm²)	x 14.223	= Pounds-force per square inch (psi; lbf/in²; lb/in²)
Pounds-force per square inch (psi; lbf/in²; lb/in²)	x 0.068	= Atmospheres (atm)	x 14.696	= Pounds-force per square inch (psi; lbf/in²; lb/in²)
Pounds-force per square inch (psi; lbf/in²; lb/in²)	x 0.069	= Bars	x 14.5	= Pounds-force per square inch (psi; lbf/in²; lb/in²)
Pounds-force per square inch (psi; lbf/in²; lb/in²)	x 6.895	= Kilopascals (kPa)	x 0.145	= Pounds-force per square inch (psi; lbf/in²; lb/in²)
Kilopascals (kPa)	x 0.01	= Kilograms-force per square centimetre (kgf/cm²; kg/cm²)	x 98.1	= Kilopascals (kPa)
Millibar (mbar)	x 100	= Pascals (Pa)	x 0.01	= Millibar (mbar)
Millibar (mbar)	x 0.0145	= Pounds-force per square inch (psi; lbf/in²; lb/in²)	x 68.947	= Millibar (mbar)
Millibar (mbar)	x 0.75	= Millimetres of mercury (mmHg)	x 1.333	= Millibar (mbar)
Millibar (mbar)	x 0.401	= Inches of water (inH₂O)	x 2.491	= Millibar (mbar)
Millimetres of mercury (mmHg)	x 0.535	= Inches of water (inH₂O)	x 1.868	= Millimetres of mercury (mmHg)
Inches of water (inH₂O)	x 0.036	= Pounds-force per square inch (psi; lbf/in²; lb/in²)	x 27.68	= Inches of water (inH₂O)

Torque (moment of force)

Pounds-force inches (lbf in; lb in)	x 1.152	= Kilograms-force centimetre (kgf cm; kg cm)	x 0.868	= Pounds-force inches (lbf in; lb in)
Pounds-force inches (lbf in; lb in)	x 0.113	= Newton metres (Nm)	x 8.85	= Pounds-force inches (lbf in; lb in)
Pounds-force inches (lbf in; lb in)	x 0.083	= Pounds-force feet (lbf ft; lb ft)	x 12	= Pounds-force inches (lbf in; lb in)
Pounds-force feet (lbf ft; lb ft)	x 0.138	= Kilograms-force metres (kgf m; kg m)	x 7.233	= Pounds-force feet (lbf ft; lb ft)
Pounds-force feet (lbf ft; lb ft)	x 1.356	= Newton metres (Nm)	x 0.738	= Pounds-force feet (lbf ft; lb ft)
Newton metres (Nm)	x 0.102	= Kilograms-force metres (kgf m; kg m)	x 9.804	= Newton metres (Nm)

Power

Horsepower (hp)	x 745.7	= Watts (W)	x 0.0013	= Horsepower (hp)

Velocity (speed)

Miles per hour (miles/hr; mph)	x 1.609	= Kilometres per hour (km/hr; kph)	x 0.621	= Miles per hour (miles/hr; mph)

Fuel consumption*

Miles per gallon, Imperial (mpg)	x 0.354	= Kilometres per litre (km/l)	x 2.825	= Miles per gallon, Imperial (mpg)
Miles per gallon, US (mpg)	x 0.425	= Kilometres per litre (km/l)	x 2.352	= Miles per gallon, US (mpg)

Temperature

Degrees Fahrenheit = (°C x 1.8) + 32   Degrees Celsius (Degrees Centigrade; °C) = (°F - 32) x 0.56

It is common practice to convert from miles per gallon (mpg) to litres/100 kilometres (l/100km), where mpg x l/100 km = 282

Notes

Chapter 1
An introduction to the diesel engine

Contents

1

1 History

Rudolf Diesel invented the first commercially-successful compression-ignition engine at the end of the 19th century. Compared with the spark ignition engine, the diesel had the advantages of lower fuel consumption, the ability to use cheaper fuel, and the potential for much higher power outputs. Over the following two or three decades, such engines were widely adopted for stationary and marine applications, but the fuel injection systems used were not capable of high-speed operation. This speed limitation, and the considerable weight of the air compressor needed to operate the injection equipment, made the first diesel engines unsuitable for use in road-going vehicles.

In the 1920s, the German engineer Robert Bosch developed the in-line injection pump, a device which is still in extensive use today. The use of hydraulic systems to pressurise and inject the fuel did away with the need for a separate air compressor, and made possible much higher operating speeds. The so called 'high-speed' diesel engine became increasingly popular as a power source for goods and public transport vehicles, but for a number of reasons (including specific power output, flexibility and cheapness of manufacture), the spark-ignition engine continued to dominate the passenger car and light commercial market.

In the 1950s and 60s, diesel engines became increasingly popular for use in taxis and vans, but it was not until the sharp rises in oil prices in the 1970s that serious attention was paid to the small passenger car market. VW's introduction of the diesel-powered Golf at the end of 1977 marked the arrival of the first 'user/friendly' diesel car, designed specifically to be acceptable to drivers who would not previously have considered abandoning the petrol engine. The diesel engine fitted to the Golf used indirect injection and a distributor type pump, and was comparable in performance to the smaller petrol engines fitted to the range.

Subsequent years have seen the growing popularity of the small diesel engine in cars and light commercial vehicles, not only for reasons of fuel economy and longevity, but also for environmental reasons. Every major European car manufacturer now offers at least one diesel-engined model. The diesel's penetration of the UK market has been relatively slow (due in part to the lack of any considerable fuel price differential in favour of diesel which exists in other parts of Europe), but it has now gained widespread acceptance, and this trend looks set to continue.

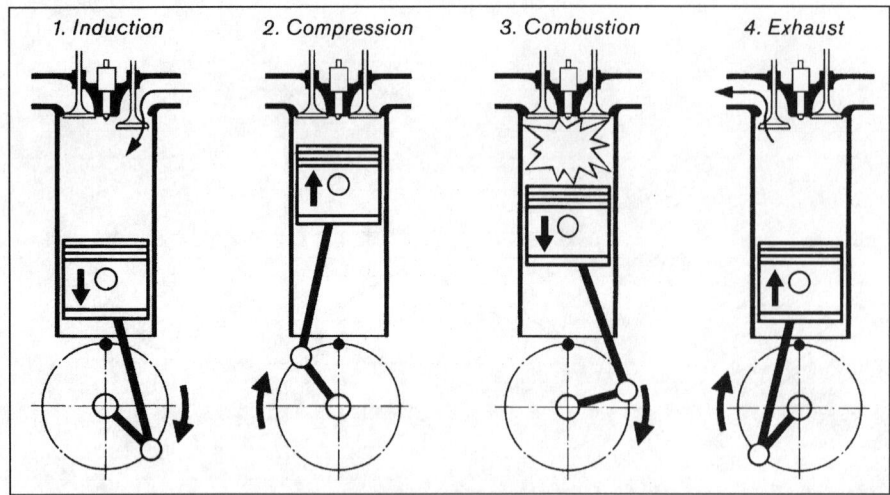

2.1 Four-stroke diesel cycle
© Robert Bosch Limited

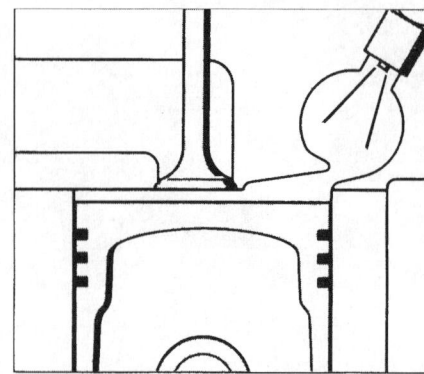

Indirect injection into a swirl chamber

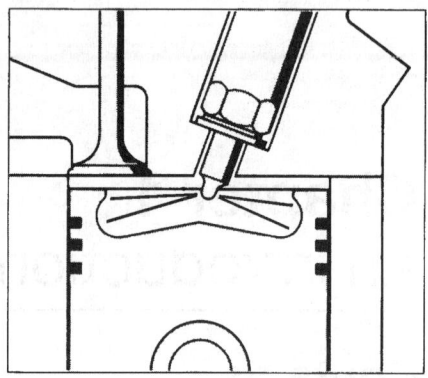

Direct injection

2.6 Indirect and direct injection
© Robert Bosch Limited

2 Principles of operation

1 All the diesel engines covered in this book operate on the familiar four-stroke cycle of induction, compression, combustion and exhaust **(see illustration)**. (Two-stroke diesels do exist, and may in future become important, but they are used in few light vehicles at present.) Most have four cylinders, some larger engines have six, and five- and three-cylinder engines also exist.

Induction and ignition

2 The main difference between diesel and petrol engines is the means by which the fuel/air mixture is introduced into the cylinder and then ignited. In the petrol engine, the fuel is mixed with the incoming air before it enters the cylinder, and the mixture is then ignited at the appropriate moment by a spark plug. At all conditions except full-throttle, the throttle butterfly restricts the airflow, and cylinder filling is incomplete.

3 In the diesel engine, air alone is drawn into the cylinder and then compressed. Because of the diesel's high compression ratio (typically 20:1), the air gets very hot when compressed – up to 750°C (1382°F). As the piston approaches the end of the compression stroke, fuel is injected into the combustion chamber under very high pressure, in the form of a finely-atomised spray. The temperature of the air is high enough to ignite the injected fuel as it mixes with the air. The mixture then burns and provides the energy which drives the piston downwards on the combustion (power) stroke.

4 When starting the engine from cold, the temperature of the compressed air in the cylinders may not be high enough to ignite the

fuel. The preheating system overcomes this problem. Most modern engines have automatically-controlled preheating systems, using electric heater plugs (glow plugs) which heat the air in the combustion chamber just before and during start-up. Full details of these systems are given in Chapter 3.

5 On most diesel engines there is no throttle valve in the inlet tract; exceptions to this are those few engines which use a pneumatic governor, which depends on a manifold depression being created. Even more rarely, a throttle valve may be used to create manifold depression for the operation of a brake servo, though it is more usual for a separate vacuum pump to be fitted for this purpose.

Direct and indirect injection

6 In practise, it is difficult to achieve smooth combustion in a small-displacement engine by injecting the fuel directly into the combustion chamber. To get around this problem, the technique of indirect injection is widely used. With indirect injection, the fuel is injected into a pre-combustion or 'swirl' chamber in the cylinder head, alongside the main combustion chamber. During the compression stroke the compressed hot air is forced into the swirl chamber where it enters a rapid swirling motion; fuel is injected into the swirl chamber, where it mixes with the rapidly moving air, enabling smoother combustion in the main combustion chamber **(see illustration)**.

7 Generally speaking, indirect injection engines are less efficient than direct injection engines, and also require more preheating when starting from cold, but these disadvantages are offset by smoother and quieter operation. Until recently, direct injection engines were mostly fitted to light commercial vehicles, where increased noise and harshness are considered acceptable trade-offs for improved fuel economy.

Recently, the use of electronic diesel engine control systems has allowed the development of more refined direct injection engines, and their use in passenger vehicles is becoming more widespread.

Mechanical construction

8 Due to the high compression ratio required in a diesel engine, and the combustion characteristics, it is necessary to ensure that the lower face of the cylinder head is flat. This is achieved by positioning the valves vertically in the cylinder head (ie, with their stems at right-angles to the cylinder head lower face), and machining the combustion chambers directly into the tops of the pistons. Locating the combustion chambers in the pistons also enables the combustion process to be contained, and allows fine control of the combustion chamber size and shape during manufacture (all the combustion chambers in a diesel engine must be of similar size and shape).

9 The pistons, crankshaft and bearings of a diesel engine are generally of more robust construction than in a petrol engine of comparable size, because of the greater loads imposed by the higher compression ratio and

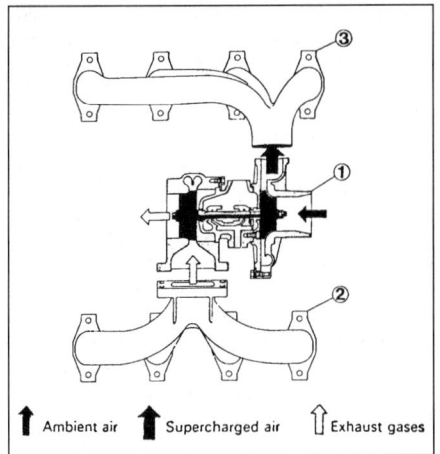

2.10 Principle of turbocharging

1 Turbocharger
2 Exhaust manifold
3 Inlet manifold

the nature of the combustion process. This is one reason for the diesel engine's longer life. Other reasons include the lubricating qualities of diesel fuel on the cylinder bores, and the fact that the diesel engine is generally lower-revving than its petrol counterpart, having much better low-speed torque characteristics and a lower maximum speed.

Turbocharging

10 Turbochargers have long been used on large diesel engines, and are becoming common on small ones. The turbocharger uses the energy of the escaping exhaust gas to drive a turbine which pressurises the air in the inlet manifold. The air is forced into the cylinders instead of being simply sucked in. If more air is present, more fuel can be burnt and more power can be developed from the same size engine **(see illustration)**.

11 Greater benefit can be gained from turbocharging if the pressurised air is cooled before it enters the engine. This is done using an air-to-air heat exchanger called an intercooler. The cooled air is denser and contains more oxygen in a given volume than warm air straight from the turbocharger **(see illustrations)**.

Exhaust emissions

12 Because combustion in a correctly-functioning diesel engine nearly always occurs in conditions of excess oxygen, there is little or no carbon monoxide (CO) present in the exhaust gas. A further environmental benefit is that there is no added lead in diesel fuel.

13 For many years, there was no need for complicated emission control systems on diesel engines. In the last few years however, simple catalytic converters, and exhaust gas recirculation systems, have become standard on most diesel engines in order to meet the increasingly stringent emission regulations.

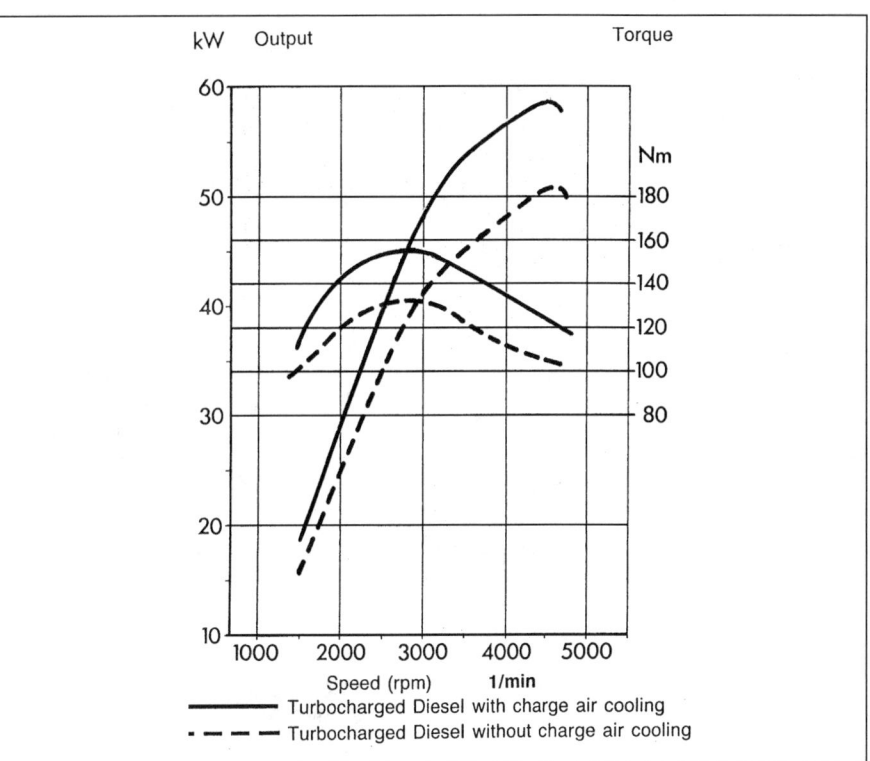

2.11a Power and torque outputs from a turbocharged engine with and without charge air cooling

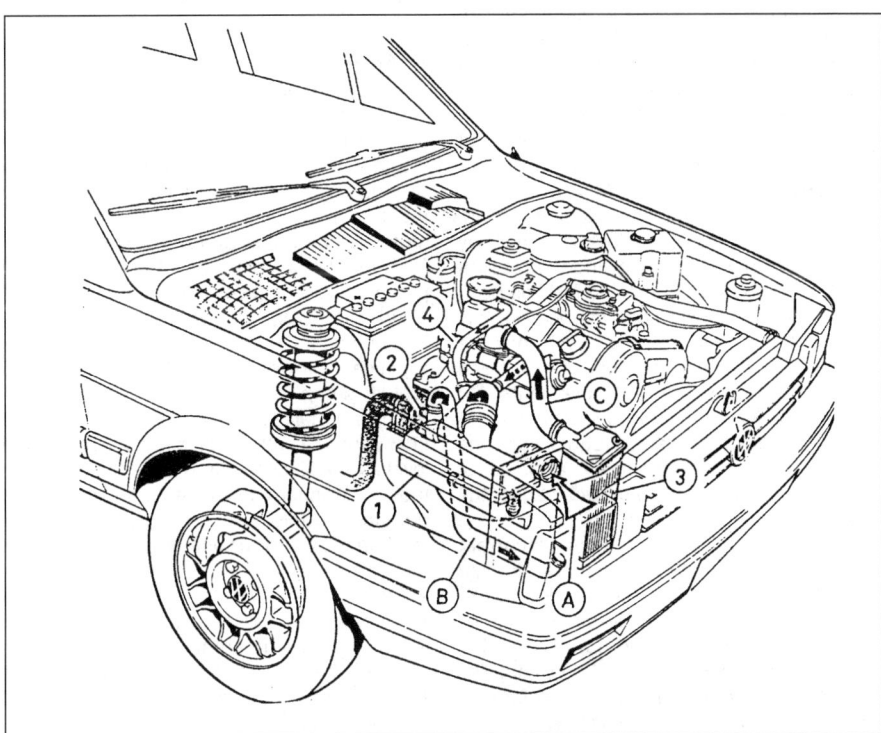

2.11b Induction airflow in a turbocharged engine with charge air cooling

1 Air cleaner
2 Turbocharger
3 Intercooler
4 Inlet manifold

A Inducted air
B Compressed air before cooling
C Compressed air after cooling

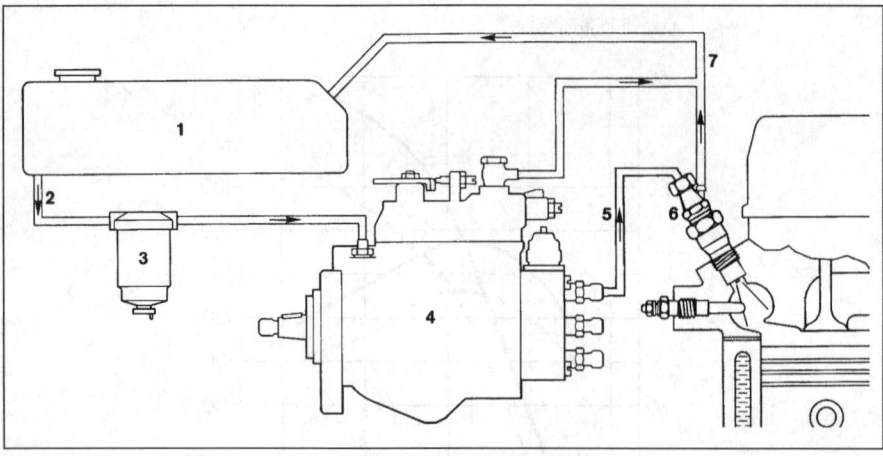

3.2 Fuel circulation – typical passenger car system

© Robert Bosch Limited

1 Fuel tank	4 Injection pump with	6 Injector
2 Fuel feed line	integral supply pump	7 Fuel return (leak-off) line
3 Fuel filter/water trap	5 Injector pipe	

The advent of electronic diesel engine control systems has also helped to improve diesel engine emissions.

Knock and smoke

14 The image of the diesel engine for many years was of a noisy, smoky machine, and to some extent this was justified. It is worth examining the causes of knock and smoke, both to see how they have been reduced in modern engines, and to understand what causes them to get worse.

15 There is inevitably a small delay (typically 0.001 to 0.002 second) between the start of fuel injection and the beginning of combustion. This delay, known as ignition lag, is greatest when the engine is cold and idling. The characteristic diesel knock is caused by the sudden increase in cylinder pressure which occurs when the injected fuel has been mixed with the hot air and starts burning. It is therefore an unavoidable part of the combustion process, though it has been greatly reduced by improvements in combustion chamber and injection system design. A defective injector (particularly one which is not atomising the fuel as it should for optimum combustion) will also cause the engine to knock.

16 Smoke is caused by incorrect combustion, but unlike knock it is more or less preventable. During start-up and warm-up a certain amount of white or blue smoke may be seen, but under normal running conditions the exhaust should be clean. The thick black smoke which is all too familiar from old or badly-maintained vehicles is caused by a lack of air for combustion, either because the air intake is restricted (clogged air cleaner), or because too much fuel is being injected (defective injectors or pump). Causes of smoke are examined in more detail in Chapter 9.

3 Fuel supply and injection systems

1 Fuel injection systems and components are covered in detail in Chapter 2. This Section gives an overview of the systems used, and their basic principles of operation.

Fuel supply

2 The fuel supply system is concerned with delivering clean fuel, free of air, water or other contaminants, to the injection pump. It always includes a fuel filter and a water trap (which may be combined in one unit), a fuel tank, and the associated pipework. Some arrangement must also be made for returning excess fuel from the fuel injectors and the fuel injection pump to the tank **(see illustration)**.

3 On older vehicles which use an in-line injection pump, or where the fuel tank outlet is significantly lower than the injection pump, a fuel lift pump is used between the tank and the filter. When a distributor injection pump is fitted, and the tank outlet is at about the same level as the injection pump (as is the case with most passenger cars), a separate lift pump is not fitted. In this case, a hand-priming pump is often provided for use when bleeding the fuel system.

4 Additional refinements may be encountered. These include a fuel heater, which may be integral with the filter, or between the tank and the filter, to prevent the formation of wax crystals in the fuel in cold weather. On some vehicles, a 'water-in-fuel' warning light may be illuminated by a device in the water trap when the water reaches a certain level.

5 The water trap and fuel filter are vital for satisfactory operation of the fuel injection system. On some vehicles, the water trap may have a glass bowl, in which case water build-up can be seen, or it may as already mentioned have some electrical device for alerting the driver to the presence of water. Whether or not these features are present, the trap must be drained at specific intervals, or more frequently if experience shows this to be necessary. If water enters the injection pump it can cause rapid corrosion, especially if the vehicle is left standing for any length of time.

6 The fuel filter may be of the disposable cartridge type, or it may consist of a renewable element inside a metal bowl **(see illustration)**. Sometimes a coarser pre-filter is fitted upstream of the main filter. Whatever the type, it must be renewed at the specified intervals. Considering the damage which can be caused to the injection equipment by the entry of even small particles of dirt, it is not worth using cheap replacement filters, which may not be of the same quality as those of reputable manufacture.

Fuel injection pump

7 In a conventional diesel injection system, the pump is a mechanical device attached to the engine, driven at half-engine speed by a chain, gears or toothed belt. Its function is to supply fuel to the injectors at the correct pressure, at the correct moment in the combustion cycle, and for the length of time necessary to ensure efficient combustion. The pump responds to depression of the accelerator pedal by increasing fuel delivery, within the limits allowed by the governor. It is

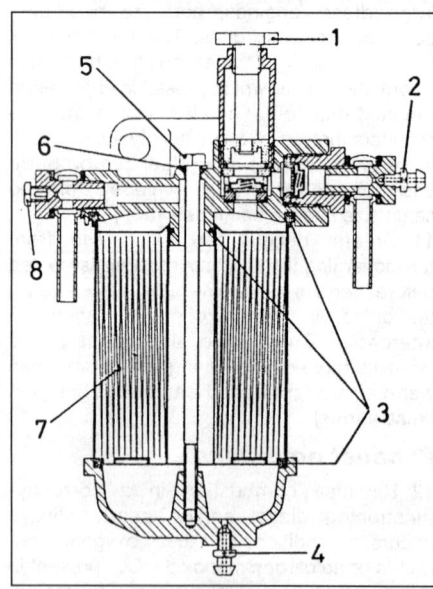

3.6 Sectional view of a typical fuel filter

1	Hand-priming plunger
2	Fuel bleed screw (on outlet union)
3	Seals
4	Water drain tap
5	Through-bolt
6	Through-bolt seal
7	Filter element
8	Air bleed screw (on inlet union)

also provided with some means of cutting off fuel delivery when it is wished to stop the engine.

8 There are two basic types of pump; the in-line pump, generally fitted to larger engines, and the distributor pump, commonly fitted to passenger car engines. The in-line pump has one pump plunger per engine cylinder. The distributor pump, as its name implies, has a single pump plunger and directs its output to each cylinder in turn **(see illustrations)**.

9 Some kind of governor is associated with the injection pump, either integral with it or attached to it. All vehicle engine governors regulate the fuel delivery to control idle speed and maximum speed; the variable-speed governor also regulates the intermediate speeds. Operation of the governor may be mechanical or hydraulic, or it may be controlled by manifold depression.

10 Other devices in, or attached to, the pump include cold start injection advance or fast idle units, turbo boost pressure sensors, and anti-stall mechanisms.

11 Fuel injection pumps are normally very reliable. If they are not damaged by dirt, water

or unskilled adjustment, they may well outlast the engine to which they are fitted.

12 Some modern electronically-controlled diesel injection systems use alternatives to the conventional in-line or distributor fuel pumps – brief details are given in Section 4, with a more detailed description in Chapter 2.

Fuel injectors

13 One fuel injector is fitted to each cylinder. The function of the injector is to spray an evenly-atomised quantity of fuel into the combustion or pre-combustion chamber when the fuel pressure exceeds a certain value, and to stop the flow of fuel cleanly when the pressure drops. Atomisation is achieved by a spring-loaded needle which vibrates rapidly against its seat when fuel under pressure passes it. The needle and seat assembly together are known as the injector nozzle.

14 Injectors in direct injection engines are

usually of the multi-hole type, while those in indirect engines are of the pintle type. The 'throttled pintle' injector gives a progressive build-up of injection, which is valuable for achieving smooth combustion **(see illustrations)**.

15 The injector tips are exposed to the temperatures and pressures of combustion, so not surprisingly they will in time suffer from carbon deposits and ultimately from erosion and burning. Service life will vary according to factors such as fuel quality and operating conditions, but typically one could expect to clean and recalibrate a set of injectors after about 50 000 miles (80 000 km), and perhaps to renew them or have them reconditioned after 100 000 miles (161 000 km).

16 Some modern electronically-controlled diesel injection systems use electronically-controlled injectors – brief details are given in Section 4, with a more detailed description in Chapter 2.

Injector pipes

17 The injector pipes are an important part of the system, and must not be overlooked. The dimensions of the pipes are important, and it should not be assumed that, just because the end fittings are the same, a pipe from a

3.8a Bosch PE in-line injection pump and associated components

© Robert Bosch Limited

1   Pump
2   Governor housing
3   Lift pump
4   Drivegear and advance mechanism

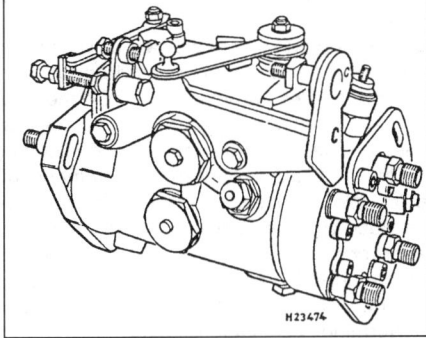

3.8b Lucas/CAV DPC-type distributor injection pump

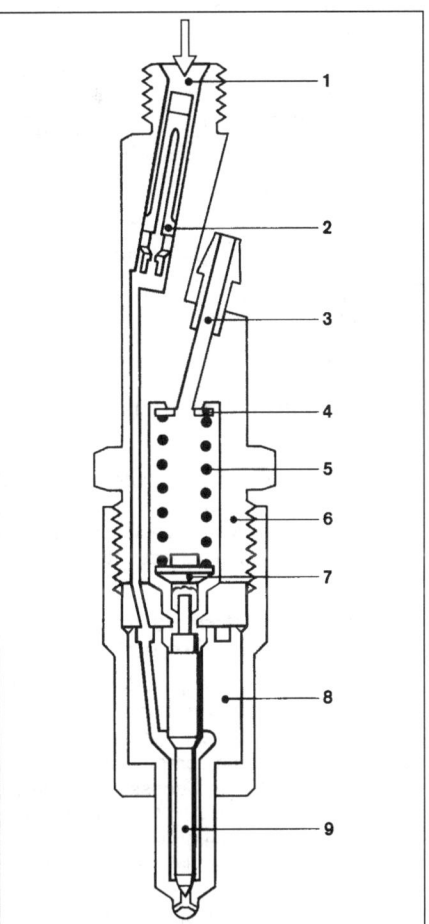

3.14a Sectional view of a multi-hole injector

© Robert Bosch Limited

1	Fuel inlet	5	Spring
2	Integral filter	6	Body
3	Fuel return	7	Spindle
4	Pressure	8	Nozzle body
	adjusting shim	9	Nozzle needle

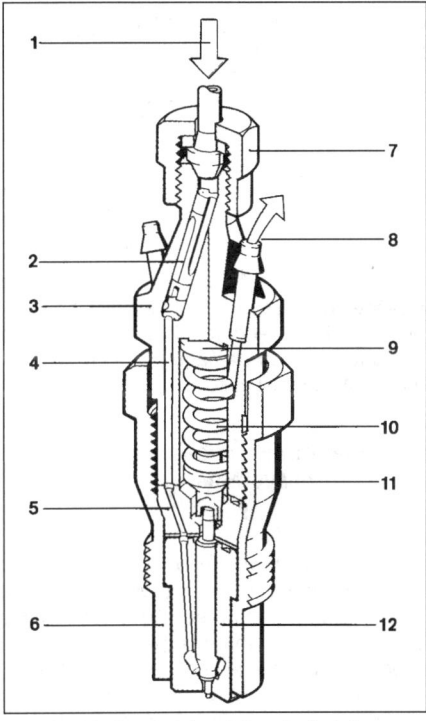

3.14b Cutaway view of a pintle injector

© Robert Bosch Limited

1	Fuel inlet	8	Fuel return
2	Integral filter	9	Pressure
3	Body		adjusting shim
4	Pressure passage	10	Spring
5	Sleeve	11	Spindle
6	Nozzle retainer	12	Nozzle
7	Union nut		

1

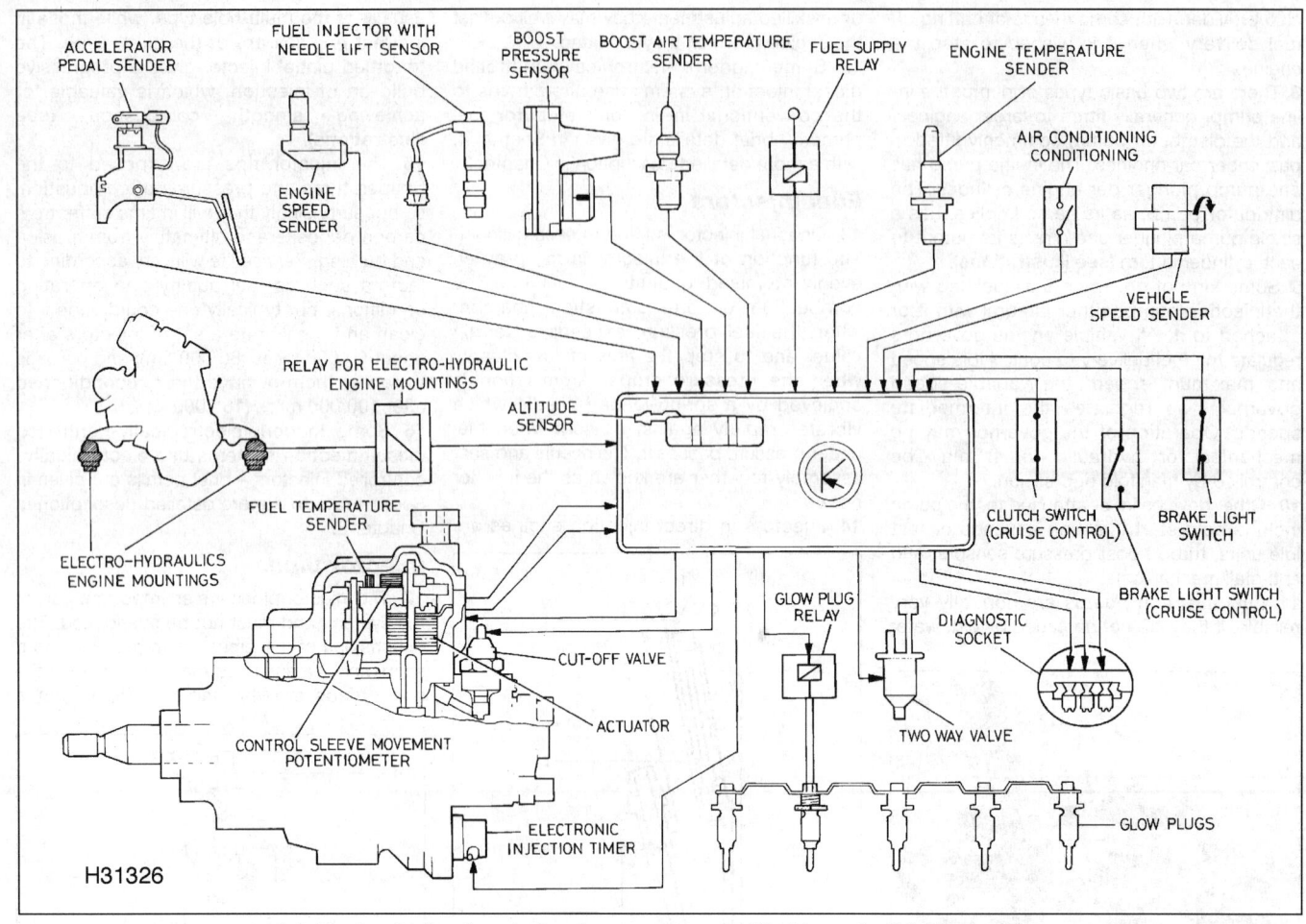

ACCELERATOR
PEDAL SENDER

FUEL INJECTOR WITH
NEEDLE LIFT SENSOR

BOOST
PRESSURE
SENSOR

BOOST AIR TEMPERATURE
SENDER

FUEL SUPPLY
RELAY

ENGINE TEMPERATURE
SENDER

AIR CONDITIONING
CONDITIONING

ENGINE
SPEED
SENDER

VEHICLE
SPEED SENDER

RELAY FOR ELECTRO-HYDRAULIC
ENGINE MOUNTINGS

ALTITUDE
SENSOR

CLUTCH SWITCH
(CRUISE CONTROL)

BRAKE LIGHT
SWITCH

FUEL TEMPERATURE
SENDER

BRAKE LIGHT SWITCH
(CRUISE CONTROL)

ELECTRO-HYDRAULICS
ENGINE MOUNTINGS

GLOW PLUG
RELAY

DIAGNOSTIC
SOCKET

CUT-OFF VALVE

CONTROL SLEEVE MOVEMENT
POTENTIOMETER

ACTUATOR

TWO WAY VALVE

ELECTRONIC
INJECTION TIMER

GLOW PLUGS

H31326

4.4 Electronic diesel control system components fitted to an Audi 2.5 litre engine

different engine can be used as a replacement. Securing clips must be kept tight, and the engine should not be run without them, as damage from vibration or fuel cavitation may result.

4 Electronic diesel engine control systems

1 Development of the diesel engine, and particularly the fuel injection system, has been relatively slow compared with the advances which have been made in petrol engine fuel injection and management systems. However, in recent years, electronic diesel engine control systems have been developed to improve diesel engine efficiency and to reduce exhaust emissions. Almost all modern engines use some form of electronic engine control system.

2 For a diesel engine to operate efficiently, it is essential that the correct amount of fuel is injected at the correct pressure, and at exactly the right time. Even small deviations can cause increased exhaust emissions,

increased noise, and increased fuel consumption. In a typical diesel engine, the injection process takes only a thousandth of a second, and only a minute quantity of fuel is injected.

Electronic control using a conventional fuel injection pump

3 As we've already seen, the function of the fuel injection pump is to supply fuel to the injectors at the correct pressure, at the correct moment in the combustion cycle, and for the length of time necessary to ensure efficient combustion. A conventional (mechanically-controlled) fuel injection pump uses an accelerator cable (connected to the driver's accelerator pedal), and various mechanical add-on devices (such as cold start injection advance, fast idle units, turbo boost pressure sensors, etc) to provide control of the fuel injection timing and the quantity of fuel injected. Even with these add-on devices, it has become increasingly difficult for a mechanical diesel control system to keep pace with modern demands on engine refinement and exhaust emission control.

4 Many electronic diesel engine control systems use a conventional in-line or distributor fuel injection pump, but the injection pump timing and the quantity of fuel injected are controlled electronically instead of mechanically. Various electronic sensors are used to measure variables such as accelerator pedal position, engine crankshaft speed, engine camshaft position, the mass of air passing into the engine, turbocharger boost pressure, engine coolant temperature, ambient air temperature, etc **(see illustration)**.

5 The information from the various sensors is passed to an electronic control unit (ECU), which evaluates the signals. The ECU memory contains a series of mapped values for injected fuel quantity, and start-of-injection point. The ECU performs a number of calculations based on the information provided by the sensors, and selects the most appropriate values for the fuel quantity and start-of-injection point from its stored values. The ECU is capable of analysing the data and performing calculations many times per second, which allows very accurate control over the operation of the engine.

Common rail diesel injection systems

6 A recent development in electronic diesel engine control is the Common Rail diesel injection system. Common rail injection systems have been developed by several manufacturers for use on direct injection diesel engines.

7 The common rail system derives its name from the fact that a common rail, or fuel reservoir, is used to supply fuel to all the fuel injectors.

8 Common rail systems give very fine control of the engine, and allow the problems of increased noise and harshness often associated with direct injection engines to be greatly reduced. Further details of common rail injection systems can be found in Chapter 2, Section 9.

'Pump injector' diesel injection systems

9 The 'pump injector' system uses a combined injection pump/injector for each engine cylinder. This eliminates the need for a separate high pressure injection pump, and the associated high pressure fuel lines.

10 As with common rail systems, pump injector systems give very fine control of the engine. Further details of pump injector systems can be found in Chapter 2, Section 10.

Future developments

11 Development of the diesel engine, and particularly the fuel injection system, has been relatively slow compared with the advances which have been made in petrol engine fuel injection and engine management systems. However, recent developments in electronic diesel control systems and the introduction of common rail and pump injector fuel systems have closed the performance and refinement gap between diesel and petrol engines.

12 There can be no doubt that the current combination of high fuel prices and increased environmental awareness, along with the increase in diesel engine vehicle sales, will provide the stimuli for further improvements in the future.

1

Notes

Chapter 2
Fuel injection systems

Contents

2

1 Introduction

A brief description of the fuel supply and injection systems in use on diesel engines was given in Chapter 1. The various systems which may be encountered will now be described in more detail. Later Chapters deal with maintenance, testing, adjustment and overhaul procedures.

2 Diesel fuel

Unlike petrol, diesel fuel must be capable of igniting under high pressure in hot air, without the aid of an electrical spark.

Diesel fuel is distilled from crude oil, and contains a percentage of oil-based compounds which are not present in petrol. The oil-based compounds give diesel fuel lubricating qualities, which eliminates the need for separate lubricating oil to protect the moving components in a fuel injection pump. Diesel fuel is less volatile than petrol, which reduces the risk of fire during storage and handling.

There are however a few disadvantages associated with diesel fuel. Firstly, the high sulphur content of the fuel makes a diesel engine more prone to deposits of carbon and other combustion products on the piston rings and contact surfaces of the pistons; this is usually combated by using engine lubricating oils with detergent additives. Secondly, diesel fuel is prone to 'waxing' in cold weather, as some of the large hydrocarbon molecules join together to form

paraffin wax compounds; if the temperature falls sufficiently, the fuel becomes so full of wax that it will no longer flow. To reduce the likelihood of waxing, fuel companies normally change the composition of diesel fuel during the winter months, and vehicle manufacturers often fit a fuel heater in the fuel line between the fuel tank and the filter.

Although diesel engine vehicles have less of a direct impact on the environment than their petrol engine counterparts, the fact that diesel fuel is distilled from the same raw crude oil as petrol means that as the reserves of natural crude oil are exhausted, both petrol and diesel fuel supplies will run out. However, currently a number of manufacturers are carrying out research into adapting diesel engines to run on alternative fuels (modified diesel engines have been run successfully using alcohol, and even rape-seed oil-based fuels).

3 Fuel supply system

Fuel lift pumps

Engines with distributor fuel injection pump

1 As mentioned in Chapter 1, a separate fuel lift pump is not normally fitted to passenger cars using a distributor type fuel injection pump. The transfer pump built into the injection pump provides enough suction to draw the fuel from the tank.

2 One effect of not using a separate lift pump is that the fuel supply lines are always under negative pressure while the engine is running. The slightest leak at any unions or seals will allow air to be drawn in, with adverse effects on the running of the engine.

3 The transfer pump also has to overcome the resistance to flow caused by the fuel filter. For this reason, the effects of filter clogging are more pronounced, and filter life is generally less than when a separate lift pump is used.

Engines with in-line fuel injection pump

4 The in-line injection pump requires fuel to be supplied at a pressure of approximately 1 bar, so a separate lift pump is always used **(see illustration)**. The lift pump may be of the diaphragm or piston type; it may be driven from the injection pump or directly from the engine. In any case, it is fitted between the fuel tank and the fuel filter. The fuel supply lines between the lift pump and the injection pump are under low pressure while the engine is running, so leakage will not result in air being drawn in, but in fuel leaking out.

Engines with common rail fuel injection system

5 In a common rail fuel injection system, a separate lift pump is used to pump fuel from the tank to the high-pressure pump, via the fuel filter. There are several different types of fuel lift pump in use on common rail systems, but the two most common are the roller-cell electric type (often submerged in the fuel tank), and the engine-driven gear type. The fuel supply lines between the lift pump and the high-pressure pump are under low pressure while the engine is running, so leakage will not result in air being drawn in, but in fuel leaking out.

Engines with 'pump injector' injection system

6 In a 'pump injector' fuel injection system, a separate lift pump is used to pump fuel from the tank to the injectors, via a fuel filter. The VW system uses a vane-type fuel lift pump, driven from the engine camshaft. The fuel supply lines between the lift pump and the injectors are under low pressure while the engine is running, so leakage will not result in air being drawn in, but in fuel leaking out.

Fuel filter and water trap

7 On most passenger vehicles, the fuel filter and water trap are combined into one unit. Some light commercial and many off-road vehicles have a separate water trap and/or pre-filter **(see illustration)**.

8 The water trap is necessary because water vapour from the atmosphere can condense in diesel fuel storage tanks (or even in the vehicle's fuel tank if it is left standing for a long time). This does not happen to the same extent in petrol storage tanks because of the higher vapour pressure of petrol.

9 In practise, water may rarely be found in the trap of a vehicle in regular use, as long as fuel is being purchased from a reputable source with a good turnover. This should not be taken as a reason to neglect periodic draining of the trap, unless a 'water-in-fuel' warning light is fitted.

10 The fuel filter is vital to the correct functioning and long life of the fuel injection system components. A clogged filter will reduce performance, and (through its effect on system pressure) can cause smoking and erratic running on some engines. A filter which allows particles of dirt to pass will cause early failure of an injection pump.

Bleeding air from the fuel supply system

11 On all systems, it is necessary to stop air entering the fuel supply lines, and to bleed out air which enters during maintenance and repair work. Methods of bleeding are described in Chapter 8.

12 On systems using a separate lift pump, a self-bleeding facility can be provided by the fuel return system, using overflow valves or restrictors on the injection pump and/or filter **(see illustration)**. When the pump is operating, a proportion of the fuel pumped is returned to the tank by the valves or restrictors; because they are at the highest point of the filter head or pump, air or fuel vapour bubbles automatically go into the return line to the tank.

13 The distributor type pump is also self-bleeding via its fuel return orifice, but the orifice is relatively small, and cannot handle large quantities of air in the fuel. Most common rail and 'pump injector' systems are self-bleeding.

Low temperature operation

14 As mentioned in Section 2, when diesel fuel is subjected to low temperatures, some of the hydrocarbon molecules join together to form

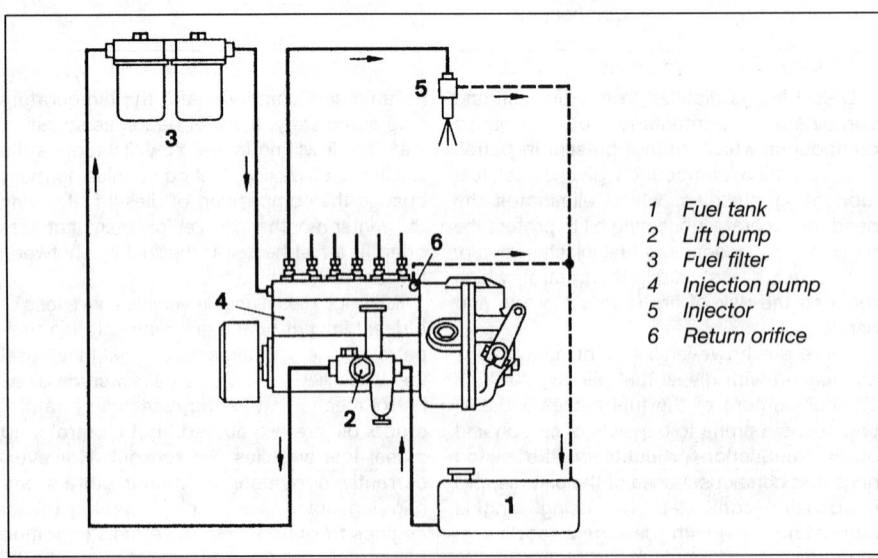

1 Fuel tank
2 Lift pump
3 Fuel filter
4 Injection pump
5 Injector
6 Return orifice

3.4 Fuel supply and return lines – in-line pump with lift pump

© Robert Bosch Limited

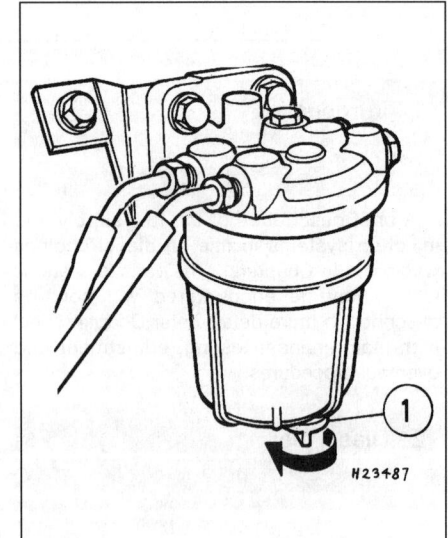

3.7 Water trap fitted to a Range Rover

1 Drain screw

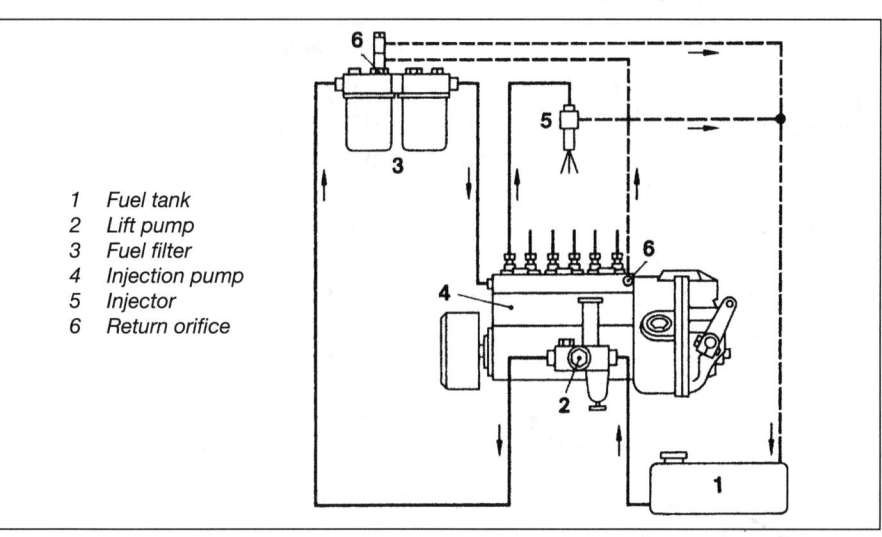

1 Fuel tank
2 Lift pump
3 Fuel filter
4 Injection pump
5 Injector
6 Return orifice

3.12 Fuel supply and return lines – in-line pump with self-bleeding fuel system

© Robert Bosch Limited

paraffin wax compounds. This causes the fuel to become cloudy in appearance. The temperature at which this happens is known as the *cloud point*; the fuel will still flow satisfactorily, and the engine will still run normally.

15 If the temperature falls further, the wax compounds combine, and present an increased resistance to flow. Eventually, the fuel will no longer flow, having a consistency similar to cold custard. The temperature at which this happens is known as the *pour point*. Obviously, if the fuel will not flow from the tank, the engine will not run at all. To get rid of these wax compounds completely, the fuel has to be raised to a much higher temperature than that at which the compounds formed.

16 In practise, problems start at a temperature between the cloud point and the pour point, when the wax compounds are not large enough to stop the fuel flowing from the tank, but are large enough to block the fine pores of the fuel filter element. This temperature is known as the *cold filter plugging point (CFPP)*.

17 When the cold filter plugging point has been reached, the engine may start normally but then stop almost immediately, or idle normally but refuse to speed up. If the filter can be warmed sufficiently to allow the fuel to flow freely, this will cure the problem, at least until the next cold start (**NEVER** use a naked flame for this – use a hair dryer or fan heater). In bad cases, it will be necessary to fit a new filter element. Once the engine has warmed up, the under-bonnet temperature is high enough to stop wax forming in the filter.

18 The oil companies alter the composition of the fuel in winter to reduce low-temperature problems. Winter-grade diesel fuel should be useable at temperatures well below freezing without extra additives or heating (the composition of winter fuel will vary depending on the country in which it is being used, and the temperatures likely to be encountered).

19 Fuel waxing obviously has a worse effect if the fuel filter element is not in good condition, and some manufacturers specify the renewal of the filter element at the beginning of every winter, regardless of the mileage covered since the previous renewal.

Anti-waxing or anti-gel fuel additives

20 Fuel additives are available to lower the temperature at which filter plugging occurs, but not all vehicle manufacturers approve their use. The additive (which is added to the fuel tank) has to mix with the fuel *before* waxing occurs – it is no good trying to avoid it once the problem has already begun.

21 Protection against waxing can also be achieved by adding a proportion (typically 10 to 20%) of petrol or paraffin to the fuel. Some vehicle manufacturers recommend or tolerate this, while others forbid it. There is no doubt that the practise works, but there are various objections to it:

a) A petrol/diesel mixture is more inflammable than plain diesel fuel, and gives off explosive vapour. If too much petrol is added, it will not mix with the diesel, but will float on top of it.

b) Combustion of a petrol/diesel mixture is not as smooth as combustion of pure diesel.

c) Petrol or paraffin will reduce the lubricating properties of the fuel, and perhaps cause accelerated injection pump wear.

d) Paraffin in the UK is not taxed as a road fuel, so its use in vehicle fuel tanks is an offence.

Fuel heaters

22 An alternative approach to fuel waxing problems is to heat the fuel before it arrives at the filter. Some vehicle manufacturers fit a fuel heater as standard, and heater kits are available from various manufacturers for aftermarket fitting.

23 Some fuel heaters are electrically-operated, forming part of the filter head, or positioned between the fuel tank and filter **(see illustrations)**. A thermostatic switch

2

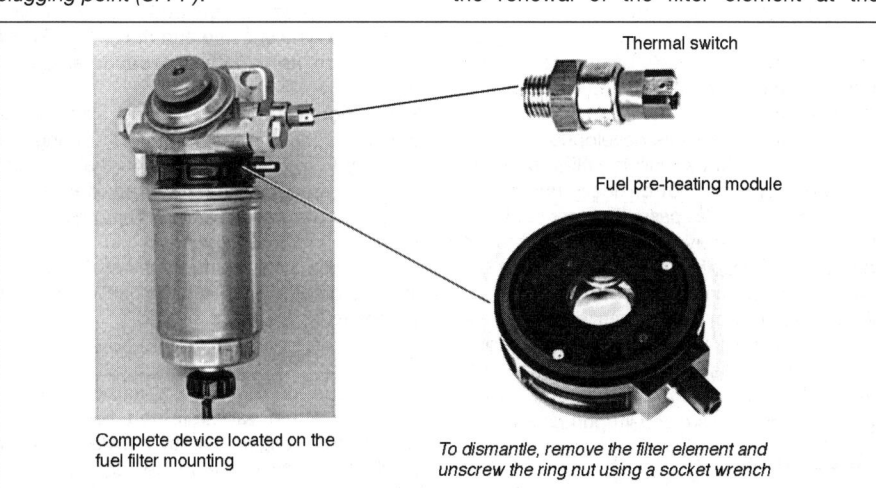

Thermal switch

Fuel pre-heating module

Complete device located on the
fuel filter mounting

*To dismantle, remove the filter element and
unscrew the ring nut using a socket wrench*

3.23a Electrically-operated fuel heater incorporated in the filter head

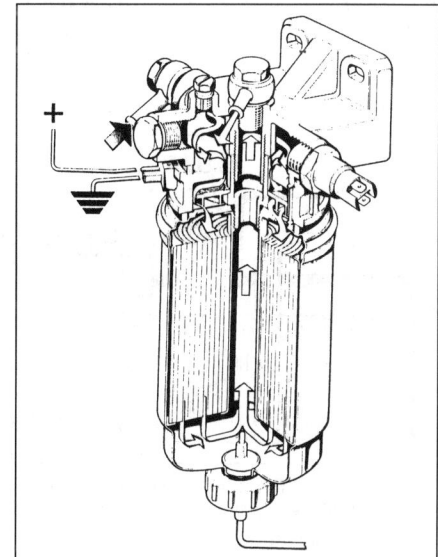

**3.23b Fuel flow through the filter and
heater element**

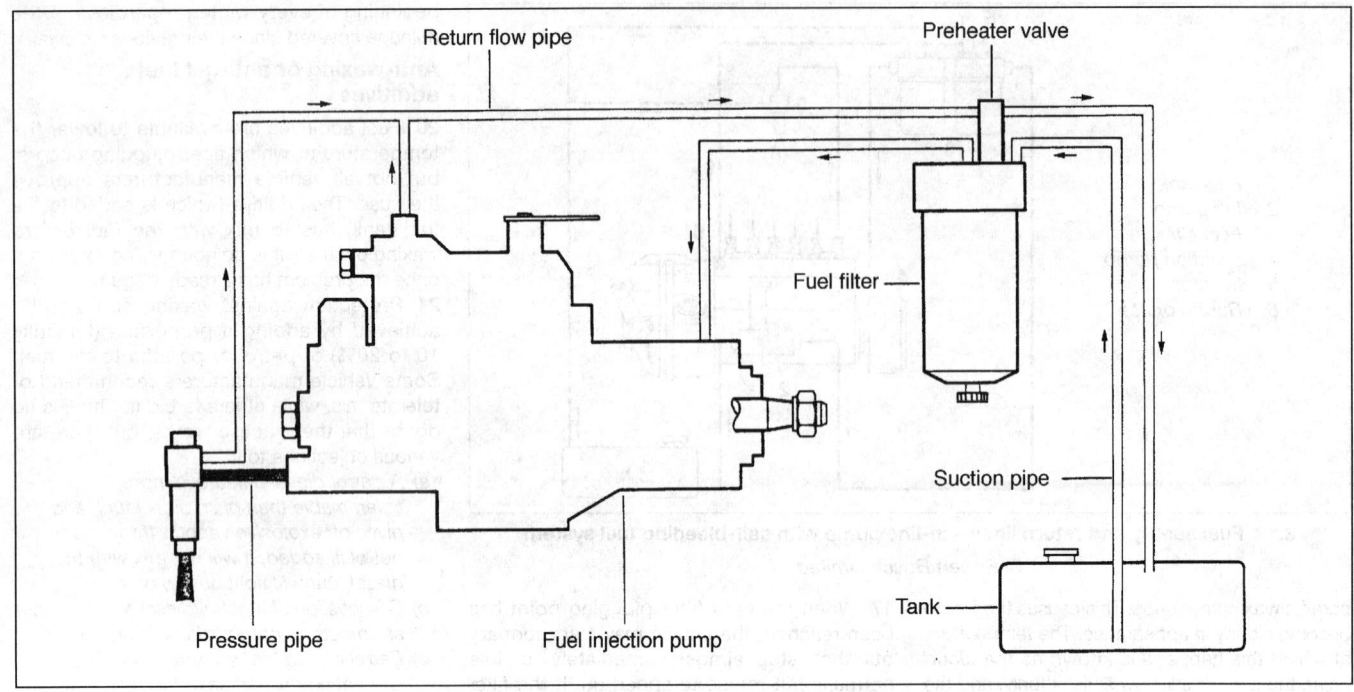

3.25a Fuel supply and return lines – VW self-heating system

ensures that the heater is only energised when the temperature is low. Generally, the heater is not energised until the ignition is switched on.

24 Other types of fuel heater use engine coolant as the heating medium – this will obviously not be effective if the temperature is already so low that the engine will not start.

25 The fuel heater fitted to later VW models diverts fuel from the return line back into the filter when the temperature is low **(see illustrations)**. The fuel returning from the pump and injectors warms up rapidly once the engine is running, so provided the engine will start, plugging of the filter should not occur. VW claim that the use of this type of heater, in conjunction with appropriate winter-grade fuel, dispenses with the need for anti-

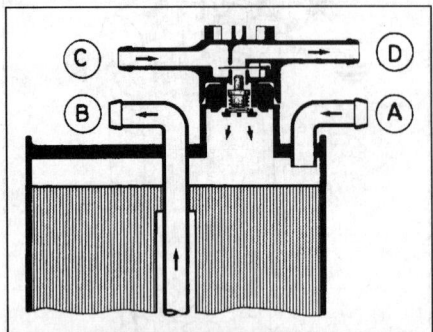

3.25b Fuel flow through filter at low temperature – VW self-heating system

A From fuel tank
B To injection pump
C Return from pump and injectors
D Return to tank

waxing additives down to -24ºC. Some other manufacturers use a similar system on vehicles destined for cold climates.

4 Fuel injection pumps - general

1 The function of an injection pump is to supply fuel to the injectors at the correct pressure, at the correct moment in the combustion cycle, and for the length of time necessary to ensure efficient combustion.

2 Pump characteristics are carefully matched to a particular engine. This matching extends to the delivery valves, injector pipes and the injectors themselves. Substitution of any of these items with those from another engine or pump will not necessarily be successful, even if the substitute items have the same external dimensions.

3 The in-line pump was developed by Robert Bosch in the 1920s, and is still in use (with some refinements, but basically unchanged) to this day. On small passenger car engines, it has been largely superseded by the distributor pump, but it may still be found on some light commercial vehicles.

4 The distributor pump has a single high-pressure pump element. As the name implies, fuel under pressure is distributed to each injector in turn. The pump contains its own governor mechanism and lift pump.

Governors

5 All diesel fuel injection pumps have a governor, either mounted externally (in-line

pump) or built into the pump housing (distributor pump). The governor on a road-going vehicle engine always regulates idle speed and maximum speed.

6 Regulation is necessary because the fuel delivery from an ungoverned fuel injection pump varies more or less directly with engine speed. Engine speed, on the other hand, varies according to load and fuel delivery. At idle, an ungoverned engine would either slow down, so reducing fuel delivery, and stall; or it would speed up, increasing fuel delivery, and carry on doing so until a mechanical failure occurred.

7 Maximum speed regulation prevents the engine from being damaged by over-revving if its load is reduced or suddenly removed.

8 Some types of governor also control speeds in between idle and maximum. These are known as *variable-speed* or *all-speed* governors.

9 Additional features found on some governors include the control of starting fuel delivery (which may exceed normal full-load delivery) and varying full-load delivery in proportion to engine speed. This last feature is known as *torque control*.

10 The various governor types are described in more detail in the Sections dealing with the pumps to which they are fitted.

Torque control

11 Torque control is the name given to systems which vary full-load fuel delivery according to engine speed, in order to match more closely the optimum fuel requirement.

12 The fuel requirement of a non-turbocharged engine reduces in its upper

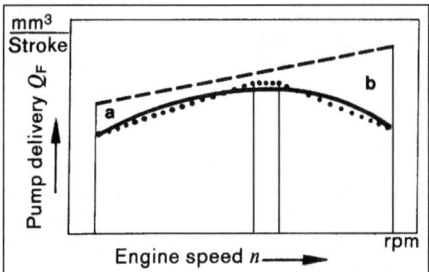

4.16 Injection pump delivery with and without torque control – curve shows actual fuel requirement

© Robert Bosch Limited

a   *Negative torque control*
b   *Positive torque control*
—   *Pump delivery without torque control*
... *Pump delivery with torque control*

speed ranges, but fuel delivery from an injection system without torque control increases with speed.

13 If the pump is set to deliver the optimum amount of fuel at low speeds, too much will be delivered at high speeds, and the engine

will smoke or overheat. On the other hand, if the pump is set only to deliver the optimum amount of fuel for high speeds, less fuel will be delivered at low speeds than could be burnt, and the engine will not produce as much torque as it could.

14 To correct this situation, a system is needed which will reduce fuel delivery with increasing speed. This is known as *positive torque control*.

15 With turbocharged engines, the opposite situation can arise: at higher speeds, the fuel requirement rises faster than the pump output. In this case, fuel delivery must be increased with speed, and this is known as *negative torque control*.

16 In practise, an engine may need to be the subject of both positive and negative torque control in order to produce its optimum output over all speed ranges **(see illustration)**.

17 Positive torque control can take place in the governor, or it can be a feature of pump delivery valve design. Negative torque control can only take place in certain types of governor. Both systems are described in more detail in the Sections dealing with the pumps to which they are fitted.

5 In-line injection pumps and governors

1 The function of a fuel injection pump is to supply fuel to the injectors at the correct pressure, at the correct moment in the combustion cycle, and for the length of time necessary to ensure efficient combustion.

2 Pump characteristics are carefully matched to a particular engine. This matching extends to the delivery valves, injector pipes and the injectors themselves. Substitution of any of these items with those from another engine or pump will not necessarily be successful, even if the substitute items have the same external dimensions.

BOSCH PUMPS AND GOVERNORS

Bosch PE pump

Basic principles

3 The in-line pump is a series of separate single-plunger pumps (one per engine cylinder) contained in one housing, and with a shared control system **(see illustration)**. A camshaft inside the pump is driven at half-engine speed by timing gears, a chain or a toothed belt. The cams operate the injection pump plungers via roller tappets. Spring-loading between the plungers and tappets keeps the latter in contact with the cams.

4 Each pump plunger works in a barrel **(see illustration)**. Each barrel has at least one fuel delivery port. At the top of the barrel is a fuel delivery valve. The delivery valve is connected to a fuel pipe, the other end of which is connected to the relevant fuel injector.

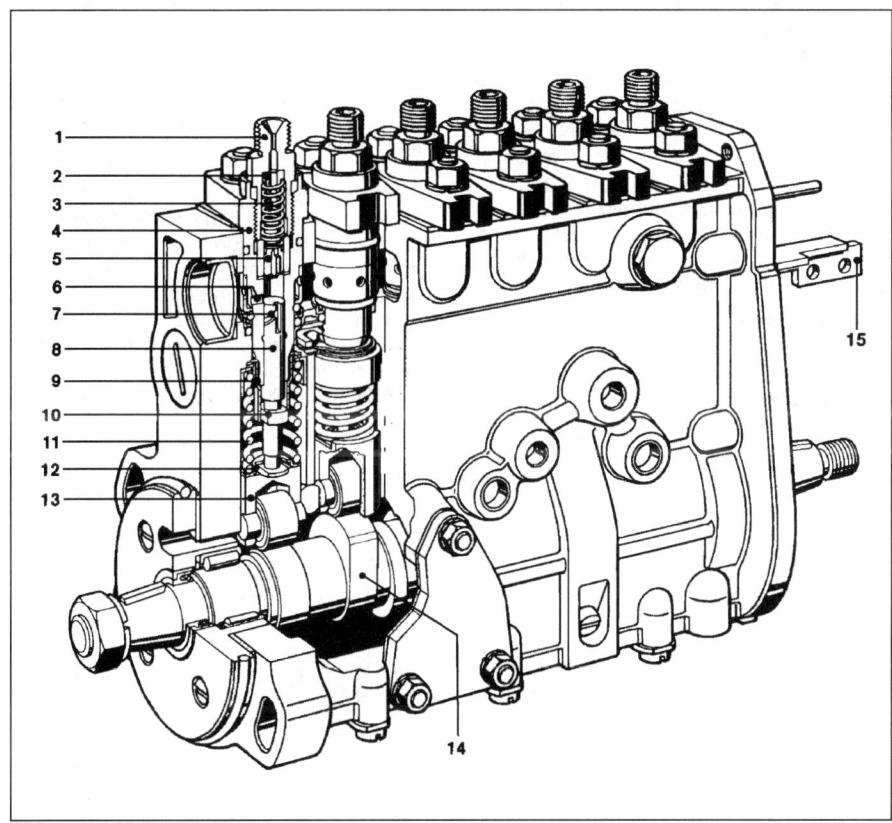

5.3 Cutaway view of a Bosch PES in-line fuel injection pump

© Robert Bosch Limited

1 Delivery valve holder	6 Inlet/spill port
2 Spacer	7 Helix
3 Spring	8 Pump plunger
4 Pump barrel	9 Control sleeve
5 Delivery valve	10 Control arm

11 Plunger return spring	
12 Spring seat	
13 Roller tappet	
14 Cam	
15 Control rod	

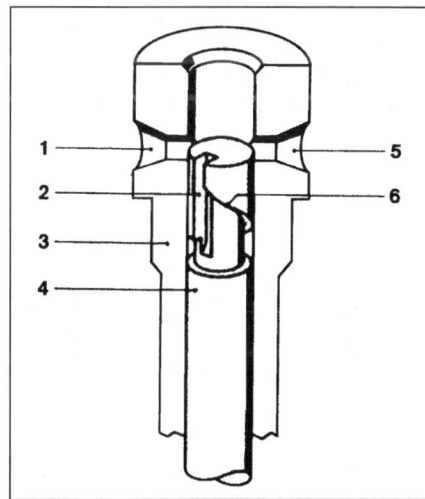

5.4 Bosch PE pump plunger and two-port barrel

© Robert Bosch Limited

1 Inlet port		4 Plunger	
2 Vertical groove		5 Inlet/spill port	
3 Barrel		6 Helix	

2

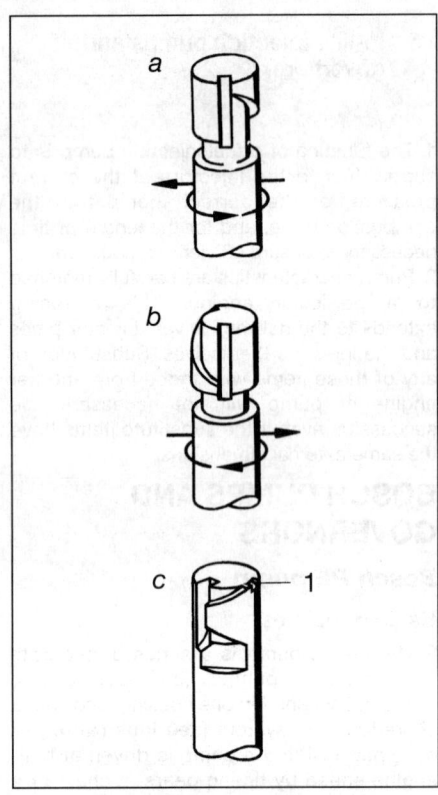

5.10 Bosch PE pump plunger versions

© Robert Bosch Limited

a   *Lower helix*
b   *Upper helix*
c   *Upper and lower helix*
1   *Starting groove*

Downward movement of the plunger allows the barrel to fill with fuel. Upward movement forces fuel under pressure out of the delivery valve and so into the injector.

5 The pump cams are arranged so that each plunger delivers fuel to the appropriate injector in the engine firing order. The cam profiles are also important in determining the point at which injection starts, and its duration.

6 Lubrication of the camshaft, bearings and roller tappets is by connection to the engine's lubrication system. Lubrication of the high-pressure components is carried out by the fuel.

Plunger and barrel construction

7 The plunger and barrel are a matched assembly, made to very fine tolerances. Some leakage of fuel takes place between them, but this is an intentional feature of the design, and is necessary for lubrication. (This leaked fuel is returned to the suction side of the pump by a drilling in the barrel, so that it does not enter the engine oil and dilute it.) The barrel has one or two ports for fuel delivery and return (spill).

8 The bottom of the plunger is connected to a control sleeve. This sleeve can rotate the plunger while still allowing it to go up and down. All the control sleeves in the pump are connected to one control rod.

9 The plunger has a vertical groove machined on one side and extending to the top. In the simplest design, a single helical groove meets the vertical groove. When the plunger is rotated in the barrel, this changes the point in the plunger stroke at which the helical groove covers and uncovers the port.

10 Practical requirements in areas such as noise reduction, exhaust emission control and cold start characteristics mean that plungers typically have more that the two grooves just described, but the basic principle remains the same: by rotating the plunger in the barrel, both the amount of fuel injected and the moment at which injection occurs can be altered **(see illustration)**.

The pump working cycle

11 The pump has a defined working cycle. The phases of the cycle are clearly shown **(see illustration)**.

12 The first phase of the cycle is the *fuel inlet*. With the plunger at BDC, both inlet ports are open, and fuel flows from the pump suction gallery into the barrel.

13 As the plunger rises in the barrel, the phase known as *prestroke* begins. Prestroke lasts until the plunger has closed off both inlet ports.

14 The plunger continues to rise, and the *retraction stroke* begins. During this phase, the fuel pressure is rising, but the delivery valve has not opened because the pressure is not great enough.

15 As soon as the delivery valve opens, the *effective stroke* begins. Movement of the plunger now forces fuel up the pipe to the injector, and fuel injection takes place. The amount of fuel delivered depends on the length of the effective stroke.

16 The end of the effective stroke occurs before TDC, when the helix uncovers an inlet port. The fuel under pressure spills out of the port, the delivery valve closes, and injection stops. The plunger carries on to TDC, and this is known as the *residual stroke*.

17 The plunger now returns to BDC, and the cycle begins again. On the way down, the plunger again covers both inlet ports, and during this phase of operation, a vacuum will be produced in the barrel until the inlet ports are uncovered and fuel can flow in.

Delivery valves

18 The delivery valve separates the pump barrel from the injector pipe. The valve only opens when the pressure in the barrel exceeds a certain value, and closes as soon as pressure begins to drop. This contributes to precise beginning and ending of injection.

19 In its simplest form, the delivery valve consists of a spring-loaded plunger in a holder **(see illustration)**. The size and shape of the plunger are matched to the length and diameter of the injector pipe. When the valve closes, the retraction of the plunger increases the effective volume of the fuel pipe by a

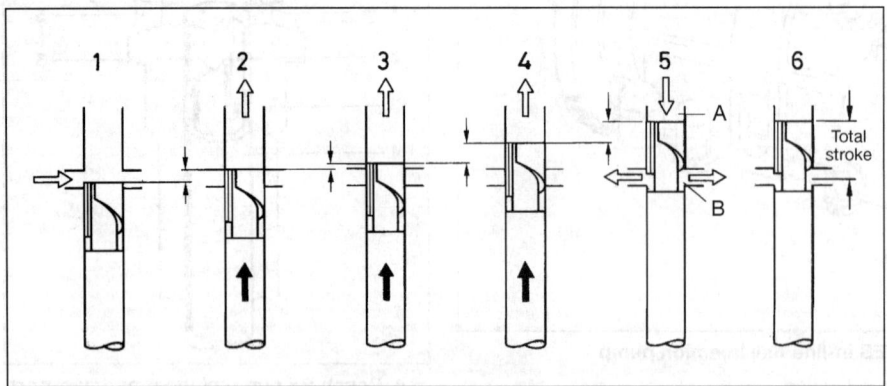

5.11 Bosch PE pump working cycle

© Robert Bosch Limited

1	BDC (fuel inlet)	4	Effective stroke	A	Plunger chamber
2	Prestroke	5	Residual stroke	B	Suction gallery
3	Retraction stroke	6	TDC (end of residual stroke)		

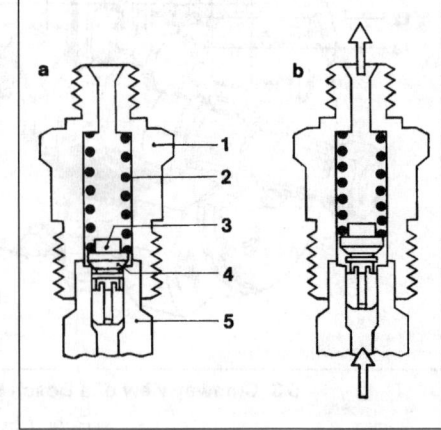

5.19 Delivery valve closed (a) and open (b)

© Robert Bosch Limited

1	Valve body	4	Valve seat
2	Spring	5	Valve holder
3	Plunger		

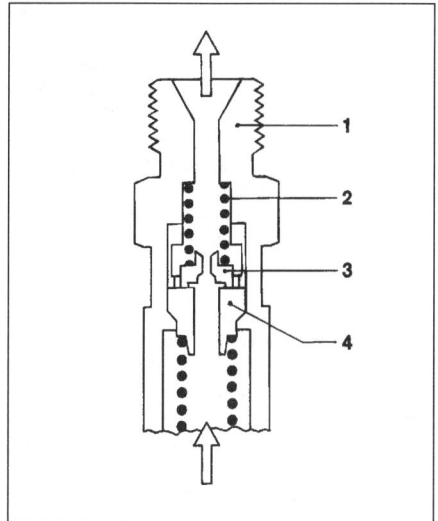

5.21 Delivery valve with return-flow restrictor

© Robert Bosch Limited

1   Valve body
2   Spring
3   Plate with return-flow restrictor
4   Valve holder

calculated amount. This relieves the pressure in the pipe and makes the injector shut cleanly.

20 The delivery valve can also be part of a torque control system. (For an explanation of torque control, see Section 4.) This is done by profiling the plunger so that it offers a greater resistance to fuel flow at higher velocities.

21 A further refinement, known as *return-flow restriction*, allows the delivery valve to play an important part in reducing turbulence and cavitation in the injector pipe. The return-flow restrictor valve sits on top of the delivery valve. It offers little resistance to fuel flow in the forward direction, but damps out the reverse-pulse which occurs when the injector needle snaps shut **(see illustration)**.

Pump control

22 As mentioned earlier, the amount of fuel delivered is controlled by rotating the plunger in the barrel, so varying its effective stroke. The plunger is turned by a control sleeve. All the control sleeves are connected to one control rod, which is in turn connected to the governor. The rod may be connected to the sleeves by gear teeth (rack and pinion), or by slots and levers **(see illustration)**.

23 If the plunger is turned so that its vertical groove is in line with a delivery port, the port will be exposed for the whole cycle, and no fuel delivery will take place. This position is selected when it is desired to stop the engine.

24 An adjustable stop may be fitted to the control rod, to limit its travel in the 'increased-delivery' direction, so limiting the maximum fuel delivery of the pump. Some engines require more fuel at start-up than at full-load,

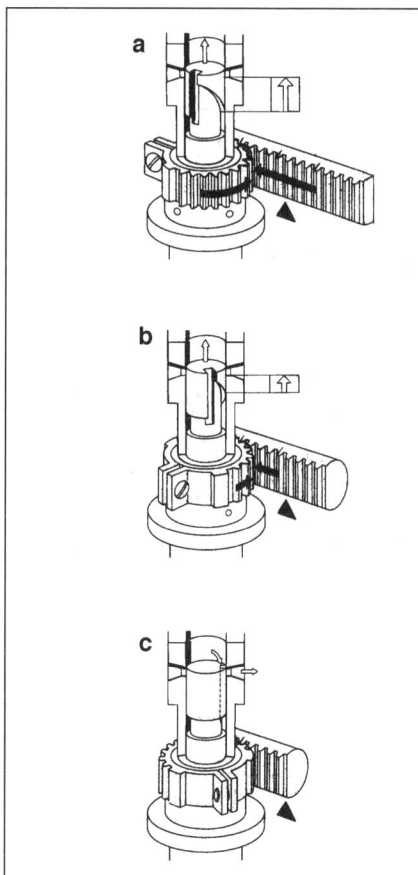

5.22 Variation of fuel delivery

© Robert Bosch Limited

a   Maximum delivery
b   Partial delivery
c   Zero delivery

and in this case, the stop is temporarily bypassed.

Governor types

25 A centrifugal minimum-maximum type governor is usually fitted to this type of pump, though occasionally a pneumatic governor will be found. These are described later in this Section.

Timing mechanisms

26 A centrifugal advance device is incorporated in the pump drive, in order to advance injection timing with increased speed **(see illustration)**.

Identification and detail differences

27 Pumps prefixed PE have a flat base or cradle mounting while those prefixed PES have an end flange mounting. Most pumps found on small and middle-sized engines will be type M (the smallest) or type A. Type MW and P pumps are of heavier construction, and capable of higher maximum pressures, but the principles of operation are the same.

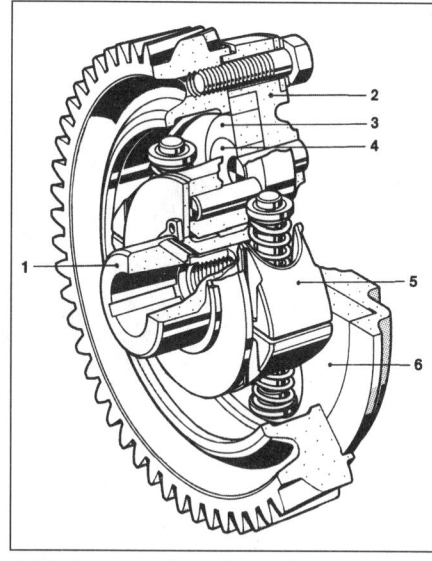

5.26 Cutaway view of centrifugal advance coupling

© Robert Bosch Limited

1   Hub
2   Housing
3   Adjusting eccentric
4   Compensating eccentric
5   Flyweight
6   Adjusting disc

Governors fitted to Bosch PE pumps

Introduction

28 One of the following types of governor will be found fitted to every Bosch PE pump. The governors divide into two basic types: centrifugal and pneumatic. The centrifugal governors are all further subdivided according to their spring arrangements (incorporated into the weight assemblies, or remote), and according to their speed regulation range (idle/maximum or variable).

Maintenance

29 Governor lubrication is almost always integral with that of the injection pump. When a separate oil filler/level plug or dipstick is provided, the oil level should be checked every 1000 miles (1500 km) or 50 hours. The oil level may rise due to fuel dilution, and in this case the excess must be drained off when checking. Every 10 000 miles (15 000 km) or 500 hours, the oil should be renewed.

RQ centrifugal governor

30 The RQ governor is a centrifugal idle/maximum speed governor. It is fitted to pump types A and P.

31 The governor is mounted on the end of the pump. The governor hub is driven, via a vibration damper, from the pump camshaft. The hub carries two flyweights, each weight with its own spring set. Bellcranks convert the radial movement of the weights into axial movement of a slider. Movement of the slider

2

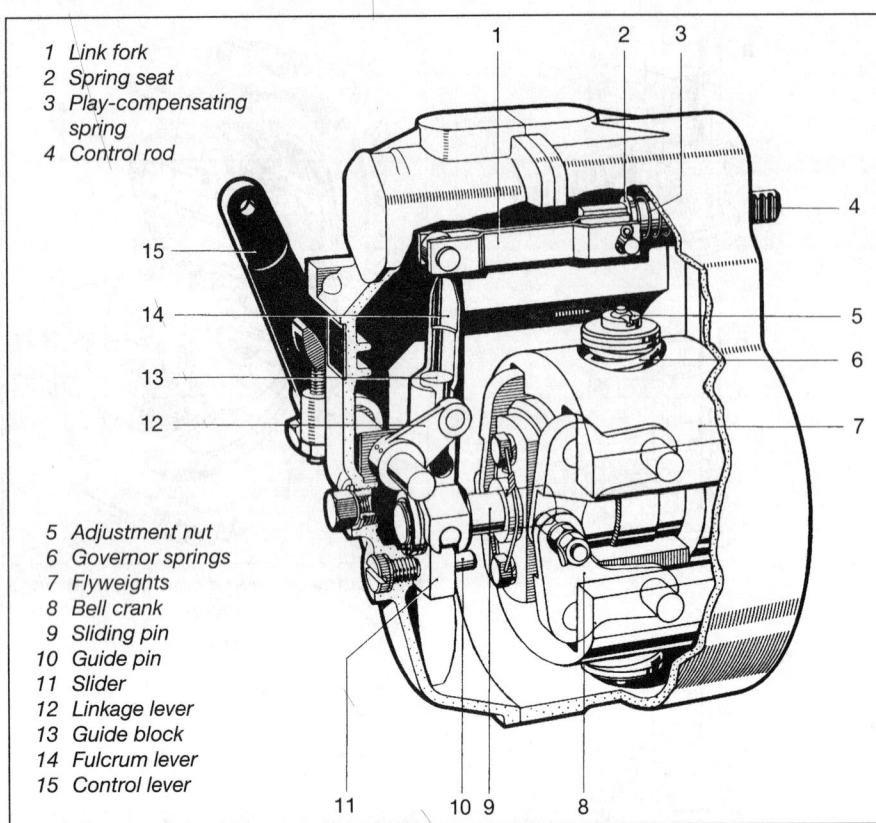

1 Link fork
2 Spring seat
3 Play-compensating spring
4 Control rod

5 Adjustment nut
6 Governor springs
7 Flyweights
8 Bell crank
9 Sliding pin
10 Guide pin
11 Slider
12 Linkage lever
13 Guide block
14 Fulcrum lever
15 Control lever

5.31a Cutaway view of the RQ centrifugal governor

© Robert Bosch Limited

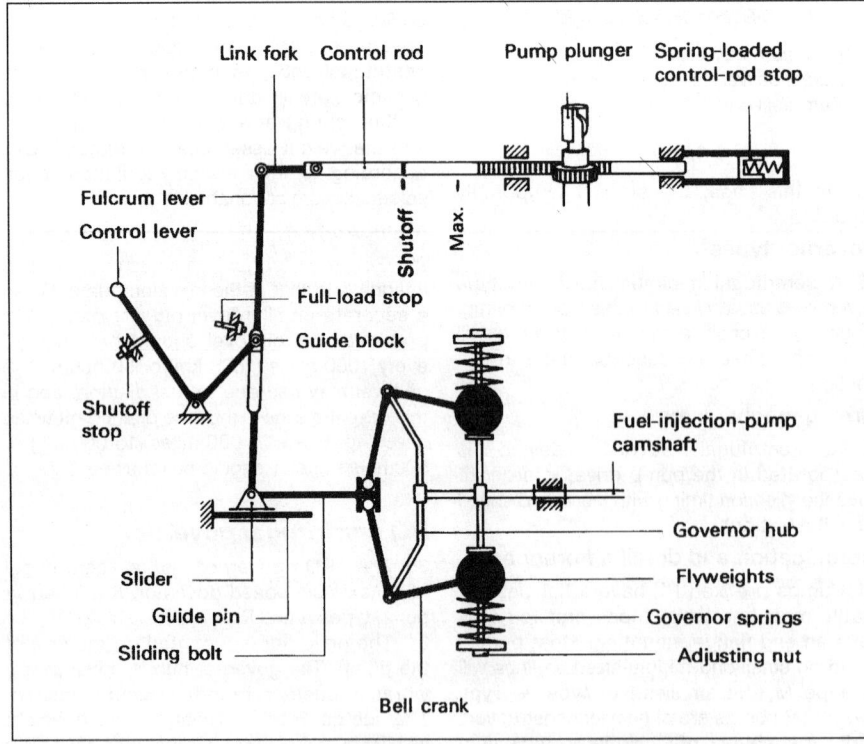

5.31b RQ centrifugal governor schematic diagram

© Robert Bosch Limited

is transmitted to the pump control rod via a fulcrum lever **(see illustrations)**.

32 The engine speed control lever is also connected to the fulcrum lever, via a guide block which slides up and down inside a channel in the lever. This sliding linkage alters the effective pivot point of the fulcrum lever for the governor, and ensures that even at low speeds, the centrifugal force of the weights is sufficient to move the control rod **(see illustration)**.

33 The spring set in each weight is made up of three springs arranged concentrically. The outer spring bears directly on the flyweight, while the inner springs bear on an inner seat. The outer ends of all the springs bear on an outer seat, the position of which is adjustable by a nut.

34 At low speeds, only the outer spring is effective, the force of the other two being taken on the inner seat. When the outer spring is compressed far enough by the weight, the weight bears on the inner seat. This marks the end of idle speed control.

35 Further increase in speed increases the force of the flyweight until it is great enough to compress all three springs. At this point, which marks the beginning of maximum-speed control, the inner seat moves with the weight **(see illustration)**.

Start-up

36 The accelerator pedal is fully depressed, which moves the control lever to the full-load position. The pump control rod overcomes the resistance of the spring-loaded full-load stop, and moves into the position for starting fuel delivery **(see illustration)**.

37 When the engine begins to operate, the driver releases the accelerator, and idle regulation begins.

Idling

38 The control lever rests against its idle stop. If engine speed reduces, the governor weights move inwards, and the control rod is moved in the direction of increased fuel delivery **(see illustration)**. If engine speed rises, the reverse occurs.

On the road

39 As soon as the accelerator pedal is depressed and the engine speed rises above idle, the governor weights compress the idle speed control springs fully and rest on the

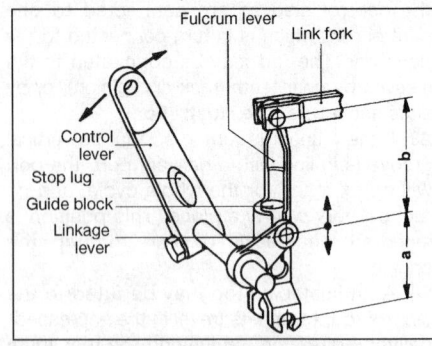

5.32 Variable fulcrum point in RQ governor

© Robert Bosch Limited

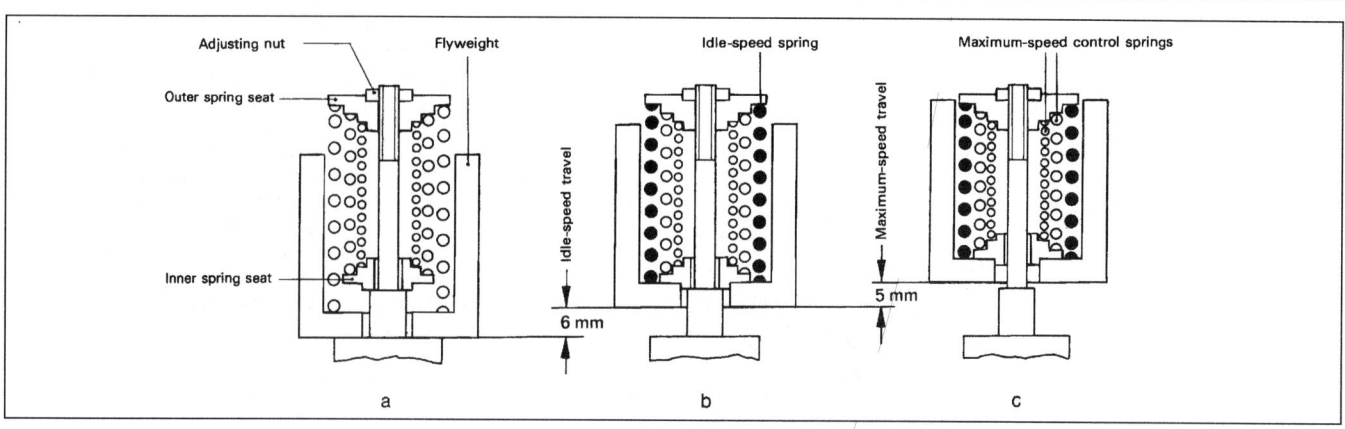

5.35 RQ governor flyweights and springs at various speeds

© Robert Bosch Limited

a Stationary   b Idling   c Maximum speed

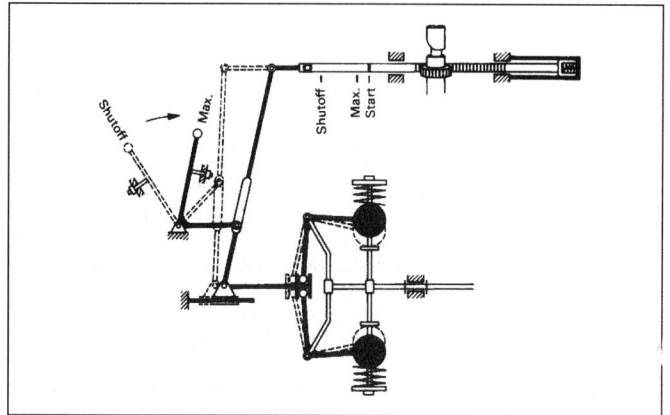

5.36 RQ governor in "start" position

© Robert Bosch Limited

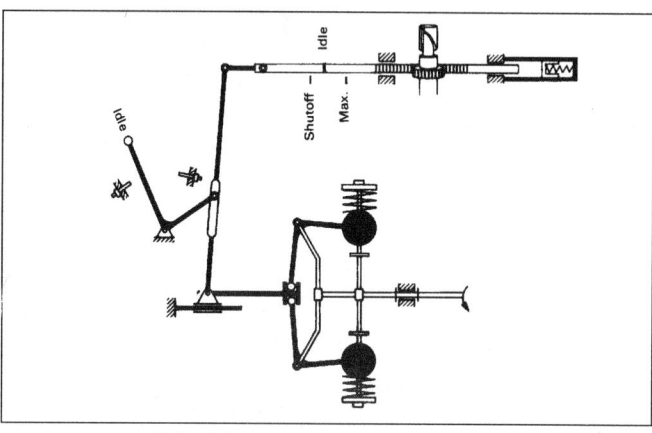

5.38 RQ governor in "idle" position

© Robert Bosch Limited

inner spring seats. The position of the weights does not alter until maximum speed is approached. At intermediate speeds, fuel delivery is determined by the position of the accelerator.

Maximum speed

40 When the flyweights begin to compress the maximum-speed control springs, the fulcrum lever moves the control rod to reduce fuel delivery **(see illustration)**. The position of the control rod is thus dependent both on the governor and on the position of the accelerator. If engine speed continues to rise, the effect of the governor exceeds that of the accelerator, to the point where fuel delivery may be cut off altogether.

Torque control

41 Positive torque control is achieved by introducing an extra spring in series with the maximum-speed control springs **(see illustration)**. The travel which this spring

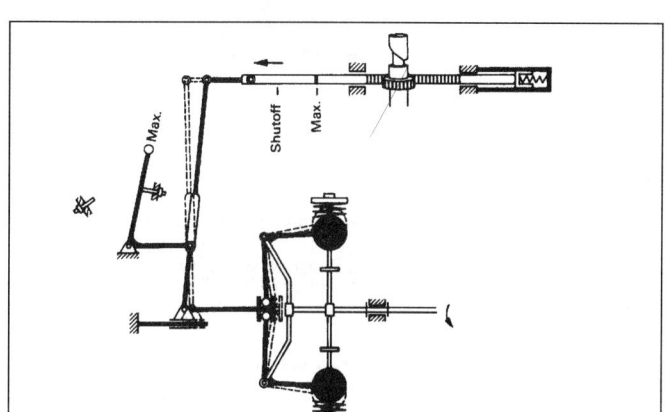

5.40 RQ governor in maximum-speed/maximum-load position

© Robert Bosch Limited

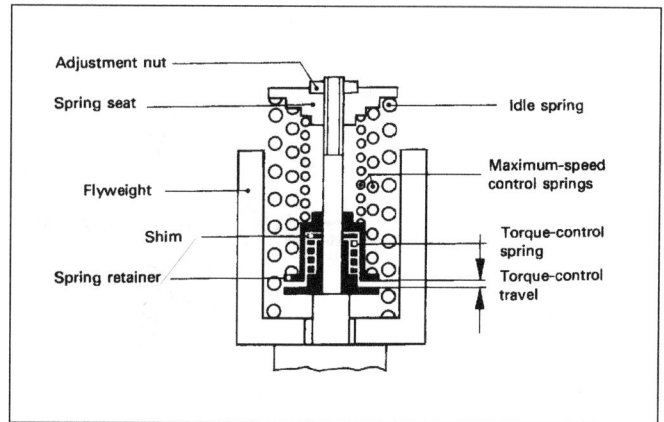

5.41 Torque control springs in the RQ governor

© Robert Bosch Limited

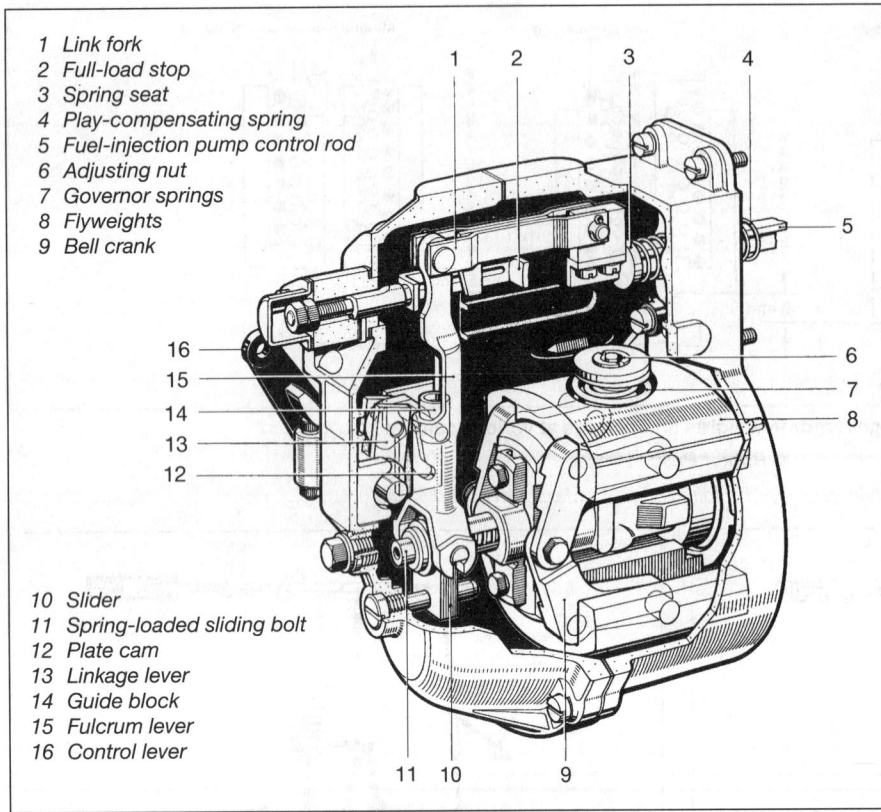

1 Link fork
2 Full-load stop
3 Spring seat
4 Play-compensating spring
5 Fuel-injection pump control rod
6 Adjusting nut
7 Governor springs
8 Flyweights
9 Bell crank

10 Slider
11 Spring-loaded sliding bolt
12 Plate cam
13 Linkage lever
14 Guide block
15 Fulcrum lever
16 Control lever

5.42 Cutaway view of Bosch RQV governor

© Robert Bosch Limited

allows can be varied with shims. The weights begin to compress the torque control springs at some speed in between idle and maximum speed. Fuel delivery is then reduced with increasing speed, until the torque control travel is taken up.

RQV centrifugal governor

42 The RQV governor closely resembles the type RQ just described, but it also incorporates a variable-speed control function **(see illustrations)**. (A variant known as a *combination governor* has control over a part of the intermediate speed range, but has an uncontrolled speed range as well.) It is fitted to pump types A and P.

43 The flyweight springs on this governor type are graded so that there is continuous movement of the weights over the entire speed range, instead of just around idle and maximum speed.

44 The connection between the governor and the fulcrum lever is spring-loaded by a spring known as a *drag spring* **(see illustration)**.

Start-up

45 The control lever is moved to the full-load position (accelerator pedal fully depressed). The pump control rod moves past the automatic full-load stop, which is unlatched, and into the position for starting fuel delivery.

Idling

46 The governor functions in the same way as the RQ governor.

On the road

47 For each position of the accelerator pedal, there is an engine speed at which regulation begins. When the pedal is initially depressed, the variable ratio of the fulcrum lever causes the pump control lever to take up the 'maximum-delivery' position, even if full throttle has not been applied. Further depression of the accelerator will tension the drag spring.

48 As the engine speed rises, the flyweights move outwards, but they have no effect on the control rod until the tension in the drag spring has been released. Further movement of the weights then results in decreased fuel delivery.

49 Increasing load will tend to reduce engine speed, and the flyweights will move inwards. At first, this movement will result in control rod movement to increase fuel delivery. When the control rod full-load stop is reached, further movement of the weights inwards will tension the drag spring **(see illustration)**.

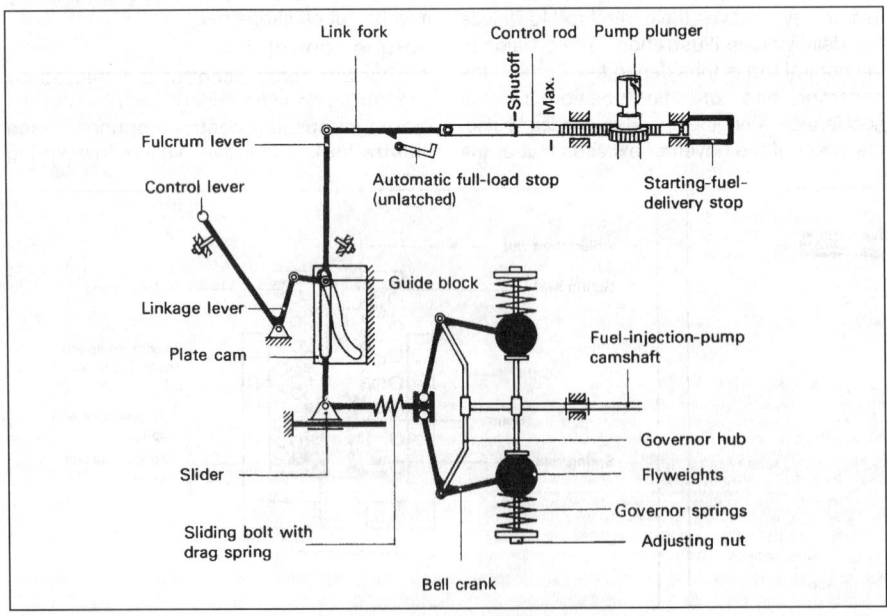

5.44 RQV governor schematic diagram

© Robert Bosch Limited

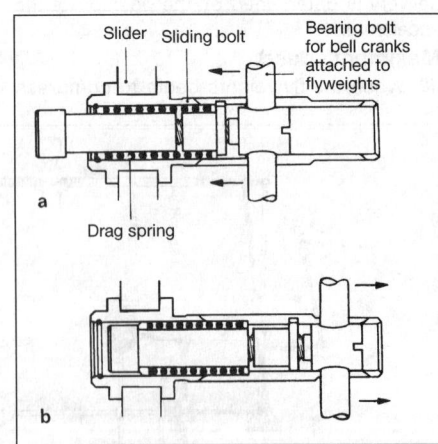

5.49 Drag spring and sliding bolt – RQV governor

© Robert Bosch Limited

a   Acceleration (control rod at full-load stop)
b   Maximum speed on overrun (control rod at shut-off stop)

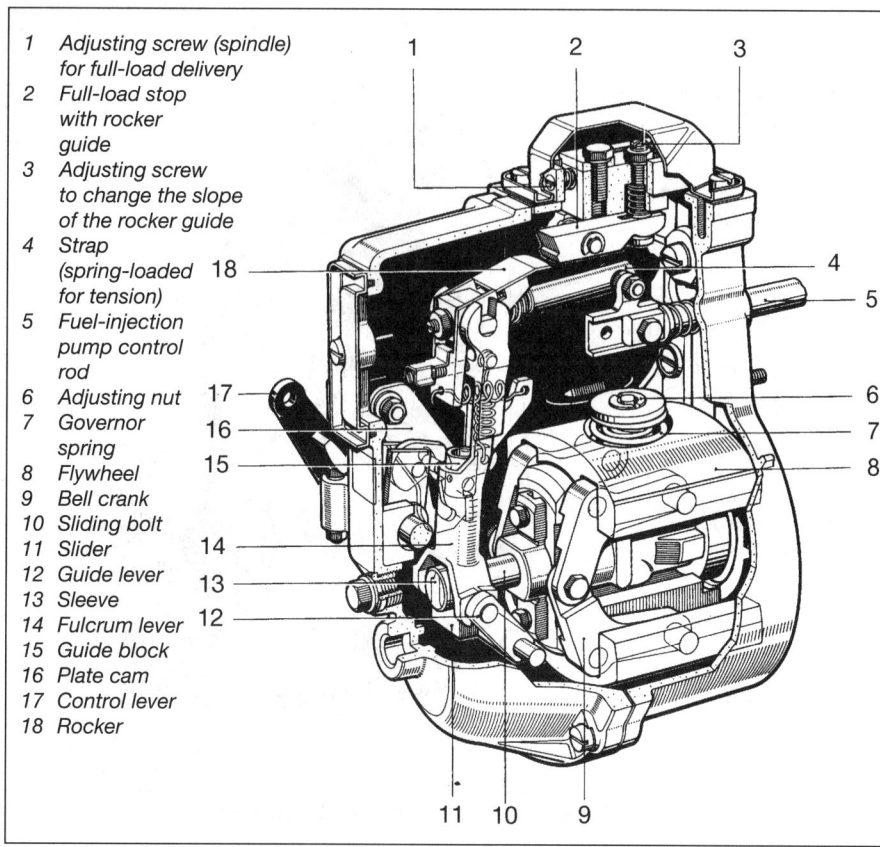

1. Adjusting screw (spindle) for full-load delivery
2. Full-load stop with rocker guide
3. Adjusting screw to change the slope of the rocker guide
4. Strap (spring-loaded for tension)
5. Fuel-injection pump control rod
6. Adjusting nut
7. Governor spring
8. Flywheel
9. Bell crank
10. Sliding bolt
11. Slider
12. Guide lever
13. Sleeve
14. Fulcrum lever
15. Guide block
16. Plate cam
17. Control lever
18. Rocker

5.52 Cutaway view of Bosch RQV-K governor

© Robert Bosch Limited

50 Reducing load and increasing speed will move the flyweights outwards, reducing fuel delivery. Once the control rod reaches the shut-off position, further movement of the weights will compress the drag spring.

Torque control

51 Positive torque control is applied using the drag spring and a spring-loaded control rod stop.

RQV-K centrifugal governor

52 The RQV-K governor has the same type of weight and spring assembly as the type RQV. It differs from the RQV (and from all other in-line pump governors) in the system used for torque control, and in the fact that negative as well as positive torque control can be applied **(see illustration)**. It is fitted to pump types A and P.

53 The fulcrum lever is connected to the pump control rod by a spring-loaded strap. The fulcrum lever and strap meet at a rocker which runs along a curved track (rocker guide) at the full-load stop **(see illustration)**.

54 The fulcrum lever guide block runs in a track on a plate cam. The cam is pivoted at one corner, and attached to a return spring.

Start-up

55 The control lever is moved to the maximum-speed position (accelerator pedal fully depressed). The rocker moves under the full-load stop, and the pump control rod is moved to the starting delivery position.

Idling

56 The governor functions in the same way as the RQ governor.

On the road

57 As with the RQV governor, for each position of the accelerator pedal, there is an engine speed at which regulation begins. Speed regulation is still carried out by the flyweights, but the full-load delivery is affected by the torque control systems.

Torque control

58 If the accelerator pedal is fully depressed at low speed, the fulcrum lever guide block moves downwards in the plate cam track and in the fulcrum lever channel **(see illustration)**. The plate cam is lifted off its stop against the

2

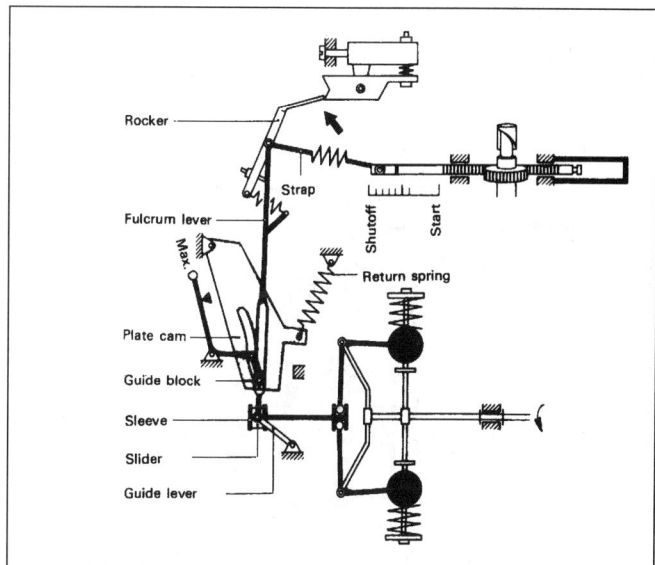

5.53 RQV-K governor schematic diagram

© Robert Bosch Limited

5.58 RQV-K governor – full-load, low-speed position

© Robert Bosch Limited

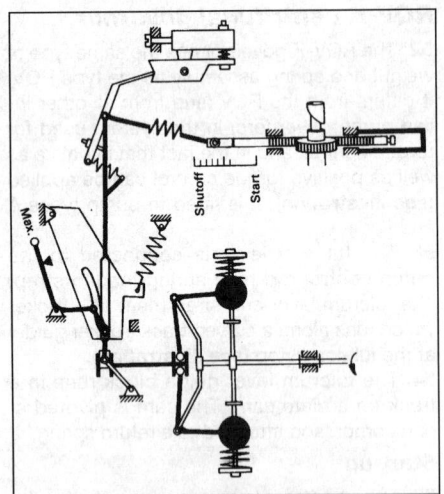

5.59 RQV-K governor – full-load, medium-speed position

© Robert Bosch Limited

1 Governor housing
2 Starting spring
3 Control lever
4 Governor cover
5 Shutoff or idle stop
6 Tensioning lever
7 Guide lever
8 Governor spring
9 Auxiliary idle-speed spring
10 Torque-control and idle speed helical compression spring
11 Full-load stop (fuel delivery)
12 Fulcrum lever
13 Guide busing
14 Flyweight
15 Swivelling lever
16 Rocker
17 Stop
18 Control rod

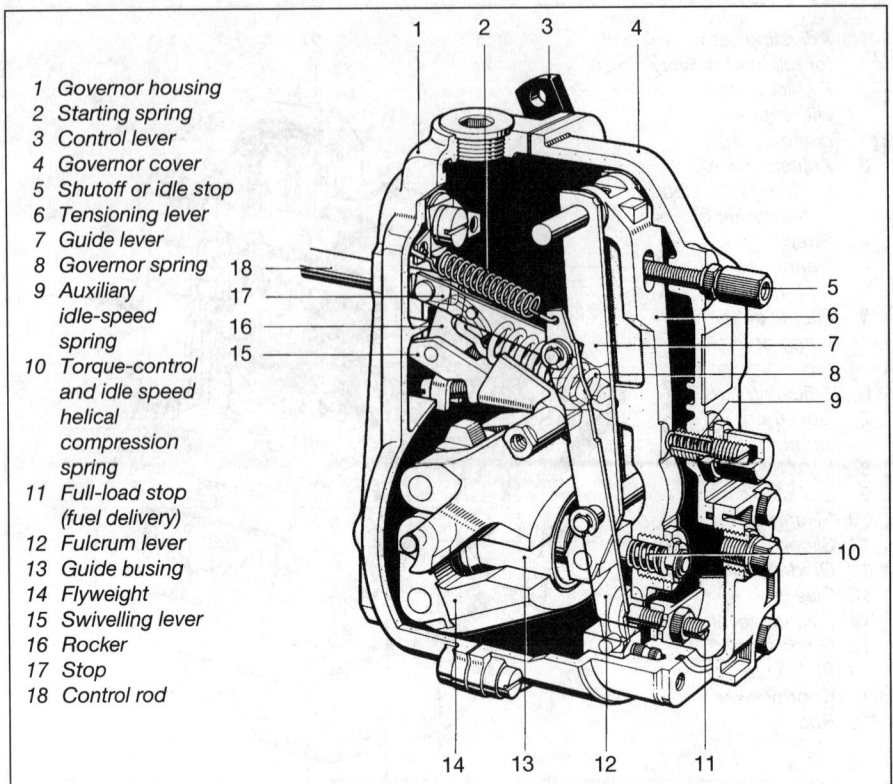

5.61a Cutaway view of Bosch EP/RSV governor

© Robert Bosch Limited

effort of the return spring. The fulcrum lever moves the control rod into the maximum-delivery position. This position is determined by the profile of the rocker guide; at low speed, the rocker arm contacts the bottom of the guide.

59 As speed rises, flyweight movement is transmitted to the fulcrum lever. The guide lever and the fulcrum lever are both raised, so the rocker arm moves further up the guide. With the 'fishtail' shape guide shown in the illustrations, movement from the bottom of the guide towards the middle section increases fuel delivery (*negative torque control*) **(see illustration)**.

60 Further increase in speed takes the rocker arm up onto the reverse slope of the fishtail. Maximum delivery is reduced with increasing speed (*positive torque control*).

EP/RSV centrifugal governor

61 The EP/RSV governor is a variable-speed type. Unlike the RQ series governors, it has only one governor spring, which is separate from the flyweights. The governor spring is attached to a lever known as the *tensioning lever*. Additional springs are used for starting, idling, and torque control **(see illustrations)**.

Start-up

62 With the engine stopped, the starting spring acts unopposed on the fulcrum lever, which moves the pump control rod into the position for starting fuel delivery **(see illustration)**. This happens without the accelerator pedal being depressed.

63 As soon as the engine turns, the flyweights begin to swing out. The sliding bolt

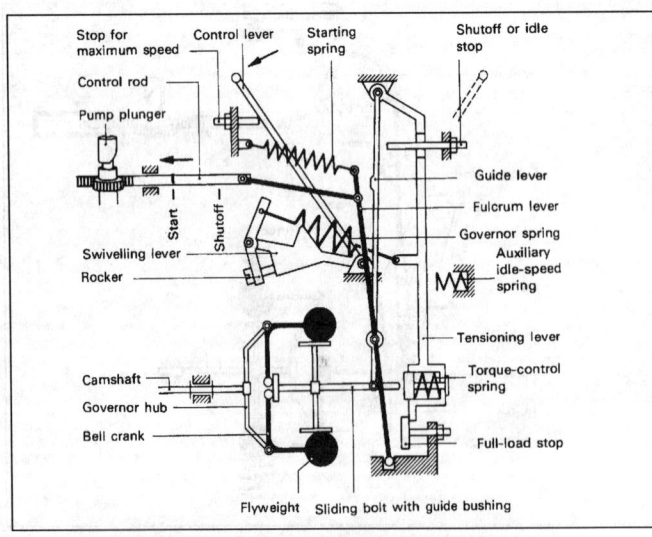

5.61b EP/RSV governor schematic diagram

© Robert Bosch Limited

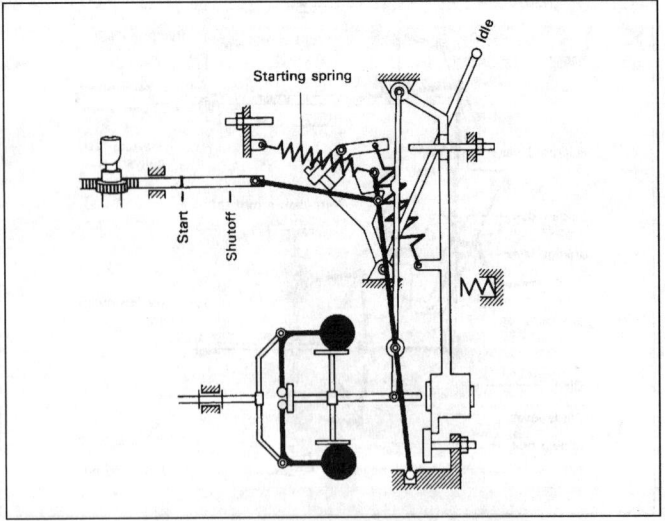

5.62 EP/RSV governor in starting position

© Robert Bosch Limited

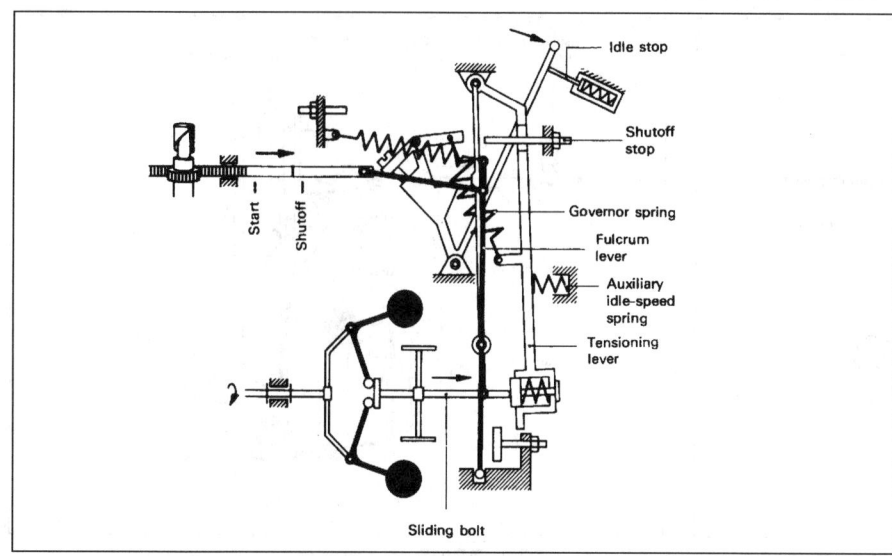

5.65 EP/RSV governor in idling position

© Robert Bosch Limited

moves the guide lever (and through it, the fulcrum lever), overcoming the force of the starting spring, and the pump control rod is moved into the idling position.

Idling

64 With the speed control lever on the idle stop, the governor spring is almost slack. This lack of tension allows the flyweights to move outwards, even at low speed. Outward movement of the weights with increased speed moves the guide lever and fulcrum lever in the direction to reduce fuel delivery, and *vice-versa*.

65 In this phase of operation, the tensioning lever is resting on the auxiliary idle spring. The auxiliary spring acts against the force transmitted by the flyweights, and gives greater smoothness of control **(see illustration)**.

On the road

66 When the accelerator pedal is depressed, the speed control lever moves off the idle stop. Movement of the swivelling lever by the control lever increases the tension of the governor spring. This increased spring force overcomes that of the flyweights. The tensioning lever is drawn against the full-load stop, and the control rod is moved to maximum delivery.

67 As engine speed rises, the force exerted by the flyweights comes back into balance with the governor spring force. The control rod is moved back towards reduced fuel delivery until the forces are in balance, and engine speed stabilises at this level.

68 If the accelerator is depressed as far as possible, the speed control lever moves up against the maximum-speed stop. The swivelling lever applies the maximum tension to the governor spring. As long as the spring force exceeds the force of the flyweights, fuel delivery is at maximum.

69 As engine speed approaches the governed maximum, the force of the flyweights once more prevails. The tensioning lever is moved away from the spring, and the fulcrum lever moves the control rod towards the shut-off position.

Torque control

70 Positive torque control is achieved using a spring in the tensioning lever. With the tensioning lever resting on the full-load stop, rising speed will cause the torque control spring to be compressed before the tensioning lever moves. Compression of the spring allows the guide lever, the fulcrum lever and the control rod to move towards reducing fuel delivery **(see illustrations)**.

Shut-off

71 Shut-off may be carried out by moving the speed control lever to a shut-off position, when lugs on the swivelling lever press on the guide lever. The guide lever moves the fulcrum lever, which brings the pump control rod into the shut-off position.

72 Sometimes a separate stop lever is used. This achieves the same result by acting on the fulcrum lever.

EP/RS centrifugal governor

73 The EP/RS governor is an idle/maximum speed unit based on the EP/RSV governor described previously. The swivelling lever which alters governor spring tension in the EP/RSV governor is locked into the maximum-speed position. Speed control is carried out by a lever acting on the fulcrum lever. The idle and torque control springs are

2

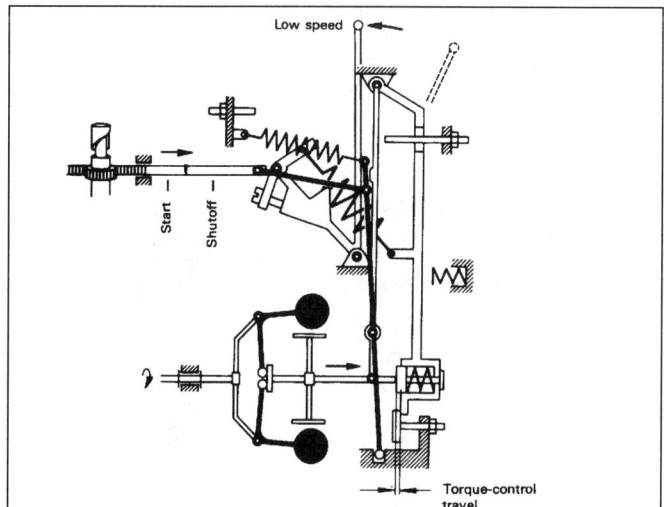

5.70a EP/RSV governor at full load and low speed; beginning of torque control

© Robert Bosch Limited

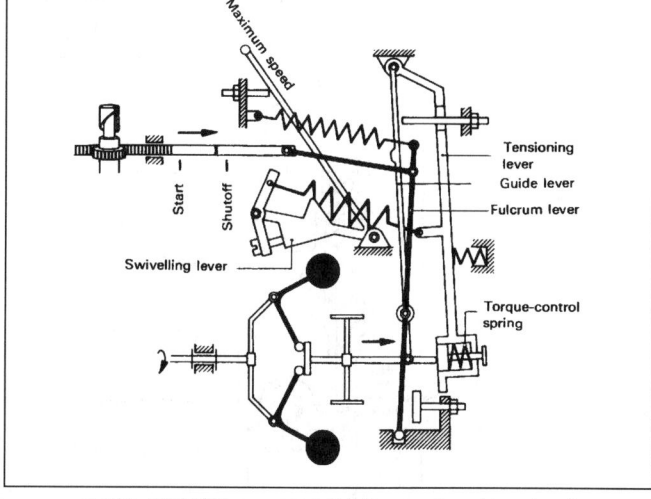

5.70b EP/RSV governor at full speed and low load

© Robert Bosch Limited

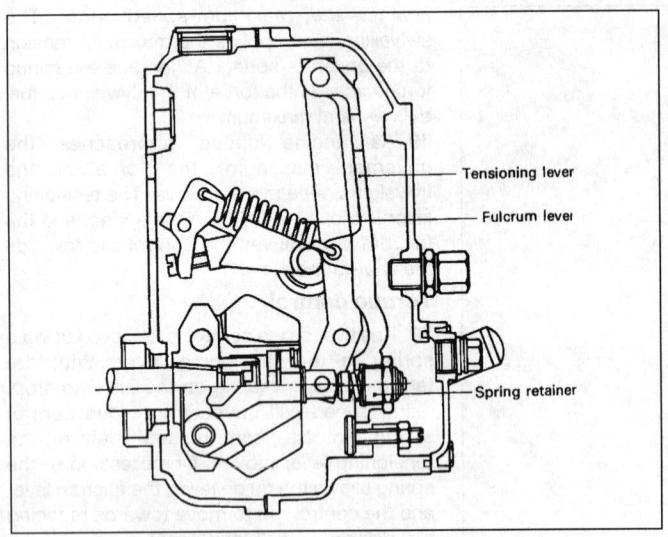

5.73a Sectional view of Bosch EP/RS governor

© Robert Bosch Limited

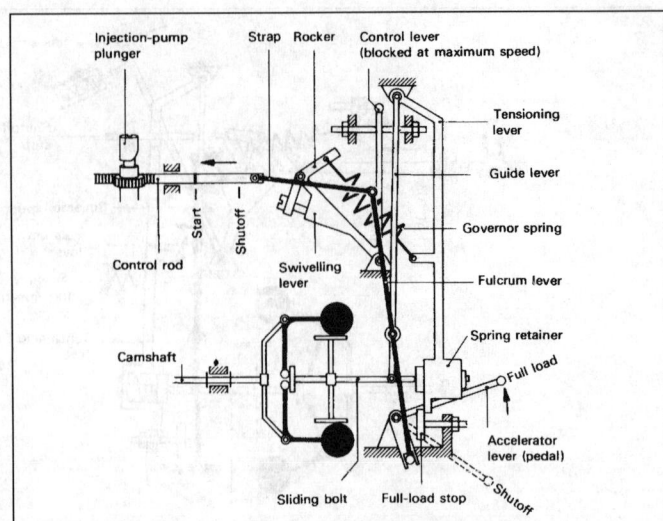

5.73b EP/RS governor schematic diagram

© Robert Bosch Limited

housed in a retainer in the tensioning lever **(see illustrations)**.

Start-up

74 The accelerator pedal must be fully depressed. Pressure of the idle spring on the sliding bolt moves the lever assembly and pump control rod to the starting fuel delivery position.

75 When the engine starts, the accelerator pedal is released. The flyweights move outwards, and the sliding bolt moves the lever assembly into the idling position.

Idling

76 The force exerted by the flyweights balances that of the idling spring. A reduction in speed reduces the flyweight force, and moves the lever assembly to increase fuel delivery, and *vice-versa*.

On the road

77 Speeds between idle and maximum are not controlled by the governor (except for torque control, as described later). The position of the accelerator pedal determines fuel delivery.

78 At maximum speed, the flyweight force exceeds the governor spring force, and the sliding bolt moves the lever assembly to reduce fuel delivery.

Torque control

79 Torque control is achieved in the same way as on the EP/RSV governor.

Centrifugal governor control lever stops

80 Stops are provided to limit control lever travel in the positions for shut-off and maximum fuel delivery. Depending on governor type and application, these stops may be fixed, variable or spring-loaded, and additional stops may be provided.

Spring-loaded idle speed stop

81 The spring-loaded idle speed stop bears on a stop lever on the end of the control lever shaft. In normal use, the spring-loading keeps the shaft in the idle position **(see illustration)**.

82 When it is wished to stop the engine, operation of the stop control overcomes the resistance of the idle speed stop spring. The control lever is moved into the stop position. When the engine has stopped and the control is released, the spring moves the lever back into the idle position.

Reduced-delivery stop

83 This stop limits control lever shaft travel towards maximum speed or maximum delivery. It may be used for setting a fixed intermediate speed.

84 The stop is spring-loaded, and is controlled by a shaft with a flat on it. When the shaft is turned, the flat raises the stop out of the way of a stop lever on the control lever shaft **(see illustration)**. Maximum speed or delivery is then available.

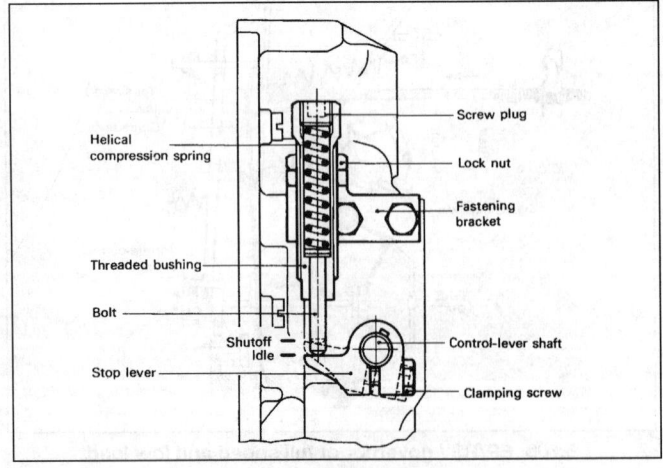

5.81 Spring-loaded idle speed stop

© Robert Bosch Limited

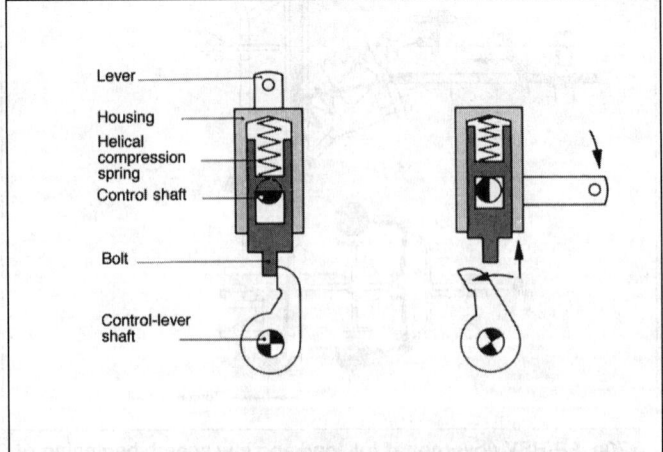

5.84 Sectional views of reduced delivery stop

© Robert Bosch Limited

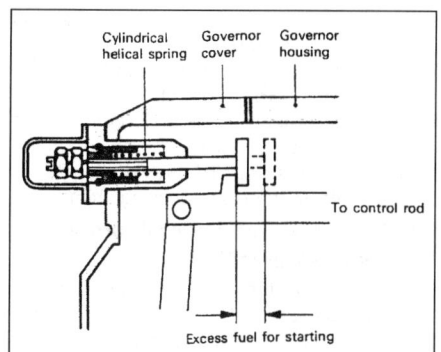

5.88 Spring-loaded excess fuel stop

© Robert Bosch Limited

Centrifugal governor control rod stops

85 Control rod stops are mainly concerned with releasing or blocking the excess fuel delivery required for start-up, and with torque control. They may also be used as part of a manifold pressure compensation system.

Rigid excess fuel stop

86 This type of stop is fitted to RQ governors. When the accelerator pedal is depressed for start-up, the stop limits the travel of the control rod to that required for starting fuel delivery. When the engine is running, the initial movement of the flyweights draws the control rod away from the stop.

87 The position of the stop is adjustable by means of a locknut and a threaded section.

Spring-loaded excess fuel stop

88 This is a development of the rigid stop. The accelerator pedal has to be depressed to overcome the spring-loading of the stop during initial start-up. As soon as the pedal is released, the spring moves the control rod back **(see illustration)**.

Automatic full-load stop

89 The components of this type of stop are a spring-loaded rocker arm and a full-load strap. When the engine is not running, pressure of the sliding bolt on the rocker arm moves the full-load strap downwards. This allows the control rod to move past the normal full-load position for start-up **(see illustration)**.

90 With the engine running, the sliding bolt moves away from the rocker arm. The rocker arm lifts the full-load strap. The stop on the strap limits control rod movement to normal full-load.

Full-load stop with external torque control (RQV governor)

91 This stop has two functions. When an external draw lever is operated, a locking bolt allows movement of the stop to the start-up fuel delivery position. The other position of the draw lever puts the stop in the normal full-load position.

92 With the engine running, the balance of forces between the torque control spring and

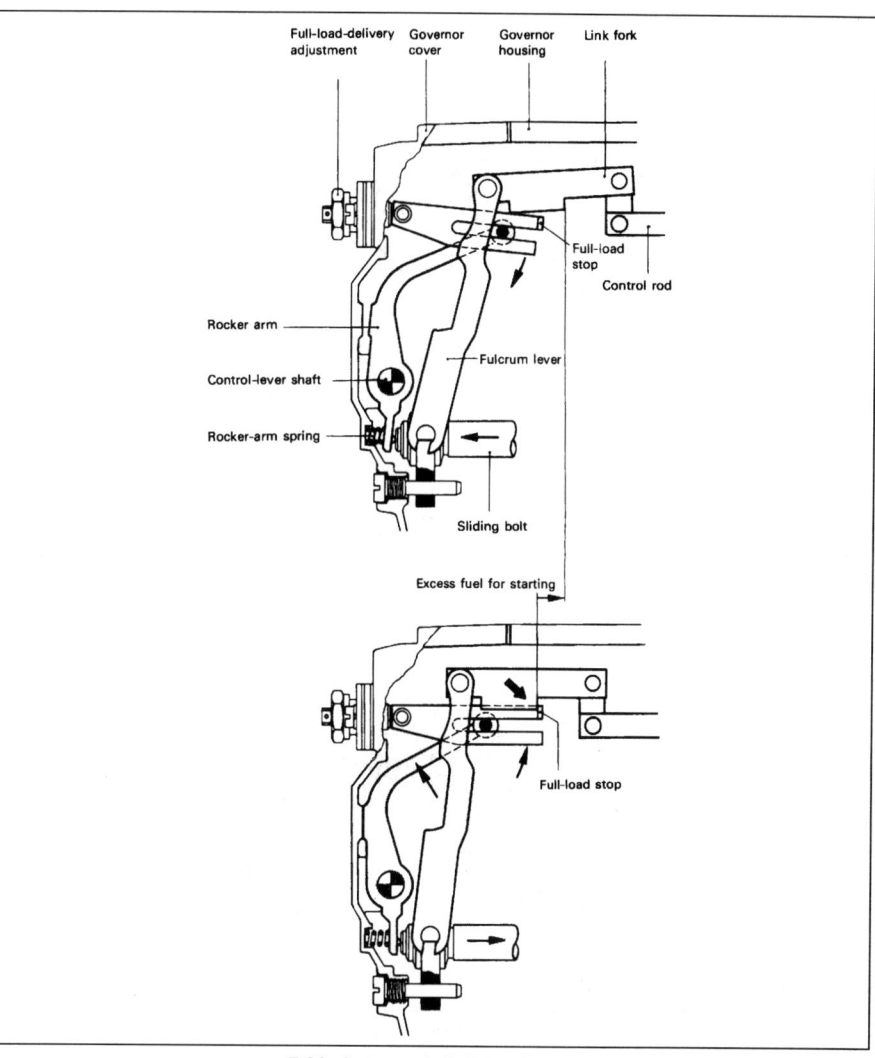

5.89 Automatic full-load stop

© Robert Bosch Limited

2

the governor drag spring provides a torque control function. When the torque control lever is moved to increase speed, the greater tension on the drag spring moves the control rod into an increased-delivery position **(see illustration)**. As speed increases, the drag

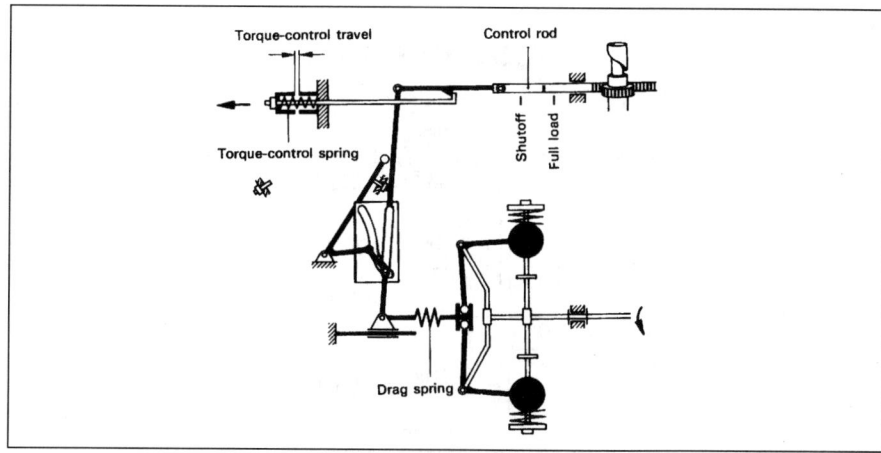

5.92 Full-load stop with external torque control

© Robert Bosch Limited

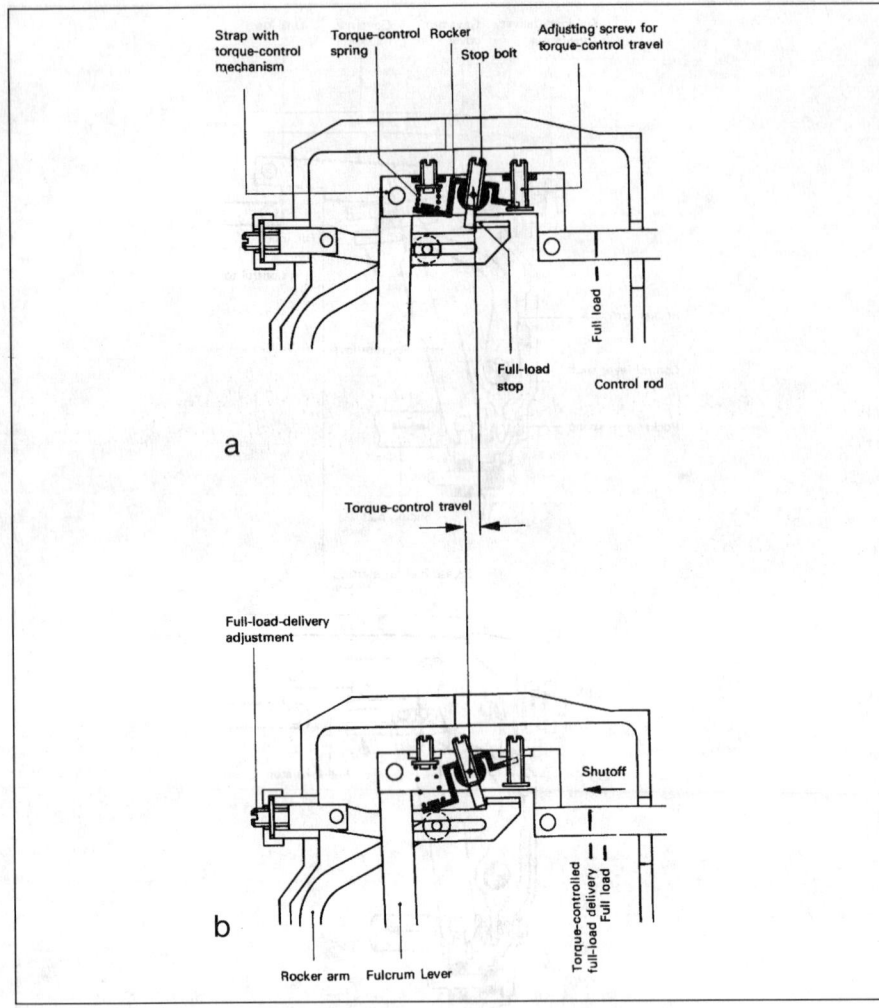

5.94 RQV governor with internal torque control mechanism

© Robert Bosch Limited

a Beginning of torque control   *b End of torque control*

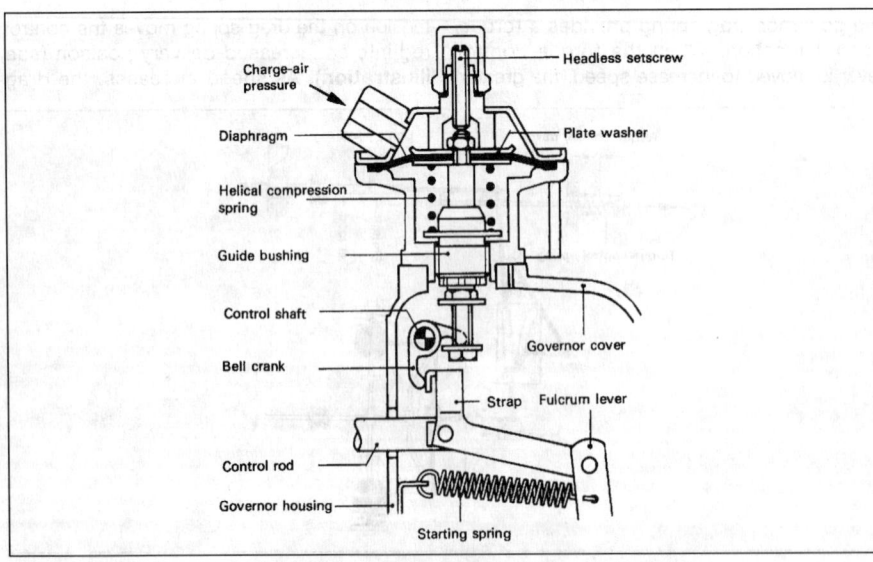

5.97 Sectional view of manifold pressure compensator

© Robert Bosch Limited

spring relaxes, and the torque control spring moves the control rod to reduce fuel delivery. Thus fuel delivery is reduced with increasing speed (positive torque control).

93 The speed at which torque control begins can be adjusted by altering the tension of the torque control spring. The travel allowed for torque control can be varied using shims.

Full-load stop with internal torque control (RQV governor)

94 This type of stop uses a pivoting stop bolt attached to a rocker. One end of the rocker is under pressure from the torque control spring. At low engine speeds, the force of the drag spring is greater than that of the torque control spring, and the stop is in the normal full-load delivery position. As engine speed rises, the drag spring relaxes, allowing the torque control spring to move the rocker. The stop bolt pivots with the rocker, and sets a reduced full-load delivery position **(see illustration)**.

95 The tension of the torque control spring can be adjusted by a screw which bears on the spring. Torque control travel is adjusted by another screw at the other end of the rocker.

Manifold pressure compensation (LDA)

96 Manifold pressure compensation is applied to turbocharged engines. It is explained fully in Section 6 – see *Bosch VE pump*.

97 The manifold pressure compensator is mounted on top of the pump. It contains a flexible diaphragm, which is subject to spring (and atmospheric) pressure on one side, and to inlet manifold pressure on the other. Movement of the diaphragm is transmitted by a threaded pin to a bellcrank. The other end of the bellcrank bears on the control rod strap **(see illustration)**.

98 When manifold pressure is at its maximum (full turbo-boost), the bellcrank moves in the direction to permit maximum fuel delivery. At reduced pressures, the bellcrank limits control rod movement. Adjustments are possible to determine the travel of the bellcrank, and also the manifold pressure at the beginning and end of travel.

99 The bellcrank can be moved out of engagement with the control rod strap by means of a shaft. This is necessary when starting fuel delivery is required. The disengagement may be controlled mechanically (with a cable) or electrically (with a solenoid).

Altitude compensation (ADA)

100 Altitude compensation involves reducing the full-load fuel delivery with increasing altitudes, to compensate for the reduced density of the air. It is not normally found on UK models, though it is used elsewhere in Europe.

101 Operation of the altitude compensator is similar to that of the manifold pressure

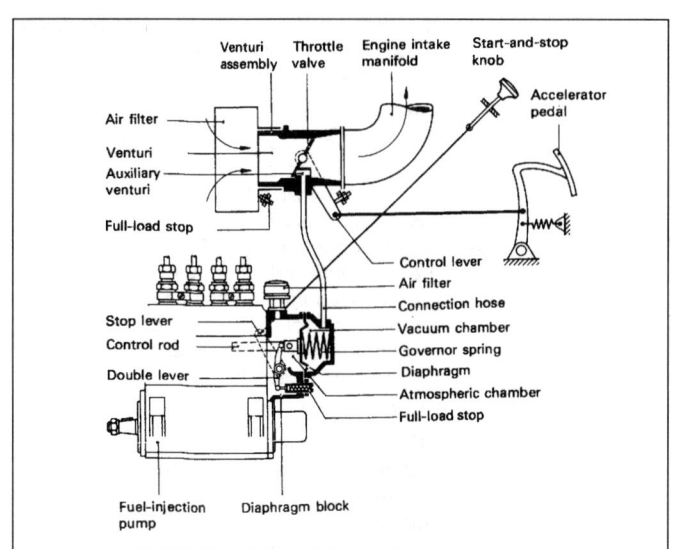

5.103 EP/M pneumatic governor schematic diagram

© Robert Bosch Limited

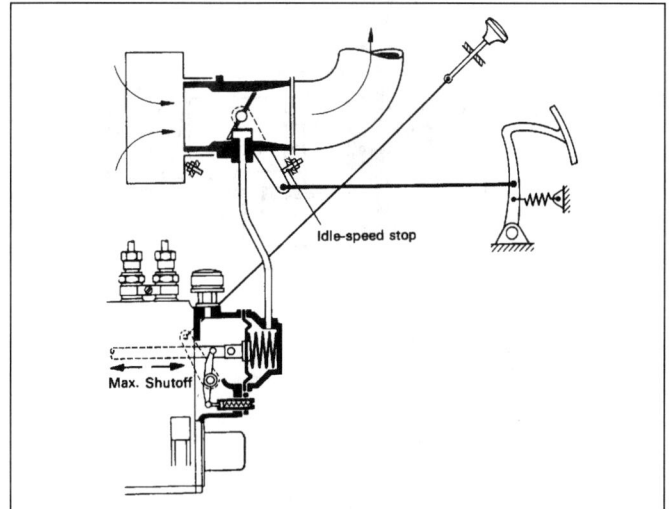

compensator. The altitude compensator contains a vacuum (aneroid) capsule, which moves a sliding bolt. The lower end of the bolt is connected to a cam plate, which acts on the full-load stop strap.

102 As air density reduces, the aneroid capsule expands and the sliding bolt moves downwards. This movement is transmitted to the cam plate, which draws the full-load stop in the direction of reduced delivery.

EP/M pneumatic governor

103 The EP/M governor is a variable-speed type, and is found on pump types A and M. Unlike the other governors fitted to the PE pump, its operation is pneumatic, and depends on the provision or a throttle valve and a venturi in the inlet manifold. The main components of the governor are a diaphragm,

a vacuum chamber and a spring **(see illustration)**.

104 The governor spring tends to force the pump control rod in the direction of increased fuel delivery. A full-load stop in the governor limits this travel. Vacuum produced in the venturi acts on the diaphragm in opposition to the spring, and pulls the control rod towards the shut-off position.

Start-up

105 An excess-fuel-delivery stop is released by the start-and-stop lever, which is connected to a start control. The governor spring moves the control rod into the starting fuel delivery position **(see illustration)**.

106 When the start control is released, spring-loading on the stop restores the normal full-load stop position.

Idling

107 The throttle valve is almost closed, and vacuum in the venturi is high. The vacuum acts on the governor diaphragm, and pulls the control rod towards the shut-off position **(see illustration)**.

108 If engine speed falls, the vacuum in the venturi will also fall, and the diaphragm will move to increase fuel delivery. If the engine speed rises, the vacuum will increase, and fuel delivery will be reduced. In this way, the idle speed is stabilised.

109 On some designs of governor, an auxiliary idle speed spring is used **(see illustration)**. This spring is stiffer than the main governor spring, and gives better regulation of idle speed. A switching cam may be used to unload the spring when the accelerator is moved out of the idle position.

2

5.105 EP/M governor in starting position

© Robert Bosch Limited

5.107 EP/M governor in idling position

© Robert Bosch Limited

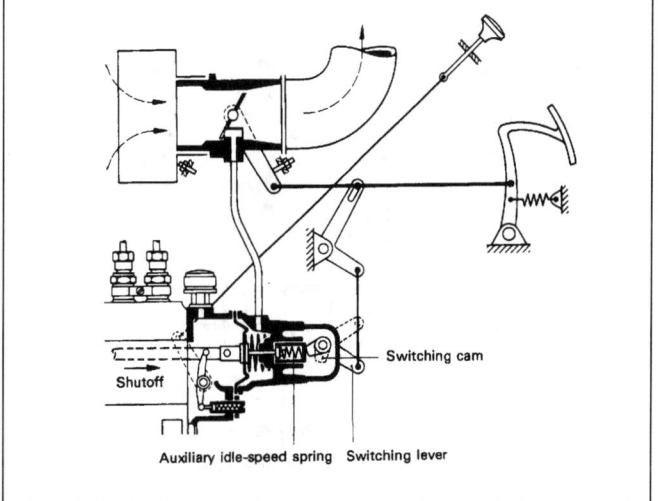

5.109 EP/M governor with auxiliary idle speed setting

© Robert Bosch Limited

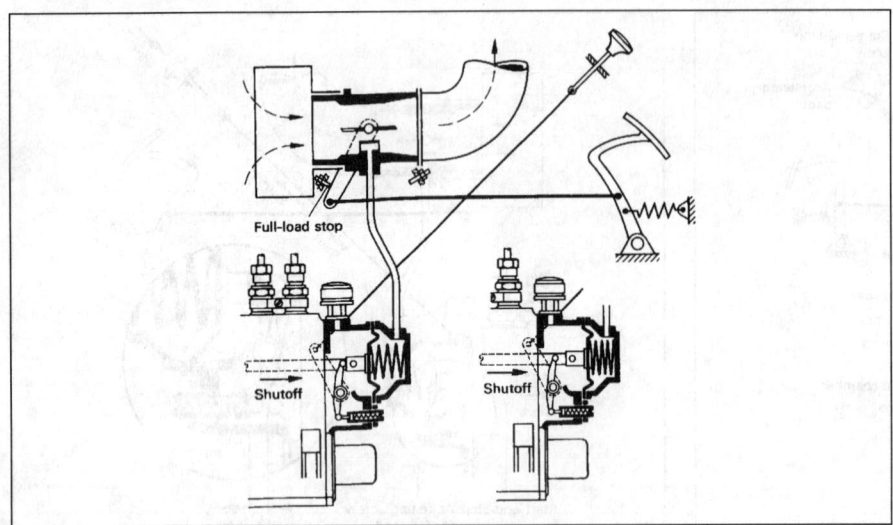

5.112 EP/M governor at full speed – left, at full load; right, unloaded

© Robert Bosch Limited

110 A restrictor is sometimes fitted in the governor vacuum line, to damp out fluctuations in vacuum which occur at idle on engines with four or fewer cylinders. Another way of reducing the effect of fluctuations is to have some slack in the coupling between the diaphragm and the control rod, so that small pulsations of the diaphragm are not transmitted to the rod.

On the road

111 Depression of the accelerator pedal opens the throttle and reduces manifold vacuum. The governor spring moves the control rod to increase fuel delivery. Engine speed will rise, and with it manifold vacuum, so that with increasing speed, the diaphragm moves the control rod back to reduce delivery.

112 For any given throttle position, there is an engine speed at which the vacuum and spring forces will be in balance. An increase in load will tend to decrease engine speed, reducing the vacuum and so increasing fuel delivery. A decrease in load has the reverse effect **(see illustration)**.

113 The dimensions of the venturi are calculated so that at the maximum no-load speed, even with the throttle wide open, the vacuum will be sufficient to move the control rod into the shut-off position.

Torque control

114 Positive torque control is achieved using a spring between the diaphragm and the control rod. When full throttle is applied at low rpm, vacuum is low or even negative, and the torque control spring is fully compressed by the governor spring. As the engine speeds up, the vacuum increases slightly, and the torque control spring can move the diaphragm and control rod to reduce fuel delivery **(see illustration)**.

Shut-off

115 A stop control or actuator moves the start-and-stop lever into the shut-off position. The lever moves the control rod to stop fuel delivery.

Reverse-operation protection

116 If an engine with a pneumatic governor starts to run backwards – for instance because of kicking-back during starting – there is a danger of it running away, since the exhaust gases will be entering the inlet manifold, and the governor diaphragm will be under pressure. The pressure on the diaphragm and the governor spring will act in the same direction, to increase fuel delivery, and the engine will run backwards out of control. In the worst case, damage will be caused to the engine bearings by inadequate lubrication, and the air cleaner will catch fire because of the exhaust gases passing through it.

117 To protect the engine against this runaway, an auxiliary venturi is provided

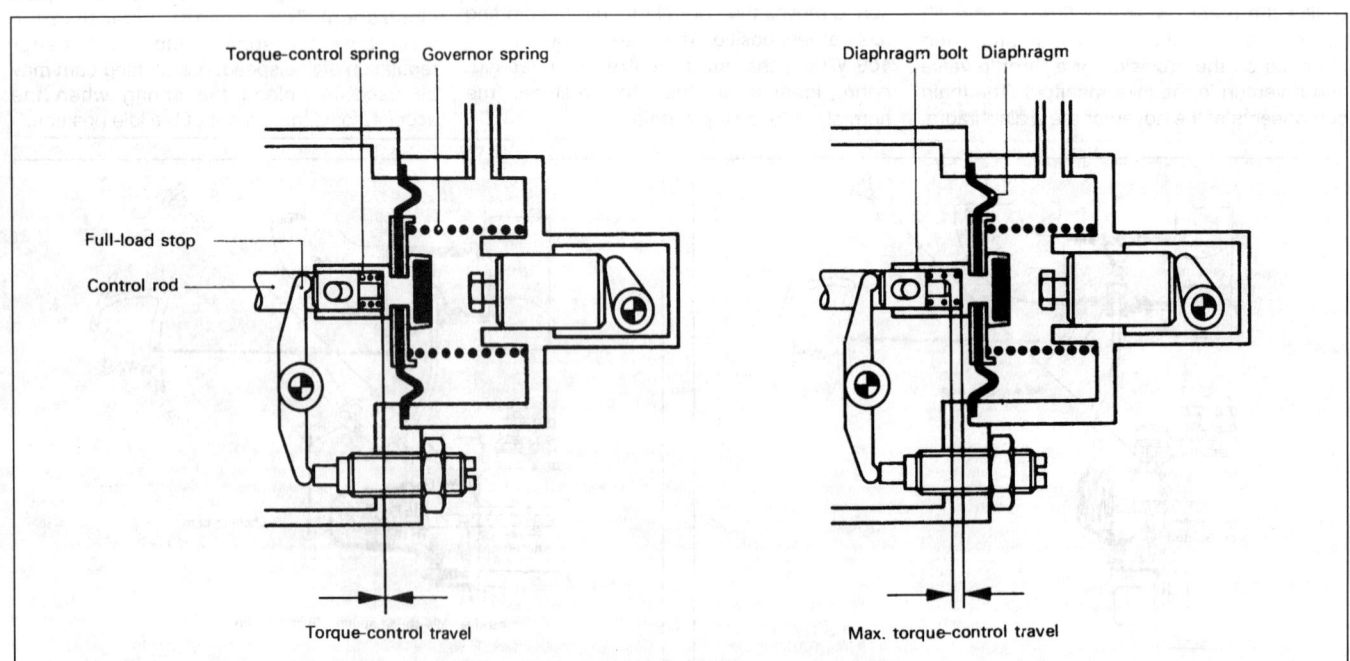

5.114 EP/M governor with torque control

© Robert Bosch Limited

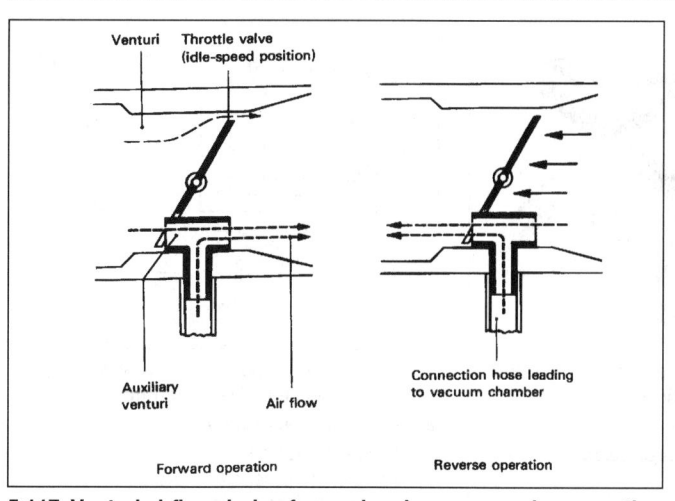

5.117 Venturi airflow during forward and reverse engine operation

© Robert Bosch Limited

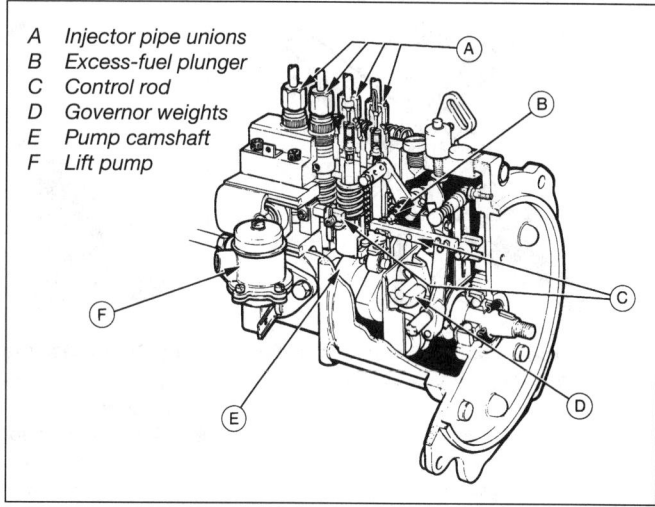

A Injector pipe unions
B Excess-fuel plunger
C Control rod
D Governor weights
E Pump camshaft
F Lift pump

5.118 Cutaway view of a CAV Minimec pump

where the governor vacuum hose taps into the inlet manifold. In the event of reverse operation, the auxiliary venturi ensures that there is still a vacuum in the governor hose **(see illustration)**. The engine can then be stopped by stalling it in gear, or by operating the stop control and fully depressing the accelerator.

CAV PUMP AND GOVERNOR

CAV Minimec pump

Basic principles

118 The CAV Minimec pump is similar in many respects to the Bosch PE pump previously described, being an in-line pump with an integral camshaft and a separate pumping element for each cylinder **(see illustration)**. The version described here has a fuel lift pump, attached to the injection pump and driven by the injection pump camshaft. The low-pressure side of the pump is connected to the engine lubrication system; the high-pressure side is lubricated by the fuel passing through it.

Fuel lift pump

119 The fuel lift pump is of the diaphragm type, and has a hand-priming lever for use

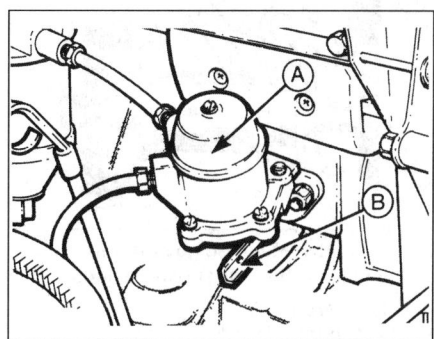

5.119 Fuel lift pump (A) and hand-priming lever (B)

when bleeding the fuel system **(see illustration)**. The pump cover can be removed for attention to the diaphragm or valves.

High-pressure pumping elements

120 Each pumping element contains a plunger and barrel, which are matched to each other in production. At the top of each barrel is a delivery valve and a pressure control spring. Each barrel contains two inlet ports, through which fuel is supplied at lift pump pressure. The side of the plunger carries vertical and helical grooves; rotation of the plunger will alter the point in the stroke at which the helical groove uncovers an inlet port.

The pump working cycle

121 With the pump plunger at BDC, fuel at lift pump pressure flows into the barrel through the inlet ports. The pressure control spring holds the delivery valve closed **(see illustration)**.

122 The pump camshaft turns and lifts the plunger. The plunger closes the delivery ports, and the pressure of fuel in the barrel begins to increase.

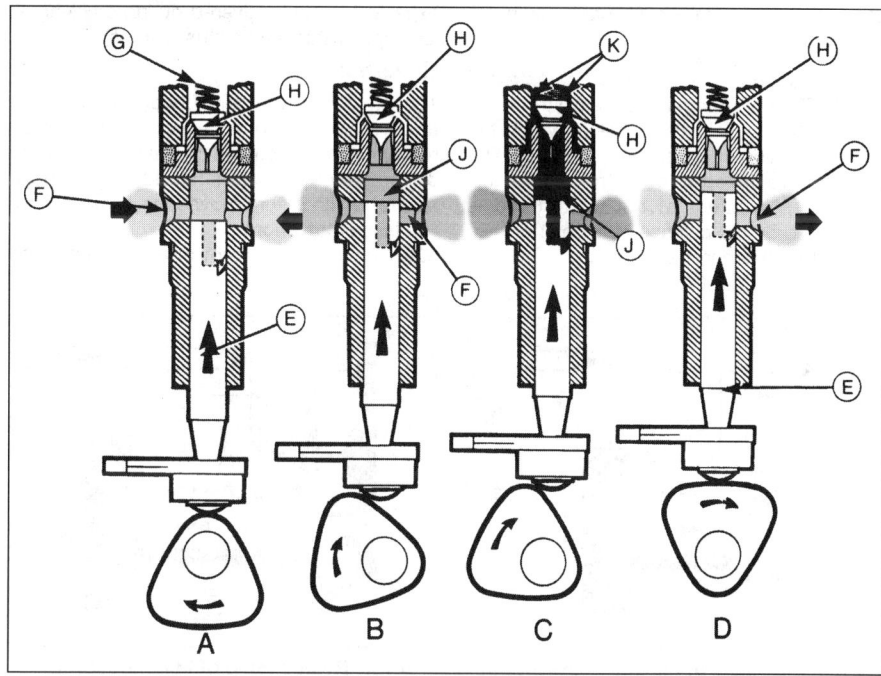

5.121 Minimec pump working cycle

A Fuel inlet
B Fuel pressurisation
C Injection
D End of injection
E Plunger
F Inlet ports
G Pressure control spring
H Delivery valve
J Fuel charge under pressure
K Outlet ports

2

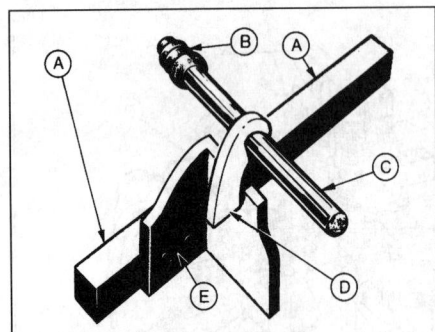

5.127a Excess-fuel device in normal (full-throttle) running position

A   Control rod
B   Excess-fuel button
C   Excess-fuel shaft
D   Excess-fuel stop
E   Control rod stop

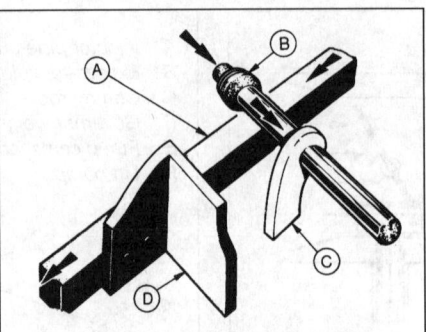

5.127b Excess fuel device in cold-start position

A   Control rod
B   Excess-fuel button depressed
C   Excess-fuel stop
D   Control rod stop

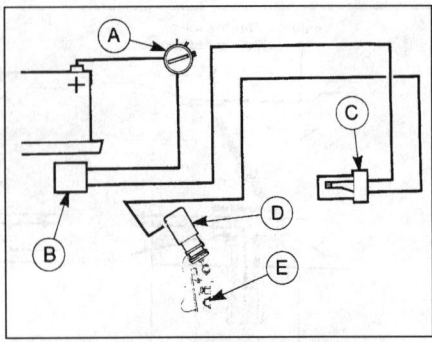

5.127c Simplified wiring diagram of excess fuel solenoid

A   Ignition/starter switch
B   Starter motor relay
C   Temperature-sensitive switch
D   Excess-fuel solenoid
E   Excess-fuel button

123 Further movement of the plunger increases fuel pressure to the point where it overcomes the pressure control spring and opens the delivery valve. Fuel under pressure reaches the injector, and injection takes place. The amount of fuel injected depends on the duration of this stroke.

124 The end of the injection stroke occurs when the helical groove in the plunger uncovers one of the inlet ports in the barrel. Pressure in the barrel falls again to lift pump pressure, the delivery valve closes, and injection ceases. The plunger passes TDC, returns to BDC, and the cycle starts again.

Pump control

125 The amount of fuel injected is controlled by rotating the plungers in their barrels, so varying the point in the cycle at which the helical grooves uncover the inlet ports. The plungers are linked to a control rod for this purpose. Movement of the control rod is determined by the balance of forces between the throttle lever and the governor.

126 If the plungers are turned so that their vertical grooves are in line with the delivery ports, the ports will be exposed for the whole cycle, and no fuel delivery will take place. This position is selected when it is desired to stop the engine.

127 Excess fuel for start-up is provided by temporarily releasing a control rod stop. On the version shown here, the excess-fuel button, which releases the stop, is electrically-actuated by a solenoid. The solenoid is connected to the starter motor relay via a temperature-sensitive switch, so ensuring that extra fuel is only delivered during cranking at low temperatures **(see illustrations)**.

Governor

128 An all-speed centrifugal governor is fitted. It consists of a pair of weights which revolve at pump camshaft speed. Movement of the weights is transmitted to a control sleeve, which in turn moves the pump control rod **(see illustration)**.

129 Outward movement of the weights with increasing speed moves the control rod towards the reduced-delivery position. This movement is opposed by a leaf spring which bears on the control sleeve; leaf spring tension is determined by throttle lever position.

130 At speeds below maximum, there is a balance between the force exerted by the governor weights in one direction and the leaf spring in the other. Increasing speed will tend to reduce fuel delivery, and *vice-versa*, so keeping engine speed constant within certain limits of load change **(see illustration)**.

131 At maximum governed speed, the force of the weights will overcome the force of the leaf spring, and the control rod will be moved to shut off fuel delivery, regardless of throttle lever position **(see illustration)**.

Timing mechanism

132 A centrifugal advance device is incorporated in the pump drive, in order to advance injection timing with increasing speed.

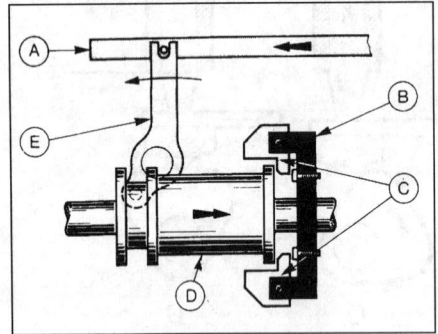

5.128 Side view of Minimec governor at low rpm

A   Control rod   D   Control sleeve
B   Weight carrier   E   Operating link
C   Weights

5.130 Relationship of Minimec throttle lever and governor (full throttle, low rpm)

A   Control rod
B   Control sleeve
C   Leaf spring
D   Throttle lever roller

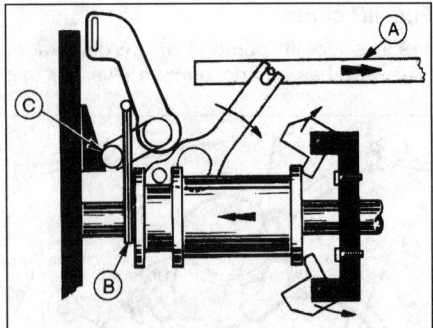

5.131 Minimec governor in full-throttle/full-speed position

A   Control rod
B   Leaf spring (closed)
C   Throttle lever roller

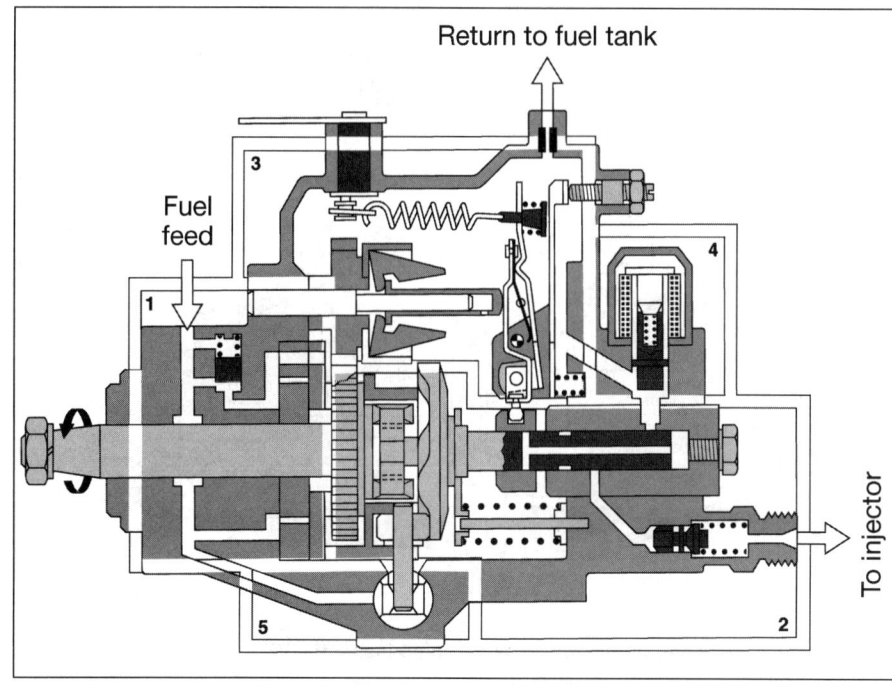

6.3 Sectional view of Bosch VE distributor injection pump

© Robert Bosch Limited

1	Supply pump	4	Shut-off solenoid
2	High-pressure pump and distributor	5	Injection timing control
3	Governor		

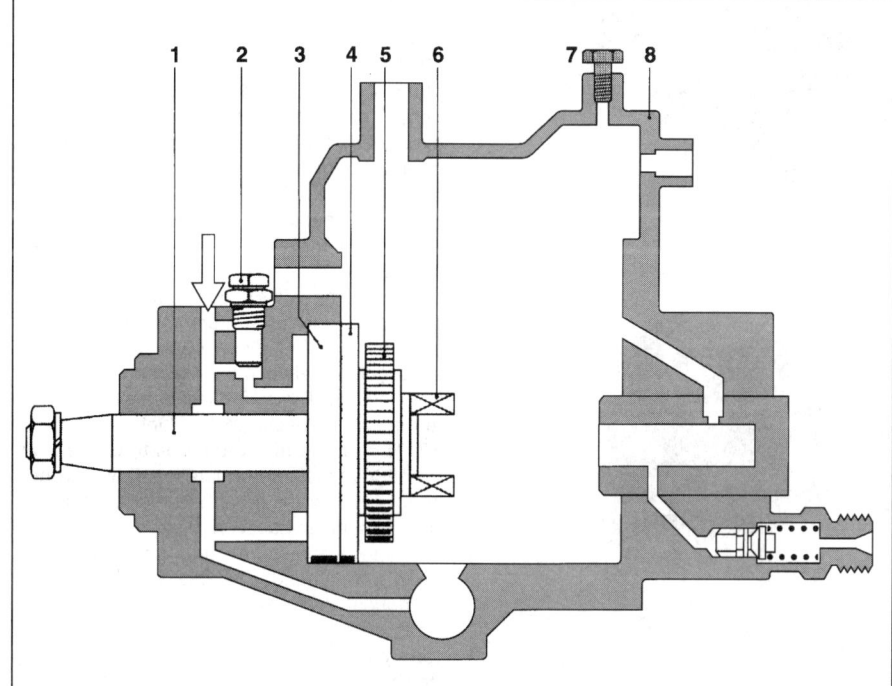

6.5 Supply pump and associated components – Bosch VE pump

© Robert Bosch Limited

1	Input shaft	4	Support ring	7	Overflow restrictor
2	Pressure control valve	5	Governor drivegear	8	Housing
3	Eccentric ring	6	Drive dogs		

6 Distributor injection pumps and governors

1 The function of a fuel injection pump is to supply fuel to the injectors at the correct pressure, at the correct moment in the combustion cycle, and for the length of time necessary to ensure efficient combustion.

2 Pump characteristics are carefully matched to a particular engine. This matching extends to the delivery valves, injector pipes and the injectors themselves. Substitution of any of these items with those from another engine or pump will not necessarily be successful, even if the substitute items have the same external dimensions.

BOSCH PUMP AND GOVERNORS

Bosch VE pump

3 The Bosch VE pump is a modern distributor pump, with an integral supply pump, timing mechanism and governor **(see illustration)**. It is widely used on small and medium-sized engines. It is made in Germany, and under licence in countries such as Japan.

Basic principles

4 The pump draws in fuel by means of its supply pump. A high-pressure pump element produces the pressure necessary for injection. Fuel under pressure is directed to each delivery port in turn by the pump plunger rotating in the distributor head.

5 Pump drive is most commonly by toothed belt, though gear or chain drive is also found. The pump driveshaft operates the supply pump directly, and a gearwheel on the shaft drives the governor. Dogs on the end of the shaft drive a cam plate (sometimes called a swash plate), which rides on a roller ring. The cam plate both rotates, thanks to the shaft, and moves back and forth, thanks to the rollers. This combined rotary/back-and-forth movement is transmitted to the pump plunger **(see illustration)**.

6 Lubrication of the pump is carried out entirely by the fuel.

Supply pump and low-pressure control

7 The supply pump impeller is keyed to the driveshaft, and runs inside an eccentric housing fixed to the body of the pump. Vanes in the impeller draw fuel in through the inlet port, and force it into the pump cavity **(see illustration overleaf)**.

8 Part of the supply pump's output passes to a pressure control valve, which opens a bypass and returns excess fuel to the suction (inlet) side when the pressure exceeds a certain level. A further part leaves the pump cavity via an overflow restrictor, and is returned to the fuel tank. The overflow restrictor is at the highest point of the pump, so any air bubbles will be bled off in the returned fuel.

2

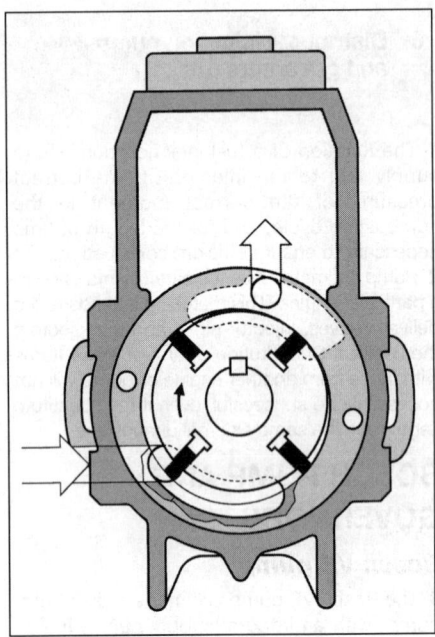

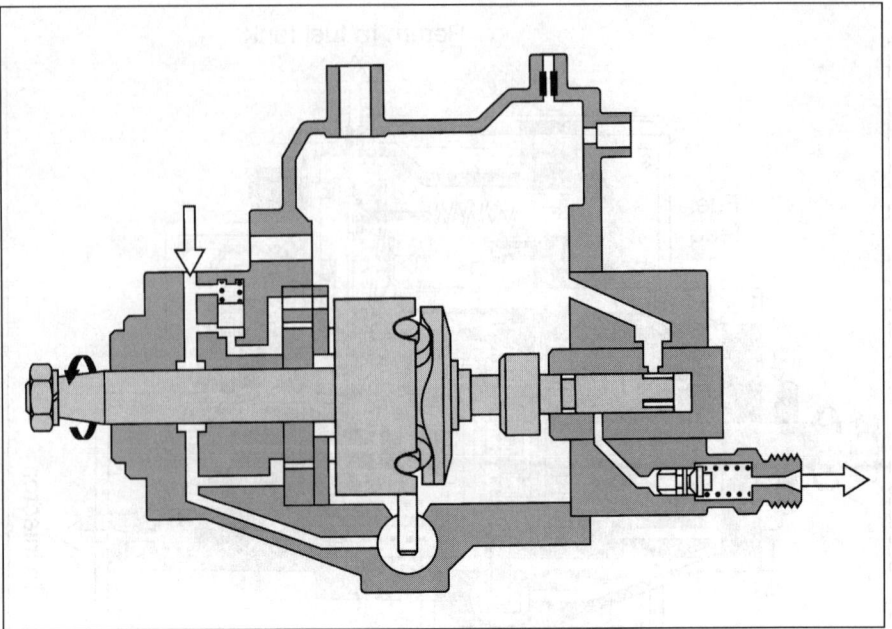

6.7 Supply pump operation – Bosch VE pump

© Robert Bosch Limited

6.10a Location of high-pressure pump inside the VE pump cavity

© Robert Bosch Limited

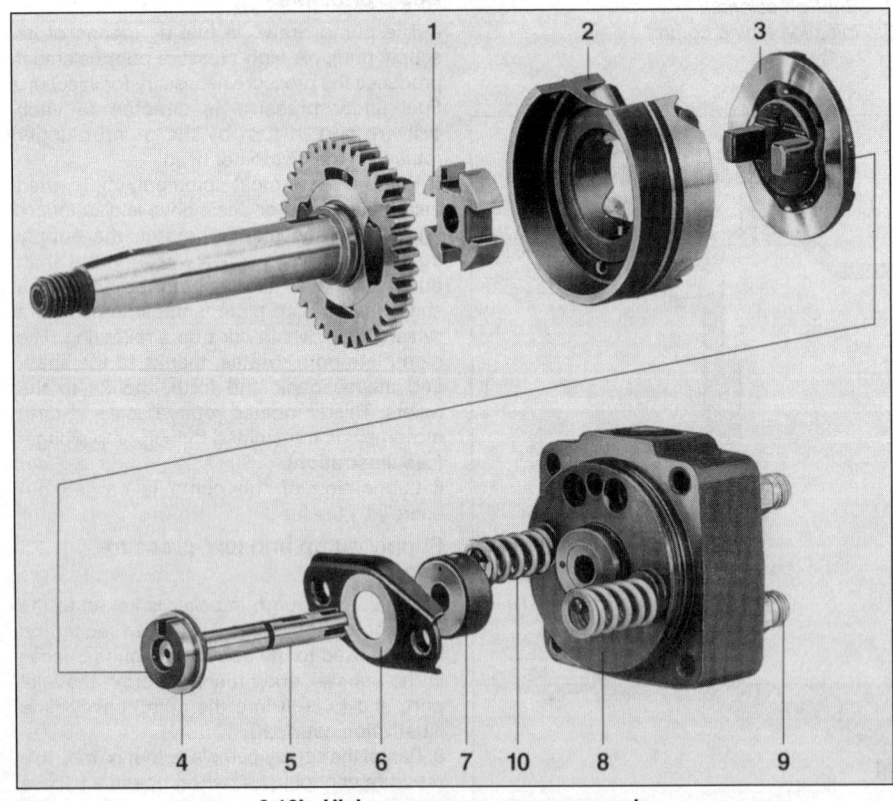

6.10b High-pressure pump components

© Robert Bosch Limited

1	Yoke	5	Pump plunger	8	Distributor head
2	Roller ring	6	Spring link	9	Delivery tube
3	Cam plate	7	Control spool	10	Return spring
4	Flange				

9 The pressure control valve and the overflow restrictor between them determine pump cavity pressure, which varies in proportion to speed, and is used to control injection timing.

High-pressure pump

10 The high-pressure pump consists of a plunger working in a bore in the distributor head **(see illustrations)**. As previously explained, the plunger rotates and moves back-and-forth. The back-and-forth movement provides the pumping effect. At the same time, the rotation brings a distributor slit in the plunger in line with each outlet bore in turn.

11 The plunger is forced towards TDC by the cam plate riding up on the rollers. It is returned to BDC by two springs which are sandwiched between the distributor head and a spring link. The spring link bears on a flange at the cam plate end of the plunger.

12 The profile of the cam plate determines injection pressure and duration. It is designed specifically for the combustion characteristics of the engine in question.

13 Where the pump plunger enters the distributor head, it passes through a cylindrical sleeve known as the *control spool*. The control spool can be moved back-and-forth by the governor. The point at which the control spool uncovers a cut-off bore in the pump plunger determines the end of injection.

14 The pump plunger, distributor head and control spool are precisely matched in production, and may only be renewed as an assembly.

The pump working cycle

15 The phases of the pump working cycle are shown in the accompanying illustration **(see illustration)**.

16 The cycle starts with fuel entry. The pump plunger is at BDC, and fuel at pump cavity pressure flows through the inlet passages, along one of the metering slits in the end of the plunger, and into the high-pressure chamber.

17 As the plunger rotates and moves towards TDC, it closes the inlet passage, and pressurises the fuel in the high-pressure chamber. Further rotation brings the distributor slit in the plunger in line with one of the outlet bores in the distributor head. Fuel under pressure opens the delivery valve at the end of the outlet bore, and fuel delivery begins.

18 Continued movement of the plunger towards TDC moves the plunger cut-off bore out of the control spool, causing the end of delivery. Fuel flows out of the high-pressure chamber via the cut-off bore, the pressure drops, and the delivery valve closes. The period between the beginning and end of fuel delivery is known as the *working stroke*.

19 The plunger completes its travel towards TDC, and fuel continues to flow from the cut-off bore. This is known as the *residual stroke*.

20 The plunger continues to rotate, and begins to return towards BDC. As the cut-off bore re-enters the control sleeve, the next metering slit comes into line with the inlet passage, and fuel entry begins again.

21 For each complete cycle, the plunger rotates one-quarter of a turn for a four-cylinder engine, one-sixth of a turn for a six-cylinder engine, and so on.

Delivery valves

22 The delivery valve separates the outlet bore from the injector pipe. The valve only opens when the pressure in the bore exceeds a certain value, and closes as soon as pressure begins to drop. This contributes to precise beginning and ending of injection.

23 In its simplest form, the delivery valve consists of a spring-loaded plunger in a holder **(see illustration)**. The size and shape of the plunger are matched to the length and diameter of the injector pipe. When the valve closes, the retraction of the plunger increases the effective volume of the fuel pipe by a

6.15 Bosch VE pump working cycle

© Robert Bosch Limited

1	Pump plunger	5	High-pressure chamber (under pressure)
2	Inlet passage	6	Distributor slit
3	Metering slit	7	Outlet bore
4	High-pressure chamber (filling)	8	Control spool
		9	Cut-off bore
		A	Fuel entry
		B	Working stroke
		C	Residual stroke
		D	Fuel entry

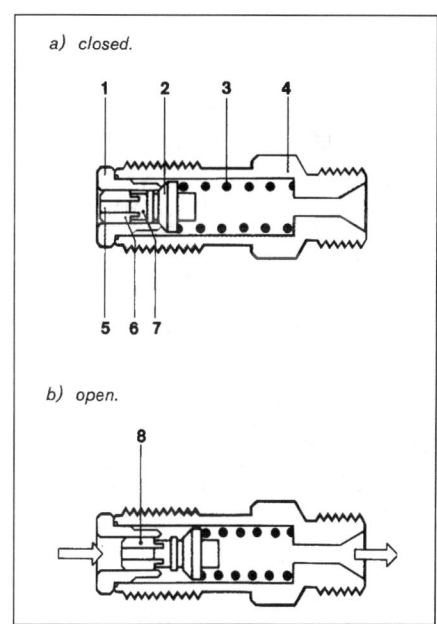

a) closed.

b) open.

6.23 Delivery valve details – Bosch VE pump

© Robert Bosch Limited

1	Valve holder	6	Relief piston
2	Plunger head	7	Ring groove
3	Spring	8	Longitudinal groove
4	Body		
5	Plunger shaft		

2

calculated amount. This relieves the pressure in the pipe, and makes the injector shut cleanly.

24 The delivery valve can also be part of a torque control system. (For an explanation of torque control, see Section 4.) This is done by profiling the plunger so that it offers a greater resistance to fuel flow at higher velocities.

25 A further refinement, known as *return-flow restriction*, allows the delivery valve to play an important part in reducing turbulence and cavitation in the injector pipe. The return-flow restrictor valve sits on top of the delivery valve. It offers little resistance to fuel flow in the forward direction, but damps out the reverse-pulse which occurs when the injector needle snaps shut.

Timing advance device

26 Injection timing advances in response to increased pump cavity pressure, which is itself increased with speed.

27 The timing advance device consists of a piston in the bottom of the pump, at right-angles to the pump driveshaft, and below the roller ring **(see illustration)**. The piston is subject to spring pressure on one side, and to pump cavity fuel pressure on the other. A sliding block and pin connect the piston with the roller ring. Movement of the piston rotates the roller ring.

28 Once the engine is running, spring pressure and pump cavity fuel pressure determine the position of the piston. Rising engine speed raises the pressure; the piston moves, and turns the roller ring so as to advance the timing. With falling engine speed, the reverse happens.

Additional features

29 So far, all the items described have been basic to the operation of the pump. The items in the following paragraphs will not be found on all pumps.

Positive torque control

30 Positive torque control can be achieved in two ways: by the delivery valves, or by the governor.

31 Positive torque control via the delivery valves is achieved by having an extra collar on the valve plungers **(see illustration)**. The collar has flats ground on it. These flats offer little resistance to fuel flow at low speeds, but with increasing flow, their effect is greater and delivery is reduced.

32 Positive torque control via the governor is achieved using an extra lever and spring **(see illustration)**. Above a certain speed, the sliding sleeve force begins to overcome the torque control spring force. The torque control lever then moves relative to the starting lever, with which it shares a pivot, and the tensioning lever moves in the direction of reduced delivery.

Negative torque control

33 Negative torque control is only achieved via the governor. As with positive torque

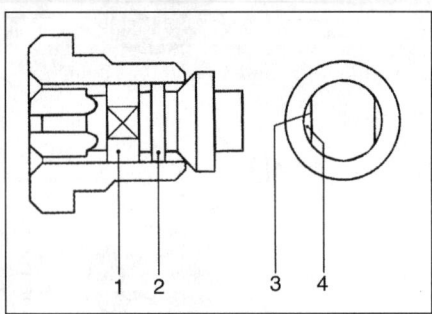

6.31 Bosch VE pump delivery valve with positive torque control

© Robert Bosch Limited

1   Relief collar
2   Torque control collar
3   Flat
4   Cross-section of restriction

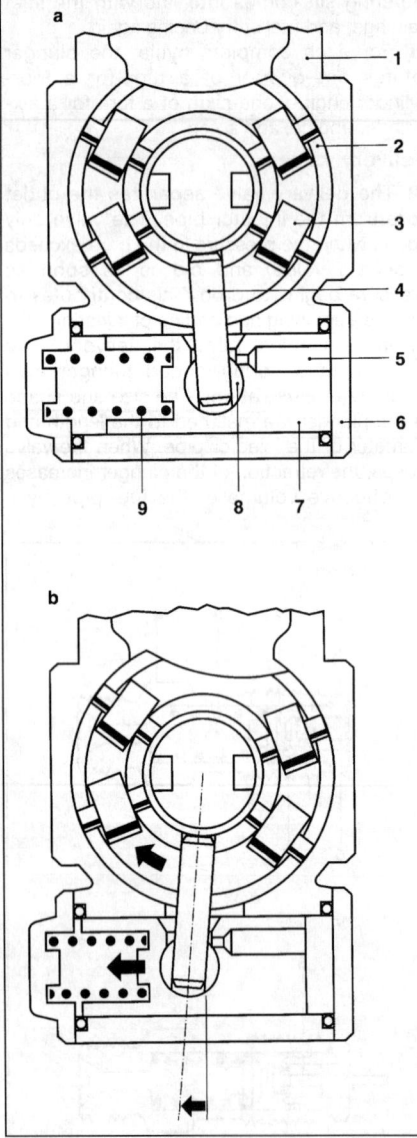

6.27 Bosch VE pump timing advance device

© Robert Bosch Limited

a   At rest
b   Running
1   Pump housing
2   Roller ring
3   Rollers
4   Pin
5   Bore in timing position
6   Cover
7   Timing piston
8   Sliding block
9   Spring

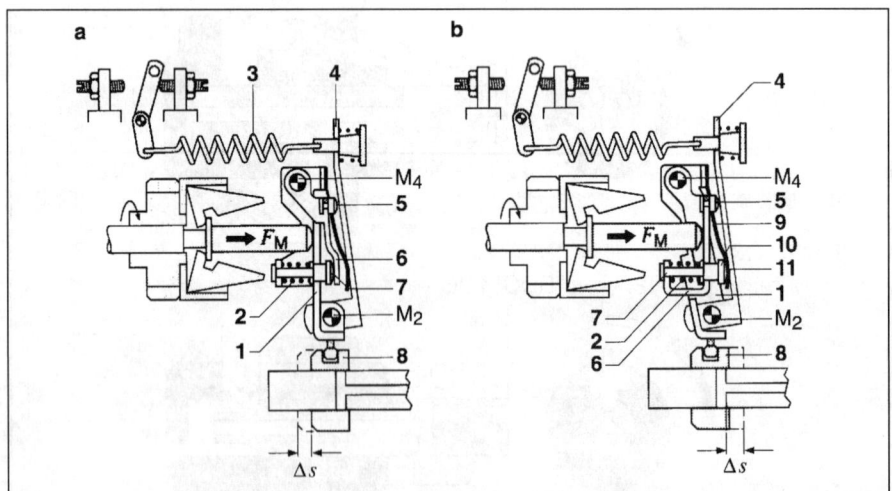

6.32 Bosch VE pump torque control by means of the governor

© Robert Bosch Limited

a	Positive torque control	6	Torque control lever
b	Negative torque control	7	Torque control pin
1	Starting lever	8	Control spool
2	Torque control spring	9	Starting spring
3	Governor spring	10	Pin collar
4	Tensioning lever	11	Stop point
5	Lug		

M_2 Pivot point for starting and tensioning levers
M_4 Pivot point for starting and torque control levers
F_M Sleeve force
Δ_s Control spool travel

control, an extra lever and spring are fitted, but the arrangement is such that when the spring pressure is overcome, the lever assembly moves to increase delivery.

Manifold pressure compensation (LDA)

34 Manifold pressure compensation is necessary on turbocharged engines, because fuel requirement varies not only with speed, but also in proportion to inlet manifold pressure (boost).

35 In practise, the pump is set to deliver the optimum amount of fuel for maximum boost; the manifold pressure compensator reduces fuel delivery at lower boost pressures. For this reason, it is sometimes known as a *smoke limiter*, since without it the engine would smoke excessively under part-load conditions.

36 The manifold pressure compensator is mounted on top of the pump **(see illustration)**. It contains a flexible diaphragm, which is subject to spring (and atmospheric) pressure on one side, and to inlet manifold pressure on the other. Movement of the diaphragm is transmitted by a sliding pin and a guide pin to a stop lever, which alters the full-load stop position of the governor.

37 When the engine speed is low, the inlet manifold pressure is not sufficient to move the diaphragm against spring force. The sliding pin is in the high position, displacing the guide pin and the stop lever towards reduced fuel delivery.

38 As speed and manifold pressure rise, the diaphragm is forced downwards. The guide pin moves into the waisted section of the sliding pin, and the stop lever moved to increase fuel delivery.

39 If the turbocharger fails, boost pressure will be zero, and the diaphragm will resume its low-pressure position. The engine will operate normally, whereas (without the pressure compensation) it would receive too much fuel, and produce smoke.

Load-dependent injection timing (LFB)

40 Smoother engine operation can be achieved if injection timing is retarded when load is reduced, and advanced when it is increased. This function can be carried out by the governor as follows.

41 A transverse bore in the governor sliding sleeve opens and closes a similar bore in the governor shaft **(see illustration)**. The governor shaft bore communicates with the inlet (suction) side of the supply pump. When the sleeve and shaft bores coincide, fuel flows out, and pump cavity pressure is reduced. Since pump cavity pressure controls injection advance, reducing the pressure retards the injection timing.

42 When the engine is running under a load which reduces without the speed control lever being moved, engine speed will tend to increase. The governor weights will move out, and the sliding sleeve will move, uncovering

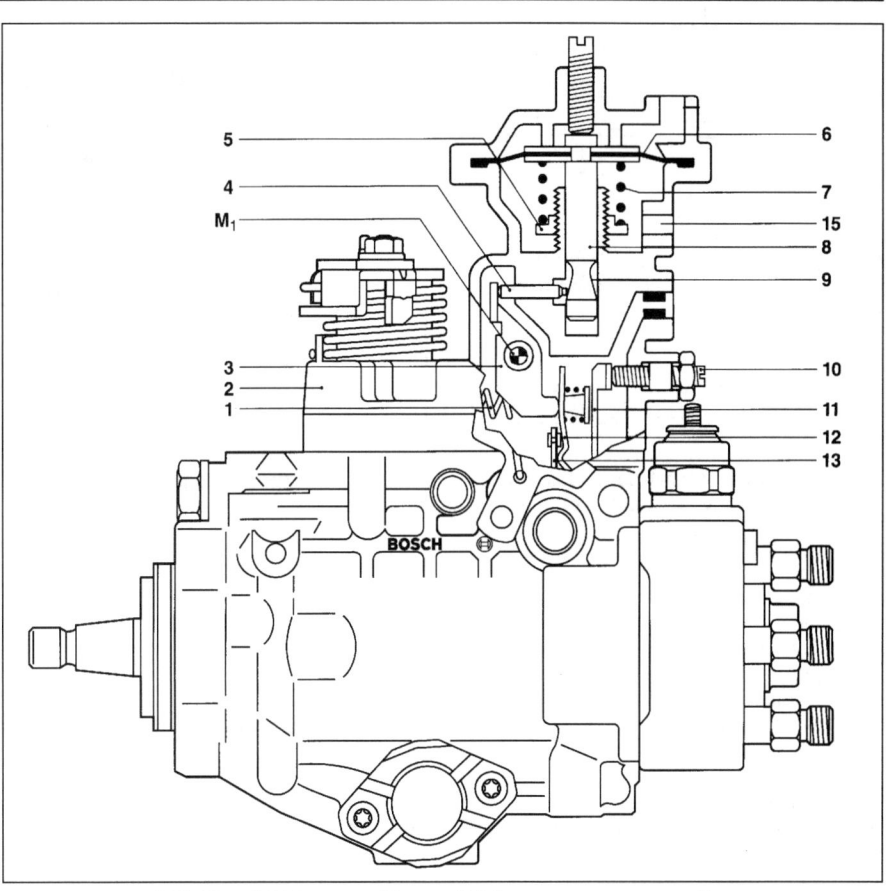

6.36 Bosch VE pump with manifold pressure compensation

© Robert Bosch Limited	5	Adjusting nut	10	Full-load adjusting screw
1 Governor spring	6	Diaphragm	11	Adjusting lever
2 Governor cover	7	Spring	12	Tensioning lever
3 Stop lever	8	Sliding pin	13	Starting lever
4 Guide pin	9	Waisted section	M_1	Pivot for stop lever

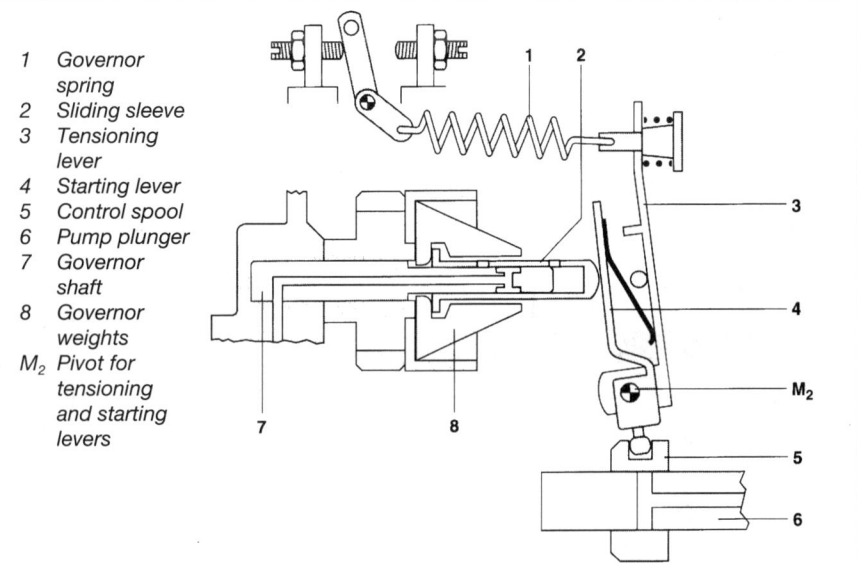

1 Governor spring
2 Sliding sleeve
3 Tensioning lever
4 Starting lever
5 Control spool
6 Pump plunger
7 Governor shaft
8 Governor weights
M_2 Pivot for tensioning and starting levers

6.41 Sectional view of Bosch VE pump load-dependant injection timing device

© Robert Bosch Limited

2

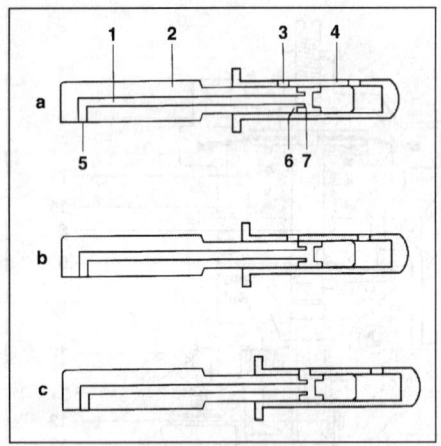

6.42 Positions of load-dependant device sliding sleeve – Bosch VE pump

© Robert Bosch Limited

a	Initial	4	Sliding sleeve
b	About to open	5	Transverse bore
c	Open		in governor shaft
1	Longitudinal bore	6	Governor shaft
	in governor shaft		port
2	Governor shaft	7	Transverse bore
3	Transverse bore		in governor shaft
	in sliding sleeve		

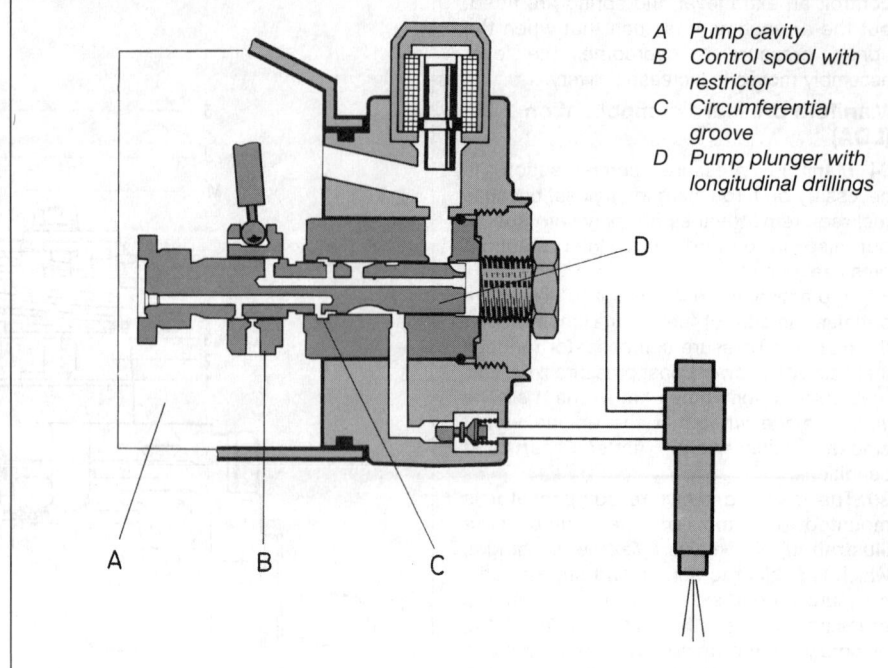

A Pump cavity
B Control spool with restrictor
C Circumferential groove
D Pump plunger with longitudinal drillings

6.47 Sectional view of pump plunger and associated components – Bosch VE pump with integral quiet-running device

the bore in the governor shaft **(see illustration)**. Pump cavity pressure will reduce, and the injection timing will be retarded.

43 If the load increases and the speed control lever is not moved, engine speed will tend to fall. As the weights move inwards and the sleeve retracts, the bores will no longer coincide. Fuel ceases to flow out of the pump cavity, pressure will increase, and the injection timing will be advanced.

Altitude compensation (ADA)

44 Altitude compensation involves reducing the full-load delivery with increasing altitudes, to compensate for the reduced density of the air. It is not normally found on UK models, though it is used elsewhere in Europe.

45 Operation of the altitude compensator is similar to that of the manifold pressure compensator described earlier. The altitude compensator contains a vacuum (aneroid) capsule, which moves a sliding pin under the influence of changing atmospheric pressure. The sliding pin is connected to a stop lever, positioned so that a reduction in atmospheric pressure reduces maximum fuel delivery.

Integral quiet-running device

46 This device is fitted to some VW models. Its aim is to reduce engine noise at idle. It does this by lengthening the period over which the fuel is injected.

47 Pumps having this device are provided with an extra longitudinal drilling in the pump plunger, a restricted outlet bore in the control spool, and a circumferential groove in the distributor head **(see illustration)**.

48 At idle, a proportion of the fuel which would normally be injected is instead discharged back into the pump cavity via the restricted outlet bore in the control spool **(see illustration)**. This delays the rise in injection pressure. The discharge only takes place while the two longitudinal drillings are in communication via the circumferential groove. Movement of the pump plunger towards TDC breaks this communication, and the later phase of injection takes place as normal.

49 The device has no effect when the engine is under load, because movement of the control spool to increase delivery results in the two drillings no longer being in communication **(see illustration)**.

Cold start devices

50 Cold start devices all advance injection timing, either automatically or in response to a mechanical cold start control. Advance may be achieved mechanically (by moving the

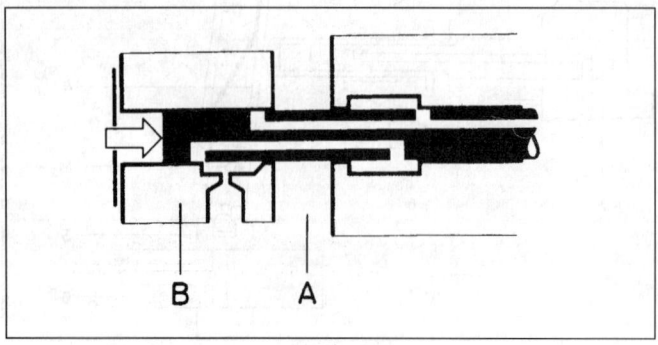

6.48 Integral quiet-running device operation at idle – Bosch VE pump

A Pump cavity   B Control spool with restrictor

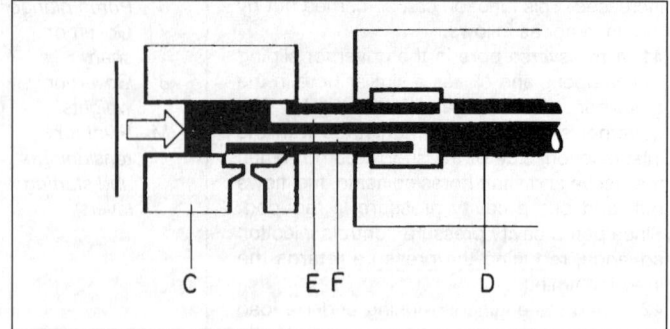

6.49 Integral quiet-running device under load – Bosch VE pump

C Control spool with restrictor   E Longitudinal drilling 1
D Pump plunger   F Longitudinal drilling 2

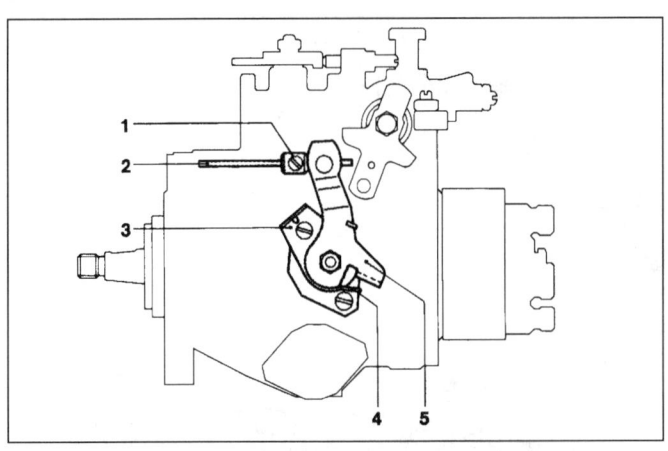

**6.52a Manually-operated cold start injection advance –
Bosch VE pump**

© Robert Bosch Limited

1	Screw	3	Stop	5	Advance lever
2	Cable	4	Spring		

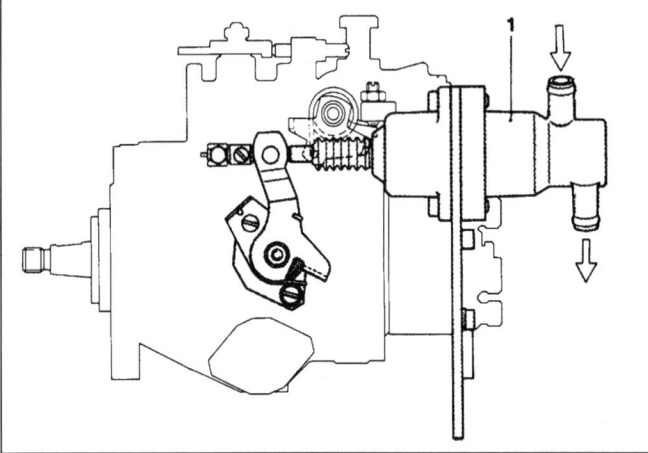

**6.52b Automatically-operated cold start advance –
Bosch VE pump**

© Robert Bosch Limited

1 Actuator

roller ring or the timing advance piston) or hydraulically (by increasing pump cavity pressure).

51 Additional devices associated with the basic cold start device may inhibit the starting delivery of fuel when the engine is warm, and increase idle speed when it is cold.

Mechanical cold start injection advance (KSB)

52 Actuation of the mechanical advance system may be automatic (by connection to a temperature-sensitive control unit plumbed into the cooling system) or manual (dashboard cold start control). In either case, the mechanism inside the pump is the same **(see illustrations)**.

53 An advance lever on the outside of the pump turns a shaft which enters the pump cavity. The inner end of the shaft carries an eccentrically-mounted ball-pin, which engages with a slot in the roller ring or the timing advance piston **(see illustration)**.

54 Movement of the advance lever to the cold start position causes a movement of the roller ring or timing advance piston in the direction of increasing advance. The ball-pin and slot allow normal movement of the roller ring to take place as engine speed rises.

55 As the engine warms up, the advance lever is returned to the 'hot' position by the control unit element expanding, or by the driver pushing home the cold start control.

Temperature-dependent starting delivery control (TAS)

56 This system can be combined with mechanical cold-start injection advance. A linkage from the cold start advance mechanism changes the position of a stop lever so that maximum starting delivery of fuel only occurs on a cold engine, when it is required **(see illustration)**.

Temperature-dependent fast idle (TLA)

57 This system is also linked with mechanical cold start injection advance. A linkage from the cold start advance mechanism lifts the engine speed control lever off the idle stop when the mechanism is in the 'cold' position. When the engine warms up and the

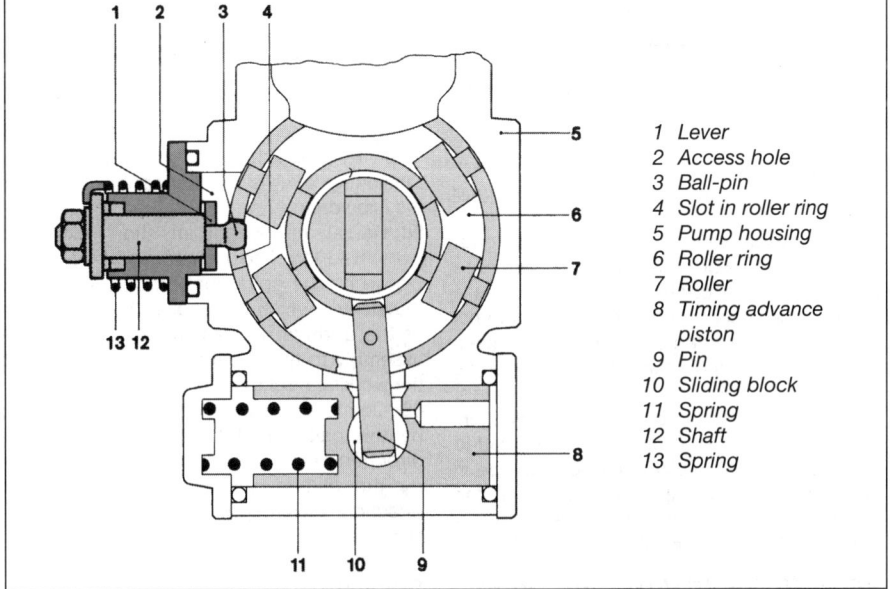

1	Lever
2	Access hole
3	Ball-pin
4	Slot in roller ring
5	Pump housing
6	Roller ring
7	Roller
8	Timing advance piston
9	Pin
10	Sliding block
11	Spring
12	Shaft
13	Spring

**6.53 Sectional view of cold start advance mechanism –
Bosch VE pump**

© Robert Bosch Limited

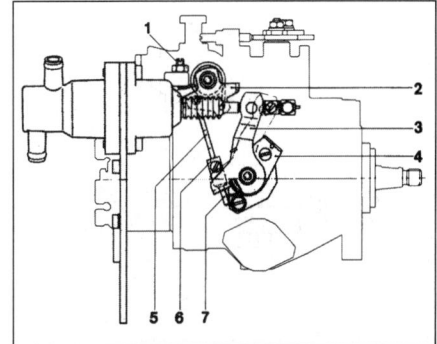

**6.56 Temperature-dependant starting
delivery control – Bosch VE pump**

© Robert Bosch Limited

1	Stop screw	5	Reduction lever
2	Outer stop screw	6	Retaining pin
3	Advance lever	7	Locating pin
4	Stop		

2

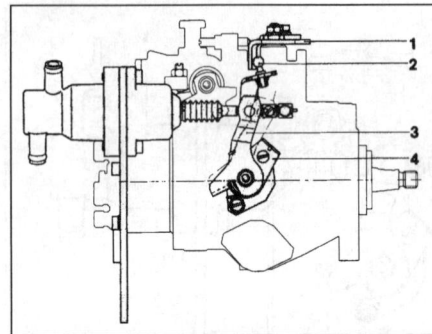

**6.57 Temperature-dependant fast idle –
Bosch VE pump**

© Robert Bosch Limited

1   Pump control lever
2   Ball-pin
3   Advance lever
4   Stop

mechanism is returned to the 'hot' position, normal idle speed is restored **(see illustration)**.

Hydraulic cold start injection advance (KSB)

58 Hydraulic cold start injection advance works by increasing pump cavity pressure when the engine is cold. This is achieved by a pressure-holding valve, which is opened by the warming up of an electrically-heated expansion element **(see illustration)**.

59 When the engine is started from cold, the expansion element is cold and the pressure-holding valve is closed. A restriction bore in the pump pressure control valve raises the pump cavity pressure and injection timing is advanced.

60 When the engine is running, the expansion element is heated electrically. When it has warmed up sufficiently, it opens the pressure-holding valve which relieves the

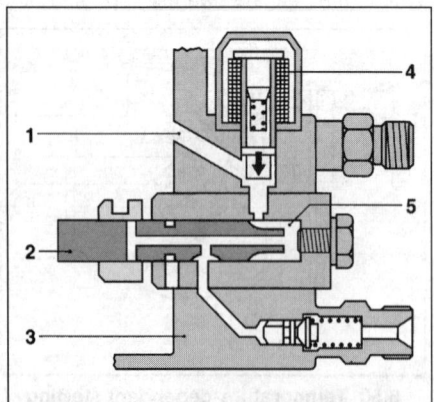

**6.62 Sectional view of shut-off solenoid –
Bosch VE pump**

© Robert Bosch Limited

1	Fuel inlet	4	Shut-off solenoid
2	Pump plunger	5	High-pressure
3	Distributor head		chamber

restriction in the pressure control valve. Pump cavity pressure returns to normal.

Engine shut-off

61 A diesel engine is shut off by physically cutting off the fuel supply. This can be done mechanically or electrically. Electrical control is preferred for passenger cars, because it can be operated through the 'ignition' switch instead of using a separate stop control.

62 Electric shut-off is achieved using a solenoid valve which opens the inlet bore to the high-pressure chamber when energised (ignition on) **(see illustration)**. When the electrical supply is removed, the valve closes the inlet so that fuel cannot reach the pump plunger.

63 Mechanical shut-off is achieved using an external stop lever which is cable-actuated by a dashboard control. The lever turns a shaft, at the other end of which is an internal stop lever. The internal stop lever presses against the governor starting lever and forces the control spool into the 'stop' position, opening the cut-off bore in the pump plunger.

64 A mechanical shut-off lever is still fitted to some pumps with electric shut-off, for use in case the solenoid valve jams open. (If the valve jams shut, or its electrical supply is interrupted, the engine will not run at all.)

Variable-speed governor (Bosch VE pump)

65 The governor assembly is driven by a gear on the pump input shaft. It consists of a pair

of centrifugal weights on a shaft, several springs, and a lever assembly.

66 The lever assembly consists of two levers: the tensioning lever and the starting lever. The levers have a common pivot. The bottom of the starting lever engages with the control spool. Between the two levers there is a spring (the starting spring) which tends to force them apart.

67 A pin passes through the top of the tensioning lever. One end of the governor spring is connected to the pin, the other end being connected to the engine speed control lever. Between the head of the pin and the tensioning lever is another spring, known as the *idle spring*.

68 Radial movement of the weights is converted into linear movement of a sliding sleeve on the governor shaft. Movement of the sleeve, which is opposed by the tension of the springs, is transmitted by the lever assembly to the control spool, and so varies the working stroke of the pump.

69 Control spool position is thus determined by the balance of forces exerted in one direction (towards reduced delivery) by the sliding sleeve, and in the other direction (towards increased delivery) by the springs.

Starting

70 With the centrifugal weights stationary, the sliding sleeve is in its rest position (fully retracted). The starting spring forces

6.58 Hydraulic cold start injection advance – Bosch VE pump

© Robert Bosch Limited

1	Pressure control valve	4	Pump cavity pressure	7	Ball valve
2	Piston	5	Pressure-holding valve	8	Fuel drain
3	Restrictor	6	Expansion element	+	Electrical feed

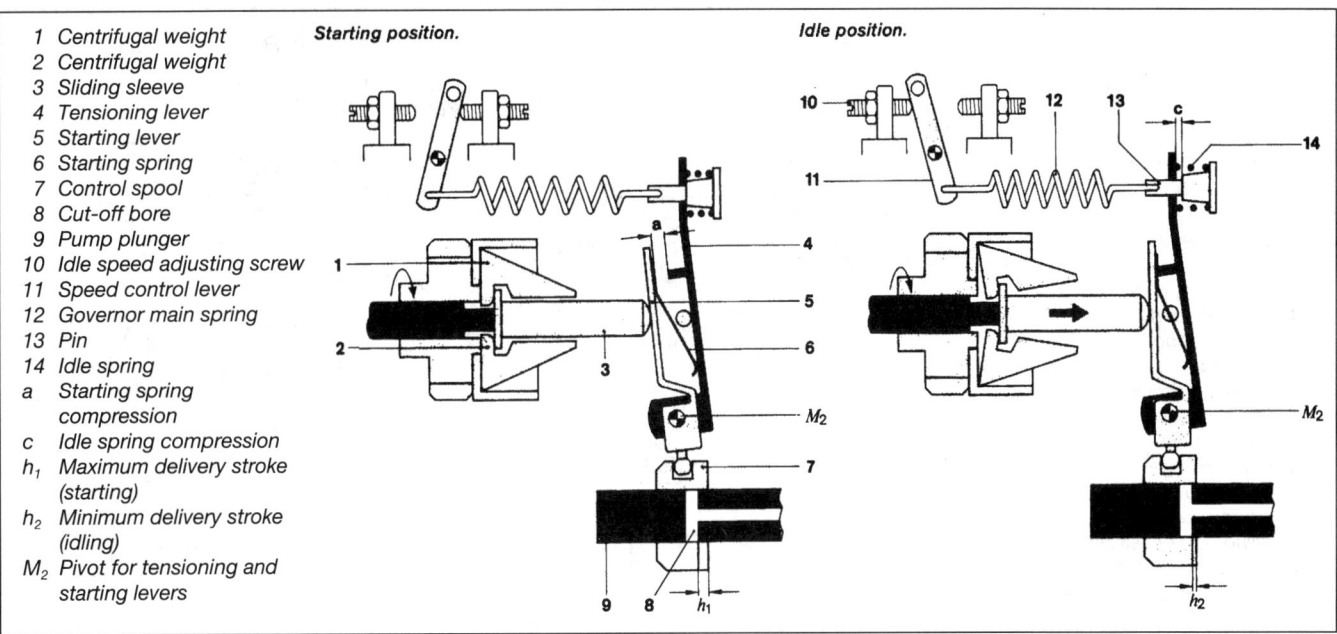

1 Centrifugal weight
2 Centrifugal weight
3 Sliding sleeve
4 Tensioning lever
5 Starting lever
6 Starting spring
7 Control spool
8 Cut-off bore
9 Pump plunger
10 Idle speed adjusting screw
11 Speed control lever
12 Governor main spring
13 Pin
14 Idle spring
a Starting spring
  compression
c Idle spring compression
h_1 Maximum delivery stroke
  (starting)
h_2 Minimum delivery stroke
  (idling)
M_2 Pivot for tensioning and
  starting levers

6.70 Starting and idling positions of Bosch VE pump variable-speed governor
© Robert Bosch Limited

the starting lever away from the tensioning lever, and the control spool is moved to the maximum-delivery position (see illustration).

71 As soon as the engine is turning, the weights start to move outwards. The sliding sleeve presses on the starting lever and overcomes the resistance of the starting spring, forcing the starting lever back into contact with the tensioning lever. The movement of the starting lever is transmitted to the control spool, which is moved into a reduced-delivery position. From now on, the starting lever and the tensioning lever move as one.

Idling

72 With the engine running and the accelerator pedal released, the engine speed control lever rests against the end of the idle speed control screw. The idle spring acts to move the tensioning lever in the direction of increased delivery, in opposition to the force exerted by the sliding sleeve.

73 If the engine speed increases, the weights will move outwards, and the sliding sleeve will move the tensioning lever in the direction of reduced fuel delivery. If the speed falls, the weights will move inwards, and the idle spring

will move the lever in the direction of increased delivery.

On the road

74 When the accelerator pedal is depressed, the engine speed control lever pulls on the governor spring. This overcomes the force of the idle spring, and also increases the spring tension on the lever assembly. The lever assembly moves the control spool to increase fuel delivery, and engine speed rises.

75 The rising engine speed causes the governor weights to move further outwards (see illustration). The sliding sleeve begins to

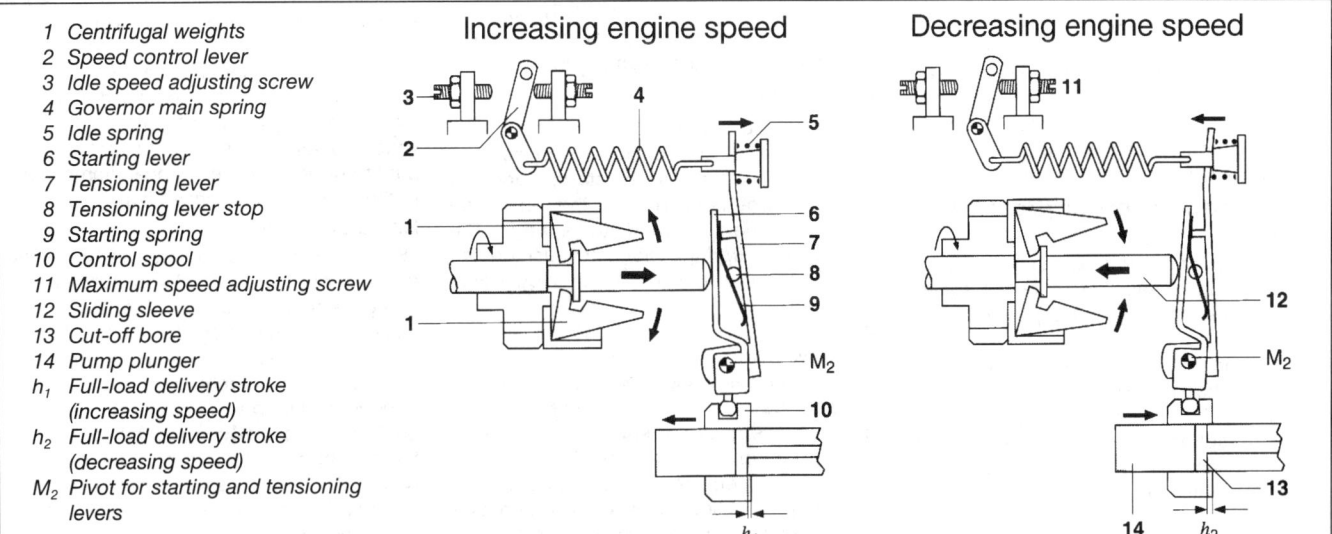

1 Centrifugal weights
2 Speed control lever
3 Idle speed adjusting screw
4 Governor main spring
5 Idle spring
6 Starting lever
7 Tensioning lever
8 Tensioning lever stop
9 Starting spring
10 Control spool
11 Maximum speed adjusting screw
12 Sliding sleeve
13 Cut-off bore
14 Pump plunger
h_1 Full-load delivery stroke
  (increasing speed)
h_2 Full-load delivery stroke
  (decreasing speed)
M_2 Pivot for starting and tensioning
  levers

6.75 Bosch VE pump variable speed governor action with increasing and decreasing speed
© Robert Bosch Limited

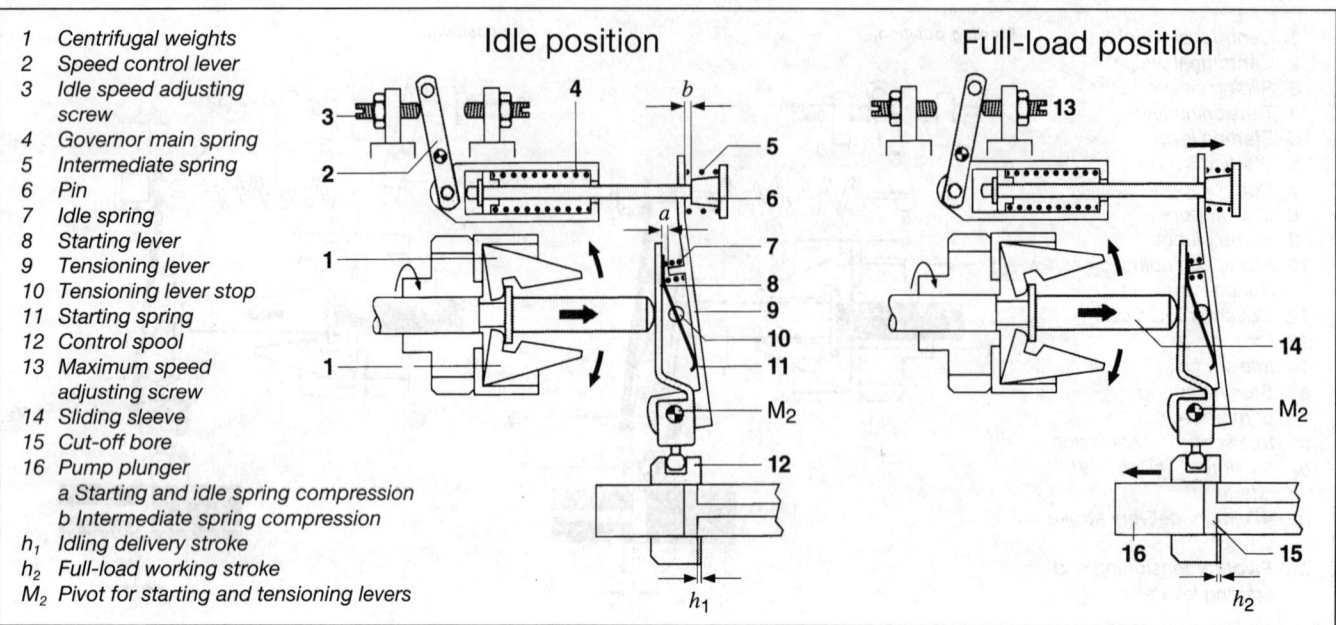

1 Centrifugal weights
2 Speed control lever
3 Idle speed adjusting screw
4 Governor main spring
5 Intermediate spring
6 Pin
7 Idle spring
8 Starting lever
9 Tensioning lever
10 Tensioning lever stop
11 Starting spring
12 Control spool
13 Maximum speed adjusting screw
14 Sliding sleeve
15 Cut-off bore
16 Pump plunger
  a Starting and idle spring compression
  b Intermediate spring compression
h_1 Idling delivery stroke
h_2 Full-load working stroke
M_2 Pivot for starting and tensioning levers

6.82 Bosch VE pump 2-speed governor at idle and under load

© Robert Bosch Limited

oppose the increased spring tension. The lever assembly is moved back towards reduced fuel delivery, until the force of the sliding sleeve and the force of the governor spring are once more in balance. If the load is constant, engine speed and fuel delivery will stabilise at this new level.

76 If the load on the engine is now reduced – for instance, because the top of a hill has been reached – the speed will begin to increase, because the same amount of fuel is being delivered. The increased speed will move the weights outwards, and the lever assembly will be moved further towards reduced delivery until the balance of forces is restored.

77 If the engine load is removed completely – for instance when going downhill and using the engine as a brake – the governor spring tension is at a minimum. The force of the sliding sleeve predominates, and moves the lever assembly into the minimum-delivery position. If the engine speed approaches the maximum no-load speed, delivery is reduced to zero.

Two-speed governor (Bosch VE pump)

78 This type of governor is only active at idle and maximum speeds. Intermediate speeds are controlled directly by the position of the accelerator pedal.

79 The construction of the governor is very similar to the variable-speed governor just described. The main difference is in the governor spring, which is a compression spring, working inside a guide. The pin at the top of the tensioning lever extends into the spring guide.

Starting

80 With the centrifugal weights stationary, the sliding sleeve is in its rest position (fully retracted). The starting spring forces the starting lever away from the tensioning lever, and the control spool is moved to the maximum-delivery position.

81 Once the engine is turning and the accelerator pedal is released, the centrifugal weights move outwards, and the sliding sleeve moves to overcome the force of the starting spring. The control spool is moved to the idle-delivery position.

Idling

82 The idle spring on this governor is located between the tension lever and the starting lever. The principle of operation remains the same: spring force and sliding force are opposed and in balance at the set idle speed. The starting spring is also contributing something to the total spring tension **(see illustration)**.

On the road

83 As the accelerator pedal is depressed, the engine speed control lever moves and overcomes the force of the starting and idle springs. Force is exerted on the tensioning lever via the intermediate spring. The intermediate spring serves to cushion the transition from governed to ungoverned operation.

84 Further depression of the accelerator pedal overcomes the force of the intermediate spring. The lever assembly moves directly in accordance with the accelerator pedal position. The governor spring and guide act as a rigid coupling at intermediate speeds.

85 As the engine speed approaches maximum, the sliding sleeve force increases enough to be able to overcome the force of the governor spring. The spring is compressed, and the lever assembly moves to reduce fuel delivery.

LUCAS/CAV DP PUMPS AND GOVERNORS

Lucas/CAV DP pumps

Source and maker's name

86 DP series pumps are made in various European countries by the Lucas group of companies. Depending on age and source, pumps will be found bearing the following makers' names, sometimes in combination:

  CAV
  ConDiesel
  Lucas
  RotoDiesel

87 Various different types of pumps may be encountered, but the types which will be found most commonly on passenger and light commercial vehicles are the DPA, DPS and DPC, and these will be described in detail in this Section **(see illustration)**. There are various differences between the three types, but the basic principles of operation common to them all will be described first.

Basic principles

88 The DP series pumps are self-contained distributor pumps, with integral transfer pump and governor, suitable for high-speed diesel engines. The pumps are flange-mounted, and may be driven by gear or sprocket. Lubrication is entirely by the fuel passing through the pump.

89 The pump draws in fuel by means of its transfer pump, which is of the sliding vane type. A high-pressure pump element produces the pressure necessary for injection. Fuel at high pressure is directed to each outlet port in turn by the movement of the pump rotor in the hydraulic head.

90 A proportion of the fuel is not injected, but

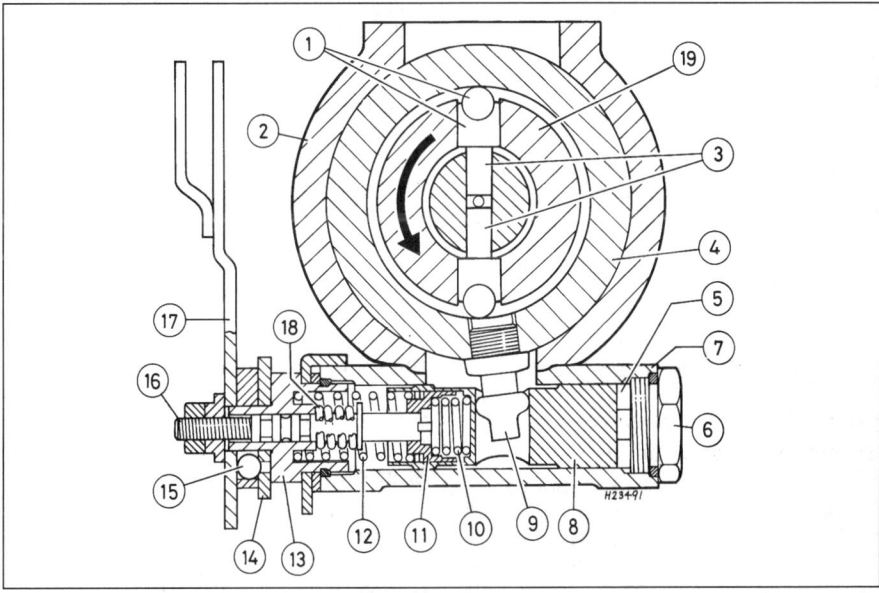

1 Fuel inlet
2 Transfer pressure regulating valve
3 Transfer pump
4 Metering valve
5 High-pressure pump
6 Injection timing piston
7 Damper
8 Governor weights
9 Governor linkage
10 Speed control (throttle) lever
11 Stop lever
12 Drivegear
13 Fuel return
14 Hydraulic head
15 High-pressure outlet

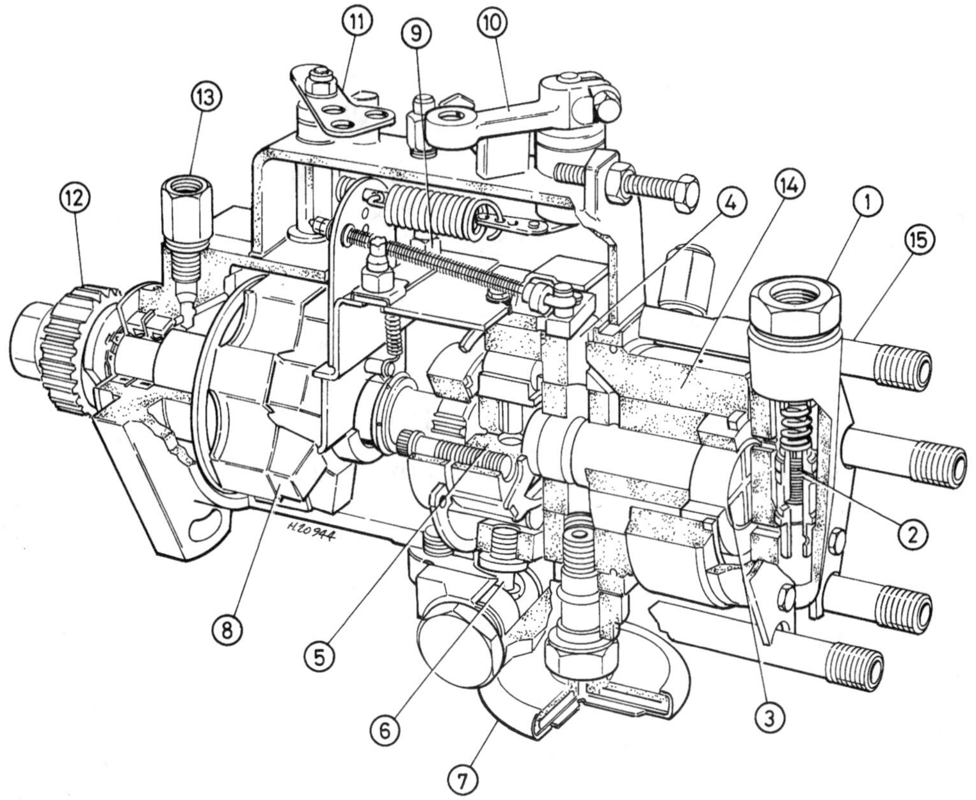

6.87 Cutaway view of a Lucas/CAV DPA pump with mechanical governor

returns to the fuel tank along with leaked-off fuel from the injectors. A spring-loaded valve maintains a residual pressure (known as *internal pressure* or *cam box pressure*) in the fuel destined to be returned; this pressure serves to keep air, dirt and water out of the pump, and contributes to smooth operation of the governor.

Transfer pump

91 Fuel is drawn in by the transfer pump, and pressurised to a value determined by a regulating valve on the pump endplate. This value, known as *transfer pressure*, varies in relation to pump speed.

Metering valve

92 Fuel at transfer pressure is passed to the high-pressure pump via a metering valve. The metering valve opens and closes in response to control lever position and to the governor. Between the metering valve and the high-pressure pump, the fuel is said to be at *metering pressure*.

High-pressure pump

93 The high-pressure pumping element consists of a pair of opposed plungers working in a common bore contained in the pump rotor **(see illustration)**. The outer ends of the plungers rest on rollers running in a cam ring. As the rotor is turned by the pump driveshaft, the cam lobes act on the rollers, and the plungers are forced towards each other.

6.93 High-pressure pump and automatic advance mechanism with manual advance lever – Lucas/CAV DPA pump

1 Roller and shoe	6 Plug	13 End cap
2 Pump housing	7 Advance piston housing	14 Detent plates
3 High-pressure pump plungers	8 Advance piston	15 Ball
4 Cam ring	9 Advance screw	16 Spindle
5 Transfer pressure chamber	10 Retard spring	17 Manual advance lever
	11 Plunger	18 Spindle spring
	12 Advance spring	19 Rotor

2

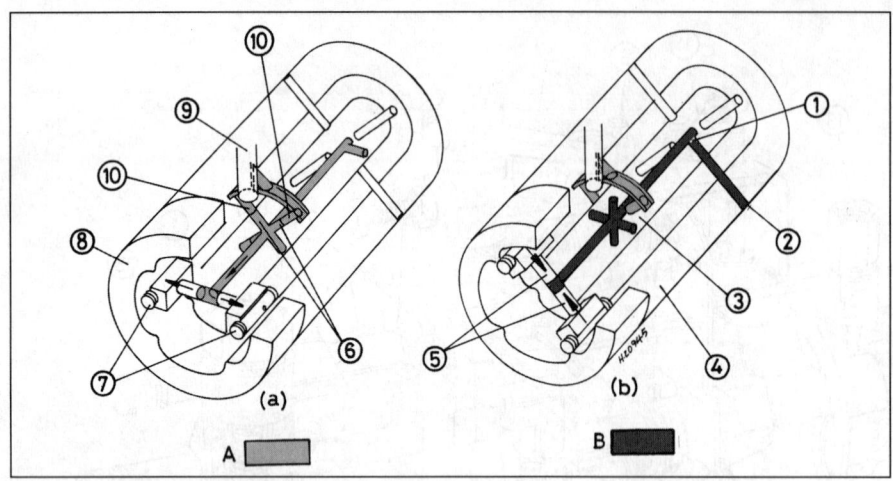

6.100 Fuel charging (a) and injection (b) – Lucas/CAV DP series pump

A	Metering pressure	3	Rotor	7	Rollers
B	Injection pressure	4	Hydraulic head	8	Cam ring
1	Delivery port	5	Pump plungers	9	Metering valves
2	Outlet port	6	Inlet ports	10	Metering ports

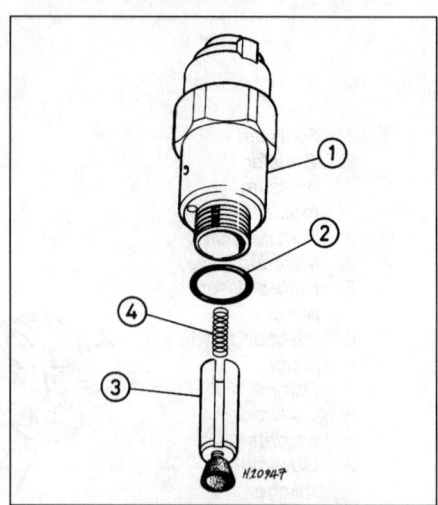

6.104 Fuel shut-off solenoid –
Lucas/CAV DP series pump

1	Valve body	3	Plunger
2	O-ring	4	Spring

Rotor and hydraulic head

94 As well as the high-pressure pump, the rotor contains a number of fuel inlet ports, and a single distributor port. The hydraulic head has a single metering port, and as many outlet ports as there are cylinders. As the rotor turns, the inlet ports align with the metering port, and the distributor port aligns with the outlet ports.

Control of fuel delivery

95 The quantity of fuel delivered per pumping stroke depends on three inter-related factors:
  a) *Metering pressure (varies with transfer pressure and metering valve opening).*
  b) *Transfer pressure (increases with pump speed).*
  c) *Duration of alignment of inlet port with metering port (reduces pump speed).*
96 On some versions of the pump, an auxiliary pair of plungers supplies the excess fuel delivery needed during start-up. Fuel supply to the auxiliary plungers is controlled by an excess-fuel delivery valve, which is closed by the increase in transfer pressure when the engine starts. On other versions, the excess fuel is provided by allowing the plungers to move further apart than normal.

Control of injection timing

97 The cam ring is not fixed, but can be rotated within certain limits by a piston. Movement of the piston is determined by spring pressure on one side, and by fuel transfer pressure on the other side. Increasing transfer pressure moves the cam ring against the direction of pump rotation, so advancing injection timing with increasing speed.
98 This device can only used to retard the injection timing during start-up, taking advantage of the fact that transfer pressure is zero before start-up, and low at cranking speeds.
99 The timing advance device can be fitted

with an external manual advance control, for improvement of cold idling.

The pump working cycle

100 The movement of the pump rotor brings one or two inlet ports into line with the corresponding metering port(s). Fuel at metering pressure flows into the rotor, and forces the pump plungers apart. The higher the metering pressure, the further the plungers will be moved, and the greater the fuel charge for the forthcoming injection. Except at maximum delivery, the pump plungers will not move out to the limit of their travel before commencing the injection stroke **(see illustration)**.
101 Continued movement of the rotor closes the inlet port. The distributor port comes into line with one of the outlet ports in the hydraulic head; at the same time, the plungers are forced towards each other by the movement of the rollers riding up on the cam ring lobes. The fuel is then pressurised, and injection takes place.
102 Further movement of the rotor takes the distributor port out of alignment with the outlet port. The next inlet port comes into line with the metering port, and the cycle beings again.

Shut-off

103 Mechanical shut-off (when provided) is a function of the governor, and is described later. Electrical shut-off by means of a solenoid is almost universal on modern vehicles.
104 The shut-off solenoid is screwed into the top of the hydraulic head. It contains a spring-loaded plunger, which in its extended position interrupts the fuel passage between the transfer pump and the metering valve **(see illustration)**.
105 When the solenoid is energised, the plunger is retracted, and the fuel at transfer pressure passes to the metering valve. When the 'ignition' is switched off, the spring forces

the plunger out. The fuel passage is interrupted, and the engine stops.

Lucas DPA pump

Description

106 The DPA pump is the earliest and simplest of the DP series. Operation is as described in the preceding paragraphs.

Air venting

107 An external vent screw is fitted, either on the pump top cover (mechanical governor) or on the pump body (hydraulic governor). This screw must be opened when bleeding the fuel system.

Lucas DPS pump

Description

108 The DPS pump is a development of the DPA type, intended specifically for passenger car and light commercial applications **(see illustration)**.
109 The transfer pump and high-pressure pump work on the same principle as those in the DPA pump, but there are detail differences. These are described in the following paragraphs.

Air venting

110 Air is automatically bled from the pump at cranking speed. This is achieved by two valves, known as the *latch valve* and the *rotor switch valve*. At cranking speed, the latch valve is closed by a combination of spring pressure and differential fuel pressure. The rotor switch valve is open, allowing fuel and air from the rotor to vent into the pump cam box.
111 With increasing speed, transfer pressure rises and opens the latch valve. The opened latch valve applies transfer pressure to the rotor switch valve, closing it.

Fuel pressure circuits

112 These are shown in the accompanying illustration **(see illustration)**. It will be seen

1   Idle lever
2   Governor main spring
3   Maximum fuel adjuster
4   Speed control (throttle) lever
5   Excess-fuel shaft
6   Metering valve
7   Shut-off valve
8   Hydraulic head
9   Venting orifice
10 Transfer pressure regulating valve
11 Transfer pump
12 Rotor
13 Head-locating fitting and damper
14 Roller and shoe
15 Automatic advance unit
16 Manual advance lever
17 Cam ring
18 Driveshaft rear bearing
19 Governor weight retainer
20 Driveshaft front bearing
21 Driveshaft
22 Governor thrust sleeve
23 Scroll plates
24 Idling spring
25 Fuel return union and residual pressure valve

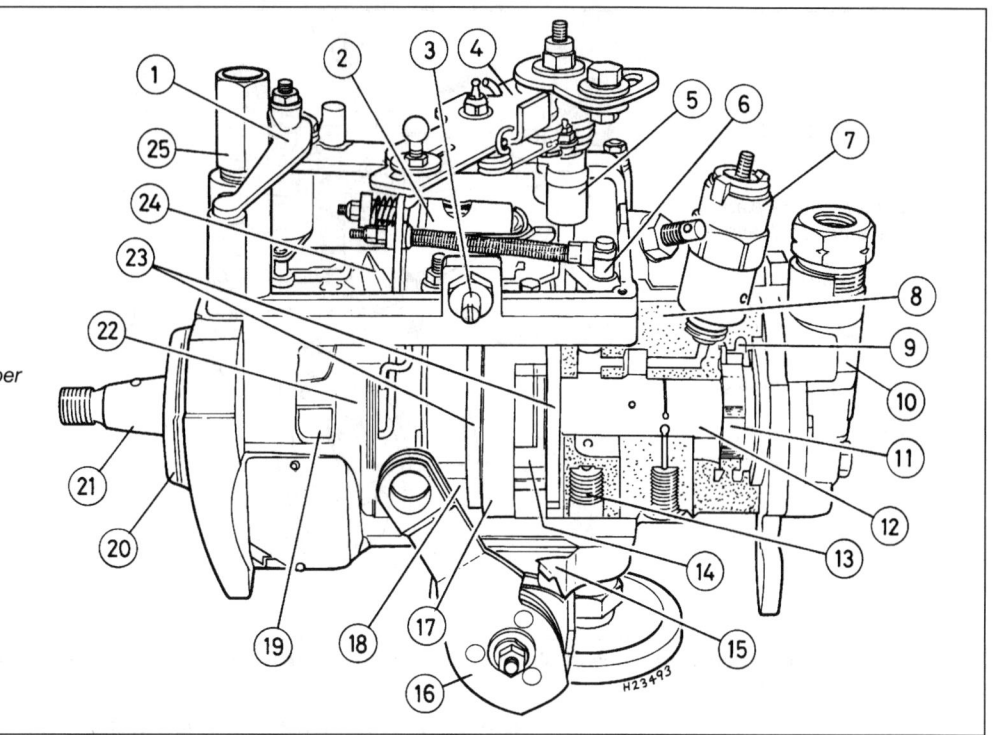

6.108 Cutaway view of Lucas/CAV DPS pump

A   Injection pressure
B   Transfer pressure
C   Metering pressure
D   Differential pressure
E   Feed pressure
F   Cam box (residual) pressure
G   Back leakage (return) pressure
1   Residual pressure valve
2   Fuel tank
3   Control (throttle) lever shaft
4   Water trap
5   Metering valve
6   Shut-off solenoid
7   Vent orifice
8   Fuel filter
9a Lift pump
9b Hand-priming pump
10 Transfer pressure regulator
11 Transfer pump
12 Hydraulic head and rotor
13 Latch valve
14 Manual advance lever
15 Advance piston
16 Head-locating fitting
17 Injector
18 Rotor vent switch valve
19 Governor
20 Cam box
21 Idle lever shaft

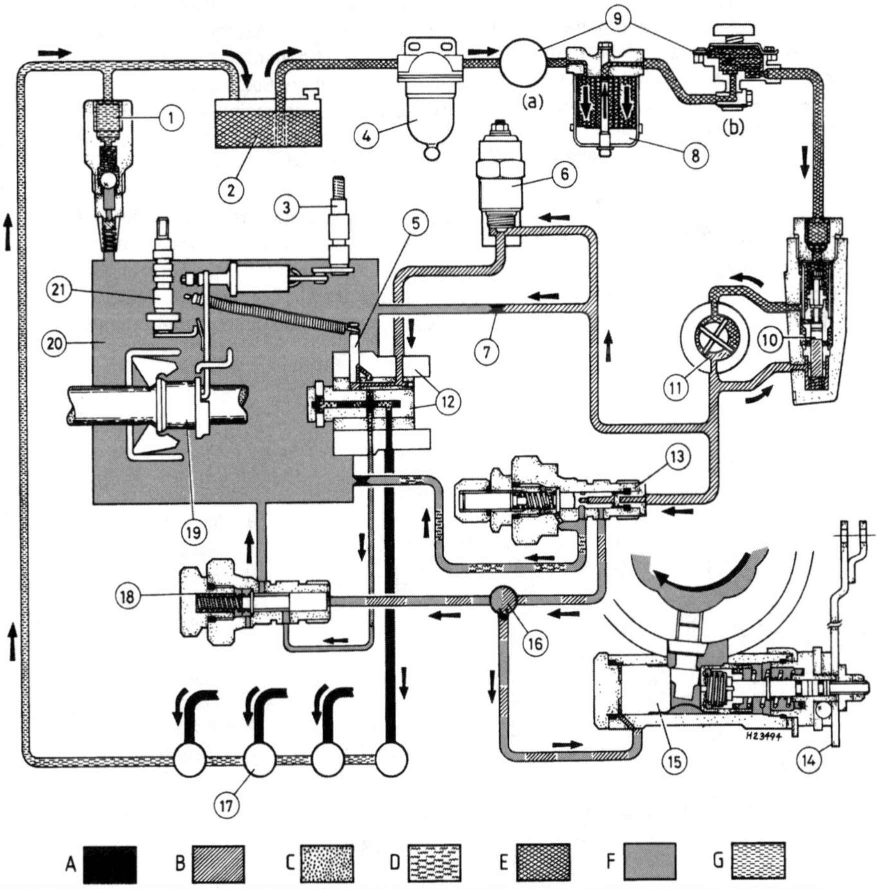

A ◼   B ▨   C ▨   D ▨   E ▨   F ▨   G ▨

6.112 Fuel hydraulic pressure diagram – Lucas/CAV DPS pump

2

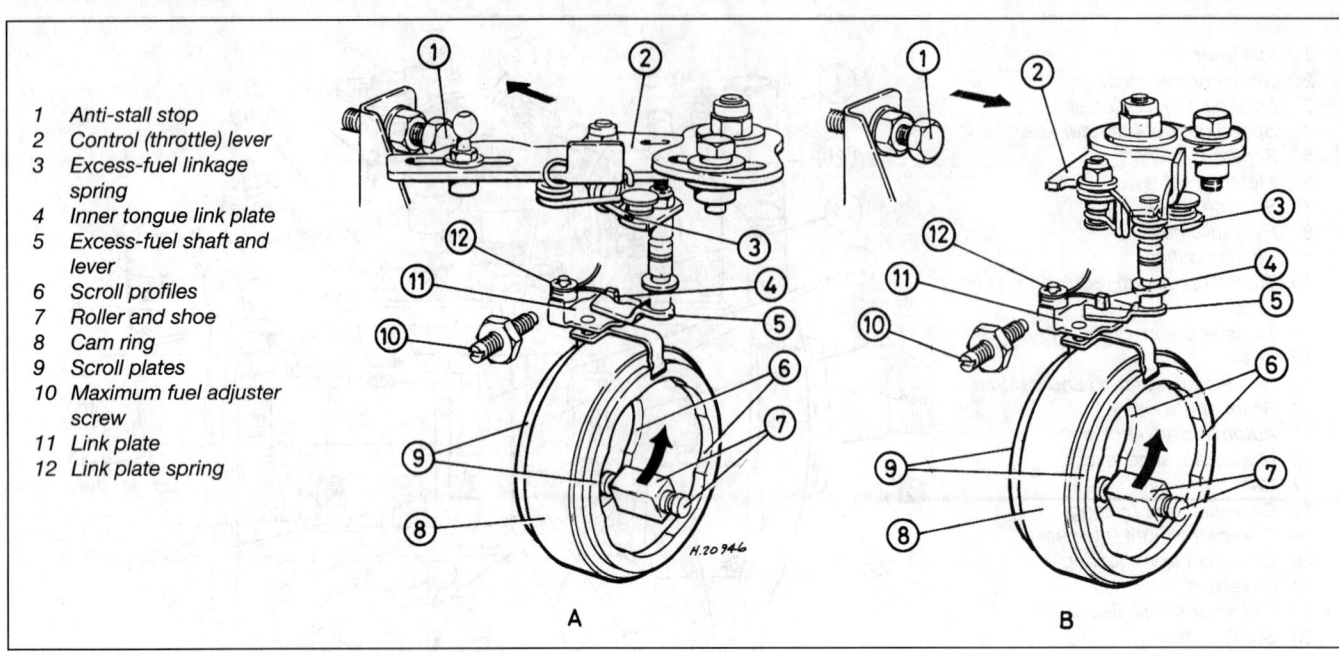

1 Anti-stall stop
2 Control (throttle) lever
3 Excess-fuel linkage spring
4 Inner tongue link plate
5 Excess-fuel shaft and lever
6 Scroll profiles
7 Roller and shoe
8 Cam ring
9 Scroll plates
10 Maximum fuel adjuster screw
11 Link plate
12 Link plate spring

6.114 Scroll plate mechanism (Lucas/CAV DPS pump) in excess-delivery position (A) and normal maximum-delivery position (B)

that the opening of the latch valve is used to apply transfer pressure to the automatic advance piston. In this way, injection timing is fully retarded during start-up.

113 Differential fuel pressure is generated at cranking speed by the effect of two restrictors in the latch valve hydraulic circuit. Its effect is to make the speed at which the valve opens higher than that at which it closes. This is necessary to prevent the latch valve closing during rapid deceleration, so retarding the timing.

Excess and maximum fuel delivery

114 Two scroll plates, one on each side of the cam ring, control automatic delivery of excess fuel during start-up, and limit maximum fuel

delivery during normal running **(see illustration)**.

115 At start-up with the accelerator pedal released, the scroll plates are positioned so as to allow the pump plunger rollers to move further apart. The metering valve is open, and excess fuel delivery takes place.

116 As soon as the engine starts, the governor closes the metering valve to the idling position, and excess fuel delivery ceases.

117 During normal running with the accelerator pedal fully depressed, the position of the scroll plates determines the maximum outward movement of the pump plunger rollers, and thus maximum fuel delivery. This position of the scroll plates is determined by a

link plate coming up against the maximum-fuel adjuster screw.

Timing advance device and head-locating fixing

118 The timing advance device is secured to the underside of the pump by a fitting which also serves to locate the hydraulic head in the pump housing **(see illustration)**. This head-locating fitting contains the passages which transmit hydraulic pressure from the head to the advance device.

119 The head-locating fitting may also incorporate a damper assembly, to even out fluctuations in transfer pressure.

Delivery valves

120 The delivery valves are fitted around the hydraulic head; each valve connects a high-pressure outlet to its injector pipe **(see illustration)**. The function of the delivery valve

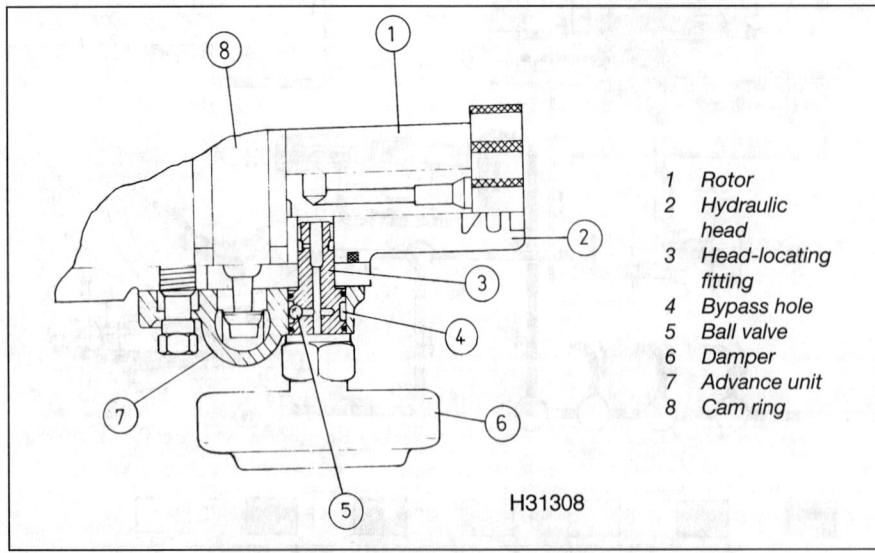

1 Rotor
2 Hydraulic head
3 Head-locating fitting
4 Bypass hole
5 Ball valve
6 Damper
7 Advance unit
8 Cam ring

H31308

6.118 Head-locating fitting – Lucas/CAV DPS pump

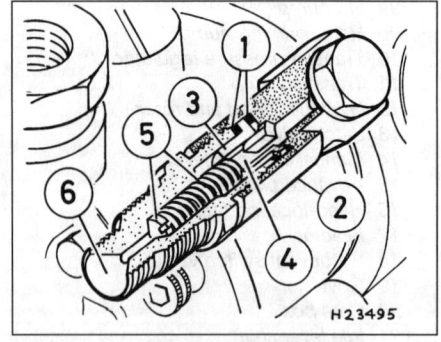

6.120 Cutaway view of a delivery valve – Lucas/CAV DPS pump

1 Washers
2 Banjo connection
3 Valve body
4 Valve piston
5 Spring and peg
6 Valve holder

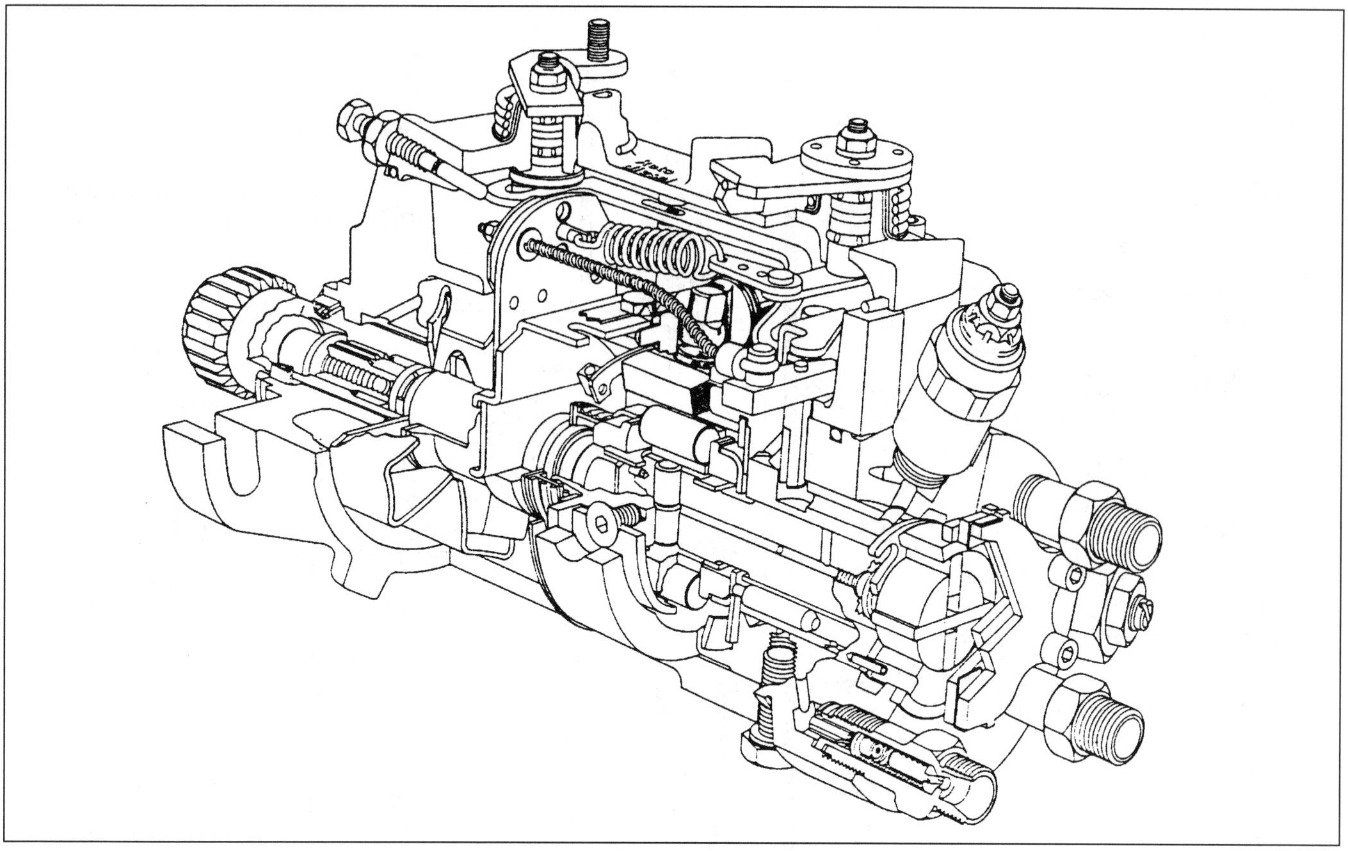

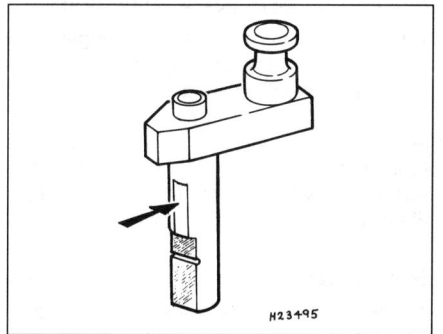

6.122 Cutaway view of a Lucas DPC pump

is to ensure a clean cut-off at the end of injection, minimising dribble and smoke. It also maintains a residual pressure in the injector pipes between injections, and protects the pipes against cavitation.

Lucas DPC pump

Description

121 The DPC pump differs from the DPS mainly in the means by which excess fuel delivery is obtained for start-up. If also has a greater variety of governors and advance mechanisms. Versions are available which are suitable for use with turbochargers.

122 Externally, the DPC pump is easily identified by having axial (in-line) high-pressure outlets, as opposed to the radial outlets found on the DPA and DPS pumps **(see illustration)**.

Air venting

123 Three different self-venting systems exist for this pump. Which one is used will depend on the fitted position of the pump, since air drawn in will accumulate at the highest point, and it is here that venting must take place.

The three venting points are **(see illustrations)**:

a) *At the metering valve, by means of a flat on the valve shaft.*
b) *Through a jet and flat at the hydraulic head securing screw.*
c) *At the excess-fuel delivery valve.*

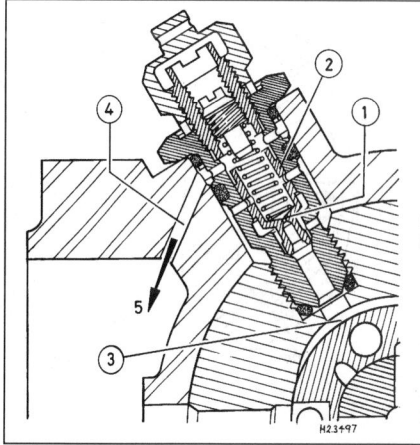

6.123c Air venting via excess-fuel delivery valve – Lucas/CAV DPC pump

1	Calibrated orifice	3	Transfer pressure groove
2	Valve stem	4	Passage
		5	Back-leakage

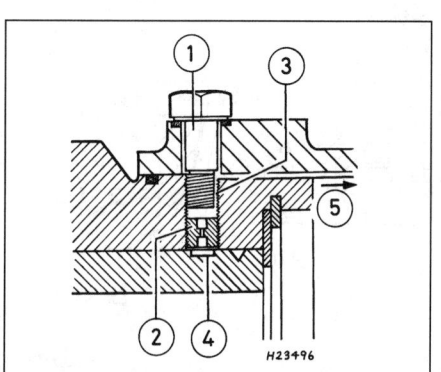

6.123b Venting jet at hydraulic head securing screw – Lucas/CAV DPC pump

1	Securing screw	4	Transfer pressure groove
2	Venting orifice		
3	Flat	5	Back-leakage

6.123a Venting flat (arrowed) on Lucas DPC pump delivery valve shaft

2

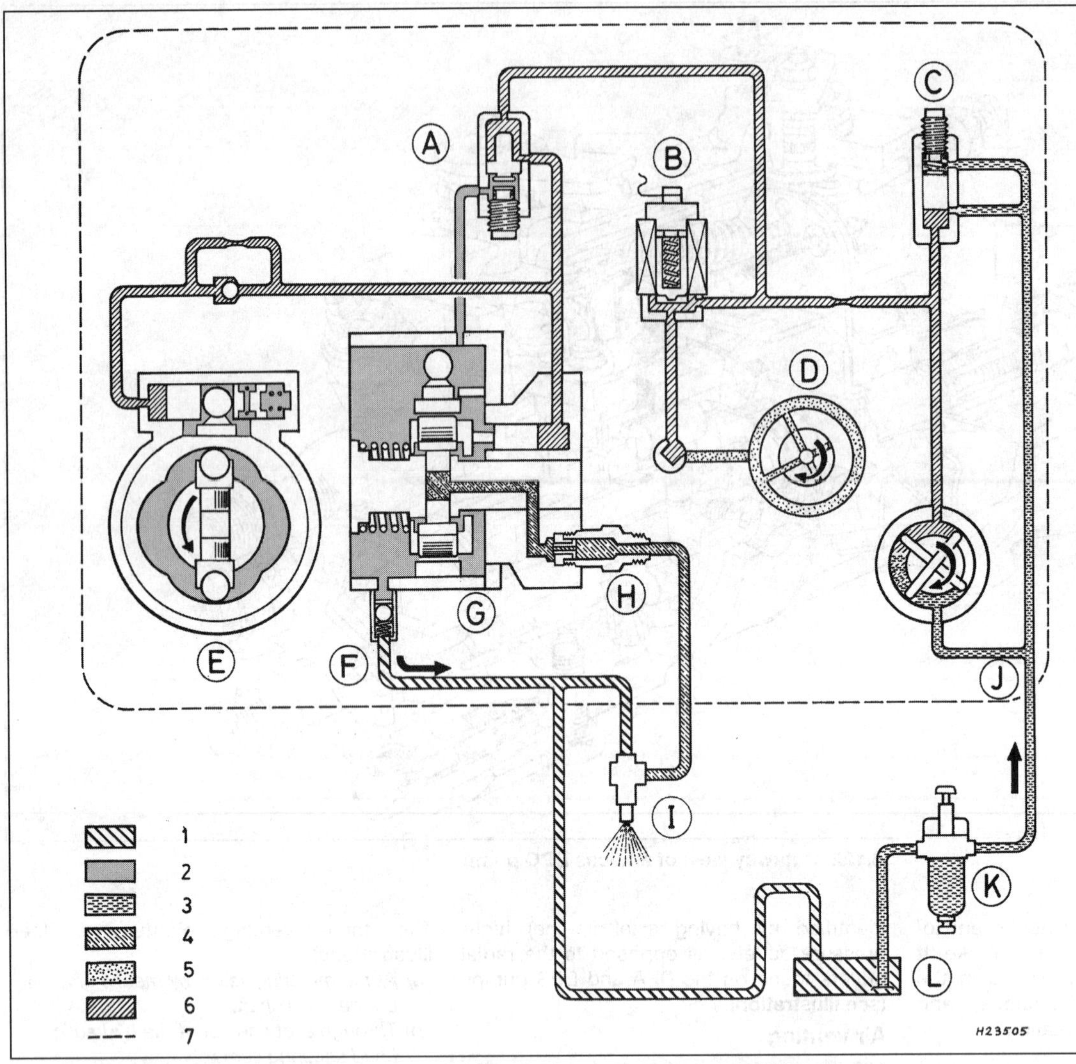

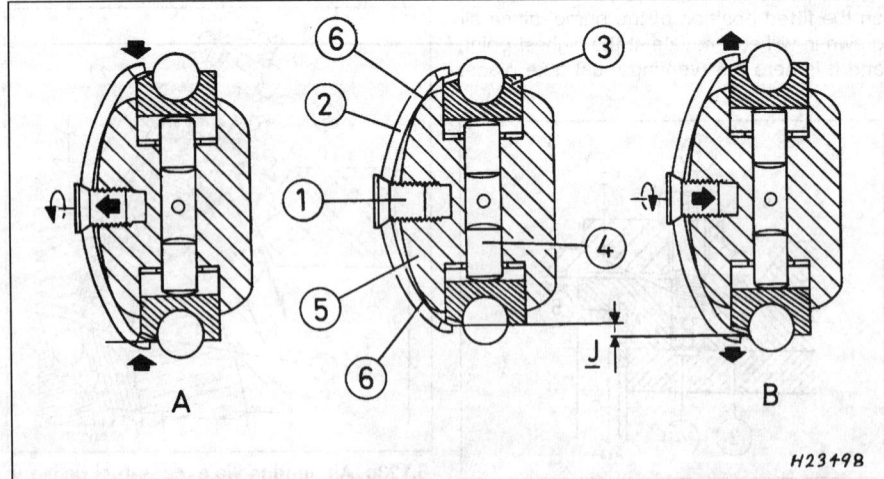

6.124 Fuel pressure circuits – Lucas DPC pump

A Differential valve
B Stop solenoid (shut-off valve)
C Transfer pressure regulator
D Inlet to hydraulic head
E Automatic advance
F Pump leak-back
G Hydraulic head
H Metering valve
I Injector
J Pump inlet/filter
K Main filter
L Fuel tank
1 Return pressure
2 Internal pressure
3 Feed pressure
4 Injection pressure
5 Metering pressure
6 Transfer pressure
7 External limits of pump

H23505

6.125 Maximum fuel delivery adjustment – Lucas DPC pump

1 Adjustment screw
2 Maximum-fuel adjustment plate
3 Shoe
4 Pump plunger
5 Rotor
6 Plate-to-rotor contact points
A Adjustment in direction of reduced delivery
B Adjustment in direction of increased delivery

H23498

Fuel pressure circuits

124 The only component unique to the DPC pump is the *differential valve* (also known as the *excess-fuel delivery valve*). This valve controls the operation of the excess-fuel delivery system, as described later **(see illustration)**.

Maximum fuel delivery regulation

125 The outward travel of the plunger roller shoes, and thus the maximum fuel delivery, is limited by the position of a spring strip known as the *maximum fuel adjustment plate*. An adjustment screw increases or decreases the tension of this strip, so moving the ends further apart or closer together **(see illustration)**.

Excess fuel delivery

126 Excess fuel for start-up is obtained by allowing the pump plungers to move further apart. This is achieved by the use of slots and castellations in the plunger roller shoes and

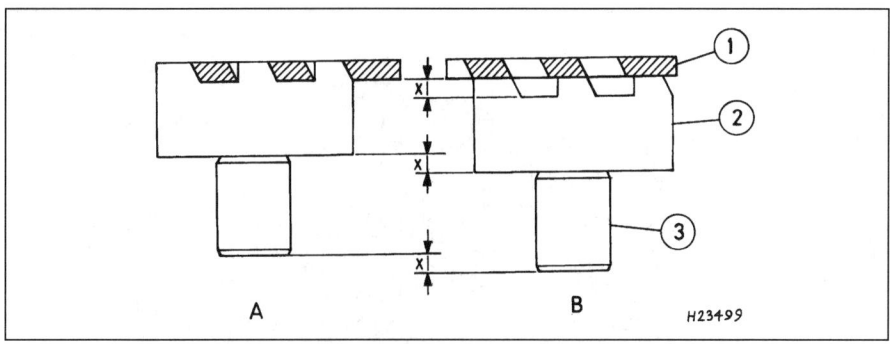

6.126 Excess-fuel delivery slots and castellations – Lucas DPC pump

1 Maximum-fuel adjustment plate	2 Shoe	A Excess delivery position
	3 Plunger	B Normal delivery position
		X Difference in stroke

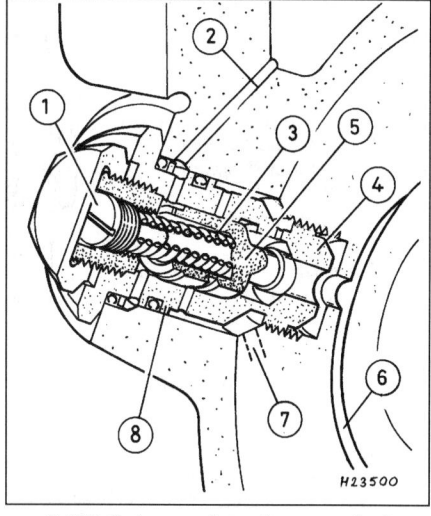

6.127 Cutaway view of excess-fuel delivery valve – Lucas DPC pump

1 Adjusting screw	6 Transfer pressure groove
2 Fuel port	7 Port connecting excess-fuel pistons
3 Return spring	8 Port connecting hollow stud
4 Hollow stud	
5 Valve stem	

the maximum fuel delivery plate **(see illustration)**. If the slots and castellations mesh, the plungers can move further out than if the castellations are in alignment.

127 Meshing of the slots and castellations is controlled by the excess-fuel delivery carriage. This is a mechanism which moves the plunger shoes in or out of mesh, in response to spring pressure on one side of the carriage, and hydraulic pressure on the other side. The application of hydraulic pressure is controlled by the excess-fuel delivery valve **(see illustration)**.

128 With the engine stopped, there is no hydraulic pressure, and the spring holds the carriage in the excess-fuel delivery position. At cranking speed, transfer pressure is too low to open the excess-fuel delivery valve, so the carriage stays in position and excess fuel is delivered **(see illustration)**.

129 When the engine starts, transfer pressure rises to a high enough level to open the excess-fuel delivery valve. This allows transfer pressure to be applied to the excess-fuel delivery pistons, which move the carriage

against spring pressure into the normal delivery position **(see illustration)**.

Control of injection timing

130 The basic automatic advance device is similar to that fitted to the DPS pump, except that the application of transfer pressure is determined by the opening of the excess-fuel delivery valve.

131 Refinements of the basic automatic advance system are described in the following paragraphs.

Manual advance override

132 This is a mechanical device which acts on the spring side of the automatic advance piston. The driver operates a 'cold idling' control which moves a lever on the side of the pump. Movement of the lever pulls the advance piston spring away from the piston, allowing the piston to move further in the advance direction.

Automatic advance override

133 This system is only fitted in conjunction with the low-load advance system. A

mechanical link between the idle lever and the low-load advance piston allows a spigot on the piston to enter a hole when the lever is in the fast idle (cold start) position. The piston thus moves to reduce spring tension, and allows greater movement of the automatic advance piston **(see illustrations)**.

134 When the accelerator pedal is depressed, load pressure increases, and the low-load advance piston moves off its seat. If the engine is now warm, the idle lever moves away from the fast idle position, and the hole is no longer in alignment with the piston spigot. The advance override is cancelled until the next cold start.

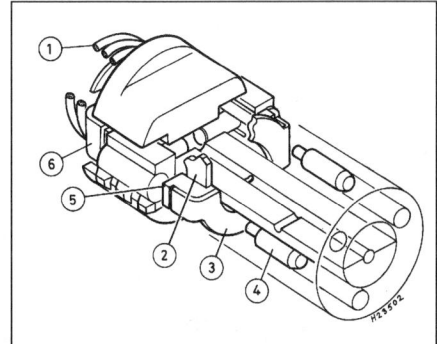

6.128 Excess-fuel delivery carriage in excess-fuel delivery position – Lucas DPC pump

1 Spring	4 Excess-fuel delivery piston
2 Spacer	5 Thrust pad
3 Rear retaining plate	6 Front retraining plate

6.129 Excess-fuel delivery carriage in normal delivery position – Lucas DPC pump

1 Spring	4 Excess-fuel delivery piston
2 Spacer	5 Thrust pad
3 Rear retaining plate	6 Front retraining plate

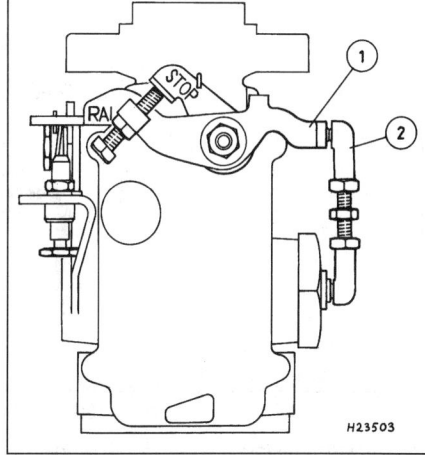

6.133a Automatic advance override external components – Lucas DPC pump

1 Idling lever	2 Connecting link

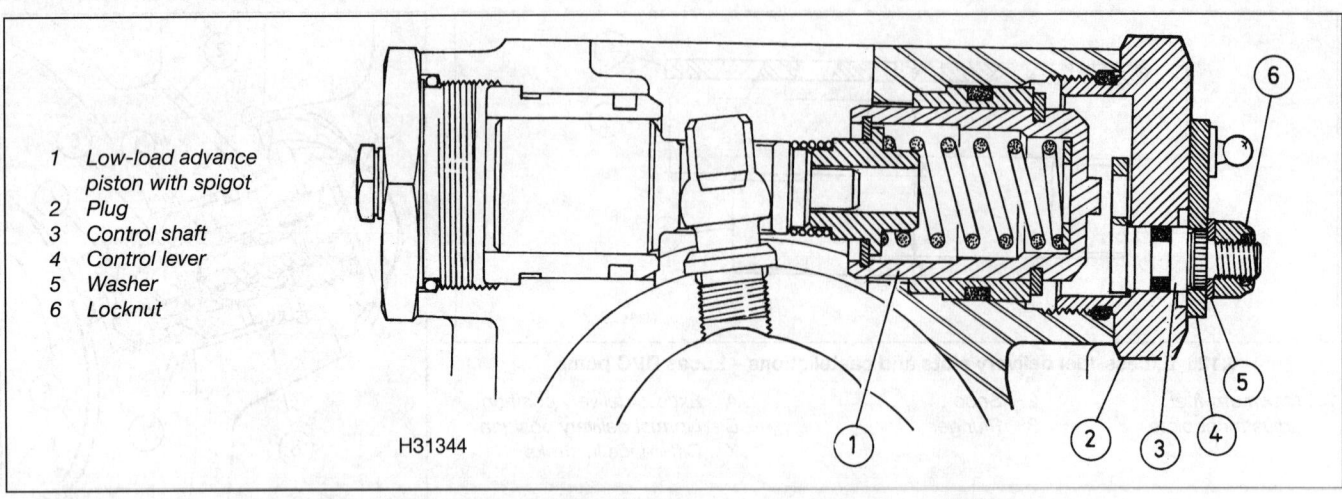

1 Low-load advance
   piston with spigot
2 Plug
3 Control shaft
4 Control lever
5 Washer
6 Locknut

H31344

6.133b Automatic advance override internal components – Lucas DPC pump

Low-load advance

135 This system advances injection timing at light loads. It works by applying a variable hydraulic pressure, known as *load pressure*, to a low-load advance piston positioned on the spring (retard) side of the automatic advance piston **(see illustration)**.

136 Load pressure has a value somewhere between transfer pressure and residual (cam box) pressure. It is controlled by a low-load advance valve and by a variable orifice. The orifice may be part of the metering valve, or

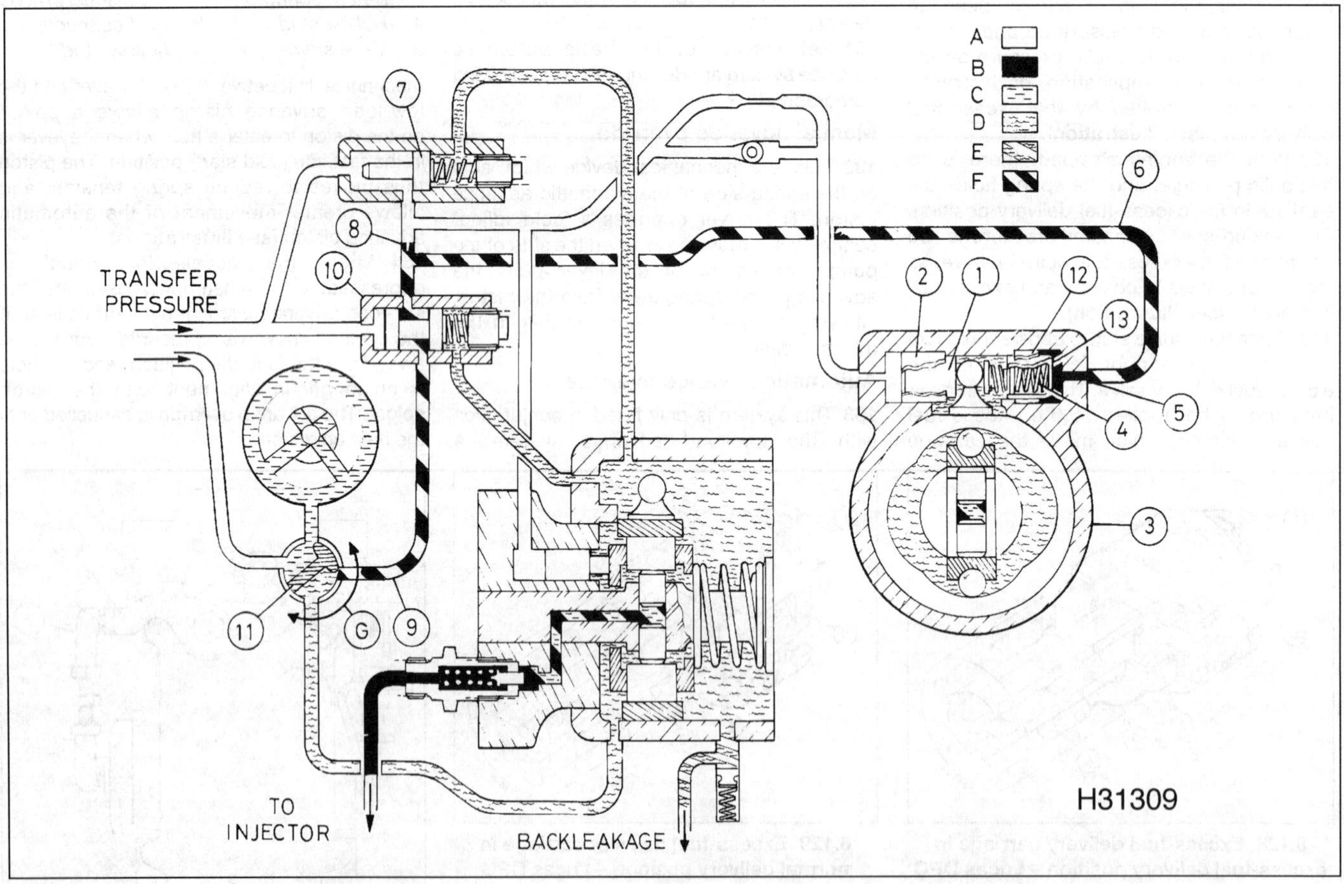

TRANSFER PRESSURE

TO INJECTOR

BACKLEAKAGE

H31309

6.135 Load pressure circuit – low-load advance system (Lucas DPC pump)

1	Advance piston	5	Low-load advance	8	Calibrated orifice	12 Low-load advance
2	Sleeve		piston	9	Duct	sleeve
3	Pump housing	6	Load pressure circuit	10	Low-load advance	13 Circlip
4	Advance spring	7	Excess-fuel delivery		valve	A Transfer pressure
			valve	11	Metering valve	B Injection pressure

C Metering pressure
D Cambox pressure
E Back-leakage pressure
F Load pressure
G Variable orifice

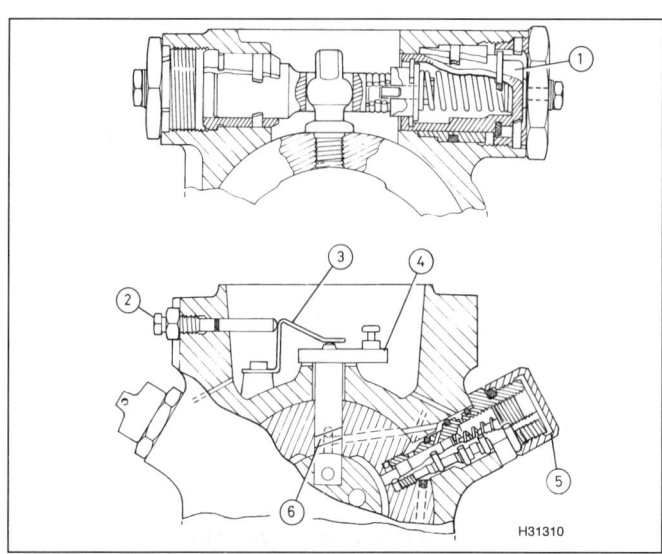

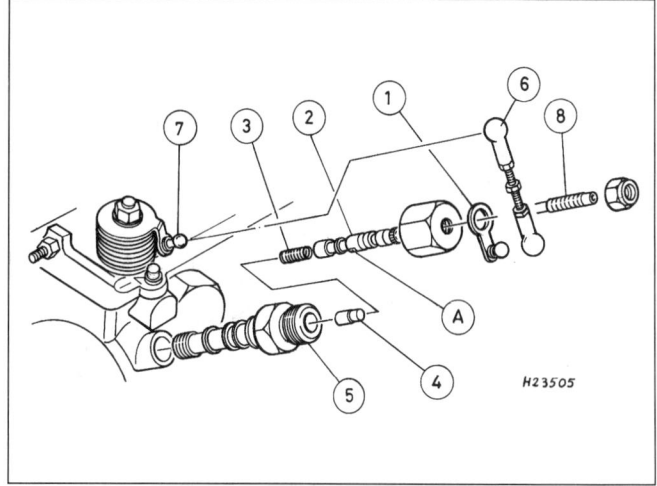

6.136a Low-load advance components with orifice on metering valve – Lucas DPC pump

1 Low-load advance piston
2 Adjuster screw
3 Metering valve stop
4 Metering valve
5 Low-load advance valve
6 Variable orifice

6.136b Low-load advance valve with external connection to control lever – Lucas DPC pump

1 Lever
2 Shaft
3 Spring
4 Cut-off piston
5 Low-load advance valve
6 Link
7 Control (throttle) lever
8 Adjuster screw
A Variable orifice

part of the low-load advance valve. In the latter case, the low-load advance valve has an external connection to the control lever **(see illustrations)**.

137 At low speed and load (metering valve opening or control lever movement small), the variable orifice is open, and load pressure is low. Transfer pressure acts on the advance piston unopposed by the low-load advance piston; a large degree of advance takes place.

138 Movement of the metering valve or control lever with increasing load reduces the variable orifice in the load pressure circuit, and causes load pressure to rise. The increased pressure moves the low-load advance piston off its seat and compresses the spring, reducing the degree of advance.

139 On some versions, low-load advance is not applied above a certain speed, as determined by transfer pressure.

Turbo boost controller

140 The function of the turbo boost controller is to increase maximum fuel delivery in proportion to turbo boost pressure.

141 Besides the boost controller itself, the system makes use of a modified excess-fuel delivery carriage **(see illustration)**. Instead of the slots and castellations machined in the plunger roller shoes and the maximum-fuel delivery plate, there is an inclined plane. Excess delivery is thus continuously variable, instead of being an 'on/off' function.

142 Increasing boost pressure acts on the diaphragms in the controller, and moves a piston **(see illustration)**. The piston uncovers a passage in the hydraulic circuit, which leads to the excess-fuel delivery pistons. Pressure in the hydraulic circuit is reduced, and the spring moves the excess-fuel delivery carriage to a position where maximum delivery is increased.

Hydraulic governor (Lucas DPA pump)

143 This governor is of the all-speed type. Its operation depends on the fact that pump transfer pressure varies with pump speed.

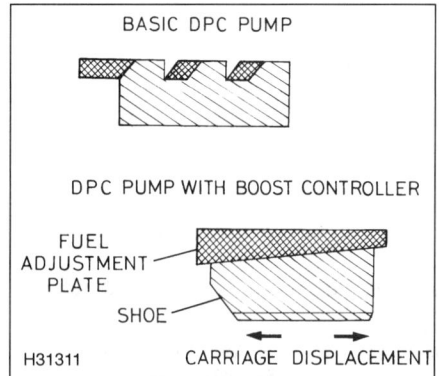

6.141 Modified excess-fuel delivery carriage used with turbo boost controller – Lucas DPC pump

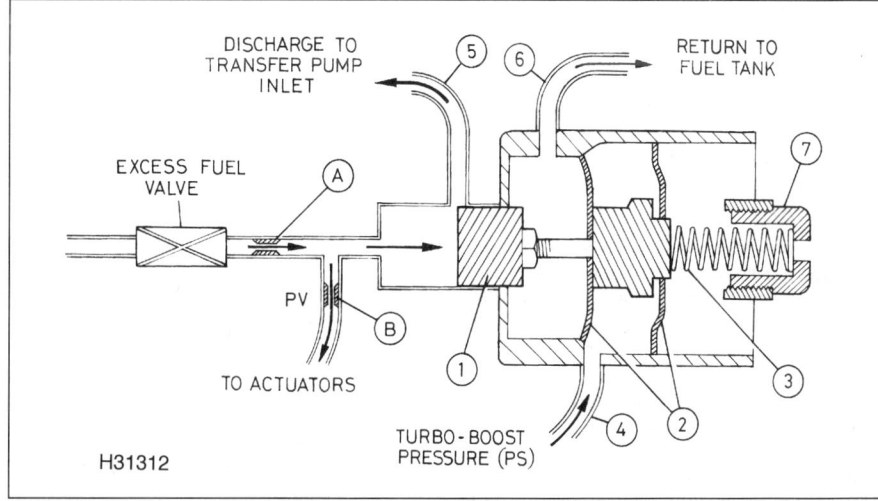

6.142 Action of turbo boost controller with increasing pressure – Lucas DPC pump

1 Piston
2 Diaphragms
3 Spring
4 Boost pressure correction
5 Fuel discharge passage
6 Fuel return passage
7 Adjuster
A Orifice
B Orifice
Pv Actuator pressure

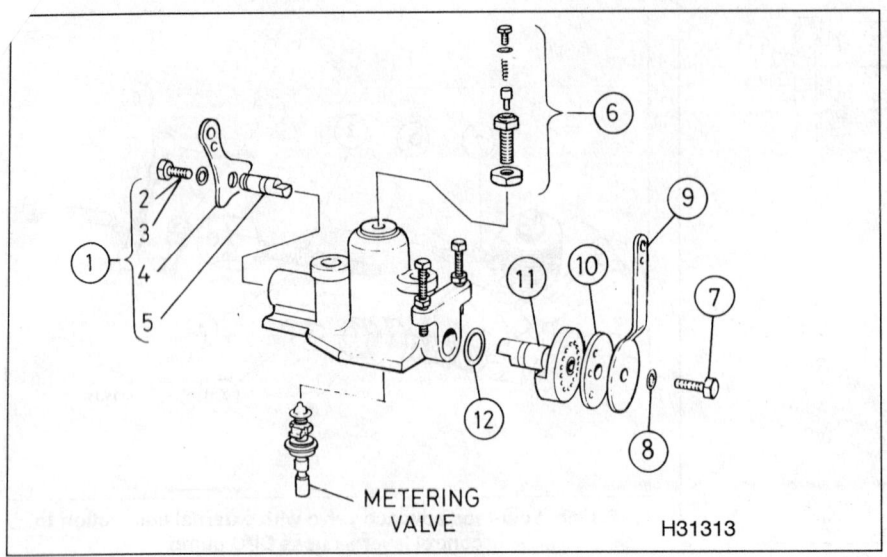

METERING VALVE

H31313

6.144a Hydraulic governor components – Lucas DPA pump

1 Shut-off control assembly	5 Shaft
2 Screw	6 Anti-stall stop
3 Lockwasher	7 Screw
4 Lever	8 Lockwasher

9 Control (throttle) lever
10 Plate
11 Control plate and shaft
12 O-ring

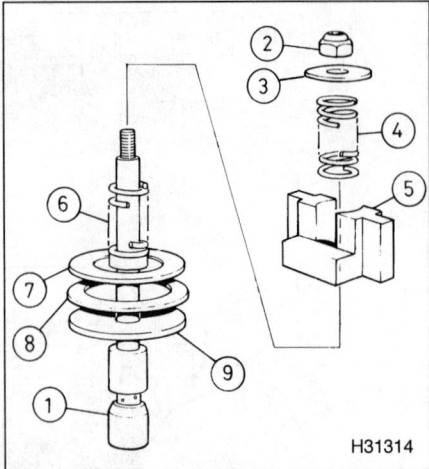

H31314

6.144b Metering valve and control sleeve components – Lucas DPA pump hydraulic governor

1 Metering valve	6 Metering valve
2 Nut	spring
3 Shut-off washer	7 Damper washer
4 Idle spring	8 Floating washer
5 Control sleeve	9 Sealing washer

144 The opening of the metering valve is determined by the balance between transfer pressure on one side, and spring pressure on the other side. Spring pressure is varied by movement of the pump control (throttle) lever, which is transmitted by means of a shaft with a flat on it and a control sleeve (see illustrations).

Idling

145 The resting position of the control lever is set by the idle speed adjusting screw. The flat on the governor shaft depresses the control sleeve, tending to open the metering valve. Transfer pressure and an idle speed spring work in the opposite direction, tending to close the valve.

146 If engine speed rises, increasing transfer pressure will lift the metering valve. Fuel delivery will be reduced, transfer pressure will fall, and the metering valve will open again. In practise, idle speed will stabilise at the point where the various forces are in balance.

On the road

147 Depression of the accelerator pedal turns the governor shaft and depresses the control sleeve further, against the metering valve spring. The metering valve will open until transfer pressure and spring pressure forces are in balance.

148 If engine load increases, speed will fall. Transfer pressure will reduce, and the metering valve will open further. If load decreases, the reverse will occur.

149 At maximum load, the control lever is against its stop, and the downward pressure on the metering valve is at its highest. An increase in engine speed to the governed maximum will lift the metering valve, and reduce or stop fuel delivery.

Anti-stall

150 An adjustable stop in the top of the governor housing limits upward movement of the metering valve during rapid deceleration, keeping fuel delivery high enough to prevent stalling.

Shut-off

151 Mechanical shut-off is achieved using a separate rod and lever, Movement of the lever turns the rod and lifts the control sleeve, closing the delivery valve.

Centrifugal all-speed governor (Lucas DPA and DPC pumps)

152 A set of centrifugal weights is mounted on the pump driveshaft. Movement of the weights is transmitted by an arm and spring link to the fuel metering valve. The force exerted by the weight is opposed by the governor main spring, the tension of which is varied by movement of the control (throttle) lever (see illustration).

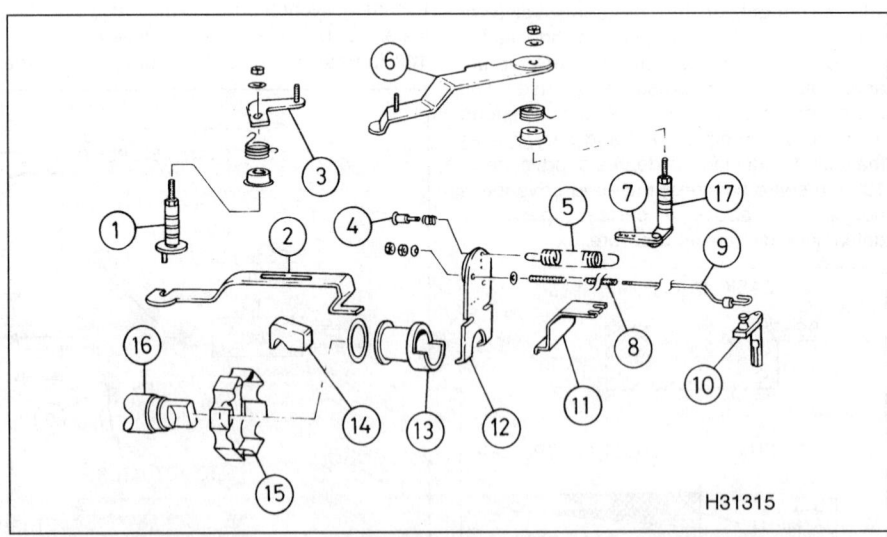

H31315

6.152 Centrifugal all-speed governor and shut-off mechanisms – Lucas DPC pump

1 Shut-off shaft	7 Governor spring link	13 Thrust sleeve
2 Shut-off bar	8 Linkage hook spring	14 Centrifugal weight
3 Shut-off lever	9 Linkage hook	15 Weight retainer
4 Idle spring guide	10 Metering valve	16 Driveshaft
5 Governor main spring	11 Control bracket	17 Control shaft
6 Control (throttle) lever	12 Governor arm	

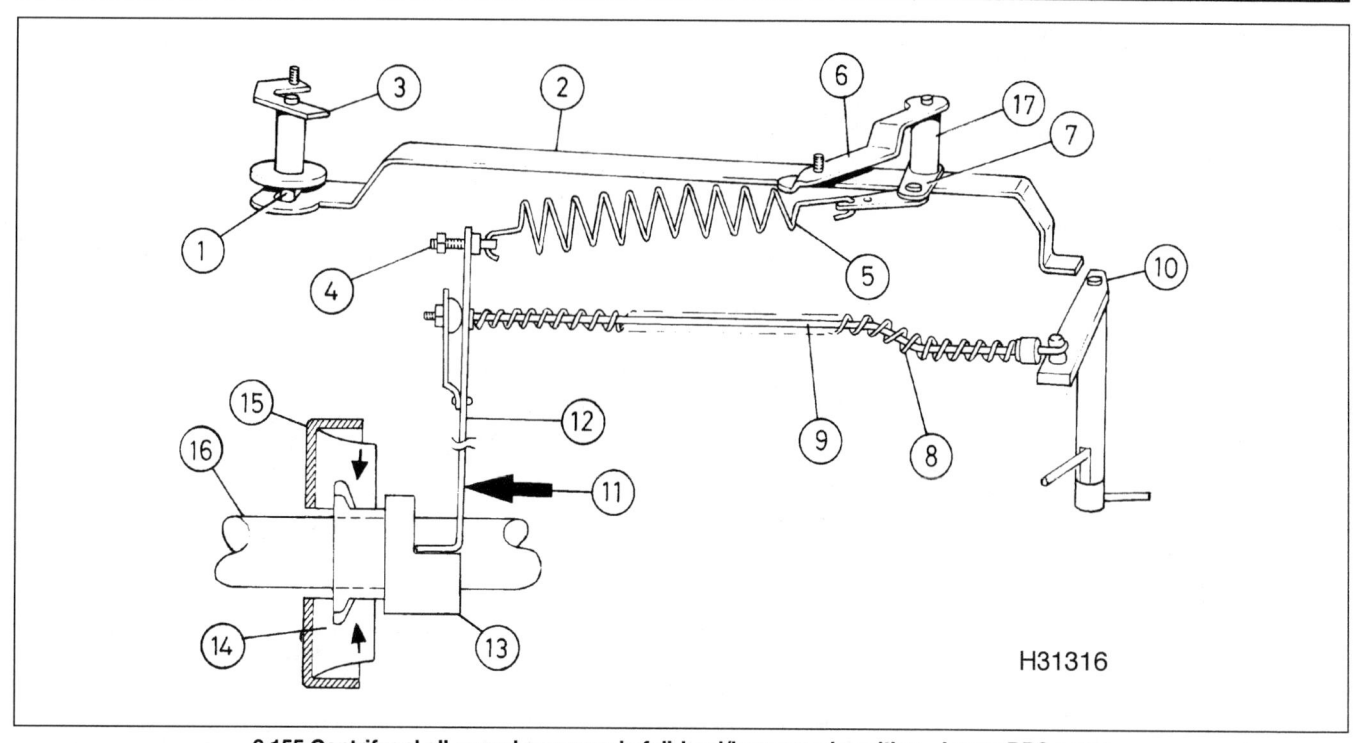

6.155 Centrifugal all-speed governor in full-load/low-speed position – Lucas DPC pump

1 *Shut-off shaft*	6 *Control (throttle) lever*	10 *Metering valve*	14 *Centrifugal weight*
2 *Shut-off bar*	7 *Governor spring link*	11 *Control bracket*	15 *Weight retainer*
3 *Shut-off lever*	8 *Linkage hook spring*	12 *Governor arm*	16 *Driveshaft*
4 *Idle spring guide*	9 *Linkage hook*	13 *Thrust sleeve*	17 *Control shaft*
5 *Governor main spring*			

Idling

153 The resting position of the control lever is set by the idle speed adjusting screw. The control lever pulls on the governor arm in the direction of increased fuel delivery; outward movement of the centrifugal weights with increasing speed creates a force in the opposite direction.

154 For greater sensitivity in the idle speed range, the governor main spring is attached to the governor arm by a pin and an idle speed spring. It is the idle speed spring which determines the balance of forces.

On the road

155 When the accelerator pedal is depressed, the control lever is moved to increase governor spring tension (**see illustration**). The governor arm moves to open the metering valve, so increasing the quantity of fuel delivered. The idle speed spring is compressed, and has no more effect.

156 As engine speed rises, the force exerted by the weights begins to overcome the spring tension. The governor arm moves to close the metering valve, reducing fuel delivery.

157 At maximum governed speed, the force of the weights is sufficient to close the metering valve completely (**see illustration**).

Anti-stall

158 An anti-stall adjustment screw limits the movement of the governor arm during rapid deceleration, keeping fuel delivery high enough to prevent stalling.

Shut-off

159 Mechanical shut-off is achieved by a separate lever, which when operated compresses the governor spring link and closes the metering valve.

Centrifugal two-speed governor (Lucas DPC and DPS pumps)

160 This governor is similar to the all-speed mechanical governor described previously,

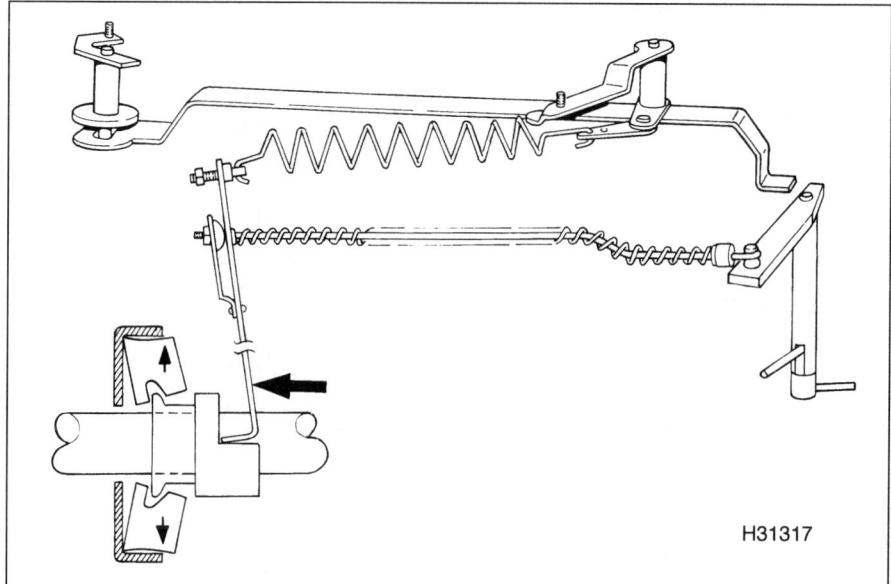

6.157 Centrifugal all-speed governor in maximum-speed position

2

ut it is only effective at idle and maximum speeds; at intermediate speeds, control lever movement is transmitted directly to the metering valve **(see illustration)**. This gives an accelerator response similar to that obtained from petrol engines.

Idling

161 At idle, the balance of forces is between the governor weights and the idle leaf spring. The tension of this spring is determined by the position of the idle actuator, attached to the idle lever shaft. In this way, idle speed adjustment is independent of control lever resting position.

On the road

162 At intermediate speeds, the governor has no effect. The main governor spring acts as a rigid link between the control lever and the metering valve.

163 As maximum speed is approached, the force of the governor weights overcomes that of the main spring, and the metering valve is moved in the direction of reduced fuel delivery.

Anti-stall

164 An anti-stall device between the governor arm and the end of the governor spring limits governor arm movement during sudden deceleration, keeping fuel delivery high enough to prevent stalling. Adjustment of the anti-stall function is effectively determined by the control lever resting position.

Shut-off

165 When fitted, manual shut-off may be controlled by a separate lever on the idling lever shaft, or by a 'stop' position of the existing idling lever.

7 Injectors

Note: *The information given in this Section refers to conventional diesel fuel injectors. For details of the injectors used in common rail and 'pump injector' systems, refer to Sections 9 and 10 respectively.*

1 One fuel injector is fitted to each cylinder. The function of the injector is to spray an evenly-atomised quantity of fuel into the combustion or pre-combustion chamber when the fuel pressure exceeds a certain value, and to stop the flow of fuel cleanly when the pressure drops. It must make a gas-tight seal where it enters the cylinder head, and a pressure-tight seal where the fuel injector pipe is connected to it. It must also have a fuel return connection, where fuel which leaks past the injector needle can be collected for return to the tank.

2 Atomisation is achieved by a spring-loaded needle, which vibrates rapidly against its seat when fuel under pressure passes it. The

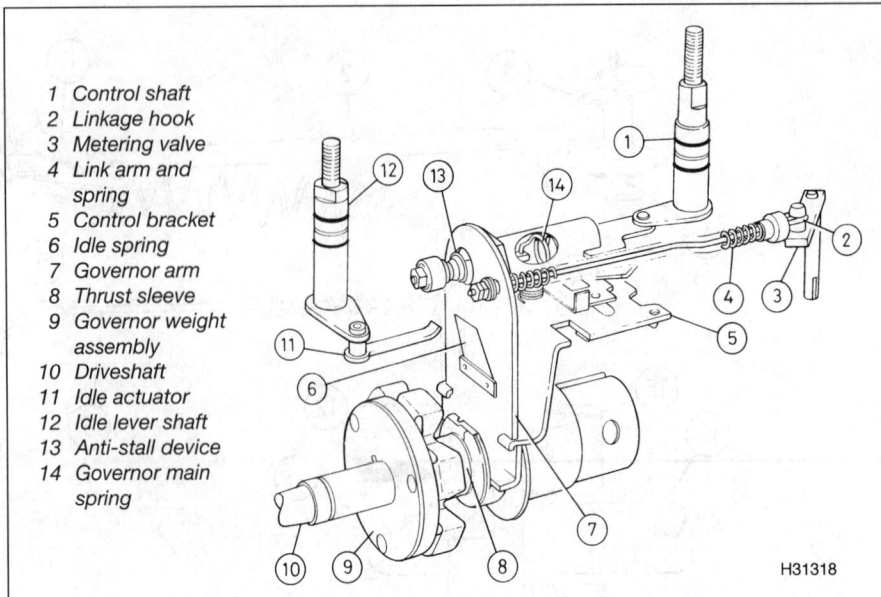

1 Control shaft
2 Linkage hook
3 Metering valve
4 Link arm and spring
5 Control bracket
6 Idle spring
7 Governor arm
8 Thrust sleeve
9 Governor weight assembly
10 Driveshaft
11 Idle actuator
12 Idle lever shaft
13 Anti-stall device
14 Governor main spring

H31318

6.160 Lucas DPS pump two-speed governor

needle and seat assembly together are known as the injector nozzle. They are finely matched in production, and may only be renewed together.

3 The pressure at which the nozzle opens is determined by the spring pressure. This is adjustable, usually by shim thickness, but sometimes by means of an adjusting screw.

4 The direction, penetration and quality of fuel spray are all important features of injector design. Each type of engine and combustion chamber has particular requirements which its injectors are designed to meet. Injectors are not necessarily interchangeable between engines, even though their external dimensions and appearance may be identical.

5 There are two basic types of injector; hole (orifice) and pintle. Hole types are usually fitted to direct injection engines, and pintle types to indirect injection units.

6 Injectors may be screwed directly into the cylinder head, or they may be secured by a clamp or flange. Clamp fitting enables the fitted position of the injector to be controlled, for example to ensure that an asymmetrical hole pattern is correctly positioned with regard to the combustion chamber.

Hole injectors

7 The nozzle seat and the injector tip are conical. The atomised fuel is discharged into a so-called blind hole, which is in fact opened by one or more injection holes. Multi-hole arrangements are common.

8 A single-hole injector may have the hole positioned either centrally or offset. The holes in a multi-hole injector are not necessarily arranged symmetrically, nor need they all be the same size. The Pintaux type nozzle has two holes; a main one, centrally positioned, and an auxiliary one offset to one side.

9 A two-stage hole injector is used in certain Rover and VW engines. Each injector contains two springs. **(see illustration)** The weaker spring allows a small 'pre-injection' spray to

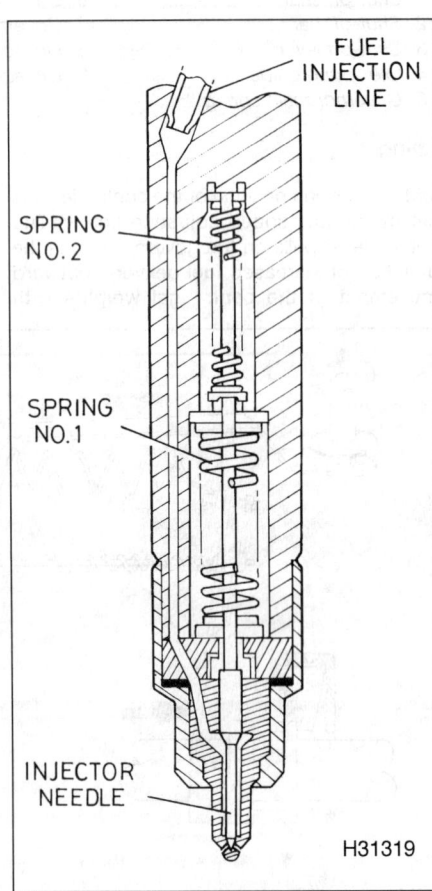

FUEL INJECTION LINE

SPRING NO. 2

SPRING NO. 1

INJECTOR NEEDLE

H31319

7.9 Sectional view of a two-stage hole injector

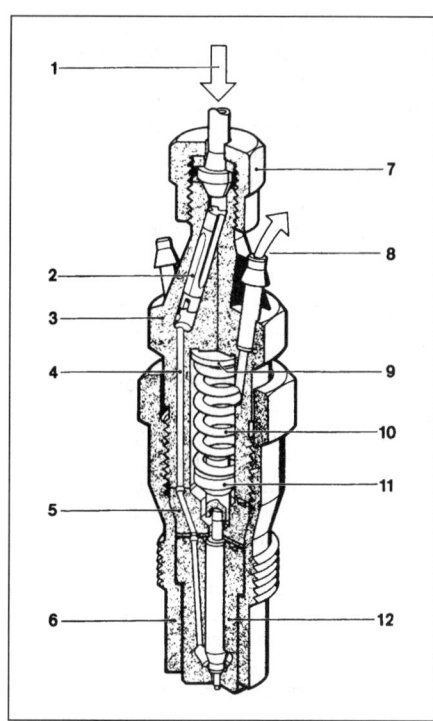

7.12a Cutaway view of a pintle injector

© Robert Bosch Limited

1	Fuel inlet	7	Union nut
2	Integral filter	8	Fuel return
3	Body	9	Pressure
4	Pressure		adjusting shim
	passage	10	Spring
5	Sleeve	11	Spindle
6	Nozzle retainer	12	Nozzle

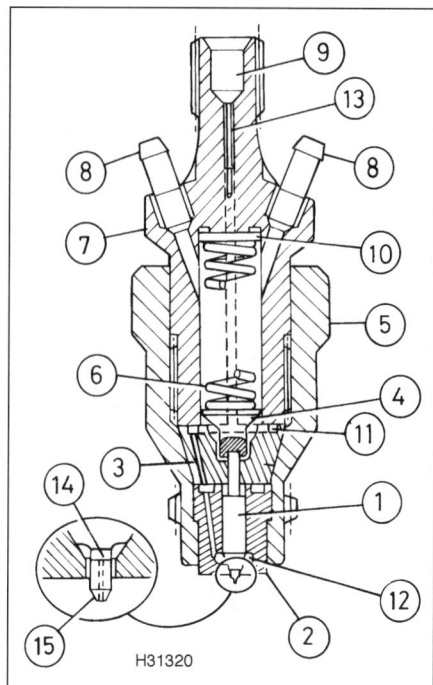

H31320

7.12b Sectional view of a perforated pintle injector

1	Nozzle needle	9	Fuel inlet
2	Nozzle body	10	Pressure-
3	Sleeve		adjusting shim
4	Spindle	11	Ring groove
5	Nozzle retainer	12	Pressure chamber
6	Spring	13	Integral filter
7	Body	14	Transverse bore
8	Fuel return	15	Longitudinal bore

occur before the main injection; as with pintle injectors, this gives a softer and quieter combustion.

10 The nozzle holes on a brand-new injector are slightly larger than they ultimately need to be. This is to allow for the carbon coating which will inevitably accumulate in service. While the holes are still too large, combustion may be harsher and noisier. This should be borne in mind when judging the noise produced by a new engine, or after fitting new injectors.

11 Opening pressures for hole type injectors are typically between 150 and 250 bar.

Pintle injectors

12 Throttling pintle injectors as they are properly known, have a nozzle seat shaped to match the pintle on the tip of the nozzle needle. The seat and pintle are shaped so that initial opening produces a small 'pre-injection' spray, which builds up relatively gradually into the main injection spray. Spray pattern and penetration will be determined by the shape of the needle and seat, and by spring rate. Sometimes the needle tip has longitudinal and transverse bores (*perforated pintle*) **(see illustrations)**.

13 The progressive injection characteristics of this type of injector, in conjunction with swirl or pre-combustion chamber design, gives smooth combustion.

14 Opening pressures for pintle-type injectors are typically between 110 and 140 bar.

8 Electronic diesel engine control systems

Description

1 In recent years, the development of the automotive diesel engine has become governed by the requirement for low exhaust emissions, and improved fuel economy, performance and driveability. These requirements place increasingly heavy demands on the fuel injection system due to the need for improved accuracy and sensitivity of fuel delivery and injection timing controls.

2 In the diesel engine, combustion, and therefore the operating characteristics are influenced by:

The quantity of fuel injected.
The start-of-injection point.
Exhaust gas recirculation.
Intake (charge) air pressure.

To ensure maximum efficiency, these variables must be adjusted to the optimum values for the prevailing operating conditions. The use of an electronic control system allows this, and also provides improved reliability compared to a mechanically-controlled system.

3 An electronic diesel control system uses various sensors (many similar to those used on a petrol injection system), an electronic control unit, and various actuators. The information from the various sensors is passed to an electronic control unit (ECU), which evaluates the signals. The ECU contains electronic 'maps' which enable it to calculate the optimum quantity of fuel to inject, and the appropriate injection timing, for any given condition of engine operation. The maps can take into account particular requirements to optimise factors such as noise, smoke emission, fuel consumption or power output. In order to perform the necessary calculations, the ECU receives signals from sensors or transducers providing the following information (not all sensors will be found on all systems):

Crankshaft speed and position.
Accelerator pedal position.
Gear lever or selector position.
Injection start point.
Pump control rod/spool position.
Intake airflow.
Road speed.
Air temperature.
Coolant temperature.
Fuel temperature.
Atmospheric pressure.
Inlet manifold pressure.
Brake/clutch switches.

4 Besides controlling injection pump delivery and injection timing, the ECU can also be used to control the preheating system, exhaust gas recirculation and, where applicable, a cruise control system **(see illustration overleaf)**. 'Limp-home' and emergency functions enable the system to keep functioning (albeit at reduced efficiency) if a sensor fails. A self-diagnosis function illuminates an instrument panel warning light in case of failure; using appropriate test equipment, the ECU can be interrogated to locate the problem rapidly.

Injection pumps

5 Electronic diesel control systems are used on both in-line and distributor injection pumps, and the fine degree of control available has also allowed the development of common rail and 'pump-injector' systems, both of which are described in more detail in Sections 9 and 10 of this Chapter.

In-line injection pumps

6 In an electronically-controlled in-line

2

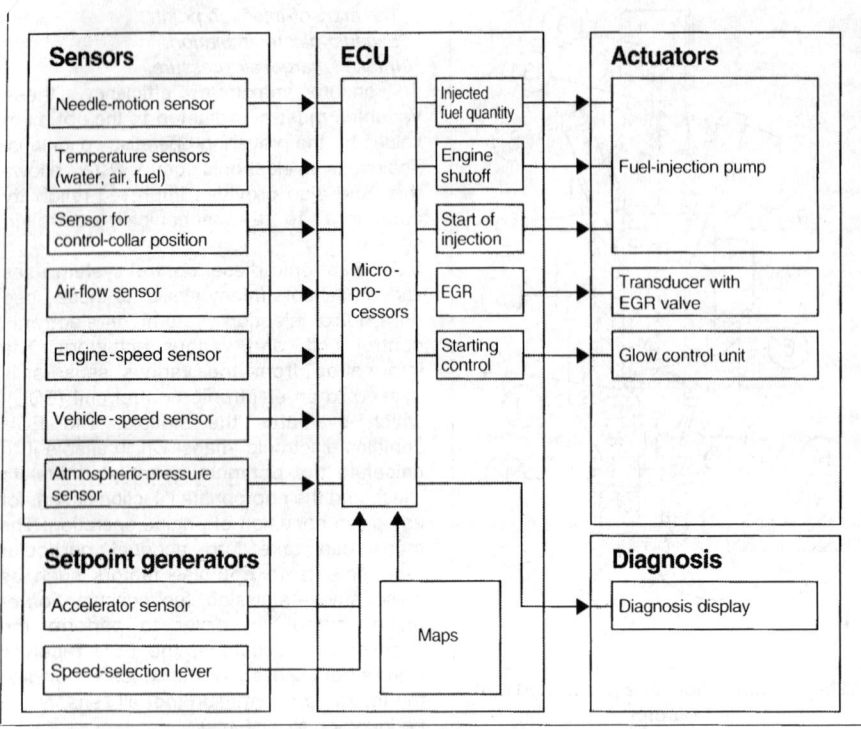

8.4 Bosch electronic diesel control system block diagram
© Robert Bosch Limited

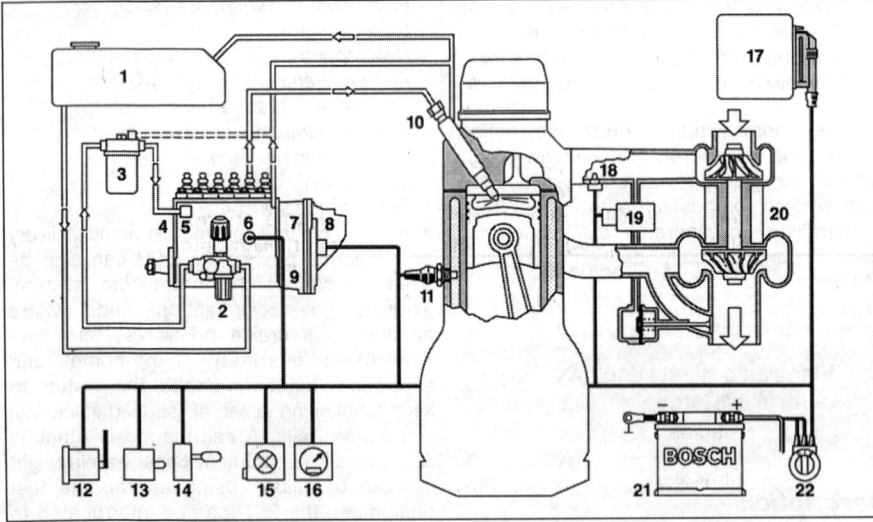

8.7 Bosch electronic diesel control system with in-line injection pump
© Robert Bosch Limited

1 Fuel tank	10 Injector	16 Vehicle speed sensor
2 Lift pump	11 Coolant temperature	17 Electronic control unit
3 Fuel filter	sensor	18 Air temperature sensor
4 In-line injection pump	12 Accelerator pedal	19 Charge-air pressure
5 Electrical shut-off device	position sensor	sensor
6 Fuel temperature sensor	13 Brake and clutch	20 Turbocharger
7 Control rod position	switches	21 Battery
sensor	14 Cruise control switch	22 Glow plug and starter
8 Linear solenoid	15 Warning lamp and	switch
9 Pump speed sensor	diagnostic connection	

injection pump, a control solenoid is fitted in place of the mechanical governor. The linear solenoid moves the pump control rod in response to signals from the ECU. The injected fuel quantity is a function of the control rod position and pump speed. The control solenoid is connected directly to the pump, and its linear movement moves the pump control rod according to the operating current supplied by the ECU. When the solenoid is de-energised, a spring forces the control rod to the 'stop' position and prevents the delivery of fuel to the engine. When the solenoid is energised, it exerts a force on the control rod which opposes the spring force; this force increases as the current in the solenoid rises, and moves the control rod in the 'increased-delivery' direction. This enables continual adjustment of the fuel quantity between zero and maximum, according to the current supplied to the solenoid.

7 A control rod position sensor and a pump speed sensor are incorporated into the pump to supply appropriate information to the ECU. The ECU selects the appropriate map for injected fuel quantity and injection timing according to the information received from the various system sensors **(see illustration)**.

Distributor injection pumps

8 One of the commonest electronically-controlled distributor injection pumps likely to be encountered is based on the Bosch VE pump **(see illustration)**. Although there are other types of electronically-controlled pump in use (eg, Lucas DP series), we will use the Bosch type as a typical example to explain the principles involved.

9 Compared with the earlier VE pumps, the electronically-controlled pump is considerably simplified. The mechanical governor and many of the add-on pump units previously used are discarded in favour of a single electronic actuator, which moves the control spool in response to signals from an electronic control unit (ECU). The actuator also moves a potentiometer which is used by the ECU to monitor the control spool position. For safety reasons, the actuator is spring-loaded so that if there is no current across its terminals, the control spool is moved into the 'stop' or zero-delivery position.

10 Internally, the injection pump resembles earlier VE pumps, as far as the supply pump and high-pressure pump are concerned. Pump cavity pressure is still used to determine injection advance, but the pressure is modulated by a solenoid valve, which is opened and closed in a rapid cycle by the ECU. The ratio of open- to closed-time determines cavity pressure, which is highest when the valve is closed.

11 A stop solenoid is still fitted to interrupt the fuel supply to the high-pressure pump, but this is a safety measure, since normally the

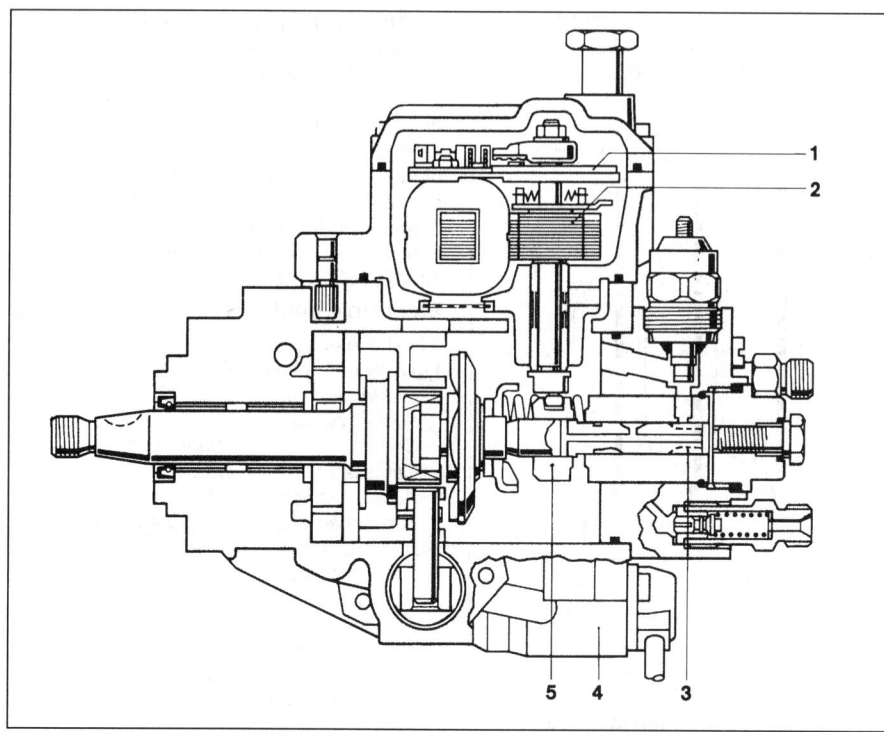

8.8 Bosch VE injection pump with electronic diesel control

© Robert Bosch Limited

1	Control spool position sensor	2	Electrical actuator
		3	Pump plunger
4	Cavity pressure control solenoid		
5	Control spool		

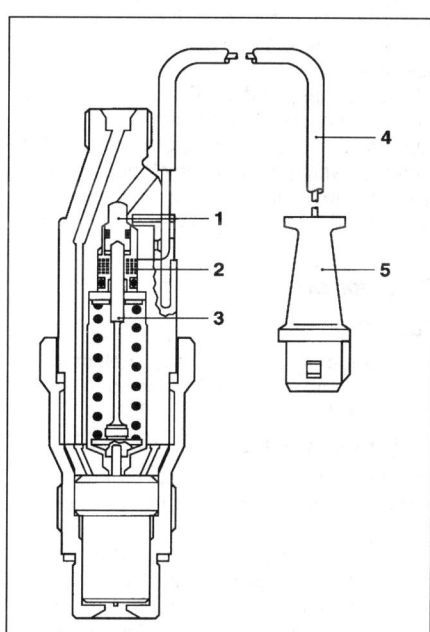

8.12 Fuel injector with needle motion sensor for electronic diesel control

© Robert Bosch Limited

1	Adjusting pin	4	Cable
2	Sensor winding	5	Plug
3	Pressure pin		

electrical actuator will stop fuel delivery when the 'ignition' is switched off.

12 Most of the sensors used in the system will be recognised as being similar to those used in petrol injection systems, but the injection start-point sensor is unique to the distributor pump-equipped diesel engine. The sensor is incorporated into one of the injectors, and senses the movement of the injector needle **(see illustration)**.

Future developments

13 Recent developments in electronic diesel control systems and the introduction of 'common rail' and 'pump injector' fuel systems have closed the performance and refinement gap between diesel and petrol engines. In spite of this, there is still a need to improve the performance, driveability and emissions standards of diesel engines.

14 The current trend in diesel engine technology is to develop direct injection engines, using high injection pressures. Higher injection pressures enable finer control of the start and end of injection points, and improved fuel atomisation, both of which improve engine efficiency and refinement. This trend has led to the development of common rail and 'pump injector' systems, which dispense with the need for high-pressure fuel lines to connect a pump to the injectors. Common rail and 'pump injector'

systems are described in greater detail in the following sections.

15 In the future it is likely that the use of these high-pressure systems will become more widespread, and other similar systems will be developed, as the narrowing gap between diesel and petrol engine refinement is closed still further.

9 Common rail systems

Description

1 The most widespread common rail system in current use is the Bosch system. Although there are other types of common rail system (eg, Caterpillar system), we will use the Bosch type as a typical example to explain the principles involved.

2 The common rail system derives its name from the fact that a common rail, or fuel reservoir, is used to supply fuel to all the fuel injectors. Instead of an in-line or distributor fuel pump, which distributes the fuel directly to each injector, a high-pressure pump is used, which generates a very high fuel pressure (up to 1350 bar on some systems) in the accumulator rail. The accumulator rail stores fuel, and maintains a constant fuel pressure, with the aid of a pressure control valve. Each injector is supplied with high-pressure fuel from the accumulator rail, and the injectors are individually controlled via signals from the system electronic control unit. The injectors are electromagnetically-operated.

3 In addition to the various sensors used on models with a conventional fuel injection pump, common rail systems also have a fuel pressure sensor. The fuel pressure sensor allows the electronic control unit to maintain the required fuel pressure, via the pressure control valve.

Operation

4 For the purposes of describing the operation of a common rail injection system, the components can be divided into three sub-systems; the low-pressure fuel system, the high-pressure fuel system and the electronic control system.

Low-pressure fuel system

5 The low-pressure fuel system consists of the following components:

Fuel tank.
Fuel lift pump.
Fuel filter/water trap.
Low-pressure fuel lines.

6 The low-pressure system (fuel supply system) is responsible for supplying clean fuel to the high-pressure fuel circuit. Refer to Section 3 of this Chapter for further details of the fuel supply components.

2

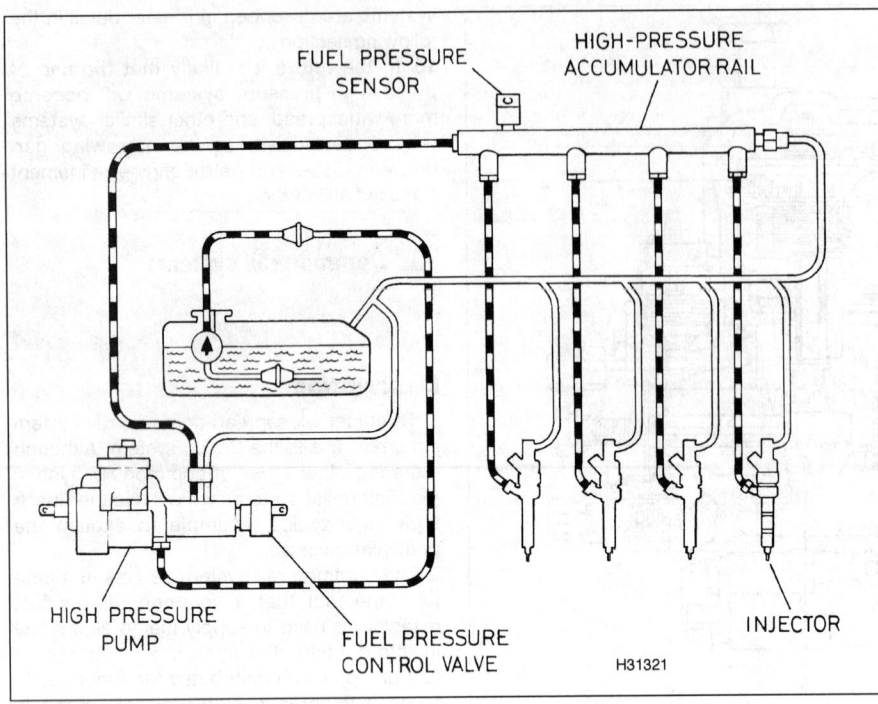

9.7 Schematic view of high-pressure fuel system – Bosch common rail system

High-pressure fuel system

7 The high-pressure fuel system consists of the following components **(see illustration)**:

High-pressure fuel pump with pressure control valve.
High-pressure accumulator rail with fuel pressure regulator.
Fuel injectors.
High-pressure fuel lines.

8 After passing through the fuel filter, the fuel reaches the high-pressure pump, which forces it into the accumulator rail, generating pressures of up to 1350 bar. As diesel fuel has

a certain elasticity, the pressure in the accumulator rail remains constant, even though fuel leaves the rail each time one of the injectors operates: additionally, a pressure control valve mounted on the high-pressure pump ensures that the fuel pressure is maintained within pre-set limits.

9 The *pressure control valve* is operated by the ECU. When the valve is opened, fuel is returned from the high-pressure pump to the tank, via the fuel return lines, and the pressure in the accumulator rail falls. To enable the ECU to trigger the pressure control valve correctly, the pressure in the accumulator rail is measured by a *fuel pressure sensor*.

10 The electromagnetically-controlled fuel injectors are operated individually, via signals from the ECU, and each injector injects fuel directly into the relevant combustion chamber. The fact that high fuel pressure is always available allows very precise and highly flexible injection in comparison to a conventional injection pump: for example combustion during the main injection process can be improved considerably by the pre-injection of a very small quantity of fuel.

Electronic control system

11 The electronic control system consists typically of the following components:

Electronic control unit (ECU).
Fuel lift pump.
Crankshaft speed/position sensor.
Camshaft position sensor.
Accelerator pedal position sensor.
Turbocharger boost pressure sensor.
Air temperature sensor.
Coolant temperature sensor.
Air mass meter.
Fuel pressure sensor.
Fuel injectors.
Fuel pressure control valve.
Preheating control circuit.
EGR valve actuator.

12 As mentioned in Section 8, the information from the various sensors is passed to the ECU, which evaluates the signals. The ECU contains electronic 'maps' which enable it to calculate the optimum quantity of fuel to inject, the appropriate start of injection, and even pre- and post injection fuel quantities, for each individual engine cylinder under any given condition of engine operation.

13 Additionally, the ECU carries out monitoring and self-diagnostic functions. Any faults in the system are stored in the ECU memory, which enables quick and accurate fault diagnosis using appropriate diagnostic equipment (such as a suitable fault code reader).

Components

Fuel lift pump

14 The fuel lift pump is usually electrically-operated, and is normally mounted in or near the fuel tank.

High-pressure pump

15 The high-pressure pump is most often mounted on the engine in the position normally occupied by a conventional distributor fuel injection pump. The pump is driven at half engine speed by the engine timing belt, timing chain, or possibly by gears, depending on application. The pump is lubricated by the fuel which is pumps.

16 The fuel lift pump forces the fuel into the high-pressure pump chamber, via a safety valve.

17 The high-pressure pump consists of a number of radially-mounted pistons and cylinders **(see illustration)**. The pistons are

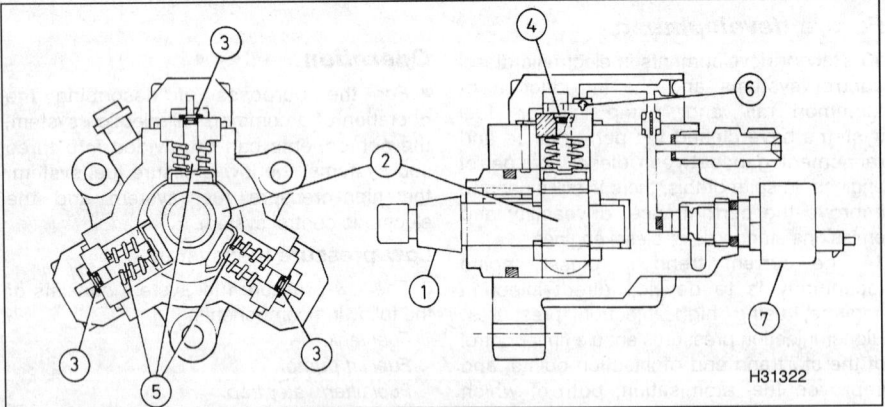

9.17 Cutaway view of high pressure pump – Bosch common rail system

1	*Driveshaft*	*4*	*Inlet valve*	*6*	*Outlet valve*
2	*Eccentric cam*	*5*	*Pistons*	*7*	*Pressure control valve*
3	*Pumping cylinders*				

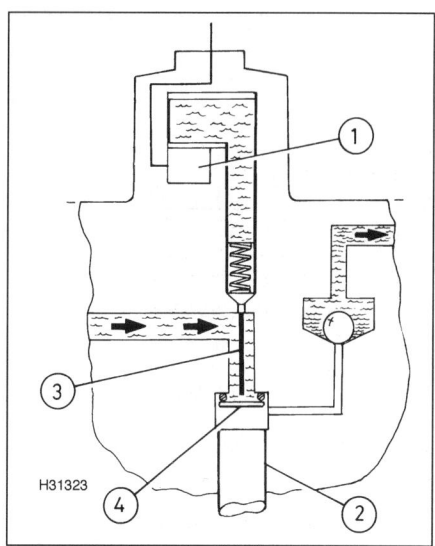

9.19 Cutaway view of high pressure pump cylinder 'switch-off' mechanism – Bosch common rail system

1	Electromagnet	3	Needle
2	Pumping cylinder	4	Inlet valve

operated by an eccentric cam mounted on the pump drive spindle. As a piston moves down, fuel enters the cylinder through an inlet valve. When the piston reaches bottom-dead-centre (BDC), the inlet valve closes, and as the piston moves back up the cylinder, the fuel is compressed. When the pressure in the cylinder reaches the pressure in the accumulator rail, an outlet valve opens, and fuel is forced into the accumulator rail. When

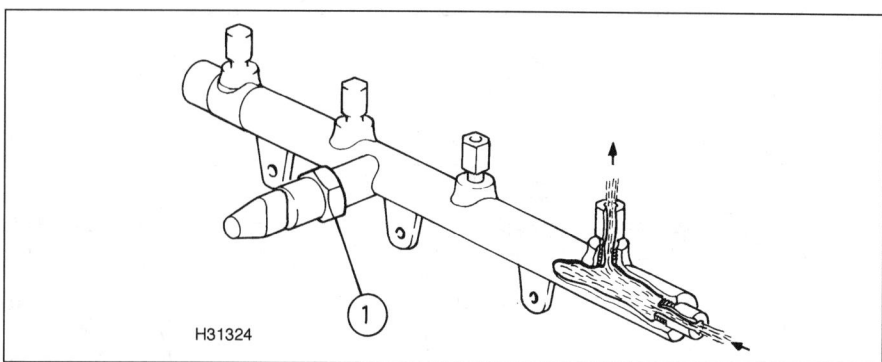

9.20 Accumulator rail – Bosch common rail system

1 Fuel pressure sensor

the piston reaches top-dead-centre (TDC), the outlet valve closes, due to the pressure drop, and the pumping cycle is repeated. The use of multiple cylinders (usually three) provides a steady flow of fuel, minimising pulses and pressure fluctuations.

18 As the pump needs to be able to supply sufficient fuel under full-load conditions, it will supply excess fuel during idle and part-load conditions. This excess fuel is returned from the high-pressure circuit to the low-pressure circuit (to the tank) via the pressure control valve.

19 The pump incorporates a facility to effectively switch off one of the cylinders to improve efficiency and reduce fuel consumption when maximum pumping capacity is not required **(see illustration)**. When this facility is operated, a solenoid-operated needle holds the inlet valve in the relevant cylinder open during the delivery

stroke, preventing the fuel from being compressed.

Accumulator rail

20 As its name suggests, the accumulator rail acts as an accumulator, storing fuel and preventing pressure fluctuations **(see illustration)**. Fuel enters the rail from the high-pressure pump, and each injector has its own connection to the rail. The fuel pressure sensor is mounted in the rail, and the rail also has a connection to the fuel pressure control valve on the pump.

Pressure control valve

21 The pressure control valve is operated by the ECU, and controls the system pressure. The valve may be mounted either on the high-pressure pump, or on the accumulator rail.

22 If the fuel pressure is excessive, the valve opens, and fuel flows back to the tank. If the pressure is too low, the valve closes, enabling the high-pressure pump to increase the pressure.

23 The valve is an electromagnetically-operated ball valve **(see illustration)**. The ball is forced against its seat, against the fuel pressure, by a powerful spring, and also by the force provided by the electromagnet. The force generated by the electromagnet is directly proportional to the current applied to it by the ECU. The desired pressure can therefore be set by varying the current applied to the electromagnet. Any pressure fluctuations are damped by the spring.

Fuel pressure sensor

24 The fuel pressure sensor is mounted in the accumulator rail, and provides very precise information on the fuel pressure to the ECU.

Injector

25 The injectors are mounted on the engine in a similar manner to conventional diesel fuel injectors. The injectors are electro-magnetically-operated via signals from the ECU, and fuel is injected at the pressure existing in the accumulator rail. The injectors are high-precision instruments and are manufactured to very high tolerances.

2

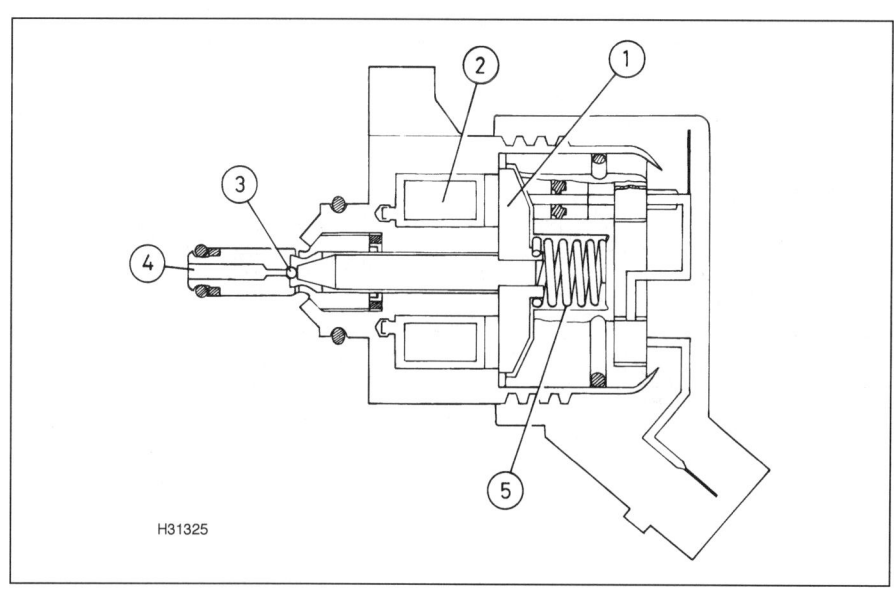

9.23 Cutaway view of pressure control valve – Bosch common rail system

1	Armature	3	Ball valve	5	Spring
2	Electromagnet	4	High pressure inlet		

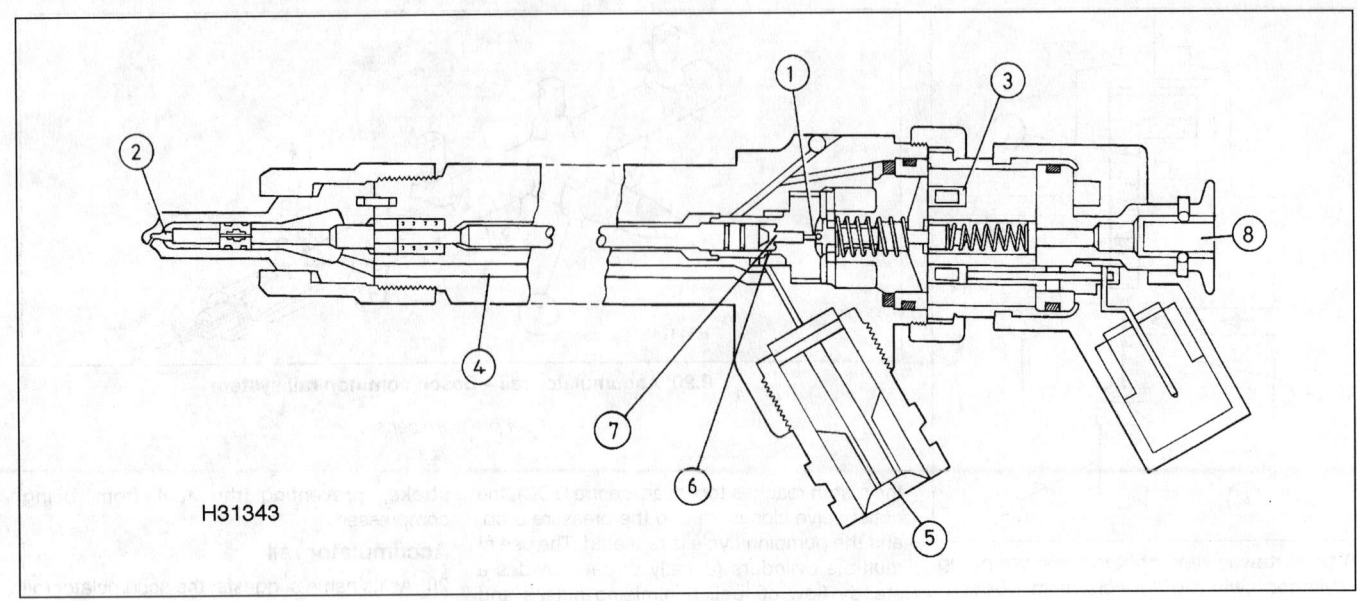

9.26 Cutaway view of injector – Bosch common rail system

1	Valve	3	Electromagnet	5	High-pressure inlet	7	Valve control chamber
2	Nozzle	4	Injector body	6	Input throttle	8	Fuel return

26 Fuel flows into the injector from the accumulator rail, via an inlet valve and an inlet throttle, and an electromagnet causes the injector nozzle to lift from its seat, allowing injection **(see illustration)**. Excess fuel is returned from the injectors to the tank via a return line. The injector operates on a hydraulic servo principle: the forces resulting inside the injector due to the fuel pressure effectively amplify the effects of the electromagnet, which does not provide sufficient force to open the injector nozzle directly. The injector functions as follows.

27 Five separate forces are essential to the operation of the injector **(see illustration)**.

a) A nozzle spring forces the nozzle needle against the nozzle seat at the bottom of the injector, preventing fuel from entering the combustion chamber.

b) In the valve at the top of the injector, the valve spring forces the valve ball against the opening to the valve control chamber. The fuel in the chamber is unable to escape through the fuel return.

c) When triggered, the electromagnet exerts a force which overcomes the valve spring force, and moves the valve ball away from

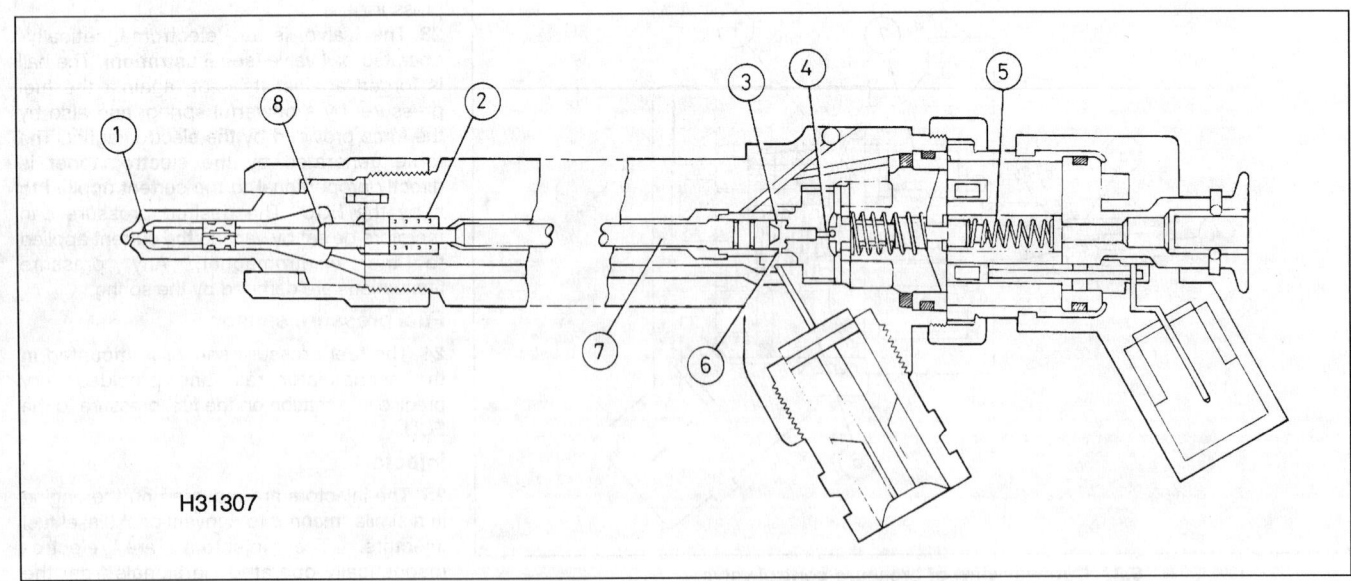

9.27 Injector components – Bosch common rail system

1	Nozzle seat	3	Valve control chamber	5	Valve spring	7	Valve control plunger
2	Nozzle spring	4	Valve ball	6	Input throttle	8	Nozzle needle chamfer

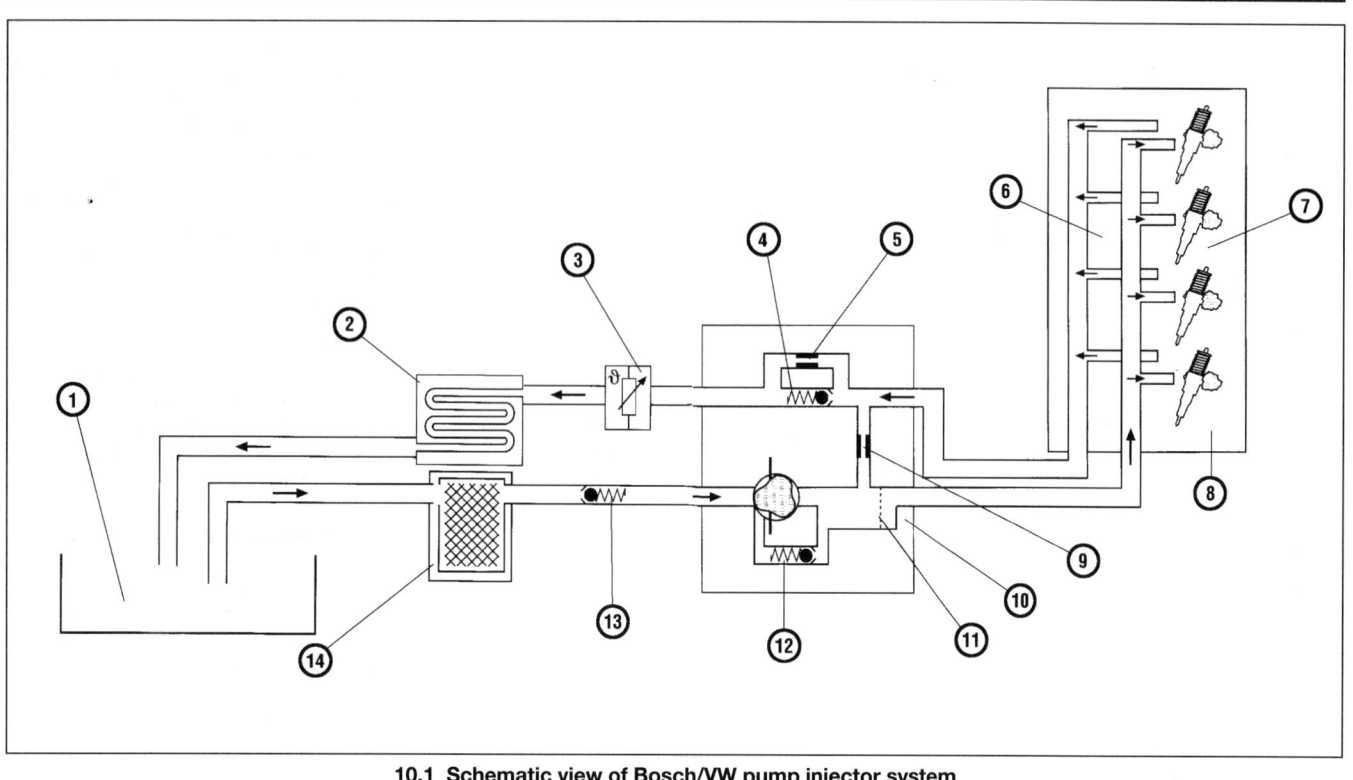

10.1 Schematic view of Bosch/VW pump injector system

1 *Fuel tank*	5 *Bypass*	9 *Restrictor*
2 *Fuel cooler*	6 *Fuel distributor pipe*	10 *Fuel pump*
3 *Fuel temperature sensor*	7 *Pump injectors*	11 *Strainer*
4 *Pressure limiting valve*	8 *Cylinder head*	

12 *Pressure limiting valve*	
13 *Non-return valve*	
14 *Fuel filter*	

its seat. This is the triggering force for the start of injection. When the valve ball moves off its seat, fuel enters the valve control chamber.

d) The pressure of the fuel in the valve control chamber exerts a force on the valve control plunger, which is added to the nozzle spring force.

e) A slight chamfer towards the lower end of the nozzle needle causes the fuel in the control chamber to exert a force on the nozzle needle.

28 When these forces are in equilibrium, the injector is in its rest (idle) state, but when a voltage is applied to the electromagnet, the forces work to lift the nozzle needle, injecting fuel into the combustion chamber. There are four phases of injector operation as follows:

a) Rest (idle) state – all forces are in equilibrium. The nozzle needle closes off the nozzle opening, and the valve spring forces the valve ball against its seat.

b) Opening - the electromagnet is triggered which opens the nozzle and triggers the injection process. The force from the electromagnet allows the valve ball to leave its seat. The fuel from the valve control chamber flows back to the tank via the fuel return line. When the valve opens, the pressure in the valve control

chamber drops, and the force on the valve plunger is reduced. However, due to the effect of the input throttle, the pressure on the nozzle needle remains unchanged. The resulting force in the valve control chamber is sufficient to lift the nozzle from its seat, and the injection process begins.

c) Injection - within a few milliseconds, the triggering current in the electromagnet is reduced to a lower holding current. The nozzle is now fully open, and fuel is injected into the combustion chamber at the pressure present in the accumulator rail.

d) Closing - the electromagnet is switched off, at which point the valve spring forces the valve ball firmly against its seat, and in the valve control chamber, the pressure is the same as that at the nozzle needle. The force at the valve plunger increases, and the nozzle needle closes the nozzle opening. The forces are now in equilibrium once more, and the injector is once more in the idle state, awaiting the next injection sequence.

ECU and sensors

29 The ECU and sensors are described earlier in this Section – see *Description*.

10 'Pump injector' systems

Description

1 The 'pump injector' system has been in use in basic form for some years on larger direct injection diesel engines **(see illustration)**. Recent developments in electronic engine control systems have enabled the system to be refined for use on smaller car and light commercial engines, and at the time of writing VW/Audi, and Land Rover were among the major manufacturers selling vehicles equipped with this system. Although there are other types of pump injector system (eg, Lucas EUI), we will use the Bosch type as a typical example to explain the principles involved.

2 As its name implies, a 'pump injector' consists of a fuel injection pump, combined with a fuel injector. Each cylinder of the engine has its own pump injector, which eliminates the need for a separate high-pressure fuel pump, and the associated high-pressure fuel lines.

3 The pump injectors are operated by the engine camshaft, and are able to generate

2

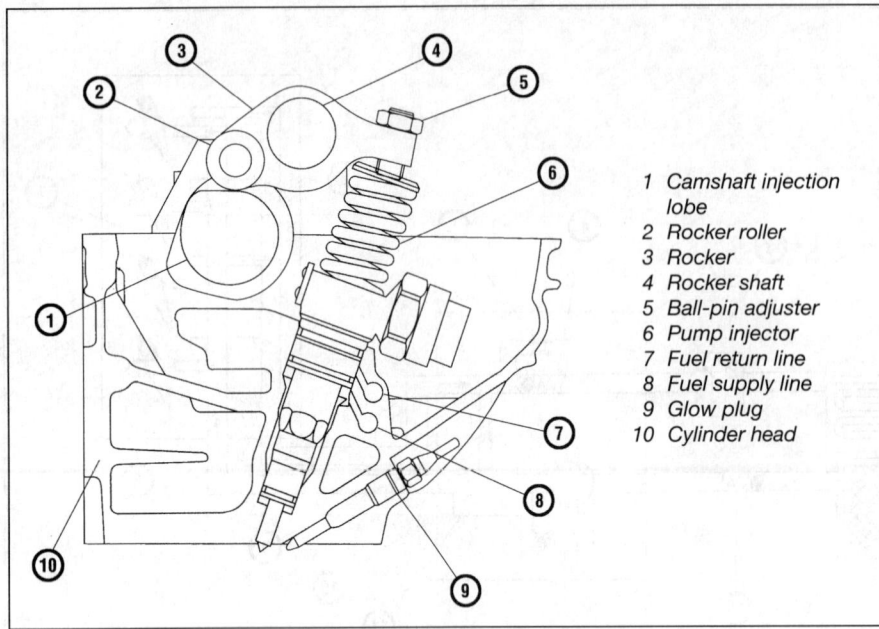

1 Camshaft injection lobe
2 Rocker roller
3 Rocker
4 Rocker shaft
5 Ball-pin adjuster
6 Pump injector
7 Fuel return line
8 Fuel supply line
9 Glow plug
10 Cylinder head

10.3 Pump injector installation - Bosch/VW pump injector system

extremely high fuel pressures (up to 2000 bar on some systems) **(see illustration)**. The pump injectors are mounted in the cylinder head, and are supplied with fuel via a distributor pipe mounted in the cylinder head. A fuel lift pump pumps fuel from the fuel tank to the distributor pipe. Each pump injector is individually controlled via signals from the system electronic control unit. The pump injectors are electromagnetically-operated.
4 Pressure limiting valves in the fuel supply and return lines maintain constant fuel pressures in the fuel supply and return lines.
5 Because of the extremely high fuel injection pressure, the fuel in the return line becomes very hot, and a fuel cooling system is used to cool the excess fuel before it is returned to the tank. Besides the obvious effect on safety, if the fuel was not cooled, the fuel temperature in the tank would rise, which means that the temperature of the fuel supplied to the injectors would also rise. Under high-pressure injection conditions, hot fuel reduces fuel delivery from the injectors; although the ECU can compensate to a reasonable degree for fuel temperature variations, cool fuel gives improved combustion and hence improved engine efficiency.

Operation

6 For the purposes of describing the operation of a pump injector system, the components can be divided as follows; the low-pressure fuel system, the fuel cooling system, the pump injectors and the electronic control system.

Low-pressure fuel system

7 The low-pressure fuel system consists of the following components:
   Fuel tank.
   Low-pressure fuel lines.
   Fuel filter/water trap.
   Fuel lift pump (incorporating pressure limiting valve).
   Fuel distributor pipe (mounted in cylinder head).
8 The low-pressure system (fuel supply system) is responsible for supplying clean, cool fuel to the pump injectors. Refer to Section 3 of this Chapter for further details of the fuel filter/water trap.
9 After passing through the filter, the fuel reaches the fuel lift pump, which supplies fuel to the fuel distributor pipe, via passages drilled in the cylinder head.
10 Any excess fuel is returned from the distributor pipe to the fuel tank, via a the fuel cooling system.

Fuel cooling system

11 The fuel cooling system is separate from the engine cooling circuit, because the temperature of the engine coolant is too high to cool the fuel when the engine is at operating temperature. In most cases, the fuel coolant circuit is connected to the main coolant expansion tank, but in such a way that the hotter engine coolant circuit has no adverse effect on the fuel coolant circuit. The connection to the expansion tank allows the system to be filled, and also allows for expansion of the coolant with varying temperature.
12 A fuel cooler is mounted on the fuel filter head **(see illustration)**. The fuel cooler is basically a fuel/coolant heat exchanger. Cold coolant is pumped through the cooler by an electric pump, controlled by the engine ECU. As the coolant passes through the cooler, it absorbs heat from the fuel. The cooled fuel then passes to the fuel tank, while the warm coolant passes to a radiator at the front of the vehicle. The radiator, which is separate from the engine cooling system radiator, is cooled by the air passing through it due to the

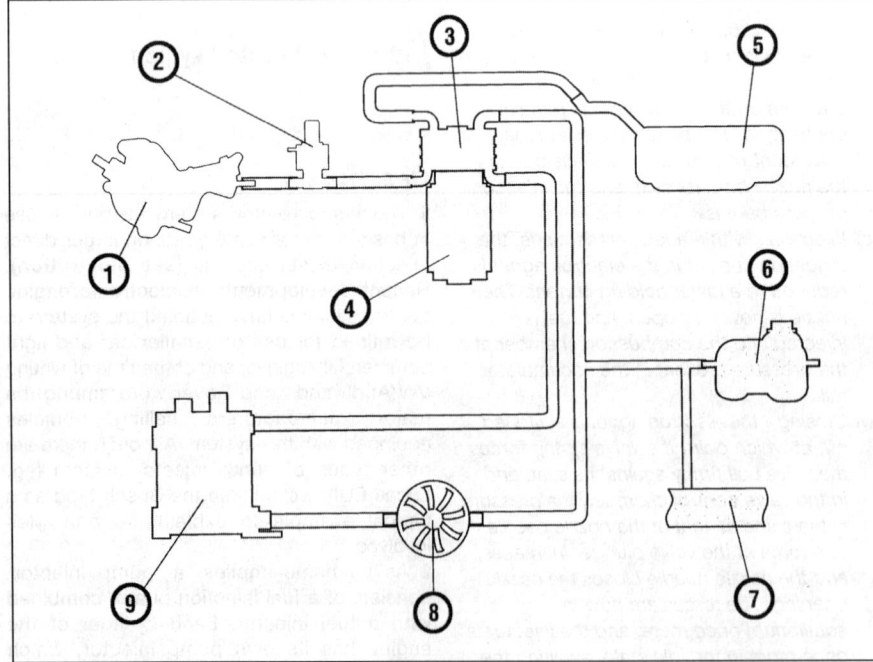

10.12 Schematic view of fuel cooling circuit – Bosch/VW pump injector system

1	Fuel pump	4	Fuel filter	7	Engine cooling system
2	Fuel temperature sensor	5	Fuel tank	8	Electric coolant pump
3	Fuel cooler	6	Coolant expansion tank	9	Coolant radiator

forward motion of the vehicle, supplemented by air from the engine cooling fan(s) when necessary. The cold coolant then passes to the coolant pump, and the cycle starts again.

Pump injectors

13 After passing through the fuel distributor pipe, the fuel reaches the pump injectors **(see illustration)**.

14 The electromagnetically-controlled pump injectors are operated individually, via signals from the ECU, and each injector injects fuel directly into the relevant combustion chamber. The fact that very high fuel pressure is always available allows very precise and highly flexible injection in comparison to a conventional injection pump: for example combustion during the main injection process can be improved considerably by the pre-injection of a very small quantity of fuel. On some systems, the individual control of the injectors also allows individual engine cylinders to be 'switched off' during part-load conditions, to improved fuel economy.

Electronic control system

15 The electronic control system consists typically of the following components:
Electronic control unit (ECU).
Fuel coolant pump.
Crankshaft speed/position sensor.
Camshaft position sensor.
Accelerator pedal position sensor.
Air temperature sensor.
Coolant temperature sensor.
Air mass meter.
Inlet manifold pressure sensor.
Fuel temperature sensor.
Clutch and brake pedal switches.
Fuel injectors.
Preheating control circuit.
EGR valve actuator.

16 As mentioned in Section 8, the information from the various sensors is

1 *Ball-pin*
2 *Pump piston*
3 *Piston spring*
4 *Solenoid valve needle*
5 *Injector solenoid valve*
6 *Fuel return line*
7 *Retraction piston*
8 *Fuel supply line*
9 *Injector spring*
10 *Injector needle damping element*
11 *Injector needle*
12 *Heat insulating seal*
13 *O-rings*

10.13 Cutaway view of pump injector – Bosch/VW pump injector system

passed to the ECU, which evaluates the signals. The ECU contains electronic 'maps' which enable it to calculate the optimum quantity of fuel to inject, the appropriate start of injection, and even pre- and post injection fuel quantities, for each individual engine cylinder under any given condition of engine operation.

17 Additionally, the ECU carries out monitoring and self-diagnostic functions. Any faults in the system are stored in the ECU memory, which enables quick and accurate fault diagnosis using appropriate diagnostic equipment (such as a suitable fault code reader).

Components

Fuel lift pump

18 The fuel lift pump is normally driven from the end of the engine camshaft. The pump is mounted directly on the cylinder head, and pumps fuel to the distributor pipe via drillings in the cylinder head. The pump incorporates a pressure-limiting valve, which keeps the fuel pressure in the distributor rail constant (approximately 7.5 bar).

19 The fuel pump is of the blocking vane-cell type, which delivers fuel even at low engine speeds. Fuel is drawn and pumped into two chambers; the intake chamber and feed chamber are separated from each other by spring-loaded blocking vanes. As the pump rotor rotates, it draws fuel from the intake chamber, compresses the fuel, and pumps it out into the feed chamber. As the rotor turns, the blocking vanes slide to maintain the separation between the intake and feed chambers **(see illustration)**.

2

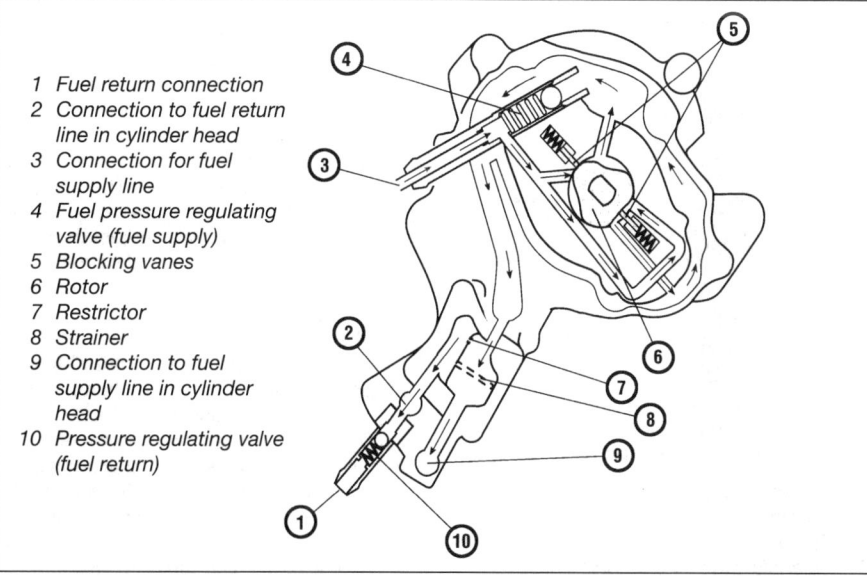

1 *Fuel return connection*
2 *Connection to fuel return line in cylinder head*
3 *Connection for fuel supply line*
4 *Fuel pressure regulating valve (fuel supply)*
5 *Blocking vanes*
6 *Rotor*
7 *Restrictor*
8 *Strainer*
9 *Connection to fuel supply line in cylinder head*
10 *Pressure regulating valve (fuel return)*

10.19 Cutaway view of fuel lift pump – Bosch/VW pump injector system

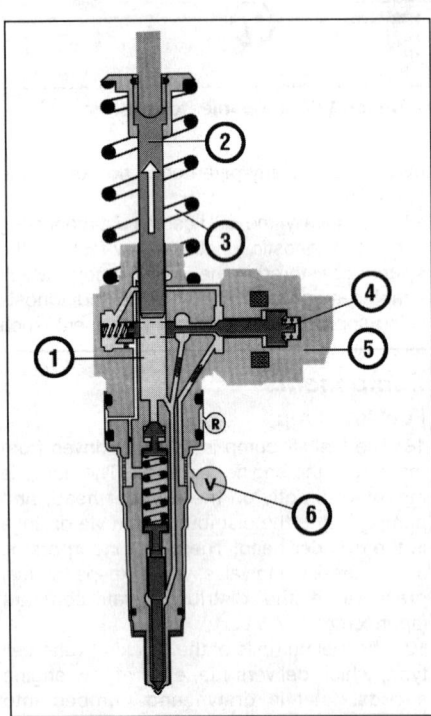

1 Annular gap
2 Cross-drillings
3 Cylinder head
4 Distributor pipe
5 Fuel from pump
   injector
6 Fuel to pump injector
7 Mixing of fuel in
   annular gap

10.20 Fuel distributor pipe – Bosch/VW pump injector system

Fuel distributor pipe

20 The purpose of the distributor pipe is to distribute fuel evenly to the pump injectors. The pipe is fitted to a passage in the cylinder head, and has a cross-drilling for each pump injector. Fuel supplied by the pump can flow out from the pipe, through the cross-drillings, and into the cylinder head passage which surrounds the pipe. Here, the cool supply fuel mixes with the hot excess (return) fuel forced back into the supply line by the pump injectors **(see illustration)**. This results in the fuel in the distributor pipe being at an even temperature all along the pipe, supplying each pump injector with fuel at the same temperature.

21 If a distributor pipe was not used, and the fuel was pumped to the pump injectors through a closed pipe, the fuel temperature would rise progressively along the pipe: the hot fuel returned by the pump injectors would be forced towards the end of the pipe by the fuel flowing into the pipe from the lift pump. If the temperature of the fuel varied between pump injectors, differing masses of fuel would be injected by each injector, causing unnecessary engine vibration and stress, and excessively high temperatures in the cylinders receiving 'hot' fuel.

Pump injector

22 The pump injectors are mounted directly in the cylinder head. The pumping action is provided by pistons which are operated by additional lobes on the engine camshaft (one lobe for each pump injector), via rocker arms. The injection action is solenoid-activated via signals from the ECU.

23 The injection cam lobe profile can be divided into two sections:

  a) *Steep leading edge - results in the pump injector piston being pushed down at high velocity, achieving a high fuel pressure very quickly.*

  b) *Flat trailing edge – allows the piston to move up and down slowly and evenly, enabling fuel to flow smoothly into the pump injector high-pressure chamber, free from air bubbles.*

24 As the cam lobe turns, it operates the rocker, which in turn operates the pump injector piston, via a ball-pin. The ball-pin is adjustable (in a similar manner to the rocker arms on an engine with adjustable valve clearances), to allow the adjustment of a minimum clearance between the bottom of the pump injector high-pressure chamber and the pump piston. This clearance must exist to prevent the piston from hitting the bottom of the high-pressure chamber when the components expand due to heat.

25 The pump injector injection cycle can be divided into six stages **(see illustrations)**:

  a) *Filling of the high-pressure chamber – the pump piston moves upwards under the force of the piston spring, increasing the volume of the high-pressure chamber. The injector solenoid valve is not*

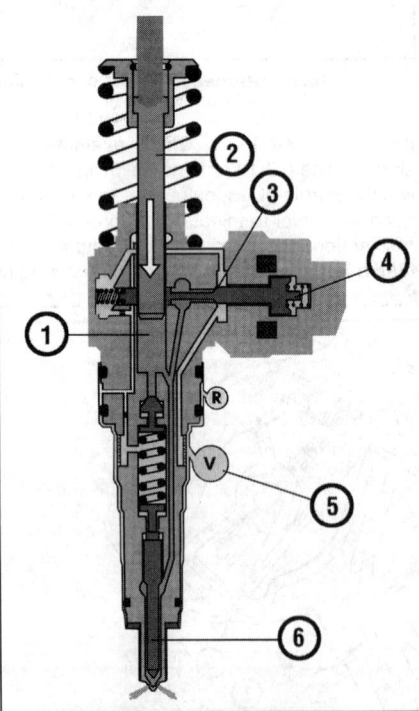

10.25a Filling of the high-pressure chamber - Bosch/VW pump injector system

1	High-pressure chamber	4	Solenoid valve needle
2	Pump piston	5	Injector solenoid valve
3	Piston spring	6	Fuel supply line

10.25b Pre-injection commences - Bosch/VW pump injector system

1	High-pressure chamber	4	Solenoid valve
2	Pump piston	5	Fuel supply line
3	Solenoid valve seat	6	Injector needle

activated at this time. The solenoid valve
needle is in its rest position, and the path
from the fuel supply line to the high-
pressure chamber is open, allowing fuel
to flow into the high-pressure chamber.

b) *Pre-injection commences* – the pump
piston moves down under the force of the
rocker arm, and pushes fuel out of the
high-pressure chamber into the fuel
supply line. The ECU activates the injector
solenoid valve. The solenoid valve needle
is pressed down onto the valve seat, and
closes off the path from the high-pressure
chamber to the fuel supply line. This
causes a build-up of pressure in the high-
pressure chamber. When the pressure
reaches around 180 bar, it overcomes the
pressure of the injector spring, which
allows the injector needle to lift. The
pressure in the high-pressure chamber
forces fuel out through the injector nozzle,
and pre-injection commences.

c) *Pre-injection damping* – during the pre-
injection stage, the movement of the
injector needle is damped by a hydraulic
'cushion', which allows the quantity of
fuel injected to be metered exactly. In the
first third of the needle stroke there is no
damping, and a quantity of fuel is injected
into the combustion chamber. As soon
as the damping piston (fitted to the top of
the injector needle) moves up into the
bore at the bottom of the injector spring
chamber, the fuel above the injector
needle can only move into injector spring
chamber through a small metered gap.
This creates a hydraulic 'cushion' which
limits the injector needle stroke during the
pre-injection cycle.

d) *Pre-injection ends* – the pre-injection
cycle ends straight after the injector
needle opens. The increasing pressure
causes the retraction piston to move
downwards, increasing the volume of the
high-pressure chamber. As a result, the
pressure drops momentarily, and the
injector needle closes. The pre-injection
cycle now ends. The downward
movement of the retraction piston
compresses the injector spring,
increasing the downward load on the
injector needle. This means that to lift
the injector needle from its seat
during the main injection cycle, the fuel
pressure must be higher than during the
pre-injection cycle.

e) *Main injection commences* – the pressure
in the high-pressure chamber increases
shortly after the injector needle closes.
The injector solenoid valve remains
closed, and the pump piston moves
downwards. When the pressure reaches
approximately 300 bar, the fuel pressure
overcomes the force exerted by the pre-
loaded injector spring. The injector needle
lifts, and the main injection quantity of fuel
is injected. The fuel pressure rises to
approximately 2050 bar, because more

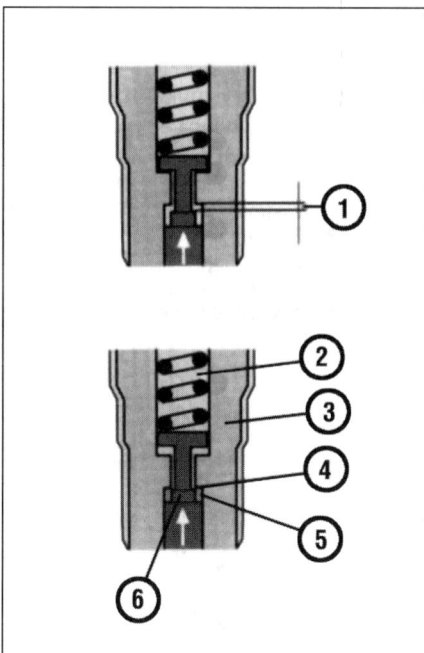

**10.25c Pre-injection damping - Bosch/VW
pump injector system**

1 Undamped stroke
2 Injector spring
   chamber
3 Injector housing

4 Leakage gap
5 Hydraulic
   cushion
6 Damping piston

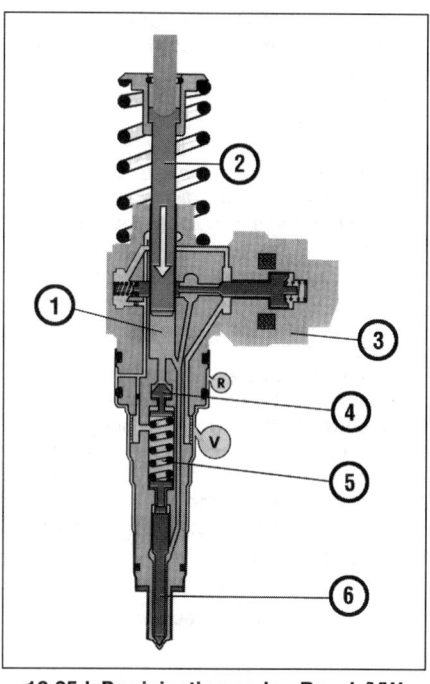

**10.25d Pre-injection ends - Bosch/VW
pump injector system**

1 Undamped stroke
2 Pump piston
3 Injector solenoid
   valve

4 Retraction
   piston
5 Injector spring
6 Injector needle

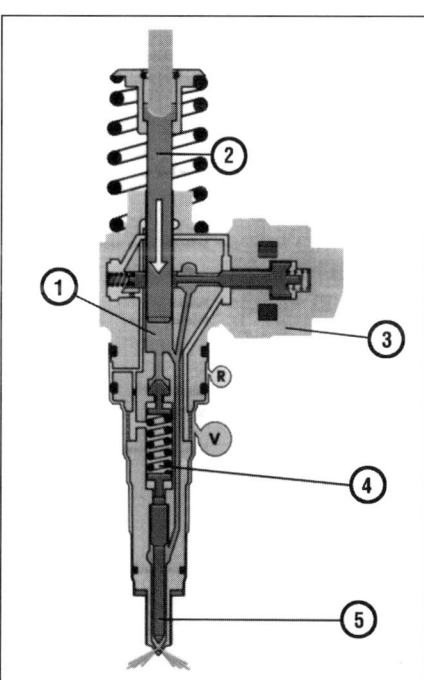

**10.25e Main injection commences -
Bosch/VW pump injector system**

1 High-pressure chamber
2 Pump piston
3 Injector solenoid valve
4 Injector spring
5 Injector needle

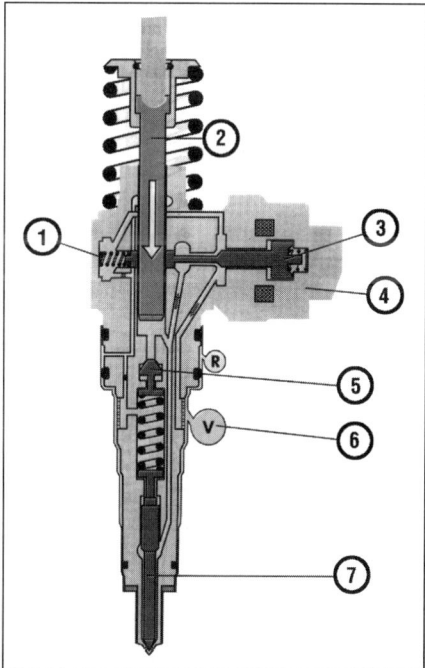

**10.25f Main injection cycle ends -
Bosch/VW pump injector system**

1 Solenoid valve
   spring
2 Pump piston
3 Solenoid valve
   needle

4 Injector solenoid
   valve
5 Retraction piston
6 Fuel supply line
7 Injector needle

2

fuel is displaced in the high-pressure chamber than can escape through the nozzle holes.

f) *Main injection cycle ends – the injection cycle ends when the ECU de-activates the injector solenoid valve. The solenoid valve spring opens the solenoid valve needle, and the fuel displaced by the pump piston enters the fuel supply line, allowing the pressure to drop. The injector needle closes, and the injector spring forces the retraction piston into its rest position. The main injection cycle now ends.*

26 When the engine is running, a quantity of fuel is flushed from the fuel supply line, through ducts in the injector body, into the fuel return line, to cool the injector. The flushing action is provided by fuel pressure, and restrictors in the ducts prevent a significant pressure loss **(see illustration)**.

ECU and sensors

27 The ECU and sensors are described earlier in this Section – see *Description*.

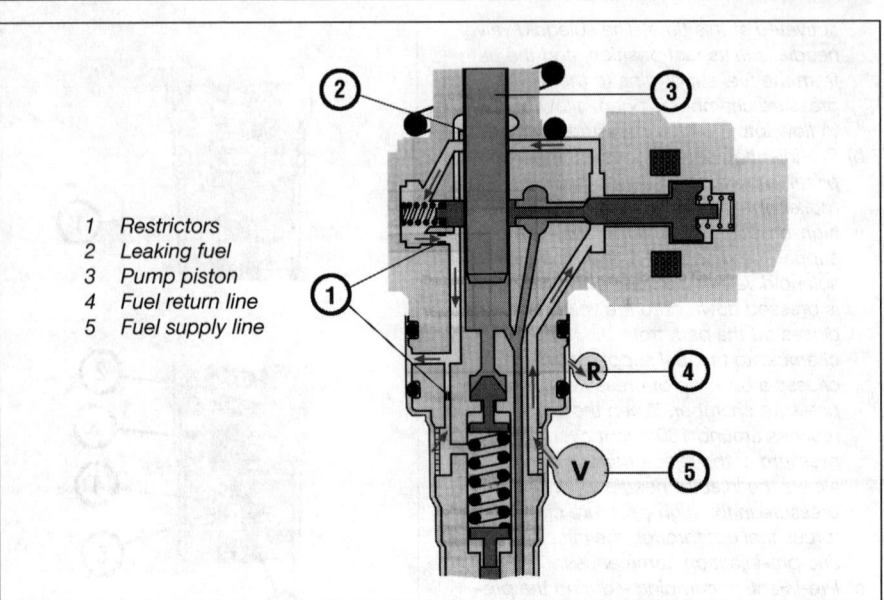

1 Restrictors
2 Leaking fuel
3 Pump piston
4 Fuel return line
5 Fuel supply line

10.26 Injector cooling action - Bosch/VW pump injector system

Chapter 3
Engine electrical systems

Contents

3

1 Introduction

1 By far the majority of faults on small diesel engines are electrical in origin, and can be attributed either to the battery or the preheating system. An understanding of how the electrical systems work and how to test them is important if mistakes in diagnosis are to be avoided. In particular, the injection pump and injectors can receive costly and unnecessary attention when in fact it is the preheating system which is at fault.

2 Apart from the absence of an ignition system, the electrical systems of the small diesel engine differ from those of the petrol engine in the following areas.

Battery

3 The battery is of larger capacity (expressed as reserve capacity – RC - or amp-hours) than on an equivalent-sized petrol engine. It is also capable of delivering more current (cold cranking capacity). Both these functions are necessary to cope with the higher demands imposed by the starter motor and the preheating system.

4 For satisfactory starting, especially in winter, it is important that the correct type of battery is fitted, and that it is in good condition and well-charged. When fitting a new battery, both the reserve capacity and the cold cranking capacity must be as specified. Battery testing is covered later in this Chapter.

Alternator

5 Alternator rated output tends to be higher on the diesel engine, in order to ensure that the battery is recharged in a sufficiently short time after the demands made on it during start-up.

Starter motor

6 Although the cranking speeds required to start petrol and diesel engines are similar, the diesel engine's starter motor needs to provide a higher torque to overcome the greater resistance caused by the high compression ratio. This may be achieved by simply increasing the physical size of the motor, or

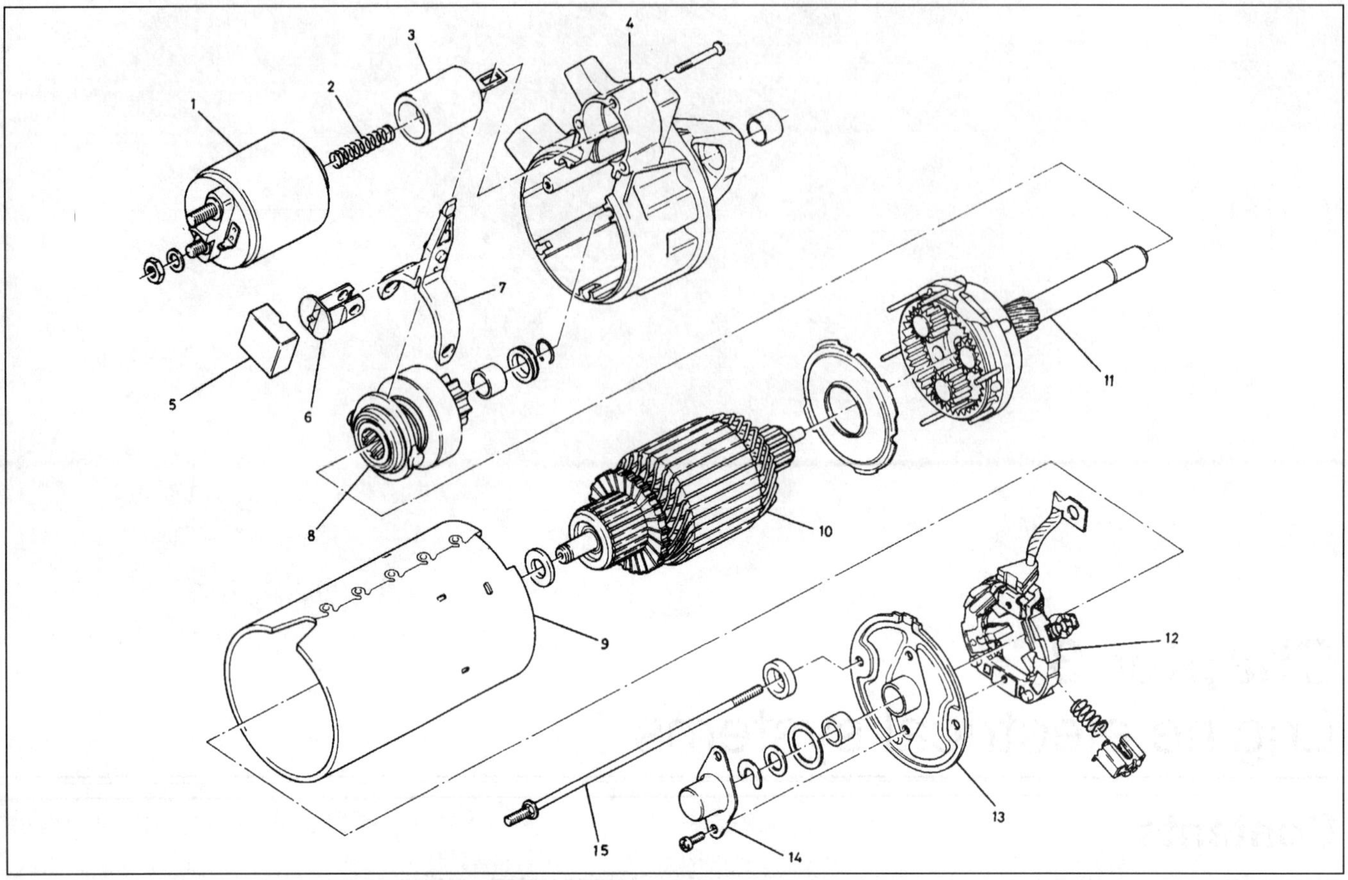

1.6 Exploded view of reduction gear starter motor with permanent magnetic field

1	Solenoid body	5	Rubber wedge	9	Field frame	12	Brushgear
2	Spring	6	Operating arm pivot	10	Armature	13	Commutator end shield
3	Solenoid plunger	7	Operating arm	11	Reduction gear and output	14	Commutator end cap
4	Drive end housing	8	Clutch-pinion assembly		shaft	15	Through-bolt

by the use of reduction gears **(see illustration)**.

7 Some small diesel engine starter motors

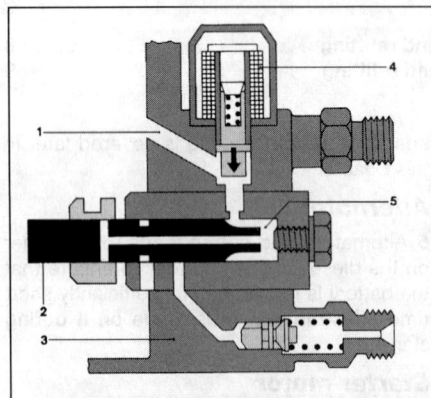

1.8 Stop solenoid fitted to Bosch VE distributor injection pump

© Robert Bosch Limited

1	Fuel inlet	4	Shut-off solenoid
2	Pump plunger	5	High-pressure
3	Distributor head		chamber

use permanent magnets instead of field coils. This, in conjunction with reduction gears, keeps the size and weight of the starter motor down.

Fuel injection pump

8 Most fuel injection pumps have an electrically-operated fuel shut-off valve, usually known as the stop solenoid. The valve opens the inlet bore to the high-pressure chamber when energised (ignition on) **(see illustration)**. If it sticks shut, or if its voltage supply is interrupted, fuel cannot reach the pump plunger, and the engine will neither start nor run (see Chapter 9).

Fuel heaters

9 Electrically-operated fuel heaters are fitted as original equipment to some vehicles, or may be encountered as an aftermarket accessory. They are fitted either upstream of the fuel filter, as close as possible, or (with units such as the Lucas/CAV D-Wax) form an integral part of the filter head. See Chapter 2, Section 3 for more details.

10 Fuel heaters are normally thermostatically-controlled, and energised only when the

ignition is switched on. Power consumption is of the order of 100 to 200 watts.

Preheating system

11 This is one of the most important engine electrical systems. It is considered in detail in the following Sections.

2 Preheating system -
description

1 The temperatures reached in the cylinders of the diesel engine at cranking speed when cold can be insufficient to ignite the fuel vapour. One way round this is to increase cranking speed, but this wound entail the use of much larger (and heavier) batteries and starter motors. A more satisfactory solution is to provide an external source of extra heat just before and during start-up. There are two ways of providing this heat: electrically, by the use of heater plugs (usually called glow plugs), or by a combination of electrical and fuel energy, in the so-called 'flame plug'.

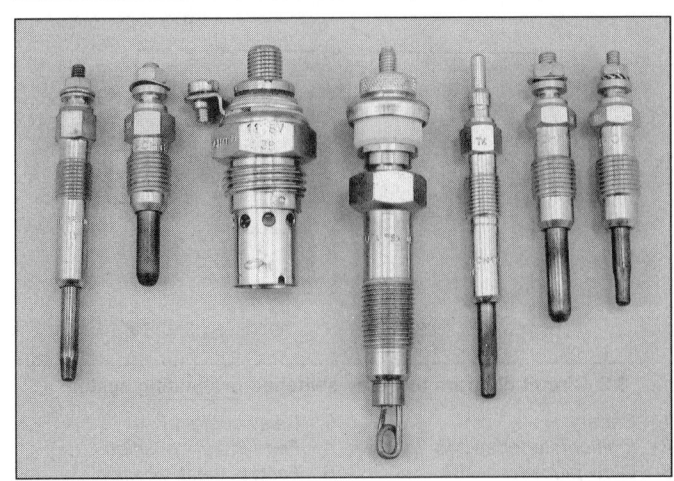

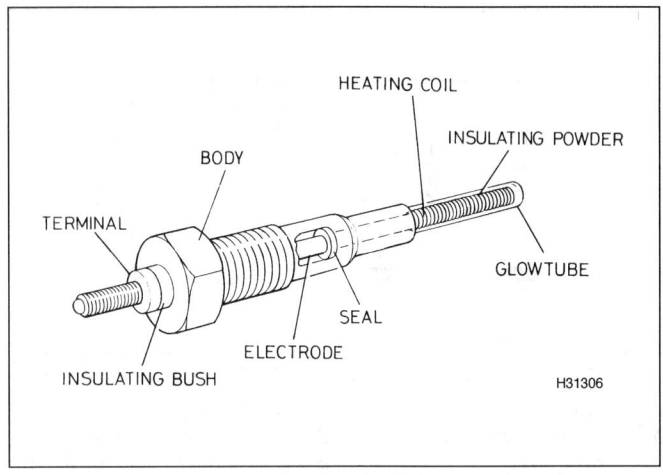

2.2a A selection of typical glow plug types

2.2b Construction of a sheathed element glow plug

Glow plugs

2 All indirect injection engines, and most direct injection engines, use glow plugs to assist cold starting. Usually, one plug is used per cylinder (sometimes, one cylinder may not be fitted with a plug due to space restrictions). Some engines still use plugs with an exposed element, but modern engines normally use the sheathed-element plug, where the heating element is enclosed and protected by a metal sheath **(see illustrations)**. On all engines in current production, the plugs are wired in parallel so that failure of one plug will not affect the operation of the others.

3 The tips of the glow plugs, which protrude into the pre-combustion or combustion chambers, become red-hot after a few seconds when battery voltage is applied to them. When the incoming fuel spray swirls into contact with the hot plug, the fuel ignites.

4 The time for which the glow plugs need to be energised before the engine can be started will depend on engine temperature and ambient air temperature. The glow plugs are also energised during starter motor operation, and sometimes for a few seconds after start-up. This last phase of operation ('afterglow' or 'post heating') contributes to smooth and smoke-free operation of the engine immediately after starting. Glow plug control circuits are described in the next Section.

5 Glow plugs may last the life of the engine, or one may fail after only a few thousand miles. A four-cylinder engine on which one glow plug has failed can normally still be started without too much trouble, except in very cold weather. With two glow plugs failed, cold starting will be difficult or impossible. It is therefore worth checking the glow plugs at major service intervals, in case one has failed unnoticed.

Flame plugs

6 Flame plug systems are used on some older types of direct injection engines. Instead of relying on electrical energy for the entire preheating effect, the flame plug uses electricity to vaporise and ignite fuel fed to it. It is located in the inlet manifold **(see illustration)**.

7 The fuel feed to the flame plug is obtained from the injector return line. Returned fuel fills a small reservoir, located at a higher level than the flame plug, so that when required the fuel will flow to the plug by gravity. The flow of fuel is controlled by an external solenoid valve, or by a temperature-sensitive valve in the plug itself.

8 The control circuitry for the electrical part of the flame plug is similar to that used for glow plugs.

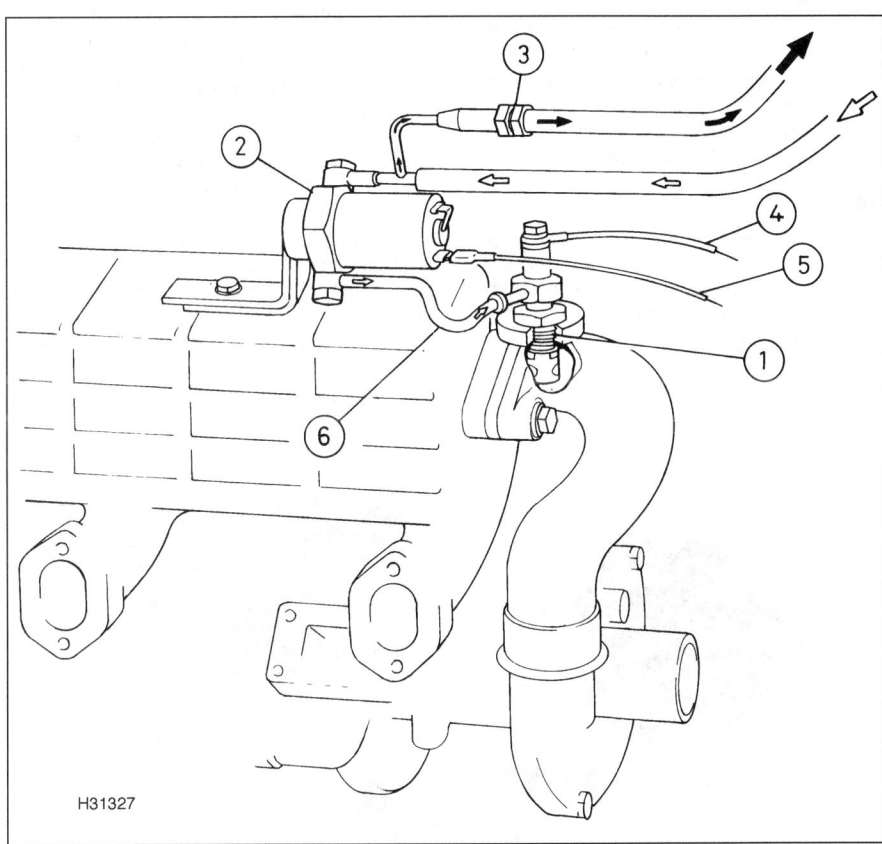

2.6 Flame plug and associated components on a Fiat engine

1	Flame plug	4	Flame plug heating
2	Solenoid valve		element wire
3	Pressure-regulating	5	Solenoid supply wire
	valve	6	Flame plug fuel supply hose

*Light arrows – fuel supply
from injection pump return
Dark arrows – fuel return
to tank*

3

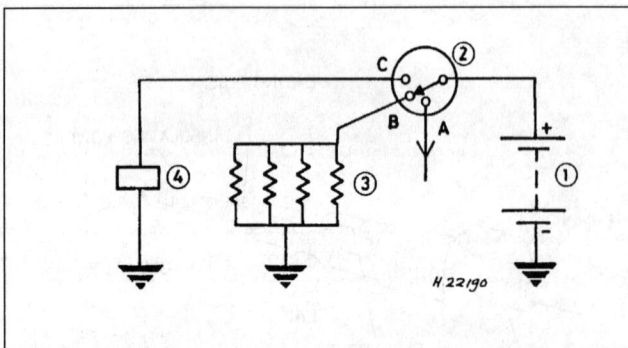

3.1 Circuit diagram for manually-controlled preheating system

1	Battery	A	Feed in 'run' position
2	Ignition/starter switch	B	Feed in 'heat' position
3	Glow plugs	B + C	Feed in 'start' position
4	Starter motor solenoid		

3.2 Circuit diagram for relay-switched preheating system

1	Battery	6	Relay contact
2	Ignition/starter switch	A	Feed in 'run' position
3	Glow plugs	B	Feed in 'heat' position
4	Starter motor solenoid	B + C	Feed in 'start' position
5	Relay control winding		

3 Preheating system control circuits

1 In the simplest form of preheating system, preheating time is controlled by the driver. There is a 're-heat' position on the ignition/starter switch, between the 'run' and 'start' positions, in which the key must be held for a few seconds (up to 30 seconds in very cold conditions) before starting. Once the engine has started and the key is released, the power supply to the glow plugs is switched off **(see illustration)**.

Relays

2 The heavy current drawn by the glow plugs – typically 9 to 12 amps each, with an initial surge two to three times greater – means that most systems make use of a relay to switch the current. A relay is an electrically-operated switch, which in this case is being used to control a heavy current at a distance from the ignition/starter switch. This eliminates the need to run thick electrical cable to and from the switch, and means that the switch itself can be of lighter construction **(see illustration)**.

Automatic control

3 In the simplest relay systems, preheating time is still controlled by the driver, but automatic control of preheating time is now virtually universal in passenger cars and light commercial vehicles. On vehicles with electronic diesel engine control, the preheating system is controlled by the engine electronic control unit (in this case the system usually operates in a similar manner to that described for *Advanced systems*, later in this Section). Basic automatic systems incorporate a timer or control unit (which may include the relay) and a preheating warning lamp. Some systems also have a 'ready to start' lamp **(see illustrations)**.

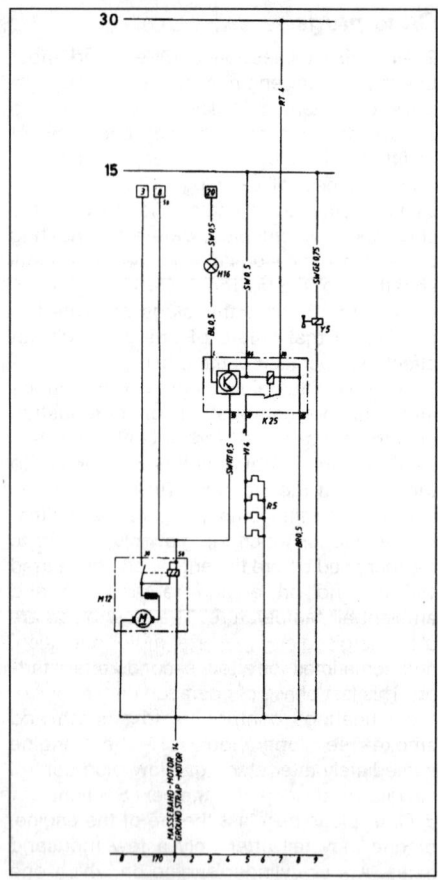

3.3b Circuit diagram for simple automatic preheating system

H16	Warning light	8	From starter switch
K25	Glow plug relay	15	Live rail (ignition-controlled)
M12	Starter motor		
R5	Glow plugs		
Y5	Stop solenoid	30	Live rail (permanent)
3	From battery		

Letters and numbers on wires relate to wire colours and thicknesses

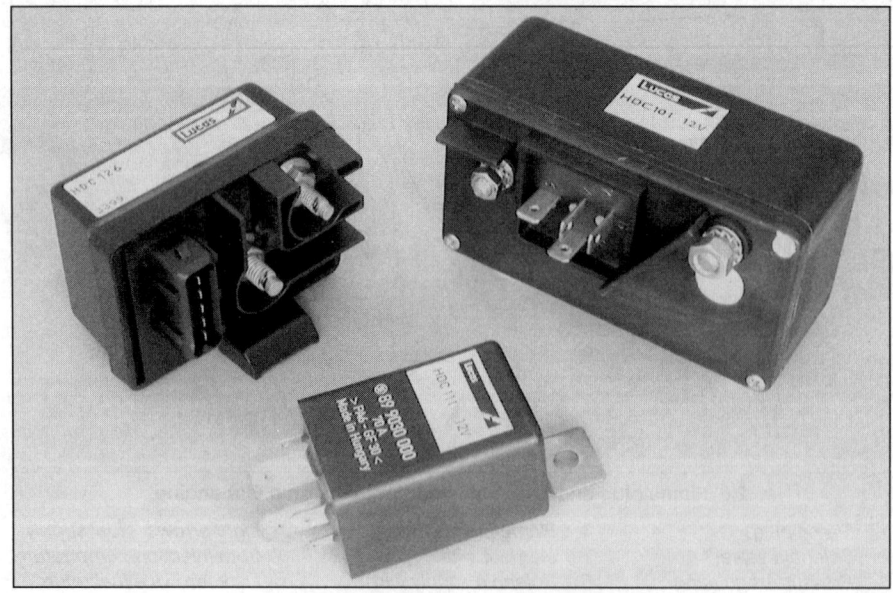

3.3a Typical preheating control units

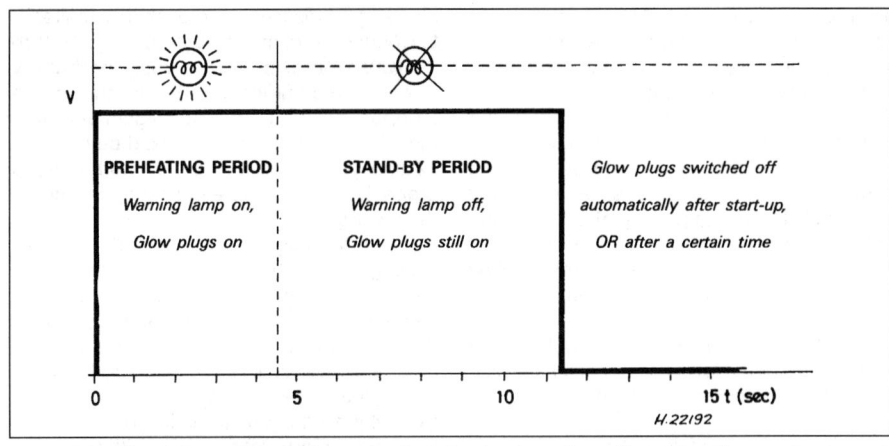

3.5 Glow plug voltage (V) versus time for basic automatic preheating system

4 The control unit responds to under-bonnet air temperature or coolant temperature to determine the preheating time required.

Additional control unit features may include protection against short-circuits, and automatic switching-off of the preheating circuit if the engine is not started within 30 seconds or so.

5 The function of the preheating system warning lamp is to advise the driver that the first stage of preheating is taking place. When the lamp goes out, the engine is ready to be started. When testing the preheating system, it is important to understand that the glow plugs are normally still energised after the warning lamp goes out, and remain energised until after start-up, or until a predetermined time has elapsed **(see illustration)**. This is known as the 'stand-by' or 'hidden glow' period.

Advanced systems

6 The advanced preheating system shown in the illustration and described in the following paragraphs is fitted to some VW and Audi models from 1987 **(see illustration)**. Similar systems are used by many other manufacturers, and on models fitted with

3

3.6 Circuit diagram for advanced preheating system

A	Battery
B	Starter motor
D	Ignition/starter motor
F60	Idling switch
G27	Coolant temperature sensor
J52	Glow plug relay and resistor
J179	Control unit
K29	Preheating warning lamp
N43	Stop solenoid
Q6	Glow plugs
S9	Fuse (in fusebox)
S39	Glow plug fuse (in relay)

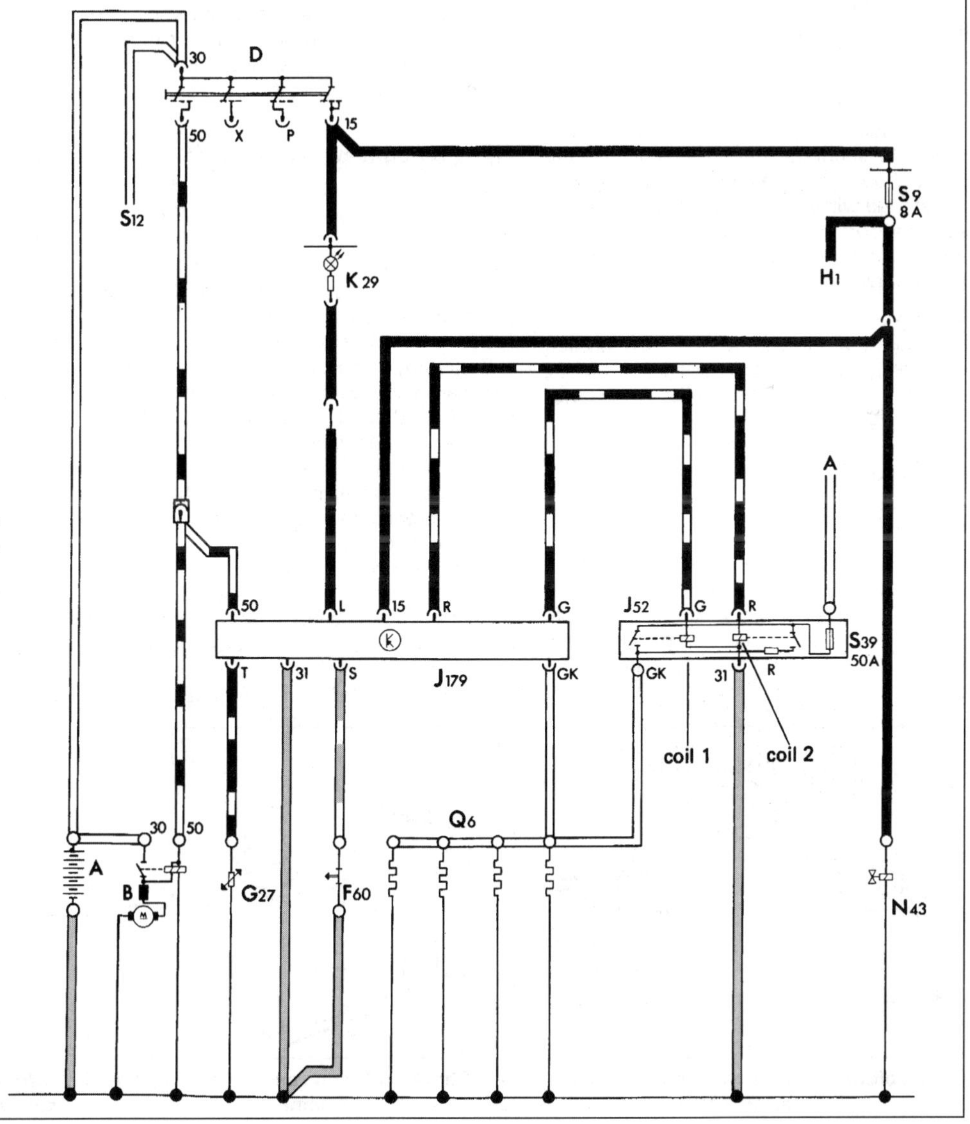

electronic engine control systems, a similar system is usually incorporated within the engine control system. In the example shown, besides the glow plugs themselves, the components of the system are:

a) *An automatic glow period control unit.*
b) *A relay, incorporating a fuse, a ballast resistor and a heat sink.*
c) *An idling switch which opens when the accelerator pedal is depressed.*
d) *A coolant temperature sensor.*
e) *A preheating warning lamp.*

Preheating

7 When the engine is warm, no preheating takes place. Operation of the system when the engine is cold (coolant temperature below 50ºC/122ºF) is as follows.

8 The driver turns the ignition switch to position II (run/preheat). The control unit activates coil No 1 in the relay, and the relay contacts close, to supply full battery voltage to the glow plugs. At the same time, the warning lamp comes on, warning the driver that the engine is not yet ready to be started.

9 After the necessary interval, the control unit extinguishes the warning lamp. The length of time taken will depend on coolant temperature. Even though the lamp has gone out, voltage is supplied to the glow plugs for a further 5 seconds, during which time the driver should start the engine. If the engine is not started within this period, the control unit switches off the supply to the relay, so preventing battery drain and overheating of the glow plugs.

Start-up

10 When the driver turns the ignition switch to position III (start), the control unit receives a signal from the wire which feeds the starter motor solenoid. As long as the starter motor is operating, the control unit activates relay coil No 1, and battery voltage is supplied to the glow plugs.

Post-heat

11 With the engine running, the post-heat period begins. For the first 5 seconds after start-up, the control unit continues to activate relay coil No 1, and the glow plugs continue to receive full battery voltage. This is to compensate for the cooling which will have occurred during starter motor operation, when battery voltage is considerably reduced.

12 After the initial 5 seconds, the control unit monitors the voltage at the glow plugs. When this rises above 11 volts, the control unit activates relay coil No 2. This causes a ballast resistor inside the relay to be switched in series with the glow plugs, which could otherwise be damaged by the increased voltage available with the engine running. If the voltage at the glow plugs drops below 10.3 volts, the control unit switches the resistor out of the circuit again.

13 Post-heat continues to be applied until one or more of the following conditions is met:

a) *Coolant temperature exceeds 50ºC/122ºF.*
b) *Post-heating has taken place continuously for 120 seconds.*
c) *The accelerator pedal is depressed for more than 10 seconds.*

In the last case, post-heating will be re-applied if the engine returns to idle while it is still cold. Full battery voltage will be applied for the first 5 seconds of the renewed post-heating period, in order to warm the glow plugs quickly. After 5 seconds, the resistor is again switched into circuit.

4 Charging system - testing

Note: *Thorough testing of the charging system is beyond the scope of this book, but the following tests will serve to show if the system is functioning or not.*

1 Begin by checking that the battery terminals and their connections are clean and tight. Also check the alternator connections and the alternator drivebelt.

2 If an ammeter with an inductive clamp connection is available, this may be used in the test procedure in addition to a voltmeter. Connect the clamp round the battery positive lead to measure the total charge or discharge current. The use of a series-connected ammeter is not recommended, partly because of the difficulty of avoiding it having to pass starter motor current, and partly because of the risk of alternator damage if it is accidentally disconnected or short-circuited.

3 Connect a voltmeter (with a range of 0 to 20 volts approximately) across the battery terminals **(see illustration)**. With the engine stopped, switch on the headlights and note the battery voltage, which should be between 12 and 13 volts. If it is lower, the battery is probably in a poor state of charge. Switch off the headlights.

4 Leave the voltmeter connected, and start the engine. Once the engine is running, voltage should rise to around 14 volts. If it does not, the alternator or voltage regulator is probably defective. (Consult the vehicle wiring diagram, or an electrical specialist, to establish if there are any particular features in the system under test which can give a similar result. Some systems incorporate fuses, either in the alternator itself or in the feed to the field windings; others will not work at low rpm if the no-charge warning light bulb has blown.)

5 Increase the engine speed to around 2000 rpm. Voltage may rise a little, but should not exceed 15 volts. If it does, the voltage regulator is defective.

6 With the engine still running at 2000 rpm, switch on as many electrical consumers as possible (headlights, heated rear window, heater blower, etc). Voltage may fall a little, but should not go below 14 volts. If voltage falls much below this, and raising the engine speed will not bring it back up, this shows that the alternator is not producing an adequate output, and should be investigated further. If an engine analyser with an oscilloscope is available, this may be used to examine the alternator output waveform to see if all three phases are present.

7 Switch off electrical consumers, and stop the engine. Disconnect the voltmeter, then disconnect the battery earth lead. Use a test lamp or an ammeter connected between the earth terminal and the earth lead to verify that with everything switched off, no current drain is taking place **(see illustration)**. (Items such as clocks and alarm systems should draw only a negligible current – certainly not enough to light a bulb.) If a drain appears to be taking place, remove the fuses one at a time to see which circuit is responsible.

8 If the current drain does not stop in relation to any fuse, try the effect of disconnecting the alternator. If this stops the drain, leakage is taking place through the alternator diodes; the diodes or the alternator itself must be renewed.

5 Battery - testing

⚠️ *Warning: Battery acid is poisonous and corrosive. Take care not to get it on clothing or skin, or in the eyes. In the case of accidental contact, wash off at once with plenty of cold water. If acid has splashed into the eyes, seek immediate*

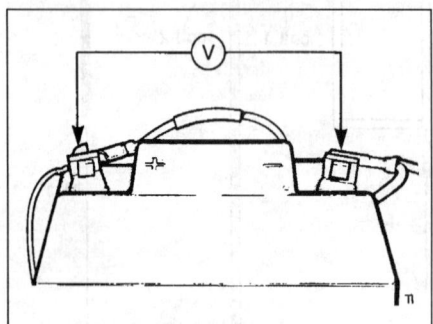

4.3 Voltmeter connected across battery terminals

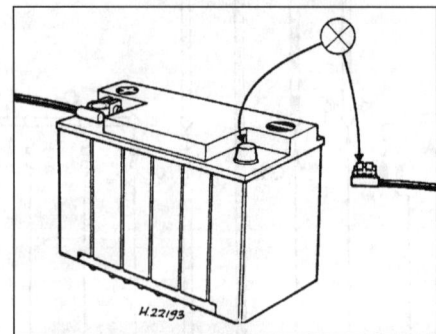

4.7 Checking for current drain using a test lamp

medical assistance as well. The gases given off by the battery during charging and testing are explosive. Do not smoke or cause sparks, or allow naked lights near the battery.

1 If battery problems are suspected and the charging system is known to be in order (Section 4), first make sure that the battery terminals and their connections are clean and tight. Also clean the top of the battery – discard the rag used for cleaning.

2 If the cell covers or plugs can be removed, check the electrolyte level and top up with distilled water if necessary. If one cell requires much more topping up than the others, this should be viewed with suspicion.

3 Check the specific gravity of the electrolyte with a hydrometer. This will give an idea of the overall state of charge, but more importantly will show up weak or failing cells, which will have a reading noticeably different from the good cells.

4 The most useful test of battery condition is to measure its terminal voltage during heavy discharge. It is this value which determines whether or not the battery can provide satisfactory cold starting performance. As a simple test, the terminal voltage can be measured while an assistant operates the starter motor. (Disconnect the stop solenoid, or manually operate the stop lever on the pump to prevent the engine starting.) If the battery is fully charged, but the voltage falls below 9.6 volts during cranking, it is almost certainly defective.

5 A purpose-made battery tester is a more satisfactory means of determining battery condition.

Warning: Older types of battery tester (consisting of a basic resistor, voltmeter and connecting tongs or leads) can

produce sparks at the battery terminals – take care as this can cause the battery to explode.

6 Modern electronic battery testers are safe to use, and will not produce sparks around the battery terminals.

6 Starter motor circuit - testing

Starter motor turns slowly

1 If the battery is fully charged and in good condition, but the starter motor turns slowly, there are three possibilities:
  a) *There is an excessive electrical resistance in the starter motor circuit.*
  b) *There is a fault in the starter motor itself.*
  c) *The engine is abnormally stiff to turn (mechanical problem or oil viscosity high).*

2 This Section deals with finding whether the problem is in the starter motor circuit or in the starter motor itself, and locating the problem if it is in the circuit.

3 Because the currents involved are very high, even a small resistance will have a large effect on the circuit. (Any point in the circuit where the resistance is higher than it should be will get hot when the starter motor is operated, and it may be possible to locate dirty or corroded connections by feeling them to detect this heating.) Such small levels of resistance cannot be measured with a conventional ohmmeter, so voltage drop tests are carried out with the starter motor operating.

4 To carry out the tests, an accurate voltmeter capable of reading fractions of one volt will be required. A high-range ammeter (0 to 400 amps or so) with an inductive clamp

pick-up would be useful, but is not essential. Some engine analysers incorporate suitable voltage and current measuring facilities.

5 If the ammeter is available, position the clamp on the battery-to-starter motor lead, and read the current drawn when the starter motor is operated. (Disconnect the fuel cut-off solenoid, or operate the pump stop lever, to prevent the engine from running.) A current draw of around 350 amps is typical. If it is higher, the circuit is in good condition, and there is a problem in the starter motor or the engine. If it is lower, carry on as follows.

Positive-side checks – solenoid mounted on starter motor

6 Measure the voltage between the battery positive terminal and the starter motor positive terminal (the output side of the solenoid), while an assistant operates the starter motor **(see illustration)**. The value obtained is the total voltage lost in the positive side of the starter motor circuit. The maximum acceptable loss is 0.5 volt; on a new system in good condition, 0.25 volt would be expected.

7 Repeat the test, but this time measure the voltage between the battery positive terminal and the input (battery) side of the solenoid. The difference between the first and second measurements is the voltage lost in the solenoid.

Positive-side checks – solenoid remote from starter motor

8 Take a reading between the battery positive terminal and the starter motor positive terminal while the starter motor is operated, as just described above **(see illustration)**.

9 Take a second reading between the battery positive terminal and the solenoid output terminal. The difference between the first and second readings is the voltage lost in the

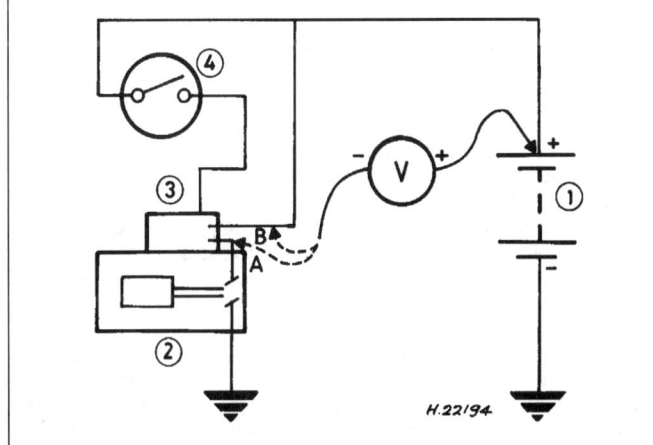

6.6 Starter motor voltage drop test points on positive side (solenoid on motor)

1	Battery	4	Ignition/starter switch
2	Starter motor	A	Solenoid output/motor input terminal
3	Solenoid	B	Solenoid input terminal

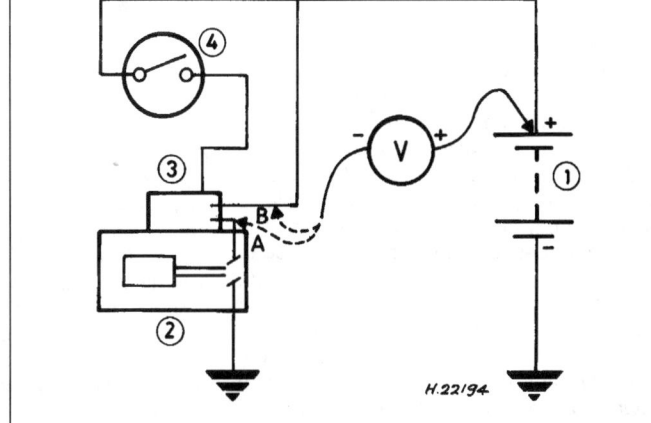

6.8 Starter motor voltage drop test points on positive side (remote solenoid)

1	Battery	A	Motor positive terminal
2	Starter motor	B	Solenoid output terminal
3	Solenoid	C	Solenoid input terminal
4	Ignition/starter switch		

3

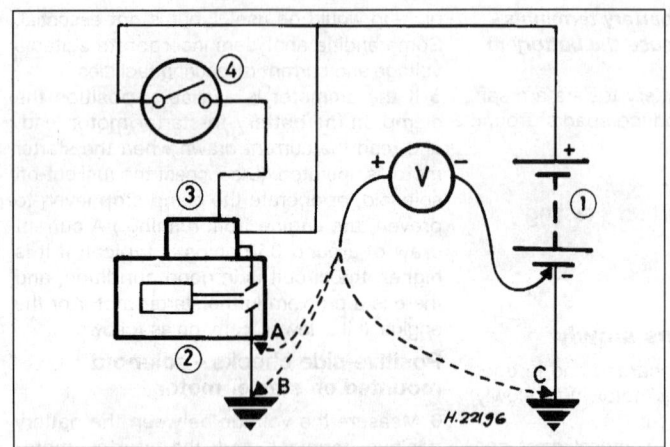

6.13 Starter motor voltage drop test points on negative side

1 Battery	A Starter motor body
2 Starter motor	B Engine earth strap
3 Solenoid	C Battery negative lead
4 Ignition/starter switch	

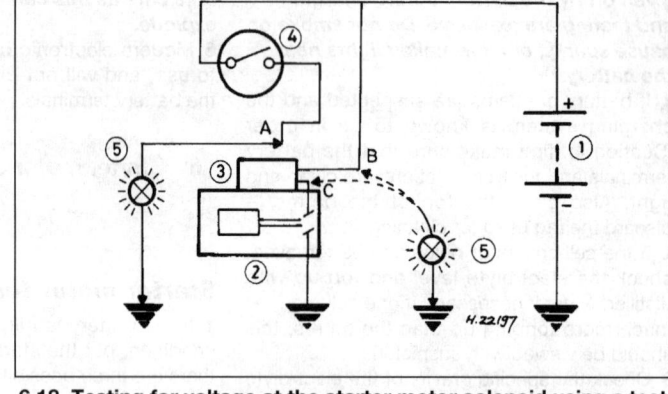

6.18 Testing for voltage at the starter motor solenoid using a test lamp

1 Battery	5 Test lamp
2 Starter motor	A Command lead
3 Solenoid	B Solenoid input terminal
4 Ignition/starter switch	C Solenoid output terminal

solenoid-to-starter motor cable. If the loss is more than 0.25 volt, clean and tighten the cable connections. If no improvement results, renew the cable.

10 Take a third reading between the battery positive terminal and the solenoid input terminal. The difference between the second and third readings is the voltage drop in the solenoid.

All solenoid types

11 If the loss in the solenoid is greater than 0.25 volt, remove it for cleaning or renewal.
12 If the loss in the battery-to-solenoid cable is greater than 0.25 volt, clean and tighten its connections. If no improvement results, renew the cable.

Negative-side checks

13 Measure the voltage between the battery negative terminal and the starter motor body, with the starter motor operating **(see illustration)**. The value obtained is the voltage lost in the whole of the negative side of the circuit. It should not exceed 0.25 volt.
14 Take a second measurement between the battery negative terminal and the body or chassis end of the earth strap, if this is separate from the battery negative lead. The difference between the first and second measurements is the voltage lost in the earth strap. If it is greater than 0.25 volt, clean and tighten the earth strap connections. If no improvement results, renew the strap. (Note that if the earth strap is loose or missing, earth return may take place through control cables, the speedometer cable, wheel bearings or similar, sometimes with bizarre results. Cases have been known of wheel bearings being damaged by high current flowing through them.)
15 Take a third measurement between the battery negative terminal and the far end

(engine or body attachment) of the battery negative lead. The value obtained is the voltage lost in the battery negative cable. If it is greater than 0.25 volt, clean and tighten cable connections. If no improvement results, renew the cable.

Starter motor does not turn at all

16 Apart from total seizure of the engine (and always assuming that the battery is in good condition), there are three possible reasons why the starter motor will not turn:
  a) The solenoid is not completing the circuit.
  b) There is a high electrical resistance in the circuit.
  c) There is a fault in the starter motor.
17 To make the following tests, an ordinary voltmeter (0 to 20 volts) or a 12-volt test lamp will suffice.
18 Disconnect the command (small) lead from the solenoid. Connect the voltmeter or test lamp between the lead and earth (vehicle metal) **(see illustration)**. Have an assistant operate the starter switch: voltage should be present when the switch is operated, and should disappear when it is released. If not, there is a break in the wiring between the switch and the solenoid, or a fault in the switch. (On automatic transmission models, it is also possible that the starter inhibitor switch is defective or mal-adjusted.)
19 Reconnect the command lead to the solenoid. Connect the voltmeter or test lamp between the input (battery side) terminal of the solenoid and earth. Voltage should be present here all the time, regardless of the position of the starter switch. If not, there is a break or a bad connection in the battery-to-solenoid lead.
20 Connect the voltmeter or test lamp between the output (motor side) terminal of the solenoid and earth. Have the assistant

operate the starter switch again: full battery voltage should be present when the switch is operated, and should disappear when it is released. If not, the solenoid is faulty.
21 If the solenoid is remote from the starter motor, connect the voltmeter or test lamp between the starter motor positive terminal and earth. Have the assistant operate the starter switch again. If voltage was present at the solenoid output terminal but does not appear at the motor, the solenoid-to-motor cable is broken or loose.
22 If it is established that full battery voltage is reaching the starter motor, but the motor is not turning (and assuming that the negative earth strap is intact and securely connected), there is a fault in the starter motor. Remove it for examination or renewal.

7 Preheating system - testing

1 Testing of the preheating system falls into three areas:
  a) Voltage supply checks.
  b) Glow plug or flame plug checks.
  c) Control circuitry checks.
2 The minimum equipment required is a 12-volt test lamp. More detailed testing will require a multimeter with the appropriate voltage and current ranges. For the professional, there exists dedicated test equipment such as the glow plug and circuit tester described later in this Section. On models with an electronic engine control system, if the system is equipped with a self-diagnostic facility, the system will store an appropriate fault code in the electronic control unit (ECU) memory if there is a fault in the preheating system (see Chapter 2, Section 8, and Chapter 9).

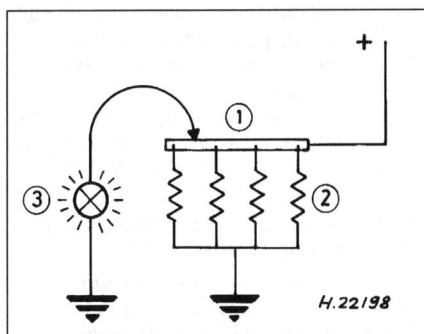

7.4 Checking for glow plug supply voltage with a test lamp

1 Glow plug supply 2 Glow plugs
  wire   3 Test lamp

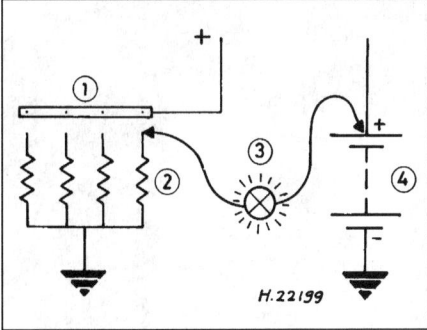

7.8 Checking glow plug continuity using a test lamp

1 Glow plug supply   2 Glow plugs
  wire   3 Test lamp
  (disconnected)   4 Battery

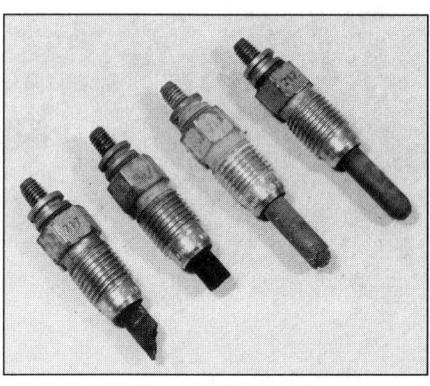

7.13 Damaged glow plugs

3 Test procedures given here are of a general nature.

Voltage supply checks with a test lamp or voltmeter

4 Connect the test lamp or voltmeter between the glow plug or flame plug supply wire and earth (vehicle metal) **(see illustration)**. Do not let the live side connections touch earth. Have an assistant energise the preheating system. The test lamp should light brightly, or the voltmeter should read at least 10 volts. (A few Japanese engines use 5-volt glow plugs; on these, a reading of 5 or 6 volts is to be expected.)

5 If there is no voltage at all, this suggests a fault such as a blown fuse, a disconnected wire, a defective relay, or a defective switch (depending on the type of system – refer to the circuit diagram). A blown fuse may only be a pointer to some underlying fault, such as a short-circuit in the wiring or a glow plug which has failed so as to cause a short. The fuse itself may be incorporated in the relay, or it may take the form of a fusible link in the feed wire near the battery.

6 If the voltage is low and the battery is in good condition, this suggests a bad connection somewhere in the wiring, or possibly a faulty relay.

7 On systems which control preheating time automatically, a check should now be made of the duration for which the voltage is applied. Remember that on most systems, preheating voltage continues to be applied after the warning lamp has gone out.

Glow plug checks with a test lamp or multimeter

8 A simple continuity check can be made by disconnecting the wire or metal strap which links the glow plug terminals (ignition off), then connecting the test lamp between the battery positive terminal and each glow plug terminal in turn **(see illustration)**. Alternatively, measure the resistance between each glow plug terminal and earth. The lamp should light brightly, or the meter read a very low resistance (typically 1 ohm or less).

9 If the lamp does not light or the meter shows a high resistance, the glow plug has failed open-circuit, and must be renewed.

10 The above is only a rough test, and will not detect a glow plug which has failed so as to cause a short-circuit, or one which is no longer heating properly, even though its resistance is still more or less correct. More accurate testing requires the use of an ammeter, of range 0 to 25, or 0 to 30 amps. It should incorporate some kind of overload protection, either in the instrument itself, or by means of a fuse in its lead.

Note: The procedure which follows applies to glow plugs which operate at full battery voltage, as is the case with most models. It does not apply to the 5-volt glow plugs fitted to some Japanese engines. For testing such plugs, a 6-volt source can safely be used.

11 Connect the ammeter between the battery positive terminal and one of the glow plugs. (The glow plugs must still be disconnected from each other.) Note the current draw over a period of 20 seconds or so. Typically, an initial surge of 20 amps or more will fall over 10 to 15 seconds to a steady draw of 9 or 10 amps. A very high draw indicates a short-circuit; a very low draw indicates an open-circuit.

12 Repeat the current draw check on the other glow plugs, and compare the results. Obvious differences such as a very high or very low draw will not be hard to spot. A difference in the rate at which the current falls off is also significant, and may indicate that the glow plug in question is no longer heating at the tip first.

13 So far, the tests have concentrated on the electrical condition of the glow plugs. Their physical condition is also important. To establish this, they must be removed (Section 8) and inspected for burning or erosion. Damage can be caused by a fault resulting in too long a post-heating time, but it is more often due to an injector fault **(see illustration)**. If damaged plugs are found, the injectors in the cylinders in question should be removed and checked for spray pattern and calibration.

14 As a final check, the glow plugs can be energised while they are out of the engine, and inspected to see that they heat evenly. The tip should glow first, with no local hot or cold spots. Some means of supporting the plug while it is being tested must be devised, and the power supply lead should be fused, or should incorporate some other overload protection. Ideally, a purpose-made glow plug tester with a hot test chamber should be used.

15 Any plug which takes much longer than specified for its tip to glow red, or which shows uneven heating, should be renewed.

Glow plug check using a glow plug tester

16 The following paragraphs describe the use of a typical glow plug tester, the Dieseltune DX 900. For other makes of tester, refer to the manufacturer's instructions.

Note: For testing the 5-volt glow plugs fitted to some Japanese engines, a resistor must be used to reduce battery voltage. A suitable resistor is available from the test equipment manufacturers.

Voltage supply

17 Disconnect the connecting wire or strap from the glow plugs (or from all but one, if preferred – but make sure that the bare ends cannot touch earth). Disconnect the feed wire from the connecting wire or strap.

18 Connect the tester leads as follows:
  Black to earth or battery negative.
  Red to the feed wire (from the relay or timer).
  Yellow to one glow plug terminal.

19 Sit in the driver's seat and hold the tester. Press and hold the 'test' button, and energise the pre-heating system (by switching on the ignition, on most models). The ammeter will show the current being drawn by one glow plug as long as the system is energised.

20 Check that current is being drawn while the preheating warning lamp is lit, and (on most systems) for some seconds after the lamp goes out.

3

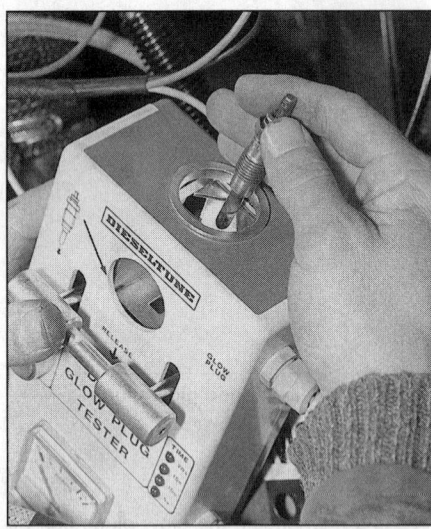

7.25 Placing a glow plug into the hot test chamber

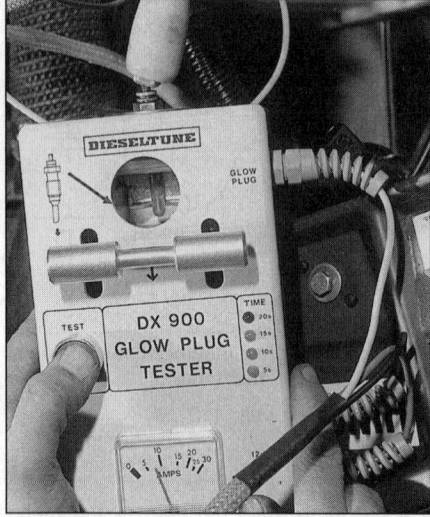

7.26 Hot testing a glow plug

Glow plugs – testing in the engine

21 With the glow plugs still disconnected from each other, connect the tester leads as follows:

Black to earth or battery negative.
Red to battery positive.
Yellow to one glow plug terminal.

22 Press and hold the 'test' button, and note the current draw shown on the ammeter. The timer LEDs on the tester will illuminate at 5-second intervals; note how long it takes for the current draw to stabilise, and what the final draw is. (If the glow plug is open-circuit, there will be no draw at all. If there is a short-circuit, the overload cut-out on the tester will trip.)

23 Release the 'test' button. Transfer the yellow lead to the next glow plug terminal, and repeat the test. Carry on until all the glow plugs have been checked.

24 Any glow plugs which show abnormal current draw characteristics should be removed (Section 8) for hot testing as follows.

Glow plugs – hot testing

25 Place the glow plug in the hot test chamber, clamping it in place with the spring-loaded bar **(see illustration)**. Connect the yellow lead to the glow plug terminal, leaving the red and black leads connected to the battery.

26 Press and hold the 'test' button, and observe the glow plug through the test chamber window **(see illustration)**. Note the time taken (as indicated by the timer LEDs) for the tip of the plug to glow red hot. If the plug takes longer than specified to glow, or if heating is uneven or starts elsewhere than at the tip, it should be renewed.

27 Release the test button and disconnect the yellow lead from the glow plug.

 Warning: The glow plug will still be very hot. Allow it to cool before removing it from the tester.

28 When testing is complete, refit the old glow plugs or fit new ones, as necessary, and remake the original electrical connections.

Control circuitry checks using a test lamp or multimeter

⚠ **Warning: On vehicles with advanced preheating systems controlled by an electronic diesel control system, it is advisable NOT to attempt to test the control circuitry using a test lamp or multimeter. Such test equipment can cause serious damage to ECUs and delicate electronic components if mis-used.**

29 The procedures given here are intended as a general guide. Refer to the manufacturer's information or to the relevant Haynes *Service and Repair Manual* for specific details.

Relays

30 The simplest type of relay is an electrically-operated on-off switch. It has four connections, two for the control winding, and two for the circuit being switched. These are shown in the accompanying illustration **(see illustration)**. The terminal numbers used correspond to the widely-used German standard (DIN); vehicles manufactured in other countries may use a different numbering system. Terminal 85 is the control winding negative (earth) side, while terminal 86 is the winding positive side. Terminal 88 is the live feed for the switched contact, and terminal 88a is the switched output.

31 To test the relay and its circuit, unplug the relay from its socket. Connect a voltmeter (0 to 20 volts approx) or a 12-volt test lamp between socket terminal 86 and earth. Energise the system (eg, by selecting the pre-heat position of the ignition/starter switch, if this is the circuit being tested). The voltmeter should indicate battery voltage, or the test lamp should light: if not, there is a fault in the control circuit or switch.

32 Measure the voltage between socket terminal 88 and earth. Again, battery voltage should be indicated, or the test lamp should light: if not, there is a fault in the supply circuit (perhaps a blown fuse, if applicable).

33 Using a multimeter set to measure resistance, or a self-powered test lamp, check for continuity between socket terminal 85 and earth. If there is no continuity, there is a fault in the wiring or the earth connection.

34 Turning to the relay itself, check for continuity between the winding terminals (85 and 86). If there is no continuity, the relay is certainly defective.

35 Check for continuity between the switched contact terminals (88 and 88a). There should be no continuity. Leave the

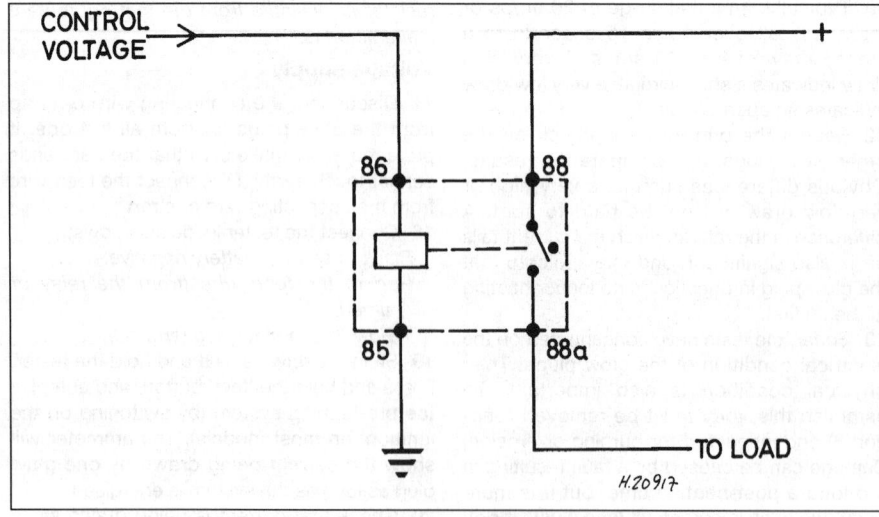

7.30 Preheating control relay schematic

85 *Control winding negative side*
86 *Control winding positive side*

88 *Live feed for switched contact*
88a *Switched output (live when energised)*

continuity tester connected, and apply battery voltage to the winding terminals. There should be an audible click, and continuity should be displayed: if not, the relay is defective. (Note that the 'click' alone is not a guarantee of correct operation, since the contacts may be defective while still producing a noise.)

36 A slightly more complicated relay which may be encountered is the changeover type **(see illustration)**. This works in the same way as the on-off relay, but it has an extra switch contact (87a), which is closed until the relay is energised. When testing this relay, continuity should be displayed between terminals 87a and 88 when the winding is not energised, and between terminals 88 and 88a when it is.

Control units

37 Refer to the manufacturer's information or to the relevant Haynes *Service and Repair Manual* for specific details of testing control units. Note that on control units incorporating semiconductors, care should be taken not to apply incorrect loads or voltages. In particular, incorrect use of a test lamp may cause damage to such units.

Temperature sensors

38 When a control unit receives a signal from a coolant temperature sensor, a quick way of testing the operation of the sensor is to disconnect it and observe the effect on preheating time. On most systems, disconnecting the sensor will result in an increased preheating time, since sensor resistance rises as temperature falls.

Flame plug checks

39 The following procedure applies to the Ford Transit 2.5 DI, and is a typical example of flame plug checking procedures. Refer to the manufacturer's information for specific details.

40 Bridge the system test connector. Have an assistant switch on the ignition **(see illustration)**. The pre-heating warning lamp should illuminate for approximately 5 seconds, and then go out. While the light is lit, and for up to 20 seconds after it goes out, the flame plug should operate. Confirm this by

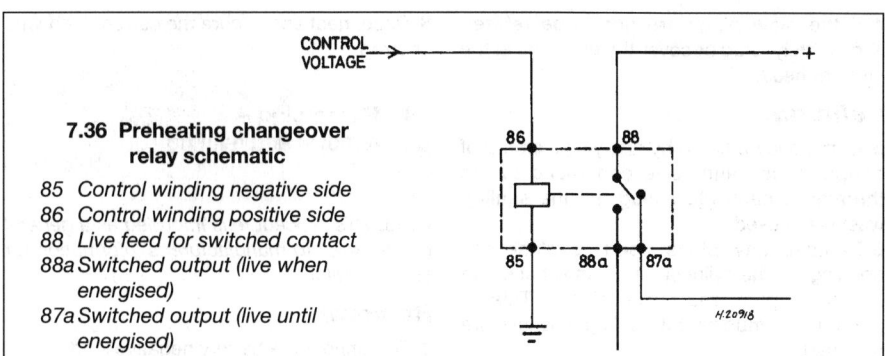

7.36 Preheating changeover relay schematic

85 *Control winding negative side*
86 *Control winding positive side*
88 *Live feed for switched contact*
88a *Switched output (live when energised)*
87a *Switched output (live until energised)*

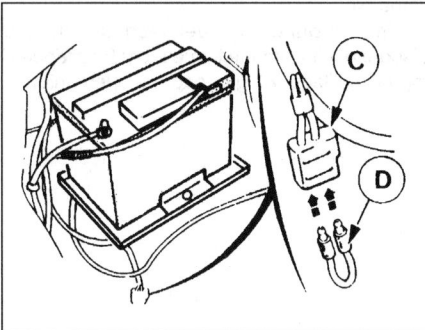

7.40 Flame plug test connector (C) and bridge wire (D)

carefully feeling for heat around the plug. Switch off the ignition.
Caution: The plug will get very hot.
41 If the system does not operate, disconnect the fuel pipe from the flame plug, and place the end of the pipe in a small container **(see illustration)**. Disconnect the electrical connector from the flame plug. Connect a 12-volt test lamp between the electrical feed and earth.
42 Switch on the ignition again. The voltmeter should read battery voltage, or the test lamp light, while the warning light is on, and for up to 20 seconds after it has gone out. While voltage is present, fuel should flow from the pipe.
43 If no voltage is present at the flame plug electrical feed, there is a fault in the wiring or the control unit. If voltage is present, but fuel

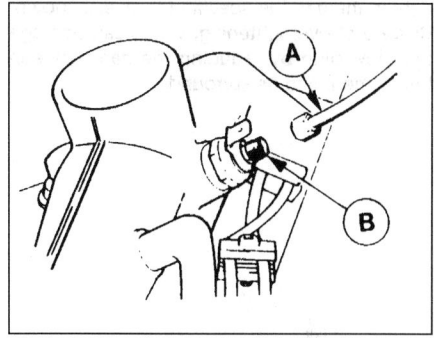

7.41 Fuel pipe (A) disconnected from flame plug (B)

does not flow even through the reservoir is full, there is a fault in the wiring to the solenoid valve, or in the valve itself.
44 If voltage and fuel are both present, but the flame plug does not operate, remove the plug and apply 12 volts to its electrical terminal. Earth the body of the plug. The heating element inside the plug should be seen to glow red. If the element does not glow, renew the plug.
Caution: The plug will get very hot.

8 Glow plugs – removal and refitting

Note: *This is general procedure. Refer to manufacturer's information or to the relevant Haynes Service and Repair Manual for specific details.*

⚠ **Warning: If the glow plugs have just been tested in the engine, or if the engine has just been running, they may be very hot.**

Removal

1 Make sure that the ignition is switched off, then disconnect the current feed wire or strap from the glow plug terminals **(see illustration)**. Recover any terminal nut washers.
2 Brush or blow away any debris from around the glow plug seats.
3 Unscrew the glow plugs from the cylinder head **(see illustration)**. Recover the sealing washers (if used); new washers should be used on reassembly.

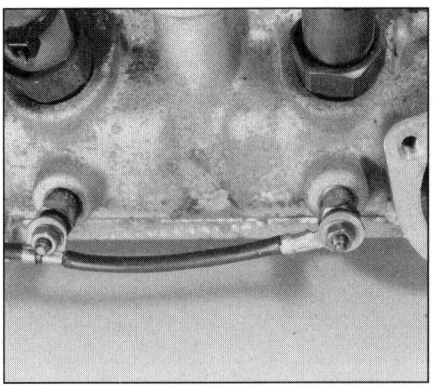

8.1 Feed wire connected to glow plug terminals

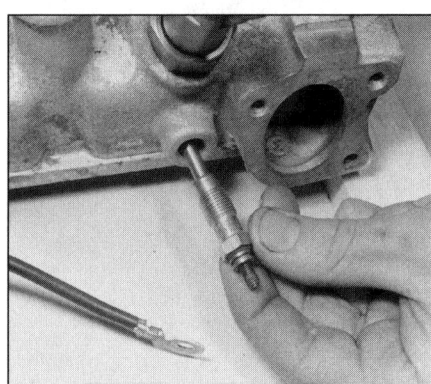

8.3 Removing a glow plug

4 If the glow plugs are not to be refitted immediately, plug or cover the openings in the cylinder head.

Refitting

5 Commence refitting by applying a smear of copper-based anti-seize compound to the threads of each glow plug. Fit the sealing washers, if used.

6 Remove any plugs used to close the openings in the cylinder head, and check that the glow plug seats are clean. (This is particularly important if sealing washers are not used.)

7 Insert the glow plugs into their holes, and tighten them to the specified torque, if known. Beware of overtightening, which can damage the glow plug by reducing the gap between the element and its surround.

8 Reconnect and secure the current feed wire or strap.

9 Flame plug –
removal and refitting

Note: *This procedure is intended as a general guide. Refer to manufacturer's information for specific details.*

Removal

1 Disconnect the battery negative lead.

2 Disconnect the electrical feed from the flame plug.

3 Clean around the fuel feed union, and disconnect it from the flame plug; be prepared for fuel spillage. Plug or cap the open unions.

4 Clean any debris from around the flame plug seat, and unscrew the flame plug from the inlet manifold. Recover the sealing washer (if used); a new washer should be used on refitting.

5 If the flame plug is not to be refitted immediately, plug the hole in the inlet manifold.

Refitting

6 Commence refitting by applying a smear of copper-based anti-seize compound to the threads of the flame plug. Fit the sealing washer, if used.

7 Where applicable, remove the plug from the hole in the inlet manifold, then insert the flame plug and tighten it.

8 Reconnect the fuel feed union, and the electrical feed.

9 Reconnect the battery negative lead.

Chapter 4
Ancillary components

Contents

1 Introduction

Besides the usual ancillaries found on vehicle engines, such as starter motors, alternators, power steering pumps, etc, there are two additional components which are very common on diesel engines; the vacuum pump, and the turbocharger.

Most light diesel engine vehicles rely on an engine-driven pump for brake vacuum servo operation, since there is insufficient depression created in a diesel engine's unthrottled inlet manifold to operate a servo. A few manufacturers also use vacuum to actuate items such as the injection pump stop control.

The turbocharger increases engine efficiency by raising the pressure in the inlet manifold above atmospheric pressure. Instead of the air supply simply being sucked into the cylinders, it is forced in. Additional air in the cylinders means that more fuel can be burnt and more power produced.

2 Vacuum pumps

Description

1 Vacuum pumps are usually rotary vane or diaphragm types. A reservoir may be fitted between the pump and the servo to provide enough vacuum for several applications of the brake with the engine stopped. On some commercial vehicles, loss of vacuum may be detected by a switch which controls a warning light or buzzer.

2 Pump drive may be by pulley and belt, by direct or indirect drive from the camshaft or an auxiliary shaft, or from some other ancillary component such as the alternator.

Maintenance

3 On belt-driven pumps, the tension and condition of the pump drivebelt should be checked periodically. Adjustment is carried out by pivoting the pump (or sometimes an idler pulley, or some other belt-driven component) on its mountings.

4 When an oil filler/level plug is fitted, this should be removed periodically to verify that oil is up to the level of the plug hole **(see illustration).** (On some types of pump, the pulley must be turned to align a mark on the pulley shoulder with a mark on the pump before checking the level.) Top up if necessary with the specified oil, and refit the level plug.

4

**2.4 Vacuum pump oil filler/level plug
(arrowed)**

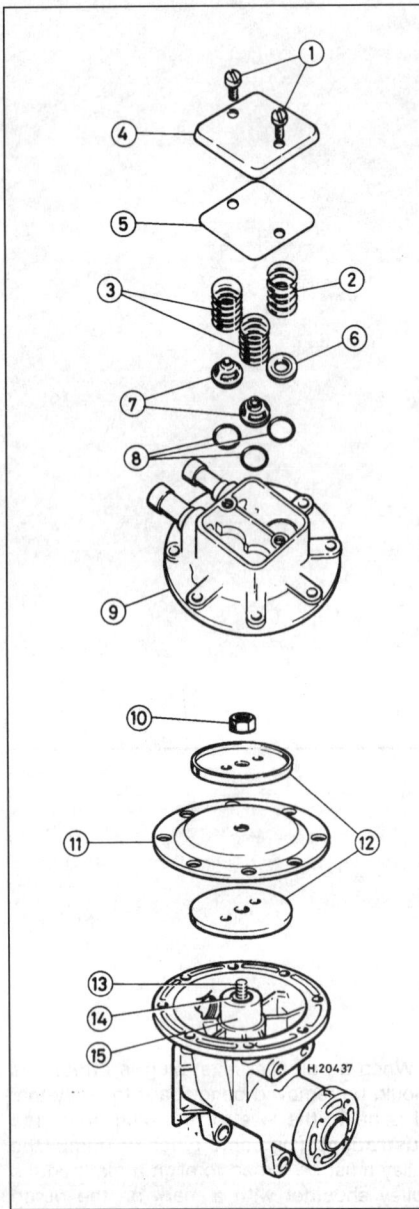

2.13 Exploded view of a diaphragm type vacuum pump

1 Cover screws	8 Seals
2 Inlet valve spring	9 Diaphragm cover
3 Outlet valve	10 Nut
springs	11 Diaphragm
4 Cover	12 Support plates
5 Gasket	13 Screw
6 Inlet valve	14 O-ring
7 Outlet valves	15 Piston

Testing

5 Disconnect the pump-to-servo (or reservoir) hose. Connect a vacuum gauge to the pump inlet port.

6 Start the engine, and allow it to idle; note the rate at which the vacuum increases. Hard-and-fast values are not often specified, but as a guide, the pump should produce a vacuum of at least 0.67 bar (500 mm Hg) within one minute. If

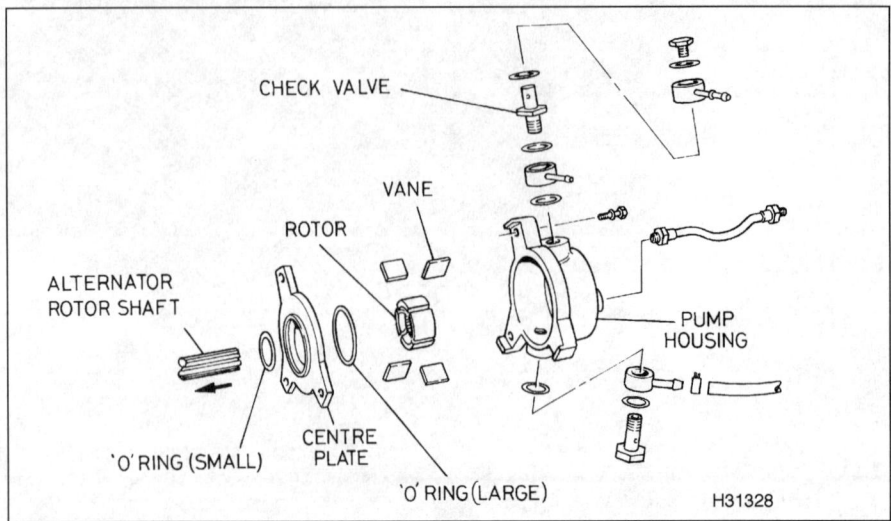

2.21 Exploded view of a vane type vacuum pump

it does not, either the drivebelt is slipping (when applicable), or the pump is faulty.

Removal and refitting

7 Removal is generally straightforward, and is typically carried out as follows.

8 Disconnect the hose(s) from the pump, noting which is inlet and which (if any) is exhaust. If separate oil feed and return hoses are fitted, disconnect them too, being prepared for oil spillage.

9 If the pump is belt-driven, slacken the pump mountings and remove the drivebelt.

10 Remove the pump mounting nuts and bolts, and remove the pump. Recover any seals or gaskets. In the case of a cam-operated pump, recover the operating plunger or pushrod if it is loose.

11 Refitting is the reverse of the removal procedure. Renew any seals or gaskets if necessary, and on belt-driven pumps, adjust the drivebelt tension.

Overhaul

12 Most modern pumps are sealed units, and no spare parts are available. If spares are available, the following procedures can be used for guidance.

Diaphragm type pump

13 Remove the screws which secure the valve cover. Lift off the cover and gasket, followed by the springs, valves and seals **(see illustration)**.

14 Mark the diaphragm cover in relation to the body of the pump. Remove the screws which secure the cover, and lift off the cover. Remove the central nut or screw, and remove the diaphragm and support plates from the piston. Recover the O-ring, if fitted, from the recess in the top of the piston.

15 Turn the pulley to bring the pump piston to the top of its stroke. In this position, check for wear by trying to move the piston from side to side. If excessive wear is present, renew the pump.

16 Clean all components. Commence

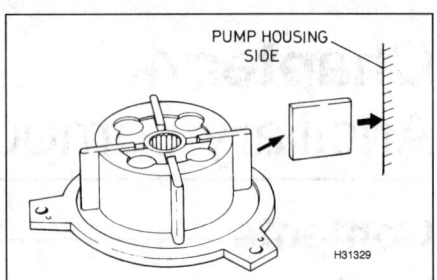

2.22 Correct fitting of vanes in rotor – vane type vacuum pump

reassembly by fitting a new O-ring, if applicable, to the top of the piston.

17 Fit a new diaphragm and the support plates to the piston, making sure that the support plates are the right way round. Apply thread-locking fluid to the threads of the central nut or screw, then fit and tighten it.

18 Refit and secure the diaphragm cover, observing the previously-made alignment marks.

19 Fit new seals, valves and springs into the valve chest. Fit a new gasket, then refit and secure the cover.

20 Top up the pump oil level before refitting.

Rotary vane type pump

21 Remove the check valve and the oil feed and return pipe unions from the pump **(see illustration)**.

22 Separate the centre plate from the pump housing. Remove the rotor and vanes, noting which way round the vanes are fitted **(see illustration)**.

23 Renew the O-rings, sealing washers, etc, as a matter of course. Renew the vanes if they are worn or damaged. If the rotor or the pump housing is worn, a complete new pump will probably be needed.

24 Test the check valve with low air pressure applied to each side. It should only pass air in one direction.

25 Reassemble the pump, coating the various parts with clean engine oil.

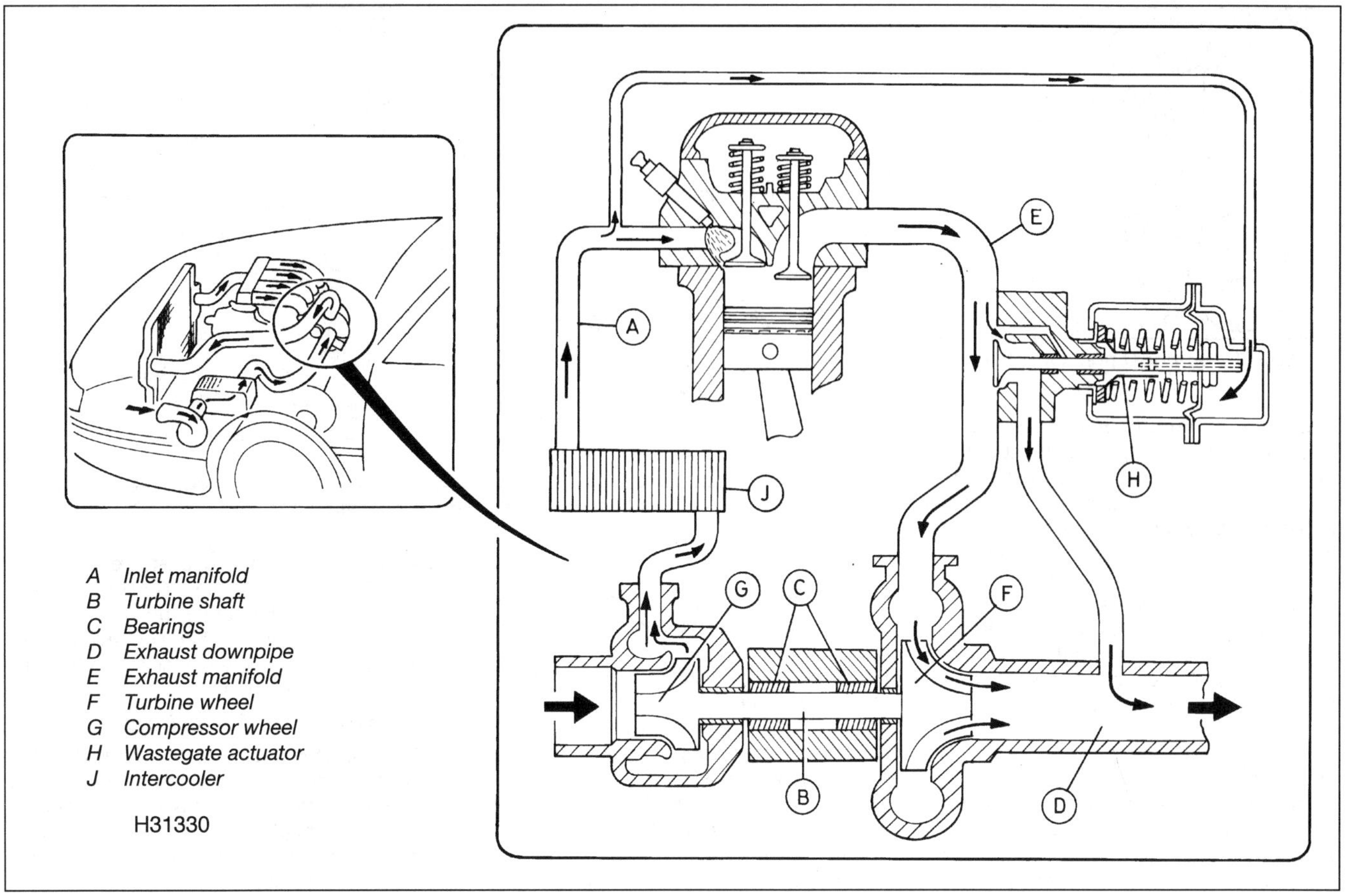

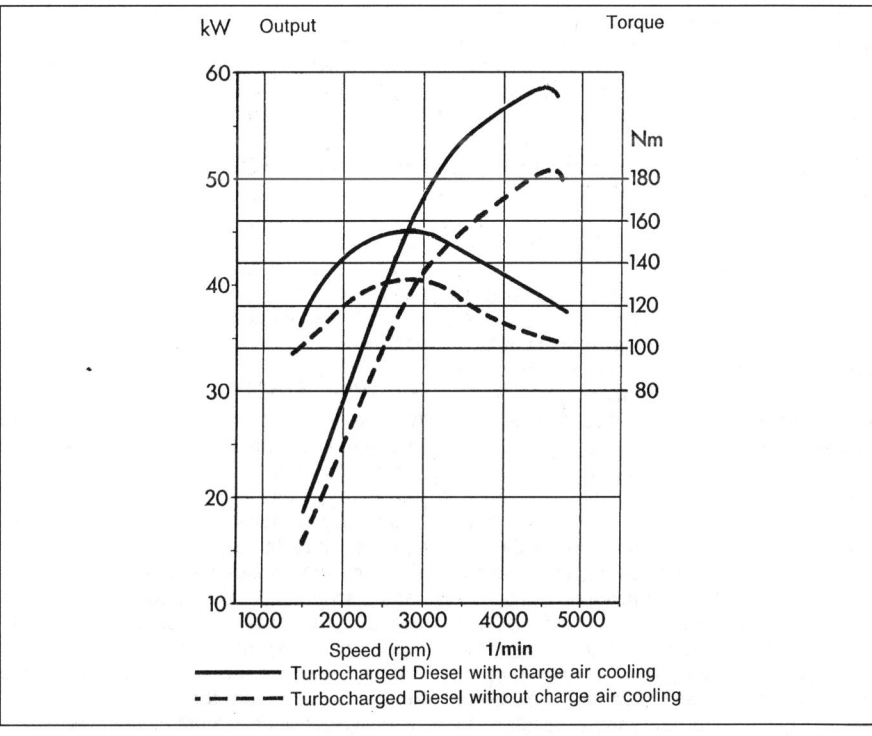

A   Inlet manifold
B   Turbine shaft
C   Bearings
D   Exhaust downpipe
E   Exhaust manifold
F   Turbine wheel
G   Compressor wheel
H   Wastegate actuator
J   Intercooler

H31330

3.1 Turbocharger location and schematic view of operation – Citroën XM

3 Turbochargers

Description

1 Energy for the operation of the turbocharger comes from the exhaust gas. The gas flows through a specially-shaped housing (the *turbine housing*), and in so doing spins the turbine wheel. The turbine wheel is attached to a shaft, at the end of which is another vaned wheel known as the *compressor wheel*. The compressor wheel spins in its own housing, and compresses the inducted air on the way to the inlet manifold **(see illustration)**.

2 On some models, the compressed air passes through an *intercooler* before arriving at the manifold. The intercooler is an air-to-air heat exchanger, which removes from the inducted air some of the heat it gained in being compressed. Under full-load conditions, the temperature of air leaving the turbocharger can be over 100°C (212°F). The intercooler in these conditions reduces the temperature by about 50°C. Cooling the air makes it denser and increases engine efficiency **(see illustration)**.

kW   Output   Torque

60

50   Nm
                                    180
                                    160
40   140
                                    120
30   100
                                    80
20

10
   1000 2000 3000 4000 5000
        Speed (rpm)   1/min

——— Turbocharged Diesel with charge air cooling
- - - Turbocharged Diesel without charge air cooling

3.2 Engine power and torque with and without charge air cooling (intercooling)

4

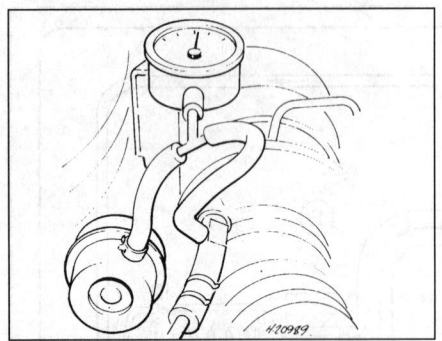

3.13 Turbo-boost pressure check

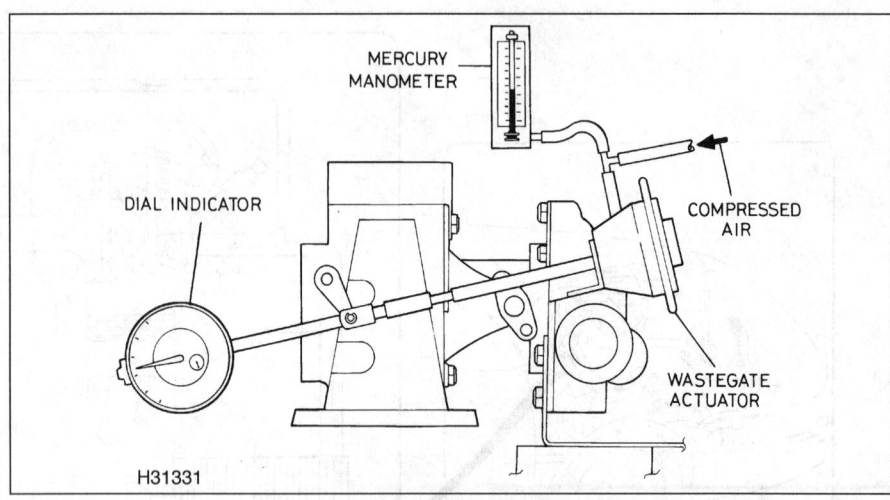

3.21 Wastegate actuator operation check

3 Boost pressure (the pressure in the inlet manifold) is limited by a wastegate, which diverts the exhaust gas away from the turbine wheel in response to a pressure-sensitive actuator. A blow-off or dump valve is often fitted in the inlet manifold, as secondary protection against excessive boost pressure.

4 The turbine shaft is pressure-lubricated by a feed from the main oil gallery. When the engine is running, the shaft 'floats' on a cushion of oil. A drain pipe returns the oil to the sump. Some models also circulate coolant round the turbo housing.

Precautions

5 The turbocharger operates at extremely high speeds and temperatures. Certain precautions must be observed to avoid premature failure of the turbo, or injury to the operator.

6 Do not operate the turbo with any parts exposed. Foreign objects falling onto the rotating vanes could cause damage and (if ejected) injury. Loose rags, small tools or other objects could also be drawn into the turbo air intake when the engine is running.

7 Do not run the engine without an air cleaner element fitted. Even if no large foreign objects pass through and damage the turbo, dust and grit in the air can damage the turbo at operating speeds.

8 Do not race the engine immediately after start-up, especially if it is cold. Give the oil a few seconds to circulate. This is particularly important after an oil change. If the turbo oil feed has drained, it is advisable to prime it by cranking the engine on the starter motor (with the pump stop solenoid disconnected, or the stop lever actuated) until the oil pressure warning light goes out.

9 Do not switch off the engine before it has returned to idle speed. After a high-speed run, allow the engine to idle for a minute or so before switching off. Do not blip the throttle and then stop the engine, as this will leave the turbo spinning without lubrication.

10 Observe the specified intervals for oil and filter changing, and use a reputable oil of the specified quality. Neglect of oil changing, or use of inferior oil, can cause the formation of carbon on the turbine shaft, leading to premature failure.

Testing

11 As a quick check of turbo function, open the bonnet and run the engine, accelerating it briskly. It should be possible to hear the turbocharger whistling or whining.

12 For accurate checking of boost pressure, a pressure gauge (range approximately 0 to 2 bar) will be needed, positioned so that it can be viewed when driving.

13 Connect the gauge into a suitable hose which conveys inlet manifold pressure – for instance, the hose which supplies this pressure to the injection pump **(see illustration)**.

14 Drive the vehicle, and briefly create the maximum possible boost by applying both brake and throttle at approximately 3000 rpm in 3rd gear. Note the boost pressure developed.

Caution: Do not allow boost pressure to reach levels in excess of those specified by the manufacturer.

Typical figures are:

Maximum working boost	Warning light/dump valve operates at
0.7 to 1.0 bar	0.9 to 1.2 bar

15 Low boost pressure is not necessarily due to a fault in the turbocharger – see Chapter 9. Excessive boost pressure can only be due to a fault associated with the wastegate actuator.

Removal and refitting

16 Removal involves disconnection of the induction and boost air pipes and hoses, the exhaust front section, and the oil feed and return pipes. Additionally, some turbochargers have coolant feed and return pipes which must be disconnected.

17 Once the various pipes and hoses have been disconnected, the turbocharger can be unbolted from the exhaust manifold, or (if access dictates) removed with the manifold and separated on the bench. Lockplates are frequently used on the turbo-to-manifold fastenings, and these must be renewed.

18 Refit by reversing the removal operations,

using new oil seals, gaskets, etc. Initially prime the turbo lubrication circuit by injecting clean engine oil into the turbocharger and/or oil feed pipe. Carry out further priming before start-up, by cranking the engine with the stop solenoid disconnected.

Inspection and overhaul

19 Inspect the turbine and compressor wheel vanes for cracks and chips. Spin them to verify that the shaft is intact, and check for shake or roughness. Some free-play is normal; this is taken up in use by the pressurised oil feed. Typical values are:

Axial (end-to-end) play	Radial (side-to-side) play
0.005 to 0.10 mm	0.30 to 0.60 mm

20 If turbine shaft oil seal failure has occurred, this will be evident by heavy deposits of oil or carbon. The induction passages and intercooler may also be contaminated; if this has happened, the intercooler should be flushed out.

21 Check the operation of the wastegate actuator by applying low air pressure (typically up to 1 bar) to it. The actuator rod must move in and out smoothly when pressure is applied and released. Vehicle manufacturers often specify travel for the wastegate actuator lever under given pressures; in this case, a pressure gauge and a dial test indicator (DTI) may be required for testing **(see illustration)**.

22 No overhaul of the turbocharger is generally possible, although it may be possible to renew the wastegate actuator separately. The old unit may be accepted in part exchange for a new or reconditioned one.

Aftermarket fitting

23 Various specialist firms offer aftermarket fitting of turbochargers. To have any useful effect, such fitting must include recalibration or renewal of the fuel injection pump. Such work will probably invalidate any manufacturer's warranty on the engine and fuel injection system; reputable firms may take on this warranty risk themselves.

Chapter 5
Emission control

Contents

1 Introduction

Legislation exists in most developed countries to reduce vehicle emissions (both diesel and petrol) in order to protect the population's health, and the environment. The permitted levels of pollution are being progressively reduced as time goes on, and vehicle manufacturers are required to invest heavily to improve engine efficiency and reduce harmful emissions.

All modern diesel engines are equipped with features designed to reduce the emissions of harmful by-products of the combustion process into the atmosphere.

When diesel fuel is burnt, a variety of combustion products are formed. The products formed depend on the fuel quality, engine design, the power output, and the working load on the engine. The major by-products of the combustion process are water (H_2O) and carbon dioxide (CO_2 - a 'greenhouse' gas, although not toxic).

Besides these two main products, the following substances are also produced in relatively low concentrations:

Carbon monoxide (CO).
Unburnt hydrocarbons (HC).
Nitrogen oxides (NOx).
Sulphur dioxide (SO_2) and sulphuric acid (H_2SO_4).
Soot (carbon) particles.

Complete combustion of the fuel leads to major reductions in the formation of toxic substances, and so one of the most important design parameters for a modern diesel engine is to achieve near-complete combustion under all engine operating conditions.

The following engine and fuel system features help to reduce exhaust gas emissions, and have the added benefit of improving fuel consumption:

Accurate start of injection timing.
Precision-manufactured injection nozzles.
Precise fuel metering.
Precisely-defined fuel-spray geometry.
High injection pressures.
Combustion chamber design.

As an example of the importance of accurate control of the injection process, a deviation of the start of injection by 1° of crankshaft rotation can cause increases in NOx and CO emissions of 5% and 15% respectively.

Diesel fuel contains sulphur, which as mentioned previously, is converted into sulphur dioxide and sulphuric acid during combustion. To reduce the levels of these harmful by-products, the permitted sulphur content in diesel fuel has been reduced in stages over recent years, and according to current standards must not exceed 0.2% by weight. This level is likely to be reduced further in the future.

Although the measures mentioned previously reduce harmful emissions significantly, additional devices and systems can be used to reduce emissions even further. The most-commonly used systems on modern diesel engines are:

Crankcase emissions control system.
Exhaust emissions control system (catalytic converter and/or soot burn-off filter).
Exhaust gas recirculation system.

These systems are described in the following Sections.

5

2 Crankcase emissions control systems

1 To reduce the emission of unburned hydrocarbons from the engine crankcase into the atmosphere, the engine is sealed. Piston blow-by gases (combustion gases which have passed by the piston rings) and oil vapour are drawn from the crankcase and the cylinder head cover, through an oil separator, into the inlet tract **(see illustration)**. The gases are then drawn into the engine, together with fresh air/fuel mixture, to be burned by the engine during normal combustion. Condensed oil vapour is returned from the oil separator to the engine sump.

3 Exhaust emissions control systems

Catalytic converter

1 To minimise the level of exhaust gas pollutants released into the atmosphere, a catalytic converter is fitted, located in the exhaust system.
2 The catalytic converter consists of a canister containing a fine mesh impregnated with a catalyst material, over which the exhaust gases pass. The catalyst speeds up the oxidation of harmful carbon monoxide and unburnt hydrocarbons, effectively reducing the quantity of harmful products reaching the atmosphere **(see illustration)**. As unburnt hydrocarbons contribute to particle emission, this can also be reduced to a limited extent by a catalytic converter.
3 A closed-loop catalytic converter system using an oxygen sensor, similar to that used on petrol engines, cannot be used on a diesel engine because a diesel engine always operates with excess air, and hence oxygen, in the exhaust gas.

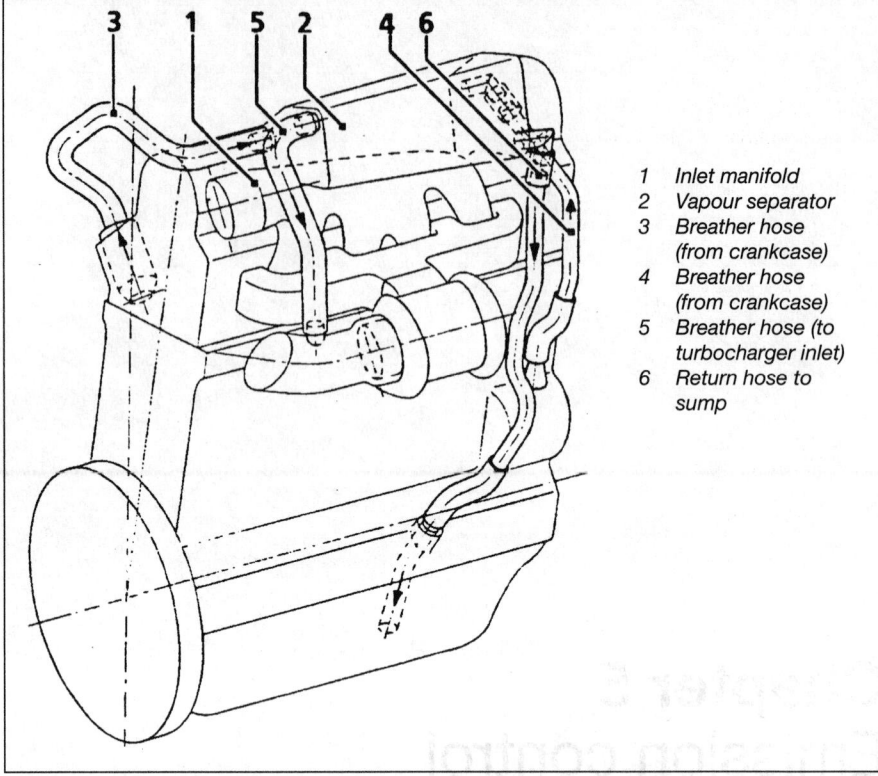

1 Inlet manifold
2 Vapour separator
3 Breather hose (from crankcase)
4 Breather hose (from crankcase)
5 Breather hose (to turbocharger inlet)
6 Return hose to sump

2.1 Typical crankcase emission control system

Particle filters

4 Particle filters are designed to reduce the level of smoke particles released into the atmosphere from a diesel engine exhaust. These devices are in the early stages of development, and various alternative systems are used by different manufacturers.
5 Particle filters work by filtering out the tiny soot particles (carbon particles coated with unburnt hydrocarbons, lubricating oil, sulphates and water) before they reach the exhaust tailpipe. Such filters are usually located in the exhaust system in the position normally occupied by a catalytic converter.

Particle filters are often combined with a catalytic converter in a single unit.
6 At the time of writing, the two most commonly used types of particle filter are the soot burn-off filter and the plasma particle filter. The two types of filter are described briefly in the following paragraphs.

Soot burn-off filter

7 Several different types of soot burn-off filter have been developed, but the following paragraphs describe a typical filter system being developed for fitment to some Peugeot/Citroën models **(see illustration)**.
8 In a soot burn off filter, the soot particles are trapped in a block of filter material, usually silicon carbide. The filter has to be carefully designed to allow the exhaust gases to flow through it, whilst trapping the soot particles; however there obviously comes a point when the filter will become blocked due to the large number of particles trapped (after, say, 200 to 300 miles of driving). As the trapped particles in the filter build up, there will be a resistance to the exhaust gas flow, and the filter must be cleaned to prevent a build-up of exhaust back-pressure.
9 Because the diesel engine always operates with excess air, the exhaust gas contains enough oxygen that at temperatures above approximately 550°C, soot will burn off of its own accord. The exhaust gas temperature in a diesel engine is normally between 150°C and 200°C, which is not high enough to burn

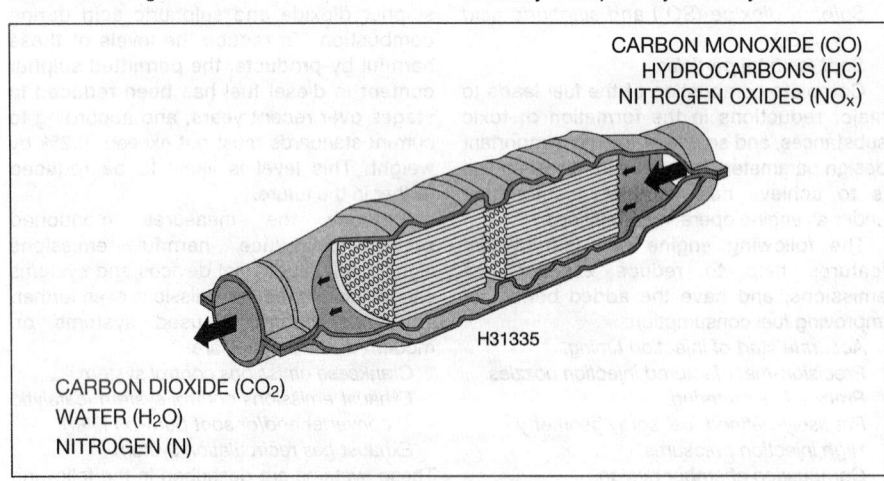

CARBON MONOXIDE (CO)
HYDROCARBONS (HC)
NITROGEN OXIDES (NO_x)

CARBON DIOXIDE (CO_2)
WATER (H_2O)
NITROGEN (N)

H31335

3.2 Cross-section of a typical catalytic converter

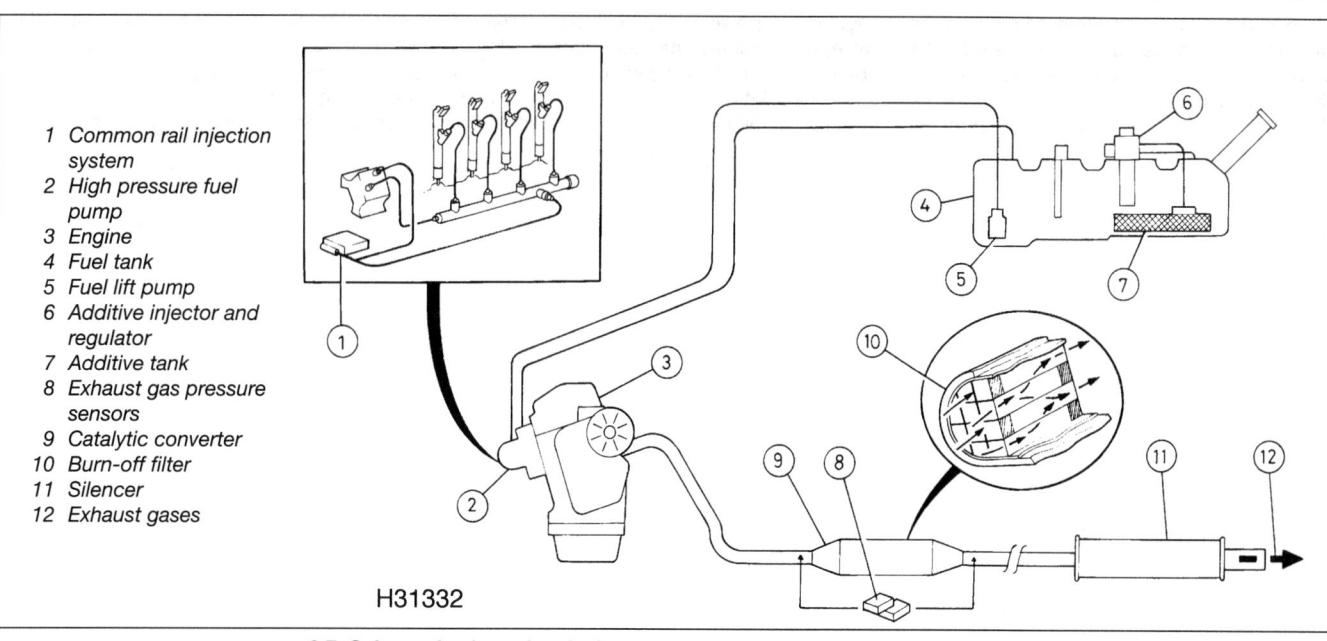

1 Common rail injection system
2 High pressure fuel pump
3 Engine
4 Fuel tank
5 Fuel lift pump
6 Additive injector and regulator
7 Additive tank
8 Exhaust gas pressure sensors
9 Catalytic converter
10 Burn-off filter
11 Silencer
12 Exhaust gases

H31332

3.7 Schematic view of emission control system using soot burn-off filter

off soot. The exhaust gas temperature can be raised by injecting extra fuel and air. Additionally, the reaction in a catalytic converter will raise the exhaust gas temperature, so soot burn-off filters are often combined with catalytic converters in a single unit.

10 Even with the extra temperature provided by additional fuel/air and the catalytic converter reaction, additional heat is still required to burn the soot. This is provided by the injection of an additive into the diesel fuel. Injection of the additive is controlled by the electronic diesel control ECU. An exhaust gas pressure sensor (or sensors) in the exhaust system provide the ECU with information which allows it to calculate when the filter is becoming blocked. When the exhaust gas pressure falls to a predetermined level, a 'shot' of additive is injected into the fuel, which raises the temperature inside the filter, and burns off the soot, cleaning the filter in the process. The additive is usually contained in a separate tank on the vehicle.

Plasma particle filter

11 This type of filter is in the early stages of development, and is designed to reduce both particle and NOx emissions. The filter contains a filter material which has a dielectric compound embedded in it. A high-voltage feed creates a plasma within the filter. Particles adhering to the filter are exposed to the plasma which causes partial oxidation of the soot particles and NOx compounds.' A remote or integral catalytic converter enables further oxidation before the exhaust gases are released into the atmosphere.

4 Exhaust gas recirculation system

1 An exhaust gas recirculation (EGR) system is designed to recirculate small quantities of exhaust gas into the inlet tract, and therefore into the combustion process. This process reduces the level of oxides of nitrogen present in the final exhaust gas which is released into the atmosphere, and also lowers the combustion temperature.

2 The volume of exhaust gas recirculated is controlled by vacuum, via a solenoid valve. The solenoid valve is controlled by a fuel injection pump-mounted sensor on models with a conventional injection pump, or by the electronic diesel engine control unit (ECU) on models with an electronically-controlled injection system (see illustration).

3 A vacuum-operated recirculation valve is

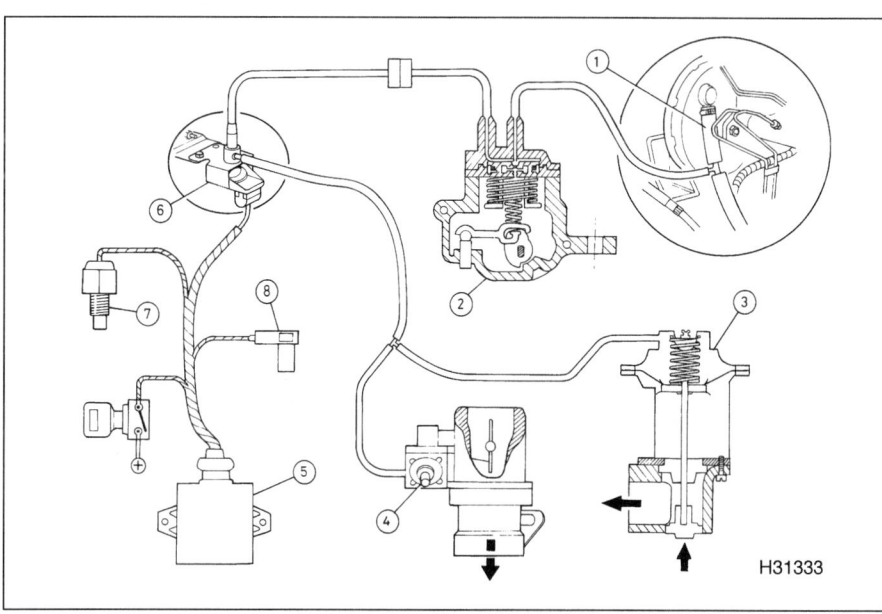

H31333

4.2 Schematic view of a typical exhaust gas recirculation system

1 Brake servo vacuum hose	4 Flow valve/butterfly housing (fitted to inlet manifold)	7 Coolant temperature sensor
2 Vacuum converter (fitted to fuel injection pump)	5 Electronic control unit	8 Crankshaft speed/position sensor
3 Recirculation valve (fitted to exhaust manifold)	6 Solenoid valve	

5

fitted to the exhaust manifold, to regulate the quantity of exhaust gas recirculated. The valve is operated by the vacuum supplied via the solenoid valve.

4 Between idle speed and a pre-determined engine load, power is supplied to the solenoid valve, which allows the recirculation valve to open. Under full-load conditions, the exhaust gas recirculation is cut off. On most EGR systems, additional control is provided by the engine temperature sensor, which cuts off the vacuum supply until the engine coolant temperature reaches a pre-determined level, preventing the recirculation valve from opening during the engine warm-up phase.

Chapter 6
Tools and equipment

Contents

1 Normal workshop tools

1 The decision as to what range of tools is necessary will depend on the work to be done, the range of vehicles which it is expected to encounter, and (not least) the financial resources available. The tools in the following list, with additions as necessary from the various categories of diesel-specific tools described later, should be sufficient for carrying out most routine maintenance and repair operations.

Combination spanners (see text)
Socket spanners (see text)
Ratchet, extension piece and universal joint (for use with sockets)
Torque wrench
Angle-tightening indicator (see text)
Adjustable spanner
Set of sump drain plug keys
Strap or chain wrench (for fuel and oil filters)
Oil drain tray

Feeler gauges
Pliers
Long-nosed pliers
Self-locking pliers ('Mole' wrench)
Screwdrivers (large and small, flat blade and cross-blade)
Set of Allen keys
Set of splined and 'Torx' keys and sockets (see text)
Ball pein hammer
Soft-faced hammer
Puller (universal type with interchangeable jaws)
Cold chisel
Scriber
Scraper
Centre-punch
Hacksaw
File
Steel straight-edge
Axle stands and/or ramps
Trolley jack
Inspection light
Inspection mirror
Telescopic magnet/pick-up tool

Socket and spanner sizes

2 A good range of open-ended and ring spanners will be required. Most modern vehicles use metric size fastenings throughout, but some early UK-built vehicles may have Imperial fastenings – or a mixture of both.
3 Split ring spanners (also known as flare nut spanners) are particularly useful for dealing with fuel pipe unions, on which a conventional ring spanner or socket cannot be used because the pipe is in the way. The most common fuel union sizes are 17 mm and 19 mm on metric systems, and 5/8 in and 3/4 in on Imperial systems.
4 Sockets are available in various drive sizes. The half-inch square drive size is most widely used, and can be used with most torque wrenches. The 3/8 in square drive is also useful for lower torque applications, especially in confined spaces, and 1/4 in and 3/4 in drive tools are also available.
5 Box spanners should not be overlooked. Box spanners are cheap, and will sometimes serve as a substitute for a deep socket,

though they cannot be used with a torque wrench, and are easily deformed.

Angle-tightening

6 For fastenings such as cylinder head bolts, many manufacturers specify tightening in terms of angular rotation rather than an absolute torque. After an initial 'pre-tightening' torque wrench setting, subsequent tightening stages are specified as angles through which each bolt must be turned. Variations in tightening torque which could be caused by the presence or absence of dirt, oil, etc, on the bolt threads are eliminated. A further benefit is that there is no need for a high-range torque wrench.

7 The owner/mechanic who expects to use this method of tightening only once or twice in the life of the vehicle may be content to make up a cardboard template or mark the bolt heads with paint spots, to indicate the angle required. Greater speed and accuracy will result from using one of the many angle-tightening indicators commercially available **(see illustration)**.

Splined and 'Torx' bolt heads

8 The conventional hexagon head bolt is being replaced in many areas by the 'splined' or 'Torx' head bolt. This type of bolt has multiple splines in place of the hexagon: splined bolts generally have 12 splines, and Torx bolts have six splines. A set of splined or Torx keys will be needed to deal with female fixings. Torx bolts with male heads also exist, and for these Torx sockets will be needed.

---

2 Diesel-specific tools

Basic tune-up and service

1 Besides the normal range of spanners, screwdrivers and so on, the following tools and equipment will be needed for basic tune-up and service operations on most models:

  Deep socket for removing and tightening screw-in injectors
  Injector puller for removing clamp-type injectors
  Optical or pulse-sensitive tachometer
  Electrical multimeter, or dedicated glow plug tester
  Compression or leakdown tester
  Vacuum pump and/or gauge

Injector socket

2 The size most commonly required is 27 mm AF; some Japanese injectors require 22 mm AF. The socket needs to be deep in order not to foul the injector body, and on some engines it also needs to be thin-walled.

Injector puller

3 This consists of a small slide hammer, with a range of adapters to screw into the various

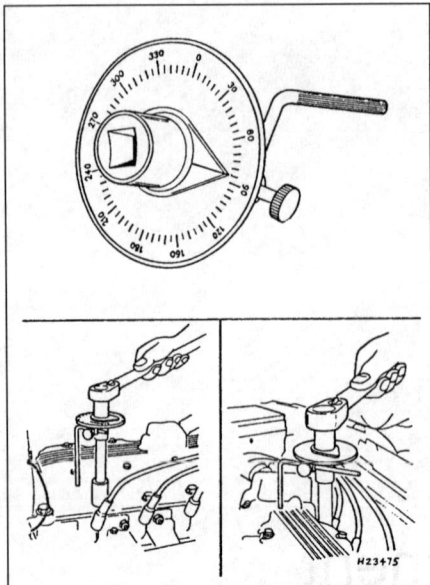

1.7 Typical angle-tightening indicator

end fittings which will be encountered **(see illustration)**. Once secured, the slide hammer is used to draw the injector out of its bore in the cylinder head, in which it may be a very tight fit.

Tachometer

4 The type of tachometer which senses ignition system HT pulses via an inductive pick-up cannot be used on diesel engines, unless a suitable timing light adapter is available. If the engine is fitted with a TDC sensor and a diagnostic socket, an electronic tachometer which reads the signals from the TDC sensor can be used.

5 Not all engines have TDC sensors; on those which do not, the use of an optical or pulse-sensitive tachometer is necessary.

6 The optical tachometer registers the passage of a paint mark or (more usually) a strip of reflective foil placed on the crankshaft pulley. It is not so convenient to use as the electronic or pulse-sensitive types, since it has to be held so that it can 'see' the pulley,

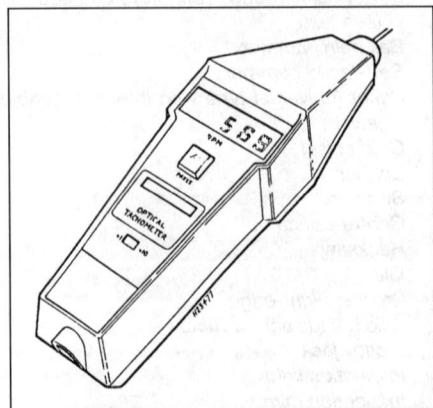

2.6 Typical optical tachometer

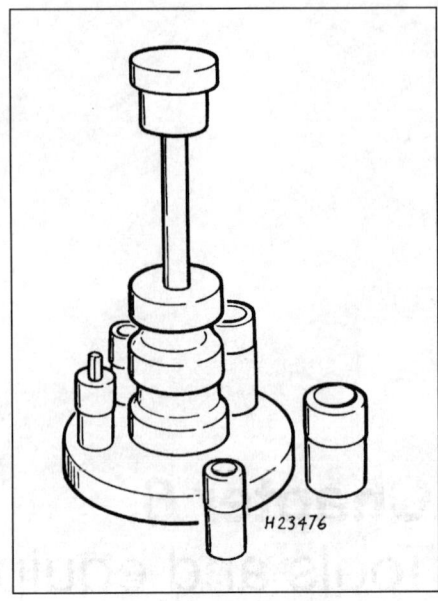

2.3 Typical injector puller set

but it has the advantage that it can be used on any engine, petrol or diesel, with or without a diagnostic socket **(see illustration)**.

7 The pulse-sensitive tachometer uses a transducer similar to that needed for a timing light. The transducer converts hydraulic or mechanical impulses in an injector pipe into electrical signals, which are displayed on the tachometer as engine speed.

8 Some dynamic timing equipment for diesel engines incorporates a means of displaying engine speed. If this equipment is available, a separate tachometer will not be required.

Electrical multimeter or glow plug tester

9 It is possible to test glow plugs and their control circuitry with a multimeter, or even (to a limited extent) with a 12-volt test lamp. A purpose-made glow plug tester will do the job faster, and is much easier to use **(see illustration)**.

10 If it is decided to purchase a multimeter, make sure that it has a high current range – ideally 0 to 100 amps – for checking glow plug

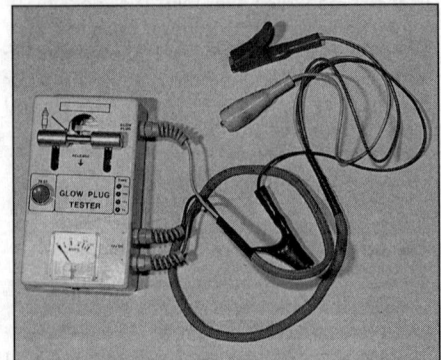

2.9 Typical glow plug tester

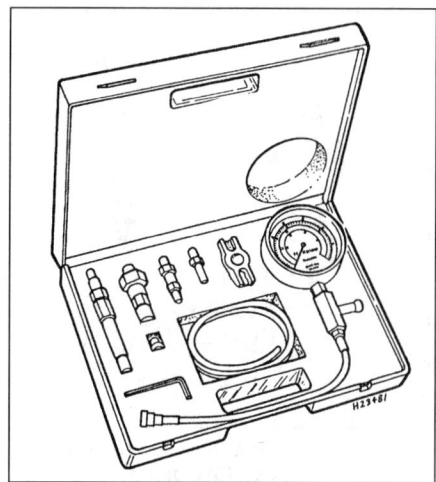

2.11 Typical diesel engine compression tester kit

current draw. Some meters require an external shunt to be fitted for this. An inductive clamp connection is preferred for high current measurement, since it can be used without breaking into the circuit. Other ranges required are dc voltage (0 to 20 or 30 volts is suitable for most applications) and resistance. Some meters have a continuity buzzer in addition to a resistance scale.

Compression tester

11 A tester specifically intended for diesel engines must be used **(see illustration)**. The push-in connectors used with some petrol engine compression testers cannot be used for diesel engines because of the higher pressures involved. Instead, the diesel engine compression tester screws into an injector or glow plug hole, using one of the adapters supplied with the tester.

12 Most compression testers are used while cranking the engine on the starter motor. A few can be used with the engine idling; this

2.13 Typical leak-down tester kit

gives more reliable results, since it is hard to guarantee that cranking speed will not fall in the course of testing all cylinders, whereas the idle speed should remain constant.

Leak-down tester

13 The leak-down tester measures the rate at which air pressure is lost from each cylinder, and can also be used to pinpoint the source of pressure loss (valves, head gasket or bores). Its use depends on the availability of a supply of compressed air, typically at 5 to 10 bar. The same tester (with different adapters) can be used on both petrol and diesel engines **(see illustration)**.

14 In use, the tester is connected to an air line and to an adapter screwed into the injector or glow plug hole, with the piston concerned at TDC on the compression stroke. The procedure is described in Chapter 9, Section 4.

Vacuum pump and/or gauge

15 A vacuum gauge with suitable adapters is useful for locating blockages or inlet air leaks in the supply side of the fuel system. A simple gauge is used with the engine running to create a vacuum in the supply lines. A hand-held vacuum pump with its own gauge can be

used without running the engine, and is also useful for bleeding the fuel system when a hand-priming pump is not fitted **(see illustration)**. Test procedures are given in Chapter 9, Section 3.

3   Injection pump timing tools

1 If work is undertaken which disturbs the position of the fuel injection pump, certain tools will be required to check the injection pump timing on reassembly. This also applies if the pump drive is disturbed – including renewal of the timing belt on some models. Checking of the timing is also a necessary part of fault diagnosis when investigating complaints such as power loss, knock and smoke.

Static timing tools

2 Static timing is still the most widely-used method of setting diesel injection pump, but it is time-consuming and sometimes messy. Precision measuring instruments are often needed for dealing with distributor pumps, and good results depend on the skill and patience of the operator.

3 The owner-mechanic who will only be dealing with one engine should refer to the manufacturer's information or to the relevant Haynes Service and Repair Manual to find out what tools will be required. The diesel engine specialist will typically need the following:

> *Two dial test indicators (DTI) with magnetic stands.*
> *DTI adapters and probes for Bosch and CAV distributor pumps.*
> *Spill tube for in-line pumps.*
> *Timing gear pins and pegs.*
> *Crankshaft or flywheel locking pins.*

Dial test indicator and magnetic stand

4 This is a useful workshop tool for many operations besides timing: it is, for example, the most accurate means of checking the protrusion or recess of swirl chambers, pistons and liners when renewing cylinder head gaskets **(see illustration)**. If major overhauls are undertaken, it can also be used for measuring values such as crankshaft endfloat.

5 Two DTIs are needed for setting the timing

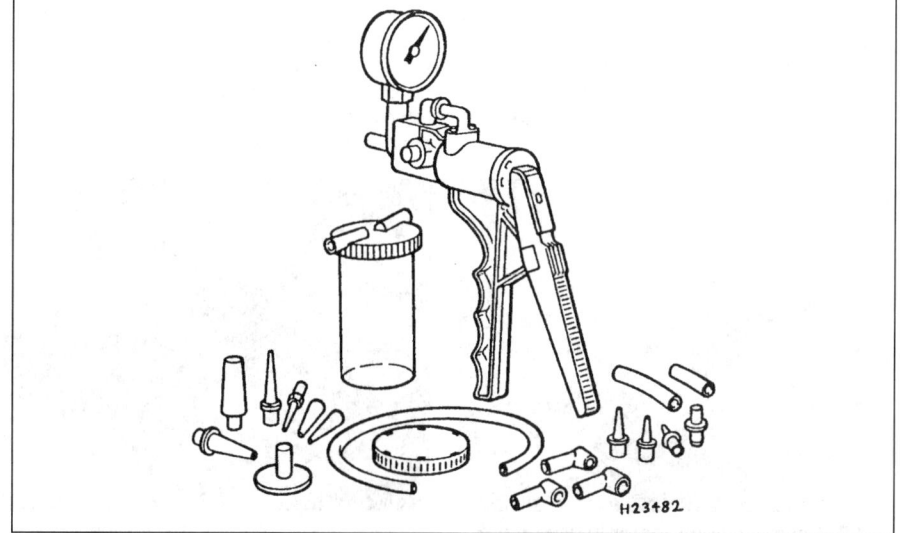

2.15 Typical hand-held vacuum pump with gauge

3.4 Dial test indicator and stand being used to check swirl chamber protrusion

6

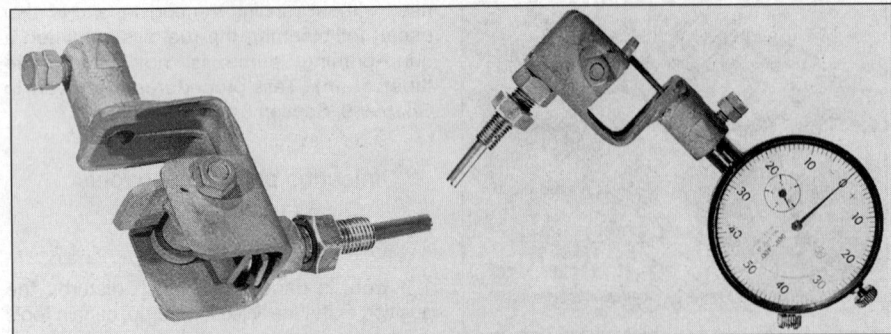

3.6 DTI and fabricated bellcrank adapter for timing a Bosch VE pump

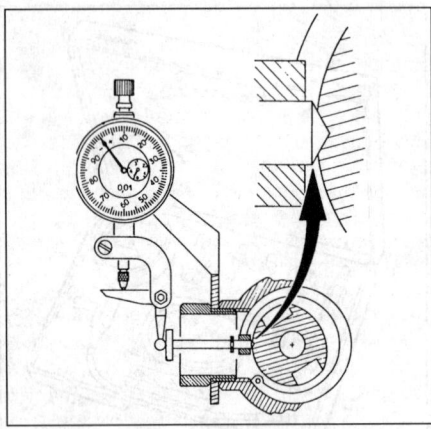

3.7 DTI and adapter used for timing Lucas/CAV pump

on some engines (for instance, the early Peugeot/Citroen XUD series): one to measure the pump plunger or rotor movement, and one to measure engine piston protrusion.

DTI adapters

6 Adapters and probes for fitting the DTI to the distributor pump are of various patterns, due partly to the need to be able to use them in conditions of poor access on the vehicle (see illustration). This means that the same adapter cannot necessarily be used on the same type of pump and engine if the under-bonnet layout is different. On the bench, it is often possible to use similar equipment.

7 A spring-loaded probe is required on some CAV/RotoDiesel pumps to find the timing groove in the pump rotor (see illustration).

Spill tube

8 This is a relatively cheap and simple piece of equipment, used for timing many in-line pumps. The tube is fitted in place of one of the pump delivery valves. The traditional form of tube has a 'swan neck' shape; more modern versions have a transparent vertical tube with a calibrated line. A spill tube can easily be made in the workshop using an old injector pipe (see illustration).

Timing gear pins or pegs

9 Pins or pegs are used on some engines to lock the pump and/or the camshaft in a particular position. They are generally specific to a particular engine or manufacturer. It is sometimes possible to use suitably-sized dowel rods, drill shanks or bolts instead.

Crankshaft or flywheel locking pins

10 These are used for locking the crankshaft at TDC (or at the injection point on some models). The crankshaft locking pin is inserted through a hole in the side of the crankcase after removal of a plug, and enters a slot in a crankshaft counterweight or web. The flywheel pin passes through a hole in the flywheel end of the crankcase, and enters a hole in the flywheel. Again, suitably-sized rods or bolts can sometimes be used instead.

Dynamic timing tools

11 Dynamic timing on diesel engines is not as widespread as static timing, due partly to the relatively expensive equipment required. Additionally, not all vehicle manufacturers provide dynamic timing values.

12 Most dynamic timing equipment depends on converting mechanical or hydraulic impulses in the injection system into electrical signals. An alternative approach is to use an optical-to-electrical conversion, with a sensor which screws into a glow plug hole and 'sees' the light of combustion.

13 Not all diesel engines have ready-made timing marks. If the engine has a TDC sensor, and the timing equipment can read the sensor output, this is not a problem. Some engines have neither timing marks nor TDC sensors; in such cases, there is no choice but to establish TDC accurately, and make suitable marks on the flywheel or crankshaft pulley.

14 For these reasons, dynamic timing methods and the tools required are not described in this book.

4 Injector testing equipment

⚠️ *Warning: never expose the hands, face or any other part of the body to injector spray. The high working pressure can penetrate the skin, with potentially fatal results. When possible, use injector test fluid rather than fuel for testing. Take precautions to avoid inhaling the vaporised fuel or injector test fluid. Remember that even diesel fuel is inflammable when vaporised.*

1 Some kind of injector tester will be needed if it is wished to identify defective injectors, or to test them after cleaning or prolonged storage. Various makes and models are available, but the essential components of all of them are a high-pressure hand-operated pump and a pressure gauge.

2 For safety reasons, injector test or calibration fluid should be used for bench-testing rather than diesel fuel or paraffin (see illustration). Use the fluid specified by the manufacturer of the test equipment if possible.

3 Some of the simpler testers have the advantage that they can be used to test opening pressure and back-leakage without removing the injectors from the engine (see illustration). A small reservoir may make such

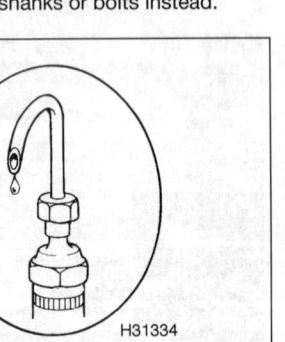

3.8 Simple spill tube

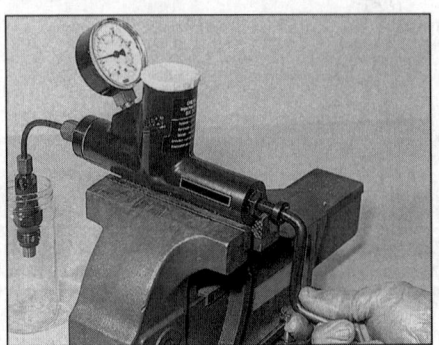

4.2 Typical injector tester in use on the bench . . .

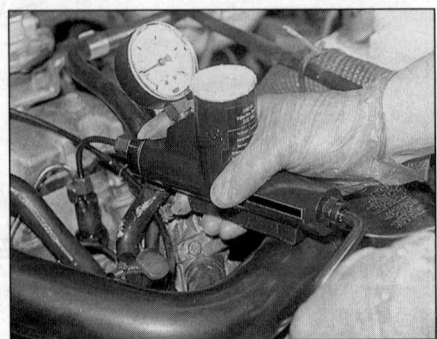

4.3 . . . and on the engine

testers of limited use for bench-testing, but good results can be obtained with practice.

4 Another method of testing injectors on the engine is to connect a pressure gauge into the line between the injection pump and the injector. This test can also detect faults caused by the injection pump high-pressure piston or delivery valve.

5 The workshop which tests or calibrates injectors regularly will need a bench-mounted tester. These testers have a lever-operated pump, and a larger fluid reservoir than the hand-held tester. The best models also incorporate a transparent chamber for safe viewing of the injector spray pattern, and perhaps a test fluid recirculation system.

6 Some means of extracting the vapour produced when testing, such as a hood connected to the workshop's fume extraction system, is desirable. Although injector test fluid is relatively non-toxic, its vapour is not particularly pleasant to inhale.

5 Injection pump testing and calibration equipment

The equipment needed for testing and calibration of injection pumps is beyond the scope of this book. Any such work should be entrusted to the pump manufacturer's agent – though the opportunity is taken to say yet again that the injection pump is often blamed for faults, when in fact the trouble lies elsewhere.

6 Smoke testing equipment

In most European countries, smoke emission testing is mandatory for heavy goods vehicles, and for passenger vehicles as part of the annual roadworthiness test.

Smoke testing equipment falls into two categories : indirect and direct reading. With the indirect systems, a sample of exhaust gas is passed over a filter paper, and the change in opacity of the paper is measured using a separate machine. With the direct systems, an optically-sensitive probe measures the opacity of the exhaust gas, and an immediate read-out is available.

7 Electronic fault code readers

Like their petrol engine counterparts, many of the modern electronic diesel engine control systems have a self-diagnostic function, which continually monitors the operation of the system.

The self-diagnostic system is able to detect system faults such as a faulty sensor or actuator, and can allocate a fault code to identify the source of the problem. The system stores any fault codes in the electronic control unit (ECU) memory, and if a fault is present, a warning light will normally be illuminated on the instrument panel to inform the driver.

Fault codes can be read using a suitable electronic fault code reader. Most vehicle manufacturer's produce their own dedicated diagnostic equipment, but aftermarket fault code readers are also available from various manufacturers.

Notes

Chapter 7
Adjustments and checks

Contents

1 Introduction

On most conventional fuel injection pumps (ie, pumps without electronic control), it is normally possible to adjust the following settings:

a) Idle speed.
b) Anti-stall controls.
c) External controls (eg, cold idle mechanisms).
d) Maximum no-load speed.
e) Injection timing.

Routine adjustments to injection pumps are normally confined to idle speed, anti-stall and external controls, which may include cold idle mechanisms. Checking injection timing is not as routine an operation as checking the ignition timing on a petrol engine. It is necessary when investigating complaints of poor performance, knock and smoke, and whenever the pump or its drive has been disturbed. This last case often includes timing belt renewal.

Some manufacturers also specify a routine check of maximum no-load speed. The screw which controls this speed is always tamperproofed in production, using a locking wire and seal, paint or a sealing cap. *Breaking or removing this tamperproof device may invalidate any manufacturer's warranty.*

Other adjustment screws may be externally accessible, either directly, or through access plugs; they control functions such as maximum-fuelling, excess-fuelling and transfer pressure. Often these screws or plugs are also tamperproofed. Do not attempt haphazard adjustment of such screws. Normally a pump test bench is needed to set (or reset) them correctly.

The following Sections give general procedures. Details specific to particular models may vary; consult manufacturer's information or the appropriate Haynes *Service and Repair Manual* for further information. Adjustment specifications can be found in Chapter 10.

2 Engine speed adjustments

Idle speed

1 Bring the engine to normal operating temperature, and connect a tachometer to it. (For details of tachometers, see Chapter 6.) If a fast idle device is fitted, make sure that it is not holding the pump control lever or idling lever off its stop.
2 Allow the engine to idle, and check the speed against that specified (Chapter 10).
3 If adjustment is necessary, slacken the locknut and turn the idle speed adjusting screw until the speed is correct. If some tolerance is allowed, adjust the speed to the

7

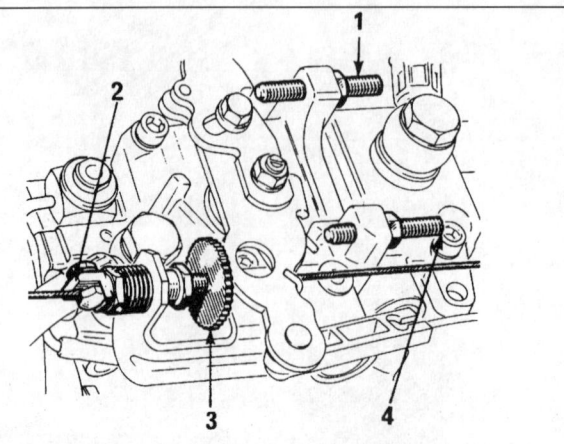

1 Idle speed adjustment screw
2 Fast idle cable and stop (when fitted)
3 Fast idle speed adjustment screw (when fitted)
4 Maximum speed adjustment screw

2.3a Typical injection pump adjustment points – Early Bosch VE pump

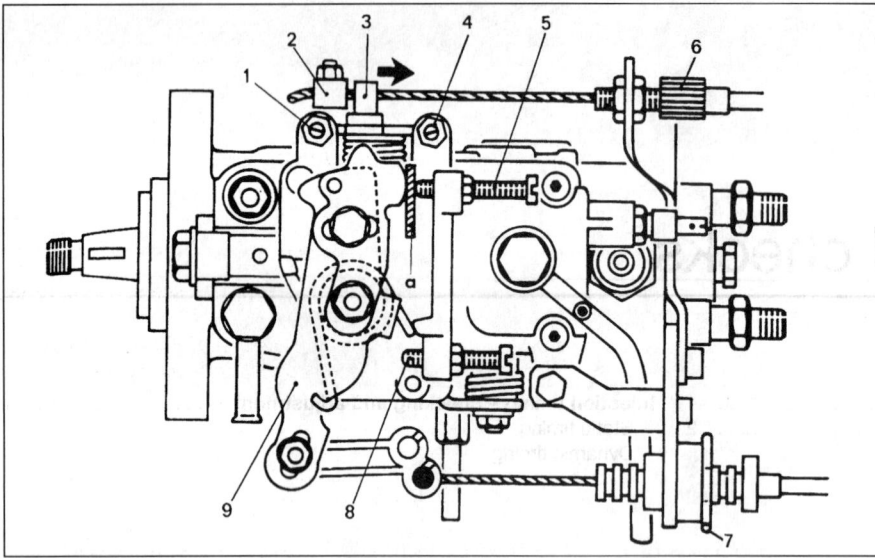

2.3b Typical injection pump adjustment points – Bosch VE pump with separate idle lever

1 Fast idle speed adjustment screw	5 Anti-stall (residual capacity) adjustment screw	8 Maximum speed adjustment screw
2 Fast idle cable end stop	6 Fast idle cable adjuster	9 Control lever
3 Idle lever	7 Accelerator cable adjuster	a Shim for anti-stall adjustment
4 Idle speed adjustment screw		

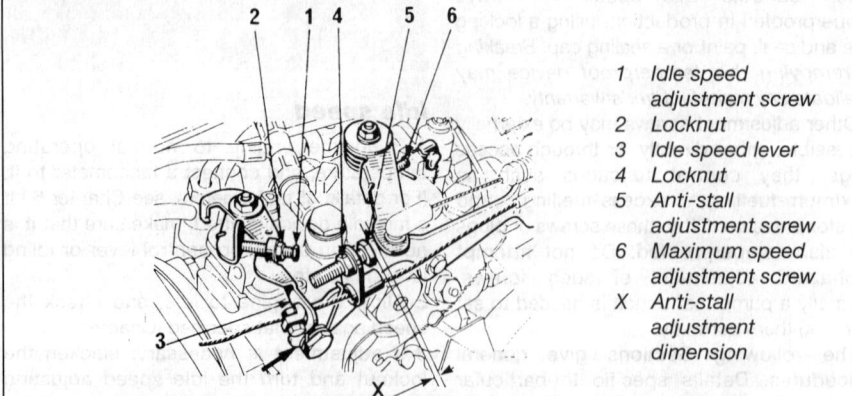

1 Idle speed adjustment screw
2 Locknut
3 Idle speed lever
4 Locknut
5 Anti-stall adjustment screw
6 Maximum speed adjustment screw
X Anti-stall adjustment dimension

2.3c Typical injection pump adjustment points – Lucas/CAV DPC pump

value within the specified range where the engine runs most smoothly. Tighten the locknut when adjustment is correct **(see illustrations)**.

4 On some pumps, it is necessary to check the anti-stall adjustment if the idle speed is altered.

Maximum speed

⚠ **Warning: Breaking or removing this tamperproof device may invalidate any manufacturer's warranty.**

5 With the engine warmed up and a tachometer connected, accelerate the engine to maximum speed for a few seconds. Note the speed reached, and compare it with that specified in Chapter 10. Do not hold maximum speed for any longer than is necessary.

6 If adjustment is necessary, remove the tamperproofing, slacken the locknut, and turn the adjustment screw. Repeat the check; tighten the locknut and fit a new tamperproof device when adjustment is correct.

Anti-stall (residual capacity)

7 Anti-stall or residual capacity adjustment determines how quickly engine speed falls off when the accelerator is suddenly released. If the adjustment is incorrect, the engine will either tend to stall after sudden deceleration, or it will 'hang' (fail to lose its speed quickly enough).

8 All CAV DP series pumps have some kind of external anti-stall adjustment facility, but most Bosch VE pumps do not **(see illustration)**.

9 When the anti-stall adjustment screw determines the resting position of the pump

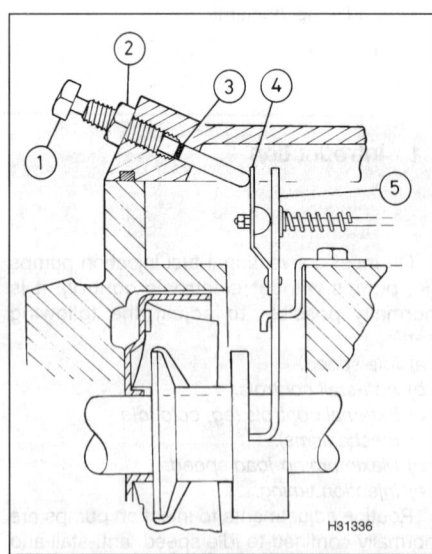

2.8 Separate anti-stall adjustment screw (1) on Lucas/CAV DPC pump with all-speed mechanical governor

2 Locknut
3 Seal
4 Spring
5 Governor arm

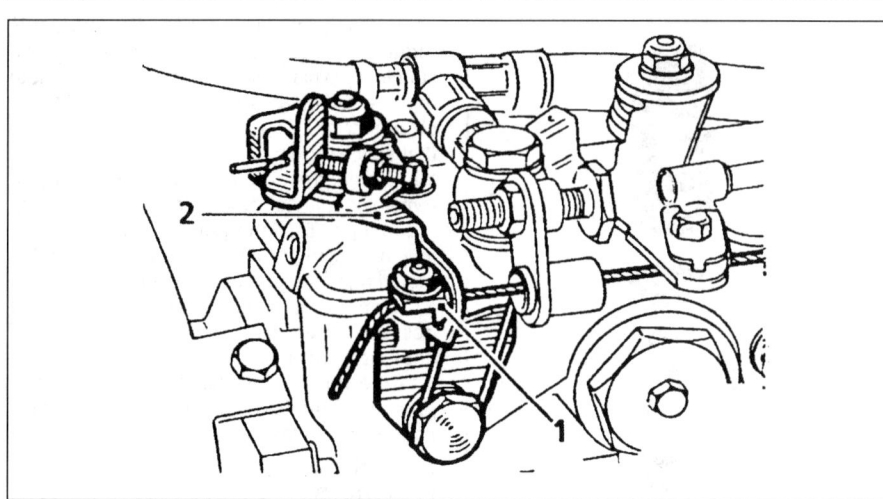

2.13a Cold idle adjustment – Lucas/CAV DPC pump with remote thermostatic capsule

When cold, adjust cable clamp (1) to hold idle lever (2) against its stop

control lever, the adjustment procedure usually consists of inserting a specified thickness of shim (or feeler gauge) between the screw and the lever, and noting the effect on idle speed. Idle speed and anti-stall adjustments are connected; if one is adjusted, it will normally be necessary to check the other.

10 When the anti-stall screw is separate, adjustment is normally on a trial-and-error basis, moving the screw by a quarter-of-a-turn at a time. Turning the screw inwards will reduce the tendency to stall; turning it outwards will reduce the tendency to 'hang'.

The effect of a change in adjustment is judged by accelerating the engine to maximum no-load speed, and then releasing the accelerator. The engine must return to idle speed within a specified number of seconds (typically 5 seconds) without stalling.

Cold-idle mechanisms

11 Cold-idle mechanisms may be automatically operated, or they may be operated by the driver using a hand control. When in operation, they may affect injection timing, idle speed or both.

12 Adjustment of the manually-operated mechanism is normally confined to checking that operation of the hand control produces the appropriate movement of the control lever on the pump, and adjusting the cable if necessary.

13 Automatic mechanisms which rely on the movement of a lever by a thermostatic capsule and a cable may also require the cable to be adjusted. If the mechanism alters both timing and idle speed, the relationship between the two functions must also be checked **(see illustrations)**.

14 Other types of automatic cold-idle mechanism alter the injection pump timing by raising the transfer pressure. Typically, this is done by an electrically-heated thermostatic capsule, which opens a valve as it warms up; no adjustment is possible.

3 Injection timing –
checking and adjustment

Static timing

1 Static timing is still the most widely-used method of setting diesel injection pumps. Precision measuring instruments are often needed for dealing with distributor pumps. Good results depend on the skill and patience of the operator.

Caution: Be careful not to introduce dirt into the injection pump during the following procedures.

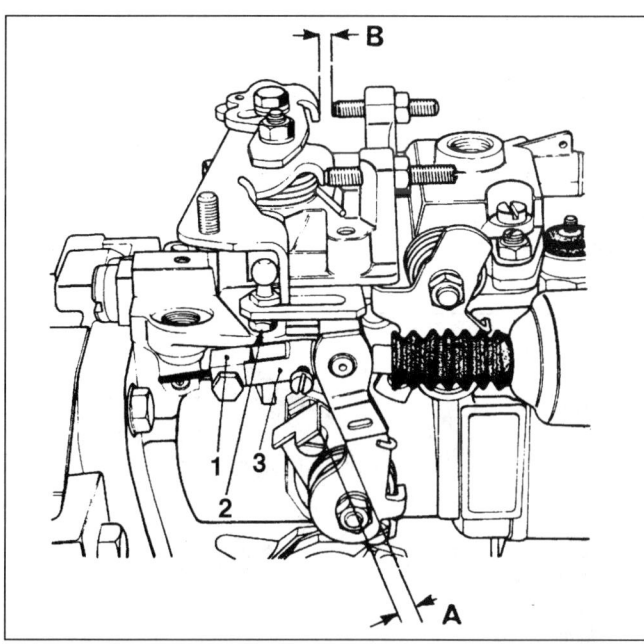

2.13b Cold idle adjustment check points – Bosch VE pump with thermostatic capsule

1	Cable end stop	A	Advance lever gap
2	Ball-pin adjusting nut	B	Fast idle gap
3	Clevis		

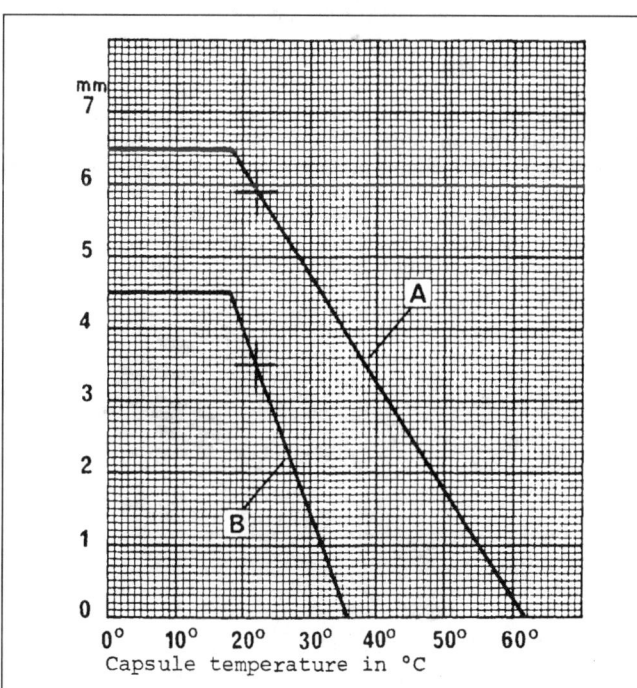

2.13c Relationship between advance lever gap (A) and fast idle gap (B) varies with temperature – Bosch VE pump with thermostatic capsule

7

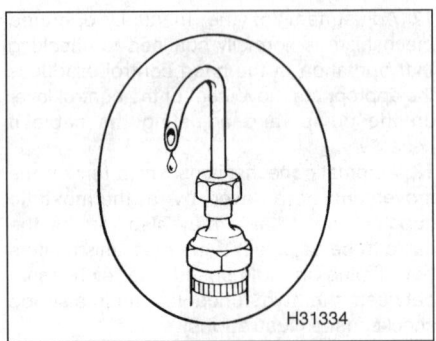

3.2 Typical spill tube

Spill timing (in-line pumps) – low-pressure method

2 This is a simple method of timing, albeit messy. The only special equipment required is a 'spill' or 'swan-neck' tube; this can be made in the workshop using part of an old injector pipe if wished **(see illustration)**.

3 The procedure finds the point in the pump cycle when one plunger covers its inlet port. This corresponds to the beginning of the retraction stroke (Chapter 2, Section 5). It can be accurately related to the engine cycle, but does not necessarily correspond to the actual beginning of injection.

4 Bring the engine to TDC, No 1 piston firing. Clean around the injector pipe unions and the connections on the pump, then remove No 1 cylinder injector pipe.

5 Unscrew No 1 cylinder pipe connection from the pump. Remove the delivery valve plunger and spring, noting which way up the plunger is fitted **(see illustration)**. **Do not** remove the valve carrier. Refit the connector and sealing washer, and fit the spill tube to the holder.

6 Make sure that the stop control is in the 'run' position. In the case of a vacuum-operated stop control, disconnect the vacuum hose. Fix the pump control lever in the maximum-speed position.

7 The fuel in the pump must now be subject to a small head of fuel pressure. On systems

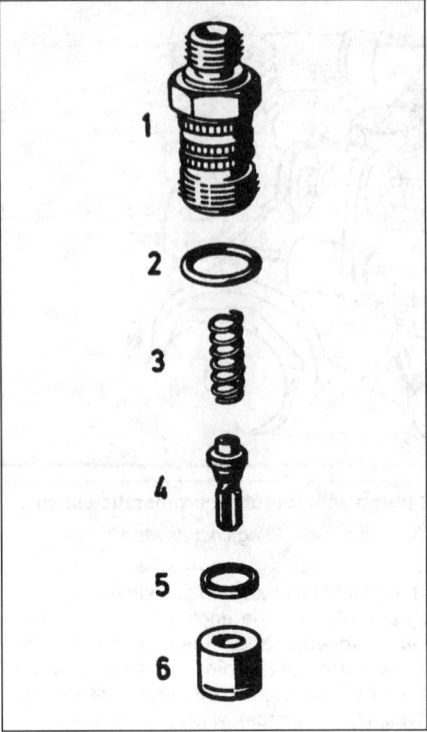

3.5 Pipe connector and delivery valve components

1	Connector	4	Plunger
2	O-ring	5	Sealing washer
3	Spring	6	Valve carrier

where the fuel filter is higher than the pump, the filter will serve as a header tank. Open the filter bleed screw, and operate the hand-priming pump until fuel emerges.

8 On systems where the fuel filter is lower than the pump, the necessary pressure can be provided by disconnecting the fuel inlet from the pump, and substituting a feed from a small reservoir of **clean** fuel positioned higher than the pump.

9 Turn the engine approximately a quarter-turn backwards. Fuel will begin to flow from the tube. Slowly, turn the engine forwards

again towards TDC, until the flow of fuel is reduced to one drop per second (or as specified). This is the spill timing point. Note the crankshaft position (degrees BTDC, or alignment of a peg hole), and compare it with that specified.

10 If adjustment is necessary, slacken the remaining injector pipe unions and the pump mountings. Turn the pump as necessary to advance or retard the timing, then tighten the mountings and repeat the check.

11 Disconnect the spill tube. Reassemble the delivery valve, using new sealing washers if necessary, being careful not to introduce dirt into the pump. Refit No 1 cylinder injector pipe, and remake the original fuel supply connections if they were disturbed.

Spill timing (in-line pumps) – high-pressure method

12 If suitable equipment is available, the fuel in the injection pump can be pressurised sufficiently to pass the delivery valve. There is thus no need to dismantle the delivery valve, with a consequent saving in time, although it will be necessary to block off the pump fuel return.

13 Because of the higher pressure involved, fuel flow from the spill tube will be much faster. The spill timing point is typically defined as the point where the jet of fuel from the spill tube turns into a chain of drops.

14 Apart from the points just noted, the procedure is the same as for the low-pressure method.

Peg systems (all pump types, when applicable)

15 Determine the location of the peg holes for the injection pump and the crankshaft or flywheel (as applicable), and the sizes of peg required.

16 Turn the engine until the crankshaft or flywheel timing peg can be inserted cleanly. With this peg in position, it must be possible to insert the injection pump peg **(see illustrations)**.

17 If adjustment is necessary, this may either be carried out by slackening the injector pipe

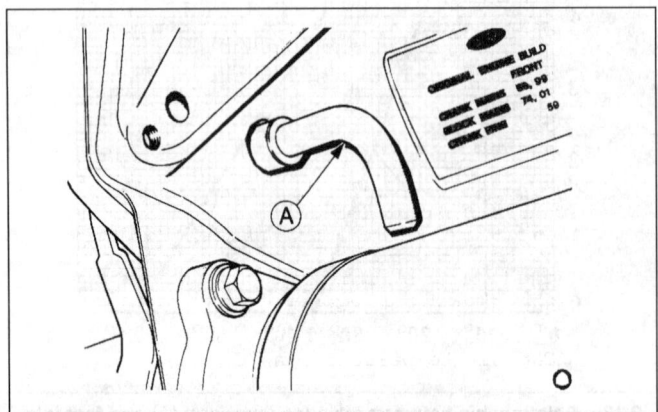

3.16a Flywheel timing peg (A) inserted – Ford 2.5 DI engine

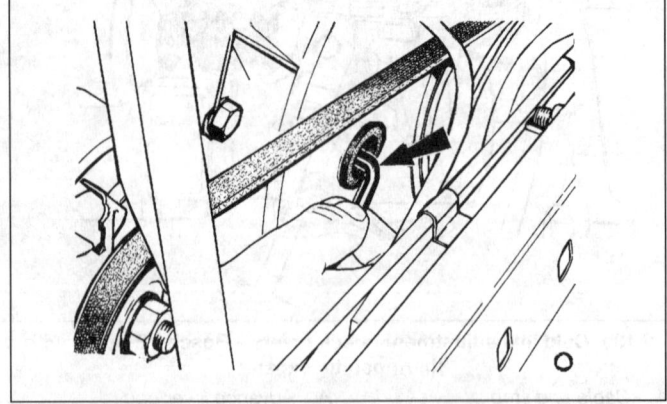

3.16b Inserting the pump timing peg (arrowed) through the timing cover access hole – Ford 2.5 DI engine

3.17 Altering the pump flange-to-sprocket relationship so that the peg can be inserted

Bolts must be slackened first!

unions and the pump mountings, and turning the pump, or by altering the relationship of the pump drive flange to its gear or sprocket **(see illustration)**. Refer to the manufacturer's information or to the relevant Haynes Service and Repair Manual for the appropriate method. Repeat the timing check from the beginning after adjustment.

CAV DP side-entry

18 Bring the engine to TDC, No 1 piston firing. Remove the access plug from the side of the injection pump – be prepared for fuel spillage.

19 Fit a dial test indicator (DTI) and probe so that the probe enters the access hole, passes through the hole in the circlip, and rests on the pump rotor. Slowly turn the crankshaft anti-clockwise to find the DTI minimum reading. In this position, the probe is resting in the bottom of the timing groove in the rotor **(see illustration)**.

20 Turn the crankshaft clockwise to bring the engine to the specified timing point. This may be TDC, or it may be a specified point before or after TDC – see Chapter 10. (If the timing point is overshot, return to the zero position established in the previous paragraph, and start again.)

21 Read the probe movement displayed on the DTI, and compare it with the value given in Chapter 10. If adjustment is necessary, refer to paragraph 17.

CAV DP top-entry

22 Bring the engine to TDC, No 1 piston firing. Remove the access plug from the top of the injection pump.

23 Insert a probe of the specified length into the access plug hole, so that the tip of the probe rests on the rotor timing piece. Position a dial test indicator to read the movement of the probe.

24 Turn the crankshaft approximately a quarter-turn backwards, and zero the DTI.

25 Turn the crankshaft clockwise to bring the engine to the specified timing point (see paragraph 20). Read the probe movement displayed on the DTI, and compare it with the value engraved on the plastic disc or tag

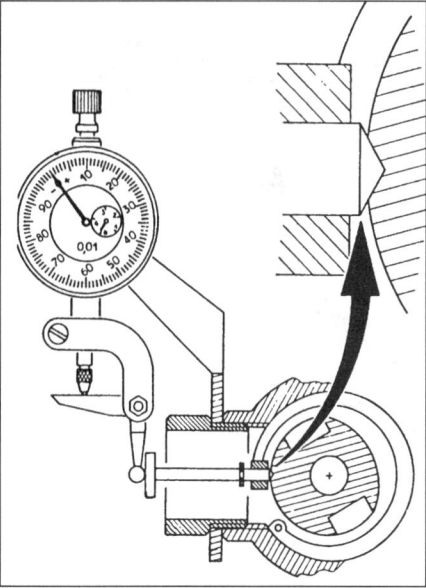

3.19 Side-entry timing of Lucas/CAV DP series pump

somewhere on the pump **(see illustrations)**.

26 If adjustment is necessary, refer to paragraph 17.

Bosch VE rear-entry

27 Bring the engine to TDC, No 1 piston firing. Remove the access plug from the rear of the injection pump – be prepared for fuel spillage **(see illustration)**.

28 Fit a dial test indicator (DTI), adapter and probe, so that the probe enters the access hole and the DTI displays movement of the pump plunger. Removing the injector pipes will improve access. On some pumps it may be necessary to use a right-angle adapter to allow the DTI to fit in the available space at the rear of the pump **(see illustrations)**.

29 Slowly turn the crankshaft anti-clockwise until the DTI reading reaches a minimum (pump plunger BDC), and zero the DTI at this point.

30 Turn the crankshaft clockwise to bring the engine to the specified timing point (see paragraph 20). Read the plunger movement displayed on the DTI, and compare it with the value given in Chapter 10.

3.27 Removing the access plug from the rear of a Bosch VE pump

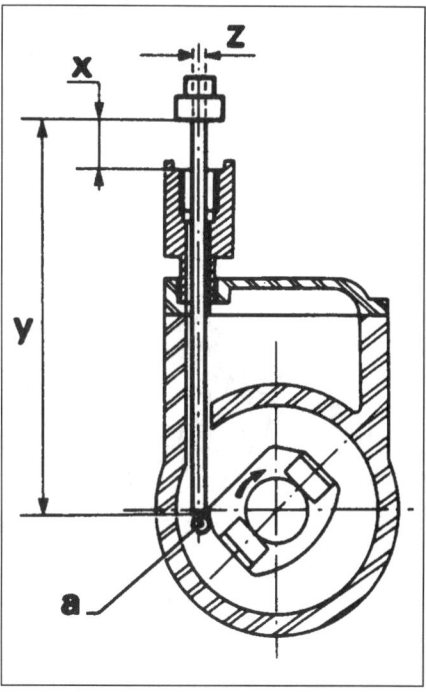

3.25a Sectional view of probe and pump – Lucas/CAV top-entry timing

a	*Timing piece*	*y*	*95.5 ± 0.01 mm*
x	*Specified timing*		*length*
	value	*z*	*7 mm diameter*

3.25b Reading probe movement – Lucas/CAV top-entry timing

3.28a Dial test indicator and adapter positioned with probe reading plunger movement – Bosch VE pump rear-entry timing

7

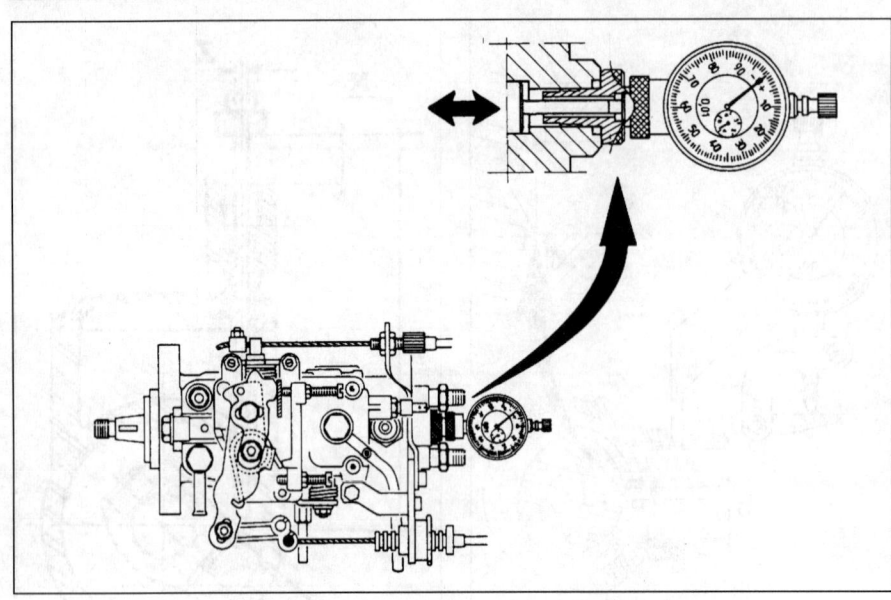

3.28b Rear-entry timing of Bosch VE pump

31 If adjustment is necessary, see paragraph 17.

Dynamic timing

32 As the name implies, dynamic timing is carried out with the engine running. Special equipment is required to carry out dynamic timing accurately, and this is unlikely to be available to the home mechanic. The equipment works by converting pressure pulses in an injector pipe into electrical signals. If such equipment is available, use it in accordance with its manufacturer's instructions.

33 Although the pump timing is checked with the engine running, any adjustment is usually carried out with the engine stopped.

34 An additional problem is that few manufacturers specify figures for dynamic timing.

35 For these reasons, static timing is generally an easier method of timing than dynamic timing, and we have chosen not to include specific details of dynamic timing in this book.

Chapter 8
Maintenance and repair operations

Contents

1 Introduction

Maintenance procedures

Due to the high working pressures, loads and temperatures found in a diesel engine, the recommended service intervals (especially oil change intervals) are generally more frequent than those for a comparable petrol engine. Frequent oil changes are particularly important for a diesel engine, as dirt or soot builds up in the oil during normal operation, leading to the deterioration of the oil's lubricating qualities.

The vehicle manufacturer's service schedule should always be followed, and it is important to use good quality lubricants and filters, which meet the manufacturer's recommendations.

This Chapter does not provide specific procedures for maintenance operations, its purpose is to provide a general guide to operations which are particularly important, or unique to diesel engines. Examples of such operations are:

   Engine oil and filter renewal.
   Draining water from the fuel filter/water separator.
   Fuel filter renewal.
   Fuel system bleeding.
   Auxiliary drivebelt checks.
   Fuel injection pump checks and adjustments.
   Fuel injector checks.
   Exhaust emissions check.
   Timing belt renewal.
   Turbocharger boost pressure check.

Repair procedures

When overhauling a diesel engine, there may be various unique features which need to be taken into account.

Always consult with the manufacturer's engine overhaul information, especially when carrying out work on the cylinder head; for example when renewing the cylinder head gasket, it may be necessary to measure the distance by which the pistons protrude from the cylinder block, in order to determine the correct thickness of cylinder head gasket.

Typical examples of such features are given in the following Sections.

2 Maintenance procedures

Note: *This Section does not provide an exhaustive list of diesel engine maintenance procedures, it gives basic information and advice on tasks which are especially important or unique to diesel engines. Always refer to the manufacturer's information for a detailed description of maintenance operations.*

Engine oil and filter renewal

1 Generally, the recommended engine oil and filter change intervals will be more frequent than those for petrol engines.

8

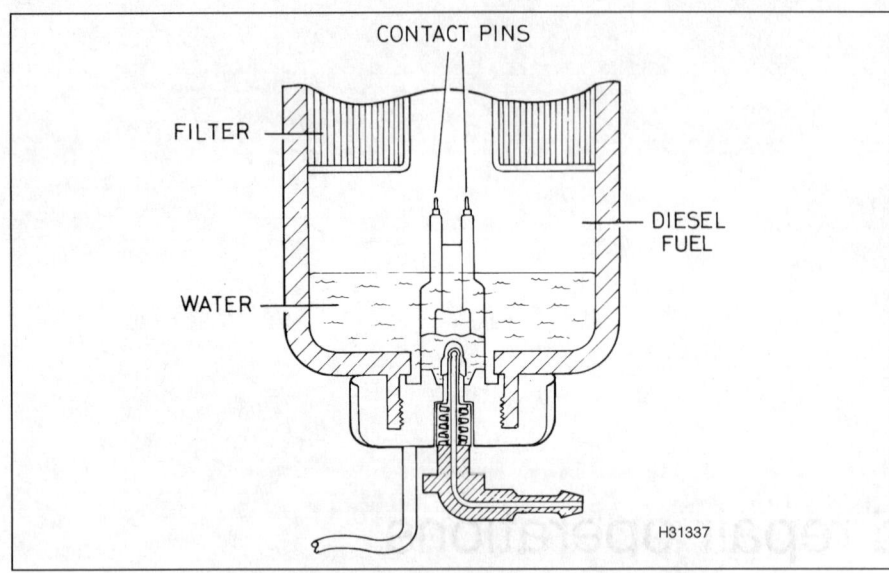

2.3 Sectional view of fuel filter with water sensor

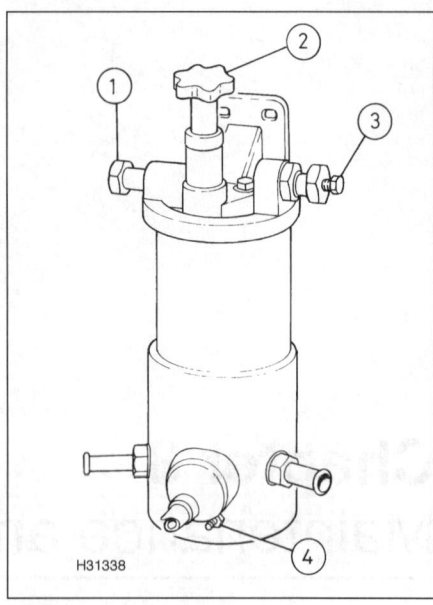

2.7 Fuel filter drain screw (4) – Lucas/CAV filter with coolant-heater base

1 Inlet union	3 Fuel bleed screw
2 Hand-priming pump	4 Water drain screw

2 It is important to use a good quality oil which is recommended for use in diesel engines.

Draining water from the fuel filter/water trap

3 If a glass bowl water trap is fitted, it is easy to see when water is accumulating in the bottom of it. When the water trap is incorporated into the filter base, the water cannot be seen. Sometimes a sensor is fitted, which illuminates a warning light to alert the driver to the presence of water **(see illustration)**.

4 Manufacturer's recommendations for the intervals at which the water trap should be drained vary widely. Obviously, operating conditions and fuel quality will determine the rate at which water accumulates, but it is better to err on the side of safety and drain the trap frequently. If water gets to through to the pump and injectors, it can cause serious damage.

5 When draining the water trap, place a small container under it to catch the fuel. It is important that fuel is not allowed to spill onto the coolant hoses, alternator, starter motor or engine mountings. Protect them with plastic sheet if necessary. On some models, the trap or filter is awkwardly placed; in such cases, it may be easier to fit a length of hose to the trap outlet.

6 On certain popular models, the filter is notoriously difficult to drain, because the brake servo or some other component effectively blocks access to the base of the filter. In these cases, it is necessary to unbolt the filter head from the bulkhead, or to make up a tool to slacken the drain screw *in situ*.

7 When the drain screw is opened, it may be found that no fuel emerges because the system is under negative pressure. Slacken the bleed screw or the inlet union on the filter head, or operate the hand-priming pump, until fuel flows **(see illustration)**.

8 When clean fuel, free of water droplets, flows out, tighten the drain screw and the bleed screw.

9 Dispose of the drained fuel and water safely, in the same way as used engine oil.

Fuel filter renewal

10 The main filter element must be renewed at the intervals specified by the manufacturer, or more often if experience of particular problems show it to be necessary. Some manufacturers specify renewal at the beginning of every winter, even if little mileage has been covered, to reduce the risk of waxing problems.

11 Filters are basically of two types: cartridge and canister **(see illustrations)**. Cartridge

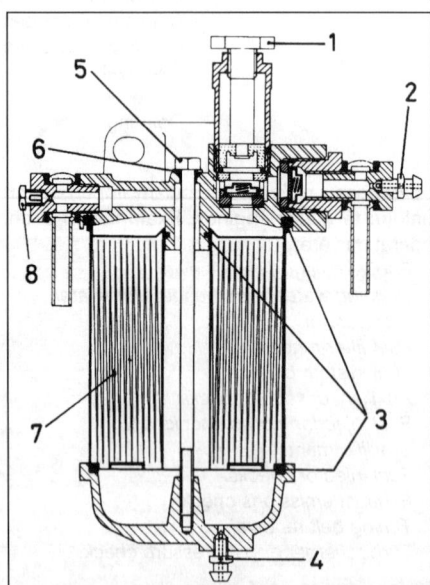

2.11a Sectional view of cartridge fuel filter with separate bowl

1 Hand-priming pump	4 Water drain screw
2 Fuel bleed screw	5 Through-bolt
	6 Through-bolt seal
3 Seals	7 Filter element
	8 Air bleed screw

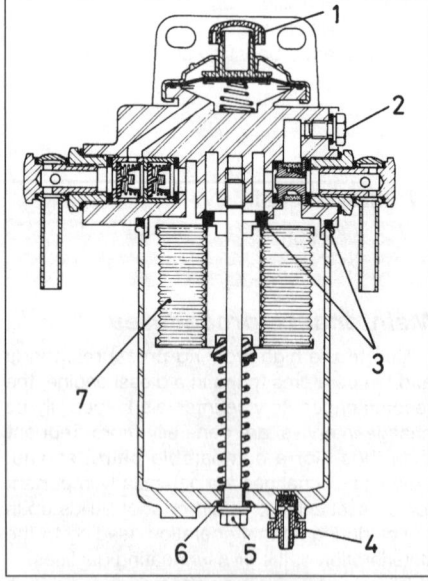

2.11b Sectional view of canister fuel filter

1 Hand-priming pump	4 Water drain screw
2 Fuel bleed screw	5 Through-bolt
3 Seals	6 Through-bolt seal
	7 Filter element

filters can be subdivided into 'spin-on' type, similar to a modern engine oil filter, 'clamp' type retained by a clamping strap or band, and 'through-bolt' type, retained by a bolt running from the filter head to a separate bowl. Canister filters are totally enclosed in the filter bowl.

12 It is best to drain the filter before removal if possible. The filter is then unscrewed with a strap or chain wrench ('spin-on' type), or the through-bolt or clamp bolt removed, according to type. Make sure that the old seals are recovered: some filter heads have a seal in a groove which is easy to overlook. The through-bolt, where fitted, may have an O-ring seal under its head. Any imperfect seals can allow air to be drawn into the system if there is no lift pump, or fuel to be forced out if there is.

13 Wipe out the filter bowl or canister, if applicable, finishing off with a **clean** non-fluffy cloth, or (if available) compressed air. *Caution: It is important that no dirt is introduced into the system.*

14 Smear the new seals with a little clean fuel. If a central seal retainer is fitted, make sure it is secure; in the case of the canister filter, make sure the seal is snugly in its groove.

15 Fit and secure the new filter or element, then bleed the fuel system if necessary.

16 If a separate water trap or pre-filter is fitted, this may incorporate a gauze screen, which should be removed for cleaning at the specified intervals.

Fuel system bleeding

17 Bleeding of the fuel system is necessary after operations in which air has been allowed to enter fuel lines, and after running out of fuel. General procedures are given here: refer to manufacturers information or to the relevant Haynes *Service and Repair Manual* for specific details.

18 Modern fuel systems are of the self-bleeding type. If no hand-priming pump is fitted, the normal way of bleeding such a system is by cranking the engine on the starter motor in 10-second bursts. If a hand-operated vacuum pump is available, this can be connected to the injection pump fuel return

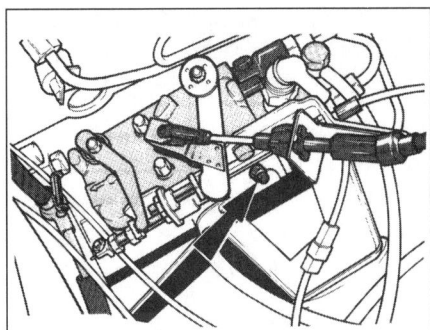

2.20 Injection pump bleed screw (arrowed) – pre-1985 Ford 2.1 litre engine

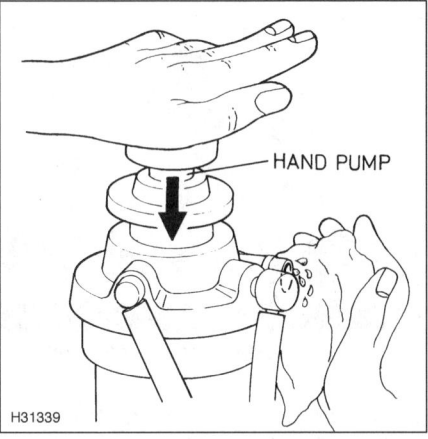

HAND PUMP

H31339

2.19 Bleeding the fuel system at the filter head

connection and used to suck fuel through the supply lines and filter; this will obviously save the battery a good deal of work.

19 When a hand-priming pump is fitted, this is operated first, with the bleed screw on the filter head open. When fuel free from air bubbles emerges, tighten the bleed screw **(see illustration)**. Carry on pumping until increased resistance is felt. Alternatively, use a vacuum pump as just described; this avoids any risk of splitting the diaphragm on the hand-priming pump, an occurrence which is not unknown on older vehicles.

20 If air has reached the injection pump, this may be bled out at a specific bleed screw if fitted, or (more usually) at the fuel return union **(see illustration)**.

21 On engines fitted with an in-line or distributor injection pump, if air has entered the injector pipes, slacken the injector unions, and crank the engine on the starter motor. When fuel emerges, tighten the unions and mop up spilt fuel.

22 When a separate fuel lift pump is fitted, this usually has a hand-priming lever for use when bleeding **(see illustration)**. If the engine has stopped with the lift pump operating arm

2.22 Hand-priming lever (1) on separate fuel lift pump

on top of its cam, it will be necessary to turn the engine before the hand-priming lever can be used.

Auxiliary drivebelt checks

23 On some diesel engines, besides the usual auxiliary drivebelt(s) used to drive ancillary units such as the alternator, power steering pump, etc, an additional drivebelt may be used to drive the brake vacuum pump, and in some cases the injection pump (most conventional injection pumps are driven by the engine timing belt).

24 Where applicable, checking of the vacuum pump and/or injection pump drivebelts should not be overlooked when carrying out routine maintenance. Always renew a drivebelt if there is any doubt about its condition.

Fuel injection pump checks and adjustments

25 On all models, the idle and maximum speeds should be checked at the manufacturer's specified intervals – see Chapter 7, Section 2 for general procedures.

26 Depending on the type of injection pump and the control systems fitted to it, there may also be a need to check the operation of the anti-stall and cold start devices. Again, general procedures are described in Chapter 7.

27 Inspect the injection pump control linkages at every service interval. Lubricate them if necessary, and renew any frayed or sticking cables. Check that fully depressing the accelerator pedal produces full movement of the pump control lever.

28 At the same intervals, inspect the fuel injector high-pressure pipes and their securing clips for security and condition. Also inspect the fuel return pipes or hoses, and (when applicable) the turbo boost pressure hose which connects the inlet manifold to the injection pump. Renew any leaking or damaged components.

Fuel injector checks

29 Some manufacturers specify that the injectors should be removed and inspected periodically, but generally they are ignored unless particular problems (excessive smoke, knocking or power loss) suggest that they may be giving trouble.

30 If suitable equipment is available for on-vehicle testing (see Chapter 6), it is worth checking the injector opening pressures after the first 48 000 miles (77 000 km), and then every 24 000 miles (39 000 km) or so. Testing, cleaning and calibration procedures are given Section 4.

31 Fuel injector cleaners are available in the form of fuel additives. If used as directed they are unlikely to be harmful, and may indeed do some good; note however that some vehicle manufacturers specifically forbid their use.

8

Exhaust emissions check

32 The only emission test applicable to diesel engines is the measuring of exhaust smoke density. The test involves the use of special test equipment (see Chapter 6), and forms part of the MoT test for vehicles in the UK.

33 The test involves accelerating the engine several times to its maximum unloaded speed, and so it is vital to ensure that the engine timing belt is in good condition before the test is carried out. Refer to Chapter 9 for details of possible causes of excessive smoke.

Timing belt renewal

34 As with petrol engines, if a timing belt is fitted, it is vital to ensure that it is in good condition. On many diesel engines, the timing belt drives the injection pump as well as the camshaft.

35 The timing belt must be renewed at the manufacturer's specified intervals, or more frequently if the vehicle is used in particularly arduous conditions (eg, frequent stop-start driving or taxi work).

36 It is strongly recommended that consideration is given to renewing the timing belt every 36 000 miles (58 000 km), regardless of the manufacturer's recommended renewal intervals.

Turbocharger boost pressure check

37 Some manufacturers recommend that a turbocharger boost pressure check is carried out as part of the routine maintenance schedule. Details of this check are given in Chapter 4, Section 3.

3 Engine repair and overhaul procedures

General

1 Always refer to the manufacturer's recommended procedures and specifications when carrying out any engine repair or overhaul work.

2 Although diesel engine components are generally similar to their petrol counterparts, there may be detailed differences. For instance, piston cooling oil spray jets may be fitted in the crankcase, and the brake vacuum pump may be driven by an auxiliary shaft.

3 It is always advisable to follow the manufacturer's repair instructions in case there are any special procedures which need to be followed, or any particular specifications which must be checked before rebuilding the engine.

Cylinder head removal and refitting

Cylinder head gasket selection

4 On many diesel engines, when carrying out work which involves removal and refitting of

3.6 Measuring piston protrusion

the cylinder head, it is necessary to carry out a measurement of the protrusion of the pistons from the top surface of the cylinder block, to determine the thickness of cylinder head gasket to use on refitting.

5 The following procedure is a typical procedure for a Peugeot/Citroën XUD type engine. Refer to the manufacturer's information or the relevant Haynes Service and Repair Manual for specific details.

6 With the cylinder head removed, turn the crankshaft until pistons 1 and 4 are at TDC. Position a dial test indicator (DTI) on the cylinder block, and zero it on the block face. Transfer the probe to the centre of the flat section of No 1 piston crown (clear of the combustion chamber, where applicable), then slowly turn the crankshaft back-and-forth past TDC, noting the highest reading obtained on the DTI **(see illustration)**. Record this reading.

7 Repeat this measurement on No 4 piston, then turn the crankshaft half a turn (180º) and repeat the procedure on Nos 2 and 3 pistons.

8 Ascertain the greatest piston protrusion measurement, and use this to determine the correct cylinder head gasket from the manufacturer's information. E.g:

Piston protrusion	Gasket identification
0.54 to 0.65 mm	*1 notch*
0.65 to 0.77 mm	*2 notches*
0.77 to 0.82 mm	*3 notches*

Swirl chamber condition and protrusion/recess checking

9 On direct injection engines, it is advisable to check the condition of the swirl chambers

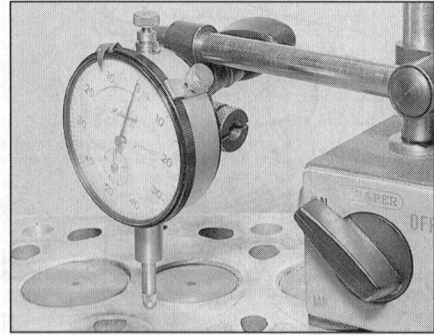

3.11a Zero the dial test indicator on the cylinder head face . . .

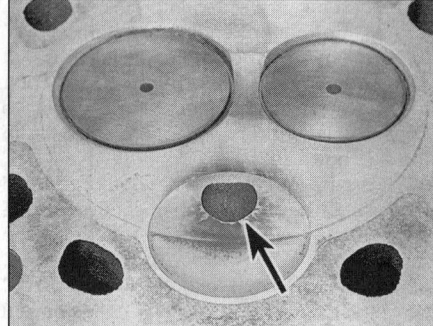

3.9 This swirl chamber shows the initial stages of cracking and burning (arrowed)

whenever the cylinder head is removed. Inspect the swirl chambers for burning or cracks **(see illustration)**. The swirl chambers can be renewed, but the work should be entrusted to a specialist.

10 On some engines it is necessary to check the protrusion or recess (as applicable) of the swirl chambers whenever the cylinder head is removed.

11 This can be done using a DTI in a similar manner to that described for the measurement of piston protrusion. Zero the DTI on the cylinder head face, then transfer the probe to the swirl chamber and measure the protrusion or recess **(see illustrations)**.

12 On some engines, the swirl chambers can be adjusted using shims; on others, the relevant swirl chamber(s) will have to be renewed if the protrusion/recess does not meet the manufacturer's specifications.

Fuel injection pump removal and refitting

13 Always ensure that the engine timing marks are in alignment when removing or refitting an injection pump. Timing marks are usually provided on the crankshaft, camshaft and injection pump sprockets. Where applicable, ensure that the timing pegs are in position in the injection pump and flywheel or crankshaft, as applicable (see Chapter 7, Section 3).

14 When removing certain types of fuel injection pump (such as those used on some Land Rover engines), the manufacturers may recommend that the pump rotor is locked in

3.11b . . . then check the swirl chamber protrusion

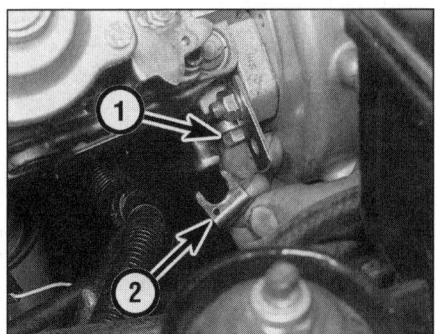

3.14 Injection pump rotor keeper plate screw (1) and keeper plate (2) – Land Rover 300 TDi engine

4.9 Disconnecting an injector leak-off pipe

4.10 Disconnecting an injector pipe

position before removal, by removing a keeper plate. The pump is then locked by refitting the keeper plate screw **(see illustration)**. This effectively locks the pump rotor in position, retaining the pump timing.

---

4 Injector testing, inspection and adjustment procedures

Note: *The information given in this Section refers to conventional diesel fuel injectors. For details of the injectors used in common rail and 'pump injector' systems, refer to Chapter 2, Sections 9 and 10 respectively.*

General

1 Refer to Chapter 2, Section 7 for a description of how typical fuel injectors function.

Testing on the vehicle

 Warning: Never expose the hands, face, or any other part of the body to injector spray; the high working pressure can penetrate the skin, with potentially fatal results.

2 A faulty injector which is causing knocking noises can be identified as follows.
3 Clean around the injector fuel pipe unions. Run the engine at a fast idle so that the knock can be heard. Using a suitable split ring spanner, slacken and retighten each injector union in turn. (Cover the union with a piece of rag to absorb the fuel which will be spilt.)
4 When the union supplying the defective injector is slackened, the knock will disappear. Stop the engine, and remove the injector for inspection. Before condemning the injector, make sure that the problem is not caused by a missing, damaged or incorrectly fitted heat shield.
5 The balance between injectors can be checked in a similar way, provided an accurate tachometer is available. With the engine idling, each union is slackened in turn, and the drop in engine rpm noted. Any injector which produces a much larger or smaller drop in rpm when its union is slackened should be viewed with suspicion.

6 With suitable equipment, it is possible to check injector opening (or 'breaking') pressures and leakage without removing the injectors from the engine. This is obviously a time-saving measure if it is simply wished to verify that the pressures are correct, or to locate a defective injector.
7 Testing spray pattern by cranking or running the engine with an injector out of its hole and connected to its fuel pipe should **not** be attempted. It is tempting to use this method because it is quick and requires no special equipment, but the risk of fire and blood poisoning from the injector spray mean that it cannot be recommended.

Removal and refitting

Note: *This is a general procedure. Refer to the manufacturer's information or to the relevant Haynes Service and Repair Manual for specific details.*

Screw-in type

8 Clean around the injector pipe unions, and around the injectors where they meet the cylinder head.
9 Disconnect the leak-off pipes from the

injectors, and plug or cap the open pipes **(see illustration)**.
10 Using a split ring spanner, disconnect the unions from the injectors. Move the pipes clear of the injectors. To do this, it is best to slacken the pipe unions at the pump as well, and to release pipe clamps or clips. Alternatively, remove the pipes completely **(see illustration)**.
Caution: Counterhold the delivery valves on the pump when slackening the unions.
11 Unscrew the injectors using a suitable deep socket, and remove them. Recover any sealing washers, heat shields, etc, and obtain new ones for reassembly. Note which way up each item is fitted **(see illustration)**.
12 Commence refitting by making sure that the injector recesses and sealing surfaces in the cylinder head are clean. Also clean the injectors themselves. On pintle type injectors, be careful not to damage the protruding needle tip.
13 Fit new heat shields, sealing washers, etc, making sure they are the right way up. Fit the injectors and tighten them to the specified torque. Be careful not to cross-thread or overtighten the injectors in an aluminium alloy cylinder head.

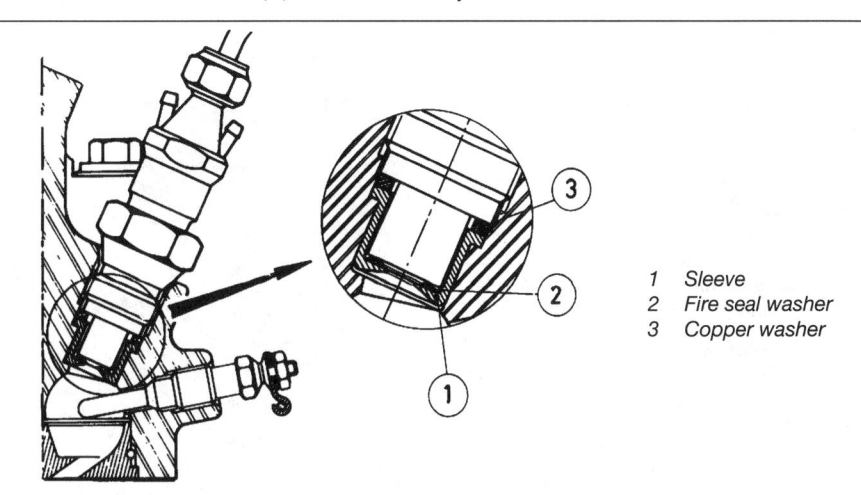

1 Sleeve
2 Fire seal washer
3 Copper washer

4.11 Sectional view of cylinder head, showing injector sealing washer detail

8

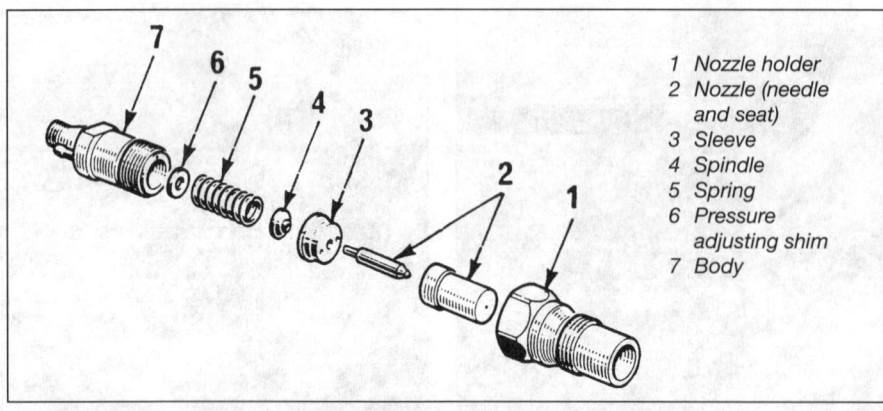

4.19a Exploded view of a Bosch screw-in pintle injector

1 Nozzle holder
2 Nozzle (needle and seat)
3 Sleeve
4 Spindle
5 Spring
6 Pressure adjusting shim
7 Body

14 Reconnect the leak-off pipes and the injector pipes. Make sure that the pipe clamps or clips are refitted to their original positions. Before tightening the unions on the injectors, bleed air from the pipes by cranking the engine until fuel emerges.

Clamp or flange type

15 The procedures are similar to the screw-in type described previously, but the injector fixing clamp nuts or bolts must be removed before the injector can be withdrawn. Carbon deposits can cause the injector to stick in its recess; in this case, an injector puller or a

small slide hammer can be used, or it may be possible to free the injector by careful levering. Do not attempt to free a sticking injector by cranking the engine: it could be ejected with enough force to cause damage or serious injury.

Cleaning and inspection

Caution: Diesel fuel is irritating to the skin and eyes. If possible use injector test fluid (which is less toxic) for cleaning and testing. Use barrier cream, and where necessary, goggles and disposable gloves, for protection. Change out of fuel-soaked clothing as soon as possible.

16 Dismantling, cleaning and calibration of injectors may not be an attractive proposition for the small garage or tune-up specialist, who will normally prefer to obtain new or reconditioned injectors on an exchange basis. The procedures are given here for the benefit of the owner-mechanic whose time is not costed at commercial rates, or for situations where replacement injectors are not available. An injector tester will be needed after reassembly, in order to check the injector opening pressure, spray pattern and leakage characteristics.

17 Injectors must only be dismantled and reassembled under conditions of near-surgical cleanliness.

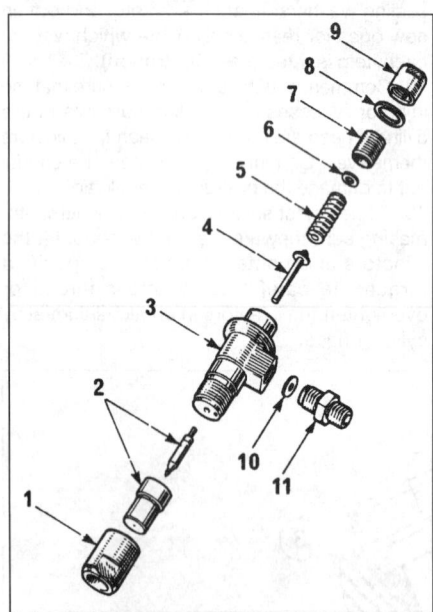

4.19c Exploded view of a RotoDiesel clamp fitting pintle injector

1 Nozzle holder
2 Nozzle (needle and seat)
3 Body
4 Spindle
5 Spring
6 Shim (not always fitted)
7 Pressure-adjusting screw
8 Washer
9 Cap nut
10 Washer
11 Injection pipe union adapter

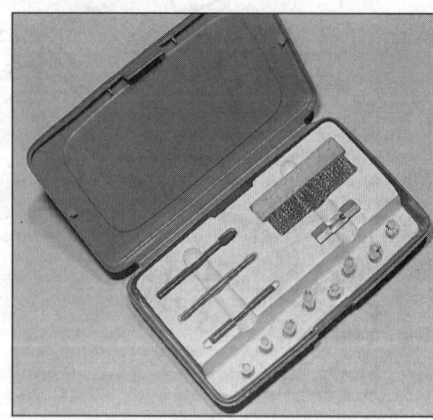

4.21 Typical injector cleaning kit

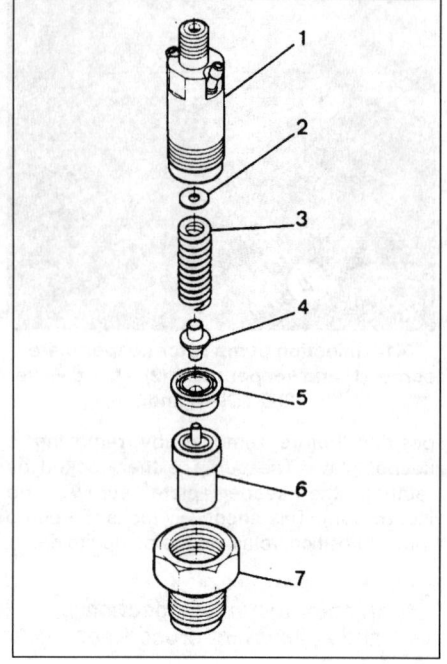

4.19b Exploded view of a RotoDiesel screw-in pintle injector

1 Body
2 Pressure adjusting shim
3 Spring
4 Spindle
5 Sleeve
6 Nozzle (needle and seat)
7 Nozzle holder

18 Clean the injector externally, using paraffin or a suitable solvent and a nylon or brass wire brush.

19 Separate the two halves of the injector body by unscrewing them. Ideally a dismantling jig should be used. If a jig is not available, **carefully** clamp the upper or central part of the injector body in a soft-jawed vice. Slacken the other part(s) using a spanner, and separate the injector body components **(see illustrations)**.

20 Remove the internal components from the injector, noting the order in which they are fitted. Do not drop the nozzle. Any copper washers or other seals must be renewed. If dismantling more than one injector, take care not to get the parts mixed up.

21 Immerse the injector components in clean diesel fuel or injector test fluid as they are removed. Clean them using a solvent such as carburettor cleaner and a wooden or plastic scraper. A nylon or brass-bristled brush, and brass wire can also be used – these are normally supplied as part of an injector cleaning kit **(see illustration)**.

22 The cleaning kit should also include brass probes or scrapers for cleaning carbon from the needle seat. When using these, work from the inside of the seat towards the outside. If a probe is inserted from the outside, there is more likelihood of damage.

23 Pay particular attention to cleaning deposits from the needle tip and seat, using a

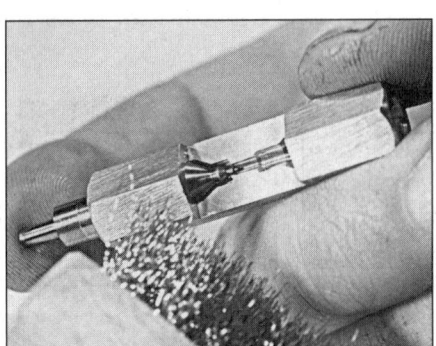

4.23 Cleaning an injector needle in a cleaning jig

cleaning jig to support the needle **(see illustration)**. Carbon build-up here can cause 'hosing' or a refusal to 'break' cleanly under test. In the case of 'perforated pintle' type

needles, clean out the drillings with the appropriate diameter brass wire.

24 When using brass wire to clean holes, be careful that it does not snap off in the hole. If this happens, its may be impossible to retrieve it.

25 Inspect all parts for obvious signs of wear or damage. Inspect the needle and its seat closely for ovality, ridging, pitting or other damage to their mating areas. The needle and seat must always be renewed as a matched set. **Do not** attempt to lap them together using grinding paste; this would damage them severely.

26 Inspect the injector body and base for corrosion and heat discoloration ('blueing'). If serious overheating has occurred, the complete injector must be renewed.

27 If the nozzle appears satisfactory after cleaning, perform a sliding test as follows. Dip

the injector needle in clean test fluid or fuel, and insert it fully into the seat, held vertically. Pull the needle back out by two-thirds of its length, then release it. When released, the needle must fall back onto the seat under its own weight – if not, the nozzle must be renewed.

28 Clean and degrease any new parts using a suitable solvent, then dip the parts in clean test fluid or diesel fuel. If possible, prime the injector body with the fluid or fuel too. Reassemble the injector and tighten the body sections together, not forgetting to fit a new sealing washer (when applicable) between the two halves **(see illustration)**.

29 Test the injector on the bench. Note that good results will not be obtained until the injector components are thoroughly coated with oil and any air has been expelled. Be prepared to make adjustments to the opening

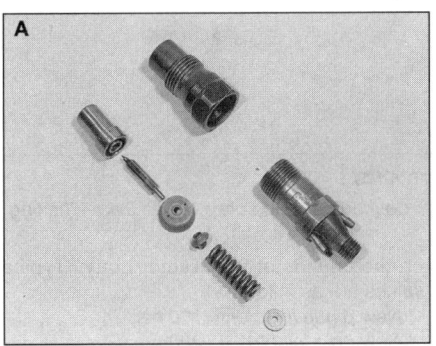 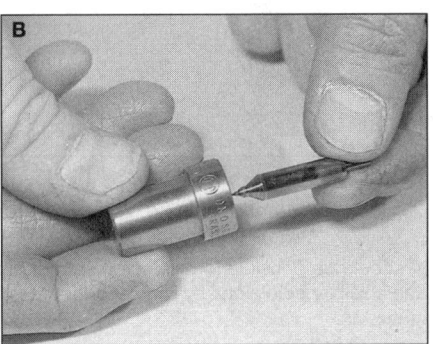 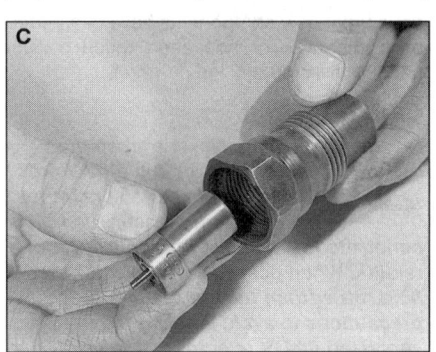
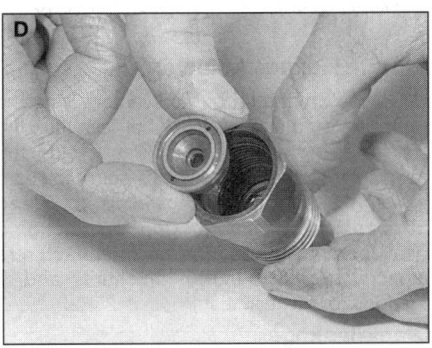 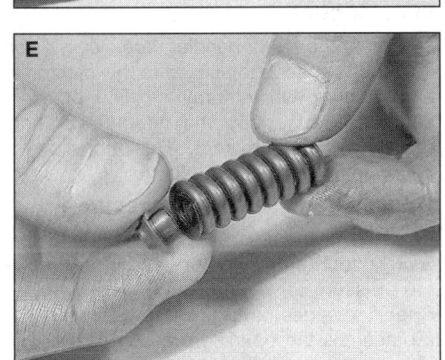 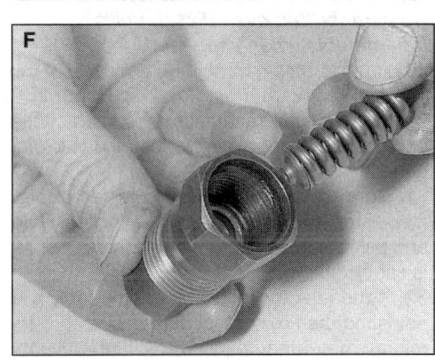
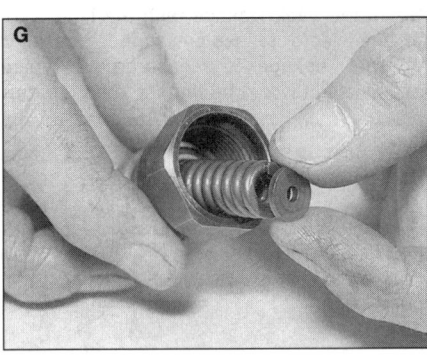 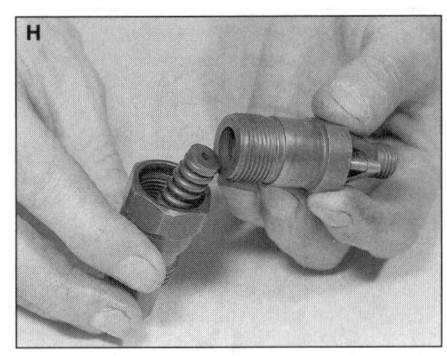 

4.28 Assembling a Bosch screw-in pintle injector

A   Injector components laid out for assembly
B   Inserting the needle into the seat
C   Fitting the nozzle into the holder
D   Fitting the sleeve
E   Fitting the spindle to the spring . . .
F   . . . the spring and spindle into the holder . . .
G   . . . and the shim on top of the spring
H   Fitting the injector body
J   Assembled injector – opening pressure (arrowed) is not always marked

8

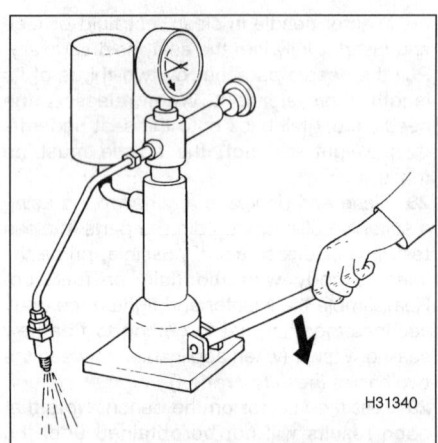

4.33a Checking injector spray pattern with an injector tester

pressure if new components have been fitted, or if the injector has seen much service (50 000 miles/80 000 km or more).

Bench-testing

 Warning: Never expose the hands, face, or any other part of the body to injector spray: the high working pressure can penetrate the skin, with potentially fatal results. When possible, use injector test fluid rather than fuel for testing. Take precautions to avoid inhaling the vaporised fuel or injector test fluid. Remember that even diesel fuel is inflammable when vaporised.

30 An injector tester will be needed for this work. These are described in Chapter 6. Follow the tester manufacturer's instructions, or if these are not available, use the following general procedure.

31 Connect the injector to the tester. Charge the tester with clean diesel fuel or test fluid, and bleed it. Make sure that the injector spray is safely directed.

32 If the injector tester has some means of adjusting the flow of test fluid, adjust it to the minimum setting which will still operate the injector satisfactorily.

Spray pattern

33 Shut off the tester pressure gauge. Pump the handle of the tester fairly rapidly (4 to 6 strokes per second) and observe the spray pattern. On pintle injectors, the spray pattern should be even and fan-shaped, free of streaks. The spray must start and stop cleanly, with no drips **(see illustrations)**.

34 On multi-hole injectors, the desired pattern is similar, but obviously there will be gaps corresponding to the intervals between the holes. Hole distribution is not always symmetrical.

35 The Pintaux type injector nozzle (with two holes, found on some Land Rover engines), is a special case. The main hole delivers a central spray of fuel, and the auxiliary one delivers a spray to one side **(see illustration)**.

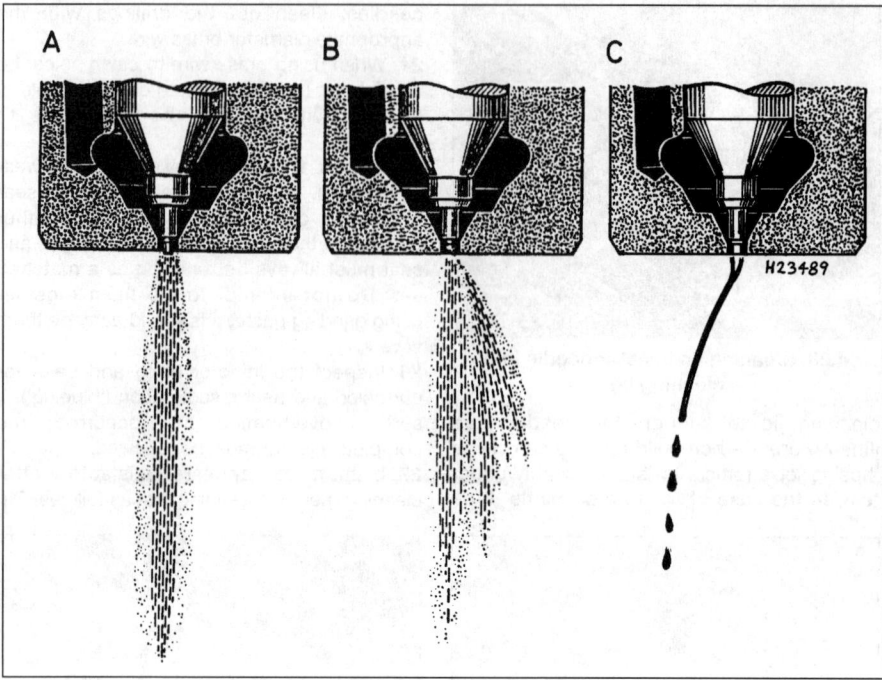

4.33b Pintle injector spray patterns

A Good – well-defined spray   B Bad – poorly-defined, ragged spray   C Bad – 'hosing'

It should be possible to produce a spray from the auxiliary hole alone by reducing the rate of pumping.

Injector buzz

36 Pump the tester handle more slowly (1 or 2 strokes per second). Listen to the injector: it should buzz while spraying. The buzzing sound is caused by the needle vibrating rapidly against its seat. This vibration is important in atomising the fuel.

Opening (or 'breaking') pressure

37 Observe the pressure gauge on the tester. Slowly depress the tester handle, and note the pressure at which injector spray begins. Specific values are given in Chapter 10. (Sometimes the opening pressure is marked

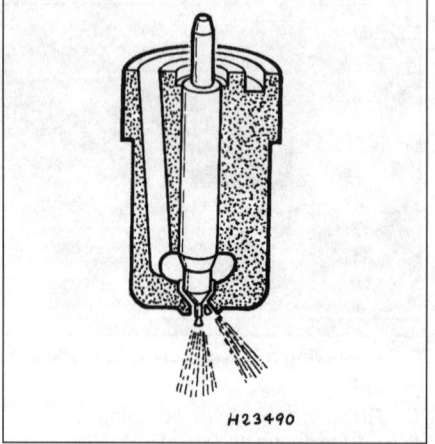

4.35 Pintaux injector spray pattern

on the outside of the injector body.) Typical values are as follows:

New pintle injector	130 bar
New hole injector	200 bar
Used injector	10 to 20 bar lower than new pressure

Nozzle leakage (dribble)

38 Wipe dry the tip of the injector. Bring the pressure on the tester to 10 bar less than the observed opening pressure. Hold the pressure at this value for 20 seconds, and observe the tip of the injector **(see illustration)**. It must not drip fuel, though it may become moist. (Some manufacturers specify slightly different pressures and/or periods for the leak test.)

Back-leakage

39 In the back-leakage test, the time taken for the injector to lose pressure as a result of internal leakage is measured. A typical specification would be that the pressure drop

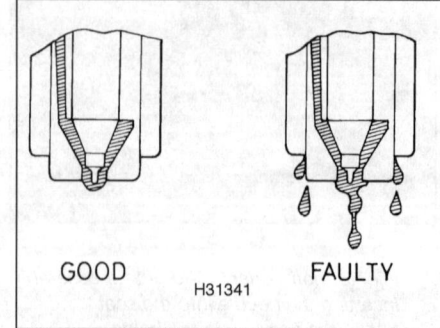

4.38 Injector nozzle leakage (dribble) test

4.51 Injector pressure-adjustment screw and cap nut

from 100 to 75 bar must take between 10 and 30 seconds. Too rapid a drop shows internal wear in the nozzle, or poor sealing between the injector body halves. (It can also show that the tester itself is worn or leaking, so do not rush to condemn an injector without establishing that the tester is OK.)

40 If there is no back-leakage at all, the needle is sticking in the nozzle, and should be removed for inspection or renewal.

41 Back-leakage can also occur in the form of a spurt of fuel from the leak-off connection at the moment of injection. If this has no effect on injector performance it can be ignored, but if it is bad enough to cause a 'kick' on the tester pressure gauge, the injector should be

stripped for inspection. This fault can cause leakage of fuel from the leak-off pipes in service.

Interpretation of results

42 If the injector has not yet been cleaned, do so now and repeat the tests.

43 A poor spray pattern or a lack of buzzing can usually be corrected by renewing the nozzle. Note, however that it is possible for an injector to perform satisfactorily in service, even though its spray pattern and noise on the bench are suspect. This is particularly true of indirect injection engines, where the swirl chamber characteristics can compensate to some extent for poor spray pattern.

44 Lack of buzzing, accompanied by an excessively wet spray pattern, can be caused by a weak or broken injector spring.

45 An incorrect opening pressure can be adjusted as described later in this Section.

46 A leaking nozzle must be renewed. A nozzle which leaks in service will rapidly become clogged up with carbon.

47 Excessive back-leakage will normally be cured by renewing the nozzle.

Adjustment of opening (or 'breaking') pressure (calibration)

48 After component renewal, or if the opening pressure is incorrect, the injector should be calibrated as follows.

49 Using the injector tester, measure the opening pressure and compare it with that specified by the manufacturer. (Opening pressure is also sometimes stamped on the injector body.)

50 Adjust the pressure to the 'new' value. This is normally done by changing the shim thickness in the injector. Obviously a range of shims will be needed. Increasing shim thickness raises the opening pressure, reducing thickness lowers it. Typically, changing the shim thickness by 0.05 mm will alter the opening pressure by approximately 5 bar.

51 A few injector types have an adjustment screw instead of a shim. After slackening a locknut or cap nut, the screw is tightened to increase opening pressure, or slackened to reduce it **(see illustration)**. Tighten the locknut or cap nut without disturbing the adjustment screw position when the opening pressure is correct.

---

5 Glow plugs – removal, inspection and refitting

Removal

1 Disconnect the battery negative lead.

2 Remove any surrounding components to allow access to the glow plugs.

3 Unscrew the nuts from the glow plug terminals, and where applicable recover the washers. Remove the interconnecting wire(s), and where applicable the feed wire from the top of the glow plugs **(see illustrations)**.

4 Unscrew the glow plugs and remove them from the cylinder head **(see illustration)**.

Inspection

5 Inspect the glow plugs for physical damage. Burnt or eroded glow plug tips can be caused by a bad injector spray pattern. Have the injectors checked if this sort of damage is found.

6 If the glow plugs are in good physical condition, check them as described in Chapter 3, Section 7.

Refitting

7 If new plugs are being fitted, ensure that the correct type plug is used, as recommended by the manufacturer.

8 Refit by reversing the removal operations. Apply a smear of copper-based anti-seize compound to the plug threads, and tighten to the specified torque. Do not overtighten, as this can damage the glow plug element.

5.3a Unscrew the nut . . .

5.3b . . . and disconnect the feed wire . . .

5.3c . . . and the interconnecting wire from the glow plug

5.4 Removing a glow plug

8

Notes

Chapter 9
Fault diagnosis

Contents

1 Introduction

The majority of starting problems on small diesel engines are electrical in origin.

When investigating difficult starting, make sure that the correct starting procedure is understood and is being followed. Some drivers are unaware of the significance of the preheating warning light; many modern engines are sufficiently forgiving for this not to matter in mild weather, but with the onset of winter, problems begin.

As a rule of thumb, if the engine is difficult to start, but runs well when it has finally got going, the problem is electrical (battery, starter motor or preheating system). If poor performance is combined with difficult starting, the problem is likely to be in the fuel system. The low-pressure (supply) side of the fuel system should be checked before suspecting the injectors and injection pump. *Normally the pump is the last item to suspect, since (unless it has been tampered with) there is no reason for it to be at fault.*

Bear in mind that most modern diesel engine vehicles with electronic control have a self-diagnosis system which will store details of any faults as fault codes in the electronic control unit memory (see Section 19). If a fault code is present, a warning light will normally be illuminated on the instrument panel to inform the driver.

9

2 Fault diagnosis charts

Engine turns but will not start (cold)

- [ ] Incorrect use of preheating system
- [ ] Preheating system fault
- [ ] Fuel waxing (in very cold weather)
- [ ] Overfuelling or cold start advance mechanism defective
- [ ] Electronic control system fault (where applicable)

Engine turns but will not start (hot or cold)

- [ ] Low cranking speed
- [ ] Poor compression
- [ ] No fuel in tank
- [ ] Air in fuel system
- [ ] Fuel feed restrictions
- [ ] Fuel contaminated
- [ ] Engine stop solenoid or mechanism defective
- [ ] Major mechanical failure
- [ ] Injection pump internal fault
- [ ] Electronic control system fault (where applicable)

Low cranking speed

- [ ] Inadequate battery capacity
- [ ] Incorrect grade of oil
- [ ] High resistance in starter motor circuit
- [ ] Starter motor internal fault

Engine is difficult to start

- [ ] Incorrect starting procedure
- [ ] Battery or starter motor fault
- [ ] Air in fuel system
- [ ] Fuel feed restriction
- [ ] Fuel lift pump defective
- [ ] Poor compression
- [ ] Valve clearances incorrect
- [ ] Valves sticking
- [ ] Blockage in exhaust system
- [ ] Valve timing incorrect
- [ ] Injector(s) faulty
- [ ] Injection pump timing incorrect
- [ ] Injection pump internal fault
- [ ] Electronic control system fault (where applicable)

Engine starts but stops again

- [ ] Fuel very low in tank
- [ ] Air in fuel system
- [ ] Idle adjustment incorrect
- [ ] Fast idle unit fault
- [ ] Fuel feed restriction
- [ ] Fuel return restriction
- [ ] Air cleaner faulty
- [ ] Blockage in induction system
- [ ] Blockage in exhaust system
- [ ] Electronic control system fault (where applicable)
- [ ] Injector(s) faulty

Engine will not stop when switched off

- [ ] Stop solenoid defective
- [ ] Stop actuator leaking or disconnected (vacuum type)
- [ ] Electronic control system fault (where applicable)

Misfiring/rough idle

- [ ] Air cleaner dirty
- [ ] Blockage in induction system
- [ ] Air in fuel system
- [ ] Fuel feed restriction
- [ ] Valve clearances incorrect
- [ ] Fuel lift pump defective (where applicable)

- [ ] Valve(s) sticking
- [ ] Valve spring(s) weak or broken
- [ ] Poor compression
- [ ] Overheating
- [ ] Injector pipe(s) wrongly connected or wrong type
- [ ] Valve timing incorrect
- [ ] Injector(s) faulty or wrong type
- [ ] Injection pump timing incorrect
- [ ] Injection pump faulty or wrong type
- [ ] Electronic control system fault (where applicable)

Lack of power

- [ ] Accelerator linkage not moving through full travel (cable slack or pedal obstructed)
- [ ] Other pump control linkages sticking or maladjusted
- [ ] Air cleaner dirty
- [ ] Blockage in induction system
- [ ] Air in fuel system
- [ ] Fuel feed restriction
- [ ] Fuel lift pump defective (where applicable)
- [ ] Valve timing incorrect
- [ ] Injection pump timing incorrect
- [ ] Blockage in exhaust system
- [ ] Turbo boost pressure inadequate (where applicable)
- [ ] Valve clearances incorrect
- [ ] Poor compression
- [ ] Injector(s) faulty or wrong type
- [ ] Injection pump faulty or wrong type
- [ ] Electronic control system fault (where applicable)

Fuel consumption excessive

- [ ] External leakage
- [ ] Fuel passing into sump
- [ ] Air cleaner dirty
- [ ] Blockage in induction system
- [ ] Valve clearances incorrect
- [ ] Valve(s) sticking
- [ ] Valve spring(s) weak
- [ ] Flame plug leaking fuel (where applicable)
- [ ] Poor compression
- [ ] Valve timing incorrect
- [ ] Injection pump timing incorrect
- [ ] Injector(s) faulty or wrong type
- [ ] Injection pump faulty or wrong type
- [ ] Electronic control system fault (where applicable)

Engine knocks

- [ ] Air in fuel system
- [ ] Fuel grade incorrect or quality poor
- [ ] Injector(s) faulty or wrong type
- [ ] Valve spring(s) weak or broken
- [ ] Valve(s) sticking
- [ ] Valve clearances incorrect
- [ ] Valve timing incorrect
- [ ] Injection pump timing incorrect
- [ ] Piston protrusion excessive/head gasket thickness inadequate (after repair)
- [ ] Valve recess incorrect (after repair)
- [ ] Piston rings broken or worn
- [ ] Pistons and/or bores worn
- [ ] Crankshaft bearings worn or damaged
- [ ] Small-end bearings worn
- [ ] Camshaft worn
- [ ] Timing gears worn
- [ ] Electronic control system fault (where applicable)

Black smoke in exhaust

- [ ] Air cleaner dirty
- [ ] Blockage in induction system
- [ ] Valve clearances incorrect
- [ ] Poor compression
- [ ] Turbo boost pressure inadequate (where applicable)
- [ ] Blockage in exhaust system
- [ ] Valve timing incorrect
- [ ] Flame plug leaking (where applicable)
- [ ] Injector(s) faulty or wrong type
- [ ] Injection pump timing incorrect
- [ ] Injection pump faulty or wrong type
- [ ] Electronic control system fault (where applicable)

Blue or white smoke in exhaust

- [ ] Engine oil incorrect grade or poor quality
- [ ] Glow plug(s) defective, or control unit faulty (smoke at start-up only)
- [ ] Flame plug leaking (where applicable)
- [ ] Overfuelling device operating after start-up (where applicable)
- [ ] Air cleaner dirty
- [ ] Blockage in induction system
- [ ] Valve timing incorrect
- [ ] Injection pump timing incorrect
- [ ] Injector(s) defective, or heat shields damaged or missing
- [ ] Engine running too cool
- [ ] Oil entering via valve stems
- [ ] Poor compression
- [ ] Head gasket blown
- [ ] Piston rings broken or worn
- [ ] Pistons and/or bores worn
- [ ] Oil consumption excessive
- [ ] External leakage (standing or running)
- [ ] New engine not yet run-in
- [ ] Engine oil incorrect grade or poor quality
- [ ] Oil level too high
- [ ] Crankcase ventilation system obstructed
- [ ] Oil leaking from oil feed pipe into fuel pipe
- [ ] Oil leakage from accessory (vacuum pump, air compressor, etc)
- [ ] Oil cooler leaking into coolant
- [ ] Oil leaking into injection pump (when applicable)
- [ ] Air cleaner dirty
- [ ] Blockage in induction system
- [ ] Cylinder bores glazed
- [ ] Piston rings broken or worn
- [ ] Pistons and/or bores worn
- [ ] Valve stems or guides worn
- [ ] Valve stem oil seals worn
- [ ] Cylinder bores glazed
- [ ] Piston rings broken or worn
- [ ] Pistons and/or bores worn
- [ ] Valve stems or guides worn
- [ ] Valve stem oil seals worn
- [ ] Electronic control system fault (where applicable)

Overheating

- [ ] Coolant leakage
- [ ] Engine oil level too high
- [ ] Electric cooling fan malfunctioning (where applicable)
- [ ] Water pump drivebelt slack or broken
- [ ] Water pump defective
- [ ] Radiator clogged externally
- [ ] Radiator clogged internally
- [ ] Hoses blocked or collapsed
- [ ] Pressure cap defective or incorrect
- [ ] Thermostat defective or incorrect
- [ ] Thermostat missing
- [ ] Air cleaner dirty
- [ ] Blockage in induction system

- [ ] Head gasket blown
- [ ] Cylinder head cracked or warped
- [ ] Valve timing incorrect
- [ ] Injection pump timing incorrect (over-advanced)
- [ ] Injector(s) faulty or wrong type
- [ ] Injection pump faulty or wrong type
- [ ] Electronic control system fault (where applicable)
- [ ] Imminent seizure (piston pick-up)

Crankcase pressure excessive (oil being blown out)

- [ ] Blockage in crankcase ventilation system
- [ ] Leakage in vacuum pump or exhauster
- [ ] Piston rings broken or sticking
- [ ] Pistons or bores worn
- [ ] Head gasket blown

Erratic running

- [ ] Operating temperature incorrect
- [ ] Stop control or accelerator linkages maladjusted or sticking
- [ ] Air cleaner dirty
- [ ] Blockage in induction system
- [ ] Air in fuel system
- [ ] Injector pipe(s) wrongly connected or wrong type
- [ ] Fuel feed restriction
- [ ] Fuel lift pump defective (where applicable)
- [ ] Valve clearances incorrect
- [ ] Valve(s) sticking
- [ ] Valve spring(s) broken or weak
- [ ] Valve timing incorrect
- [ ] Poor compression
- [ ] Injector(s) faulty or wrong type
- [ ] Injection pump mountings loose
- [ ] Injection pump timing incorrect
- [ ] Injection pump faulty or wrong type
- [ ] Electronic control system fault (where applicable)

Vibration

- [ ] Accelerator linkage sticking
- [ ] Engine mountings loose or worn
- [ ] Cooling fan damaged or loose
- [ ] Crankshaft pulley/damper damaged or loose
- [ ] Injector pipe(s) wrongly connected or wrong type
- [ ] Valve(s) sticking
- [ ] Flywheel or (where applicable) flywheel housing loose
- [ ] Poor (uneven) compression

Low oil pressure

- [ ] Oil level low
- [ ] Oil grade or quality incorrect
- [ ] Oil filter clogged
- [ ] Overheating
- [ ] Oil contaminated
- [ ] Gauge or warning light sender inaccurate
- [ ] Oil pump pick-up strainer clogged
- [ ] Oil pump suction pipe loose or cracked
- [ ] Oil pressure relief valve defective or stuck open
- [ ] Oil pump worn
- [ ] Crankshaft bearings worn

High oil pressure

- [ ] Oil grade or quality incorrect
- [ ] Gauge inaccurate
- [ ] Oil pressure relief valve stuck shut

Injector pipe(s) break or split repeatedly

- [ ] Missing or wrongly-located clamps
- [ ] Wrong type or length of pipe
- [ ] Faulty injector
- [ ] Faulty delivery valve

9

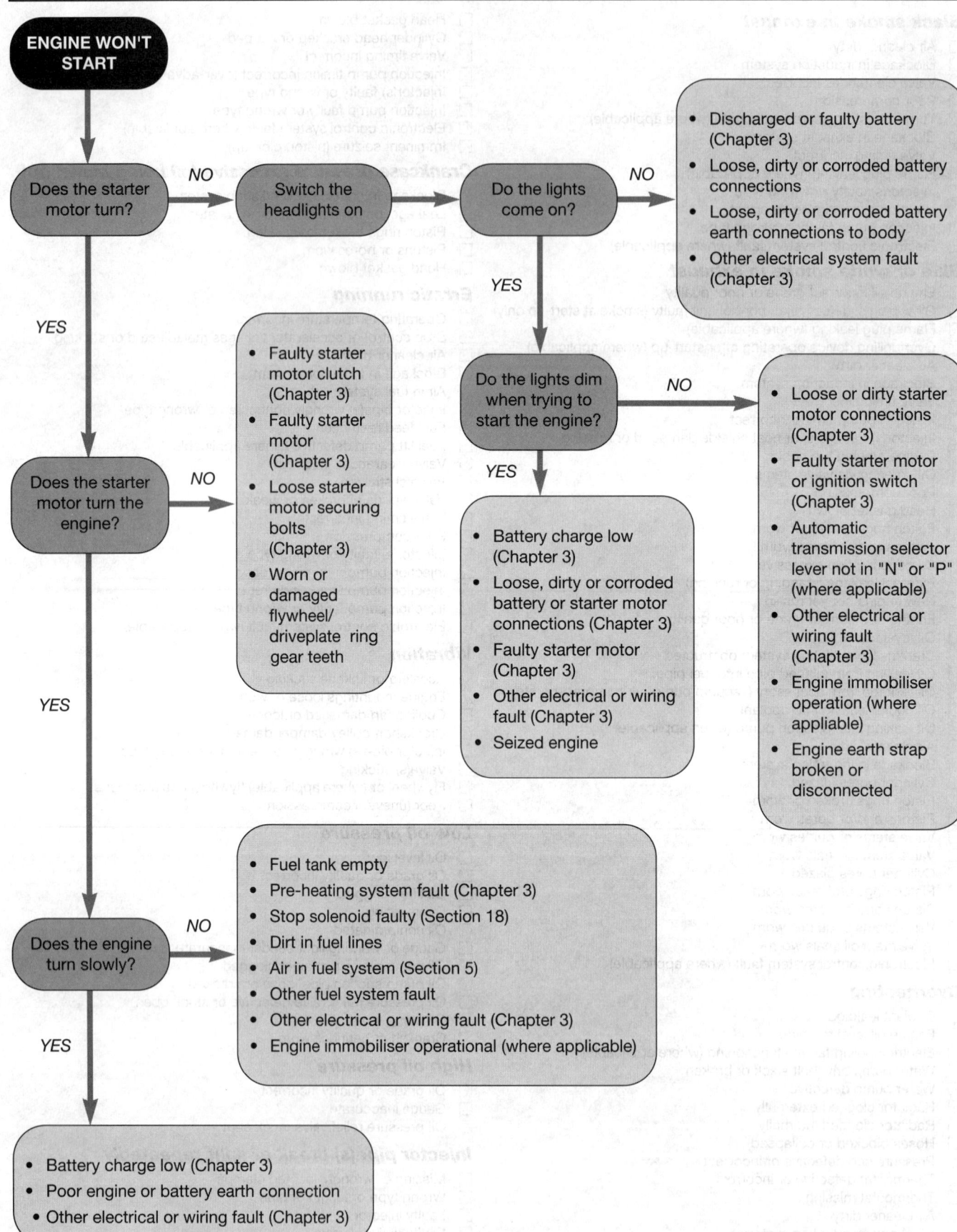

ENGINE WON'T START

Does the starter motor turn? — NO → Switch the headlights on → Do the lights come on? — NO →
- Discharged or faulty battery (Chapter 3)
- Loose, dirty or corroded battery connections
- Loose, dirty or corroded battery earth connections to body
- Other electrical system fault (Chapter 3)

Do the lights come on? — YES ↓

Do the lights dim when trying to start the engine? — NO →
- Loose or dirty starter motor connections (Chapter 3)
- Faulty starter motor or ignition switch (Chapter 3)
- Automatic transmission selector lever not in "N" or "P" (where applicable)
- Other electrical or wiring fault (Chapter 3)
- Engine immobiliser operation (where appliable)
- Engine earth strap broken or disconnected

Do the lights dim when trying to start the engine? — YES ↓
- Battery charge low (Chapter 3)
- Loose, dirty or corroded battery or starter motor connections (Chapter 3)
- Faulty starter motor (Chapter 3)
- Other electrical or wiring fault (Chapter 3)
- Seized engine

Does the starter motor turn? — YES ↓

Does the starter motor turn the engine? — NO →
- Faulty starter motor clutch (Chapter 3)
- Faulty starter motor (Chapter 3)
- Loose starter motor securing bolts (Chapter 3)
- Worn or damaged flywheel/driveplate ring gear teeth

Does the starter motor turn the engine? — YES ↓

Does the engine turn slowly? — NO →
- Fuel tank empty
- Pre-heating system fault (Chapter 3)
- Stop solenoid faulty (Section 18)
- Dirt in fuel lines
- Air in fuel system (Section 5)
- Other fuel system fault
- Other electrical or wiring fault (Chapter 3)
- Engine immobiliser operational (where applicable)

Does the engine turn slowly? — YES ↓
- Battery charge low (Chapter 3)
- Poor engine or battery earth connection
- Other electrical or wiring fault (Chapter 3)

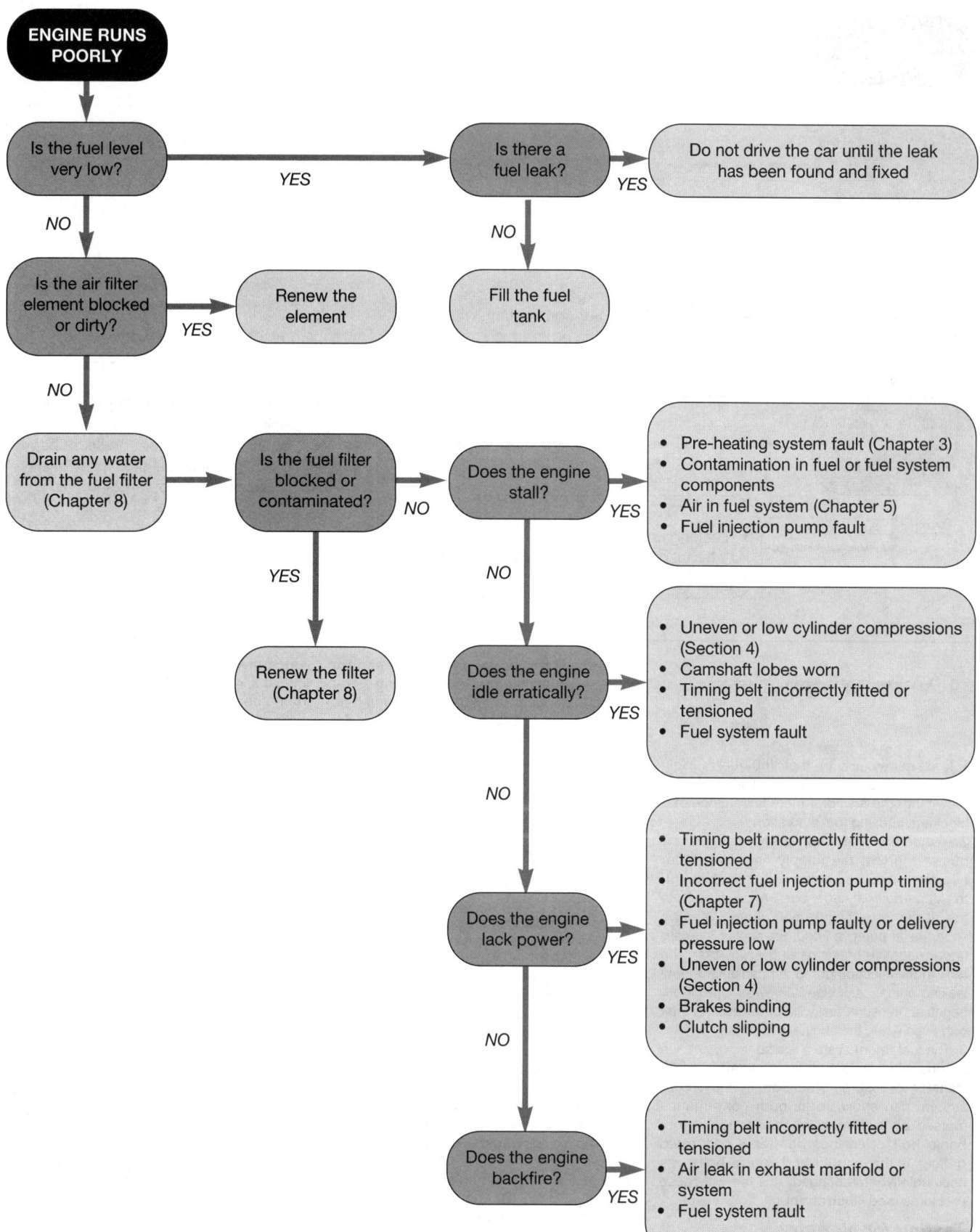

ENGINE RUNS POORLY

Is the fuel level very low? — NO → Is the air filter element blocked or dirty? — NO → Drain any water from the fuel filter (Chapter 8) → Is the fuel filter blocked or contaminated? — NO → Does the engine stall?

Is the fuel level very low? — YES → Is there a fuel leak? — YES → Do not drive the car until the leak has been found and fixed

Is there a fuel leak? — NO → Fill the fuel tank

Is the air filter element blocked or dirty? — YES → Renew the element

Is the fuel filter blocked or contaminated? — YES → Renew the filter (Chapter 8)

Does the engine stall? — YES →
- Pre-heating system fault (Chapter 3)
- Contamination in fuel or fuel system components
- Air in fuel system (Chapter 5)
- Fuel injection pump fault

Does the engine stall? — NO → Does the engine idle erratically? — YES →
- Uneven or low cylinder compressions (Section 4)
- Camshaft lobes worn
- Timing belt incorrectly fitted or tensioned
- Fuel system fault

Does the engine idle erratically? — NO → Does the engine lack power? — YES →
- Timing belt incorrectly fitted or tensioned
- Incorrect fuel injection pump timing (Chapter 7)
- Fuel injection pump faulty or delivery pressure low
- Uneven or low cylinder compressions (Section 4)
- Brakes binding
- Clutch slipping

Does the engine lack power? — NO → Does the engine backfire? — YES →
- Timing belt incorrectly fitted or tensioned
- Air leak in exhaust manifold or system
- Fuel system fault

9

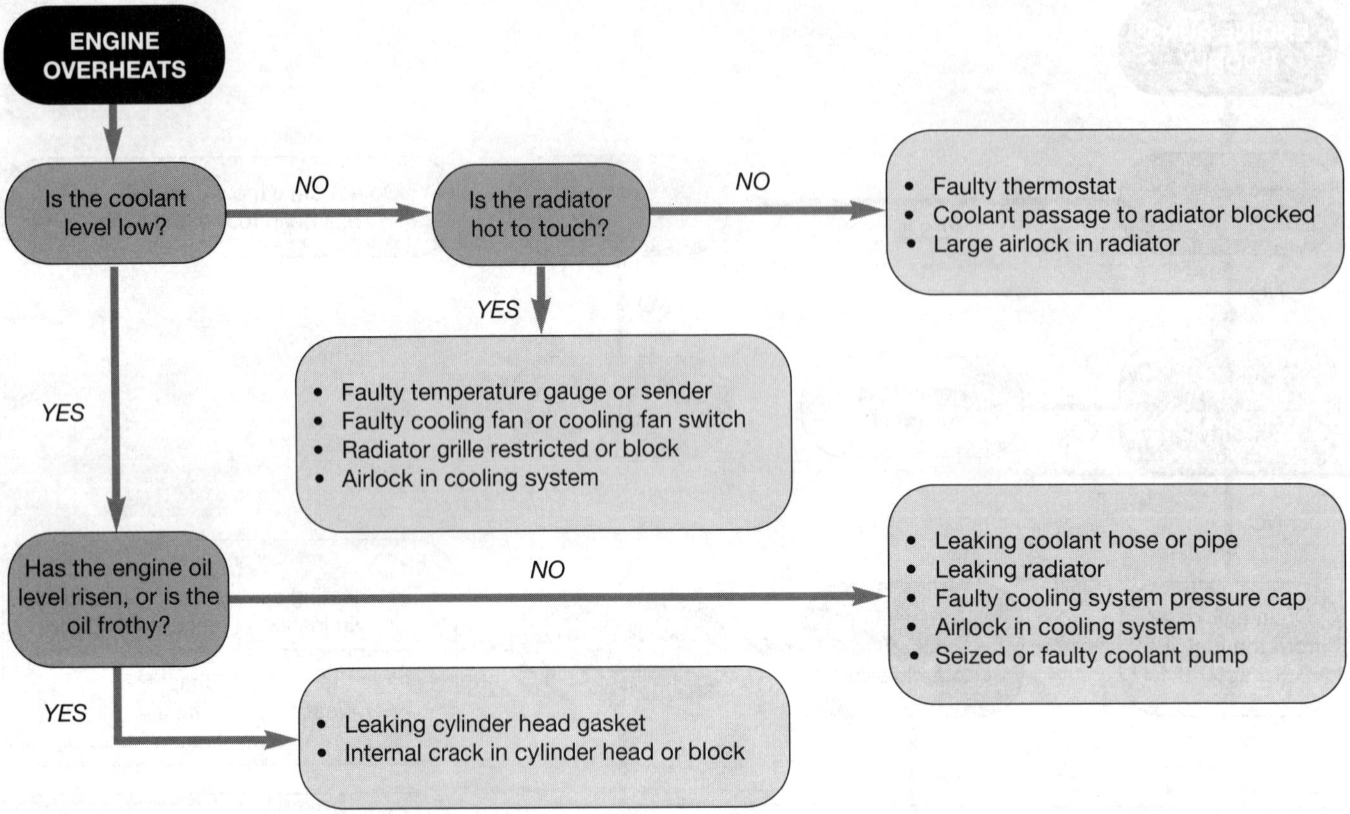

ENGINE OVERHEATS

Is the coolant level low? → **NO** → Is the radiator hot to touch? → **NO** →
- Faulty thermostat
- Coolant passage to radiator blocked
- Large airlock in radiator

Is the radiator hot to touch? **YES** →
- Faulty temperature gauge or sender
- Faulty cooling fan or cooling fan switch
- Radiator grille restricted or block
- Airlock in cooling system

Is the coolant level low? **YES** →

Has the engine oil level risen, or is the oil frothy? → **NO** →
- Leaking coolant hose or pipe
- Leaking radiator
- Faulty cooling system pressure cap
- Airlock in cooling system
- Seized or faulty coolant pump

Has the engine oil level risen, or is the oil frothy? **YES** →
- Leaking cylinder head gasket
- Internal crack in cylinder head or block

---

3 Fuel supply system - testing

1 It is necessary to test the fuel supply system if it is suspected that air is being drawn into the fuel, or if there is evidence of a blockage causing fuel starvation.

2 Remember that when a separate fuel lift pump is fitted, the supply lines are under negative pressure between the fuel tank and the lift pump, and under positive pressure from the lift pump to the injection pump. When no separate lift pump is fitted, the supply lines are under negative pressure all the way from the tank to the injection pump. Air can enter at any leaking union, seal, bleed screw or pipe under negative pressure; fuel will not necessarily leak out, even when the engine is stopped.

3 The fuel return system is also important. On pumps where injection timing is affected by transfer pressure, blockage in the return system can show up as poor performance caused by incorrect timing. The fuel return banjo bolt often incorporates a calibrated orifice; if the supply and return bolts are accidentally interchanged, this too will cause problems **(see illustration)**.

Testing for air entry

4 The presence of air in the fuel can be verified by fitting a piece of transparent hose in the injection pump fuel return line, and running the engine at 2000 to 3000 rpm. If air is being drawn in, bubbles will be visible in the returning fuel. A few bubbles are nothing to worry about, but a continuous stream means trouble. If a return line is fitted at the fuel filter, a similar check can be made there.

5 Further testing requires a hand-operated vacuum pump, with a vacuum gauge.

6 Locate the fuel supply line where it leaves the fuel tank. Either disconnect the line and plug it, or (if a flexible hose is fitted at this point) clamp it with a brake hose clamp or self-locking pliers with protected jaws.

7 Clean around the supply line where it enters the fuel injection pump (or lift pump, if applicable), and disconnect it. Connect the vacuum pump to the line.

Caution: It is important that no dirt is allowed to enter the pump.

8 Apply vacuum to the line, and watch the gauge. If the gauge falls, air is entering the line somewhere. If the gauge does not fall, air was being drawn in on the tank side of the point where the line is clamped or plugged. Check that the pick-up pipe (stack pipe) is not split.

9 Disconnect the vacuum pump, and reconnect it at the first union in the direction of the fuel tank. (On systems without a lift pump, this will be on the injection pump side of the filter.) Again, take care to clean around the union first. Reapply the vacuum and again watch the gauge. If the vacuum is held this time, the leak was in the section first tested.

10 If the gauge still falls, disconnect the pump and repeat the test one union nearer the fuel tank. Carry on until the leaking union, section or component is located, then repair and re-test. When a diaphragm type hand-priming pump is fitted, do not overlook this as a possible source of air entry, especially if it has seen much service.

11 Unplug or unclamp the supply line at the tank, and remake the original connections.

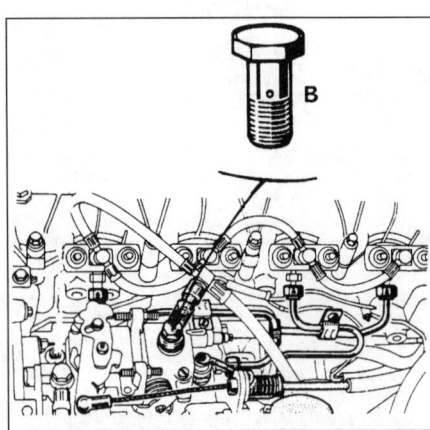

3.3 Fuel return banjo bolt (B) with calibrated orifice

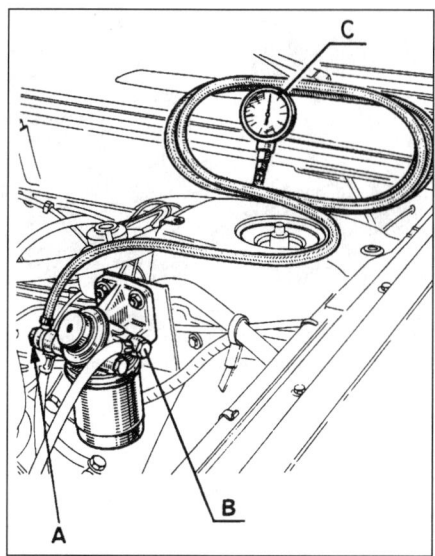

3.15 Checking the pressure drop across the fuel filter

A Inlet
B Outlet
C Pressure gauge

Testing for blockage

Systems without a separate lift pump

12 A fuel line vacuum gauge with the necessary adapters will be needed for this test. The gauge range should be approximately 0 to 1 bar.

13 Clean around the fuel pipe union on the outlet (injection pump side) of the fuel filter. Connect the gauge to this union using a T-piece.
Caution: It is important that no dirt is allowed to enter the pump.

14 Run the engine at maximum rpm, and note the gauge reading. Vacuum of 0.2 bar is acceptable. Any higher reading shows that there is a blockage.

15 Stop the engine. Remake the original fuel line connections on the pump side of the filter, and connect the vacuum gauge to the inlet (tank side) union of the filter **(see illustration)**. Run the engine at maximum rpm again, and note the gauge reading. Vacuum of up to 0.1 bar is acceptable. Any higher reading shows that there is a blockage on the tank side of the filter.

16 The difference between the two gauge readings is caused by the resistance to flow of the fuel filter. If the difference is greater than 0.15 bar, this shows that the fuel filter element is clogged, and should be renewed.

17 A blockage on the tank side of the filter may be caused by one of the following:
  a) Blocked tank vent of filler cap vent (as applicable).
  b) Clogged pick-up strainer (if fitted) in tank.
  c) Tank-to-filter pipe kinked or squashed (external damage).
  d) Tank-to-filter pipe blocked internally.

18 When testing is complete, stop the engine and remake the original fuel line connections.

Systems with a separate fuel lift pump

19 The principle of testing is the same as described previously for systems without a lift pump, with the difference being that there is negative pressure on the tank side of the lift pump, and positive pressure on the injection pump side **(see illustration)**. The gauge used must therefore have both positive and negative ranges – typically 0 to 1 bar vacuum, and 0 to 2 bars pressure.

4 Poor compression

1 Poor compression may give rise to a number of faults, including difficult starting, loss of power, misfiring or uneven running and smoke in the exhaust.

2 Before looking for mechanical reasons for compression loss, check that the problem is not on the induction side. A dirty air cleaner or some other blockage in the induction system can restrict air intake to the point where compression suffers.

3 Mechanical reasons for low compression include:
  a) Incorrect valve clearances.
  b) Sticking valves.
  c) Weak or broken valve springs.
  d) Incorrect valve timing.
  e) Worn or burnt valve heads and seats.
  f) Worn valve stems and guides.
  g) Head gasket blown.
  h) Piston rings broken or sticking.
  i) Pistons or bores worn.
  j) Head gasket thickness incorrect (after rebuild).

4 Compression loss on one cylinder alone can be due to a defective or badly-seated glow plug, or a leaking injector sealing washer. Some engines also have a cylinder head plug (for the insertion of a dial test indicator probe when determining TDC), and this should not be overlooked as a possible source of leaks.

5 Compression loss on two adjacent cylinders is almost certainly due to the head gasket blowing between them. Sometimes the fault will be corrected by renewing the gasket, but a blown gasket can also be an indication that the cylinder head itself is warped. Always check the head mating face for distortion when renewing the gasket. On wet-liner engines, also check the liner protrusion.

Compression test

6 A compression tester specifically intended for diesel engines must be used, because of the higher pressures involved compared to a petrol engine – see Chapter 6. The tester is connected to an adapter which screws into the glow plug or injector hole. Normally, sealing washers must be used on both sides of the adapter.

7 Unless specific instructions to the contrary are supplied with the tester, observe the following points:
  a) The battery must be in a good state of charge, the air cleaner element must be clean, and the engine should be at normal operating temperature.
  b) All the injectors or glow plugs should be removed before starting the test. If removing the injectors, also remove their heat shields (when fitted), otherwise they may be blown out.
  c) The stop control lever on the injection pump must be operated, or the stop solenoid disconnected, to prevent the engine from running, or fuel from being discharged.

8 There is no need to hold the accelerator pedal down during the test, because the diesel engine air inlet is not throttled. (There

9

3.19 Fuel supply checking points – systems with a separate lift pump

A Injection pump inlet/filter outlet (lift pump pressure)
B Filter inlet/lift pump outlet (lift pump pressure)
C Lift pump inlet (vacuum)
D Water separator outlet (vacuum)
E Water separator inlet (vacuum)
F Fuel tank outlet (vacuum)

4.13a Leakdown test adapter being fitted to a glow plug hole

4.13b Whistle fitted to test adapter to find TDC

4.15 Leakdown tester in use

are rare exceptions to this case, when a throttle valve is used to produce vacuum for servo or governor operation – for example, early Land Rovers.)

9 The actual compression pressures measured are not so important as the balance between cylinders. Typical values at cranking speeds are:

Good condition	25 to 30 bar
Minimum	18 bar
Maximum difference between cylinders	5 bar

10 The cause of poor compression is less easy to establish on a diesel engine than on a petrol one. The effect of introducing oil into the cylinders ('wet' testing) is not conclusive, because there is a risk that the oil will sit in the bowl in the piston crown (direct injection engines) or in the swirl chamber (indirect injection engines), instead of passing to the piston rings.

Leakdown test

11 A leakdown test measures the rate at which compressed air fed into the cylinder is lost. It is an alternative to a compression test, and in many ways is better, since it provides easy identification of where pressure loss is occurring (piston ring, valves or head gasket). However, it does require a source of compressed air.

Caution: Before beginning the test, remove the cooling system pressure cap. This is necessary because if there is a leak into the cooling system, the introduction of compressed air may damage the radiator. Similarly, it is advisable to remove the dipstick or the oil filler cap, to prevent excessive crankcase pressurisation.

12 Connect the tester to a compressed air line, and adjust the reading to 100% as instructed by the manufacturer.

13 Remove the glow plugs or injectors, and screw the appropriate adapter into a glow plug or injector hole. Fit the whistle to the adapter, and turn the crankshaft. When the whistle begins to sound, the piston in question is rising on compression. When the whistle stops, TDC has been reached **(see illustrations)**.

14 Engage a gear and apply the handbrake

to stop the engine turning. Remove the whistle and connect the tester to the adapter. Note the tester reading, which indicates the rate at which the air escapes. Repeat the test on the other cylinders.

15 The tester reading is in the form of a percentage, where 100% is perfect. Readings of 80% or better are to be expected from an engine in good condition. The actual reading is less important than the balance between cylinders, which should be within 5% **(see illustration)**.

16 The areas from which escaping air emerges show where a fault lies as follows:

Air escaping from	Probable cause
Oil filler cap or dipstick tube	Worn piston rings or cylinder bores
Exhaust pipe	Worn or burnt exhaust valve
Air cleaner/inlet manifold	Worn or burnt inlet valve
Cooling system	Blown head gasket or cracked cylinder head

17 Bear in mind that if the head gasket is blown between two adjacent cylinders, air escaping from the cylinder under test may emerge via an open valve in the cylinder adjacent.

5 Air in fuel system

1 The diesel engine will not run at all, or at best will run erratically, if there is air in the fuel lines. If the fuel tank has been allowed to run dry, or after operations in which the fuel supply lines have been opened, the fuel system must be bled before the engine will run. Methods of bleeding are given in Chapter 8, Section 2.

2 Air will also enter the fuel lines through any leaking joint or seal, since the supply side is under negative pressure all the time that the engine is running. For testing procedures, see Section 3.

6 Fuel feed restricted

1 Restriction in the fuel feed from the tank to the pump may be caused by any one of the following faults:

   a) Fuel filter blocked.
   b) Tank vent blocked.
   c) Feed pipe blocked or collapsed.
   d) Fuel waxing (in very cold weather).

2 Testing of the fuel supply system is covered in Section 3.

Fuel waxing

3 In the case of fuel waxing, the wax normally builds up first in the filter. If the filter can be warmed, this will often allow the engine to run. Only in exceptionally severe weather will waxing prevent winter-grade fuel from being pumped out of the tank. See Chapter 2 for more details.

Caution: Do not use a naked flame for this.

Microbiological contamination

4 Under certain conditions, it is possible for micro-organisms to colonise the fuel tank and supply lines. These micro-organisms produce a black sludge or slime, which can block the filter and cause corrosion on metal parts. The problem normally shows up first as unexpected blockage of the filter.

5 If such contamination is found, drain the fuel tank, and discard the drained fuel. Flush the tank and fuel lines with clean fuel, and renew the fuel filter; in bad cases, steam-clean the tank as well. If there is evidence that the contamination has passed the fuel filter, have the injection pump cleaned by a specialist.

6 Further trouble may be avoided by only using fuel from reputable outlets with a high turnover. Proprietary additives are also available to inhibit the growth of micro-organisms in storage tanks or in the vehicle fuel tank.

7 Lack of power

1 Complaints of lack of power are not always justified. If necessary, perform a road or dynamometer test to verify the condition Even if power is down, the complaint is not necessarily due to an engine or injection system fault.
2 Before commencing detailed investigation, check that the accelerator linkage is moving through its full travel. Also make sure that an apparent power loss is not caused by items such as binding brakes, under-inflated tyres, overloading of the vehicle, or some particular feature of operation.

8 Turbo-boost pressure inadequate

1 If boost pressure is low, power will be down, and too much fuel may be delivered at high engine speeds (depending on the method of pump control). Possible reasons for low boost pressure include:
   a) *Air cleaner dirty.*
   b) *Leaks in induction system.*
   c) *Blockage in exhaust system.*
   d) *Turbo control fault (wastegate or actuator).*
   e) *Turbo mechanical fault.*
2 Refer to Chapter 4, Section 3 for details of turbocharger boost pressure checking.

9 Fuel consumption excessive

Complaints of excessive fuel consumption, as with lack of power, may not mean that a fault exists. If the complaint is justified and there are no obvious fuel leaks, check the same external factors as for lack of power (Section 7) before turning to the engine and injection system.

10 Fuel in sump

1 If fuel oil is found to be diluting the engine oil in the sump, this may have arrived in one or more of the following ways:
   a) *Down the cylinder bores, especially when the engine is cold.*
   b) *Through a leaking fuel lift pump diaphragm (where applicable).*
   c) *Through leaking injection pump seals, when these communicate with the timing case.*
2 Fuel contamination of the oil can be detected by smell, and in bad cases, by an obvious reduction in viscosity.

11 Knocking caused by injector fault

1 A faulty injector which is causing knocking noises can be identified as follows.
2 Clean around the injector fuel pipe unions, then run the engine at a fast idle so that the knock can be heard. Using a suitable spanner, slacken and retighten each injector in turn. (Cover the union with a piece of rag to absorb the fuel which will spray out.)
3 When the union supplying the defective injector is slackened, the knock will disappear. Stop the engine and remove the injector for inspection (see Chapter 8, Section 4).

12 Excessive exhaust smoke

1 When investigating a complaint of excessive exhaust smoke, check first (by means of a dynamometer or road test) that the smoke is still excessive when the engine has reached normal operating temperature. A cold engine may produce some blue or white smoke until it has warmed up; this is not necessarily a fault.
2 **Black smoke** is produced by incomplete combustion of the fuel in such a way that carbon particles (soot) are formed. Incomplete combustion shows that there is a lack of oxygen, either because too much fuel is being delivered, or because not enough air is being drawn into the cylinders. A dirty air cleaner is an obvious cause of air starvation; incorrect valve clearances (where applicable) should also be considered. Combustion may also be incomplete because the injection timing is incorrect (too far retarded), or because the injector spray pattern is poor.
3 **Blue smoke** is produced either by incomplete combustion of the fuel, or by burning engine (sump) oil. This type of incomplete combustion may be caused by incorrect injection timing (too far advanced), by defective injectors, or by damaged or missing injector heat shields.
4 All engines burn a certain amount of engine oil, especially when cold, but if enough is being burnt to cause excessive exhaust smoke, this suggests that there is a significant degree of wear or some other problem.
5 White smoke (not to be confused with steam) is produced by unburnt or partially-burnt fuel appearing in the exhaust gases. Some white smoke is normal during and immediately after start-up, especially in cold conditions. Excessive amounts of white smoke can be caused by a preheating system fault, by incorrect injection pump timing, or by too much fuel being delivered by the injection pump (overfuelling device malfunctioning). The use of poor quality fuel with a low cetane number, and thus a long ignition delay, can also increase emissions of white smoke.
6 Accurate measurement of exhaust smoke requires the use of some kind of smoke meter; these are described in Chapter 6.

13 Oil entering the engine via valve stems

1 Excessive oil consumption due to oil passing down the valve stems can have three causes:
   a) *Valve stem wear.*
   b) *Valve guide wear.*
   c) *Valve stem oil seal wear.*
2 In the first two cases, the cylinder head must be removed and dismantled so that the valves and guides can be inspected and measured for wear.
3 In the case of worn valve stem oil seals, on some engines, these can be renewed without removing the head. Whether or not this is worthwhile will depend on how worn the valve stems are.

14 Oil consumption excessive

When investigating complaints of excessive oil consumption, make sure that the correct level-checking procedure is being followed. If insufficient time is allowed for the oil to drain down after stopping the engine, or if the level is checked while the vehicle is standing on a slope, a false low reading may result. The unnecessary topping-up which follows may itself cause increased oil consumption, as a result of the level being too high.

15 Cylinder bore glazing

Engines which spend long periods idling can suffer from glazing of the cylinder bores, leading to high oil consumption, even though no significant wear has taken place. The same effect can be produced by incorrect running-in procedures, or by the use of the incorrect grade of oil during running-in. The remedy is to remove the pistons, deglaze the bores with a hone or 'glaze buster' tool, and to fit new piston rings.

16 Overheating

1 Complaints of overheating should first be verified , if they are based only on gauge readings, and not on more definite symptoms. Road-test the vehicle, and use a thermometer

18.3 Stop solenoid wire secured by nut (arrowed)

18.5a Removing the stop solenoid plunger from the pump

18.5b Stop solenoid components

of known accuracy to measure the temperature of the coolant in the radiator or expansion tank when the gauge shows that overheating is taking place.

> ⚠ **Warning: Take care to avoid scalding when removing the coolant filler cap on a hot engine.**

2 Sometimes the thermostat is removed if it is suspected of being the cause of overheating. If the thermostat is of the bypass-blanking type, this will actually make matters worse, since removing the thermostat increases coolant flow through the bypass, at the expense of flow through the radiator. **Do not** run an engine without the thermostat fitted if it is of this type.

17 Oil contamination

1 Oil contamination falls into three categories: dirt, sludge and dilution.

2 **Dirt** or soot builds up in the oil in normal operation, and is not a problem if regular oil and filter changes are carried out. If it gets to the stage where it is causing low oil pressure, change the oil and filter immediately.

3 **Sludge** occurs when inferior grades of oil are used, or when regular oil changing has been neglected; it is more likely to occur on engines which rarely reach optimum operating temperature. If sludge is found when draining, a flushing oil may be used if the engine manufacturer allows it.

Caution: Some engine manufacturers – for example Renault – forbid the use of flushing oil, because it cannot all be drained afterwards.

The engine should then be refilled with fresh oil of the correct grade, and a new oil filter be fitted.

4 **Dilution** is of two kinds: fuel or water (coolant). In either case, if the dilution is bad enough, the engine oil level will appear to rise

with use. The routes by which fuel may get into the sump are explained in Section 10.

5 Coolant dilution of the oil is indicated by the 'mayonnaise' appearance of the oil-and-water mixture. Sometimes oil will also appear in the coolant. Possible reasons are:
a) Blown head gasket.
b) Cracked or porous cylinder head or block.
c) Cylinder liner seal failure (on wet-liner engines).
d) Leaking oil-to-coolant oil cooler (when fitted).

6 With either type of dilution, the cause must be dealt with, and the oil and filter changed.

18 Engine stop (fuel cut-off) solenoid – emergency repair

1 Most small diesel engines have a solenoid valve for cutting off the supply of fuel to the high-pressure side of the injection pump when the 'ignition' is switched off. If the solenoid fails electrically or mechanically so that its plunger is in the 'shut' position, the engine will not run. (One possible reason for such a failure is that the 'ignition' has been switched off while the engine speed is still high. In such a case, the plunger will be sucked onto its seat with considerable force, and may jam.)

2 Should the valve fail on the road and a spare not be immediately available, the following procedure will serve to get the engine running again.

Caution: It is important that no dirt is allowed to enter the injection pump, via the solenoid hole.

3 With the 'ignition' off, disconnect the wire from the solenoid, and thoroughly clean around the solenoid where it screws into the pump **(see illustration)**.

4 Unscrew the solenoid and remove it. If a hand-priming pump is fitted, operate the pump a few times while lifting out the

solenoid, to wash away any particles of dirt. Do not lose the sealing washer.

5 Remove the plunger from the solenoid (or from the recess in the pump, if it is stuck inside). Refit the solenoid body, making sure the sealing washer is in place, again operating the priming pump at the same time to flush away dirt **(see illustrations)**.

6 Tape up the end of the solenoid wire so that it cannot touch bare metal.

7 The engine will now start and run as usual, but it will not stop when the 'ignition' is switched off: it will be necessary to use the manual stop lever (if fitted) on the injection pump, or to stall the engine in gear.

8 Fit a new solenoid and sealing washer at the earliest opportunity.

19 Reading fault codes

Most modern diesel engines with electronic diesel engine control have a self-diagnostic system. If a fault occurs in any of the system sensors or actuators, this is recognised by the electronic control unit (ECU), which stores an appropriate fault code in its memory.

If a sensor or actuator is faulty, the ECU will usually substitute an 'emergency' value in place of the signal normally associated with the relevant sensor or actuator. This will enable the engine to carry on operating, albeit with reduced performance and efficiency. This situation is often referred to as 'limp home' mode.

If a fault code (or codes) is recorded, the ECU will usually illuminate an engine warning light on the instrument panel to inform the driver that there is a problem. The warning light will normally extinguish once the fault code has bee read.

To read fault codes, a suitable fault code reader will be required. A diagnostic connector is provided in the vehicle wiring loom, and a fault code reader can be connected via an appropriate adapter harness.

Chapter 10
Maintenance and repair data

Special notes concerning the use of this Chapter

The layout of this Chapter has been designed for the easiest possible useage, and is therefore generally self-explanatory. However, the following notes should serve to clarify some of the slightly less-obvious aspects of the data, and should be read before consulting the main part of this Chapter.

Valve clearances

These are specified with the engine COLD unless otherwise stated.

Brakes – minimum friction material thickness

Where this is not quoted by the manufacturers, it is suggested that the following is used for general guidance.

Disc brake pads should be renewed if the *friction material* thickness is less than 1.5 mm, or if it will reach this state before the next service is due.

Drum brake shoes should similarly be renewed if the friction material thickness is less than 2.0 mm for *bonded* shoe linings, or is within 1.0 mm of the rivet heads for *riveted* shoe linings.

Tyres

Note: *Pressures for compact/space saver spare tyres are not quoted.*

Tyre pressures are quoted COLD, with the vehicle UNLADEN, and are typical examples only. Space considerations prevent the inclusion of all possible vehicle type/tyre size/loading combinations. Always follow the manufacturer's recommendations for correct tyre pressures wherever possible. With regard to tyre sizes, these do not generally include the relevant speed/load rating. It is therefore important to refer to the manufacturer's or a tyre specialist's recommendations regarding this aspect, especially for light commercial vehicles.

Wheel alignment/suspension geometry

Figures are generally quoted with the vehicle UNLADEN, unless otherwise specified. Figures quoted are for both sides of the vehicle in total, unless stated otherwise. Certain vehicles, in particular those produced by Renault and Ford are set at specific vehicle ride heights, and figures quoted are therefore nominal values only.

Torque wrench settings

The correct assembly of certain cylinder heads, main bearings and wheel hubs often depends on special procedures being strictly adhered to. When using the torque wrench settings given for these items, it is recommended that reference is made to the appropriate manufacturer's information, or the relevant Haynes *Service and Repair Manual*, for further details of any such special procedures. It should be noted that new bolts or nuts should be used as specified by the manufacturer, especially where angular torque tightening methods are used. Note that in some cases, certain bolts fitted to the cylinder head are required to be tightened 'hand-tight'. This equates to a torque wrench setting of approximately 25 Nm. Take special note of any requirements regarding cylinder head bolt types, so that the appropriate torque wrench setting is used. Reference is made to *Torx, splined, hex (hexagon)* and *Allen* type bolts, and it is vital to use the correct setting according to type **(see illustration)**.

Illustrations

These are provided to clearly show the relevant cylinder head bolt/nut tightening sequence, and valve positions. The location of the engine flywheel is clearly visible to aid correct identification of the cylinder locations. The valve *head* positions are indicated, rather than the locations of the adjustment points (eg, rocker arm adjusters).

Information not available

Where the relevant technical data is not quoted by the vehicle manufacturer, this is indicated by a dash (-). A dash may also show that the particular item of data is not applicable to the model in question.

Abbreviations used in this Chapter

Most abbreviations are well-known, and will therefore be familiar to the user of this Chapter. The following list is provided to explain some of the less-familiar abbreviations, although it is stressed that this is a typical selection rather than a comprehensive list

ABS	=	Anti-lock braking system
A/C	=	Air conditioning
AT	=	Automatic transmission
DI	=	Direct injection
EDC	=	Electronic diesel control
FoG	=	Ford of Germany
IDI	=	Indirect injection
IRS	=	Independent rear suspension
L	=	Laden
MT	=	Manual transmission
MY	=	Model year
N/A	=	Not adjustable
O/D	=	Overdrive
OHB	=	Owner's handbook
PAS	=	Power-assisted steering
TWS	=	Torque wrench setting
U/L	=	Unladen
>	=	Greater than
<	=	Less than

Rover vehicles

The Section covering this range also incorporates data on vehicles originally manufactured under the AUSTIN ROVER or BRITISH LEYLAND marque names, including Austin, Morris and Rover.

Vauxhall/Opel vehicles

The section covering this range also incorporates data on vehicles originally manufactured under the BEDFORD marque name.

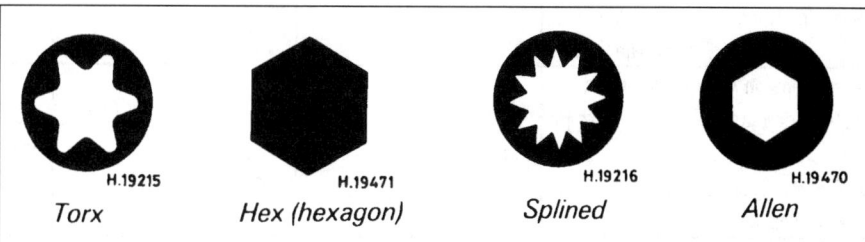

Torx   H.19215
Hex (hexagon)   H.19471
Splined   H.19216
Allen   H.19470

Cylinder head bolt types

ALFA ROMEO

	33 1.8 TD Intercooler 1990 to 1993	145 / 146 1.9 TD 1994 to 1998	75 2.0 TD 1985 to 1992	75 2.4 TD 1988 to 1992
Engine				
Engine type/code...............	VM96A OHV 62kW	AR67501 SOHC Turbo 66kW	VM80A OHV Turbo 70kW	VM81A OHV Turbo 81kW
Capacity (cm^3) / cylinders........................	1779 / 3	1929 / 4	1995 / 4	2393 / 4
Compression ration / pressure ....bar	22.2 / _	19.2 /	22.0 /	22.0 / ≥21.5
Torque output ....Nm	0	186	0	0
Oil pressure ....idle [running] bar	[4.0 to 5.5 @ 4000]	[3.5]	[3.4 to 5.9 @ 4000]	0.5 [3.4 to 5.9]
Oil temperature ....°C	80	80	80	80
Valve clearances - inlet (mm)	0.30	0.3 ± 0.05	0.30	0.30
- exhaust (mm)	0.30	0.35 ± 0.05	0.30	0.30
Injection order	1-3-2	1-3-4-2	1-3-4-2	1-3-4-2
No. 1 cylinder position	_	_	_	_
Cooling system				
Thermostat opening temperature ....°C	81 to 85	83		80
Radiator cap pressure ....bar	0.88 to 1.08	1.0 ± 0.1		1.1
Fuel system				
Idle speed ....rpm	875 to 925	900 ± 20	825 ± 25	800 to 850
Maximum (no load) speed ....rpm	4000	4870 to 5070	4800 to 4830	4800
Smoke test/opacity ....M^{-1} %	2.0	2.0	2.0	2.0
Static timing method................................	Plunger travel	Plunger travel	Plunger travel	Plunger travel
Timing dimension................................mm	1.00 to 1.03 [8]	0.8	0.97 to 0.97	0.78 to 0.80
Crankshaft position ....mm [°]	TDC	TDC	TDC	TDC
Turbo type / ref / pressure ....bar	_	IHI 1.1 to 1.2 bar	_	_
Injection pump make	Bosch	Bosch	Bosch	Bosch
Injection pump part no............................	VE3/10 FL353	VER 0460 494 304	VE 4/9	VE 4/10F 0460 404 042
Injector Make / type................................	Bosch	Bosch	Bosch	Bosch
Injector part no..	0434 250 125	0 432 217 211	0 432 297 047	0 432 297 047
Injection type...	Indirect VE	Indirect VER	Indirect	Indirect VE4/10
Injection opening pressure, New [used]...bar	150	150 to 158 [130 to 138]	155 [140]	150 [135]
Glow plugs				
Maker	Bosch/Champion	Beru/Bosch/Champion	Bosch/Champion	Bosch/Champion
Type	0250 201 012 / CH70	663 MJ / 0 250 201 005 / CH163	0 250 201 012 / CH70	0 250 201 012 / CH70
Nominal rating.................................V/A	11 /	11 / 14	11 /	11 / 13
Brakes				
minimum friction material thickness				
Front..mm	1.5	1.5	1.5	1.5
Rear...mm	1.0	1.5	1.5	1.5
Tyres - Saloon / Hatch.................Size	175/70x13	175/65x14: 185/60x14	_	185/70x14
- Estate / Van.................Size	175/70x13	_		_
Pressure - front / rear - Saloon / Hatch ...bar	2.2 / 1.8	2.3 / 2.1	_	2.1 / 2.1
- Estate / Van ......bar	2.2 / 1.8	_		_
Front suspension / wheel alignment				
Toe-in (+) / Toe-out (–)....................mm [°]	-2 ± 2	-2.5 ± 1.0	-2.0 ± 1.0	0.0 to -2.0
Camber	-1°55' ± 30'	-50' ± 20'	-0°30' ± 30'	-30' ± 30'
Castor	1° ± 30'	3°20' ± 30'	3°30' ± 30'	4° ± 30'
King pin inclination................................				
Rear suspension / wheel alignment				
Toe-in (+) / Toe-out (–)....................mm [°]	[0° ± 25']	4.0 ± 2.0	_	[0° ± 10']
Camber	0° ± 25'	-1° ± 15'		0° ± 30'

ALFA ROMEO

Torque wrench settings	33 1.8 TD Intercooler 1990 to 1993	45 / 146 1.9 TD 1994 to 1998	75 2.0 TD 1985 to 1992	75 2.4 TD 1988 to 1992
Cylinder head - stage 1 ....Nm	29 ± 5	100	29	29
- stage 2 ....Nm	+ 50°	+ 90°	+ 50°	+ 50°
- stage 3 ....Nm	+ 50°1	+ 90°	+ 50°1	+ 50°1
- stage 4 ....Nm	–	M8: 30	–	–
- stage 5 ....Nm	–	–	–	–
- stage 6 ....Nm		–		
Big-end bearings......................Nm	79.4 to 84.3 N	25 + 50° N	42, 85 N	40, 80 N
Main bearings......................Nm	44.1 to 49.0	55, 110	–	–
Crankshaft pulley bolt ....Nm	108	190	150	150
Camshaft pulley bolt ....Nm	–	110		–
Flywheel [driveplate] bolt......................Nm	107.8	130	108	108
Front hubs ....Nm	226 to 275	275	WSM	WSM
Rear hubs ....Nm	265 to 324	275	280	280
Wheel nuts / bolts ....Nm	80	100	98	98
Glow plugs ....Nm	22.5 to 24.5	15	–	–
Clutch pressure plate bolts ....Nm	20	22	–	–
Injection pump sprocket......................Nm	88.2	50	88	88
Injectors......................Nm	24.5 to 29.4	50	27	27
Injection pump mounting bolts ....Nm	29.4 to 31.3	25	–	–
Injector pipe unions......................Nm	14.7 to 19.6	30	20	20
Capacities				
Engine oil & filter......................litres	5.25	5.0	6.0	6.0
Gearbox......................litres	2.6	2.4	2.1	2.1
Automatic transmission ....litres	–		–	–
Final drive......................litres	WT	WT	WT	WT
Cooling system......................litres	8.8	8.9	–	10.0
Fuel tank......................litres	50			50

Notes

33 1.8 TD Intercooler 1990 to 1993
1Warm up, allow to cool, + 30° .
outside rows [b]: + 88 Nm
145 / 146 1.9 TD 1994 to 1998
75 2.0 TD 1985 to 1992
1Warm up, allow to cool, + 30°
outside row [b]: + 80 Nm
75 2.4 TD 1988 to 1992
1Warm up, allow to cool, + 30°
outside row [b]: + 80 Nm

– Not applicable, or information not available

1779 cm³

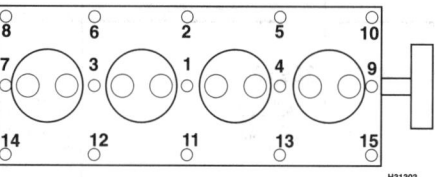

1929 cm³

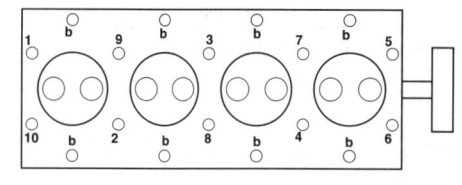

1995 cm³ / 2393 cm³

ALFA ROMEO

	155 2.0 TD 1993 to 1997	156 2.4 JTD 1998 to 2000	164 2.5 TD 1990 to 1994	
Engine				
Engine type/code........................	AR67501 SOHC Turbo 66kW	AR32501 SOHC 100kW	VM84A Turbo 86kW	
Capacity (cm³) / cylinders..................	1929 / 4	2387 / 5	2495 / 4	
Compression ration / pressure ....bar	19.2 /	18.45 /	22.0 /	
Torque output ....Nm	0	224	258	
Oil pressure ....idle [running] bar	[3.4 to 3.9]	0.2 to 0.4 [2.9 to 3.1]	0.98	
Oil temperature ....°C	80	80	90	
Valve clearances - inlet (mm)	0.30 ± 0.05	0.30 ± 0.05	0.30	
- exhaust (mm)	0.35 ± 0.05	0.35 ± 0.05	0.30	
Injection order	1-3-4-2	1-2-4-5-3	1-3-4-2	
No. 1 cylinder position	_	TCE	_	
Cooling system				
Thermostat opening temperature ....°C	_	85 to 89	81 to 85	
Radiator cap pressure ....bar	_	0.98 ± 0.1	0.6 to 0.8	
Fuel system				
Idle speed ....rpm	850 to 950	830 ± 30	875 to 925	
Maximum (no load) speed ....rpm	4950 to 5050	_	4200	
Smoke test/opacity ....M⁻¹ %	1.82	2.5	2.5	
Static timing method...............................	Plunger travel	Refer to wsm	Dial gauge	
Timing dimension..................................mm	_	_	1.0	
Crankshaft position ....mm [°]	_	_	TDC	
Turbo type / ref / pressure ....bar	0.8 bar @ 2400 rpm	Garrett VNT25	_	
Injection pump make	Bosch	Bosch	Bosch	
Injection pump part no...........................	VER 466 0460 494 304	_	VE4/10F2100L269	
Injector Make / type	Bosch	_	Bosch	
Injector part no................................	0 432 217 211	_	KB258 S4/4	
Injection type...............................	Indirect VER	EDC 15C Common rail	VE	
Injection opening pressure, New [used]...bar	_	1350	147 155 [127]	
Glow plugs				
Maker	Bosch/Champion	Champion	Bosch	
Type	0 250 201 019 / CH68	CH166	0250 201 012	
Nominal rating.............................V/A	11 / 13	12 /	11 / 13	
Brakes				
minimum friction material thickness				
Front.........................mm	1.5	_	1.0	
Rear...........................mm	1.5	_	1.0	
Tyres - Saloon / Hatch.......................Size	_	205/55x16	185/65x15: 195/60x15:195/65x15	
- Estate / Van.......................Size	_			
Pressure - front / rear - Saloon / Hatch ...bar	_	2.2 / 2.2	2.2 / 2.0	
- Estate / Van ......bar	_	_	_	
Front suspension / wheel alignment				
Toe-in (+) / Toe-out (–).....................mm [°]	0.0 + 1.0	2.0 ± 1.0	0 ± 1.0	
Camber	-0°20'	-0°42' ± 20'	-1°40' ± 20'	
Castor	3°30'	3°55' ± 30'	2°30' ± 20'	
King pin inclination................................	11°30'	_	_	
Rear suspension / wheel alignment				
Toe-in (+) / Toe-out (–).....................mm [°]	_	3.1 ± 1.0	3.0 ± 1.0	
Camber	_	-0°53' ± 20'	-15' ± 20'	

ALFA ROMEO

	155 2.0 TD 1993 to 1997	156 2.4 JTD 1998 to 2000	164 2.5 TD 1990 to 1994	
Torque wrench settings				
Cylinder head - stage 1 ....Nm	50	65 N	29	
- stage 2 ....Nm	+ 90°	+ 90°	+ 50°	
- stage 3 ....Nm	+ 90°[1]	+ 90°	+ 50°[1]	
- stage 4 ....Nm	–	+ 90°	–	
- stage 5 ....Nm	–	–	–	
- stage 6 ....Nm	–	–	–	
Big-end bearings......................Nm	25 + 50° N	25 + 60° N	79 to 84 N	
Main bearings......................Nm	55, 110	25 + 100° N	43 to 48	
Crankshaft pulley bolt ....Nm	190	306 to 378 Left-hand	176	
Camshaft pulley bolt ....Nm	–	102 to 126	–	
Flywheel [driveplate] bolt......................Nm	140	168	108	
Front hubs ....Nm	WSM	235	356 to 400	
Rear hubs ....Nm	320	284	250 to 300	
Wheel nuts / bolts ....Nm	85 Alloy: 95	88 to 108	95 to 105	
Glow plugs ....Nm	17	28	23	
Clutch pressure plate bolts ....Nm	35	21	19 to 25	
Injection pump sprocket......................Nm	50	–	88	
Injectors......................Nm	78	–	30	
Injection pump mounting bolts ....Nm	29	–	–	
Injector pipe unions......................Nm	23	–	20	
Capacities				
Engine oil & filter......................litres	5.0	5.0	6.7	
Gearbox......................litres	1.4	2.0	1.55	
Automatic transmission ....litres	–	–	–	
Final drive......................litres	WT	WT	WT	
Cooling system......................litres	–	6.8	10.0 Refill: 8.0	
Fuel tank......................litres	–	63	70	

Notes

155 2.0 TD 1993 to 1997
[1]Warm up, allow to cool, + 90° M8: 30, Warm up, allow to cool, + 30 Nm
156 2.4 JTD 1998 to 2000
164 2.5 TD 1990 to 1994
[1]run for 20 mins, wait 2.5 hrs, + 30° outside rows: + 88 Nm

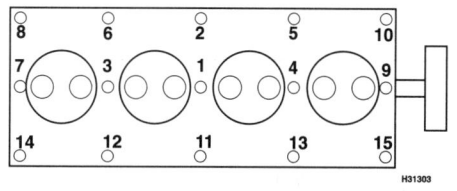

1929 cm³

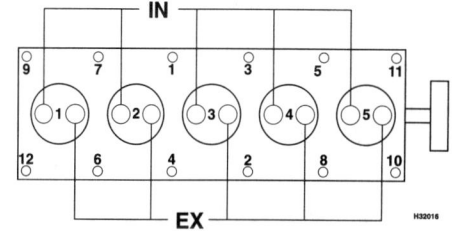

2387 cm³

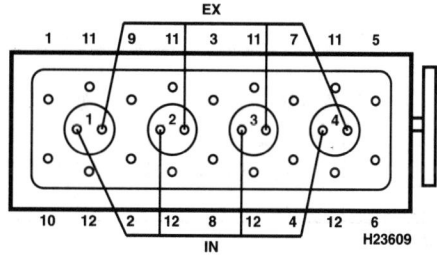

2495 cm³

– Not applicable, or information not available

AUDI

	A3 1.9 TDi 1996 to 1998	A3 1.9 TDi 110 1998 to 1999	80, 90 1.6 Turbo 1986 to 1991	80 1.9 1989 to 1991
Engine				
Engine type/code........................	AGR SOHC Turbo 66kW	AHF SOHC Turbo 81kW	SB SOHC Turbo 59kW	1Y SOHC 50kW
Capacity (cm³) / cylinders......................	1896 / 4	1896 / 4	1588 / 4	1896 / 4
Compression ration / pressure ....bar	19.5 / 19 to 31	19.5 / 25 to 31	23.0 / ≥26.0	23.0 / ≥26.0
Torque output ....Nm	210	235	0	0
Oil pressure ....idle [running] bar	[2.0 @ 2000]	[2.0 @ 2000]	[2.0 @ 2000]	[2.0 @ 2000]
Oil temperature ....°C	80	80	80	80
Valve clearances - inlet (mm)	0: Hyd.	0: Hyd.	0: Hyd.	0: Hyd.
- exhaust (mm)	0: Hyd.	0: Hyd.	0: Hyd.	0: Hyd.
Injection order	1-3-4-2	1-3-4-2	1-3-4-2	1-3-4-2
No. 1 cylinder position	TBE	TBE	TBE	TBE
Cooling system				
Thermostat opening temperature ....°C	87	87	85	85
Radiator cap pressure ....bar	1.2 to 1.5	1.2	1.2 to 1.5	1.2 to 1.5
Fuel system				
Idle speed ....rpm	860 to 940	860 to 940	900 ± 30	900 ± 30
Maximum (no load) speed ....rpm	4800 to 5200	4800 to 5200	5100 ± 50	5050 ± 100
Smoke test/opacity ....M⁻¹ %	2.0	2.0	2.0	2.0
Static timing method...........................	Refer to wsm	Refer to wsm	Plunger travel	Plunger travel
Timing dimension........................mm	_	_	0.9 ± 0.02	1.0 ± 0.02
Crankshaft position ....mm [°]	_	_	TDC	[0] TDC
Turbo type / ref / pressure ....bar	1.5 to 1.7 bar @ 3000 rpm	_	0.63 to 0.83 bar	_
Injection pump make	Bosch	Bosch	Bosch	Bosch
Injection pump part no.......................	_	VE	VE 4/9	
Injector Make / type	Bosch	_	Bosch	Bosch
Injector part no.......................	_	_	0 432 217 165	_
Injection type...........................	EDC	Direct VE	VE	VE
Injection opening pressure, New [used]...bar	190 to 200 [170]	190 to 200 [170]	≥155 [≥140]	≥130 [≥120]
Glow plugs				
Maker	Bosch	Champion	Bosch/Champion	Champion
Type	0 250 202 022	CH171	0250 201 032 / CH69	CH69
Nominal rating.......................V/A	11.5 /	_	11 / 13	11 / 13
Brakes				
minimum friction material thickness				
Front...........................mm	7.0 with backing	7.0 with backing	2.0	2.0
Rear............................mm	7.0 with backing	7.0 with backing	2.5	2.5
Tyres - Saloon / Hatch.....................Size	195/65x15	195/65x15	195/70x14	175/70x14
- Estate / Van.....................Size	_	_	_	_
Pressure - front / rear - Saloon / Hatch ...bar	1.8 / 1.8	Refer to vehicle	1.9 / 1.9	1.9 / 1.9
- Estate / Van ......bar	_	_	_	_
Front suspension / wheel alignment				
Toe-in (+) / Toe-out (–)......................mm [°]	[0° ± 10']	[0° ± 10']	[10' ± 10']	[10' ± 10']
Camber	-0°30' ± 30' Sport: -33'±30'	-30' ± 30' Sport: -33' ± 30'	-45' ± 30'	-45' ± 30'
Castor	7°40' ± 30' Sport: 7°0' ± 30'	7°40' ± 30' Sport: 7°0' ± 30'	1°15' ± 30'	1°15' ± 30'
King pin inclination...........................				
Rear suspension / wheel alignment				
Toe-in (+) / Toe-out (–)......................mm [°]	[0°25' ± 15']	[0°25' ± 15']	[20' ± 20']	[20' ± 20']
Camber	-1°36' ± 20'	-1°36' ± 20'	-1° ± 20'	-1° ± 20'

AUDI

Torque wrench settings	A3 1.9 TDi 1996 to 1998	A3 1.9 TDi 110 1998 to 1999	80, 90 1.6 Turbo 1986 to 1991	80 1.9 1989 to 1991
Cylinder head - stage 1 ....Nm	40 N	40 N	40	40
- stage 2 ....Nm	+ 90°	+ 90°	60	60
- stage 3 ....Nm	+ 90°	+ 90°	+ 180°	+ 180°
- stage 4 ....Nm	–	–	Warm engine, + 90°	Warm engine
- stage 5 ....Nm	–	–	After 1000km,	+ 90°
- stage 6 ....Nm	–	–	+ 90°	–
Big-end bearings........................Nm	30 + 90° oiled N	30 + 90° oiled N	30 + 90° N	30 + 90° N
Main bearings............................Nm	40 + 90° N	40 + 90° N	65	65
Crankshaft pulley bolt ....Nm	90 + 90°	90 + 90°	Hexagon head: 180¹	90 + 180°
Camshaft pulley bolt ....Nm	100	100	45	45
Flywheel [driveplate] bolt...........Nm	60 + 90° N	60 + 90° N	30 + 90°	30 + 90°
Front hubs ....Nm	¹	¹	265 LkC²	265 LkC¹
Rear hubs ....Nm	175	175	WSM	WSM
Wheel nuts / bolts ....Nm	120	120	110	110
Glow plugs ....Nm	15	15	30	30
Clutch pressure plate bolts ....Nm	25	13	25	25
Injection pump sprocket.............Nm	20 + 90° N	20 + 90° N	45	45
Injectors....................................Nm	20	20	70	70
Injection pump mounting bolts ....Nm	25	25	25	25
Injector pipe unions...................Nm	15	25	25	25
Capacities				
Engine oil & filter.......................litres	4.5	4.5	3.5 3.88 ▶: 4.0	4.5
Gearbox....................................litres	1.9	1.9	2.4	2.4
Automatic transmission ....litres	5.3	5.3	–	–
Final drive ....litres	WT AT: 0.75	WT AT: 0.75	WT	WT
Cooling system..........................litres	5.0	5.0	7.0	7.0
Fuel tank...................................litres	–	55	68	68

Notes

A3 1.9 TDi 1996 to 1998
¹300, -360°, 50 + 30°
A3 1.9 TDi 110 1998 to 1999
¹300, -360°, 50 + 30°
80, 90 1.6 Turbo 1986 to 1991
¹Double hexagon head: 90 + 180°
²PAS with M14 bolts: 120 + 90°.
PAS with M16: 200 + 90°
80 1.9 1989 to 1991
¹PAS with M14 bolts: 120 + 90°.
PAS with M16: 200 + 90°

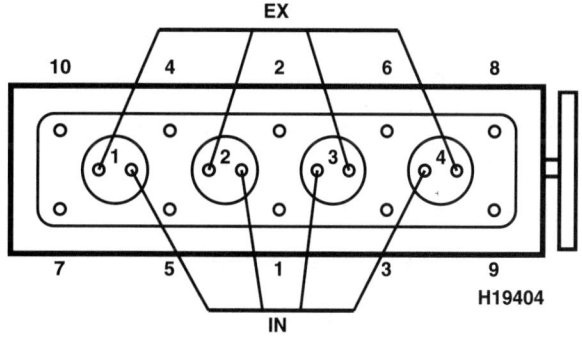

1588 cm³ / 1896 cm³

– Not applicable, or information not available

AUDI

	80 1.9 Turbo CAT 1992 to 1994	80 1.9 TDi 1992 to 1996	A4 1.9 TDi CAT 1995 to 1996	A4 1.9 TDi CAT 1994 to 1999
Engine				
Engine type/code	AAZ SOHC Turbo 55kW	1Z SOHC 8V Turbo 66kW	1Z SOHC Turbo 66kW	AFN SOHC 8V Turbo 81kW
Capacity (cm^3) / cylinders........................	1896 / 4	1896 / 4	1896 / 4	1896 / 4
Compression ration / pressure ....bar	22.5 / ≥26.0	19.5 / ≥26.0	19.5 / ≥19.0	19.5 / ≥19.0
Torque output ....Nm	140	202	202	225
Oil pressure ....idle [running] bar	[2.0 @ 2000]	[2.0 @ 2000]	1.0 to 2.5 [3.0 to 5.0]	1.0 to 2.5 [3.0 to 5.0]
Oil temperature ....°C	80	80	80	80
Valve clearances - inlet (mm)	0: Hyd.	0: Hyd.	0: Hyd.	0: Hyd.
- exhaust (mm)	0: Hyd.	0: Hyd.	0: Hyd.	0: Hyd.
Injection order......................................	1-3-4-2	1-3-4-2	1-3-4-2	1-3-4-2
No. 1 cylinder position	TBE	TBE	TBE	TBE
Cooling system				
Thermostat opening temperature ....°C	87	87	87	87
Radiator cap pressure ....bar	1.2 to 1.5	1.2 to 1.5	1.2 to 1.5	1.2 to 1.5
Fuel system				
Idle speed ....rpm	900 ± 30	900 ± 30	840 ± 60	780 to 900
Maximum (no load) speed ....rpm	5200 ± 100	5200 ± 100	5200 ± 100	5200 ± 100
Smoke test/opacity ....M^{-1} %	2.0	2.0	2.0	2.0
Static timing method.............................	Plunger travel	Plunger travel	Plunger travel	Plunger travel
Timing dimension...........................mm	0.9 ± 0.02	0.9 ± 0.02	0.7 ± 0.02	0.7 ± 0.02
Crankshaft position ....mm [°]	[0] TDC	[0] TDC	[0] TDC	[0] TDC
Turbo type / ref / pressure ....bar	1.6 to 1.8 bar	1.5 to 1.7 bar @ 3000 rpm	1.5 to 1.7 bar @ 3000 rpm	1.8 to 2.05 bar @ 3000
Injection pump make	Bosch	Bosch	Bosch	Bosch
Injection pump part no.............................	VE 4/9	VE4/10	VE R510	VE R638
Injector Make / type	Bosch	Bosch	Bosch	Bosch
Injector part no......................................	–	0 432 193 850	0 432 193 799	–
Injection type..	VE	VE	Direct VE	Direct VE
Injection opening pressure, New [used]...bar	≥155 [≥140]	≥155 [≥140]	190 to 200 [170]	190 to 200 [170]
Glow plugs				
Maker	Bosch	Bosch	Bosch	Bosch
Type	0 250 201 032	0 250 202 009	0 250 202 009	0 250 202 022
Nominal rating....................................V/A	11 / 12	11 / 12	11 / 12	11 / 12
Brakes				
minimum friction material thickness				
Front...mm	7.0 with backing	7.0 with backing	7.0 with backing	7.0 with backing
Rear...mm	2.5¹	2.5¹	7.0 with backing	7.0 with backing
Tyres - Saloon / Hatch.....................Size	195/65x15: 205/60x15	195/65x15: 205/60x15	195/65x15: 205/60x16:205/55x16	195/65x15: 205/60x16:205/55x16
- Estate / Van...........................Size	–	–	–	–
Pressure - front / rear - Saloon / Hatch ...bar	1.8 / 1.8	1.8 / 1.8	2.0 / 2.0 205/55: 2.2 / 2.2	2.0 / 2.0. 205/55: 2.2 / 2.2
- Estate / Van ......bar	–	–	–	–
Front suspension / wheel alignment				
Toe-in (+) / Toe-out (–).....................mm [°]	[0°10' ± 10']	[0°10' ± 10']	[0°10' ± 5']	[0°10' ± 5']
Camber	-0°45' ± 30'	-0°45' ± 30'	-0°20' ± 20'	-0°20' ± 20'
Castor	2°10' ± 30'	2°10' ± 30'	–	–
King pin inclination.................................	–	–	–	–
Rear suspension / wheel alignment				
Toe-in (+) / Toe-out (–)....................mm [°]	[0°15' ± 10']	[0°15' ± 10'] PW	[0°10' ± 5']	[0°10' ± 5']
Camber	-1°30' ± 20'	-1°30' ± 20'	-1°30' ± 20'	-1°30' ± 20'

AUDI

	80 1.9 Turbo CAT 1992 to 1994	80 1.9 TDi 1992 to 1996	A4 1.9 TDi CAT 1995 to 1996	A4 1.9 TDi CAT 1994 to 1999
Torque wrench settings				
Cylinder head - stage 1 ....Nm	40 N	40 N	40 N	40 N
- stage 2 ....Nm	60	60	60	60
- stage 3 ....Nm	+ 180°	+ 180°	+ 90°	+ 90°
- stage 4 ....Nm	Warm engine	Warm engine	+ 90°	+ 90°
- stage 5 ....Nm	+ 90°	+ 90°	–	–
- stage 6 ....Nm	–	–	–	–
Big-end bearings................................Nm	30 + 90° N	30 + 90° N	30 + 90° N	30 + 90° N
Main bearings.....................................Nm	65	65	65 + 90° N	65 + 90° N
Crankshaft pulley bolt ....Nm	90 + 90°	90 + 90°	90 + 90°	90 + 90°
Camshaft pulley bolt ....Nm	45	45	45	45
Flywheel [driveplate] bolt...................Nm	30 + 90°	30 + 90°	60 + 90° N	60 + 90° N
Front hubs ....Nm	M14: 120 + 90°²	M14: 120 + 90°²	115 + 90° N M16: 190 + 90° N	120 + 90° N M16: 200 + 90° N
Rear hubs ....Nm	WSM	WSM	WSM	WSM
Wheel nuts / bolts ....Nm	110	110	120	120
Glow plugs ....Nm	30	30	15	15
Clutch pressure plate bolts ....Nm	25	25	25	25
Injection pump sprocket.....................Nm	45	45	45	45
Injectors..Nm	70	70	20	20
Injection pump mounting bolts ....Nm	25	25	25	25
Injector pipe unions...........................Nm	25	25	25	25
Capacities				
Engine oil & filter................................litres	4.5	4.5	3.5	3.5
Gearbox...litres	2.35	2.35	2.25	2.25
Automatic transmission ....litres	–	3.5	–	–
Final drive ....litres	WT	WT AT: 0.75	WT	WT
Cooling system...................................litres	6.5	6.5	7.5	7.5
Fuel tank..litres	66	66	–	62

Notes

80 1.9 Turbo CAT 1992 to 1994
¹Discs: 7.0 with backing
²M16: 200 + 90°
80 1.9 TDi 1992 to 1996
¹Discs: 7.0 with backing
²M16: 200 + 90°

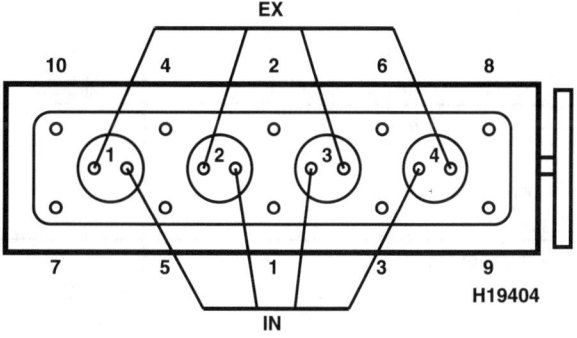

1588 cm³ / 1896 cm³

– Not applicable, or information not
  available

AUDI

	A4 2.5 TDi 1998 to 2000	A4 2.5 TDi 1998 to 2000	100 & A6 1.9 TDi 1995 to 1997	100 & Avant 2.0 D 1983 to 1991
Engine				
Engine type/code...............................	AFB DOHC 24V V6 110kW	AKN DOHC 24V V6 110kW	1Z / AHU SOHC Turbo 66kW	CN SOHC 51kW
Capacity (cm^3) / cylinders......................	2497 / 6	2497 / 6	1896 / 4	1986 / 5
Compression ration / pressure ....bar	19.5 / ≥24.0	19.5 / ≥24.0	19.5 / ≥19	23.0 / ≥26.0
Torque output ....Nm	310	310	202	123
Oil pressure ....idle [running] bar	0.8 [2.0 @ 2000]	0.8 [2.0 @ 2000]	[2.0 @ 2000]	[2.0 @ 2000]
Oil temperature ....°C	80	80	80	80
Valve clearances - inlet (mm)	0: Hyd.	0: Hyd.	0: Hyd.	0.15 to 0.25
- exhaust (mm)	0: Hyd.	0: Hyd.	0: Hyd.	0.35 to 0.45
Injection order	1-4-3-6-2-5	1-4-3-6-2-5	1-3-4-2	1-2-4-5-3
No. 1 cylinder position	–	–	TBE	TBE
Cooling system				
Thermostat opening temperature ....°C	87	87	87	87
Radiator cap pressure ....bar	1.4 to 1.6	1.4 to 1.6	1.2 to 1.5	1.2 to 1.5
Fuel system				
Idle speed ....rpm	680 to 860	680 to 860	800 to 840 N/A	750 ± 50
Maximum (no load) speed ....rpm	4500 to 5500	4500 to 5500	5000 to 5200	5400 ± 50
Smoke test/opacity ....M^{-1} %	2.5	2.5	2.0	2.0
Static timing method................................	Refer to wsm	Refer to wsm	Plunger travel	Plunger travel
Timing dimension..............................mm	–	–	0.7 ± 0.02	0.85 ± 0.02
Crankshaft position ....mm [°]	–	–	[0] TDC	[0] TDC
Turbo type / ref / pressure ....bar	–	–	1.8 to 2.05 bar @ 3000 rpm	–
Injection pump make	Bosch	Bosch	Bosch	Bosch
Injection pump part no.............................	VP44		VE R440-1	VE L 35/137
Injector Make / type	–	–	Bosch	Bosch
Injector part no..			0 432 193 838	0 432 217 059
Injection type...	Direct VP	Direct	Direct VE EDC	Indirect VE
Injection opening pressure, New [used]...bar	180 to 190 [160]	180 to 190 [160]	190 to 200 [170]	≥130 [≥120]
Glow plugs				
Maker	Beru/Champion	Champion	Bosch/Champion	Beru/Champion
Type	GN855 / CH171	CH171	0 250 202 009 / CH171	N19 1001 / CH69
Nominal rating.................................V/A	11.5 /	11 /	11 / 12	11 / 13
Brakes				
minimum friction material thickness				
Front..mm	7.0 with backing	7.0 with backing	2.0	7.0 with backing
Rear...mm	7.0 with backing	7.0 with backing	2.5 Disc: 2.0	2.5
Tyres - Saloon / Hatch......................Size	–	–	195/65x15:205/60x15:215/60x15	185/70x14
- Estate / Van............................Size			–	185/70x14
Pressure - front / rear - Saloon / Hatch ...bar	–		2.3 / 2.3'	1.7 / 1.7
- Estate / Van ......bar	–		–	1.7 / 1.7
Front suspension / wheel alignment				
Toe-in (+) / Toe-out (–)....................mm [°]	[0°20' ± 5']	[0°20' ± 15']	[15' +5' -10']	[0 +5' -10']
Camber	-0°25' ± 25'¹	-0°25' ± 25'¹	-40' +35' -25'	-30' ± 30'
Castor	–	–	1°10' ± 40'	-15' ± 40'¹ PAS: 50' ± 40'
King pin inclination...................................	–	–	–	–
Rear suspension / wheel alignment				
Toe-in (+) / Toe-out (–)....................mm [°]	[0°20' +15' -10']²	[0°20' +15' -10']²	[10' ± 5'] PW	[15' ± 10']²
Camber	-1°30' ± 20'¹³	-1°30' ± 20'¹³	-50' ± 30'	-40' ± 30'¹³

AUDI

Torque wrench settings	A4 2.5 TDi 1998 to 2000	A4 2.5 TDi 1998 to 2000	100 & A6 1.9 TDi 1995 to 1997	100 & Avant 2.0 D 1983 to 1991
Cylinder head - stage 1 ....Nm	35 N	35 N	40 N	40 N
- stage 2 ....Nm	60	60	60	60
- stage 3 ....Nm	+ 90°	+ 90°	+ 90°	+ 180°
- stage 4 ....Nm	+ 90°	+ 90°	+ 90°	Warm up, + 90°
- stage 5 ....Nm	–	–	–	After 1000km,
- stage 6 ....Nm	–	–	–	+ 90°
Big-end bearings........................Nm	30 + 90° N	30 + 90° N	30 + 90° N	30 + 180° N
Main bearings........................Nm	WSM	WSM	65 + 90° N	65
Crankshaft pulley bolt ....Nm	200 + 180° N	200 + 180° N	90 + 90° N	350
Camshaft pulley bolt ....Nm	75	75	45	45 Pump end: 100
Flywheel [driveplate] bolt........................Nm	60 + 180° N [60 + 90° N]	60 + 180° N [60 + 90° N]	60 + 90° N [60 + 90° N]	30 + 90°
Front hubs ....Nm	115 + 180° N M16: 190 + 180°N	115 + 180° N M16: 190 + 180° N	200 + 90°	280
Rear hubs ....Nm	4x4: 115 + 90° N	4x4: 115 + 90° N	WSM	WSM
Wheel nuts / bolts ....Nm	120	–	110	110
Glow plugs ....Nm	15	15	15	30
Clutch pressure plate bolts ....Nm	25	25	25	25
Injection pump sprocket........................Nm	–	–	45	45
Injectors........................Nm	10	–	20	70
Injection pump mounting bolts ....Nm	20	–	25	25
Injector pipe unions........................Nm	30	30	25	25
Capacities				
Engine oil & filter........................litres	5.4	5.4	3.5	5.0
Gearbox........................litres	2.4	2.4	2.4	2.6
Automatic transmission ....litres	–	–	3.0	3.0
Final drive ....litres	WT	WT	AT: 1.0	AT: 1.0
Cooling system........................litres	10.0	10.0	6.5	8.0
Fuel tank........................litres	62	62	80	80

Notes

A4 2.5 TDi 1998 to 2000
[1] Sports: -0°40' ± 25'
[2] Sports: [0°28' +15' -10'] 4x4: [0°15' ± 5']
[3] 4x4: -0°40' ± 30'
A4 2.5 TDi 1998 to 2000
[1] Sports: -0°40' ± 25'
[2] Sports: [0°28' +15' -10'] 4x4: [0°15' ± 5']
[3] 4x4: -0°40' ± 30'
100 & A6 1.9 TDi 1995 to 1997
[2] 215/60x15: 2.1 / 2.1
100 & Avant 2.0 D 1983 to 1991
[1] From chassis no. 44GA 024 419: 50' ± 40'
[2] From chassis no. EA 085 288 (Discs: EA 082 449):[10' ±50'
[3] From chassis no. 44 JN 201 031: -50' ± 30'

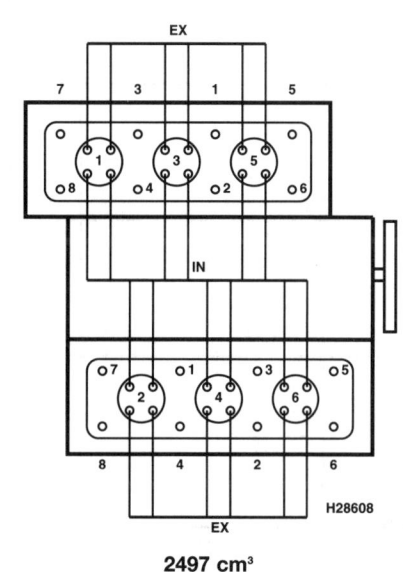

2497 cm³

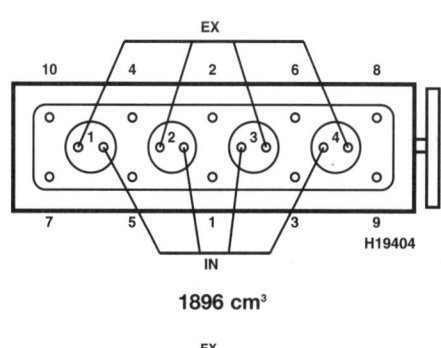

1896 cm³

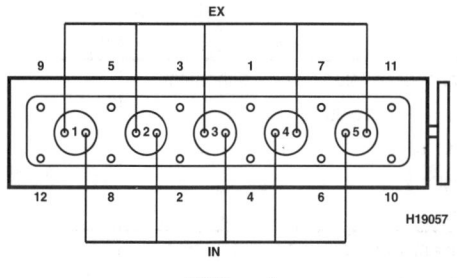

1986 cm³

– Not applicable, or information not
available

AUDI

Haynes THE BOOK ®

	100 & Avant 2.0 TD 1988 to 1991	100 & Avant 2.4 D 1989 to 1991	100 & Avant 2.5 TDi 1990 to 1991	100 & A6 2.5 TDi 1991 to 1997
Engine				
Engine type/code....................	NC SOHC Turbo 74kW	3D SOHC 60kW	1T SOHC Turbo 88kW	ABP SOHC Turbo 85kW
Capacity (cm^3) / cylinders......................	1986 / 5	2370 / 5	2461 / 5	2461 / 5
Compression ration / pressure ....bar	23.0 / ≥26.0	22.5 / ≥26.0	20.5 / ≥24.0	20.5 / ≥24.0
Torque output ....Nm	192	164	0	0
Oil pressure ....idle [running] bar	[2.0 @ 2000]	[2.0 @ 2000]	[2.0 @ 2000]	[2.0 @ 2000]
Oil temperature ....°C	80	80	80	80
Valve clearances - inlet (mm)	0: Hyd.	0: Hyd.	0: Hyd.	0: Hyd.
- exhaust (mm)	0: Hyd.	0: Hyd.	0: Hyd.	0: Hyd.
Injection order	1-2-4-5-3	1-2-4-5-3	1-2-4-5-3	1-2-4-5-3
No. 1 cylinder position	TBE	TBE	TBE	TBE
Cooling system				
Thermostat opening temperature ....°C	87	87	87	87
Radiator cap pressure ....bar	1.2 to 1.5	1.2 to 1.5	1.2 to 1.5	1.2 to 1.5
Fuel system				
Idle speed ....rpm	830 ± 30	830 ± 30	800 to 840	800 to 840 N/A
Maximum (no load) speed ....rpm	5100 ± 50	5000 ± 50	5200 ± 50	4500 to 5500
Smoke test/opacity ....M^{-1} %	2.0	2.0	2.5	2.5
Static timing method............................	Plunger travel	Plunger travel	Plunger travel	Plunger travel
Timing dimension.............................mm	0.9 ± 0.02	1.0 ± 0.02	0.35 ± 0.02	0.35 ± 0.02
Crankshaft position ....mm [°]	[0] TDC	[0] TDC	[0] TDC	[0] TDC
Turbo type / ref / pressure ....bar	1.6 to 1.7 bar	–	0.8 to 0.9 bar	1.8 to 1.9 bar
Injection pump make	Bosch	Bosch	Bosch	Bosch
Injection pump part no.............................	VE 5.10	–	VE5 11E	VE 5/11
Injector Make / type	Bosch	Bosch	Bosch	Bosch
Injector part no............................	0 432 217 165	0 432 217 206	0 432 193 838	0 432 193 877
Injection type............................	Indirect VE	Indirect VE	Direct EDC	Direct VE EDC
Injection opening pressure, New [used]...bar	≥155 [≥140]	≥130 [≥120]	≥180 [≥160]	≥180 [≥160]
Glow plugs				
Maker	Champion	Champion	Bosch/Champion	Champion
Type	CH69	CH69	0 250 202 009 / CH69	HDS342
Nominal rating.................................V/A	10 / 12	11 / 13	11 / 13	11 / 12
Brakes				
minimum friction material thickness				
Front...mm	7.0 with backing	7.0 with backing	7.0 with backing	2.0
Rear...mm	2.5	2.5	2.5	2.5 Disc: 2.0
Tyres - Saloon / Hatch......................Size	185/70x14	185/70x14	185/70x14	195/65x15:205/60x15:215/60x15
- Estate / Van......................Size	185/70x14	185/70x14	185/70x14	–
Pressure - front / rear - Saloon / Hatch ...bar	1.7 / 1.7	1.7 / 1.7	1.7 / 1.7	2.3 / 2.3¹
- Estate / Van .....bar	1.7 / 1.7	1.7 / 1.7	1.7 / 1.7	–
Front suspension / wheel alignment				
Toe-in (+) / Toe-out (–)....................mm [°]	[0 +5' -10']	[0° +5' -10']	[0° +5' -10']	[15' +5' -10']
Camber	-30' ± 30'	-30' ± 30'	-30' ± 30'	-40' +35' -25'
Castor	-15' ± 40'¹¹ PAS: 50' ± 40'	50' ± 40'	50' ± 40'	1°10' ± 40'
King pin inclination........................	–	–	–	–
Rear suspension / wheel alignment				
Toe-in (+) / Toe-out (–)....................mm [°]	[15' ± 10']²	[10' ± 5']	[10' ± 5']	[10' ± 5'] PW
Camber	-40' ± 30'³	-50' ± 30'	-50' ± 30'	-50' ± 30'

AUDI

Torque wrench settings	100 & Avant 2.0 TD 1988 to 1991	100 & Avant 2.4 D 1989 to 1991	100 & Avant 2.5 TDi 1990 to 1991	100 & A6 2.5 TDi 1991 to 1997
Cylinder head - stage 1 ....Nm	40 N	40 N	40 N	40 N
- stage 2 ....Nm	60	60	60	60
- stage 3 ....Nm	+ 180°	+ 180°	+ 180°	+ 180°
- stage 4 ....Nm	Warm up, + 90°	Warm up, + 90°	Warm up, + 90°	Warm up, + 90°
- stage 5 ....Nm	After 1000km,	After 1000km,	After 1000km,	After 1000km,
- stage 6 ....Nm	+ 90°	+ 90°	+ 90°	+ 90°
Big-end bearings.................................Nm	30 + 180° N	30 + 90° N	30 + 90° N	30 + 90° N
Main bearings.....................................Nm	65	65	65	30 + 90°
Crankshaft pulley bolt ....Nm	350	350	100 + 180°	160 + 180°
Camshaft pulley bolt ....Nm	45 Pump end: 100	45 Pump end: 100	120 Pump end: 160	30 + 90°
Flywheel [driveplate] bolt...................Nm	30 + 90°	30 + 90°	30 + 90°	30 + 90°
Front hubs ....Nm	280	280	280	200 + 90°
Rear hubs ....Nm	WSM	WSM	WSM	WSM
Wheel nuts / bolts ....Nm	110	110	110	110
Glow plugs ....Nm	30	30	15	15
Clutch pressure plate bolts ....Nm	25	25	25	25
Injection pump sprocket.....................Nm	45	45	45	160
Injectors..Nm	70	70	20	20
Injection pump mounting bolts ....Nm	25	25	20	25
Injector pipe unions...........................Nm	25	25	25	25
Capacities				
Engine oil & filter................................litres	5.0	5.0	4.5	5.0
Gearbox...litres	2.6	2.6	2.6	2.4
Automatic transmission ....litres	3.0	3.0	3.0	3.0
Final drive ....litres	AT: 1.0	AT: 1.0	AT: 1.0	AT: 1.0
Cooling system..................................litres	9.4	8.0	8.5	6.5
Fuel tank..litres	80	80	80	80

Notes

100 & Avant 2.0 TD 1988 to 1991
[1]From chassis no. 44 GA 024 419: 50' ± 40'
[2]From chassis no. EA 085 288(Discs: EA 082 449):
[10' ± 5'
[3]From chassis no. 44 JN 201 031: -50' ± 30'
100 & A6 2.5 TDi 1991 to 1997
[2]215/60x15: 2.1 / 2.1

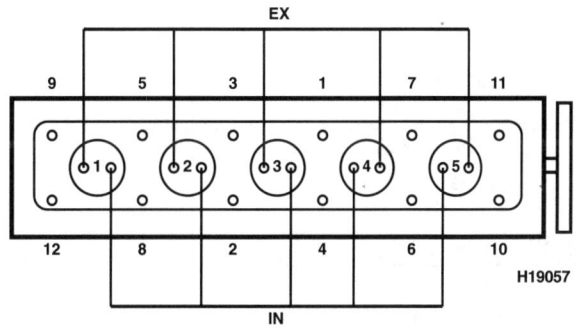

1986 cm³ / 2370 cm³ / 2461 cm³

– Not applicable, or information not available

AUDI

	100 & A6 2.5 TDi 1991 to 1997	A6 1.9 TDi 1994 to 1997	A6 1.9 TDi 1997 to 2000	A6 2.5 TDi 1994 to 1997
Engine				
Engine type/code	AAT SOHC Turbo 85kW	1Z SOHC Turbo 66kW	AFN SOHC Turbo 81kW	AEL SOHC Turbo 103kW
Capacity (cm^3) / cylinders........................	2461 / 5	1896 / 4	1896 / 4	2461 / 5
Compression ration / pressure ....bar	20.5 / ≥24.0	19.5 /	19.5 / _	20.5 / ≥24.0
Torque output ....Nm	0	202	230	290
Oil pressure ....idle [running] bar	[2.0 @ 2000]	[2.0 @ 2000]	[2.0 @ 2000]	[2.0 @ 2000]
Oil temperature ....°C	80	80	80	80
Valve clearances - inlet (mm)	0: Hyd.	0: Hyd.	0: Hyd.	0: Hyd.
- exhaust (mm)	0: Hyd.	0: Hyd.	0: Hyd.	0: Hyd.
Injection order	1-2-4-5-3	1-3-4-2	1-3-4-2	1-2-4-5-3
No. 1 cylinder position	TBE	TBE	TBE	TBE
Cooling system				
Thermostat opening temperature ....°C	87	87	87	87
Radiator cap pressure ....bar	1.2 to 1.5	1.2 to 1.5	1.2 to 1.5	1.2 to 1.5
Fuel system				
Idle speed ....rpm	750 to 810	800 to 900	820 to 900	780 ± 30
Maximum (no load) speed ....rpm	4500 to 5500	5000 to 5200	4800 to 5200	4500 to 4800
Smoke test/opacity ....M^{-1} %	2.5	2.5	2.0	2.0
Static timing method...............................	Plunger travel	Plunger travel	Plunger travel	Plunger travel
Timing dimension.................................mm	0.35 ± 0.02	0.7 ± 0.02	0.7 ± 0.02	0.35 ± 0.02
Crankshaft position ....mm [°]	[0] TDC	TDC [0]	TDC [0]	[0] TDC
Turbo type / ref / pressure ....bar	1.8 to 1.9 bar	_	1.5 to 1.7 bar @ 3000 rpm	_
Injection pump make	Bosch	_	Bosch	Bosch
Injection pump part no............................	VEL 400/460-1	_	VP 37	_
Injector Make / type	Bosch	_	_	Bosch
Injector part no.......................................	0 432 193 858/6	_	_	0 432 193 786
Injection type...	Direct VE EDC	Direct	Direct VP	Direct VE EDC
Injection opening pressure, New [used]...bar	≥180 [≥160]	190 to 200 [170]	190 to 200 [170]	≥180 [≥160]
Glow plugs				
Maker	Champion	Champion	Bosch/Champion	Bosch
Type	HDS342	HDS342	0 250 202 022 / HDS342	0 250 202 009
Nominal rating....................................V/A	11 / 12	_	11 / 12	11 / 12
Brakes				
minimum friction material thickness				
Front........................mm	2.0	2.0	2.0	2.0
Rear..........................mm	2.5 Disc: 2.0	2.5 Disc: 2.0	2.5 Disc: 2.0	2.5 Disc: 2.0
Tyres - Saloon / Hatch......................Size	195/65x15: 205/60x15[1]	195/65x15: 205/60x15	195/65x15: 205/60x15	195/65x15: 205/60x15[1]
- Estate / Van............................Size	_	195/65x15: 205/60x15	195/65x15: 205/60x15	_
Pressure - front / rear - Saloon / Hatch...bar	1.9 / 1.9[2]	1.9 / 1.9	1.9 / 1.9	1.9 / 1.9[2]
- Estate / Van ......bar	_	1.9 / 1.9	1.9 / 1.9	_
Front suspension / wheel alignment				
Toe-in (+) / Toe-out (−)......................mm [°]	[15' +5' -10']	[15' ± 5']	[15' ± 5']	[15' +5' -10']
Camber	-40' +35' -25'	-40' +35' -25' Sports: -50'	-40' +35' -25' Sports: -50'	-40' +35' -25'
Castor	1°10' ± 40'	1°10' ± 40'	1°10' ± 40'	1°10' ± 40'
King pin inclination..................................	_	_	_	_
Rear suspension / wheel alignment				
Toe-in (+) / Toe-out (−)......................mm [°]	[10' ± 5'] PW	[20' ± 10']	[20' ± 10']	[10' ± 5'] PW
Camber	-50' ± 30'	-50' ± 30'	-50' ± 30'	-50' ± 30'

AUDI

Torque wrench settings	100 & A6 2.5 TDi 1991 to 1997	A6 1.9 TDi 1994 to 1997	A6 1.9 TDi 1997 to 2000	A6 2.5 TDi 1994 to 1997
Cylinder head - stage 1 ....Nm	40 N	40 N	40 N	40 N
- stage 2 ....Nm	60	60	60	60
- stage 3 ....Nm	+ 180°	+ 90°	+ 90° N	+ 180°
- stage 4 ....Nm	Warm up, + 90°	+ 90°	+ 90° N	Warm up, + 90°
- stage 5 ....Nm	After 1000km,	–	–	After 1000km,
- stage 6 ....Nm	+ 90°	–	–	+ 90°
Big-end bearings.........................Nm	30 + 90° N	30 + 90° N	30 + 90° N	30 + 90° N
Main bearings...........................Nm	30 + 90°	65 + 90° N	65 + 90° N	30 + 90°
Crankshaft pulley bolt ....Nm	160 + 180°	90 + 90° N	90 + 90° N	160 + 180°
Camshaft pulley bolt ....Nm	30 + 90°	20	45	30 + 90°
Flywheel [driveplate] bolt..................Nm	30 + 90°	60 + 90° N	60 + 90° N	30 + 90°
Front hubs ....Nm	200 + 90°	200 + 90°	200 + 90°	200 + 90°
Rear hubs ....Nm	WSM	WSM	WSM	WSM
Wheel nuts / bolts ....Nm	110	110	110	110
Glow plugs ....Nm	15	15	15	15
Clutch pressure plate bolts ....Nm	25	25	25	25
Injection pump sprocket.....................Nm	50 Gear: 160	45	45	50 Gear: 160
Injectors.............................Nm	20	20	20	20
Injection pump mounting bolts ....Nm	25	25	25	25
Injector pipe unions...........................Nm	25	25	25	25
Capacities				
Engine oil & filter................................litres	5.0	3.5	3.0	5.0
Gearbox.......................................litres	2.4	2.4	2.4	2.4
Automatic transmission ....litres	3.0	3.0	3.0	3.0
Final drive ....litres	AT: 1.0	AT: 1.0	AT: 1.0	AT: 1.0
Cooling system..................................litres	6.5	7.5	7.0	6.5
Fuel tank.............................litres	80	80	80	80

Notes

100 & A6 2.5 TDi 1991 to 1997
¹Also 215/60x15
²215/60x15: 1.8 / 1.8
A6 2.5 TDi 1994 to 1997
¹Also 215/60x15
²215/60x15: 1.8 / 1.8

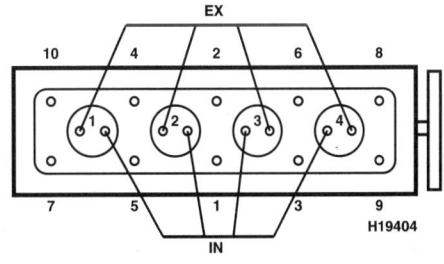

1896 cm³

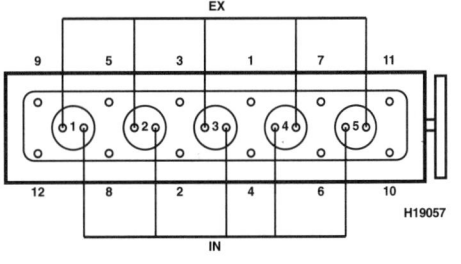

2461 cm³

– Not applicable, or information not available

AUDI

	A6 2.5 TDi 1997 to 2000	A6 2.5 TDi 1998 to 2000	A8 2.5 TDi 1996 to 1998	A8 2.5 TDi 1999 to 2000
Engine				
Engine type/code	AFB DOHC 24V Turbo 110kW	AKN DOHC 24V Turbo 110kW	AFB DOHC 24V Turbo 110kW	AKN DOHC 24V Turbo 110kW
Capacity (cm³) / cylinders	2496 / 6	2496 / 6	2496 / 6	2496 / 6
Compression ration / pressure .....bar	19.5 / _	19.5 / ≥24.0	19.5 /	19.5 / ≥24.0
Torque output .....Nm	310	310	310	310
Oil pressure .....idle [running] bar	[2.0 @ 2000]	[2.0 @ 2000]	[2.0 @ 2000]	[2.0 @ 2000]
Oil temperature .....°C	80	80	80	80
Valve clearances - inlet (mm)	0: Hyd.	0: Hyd.	0: Hyd.	0: Hyd.
- exhaust (mm)	0: Hyd.	0: Hyd.	0: Hyd.	0: Hyd.
Injection order	_	_	1-4-3-6-2-5	1-4-3-6-2-5
No. 1 cylinder position	TBE	TBE	PER	PER
Cooling system				
Thermostat opening temperature .....°C	87	87	87 to 102	87
Radiator cap pressure .....bar	1.2 to 1.5	1.2 to 1.5	1.2 to 1.5	1.2 to 1.5
Fuel system				
Idle speed .....rpm	780 ± 40	680 to 860	780 ± 40	680 to 860
Maximum (no load) speed .....rpm	4500 to 5500	4500 to 5500	4500 to 5500	4500 to 5500
Smoke test/opacity .....M⁻¹ %	2.0	2.0	2.0	2.0
Static timing method	Plunger travel	Plunger travel	Refer to wsm	Refer to wsm
Timing dimension .....mm	0.7 ± 0.02	0.7 ± 0.02	_	_
Crankshaft position .....mm [°]	TDC [0]	TDC [0]	_	_
Turbo type / ref / pressure .....bar	_	_		
Injection pump make	Bosch	_	Bosch	Bosch
Injection pump part no	VP 44	_	VP44	
Injector Make / type	_	_		
Injector part no	_	_		
Injection type	Direct VP	Direct	Direct VP	Direct
Injection opening pressure, New [used].....bar	190 to 200 [170]	190 to 200 [170]	190 to 200 [170]	190 to 200 [160]
Glow plugs				
Maker	Beru	_	Beru	Champion
Type	GN855	_	GN855	CH171
Nominal rating .....V/A	11.5 /	_	11.5 /	
Brakes				
minimum friction material thickness				
Front .....mm	2.0	2.0	3.0	3.0
Rear .....mm	2.5 Disc: 2.0	2.5 Disc: 2.0	7.0 with backing	7.0 with backing
Tyres - Saloon / Hatch .....Size	195/65x15: 205/60x15	195/65x15: 205/60x15	225/60x16: 225/55x17	225/60x16: 225/55x17
- Estate / Van .....Size	195/65x15: 205/60x15	195/65x15: 205/60x15	_	_
Pressure - front / rear - Saloon / Hatch ...bar	1.9 / 1.9	1.9 / 1.9	2.0 / 2.0	2.0 / 2.0
- Estate / Van .....bar	1.9 / 1.9	1.9 / 1.9		
Front suspension / wheel alignment				
Toe-in (+) / Toe-out (–) .....mm [°]	[15' ± 5']	[15' ± 5']	[15' ± 5'] PW	[15' ± 5'] PW
Camber	-40' +35' -25' Sports: -50'	-40' + 35' -25' Sports: -50'	-30' ± 30' PW	-30' ± 30' PW
Castor	1°10' ± 40'	1°10' ± 40'		
King pin inclination	_	_		
Rear suspension / wheel alignment				
Toe-in (+) / Toe-out (–) .....mm [°]	[20' ± 10']	[20' ± 10']	[6' ± 5'] PW	[6' ± 5'] PW
Camber	-50' ± 30'	-50' ± 30'	-40' ± 30' PW	-40' ± 30' PW

AUDI

Torque wrench settings	A6 2.5 TDi 1997 to 2000	A6 2.5 TDi 1998 to 2000	A8 2.5 TDi 1996 to 1998	A8 2.5 TDi 1999 to 2000
Cylinder head - stage 1 ....Nm	35 N	35 N	35 N	35 N
- stage 2 ....Nm	60	60	60	60
- stage 3 ....Nm	+ 90°	+ 90°	+ 90°	+ 90°
- stage 4 ....Nm	+ 90°	+ 90°	+ 90°	+ 90°
- stage 5 ....Nm	–	–	–	–
- stage 6 ....Nm	–	–	–	–
Big-end bearings.............................Nm	30 + 90° N	30 + 90° N	30 + 90° N	30 + 90° N
Main bearings.................................Nm	60 + 90° N	WSM	60 + 90° N	WSM
Crankshaft pulley bolt ....Nm	200 + 180° N	200 + 180° N	200 + 180° N	200 + 180° N
Camshaft pulley bolt ....Nm	80	75	80	75
Flywheel [driveplate] bolt..................Nm	60 + 180° N [60 + 90° N]	60 + 180° N [60 + 90° N]	60 + 90° N Dual mass: 40 +180°	60 + 180° N [60 + 90° N]
Front hubs ....Nm	200 + 90°	200 + 90°	190 + 180° N	190 + 180° N
Rear hubs ....Nm	WSM	WSM	190 + 180° N	190 + 180° N
Wheel nuts / bolts ....Nm	110	110	120	120
Glow plugs....................................Nm	15	15	15	15
Clutch pressure plate bolts ....Nm	25	25	25	25
Injection pump sprocket....................Nm	45	45	–	45
Injectors.......................................Nm	20	20	10	20
Injection pump mounting bolts ....Nm	25	25	20	25
Injector pipe unions..........................Nm	25	25	30	25
Capacities				
Engine oil & filter...............................litres	5.4	5.4	5.4	5.4
Gearbox...litres	2.4	2.4	2.4	2.4
Automatic transmission ....litres	3.0	3.0	2.4	2.4
Final drive ....litres	AT: 1.0	AT: 1.0	WT	WT
Cooling system.................................litres	10.0	10.0	10.0	10.0
Fuel tank..litres	80	80	70	–

Notes

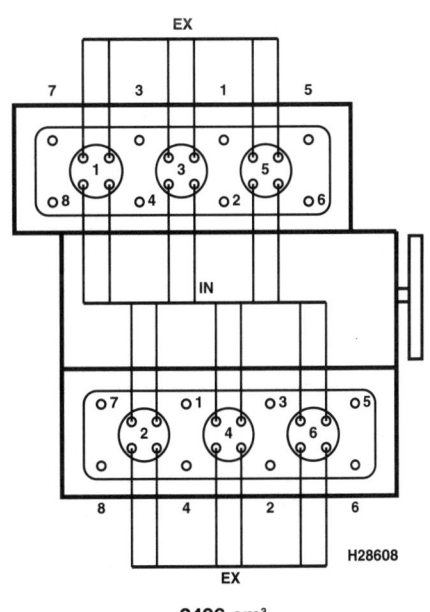

2496 cm³

– Not applicable, or information not
 available

	318tds Compact 1998 to 2000	318tds Touring 1998 to 2000	325td 1995 to 1998	325tds 1995 to 1998
Engine				
Engine type/code....................	M41 SOHC Turbo 66kW	M41 SOHC Turbo 66kW	M51/D25 SOHC Turbo 85kW	M51/D25 SOHC Turbo 105kW[1]
Capacity (cm^3) / cylinders........................	1665 / 4	1665 / 4	2498 / 6	2498 / 6
Compression ration / pressure ....bar	22.0 /	22.0 /	22 / ≥20.0	22 / ≥20.0
Torque output ....Nm	190	190	222	260
Oil pressure ....idle [running] bar	≥0.5	≥0.5	2.0	2.0
Oil temperature ....°C	80	80	80	80
Valve clearances - inlet (mm)	0: Hyd.	0: Hyd.	0: Hyd.	0: Hyd.
- exhaust (mm)	0: Hyd.	0: Hyd.	0: Hyd.	0: Hyd.
Injection order..	1-3-4-2	1-3-4-2	1-5-3-6-2-4	1-5-3-6-2-4
No. 1 cylinder position	TCE	TCE	TCE	TCE
Cooling system				
Thermostat opening temperature ....°C	80	80	80	80
Radiator cap pressure ....bar	1.4	1.4	1.2 to 1.6	1.2 to 1.6
Fuel system				
Idle speed ....rpm	820 ± 50	820 ± 50	750 ± 50	750 ± 50
Maximum (no load) speed ....rpm	5200 to 5300	5250	5300 ± 100	5300 ± 100
Smoke test/opacity ....M^{-1} %	2.5	1.9	2.5	2.1
Static timing method................................	Dial gauge	Dial gauge	Plunger travel	Plunger travel
Timing dimension.............................mm	0.6 ± 0.03	0.6	0.95 ± 0.02	0.95 ± 0.02
Crankshaft position ....mm [°]	TDC	TDC	TDC	TDC
Turbo type / ref / pressure ....bar	–	–	–	–
Injection pump make	Bosch	Bosch	Bosch	Bosch
Injection pump part no............................	VP36	VP36	VE	VP36
Injector Make / type	–	Bosch	Bosch	Bosch
Injector part no.......................................	DNOSD 318	DNOSD 318	0 432 217 223	0 432 217 223
Injection type..	Indirect VP	Indirect VP	Indirect VE	Indirect VP
Injection opening pressure, New [used]...bar	140	140	150 to 158 [140]	150 to 158 [140]
Glow plugs				
Maker	Bosch	Bosch/Champion	Bosch/Champion	Bosch/Champion
Type	0 250 201 033	0 250 201 033 / CH159	0 250 201 033 / CH159	0 250 201 033 / CH159
Nominal rating.....................................V/A	12 / 14	12 / 14	11.5 / 13	11.5 / 13
Brakes				
minimum friction material thickness				
Front.............................mm	2.0	2.0	2.0	2.0
Rear..............................mm	1.5	1.5	2.0	2.0
Tyres - Saloon / Hatch......................Size	185/65x15: 205/55x16	–	185/65x15	205/60x15
- Estate / Van............................Size	–	185/65x15: 205/60x15	–	–
Pressure - front / rear - Saloon / Hatch....bar	1.8 / 2.2 : 1.8 / 2.0	–	1.8 / 2.1	2.0 / 2.3
- Estate / Van ......bar	–	2.0 / 2.3: 1.8 / 2.0	–	–
Front suspension / wheel alignment				
Toe-in (+) / Toe-out (–)....................mm [°]	[0°18' ± 5']	[0°18' ± 5']	[18' ± 8']	[18' ± 8']
Camber	-0°20' ± 30' Sport: -51'±30'	-0°20' ± 30' Sport: -51' ± 30'	-0°30 ± 30'	-0°30 ± 30'
Castor	3°52' ± 30' Sport: 3°57' ± 30'	3°52' ± 30' Sport: 3°57' ± 30'	3°52' ± 30'	3°52' ± 30'
King pin inclination..................................	–	–	–	–
Rear suspension / wheel alignment				
Toe-in (+) / Toe-out (–)....................mm [°]	[30' ± 13'] Sport: [36'±13']	[30' ± 13'] Sport: [36'±13']	[24' ± 6']	[24' ± 6']
Camber	-2° ± 30' Sport: -2°30'±30'	-2° ± 30' Sport: -2°30'±30'	-1°40' ± 15'	-1°40' ± 15'

BMW

Torque wrench settings	318tds Compact 1998 to 2000	318tds Touring 1998 to 2000	325td 1995 to 1998	325tds 1995 to 1998
Cylinder head - stage 1 ....Nm	80, slacken 180° N	80, slacken 180° N	80, slacken	80, slacken
- stage 2 ....Nm	50 + 90°	50 + 90°	50	50
- stage 3 ....Nm	+ 90°	+ 90°	+ 90°	+ 90°
- stage 4 ....Nm	Warm up	Warm up	+ 90°	+ 90°
- stage 5 ....Nm	+ 90°	+ 90°	warm up	warm up
- stage 6 ....Nm	–	–	+ 90°	+ 90°
Big-end bearings.................................Nm	20 + 70° N	20 + 70° N	23 + 70° N	23 + 70° N
Main bearings......................................Nm	20 + 50° N	20 + 50° N	23 + 50°	23 + 50°
Crankshaft pulley bolt ....Nm	330	330	100 + 150°	100 + 150°
Camshaft pulley bolt ....Nm	20 + 35°	20 + 35°	20 + 35°	20 + 35°
Flywheel [driveplate] bolt....................Nm	120 N	120 N	120	105 [120]
Front hubs ....Nm	290	290	290	290
Rear hubs ....Nm	200	200	248	248
Wheel nuts / bolts ....Nm	100 ± 10	100 ± 10	100	100
Glow plugs ....Nm	20	20	20	20
Clutch pressure plate bolts ....Nm	WSM	WSM	24	24
Injection pump sprocket.....................Nm	50	50	48	48
Injectors..Nm	65	65	65	65
Injection pump mounting bolts ....Nm	–	–	22	22
Injector pipe unions............................Nm	20	20	22	22
Capacities				
Engine oil & filter................................litres	5.5	5.5	6.5	6.5
Gearbox..litres	1.0	1.0	1.3	1.3
Automatic transmission ....litres	–	–	3.0	3.0
Final drive ....litres	1.1	1.1	1.7	1.7
Cooling system....................................litres	7.5	7.5	8.75	8.75
Fuel tank...litres	52	62	65	65

Notes

325tds 1995 to 1998
¹with intercooler

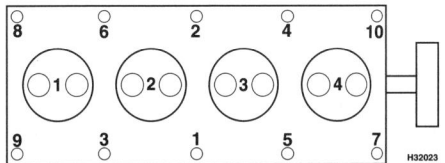

1665 cm³

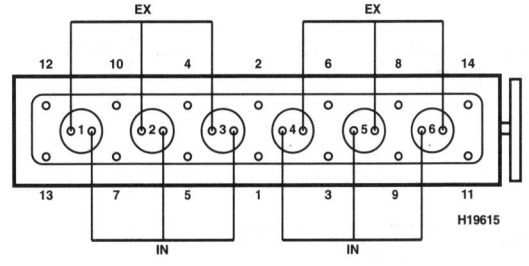

2499 cm³

– Not applicable, or information not available

	325tds Touring 1998 to 2000	525td 1993 to 1996	525tds 1993 to 1997	525tds 1996 to 1997
Engine				
Engine type/code.........................	M51/D25 SOHC 105kW	M51/D25 SOHC Turbo 85kW	M51/D25 SOHC Turbo 105kW	M51/D25 SOHC 105kW
Capacity (cm³) / cylinders..................	2498 / 6	2498 / 6	2498 / 6	2498 / 6
Compression ration / pressure ....bar	22.0 /	22.0 / _	22.0 / _	22.0 / _
Torque output ....Nm	260	222	260	260
Oil pressure ....idle [running] bar	≥0.5	≥0.5 / 4.0	≥0.5 / 4.0	≥0.5 / 4.0
Oil temperature ....°C	80	80	80	80
Valve clearances - inlet (mm)	0: Hyd.	0: Hyd.	0: Hyd.	0: Hyd.
- exhaust (mm)	0: Hyd.	0: Hyd.	0: Hyd.	0: Hyd.
Injection order	1-5-3-6-2-4	1-5-3-6-2-4	1-5-3-6-2-4	1-5-3-6-2-4
No. 1 cylinder position	TCE	TCE	TCE	TCE
Cooling system				
Thermostat opening temperature ....°C	80	82	82	82
Radiator cap pressure ....bar	1.4	1.2 to 1.6	1.2 to 1.6	1.2 to 1.6
Fuel system				
Idle speed ....rpm	770 ± 50	750 ± 50	750 ± 50	750 ± 50
Maximum (no load) speed ....rpm	5300	5250 ± 50	5250 ± 50	5250
Smoke test/opacity ....M⁻¹ %	2.5	2.4	2.1	2.5
Static timing method................................	Dial gauge	Rotor lift	Rotor lift	Rotor lift
Timing dimension.................................mm	0.95	0.95 ± 0.02	0.95 ± 0.02	0.95 ± 0.02
Crankshaft position ....mm [°]	0 TDC	TDC	TDC	TDC
Turbo type / ref / pressure ....bar	_	_	_	_
Injection pump make	Bosch	Bosch	Bosch	Bosch
Injection pump part no............................	VP36	VP15	VP15	VP 36 VE6/10E 2400 R515
Injector Make / type	_	_	_	_
Injector part no.......................................	DNOSD 300	_	_	_
Injection type..	Indirect VP36	Indirect VP	Indirect VP	DDE 2.1
Injection opening pressure, New [used]...bar	140	140 to 160	140 to 160	140 to 160
Glow plugs				
Maker	Champion	Bosch/Champion	Bosch/Champion	Bosch/Beru
Type	CH159	0 250 201 033 / CH159	0 250 201 033 / CH159	0250 201 033 / 719 MJ
Nominal rating....................................V/A	11 /	11.5 / 13	11.5 / 13	12 / 14
Brakes				
minimum friction material thickness				
Front..mm	2.0	2.0	2.0	3.0
Rear..mm	1.5	2.0	2.0	3.0
Tyres - Saloon / Hatch......................Size	_	205/65x15	205/65x15	205/65x15
- Estate / Van......................Size	205/60x15	225/60x15	225/60x15	_
Pressure - front / rear - Saloon / Hatch ...bar	_	2.0 / 2.1	2.0 / 2.1	Refer to vehicle
- Estate / Van ......bar	2.0 / 2.3	2.3 / 2.6	2.3 / 2.6	
Front suspension / wheel alignment				
Toe-in (+) / Toe-out (–)....................mm [°]	[0°18' ± 5']	[18' ± 5']	[18' ± 5']	[18' ± 10']
Camber	-30' ± 30' Sport: -51' ± 30'	-0°13' ± 30'	-0°13' ± 30'	-0°30' ± 30'
Castor	3°52' ± 30' Sport: 3°57' ± 30'	8°08' ± 30'	8°08' ± 30'	3°52' ± 30'
King pin inclination..................................	_	_		
Rear suspension / wheel alignment				
Toe-in (+) / Toe-out (–)....................mm [°]	[30' ± 13'] Sport: [36' ± 13']	[25' ± 7']	[25' ± 7']	[24' ± 10']
Camber	-2° ± 30' Sport: -2°30' ± 30'	-2°20' ± 30'	-2°20' ± 30'	-1°40' ± 20'

BMW

	325tds Touring 1998 to 2000	525td 1993 to 1996	525tds 1993 to 1997	525tds 1996 to 1997
Torque wrench settings				
Cylinder head - stage 1 ....Nm	80, slacken 180° N	80 N	80 N	80 N
- stage 2 ....Nm	50 + 90°	Slacken 180°	Slacken 180°	Slacken 180°
- stage 3 ....Nm	+ 90°	50	50	50
- stage 4 ....Nm	Warm up	+ 90°	+ 90°	+ 90°
- stage 5 ....Nm	+ 90°	+ 90°	+ 90°	+ 90°
- stage 6 ....Nm	–	Warm-up + 90°	Warm-up + 90°	Warm-up + 90°
Big-end bearings.................................Nm	20 + 70° N	20 + 70° N	20 + 70° N	20 + 70° N
Main bearings.....................................Nm	20 + 50° N	20 + 50° N	20 + 50° N	20 + 50° N
Crankshaft pulley bolt ....Nm	330	100 + 150°	100 + 150°	100 + 150°
Camshaft pulley bolt ....Nm	–	20 + 35°	20 + 35°	20 + 35°
Flywheel [driveplate] bolt.....................Nm	105 [120] N	105 N [120 N]	105 N [120 N]	105 N [120 N]
Front hubs ....Nm	290	290	290	290
Rear hubs ....Nm	250	250	250	250
Wheel nuts / bolts ....Nm	100 ± 10	100 ± 10	100 ± 10	100 ± 10
Glow plugs ....Nm	24	20	20	20
Clutch pressure plate bolts ....Nm	WSM	24	24	24
Injection pump sprocket.......................Nm	–	50	50	45 to 50
Injectors...Nm	65	65	65	65
Injection pump mounting bolts ....Nm	–	22	22	25
Injector pipe unions.............................Nm	–	22	22	20 to 25
Capacities				
Engine oil & filter................................litres	7.0	6.75	6.75	6.75
Gearbox...litres	1.25	1.2	1.2	1.2
Automatic transmission ....litres	3.0	3.3	3.3	–
Final drive...litres	1.7	1.7	1.7	1.0
Cooling system....................................litres	8.75	9.75	9.75	9.75
Fuel tank...litres	65	80	80	70

Notes

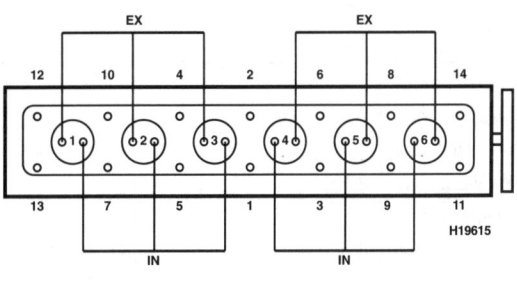

2498 cm³

– Not applicable, or information not
available

BMW 5

Haynes THE BOOK ®

BMW

	525tds Touring 1997 to 1998	725tds 1996 to 1997		
Engine				
Engine type/code....................	M51/D25 SOHC Turbo 110kW	M51/D25 SOHC 105kW		
Capacity (cm^3) / cylinders..................	2498 / 6	2498 / 6		
Compression ration / pressure ....bar	22.0 /	22.0 / _		
Torque output ....Nm	260	260		
Oil pressure ....idle [running] bar	≥0.5	≥0.5 [4.0]		
Oil temperature ....°C	80	80		
Valve clearances - inlet (mm)	0: Hyd.	0: Hyd.		
- exhaust (mm)	0: Hyd.	0: Hyd.		
Injection order	1-5-3-6-2-4	1-5-3-6-2-4		
No. 1 cylinder position	TCE	TCE		
Cooling system				
Thermostat opening temperature ....°C	80	92		
Radiator cap pressure ....bar	1.4	_		
Fuel system				
Idle speed ....rpm	770 ± 50	750 ± 50		
Maximum (no load) speed ....rpm	5300	_		
Smoke test/opacity ....M^{-1} %	2.1	2.5		
Static timing method................................	Dial gauge	Refer to wsm		
Timing dimension................................mm	0.95	_		
Crankshaft position ....mm [°]	0 TDC	_		
Turbo type / ref / pressure ....bar	_	_		
Injection pump make	Bosch	Bosch		
Injection pump part no............................	VP36	VP36 VE 6/10E 2400 R515		
Injector Make / type	_	_		
Injector part no..	DNOSD 300	_		
Injection type...	Indirect VP36	DDE 2.1		
Injection opening pressure, New [used]...bar	140	140 to 160		
Glow plugs				
Maker	Bosch/Champion	Bosch		
Type	0 250 201 033 / CH159	0250 201 033		
Nominal rating.................................V/A	12 / 14	12 / 14		
Brakes				
minimum friction material thickness				
Front..mm	2.0	3.0		
Rear...mm	1.5	3.0		
Tyres - Saloon / Hatch.....................Size	_	215/65x16		
- Estate / Van.........................Size	205/60x15	_		
Pressure - front / rear - Saloon / Hatch ...bar	_	Refer to vehicle		
- Estate / Van ......bar	2.0 / 2.3	_		
Front suspension / wheel alignment				
Toe-in (+) / Toe-out (–)....................mm [°]	[0°5' ± 5']	[0°14' ± 10']		
Camber	-0°13' ± 30' Sport: -35'±30'	-0°13' ± 30'		
Castor	6°42' ± 30' Sport: 6°55'±30'	6°06' ± 30'		
King pin inclination..................................	_	_		
Rear suspension / wheel alignment				
Toe-in (+) / Toe-out (–)....................mm [°]	[0°16' ± 10']	[0°18' ± 10']		
Camber	-2°10' ± 20'	-1°30' ± 20'		

BMW

	525tds Touring 1997 to 1998	725tds 1996 to 1997		
Torque wrench settings				
Cylinder head - stage 1 ....Nm	80, slacken 180° N	80 N		
- stage 2 ....Nm	50 + 90°	Slacken 180°		
- stage 3 ....Nm	+ 90°	50		
- stage 4 ....Nm	Warm up	+ 90°		
- stage 5 ....Nm	+ 90°	+ 90°		
- stage 6 ....Nm	–	warm-up + 90°		
Big-end bearings.................................Nm	20 + 70° N	20 + 70° N		
Main bearings......................................Nm	20 + 50° N	20 + 50° N		
Crankshaft pulley bolt ....Nm	330	100 + 150°		
Camshaft pulley bolt ....Nm	20 + 35°	20 + 35°		
Flywheel [driveplate] bolt....................Nm	105 [120] N	105		
Front hubs ....Nm	290	–		
Rear hubs ....Nm	300	300		
Wheel nuts / bolts ....Nm	100 ± 10	100 ± 10		
Glow plugs ....Nm	20	20		
Clutch pressure plate bolts ....Nm	WSM	24		
Injection pump sprocket......................Nm	45	–		
Injectors...Nm	65	50		
Injection pump mounting bolts ....Nm	25	–		
Injector pipe unions............................Nm	22	–		
Capacities				
Engine oil & filter.................................litres	6.75	7.0		
Gearbox..litres	1.25	1.3		
Automatic transmission ....litres	3.0	–		
Final drive...litres	1.0	1.2		
Cooling system....................................litres	10.5	10.5		
Fuel tank...litres	70	85		

Notes

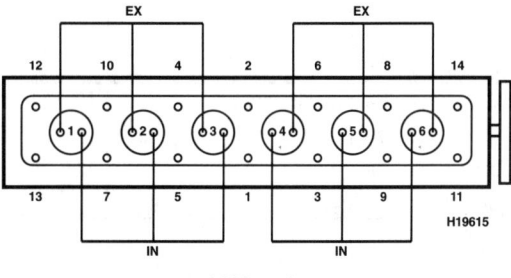

2498 cm³

– Not applicable, or information not available

CARBODIES LTI

	FX4S 2.5 1985 to 1991	FX4S 2.7 1989 to 1991		
Engine				
Engine type/code.....................	Land Rover OHV	Nissan TD27 OHV		
Capacity (cm^3) / cylinders.........................	2495 / 4	2663 / 4		
Compression ration / pressure ....bar	21.0 / _	21.8 / ≥24.5		
Torque output ....Nm	0	0		
Oil pressure ....idle [running] bar	[2.4 to 4.5]	[2.9 to 3.9]		
Oil temperature ....°C	80	80		
Valve clearances - inlet (mm)	0.25	0.35 H		
- exhaust (mm)	0.25	0.35 H		
Injection order	1-3-4-2	1-3-4-2		
No. 1 cylinder position	F	F		
Cooling system				
Thermostat opening temperature ....°C	82	82		
Radiator cap pressure ....bar	0.62	0.78 to 0.98		
Fuel system				
Idle speed ....rpm	650 ± 20	700 ± 50		
Maximum (no load) speed ....rpm	4400 ± 80	5100 +50 -150		
Smoke test/opacity ....M^{-1} %	2.5	2.5		
Static timing method.................................	_	Plunger travel		
Timing dimension.................................mm	_	0.65 ± 0.02 [5]		
Crankshaft position ....mm [°]	_	[0] TDC		
Turbo type / ref / pressure ....bar	_	_		
Injection pump make	CAV	Diesel Kiki		
Injection pump part no.............................	_	16700-43G13		
Injector Make / type	CAV	Diesel Kiki		
Injector part no..	BDNO SPC 6209	16600-43G02		
Injection type...	Indirect DPS	Indirect VE		
Injection opening pressure, New [used]...bar	≥135	103 to 113 [98 to 103]		
Glow plugs				
Maker	Champion	Champion		
Type	CH63	CH63		
Nominal rating....................................V/A	12.0 / _	_		
Brakes				
minimum friction material thickness				
Front...mm	_	_		
Rear...mm				
Tyres - Saloon / Hatch.....................Size	175x16: 185x16	175x16: 185x16		
- Estate / Van...........................Size	_	_		
Pressure - front / rear - Saloon / Hatch ...bar	2.4 / 2.2^1	2.4 / 2.2^1		
- Estate / Van ......bar	_	_		
Front suspension / wheel alignment				
Toe-in (+) / Toe-out (–)....................mm [°]	1.6	1.6		
Camber	1°	1°		
Castor	3°	3°		
King pin inclination..................................	6.5°	6.5°		
Rear suspension / wheel alignment				
Toe-in (+) / Toe-out (–)....................mm [°]	_	_		
Camber				

CARBODIES LTI

	FX4S 2.5 1985 to 1991	FX4S 2.7 1989 to 1991		
Torque wrench settings				
Cylinder head - stage 1 ....Nm	115 to 130	39 to 44		
- stage 2 ....Nm	–	54 to 59		
- stage 3 ....Nm	–	+ 90 ± 10°		
- stage 4 ....Nm	–	–		
- stage 5 ....Nm	–	–		
- stage 6 ....Nm	–	–		
Big-end bearings.......................Nm	34 to 46	78 to 83 N		
Main bearings.......................Nm	130 to 136	167 to 177		
Crankshaft pulley bolt ....Nm	260 to 280	294 to 324		
Camshaft pulley bolt ....Nm	40 to 50	44 to 49		
Flywheel [driveplate] bolt.......................Nm	130 to 143	147 to 167		
Front hubs ....Nm	–	–		
Rear hubs ....Nm	–	–		
Wheel nuts / bolts ....Nm	88	88		
Glow plugs ....Nm	34	15 to 20		
Clutch pressure plate bolts ....Nm	25 to 34	25 to 34		
Injection pump sprocket.......................Nm	–	59 to 69		
Injectors.......................Nm	22 to 28	54 to 64		
Injection pump mounting bolts ....Nm	22 to 28	30 to 41		
Injector pipe unions.......................Nm	–	20 to 25		
Capacities				
Engine oil & filter.......................litres	6.9	5.9		
Gearbox.......................litres	2.0	2.0		
Automatic transmission ....litres	Type 35: 6.4²	Type 35: 6.4²		
Final drive ....litres	2.0	2.0		
Cooling system.......................litres	10.3	10.3		
Fuel tank.......................litres	58.2	58.2		

Notes

FX4S 2.5 1985 to 1991
¹185x16: 2.5 / 2.5
²Type 65: 7.0
FX4S 2.7 1989 to 1991
¹185x16: 2.5 / 2.5
²Type 65: 7.0

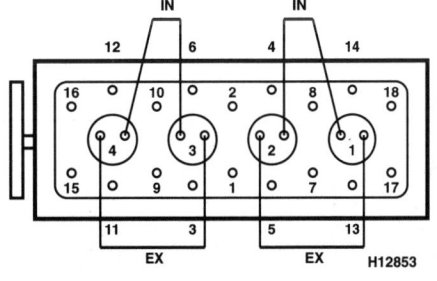

2495 cm³

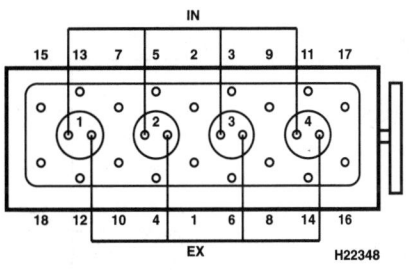

2663 cm³

– Not applicable, or information not
  available

CHRYSLER JEEP

	Voyager 2.5 TD 1998 to 2000	Cherokee 2.5 TD 1995 to 2000	Grand Cherokee 2.5 TD 1997 to 2000	
Engine				
Engine type/code..........................	HR 425 OHV Turbo 87kW	09B OHV Turbo 84kW	09B OHV Turbo 84kW	
Capacity (cm^3) / cylinders........................	2499 / 4	2499 / 4	2499 / 4	
Compression ration / pressure ....bar	21.0 /	21.0 /	21.0 /	
Torque output ....Nm	0	300	300	
Oil pressure ....idle [running] bar	_	[3.5 to 5.0]	[3.5 to 5.0]	
Oil temperature ....°C	_	80	80	
Valve clearances - inlet (mm)	0.30	0: Hyd.	0: Hyd.	
- exhaust (mm)	0.30	0: Hyd.	0: Hyd.	
Injection order	1-3-4-2	1-3-4-2	1-3-4-2	
No. 1 cylinder position	TGF	TGF	TGF	
Cooling system				
Thermostat opening temperature ....°C	_	80	80	
Radiator cap pressure ....bar	_	1.2	1.2	
Fuel system				
Idle speed ....rpm	850 ± 100	900 ± 25	900 ± 25	
Maximum (no load) speed ....rpm	4700 ± 100	4750	4750	
Smoke test/opacity ....M^{-1} %	2.5	2.5	2.5	
Static timing method...............................	Plunger travel	Plunger travel	Plunger travel	
Timing dimension.............................mm	0.82 ± 0.02	0.65	0.65	
Crankshaft position ....mm [°]	TDC	TDC	TDC	
Turbo type / ref / pressure ....bar	_	_	_	
Injection pump make	Bosch	Bosch	Bosch	
Injection pump part no...........................	VE	VP36	VP36	
Injector Make / type	Bosch	Bosch	Bosch	
Injector part no......................................	0 460 404 073	_	_	
Injection type...	Indirect VE	Indirect VP	Indirect VP	
Injection opening pressure, New [used]...bar	125 to 138	150 to 158	150 to 158	
Glow plugs				
Maker	Bosch	Beru	Beru	
Type	0 250 201 033	642 MJ	642 MJ	
Nominal rating.................................V/A	_	11 / 12	11 / 12	
Brakes				
minimum friction material thickness				
Front............................mm	7.95 with backing	_	_	
Rear............................mm	_		_	
Tyres - Saloon / Hatch.................Size	_	225/75x15	_	
- Estate / Van..........................Size	215/65x15	_	215/75x15	
Pressure - front / rear - Saloon / Hatch ...bar	_	2.1 / 2.1	_	
- Estate / Van ......bar	Refer to vehicle	_	2.3 / 2.3	
Front suspension / wheel alignment				
Toe-in (+) / Toe-out (–)....................mm [°]	[6' ± 12']	0.0	-1.8	
Camber	8' ± 25'	-0° 15'	-0°25'	
Castor	1°25' ± 1°	7°	7°	
King pin inclination.................................	_	_	_	
Rear suspension / wheel alignment				
Toe-in (+) / Toe-out (–)....................mm [°]	[0° ± 25']	_	_	
Camber	0° ± 15'	_	_	

CHRYSLER JEEP

	Voyager 2.5 TD 1998 to 2000	Cherokee 2.5 TD 1995 to 2000	Grand Cherokee 2.5 TD 1997 to 2000	
Torque wrench settings				
Cylinder head - stage 1 ......Nm	30	WSM	WSM	
- stage 2 ......Nm	+ 70°	–	–	
- stage 3 ......Nm	Warm up, allow to cool	–	–	
- stage 4 ......Nm	Slacken	–	–	
- stage 5 ......Nm	30	–	–	
- stage 6 ......Nm	+ 65°	–	–	
Big-end bearings......Nm	30 + 60° N	30 + 60° N	30 + 60° N	
Main bearings......Nm	20, 42	54, 95, 108	54, 95, 108	
Crankshaft pulley bolt ......Nm	–	160	160	
Camshaft pulley bolt ......Nm	–	–	–	
Flywheel [driveplate] bolt......Nm	130	68 + 60°	68 + 60°	
Front hubs ......Nm	244	237	237	
Rear hubs ......Nm	244	–	–	
Wheel nuts / bolts ......Nm	120	120	120	
Glow plugs ......Nm	25	23	23	
Clutch pressure plate bolts ......Nm	–	31	31	
Injection pump sprocket......Nm	–	86	86	
Injectors......Nm	–	70	70	
Injection pump mounting bolts ......Nm	–	28	28	
Injector pipe unions......Nm	–	25	25	
Capacities				
Engine oil & filter......litres	6.0	6.5	6.5	
Gearbox......litres	2.4	3.3 Transfer: 1.0	3.3 Transfer: 1.0	
Automatic transmission ......litres	–	–	–	
Final drive......litres	–	1.49 Rear: 1.6	1.49 Rear: 1.6	
Cooling system......litres	–	9.5	9.5	
Fuel tank......litres	–	76	87	

Notes

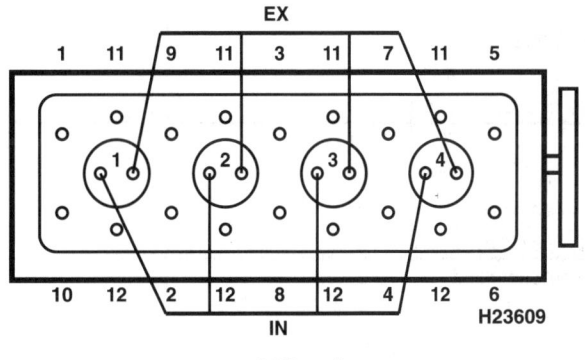

2499 cm³

– Not applicable, or information not available

CITROEN

	AX 14 D 1988 to 1991	AX 1.4 D 1989 to 1997	AX 1.5 D 1994 to 1997	Saxo 1.5 (Bosch) 1996 to 2000
Engine				
Engine type/code..........................	TUD3 K9A SOHC 38kW	TUD3/Y/L K9Y SOHC 37kW	TUD5/L/Y/L3 VJZ/Y/X SOHC 42kW	TUD5/L/L3 VJZ/VJY 42/40kW
Capacity (cm³) / cylinders........................	1360 / 4	1360 / 4	1527 / 4	1527 / 4
Compression ration / pressure ....bar	22.0 / _	22.0 / _	23.0 / _	23.0 / _
Torque output ....Nm	0	0	95	95
Oil pressure ....idle [running] bar	[4.0]	[4.0]	[4.0]	[≥4.0]
Oil temperature ....°C	90	90	90	90
Valve clearances - inlet (mm)	0.15	0.15 ± 0.08	0.15 ± 0.08	0.15
- exhaust (mm)	0.30	0.30 ± 0.08	0.30 ± 0.08	0.30
Injection order..................................	1-3-4-2	1-3-4-2	1-3-4-2	1-3-4-2
No. 1 cylinder position	FE	FE	FE	FE
Cooling system				
Thermostat opening temperature ....°C	88	88	89	88
Radiator cap pressure ....bar	1.4	1.4	1.4	1.4
Fuel system				
Idle speed ....rpm	775 ± 25	775 ± 25	775 ± 25	800 ± 25
Maximum (no load) speed ....rpm	5500	5500	5450 ± 125	5450 ± 125
Smoke test/opacity ....M⁻¹ %	2.5	2.5	2.5	2.5
Static timing method................................	Rotor lift	Rotor lift	_	_
Timing dimension...............................mm	0.8	0.03	_	_
Crankshaft position ....mm [°]	TDC	TDC	[14° BTDC[	[14° BTDC]
Turbo type / ref / pressure ....bar	_	_	_	_
Injection pump make	Roto Diesel	Lucas	Lucas	Bosch
Injection pump part no..............................	R8443B460A 91▶: R8443B481	D66 R8443B971B	070 R8444B421B	VE4 / 8F2500 R611
Injector Make / type	_	CAV	CAV	Bosch
Injector part no..................................	RDNOSDC 6863C	RDNOSDC 688DC	RDN 12 SDC 6849D	DNOSD 299-A / DNOSD 3026
Injection type......................................	DPC TU D 100	DPC TU D 100	VJZ: R8444B421. VJY: R8444B480	Indirect VE
Injection opening pressure, New [used]...bar	125 + 5	125 + 5	130 + 5-0	120.0 ± 5.0
Glow plugs				
Maker	Bosch/Champion	Beru/Bosch	Beru/Bosch	Bosch/Beru
Type	B250 601 382 / CH147	0100 221 144 / 0250 202 001	0 100 226 188 / 0250 202 020	0 250 201 033 / 0 100 226 188
Nominal rating....................................V/A	_	_	11.0 / _	11.0 / _
Brakes				
minimum friction material thickness				
Front...mm	2.0	2.0	2.0	2.0
Rear..mm	1.0	1.0	1.0	1.0
Tyres - Saloon / Hatch.......................Size	145/70x13: 155/70x13	145/70x13: 155/70x13	155/70x13	165/70x13: 165/65x14
- Estate / Van.........................Size	_	_	_	_
Pressure - front / rear - Saloon / Hatch ...bar	1.9 / 1.9	1.9 / 1.9	2.1 / 2.1	2.2 / 2.0
- Estate / Van ......bar	_	_	_	_
Front suspension / wheel alignment				
Toe-in (+) / Toe-out (–)...................mm [°]	-2.0 ± 1.5	-2.0 ± 1.5	-0.5 to -3.5	-1.0 to -3.0 PAS: 1.0 to 3.0
Camber	57' + 30'	-30'	-30'	-40' ± 30'
Castor	1°5' ± 30'	2°15' ± 1°	2°15' ± 1°	2°13' ± 30' PAS: 3°14'±30'
King pin inclination.................................	12.58°	_	_	12°41' ± 40' PAS: 12°42'±40'
Rear suspension / wheel alignment				
Toe-in (+) / Toe-out (–)...................mm [°]	-1.0 ± 1.0	1.8 to 5.4	1.8 to 5.4	1.49 to 6.39
Camber	-1° ± 20'	-1°15'	-1°15'	-59' ± 20'

CITROEN

	AX 14 D 1988 to 1991	AX 1.4 D 1989 to 1997	AX 1.5 D 1994 to 1997	Saxo 1.5 (Bosch) 1996 to 2000
Torque wrench settings				
Cylinder head - stage 1 ....Nm	60 N oiled threads	20 N	Bolt length: ≤197.5mm	Bolts ≤184.4 mm below head
- stage 2 ....Nm	Slacken	+ 160°	40	20[1]
- stage 3 ....Nm	20	+ 160°	+ 260°	+ 260° ± 5°
- stage 4 ....Nm	+ 160°	–	–	–
- stage 5 ....Nm	+ 160°	–	–	–
- stage 6 ....Nm	–	–	–	–
Big-end bearings.............................Nm	40 N	40 N	40 N	40 N
Main bearings..Nm	20 + 45°	20 + 45°	20 + 45°	20 + 45°
Crankshaft pulley bolt ....Nm	90 to 130	90 to 130	110	70 + 45°
Camshaft pulley bolt ....Nm	40 to 50	40 to 50	–	40 + 20°
Flywheel [driveplate] bolt.....................Nm	70 LkC	70 LkC	66	65 LkC
Front hubs ....Nm	250	250	250	250
Rear hubs ....Nm	140	140	140	140
Wheel nuts / bolts ....Nm	90	90	90	85
Glow plugs ....Nm	22	22	20	25
Clutch pressure plate bolts ....Nm	13	13	13	13
Injection pump sprocket.......................Nm	–	–	–	23
Injectors...Nm	70	70	55	70
Injection pump mounting bolts ....Nm	–	–	–	25
Injector pipe unions.............................Nm	20	20	20	20
Capacities				
Engine oil & filter..................................litres	3.75	3.75	4.75	4.75
Gearbox...litres	2.0	2.0	2.0	2.0
Automatic transmission ....litres	–	–	–	–
Final drive ....litres	WT	WT	WT	WT
Cooling system.....................................litres	4.8	4.8	6.0	4.8
Fuel tank..litres	43	43	43	45

Notes

Saxo 1.5 (Bosch) 1996 to 2000
[1] 01.99 on: 40 Nm + 300° ± 5°

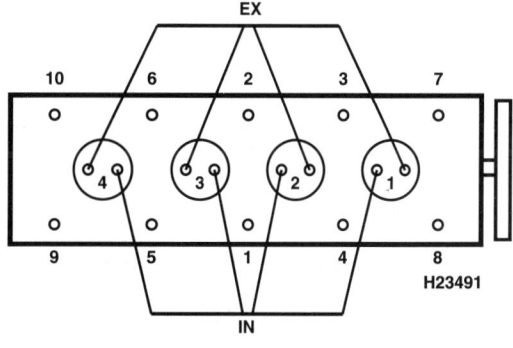

1360 cm³ / 1527 cm³

– Not applicable, or information not
  available

CITROEN

	Saxo 1.5 (Lucas) 1996 to 2000	Picasso 2.0 HDi 2000	ZX 1.4D (Lucas) 1995 to 1998	ZX 1.4D (Bosch) 1995 to 1998
Engine				
Engine type/code	TUD5/L/L3 VJZ/VJY 42/40kW	DW10TD/L3 RHY 66kW	TUD3Y/L K9Y SOHC 37kW	TUD3Y/L K9Y SOHC 37kW
Capacity (cm³) / cylinders	1527 / 4	1997 / 4	1360 / 4	1360 / 4
Compression ration / pressure ....bar	23.0 / _	17.6 /	22.5 / _	22.5 / _
Torque output ....Nm	95	0	0	0
Oil pressure ....idle [running] bar	[≥4.0]	[4.0]	[3.5]	[3.5]
Oil temperature ....°C	90	90	80	80
Valve clearances - inlet (mm)	0.15	_	0.15 ± 0.08	0.15 ± 0.08
- exhaust (mm)	0.30	_	0.30 ± 0.08	0.30 ± 0.08
Injection order	1-3-4-2	_	1-3-4-2	1-3-4-2
No. 1 cylinder position	FE	_	FE	FE
Cooling system				
Thermostat opening temperature ....°C	88	_	88	88
Radiator cap pressure ....bar	1.4	_	1.4	1.4
Fuel system				
Idle speed ....rpm	800 ± 25	_	800 ± 50	800 ± 50
Maximum (no load) speed ....rpm	5450 ± 125	_	5100 ± 125	5100 ± 125
Smoke test/opacity ....M⁻¹ %	2.5	2.5	2.5	2.5
Static timing method................................	Dial gauge	_	_	_
Timing dimension................................mm	Dimension on pump	_	Dimension on pump	Dimension on pump
Crankshaft position ....mm [°]	_	_		
Turbo type / ref / pressure ....bar	_	_		
Injection pump make	Lucas	_	Lucas	Bosch
Injection pump part no..............................	DPC / 8444 B425/482/660	_	TUD100 DPC 8443	TUD200 VER311
Injector Make / type	CAV	_	_	_
Injector part no..	RDN 12SDC 6849	_	_	_
Injection type..	Indirect DPC	_	DPC	VER
Injection opening pressure, New [used]...bar	130.0 ± 5.0	_	125 ± 5	_
Glow plugs				
Maker	Bosch/Beru	_	Beru/Bosch	Beru/Bosch
Type	0 250 201 033 / 0 100 226 188	_	0100 221 144 / 0250 202 001	0100 221 144 / 0250 202 001
Nominal rating.....................................V/A	11.0 / _			
Brakes				
minimum friction material thickness				
Front.............................mm	2.0	_	2.0	2.0
Rear.............................mm	1.0	_	1.0	1.0
Tyres - Saloon / Hatch ....Size	165/70x13: 165/65x14	_	_	_
- Estate / Van.............................Size	_	_	_	_
Pressure - front / rear - Saloon / Hatch ...bar	2.2 / 2.0	_	_	_
- Estate / Van ......bar		_	_	_
Front suspension / wheel alignment				
Toe-in (+) / Toe-out (–)....................mm [°]	-1.0 to -3.0 PAS: 1.0 to 3.0	_	-1.0 to -3.0 PAS: 1.0 to 3.0	-1.0 to -3.0 PAS: 1.0 to 3.0
Camber	-40' ± 30'	_	0 ± 40'	0° ± 40'
Castor	2°13' ± 30' PAS: 3°14'±30'	_	1°30' ± 40' PAS: 3°±40'	1°30' ± 40' PAS: 3°±40'
King pin inclination...................................	12°41' ± 40' PAS: 12°42'±40	_	10°45' ± 40'	10°45' ± 40'
Rear suspension / wheel alignment				
Toe-in (+) / Toe-out (–)....................mm [°]	1.49 to 6.39	_	2.5 to 6.0	2.5 to 6.0
Camber	-59' ± 20'	_	-1° ± 40'	-1° ± 40'

CITROEN

	Saxo 1.5 (Lucas) 1996 to 2000	Picasso 2.0 HDi 2000	ZX 1.4D (Lucas) 1995 to 1998	ZX 1.4D (Bosch) 1995 to 1998
Torque wrench settings				
Cylinder head - stage 1 ....Nm	Bolts ≤184.4 mm below head	20	20[1]	20[1]
- stage 2 ....Nm	20[1]	60	60	60
- stage 3 ....Nm	+ 260° ± 5°	220° ± 5°	+ 180°	+ 180°
- stage 4 ....Nm	–	–	–	–
- stage 5 ....Nm	–	–	–	–
- stage 6 ....Nm	–	–	–	–
Big-end bearings.................................Nm	40 N	–	40 N	40 N
Main bearings.....................................Nm	20 + 45°	–	20 + 45°	20 + 45°
Crankshaft pulley bolt ....Nm	70 + 45°	–	90 to 130	90 to 130
Camshaft pulley bolt ....Nm	40 + 20°	–	40 to 50	40 to 50
Flywheel [driveplate] bolt....................Nm	65 LkC	–	70 LkC	70 LkC
Front hubs ....Nm	250	–	–	–
Rear hubs..Nm	140	–	–	–
Wheel nuts / bolts ....Nm	85	–	–	–
Glow plugs ....Nm	25	–	22	22
Clutch pressure plate bolts ....Nm	13	–	13	13
Injection pump sprocket......................Nm	23	–	–	–
Injectors..Nm	70	–	–	–
Injection pump mounting bolts ....Nm	25	–	–	–
Injector pipe unions............................Nm	20	–	–	–
Capacities				
Engine oil & filter.................................litres	4.75	4.5	4.25 A/C: 4.0	4.25 A/C: 4.0
Gearbox...litres	2.0	–	1.9	1.9
Automatic transmission ....litres	–	–	2.4	2.4
Final drive ....litres	WT	–	WT	WT
Cooling system...................................litres	6.0	11.0	9.0	9.0
Fuel tank..litres	45	60	56	56

Notes

Saxo 1.5 (Lucas) 1996 to 2000
[1] 01.99 on: 40 Nm + 300° ± 5°
ZX 1.4D (Lucas) 1995 to 1998
[1] Bolt length below head: ≤121.5 mm Bolt with guide boss: ≤125.5 mm
ZX 1.4D (Bosch) 1995 to 1998
[1] Bolt length below head: ≤121.5 mm Bolt with guide boss: ≤125.5 mm

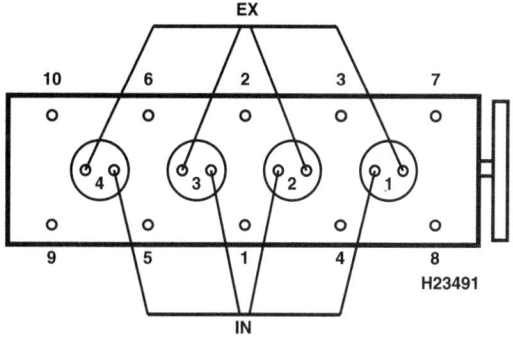

1360 cm³ / 1527 cm³

– Not applicable, or information not available

CITROEN

	ZX 1.8D (Lucas) 1993 to 1996	ZX 1.8D (Bosch) 1993 to 1996	ZX 1.9D (Lucas) 1992 to 1996	ZX 1.9D (Bosch) 1992 to 1996
Engine				
Engine type/code	XUD7/L A9A SOHC 43.5kW	XUD7/L A9A SOHC 43.5kW	XUD9A/L D9B SOHC 51kW	XUD9A/L D9B SOHC 51kW
Capacity (cm³) / cylinders........................	1769 / 4	1769 / 4	1905 / 4	1905 / 4
Compression ration / pressure ....bar	23.0 / _	23.0 / _	23.0 / _	23.0 / _
Torque output ....Nm	90	90	0	90
Oil pressure ....idle [running] bar	[3.5]	[3.5]	[3.5]	[3.5]
Oil temperature ....°C	80	80	80	80
Valve clearances - inlet (mm)	0.15 ± 0.08	0.15 ± 0.08	0.15 ± 0.08	0.15 ± 0.08
- exhaust (mm)	0.30 ± 0.08	0.30 ± 0.08	0.30 ± 0.08	0.30 ± 0.08
Injection order	1-3-4-2	1-3-4-2	1-3-4-2	1-3-4-2
No. 1 cylinder position	FE	FE	FE	FE
Cooling system				
Thermostat opening temperature ....°C	89	89	89	89
Radiator cap pressure ....bar	1.4	1.4	1.4	1.4
Fuel system				
Idle speed ....rpm	800 ± 50	800 ± 50	800 ± 50	800 ± 50
Maximum (no load) speed ....rpm	5100 ± 125	5100 ± 125	5150 ± 125	5100 ± 125
Smoke test/opacity ....M⁻¹ %	2.5	2.5	2.5	2.5
Static timing method..............................	_	Plunger travel	_	_
Timing dimension...............................mm	Dimension on pump	0.9	Dimension on pump	1.07 [AT: 0.98]
Crankshaft position ....mm [°]	_	TDC [0]	_	TDC [0]
Turbo type / ref / pressure ....bar	_	_	_	_
Injection pump make	Lucas	Bosch	Lucas	Bosch
Injection pump part no..............................	0523 B930A	523 VER171/2	XUD101 8443 B952B. 95▶: B953C	VER 425/1 XUD 201
Injector Make / type	CAV	Bosch	CAV	Bosch
Injector part no.......................................	RDNOSDC 6850D	DNOSD 256/	RDNOSDC 6751D	DNOSD +299A
Injection type..	Indirect DPC	Indirect VER	Indirect DPC	Indirect VER
Injection opening pressure, New [used]...bar	130.0 ± 5.0	130.0 ± 5.0	125.0 ± 5.0	130 +0 -5.0
Glow plugs				
Maker	Beru/Bosch	Beru/Bosch	Beru/Bosch	Beru/Bosch
Type	0100 221 133 / 0250 201 019	0100 221 133 / 0250 201 019	0100 221 133 / 0250 201 019	0100 221 133 / 0250 201 019
Nominal rating.....................................V/A	11 / 13	11 / 13	11 / 13	11 / 13
Brakes				
minimum friction material thickness				
Front..mm	2.0	2.0	2.0	2.0
Rear..mm	1.0	1.0	1.0	1.0
Tyres - Saloon / Hatch......................Size	175/65x14	175/65x14	175/65x15	175/65x14
- Estate / Van............................Size	_	_	_	_
Pressure - front / rear - Saloon / Hatch ...bar	2.2 / 2.1	2.1 2.2	2.2 / 2.1	2.2 / 2.1
- Estate / Van ......bar	_	_	_	_
Front suspension / wheel alignment				
Toe-in (+) / Toe-out (−)....................mm [°]	-1.0 to -3.0 PAS: 1.0 to 3.0	-1.0 to -3.0 PAS: 1.0 to 3.0	-1.0 to -3.0 PAS: 1.0 to 3.0	-1.0 to -3.0 PAS: 1.0 to 3.0
Camber	0° ± 40'	0 ± 40'	0° ± 40'	0 ± 40'
Castor	1°30' ± 40' PAS: 3° ± 40'	1°30' ± 40' PAS: 3° ± 40'	1°30' ± 40' PAS: 3° ± 40'	1°30' ± 40' PAS: 3° ± 40'
King pin inclination..................................	10°45' ± 40'	10°45' ± 40'	10°45' ± 40'	10°45' ± 40'
Rear suspension / wheel alignment				
Toe-in (+) / Toe-out (−).....................mm [°]	2.5 to 6.0	2.5 to 6.0	2.5 to 6.0	2.5 to 6.0
Camber	-1° ± 40'	-1° ± 40'	-1° ± 40'	-1° ± 40'

CITROEN

Torque wrench settings	ZX 1.8D (Lucas) 1993 to 1996	ZX 1.8D (Bosch) 1993 to 1996	ZX 1.9D (Lucas) 1992 to 1996	ZX 1.9D (Bosch) 1992 to 1996
Cylinder head - stage 1 ....Nm	20[1]	20[1]	20[1]	20[1]
- stage 2 ....Nm	60	60	60	70
- stage 3 ....Nm	+ 180°	+ 180°	+ 180°	+ 140°
- stage 4 ....Nm	–	–	–	–
- stage 5 ....Nm	–	–	–	–
- stage 6 ....Nm	–	–	–	–
Big-end bearings.................................Nm	20 + 70° N	20 + 70° N	20 + 70° N	20 + 70° N
Main bearings.....................................Nm	35, 70	35, 70	70	70
Crankshaft pulley bolt ....Nm	40 + 60°	40 + 60°	40 + 60°	40 + 60°
Camshaft pulley bolt ....Nm	35	35	35	35
Flywheel [driveplate] bolt....................Nm	50	50	50	50
Front hubs ....Nm	320	320	320	320
Rear hubs ....Nm	190	190	190	190
Wheel nuts / bolts ....Nm	90	90	90	90
Glow plugs ....Nm	22	22	22	22
Clutch pressure plate bolts ....Nm	20	20	20	20
Injection pump sprocket.......................Nm	50	50	50	50
Injectors..Nm	90	90	90	90
Injection pump mounting bolts ....Nm	–	–	–	–
Injector pipe unions............................Nm	20	20	20	20
Capacities				
Engine oil & filter.................................litres	4.25 A/C: 4.0	4.25 A/C: 4.0	4.25 A/C: 4.0	4.25 A/C: 4.0
Gearbox..litres	1.9	1.9	1.9	1.9
Automatic transmission ....litres	2.4	2.4	2.4	2.4
Final drive ....litres	WT	WT	WT	WT
Cooling system....................................litres	9.0	9.0	8.5	8.5
Fuel tank..litres	56	56	56	56

Notes

ZX 1.8D (Lucas) 1993 to 1996
[1]Bolt length below head: ≤121.5 mm Bolt with guide boss: ≤125.5 mm
ZX 1.8D (Bosch) 1993 to 1996
[1]Bolt length below head: ≤121.5 mm Bolt with guide boss: ≤125.5 mm
ZX 1.9D (Lucas) 1992 to 1996
[1]Bolt length below head: ≤121.5 mm Bolt with guide boss: ≤125.5 mm
ZX 1.9D (Bosch) 1992 to 1996
[1]Bolt length below head: ≤121.5 mm Bolt with guide boss: ≤125.5 mm

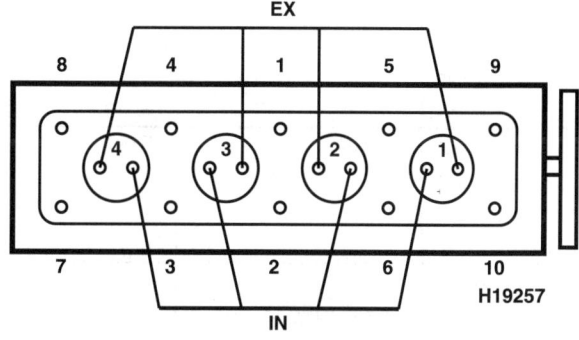

1769 cm³ / 1905 cm³

– Not applicable, or information not available

CITROEN

	ZX 1.9D (Lucas) 1995 to 1998	ZX 1.9D (Bosch) 1995 to 1996	ZX 1.9D Turbo (Bosch) 1992 to 1996	ZX 1.9D Turbo (Lucas) 1992 to 1998
Engine				
Engine type/code............................	XUD9A/L D9B SOHC 51kW	XUD9AY/L3 DJY SOHC 50kW	XUD9TE/L D8A Turbo 67.5kW	XUD9T/L D8A Turbo 67.5kW
Capacity (cm³) / cylinders......................	1905 / 4	1905 / 4	1905 / 4	1905 / 4
Compression ration / pressure ....bar	23.0 / _	23.0 / _	21.8 / _	21.8 / _
Torque output ....Nm	90	90	202	202
Oil pressure ....idle [running] bar	[3.5]	[3.5]	[4.9]	[4.9]
Oil temperature ....°C	80	80	80	80
Valve clearances - inlet (mm)	0.15 ± 0.08	0.15 ± 0.08	0.15 ± 0.08	0.15 ± 0.08
- exhaust (mm)	0.30 ± 0.08	0.30 ± 0.08	0.30 ± 0.08	0.30 ± 0.08
Injection order	1-3-4-2	1-3-4-2	1-3-4-2	1-3-4-2
No. 1 cylinder position	FE	FE	FE	FE
Cooling system				
Thermostat opening temperature ....°C	89	89	83	83
Radiator cap pressure ....bar	1.4	1.4	1.4	1.4
Fuel system				
Idle speed ....rpm	800 ± 50	800 ± 50	800 ± 50	800 ± 50
Maximum (no load) speed ....rpm	5150 ± 125	5150 ± 125	5050 ± 125	5050 ± 125
Smoke test/opacity ....M⁻¹ %	2.5	2.5	2.5	2.5
Static timing method................................	Dial gauge	Plunger travel	Plunger travel	Dial gauge
Timing dimension...............................mm	Dimension on pump	0.9	0.66	Dimension on pump
Crankshaft position ....mm [°]	_	TDC [0]	TDC [0]	TDC [0]
Turbo type / ref / pressure ....bar	_	_	KKK K14 / Garrett T2'	KKK K14 / Garrett T2'
Injection pump make	Lucas	Bosch	Bosch	Lucas
Injection pump part no.............................	XUD101 8443 3B980A	VER 425/6 XUD 211	VER 445 XUD 203	XUD 103
Injector Make / type................................	CAV	Bosch	Bosch	CAV
Injector part no.......................................	RDNOSDC 6751D	DNOSD 299A	DNOSD 299	_
Injection type..	Indirect DPC	Indirect VER	Indirect VER	Indirect
Injection opening pressure, New [used]...bar	125.0 ± 5.0	130.0 +0 -5.0	175.0 +5.0 -0	175.0 +5.0 -0
Glow plugs				
Maker	Beru/Bosch	Beru/Bosch	Beru/Bosch	Beru/Bosch
Type	0100 221 133 / 0250 201 019	0100 221 133 / 0250 201 019	0100 226 186 / 0250 201 039	0100 226 186 / 0250 201 039
Nominal rating...................................V/A	11 / 12	11 / 12	11 / 12	11 / 12
Brakes				
minimum friction material thickness				
Front.............................mm	2.0	2.0	2.0	2.0
Rear..............................mm	1.0	1.0	1.0	1.0
Tyres - Saloon / Hatch................Size	175/65x14	175/65x14	175/65x14	175/70x14
- Estate / Van.............................Size	_	_	_	175/70x14
Pressure - front / rear - Saloon / Hatch...bar	2.2 / 2.1	2.2 / 2.1	2.4 / 2.2	2.3 / 2.1
- Estate / Van ......bar	_	_	_	_
Front suspension / wheel alignment				
Toe-in (+) / Toe-out (−).....................mm [°]	-1.0 to -3.0 PAS: 1.0 to 3.0	-1.0 to -3.0 PAS: 1.0 to 3.0	-1.0 to -3.0 PAS: 1.0 to 3.0	-1.0 to -3.0 PAS: 1.0 to 3.0
Camber	0 ± 40'	0° ± 40'	0° ± 40'	0° ± 40'
Castor	1°30' ± 40' PAS: 3° ± 40'	1°30' ± 40' PAS: 3° ± 40'	1°30' ± 40' PAS: 3° ± 40'	1°30' ± 40' PAS: 3° ± 40'
King pin inclination................................	10°45' ± 40'	10°45' ± 40'	10°45' ± 40'	10°45' ± 40'
Rear suspension / wheel alignment				
Toe-in (+) / Toe-out (−)....................mm [°]	2.5 to 6.0	2.5 to 6.0	2.5 to 6.0	2.5 to 6.0
Camber	-1° ± 40'	-1° ± 40'	-1° ± 40'	-1° ± 40'

CITROEN

	ZX 1.9D (Lucas) 1995 to 1998	ZX 1.9D (Bosch) 1995 to 1996	ZX 1.9D Turbo (Bosch) 1992 to 1996	ZX 1.9D Turbo (Lucas) 1992 to 1998
Torque wrench settings				
Cylinder head - stage 1 ....Nm	20[1]	20[1]	20[2]	20[2]
- stage 2 ....Nm	70	70	60	60
- stage 3 ....Nm	+ 140°	+ 140°	20 (oiled)	+ 220°
- stage 4 ....Nm	–	–	+ 220°	–
- stage 5 ....Nm	–	–	–	–
- stage 6 ....Nm	–	–	–	–
Big-end bearings.............................Nm	20 + 70° N	20 + 70° N	20 + 70° N	20 + 70° N
Main bearings..................................Nm	70	70	70	70
Crankshaft pulley bolt ....Nm	40 + 60°	40 + 60°	40 + 51°	40 + 50°
Camshaft pulley bolt ....Nm	35	35	35	35
Flywheel [driveplate] bolt....................Nm	50	50	50	50
Front hubs ....Nm	320	320	320	320
Rear hubs ....Nm	190	190	190	190
Wheel nuts / bolts ....Nm	90	90	90	90
Glow plugs ....Nm	22	22	22	22
Clutch pressure plate bolts ....Nm	20	20	20	20
Injection pump sprocket.....................Nm	50	50	50	50
Injectors..Nm	90	90	90	90
Injection pump mounting bolts ....Nm	–	–	22	22
Injector pipe unions...........................Nm	20	20	20	20
Capacities				
Engine oil & filter...............................litres	4.25 A/C: 4.0	4.25 A/C: 4.0	4.25 A/C: 4.0	4.25 A/C: 4.0
Gearbox..litres	1.9	1.9	1.9	1.9
Automatic transmission ....litres	2.4	2.4	–	–
Final drive ....litres	WT	WT	WT	WT
Cooling system..................................litres	8.5	8.5	9.0	9.0
Fuel tank...litres	56	56	56	56

Notes

ZX 1.9D (Lucas) 1995 to 1998
[1]Bolt length below head: ≤121.5 mm Bolt with guide boss: ≤125.5 mm
ZX 1.9D (Bosch) 1995 to 1996
[1]Bolt length below head: ≤121.5 mm Bolt with guide boss: ≤125.5 mm
ZX 1.9D Turbo (Bosch) 1992 to 1996
[1]0.66 to 0.76 bar @ 2000 rpm
[2]Bolt length below head: ≤146.5 mm Bolt with guide boss: ≤150.5 mm
ZX 1.9D Turbo (Lucas) 1992 to 1998
[1]0.66 to 0.76 bar @ 2000 rpm
[2]Bolt length below head: ≤146.5 mm Bolt with guide boss: ≤150.5 mm

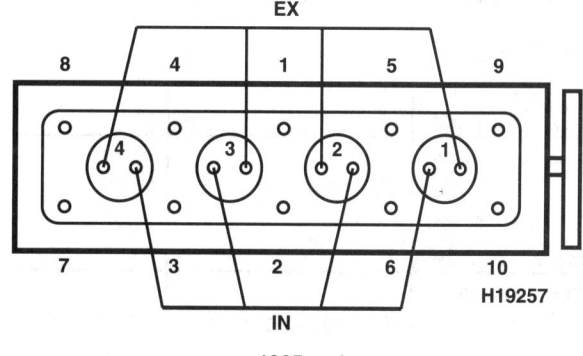

1905 cm³

– Not applicable, or information not available

CITROEN

	ZX 1.9D Turbo (Bosch) 1995 to 1998	ZX 1.9D Turbo (Lucas) 1997 to 1998	Xsara 1.5 D (Lucas) 1998 to 2000	Xsara 1.5 D (Bosch) 1998 to 2000
Engine				
Engine type/code...............................	XUD9TEY/L3 DHY Turbo 66kW	XUD9TEY/L3 DHY Turbo 66kW	TUD5 L3 VJZ/Y SOHC 42/40kW	TUD5 L3 VJZ/Y SOHC 42/40kW
Capacity (cm^3) / cylinders.......................	1905 / 4	1905 / 4	1527 / 4	1527 / 4
Compression ration / pressure ....bar	21.8 / _	21.8 / _	23.0 /_	23.0 /_
Torque output ....Nm	202	202	95	95
Oil pressure ....idle [running] bar	[4.9]	[4.9]	[4.0 @ 4000]	[4.0 @ 4000]
Oil temperature ....°C	80	80	90	90
Valve clearances - inlet (mm)	0.15 ± 0.08	0.15 ± 0.08	0.15	0.15
- exhaust (mm)	0.30 ± 0.08	0.30 ± 0.08	0.30	0.30
Injection order	1-3-4-2	1-3-4-2	1-3-4-2	1-3-4-2
No. 1 cylinder position	FE	FE	FE	FE
Cooling system				
Thermostat opening temperature ....°C	83	83	88	88
Radiator cap pressure ....bar	1.4	1.4	1.4	1.4
Fuel system				
Idle speed ....rpm	800 ± 50	800 ± 50	800 ± 50[2]	800 ± 50[2]
Maximum (no load) speed ....rpm	5050 ± 125	5050 ± 125	5450 ± 125	5450 ± 125
Smoke test/opacity ....M^{-1} %	2.5	2.5	2.5	2.5
Static timing method................................	Plunger travel	Plunger travel	Refer to wsm	_
Timing dimension................................mm	0.66	0.66	_	_
Crankshaft position ....mm [°]	TDC [0]	TDC [0]	_	_
Turbo type / ref / pressure ....bar	_	KKK K14 /Garrett T2[1]	_	_
Injection pump make	Bosch	Lucas	Lucas	Bosch
Injection pump part no............................	VER 445 XUD 210	XUD 110	R8444B662B	VE4 537
Injector Make / type	Bosch	_		
Injector part no.......................................	DNOSD 299	_	RDN 12SDC 6849 E	_
Injection type..	Indirect VER	Indirect	Indirect DPC	Indirect VER
Injection opening pressure, New [used]...bar	175.0 +5.0 -0	175	130 to 135	115 to 125
Glow plugs				
Maker	Beru/Bosch	Beru/Bosch	Beru/Champion	Bosch
Type	0100 226 186 / 0250 201 039	0100 226 186 / 0250 201 039	A 0 100 226 188 / C625 037 00	0 250 201 033
Nominal rating.................................V/A	11 / 12	11 / 12	11 / 9	11 / 9
Brakes				
minimum friction material thickness				
Front...............................mm	2.0	2.0	2.0	2.0
Rear.................................mm	1.0	1.0	2.0	2.0
Tyres - Saloon / Hatch.......................Size	175/70x14	175/70x14	175/65x14: 185/65x14	175/65x14: 185/65x14
- Estate / Van...........................Size	175/70x14	175/70x14	185/65x14	185/65x14
Pressure - front / rear - Saloon / Hatch ...bar	2.3 / 2.1	2.3 / 2.1	2.2 / 2.2: 2.2 / 2.3	2.2 / 2.2: 2.3 / 2.2
- Estate / Van ......bar	_	_	2.2 / 2.3	2.3 / 2.2
Front suspension / wheel alignment				
Toe-in (+) / Toe-out (–).....................mm [°]	-1.0 to -3.0 PAS: 1.0 to 3.0	-1.0 to -3.0 PAS: 1.0 to 3.0	-1.5 ± 1.0	-1.5 ± 1.0
Camber	0° ± 40'	0° ± 40'	0° ± 1°	0° ± 1°
Castor	1°30' ± 40' PAS: 3° ± 40'	1°30' ± 40' PAS: 3° ± 40'	1°30' ± 40'	1°30' ± 40'
King pin inclination................................	10°45' ± 40'	10°45' ± 40'	10°50' ± 1'	10°50' ± 1'
Rear suspension / wheel alignment				
Toe-in (+) / Toe-out (–).....................mm [°]	2.5 to 6.0	2.5 to 6.0	4.5 ± 1.0	4.5 ± 1.0
Camber	-1° ± 40'	-1° ± 40'	-1°20' ± 40'	-1°20' ± 40'

CITROEN

	ZX 1.9D Turbo (Bosch) 1995 to 1998	ZX 1.9D Turbo (Lucas) 1997 to 1998	Xsara 1.5 D (Lucas) 1998 to 2000	Xsara 1.5 D (Bosch) 1998 to 2000
Torque wrench settings				
Cylinder head - stage 1 ....Nm	20[1]	20[2]	40[1]	40[1]
- stage 2 ....Nm	60	60	+ 260°	+ 260°
- stage 3 ....Nm	+ 220°	+ 220°	–	–
- stage 4 ....Nm	–	–	–	–
- stage 5 ....Nm	–	–	–	–
- stage 6 ....Nm	–	–	–	–
Big-end bearings.................................Nm	20 + 70° N	20 + 70° N	40 N	40 N
Main bearings.....................................Nm	70	70	20 + 45°	20 + 45°
Crankshaft pulley bolt ....Nm	40 + 50°	40 + 50°	110	110
Camshaft pulley bolt ....Nm	35	35	80	80
Flywheel [driveplate] bolt....................Nm	50	50	48	48
Front hubs ....Nm	320	320	250 N	250 N
Rear hubs ....Nm	190	190	WSM	WSM
Wheel nuts / bolts ....Nm	90	90	85	85
Glow plugs ....Nm	22	22	20[3]	25
Clutch pressure plate bolts ....Nm	20	20	15	15
Injection pump sprocket.....................Nm	50	50	23	23
Injectors...Nm	90	90	70	70
Injection pump mounting bolts ....Nm	22	22	25	25
Injector pipe unions............................Nm	20	20	20	20
Capacities				
Engine oil & filter................................litres	4.25 A/C: 4.0	4.25 A/C: 4.0	4.75	4.75
Gearbox..litres	1.9	1.9	2.0	2.0
Automatic transmission ....litres	–	–	–	–
Final drive...litres	WT	WT	WT	WT
Cooling system...................................litres	9.0	9.0	8.8	8.8
Fuel tank...litres	56	56	54	54

Notes

ZX 1.9D Turbo (Bosch) 1995 to 1998
[1]Bolt length below head: ≤146.5 mm Bolt with guide boss: ≤150.5 mm
ZX 1.9D Turbo (Lucas) 1997 to 1998
[1]0.66 to 0.76 bar @ 2000 rpm
[2]Bolt length below head: ≤146.5 mm Bolt with guide boss: ≤150.5 mm
Xsara 1.5 D (Lucas) 1998 to 2000
[1] Bolt length below head ≤184.4 mm
[2] With A/C: 825 ± 25 rpm
[3] Champion: 12 Nm
Xsara 1.5 D (Bosch) 1998 to 2000
[1]Bolt length ≤184.4 mm below head
[2] With A/C: 825 ± 25 rpm

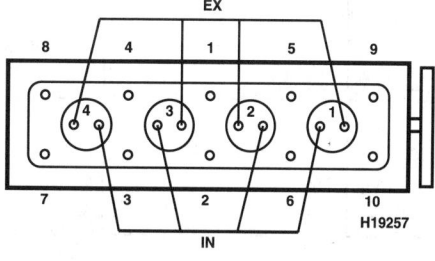

1905 cm³

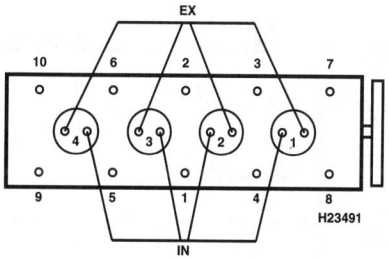

1527 cm³

– Not applicable, or information not available

CITROEN

Haynes
THE BOOK ®

	Xsara 1.8 D 1998 to 1999	Xsara 1.9 D (Bosch) 1998 to 2000	Xsara 1.9 D (Lucas) 1998 to 2000	Xsara 1.9 SD 1998 to 2000
Engine				
Engine type/code	XUD7 L3 A9A SOHC 43kW	XUD9A L3 DJY SOHC 50kW	XUD9A L3 DJY SOHC 50kW	XUD9BSD L4 DHV Turbo 55kW
Capacity (cm^3) / cylinders......................	1769 / 4	1905 / 4	1905 / 4	1905 / 4
Compression ration / pressure ....bar	23.0 /_	23.0 /_	23.0 /_	21.1 /_
Torque output ....Nm	100	110	110	135
Oil pressure ....idle [running] bar	[4.0]	[4.0]	[4.0]	[4.0]
Oil temperature ....°C	90	90	90	90
Valve clearances - inlet (mm)	0.20 ± 0.5	0.15 ± 0.5	0.15 ± 0.5	0.15 ± 0.8
- exhaust (mm)	0.40 ± 0.5	0.40 ± 0.5	0.40 ± 0.5	0.30 ± 0.8
Injection order	1-3-4-2	1-3-4-2	1-3-4-2	1-3-4-2
No. 1 cylinder position	FE	FE	FE	FE
Cooling system				
Thermostat opening temperature ....°C	89 to 95	89 to 95	89 to 95	89 to 95
Radiator cap pressure ....bar	1.4	1.4	1.4	1.4
Fuel system				
Idle speed ....rpm	850 ± 50	850 ± 50	850 ± 50	875 ± 25 N/A
Maximum (no load) speed ....rpm	5100 ± 125	5150 ± 125	5150 ± 125	5100 ± 125
Smoke test/opacity ....M^{-1} %	2.5	2.5	2.5	3.0
Static timing method..............................	Dial gauge	Plunger travel	Peg	Rotor lift
Timing dimension.............................mm	Dimension on pump	0.9	Dimension on pump	0.74
Crankshaft position ....mm [°]	TDC	TDC	TDC	TDC
Turbo type / ref / pressure ....bar	_	_	_	KKK 0.3 to 0.4 bar
Injection pump make	Lucas	Bosch	Lucas	Bosch
Injection pump part no.............................	683B	XUD LP05 R753	XUD 211	VE 4/8F 2300 R 753
Injector Make / type	_	_	_	_
Injector part no..	_	_	_	KCA 17 S 42
Injection type......................................	XUD LP04	Indirect	XUD LP05	VE4/8F 2300 R753 ECU control
Injection opening pressure, New [used]...bar	_	140	140	165 ± 5
Glow plugs				
Maker	_	Bosch	Bosch	Beru
Type	_	0 502 010 33	0 502 010 33	0100226326
Nominal rating...............................V/A	_	_	_	11 / 7.5
Brakes				
minimum friction material thickness				
Front.......................................mm	2.0	2.0	2.0	2.0
Rear.......................................mm	2.0	2.0	2.0	2.0
Tyres - Saloon / Hatch......................Size	175/65x14	175/65x14	175/65x14	175/65x14
- Estate / Van......................Size	185/65x14	185/65x14	185/65x14	185/65x14
Pressure - front / rear - Saloon / Hatch ...bar	2.2 / 2.2	2.2 / 2.2	2.2 / 2.2	2.2 / 2.2
- Estate / Van ......bar	2.3 / 2.3	2.3 / 2.3	2.3 / 2.3	2.3 / 2.3
Front suspension / wheel alignment				
Toe-in (+) / Toe-out (–)....................mm [°]	-1.5 ± 1.0	-1.5 ± 1.0	-1.5 ± 1.0	-1.5 ± 1.0
Camber	0° ± 1°	0° ± 1°	0° ± 1°	0° ± 1°
Castor	1°30' ± 40'	1°30' ± 40'	1°30' ± 40'	1°30' ± 40'
King pin inclination..............................	10°50' ± 1'	10°50' ± 1'	10°50' ± 1'	10°50' ± 1'
Rear suspension / wheel alignment				
Toe-in (+) / Toe-out (–)....................mm [°]	4.5 ± 1.0	4.5 ± 1.0	4.5 ± 1.0	4.5 ± 1.0
Camber	-1°20' ± 40'	-1°20' ± 40'	-1°20' ± 40'	-1°20' ± 40'

CITROEN

	Xsara 1.8 D 1998 to 1999	Xsara 1.9 D (Bosch) 1998 to 2000	Xsara 1.9 D (Lucas) 1998 to 2000	Xsara 1.9 SD 1998 to 2000
Torque wrench settings				
Cylinder head - stage 1 ....Nm	20¹	20¹	20¹	20¹
- stage 2 ....Nm	60	60	60	60
- stage 3 ....Nm	+ 180°	+ 180°	+ 180°	+ 180°
- stage 4 ....Nm	–	–	–	–
- stage 5 ....Nm	–	–	–	–
- stage 6 ....Nm		–	–	–
Big-end bearings......................Nm	20 + 70° N	20 + 70° N	20 + 70° N	20 + 70° N
Main bearings......................Nm	70	70	70	15 + 60°
Crankshaft pulley bolt ....Nm	40 + 51°	40 + 51°	40 + 51°	40 + 51°
Camshaft pulley bolt ....Nm	45	45	45	45
Flywheel [driveplate] bolt......................Nm	48	48	48	48
Front hubs ....Nm	M24: 320 M20: 250	M24: 320 M20: 250	M24: 320 M20: 250	M24: 320 M20: 250
Rear hubs ....Nm	WSM	WSM	WSM	WSM
Wheel nuts / bolts ....Nm	85	85	85	85
Glow plugs ....Nm	25	25	25	22
Clutch pressure plate bolts ....Nm	15	15	15	15
Injection pump sprocket......................Nm	–	50	50	50
Injectors......................Nm	90	90	90	90
Injection pump mounting bolts ....Nm	25	–	–	–
Injector pipe unions......................Nm	–	25	25	20
Capacities				
Engine oil & filter......................litres	4.75	4.25	4.25	4.25 A/C : 4.75
Gearbox......................litres	2.0	2.0	2.0	2.0
Automatic transmission ....litres	–	–	–	–
Final drive......................litres	WT	WT	WT	WT
Cooling system......................litres	6.5	6.5	6.5	6.5
Fuel tank......................litres	54	54	54	54

Notes

Xsara 1.8 D 1998 to 1999
¹Bolt length below head: ≤125.5 mm Bolt without guide boss: ≤121.5 mm

Xsara 1.9 D (Bosch) 1998 to 2000
¹Bolt length below head: ≤121.5 mm Bolt with guide boss: ≤125.5 mm

Xsara 1.9 D (Lucas) 1998 to 2000
¹Bolt length below head: ≤121.5 mm Bolt with guide boss: ≤125.5 mm

Xsara 1.9 SD 1998 to 2000
¹ Bolt length below head ≤125.5 mm

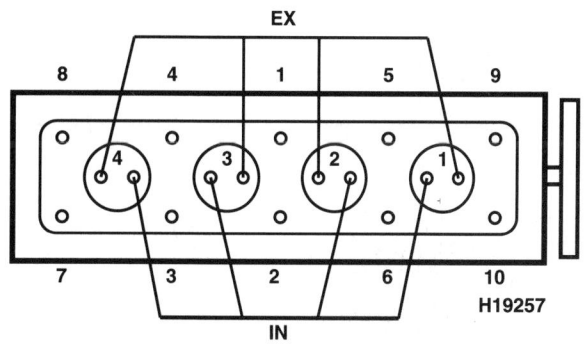

1769 cm³ / 1905 cm³

– Not applicable, or information not available

CITROEN

	Xsara 1.9 TD 1998 to 2000	Xsara 1.9 D 1998 to 2000	Xsara 2.0 HDi 1999 to 2000	BX 17 1987 to 1993
Engine				
Engine type/code................................	XUD9TE L3 DHY Turbo 66kW	DW8 WJZ SOHC EGR 51kW	DW10TD/L3 RHY SOHC 66kW	XUD7 161A SOHC 44kW
Capacity (cm³) / cylinders.....................	1905 / 4	1868 / 4	1997 / 4	1769 / 4
Compression ration / pressure ....bar	21.8 /_	23.0 /_	17.6 / ≥30 ± 5	23.0 / _
Torque output ....Nm	196	125	205	0
Oil pressure ....idle [running] bar	2.5 [4.9 @ 4000]	1.8 [3.7 @ 2000]	[3.8]	[3.5]
Oil temperature ....°C	90	80	80	80
Valve clearances - inlet (mm)	0.20 ± 0.5	0.15 ± 0.05	0: Hyd.	0.15 ± 0.08
- exhaust (mm)	0.40 ± 0.5	0.30 ± 0.05	0: Hyd.	0.30 ± 0.08
Injection order	1-3-4-2	1-3-4-2	1-3-4-2	1-3-4-2
No. 1 cylinder position	FE	FE	FE	FE
Cooling system				
Thermostat opening temperature ....°C	89 to 95	83	83	81 4.87 ▶: 88
Radiator cap pressure ....bar	1.4	1.4	1.4	1.0
Fuel system				
Idle speed ....rpm	850 ± 50	800 ± 25[2]	_	750 ± 50
Maximum (no load) speed ....rpm	5150 ± 125	5350 ± 125	_	5100 ± 125
Smoke test/opacity ....M⁻¹ %	3.0	3.0	2.0	2.5
Static timing method............................	Peg	Peg	_	Dial gauge
Timing dimension.................................mm	Dimension on pump	0.58 ± 0.04[6]	_	Dimension on pump
Crankshaft position ....mm [°]	TDC	_	_	[TDC hole]
Turbo type / ref / pressure ....bar	KKK 03 / Garrett GT15[2]	_	_	_
Injection pump make	Lucas	Bosch	Bosch	Roto Diesel
Injection pump part no...........................	XUD 110	[1]	EDC 15C2	R8443B264F
Injector Make / type	_	_	Bosch	Roto Diesel
Injector part no....................................	_	KCA 20S106[5]		RDNOSDC 6850 C
Injection type......................................	R8445B 081A	VE4 / 537	Direct EDC 15C2	Indirect DPC 052
Injection opening pressure, New [used]...bar	150	120[9]	200 to 1350	115 ± 5
Glow plugs				
Maker	Bosch	Beru/Champion	Bosch/Champion	Bosch
Type	0 100 226 186	0100226371 / CH185	0250 202 032 / CH170	0 250 201 019
Nominal rating....................................V/A	_	_	_	_
Brakes				
minimum friction material thickness				
Front.......................................mm	2.0	2.0	2.0	2.0
Rear.......................................mm	2.0	2.0	2.0	2.0
Tyres - Saloon / Hatch....................Size	175/65x14	175/65x14	175/65x14	165/70x14
- Estate / Van..........................Size	185/65x14	185/65x14	185/65x14	165/70x14
Pressure - front / rear - Saloon / Hatch ...bar	2.2 / 2.2	2.2 / 2.2	2.2 / 2.2	2.0 / 2.0
- Estate / Van ......bar	2.3 / 2.3	2.3 / 2.3	2.3 / 2.3	2.3 / 2.5
Front suspension / wheel alignment				
Toe-in (+) / Toe-out (–)....................mm [°]	-1.5 ± 1.0	-1.5 ± 1.0	-1.5 ± 1.0	0 to -3.0
Camber	0° ± 1°	0° ± 1°	0° ± 1°	0° ± 30'
Castor	1°30' ± 40'	1°30' ± 40'	1°30' ± 40'	2° ± 35'
King pin inclination................................	10°50' ± 1'	10°50' ± 1'	10°50' ± 1°	11°58'
Rear suspension / wheel alignment				
Toe-in (+) / Toe-out (–)....................mm [°]	4.5 ± 1.0	4.5 ± 1.0	4.5 ± 1.0	1.6 to 5.0
Camber	-1°20' ± 40'	-1°20' ± 40'	-1°20' ± 40	-9' ± 20' 88 ▶: -1° ± 20'

CITROEN

	Xsara 1.9 TD 1998 to 2000	Xsara 1.9 D 1998 to 2000	Xsara 2.0 HDi 1999 to 2000	BX 17 1987 to 1993
Torque wrench settings				
Cylinder head - stage 1 ....Nm	20[1]	20[4]	20[1]	30 N
- stage 2 ....Nm	60	60	60	70
- stage 3 ....Nm	+ 220°	+ 180°	+ 220°	+ 120°
- stage 4 ....Nm	–	–	–	–
- stage 5 ....Nm	–	–	–	–
- stage 6 ....Nm	–	–	–	–
Big-end bearings...............................Nm	20 + 70° N	20 + 70° N	20 + 70° N	50 N
Main bearings...................................Nm	15 + 60°	70	25 + 60°	70
Crankshaft pulley bolt ....Nm	40 + 51°	40 + 60°	197	150 LkC
Camshaft pulley bolt ....Nm	45	43	43	35
Flywheel [driveplate] bolt....................Nm	48	48 N	48	50 LkC
Front hubs ....Nm	M24: 320 M20: 250	M24: 320 M20: 250	320 M20: 250	270
Rear hubs ....Nm	WSM	WSM	WSM	270
Wheel nuts / bolts ....Nm	85	85	85	80 Alloy: 90
Glow plugs ....Nm	25	25	25	22
Clutch pressure plate bolts ....Nm	15	20	–	22
Injection pump sprocket......................Nm	50	23	–	50
Injectors...Nm	90	90	–	90
Injection pump mounting bolts ....Nm	22	20	–	18
Injector pipe unions............................Nm	20	25	–	20
Capacities				
Engine oil & filter................................litres	4.25 A/C : 4.0	4.75	4.5 A/C : 4.5	5.0
Gearbox..litres	2.0	2.0	2.0	2.0
Automatic transmission ....litres	–	–	–	–
Final drive...litres	WT	WT	WT	WT
Cooling system..................................litres	6.5	9.0	7.0	6.5
Fuel tank...litres	54	54	54	52

Notes

Xsara 1.9 TD 1998 to 2000
[1]Bolt length below head: ≤146.5 mm Bolt with guide boss: ≤150.5 mm
[2]0.66 to 0.76 bar @ 2000 rpm
Xsara 1.9 D 1998 to 2000
[1]Lucas Varity DPC pump available as alternative
[2]With A/C: 875 ± 25
[3]Lucas: 142 to 149
[4]Bolt length ≤125.5 mm below head
[5]or Lucas LRC6736001
[6]Lucas: Dimension on pump
Xsara 2.0 HDi 1999 to 2000
[1] Bolt length ≤133.3 mm below head

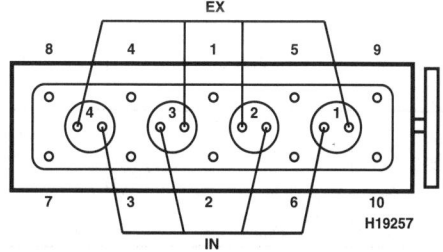

1769 cm³ / 1905 cm³

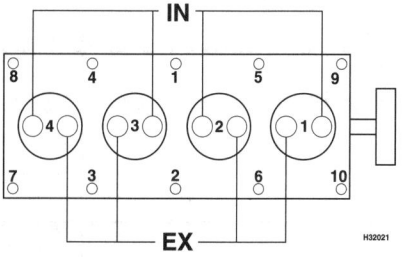

1868 cm³ / 1997 cm³

– Not applicable, or information not available

CITROEN

	BX 17 1987 to 1993	BX 17 Turbo 1988 to 1993	BX 17 Turbo 1988 to 1993	BX 19 1987 to 1994
Engine				
Engine type/code	XUD7 161A SOHC 44kW	XUD7TE A8A SOHC Turbo 66kW	XUD7TE A8A SOHC Turbo 66kW	XUD9A D9B SOHC 51kW
Capacity (cm³) / cylinders	1769 / 4	1769 / 4	1769 / 4	1905 / 4
Compression ration / pressure ......bar	23.0 / _	22.0 / _	22.0 / _	23.0 / _
Torque output ....Nm	0	0	0	0
Oil pressure ....idle [running] bar	[3.5]	[3.4]	[3.4]	[3.5]
Oil temperature ....°C	80	80	80	80
Valve clearances - inlet (mm)	0.15 ± 0.08	0.15 ± 0.08	0.15 ± 0.08	0.15 ± 0.08
- exhaust (mm)	0.30 ± 0.08	0.30 ± 0.08	0.30 ± 0.08	0.30 ± 0.08
Injection order	1-3-4-2	1-3-4-2	1-3-4-2	1-3-4-2
No. 1 cylinder position	FE	FE	FE	FE
Cooling system				
Thermostat opening temperature ....°C	81	88	88	88
Radiator cap pressure ....bar	1.0	1.0	1.0	1.0
Fuel system				
Idle speed ....rpm	750 ± 50	750 ± 50	750 ± 50	750 ± 50
Maximum (no load) speed ....rpm	5100 ± 150	4800 ± 100	_	5100 ± 150
Smoke test/opacity ....M⁻¹ %	2.5	2.5	2.5	2.5
Static timing method....................	Dial gauge	Dial gauge	Dial gauge	Dial gauge
Timing dimension......................mm	0.9 ABDC	Dimension on pump	0.8 ABDC	Dimension on pump
Crankshaft position ....mm [°]	TDC peg	[TDC hole]	TDC peg	[TDC hole]
Turbo type / ref / pressure ....bar	_	_	_	_
Injection pump make	Bosch	Roto Diesel	Bosch	Roto Diesel
Injection pump part no.	VE R171.1	R8443B451B	R8443B451B	R8443B380A
Injector Make / type	Bosch	Roto Diesel	Bosch	Roto Diesel
Injector part no.......................	DNOSD 256	RDNOSDC 6862 C	DNOSD 289	RDNOSDC 6751 C
Injection type..........................	Indirect VER	Indirect DPC 058	Indirect VE R 316	Indirect DPC 057
Injection opening pressure, New [used]...bar	130 +5 -0	130 ± 5	130 ± 5	125 ± 5
Glow plugs				
Maker	Bosch	Bosch	Bosch	Bosch
Type	0 250 201 019	0 250 201 019	0 250 201 019	0 250 201 019
Nominal rating.......................V/A	_	_	_	_
Brakes				
minimum friction material thickness				
Front......................mm	2.0	2.0	2.0	2.0
Rear......................mm	1.0	2.0	2.0	2.0
Tyres - Saloon / Hatch ....Size	165/70x14	165/70x14	165/70x14	165/70x14
- Estate / Van..............Size	165/70x14	165/70x14	165/70x14	165/70x14
Pressure - front / rear - Saloon / Hatch ...bar	2.0 / 2.0	2.0 / 2.0	2.0 / 2.0	2.0 / 2.0
- Estate / Van ......bar	2.3 / 2.5	2.3 / 2.5	2.3 / 2.5	2.3 / 2.5
Front suspension / wheel alignment				
Toe-in (+) / Toe-out (–)................mm [°]	0 to -3.0	0 to -3.0	0 to -3.0	0 to -3.0
Camber	0° ± 30'	0° ± 30'	0° ± 30'	0° ± 30'
Castor	2° ± 35'	2° ± 35'	2° ± 35'	2° ± 35'
King pin inclination....................	11°58'	11°58'	11°58'	11°58'
Rear suspension / wheel alignment				
Toe-in (+) / Toe-out (–)................mm [°]	1.6 to 5.0	1.6 to 5.0	1.6 to 5.0	1.6 to 5.0
Camber	-9' ± 20' 88 ▶: -1° ± 20'	-1° ± 20'	-1° ± 20'	-1° ± 20'

CITROEN

Torque wrench settings	BX 17 1987 to 1993	BX 17 Turbo 1988 to 1993	BX 17 Turbo 1988 to 1993	BX 19 1987 to 1994
Cylinder head - stage 1 ....Nm	30 N	30 N	30 N	30 N
- stage 2 ....Nm	70	70	70	70
- stage 3 ....Nm	+ 120°	+ 120°	+ 120°	+ 120°
- stage 4 ....Nm	–	–	–	–
- stage 5 ....Nm	–	–	–	–
- stage 6 ....Nm	–	–	–	–
Big-end bearings.................................Nm	50 N	50 N	50 N	50 N
Main bearings.....................................Nm	70	70	70	70
Crankshaft pulley bolt ....Nm	150 LkC	150 LkC	150 LkC	150 LkC
Camshaft pulley bolt ....Nm	35	35	35	35
Flywheel [driveplate] bolt....................Nm	50 LkC	50 LkC	50 LkC	50 LkC
Front hubs ....Nm	270	270	270	270
Rear hubs ....Nm	270	270	270	270
Wheel nuts / bolts ....Nm	80 Alloy: 90	80 Alloy: 90	80 Alloy: 90	80 Alloy: 90
Glow plugs ....Nm	22	22	22	22
Clutch pressure plate bolts ....Nm	22	22	22	22
Injection pump sprocket......................Nm	50	50	50	50
Injectors...Nm	90	90	90	90
Injection pump mounting bolts ....Nm	18	18	18	18
Injector pipe unions............................Nm	20	20	20	20
Capacities				
Engine oil & filter................................litres	5.0	5.0	5.0	5.0
Gearbox...litres	2.0	2.0	2.0	2.0
Automatic transmission ....litres	–	–	–	6.0
Final drive ....litres	WT	WT	WT	WT
Cooling system...................................litres	6.5	6.5	6.5	6.5
Fuel tank..litres	52	66	66	52

Notes

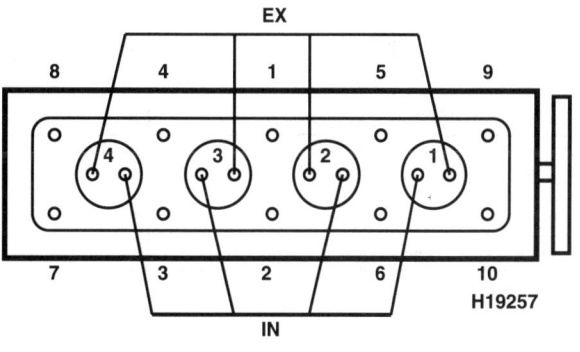

1769 cm³ / 1905 cm³

– Not applicable, or information not
  available

CITROEN

Haynes THE BOOK ®

	BX 19 1988 to 1994	Xantia 1.9D (Lucas) 1993 to 1996	Xantia 1.9D (Bosch) 1993 to 1996	Xantia 1.9D (Bosch) 1995 to 1996
Engine				
Engine type/code..................................	XUD9A D9B SOHC 51kW	XUD9A/L D9B SOHC 51kW	XUD9A/L D9B SOHC 51kW	XUD9A/Y DJY SOHC 50kW
Capacity (cm³) / cylinders........................	1905 / 4	1905 / 4	1905 / 4	1905 / 4
Compression ration / pressure ....bar	23.0 / _	23.0 / _	23.0 / _	23.0 / _
Torque output ....Nm	0	120	120	120
Oil pressure ....idle [running] bar	[3.5]	[3.5]	[3.5]	[3.5]
Oil temperature ....°C	80	80	80	80
Valve clearances - inlet (mm)	0.15 ± 0.08	0.15 ± 0.08	0.15 ± 0.08	0.15 ± 0.08
- exhaust (mm)	0.30 ± 0.08	0.30 ± 0.08	0.30 ± 0.08	0.30 ± 0.08
Injection order	1-3-4-2	1-3-4-2	1-3-4-2	1-3-4-2
No. 1 cylinder position	FE	FE	FE	FE
Cooling system				
Thermostat opening temperature ....°C	88	89	89	89
Radiator cap pressure ....bar	1.0	1.4	1.4	1.4
Fuel system				
Idle speed ....rpm	750 ± 50	800 ± 50	800 ± 50	800 ± 50
Maximum (no load) speed ....rpm	5100 ± 150	5150 ± 125	5100 ± 125	5100 ± 125
Smoke test/opacity ....M⁻¹ %	2.5	2.5	2.5	2.5
Static timing method...............................	Dial gauge	Dial gauge	Plunger travel	Plunger travel
Timing dimension................................mm	0.9 ABDC	Dimension on pump	1.07 [AT: 0.98]	1.07 [AT: 0.98]
Crankshaft position ....mm [°]	TDC peg	TDC [0]	TDC [0]	TDC [0]
Turbo type / ref / pressure ....bar	_	_	_	_
Injection pump make	Bosch	Lucas	Bosch	Bosch
Injection pump part no.............................	_	XUD101 8443 B952B. 95▶: B953C	VER 425/ XUD 201	VER 425/ XUD 211
Injector Make / type	Bosch	CAV	Bosch	Bosch
Injector part no......................................	DNOSD 287 +	RDNOSDC 6751D	DNOSD +299A	DNOSD +299A
Injection type...	Indirect VER 272.2	Indirect	Indirect	Indirect
Injection opening pressure, New [used]...bar	130 +5 -0	125.0 ± 5.0	130.0 +5.0 -0	130.0 +5.0 -0
Glow plugs				
Maker	Bosch	Beru/Bosch	Beru/Bosch	Beru/Bosch
Type	0 250 201 019	0100 221 133 / 0250 201 019'	0100 221 133 / 0250 201 019	0100 221 133 / 0250 201 019
Nominal rating ....V/A	_	11 /	11 /	11 /
Brakes				
minimum friction material thickness				
Front...mm	2.0	3.0	3.0	3.0
Rear..mm	2.0	2.0	2.0	2.0
Tyres - Saloon / Hatch......................Size	165/70x14	175/70x14	175/70x14	175/70x14
- Estate / Van............................Size	165/70x14	_	_	_
Pressure - front / rear - Saloon / Hatch ...bar	2.0 / 2.0	2.3 / 2.1	2.3 / 2.1	2.3 / 2.1
- Estate / Van ......bar	2.3 / 2.5	_		
Front suspension / wheel alignment				
Toe-in (+) / Toe-out (–)....................mm [°]	0 to -3.0	0 to -3.0	0 to -3.0	0 to -3.0
Camber...	0° ± 30'	0° ± 30'	0° ± 30'	0° ± 30'
Castor...	2° ± 35'	1° ± 30' PAS/96▶: 3° ± 30'	1° ± 30' PAS/96▶: 3° ± 30'	1° ± 30' PAS/96▶: 3° ± 30'
King pin inclination..................................	11°58'	13°20' ± 30' 96▶: 13°15' ± 35'	13°20' ± 30' 96▶: 13°15' ±35'	13°20' ± 30' 96▶: 13°15' ± 35'
Rear suspension / wheel alignment				
Toe-in (+) / Toe-out (–)....................mm [°]	1.6 to 5.0	1.0 to 6.0	1.0 to 6.0	1.0 to 6.0
Camber..	-1° ± 20'	-1°15' ± 20'	-1°15' ± 20'	-1°15' ± 20'

CITROEN

Torque wrench settings	BX 19 1988 to 1994	Xantia 1.9D (Lucas) 1993 to 1996	Xantia 1.9D (Bosch) 1993 to 1996	Xantia 1.9D (Bosch) 1995 to 1996
Cylinder head - stage 1 ....Nm	30 N	20[2]	20[1]	20[1]
- stage 2 ....Nm	70	60	60	60
- stage 3 ....Nm	+ 120°	+ 180°	+ 180°	+ 180°
- stage 4 ....Nm	–	–	–	–
- stage 5 ....Nm	–	–	–	–
- stage 6 ....Nm	–	–	–	–
Big-end bearings.............................Nm	50 N	20 + 70° N	20 + 70° N	20 + 70° N
Main bearings................................Nm	70	70	70	70
Crankshaft pulley bolt ....Nm	150 LkC	40 + 60°	40 + 60°	40 + 60°
Camshaft pulley bolt ....Nm	35	35	35	35
Flywheel [driveplate] bolt....................Nm	50 LkC	50	50	50
Front hubs....................................Nm	270	320	320	320
Rear hubs....................................Nm	270	280	280	280
Wheel nuts / bolts ....Nm	80 Alloy: 90	90	90	90
Glow plugs ....Nm	22	Bosch: 25 Beru: 20	Bosch: 25 Beru: 20	Bosch: 25 Beru: 20
Clutch pressure plate bolts ....Nm	22	20	20	20
Injection pump sprocket....................Nm	50	50	50	50
Injectors....................................Nm	90	90	90	90
Injection pump mounting bolts ....Nm	18	22	22	22
Injector pipe unions..........................Nm	20	25	20	20
Capacities				
Engine oil & filter................................litres	5.0	4.2 A/C: 4.0	4.2 A/C: 4.0	4.25 A/C: 4.0
Gearbox..litres	2.0	1.9	1.9	1.9
Automatic transmission ....litres	6.0	2.4	2.4	2.4
Final drive.....................................litres	WT	WT	WT	WT
Cooling system................................litres	6.5	7.5 A/C: 8.5	7.5 A/C: 8.5	7.5 A/C: 8.5
Fuel tank......................................litres	52	65	65	65

Notes

Xantia 1.9D (Lucas) 1993 to 1996
[1]AT: 0100 226 186 / 0250 201 039
[2]Bolt length below head: ≤121.5 mm Bolt with guide boss: ≤125.5 mm
Xantia 1.9D (Bosch) 1993 to 1996
[1]Bolt length below head: ≤121.5 mm Bolt with guide boss: ≤125.5 mm
Xantia 1.9D (Bosch) 1995 to 1996
[1]Bolt length below head: ≤121.5 mm Bolt with guide boss: ≤125.5 mm

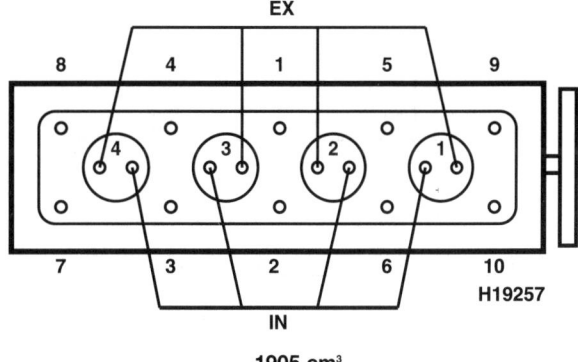

1905 cm³

– Not applicable, or information not available

CITROEN

	Xantia 1.9D Turbo 1993 to 1996	Xantia 1.9D Turbo 1995 to 1996	Xantia 1.9D Turbo 1996 to 2000	Xantia 1.9D LPT Turbo 1997 to 2000
Engine				
Engine type/code....................	XUD9TF/L D8B OHC Turbo 67kW	XUD9TF/Y DHX OHC Turbo 66kW	XUD9BTF/Y/L3 DHX Turbo 65kW	XUD9SD/L3 DHW LP Turbo 55kW
Capacity (cm³) / cylinders......................	1905 / 4	1905 / 4	1905 / 4	1905 / 4
Compression ration / pressure ....bar	21.8 / _	21.8 / _	23.0 / _	21.1 / _
Torque output ....Nm	202	202	197	135
Oil pressure ....idle [running] bar	[4.9]	[4.9]	[4.9]	[4.0 @ 4000]
Oil temperature ....°C	80	80	80	80
Valve clearances - inlet (mm)	0.15 ± 0.08	0.15 ± 0.08	0.15 ± 0.08	0.15 ± 0.08
- exhaust (mm)	0.30 ± 0.08	0.30 ± 0.08	0.30 ± 0.08	0.30 ± 0.08
Injection order	1-3-4-2	1-3-4-2	1-3-4-2	1-3-4-2
No. 1 cylinder position	FE	FE	FE	FE
Cooling system				
Thermostat opening temperature ....°C	89	89	89	83
Radiator cap pressure ....bar	1.4	1.4	1.4	1.40
Fuel system				
Idle speed ....rpm	800 ± 50	800 ± 50	800 ± 50	800 ± 50 A/C: 850 ± 50
Maximum (no load) speed ....rpm	5100 ± 80	5100 ± 80	5100 ± 80	5150 ± 125
Smoke test/opacity ....M⁻¹ %	2.5	2.5	2.5	2.5
Static timing method.............................	Plunger travel	Plunger travel	Plunger travel	Dial gauge
Timing dimension............................mm	0.66	0.66	0.66	Dimension on pump
Crankshaft position ....mm [°]	TDC [0]	TDC [0]	TDC [0]	TDC [0]
Turbo type / ref / pressure ....bar	KKK K14 / Garrett T2'	KKK K14 / Garrett T2'	KKK K14 / Garrett T2'	KKK 0.6 bar
Injection pump make	Bosch	Bosch	Bosch	Lucas
Injection pump part no.............................	VER 445/2 XUD 203	VER 445/2 XUD 203	VP20 AS3.1 XUD BC02	XUDLP06
Injector Make / type	Bosch	Bosch	Bosch	CAV
Injector part no.............................	DNOSD 299	DNOSD 299	_	RDN12 3DC 6895
Injection type.............................	VER	VER	VP20	XUD LP06 791B / 711B
Injection opening pressure, New [used]...bar	175.0 +5.0 -0	175.0 +5.0 -0	175	135.0
Glow plugs				
Maker	Beru/Bosch	Beru/Bosch	Beru/Bosch	Beru/Bosch
Type	0100 226 186 / 0250 201 039	0100 226 186 / 0250 201 039	0100 226 186 / 0250 201 039	0100 226 186 / 0250 201 033
Nominal rating............................V/A	11 /	_		
Brakes				
minimum friction material thickness				
Front.......................................mm	3.0 Break: 2.0	3.0 Break: 2.0	3.0 Break: 2.0	3.0 Break: 2.0
Rear.......................................mm	2.0	2.0	2.0	2.0
Tyres - Saloon / Hatch.......................Size	185/65x14: 195/55x15	185/65x14: 195/55x15	185/65x14: 195/55x15	185/65x14
- Estate / Van.........................Size	185/65x15	185/65x15	185/65x15	185/65x14
Pressure - front / rear - Saloon / Hatch ...bar	Refer to car	Refer to car	Refer to car	2.3 / 2.1
- Estate / Van ......bar	Refer to car	Refer to car	Refer to car	Refer to car
Front suspension / wheel alignment				
Toe-in (+) / Toe-out (–).....................mm [°]	0 to -3.0	0 to -3.0	0 to -3.0	0 to -3.0
Camber	0° ± 30'	0° ± 30'	0° ± 30'	0° ± 30'
Castor	3° ± 30'	3° ± 30'	3° ± 30'	3° ± 30'
King pin inclination............................	13°20' ± 30' 96▶: 13°15'±35'	13°20' ± 30' 96▶: 13°15'±35'	13°20' ± 30' 96▶: 13°15'±35'	13°20' ± 30' 96▶: 13°15'±35'
Rear suspension / wheel alignment				
Toe-in (+) / Toe-out (–).....................mm [°]	1.0 to 6.0	1.0 to 6.0	1.0 to 6.0	1.0 to 6.0
Camber	-1°15' ± 20'	-1°15' ± 20'	-1°15' ± 20'	-1°15' ± 20'

CITROEN

	Xantia 1.9D Turbo 1993 to 1996	Xantia 1.9D Turbo 1995 to 1996	Xantia 1.9D Turbo 1996 to 2000	Xantia 1.9D LPT Turbo 1997 to 2000
Torque wrench settings				
Cylinder head - stage 1 ....Nm	20[2]	20[2]	20[2]	20[1]
- stage 2 ....Nm	60	60	60	60
- stage 3 ....Nm	+ 220°	+ 220°	+ 220°	+ 180°
- stage 4 ....Nm	–	–	–	–
- stage 5 ....Nm	–	–	–	–
- stage 6 ....Nm	–	–	–	–
Big-end bearings.............................Nm	20 + 70° N	20 + 70° N	20 + 70° N	20 + 70° N
Main bearings.................................Nm	70	70	70	70
Crankshaft pulley bolt ....Nm	40 + 60°	40 + 60°	40 + 51°	40 + 60°
Camshaft pulley bolt ....Nm	–	35	35	35
Flywheel [driveplate] bolt....................Nm	50	50	50	50
Front hubs ....Nm	320	320	320	320
Rear hubs ....Nm	280	280	280	280
Wheel nuts / bolts ....Nm	90	90	90	90
Glow plugs ....Nm	Bosch: 25 Beru: 20	Bosch: 25 Beru: 20	Bosch: 25 Beru: 20	Bosch: 25 Beru: 20
Clutch pressure plate bolts ....Nm	20	20	20	20
Injection pump sprocket.....................Nm	50	50	50	50
Injectors...Nm	90	90	90	90
Injection pump mounting bolts ....Nm	22	22	22	22
Injector pipe unions...........................Nm	20	20	20	20
Capacities				
Engine oil & filter.............................litres	4.25 A/C: 4.0	4.25 A/C: 4.0	4.25 A/C: 4.0	4.25 A/C: 4.0
Gearbox...litres	1.9	1.9	1.9	1.9
Automatic transmission ....litres	–	–	–	–
Final drive ....litres	WT	WT	WT	WT
Cooling system................................litres	9.0	9.0	9.0	9.0
Fuel tank..litres	65	65	65	65

Notes

Xantia 1.9D Turbo 1993 to 1996
[1] 0.66 to 0.76 bar @ 2000 rpm
[2] Bolt length below head: ≤146.5 mm Bolt with guide boss: ≤150.5 mm

Xantia 1.9D Turbo 1995 to 1996
[1] 0.66 to 0.76 bar @ 2000 rpm
[2] Bolt length below head: ≤146.5 mm Bolt with guide boss: ≤150.5 mm

Xantia 1.9D Turbo 1996 to 2000
[1] 0.66 to 0.76 bar @ 2000 rpm
[2] Bolt length below head: ≤146.5 mm Bolt with guide boss: ≤150.5 mm

Xantia 1.9D LPT Turbo 1997 to 2000
[1] Bolt length below head: ≤125.5 mm

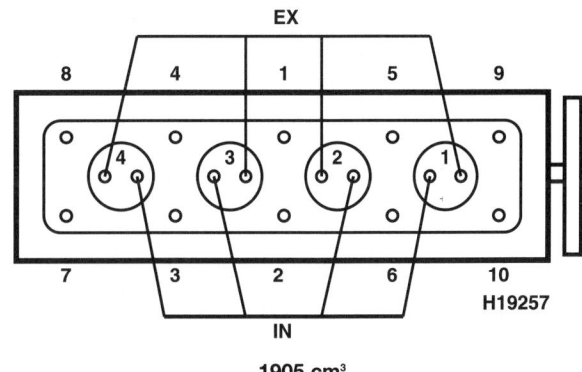

H19257

1905 cm³

– Not applicable, or information not available

CITROEN

	Xantia 2.0 HDi 1998 to 2000	Xantia 2.0 HDi 1999 to 2000	Xantia 2.1D Turbo 1996 to 1999	XM 2.1D Turbo 1989 to 1995
Engine				
Engine type/code................	DW10ATED RHZ Turbo 80kW	DW10TD RHY SOHC 66kW	XUD11BTE/Y/L3 P8C SOHC 80kW	XUD11ATE/L P8A Turbo 80kW
Capacity (cm^3) / cylinders..................	1997 / 4	1997 / 4	2088 / 4	2088 / 4
Compression ration / pressure ....bar	17.6 / ≥30 ± 5	17.6 / ≥30 ± 5	21.8 / _	21.5 / _
Torque output ....Nm	250	205	250	250
Oil pressure ....idle [running] bar	[3.8]	[3.8]	[4.9]	[4.0]
Oil temperature ....°C	80	80	80	80
Valve clearances - inlet (mm)	0: Hyd.	0: Hyd.	0: Hyd.	0: Hyd.
- exhaust (mm)	0: Hyd.	0: Hyd.	0: Hyd.	0: Hyd.
Injection order	1-3-4-2	1-3-4-2	1-3-4-2	1-3-4-2
No. 1 cylinder position	FE	FE	FE	FE
Cooling system				
Thermostat opening temperature.........°C	83	83	89	85 AT: 80
Radiator cap pressure ....bar	1.4	1.4	1.4	1.4
Fuel system				
Idle speed ....rpm	800 ± 50	800 ± 50	800 ± 50	700 ± 50
Maximum (no load) speed ....rpm	5100 ± 80	5100 ± 80	5100 ± 80	5150 ± 125
Smoke test/opacity ....M^{-1} %	2.0	2.0	2.5	2.5
Static timing method..............................	Refer to wsm	Refer to wsm	_	Dial gauge
Timing dimension..............................mm	Computer controlled	Computer controlled	ECU controlled	Dimension on pump
Crankshaft position ....mm [°]	_	_	_	[TDC hole]
Turbo type / ref / pressure ....bar	_	_	Garrett TB0251[1]	_
Injection pump make	Bosch	Bosch	Lucas	Roto Diesel
Injection pump part no...........................	_	EDC 15C2	EPIC XUD LC01	R8443 B740A[2]
Injector Make / type	_	Bosch		Roto Diesel
Injector part no.................................	_	9625542580	_	RDN 12 SDC 6874C
Injection type.................................	EDC 15C2 Direct	Direct EDC 15C2	EPIC	Indirect DPC 062
Injection opening pressure, New [used]...bar	200 to 1350	200 to 1350	150	150 ± 5
Glow plugs				
Maker	Bosch/Champion	Bosch/Champion	Beru/Bosch	Bosch
Type	0250 202 032 / CH170	0250202032 / CH170	0100 226 186 / 0250 201 039	0250 201 019
Nominal rating....................................V/A	_	_	_	
Brakes				
minimum friction material thickness				
Front.............................mm	3.0 Break: 2.0	3.0	3.0 Break: 2.0	3.0
Rear.............................mm	2.0	2.0	2.0	2.0
Tyres - Saloon / Hatch......................Size	205/60x15	205/60x15	205/60x15	195/65x15: 205/65x15
- Estate / Van...........................Size	205/60x15	205/60x15	205/60x15	195/65x15: 205/65x15
Pressure - front / rear - Saloon / Hatch ...bar	Refer to car	Refer to vehicle	Refer to car	2.2 / 1.9 95▶: 2.3 / 1.9
- Estate / Van ......bar	Refer to car	Refer to vehicle	Refer to car	2.4 / 2.4 AT: 2.3 / 2.3
Front suspension / wheel alignment				
Toe-in (+) / Toe-out (–)....................mm [°]	0 to -3.0	0 to -3.0	0 to -3.0	0 to -3.0
Camber	0° ± 30'	0° ± 30'	0° ± 30'	0° ± 30'[3]
Castor	3° ± 30'	3° ± 30'	3° ± 30'	2°30' ± 30'
King pin inclination...........................	13°20' ± 30' 96▶:13°15'±35'	13°20' ± 30'	13°20' ± 30' 96▶: 13°15'±35'	13°14'[4]
Rear suspension / wheel alignment				
Toe-in (+) / Toe-out (–)....................mm [°]	1.0 to 6.0	1.0 to 6.0	1.0 to 6.0	0.5 to 6.5
Camber	-1°15' ± 20'	-1°15' ± 20'	-1°15' ± 20'	50' ± 20'

CITROEN

Torque wrench settings	Xantia 2.0 HDi 1998 to 2000	Xantia 2.0 HDi 1999 to 2000	Xantia 2.1D Turbo 1996 to 1999	XM 2.1D Turbo 1989 to 1995
Cylinder head - stage 1 ....Nm	20[1]	20[1]	Bolt length: ≤146.5mm OR...	70[5]
- stage 2 ....Nm	60	60	Pilot section type: ≤151.5mm	+ 140°
- stage 3 ....Nm	+ 220° ± 5°	+ 220°	60	Warm up, allow to cool
- stage 4 ....Nm	–	–	+ 180°	Slacken
- stage 5 ....Nm	–	–	–	70
- stage 6 ....Nm	–	–	–	+ 140°
Big-end bearings................................Nm	20 + 70° N	20 + 70° N	20 + 70° N	20 + 70° N
Main bearings.....................................Nm	25 + 60°	25 + 60°	15 + 60°	15 + 60°
Crankshaft pulley bolt ....Nm	–	–	70 + 60°	70 + 60°
Camshaft pulley bolt ....Nm	43	43	43	43
Flywheel [driveplate] bolt....................Nm	48	48	48 N [43]	50 N
Front hubs ....Nm	320	320	320	320
Rear hubs ....Nm	280	280	280	275
Wheel nuts / bolts ....Nm	90	90	90	90
Glow plugs ....Nm	Bosch: 25 Beru: 20	20	Bosch: 25 Beru: 20	22
Clutch pressure plate bolts ....Nm	20	20	20	20
Injection pump sprocket.....................Nm	50	50	50	50
Injectors..Nm	30	30	90	90
Injection pump mounting bolts ....Nm	22.5	25	20	18
Injector pipe unions...........................Nm	20	20	25	25
Capacities				
Engine oil & filter................................litres	4.5 A/C: 4.25	4.5 A/C: 4.25	5.25 A/C: 4.75	6.0
Gearbox...litres	1.8	1.8	1.8	1.85
Automatic transmission ....litres	–	–	–	2.0
Final drive ....litres	WT	WT	WT	WT
Cooling system..................................litres	8.5 A/C: 11.0	8.5 A/C: 11.0	9.0	10.0
Fuel tank..litres	65	65	65	80

Notes

Xantia 2.0 HDi 1998 to 2000
[1]Bolt length below head: ≤133.3 mm
Xantia 2.0 HDi 1999 to 2000
[1]Bolt length ≤133.3 mm below head
Xantia 2.1D Turbo 1996 to 1999
[1]0.9 ± 0.1 bar @ 2500 rpm
XM 2.1D Turbo 1989 to 1995
[1]Or Condiesel
[2]Or 742A / 744B / 745B / 748E. 91▶: B640
[3]Hydractive suspension: 15' ± 30'
[4]Hydractive suspension: 13°28'
[5] Bolt length ≤146.8 mm below head

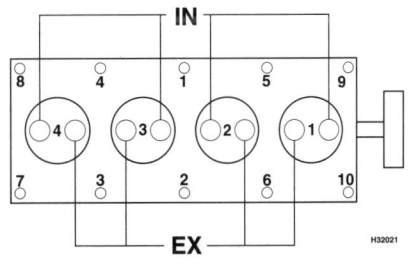

1997 cm³

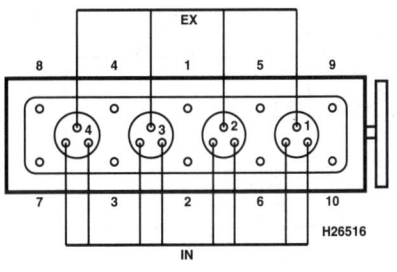

2088 cm³

– Not applicable, or information not
available

CITROEN

	XM 2.1D 1989 to 1991	XM 2.1D Turbo 1994 to 1996	XM 2.1D Turbo 1995 to 1999	XM 2.1D Turbo 1995 to 1996
Engine				
Engine type/code..........................	XUD11A P9A SOHC 12V 60kW	XUD11ATE/L P8B Turbo 80kW	XUD11ATE/Y PHZ Turbo 80kW	XUD11BTE/L P8C Turbo 80kW
Capacity (cm³) / cylinders........................	2138 / 4	2088 / 4	2088 / 4	2088 / 4
Compression ration / pressure ....bar	22.5 / _	21.5 / _	21.5 / _	21.5 / _
Torque output ....Nm	0	250	250	250
Oil pressure ....idle [running] bar	[4.0]	[2.5]	[2.5]	[2.5]
Oil temperature ....°C	80	100	100	100
Valve clearances - inlet (mm)	0: Hyd.	0: Hyd.	0: Hyd.	0: Hyd.
- exhaust (mm)	0: Hyd.	0: Hyd.	0: Hyd.	0: Hyd.
Injection order	1-3-4-2	1-3-4-2	1-3-4-2	1-3-4-2
No. 1 cylinder position	FE	FE	FE	FE
Cooling system				
Thermostat opening temperature ....°C	85	85 AT: 80	85 AT: 80	85 AT: 80
Radiator cap pressure ....bar	1.4	1.4	1.4	1.4
Fuel system				
Idle speed ....rpm	675 ± 25	775 ± 25	775 ± 25	750 ± 25
Maximum (no load) speed ....rpm	5150 ± 125	5150 ± 125	5150 ± 125	5100 ± 75
Smoke test/opacity ....M⁻¹ %	2.5	2.5	2.5	2.5
Static timing method............................	Dial gauge	Plunger travel	Plunger travel	Plunger travel
Timing dimension...............................mm	Dimension on pump	0.88	0.88	0.88
Crankshaft position ....mm [°]	[TDC hole]	TDC [0]	TDC [0]	TDC [0]
Turbo type / ref / pressure ....bar	_	_	0.7 to 0.9 bar @ 3500 rpm	Mitsubishi TD04-11B⁴
Injection pump make	Roto Diesel	Bosch	Bosch	Lucas
Injection pump part no..........................	R8443 B730A²	VE4/9F 2150 R474	VE4/9F 2150 R474	XUD 11P01 R8640 A042A
Injector Make / type	Roto Diesel	Bosch	Bosch	CAV
Injector part no.	RDN 12 SDC 6872C	DNOSD 299	DNOSD 299	RDNOSD 6751H¹
Injection type......................................	Indirect DPC 061	VE4/9F	VE4/9F	XUD
Injection opening pressure, New [used]...bar	130 ± 5	175.0 ± 5.0	175.0 ± 5.0	150.0 ± 5.0
Glow plugs				
Maker	Bosch	Beru/Bosch	Beru/Bosch	Beru/Bosch
Type	0250 201 019	0100 221 133 / 0250 202 019	0100 221 133 / 0250 202 019	0100 221 133 / 0250 202 019
Nominal rating.................................V/A	_	_	11 / 12	_
Brakes				
minimum friction material thickness				
Front......................................mm	3.0	3.0	3.0	3.0
Rear......................................mm	2.0	2.0	2.0	2.0
Tyres - Saloon / Hatch......................Size	175/70x15: 185/65x15	195/65x15: 205/65x15	195/65x15: 205/65x15	195/65x15: 205/65x15
- Estate / Van.....................Size	195/65x15	195/65x15: 205/65x15	195/65x15: 205/65x15	195/65x15: 205/65x15
Pressure - front / rear - Saloon / Hatch...bar	2.4 / 1.9 185/65: 2.3 / 1.9	2.2 / 1.9 95▶: 2.3 / 1.9	2.3 / 1.9	2.2 / 1.9 95▶: 2.3 / 1.9
- Estate / Van ......bar	Refer to car	2.4 / 2.4 AT: 2.3 / 2.3	2.4 / 2.4 AT: 2.3 / 2.3	2.4 / 2.4 AT: 2.3 / 2.3
Front suspension / wheel alignment				
Toe-in (+) / Toe-out (−)....................mm [°]	0 to -3.0	0 to -3.0	0 to -3.0	0 to -3.0
Camber	0° ± 30'³	0° ± 30'¹	0° ± 30'¹	0° ± 30'²
Castor	2°30' ± 30'	2°30' ± 30'	2°30' ± 30'	2°30' ± 30'
King pin inclination................................	13°14'⁴	13°14'²	13°14'²	13°14'³
Rear suspension / wheel alignment				
Toe-in (+) / Toe-out (−)....................mm [°]	0.5 to 6.5	0.5 to 6.5	0.5 to 6.5	0.5 to 6.5
Camber	50' ± 20'	50' ± 20'	50' ± 20'	50' ± 20'

CITROEN

Torque wrench settings	XM 2.1D 1989 to 1991	XM 2.1D Turbo 1994 to 1996	XM 2.1D Turbo 1995 to 1999	XM 2.1D Turbo 1995 to 1996
Cylinder head - stage 1 .....Nm	70 N	Bolt length ≤146.5 mm[4]	Bolt length ≤146.5 mm[3]	Bolt length ≤151.5 mm
- stage 2 .....Nm	+ 140°	70	70[4]	60
- stage 3 .....Nm	–	+ 150°	+ 150°	+ 180°
- stage 4 .....Nm	–	–	warm-up, allow to cool	–
- stage 5 .....Nm	–	–	- 90°	–
- stage 6 .....Nm	–	–	70 + 150°	–
Big-end bearings.....Nm	20 + 70° N	20 + 70° N	20 + 70° N	20 + 70° N
Main bearings.....Nm	70	70	15 + 60°	15 + 60°
Crankshaft pulley bolt .....Nm	70 + 60°	70 + 60°	70 + 60°	70 + 60°
Camshaft pulley bolt .....Nm	43	43	43	43
Flywheel [driveplate] bolt.....Nm	50	48 N	48 N	48 N
Front hubs .....Nm	320	320	320	320
Rear hubs .....Nm	275	275	275	275
Wheel nuts / bolts .....Nm	90	90	90	90
Glow plugs .....Nm	22	22	22	22
Clutch pressure plate bolts .....Nm	20	20	20	20
Injection pump sprocket.....Nm	50	50	50	50
Injectors.....Nm	90	90	90	90
Injection pump mounting bolts .....Nm	18	18	18	18
Injector pipe unions.....Nm	25	25	25	25
Capacities				
Engine oil & filter.....litres	6.0	5.75 A/C: 5.5[3]	5.75 A/C: 5.5	5.0 A/C: 4.5
Gearbox.....litres	1.8	1.85	1.85	1.85
Automatic transmission .....litres	–	2.7	2.7	2.7
Final drive.....litres	WT	WT	WT	WT
Cooling system.....litres	9.6	11.4 to 12.0	11.4 to 12.0	11.4 to 12.0
Fuel tank.....litres	80	80	80	80

Notes

XM 2.1D 1989 to 1991
[1] Or Condiesel
[2] Or B731A / B732B / B733C / B735D. 91▶: B631
[3] Hydractive suspension: 15' ± 30'
[4] Hydractive suspension: 13°28'
XM 2.1D Turbo 1994 to 1996
[1] Hydractive suspension: 15' ± 30'
[2] Hydractive suspension: 13°28'
[3] 96▶: 4.25 A/C: 4.0
[4] Bolt with guide boss: ≤151.5 mm
XM 2.1D Turbo 1995 to 1999
[1] Hydractive suspension: 15' ± 30'
[2] Hydractive suspension: 13°28'
[3] Bolts with guide boss: 151.5 mm
[4] Bolts with guide boss: 60 + 180°
XM 2.1D Turbo 1995 to 1996
[1] No 4: RDNOSDC 6751H
[2] Hydractive suspension: 15' ± 30'
[3] Hydractive suspension: 13°28'
[4] 0.9 ± 0.1 bar : 3000 rpm

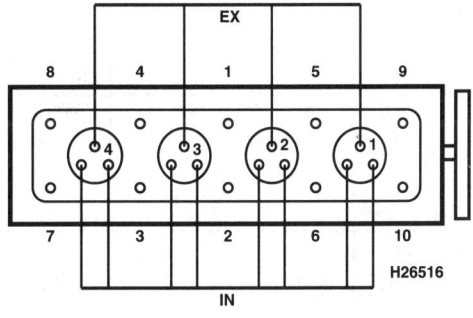

2088 cm³ / 2138 cm³

– Not applicable, or information not available

CITROEN

	XM 2.1D Turbo 1996 to 1999	XM 2.1D 1994 to 1996	XM 2.1D Turbo 1998 to 1999	XM 2.5D Turbo 1994 to 1996
Engine				
Engine type/code........................	XUD11BTE/L3 P8C Turbo 80kW	XUD11A/L PJZ SOHC 12V 60kW	XUD11BTE/L4 P8C Turbo 80kW	DK5ATE/L THY Turbo 94kW
Capacity (cm³) / cylinders.........................	2088 / 4	2138 / 4	2088 / 4	2446 / 4
Compression ration / pressure ....bar	21.5 / _	22.5 / _	21.5 / _	21.0 / _
Torque output ....Nm	250	0	250	217
Oil pressure ....idle [running] bar	[2.5]	[2.5]	[2.5]	[3.0]
Oil temperature ....°C	100	100	100	90
Valve clearances - inlet (mm)	0: Hyd.	0: Hyd.	0: Hyd.	0: Hyd.
- exhaust (mm)	0: Hyd.	0: Hyd.	0: Hyd.	0: Hyd.
Injection order	1-3-4-2	1-3-4-2	1-3-4-2	1-3-4-2
No. 1 cylinder position	FE	FE	FE	FE
Cooling system				
Thermostat opening temperature ....°C	85 AT: 80	85	85 AT: 80	85 AT: 80
Radiator cap pressure ....bar	1.4	1.4	1.4	1.4
Fuel system				
Idle speed ....rpm	750 ± 25	725 ± 25	750 ± 25	750 ± 50
Maximum (no load) speed ....rpm	5100 ± 75	5150 ± 125	5100 ± 75	5150
Smoke test/opacity ....M⁻¹ %	2.5	2.5	2.5	3.0
Static timing method................................	Plunger travel	Dial gauge	Plunger travel	Refer to wsm
Timing dimension.............................mm	0.88	Dimension on pump	0.88	_
Crankshaft position ....mm [°]	TDC [0]	[TDC hole]	TDC [0]	_
Turbo type / ref / pressure ....bar	Mitsubishi TD04-11B⁴	_	Mitsubishi TD04-11B¹	Garrett T2 0.3 bar @ 3000 rpm
Injection pump make	Lucas	Lucas	Lucas	Bosch
Injection pump part no............................	EPIC XUD 11CO1 R04010011B	DPC 061 R8443 B962A	EPIC XUD 11CO1 R04010011B	96: MSA11 3.6 0281001212/3
Injector Make / type	CAV	CAV	_	Bosch
Injector part no.......................................	RDNOSD 6751H¹	RDNOSDC 6872D	_	DNOSD 312. No 3: 316
Injection type..	EPIC	DPC 061	EPIC	VP36 VER520 type 535
Injection opening pressure, New [used]...bar	150.0 ± 5.0	130.0 ± 5.0	150.0 ± 5.0	170.0 +5.0 -0
Glow plugs				
Maker	Beru/Bosch	Beru/Bosch	Beru/Bosch	Beru/Bosch
Type	0100 221 133 / 0250 202 019	0100 221 133 / 0250 202 019	0100 221 133 / 0250 202 019	0100 226 186 / 0250 201 033
Nominal rating ....V/A	_	_	_	_
Brakes				
minimum friction material thickness				
Front..........................mm	3.0	3.0	3.0	3.0
Rear..........................mm	2.0	2.0	2.0	2.0
Tyres - Saloon / Hatch.....................Size	195/65x15: 205/65x15	195/65x15	195/65x15: 205/65x15	205/65x15
- Estate / Van............................Size	195/65x15: 205/65x15	195/65x15	195/65x15: 205/65x15	205/65x15
Pressure - front / rear - Saloon / Hatch ...bar	2.2 / 1.9 95▶: 2.3 / 1.9	2.2 / 2.0	2.3 / 1.9	2.3 / 1.9
- Estate / Van ......bar	2.4 / 2.4 AT: 2.3 / 2.3	2.4 / 2.4	2.4 / 2.4 AT: 2.3 / 2.3	2.3 / 2.3
Front suspension / wheel alignment				
Toe-in (+) / Toe-out (–)....................mm [°]	0 to -3.0	0 to -3.0	0 to -3.0	0 to -3.0
Camber	0° ± 30'²	0° ± 30'¹	0° ± 30'	0° ± 30'¹
Castor	2°30' ± 30'	2°30' ± 30'	2°30' ± 30'	2°30' ± 30'
King pin inclination.................................	13°14'³	13°14'²	13°14'	13°14'²
Rear suspension / wheel alignment				
Toe-in (+) / Toe-out (–)....................mm [°]	0.5 to 6.5	0.5 to 6.5	0.5 to 6.5	0.5 to 6.5
Camber	50' ± 20'	50' ± 20'	50' ± 20'	50' ± 20'

CITROEN

Torque wrench settings	XM 2.1D Turbo 1996 to 1999	XM 2.1D 1994 to 1996	XM 2.1D Turbo 1998 to 1999	XM 2.5D Turbo 1994 to 1996
Cylinder head - stage 1 ......Nm	Bolt length ≤151.5 mm	70[3]	Bolt length ≤151.5 mm	Bolt length no 1 to 14 ≤153.5
- stage 2 ......Nm	60	+ 150°	60	No 15 to 22: ≤162.5 mm
- stage 3 ......Nm	+ 180°	–	+ 180°	No 1 to 14: 50
- stage 4 ......Nm	–	–	–	No 15 to 22: 35
- stage 5 ......Nm	–	–	–	+ 120 ± 5°
- stage 6 ......Nm	–	–	–	–
Big-end bearings......Nm	20 + 70° N	20 + 70° N	20 + 70° N	20 + 70° N
Main bearings......Nm	15 + 60°	70	15 + 60°	15 + 60°
Crankshaft pulley bolt ......Nm	70 + 60°	70 + 60°	70 + 60°	70 + 60°
Camshaft pulley bolt ......Nm	43	43	43	43
Flywheel [driveplate] bolt......Nm	48 N	50 N	50 N	50
Front hubs......Nm	320	320	320	320
Rear hubs......Nm	275	275	275	275
Wheel nuts / bolts ......Nm	90	90	90	90
Glow plugs ......Nm	22	22	22	Beru: 20 Bosch: 25
Clutch pressure plate bolts ......Nm	20	20	20	20
Injection pump sprocket......Nm	50	50	50	50
Injectors......Nm	90	90	90	90
Injection pump mounting bolts ......Nm	18	18	18	18
Injector pipe unions......Nm	25	25	25	25
Capacities				
Engine oil & filter......litres	5.0 A/C: 4.5	5.75 A/C: 5.5	5.0 A/C: 4.5	8.0
Gearbox......litres	1.85	1.9	1.85	2.2
Automatic transmission ......litres	2.7	–	2.7	–
Final drive......litres	WT	WT	WT	WT
Cooling system......litres	11.4 to 12.0	9.8	11.4 to 12.0	13.2
Fuel tank......litres	80	80	80	80

Notes

XM 2.1D Turbo 1996 to 1999
[1]No 4: RDNOSDC 6751H
[2]Hydractive suspension: 15' ± 30'
[3]Hydractive suspension: 13°28'
[4]0.9 ± 0.1 bar @ 3000 rpm
XM 2.1D 1994 to 1996
[1]Hydractive suspension: 15' ± 30'
[2]Hydractive suspension: 13°28'
[3]Bolt length below head: ≤146.5 mm Bolt with guide boss: ≤150.5 mm
XM 2.1D Turbo 1998 to 1999
[1]0.9 ± 0.1 bar @ 3000 rpm
XM 2.5D Turbo 1994 to 1996
[1]Hydractive suspension: 15' ± 30'
[2]Hydractive suspension: 13°28'

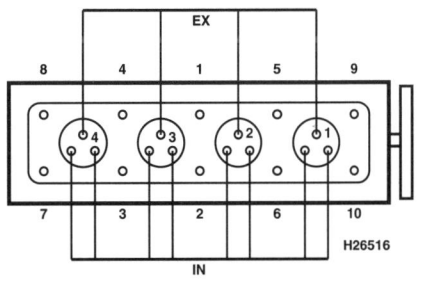

2088 cm³ / 2138 cm³

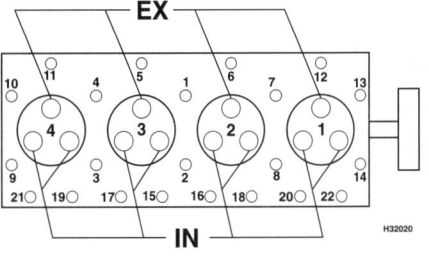

2446 cm³

– Not applicable, or information not available

CITROEN

	XM 2.5D Turbo 1996 to 1999	Evasion/Synergie 1.9TD 1994 to 2000	Evasion/Synergie 1.9TD 1994 to 2000	Evasion/Synergie 2.0 HDi 1999 to 2000
Engine				
Engine type/code...............................	DK5ATE/L3 THY Turbo 94kW	XUD9TF/L/W2 D8B Turbo 67kW	XUD9BTF/Y/L3 DHX Turbo 66kW	DW10ATED RHZ SOHC 80kW
Capacity (cm³) / cylinders......................	2446 / 4	1905 / 4	1905 / 4	1997 / 4
Compression ration / pressure ....bar	21.0 / _	21.8 / _	21.8 / _	17.6 / ≥30 ± 5
Torque output ....Nm	217	196	196	250
Oil pressure ....idle [running] bar	[3.0]	[4.9]	[4.9]	[3.8]
Oil temperature ....°C	90	80	80	80
Valve clearances - inlet (mm)	0: Hyd.	0.15 ± 0.08	0.15 ± 0.08	0: Hyd.
- exhaust (mm)	0: Hyd.	0.30 ± 0.08	0.30 ± 0.08	0: Hyd.
Injection order	1-3-4-2	1-3-4-2	1-3-4-2	1-3-4-2
No. 1 cylinder position	FE	FE	FE	FE
Cooling system				
Thermostat opening temperature ....°C	85 AT: 80	89	89	83
Radiator cap pressure ....bar	1.4	1.4	1.4	1.4
Fuel system				
Idle speed ....rpm	750 ± 50	800 ± 50	800 ± 50 A/C: 850 ± 50	800 ± 50 A/C: 850 ± 50
Maximum (no load) speed ....rpm	5150	5100 ± 80	5100 ± 80	5100 ± 80
Smoke test/opacity ....M⁻¹ %	3.0	2.5	2.5	2.5
Static timing method.............................	Refer to wsm	Plunger travel	Plunger travel	Plunger travel
Timing dimension............................mm	_	0.66	0.57	0.57
Crankshaft position ....mm [°]	_	TDC	TDC	TDC
Turbo type / ref / pressure ....bar	Garrett T2 0.3 bar @ 3000rpm	KKK K14 / Garrett T2¹	KKK K14 / Garrett T2¹	_
Injection pump make	Bosch	Bosch	Bosch	Bosch
Injection pump part no...........................	96: MSA11 7.6 0281001336	VER 445/2 XUD203	XUD BP02 R601/1/2	EDC 15C2
Injector Make / type	Bosch	Bosch	Bosch	Bosch
Injector part no...................................	DNOSD 312. No 3: 316	DNDOSD 299	DNDOSD 299	_
Injection type....................................	VP36 VER520 type 535	Indirect	VP20	EDC 15C2 Direct
Injection opening pressure, New [used]...bar	170.0 +5.0 -0	175 +5 -0	175 +5 -0	_
Glow plugs				
Maker	Beru/Bosch	Bosch/Beru	Bosch/Beru	Bosch/Beru/Champion
Type	0100 226 186 / 0250 201 033	0 250 201 039 / 0 100 226 186	0 250 201 039 / 0 100 226 186	0250 202 032 / CH170
Nominal rating ....V/A	_	11 / 9	11 / 9	_
Brakes				
minimum friction material thickness				
Front..mm	3.0	8.5 with backing	8.5 with backing	3.0 Break: 2.0
Rear..mm	2.0	_	_	2.0
Tyres - Saloon / Hatch ....Size	205/65x15	_	_	205/65x15
- Estate / Van......................Size	205/65x15	205/65x15	205/65x15	205/65x15
Pressure - front / rear - Saloon / Hatch ...bar	2.3 / 1.9	_	_	Refer to car
- Estate / Van ......bar	2.3 / 2.3	2.3 / 2.3	2.3 / 2.3	Refer to car
Front suspension / wheel alignment				
Toe-in (+) / Toe-out (–)....................mm [°]	0 to -3.0	0.5 to 1.5	0.5 to 1.5	0 to -3.0
Camber	0° ± 30'¹	0° ± 30'	0° ± 30'	0° ± 30'
Castor	2°30' ± 30'	3°30' ± 30'	3°30' ± 30'	3° ± 30'
King pin inclination...............................	13°14'²	11°30' ± 40'	11°30' ± 40'	13°20' ± 30'
Rear suspension / wheel alignment				
Toe-in (+) / Toe-out (–)....................mm [°]	0.5 to 6.5	0.5 to 1.5	0.5 to 1.5	1.0 to 6.0
Camber	50' ± 20'	1° ± 30'	1° ± 30'	1°15' ± 20'

CITROEN

	XM 2.5D Turbo 1996 to 1999	Evasion/Synergie 1.9TD 1994 to 2000	Evasion/Synergie 1.9TD 1994 to 2000	Evasion/Synergie 2.0 HDi 1999 to 2000
Torque wrench settings				
Cylinder head - stage 1 ....Nm	Bolt length no 1 to 14 ≤153.5	20[2]	20[2]	20[1]
- stage 2 ....Nm	No 15 to 22: ≤162.5 mm	60	60	60
- stage 3 ....Nm	No 1 to 14: 50	+ 220°	+ 220°	+ 220° ± 5°
- stage 4 ....Nm	No 15 to 22: 35	–	–	–
- stage 5 ....Nm	+ 120 ± 5°	–	–	–
- stage 6 ....Nm	–	–	–	–
Big-end bearings.............................Nm	20 + 70° N	20 + 70° N	20 + 70° N	20 + 70° N
Main bearings..................................Nm	15 + 60°	15 + 60°	70	25 + 60°
Crankshaft pulley bolt ....Nm	70 + 60°	40 + 60°	40 + 60°	–
Camshaft pulley bolt ....Nm	43	45	45	43
Flywheel [driveplate] bolt................Nm	50	50	50	48
Front hubs.......................................Nm	320	100 + 60°	100 + 60°	320
Rear hubs..Nm	275	100 + 60°	100 + 60°	280
Wheel nuts / bolts ....Nm	90	100	100	90
Glow plugs ....Nm	Beru: 20 Bosch: 25	22	22	Bosch: 25 Beru: 20
Clutch pressure plate bolts ....Nm	20	22	20	–
Injection pump sprocket...................Nm	50	50	90	50
Injectors..Nm	90	90	90	30
Injection pump mounting bolts ....Nm	18	22	22	22.5
Injector pipe unions.........................Nm	25	20	20	20
Capacities				
Engine oil & filter.............................litres	8.0	4.25 A/C: 4.0	4.75 A/C: 4.0	4.5 A/C: 4.25
Gearbox...litres	2.2	1.85	1.85	1.85
Automatic transmission ....litres	–	–	–	–
Final drive.......................................litres	WT	WT	WT	WT
Cooling system................................litres	13.2	9.0	9.0	8.5 A/C: 11.0
Fuel tank..litres	80	80	80	80

Notes

XM 2.5D Turbo 1996 to 1999
[1]Hydractive suspension: 15' ± 30'
[2]Hydractive suspension: 13°28'
Evasion/Synergie 1.9TD 1994 to 2000
[1]0.66 to 0.76 bar @ 2000 rpm
[2]Bolt length below head: ≤146.5 mm Bolt with guide boss: ≤150.5 mm
Evasion/Synergie 1.9TD 1994 to 2000
[1]0.66 to 0.76 bar @ 2000 rpm
[2]Bolt length below head: ≤146.5 mm Bolt with guide boss: ≤150.5 mm
Evasion/Synergie 2.0 HDi 1999 to 2000
[1]Bolt length below head: ≤133.3 mm

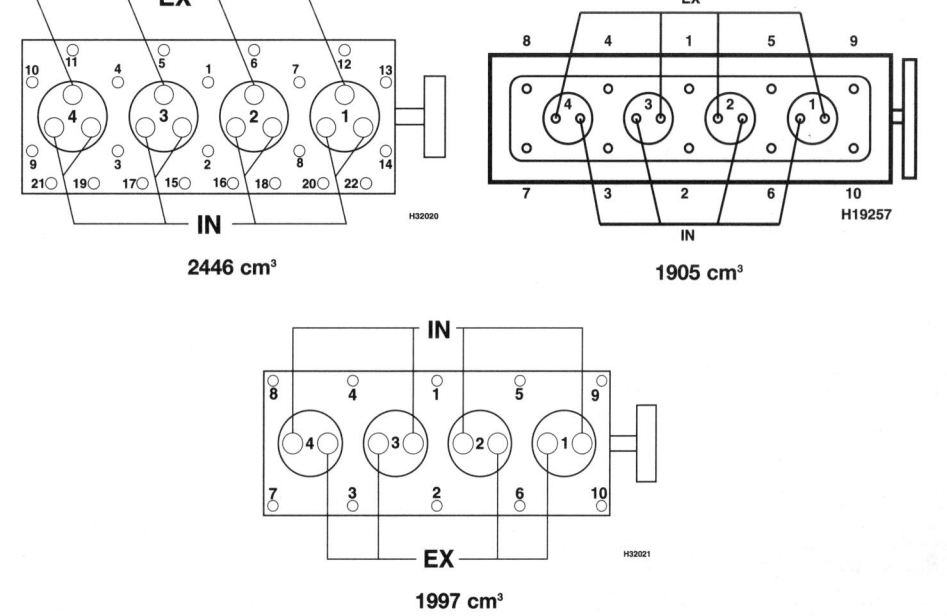

2446 cm³

1905 cm³

1997 cm³

– Not applicable, or information not available

CITROEN

	Evasion/Synergie 2.1TD 1996 to 1999	C15D Van 1984 to 1992	C15D Van 1987 to 1992	C15D Van 1993 to 1999
Engine				
Engine type/code...................	XUD11BTE/L3 P8C Turbo 80kW	XUD7 161A SOHC 44kW	XUD7 161A SOHC 44kW	XUD7/W2 161A SOHC 43kW
Capacity (cm³) / cylinders...................	2088 / 4	1769 / 4	1769 / 4	1769 / 4
Compression ration / pressure ....bar	21.5 / _	23.0 / _	23.0 / _	23.0 / _
Torque output ....Nm	250	110	110	110
Oil pressure ....idle [running] bar	[4.9]	[3.0]	[3.0]	[3.0]
Oil temperature ....°C	80	80	80	80
Valve clearances - inlet (mm)	0: Hyd.	0.15 ± 0.08	0.15 ± 0.08	0.15 ± 0.08
- exhaust (mm)	0: Hyd.	0.30 ± 0.08	0.30 ± 0.08	0.30 ± 0.08
Injection order	1-3-4-2	1-3-4-2	1-3-4-2	1-3-4-2
No. 1 cylinder position	FE	FE	FE	FE
Cooling system				
Thermostat opening temperature ....°C	89	79 to 82	79 to 82	79 to 82
Radiator cap pressure ....bar	1.4	1.0	1.0	1.0
Fuel system				
Idle speed ....rpm	750	750 ± 50	750 ± 50	750 + 50
Maximum (no load) speed ....rpm	5100 ± 75	5150 ± 125	5100 ± 125	5100 ± 125
Smoke test/opacity ....M⁻¹ %	2.5	2.5	2.5	2.5
Static timing method............................	Plunger travel	Dial gauge	Dial gauge	Dial gauge
Timing dimension............................mm	0.66	0.3 ABDC	Dimension on pump	Dimension on pump
Crankshaft position ....mm [°]	TDC	0.8 ± 0.03 [9.5]	[TDC hole]	[TDC hole]
Turbo type / ref / pressure ....bar	_	_	_	_
Injection pump make	Lucas	Bosch	Roto Diesel	Roto Diesel
Injection pump part no...........................	R040 0A 100A	_	R8443B264F	R8443B264F
Injector Make / type	_	Bosch	Roto Diesel	Roto Diesel
Injector part no......................................	_	RDNOSD 256	RDNOSDC 6850C	RDNOSDC 6850C
Injection type...	Diesel EPIC	VER 171 or 171.1	DPC	DPC
Injection opening pressure, New [used]...bar	175 +5 -0	130 +5 -0	125 ± 5	125 ± 5
Glow plugs				
Maker	Bosch/Beru	Bosch	Bosch	Bosch
Type	0 250 201 039 / 0 100 226 186	0250 201 005	0250 201 005	0250 201 005'
Nominal rating...................................V/A		11 / 13	11 / 13	11 / 13
Brakes				
minimum friction material thickness				
Front.....................................mm	8.5 with backing	2.0	2.0	2.0
Rear.......................................mm	_	1.0	1.0	1.0
Tyres - Saloon / Hatch.....................Size	_	_	_	_
- Estate / Van.....................Size	205/65x15	155x13	155x13	155x13
Pressure - front / rear - Saloon / Hatch ...bar	_	_	_	_
- Estate / Van ......bar	2.3 / 2.3	2.3 / 2.6	2.3 / 2.6	2.3 / 2.6
Front suspension / wheel alignment				
Toe-in (+) / Toe-out (–)....................mm [°]	0.5 to 1.5	0 to 2.0	0 to 2.0	0 to 2.0 PAS: 3.0
Camber	0° ± 30'	30' ± 30' 765kg: 41'±30'	30' ± 30' 765kg: 41'±30'	30' ± 30' 765: 41' ± 30'²
Castor	3°30' ± 30'	55' ± 30' 765kg: 1°47'±30'	55' ± 30' 765kg: 1°47'±30'	55' ± 30' 765: 1°47' ± 30'³
King pin inclination................................	11°30' ± 40'	8°50' ± 40' 765kg: 8°34'±40'	8°50' ± 40' 765kg: 8°34'±40'	8°50' ± 40' 765: 8°34' ± 40'⁴
Rear suspension / wheel alignment				
Toe-in (+) / Toe-out (–)....................mm [°]	0.5 to 1.5	1.6 to 5.0	1.6 to 5.0	1.6 to 5.0
Camber	1° ± 30'	-9' ± 20'	-9' ± 20'	-9' ± 20'

CITROEN

Torque wrench settings	Evasion/Synergie 2.1TD 1996 to 1999	C15D Van 1984 to 1992	C15D Van 1987 to 1992	C15D Van 1993 to 1999
Cylinder head - stage 1 ....Nm	Bolt length: ≤146.5mm or....	30 N¹	30 N¹	30 N
- stage 2 ....Nm	Pilot type: ≤150.5mm	60	60	70
- stage 3 ....Nm	20	Slacken	Slacken	+ 120°
- stage 4 ....Nm	60	60	60	–
- stage 5 ....Nm	+ 180°	Warm up, allow to cool	Warm up, allow to cool	–
- stage 6 ....Nm	–	Slacken, 70	Slacken, 70	–
Big-end bearings.................................Nm	20 + 70° N	50 N	50 N	50 N
Main bearings.....................................Nm	15 + 60°	70	70	70
Crankshaft pulley bolt ....Nm	70 + 60°	150 LkC	150 LkC	150 LkC
Camshaft pulley bolt ....Nm	43	35	35	35
Flywheel [driveplate] bolt....................Nm	50	50 LkC	50 LkC	50 LkC
Front hubs ....Nm	100 + 60°	250	250	250
Rear hubs ....Nm	100 + 60°	275	275	275
Wheel nuts / bolts ....Nm	100	60 to 80	60 to 80	60 to 80
Glow plugs ....Nm	22	22	22	22
Clutch pressure plate bolts ....Nm	–	22	22	22
Injection pump sprocket......................Nm	50	50	50	50
Injectors..Nm	90	90	90	90
Injection pump mounting bolts ....Nm	–	–	18	18
Injector pipe unions............................Nm	20	20	20	20
Capacities				
Engine oil & filter.................................litres	4.5 A/C: 4.25¹	5.0	5.0	4.25
Gearbox...litres	1.85	2.0	2.0	2.0
Automatic transmission ....litres	–	–	–	–
Final drive ....litres	WT	WT	WT	WT
Cooling system....................................litres	9.0	8.0	8.0	8.0
Fuel tank..litres	80	48	48	48

Notes

Evasion/Synergie 2.1TD 1996 to 1999
¹Alloy sump: 4.0 l
C15D Van 1984 to 1992
¹9.86 ▶: 30, 70 + 120°
C15D Van 1993 to 1999
¹96▶: Beru 0100 226 186 / Bosch 0250 201 033
²PAS: 30' ± 30'
³PAS: 55' ± 30'
⁴PAS: 8°50' ± 40'

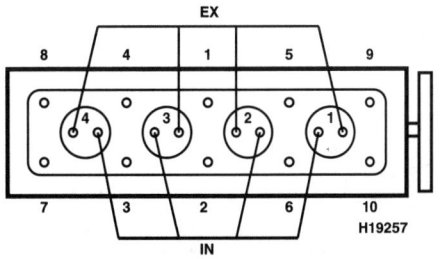

1769 cm³

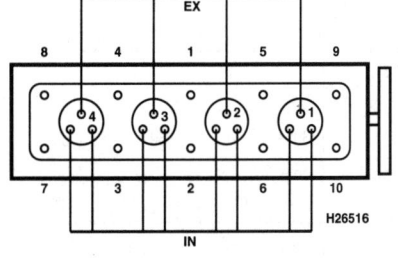

2088 cm³

– Not applicable, or information not
 available

CITROEN

	C15D Van 1993 to 1999	C15D Van 1999 to 2000	Dispatch / Jumpy 1.9 D 1995 to 1999	Dispatch / Jumpy 1.9 TD 1995 to 1999
Engine				
Engine type/code	XUD7/L 161A SOHC 43kW	DW8 WJZ/C SOHC EGR 51kW	XUD9A/W2 D9B SOHC 52kW	XUD9TF/W2 D8B Turbo 67kW
Capacity (cm^3) / cylinders	1769 / 4	1868 / 4	1905 / 4	1905 / 4
Compression ration / pressure ....bar	23.0 / ≥20	23.0 / _	23.0 /	21.8 /
Torque output ....Nm	110	125	120	196
Oil pressure ....idle [running] bar	2.5 [3.0 to 5.0 @ 4000]	1.8 [3.7 @ 2000]	[5.0 @ 4000]	[5.0 @ 4000]
Oil temperature ....°C	80	80	80	80
Valve clearances - inlet (mm)	0.15 ± 0.08	0.15 ± 0.08	0.15 ± 0.05	0.15 ± 0.05
- exhaust (mm)	0.30 ± 0.08	0.30 ± 0.08	0.30 ± 0.05	0.30 ± 0.05
Injection order	1-3-4-2	1-3-4-2	1-3-4-2	1-3-4-2
No. 1 cylinder position	FE	FE	FE	FE
Cooling system				
Thermostat opening temperature ....°C	79 to 82	83	83	83
Radiator cap pressure ....bar	1.0	1.4	1.1	1.1
Fuel system				
Idle speed ....rpm	800 ± 50	800 ± 25	800 to 850	800 ± 50
Maximum (no load) speed ....rpm	5100 ± 125	5350 ± 125	5150 ± 125	5150 ± 80
Smoke test/opacity ....M^{-1} %	2.5	2.5	3.0	2.5
Static timing method...............................	Dial gauge	Plunger travel	Plunger travel	Dial gauge
Timing dimension................................mm	Dimension on pump	0.58 ± 0.04	1.07 ± 0.03	0.66
Crankshaft position ....mm [°]	[TDC hole]	TDC	TDC	TDC
Turbo type / ref / pressure ....bar	–	–	–	KKK K14 / Garrett T2^2
Injection pump make	Roto Diesel	Bosch	Bosch	Bosch
Injection pump part no............................	DPC R8443B930A	VE4 / 537[1]	VER 425 10/11	R445
Injector Make / type	Roto Diesel	–	–	Bosch
Injector part no......................................	RDNOSDC 6850C	KCA 20S106	DNOSD 299 A	KCA 17S42
Injection type..	Indirect DPC	Indirect	Indirect	Indirect
Injection opening pressure, New [used]...bar	142 to 147	120[2]	130	175 to 180
Glow plugs				
Maker	Bosch	Beru/Champion	Bosch	Bosch
Type	0250 201 019	0100226371 / CH185	0 250 201 033	0 250 201 033
Nominal rating...............................V/A	11 / 13	11	11 / 18	11 / 18
Brakes				
minimum friction material thickness				
Front.......................................mm	2.0	2.0	8.5 with backing	8.5 with backing
Rear..mm	1.0	1.0	1.0	1.0
Tyres - Saloon / Hatch.....................Size	–	–	–	–
- Estate / Van.......................Size	155x13	155x13	195/70x14	195/70x14
Pressure - front / rear - Saloon / Hatch ...bar	–	–	–	–
- Estate / Van ......bar	2.3 / 2.6	2.3 / 2.6	2.5 / 2.5	3.0 / 3.0
Front suspension / wheel alignment				
Toe-in (+) / Toe-out (–)....................mm [°]	0 to 2.0 PAS: 3.0	0 to 2.0 PAS: 3.0	1.0 to 3.0	1.0 to 3.0
Camber	30' ± 30' 765: 41' ± 30'[2]	30' ± 30' 765: 41' ± 30'[2]	0° ± 30'	0° ± 30'
Castor	55' ± 30' 765: 1°47' ± 30'[3]	55' ± 30' 765: 1°47' ± 30'[3]	1° ± 30' PAS: 3°30' ± 30'	1° ± 30' PAS: 3°30'±30'
King pin inclination.................................	8°50' ± 40' 765: 8°34' ± 40'[4]	8°50' ± 40' 765: 8°34' ± 40'[4]	11°28' ± 40'	11°28' ± 40'
Rear suspension / wheel alignment				
Toe-in (+) / Toe-out (–)....................mm [°]	1.6 to 5.0	1.6 to 5.0	1.0 to 3.0	1.0 to 3.0
Camber	-9' ± 20'	-9' ± 20'	-1° ± 30'	-1° ± 30'

CITROEN

Torque wrench settings	C15D Van 1993 to 1999	C15D Van 1999 to 2000	Dispatch / Jumpy 1.9 D 1995 to 1999	Dispatch / Jumpy 1.9 TD 1995 to 1999
Cylinder head - stage 1 ....Nm	20 N	Bolt ≤125.5mm below head	20[1]	20[1]
- stage 2 ....Nm	60	20	60	60
- stage 3 ....Nm	+ 180°	60	+ 180°	+ 220°
- stage 4 ....Nm	–	+ 180°	–	–
- stage 5 ....Nm	–	–	–	–
- stage 6 ....Nm	–	–	–	–
Big-end bearings.....................Nm	20 + 70° N	20 + 70° N	40, slacken, 20 + 70° N	20 + 70° N
Main bearings.....................Nm	70	70	70	70
Crankshaft pulley bolt ....Nm	150 LkC	40 + 60°	40 + 51°	40 + 50°
Camshaft pulley bolt.....................Nm	35	25 Hub: 43	45	35
Flywheel [driveplate] bolt.....................Nm	50 LkC	48 N	50 N LkC	50
Front hubs.....................Nm	250	250	100 + 60° oiled	100 + 60° oiled
Rear hubs.....................Nm	275	275	100 + 60°	100 + 60°
Wheel nuts / bolts ....Nm	60 to 80	60 to 80	100	100
Glow plugs ....Nm	22	25	18	22
Clutch pressure plate bolts ....Nm	22	20	20	20
Injection pump sprocket.....................Nm	50	23	50	50
Injectors.....................Nm	90	90	90	90
Injection pump mounting bolts ....Nm	18	20	–	–
Injector pipe unions.............................Nm	20	25	25	25
Capacities				
Engine oil & filter.....................litres	5.0	4.75	4.2	4.25
Gearbox.....................litres	2.0	2.0	1.8	1.8
Automatic transmission ....litres	–	–	–	–
Final drive.....................litres	WT	WT	WT	WT
Cooling system.....................litres	7.5	10.5	9.0	9.0
Fuel tank.....................litres	48	48	80	80

Notes

C15D Van 1993 to 1999
[2]PAS: 30' ± 30'
[3]PAS: 55' ± 30'
[4]PAS: 8°50'± 40'
C15D Van 1999 to 2000
[1]Lucas Varity DPC pump available as alternative
[2]Lucas: 142 to 149
[3]PAS: 55' ± 30'
[4]PAS: 8°50'± 40'
Dispatch / Jumpy 1.9 D 1995 to 1999
[1]Bolt length below head: ≤124.5 mm
Dispatch / Jumpy 1.9 TD 1995 to 1999
[1]Bolt length below head: ≤150.5 mm
[2]0.66 to 0.76 bar @ 2000 rpm

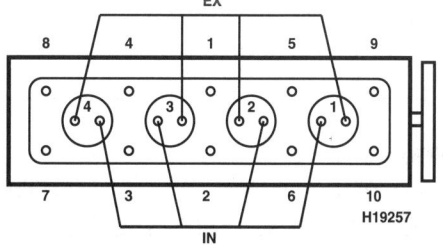

1769 cm³ / 1905 cm³

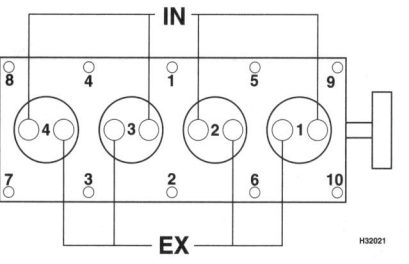

1868 cm³

– Not applicable, or information not available

CITROEN

	Dispatch / Jumpy 1.9 TD 1997 to 2000	Dispatch / Jumpy 1.9 D 1999 to 2000	Berlingo 1.8D 1996 to 1999	Berlingo 1.8D 1998 to 2000
Engine				
Engine type/code.....................	XUD9BTF/L3 DHX Turbo 66kW	DW8 WJZ EGR SOHC 51kW	XUD7/L3 A9A SOHC 44kW	DW8 WJZ/C SOHC EGR 51kW
Capacity (cm³) / cylinders......................	1905 / 4	1868 / 4	1769 / 4	1868 / 4
Compression ration / pressure ....bar	21.5 / _	23.0 /	23.0 /	23.0 /
Torque output ....Nm	196	125	110	125
Oil pressure ....idle [running] bar	[4.1 @ 2000]	1.8 [3.7]	2.5	1.8 [3.7 @ 2000]
Oil temperature ....°C	80	90	80	80
Valve clearances - inlet (mm)	0.15 ± 0.05	0.15 ± 0.05	0.15 ± 0.08	0.15 ± 0.05
- exhaust (mm)	0.30 ± 0.05	0.30 ± 0.05	0.30 ± 0.08	0.30 ± 0.05
Injection order...	1-3-4-2	1-3-4-2	1-3-4-2	1-3-4-2
No. 1 cylinder position	FE	FE	FE	FE
Cooling system				
Thermostat opening temperature ....°C	83	89 to 95	83	83
Radiator cap pressure ....bar	1.1	1.4	1.4	1.4
Fuel system				
Idle speed ....rpm	800 ± 50	800 ± 25	800 ± 50	800 ± 25 A/C: 875 ± 25
Maximum (no load) speed ....rpm	5100 ± 80	5350 ± 225	5150 ± 125	5350 ± 125
Smoke test/opacity ....M⁻¹ %	2.5	3.0	1.56	2.5
Static timing method...............................	Peg	Refer to wsm	_	Plunger travel
Timing dimension............................mm	0.57		Dimension on pump	0.58 ± 0.04
Crankshaft position ....mm [°]	[0] TDC	_	_	TDC
Turbo type / ref / pressure ....bar	KKK K14 to Garrett T2	_	_	_
Injection pump make	Bosch	Bosch	Lucas	Bosch
Injection pump part no.............................	R601	VE4 / 537	DPC 8443B930A	VE4 / 537¹
Injector Make / type	Bosch	_	CAV	
Injector part no..	DN0SD 299	_	LCR6730 75D	KCA 20S106
Injection type...	Indirect	Bosch VE	Indirect DPC	Indirect
Injection opening pressure, New [used]...bar	175	120²	140	120²
Glow plugs				
Maker	Bosch	Beru/Champion	Bosch	Beru/Champion
Type	0 250 201 033	0100226371 / CH185	0 502 010 33	0100226371 / CH185
Nominal rating...................................V/A	11 / 9	_	11 / 12	11
Brakes				
minimum friction material thickness				
Front...mm	8.5 with backing	8.5 with backing	1.0	1.5
Rear...mm	1.0	1.0	1.0	1.0
Tyres - Saloon / Hatch......................Size	_	_	_	_
- Estate / Van..........................Size	195/70x14	195/70x14	165/70x14	165/70x14: 175/65x14
Pressure - front / rear - Saloon / Hatch ...bar	_	_	_	_
- Estate / Van ......bar	2.5 / 2.5	2.5 / 2.5	2.5 / 3.7	2.5 / 3.2
Front suspension / wheel alignment				
Toe-in (+) / Toe-out (−)...................mm [°]	1.0 to 3.0	1.0 to 3.0	-1.0 ± 1.0 PAS: 3.5 ± 1.0	-1.0 ± 1.0 PAS: 3.5 ± 1.0
Camber	0 ± 30'	0° ± 30'	0° ± 1°	0° ± 1°
Castor	1° ± 30' PAS: 3°30' ± 30'	1° ± 30' PAS: 3°30' ± 30'	30' ± 40' PAS: 2° ± 40'	30' ± 40' PAS: 2° ± 40'
King pin inclination..................................	11°28' ± 40'	11°28' ± 40'	10°8' ± 1°	10°50' ± 1°
Rear suspension / wheel alignment				
Toe-in (+) / Toe-out (−)...................mm [°]	1.0 to 3.0	1.0 to 3.0	-2.5 ± 1.0	-2.5 ± 1.0
Camber	-1° ± 30'	-1° ± 30'	1°15' ± 30'	-1°15' ± 1°

CITROEN

Torque wrench settings	Dispatch / Jumpy 1.9 TD 1997 to 2000	Dispatch / Jumpy 1.9 D 1999 to 2000	Berlingo 1.8D 1996 to 1999	Berlingo 1.8D 1998 to 2000
Cylinder head - stage 1 ....Nm	20[1]	20[1]	20[1]	20[3]
- stage 2 ....Nm	60	60	60	60
- stage 3 ....Nm	+ 220°	+ 180°	+ 180°	+ 180°
- stage 4 ....Nm	–	–	–	–
- stage 5 ....Nm	–	–	–	–
- stage 6 ....Nm	–	–	–	–
Big-end bearings.............................Nm	20 + 70° N	20 + 70° N	20 + 70° N	20 + 70° N
Main bearings................................Nm	15 + 60°	70	70	70
Crankshaft pulley bolt ....Nm	40 + 50°	40 + 60°	40 + 60°	40 + 60°
Camshaft pulley bolt ....Nm	35	43	43	25 Hub: 43
Flywheel [driveplate] bolt....................Nm	50	48 N	75	48 N
Front hubs ....Nm	100 + 60° oiled	100 + 60° oiled	320	320
Rear hubs ....Nm	100 + 60°	100 + 60°	275	275
Wheel nuts / bolts ....Nm	100	100	85	–
Glow plugs ....Nm	22	25	22	25
Clutch pressure plate bolts ....Nm	20	20	20	20
Injection pump sprocket.....................Nm	50	23	50	23
Injectors......................................Nm	90	90	90	90
Injection pump mounting bolts ....Nm	20	20	–	20
Injector pipe unions...........................Nm	20	25	25	25
Capacities				
Engine oil & filter................................litres	4.25	4.5	4.25	4.75
Gearbox...litres	1.8	1.8	1.8	2.0
Automatic transmission ....litres	–	–	–	–
Final drive ....litres	WT	WT	WT	WT
Cooling system.................................litres	9.0	9.0	8.0	10.5
Fuel tank..litres	80	80	60	55

Notes

Dispatch / Jumpy 1.9 TD 1997 to 2000
[1]Bolt length below head: ≤150.5 mm
Dispatch / Jumpy 1.9 D 1999 to 2000
[1]Bolt length below head: ≤125.5 mm
[2]Lucas: 142 to 149
Berlingo 1.8D 1996 to 1999
[1]Bolt length below head: 125.5 mm Without guide boss: 121.5 mm
Berlingo 1.8D 1998 to 2000
[1]lucary Varity DPC pump available as alternative
[2]Lucas: 142 to 149
[3]Bolt length below head: ≤125.5 mm

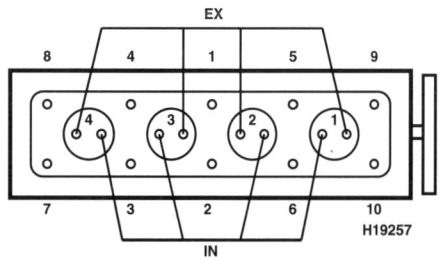

1769 cm³ / 1905 cm³

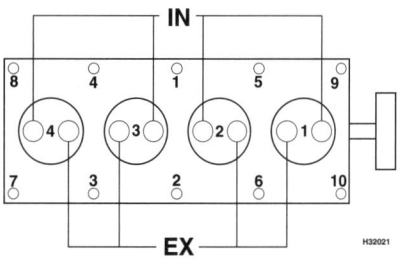

1868 cm³

– Not applicable, or information not available

CITROEN

	Berlingo 1.9D 1998 to 1999	C25D 1.9 1987 to 1991	C25D 1.9 1987 to 1991	C25D 2.5 1987 to 1991
Engine				
Engine type/code....................	XUD9A/L3 D9B SOHC 52kW	XUD9A D9B SOHC 51kW	XUD9A D9B SOHC 51kW	U25/661 OHV 54kW
Capacity (cm³) / cylinders........................	1905 / 4	1905 / 4	1905 / 4	2500 / 4
Compression ration / pressure ....bar	23.0 /	23.0 / _	23.0 / _	22.25 / _
Torque output ....Nm	120	0	0	0
Oil pressure ....idle [running] bar	[3.5 @ 4000]	[3.5]	[3.5]	[4.5 to 5.0]
Oil temperature ....°C	80	80	80	95
Valve clearances - inlet (mm)	0.15 ± 0.05	0.15 ± 0.08	0.15 ± 0.08	0.30
- exhaust (mm)	0.30 ± 0.05	0.30 ± 0.08	0.30 ± 0.08	0.20
Injection order	1-3-4-2	1-3-4-2	1-3-4-2	1-3-4-2
No. 1 cylinder position	FE	FE	FE	FE
Cooling system				
Thermostat opening temperature ....°C	83	88	88	83 to 86
Radiator cap pressure ....bar	1.4	1.0	1.0	0.9
Fuel system				
Idle speed ....rpm	800 ± 50	750 ± 50	750 ± 50	725 to 775
Maximum (no load) speed ....rpm	5150 ± 125	5150 ± 125	5100 ± 100	4375 to 4525
Smoke test/opacity ....M⁻¹ %	2.5	2.5	2.5	2.5
Static timing method...............................		Plunger travel	Dial gauge	Refer to wsm
Timing dimension.............................mm	Refer to pump	0.9	Dimension on pump	_
Crankshaft position ....mm [°]	_	TDC peg	[TDC hole]	4.71 ± 0.05
Turbo type / ref / pressure ....bar		_	_	
Injection pump make	Lucas	Bosch	Roto Diesel	Roto Diesel
Injection pump part no............................	1018 445B010A	_	R8443B380A	R8443B111A
Injector Make / type	_	Bosch	Roto Diesel	Roto Diesel
Injector part no.......................................	Lucas 6887	DNOSD 287 or 287 +	RDNOSDC 6751 C	RDNOSDC 6577B
Injection type...	Indirect	VER 272.1 or 272.2	DPC 057	Indirect DPC MA260
Injection opening pressure, New [used]...bar	140	130 +5 -0	125 ± 5	112 +5 -0
Glow plugs				
Maker	Bosch	Bosch	Bosch	Bosch
Type	0 250 201 039	91 521 918	0 250 201 019	0250 200 059
Nominal rating.................................V/A	_	11 / 13	11 / 13	11 / 8.5
Brakes				
minimum friction material thickness				
Front..............................mm	1.0	2.0	2.0	2.0
Rear...............................mm	2.0	1.0	1.0	1.0
Tyres - Saloon / Hatch.......................Size	_	_	_	_
- Estate / Van............................Size	165/70x14	185x14: 185/75x14	185x14: 185/75x14	185x14: 185/75x14
Pressure - front / rear - Saloon / Hatch...bar	_	_	_	_
- Estate / Van ......bar	2.3 / 3.7	Refer to manufacturer/handbook	Refer to manufacturer/handbook	Refer to manufacturer/handbook
Front suspension / wheel alignment				
Toe-in (+) / Toe-out (–).....................mm [°]	1.0 ± 1.0 PAS: 3.5 ± 1.0	0.5 ± 1.0	0.5 ± 1.0	0.5 ± 1.0
Camber	0° ± 1°	1°50' to 2°50'	1°50' to 2°50'	1°50' to 2°50'
Castor	30' ± 40' PAS: 2° ± 40'	30' ± 30' 91 ▶: 45' ± 30'	30' ± 30' 91 ▶: 45' ± 30'	0° to 1°
King pin inclination...................................	10° 08' ± 1°	_	_	_
Rear suspension / wheel alignment				
Toe-in (+) / Toe-out (–)....................mm [°]	2.5 ± 1.0	0 ± 1.0	0 ± 1.0	0 ± 1.0
Camber	1°15' ± 30'	_	_	_

CITROEN

Torque wrench settings	Berlingo 1.9D 1998 to 1999	C25D 1.9 1987 to 1991	C25D 1.9 1987 to 1991	C25D 2.5 1987 to 1991
Cylinder head - stage 1 ....Nm	20[1]	30 N	30 N	40
- stage 2 ....Nm	60	70	70	+ 100°
- stage 3 ....Nm	+ 180°	+ 120°	+ 120°	+ 100°
- stage 4 ....Nm	–	–	–	Warm up, allow to cool
- stage 5 ....Nm	–	–	–	+ 45°
- stage 6 ....Nm	–	–	–	–
Big-end bearings..................................Nm	50 N	50 N	50	90 N
Main bearings......................................Nm	70	70	70	95
Crankshaft pulley bolt ....Nm	40 + 60°	150 LkC	150 LkC	–
Camshaft pulley bolt ....Nm	35	35	35	32
Flywheel [driveplate] bolt.....................Nm	65	50 LkC	50 LkC	90 LkC
Front hubs ....Nm	320	500 LkC	500 LkC	500 LkC
Rear hubs ....Nm	275	170	170	105 6.86 ▶: 170
Wheel nuts / bolts ....Nm	85	180	180	180
Glow plugs ....Nm	22	22	22	28
Clutch pressure plate bolts ....Nm	20	22	22	35
Injection pump sprocket.......................Nm	50	–	50	–
Injectors..Nm	90	90	90	25
Injection pump mounting bolts ....Nm	–	–	–	–
Injector pipe unions.............................Nm	25	20	20	9
Capacities				
Engine oil & filter................................litres	4.25	5.0	5.0	4.9
Gearbox..litres	1.8	1.25	1.25	1.25
Automatic transmission ....litres	–	–	–	–
Final drive ....litres	WT	WT	WT	WT
Cooling system...................................litres	8.0	6.5	6.5	10.6
Fuel tank...litres	60	70	70	70

Notes

Berlingo 1.9D 1998 to 1999
[1]Bolt length below head: ≤121.5 mm

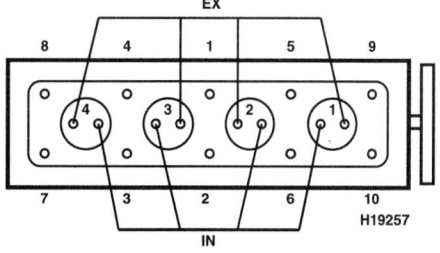

1905 cm³

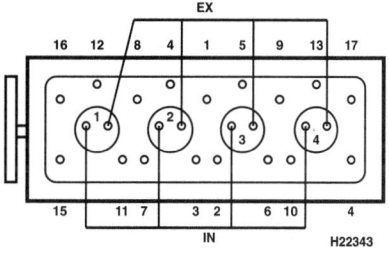

2500 cm³

– Not applicable, or information not available

CITROEN

	Jumper/Relay 1.9 D 1996 to 1999	Jumper/Relay 1.9 TD 1996 to 1999	Jumper/Relay 1.9 TD 1995 to 1996	Jumper/Relay 1.9 TD 1999 to 2000
Engine				
Engine type/code...............	XUD9AU/W2 D9B SOHC 51kW	XUD9UTF/W2 D8C Turbo 67kW	XUD9UTF/X3 DHX Turbo 66kW	XUD9TFU/W3 DHX Turbo 66kW
Capacity (cm^3) / cylinders........................	1905 / 4	1905 / 4	1905 / 4	1905 / 4
Compression ration / pressure ....bar	23.0 / _	21.8 / _	21.8 / _	21.8 / _
Torque output ....Nm	120	196	196	196
Oil pressure ....idle [running] bar	2.2 [4.6]	2.4 [4.8]	2.4 [4.8]	2.4 [4.8]
Oil temperature ....°C	80	80	80	80
Valve clearances - inlet (mm)	0.15 ± 0.05	0.15 ± 0.08	0.15 ± 0.08	0.15 ± 0.08
- exhaust (mm)	0.30 ± 0.05	0.30 ± 0.08	0.30 ± 0.08	0.30 ± 0.08
Injection order	1-3-4-2	1-3-4-2	1-3-4-2	1-3-4-2
No. 1 cylinder position	FE	FE	FE	FE
Cooling system				
Thermostat opening temperature ....°C	83	83	83	83
Radiator cap pressure ....bar	1.0	1.0	1.0	1.0
Fuel system				
Idle speed ....rpm	750 to 800	800 ± 50 A/C: 850 ± 50	800 ± 50 A/C: 850 ± 50	800 ± 50 A/C: 850 ± 50
Maximum (no load) speed ....rpm	5150 ± 125	5050 ± 125	5050 ± 125	5050 ± 125
Smoke test/opacity ....M^{-1} %	2.5	2.5	2.5	2.5
Static timing method..............................	Plunger travel	Plunger travel	Plunger travel	Plunger travel
Timing dimension.........................mm	1.07 ± 0.01	0.66 ± 0.01	0.66 ± 0.01	0.66 ± 0.01
Crankshaft position ....mm [°]	TDC [0]	TDC [0]	TDC [0]	TDC [0]
Turbo type / ref / pressure ....bar	_	_	_	_
Injection pump make	Bosch	Bosch	Bosch	Bosch
Injection pump part no............................	VER 425 / XUD201	VER 445 / XUD203	XUD 212	VER 445 / XUD203
Injector Make / type	Bosch	Bosch	Bosch	Bosch
Injector part no......................................	299A	299	299	299
Injection type...	Indirect VER	Indirect VER	Indirect	Indirect VER
Injection opening pressure, New [used]...bar	130	175.0	175.0	175
Glow plugs				
Maker	Beru/Bosch	Beru/Bosch	Beru/Bosch	Beru/Bosch
Type	0100 221 133 / 0250 201 019	0100 226 186 / 0250 201 033	0100 226 186 / 0250 201 033	0100 226 186 / 0250 201 033
Nominal rating...................................V/A	11 /	11 /	11 /	11 /
Brakes				
minimum friction material thickness				
Front............................mm	2.0	2.0	2.0	2.0
Rear............................mm	1.0	1.0	1.0	1.0
Tyres - Saloon / Hatch.....................Size	_	_	_	_
- Estate / Van.............................Size	195/70x15	195/70x15: 205/70x15	195/70x15: 205/70x15	195/70x15: 205/70x15
Pressure - front / rear - Saloon / Hatch ...bar	_	_	_	_
- Estate / Van ......bar	3.0 / 3.0	3.0 / 3.0	3.0 / 3.0	3.0 / 3.0
Front suspension / wheel alignment				
Toe-in (+) / Toe-out (–).....................mm [°]	0 ± 1.0	0 ± 1.0	0 ± 1.0	0 ± 1.0
Camber	0° ±30'	0° ± 30'	0° ± 30'	0° ± 30'
Castor	1° ± 30'	1° ± 30'	1° ± 30'	1° ± 30'
King pin inclination............................	_	_	_	_
Rear suspension / wheel alignment				
Toe-in (+) / Toe-out (–)....................mm [°]	0 ± 1.0	0 ± 1.0	0 ± 1.0	0 ± 1.0
Camber	0° ± 10'	0° ± 10'	0° ± 10'	0° ± 10'

CITROEN

	Jumper/Relay 1.9 D 1996 to 1999	Jumper/Relay 1.9 TD 1996 to 1999	Jumper/Relay 1.9 TD 1995 to 1996	Jumper/Relay 1.9 TD 1999 to 2000
Torque wrench settings				
Cylinder head - stage 1 ....Nm	20[1]	20[1]	20[1]	20[1]
- stage 2 ....Nm	60	60	60	60
- stage 3 ....Nm	+ 180°	+ 220°	+ 220°	+ 220°
- stage 4 ....Nm	–	–	–	–
- stage 5 ....Nm	–	–	–	–
- stage 6 ....Nm		–		–
Big-end bearings..............................Nm	40, slacken, 20 + 70° N	40, slacken, 20 + 70° N	40, slacken, 20 + 70° N	40, slacken, 20 + 70° N
Main bearings...................................Nm	70	70	70	70
Crankshaft pulley bolt ....Nm	40 + 51°	40 + 51°	40 + 51°	40 + 51°
Camshaft pulley bolt ....Nm	45	45	45	45
Flywheel [driveplate] bolt....................Nm	50 N LkC	50 N LkC	50 N LkC	50 N LkC
Front hubs ....Nm	450 18Q model: 500	450 18Q model: 500	450 18Q model: 500	450 18Q model: 500
Rear hubs...Nm	WSM	WSM	WSM	WSM
Wheel nuts / bolts ....Nm	160 18Q model: 180	160 18Q model: 180	160 18Q model: 180	160 18Q model: 180
Glow plugs ....Nm	18	18	18	18
Clutch pressure plate bolts ....Nm	20	20	20	20
Injection pump sprocket......................Nm	–	–	–	–
Injectors...Nm	90	90	90	90
Injection pump mounting bolts ....Nm	–	–	–	–
Injector pipe unions............................Nm	–	–	–	–
Capacities				
Engine oil & filter................................litres	5.0	6.3	6.3	6.3
Gearbox...litres	1.85	1.85	1.85	1.85
Automatic transmission ....litres				–
Final drive ....litres	WT	WT	WT	WT
Cooling system..................................litres	9.5	12.0	12.0	12.0
Fuel tank...litres	80	–	80	80

Notes

Jumper/Relay 1.9 D 1996 to 1999
[1]Bolt length below head: ≤121.5 mm Bolt with guide boss: ≤124.5 mm
Jumper/Relay 1.9 TD 1996 to 1999
[1]Bolt length below head: ≤146.5 mm Bolt with guide boss: ≤150.5 mm
Jumper/Relay 1.9 TD 1995 to 1996
[1]Bolt length below head: ≤146.5 mm Bolt with guide boss: ≤150.5 mm
Jumper/Relay 1.9 TD 1999 to 2000
[1]Bolt length below head: ≤146.5 mm Bolt with guide boss: ≤150.5 mm

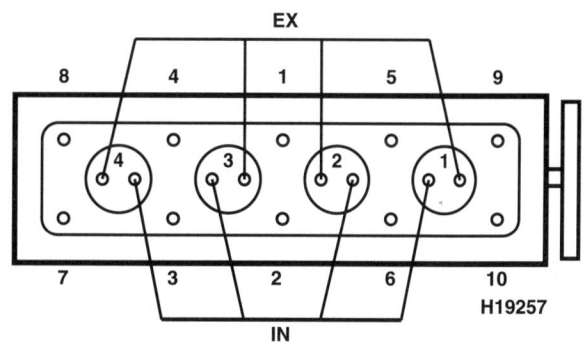

1905 cm³

– Not applicable, or information not available

CITROEN

Haynes
THE BOOK ®

	Jumper/Relay 1.9 D 1999 to 2000	Jumper/Relay 2.5 D 1996 to 2000	Jumper/Relay 2.5D Turbo 1996 to 2000	Jumper/Relay 2.5 iDi Turbo 1995 to 1996
Engine				
Engine type/code.....................................	XUD9AU/W3 DJY 50kW	DJ5/W2 T9A SOHC 12V 63kW	DJ5T/W2 T8A Turbo 76kW	DJ5T/X3 THZ Turbo 12V 76kW
Capacity (cm³) / cylinders........................	1905 / 4	2446 / 4	2446 / 4	2446 / 4
Compression ration / pressure ....bar	23.0 / _	23.0 / _	23.0 / _	21.0 / _
Torque output ....Nm	120	153	230	230
Oil pressure ....idle [running] bar	2.2 [4.6]	2.2 [4.6]	2.2 [4.6]	2.2 [4.6]
Oil temperature ....°C	80	80	80	80
Valve clearances - inlet (mm)	0.15 ± 0.08	0: Hyd.	0: Hyd.	0: Hyd.
- exhaust (mm)	0.30 ± 0.08	0: Hyd.	0: Hyd.	0: Hyd.
Injection order	1-3-4-2	1-3-4-2	1-3-4-2	1-3-4-2
No. 1 cylinder position	FE	FE	FE	FE
Cooling system				
Thermostat opening temperature ....°C	83	83	83	83
Radiator cap pressure ....bar	1.00	1.0	1.0	1.0
Fuel system				
Idle speed ....rpm	800 ± 50 A/C: 850 ± 50	800 ± 50	800 ± 50	725 ± 25 A/C: 775 ± 25
Maximum (no load) speed ....rpm	5150 ± 125	5150 ± 125	5150 ± 125	4900 ± 125
Smoke test/opacity ....M⁻¹ %	2.5	3.0	3.0	2.0
Static timing method..............................	Plunger travel	Plunger travel	Refer to wsm	Refer to wsm
Timing dimension...............................mm	1.07 ± 0.01	1.07 ± 0.01	_	_
Crankshaft position ....mm [°]	TDC [0]	TDC [0]	_	_
Turbo type / ref / pressure ....bar	_	_	KKK K14	_
Injection pump make	Bosch	Bosch	Bosch	Lucas
Injection pump part no............................	VER 425/* XUD201	VER 4/9F 2175 R504	VER 425/* XUD201	DPC R8444B281B
Injector Make / type	Bosch	Bosch	Bosch	CAV
Injector part no.......................................	299A	DNOSD 312	299A	LDC003R
Injection type...	Indirect VER	Indirect VER	Indirect VER	Direct DPC
Injection opening pressure, New [used]...bar	130.0	140 to 145	130.0	168 to 173
Glow plugs				
Maker	Beru/Bosch	Beru/Bosch	Beru/Bosch	Beru/Bosch
Type	0100 221 133 / 0250 201 019	0100 221 133 / 0250 201 019	0100 221 133 / 0250 201 019	0100 221 133 / 0250 201 019
Nominal rating.....................................V/A	11 /	11 /	11 /	11 / 9
Brakes				
minimum friction material thickness				
Front..mm	2.0	2.0	2.0	2.0
Rear...mm	1.0	1.0	1.0	1.0
Tyres - Saloon / Hatch......................Size	_	_	_	_
- Estate / Van.........................Size	195/70x15: 205/70x15	195/70x15: 205/70x16	195/70x15: 205/70x16	195/70x15: 205/70x15
Pressure - front / rear - Saloon / Hatch ...bar	_	_	_	_
- Estate / Van ......bar	Refer to vehicle	3.0 / 3.0	3.0 / 3.0	Refer to vehicle
Front suspension / wheel alignment				
Toe-in (+) / Toe-out (–)....................mm [°]	0 ± 1.0	0 ± 1.0	0 ± 1.0	0 ± 1.0
Camber	0° ± 30'	0° ± 30'	0° ± 30'	0° ± 30'
Castor	1° ± 30'	1° ± 30'	1° ± 30'	1° ± 30'
King pin inclination................................	_	_	_	_
Rear suspension / wheel alignment				
Toe-in (+) / Toe-out (–)....................mm [°]	0 ± 1.0	0 ± 1.0	0 ± 1.0	0 ± 1.0
Camber	0° ± 10'	0° ± 1°	0° ± 10'	0° ± 10'

CITROEN

	Jumper/Relay 1.9 D 1999 to 2000	Jumper/Relay 2.5 D 1996 to 2000	Jumper/Relay 2.5D Turbo 1996 to 2000	Jumper/Relay 2.5 iDi Turbo 1995 to 1996
Torque wrench settings				
Cylinder head - stage 1 ....Nm	20[1]	M10: 35 M12: 50[1]	M10: 3.5 M12: 50[1]	M10: 35 M12: 50[1]
- stage 2 ....Nm	60	+ 120°	+ 150°	+ 150°
- stage 3 ....Nm	+ 180°	–	–	–
- stage 4 ....Nm	–	–	–	–
- stage 5 ....Nm	–	–	–	–
- stage 6 ....Nm	–	–	–	–
Big-end bearings................................Nm	40, slacken, 20 + 70° N	50 N	50 N	50 N
Main bearings.....................................Nm	70	70	70	70
Crankshaft pulley bolt ....Nm	40 + 51°	40 + 51°	40 + 51°	40 + 51°
Camshaft pulley bolt ....Nm	45	45	45	45
Flywheel [driveplate] bolt....................Nm	50 N LkC	50 N LkC	50 N LkC	50 N LkC
Front hubs ....Nm	450 18Q model: 500	450 18Q model: 500	450 18Q model: 500	450 18Q model: 500
Rear hubs ....Nm	WSM	WSM	WSM	WSM
Wheel nuts / bolts ....Nm	160 18Q model: 180	160 18Q model: 180	160 18Q model: 180	160 18Q model: 180
Glow plugs ....Nm	18	18	18	18
Clutch pressure plate bolts ....Nm	20	20	20	20
Injection pump sprocket......................Nm	–	25	25	25
Injectors...Nm	90	55	90	55
Injection pump mounting bolts ....Nm	–	–	–	25
Injector pipe unions............................Nm	–	20	25	25
Capacities				
Engine oil & filter................................litres	6.3	9.0	8.0	6.3
Gearbox..litres	1.85	1.85	1.85	1.85
Automatic transmission ....litres	–	–	–	–
Final drive ....litres	WT	WT	WT	WT
Cooling system...................................litres	12.0	12.0	12.0	12.0
Fuel tank...litres	–	80	80	80

Notes

Jumper/Relay 1.9 D 1999 to 2000
[1]Bolt length below head: ≤121.5 mm Bolt with guide boss: ≤125.5 mm
Jumper/Relay 2.5 D 1996 to 2000
[1] Max overall bolt length: M10 162.5 mm
M12 153.5 mm
Jumper/Relay 2.5D Turbo 1996 to 2000
[1] Max overall bolt length: M10 189.3 mm
M12 178.3 mm
Jumper/Relay 2.5 iDi Turbo 1995 to 1996
[1]Max overall bolt length: M10 189.3 mm
M12 178.3 mm

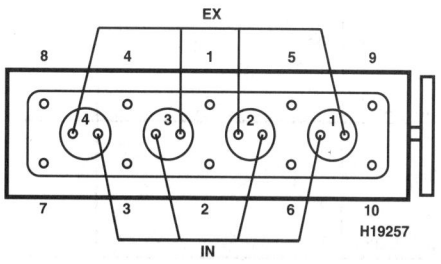

1905 cm³

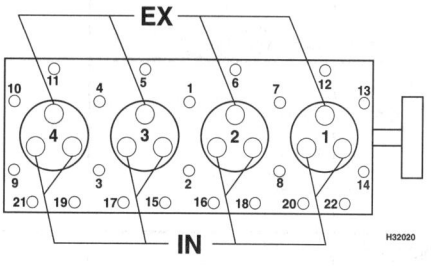

2446 cm³

– Not applicable, or information not available

CITROEN

	Jumper/Relay 2.5 TDi 1996 to 1997			
Engine				
Engine type/code............................	DJ5TED/W2 THX Turbo 12V 79kW			
Capacity (cm³) / cylinders.........................	2446 / 4			
Compression ration / pressure ....bar	20.0 / _			
Torque output ....Nm	230			
Oil pressure ....idle [running] bar	2.2 [4.6]			
Oil temperature ....°C	80			
Valve clearances - inlet (mm)	0: Hyd.			
- exhaust (mm)	0: Hyd.			
Injection order	1-3-4-2			
No. 1 cylinder position	FE			
Cooling system				
Thermostat opening temperature ....°C	83			
Radiator cap pressure ....bar	1.0			
Fuel system				
Idle speed ....rpm	850 ± 50 A/C: 900 ± 50			
Maximum (no load) speed ....rpm	4900 ± 150			
Smoke test/opacity ....M⁻¹ %	2.0			
Static timing method.................................	Refer to wsm			
Timing dimension...............................mm	_			
Crankshaft position ....mm [°]	TDC			
Turbo type / ref / pressure ....bar	KKK K14			
Injection pump make	Bosch			
Injection pump part no.............................	VER683			
Injector Make / type	Bosch			
Injector part no..	140PV3375955			
Injection type...	Direct VER			
Injection opening pressure, New [used]...bar	200			
Glow plugs				
Maker	Beru			
Type	0100 221 133			
Nominal rating.......................................V/A	_			
Brakes				
minimum friction material thickness				
Front..mm	2.0			
Rear..mm	1.0			
Tyres - Saloon / Hatch......................Size	_			
- Estate / Van............................Size	195/70x15: 205/70x15			
Pressure - front / rear - Saloon / Hatch...bar	_			
- Estate / Van ......bar	Refer to vehicle			
Front suspension / wheel alignment				
Toe-in (+) / Toe-out (−).....................mm [°]	0 ± 1.0			
Camber	0° ± 30'			
Castor	1° ± 30'			
King pin inclination...................................	_			
Rear suspension / wheel alignment				
Toe-in (+) / Toe-out (−).....................mm [°]	0 ± 1.0			
Camber	0° ± 10'			

Haynes THE BOOK ®

CITROEN

	Jumper/Relay 2.5 TDi 1996 to 1997			
Torque wrench settings				
Cylinder head - stage 1 ....Nm	M10: 35 M12: 50[1]			
- stage 2 ....Nm	+ 150°			
- stage 3 ....Nm	_			
- stage 4 ....Nm	_			
- stage 5 ....Nm	_			
- stage 6 ....Nm	_			
Big-end bearings..............................Nm	50 N			
Main bearings...................................Nm	70			
Crankshaft pulley bolt ....Nm	40 + 51°			
Camshaft pulley bolt ....Nm	45			
Flywheel [driveplate] bolt.................Nm	50 N LkC			
Front hubs ....Nm	450 18Q model: 500			
Rear hubs...Nm	WSM			
Wheel nuts / bolts ....Nm	160 18Q model: 180			
Glow plugs ....Nm	18			
Clutch pressure plate bolts ....Nm	20			
Injection pump sprocket.....................Nm	25			
Injectors..Nm	55			
Injection pump mounting bolts ....Nm	_			
Injector pipe unions...........................Nm	25			
Capacities				
Engine oil & filter.............................litres	8.0			
Gearbox...litres	1.85			
Automatic transmission ....litres	_			
Final drive ....litres	WT			
Cooling system.................................litres	12.0			
Fuel tank...litres	80			

Notes

Jumper/Relay 2.5 TDi 1996 to 1997
[1] Max overall bolt length: M10 189.3 mm
M12 178.3 mm

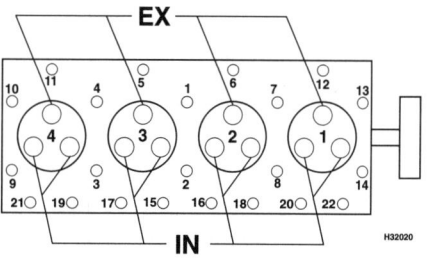

2446 cm³

– Not applicable, or information not
available

DACIA

	Duster 1989 to 1991			
Engine				
Engine type/code....................................	F8M			
Capacity (cm³) / cylinders........................	1596 / 4			
Compression ration / pressure ....bar	22.5 / ≥20.0			
Torque output ....Nm	0			
Oil pressure ....idle [running] bar	2.0 [3.5]			
Oil temperature ....°C	80			
Valve clearances - inlet (mm)	0.20			
- exhaust (mm)	0.40			
Injection order	1-3-4-2			
No. 1 cylinder position	FE			
Cooling system				
Thermostat opening temperature ....°C	83			
Radiator cap pressure ....bar	0.9 ± 0.1			
Fuel system				
Idle speed ....rpm	850 ± 25			
Maximum (no load) speed ....rpm	5300			
Smoke test/opacity ....M⁻¹ %	2.5			
Static timing method................................	Dial gauge			
Timing dimension.................................mm	1.6 ± 0.02			
Crankshaft position ....mm [°]	[0] TDC			
Turbo type / ref / pressure ....bar	–			
Injection pump make	Roto Diesel			
Injection pump part no.............................				
Injector Make / type	Roto Diesel			
Injector part no..	RDNOSDC 6843C			
Injection type...	DPCR 8443 A370A			
Injection opening pressure, New [used]...bar	118 +7 -5			
Glow plugs				
Maker	Champion			
Type	CH88, CH137			
Nominal rating.....................................V/A	–			
Brakes				
minimum friction material thickness				
Front...mm	7.0 with backing			
Rear...mm	–			
Tyres - Saloon / Hatch.......................Size	–			
- Estate / Van...........................Size	175x14			
Pressure - front / rear - Saloon / Hatch ...bar	–			
- Estate / Van ......bar	1.9 / 2.1			
Front suspension / wheel alignment				
Toe-in (+) / Toe-out (–).....................mm [°]	1.5 to 5.0			
Camber	1°30' ± 30'			
Castor	2°30' ± 30'			
King pin inclination..................................	9°30' ± 30'			
Rear suspension / wheel alignment				
Toe-in (+) / Toe-out (–).....................mm [°]	–			
Camber	–			

DACIA

	Duster 1989 to 1991			
Torque wrench settings				
Cylinder head - stage 1 ....Nm	30			
- stage 2 ....Nm	70			
- stage 3 ....Nm	Wait 3 mins			
- stage 4 ....Nm	Slacken			
- stage 5 ....Nm	20			
- stage 6 ....Nm	+ 123°			
Big-end bearings......................Nm	45 to 50 N			
Main bearings......................Nm	60 to 65			
Crankshaft pulley bolt ....Nm	90 to 100			
Camshaft pulley bolt ....Nm	50			
Flywheel [driveplate] bolt......................Nm	50 to 55			
Front hubs ....Nm	_			
Rear hubs ....Nm	_			
Wheel nuts / bolts ....Nm	80			
Glow plugs ....Nm	15 to 30			
Clutch pressure plate bolts ....Nm	25			
Injection pump sprocket......................Nm	50			
Injectors......................Nm	65 to 75			
Injection pump mounting bolts ....Nm	_			
Injector pipe unions......................Nm	25			
Capacities				
Engine oil & filter......................litres	5.5			
Gearbox......................litres	1.4[1]			
Automatic transmission ....litres	_			
Final drive ....litres	Front: 0.8 Rear:1.4			
Cooling system......................litres	7.0			
Fuel tank......................litres	46			

Notes

Duster 1989 to 1991
[1]Transfer box: 0.8

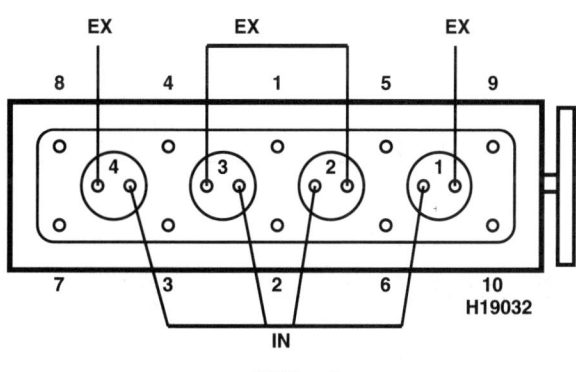

1596 cm³

– Not applicable, or information not
  available

DAEWOO

Haynes THE BOOK ®		Korando 2.9 TDi 1999 to 2000	Musso 2.9 TDi 1999 to 2000		
Engine					
Engine type/code.....................		OM662LA Turbo SOHC 87kW	OM662LA Turbo SOHC 87kW		
Capacity (cm³) / cylinders........................		2874 / 5	2874 / 5		
Compression ration / pressure ....bar		22.1 / ≥18.0	22.1 / ≥18.0		
Torque output ....Nm		250	250		
Oil pressure ....idle [running] bar		–	–		
Oil temperature ....°C					
Valve clearances - inlet (mm)		0: Hyd.	0: Hyd.		
- exhaust (mm)		0: Hyd.	0: Hyd.		
Injection order..		1-2-4-5-3	1-2-4-5-3		
No. 1 cylinder position		F	F		
Cooling system					
Thermostat opening temperature ....°C		80	80		
Radiator cap pressure ....bar		–	–		
Fuel system					
Idle speed ....rpm		770 ± 50	770 ± 50		
Maximum (no load) speed ....rpm		–	–		
Smoke test/opacity ....M⁻¹ %		2.0	2.0		
Static timing method...............................		Refer to wsm	Refer to wsm		
Timing dimension.............................mm		–	–		
Crankshaft position ....mm [°]		–	–		
Turbo type / ref / pressure ....bar		–	–		
Injection pump make		Bosch	Bosch		
Injection pump part no............................		PES 4 M55 C320 RS 191	PES 4 M55 C320 RS 191		
Injector Make / type		–	–		
Injector part no.......................................		–	–		
Injection type...		Indirect, intercooled	Indirect, intercooled		
Injection opening pressure, New [used]...bar		115 to 125 [100]	115 to 125 [100]		
Glow plugs					
Maker		–	–		
Type		–	–		
Nominal rating...................................V/A		–	–		
Brakes					
minimum friction material thickness					
Front.....................................mm		2.0	2.0		
Rear.....................................mm		1.5	1.5		
Tyres - Saloon / Hatch.....................Size		–	–		
- Estate / Van...........................Size		235/75x15	235/75x15		
Pressure - front / rear - Saloon / Hatch ...bar					
- Estate / Van ......bar		2.1 / 2.1	2.1 / 2.1		
Front suspension / wheel alignment					
Toe-in (+) / Toe-out (–)..................mm [°]		0.4	0.4		
Camber		0° ± 30'	0° ± 30'		
Castor		2° 30' ± 30'	2° 30' ± 30'		
King pin inclination.................................		12° 30'	12° 30'		
Rear suspension / wheel alignment					
Toe-in (+) / Toe-out (–)....................mm [°]		–	–		
Camber		–	–		

DAEWOO

Torque wrench settings	Korando 2.9 TDi 1999 to 2000	Musso 2.9 TDi 1999 to 2000		
Cylinder head - stage 1 ....Nm	15 N	15 N		
- stage 2 ....Nm	35	35		
- stage 3 ....Nm	+ 90°	+ 90°		
- stage 4 ....Nm	+ 90°	+ 90°		
- stage 5 ....Nm	–	–		
- stage 6 ....Nm	–	–		
Big-end bearings.............................Nm	40 + 90° N	40 + 90° N		
Main bearings...................................Nm	55 + 90° N	55 + 90° N		
Crankshaft pulley bolt ....Nm	–	–		
Camshaft pulley bolt ....Nm	–	–		
Flywheel [driveplate] bolt....................Nm	45 + 90° N	45 + 90° N		
Front hubs ....Nm	WSM	WSM		
Rear hubs ....Nm	WSM	WSM		
Wheel nuts / bolts ....Nm	90[1]	90[1]		
Glow plugs ....Nm	–	–		
Clutch pressure plate bolts ....Nm	35	35		
Injection pump sprocket.....................Nm	–	–		
Injectors..Nm	–	–		
Injection pump mounting bolts ....Nm	–	–		
Injector pipe unions...........................Nm	–	–		
Capacities				
Engine oil & filter...............................litres	8.0	8.0		
Gearbox...litres	1.8	1.8		
Automatic transmission ....litres	–	–		
Final drive ....litres	Rear: 1.9 Front: 1.4[2]	Rear: 1.9 Front: 1.4[2]		
Cooling system..................................litres	10.0	10.0		
Fuel tank..litres	80	80		

Notes

Korando 2.9 TDi 1999 to 2000
[1]Alloy: 125
[2]Transfer: 1.4
Musso 2.9 TDi 1999 to 2000
[1]Alloy: 125
[2]Transfer: 1.4

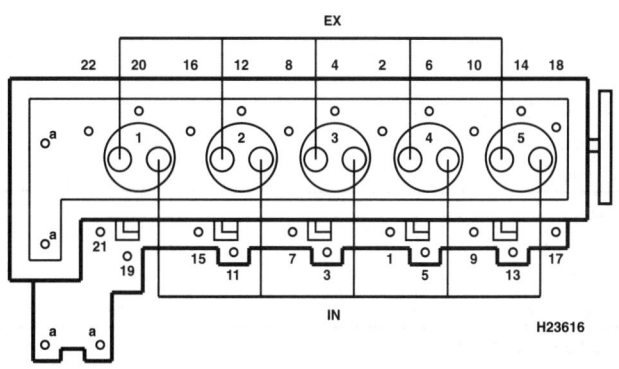

2874 cm³

– Not applicable, or information not available

DAIHATSU

	Charade 1.0 Turbo 1990 to 1993	F70 & Fourtrak 1984 to 1993	F70 & Fourtrak Turbo 1984 to 1993	F70 & Fourtrak 1989 to 1998
Engine				
Engine type/code.............................	CL61 SOHC Turbo 35kW	DL50 OHV 54kW	DL51 OHV Turbo 67kW	DL42 OHV 54kW
Capacity (cm^3) / cylinders........................	993 / 3	2765 / 4	2765 / 4	2765 / 4
Compression ration / pressure ....bar	21.5 / ≥18.0	21.5 / ≥21.6	21.5 / ≥21.6	21.2 / ≥21.6
Torque output ....Nm	0	0	0	0
Oil pressure ....idle [running] bar	[3.1 to 4.1]	[4.4 to 5.4]	[4.4 to 5.4]	[4.4 to 5.4]
Oil temperature ....°C	80	80	80	80
Valve clearances - inlet (mm)	0.25 ± 0.05	0.25 H	0.25 H	0.25 H
- exhaust (mm)	0.30 ± 0.05	0.30 H	0.35 H	0.25 H
Injection order	1-2-3	1-3-4-2	1-3-4-2	1-3-4-2
No. 1 cylinder position	TBE	TCE	TCE	TCE
Cooling system				
Thermostat opening temperature ....°C	88 ± 1.5	80 to 84	80 to 84	80 to 84
Radiator cap pressure ....bar	0.9	0.9	0.9	0.9
Fuel system				
Idle speed ....rpm	850 ± 50	650 ± 50	650 ± 50	650 ± 50
Maximum (no load) speed ....rpm	5600 +150 -100	4100 to 4175	4600	4500
Smoke test/opacity ....M^{-1} %	2.5	3.0	3.0	3.0
Static timing method...............................	Plunger travel	Plunger travel	Plunger travel	Plunger travel
Timing dimension...............................mm	0.89 ± 0.03	0.97 to 1.03	0.97 to 1.03	1.06 ± 0.03
Crankshaft position ....mm [°]	[0] TDC	[0] TDC	TDC	[1] BTDC
Turbo type / ref / pressure ....bar	0.5 to 0.7 bar	–	0.55 to 0.85 @ 3400 rpm	–
Injection pump make	Bosch	Nippon Denso	Bosch	Bosch
Injection pump part no............................	VE	VE	VE4	VE
Injector Make / type	Nippon Denso	Nippon Denso	Nippon Denso	Single hole
Injector part no..	ND-DN12SD12A	DN12S 012A	093400 - 0510	–
Injection type...	Indirect VE	Indirect VE	Indirect VE	Indirect VE
Injection opening pressure, New [used]...bar	132 to 142	108 ± 5	105 to 115	108 ± 5
Glow plugs				
Maker	Bosch	Bosch	Bosch	Bosch
Type	0250 2020 074	0250 202 051	0250 202 051	0250 202 051
Nominal rating.....................................V/A	11.0 / 6.4	8.5 / 9	–	8.5 / 10
Brakes				
minimum friction material thickness				
Front...............................mm	1.0	1.0	1.0	1.0
Rear................................mm	1.0	1.0	1.0	1.0
Tyres - Saloon / Hatch ....Size	165/70x13	–	–	–
- Estate / Van...........................Size	–	215x15	215x15	215x15
Pressure - front / rear - Saloon / Hatch ...bar	2.0 / 2.0	–	–	–
- Estate / Van ......bar	–	1.2 / 1.7	1.2 / 1.7	1.2 / 1.7
Front suspension / wheel alignment				
Toe-in (+) / Toe-out (–)...................mm [°]	1.0	0 ± 3.0	0 ± 3.0	0 ± 3.0
Camber	20' ± 1°	1° ± 1°	1° ± 1°	1° ± 1°
Castor	2°55' ± 1°	1°30' ± 1°	1°30' ± 1°	1°30' ± 1°
King pin inclination.................................	12° ± 1°	9°30' ± 1°	9°30' ± 1°	9°30' ± 1°
Rear suspension / wheel alignment				
Toe-in (+) / Toe-out (–)....................mm [°]	5.0	–	–	–
Camber	40'			

DAIHATSU

Torque wrench settings	Charade 1.0 Turbo 1990 to 1993	F70 & Fourtrak 1984 to 1993	F70 & Fourtrak Turbo 1984 to 1993	F70 & Fourtrak 1989 to 1998
Cylinder head - stage 1 ....Nm	85 to 95	108 to 127	108 to 127	108 to 127
- stage 2 ....Nm	–	–	–	–
- stage 3 ....Nm	–	–	–	–
- stage 4 ....Nm	–	–	–	–
- stage 5 ....Nm	–	–	–	–
- stage 6 ....Nm	–	–		
Big-end bearings.....................Nm	42 to 52 N	64 to 73 N	64 to 73 N	64 to 73 N
Main bearings......................Nm	54 to 66	96 to 110	96 to 110	96 to 110
Crankshaft pulley bolt ....Nm	55	240	240	240
Camshaft pulley bolt ....Nm	45	50	50	50
Flywheel [driveplate] bolt.....................Nm	50 to 60	113 to 137	113 to 137	113 to 137
Front hubs ....Nm	176 to 216	WSM	WSM	WSM
Rear hubs ....Nm	58 to 98	WSM	WSM	WSM
Wheel nuts / bolts ....Nm	88 to 118	88 to 118	88 to 118	88 to 118
Glow plugs ....Nm	14	18	18	18
Clutch pressure plate bolts ....Nm	7 to 10	15 to 24	15 to 24	15 to 24
Injection pump sprocket.....................Nm	65	–	90	90
Injectors.....................Nm	50 to 70	80	60 to 80	60 to 80
Injection pump mounting bolts ....Nm	18	30	30	–
Injector pipe unions.....................Nm	25	27	30	30
Capacities				
Engine oil & filter.....................litres	3.5	6.0	6.0	6.4
Gearbox.....................litres	2.1	2.8¹	2.8¹	2.8¹
Automatic transmission ....litres	–	–	–	–
Final drive.....................litres	WT	1.5	1.5	1.5
Cooling system.....................litres	4.4	11.0	11.0	11.0
Fuel tank.....................litres	38	60	60	60

Notes

F70 & Fourtrak 1984 to 1993
¹Transfer box: 1.3
F70 & Fourtrak Turbo 1984 to 1993
¹Transfer box: 1.3
F70 & Fourtrak 1989 to 1998
¹Transfer box: 1.3

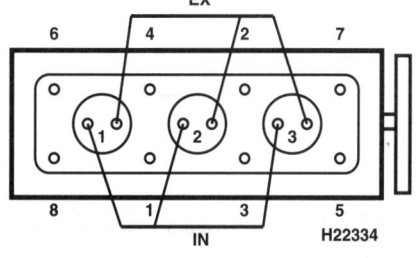

993 cm³

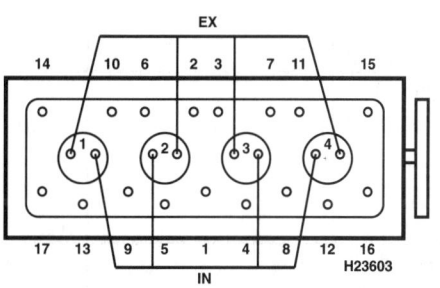

2765 cm³

– Not applicable, or information not available

DAIHATSU

	Fourtrack Independant 1993 to 2000	Hijet 1.2 D 1996 to 2000		
Engine				
Engine type/code...........................	DL52 OHV Turbo 73kW	SOHC 26kW		
Capacity (cm^3) / cylinders........................	2765 / 4	1221 / 4		
Compression ration / pressure ....bar	21.2 / ≥21.6	23.0 /		
Torque output ....Nm	245	71		
Oil pressure ....idle [running] bar	1.5 [3.0]	[3.0 @ 2200]		
Oil temperature ....°C	80	80		
Valve clearances - inlet (mm)	0.25 H	0.20		
- exhaust (mm)	0.35 H	0.20		
Injection order	1-3-4-2	1-3-4-2		
No. 1 cylinder position	TBE	TBE		
Cooling system				
Thermostat opening temperature ....°C	80 to 84	88		
Radiator cap pressure ....bar	0.9	0.9		
Fuel system				
Idle speed ....rpm	650 ± 50	900 ± 50		
Maximum (no load) speed ....rpm	4500	4900		
Smoke test/opacity ....M^{-1} %	3.0	3.0		
Static timing method.................................	Plunger travel	–		
Timing dimension.................................mm	1.06 ± 0.03	–		
Crankshaft position ....mm [°]	[1] BTDC	–		
Turbo type / ref / pressure ....bar	0.5 to 0.8 bar @ 3500	–		
Injection pump make	Bosch	–		
Injection pump part no.............................	VE	–		
Injector Make / type	Single hole	–		
Injector part no...	–			
Injection type...	Indirect VE	Indirect		
Injection opening pressure, New [used]...bar	108 ± 5 [127]	140 to 155		
Glow plugs				
Maker	Bosch	–		
Type	0250 202 051	–		
Nominal rating....................................V/A	7 / 8.5	–		
Brakes				
minimum friction material thickness				
Front....................................mm	1.0	1.0		
Rear.....................................mm	1.0	1.0		
Tyres - Saloon / Hatch......................Size	–	–		
- Estate / Van............................Size	215x15	155x12		
Pressure - front / rear - Saloon / Hatch ...bar	–			
- Estate / Van ......bar	2.0 / 1.8	Refer to vehicle		
Front suspension / wheel alignment				
Toe-in (+) / Toe-out (–)....................mm [°]	2.0 ± 1.0	1.5 ± 1.0		
Camber	0°30' ± 30'	1°40'		
Castor	2°20' ± 45'	3° 02' ± 1°		
King pin inclination..................................	10°	11°25' ± 1°		
Rear suspension / wheel alignment				
Toe-in (+) / Toe-out (–)....................mm [°]	–	–		
Camber	–	–		

DAIHATSU

	Fourtrack Independant 1993 to 2000	Hijet 1.2 D 1996 to 2000		
Torque wrench settings				
Cylinder head - stage 1 ....Nm	108 to 127	50		
- stage 2 ....Nm	–	+ 90°		
- stage 3 ....Nm	–	+ 120°		
- stage 4 ....Nm	–	–		
- stage 5 ....Nm	–	–		
- stage 6 ....Nm	–	–		
Big-end bearings..............................Nm	64 to 73 N	40 N		
Main bearings..................................Nm	96 to 110	60		
Crankshaft pulley bolt ....Nm	240	360		
Camshaft pulley bolt ....Nm	50	–		
Flywheel [driveplate] bolt....................Nm	113 to 137	80		
Front hubs ....Nm	WSM	177 to 216		
Rear hubs ....Nm	WSM	39 to 53		
Wheel nuts / bolts ....Nm	88 to 118	55		
Glow plugs ....Nm	18	20		
Clutch pressure plate bolts ....Nm	15 to 24	10		
Injection pump sprocket......................Nm	90	–		
Injectors..Nm	60 to 80	20		
Injection pump mounting bolts ....Nm	–	–		
Injector pipe unions............................Nm	30	–		
Capacities				
Engine oil & filter.................................litres	6.4	4.15		
Gearbox..litres	6.4¹	0.7		
Automatic transmission ....litres	–	–		
Final drive ....litres	1.5	1.2		
Cooling system................................litres	10.6	5.5		
Fuel tank...litres	60	35		

Notes

Fourtrack Independant 1993 to 2000
¹Transfer box: 1.3

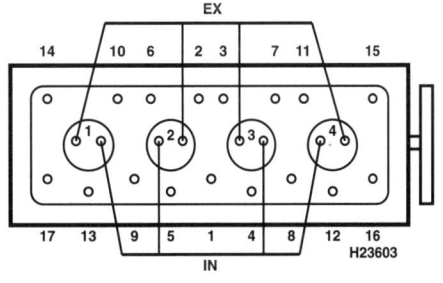

2765 cm³

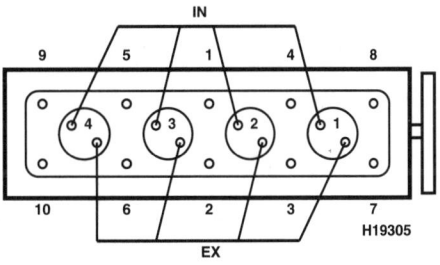

1221 cm³

– Not applicable, or information not available

FIAT

	Panda 1300 & Van 1987 to 1991	Uno 1.3 1990 to 1991	Uno 1.4 Turbo 1990 to 1994	Uno 1.7 1990 to 1995
Engine				
Engine type/code	156 A5.000 OHC 27kW	146 B1.000 OHC 33 kW	146 B3.000 OHC Turbo 52kW	146 B2.000 OHC 42kW
Capacity (cm^3) / cylinders.......................	1301 / 4	1301 / 4	1367 / 4	1697 / 4
Compression ration / pressure ....bar	20.0 / _	20.0 / _	20.0 ± 0.5 / _	20.5 ± 0.5 / _
Torque output ....Nm	0	0	0	0
Oil pressure ....idle [running] bar	[2.9 to 3.9]	[3.4 to 4.9]	[3.4 to 4.9]	[3.4 to 4.9]
Oil temperature ....°C	100	100	100	100
Valve clearances - inlet (mm)	0.35	0.35	0.30 ± 0.05	0.30 ± 0.05
- exhaust (mm)	0.40	0.40	0.35 ± 0.05	0.35 ± 0.05
Injection order	1-3-4-2	1-3-4-2	1-3-4-2	1-3-4-2
No. 1 cylinder position	TBE	TBE	TBE	TBE
Cooling system				
Thermostat opening temperature ....°C	85 to 89	78 to 82	78 to 82	78 to 82
Radiator cap pressure ....bar	0.98	0.78	0.78	0.78
Fuel system				
Idle speed ....rpm	775 ± 25	780 ± 20	800 ± 20	760 ± 20
Maximum (no load) speed ....rpm	5560 ± 50	5560 ± 50	5460 ± 50	5150 ± 50
Smoke test/opacity ....M^{-1} %	2.5	2.5	2.5	2.5
Static timing method................................	Plunger travel	Plunger travel	Plunger travel	Plunger travel
Timing dimension............................mm	0.77 [4 ± 1 ATDC]	0.82 [3 ± 1 ATDC]	1.05 [1 ± 1 ATDC]	1.0 [0 ± 1 ATDC]
Crankshaft position ....mm [°]	0 TDC	0 TDC	0 TDC	0 TDC
Turbo type / ref / pressure ....bar	_	_	_	_
Injection pump make	Bosch	Bosch	Bosch	Bosch
Injection pump part no.............................	_	_	VER 349	0 460 484 015
Injector Make / type	Bosch	Bosch	Bosch	Bosch
Injector part no..	DM 12 SD 1750	DN 12 SD 296	DN 12 SD 290	DN 12 SD 290
Injection type...	VE 4/8 F2000 R61-1	VE R 355	VE R 349	VE R 308
Injection opening pressure, New [used]...bar	125 to 133	125 to 133	150 to 158	125 to 133
Glow plugs				
Maker	Marelli/Champion	Marelli/Champion	Marelli/Champion	Marelli/Champion
Type	UX2A / CH68	UX2A / CH68	UX2A / CH68	UX2A / CH68
Nominal rating.....................................V/A	_	_	11 / 12	11 / 12
Brakes				
minimum friction material thickness				
Front..mm	1.5	1.5	1.5	1.5
Rear...mm	1.5	1.5	1.5	1.5
Tyres - Saloon / Hatch.....................Size	135x13	155/70x13	155/70x13	155/70x13
- Estate / Van...........................Size	135x13	_	_	_
Pressure - front / rear - Saloon / Hatch...bar	2.0 / 2.0	1.9 / 1.9	2.0 / 1.9	2.0 / 1.9
- Estate / Van ......bar	2.0 / 2.0	_	_	_
Front suspension / wheel alignment				
Toe-in (+) / Toe-out (–)....................mm [°]	-2.0 ± 2.0	1.0 ± 1.0	1.0 ± 1.0	1.0 ± 1.0
Camber	1° ± 30'	-30' ± 30'	-30' ± 30'	-30' ± 30'
Castor	2°30' ± 30'	2°10' ± 30'	2°10' ± 30'	2°10' ± 30'
King pin inclination..................................	_	_	_	_
Rear suspension / wheel alignment				
Toe-in (+) / Toe-out (–)....................mm [°]	0	_	_	_
Camber	0	_	_	_

FIAT

	Panda 1300 & Van 1987 to 1991	Uno 1.3 1990 to 1991	Uno 1.4 Turbo 1990 to 1994	Uno 1.7 1990 to 1995
Torque wrench settings				
Cylinder head - stage 1 ....Nm	30	30	50	50
- stage 2 ....Nm	65	65	100	100
- stage 3 ....Nm	Bolts: + 90°	Bolts: + 90°	+ 90°	+ 90°
- stage 4 ....Nm	Bolts: + 90°	Bolts: + 90°	+ 90°	+ 90°
- stage 5 ....Nm	Nuts: + 40° + 40°	Nuts: + 40° + 40°	Bolts 11 to 15: 30	Bolts 11 to 15: 30
- stage 6 ....Nm	Bolts 11 to 14: 30	Bolts 11 to 14: 30	–	–
Big-end bearings...............................Nm	51 N	51 N	25 + 50° N	25 + 50° N
Main bearings....................................Nm	80	80	113	113
Crankshaft pulley bolt ....Nm	137	137	190	190
Camshaft pulley bolt ....Nm	118	118	118	118
Flywheel [driveplate] bolt....................Nm	83	83	83	142
Front hubs ....Nm	220	294	294	294
Rear hubs ....Nm	70	216	216	216
Wheel nuts / bolts ....Nm	90	86	86	86
Glow plugs ....Nm	15	15	15	15
Clutch pressure plate bolts ....Nm	16	16	16	16
Injection pump sprocket.....................Nm	49	49	50	49
Injectors...Nm	55	55	55	55
Injection pump mounting bolts ....Nm	29	29	25	25
Injector pipe unions...........................Nm	24	24	24	24
Capacities				
Engine oil & filter................................litres	4.1	4.1	4.9	4.1
Gearbox..litres	2.4	2.4	3.3	2.0
Automatic transmission ....litres	–	–	–	–
Final drive ....litres	WT	WT	WT	WT
Cooling system...................................litres	5.6	8.0	8.9	8.9
Fuel tank...litres	32 Van: 40	42	42	42

Notes

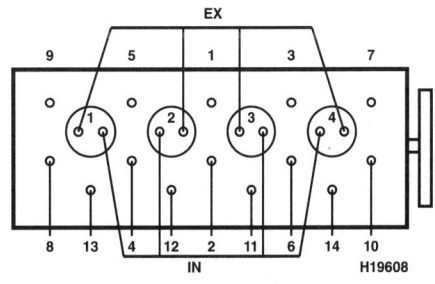

1301 cm³

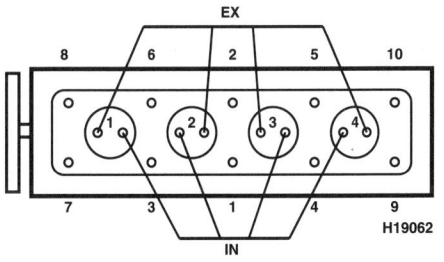

1347 cm³ / 1697 cm³

– Not applicable, or information not available

FIAT

	Uno 1.7 1990 to 1991	Punto D S 1996 to 1997	Punto TD 60 S 1996 to 1999	Punto TD 70 SX 1996 to 1999
Engine				
Engine type/code	146 B2.000 OHC 42kW	176 B3.000 SOHC 42kW	176 B7.000 Turbo 46kW	176 A3.000 Turbo 51kW
Capacity (cm³) / cylinders	1697 / 4	1698 / 4	1698 / 4	1698 / 4
Compression ration / pressure ....bar	20.5 ± 0.5 / _	20.3 /	20.3 /	20.3 /
Torque output ....Nm	0	0	118	134
Oil pressure ....idle [running] bar	[3.4 to 4.9]	[3.4]	[3.4]	[3.4]
Oil temperature ....°C	100	100	100	100
Valve clearances - inlet (mm)	0.30 ± 0.05	0.30 ± 0.05	0.30 ± 0.05	0.30 ± 0.05
- exhaust (mm)	0.35 ± 0.05	0.35 ± 0.05	0.35 ± 0.05	0.35 ± 0.05
Injection order	1-3-4-2	1-3-4-2	1-3-4-2	1-3-4-2
No. 1 cylinder position	TBE	TBE	TBE	TBE
Cooling system				
Thermostat opening temperature ....°C	78 to 82	78	78	78
Radiator cap pressure ....bar	0.78	0.98	0.98	0.98
Fuel system				
Idle speed ....rpm	810 ± 20	800 ± 50	900	900
Maximum (no load) speed ....rpm	5150 ± 50	5250	5250	5250
Smoke test/opacity ....M⁻¹ %	2.5	1.85	1.89	1.35
Static timing method................................	Refer to text	Dial gauge	Plunger travel	Plunger travel
Timing dimension........................mm	[0 ± 1]	Dimension on pump	0.94 ± 0.05	0.94 ± 0.05
Crankshaft position ....mm [°]	0 TDC	TDC	TDC	TDC
Turbo type / ref / pressure ....bar	_	_	TBO 227	0.75 bar
Injection pump make	Roto Diesel	CAV	Bosch	Bosch
Injection pump part no.........................	B 571B (FT01 E FT02)	FT08	VER 691	VER 537-1
Injector Make / type	Roto Diesel	CAV	Bosch	Bosch
Injector part no.........................	BDNOSDC6751C¹	6732 604C	DN 12 SD 290	DN 12 SD 290
Injection type.........................	DPCR 8443	Indirect FT08	Bosch VE	Indirect VER
Injection opening pressure, New [used]...bar	124 to 131²	124 to 131 [116 to 123]	150	150
Glow plugs				
Maker	Marelli/Champion	Bosch	Bosch	Bosch
Type	UX2A / CH68	0250 201 33	0250 001 538	0250 201 033
Nominal rating ....V/A	_	_	_	_
Brakes				
minimum friction material thickness				
Front..mm	1.5	1.5	1.5	1.5
Rear..mm	1.5	1.5	1.5	1.5
Tyres - Saloon / Hatch.....................Size	155/70x13	165/65x14	165/65x14	165/65x14
- Estate / Van.........................Size	_	_	_	_
Pressure - front / rear - Saloon / Hatch...bar	2.0 / 1.9	2.4 / 2.0	2.4 / 2.0	2.4 / 2.0
- Estate / Van ......bar	_	_	_	_
Front suspension / wheel alignment				
Toe-in (+) / Toe-out (–)....................mm [°]	1.0 ± 1.0	0.0 ± 1.0	0.0 ± 1.0	0.0 ± 1.0
Camber	-30' ± 30'	-0°15' ± 30'	-0°15' ± 30'	-0° 15' ± 30'
Castor	2°10' ± 30'	2°30' ± 30'	2°50' ± 30'	2° 50' ± 30'
King pin inclination..........................	_	_	_	_
Rear suspension / wheel alignment				
Toe-in (+) / Toe-out (–)....................mm [°]	_	1.3 ± 2.0	1.0 ± 2.0	1.0 ± 2.0
Camber	_	-0°15' ± 30'	0°15' ± 30'	-0° 15' ± 30'

FIAT

Torque wrench settings	Uno 1.7 1990 to 1991	Punto D S 1996 to 1997	Punto TD 60 S 1996 to 1999	Punto TD 70 SX 1996 to 1999
Cylinder head - stage 1 ....Nm	50	100	100	100
- stage 2 ....Nm	100	+ 90°	+ 90°	+ 90°
- stage 3 ....Nm	+ 90°	+ 90°	+ 90°	+ 90°
- stage 4 ....Nm	+ 90°	–	–	–
- stage 5 ....Nm	Bolts 11 to 15: 30	–	–	–
- stage 6 ....Nm	–	–	–	–
Big-end bearings..............................Nm	25 + 50° N	25 + 50° N	25 + 50° N	25 + 50° N
Main bearings....................................Nm	113	113	113	113
Crankshaft pulley bolt ....Nm	190	190	190	190
Camshaft pulley bolt ....Nm	118	120	120	120
Flywheel [driveplate] bolt....................Nm	142	142	142	142
Front hubs ....Nm	294	280 N	280 N	280 N
Rear hubs ....Nm	216	280 N	280 N	280 N
Wheel nuts / bolts ....Nm	86	85	85	85
Glow plugs ....Nm	15	15	15	15
Clutch pressure plate bolts ....Nm	16	–	–	–
Injection pump sprocket......................Nm	49	49	–	49
Injectors...Nm	55	55	55	55
Injection pump mounting bolts ....Nm	25	25	–	30
Injector pipe unions............................Nm	24	25	25	25
Capacities				
Engine oil & filter................................litres	4.1	4.8	4.8	4.84
Gearbox...litres	2.0	1.98	1.98	1.98
Automatic transmission ....litres	–	–	–	–
Final drive ....litres	WT	WT	WT	WT
Cooling system..................................litres	8.9	7.2	7.2	1.98
Fuel tank..litres	42	47	47	47

Notes

Uno 1.7 1990 to 1991
[1]Also RDNOS6751C
[2]RDNOS6751C: 116 to 123

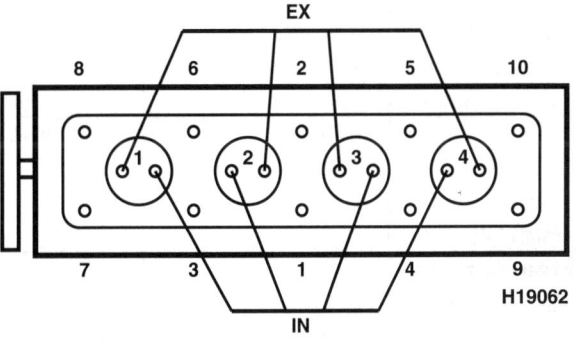

1697 cm³ / 1698 cm³

– Not applicable, or information not
available

	Tipo 1.7 1988 to 1992	Tipo 1.7 1988 to 1992	Tipo 1.7 1993 to 1995	Tipo 1.7 1993 to 1995
Engine				
Engine type/code....................	149 B4.000 SOHC 42kW	149 B4.000 SOHC 42kW	149 B4.000 SOHC 42kW	149 B4.000 SOHC 42kW
Capacity (cm³) / cylinders........................	1697 / 4	1697 / 4	1697 / 4	1697 / 4
Compression ration / pressure ....bar	20.5 ± 0.5 / _	20.5 ± 0.5 / _	20.5 ± 0.5 / _	20.5 ± 0.5 / _
Torque output ....Nm	100	100	100	100
Oil pressure ....idle [running] bar	[3.4 to 4.9]	[3.4 to 4.9]	[3.4 to 4.9]	[3.4 to 4.9]
Oil temperature ....°C	100	100	100	100
Valve clearances - inlet (mm)	0.30 ± 0.05	0.30 ± 0.05	0.30 ± 0.05	0.30 ± 0.05
- exhaust (mm)	0.35 ± 0.05	0.35 ± 0.05	0.35 ± 0.05	0.35 ± 0.05
Injection order	1-3-4-2	1-3-4-2	1-3-4-2	1-3-4-2
No. 1 cylinder position	TBE	TBE	TBE	TBE
Cooling system				
Thermostat opening temperature ....°C	78 to 82	78 to 82	78 to 82	78 to 82
Radiator cap pressure ....bar	0.78	0.78	0.78	0.78
Fuel system				
Idle speed ....rpm	790 to 840	760 ± 20	880 to 920	790 to 840
Maximum (no load) speed ....rpm	5100 to 5200	5150 ± 50	5150 ± 50	5150 ± 50
Smoke test/opacity ....M⁻¹ %	1.78	1.78	1.78	1.78
Static timing method................................	Refer to text	Plunger travel	Plunger travel	Plunger travel
Timing dimension................................mm	Dimension on pump	1.0 [2 ± 1 ATDC]	1.0 [2 ± 1 ATDC]	Dimension on pump
Crankshaft position ....mm [°]	0 TDC	0 TDC	0 TDC	0 TDC
Turbo type / ref / pressure ....bar	_	_	_	_
Injection pump make	Roto Diesel	Bosch	Bosch	CAV
Injection pump part no............................	FT01	_	0460 484 051	R8443B572C
Injector Make / type	Roto Diesel	Bosch	Bosch	CAV
Injector part no.......................................	BDNOSDC6751C	DN 12 SD 290	0432 217 195	LCR 67326
Injection type..	DPCR 8443	VE R 308	VE R 463	CAV DPC
Injection opening pressure, New [used]...bar	≥124 [116]	135 to 143	125 to 133	124 to 131 [119]
Glow plugs				
Maker	Bosch	Bosch	Bosch	Bosch
Type	0250 201 005	0250 201 005	0250 201 019	0250 201 019
Nominal rating.......................................V/A	11 / 12	11 / 12	11 / 12	11 / 12
Brakes				
minimum friction material thickness				
Front.................................mm	1.5	1.5	1.5	1.5
Rear..................................mm	1.5	1.5	1.5	1.5
Tyres - Saloon / Hatch......................Size	165/70x13	165/70x13: 165/65x14	165/70x13: 165/65x14	165/70x13: 165/65x14
- Estate / Van............................Size	_	_	_	_
Pressure - front / rear - Saloon / Hatch...bar	2.1 / 1.9	2.1 / 1.9	2.1 / 1.9	2.1 / 1.9
- Estate / Van ......bar	_	_	_	_
Front suspension / wheel alignment				
Toe-in (+) / Toe-out (–)....................mm [°]	0 ± 1.0	0 ± 1.0	0 ± 1.0	0 ± 1.0
Camber	25' ± 30'	25' ± 30'	25' ± 30'	25' ± 30'
Castor	1°40' ± 30'	1°40' ± 30'	1°40' ± 30'	1°40' ± 30'
King pin inclination...................................	_	_	_	_
Rear suspension / wheel alignment				
Toe-in (+) / Toe-out (–)....................mm [°]	0 ± 2.0	0 ± 2.0	0 ± 2.0	0 ± 2.0
Camber	-1°30' ± 15'	-1°30' ± 15'	-1°30' ± 15'	-1°30' ± 15'

FIAT

	Tipo 1.7 1988 to 1992	Tipo 1.7 1988 to 1992	Tipo 1.7 1993 to 1995	Tipo 1.7 1993 to 1995
Torque wrench settings				
Cylinder head - stage 1 ....Nm	50	50	50	50
- stage 2 ....Nm	100	100	100	100
- stage 3 ....Nm	+ 90°	+ 90°	+ 90°	+ 90°
- stage 4 ....Nm	+ 90°	+ 90°	+ 90°	+ 90°
- stage 5 ....Nm	Bolts 11 to 15: 30	Bolts 11 to 15: 30	Bolts 11 to 15: 30	Bolts 11 to 15: 30
- stage 6 ....Nm	–	–	–	–
Big-end bearings.................................Nm	74 N	74 N	25 + 50° N	25 + 50° N
Main bearings.....................................Nm	113	113	113	113
Crankshaft pulley bolt ....Nm	190	190	190	190
Camshaft pulley bolt ....Nm	118	118	118	118
Flywheel [driveplate] bolt....................Nm	142	142	142	142
Front hubs ....Nm	250[1]	250[2]	240	240
Rear hubs ....Nm	280	280	280	280
Wheel nuts / bolts ....Nm	86	86	86	86
Glow plugs ....Nm	15	15	15	15
Clutch pressure plate bolts ....Nm	M6: 17 M8: 30	M6: 17 M8: 30	38	38
Injection pump sprocket......................Nm	49	49	49	49
Injectors..Nm	55	55	55	55
Injection pump mounting bolts ....Nm	25	25	25	25
Injector pipe unions............................Nm	24	24	24	24
Capacities				
Engine oil & filter................................litres	5.0	5.0	5.0	5.0
Gearbox...litres	1.4	1.4	1.4	1.4
Automatic transmission ....litres	–	–	–	–
Final drive ....litres	WT	WT	WT	WT
Cooling system....................................litres	8.0	8.0	8.0	8.0
Fuel tank..litres	55	55	55	55

Notes

Tipo 1.7 1988 to 1992
[1]1990 ▶, with integral washers: 235
Tipo 1.7 1988 to 1992
[2]1990 ▶, with integral washers: 235

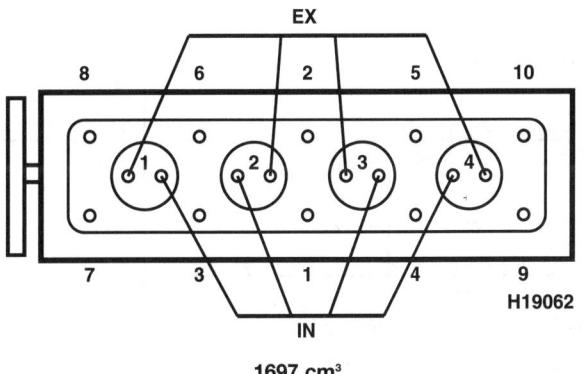

1697 cm³

– Not applicable, or information not
available

FIAT

	Tipo 1.9 Turbo 1988 to 1992	Tipo 1.9 1990 to 1992	Tipo 1.9 Turbo EGR 1990 to 1992	Tipo 1.9 Turbo 1990 to 1992
Engine				
Engine type/code..................	160 A6.000 SOHC Turbo 66kW	160 A7.000 SOHC 48kW	160 B6.046 SOHC Turbo 59kW	160 A6.046 SOHC Turbo 66kW
Capacity (cm³) / cylinders................	1929 / 4	1929 / 4	1929 / 4	1929 / 4
Compression ration / pressure ....bar	19.2 ± 0.5 / _	19.2 ± 0.5 / _	19.2 ± 0.5 / _	19.2 ± 0.5 / _
Torque output ....Nm	186	0	0	186
Oil pressure ....idle [running] bar	[3.4 to 4.9]	[3.4 to 4.9]	[3.4 to 4.9]	[3.4 to 4.9]
Oil temperature ....°C	100	100	100	100
Valve clearances - inlet (mm)	0.30 ± 0.05	0.30 ± 0.05	0.30 ± 0.05	0.30 ± 0.05
- exhaust (mm)	0.35 ± 0.05	0.35 ± 0.05	0.35 ± 0.05	0.35 ± 0.05
Injection order	1-3-4-2	1-3-4-2	1-3-4-2	1-3-4-2
No. 1 cylinder position	TBE	TBE	TBE	TBE
Cooling system				
Thermostat opening temperature ....°C	78 to 82	78 to 82	78 to 82	78 to 82
Radiator cap pressure ....bar	0.78	0.78	0.78	0.78
Fuel system				
Idle speed ....rpm	900 ± 20	740 to 780	740 to 780	830 ± 20
Maximum (no load) speed ....rpm	4860 to 4940	4860 to 4940	4860 to 4940	5000 ± 50
Smoke test/opacity ....M⁻¹ %	1.78	1.82	1.16	1.78
Static timing method................................	Plunger travel	Plunger travel	Plunger travel	Plunger travel
Timing dimension...............................mm	0.82 [3 ± 1 ATDC]	0.82	0.82	0.82
Crankshaft position ....mm [°]	0 TDC	0 TDC	0 TDC	0 TDC
Turbo type / ref / pressure ....bar	0.8 bar @ 2000 rpm	_	_	_
Injection pump make	Bosch	Bosch	Bosch	Bosch
Injection pump part no.............................	_	_	_	_
Injector Make / type	Bosch	Bosch	Bosch	Bosch
Injector part no.............................	DN 12 SD 283	DN 12 SD 283	DN 12 SD 283	DNOSD 1510
Injection type.............................	VE R 303	VE R 303	VE R 303	VE 4/R 343
Injection opening pressure, New [used]...bar	150 to 158	150 to 158	150 to 158	150 to 158
Glow plugs				
Maker	Bosch	Bosch	Bosch	Beru
Type	0250 201 005	0250 201 005	0250 201 005	0100221166
Nominal rating..................................V/A	11 / 12	11 / 12	11 / 12	_
Brakes				
minimum friction material thickness				
Front.............................mm	1.5	1.5	1.5	1.5
Rear.............................mm	1.5	1.5	1.5	1.5
Tyres - Saloon / Hatch......................Size	175/65x14	175/65x14	175/65x14	175/65x14: 185/60x14
- Estate / Van............................Size	_	_	_	_
Pressure - front / rear - Saloon / Hatch ...bar	2.2 / 2.2	2.2 / 2.2	2.2 / 2.2	2.1 / 2.1
- Estate / Van ......bar				
Front suspension / wheel alignment				
Toe-in (+) / Toe-out (−).....................mm [°]	0 ± 1.0	0 ± 1.0	0 ± 1.0	-1.0 to 1.0
Camber	25' ± 30'	25' ± 30'	25' ± 30'	-4' ± 20'
Castor	2°50' ± 30'	2°50' ± 30'	2°50' ± 30'	3°11' ± 30'
King pin inclination...............................	_	_	_	_
Rear suspension / wheel alignment				
Toe-in (+) / Toe-out (−).....................mm [°]	0 ± 2.0	0 ± 2.0	0 ± 2.0	-1.0 ± 2.0
Camber	-1°30' ± 15'	-1°30' ± 15'	-1°30' ± 15'	-1° ± 15'

FIAT

	Tipo 1.9 Turbo 1988 to 1992	Tipo 1.9 1990 to 1992	Tipo 1.9 Turbo EGR 1990 to 1992	Tipo 1.9 Turbo 1990 to 1992
Torque wrench settings				
Cylinder head - stage 1 ....Nm	50	50	50	50
- stage 2 ....Nm	100	100	100	100
- stage 3 ....Nm	+ 90°	+ 90°	+ 90°	+ 90°
- stage 4 ....Nm	+ 90°	+ 90°	+ 90°	+ 90°
- stage 5 ....Nm	Bolts 11 to 15: 30	Bolts 11 to 15: 30	Bolts 11 to 15: 30	Bolts 11 to 15: 30
- stage 6 ....Nm	–	–	–	–
Big-end bearings.................................Nm	25 + 50° N	25 + 50° N	25 + 50° N	25 + 50° N
Main bearings.....................................Nm	113	113	113	113
Crankshaft pulley bolt ....Nm	190	190	190	190
Camshaft pulley bolt ....Nm	118	118	118	118
Flywheel [driveplate] bolt....................Nm	142	142	142	142
Front hubs ....Nm	460	460	460	460
Rear hubs ....Nm	280	280	280	280
Wheel nuts / bolts ....Nm	86	86	86	86
Glow plugs ....Nm	15	15	15	15
Clutch pressure plate bolts ....Nm	M6: 17 M8: 30	M6: 17 M8: 30	M6: 17 M8: 30	M6: 17 M8: 30
Injection pump sprocket......................Nm	49	49	49	49
Injectors...Nm	55	55	55	55
Injection pump mounting bolts ....Nm	25	25	25	25
Injector pipe unions............................Nm	24	24	24	24
Capacities				
Engine oil & filter.................................litres	5.0	5.0	5.0	5.0
Gearbox...litres	1.4	1.4	1.4	1.4
Automatic transmission ....litres	–	–	–	–
Final drive ....litres	WT	WT	WT	WT
Cooling system....................................litres	8.8	8.8	8.8	8.8
Fuel tank..litres	55	55	55	55

Notes

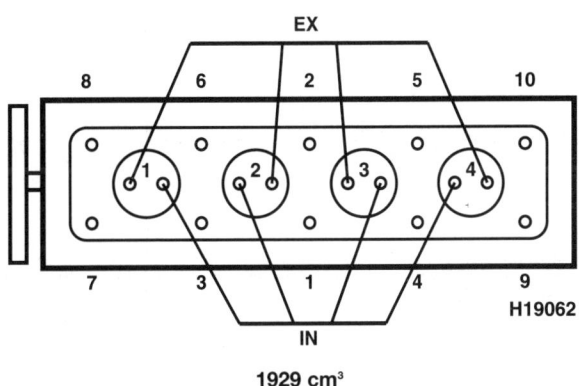

– Not applicable, or information not available

FIAT

	Tipo 1.9 Turbo 1993 to 1995	Bravo & Brava 1.9D 1996 to 2000	Bravo & Brava 1.9TD 75 1996 to 2000	Bravo & Brava 1.9TD 100 1996 to 2000
Engine				
Engine type/code	160 A6.000 SOHC Turbo 66kW	160 A7.000 SOHC 48kW	182 A8.000 SOHC Turbo 55kW	182 A7.000 SOHC Turbo 74kW
Capacity (cm^3) / cylinders.............	1929 / 4	1929 / 4	1910 / 4	1910 / 4
Compression ration / pressure ....bar	19.2 ± 0.5 / _	21.0 / _	20.7 / _	20.7 / _
Torque output ....Nm	186	119	147	200
Oil pressure ....idle [running] bar	[3.4 to 4.9]	[3.7]	[3.7]	[3.7]
Oil temperature ....°C	100	100	100	100
Valve clearances - inlet (mm)	0.30 ± 0.05	0.30 ± 0.05	0.35 ± 0.05	0.35 ± 0.05
- exhaust (mm)	0.35 ± 0.05	0.35 ± 0.05	0.35 ± 0.05	0.35 ± 0.05
Injection order	1-3-4-2	1-3-4-2	1-3-4-2	1-3-4-2
No. 1 cylinder position	TBE	TBE	TBE	TBE
Cooling system				
Thermostat opening temperature ....°C	78 to 82	81	81	81
Radiator cap pressure ....bar	0.78	0.98	0.98	0.98
Fuel system				
Idle speed ....rpm	900 ± 20	800 ± 20	850 ± 20	860 ± 10
Maximum (no load) speed ....rpm	5000	5050 ± 50	5050 ± 50	4970 ± 50
Smoke test/opacity ....M^{-1} %	1.85	2.0	2.0	2.0
Static timing method................................	Plunger travel	Plunger travel	_	_
Timing dimension........................mm	0.8 ± 0.05	0.92 ± 0.05	0.87 to 0.97	0.87 to 0.97
Crankshaft position ....mm [°]	0 TDC	TDC	WSM	WSM
Turbo type / ref / pressure ....bar	0.8 bar @ 2000 rpm	_	Garrett GT15 ≤1.05 bar	Garrett GT15 ≤1.05 bar
Injection pump make	Bosch	Lucas	Bosch	Lucas
Injection pump part no...........................	0460 494 304	FT 08	VER 679	DPC FT09
Injector Make / type	Bosch	CAV	Bosch	CAV
Injector part no....................................	0432 217 211	LCR 6734 202D	KCA 30S41	RDN 05DC6897
Injection type......................................	VE R 466	Indirect	Indirect	Indirect
Injection opening pressure, New [used]...bar	150 to 158	124 to 131 [116 to 123]	150 to 158	143 to 150
Glow plugs				
Maker	Bosch	Beru/Champion	Beru	Beru
Type	0250 001 538	0 100 221 145 / CH 68	0 100 226 249	0 100 226 249
Nominal rating..............................V/A	11 / 18	_	_	_
Brakes				
minimum friction material thickness				
Front................................mm	1.5	1.5	1.5	1.5
Rear................................mm	1.5	1.5	1.5	1.5
Tyres - Saloon / Hatch.....................Size	175/65x14	165/65x14: 175/65x14'	165/65x14: 175/65x14'	175/65x14: 185/60x14
- Estate / Van.....................Size	_	_	_	_
Pressure - front / rear - Saloon / Hatch ...bar	2.2 / 2.2	2.3 / 2.2	2.3 / 2.2	2.3 / 2.2
- Estate / Van ......bar				
Front suspension / wheel alignment				
Toe-in (+) / Toe-out (−)....................mm [°]	0 ± 1.0	0 ± 1.0	0 ± 1.0	0 ± 1.0
Camber	10' ± 30'	-7' ± 30'	-7' ± 30'	-7' ± 30'
Castor	2°0' ± 50'	2°50' ± 30'	2°42' ± 30'	2°50' ± 30'
King pin inclination...............................	_	_	_	_
Rear suspension / wheel alignment				
Toe-in (+) / Toe-out (−)....................mm [°]	-1 ± 2.0	-2.5 to 1.5	-2.5 to 1.5	-2.5 to 1.5
Camber	-1°0' ± 30'	-1° ± 30'	-1° ± 30'	-1° ± 30'

FIAT

	Tipo 1.9 Turbo 1993 to 1995	Bravo & Brava 1.9D 1996 to 2000	Bravo & Brava 1.9TD 75 1996 to 2000	Bravo & Brava 1.9TD 100 1996 to 2000
Torque wrench settings				
Cylinder head - stage 1 ....Nm	50	50	65	65
- stage 2 ....Nm	100	+ 90°	+ 90°	+ 90°
- stage 3 ....Nm	+ 90°	+ 90°	+ 90°	+ 90°
- stage 4 ....Nm	+ 90°	+ 90°	+ 90°	+ 90°
- stage 5 ....Nm	Bolts 11 to 15: 30	–	–	–
- stage 6 ....Nm	–	–	–	–
Big-end bearings................................Nm	25 + 50° N	25 + 50° N	20 + 60° N	20 + 60° N
Main bearings....................................Nm	113	20 + 100°	20 + 100°	20 + 100°
Crankshaft pulley bolt ....Nm	190	190	360 Left-hand	360 Left-hand
Camshaft pulley bolt ....Nm	118	120	120	120
Flywheel [driveplate] bolt....................Nm	142	142	160	160
Front hubs ....Nm	460	240	280	280
Rear hubs ....Nm	280	280	280	280
Wheel nuts / bolts ....Nm	86	86	86	86
Glow plugs ....Nm	15	15	15	15
Clutch pressure plate bolts ....Nm	M6: 17 M8: 30	38	38	38
Injection pump sprocket.....................Nm	49	50	50	50
Injectors..Nm	55	55	55	55
Injection pump mounting bolts ....Nm	25	25	25	25
Injector pipe unions...........................Nm	24	29	29	29
Capacities				
Engine oil & filter...............................litres	5.0	4.9	4.8	4.8
Gearbox...litres	1.4	1.98	1.98	1.98
Automatic transmission ....litres	–	–	–	–
Final drive ....litres	WT	WT	WT	WT
Cooling system..................................litres	8.8	7.4	6.3	6.3
Fuel tank..litres	55	60	60	60

Notes

Bravo & Brava 1.9D 1996 to 2000
[1]Also 185/60x14: 195/50x15
Bravo & Brava 1.9TD 75 1996 to 2000
[1]Also 185/60x14: 195/50x15

– Not applicable, or information not available

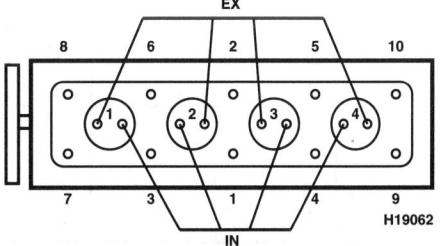

Tipo 1929 cm³

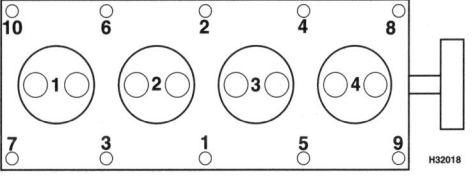

1910 cm³ / 1929 cm³

FIAT

	Bravo & Brava 1.9 JTD 105 1999 to 2000	Tempra 2.0 DS 1990 to 1996	Tempra 2.0 DS 1990 to 1996	Tempra 2.0 Turbo DS 1990 to 1996
Engine				
Engine type/code...............................	182 B4.000 SOHC Turbo 77kW	160 A7.000 SOHC 48kW	160 A7.000 SOHC 48kW	160 A6.000 SOHC Turbo 66kW
Capacity (cm³) / cylinders...................	1910 / 4	1929 / 4	1929 / 4	1929 / 4
Compression ration / pressure ....bar	18.45 / _	21.0 ± 0.5 / _	21.0 ± 0.5 / _	19.2 ± 0.5 / _
Torque output ....Nm	200	119	119	191
Oil pressure ....idle [running] bar	[3.43 to 4.9]	[3.4 to 4.9]	[3.4 to 4.9]	[3.4 to 4.9]
Oil temperature ....°C	100	100	100	100
Valve clearances - inlet (mm)	0.30 ± 0.05	0.30 ± 0.05	0.30 ± 0.05	0.30 ± 0.05
- exhaust (mm)	0.35 ± 0.05	0.35 ± 0.05	0.35 ± 0.05	0.35 ± 0.05
Injection order	1-3-4-2	1-3-4-2	1-3-4-2	1-3-4-2
No. 1 cylinder position	TBE	TBE	TBE	TBE
Cooling system				
Thermostat opening temperature ....°C	86 to 90	78 to 82	78 to 82	78 to 82
Radiator cap pressure ....bar	0.9 to 1.1	0.78	0.78	0.78
Fuel system				
Idle speed ....rpm	800 ± 30	760 to 800	740 to 780	880 to 920
Maximum (no load) speed ....rpm	ECU controlled	5100 to 5200	5120 to 5200	4950 to 5050
Smoke test/opacity ....M⁻¹ %	2.0	1.8	1.8	1.88
Static timing method................................	_	Plunger travel	Rotor groove	Plunger travel
Timing dimension...............................mm	_	0.85	Dimension on pump	0.8
Crankshaft position ....mm [°]	_	0 TDC	_	0 TDC
Turbo type / ref / pressure ....bar	_	_	_	KKK K16 ≤0.8 bar @ 2400rpm
Injection pump make	Bosch	Bosch	Roto Diesel	Bosch
Injection pump part no............................	EDC 15C	0460 484 024	_	VER 466
Injector Make / type	_	Bosch	Roto Diesel	Bosch
Injector part no..	0.445.110.002	DN 12 SD 290	RDN or BDN 12SDC 6849C	DN 12 SD 290
Injection type...	Direct common rail EDC 15C	VE R 334	FT 03	Indirect VER
Injection opening pressure, New [used]...bar	_	125 to 133	124 to 131 [119 to 131]	150 to 158
Glow plugs				
Maker	Bosch	Bosch/Champion	Bosch/Champion	Bosch/Champion
Type	0250 202 028	0250 201 005 / CH68	0250 201 005 / CH68	0250 201 005 / CH68
Nominal rating...................................V/A	_	11 / 12	11 / 12	11 / 18
Brakes				
minimum friction material thickness				
Front......................mm	1.5	1.5	1.5	1.5
Rear......................mm	1.5	1.5	1.5	1.5
Tyres - Saloon / Hatch......................Size	185/55x15: 185/60x14	175/65x14: 185/60x14	175/65x14: 185/60x14	175/65x14: 185/60x14
- Estate / Van...........................Size	_	_	175/65x14: 185/60x14	_
Pressure - front / rear - Saloon / Hatch ...bar	2.3 / 2.2	2.2 / 2.2	2.2 / 2.2	2.2 / 2.2
- Estate / Van ......bar	_	_	2.2 / 2.2	_
Front suspension / wheel alignment				
Toe-in (+) / Toe-out (–)....................mm [°]	0 ± 1.0	0 ± 1.0	0 to 1.0	0 ± 1.0
Camber	-7' ± 30'	10' ± 30'	10' ± 30'	10' ± 30'
Castor	2°50' ± 30'	2°40' ± 30'	2°40' ± 30' Est: 2°20' ± 30'	2°50' to 3°50'
King pin inclination................................	_	_	_	_
Rear suspension / wheel alignment				
Toe-in (+) / Toe-out (–)....................mm [°]	0.0 to 4.0	-3.0 to 1.0	-3.0 to 1.0	-3.0 to 1.0
Camber	-0°46' ± 30'	-1° ± 30'	-1° ± 30'	-30' to -1°30'

FIAT

Torque wrench settings	Bravo & Brava 1.9 JTD 105 1999 to 2000	Tempra 2.0 DS 1990 to 1996	Tempra 2.0 DS 1990 to 1996	Tempra 2.0 Turbo DS 1990 to 1996
Cylinder head - stage 1 ....Nm	65	50	50	50
- stage 2 ....Nm	+ 90°	100	100	100
- stage 3 ....Nm	+ 90°	+ 90°	+ 90°	+ 90°
- stage 4 ....Nm	+ 90°	+ 90°	+ 90°	+ 90°
- stage 5 ....Nm	–	Bolts 11 to 15: 30	Bolts 11 to 15: 30	Bolts 11 to 15: 30
- stage 6 ....Nm	–	–	–	–
Big-end bearings...............................Nm	25 + 60° N	25 + 50° N	25 + 50° N	25 + 50° N
Main bearings....................................Nm	20 + 100°	113	113	113
Crankshaft pulley bolt ....Nm	360 Left-hand	190	190	190
Camshaft pulley bolt ....Nm	120	118	118	118
Flywheel [driveplate] bolt......................Nm	160	142	142	142
Front hubs ....Nm	70 + 55° M24: 70 + 62°	235	235	451
Rear hubs ....Nm	284	284	284	284
Wheel nuts / bolts ....Nm	86	86	86	86
Glow plugs ....Nm	15	15	15	15
Clutch pressure plate bolts ....Nm	38	38	38	38
Injection pump sprocket......................Nm	50	49	49	49
Injectors..Nm	32	55	55	55
Injection pump mounting bolts ....Nm	32	25	25	25
Injector pipe unions............................Nm	32	24	24	24
Capacities				
Engine oil & filter................................litres	4.9	5.0	5.0	5.0
Gearbox..litres	2.0	2.0	2.0	2.0
Automatic transmission ....litres	–	–	–	–
Final drive ....litres	WT	WT	WT	WT
Cooling system..................................litres	6.3	8.0	8.0	8.9
Fuel tank..litres	60	65	65	65

Notes

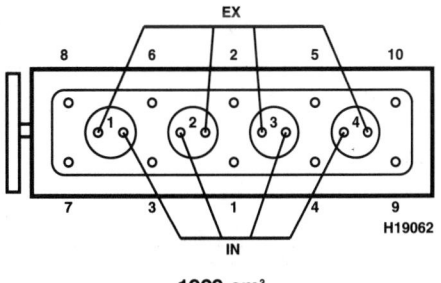

1929 cm³

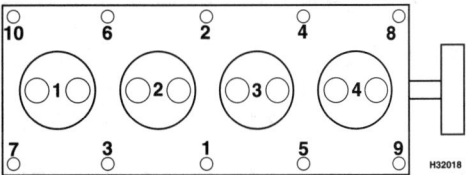

1910 cm³

– Not applicable, or information not available

FIAT

	Tempra 2.0 Turbo DS 1993 to 1996	Marea & Weekend 1.9 TD75 1997 to 1999	Marea & Weekend 1.9 TD100 1997 to 1999	Marea & Weekend JTD 105 1999 to 2000
Engine				
Engine type/code................................	160 D1.000 SOHC Turbo 66kW	182 A8.000 SOHC Turbo 55kW	182 A7.000 SOHC Turbo 74kW	182 B4.000 SOHC Turbo 77kW
Capacity (cm³) / cylinders........................	1929 / 4	1910 / 4	1910 / 4	1910 / 4
Compression ration / pressure ....bar	19.2 ± 0.5 / _	20.7 /	20.7 /	18.45 /
Torque output ....Nm	191	147	200	200
Oil pressure ....idle [running] bar	[3.4 to 4.9]	[3.7]	[3.7]	[3.43 to 4.9]
Oil temperature ....°C	100	100	100	100
Valve clearances - inlet (mm)	0.30 ± 0.05	0.30 ± 0.05	0.30 ± 0.05	0.30 ± 0.05
- exhaust (mm)	0.35 ± 0.05	0.35 ± 0.05	0.35 ± 0.05	0.35 ± 0.05
Injection order	1-3-4-2	1-3-4-2	1-3-4-2	1-3-4-2
No. 1 cylinder position	TBE	TBE	TBE	TBE
Cooling system				
Thermostat opening temperature ....°C	78 to 82	81	81	86 to 90
Radiator cap pressure ....bar	0.78	0.98	0.98	0.98
Fuel system				
Idle speed ....rpm	850 to 950	830 to 870	830 to 870	830 ± 30
Maximum (no load) speed ....rpm	4950 to 5050	5100	4920 to 5020	_
Smoke test/opacity ....M⁻¹ %	1.58	2.0	1.16	2.0
Static timing method..............................	Plunger travel	Plunger travel	Plunger travel	Refer to wsm
Timing dimension...............................mm	0.8	0.92	_	_
Crankshaft position ....mm [°]	0 TDC	TDC	_	_
Turbo type / ref / pressure ....bar	KKK K16 ≤0.8 bar @ 2400rpm	Garrett GT15 ≤1.05 bar	Garrett GT15 ≤1.05 bar	_
Injection pump make	Bosch	Bosch	Lucas	Bosch
Injection pump part no............................	VER 508	VER 679	DPC FT09	EDC 15C
Injector Make / type	Bosch	Bosch		
Injector part no.......................................	DN 12 SD 290	DNOSD321	RDN SDC6888D	
Injection type..	Indirect VER	Indirect VER	Indirect DPC	Direct common rail EDC 15C
Injection opening pressure, New [used]...bar	150 to 158	150 to 158	143 to 150	_
Glow plugs				
Maker	Bosch/Champion	Beru	Beru	Bosch
Type	0250 201 005 / CH68	0 100 226 249	0100 226 249	0250 202 028
Nominal rating....................................V/A	11 / 18	_	_	_
Brakes				
minimum friction material thickness				
Front..mm	1.5	1.5	1.5	1.5
Rear..mm	1.5	1.5	1.5	1.5
Tyres - Saloon / Hatch.......................Size	175/65x14: 185/60x14	175/70x14	165/65x14: 175/65x14'	175/70x14: 185/65x14
- Estate / Van...........................Size	_	_	165/65x14: 175/65x14'	175/70x14: 185/65x14
Pressure - front / rear - Saloon / Hatch...bar	2.2 / 2.2	2.1 2.3	2.2 / 2.1	2.2 / 2.1
- Estate / Van ......bar	_	_	2.2 / 2.2	2.2 / 2.2
Front suspension / wheel alignment				
Toe-in (+) / Toe-out (–)....................mm [°]	0 ± 1.0	0 ± 1.0	0 ± 1.0	0 ± 1.0
Camber	10' ± 30'	-7' ± 30'	-7' ± 30'	-7' ± 30'
Castor	2°50' to 3°50'	2°50' ± 30'	2°40' ± 30' Est: 2°20' ± 30'	2°42' ± 30'
King pin inclination.................................	_	_	_	_
Rear suspension / wheel alignment				
Toe-in (+) / Toe-out (–)....................mm [°]	-3.0 to 1.0	-2.5 to 1.5	-2.5 to 1.5	-2.5 to 1.5
Camber	-30' to -1°30'	-1° ± 30'	-1° ± 30'	-1° ± 30'

FIAT

Torque wrench settings	Tempra 2.0 Turbo DS 1993 to 1996	Marea & Weekend 1.9 TD75 1997 to 1999	Marea & Weekend 1.9 TD100 1997 to 1999	Marea & Weekend JTD 105 1999 to 2000
Cylinder head - stage 1 ....Nm	50	60 N	65 N	65
- stage 2 ....Nm	100	+ 90°	+ 90°	+ 90°
- stage 3 ....Nm	+ 90°	+ 90°	+ 90°	+ 90°
- stage 4 ....Nm	+ 90°	+ 90°	+ 90°	+ 90°
- stage 5 ....Nm	Bolts 11 to 15: 30	–	–	–
- stage 6 ....Nm	–	–	–	–
Big-end bearings.................................Nm	25 + 50° N	25 + 90° N	25 + 60° N	20 + 60° N
Main bearings.......................................Nm	113	20 + 100°	20 + 100°	20 + 100°
Crankshaft pulley bolt ....Nm	190	360 left-hand	360 left-hand	360 Left-hand
Camshaft pulley bolt ....Nm	118	120	120	120
Flywheel [driveplate] bolt....................Nm	142	142	142	160
Front hubs ....Nm	451	240	240	240
Rear hubs ....Nm	284	284	284	–
Wheel nuts / bolts ....Nm	86	86	86	86
Glow plugs ....Nm	15	15	15	15
Clutch pressure plate bolts ....Nm	38	38	38	38
Injection pump sprocket.....................Nm	49	50	50	50
Injectors..Nm	55	78	78	32
Injection pump mounting bolts ....Nm	25	25	25	32
Injector pipe unions............................Nm	24	29	29	32
Capacities				
Engine oil & filter................................litres	5.0	4.3	4.3	4.9
Gearbox...litres	2.0	2.0	2.0	2.0
Automatic transmission ....litres	–	–	–	–
Final drive ....litres	WT	WT	WT	WT
Cooling system...................................litres	8.9	6.0	6.0	6.3
Fuel tank..litres	65	50	50	50

Notes

Marea & Weekend 1.9 TD100 1997 to 1999
[1]Also 185/60x14: 195/50x15

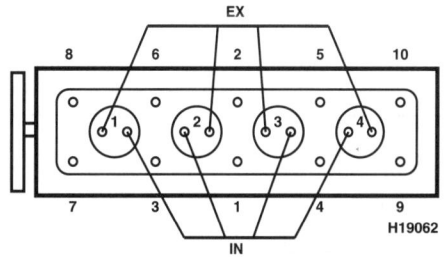

1929 cm³

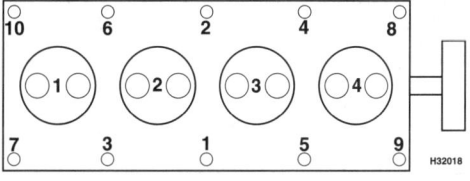

1910 cm³

– Not applicable, or information not available

FIAT

	Marea & Weekend 2.4TD 1997 to 1999	Croma 1.9 Turbo D 1988 to 1992	Croma 2500 Turbo D 1989 to 1993	Croma 2500 Turbo D 1990 to 1993
Engine				
Engine type/code............................	182 A2 000 SOHC Turbo 91kW	154 B.000 SOHC Turbo 66kW	8144.97 SOHC Turbo 85kW	8144.97R SOHC Turbo 80kW US83
Capacity (cm³) / cylinders........................	2387 / 5	1929 / 4	2499 / 4	2499 / 4
Compression ration / pressure ....bar	20.7 /	18.0 / _	21.0 / _	21.0 / _
Torque output ....Nm	265	191	0	0
Oil pressure ....idle [running] bar	2.0 [4.0]	[3.4 to 4.9]	[3.4 to 4.9]	[3.4 to 4.9]
Oil temperature ....°C	100	100	100	100
Valve clearances - inlet (mm)	0.30 ± 0.05	0.40 ± 0.05	0.50 ± 0.05	0.50 ± 0.05
- exhaust (mm)	0.35 ± 0.05	0.40 ± 0.05	0.50 ± 0.05	0.50 ± 0.05
Injection order	1-2-4-5-3	1-3-4-2	1-3-4-2	1-3-4-2
No. 1 cylinder position	TBE	TBE	TBE	TBE
Cooling system				
Thermostat opening temperature ....°C	81	78 to 82	77 to 81	77 to 81
Radiator cap pressure ....bar	0.98	0.78	0.78	0.78
Fuel system				
Idle speed ....rpm	830 to 870	740 to 780	800 to 850	740 to 780
Maximum (no load) speed ....rpm	5050 to 5150	4860 to 4940	5100 to 5200	5100 to 5200
Smoke test/opacity ....M⁻¹ %	1.13	2.5	2.5	2.0
Static timing method............................	Plunger travel	Plunger travel	Plunger travel	Plunger travel
Timing dimension.................................mm	0.75	0.82	1.0 ± 0.04	0.75 ± 0.04
Crankshaft position ....mm [°]	TDC	0 TDC	0 TDC	0 TDC
Turbo type / ref / pressure ....bar	IHI VL6 ≤1.05 bar	_	0.95 bar @ 2400rpm	_
Injection pump make	Bosch	Bosch	Bosch	Bosch
Injection pump part no............................	VER 560	0460 414 056	VER 4/10	VER 361
Injector Make / type	Bosch	Bosch	Bosch	Bosch
Injector part no.............................	DN12 SD290	DN 12 SD 283	DNOSD 259	DNOSD 259
Injection type............................	Indirect VER	VE R 303	Indirect VER	Indirect VER
Injection opening pressure, New [used]...bar	150 to 158	150 to 158	150 to 158 [135]	150 to 158
Glow plugs				
Maker	Beru	Bosch	Beru/Champion	Bosch
Type	0100 226 249	0250 201 025	0100 221 167 / CH88	0250 201 006
Nominal rating...................................V/A	_	_	11 / 12	11 / 12
Brakes				
minimum friction material thickness				
Front...........................mm	1.5	1.5	1.5	1.5
Rear............................mm	1.5	1.5	1.5	1.5
Tyres - Saloon / Hatch.....................Size	195/55x15: 205/50x15	185/65x14: 195/60x14	185/65x14: 195/60x14	185/65x14: 195/60x14
- Estate / Van.........................Size	195/55x15: 205/50x15	_	_	_
Pressure - front / rear - Saloon / Hatch ...bar	2.5 / 2.2	2.2 / 2.2	2.2 / 2.2	2.2 / 2.2
- Estate / Van ......bar	2.5 / 2.3	_	_	_
Front suspension / wheel alignment				
Toe-in (+) / Toe-out (−)....................mm [°]	0 ± 1.0	-1.0 to 1.0	-1.0 to 1.0	-1.0 to 1.0
Camber	10' ± 30'	-40' to 0'	-40' to 0'	-40' to 0'
Castor	2°40' ± 30' Est: 2°20' ± 30'	2°30' to 3°10'	2°30' to 3°10'	2°30' to 3°10'
King pin inclination..............................	_	_	_	_
Rear suspension / wheel alignment				
Toe-in (+) / Toe-out (−)....................mm [°]	-2.5 to 1.5	2.0 to 5.0	2.0 to 5.0	2.0 to 5.0
Camber	-1° ± 30'	-40' to 0'	-40' to 0'	-40' to 0'

FIAT

	Marea & Weekend 2.4TD 1997 to 1999	Croma 1.9 Turbo D 1988 to 1992	Croma 2500 Turbo D 1989 to 1993	Croma 2500 Turbo D 1990 to 1993
Torque wrench settings				
Cylinder head - stage 1 ....Nm	65	50	60	40
- stage 2 ....Nm	+ 90°	100	Slacken	Slacken
- stage 3 ....Nm	+ 90°	+ 90°	60	40
- stage 4 ....Nm	+ 90°	+ 90°	+ 180°	+ 180°
- stage 5 ....Nm	–	Bolts 11 to 15: 30	–	–
- stage 6 ....Nm	–	–	–	–
Big-end bearings......................Nm	20 + 60° N	25 + 50° N	115 N	115 N
Main bearings......................Nm	20 + 100°	113	80, 160	80, 160
Crankshaft pulley bolt ....Nm	360 Left-hand	190	205	205
Camshaft pulley bolt ....Nm	120	118	25	25
Flywheel [driveplate] bolt......................Nm	160	142	125	125
Front hubs ....Nm	235	450 LkC [1]	450 LkC	450 LkC
Rear hubs ....Nm	280	320	320	320
Wheel nuts / bolts ....Nm	86	98	98	98
Glow plugs ....Nm	15	15	37	37
Clutch pressure plate bolts ....Nm	38	M6: 17. M8: 30	20	20
Injection pump sprocket......................Nm	50	49	25	25
Injectors......................Nm	55	55	35	35
Injection pump mounting bolts ....Nm	25	25	25	25
Injector pipe unions......................Nm	–	24	–	–
Capacities				
Engine oil & filter......................litres	5.0	5.0	6.4	5.5
Gearbox......................litres	2.0	1.5 to 1.6	1.5 to 1.6	1.5 to 1.6
Automatic transmission ....litres	–	–	–	–
Final drive......................litres	WT	WT	WT	WT
Cooling system......................litres	7.4	9.0	9.0	9.0
Fuel tank......................litres	60	70	70	70

Notes

Croma 1.9 Turbo D 1988 to 1992
[1]From chassis no. 290 838: 360

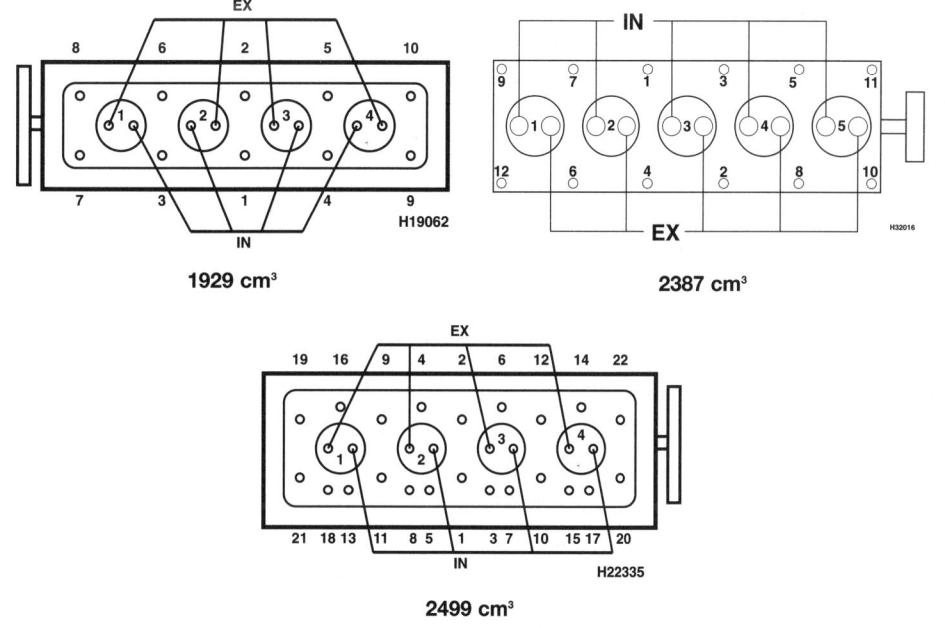

1929 cm³

2387 cm³

2499 cm³

– Not applicable, or information not available

FIAT

	Ulysse 1.9 TD S 1995 to 2000	Ulysse 1.9 TD S 1995 to 2000	Ulysse 2.1 TD S 1996 to 1999	Citivan & Duna 1.7 1988 to 1991
Engine				
Engine type/code............................	XUD9BTF DHX SOHC Turbo 66kW	XUD9BTF D8B SOHC Turbo 66kW	XUD11BTE P8C 12V Turbo 80kW	149 B3.000 OHC 44kW
Capacity (cm³) / cylinders........................	1905 / 4	1905 / 4	2088 / 4	1697 / 4
Compression ration / pressure ....bar	21.8 /	21.8 /	21.5 /	20.5 ± 0.5 /
Torque output ....Nm	196	196	250	0
Oil pressure ....idle [running] bar	[5.0]	[5.0]	[5.0]	[3.4 to 4.9]
Oil temperature ....°C	80	80	80	100
Valve clearances - inlet (mm)	0.15 ± 0.05	0.15 ± 0.05	0: Hyd.	0.30 ± 0.05
- exhaust (mm)	0.30 ± 0.05	0.30 ± 0.05	0: Hyd.	0.30 ± 0.05
Injection order	1-3-4-2	1-3-4-2	1-3-4-2	1-3-4-2
No. 1 cylinder position	TBE	TBE	TBE	TBE
Cooling system				
Thermostat opening temperature ....°C	83	83	83	80 ± 2
Radiator cap pressure ....bar	1.1	1.1	1.1	0.78
Fuel system				
Idle speed ....rpm	850 ± 20	850 ± 20	690 ± 30	760 ± 20
Maximum (no load) speed ....rpm	5070	5070	–	5150 ± 50
Smoke test/opacity ....M⁻¹ %	1.85	1.85	1.85	1.65
Static timing method................................	Plunger travel	Plunger travel	Dial gauge	Plunger travel
Timing dimension.................................mm	0.9 ± 0.05	0.9 ± 0.05	5.12	0.88
Crankshaft position ....mm [°]	TDC	TDC	[24] BTDC	[0] TDC
Turbo type / ref / pressure ....bar	KKK K14 / Garrett T2 0.8 bar	KKK K14 / Garrett T2 0.8 bar		–
Injection pump make	Bosch	Bosch	Lucas	Bosch
Injection pump part no.............................	VER 513	VER 601	EPIC	–
Injector Make / type	Bosch	Bosch		Bosch
Injector part no..	0432 217 218	0432 217 218	–	DNO SD 259
Injection type..	ECU controlled	ECU controlled	Electronic control	VE R 256
Injection opening pressure, New [used]...bar	175	175	–	125 to 133
Glow plugs				
Maker	Bosch	Bosch	Bosch	Bosch/Marelli
Type	0250 201 033	0250 201 033	0250 201 033	0250 201 005 / UX2A
Nominal rating.....................................V/A	11 / 12	11 / 12	11 / 9	
Brakes				
minimum friction material thickness				
Front..............................mm	8.5 with backing	8.5 with backing	8.5 with backing	1.5
Rear.................................mm	–	–	–	1.5
Tyres - Saloon / Hatch......................Size	–	–	–	165/70x13
- Estate / Van...........................Size	205/65x15	205/65x15	205/65x15	165/70x13
Pressure - front / rear - Saloon / Hatch...bar	–	–	–	2.2 / 1.9
- Estate / Van ......bar	2.3 / 2.3	2.3 / 2.3	2.3 / 2.3	2.2 / 1.9
Front suspension / wheel alignment				
Toe-in (+) / Toe-out (–).....................mm [°]	2.0 ± 1.0	2.0 ± 1.0	2.0 ± 1.0	-3.0 ± 1.0
Camber	-1° ± 30'	-1° ± 30'	-1° ± 30'	1°45' ± 30' Est/van: 1°55'±30
Castor	3° 30' ± 30'	3° 30' ± 30'	3° 30' ± 30'	1°5' ± 30' Est/van: 55'±30'
King pin inclination..................................	–	–	–	–
Rear suspension / wheel alignment				
Toe-in (+) / Toe-out (–).....................mm [°]	2.0 ± 1.0	2.0 ± 1.0	2.0 ± 1.0	5.0 ± 2.0
Camber	-1° ± 30'	-1° ± 30'	-1° ± 30'	-1°55' ± 30' Est/van: -1°20'

FIAT

Torque wrench settings	Ulysse 1.9 TD S 1995 to 2000	Ulysse 1.9 TD S 1995 to 2000	Ulysse 2.1 TD S 1996 to 1999	Citivan & Duna 1.7 1988 to 1991
Cylinder head - stage 1 ....Nm	20¹	20¹	Bolt length: ≤146.5mm or...	50
- stage 2 ....Nm	60	60	Pilot type: ≤150.5mm	100
- stage 3 ....Nm	+ 220°	+ 220°	20	+ 90°
- stage 4 ....Nm	–	–	60	+ 90°
- stage 5 ....Nm	–	–	+ 180°	Bolts 11 to 15: 30
- stage 6 ....Nm	–	–	–	–
Big-end bearings................................Nm	20 + 70° N	20 + 70° N	20 + 70° N	74 N
Main bearings....................................Nm	70	70	15 + 60°	113
Crankshaft pulley bolt ....Nm	40 + 60°	40 + 60°	70 + 60°	180
Camshaft pulley bolt ....Nm	45	45	45	118
Flywheel [driveplate] bolt...................Nm	50	50	50	142
Front hubs ....Nm	110	110	110	216
Rear hubs...Nm	100 + 60°	100 + 60°	100 + 60°	216
Wheel nuts / bolts ....Nm	100	100	100	86
Glow plugs ....Nm	18	18	18	15
Clutch pressure plate bolts ....Nm	20	20	20	16
Injection pump sprocket.....................Nm	50	50	50	49
Injectors...Nm	90	90	90	55
Injection pump mounting bolts ....Nm	20	20	20	25
Injector pipe unions...........................Nm	25	25	25	24
Capacities				
Engine oil & filter...............................litres	4.2	4.2	6.0	4.3
Gearbox..litres	1.9	1.9	1.9	3.0
Automatic transmission ....litres	–	–	–	–
Final drive ....litres	WT	WT	WT	WT
Cooling system...................................litres	9.0	9.0	9.0	5.0
Fuel tank...litres	80	80	80	54

Notes

Ulysse 1.9 TD S 1995 to 2000
¹Bolt length below head ≤146.8 mm
Ulysse 1.9 TD S 1995 to 2000
¹Bolt length below head ≤146.8 mm

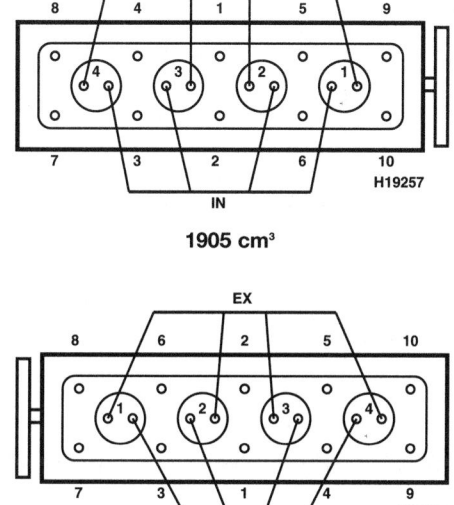

1905 cm³

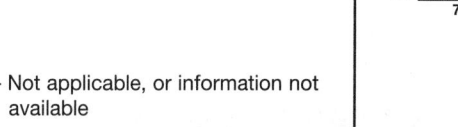

1697 cm³

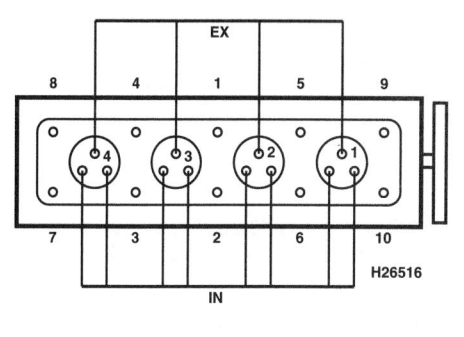

2088 cm³

– Not applicable, or information not available

Haynes THE BOOK ®

FIAT

	Fiorino 1.7 1988 to 1991	Fiorino 1.7 1991 to 1993	Fiorino 1.7 1991 to 1993	Fiorino 1.7 1994 to 1997
Engine				
Engine type/code...............	149 B3.000 OHC 44kW	146 B2.000 OHC 42kW	146 B2.000 OHC 42kW	146 B2.000 OHC 42kW
Capacity (cm³) / cylinders........................	1697 / 4	1697 / 4	1697 / 4	1697 / 4
Compression ration / pressure ....bar	20.5 ± 0.5 / _	20.5 ± 0.5 / _	20.5 ± 0.5 / _	20.5 ± 0.5 / _
Torque output ....Nm	0	0	0	0
Oil pressure ....idle [running] bar	[3.4 to 4.9]	[3.4 to 4.9]	[3.4 to 4.9]	[3.4 to 4.9]
Oil temperature ....°C	100	100	100	100
Valve clearances - inlet (mm)	0.30 ± 0.05	0.30 ± 0.05	0.30 ± 0.05	0.30 ± 0.05
- exhaust (mm)	0.30 ± 0.05	0.35 ± 0.05	0.35 ± 0.05	0.35 ± 0.05
Injection order	1-3-4-2	1-3-4-2	1-3-4-2	1-3-4-2
No. 1 cylinder position	TBE	TBE	TBE	TBE
Cooling system				
Thermostat opening temperature ....°C	80 ± 2	80 ± 2	80 ± 2	80 ± 2
Radiator cap pressure ....bar	0.78	0.78	0.78	0.78
Fuel system				
Idle speed ....rpm	760 ± 20	760 ± 20	810 ± 20	900 ± 20
Maximum (no load) speed ....rpm	5150 ± 50	5150 ± 50	5150 ± 50	5150 ± 50
Smoke test/opacity ....M⁻¹ %	1.65	1.65	1.74	1.74
Static timing method........................	Plunger travel	Plunger travel	Plunger travel	Plunger travel
Timing dimension........................mm	0.88	1.0 ± 0.05	Dimension on pump	1.0 ± 0.05
Crankshaft position ....mm [°]	[0] TDC	[0] TDC	[0] TDC	[0] TDC
Turbo type / ref / pressure ....bar	_	_	_	_
Injection pump make	Bosch	Bosch	CAV	Bosch
Injection pump part no........................	_	VER 308	DPC R8443	VER 463
Injector Make / type	Bosch	Bosch	CAV	Bosch
Injector part no........................	DNO SD 259	KCA 30S41	DNOSD6751C	0432 217 195
Injection type........................	VE R 256	Indirect VER	Indirect DPC	Indirect VER
Injection opening pressure, New [used]...bar	125 to 133	125 to 133	123	125 to 133
Glow plugs				
Maker	Bosch/Marelli	Bosch	Bosch	Bosch
Type	0250 201 005 / UX2A	0250 201 019	0250 201 019	0250 201 019
Nominal rating........................V/A	_	11 / 12	11 / 12	11 / 12
Brakes				
minimum friction material thickness				
Front........................mm	1.5	1.5	1.5	1.5
Rear........................mm	1.5	1.5	1.5	1.5
Tyres - Saloon / Hatch........................Size	_	_	_	_
- Estate / Van........................Size	165/70x13	165/70x13	165/70x13	165/70x13
Pressure - front / rear - Saloon / Hatch ...bar	_	_	_	_
- Estate / Van ......bar	1.9 / 1.9	1.9 / 1.9	1.9 / 1.9	1.9 / 1.9
Front suspension / wheel alignment				
Toe-in (+) / Toe-out (−)........................mm [°]	-3.0 ± 1.0	-3.0 ± 1.0	-3.0 ± 1.0	-3.0 ± 1.0
Camber	1°50' ± 30'	1°50' ± 30'	1°50' ± 30'	1°50' ± 30'
Castor	30' ± 30' Pick-up: 25'±30'	30' ± 30' Pick-up: 25'±30'	30' ± 30' Pick-up: 25'±30'	30' ± 30' Pick-up: 25'±30'
King pin inclination........................				
Rear suspension / wheel alignment				
Toe-in (+) / Toe-out (−)........................mm [°]	5.0 ± 2.0	5.0 ± 2.0	5.0 ± 2.0	5.0 ± 2.0
Camber	-1° ± 30' Pick-up: -40'±30'	-1° ± 30' Pick-up: -40'±30'	-1° ± 30' Pick-up: -40'±30'	-1° ± 30' Pick-up: -40'±30'

FIAT

Torque wrench settings	Fiorino 1.7 1988 to 1991	Fiorino 1.7 1991 to 1993	Fiorino 1.7 1991 to 1993	Fiorino 1.7 1994 to 1997
Cylinder head - stage 1 ....Nm	50	50	50	50
- stage 2 ....Nm	100	100	100	100
- stage 3 ....Nm	+ 90°	+ 90°	+ 90°	+ 90°
- stage 4 ....Nm	+ 90°	+ 90°	+ 90°	+ 90°
- stage 5 ....Nm	Bolts 11 to 15: 30	Bolts 11 to 15: 30	Bolts 11 to 15: 30	Bolts 11 to 15: 30
- stage 6 ....Nm	–	–	–	–
Big-end bearings..............................Nm	74 N	74 N	74 N	74 N
Main bearings.................................Nm	113	113	113	113
Crankshaft pulley bolt ....Nm	180	180	180	180
Camshaft pulley bolt ....Nm	118	118	118	118
Flywheel [driveplate] bolt....................Nm	142	142	142	142
Front hubs ....Nm	216	294	294	240
Rear hubs ....Nm	216	216	216	280
Wheel nuts / bolts ....Nm	86	86	86	86
Glow plugs ....Nm	15	15	15	15
Clutch pressure plate bolts ....Nm	16	16	16	16
Injection pump sprocket.....................Nm	49	49	49	49
Injectors..Nm	55	55	55	55
Injection pump mounting bolts ....Nm	25	25	25	25
Injector pipe unions...........................Nm	24	24	24	24
Capacities				
Engine oil & filter.................................litres	4.3	4.3	4.3	4.6
Gearbox...litres	3.0	3.0	3.0	3.0
Automatic transmission ....litres	–	–	–	–
Final drive ....litres	WT	WT	WT	WT
Cooling system................................litres	5.0	5.0	5.0	5.0
Fuel tank..litres	54	54	54	54

Notes

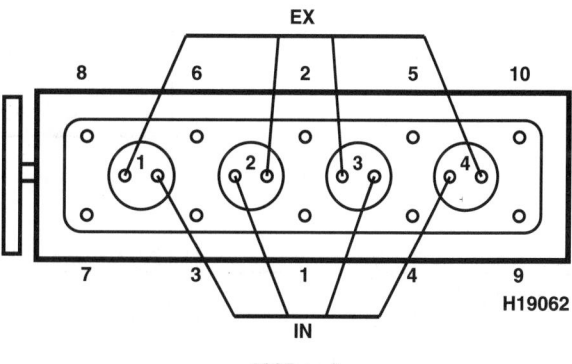

1697 cm³

– Not applicable, or information not available

FIAT

	Fiorino 1.7 1994 to 1997	Fiorino 1.7 TD 1997 to 1999	Scudo 1.9 D 1999 to 2000	Scudo 1.9 D 1996 to 1999
Engine				
Engine type/code..............................	146 B2.000 OHC 42kW	146 D7.000 SOHC Turbo 46kW	DW8 WJZ SOHC 51kW	XUD9 D9B SOHC 51kW
Capacity (cm³) / cylinders........................	1697 / 4	1697 / 4	1868 / 4	1905 / 4
Compression ration / pressure ....bar	20.5 ± 0.5 / _	20.3 /	23.0 /	23 / _
Torque output ....Nm	0	118	125	120
Oil pressure ....idle [running] bar	[3.4 to 4.9]	[3.4]	[5.0 @ 4000]	0.8 [3.4 to 4.9 @ 4000]
Oil temperature ....°C	100	100	80	80
Valve clearances - inlet (mm)	0.30 ± 0.05	0.30 ± 0.05	0.15 ± 0.05	0.15 ± 0.05
- exhaust (mm)	0.35 ± 0.05	0.35 ± 0.05	0.30 ± 0.05	0.30 ± 0.08
Injection order	1-3-4-2	1-3-4-2	1-3-4-2	1-3-4-2
No. 1 cylinder position	TBE	_	FE	FE
Cooling system				
Thermostat opening temperature ....°C	80 ± 2	78	83	87 to 98
Radiator cap pressure ....bar	0.78	0.8	1.1	1.0
Fuel system				
Idle speed ....rpm	810 ± 20	900 ± 50	750 to 800	750 to 800
Maximum (no load) speed ....rpm	5150 ± 50	4950 ± 50	5300 ± 50	5250 to 5350
Smoke test/opacity ....M⁻¹ %	1.74	1.8	2.5	1.48
Static timing method...............................	Plunger travel	Plunger travel	Plunger travel	Plunger travel
Timing dimension............................mm	Dimension on pump	1.0	0.9	0.9 ± 0.05
Crankshaft position ....mm [°]	[0] TDC	TDC	TDC	TDC
Turbo type / ref / pressure ....bar	_	Garrett T2 0.75 bar	_	_
Injection pump make	CAV	Lucas	Bosch	Bosch
Injection pump part no............................	R8444 B300A	FT11 8445	VER 425-10/11	VER 425-6
Injector Make / type	CAV			Bosch
Injector part no..	LRC 67326	Lucas RDNO 6888D	DNOSD 299A	0432 217 248
Injection type...	Indirect DPC	Indirect	Indirect VER	Indirect VER
Injection opening pressure, New [used]...bar	124 to 131 [116]	143 to 150	_	130
Glow plugs				
Maker	Bosch	Bosch		Bosch/Champion
Type	0250 201 019	0250 201 005		0250 201 039 / CH1693
Nominal rating......................................V/A	11 / 12	11 / 18		11 / 9
Brakes				
minimum friction material thickness				
Front.......................................mm	1.5	1.5	1.5	1.5
Rear..mm	1.5	1.5	1.5	1.0
Tyres - Saloon / Hatch.....................Size	_	_	_	_
- Estate / Van...........................Size	165/70x13	165/70x13	195/70x14	195/70x14
Pressure - front / rear - Saloon / Hatch ...bar	_	_	_	_
- Estate / Van ......bar	1.9 / 1.9	2.2 / 1.9	3.0 / 3.0	2.5 / 2.5
Front suspension / wheel alignment				
Toe-in (+) / Toe-out (–).....................mm [°]	-3.0 ± 1.0	-3.0 ± 1.0	1.0 ± 0.5	2.0 ± 1.0
Camber	1°50' ± 30'	-0°15' ± 20'	0° ± 30'	0° ± 30'
Castor	30' ± 30' Pick-up: 25'±30'	0°40' ± 20'¹	3°30' ± 30'	1° ± 30' PAS: 3°30'±30'
King pin inclination................................	_	_	11°28' ± 40'	11°28' ± 40'
Rear suspension / wheel alignment				
Toe-in (+) / Toe-out (–)....................mm [°]	5.0 ± 2.0	0	2.0 ± 1.0	2.0 ± 1.0
Camber	-1° ± 30' Pick-up: -40'±30'	_	-1° ± 30'	-1° ± 30'

FIAT

Torque wrench settings	Fiorino 1.7 1994 to 1997	Fiorino 1.7 TD 1997 to 1999	Scudo 1.9 D 1999 to 2000	Scudo 1.9 D 1996 to 1999
Cylinder head - stage 1 ....Nm	50	50 N	20	20 N
- stage 2 ....Nm	100	100	60	60
- stage 3 ....Nm	+ 90°	+ 90°	+ 180°	+ 90°
- stage 4 ....Nm	+ 90°	+ 90°	–	+ 90°
- stage 5 ....Nm	Bolts 11 to 15: 30	–	–	–
- stage 6 ....Nm	–	–	–	–
Big-end bearings...............................Nm	74 N	20 + 50° N	20 + 70° N	20 + 70° N
Main bearings....................................Nm	113	113	70	15 + 60°
Crankshaft pulley bolt ....Nm	180	190	–	40 + 60°
Camshaft pulley bolt ....Nm	118	118	–	45
Flywheel [driveplate] bolt....................Nm	142	142	48	50
Front hubs ....Nm	240	240	100 + 60°	100 + 60° oiled
Rear hubs ....Nm	280	280	100 ± 60°	100 + 60°
Wheel nuts / bolts ....Nm	86	86	100	100
Glow plugs ....Nm	15	15	–	20
Clutch pressure plate bolts ....Nm	16	18	20	23
Injection pump sprocket......................Nm	49	49	–	50
Injectors...Nm	55	55	–	90
Injection pump mounting bolts ....Nm	25	25	–	20
Injector pipe unions............................Nm	24	24	–	–
Capacities				
Engine oil & filter...............................litres	4.6	4.6	5.0	5.0
Gearbox...litres	3.0	3.1	1.9	1.8
Automatic transmission ....litres	–	–	–	–
Final drive ....litres	WT	WT	WT	WT
Cooling system..................................litres	5.0	5.0	8.5	8.5
Fuel tank..litres	54	64	80	80

Notes

Fiorino 1.7 TD 1997 to 1999
'Pick-up: 2° ± 20'

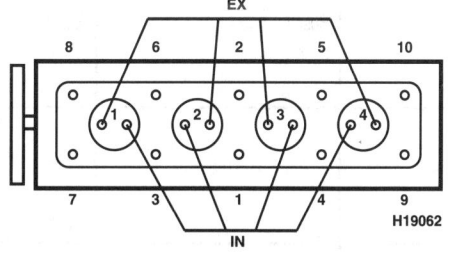

1697 cm³ / 1698 cm³

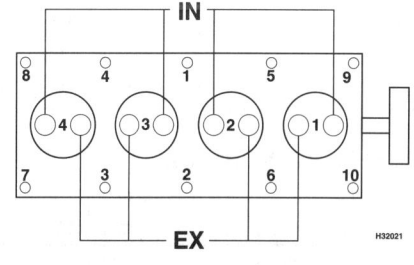

1868 cm³

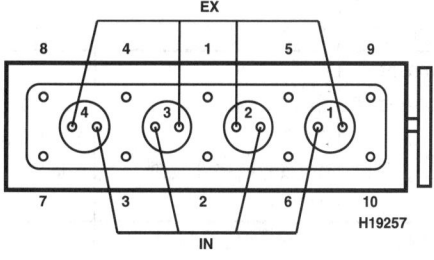

1905 cm³

– Not applicable, or information not available

FIAT

	Scudo 1.9 TD 1996 to 1999	Ducato 1.9 D 1987 to 1993	Ducato 1930 Turbo D 1986 to 1993	Ducato 1.9 DS 1994 to 1998
Engine				
Engine type/code	XUD9BTF DHX Turbo 68kW	149 B1.000 SOHC 52kW	280 A1.000 SOHC Turbo 60kW	230 A2.000 SOHC 51kW
Capacity (cm³) / cylinders........................	1905 / 4	1929 / 4	1929 / 4	1929 / 4
Compression ration / pressure ....bar	21.8 / _	21.5 ± 0.5 / _	20.0 ± 0.5 / _	21.0 /
Torque output ....Nm	196	0	0	0
Oil pressure ....idle [running] bar	0.8 [3.4 to 4.9]	[3.4 to 4.9]	[3.4 to 4.9]	[3.5 to 5.5]
Oil temperature ....°C	80	100	100	100
Valve clearances - inlet (mm)	0.15 ± 0.05	0.30 ± 0.05	0.30 ± 0.05	0.30 ± 0.05
- exhaust (mm)	0.30 ± 0.05	0.35 ± 0.05	0.35 ± 0.05	0.35 ± 0.05
Injection order	1-3-4-2	1-3-4-2	1-3-4-2	1-3-4-2
No. 1 cylinder position	FE	F	F	_
Cooling system				
Thermostat opening temperature ....°C	87 to 98	78 to 82	78 to 82	_
Radiator cap pressure ....bar	1.0	0.78	0.78	_
Fuel system				
Idle speed ....rpm	825 ± 25	740 to 780	780 ± 20	900 ± 20
Maximum (no load) speed ....rpm	5070 to 5230	5100 to 5200	4900 ± 40	5200
Smoke test/opacity ....M⁻¹ %	2.5	1.84	2.0	1.68
Static timing method...............................	Plunger travel	Plunger travel	Plunger travel	Plunger travel
Timing dimension............................mm	0.9	1.0	1.0	0.86 ± 0.05
Crankshaft position ....mm [°]	TDC	0 TDC	0 TDC	TDC
Turbo type / ref / pressure ....bar	KKK K14 or Garrett T2²	_	0.8 bar	_
Injection pump make	Bosch	Bosch	Bosch	Bosch
Injection pump part no............................	VER 513	0460 494 214	VE4/9	VE R549
Injector Make / type	Bosch	Bosch	Bosch	Bosch
Injector part no.......................................	DNOSD 299	0432 217 129	0432 217 172	0432 217 195
Injection type...	Indirect VER	VER	VE 4/9F 2100L184	Indirect VE
Injection opening pressure, New [used]...bar	175	125 to 133	150 to 158	130 to 133
Glow plugs				
Maker	Beru	Bosch	Bosch	Bosch/Champion
Type	GN909	0250 201 005	0250 201 005	0250 201 005 / CH68
Nominal rating....................................V/A	11 / 12	11 / 12	11 / 12	_
Brakes				
minimum friction material thickness				
Front...mm	1.5	2.0	2.0	1.0
Rear..mm	1.0	1.0	1.0	1.5
Tyres - Saloon / Hatch......................Size	_	_	_	_
- Estate / Van......................Size	195/70x14	185x14	185x14	195/75x15
Pressure - front / rear - Saloon / Hatch ...bar	_	_		_
- Estate / Van ......bar	2.5 / 2.5	3.5 / 3.2	3.5 / 3.2	Refer to vehicle
Front suspension / wheel alignment				
Toe-in (+) / Toe-out (–)....................mm [°]	2.0 ± 1.0	0.5 to 2.5	0.5 to 2.5	-2.0 ± 1.0
Camber	0° ± 30'	1°30' to 2°30'	1°30' to 2°30'	0° ± 30'
Castor	1° ± 30' PAS: 3°30' ± 30'	0° to 1°	08 to 1°	1° ± 30'
King pin inclination..................................	11°28' ± 40'	_	_	_
Rear suspension / wheel alignment				
Toe-in (+) / Toe-out (–)....................mm [°]	2.0 ± 1.0	_	_	0 ± 1.0
Camber	-1° ± 30'	_	_	0° ± 10'

FIAT

	Scudo 1.9 TD 1996 to 1999	Ducato 1.9 D 1987 to 1993	Ducato 1930 Turbo D 1986 to 1993	Ducato 1.9 DS 1994 to 1998
Torque wrench settings				
Cylinder head - stage 1 ....Nm	20[1]	50	50	20
- stage 2 ....Nm	60	100	100	30
- stage 3 ....Nm	+ 220°	+ 90°	+ 90°	+ 180°
- stage 4 ....Nm	–	+ 90°	+ 90°	+ 90°
- stage 5 ....Nm	–	Bolts 11 to 15: 30	Bolts 11 to 15: 30	–
- stage 6 ....Nm	–	–	–	–
Big-end bearings.............................Nm	20 + 70° N	25 + 50° N	25 + 50° N	25 + 50° N
Main bearings.................................Nm	70	113	113	113
Crankshaft pulley bolt ....Nm	40 + 60°	190	190	190
Camshaft pulley bolt ....Nm	45	118	118	118
Flywheel [driveplate] bolt.................Nm	50	142	142	142
Front hubs ....Nm	100 + 60° oiled	421 to 510	421 to 510	450
Rear hubs ....Nm	100 + 60°	WSM	WSM	WSM
Wheel nuts / bolts ....Nm	100	160	160	160
Glow plugs ....Nm	22	15	15	15
Clutch pressure plate bolts ....Nm	23	15	15	–
Injection pump sprocket...................Nm	50	49	49	49
Injectors...Nm	90	55	55	80
Injection pump mounting bolts ....Nm	22	25	25	25
Injector pipe unions.........................Nm	20	24	24	30
Capacities				
Engine oil & filter..............................litres	5.0	5.0	5.0	5.0
Gearbox...litres	1.8	1.25 89 ▶: 1.6	1.25 89 ▶: 1.6	2.0
Automatic transmission ....litres	–	–	–	–
Final drive ....litres	WT	WT	WT	WT
Cooling system.................................litres	8.5	9.0	9.0	–
Fuel tank..litres	80	70	70	80

Notes

Scudo 1.9 TD 1996 to 1999
[1]Bolt length below head: ≤121.5mm
[2]0.67 to 0.76 bar @ 2000rpm

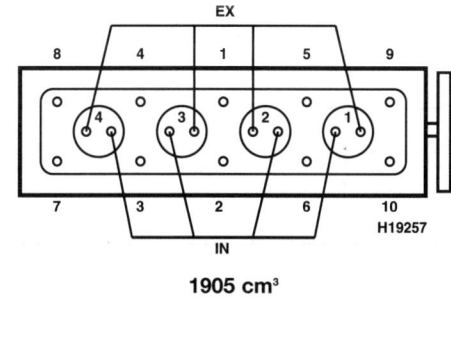

1905 cm³

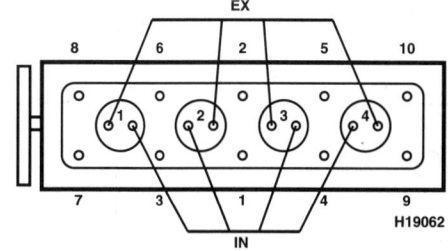

1929 cm³

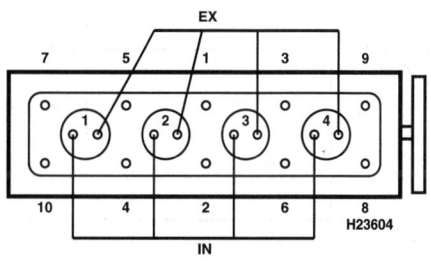

1929 cm³

– Not applicable, or information not available

FIAT

	Ducato 1.9 TDS 1994 to 1998	Ducato 1.9 TD 1999 to 2000	Ducato 2500 Turbo D & 4x4 1990 to 1994	Ducato 2500 D & 4x4 1986 to 1994
Engine				
Engine type/code..................................	230 A3.000 SOHC Turbo 60kW	XUD9BTF DHX SOHC Turbo 66kW	8140.27 SOHC Turbo 70kW	8144.67 SOHC 55kW
Capacity (cm^3) / cylinders......................	1929 / 4	1905 / 4	2445 / 4	2499 / 4
Compression ration / pressure ....bar	21.0 /	21.8 /	18.0 / _	22.0 / _
Torque output ....Nm	180	196	0	0
Oil pressure ....idle [running] bar	[3.5 to 5.5]	0.8 [3.4 to 3.9]	0.78 [3.9 @ 4000]	0.78 [3.9]
Oil temperature ....°C	100	80	100	100
Valve clearances - inlet (mm)	0.30 ± 0.05	0.15 ± 0.05	0.50	0.50
- exhaust (mm)	0.35 ± 0.05	0.30 ± 0.05	0.50	0.50
Injection order	1-3-4-2	1-3-4-2	1-3-4-2	1-3-4-2
No. 1 cylinder position	_	FE	FE	FE
Cooling system				
Thermostat opening temperature ....°C	_	87 to 98	72 to 76	77 to 81
Radiator cap pressure ....bar	_	1.0	0.78	0.78
Fuel system				
Idle speed ....rpm	900 ± 20	825 ± 25	740 to 780	740 to 780
Maximum (no load) speed ....rpm	5200	5070 to 5230	4475 ± 50	4600 to 4900
Smoke test/opacity ....M^{-1} %	1.61	2.5	2.0	2.0
Static timing method..............................	Plunger travel	Plunger travel	Plunger travel	Plunger travel
Timing dimension................................mm	0.8 ± 0.05	0.9	1.48 ± 0.04	1.0 ± 0.04
Crankshaft position ....mm [°]	TDC	TDC	0 TDC	0 TDC
Turbo type / ref / pressure ....bar	0.8 bar @ 4200	_	_	_
Injection pump make	Bosch	Bosch	Bosch	Bosch
Injection pump part no.............................	VE R545	VER 513	0460 414 081	0460 494 189
Injector Make / type	Bosch	Bosch	Bosch	Bosch
Injector part no.................................	0432 217 195	DNOSD299	0432 191 760	0432 297 034
Injection type...................................	Indirect VE	Indirect VER	Direct VER 127	VER 22/7
Injection opening pressure, New [used]...bar	130 to 133	175	222 to 238	120 to 130
Glow plugs				
Maker	Bosch/Champion	Beru	Bosch	Bosch
Type	0250 201 033 / CH68	GN909	0250 201 006	0250 201 006
Nominal rating.....................................V/A	_	11 / 12	11 / 12	11 / 12
Brakes				
minimum friction material thickness				
Front..............................mm	1.0	1.0	2.0	2.0
Rear...............................mm	1.5	1.5	1.0	1.0
Tyres - Saloon / Hatch.....................Size	_	_	_	_
- Estate / Van.....................Size	195/75x15	195/75x15	185x14; 195/75x16	185x14; 195/75x16
Pressure - front / rear - Saloon / Hatch ...bar	_	_	_	_
- Estate / Van ......bar	Refer to vehicle	Refer to vehicle	Refer to vehicle	Refer to vehicle
Front suspension / wheel alignment				
Toe-in (+) / Toe-out (–).....................mm [°]	-2.0 ± 1.0	-2.0 ± 1.0	0.5 to 2.5'	0.5 to 2.5'
Camber	0° ± 30'	0° ± 30'	1°30' to 2°30'²	1°30' to 2°30'²
Castor	1° ± 30'	1° ± 30'	0° to 1° 18: 0° to 30'	0° to 1° 18: 0° to 30'
King pin inclination................................	_	_	_	_
Rear suspension / wheel alignment				
Toe-in (+) / Toe-out (–)......................mm [°]	0 ± 1.0	0 ± 1.0	_	_
Camber	0° ± 10'	0° ± 10'		

FIAT

	Ducato 1.9 TDS 1994 to 1998	Ducato 1.9 TD 1999 to 2000	Ducato 2500 Turbo D & 4x4 1990 to 1994	Ducato 2500 D & 4x4 1986 to 1994
Torque wrench settings				
Cylinder head - stage 1 ....Nm	20	20¹	40	40
- stage 2 ....Nm	30	60	Slacken, 40	Slacken, 40
- stage 3 ....Nm	+ 180°	+ 220°	+ 180°	+ 180°
- stage 4 ....Nm	+ 90°	–	–	–
- stage 5 ....Nm	–	–	–	–
- stage 6 ....Nm	–	–	–	–
Big-end bearings...................................Nm	25 + 50° N	20 + 70° N	110 N	115 N
Main bearings..Nm	113	70	160	160
Crankshaft pulley bolt ....Nm	190	40 + 60°	205	190
Camshaft pulley bolt ....Nm	118	45	25	25
Flywheel [driveplate] bolt......................Nm	142	50	125	125
Front hubs ....Nm	450	450	421 to 510	421 to 510
Rear hubs ....Nm	WSM	WSM	WSM	WSM
Wheel nuts / bolts ....Nm	160	160	160	160
Glow plugs ....Nm	15	22	23	23
Clutch pressure plate bolts ....Nm	–	23	15	15
Injection pump sprocket.......................Nm	49	50	60	25
Injectors...Nm	80	90	35	95
Injection pump mounting bolts ....Nm	25	22	–	25
Injector pipe unions..............................Nm	30	20	–	–
Capacities				
Engine oil & filter..................................litres	5.0	5.0	6.4	5.5
Gearbox..litres	2.0	2.0	1.25 89 ▶: 1.6	1.25 89 ▶: 1.6
Automatic transmission ....litres	–	–	–	–
Final drive ....litres	WT	WT	WT	WT
Cooling system......................................litres	–	8.5	9.0	9.0
Fuel tank..litres	80	80	70	70

Notes

Ducato 1.9 TD 1999 to 2000
¹Bolt length below head: ≤121.5 mm
Ducato 2500 Turbo D & 4x4 1990 to 1994
¹Ducato 18: -1.0 to 1.0
²Ducato 18: 2° to 2°30′
Ducato 2500 D & 4x4 1986 to 1994
¹Ducato 18: -1.0 to 1.0
²Ducato 18: 2° to 2°30′

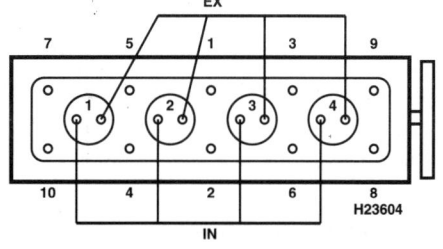

1929 cm³

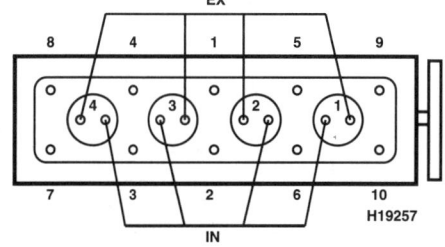

1905 cm³

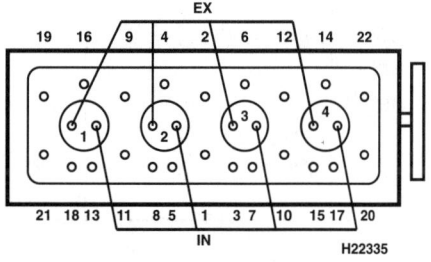

2445 cm³ / 2499 cm³

– Not applicable, or information not
available

FIAT

	Ducato 2500 D (CRD) 1987 to 1994	Ducato 2.5 D 1994 to 1999	Ducato 2.5 TD 1994 to 1999	Ducato 2.8 D 1998 to 2000
Engine				
Engine type/code...........................	U25/661 OHV 54kW	8140.67 SOHC 62kW	8140.47 SOHC Turbo 85kW	8140.67 SOHC 64kW
Capacity (cm^3) / cylinders......................	2500 / 4	2499 / 4	2499 / 4	2800 / 4
Compression ration / pressure ....bar	22.25 / _	22.5 /	18.5 /	21.7 /
Torque output ....Nm	0	164	245	180
Oil pressure ....idle [running] bar	[3.7 to 4.5 @ 3500]	0.8	0.8	0.8
Oil temperature ....°C	100	100	100	100
Valve clearances - inlet (mm)	0.30	0.50	0.50	0.50
- exhaust (mm)	0.20	0.50	0.50	0.50
Injection order	1-3-4-2	1-3-4-2	1-3-4-2	1-3-4-2
No. 1 cylinder position	FE	_	_	TBE
Cooling system				
Thermostat opening temperature ....°C	75 to 78	81	81	82
Radiator cap pressure ....bar	1.0	1.0	1.0	1.0
Fuel system				
Idle speed ....rpm	775 to 825	775 ± 25	750 ± 25	775 ± 25
Maximum (no load) speed ....rpm	4625 ± 25	4750 ± 50	4450 ± 50	4750 ± 50
Smoke test/opacity ....M^{-1} %	2.5	1.85	1.85	2.5
Static timing method..............................	Refer to wsm	Plunger travel	Rotor lift	Rotor lift
Timing dimension................................mm	_	1.0 ± 0.03	1.31 ± 0.03	1.15 ± 0.05
Crankshaft position ....mm [°]	4.32 [22]	TDC	TDC	TDC
Turbo type / ref / pressure ....bar	_	_	_	_
Injection pump make	Roto Diesel	Bosch	Bosch	Bosch
Injection pump part no............................	R3449 F 010	VER 518	VER 542	R735
Injector Make / type	Roto Diesel	Bosch	Bosch	_
Injector part no......................................	RDNOSDC 6577	_	_	DSLA 134P 604
Injection type..	MA300	Indirect VER	Direct	Direct injection
Injection opening pressure, New [used]...bar	112 to 117 [112]	120 to 128	240 to 248	240 to 252
Glow plugs				
Maker	Beru	Beru	Beru	_
Type	0100 221 106	GN908	GN927	_
Nominal rating.......................................V/A	11 / 12	_	_	_
Brakes				
minimum friction material thickness				
Front.......................................mm	2.0	1.0	1.0	4.0
Rear.......................................mm	1.0	1.5	1.5	1.0
Tyres - Saloon / Hatch......................Size	_	_	_	_
- Estate / Van............................Size	185x14	195/70x15: 205/75x16	195/70x15: 205/75x16	214/70x15
Pressure - front / rear - Saloon / Hatch ...bar	_	_	_	_
- Estate / Van ......bar	Refer to vehicle	Refer to vehicle	Refer to vehicle	5.0 / 5.0
Front suspension / wheel alignment				
Toe-in (+) / Toe-out (–)........................mm [°]	0.5 to 2.5	-2.0 ± 1.0	-2.0 ± 1.0	2.0 ± 1.0
Camber	1°30' to 2°30'	0° ± 30'	0° ± 30'	0° ± 30'
Castor	0° to 1°	1° ± 30'	1° ± 30'	1° ± 30'
King pin inclination................................	_	_	_	_
Rear suspension / wheel alignment				
Toe-in (+) / Toe-out (–)....................mm [°]	_	0 ± 1.0	0 ± 1.0	0.0 ± 1.0
Camber	_	0° ± 10'	0° ± 10'	0° ± 10'

FIAT

	Ducato 2500 D (CRD) 1987 to 1994	Ducato 2.5 D 1994 to 1999	Ducato 2.5 TD 1994 to 1999	Ducato 2.8 D 1998 to 2000
Torque wrench settings				
Cylinder head - stage 1 ....Nm	40	60	60	60 N
- stage 2 ....Nm	+ 100°	+ 180°	+ 180°	+ 180°
- stage 3 ....Nm	+ 100°	–	–	–
- stage 4 ....Nm	warm up, allow to cool	–	–	–
- stage 5 ....Nm	+ 45°	–	–	–
- stage 6 ....Nm	–	–		–
Big-end bearings............................Nm	72 to 90 N	118 N	118 N	118 N
Main bearings..............................Nm	95	160	160	WSM
Crankshaft pulley bolt ....Nm	–	200	200	–
Camshaft pulley bolt ....Nm	32	–	–	–
Flywheel [driveplate] bolt.................Nm	90 LkC	125	125	30 + 90°
Front hubs ....Nm	421 to 510	450	450	450
Rear hubs....................................Nm	WSM	WSM	WSM	WSM
Wheel nuts / bolts ....Nm	160	160 Maxi: 180	160 Maxi: 180	160
Glow plugs ....Nm	28	25	25	–
Clutch pressure plate bolts ....Nm	15	25	25	25
Injection pump sprocket...................Nm	23	95	95	–
Injectors.....................................Nm	24	50	50	–
Injection pump mounting bolts ....Nm	30	–	–	–
Injector pipe unions........................Nm	24	25	25	–
Capacities				
Engine oil & filter............................litres	4.7[1]	7.3	7.3	6.7
Gearbox......................................litres	1.25 89 ▶: 1.6	2.0 Maxi 2.45	2.0 Maxi 2.45	2.75
Automatic transmission ....litres	–	–	–	–
Final drive ....litres	WT	WT	WT	WT
Cooling system...............................litres	10.6	11.0	10.0	10.0
Fuel tank......................................litres	70	80	80	80

Notes

Ducato 2500 D (CRD) 1987 to 1994
[1]Engine no. GJ 06061862 ▶: 5.6

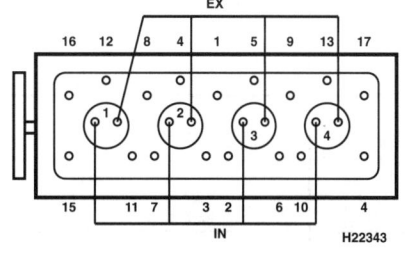

2500 cm³

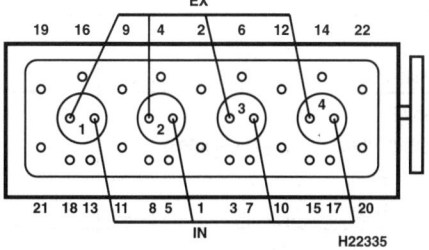

2499 cm³

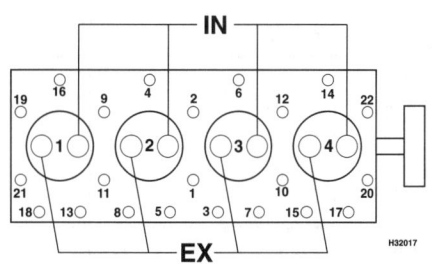

2800 cm³

– Not applicable, or information not available

FIAT

Haynes THE BOOK ®

	Ducato 2.8 TD 1998 to 2000			
Engine				
Engine type/code	8140.43 SOHC Turbo 90kW			
Capacity (cm³) / cylinders	2800 / 4			
Compression ration / pressure ....bar	19.0 /			
Torque output ......Nm	283			
Oil pressure ......idle [running] bar	0.8			
Oil temperature ......°C	100			
Valve clearances - inlet (mm)	0.50			
- exhaust (mm)	0.50			
Injection order	1-3-4-2			
No. 1 cylinder position	–			
Cooling system				
Thermostat opening temperature ......°C	82			
Radiator cap pressure ......bar	1.0			
Fuel system				
Idle speed ......rpm	750 ± 25			
Maximum (no load) speed ......rpm	4450 ± 50			
Smoke test/opacity ......M⁻¹ %	2.5			
Static timing method	Rotor lift			
Timing dimension......mm	1.1 ± 0.05			
Crankshaft position ......mm [°]	TDC			
Turbo type / ref / pressure ......bar	KKK K14 ot Mitsubishi TFO 35 M			
Injection pump make	Bosch			
Injection pump part no	R799			
Injector Make / type	Bosch			
Injector part no	DSLA 141P 405			
Injection type	Direct			
Injection opening pressure, New [used]...bar	240 to 248			
Glow plugs				
Maker	–			
Type	–			
Nominal rating ......V/A	–			
Brakes				
minimum friction material thickness				
Front......mm	4.0			
Rear......mm	1.0			
Tyres - Saloon / Hatch......Size	–			
- Estate / Van......Size	215/70x15			
Pressure - front / rear - Saloon / Hatch...bar	–			
- Estate / Van ......bar	5.0 / 5.0			
Front suspension / wheel alignment				
Toe-in (+) / Toe-out (–)......mm [°]	2.0 ± 1.0			
Camber	0° ± 30'			
Castor	1° ± 30'			
King pin inclination	–			
Rear suspension / wheel alignment				
Toe-in (+) / Toe-out (–)......mm [°]	0.0 ± 1.0			
Camber	0° ± 10'			

FIAT

	Ducato 2.8 TD 1998 to 2000			
Torque wrench settings				
Cylinder head - stage 1 ....Nm	60 N			
- stage 2 ....Nm	+ 180°			
- stage 3 ....Nm	_			
- stage 4 ....Nm	_			
- stage 5 ....Nm	_			
- stage 6 ....Nm	_			
Big-end bearings.................................Nm	118 N			
Main bearings.....................................Nm	WSM			
Crankshaft pulley bolt ....Nm	_			
Camshaft pulley bolt ....Nm	_			
Flywheel [driveplate] bolt....................Nm	30 + 90°			
Front hubs ....Nm	450			
Rear hubs ....Nm	_			
Wheel nuts / bolts ....Nm	160			
Glow plugs ....Nm	_			
Clutch pressure plate bolts ....Nm	25			
Injection pump sprocket......................Nm	_			
Injectors..Nm	_			
Injection pump mounting bolts ....Nm	_			
Injector pipe unions............................Nm	_			
Capacities				
Engine oil & filter................................litres	6.7			
Gearbox...litres	2.75			
Automatic transmission ....litres	_			
Final drive...litres	WT			
Cooling system...................................litres	10.0			
Fuel tank...litres	80			

Notes

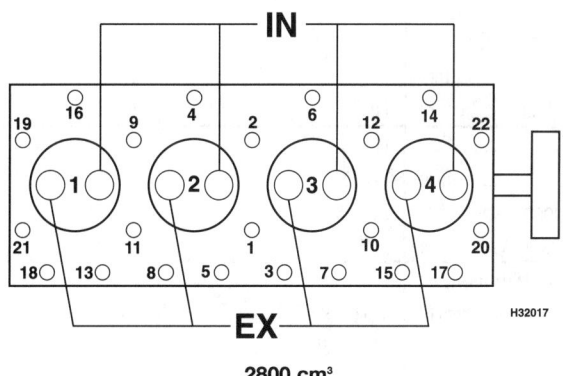

2800 cm³

– Not applicable, or information not available

FORD

Haynes THE BOOK ®	Fiesta 1.8 & Van 1989 to 1996	Fiesta 1.8 & Van 1989 to 1996	Fiesta 1.8 1995 to 1999	Escort & Orion 1.8 TD 1995 to 1997
Engine				
Engine type/code.........................	RTC/RTD SOHC 44kW	RTC/RTD SOHC 44kW	Endura-DE RTJ⁴ SOHC 43kW	RVA Endura-DE SOHC¹ 51kW
Capacity (cm³) / cylinders......................	1753 / 4	1753 / 4	1753 / 4	1753 / 4
Compression ration / pressure ....bar	21.5 / 28 to 34	21.5 / 28 to 34	21.5 / 28.0 to 34.0	21.5 / 28.0 to 34.0
Torque output ....Nm	110	110	105	135
Oil pressure ....idle [running] bar	0.75 [1.5]	0.75 [1.5]	0.75 [1.5]	0.75 [1.5]
Oil temperature ....°C	80	80	80	80
Valve clearances - inlet (mm)	0.30 to 0.40	0.30 to 0.40	0.30 to 0.40	0.35 ± 0.05
- exhaust (mm)	0.45 to 0.55	0.45 to 0.55	0.45 to 0.55	0.50 ± 0.05
Injection order	1-3-4-2	1-3-4-2	1-3-4-2	1-3-4-2
No. 1 cylinder position	TBE	TBE	TBE	TBE
Cooling system				
Thermostat opening temperature ....°C	85 to 89	85 to 89	85 to 89	85 to 89
Radiator cap pressure ....bar	1.0 to 1.3	1.0 to 1.3	1.2	1.2
Fuel system				
Idle speed ....rpm	850 ± 50	850 ± 50	920 ± 50	850 ± 50
Maximum (no load) speed ....rpm	5350 ± 50	5350 ± 50	5350 ±50	5150 ± 50
Smoke test/opacity ....M⁻¹ %	2.5	2.5	2.5	2.5
Static timing method...............................	Refer to wsm	Refer to wsm	Refer to wsm	Refer to wsm
Timing dimension.................................mm	–	–	–	–
Crankshaft position ....mm [°]	–	–	–	–
Turbo type / ref / pressure ....bar	–	–	–	–
Injection pump make	CAV	Bosch	Bosch	Lucas
Injection pump part no............................	–	–	VE	DPC
Injector Make / type	Pintle	Pintle	Bosch	CAV
Injector part no.......................................	–	–	–	Pintle
Injection type..	Indirect	Indirect	Indirect	EEC V
Injection opening pressure, New [used]...bar	120 ± 6	143 ± 7	158 to 165	120
Glow plugs				
Maker	Champion	Champion	Bosch/Champion	Motorcraft
Type	CH147	CH147	0250 202 001 / CH147	EZD34
Nominal rating.....................................V/A	11.5 / 16	11.5 / 16		11.5 / 16
Brakes				
minimum friction material thickness				
Front...mm	1.5	1.5	1.5	1.5
Rear...mm	1.0	1.0	1.5	1.0
Tyres - Saloon / Hatch....................Size	145x13: 155/70x13	145x13: 155/70x13	165/70x13	155x13: 175/70x13: 185/60x14
- Estate / Van..........................Size	145x13	145x13	165/70x13	155x13: 175/70x13: 185/60x14
Pressure - front / rear - Saloon / Hatch ...bar	2.0 / 1.8¹	2.0 / 1.8¹	2.1 / 1.8	2.0 / 1.8
- Estate / Van ......bar	2.0 / 1.8	2.0 / 1.8	2.1 / 1.8	2.0 / 1.8
Front suspension / wheel alignment				
Toe-in (+) / Toe-out (–)....................mm [°]	-3.0 to 3.0²	-3.0 to 3.0²	0 ± 1.0	2.0 ± 1.0
Camber	12' ± 1°15'³	12' ± 1°15'³	-40' ± 1°10'²	-19' ± 14'
Castor	18' ± 1°⁴	18' ± 1°⁴	1°20' ± 1°15'³	1°12' ± 1°20'
King pin inclination..................................	–	–		
Rear suspension / wheel alignment				
Toe-in (+) / Toe-out (–)....................mm [°]	2.3 ± 2.0	2.3 ± 2.0	2.5 ± 2.0	2.0 ± 2.0
Camber	-1° +30' -1°	-1° +30' -1°	-1° ± 70'	-1° ± 50'

FORD

	Fiesta 1.8 & Van 1989 to 1996	Fiesta 1.8 & Van 1989 to 1996	Fiesta 1.8 1995 to 1999	Escort & Orion 1.8 TD 1995 to 1997
Torque wrench settings				
Cylinder head - stage 1 ....Nm	20 to 30	20 to 30	10 N	10 N
- stage 2 ....Nm	76 to 92	76 to 92	100	100
- stage 3 ....Nm	Wait 2 min.	Wait 2 min.	- 180°	Wait 3 min.
- stage 4 ....Nm	+ 90°	+ 90°	70	- 180°
- stage 5 ....Nm	Bolts must not be retorqued	Bolts must not be retorqued	+ 120°	70
- stage 6 ....Nm	–	–	–	+ 120°
Big-end bearings.................................Nm	20 to 30[5] N	20 to 30[5] N	25 + 60° + 20° N	30 + 60° + 20° N
Main bearings.....................................Nm	27 + 45°	27 + 45°	25 + 75°	27 + 75°
Crankshaft pulley bolt ....Nm	180	180	150, slacken, 120 + 60°	150, slacken, 120 + 60°
Camshaft pulley bolt ....Nm	M6: 9 to 11[6]	M6: 9 to 11[6]	23	11 M8: 33
Flywheel [driveplate] bolt....................Nm	15 to 20[7]	15 to 20[7]	18 + 45° + 45°	20 + 35° + 35°
Front hubs ....Nm	205 to 235	205 to 235	270	220 M22: 235
Rear hubs ....Nm	250 to 290	250 to 290	270	260
Wheel nuts / bolts ....Nm	100	100	85	100
Glow plugs ....Nm	25 to 30	25 to 30	28	30
Clutch pressure plate bolts ....Nm	25 to 34	25 to 34	30	30
Injection pump sprocket......................Nm	20 to 25	20 to 25	23	25
Injectors...Nm	60 to 80	60 to 80	70	60 to 80
Injection pump mounting bolts ....Nm	18 to 22	18 to 22	24	25
Injector pipe unions............................Nm	15 to 25	15 to 25	25	25
Capacities				
Engine oil & filter................................litres	4.5	4.5	5.0	4.5
Gearbox..litres	3.1	3.1	2.6	2.8
Automatic transmission ....litres	–	–	–	–
Final drive ....litres	WT	WT	WT	WT
Cooling system...................................litres	8.6	8.6	9.3	9.3
Fuel tank...litres	42	42	42	55

Notes

Fiesta 1.8 & Van 1989 to 1996
[1]155/70x13: 2.2 / 1.8
[2]90 ▶: -4.5 to +0.5
[3]90 ▶: -8'±1°20'. Van: -3'±1°20'
[4]90 ▶: 53'±1°15'. Van: 51'±1°15'
[5]+ 60° + 20°
[6]M8: 27 to 33
[7]+ 45+0-10° + 45+0-10°
Fiesta 1.8 & Van 1989 to 1996
[1]155/70x13: 2.2 / 1.8
[2]90 ▶: -4.5 to +0.5
[3]90 ▶: -8'±1°20'. Van: -3'±1°20'
[4]90 ▶: +53'±1°15'. Van: +51'±1°15'
[5]+ 60° + 20°
[6]M8: 27 to 33
[7]+ 45+0-10° + 45+0-10°
Fiesta 1.8 1995 to 1999
[2] Van: 0°44' -2' +4'
[3] Van: 2°33'
[4]or RTK
Escort & Orion 1.8 TD 1995 to 1997
[1]Turbo

– Not applicable, or information not available

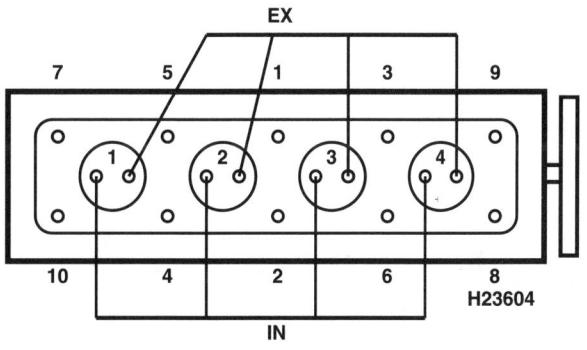

EX

7   5   1   3   9

1   2   3   4

10   4   2   6   8

H23604

IN

1753 cm³

FORD

	Escort & Orion 1.8 TD 1995 to 1999	Escort & Orion 1.8 TDi 1993 to 1999	Escort & Orion 1.8 1990 to 1998	Escort & Orion 1.8 1990 to 1998
Engine				
Engine type/code....................	RVA Endura-DE SOHC³ 51kW	RFD/RFK Endura-DE¹ 66kW	RTE SOHC 44kW	RTE/RTF/RTH SOHC 44kW¹
Capacity (cm³) / cylinders........................	1753 / 4	1753 / 4	1753 / 4	1753 / 4
Compression ration / pressure ....bar	21.5 / 28.0 to 34.0	21.5 / 28.0 to 34.0	21.5 / 28 to 34	21.5 / 28 to 34
Torque output ....Nm	135	180	110	110
Oil pressure ....idle [running] bar	0.75 [1.5]	0.75 [1.5]	0.75 [1.5]	0.75 [1.5]
Oil temperature ....°C	80	80	80	80
Valve clearances - inlet (mm)	0.35 ± 0.05	0.35 ± 0.05	0.30 to 0.40	0.30 to 0.40
- exhaust (mm)	0.50 ± 0.05	0.50 ± 0.05	0.45 to 0.55	0.45 to 0.55
Injection order	1-3-4-2	1-3-4-2	1-3-4-2	1-3-4-2
No. 1 cylinder position	TBE	TBE	TBE	TBE
Cooling system				
Thermostat opening temperature ....°C	85 to 89	85 to 89	85 to 89	85 to 89
Radiator cap pressure ....bar	1.2	1.2	1.2	1.2
Fuel system				
Idle speed ....rpm	850 ± 50	850 ± 50	850 ± 50	850 ± 50
Maximum (no load) speed ....rpm	5150 ± 50	5150 ± 50	5350 ± 50	5350 ± 50
Smoke test/opacity ....M⁻¹ %	2.5	2.0	2.5	2.5
Static timing method...............................	Refer to wsm	Refer to wsm	Refer to wsm	Refer to wsm
Timing dimension.............................mm	–	–	–	–
Crankshaft position ....mm [°]	–	–	–	–
Turbo type / ref / pressure ....bar	–	–	–	–
Injection pump make	Bosch	Lucas	CAV	Bosch
Injection pump part no.............................	VE	–	–	–
Injector Make / type	Pintle	Pintle	Pintle	Bosch
Injector part no..	–	–	–	–
Injection type..	VE	EEC V	Indirect	VE
Injection opening pressure, New [used]...bar	143	–	120 ± 6	143 ± 7
Glow plugs				
Maker	Motorcraft	Motorcraft	Motorcraft/Champion	Motorcraft/Champion
Type	EZD34	EZD39	EZD6 / CH147	EZD6 / CH147
Nominal rating............................V/A	11.5 / 16	11.5 / 16	11.5 / 16	11.5 / 16
Brakes				
minimum friction material thickness				
Front............................mm	1.5	1.5	1.5	1.5
Rear............................mm	1.0	1.0	1.0	1.0
Tyres - Saloon / Hatch.................Size	155x13: 175/70x13: 185/60x14	155x13: 175/70x13: 185/60x14	155x13: 175/70x13	155x13: 175/70x13
- Estate / Van.......................Size	155x13: 175/70x13: 185/60x14	155x13: 175/70x13: 185/60x14	155x13: 165x13: 175/70x13	155x13: 165x13: 175/70x13
Pressure - front / rear - Saloon / Hatch ...bar	2.0 / 1.8	2.0 / 1.8	2.0 / 1.8	2.0 / 1.8
- Estate / Van ......bar	2.0 / 1.8	2.0 / 1.8	2.0 / 1.8	2.0 / 1.8
Front suspension / wheel alignment				
Toe-in (+) / Toe-out (–)....................mm [°]	2.0 ± 1.0	2.0 ± 1.0	-2.0 ± 1.0	-2.0 ± 1.0
Camber	-19' ± 14'¹	-19' ± 14'	-15' ± 1°20'	-15' ± 1°20'
Castor	1°12' ± 1°20'²	1°12' ± 1°20'	0 ± 1°15'	0 ± 1°15'
King pin inclination..................................			–	–
Rear suspension / wheel alignment				
Toe-in (+) / Toe-out (–).....................mm [°]	2.0 ± 2.0	2.0 ± 2.0	2.0 ± 2.0	2.0 ± 2.0
Camber	-1° ± 50'	-1° ± 50'	-1° +30' -1°	-1° +30' -1°

FORD

	Escort & Orion 1.8 TD 1995 to 1999	Escort & Orion 1.8 TDi 1993 to 1999	Escort & Orion 1.8 1990 to 1998	Escort & Orion 1.8 1990 to 1998
Torque wrench settings				
Cylinder head - stage 1 ....Nm	10 N	10 N	20 to 30	20 to 30
- stage 2 ....Nm	100	100	76 to 92	76 to 92
- stage 3 ....Nm	Wait 3 min.	Wait 3 min.	Wait 2 min.	Wait 2 min.
- stage 4 ....Nm	- 180°	- 180°	+ 90°	+ 90°
- stage 5 ....Nm	70	70	Bolts must not be retorqued	Bolts must not be retorqued
- stage 6 ....Nm	+ 120°	+ 120°	–	–
Big-end bearings................................Nm	30 + 60° + 20° N	30 + 60° + 20° N	20 to 30[1] N	20 to 30[2] N
Main bearings....................................Nm	27 + 75°	27 + 75°	27 + 45°	27 + 45°
Crankshaft pulley bolt ....Nm	150, slacken, 120 + 60°	150, slacken, 120 + 60°	180	180
Camshaft pulley bolt ....Nm	11 M8: 33	11 M8: 33	M6: 9 to 11[2]	M6: 9 to 11[3]
Flywheel [driveplate] bolt...................Nm	20 + 35° + 35°	20 + 35° + 35°	15 to 20[3]	15 to 20[4]
Front hubs ....Nm	220 M22: 235	220 M22: 235	220 to 250	220 to 250
Rear hubs ....Nm	260	260	250 to 270	250 to 270
Wheel nuts / bolts ....Nm	100	100	70 to 100	70 to 100
Glow plugs ....Nm	30	30	25 to 30	25 to 30
Clutch pressure plate bolts ....Nm	30	30	25 to 34	25 to 34
Injection pump sprocket......................Nm	25	25	20 to 25	20 to 25
Injectors..Nm	60 to 80	60 to 80	60 to 80	60 to 80
Injection pump mounting bolts ....Nm	25	25	18 to 22	18 to 22
Injector pipe unions............................Nm	25	25	15 to 25	15 to 25
Capacities				
Engine oil & filter................................litres	4.5	4.5	4.5	4.5
Gearbox...litres	2.8	2.8	3.1	3.1
Automatic transmission ....litres	–	–	–	–
Final drive ....litres	WT	WT	WT	WT
Cooling system..................................litres	9.3	9.3	9.3	9.3
Fuel tank..litres	55	55	55	55

Notes

Escort & Orion 1.8 TD 1995 to 1999
[1] Van: -1°47' to 1°01'
[2] Van: -0°43' to 1°57'
[3]Turbo
Escort & Orion 1.8 TDi 1993 to 1999
[1]SOHC Turbo Intercooled
Escort & Orion 1.8 1990 to 1998
[1]+ 60° + 20°
[2]M8: 27 to 33
[3] 45 +0 -10° + 45+0-10°
Escort & Orion 1.8 1990 to 1998
[1]RTH: catalyst
[2]+ 60° + 20°
[3]M8: 27 to 33
[4]+ 45+0-10° + 45+0-10°

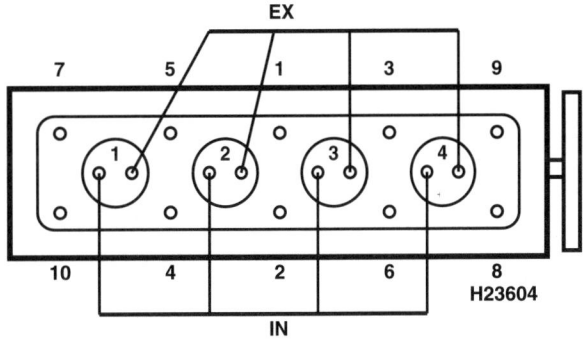

1753 cm³

– Not applicable, or information not
   available

FORD

	Focus 1.8 TDi 1999 to 2000	Sierra & Sapphire 1.8 TD 1990 to 1993	Mondeo 1.8 TD 1993 to 2000	Granada, Scorpio 2.5 Turbo 1988 to 1994
Engine				
Engine type/code...............	C9DC Endura DI² 66kW	RFA/RFB/RFL SOHC Turbo 55kW¹	RFM/RFN¹ SOHC Turbo 65kW	SFA/SFC OHV Turbo 68kW
Capacity (cm³) / cylinders...............	1753 / 4	1753 / 4	1753 / 4	2498 / 4
Compression ration / pressure ....bar	19.4 / ≥28	21.5 / 28 to 34	21.5 / 28.0 to 34.0	21.0 / 20 to 25
Torque output ....Nm	200	0	180	0
Oil pressure ....idle [running] bar	0.75 [1.5 @ 2000]	0.75 [1.5]	0.75 [1.5 @ 2000]	1.2 [3.0]
Oil temperature ....°C	80	80	80	80
Valve clearances - inlet (mm)...............	0.30 to 0.40	0.30 to 0.40	0.30 to 0.40	0.10 to 0.20
- exhaust (mm).............	0.45 to 0.55	0.45 to 0.55	0.45 to 0.55	0.20 to 0.30
Injection order...............	1-3-4-2	1-3-4-2	1-3-4-2	1-3-4-2
No. 1 cylinder position...............	TBE	TBE	TBE	FE
Cooling system				
Thermostat opening temperature.........°C	–	85 to 89	85 to 89	69 to 72
Radiator cap pressure ....bar	–	1.00 to 1.25	refer to cap	1.0 to 1.2
Fuel system				
Idle speed ....rpm	800 ± 50	850 ± 50	850 to 860	875 to 900
Maximum (no load) speed ....rpm	–	5150 ± 50	5200 to 5300	4800 ± 50
Smoke test/opacity ....M⁻¹ %	2.5	2.5	2.5	2.5
Static timing method...............	–	Refer to wsm	Refer to wsm	Dial gauge
Timing dimension............... mm	–	–	–	Dimension on pump
Crankshaft position ....mm [°]	–	–	–	–
Turbo type / ref / pressure ....bar	–	–	–	–
Injection pump make	Bosch	CAV	Lucas	CAV
Injection pump part no...............	VP30 234140	–	–	–
Injector Make / type	Multi hole	Pintle	–	Pintle
Injector part no...............	–	–	–	–
Injection type...............	EEC V Direct	Indirect	Indirect	DPCR
Injection opening pressure, New [used]...bar	≤1500	143 to 150	158 to 165	130 ± 5
Glow plugs				
Maker	Motorcraft	Motorcraft/Champion	Motorcraft	Motorcraft/Champion
Type	EZD 37	EZD6 / CH147	EZD 39	EZD5 / CH68
Nominal rating............... V/A		11.5 / 16		11 / 8.5
Brakes				
minimum friction material thickness				
Front............... mm	1.5	1.5	–	1.5
Rear............... mm	1.0	1.0	–	1.5
Tyres - Saloon / Hatch .... Size	175/70x14:185/65x14:195/55x15	165x13: 185/70x13: 195/65x14²	185/65x14	175x14: 185/70x14
- Estate / Van............... Size	185/65x14:195/55x15:195/60x15	175x13: 195/65x14: 185/65x14³	185/65x14	
Pressure - front / rear - Saloon / Hatch ...bar	2.2 / 2.2 195: 2.0 / 2.0	1.8 / 1.8⁴	2.1 / 2.1	1.8 / 1.8
- Estate / Van ......bar	2.2 / 2.2 195: 2.0 / 2.0	1.8 / 1.8	2.1 / 2.1	
Front suspension / wheel alignment				
Toe-in (+) / Toe-out (–)....................mm [°]	1.0 ± 1.0	2.0 ± 1.0	-0.5 to -3.5	2.0 ± 1.0
Camber	-0°32' Est: -0°32'	-21' ± 1° Estate: -25'±1°⁵	-32' ± 1°18' Sport:-35'±1°15'	-23' ± 1°¹
Castor	3° Est: 2°27'	1°52' Estate: 1°44'⁵	2°19' ± 1° Sport: 2°21'±1°	1°51' ± 1°²
King pin inclination...............	–	–	–	–
Rear suspension / wheel alignment				
Toe-in (+) / Toe-out (–)....................mm [°]	1.1 to 5.1 Est: 0.1 to 4.1	–	3.9 to -0.1	
Camber	-0°55' Est: -0°35'	–	-34' ± 19' Sport: -37' ± 1°	

FORD

Torque wrench settings	Focus 1.8 TDi 1999 to 2000	Sierra & Sapphire 1.8 TD 1990 to 1993	Mondeo 1.8 TD 1993 to 2000	Granada, Scorpio 2.5 Turbo 1988 to 1994
Cylinder head - stage 1 ....Nm	20	35 to 40	10 N	30
- stage 2 ....Nm	54	70 to 75	100	70
- stage 3 ....Nm	+ 160°[1]	Wait 1 min.	Wait 3 mins.	Slacken 90°
- stage 4 ....Nm	–	70 to 75	- 180°	70, warm up,
- stage 5 ....Nm	–	+ 90°	70	Allow to cool, slacken
- stage 6 ....Nm	–	–	+ 120°	70 + 120°
Big-end bearings.....................Nm	25 + 60° + 20° N	25 to 30[6] N	25 + 60° + 20° N	53 to 62 N
Main bearings.....................Nm	27 + 75° N	27 + 75°	27 + 75°	100 to 120
Crankshaft pulley bolt ....Nm	120 - 90°, 150	180	150, slacken, 120 + 60°	55 + 60°
Camshaft pulley bolt ....Nm	35 M10: 48	M6: 9 to 11[7]	35	–
Flywheel [driveplate] bolt......................Nm	20 + 45°	15 to 20[8]	18 + 45° + 45°	15, 95
Front hubs ....Nm	316	310 to 350	340	390 to 450
Rear hubs ....Nm	235	250 to 290	290	250 to 290
Wheel nuts / bolts ....Nm	85	70 to 100	85	70 to 100
Glow plugs ....Nm	15	25 to 30	20	30 to 40
Clutch pressure plate bolts ....Nm	30	25 to 33	30	13 to 17
Injection pump sprocket.....................Nm	33	20 to 25	25	–
Injectors.............................Nm	23	70 to 80	95	90
Injection pump mounting bolts ....Nm	22	18 to 22	24	15 to 20
Injector pipe unions............................Nm	25	15 to 25	–	20 to 30
Capacities				
Engine oil & filter..............................litres	5.6	5.1	5.1	5.6
Gearbox..............................litres	2.6	1.25	2.6	1.2
Automatic transmission ....litres	–	.	–	–
Final drive ....litres	WT	0.9	WT	1.3
Cooling system..............................litres	9.3	9.5	9.3	7.0
Fuel tank..............................litres	52.7	60	61.5	70

Notes

Focus 1.8 TDi 1999 to 2000
[1]Short bolt: 180°
[2]SOHC Turbo
Sierra & Sapphire 1.8 TD 1990 to 1993
[1]RFA is 15.04. RFB, RFL are US87, RFL may be fitted with a catalyst
[2]Also: 185/65x14
[3]Also: 185/70x13
[4]165X13: 2.0 / 2.0
[5]WSM
[6]+ 55 to 60° + 15 to 20°
[7]M8: 27 to 33
[8]+ 35 to 40° + 35 to 40°
Mondeo 1.8 TD 1993 to 2000
[1]With CAT
Granada, Scorpio 2.5 Turbo 1988 to 1994
[1]'90 ▶: -23'±1°. Low series: -10'±1°
[2]Self-levelling: 1°58'±1°. 90 ▶: 2°25'±1°. Low series: 2°35'±1°

– Not applicable, or information not available

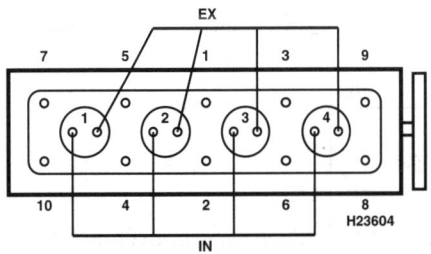

1753 cm³

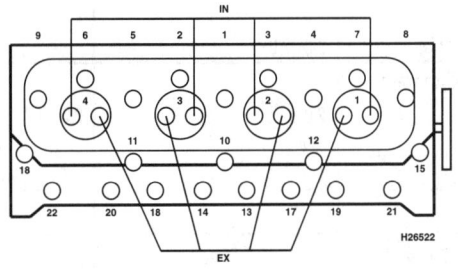

2498 cm³

Haynes THE BOOK ®

FORD

	Granada, Scorpio 2.5 TCi 1988 to 1994	Scorpio 2.5 TDi 1995 to 1998	Scorpio 2.5 TDi 1997 to 1998	Galaxy 1.9 TDi 1995 to 1999
Engine				
Engine type/code..................	SCB OHV Turbo 85kW	SCC OHV Turbo 85kW	SCD OHV Turbo 92kW	1Z/AHU SOHC Turbo 66kW
Capacity (cm^3) / cylinders........................	2498 / 4	2498 / 4	2498 / 4	1896 / 4
Compression ration / pressure ....bar	21.0 / 20 to 25	21.0 / 24 to 26	21.0 / 24 to 26	19.5 / 25 to 30
Torque output ....Nm	270	293	293	202
Oil pressure ....idle [running] bar	1.2 [3.0 @ 4000]	[4.0 @ 4000]	1.5 [4.0 @ 4000]	[2.0 @ 2000]
Oil temperature ....°C	80	80	100	80
Valve clearances - inlet (mm)	0: Hyd.	0: Hyd.	0: Hyd.	0: Hyd.
- exhaust (mm)	0: Hyd.	0: Hyd.	0: Hyd.	0: Hyd.
Injection order...	1-3-4-2	1-3-4-2	1-3-4-2	1-3-4-2
No. 1 cylinder position	TCE	TCE	TCE	TBE
Cooling system				
Thermostat opening temperature ....°C	69 to 72	–	–	77
Radiator cap pressure ....bar	1.0 to 1.2	refer to cap	refer to cap	0.78 to 0.98
Fuel system				
Idle speed ....rpm	850 ± 50	875 ± 25	875 ± 55	860 to 940
Maximum (no load) speed ....rpm	4800 ± 50	–	4750	4800 to 5100
Smoke test/opacity ....M^{-1} %	2.0	2.0	2.0	2.0
Static timing method................................	Dial gauge	Dial gauge	Dial gauge	–
Timing dimension................................mm	0.61 to 0.62	0.65 ± 0.02	0.53 to 0.57	Computer controlled
Crankshaft position ....mm [°]	TDC	TDC	TDC	–
Turbo type / ref / pressure ....bar	–	–	–	–
Injection pump make	Bosch	Bosch	Bosch	Bosch
Injection pump part no............................	VE L503	VE L503	VE4	–
Injector Make / type................................	Bosch	–	Bosch	Multi hole
Injector part no.......................................	DNSOD301	–	–	–
Injection type..	Indirect	Indirect EDC	Indirect EEC V EDC	Direct EDC
Injection opening pressure, New [used]...bar	150 to 158	–	–	–
Glow plugs				
Maker	Motorcraft/Beru	Bosch	Bosch	Motorcraft/Bosch
Type	EZD30 / 909 MJ	0250 201 033	0250 201 033	EDZ40 / 0250 201 036
Nominal rating....................................V/A	–	–	–	–
Brakes				
minimum friction material thickness				
Front..mm	1.5	1.0	1.0	7.0 with backing
Rear..mm	1.5	1.0	1.0	7.0 with backing
Tyres - Saloon / Hatch......................Size	195/65x15	195/65x15: 215/60x15	195/65x15: 215/60x15	–
- Estate / Van...........................Size	205/60x15	195/65x15: 215/60x15	195/65x15: 215/60x15	195/65x15: 205/60x15:215/60x15
Pressure - front / rear - Saloon / Hatch..bar	2.0 / 2.0	2.0 / 2.0	2.0 / 2.0	–
- Estate / Van ......bar	2.0 / 2.0	–	–	2.6 / 2.4 205: 2.7 / 2.5'
Front suspension / wheel alignment				
Toe-in (+) / Toe-out (–)...................mm [°]	2.0 ± 1.0	2.0 ± 1.0	2.0 ± 1.0	[5' ± 10']
Camber	-23' ± 1°¹	-1°38' ± 1° Est: -1°41' ± 1°	-1°38' ± 1° Est: -1°41' ± 1°	-20' ± 45'
Castor	1°51' ± 1°²	4°18' to 5° 48'	4°18' to 5° 48'	3°20' ± 40'
King pin inclination................................	–	–	–	–
Rear suspension / wheel alignment				
Toe-in (+) / Toe-out (–)...................mm [°]	–	–	–	[0 ± 25']
Camber				-20' ± 30'

FORD

Torque wrench settings	Granada, Scorpio 2.5 TCi 1988 to 1994	Scorpio 2.5 TDi 1995 to 1998	Scorpio 2.5 TDi 1997 to 1998	Galaxy 1.9 TDi 1995 to 1999
Cylinder head - stage 1 ....Nm	30	10[1]	30[1]	40 N
- stage 2 ....Nm	70	100	+ 70°	60
- stage 3 ....Nm	Slacken 90°	wait 3 mins	+ 70°, warm up	+ 90°
- stage 4 ....Nm	70, warm up	slacken	allow to cool, slacken	+ 90°
- stage 5 ....Nm	Allow to cool, slacken	70 + 120°	30	–
- stage 6 ....Nm	70 + 120°	–	+ 130°[2]	–
Big-end bearings.......................Nm	53 to 62 N	30 + 60° N	30 + 60° N	30 + 90° N
Main bearings.......................Nm	100 to 120	42	42	65 + 90° N
Crankshaft pulley bolt ....Nm	55 + 60°	152	196	90 + 90° N
Camshaft pulley bolt ....Nm	–	–	–	45
Flywheel [driveplate] bolt.......................Nm	15 then 95	20 + 60°	20 + 75°	60 + 90° N
Front hubs ....Nm	390 to 450	420	420	150 + 90°
Rear hubs ....Nm	250 to 290	270	270	200
Wheel nuts / bolts ....Nm	70 to 100	85	85	140
Glow plugs ....Nm	30 to 40	24	14	15
Clutch pressure plate bolts ....Nm	13 to 17	27	27	25
Injection pump sprocket.......................Nm	–	88	88	55
Injectors.......................Nm	90	70	70	22
Injection pump mounting bolts ....Nm	15 to 20	30	30	–
Injector pipe unions.......................Nm	20 to 30	17	17	25
Capacities				
Engine oil & filter.......................litres	5.6	6.6	6.6	4.3
Gearbox.......................litres	1.2	1.2	1.2	2.0
Automatic transmission ....litres	–	–	–	–
Final drive.......................litres	1.3	1.3	1.3	WT
Cooling system.......................litres	11.0	7.0	7.0	9.2
Fuel tank.......................litres	70	70	70	70

Notes

Granada, Scorpio 2.5 TCi 1988 to 1994
[1]'90 ▶: -23' ± 1°   Low series: -10' ± 1°
[2]Self-levelling: 1°58' ± 1°   90 ▶: 2°25' ± 1° Low series 2°35' ± 1°
Scorpio 2.5 TDi 1995 to 1998
[1] Hexagon head bolts: 30, 92, wait 3 mins, + 90°
Scorpio 2.5 TDi 1997 to 1998
[1] M12: 30 + 85°, warm-up, allow to cool, 90
[2] All bolts: + 15° after 20 000kms
Galaxy 1.9 TDi 1995 to 1999
[1]215/60x15: 2.3 / 2.1

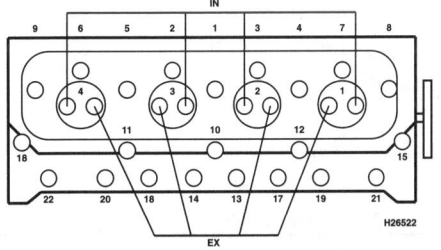

2498 cm³

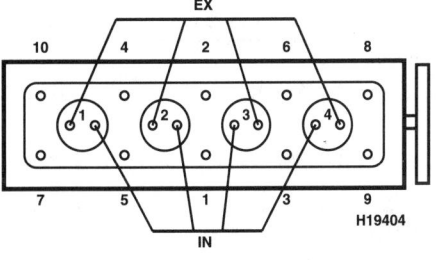

1896 cm³

– Not applicable, or information not available

FORD

	Galaxy 1.9 TDi 1996 to 2000	Maverick 2.7 TD 1993 to 1996	Maverick 2.7 TDi 1996 to 1998	P100 1.8 Turbo 1989 to 1992
Engine				
Engine type/code	AFN SOHC Turbo 81kW	TD27T VP OHV Turbo 73kW	TD27E VP OHV Turbo 92kW	RFA SOHC Turbo 55kW
Capacity (cm³) / cylinders................	1896 / 4	2663 / 4	2663 / 4	1753 / 4
Compression ration / pressure ....bar	19.5 / 25 to 30	21.9 / _	21.9 / 29	21.5 / 28 to 34
Torque output ....Nm	235	221	270	0
Oil pressure ....idle [running] bar	[2.0]	0.78 [2.9 to 3.9 @ 3000]	0.78 [2.94 to 3.92]	0.75 [1.5]
Oil temperature ....°C	80	80	80	80
Valve clearances - inlet (mm)	0: Hyd.	0.25 H	0.25 H	0.30 to 0.40
- exhaust (mm)	0: Hyd.	0.25 H	0.25 H	0.45 to 0.55
Injection order	1-3-4-2	1-3-4-2	1-3-4-2	1-3-4-2
No. 1 cylinder position	TBE	TCE	TCE	TBE
Cooling system				
Thermostat opening temperature ....°C	_	82	82	85 to 89
Radiator cap pressure ....bar	_	0.78 to 0.98	0.78 to 0.98	1.2 to 1.5
Fuel system				
Idle speed ....rpm	860 to 940	700 ± 50	700 ± 50	850 ± 50
Maximum (no load) speed ....rpm	4800 to 5200	5050	5050	5150 ± 50
Smoke test/opacity ....M⁻¹ %	2.0	2.5	2.0	2.5
Static timing method...............................	_	Plunger travel	Plunger travel	Refer to wsm
Timing dimension.............................mm	Computer controlled	0.38 ± 0.02	0.38 ± 0.02	_
Crankshaft position ....mm [°]	_	[0] TDC	[0] TDC	_
Turbo type / ref / pressure ....bar				
Injection pump make	Bosch	Diesel Kiki	Diesel Kiki	CAV
Injection pump part no........................	VER 510	VE	NP VE4	_
Injector Make / type	Multi hole	Bosch	Bosch	Pintle
Injector part no..................................	DSLA150 P 520	_	_	_
Injection type.....................................	Direct EDC	Indirect EDC	Indirect EDC	Indirect
Injection opening pressure, New [used]...bar	190 to 200 [170]	105 to 113 [98 to 103]	103 to 123 [98 to 103]	143 to 150
Glow plugs				
Maker	Motorcraft/Bosch	Beru	Motorcraft	Motorcraft/Champion
Type	EDZ40 / 0250 202 009	854 MJ	1 960 555	EZD6 / CH147
Nominal rating ....V/A	11 / 12	_	_	11.5 / 16
Brakes				
minimum friction material thickness				
Front...................................mm	7.0 with backing	2.0	2.0	1.5
Rear......................................mm	7.0 with backing	2.0	2.0	1.0
Tyres - Saloon / Hatch......................Size	_	_	_	_
- Estate / Van......................Size	195/65x15: 205/60x15:215/60x15	215/80x15: 235/75x15	215/80x15: 235/75x15	185x14
Pressure - front / rear - Saloon / Hatch ...bar	_	_	_	_
- Estate / Van ......bar	2.6 / 2.4 205: 2.7 / 2.5¹	3D: 1.8 / 2.2²	3D: 1.8 / 2.2²	1.8 / 2.5
Front suspension / wheel alignment				
Toe-in (+) / Toe-out (–)......................mm [°]	[5' ± 10']	3.0 to 5.0	3.0 to 5.0	2.0 ± 2.0
Camber	-20' ± 45'	0 to 1°	0 to 1°	30' +1° -1°30'
Castor	3°20' ± 40'	1°10' to 2°10'	1°10' to 2°10'	1°39' +1°30'-1°
King pin inclination...............................	_	7°36' to 8°36'	7°36' to 8°36'	
Rear suspension / wheel alignment				
Toe-in (+) / Toe-out (–)....................mm [°]	[0 ± 25']	_	_	
Camber	-20' ± 30'			

FORD

	Galaxy 1.9 TDi 1996 to 2000	Maverick 2.7 TD 1993 to 1996	Maverick 2.7 TDi 1996 to 1998	P100 1.8 Turbo 1989 to 1992
Torque wrench settings				
Cylinder head - stage 1 ....Nm	40 N	42[1]	42[1]	35 to 40
- stage 2 ....Nm	60	57	57	70 to 75
- stage 3 ....Nm	+ 90°	+ 90°	+ 90°	Wait 1 min.
- stage 4 ....Nm	+ 90°	–	–	70 to 75
- stage 5 ....Nm	–	–	–	+ 90°
- stage 6 ....Nm	–	–	–	–
Big-end bearings...........................Nm	30 + 90° N	63 N	63 N	25 to 30[1]
Main bearings...............................Nm	65 + 90° N	172	172	27 + 75°
Crankshaft pulley bolt ....Nm	90 + 90° N	309	309	180
Camshaft pulley bolt ....Nm	45	–	–	M6: 9 to 11[2]
Flywheel [driveplate] bolt...............Nm	60 + 90° N	152	152	15 to 20[3]
Front hubs ....Nm	150 + 90°	WSM	WSM	390 to 450
Rear hubs ....Nm	200	WSM	WSM	280 to 300
Wheel nuts / bolts ....Nm	140	123	123	85 to 90
Glow plugs ....Nm	15	18	18	25 to 30
Clutch pressure plate bolts ....Nm	25	26	26	25 to 33
Injection pump sprocket.....................Nm	55	64	64	20 to 25
Injectors......................................Nm	20	59	59	70 to 80
Injection pump mounting bolts ....Nm	25	23	23	18 to 22
Injector pipe unions............................Nm	25	23	23	15 to 25
Capacities				
Engine oil & filter............................litres	4.3	7.2	7.2	5.5
Gearbox..litres	2.0	3.5 Transfer: 2.3	3.5 Transfer: 2.3	1.25 MT75: 1.2
Automatic transmission ....litres	–	–	–	–
Final drive ....litres	WT	Rear: 2.8	Rear: 2.8	1.4
Cooling system................................litres	9.2	10.0	10.0	9.5
Fuel tank.......................................litres	70	3D: 72 5D: 80	3D: 72 5D: 80	60

Notes

Galaxy 1.9 TDi 1996 to 2000
[1]215/60x15: 2.3 / 2.1
Maverick 2.7 TD 1993 to 1996
[1]Re-use once only
[2]5D: 1.9 / 2.5
Maverick 2.7 TDi 1996 to 1998
[1]Re-use once only
[2]5D: 1.9 / 2.5
P100 1.8 Turbo 1989 to 1992
[1]+ 55 to 60° + 15 to 20°
[2]M8: 27 to 33
[3]+ 35 to 40° + 35 to 40°

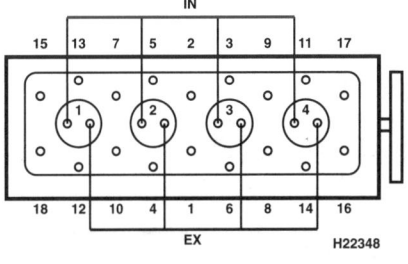

1896 cm³

2663 cm³

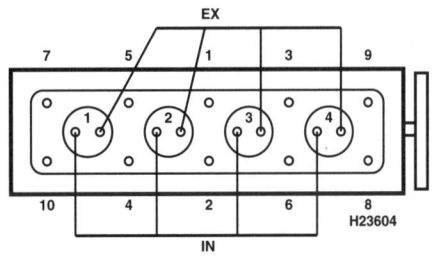

1753 cm³

– Not applicable, or information not available

FORD

	Transit 2.5 1989 to 1991	Transit 2.5 1989 to 1991	Transit 2.5 Turbo 1989 to 1994	Transit 2.5 1991 to 1993
Engine				
Engine type/code........................	4CA OHV 52kW	4CA OHV 52kW	4AE OHV Turbo 60kW	4FA OHV 52kW
Capacity (cm^3) / cylinders........................	2496 / 4	2496 / 4	2496 / 4	2498 / 4
Compression ration / pressure ....bar	19.0 / 33.8	19.0 / 33.8	19.0 / 33.8	19.0 / 33
Torque output ....Nm	0	0	0	0
Oil pressure ....idle [running] bar	[2.7 to 4.8]	[2.7 to 4.8]	[2.7 to 4.8]	1.0
Oil temperature ....°C	80	80	80	80
Valve clearances - inlet (mm)	0.20	0.20	0.20	0.20
- exhaust (mm)	0.38	0.38	0.40	0.38
Injection order	1-2-4-3	1-2-4-3	1-2-4-3	1-2-4-3
No. 1 cylinder position	–	TBE	TBE	–
Cooling system				
Thermostat opening temperature ....°C	82	82	82	88 to 102
Radiator cap pressure ....bar	0.85 to 1.10	0.85 to 1.10	0.85 to 1.10	1.0
Fuel system				
Idle speed ....rpm	800 to 850	800 to 850	825 ± 25	825 ± 25
Maximum (no load) speed ....rpm	4320 to 4480	4320 to 4560	4320 to 4560	4480
Smoke test/opacity ....M^{-1} %	2.5	2.5	2.5	2.5
Static timing method........................	Refer to wsm	Refer to wsm	Refer to wsm	Refer to wsm
Timing dimension........................mm	–	–	–	–
Crankshaft position ....mm [°]	–	[11] BTDC	[11] BTDC	–
Turbo type / ref / pressure ....bar	–	–	0.67 bar	–
Injection pump make	CAV	Bosch	Lucas	Bosch
Injection pump part no........................	–	0460 414 051	–	0460 414 083
Injector Make / type	CAV	Bosch	–	–
Injector part no........................	–	–	–	–
Injection type........................	Direct	Direct	Direct EPIC	Direct VE
Injection opening pressure, New [used]...bar	275 [241]	275 [241]	275 [241]	275 to 280 [240 to 266]
Glow plugs				
Maker	Beru	Beru	Beru	Motorcraft
Type	GF859	GF859	GF859	EZD26
Nominal rating........................V/A	–	–	–	–
Brakes				
minimum friction material thickness				
Front........................mm	1.5	1.5	1.5	1.5
Rear........................mm	1.0	1.0	1.0	1.0
Tyres - Saloon / Hatch........................Size	–	–	–	–
- Estate / Van........................Size	185x14: 195x14	185x14: 195x14	185x14: 195x14	185x14: 195x14: 215/70x15^1
Pressure - front / rear - Saloon / Hatch ...bar	–	–	–	–
- Estate / Van ......bar	Owners Handbook	Owners handbook	Owners handbook	WSM
Front suspension / wheel alignment				
Toe-in (+) / Toe-out (–)........................mm [°]	0 to 1.6	0 to 1.6	0 to 1.6	0 ± 0.5
Camber	-30' to 2°30'^{1}	-30' to 2°30'^{1}	-30' to 2°30'^{1}	WSM
Castor	15' to 4°30'^{2}	15' to 4°30'^{2}	15' to 4°30'^{2}	WSM
King pin inclination........................				
Rear suspension / wheel alignment				
Toe-in (+) / Toe-out (–)........................mm [°]	–	–	–	–
Camber	–	–	–	–

FORD

	Transit 2.5 1989 to 1991	Transit 2.5 1989 to 1991	Transit 2.5 Turbo 1989 to 1994	Transit 2.5 1991 to 1993
Torque wrench settings				
Cylinder head - stage 1 ....Nm	70	70	70	70
- stage 2 ....Nm	Slacken	Slacken	Slacken	Slacken
- stage 3 ....Nm	70	70	70	70
- stage 4 ....Nm	+ 90°	+ 90°	+ 90°	+ 90°
- stage 5 ....Nm	–	–	–	–
- stage 6 ....Nm	–	–	–	–
Big-end bearings.............................Nm	47 to 65³	47 to 65³	47 to 65³	72 N
Main bearings.................................Nm	110 to 126⁴	110 to 126⁴	110 to 126⁴	180
Crankshaft pulley bolt ....Nm	M14: 144 to 184⁵	M14: 144 to 184⁵	M14: 144 to 184⁵	320
Camshaft pulley bolt ....Nm	8 to 12 + 60°	8 to 12 + 60°	8 to 12 + 60°	12 + 60°
Flywheel [driveplate] bolt..................Nm	59 to 67	59 to 67	59 to 67	65
Front hubs ....Nm	WSM	WSM	WSM	WSM
Rear hubs ....Nm	WSM	WSM	WSM	WSM
Wheel nuts / bolts ....Nm	5 stud: 75 to 95⁶	5 stud: 75 to 95⁶	5 stud: 75 to 95⁶	5 stud: 80 6 stud: 168
Glow plugs ....Nm	–	–	–	–
Clutch pressure plate bolts ....Nm	17 to 21	17 to 21	17 to 21	19
Injection pump sprocket....................Nm	22 to 27	22 to 27	22 to 27	27
Injectors...Nm	40	40	40	40
Injection pump mounting bolts ....Nm	21 to 26	21 to 26	21 to 26	25
Injector pipe unions...........................Nm	18 to 20	18 to 20	18 to 20	20
Capacities				
Engine oil & filter..............................litres	6.2	6.2	6.2	7.0
Gearbox..litres	Type F: 1.5⁷	Type F: 1.5⁷	Type F: 1.5⁷	1.25
Automatic transmission ....litres	6.3 A4LD: 8.5	6.3 A4LD: 8.5	6.3 A4LD: 8.5	8.5
Final drive ....litres	Type G: 1.7⁸	Type G: 1.7⁸	Type G: 1.7⁸	Type G: 1.7²
Cooling system.................................litres	11.6	11.6	11.6	8.4
Fuel tank...litres	68	68	68	68

Notes

Transit 2.5 1989 to 1991
¹Beam axle: 30' to 1°30'
²Beam axle: 2° to 5°45'
³Then 65 to 72
⁴Stage 2: Centre: 140 to 155. Others: 126 to 140
⁵M24: 312 to 346
⁶6 stud: 155 to 180
⁷Type G: 2.0 [Type N: 1.5]. O/D: 3.1. MT 75: 1.2
⁸Type H axle: 2.7. Type F: 1.4
Transit 2.5 1989 to 1991
¹Beam axle: +30' to 1°30'
²Beam axle: +2° to 5°45'
³Then 65 to 72
⁴Stage 2: Centre: 140 to 155. Others: 126 to 140
⁵M24: 312 to 346
⁶6 stud: 155 to 180
⁷Type G: 2.0 [Type N: 1.5]. O/D: 3.1. MT 75: 1.2
⁸Type H axle: 2.7. Type F: 1.4
Transit 2.5 Turbo 1989 to 1994
¹Beam axle: +30' to 1°30'
²Beam axle: +2° to 5°45'
³Then 65 to 72
⁴Stage 2: Centre: 140 to 155. Others: 126 to 140
⁵M24: 312 to 346
⁶6 stud: 155 to 180
⁷Type G: 2.0 [Type N: 1.5]. O/D: 3.1. MT 75: 1.2
⁸Type H axle: 2.7. Type F: 1.4
Transit 2.5 1991 to 1993
¹185x15: 225/70x15 may be fitted
²Type H axle: 2.7

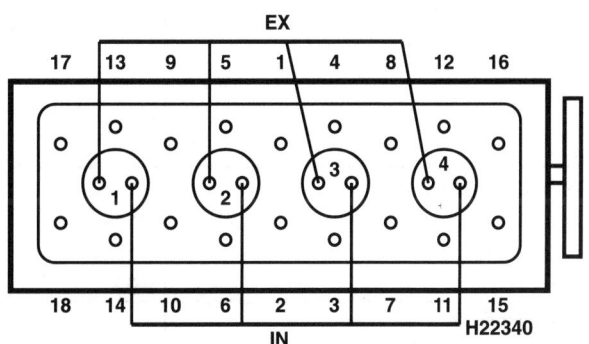

2496 cm³ / 2498 cm³

– Not applicable, or information not available

FORD

	Transit 2.5 1991 to 1993	Transit 2.5 1991 to 1993	Transit 2.5 Turbo 1991 to 1994	Transit 2.5 Turbo 1991 to 1994
Engine				
Engine type/code..........................	4DA OHV 59kW	4DA OHV 59kW	4GA OHV Turbo 63kW	4EA OHV Turbo 74kW
Capacity (cm³) / cylinders.............	2498 / 4	2498 / 4	2498 / 4	2498 / 4
Compression ration / pressure ....bar	19.0 / 33	19.0 / 33	18.3 /	18.3 / 33.8
Torque output ....Nm	0	0	200	0
Oil pressure ....idle [running] bar	1.0	1.0	1.0	1.0 [2.7 to 4.8]
Oil temperature ....°C	80	80	80	80
Valve clearances - inlet (mm)	0.20	0.20	0.20	0.20
- exhaust (mm)	0.38	0.38	0.38	0.40
Injection order	1-2-4-3	1-2-4-3	1-2-4-3	1-3-4-2
No. 1 cylinder position	–	–	–	TBE
Cooling system				
Thermostat opening temperature ....°C	88 to 102	88 to 102	88 to 102	88 to 102
Radiator cap pressure ....bar	1.0	1.0	1.0	1.0
Fuel system				
Idle speed ....rpm	825 ± 25	825 ± 25	850 ± 25	800 to 90
Maximum (no load) speed ....rpm	4480	4480	4480	4480
Smoke test/opacity ....M⁻¹ %	2.5	2.5	2.5	2.5
Static timing method..............................	Refer to wsm	Refer to wsm	Refer to wsm	Refer to wsm
Timing dimension........................mm	–	–	–	–
Crankshaft position ....mm [°]	–	–	–	–
Turbo type / ref / pressure ....bar	–	–	0.64 bar	0.67 bar
Injection pump make	Bosch	CAV	Lucas	Lucas
Injection pump part no...........................	0460 414 083	–	–	–
Injector Make / type	–	–	–	–
Injector part no................................	–	–	–	Stanadyne
Injection type.................................	Direct VE	Direct	Direct EPIC	EPIC
Injection opening pressure, New [used]...bar	275 to 280 [240 to 266]	275 to 280 [240 to 266]	280 [266]	280 [260]
Glow plugs				
Maker	Motorcraft	Motorcraft	Motorcraft	Motorcraft
Type	EZD26	EZD26	EZD31	EZD26
Nominal rating.........................V/A	–	–	–	–
Brakes				
minimum friction material thickness				
Front........................mm	1.5	1.5	1.5	1.5
Rear........................mm	1.0	1.0	1.0	1.0
Tyres - Saloon / Hatch.....................Size	–	–	–	–
- Estate / Van.....................Size	185x14: 195x14: 215/70x15'	185x14: 195x14: 215/70x15'	185x14: 195x14: 215/70x15'	185x14: 195x14: 215/70x15'
Pressure - front / rear - Saloon / Hatch ...bar	–	–	–	–
- Estate / Van ......bar	WSM	WSM	WSM	WSM
Front suspension / wheel alignment				
Toe-in (+) / Toe-out (–)....................mm [°]	0 ± 0.5	0 ± 0.5	0 ± 0.5	0 ± 0.5
Camber	WSM	WSM	WSM	WSM
Castor	WSM	WSM	WSM	WSM
King pin inclination..............................	–	–	–	–
Rear suspension / wheel alignment				
Toe-in (+) / Toe-out (–)....................mm [°]	–	–	–	–
Camber	–	–	–	–

FORD

	Transit 2.5 1991 to 1993	Transit 2.5 1991 to 1993	Transit 2.5 Turbo 1991 to 1994	Transit 2.5 Turbo 1991 to 1994
Torque wrench settings				
Cylinder head - stage 1 ....Nm	70 N	70 N	70 N	70 N
- stage 2 ....Nm	Slacken	Slacken	Slacken	Slacken
- stage 3 ....Nm	70	70	70	70
- stage 4 ....Nm	+ 90°	+ 90°	+ 90°	+ 90°
- stage 5 ....Nm	–	–	–	–
- stage 6 ....Nm	–	–	–	–
Big-end bearings......................Nm	72 N	72 N	64 N	72 N
Main bearings......................Nm	180	180	170	170
Crankshaft pulley bolt ....Nm	320	320	320	185
Camshaft pulley bolt ....Nm	12 + 60°	12 + 60°	12 + 60°	12 + 60°
Flywheel [driveplate] bolt.....................Nm	65	65	67	25 + 30°
Front hubs ....Nm	WSM	WSM	WSM	WSM
Rear hubs ....Nm	WSM	WSM	WSM	WSM
Wheel nuts / bolts ....Nm	5 stud: 80 6 stud: 168	5 stud: 80 6 stud: 168	5 stud: 80 6 stud: 168	5 stud: 80 6 stud: 168
Glow plugs ....Nm	–	–	–	21 to 27
Clutch pressure plate bolts ....Nm	19	19	19	19
Injection pump sprocket......................Nm	27	27	27	–
Injectors......................Nm	40	40	40	–
Injection pump mounting bolts ....Nm	25	25	–	42
Injector pipe unions......................Nm	20	20	20	20
Capacities				
Engine oil & filter......................litres	7.0	7.0	6.0	6.0
Gearbox......................litres	1.25	1.25	1.25	1.25
Automatic transmission ....litres	8.5	8.5	8.5	8.5
Final drive ....litres	Type G: 1.7²	Type G: 1.7²	Type G: 1.7²	Type G: 1.7²
Cooling system......................litres	8.4	8.4	8.4	8.4
Fuel tank......................litres	68	68	68	68

Notes

Transit 2.5 1991 to 1993
¹185x15: 225/70x15 may be fitted
²Type H axle: 2.7
Transit 2.5 1991 to 1993
¹185x15: 225/70x15 may be fitted
²Type H axle: 2.7
Transit 2.5 Turbo 1991 to 1994
¹185x15: 225/70x15 may be fitted
²Type H axle: 2.7
Transit 2.5 Turbo 1991 to 1994
¹185x15: 225/70x15 may be fitted
²Type H axle: 2.7

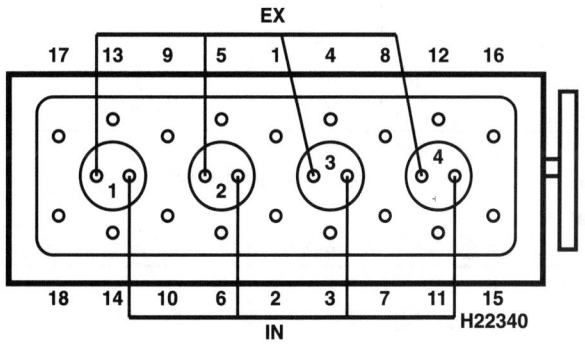

2498 cm³

– Not applicable, or information not
  available

FORD

	Transit 2.5 Turbo 1995 to 2000	Transit 2.5 TDi 1997 to 2000	Transit 2.5 TDi 1997 to 2000	Transit 2.5 Di 1993 to 2000
Engine				
Engine type/code....................................	4EB/4EC OHV Turbo 74kW	4GC/4GB OHV Turbo 63kW	4GD/4GE/4GF OHV Turbo 63kW	4FB/4FC/4FD OHV 52kW
Capacity (cm³) / cylinders	2498 / 4	2496 / 4	2496 / 4	2496 / 4
Compression ration / pressure ....bar	18.3 /	18.3 / _	19.0 / _	20.8 /
Torque output ....Nm	220	200	200	145
Oil pressure ....idle [running] bar	[2.7 to 4.8 @ 4000]	1.6 [3.1 @ 2000]	1.6 [3.1 @ 2000]	[2.7 to 4.8]
Oil temperature ....°C	80	80	80	80
Valve clearances - inlet (mm)	0.20	0.20	0.20	0.20
- exhaust (mm)	0.38	0.38	0.38	0.38
Injection order	1-2-4-3	1-2-4-3	1-2-4-3	1-2-4-3
No. 1 cylinder position	_	_	_	TBE
Cooling system				
Thermostat opening temperature ....°C	88 to 102	_	_	_
Radiator cap pressure ....bar	1.0	refer to cap	refer to cap	refer to cap
Fuel system				
Idle speed ....rpm	850 ± 25	850 ± 50	850 ± 50	850 ± 25
Maximum (no load) speed ....rpm	4400	4400	4400	4280 to 4480
Smoke test/opacity ....M⁻¹ %	2.5	2.0	2.0	2.5
Static timing method...............................	Refer to wsm	Refer to wsm	Refer to wsm	Refer to wsm
Timing dimension..............................mm	_	_	_	_
Crankshaft position ....mm [°]	_	_	_	_
Turbo type / ref / pressure ....bar	_	_	_	_
Injection pump make	Lucas	Lucas	Lucas	Bosch
Injection pump part no............................	_	_	_	VE
Injector Make / type	_	_	_	_
Injector part no..	_	_	_	_
Injection type...	Direct EPIC	Direct EPIC	Direct EPIC	Direct
Injection opening pressure, New [used]...bar	_	_	280 [260]	_
Glow plugs				
Maker	Motorcraft	Motorcraft	Motorcraft	_
Type	EZD31	EZD31	EZD31	_
Nominal rating....................................V/A	_	_	_	_
Brakes				
minimum friction material thickness				
Front..mm	1.5	_	_	_
Rear...mm	1.0	1.0	1.0	1.0
Tyres - Saloon / Hatch......................Size	_	_	_	_
- Estate / Van......................Size	185x14: 195x14: 215/70x15'	195x14	195x14	195x14
Pressure - front / rear - Saloon / Hatch ...bar				
- Estate / Van .....bar	Owners handbook	Owners handbook	Owners handbook	2.8 / 2.6
Front suspension / wheel alignment				
Toe-in (+) / Toe-out (–).....................mm [°]	0 ± 0.5	0.0 ± 1.0	0.0 ± 1.0	0.0 ± 1.0
Camber	WSM	-52' to 1°8'	-52' to 1°8'	-0°52' to 1°08'
Castor	WSM	1°9' to 4°9'	1°9' to 4°9'	1°09' to 4°09'
King pin inclination..................................	_	_	_	_
Rear suspension / wheel alignment				
Toe-in (+) / Toe-out (–)....................mm [°]	_	_	_	_
Camber				

FORD

Torque wrench settings	Transit 2.5 Turbo 1995 to 2000	Transit 2.5 TDi 1997 to 2000	Transit 2.5 TDi 1997 to 2000	Transit 2.5 Di 1993 to 2000
Cylinder head - stage 1 ....Nm	70	70	70	70
- stage 2 ....Nm	Slacken	70	70	70
- stage 3 ....Nm	70	+ 90°	+ 90°	+ 90°
- stage 4 ....Nm	+ 90°	–	–	–
- stage 5 ....Nm	–	–	–	–
- stage 6 ....Nm	–	–	–	–
Big-end bearings.....................Nm	70 N	69 N	69 N	69 N
Main bearings.....................Nm	120	84, 170	84, 170	84, 170
Crankshaft pulley bolt ....Nm	163	163	163	163
Camshaft pulley bolt ....Nm	160	–	–	10 + 60°
Flywheel [driveplate] bolt.....................Nm	21 + 30°	21 + 30°	21 + 30°	21 + 28°
Front hubs ....Nm	WSM	WSM	WSM	WSM
Rear hubs ....Nm	WSM	WSM	WSM	WSM
Wheel nuts / bolts ....Nm	5 stud: 80. 6 stud: 168	80 6 stud: 168	80 6 stud: 168	85 6 stud: 168
Glow plugs ....Nm	–	18	18	–
Clutch pressure plate bolts ....Nm	30	30	30	30
Injection pump sprocket.....................Nm	–	–	25	–
Injectors.....................Nm	–	40	40	–
Injection pump mounting bolts ....Nm	–	–	25	–
Injector pipe unions.....................Nm	–	–	19	–
Capacities				
Engine oil & filter.....................litres	6.0	6.0	6.0	6.0
Gearbox.....................litres	1.25	1.25	1.25	1.25
Automatic transmission ....litres	8.5	–	–	–
Final drive ....litres	Type G: 1.7²	1.72 Type 34: 2.70	1.72 Type 34: 2.70	1.72
Cooling system.....................litres	8.4	7.9	11.6	11.6
Fuel tank.....................litres	68	68	68	68

Notes

Transit 2.5 Turbo 1995 to 2000
¹185x15: 225/70x15 may be fitted
²Type H axle: 2.7

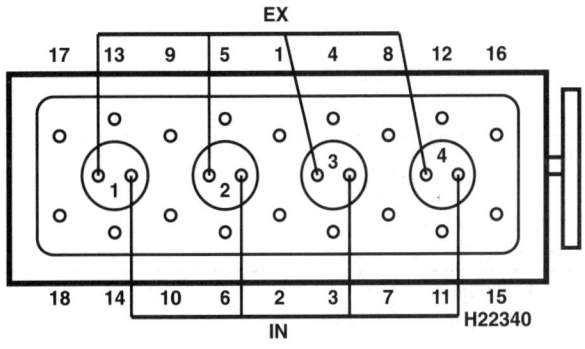

2496 cm³ / 2498 cm³

– Not applicable, or information not available

FORD

	Transit 2.5 Di 1996 to 2000	Transit 2.5 Di 1997 to 2000	Transit 2.5 TDi 1995 to 2000	Transit 2.5 TCi 1995 to 2000
Engine				
Engine type/code....................	4HB OHV 56kW	4HC OHV 56kW	4HA/4HD OHV Turbo 55kW	4ED OHV Turbo 85kW
Capacity (cm³) / cylinders.......................	2496 / 4	2496 / 4	2496 / 4	2496 / 4
Compression ration / pressure ....bar	20.6 /	20.8 /	20.6 /	20.6 /
Torque output ....Nm	168	168	145	255
Oil pressure ....idle [running] bar	2.7	2.7	2.7	2.7
Oil temperature ....°C	80	80	80	80
Valve clearances - inlet (mm)	0.20	0.20	0.20	0.20
- exhaust (mm)	0.38	0.38	0.40	0.40
Injection order..	1-2-4-3	1-2-4-3	1-2-4-3	1-2-4-3
No. 1 cylinder position	TBE	TBE	TBE	TBE
Cooling system				
Thermostat opening temperature ....°C	–	–	–	–
Radiator cap pressure ....bar	refer to cap	refer to cap	refer to cap	refer to cap
Fuel system				
Idle speed ....rpm	800 to 850	850 ± 25	850 ± 50	850 ± 50
Maximum (no load) speed ....rpm	4320 to 4480	4320 to 4480	4280 to 4480	4280 to 4480
Smoke test/opacity ....M⁻¹ %	2.5	2.5	2.0	2.0
Static timing method...............................	Refer to wsm	Refer to wsm	Refer to wsm	Refer to wsm
Timing dimension................................mm	–	–	–	–
Crankshaft position ....mm [°]	–	–	–	–
Turbo type / ref / pressure ....bar	–	–	–	–
Injection pump make	Bosch	–	Lucas	Lucas
Injection pump part no.............................	VE	–	–	–
Injector Make / type	–	–	–	–
Injector part no.......................................				
Injection type..	Direct	Direct	Direct EPIC	Direct EPIC
Injection opening pressure, New [used]...bar	–			
Glow plugs				
Maker	Motorcraft	Motorcraft	Motorcraft	Motorcraft
Type	EDZ31	EDZ31	EDZ31	EDZ31
Nominal rating......................................V/A	–	–	–	–
Brakes				
minimum friction material thickness				
Front....................................mm	1.0	–	–	1.0
Rear.....................................mm	1.5	1.0	1.0	1.0
Tyres - Saloon / Hatch......................Size	–	–	–	–
- Estate / Van.....................Size	215/70x15	215/70x15	195x14	195x14
Pressure - front / rear - Saloon / Hatch ...bar				
- Estate / Van ......bar	2.7 / 4.2	2.7 / 4.2	2.8 / 2.6	2.8 / 2.6
Front suspension / wheel alignment				
Toe-in (+) / Toe-out (–)....................mm [°]	0.0 ± 1.0	0.0 ± 1.0	0.0 ± 1.0	0.0 ± 1.0
Camber	-0°53' to 1° 06'	-0°53' to 1° 06'	-0°52' to 1°08'	-0°52' to 1°08'
Castor	-0°04' to 2° 56'	-0°04' to 2° 56'	1°09' to 4°09'	1°09' to 4°09'
King pin inclination.................................				
Rear suspension / wheel alignment				
Toe-in (+) / Toe-out (–)....................mm [°]	–	–	–	–
Camber				

FORD

	Transit 2.5 Di 1996 to 2000	Transit 2.5 Di 1997 to 2000	Transit 2.5 TDi 1995 to 2000	Transit 2.5 TCi 1995 to 2000
Torque wrench settings				
Cylinder head - stage 1 ....Nm	70	70	70	30' WSM
- stage 2 ....Nm	70	70	70	+ 70°
- stage 3 ....Nm	+ 90°	+ 90°	+ 90°	+ 70°
- stage 4 ....Nm	–	–	–	–
- stage 5 ....Nm	–	–	–	–
- stage 6 ....Nm	–	–	–	–
Big-end bearings.................................Nm	69 N	69 N	69 N	69 N
Main bearings.....................................Nm	84, 170	84, 170	84, 170	84, 170
Crankshaft pulley bolt ....Nm	163	163	163	163
Camshaft pulley bolt ....Nm	160	–	–	–
Flywheel [driveplate] bolt.....................Nm	21 + 20°	21 + 20°	21 + 28°	21 + 28°
Front hubs ....Nm	WSM	WSM	WSM	WSM
Rear hubs ....Nm	WSM	WSM	WSM	WSM
Wheel nuts / bolts ....Nm	85 6 stud: 168	85 6 stud: 168	85 6 stud: 168	85 6 stud: 168
Glow plugs ....Nm	–	–	–	–
Clutch pressure plate bolts ....Nm	30	30	30	30
Injection pump sprocket......................Nm	25	–	–	25
Injectors...Nm	40	–	–	40
Injection pump mounting bolts ....Nm	24	23	–	24
Injector pipe unions............................Nm	20	–	–	20
Capacities				
Engine oil & filter................................litres	6.0	6.0	6.0	6.0
Gearbox..litres	1.25	1.25	1.25	1.25
Automatic transmission ....litres	–	–	–	–
Final drive ....litres	1.75	1.75	1.72	1.72
Cooling system...................................litres	11.6	11.6	11.6	11.6
Fuel tank...litres	68	68	68	68

Notes

Transit 2.5 TCi 1995 to 2000
'M12 bolts: 30 + 85°

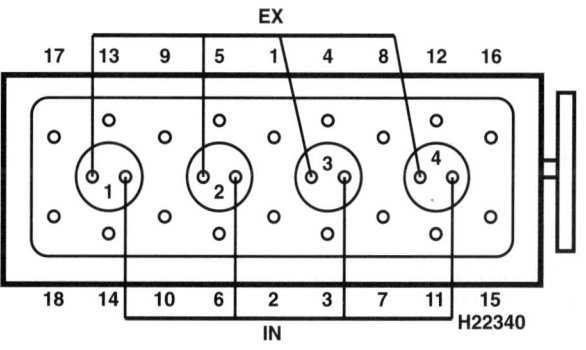

2496 cm³

– Not applicable, or information not available

FSO

	Caro 1.9D 1994 to 1999			
Engine				
Engine type/code....................	XUD9 EJ SOHC 51kW			
Capacity (cm³) / cylinders........................	1905 / 4			
Compression ration / pressure ....bar	23.0 /			
Torque output ....Nm	0			
Oil pressure ....idle [running] bar	[3.4 @ 4000]			
Oil temperature ....°C	80			
Valve clearances - inlet (mm)	0.07 to 0.23			
- exhaust (mm)	0.22 to 0.38			
Injection order	1-3-4-2			
No. 1 cylinder position	FE			
Cooling system				
Thermostat opening temperature ....°C	_			
Radiator cap pressure ....bar	_			
Fuel system				
Idle speed ....rpm	750 to 800			
Maximum (no load) speed ....rpm	5100			
Smoke test/opacity ....M⁻¹ %	2.5			
Static timing method................................	Plunger travel			
Timing dimension..............................mm	0.9			
Crankshaft position ....mm [°]	TDC			
Turbo type / ref / pressure ....bar	_			
Injection pump make	Bosch			
Injection pump part no.............................	VER 272 2			
Injector Make / type	Bosch			
Injector part no..	DNOSD287			
Injection type...	Indirect			
Injection opening pressure, New [used]...bar	125 to 135			
Glow plugs				
Maker	Bosch			
Type	0250 201 019			
Nominal rating.................................V/A	11 / 13			
Brakes				
minimum friction material thickness				
Front...mm	1.0			
Rear...mm	1.5			
Tyres - Saloon / Hatch......................Size	_			
- Estate / Van...........................Size	_			
Pressure - front / rear - Saloon / Hatch...bar	_			
- Estate / Van ......bar	_			
Front suspension / wheel alignment				
Toe-in (+) / Toe-out (–)....................mm [°]	3 ± 1			
Camber	0°30' ± 10'			
Castor	4°30' ± 30'			
King pin inclination..................................	_			
Rear suspension / wheel alignment				
Toe-in (+) / Toe-out (–)....................mm [°]	_			
Camber	_			

FSO

	Caro 1.9D 1994 to 1999			
Torque wrench settings				
Cylinder head - stage 1 ....Nm	30			
- stage 2 ....Nm	70			
- stage 3 ....Nm	+ 120°			
- stage 4 ....Nm	_			
- stage 5 ....Nm	_			
- stage 6 ....Nm	_			
Big-end bearings................................Nm	50 N			
Main bearings....................................Nm	70			
Crankshaft pulley bolt ....Nm	40 + 60°			
Camshaft pulley bolt ....Nm	35			
Flywheel [driveplate] bolt....................Nm	_			
Front hubs ....Nm	_			
Rear hubs ....Nm	_			
Wheel nuts / bolts ....Nm	85			
Glow plugs ....Nm	22			
Clutch pressure plate bolts ....Nm	22			
Injection pump sprocket.....................Nm	_			
Injectors...Nm	_			
Injection pump mounting bolts ....Nm	18			
Injector pipe unions...........................Nm	22			
Capacities				
Engine oil & filter................................litres	5.0			
Gearbox..litres	1.6			
Automatic transmission ....litres	_			
Final drive ....litres	WT			
Cooling system.................................litres	_			
Fuel tank..litres	_			

Notes

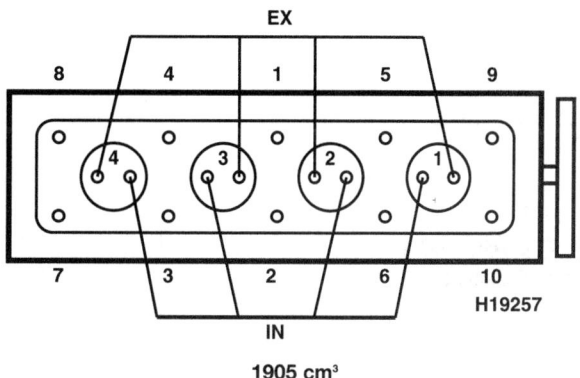

– Not applicable, or information not available

HONDA

	Civic 2.0 TDi 1997 to 2000	Accord 2.0 TDi 1996 to 1998		
Engine				
Engine type/code.....................................	20T2R/N' SOHC Turbo 64kW	20T2N SOHC Turbo 76kW		
Capacity (cm³) / cylinders......................	1994 / 4	1994 / 4		
Compression ration / pressure ....bar	19.5 /	19.5 /		
Torque output ....Nm	170	210		
Oil pressure ....idle [running] bar	0.7	0.7 [3.8 @ 3000]		
Oil temperature ....°C	80	80		
Valve clearances - inlet (mm)	0: Hyd.	0: Hyd.		
- exhaust (mm)	0: Hyd.	0: Hyd.		
Injection order	1-3-4-2	1-3-4-2		
No. 1 cylinder position	–	–		
Cooling system				
Thermostat opening temperature ....°C	82	82		
Radiator cap pressure ....bar	0.9	1.1		
Fuel system				
Idle speed ....rpm	850 ± 50	805 ± 50		
Maximum (no load) speed ....rpm	–	–		
Smoke test/opacity ....M⁻¹ %	2.0	2.0		
Static timing method...............................				
Timing dimension.................................mm	Computer controlled	Computer controlled		
Crankshaft position ....mm [°]	–	–		
Turbo type / ref / pressure ....bar	–	–		
Injection pump make	Bosch	Bosch		
Injection pump part no............................	VP21	L 580		
Injector Make / type	–	Bosch		
Injector part no..	–	KBAL 70P45		
Injection type...	ECU controlled	Direct EDC		
Injection opening pressure, New [used]...bar	200	200 to 330		
Glow plugs				
Maker	–	Bosch		
Type	–	0250 202 025		
Nominal rating.................................V/A	–	–		
Brakes				
minimum friction material thickness				
Front..mm	1.6	1.6		
Rear..mm	1.6	1.6		
Tyres - Saloon / Hatch......................Size	185/60x14	185/65x15		
- Estate / Van...........................Size	–	–		
Pressure - front / rear - Saloon / Hatch ...bar	2.5 / 2.2	2.2 / 2.2		
- Estate / Van .....bar	–	–		
Front suspension / wheel alignment				
Toe-in (+) / Toe-out (–)......................mm [°]	[0° ± 12']	0.0 ± 3.0		
Camber	0° ± 1°	0° ± 1°		
Castor	1° 10'	3° ± 1°		
King pin inclination..................................				
Rear suspension / wheel alignment				
Toe-in (+) / Toe-out (–)......................mm [°]	2.3	-2.0 ± 2.0		
Camber	-0° 50'	-0°30' ± 30'		

HONDA

Torque wrench settings	Civic 2.0 TDi 1997 to 2000	Accord 2.0 TDi 1996 to 1998		
Cylinder head - stage 1 ....Nm	30²	30		
- stage 2 ....Nm	60	65		
- stage 3 ....Nm	+ 90°	+ 90°		
- stage 4 ....Nm	+ 90°	+ 90°		
- stage 5 ....Nm	–	–		
- stage 6 ....Nm	–	–		
Big-end bearings.................................Nm	48 N Bolts: 20 N	48 N		
Main bearings.....................................Nm	112	112		
Crankshaft pulley bolt ....Nm	63	63 + 90°		
Camshaft pulley bolt ....Nm	20	20 + 90°		
Flywheel [driveplate] bolt....................Nm	103	15 + 90° N		
Front hubs..Nm	181 N	250		
Rear hubs...Nm	181 N	185		
Wheel nuts / bolts ....Nm	110	110		
Glow plugs ....Nm	20	20		
Clutch pressure plate bolts ....Nm	19	26		
Injection pump sprocket......................Nm	–	60		
Injectors...Nm	25	25		
Injection pump mounting bolts ....Nm	–	25		
Injector pipe unions............................Nm	20	20		
Capacities				
Engine oil & filter................................litres	4.0	5.0		
Gearbox..litres	2.0	2.2		
Automatic transmission ....litres	–	–		
Final drive ....litres	WT	WT		
Cooling system...................................litres	7.0	7.0		
Fuel tank...litres	55	65		

Notes

Civic 2.0 TDi 1997 to 2000
¹20T2N: Intercooled
²Bolt length: ≤ 243.41 mm

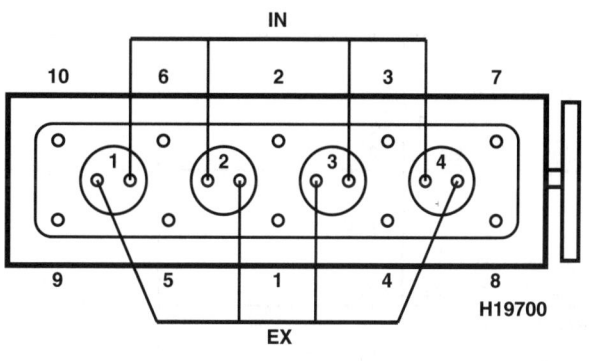

1994 cm³

– Not applicable, or information not available

HYUNDAI

	H100 2.5 D 1997 to 2000			
Engine				
Engine type/code.....................	D4BA SOHC 56kW			
Capacity (cm³) / cylinders.....................	2477 / 4			
Compression ration / pressure ....bar	21.0 / ≥19.2			
Torque output ....Nm	149			
Oil pressure ....idle [running] bar	_			
Oil temperature ....°C	_			
Valve clearances - inlet (mm)	0.25 H			
- exhaust (mm)	0.25 H			
Injection order	1-3-4-2			
No. 1 cylinder position	TBE			
Cooling system				
Thermostat opening temperature ....°C	82			
Radiator cap pressure ....bar	0.75 to 1.05			
Fuel system				
Idle speed ....rpm	750 ± 30			
Maximum (no load) speed ....rpm	5200			
Smoke test/opacity ....M⁻¹ %	2.5			
Static timing method............................	Rotor lift			
Timing dimension................................mm	1.0			
Crankshaft position ....mm [°]	[4°] ATDC			
Turbo type / ref / pressure ....bar	_			
Injection pump make	Lucas			
Injection pump part no............................				
Injector Make / type	Throttle type			
Injector part no..................................	_			
Injection type..................................	Indirect			
Injection opening pressure, New [used]...bar	120 to 130			
Glow plugs				
Maker	Beru			
Type	GN955			
Nominal rating.................................V/A	_			
Brakes				
minimum friction material thickness				
Front............................mm	2.0			
Rear.............................mm	1.0			
Tyres - Saloon / Hatch.......................Size	_			
- Estate / Van.....................Size	185x14			
Pressure - front / rear - Saloon / Hatch ...bar	_			
- Estate / Van ......bar	3.0 / 3.0			
Front suspension / wheel alignment				
Toe-in (+) / Toe-out (–).....................mm [°]	0.0 to 3.0			
Camber	0°30' ± 45'			
Castor	3° ± 1°			
King pin inclination............................	_			
Rear suspension / wheel alignment				
Toe-in (+) / Toe-out (–).....................mm [°]	_			
Camber	_			

Haynes THE BOOK ®

HYUNDAI

Torque wrench settings	H100 2.5 D 1997 to 2000			
Cylinder head - stage 1 ....Nm	105			
- stage 2 ....Nm	_			
- stage 3 ....Nm	_			
- stage 4 ....Nm	_			
- stage 5 ....Nm	_			
- stage 6 ....Nm	_			
Big-end bearings........................Nm	46 N			
Main bearings..........................Nm	80			
Crankshaft pulley bolt ....Nm	_			
Camshaft pulley bolt ....Nm	_			
Flywheel [driveplate] bolt................Nm	135			
Front hubs ....Nm	WSM			
Rear hubs ....Nm	WSM			
Wheel nuts / bolts ....Nm	130			
Glow plugs ....Nm	_			
Clutch pressure plate bolts ....Nm	20			
Injection pump sprocket................Nm	_			
Injectors............................Nm	55			
Injection pump mounting bolts ....Nm	_			
Injector pipe unions...................Nm	_			
Capacities				
Engine oil & filter......................litres	5.2			
Gearbox.............................litres	2.1			
Automatic transmission ....litres	_			
Final drive ....litres	1.2			
Cooling system.......................litres	8.7			
Fuel tank............................litres	55			

Notes

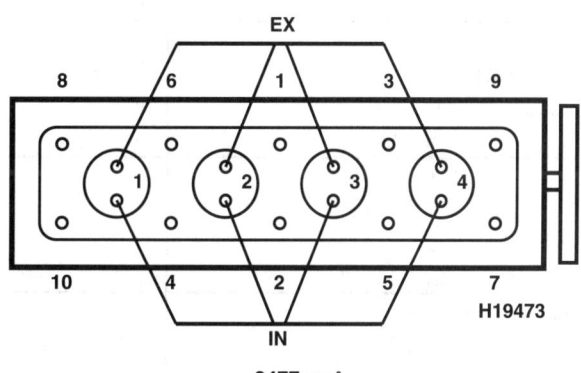

2477 cm³

– Not applicable, or information not available

ISUZU

	Trooper 2.8 (UBS55) 1988 to 1992	Trooper 2.8 Turbo (UBS55-J) 1988 to 1992	Trooper 2.8 Turbo 1989 to 1992	Trooper 3.0 TD 1998 to 2000
Engine				
Engine type/code.....................	4JB1 OHV 64kW	4JB1T OHV Turbo 71kW	4JB1TC + EGR OHV 78kW	4JX1 DOHC 16V Turbo 116kW
Capacity (cm^3) / cylinders......................	2771 / 4	2771 / 4	2771 / 4	2999 / 4
Compression ration / pressure ....bar	18.2 / ≥30.4	17.5 / ≥30.4	17.5 / ≥30.4	19.0 / ≥21.5
Torque output ....Nm	0	0	0	331
Oil pressure ....idle [running] bar	[3.9 to 4.9]	[3.9 to 4.9]	[3.9 to 4.9]	[3.9]
Oil temperature ....°C	50	50	50	80
Valve clearances - inlet (mm)	0.40	0.40	0.40	0.15
- exhaust (mm)	0.40	0.40	0.40	0.25
Injection order	1-3-4-2	1-3-4-2	1-3-4-2	1-3-4-2
No. 1 cylinder position	F	F	F	TBE
Cooling system				
Thermostat opening temperature ....°C	82	82	82	83
Radiator cap pressure ....bar	0.9 to 1.2	0.9 to 1.2	0.9 to 1.2	0.8 to 1.1
Fuel system				
Idle speed ....rpm	750 to 790 A/C: 900	750 to 790 A/C: 900	750 to 790	720 ± 30
Maximum (no load) speed ....rpm	–	4400 to 4800	4400 to 4800	–
Smoke test/opacity ....M^{-1} %	2.5	2.5	2.5	3.0
Static timing method..............................	Plunger travel	Plunger travel	Plunger travel	–
Timing dimension.............................mm	0.50	0.50	0.50	–
Crankshaft position ....mm [°]	[12]	[12] Belt drive: [10]	[12] Belt drive: [10]	
Turbo type / ref / pressure ....bar	–	–	–	IHI RHF5
Injection pump make	Diesel Kiki	Bosch	Bosch	–
Injection pump part no........................		9460 610 315	–	–
Injector Make / type	4 hole	Bosch		–
Injector part no........................	DN12SD12T or EF8511/9	DN12SD12T or EF8511/9	DN12SD12T or EF8511/9	–
Injection type..............................	Bosch VE type	Zexel VE4	Zexel Direct VE4	Direct common rail[1]
Injection opening pressure, New [used]...bar	185	185[1]	185[1]	350 to 1000
Glow plugs				
Maker	Bosch	Bosch	–	–
Type	0250 202 065	0250 202 065	–	–
Nominal rating.............................V/A	5.0 / 11.1	5.0 / 11.1	5.0 / 11.1	–
Brakes				
minimum friction material thickness				
Front...mm	1.0	1.0	1.0	–
Rear...mm	1.0	1.0	1.0	
Tyres - Saloon / Hatch......................Size	–	–	–	
- Estate / Van......................Size	215x15	215x15	215x15	245/70x16
Pressure - front / rear - Saloon / Hatch ...bar	–	–	–	–
- Estate / Van ......bar	1.8 / 1.8	1.8 / 1.8	1.8 / 1.8	–
Front suspension / wheel alignment				
Toe-in (+) / Toe-out (–)......................mm [°]	0 to 4.0	0 to 4.0	0 to 4.0	–
Camber	30' ± 30'	30' ± 30'	30' ± 30'	–
Castor	2°30' ± 45'	2°30' ± 45'	2°30' ± 45'	–
King pin inclination..............................	10° ± 1°	10° ± 1°	10° ± 1°	–
Rear suspension / wheel alignment				
Toe-in (+) / Toe-out (–)......................mm [°]	–	–	–	–
Camber	–	–	–	–

ISUZU

Torque wrench settings	Trooper 2.8 (UBS55) 1988 to 1992	Trooper 2.8 Turbo (UBS55-J) 1988 to 1992	Trooper 2.8 Turbo 1989 to 1992	Trooper 3.0 TD 1998 to 2000
Cylinder head - stage 1 ....Nm	49 ± 5	49 ± 5	49 ± 5	49
- stage 2 ....Nm	+ 60°	+ 60°	+ 60°	+ 60°
- stage 3 ....Nm	+ 60°	+ 60°	+ 60°	+ 60°
- stage 4 ....Nm	–	–	–	–
- stage 5 ....Nm	–	–	–	–
- stage 6 ....Nm	–	–	–	–
Big-end bearings.................................Nm	83 ± 5 N	83 ± 5 N	83 ± 5 N	29 + 50° N
Main bearings.................................Nm	166 ± 10	166 ± 10	166 ± 10	167
Crankshaft pulley bolt ....Nm	186 ± 17	186 ± 17	186 ± 17	WSM
Camshaft pulley bolt ....Nm	108 ± 10	108 ± 10	108 ± 10	40 + 80°
Flywheel [driveplate] bolt.....................Nm	117 ± 5	117 ± 5	117 ± 5	59 + 75°
Front hubs ....Nm	WSM	WSM	WSM	–
Rear hubs.................................Nm	WSM	WSM	WSM	–
Wheel nuts / bolts ....Nm	Steel: 78 to 118[1]	Steel: 78 to 118[2]	Steel: 78 to 118[2]	–
Glow plugs ....Nm	22 ± 2	22 ± 2	22 ± 2	15
Clutch pressure plate bolts ....Nm	16 to 20	16 to 20	16 to 20	–
Injection pump sprocket.....................Nm	64 ± 5	64 ± 5	64 ± 5	10
Injectors.................................Nm	32 to 44	32 to 44	32 to 44	30
Injection pump mounting bolts ....Nm	19 ± 5	19 ± 5	19 ± 5	–
Injector pipe unions.............................Nm	20 to 40	20 to 40	20 to 40	80
Capacities				
Engine oil & filter.................................litres	4.3	4.3	4.3	6.0
Gearbox.................................litres	2.95[2]	2.95 Transfer: 1.45	2.95 Transfer: 1.45	–
Automatic transmission ....litres	4.5[3]	4.5 Transfer: 0.8	4.5 Transfer: 0.8	–
Final drive.................................litres	Front: 1.5[4]	Front: 1.5 Rear: 1.8	Front: 1.5 Rear: 1.8	–
Cooling system.................................litres	10.3	10.3	10.3	5.8
Fuel tank.................................litres	83	83	83	85

Notes

Trooper 2.8 (UBS55) 1988 to 1992
[1]Alloy: 108 to 127
[2]Transfer box: 1.45
[3]Transfer box: 0.8
[4]Rear: 1.8
Trooper 2.8 Turbo (UBS55-J) 1988 to 1992
[1]Belt drive, two spring type: 1st pressure: 175, 2nd pressure: 275
[2]Alloy: 108 to 127
Trooper 2.8 Turbo 1989 to 1992
[1]Two spring type: 1st pressure: 175. 2nd pressure: 275
[2]Alloy: 108 to 127
Trooper 3.0 TD 1998 to 2000
[1]Caterpillar system.

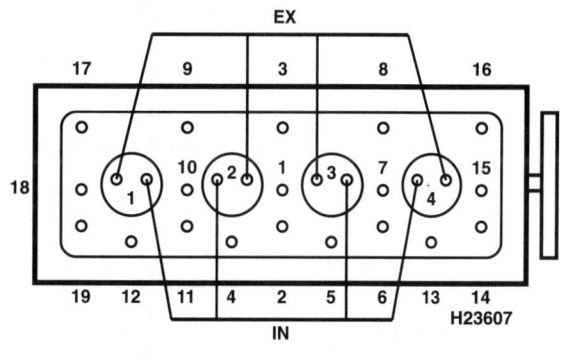

2771 cm³

– Not applicable, or information not available

ISUZU

	Trooper 3.1 TD 1992 to 1998			
Engine				
Engine type/code.............................	4JG2TC OHV Turbo 84kW			
Capacity (cm³) / cylinders.......................	3059 / 4			
Compression ration / pressure ....bar	20.0 / 30.4			
Torque output ....Nm	260			
Oil pressure ....idle [running] bar	–			
Oil temperature ....°C	–			
Valve clearances - inlet (mm)	0.40			
- exhaust (mm)	0.40			
Injection order	1-3-4-2			
No. 1 cylinder position	–			
Cooling system				
Thermostat opening temperature ....°C	74			
Radiator cap pressure ....bar	0.9 to 1.2			
Fuel system				
Idle speed ....rpm	720			
Maximum (no load) speed ....rpm	4250			
Smoke test/opacity ....M⁻¹ %	2.5			
Static timing method..............................	Plunger travel			
Timing dimension.................................mm	0.5			
Crankshaft position ....mm [°]	TDC			
Turbo type / ref / pressure ....bar	–			
Injection pump make	Bosch			
Injection pump part no...........................	VE			
Injector Make / type	–			
Injector part no......................................	–			
Injection type..	Indirect			
Injection opening pressure, New [used]...bar	147			
Glow plugs				
Maker	–			
Type	–			
Nominal rating ....V/A				
Brakes				
minimum friction material thickness				
Front...mm	1.0			
Rear..mm	1.0			
Tyres - Saloon / Hatch......................Size	–			
- Estate / Van............................Size	245/70x16			
Pressure - front / rear - Saloon / Hatch ...bar	–			
- Estate / Van ......bar	2.1 / 2.3			
Front suspension / wheel alignment				
Toe-in (+) / Toe-out (–)....................mm [°]	0.0 ± 2.0			
Camber	0° ± 30'			
Castor	2°10' ± 45'			
King pin inclination..............................	–			
Rear suspension / wheel alignment				
Toe-in (+) / Toe-out (–)....................mm [°]	0			
Camber	0			

ISUZU

	Trooper 3.1 TD 1992 to 1998			
Torque wrench settings				
Cylinder head - stage 1 ....Nm	49			
- stage 2 ....Nm	+ 75°			
- stage 3 ....Nm	+ 75°			
- stage 4 ....Nm	–			
- stage 5 ....Nm	–			
- stage 6 ....Nm	–			
Big-end bearings.................................Nm	29 + 50° N			
Main bearings.....................................Nm	167			
Crankshaft pulley bolt ....Nm	275			
Camshaft pulley bolt ....Nm	118			
Flywheel [driveplate] bolt....................Nm	118			
Front hubs ....Nm	–			
Rear hubs ....Nm	–			
Wheel nuts / bolts ....Nm	118			
Glow plugs ....Nm	23			
Clutch pressure plate bolts ....Nm	18			
Injection pump sprocket......................Nm	64			
Injectors...Nm	64			
Injection pump mounting bolts ....Nm	20			
Injector pipe unions............................Nm	29			
Capacities				
Engine oil & filter................................litres	6.0			
Gearbox..litres	2.95			
Automatic transmission ....litres	–			
Final drive ....litres	1.8 Front: 1.5[1]			
Cooling system...................................litres	8.6			
Fuel tank...litres	85			

Notes

Trooper 3.1 TD 1992 to 1998
[1]Transfer: 1.45

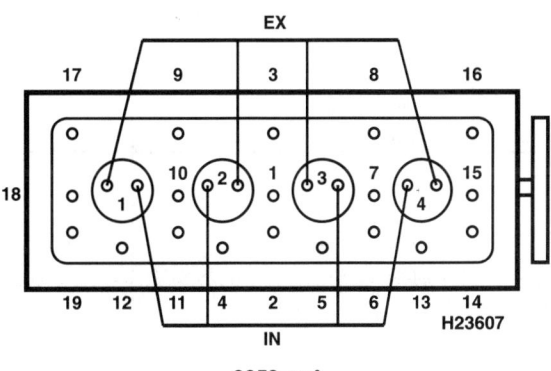

3059 cm³

– Not applicable, or information not available

LANCIA

	Delta & Prisma 1.9 Turbo 1986 to 1993	Prisma 1.9 1986 to 1991	Dedra 1.9 Turbo D 1990 to 1998	Thema Turbo DS 1989 to 1992
Engine				
Engine type/code...........................	831 D1.000 SOHC Turbo 59kW	831 D.000 48kW	835 A4.000 SOHC Turbo 66kW	8144.97 SOHC Turbo 85kW
Capacity (cm³) / cylinders........................	1929 / 4	1929 / 4	1929 / 4	2499 / 4
Compression ration / pressure ....bar	20.0±0.5 / _	21.0±0.5 / _	19.2 ± 0.5 / _	21.0 / _
Torque output ....Nm	0	0	0	0
Oil pressure ....idle [running] bar	[3.4 to 4.9]	[3.4 to 4.9]	[3.4 to 4.9]	0.78 [3.9]
Oil temperature ....°C	100	100	100	100
Valve clearances - inlet (mm)	0.30 ± 0.05	0.30 ± 0.05	0.30 ± 0.05	0.50 ± 0.05
- exhaust (mm)	0.35 ± 0.05	0.35 ± 0.05	0.35 ± 0.05	0.50 ± 0.05
Injection order	1-3-4-2	1-3-4-2	1-3-4-2	1-3-4-2
No. 1 cylinder position	TBE	TBE	TBE	TBE
Cooling system				
Thermostat opening temperature ....°C	78 to 82	78 to 82	78 to 82	77 to 81
Radiator cap pressure ....bar	0.78	0.78	0.78	0.78
Fuel system				
Idle speed ....rpm	780 ± 20	780 ± 20	830 ± 20	800 to 850
Maximum (no load) speed ....rpm	4900 ± 40	5160 ± 40	4900 ± 40	4950
Smoke test/opacity ....M⁻¹ %	2.5	2.5	2.5	2.5
Static timing method................................	Plunger travel	Plunger travel	Plunger travel	Plunger travel
Timing dimension................................mm	1.0	0.88	0.8	1.0
Crankshaft position ....mm [°]	[0] TDC	[0] TDC	[0] TDC	[0] TDC
Turbo type / ref / pressure ....bar	_		_	_
Injection pump make	Bosch	Bosch	Bosch	Bosch
Injection pump part no.............................	0460 494 163	_	0460 494 250	046 404 059
Injector Make / type	Bosch	Bosch	Bosch	Bosch
Injector part no...	DN12 SD 1750	DN12 SD 1750	DNO SD 290	DNO SD 259
Injection type..	VE 4/9F 2100 L184	VE 4/9F 2300 L157	VER 303-1	VER 318
Injection opening pressure, New [used]...bar	150 to 158	125 to 133	150 to 158	150 to 158
Glow plugs				
Maker	Marelli/Bosch	Marelli/Bosch	Marelli/Bosch	Beru
Type	UX2A / 0250 201 005	UX2A / 0.250.201.005	UX2A / 0250 201 005	0100 221 167
Nominal rating....................................V/A	11 / 12	_	11 / 12	11 / 12
Brakes				
minimum friction material thickness				
Front...........................mm	1.5	1.5	1.5	1.5
Rear...........................mm	1.5	1.5	1.5	1.5
Tyres - Saloon / Hatch......................Size	165/65x14	165/65x14	175/65x14: 185/60x14	185/70x14: 195/60x14
- Estate / Van............................Size	_	_	_	185/70x14: 195/65x14
Pressure - front / rear - Saloon / Hatch...bar	2.3 / 2.0	2.3 / 2.0	2.3 / 2.1	2.2 / 2.2
- Estate / Van ......bar	_	_		2.3 / 2.4
Front suspension / wheel alignment				
Toe-in (+) / Toe-out (–)....................mm [°]	1.0 to -2.5	1.0 to -2.5	-1.0 to 1.0	-1.0 to 1.0
Camber	10' ± 30'	10' ± 30' N/A	10' ± 30'	-40' to 0'
Castor	2°40' ± 30''	2°40' ± 30' N/A'	2°40' ± 30'	2°30' to 3°10'
King pin inclination...................................	9°18'	9°18'	_	_
Rear suspension / wheel alignment				
Toe-in (+) / Toe-out (–)....................mm [°]	2.0 to 5.0	2.0 to 5.0	-2.5 to 1.5	2.0 to 5.0
Camber	-20' ± 30'	-20'±30' N/A	-1°30' ± 30'	-40' to 0'

LANCIA

Torque wrench settings	Delta & Prisma 1.9 Turbo 1986 to 1993	Prisma 1.9 1986 to 1991	Dedra 1.9 Turbo D 1990 to 1998	Thema Turbo DS 1989 to 1992
Cylinder head - stage 1 ....Nm	50	50	50	40
- stage 2 ....Nm	100	100	100	Slacken, then 40
- stage 3 ....Nm	+ 90°	+ 90°	+ 90°	+ 180°
- stage 4 ....Nm	+ 90°	+ 90°	+ 90°	–
- stage 5 ....Nm	–	–	M8 bolts: 30	–
- stage 6 ....Nm	–	–	–	–
Big-end bearings........................Nm	25 + 50° N	25 + 50°	25 + 50° N	115 N
Main bearings........................Nm	113	113	113	78 then 157
Crankshaft pulley bot ....Nm	190	190	190	–
Camshaft pulley bolt ....Nm	118	118	118	25
Flywheel [driveplate] bolt........................Nm	142	142	142	125
Front hubs........................Nm	294	294	280	360
Rear hubs........................Nm	216	216	320	320
Wheel nuts / bolts ....Nm	86	86	86	88
Glow plugs ....Nm	15	15	15	37
Clutch pressure plate bolts ....Nm	38	38	38	20
Injection pump sprocket........................Nm	50	50	49	25
Injectors........................Nm	78	78	55	35
Injection pump mounting bolts ....Nm	25	25	25	25
Injector pipe unions........................Nm	32	32	32	–
Capacities				
Engine oil & filter........................litres	5.0	5.0	5.0	6.4
Gearbox........................litres	3.2	3.2	1.4	1.8 to 2.1
Automatic transmission ....litres	–	–	–	
Final drive........................litres	WT	WT	WT	WT
Cooling system........................litres	7.0	7.0	8.9	9.0
Fuel tank........................litres	57	57	63	70

Notes

Delta & Prisma 1.9 Turbo 1986 to 1993
¹Prisma PAS: 3°25'±30'
Prisma 1.9 1986 to 1991
¹Prisma PAS: 3°25'±30' N/A

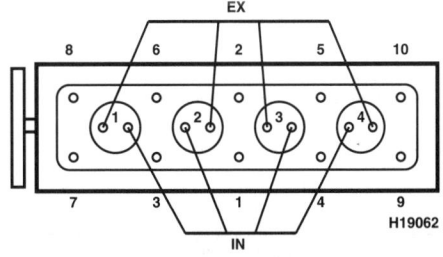

1929 cm³

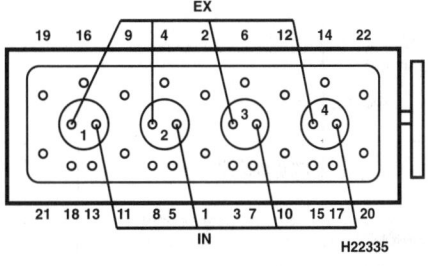

2499 cm³

– Not applicable, or information not available

LANCIA

	Kappa 2.4 TD 1995 to 1998	Zeta 2.1 TD 1996 to 1999		
Engine				
Engine type/code.............................	838 A3.000 10V 91kW	XUD11 P8C SOHC Turbo 80kW		
Capacity (cm³) / cylinders........................	2387 / 5	2088 / 4		
Compression ration / pressure ....bar	21.0 / _	21.5 /		
Torque output ....Nm	0	250		
Oil pressure ....idle [running] bar	[5.0 @ 4000]	[4.9]		
Oil temperature ....°C	100	80		
Valve clearances - inlet (mm)	0.30 ± 0.05	0: Hyd.		
- exhaust (mm)	0.35 ± 0.05	0: Hyd.		
Injection order	1-2-4-5-3	1-3-4-2		
No. 1 cylinder position	TBE	F		
Cooling system				
Thermostat opening temperature ....°C	_	_		
Radiator cap pressure ....bar	0.98	_		
Fuel system				
Idle speed ....rpm	830 to 870	690 ± 25		
Maximum (no load) speed ....rpm	5150	_		
Smoke test/opacity ....M⁻¹ %	1.6	2.5		
Static timing method.................................	Plunger travel	Refer to wsm		
Timing dimension................................mm	0.65	_		
Crankshaft position ....mm [°]	TDC	_		
Turbo type / ref / pressure ....bar	_	_		
Injection pump make	Bosch	Lucas		
Injection pump part no.............................	VER 560	EPIC		
Injector Make / type.................................	Bosch	_		
Injector part no..	KCA SV0216 138	_		
Injection type..	Indirect	EPIC		
Injection opening pressure, New [used]...bar	150 to 158	150		
Glow plugs				
Maker	Bosch	Bosch		
Type	0281 003 010	0250 201 033		
Nominal rating.......................................V/A	_	11 / 9		
Brakes				
minimum friction material thickness				
Front..mm	1.5	1.5		
Rear...mm	1.5	1.5		
Tyres - Saloon / Hatch......................Size	185/65x14: 195/60x14	_		
- Estate / Van...........................Size	185/70x14: 195/65x14	_		
Pressure - front / rear - Saloon / Hatch ...bar	2.2 / 2.2	_		
- Estate / Van ......bar	2.3 / 2.4	_		
Front suspension / wheel alignment				
Toe-in (+) / Toe-out (–)....................mm [°]	-1.0 to 1.0	_		
Camber	-40' to 0°	_		
Castor	2°30' to 3°10'	_		
King pin inclination....................................	_	_		
Rear suspension / wheel alignment				
Toe-in (+) / Toe-out (–)....................mm [°]	1.0 to 4.0'	_		
Camber	-40' to 0°	_		

LANCIA

	Kappa 2.4 TD 1995 to 1998	Zeta 2.1 TD 1996 to 1999		
Torque wrench settings				
Cylinder head - stage 1 ....Nm	60	35		
- stage 2 ....Nm	+ 90°	70 + 150°		
- stage 3 ....Nm	+ 90°	warm up, allow to cool		
- stage 4 ....Nm	+ 90°	slacken		
- stage 5 ....Nm	_	70		
- stage 6 ....Nm	_	+ 150°		
Big-end bearings.............................Nm	25 + 50° N	70 N		
Main bearings.................................Nm	113	60		
Crankshaft pulley bolt ....Nm	360	70 + 60°		
Camshaft pulley bolt ....Nm	118	45		
Flywheel [driveplate] bolt...................Nm	142	50		
Front hubs ....Nm	360	300		
Rear hubs......................................Nm	320	_		
Wheel nuts / bolts ....Nm	98	_		
Glow plugs ....Nm	15	23		
Clutch pressure plate bolts ....Nm	20	15		
Injection pump sprocket.....................Nm	50	50		
Injectors..Nm	55	90		
Injection pump mounting bolts ....Nm	25	20		
Injector pipe unions............................Nm	_	_		
Capacities				
Engine oil & filter..............................litres	5.5	4.75		
Gearbox...litres	2.3	1.85		
Automatic transmission ....litres	4.4	_		
Final drive......................................litres	WT	WT		
Cooling system................................litres	8.2	9.0		
Fuel tank.......................................litres	65	_		

Notes

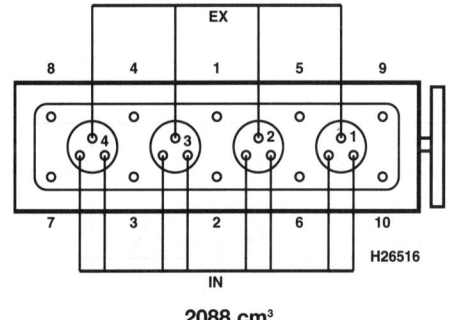

2088 cm³

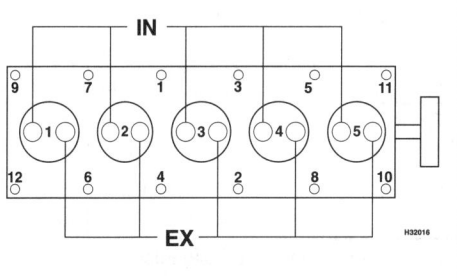

2387 cm³

– Not applicable, or information not available

LAND ROVER

	90 & 110 2.5 Turbo 1986 to 1992	Defender 200TDi 1990 to 1994	Defender 90, 110 300TDi 1994 to 2000	Freelander 2.0 TDi 1997 to 2000
Engine				
Engine type/code....................	12J OHV Turbo 63kW	200TDi OHV Turbo 79kW	300TDi OHV Turbo 82kW	L Series SOHC TDI EGR 88kW
Capacity (cm^3) / cylinders........................	2495 / 4	2495 / 4	2495 / 4	1994 / 4
Compression ration / pressure ....bar	21.0 / _	19.5 ± 0.5 / _	19.5 /	19.5 /
Torque output ....Nm	0	0	265	210
Oil pressure ....idle [running] bar	[2.4 to 4.5]	[1.7 to 3.8]	1.7 to 3.8	0.7 [3.8]
Oil temperature ....°C	80	80	80	80
Valve clearances - inlet (mm)	0.25	0.20	0.20	0: Hyd.
- exhaust (mm)	0.25	0.20	0.20	0: Hyd.
Injection order	1-3-4-2	1-3-4-2	1-3-4-2	1-3-4-2
No. 1 cylinder position	F	TBE	TBE	TBE
Cooling system				
Thermostat opening temperature ....°C	82	88	88	80 to 84
Radiator cap pressure ....bar	1.0	1.04	1.0	0.9 to 1.2
Fuel system				
Idle speed ....rpm	650 ± 20	780 to 800	720 ± 20	850 ± 50'
Maximum (no load) speed ....rpm	4400 ± 80	4100 to 4260	4490	5300
Smoke test/opacity ....M^{-1} %	2.5	2.5	2.0	2.0
Static timing method................................	Refer to wsm	Plunger travel	Plunger travel	_
Timing dimension............................mm	_	1.54	1.54	Computer controlled
Crankshaft position ....mm [°]	_	TDC	TDC	
Turbo type / ref / pressure ....bar		_	Allied signal	Garrett GT 1549
Injection pump make	CAV	Bosch	Bosch	Bosch
Injection pump part no.............................	_	1900R 347-1	R509	VP37
Injector Make / type	CAV	Bosch	Bosch	
Injector part no..	BDNO/SPC 6209	DSLA 145P208	DSLA 145P366	_
Injection type...	DPS	VE4/11F	VER	EDC Two stage
Injection opening pressure, New [used]...bar	135	203 to 284	200 to 300	200 to 320
Glow plugs				
Maker	Champion	Bosch	Bosch	Beru
Type	CH63	0250 201 012	0250 201 012	0100226 184
Nominal rating....................................V/A	11 / 8.5	11.0 / _	_	_
Brakes				
minimum friction material thickness				
Front............................mm	_	_	_	3.0
Rear............................mm	_	_	_	2.0
Tyres - Saloon / Hatch......................Size	_	_	_	_
- Estate / Van...........................Size	6.00x16: 7.50x16: 205x16	205x16: 750x16	Refer to vehicle	195/80x15: 215/65x16
Pressure - front / rear - Saloon / Hatch ...bar	_	_	_	_
- Estate / Van ......bar	Refer to manufacturer / OHB	Refer to manufacturer / OHB	Refer to vehicle	2.1 / 2.1: 2.1 / 2.1
Front suspension / wheel alignment				
Toe-in (+) / Toe-out (–)....................mm [°]	1.2 to 2.4'	1.2 to 2.4'	0.0 to -2.0	[-0°10' ± 15']
Camber	0	0	0.0	-0°15' ± 45'
Castor	3°	3°	3.0	3°30' ± 1°
King pin inclination..................................	7°	7°	_	12°18'
Rear suspension / wheel alignment				
Toe-in (+) / Toe-out (–)....................mm [°]	_	_	_	[20' ± 15']
Camber				-0°30' ± 45'

LAND ROVER

	90 & 110 2.5 Turbo 1986 to 1992	Defender 200TDi 1990 to 1994	Defender 90, 110 300TDi 1994 to 2000	Freelander 2.0 TDi 1997 to 2000
Torque wrench settings				
Cylinder head - stage 1 ....Nm	115 to 130	40	40	30
- stage 2 ....Nm	–	+ 60°	+ 60°	65
- stage 3 ....Nm	–	+ 60°	+ 60°	+ 90°
- stage 4 ....Nm	–	–	M12: + 20°	+ 90°
- stage 5 ....Nm	–	–	–	–
- stage 6 ....Nm	–	–	–	–
Big-end bearings..............................Nm	34 to 46 N	56 to 62 N	59 N	20 + 85° N
Main bearings.................................Nm	130 to 136	130 to 136	133	112
Crankshaft pulley bolt ....Nm	260 to 280	335 to 350	343	63 + 90°
Camshaft pulley bolt ....Nm	40 to 50	40 to 50	80	20 + 90°
Flywheel [driveplate] bolt.................Nm	130 to 143	139 to 153	145	15 + 90° N
Front hubs ....Nm	–	–	WSM	400 N
Rear hubs ....Nm	–	–	WSM	400 N
Wheel nuts / bolts ....Nm	108	108	108	115
Glow plugs ....Nm	34	15 to 30	20	20
Clutch pressure plate bolts ....Nm	25 to 34	30 to 38	34	25
Injection pump sprocket....................Nm	–	22 to 28	–	20 + 90°
Injectors..Nm	22 to 28	20 to 25	25	25
Injection pump mounting bolts ....Nm	22 to 28	22 to 28	25	–
Injector pipe unions...........................Nm	–	22 to 25	25	28
Capacities				
Engine oil & filter...............................litres	6.9	6.9	6.85	4.5
Gearbox...litres	2.2[2]	2.2[2]	2.6 Transfer: 28	2.0
Automatic transmission ....litres	–	–	–	–
Final drive ....litres	1.7[3]	1.7[3]	F: 1.7 R: 2.3	Rear: 0.8
Cooling system..................................litres	11.0	11.1	11.1	6.5
Fuel tank...litres	79.5[4]	79.5[4]	90: 54 110: 80	59

Notes

90 & 110 2.5 Turbo 1986 to 1992
[1]Permanent 4 wheel drive: figures are toe-out (-)
[2]Transfer box: 2.8
[3]Rear, 90: 1.7. 110: 2.3
[4]Rear tank only. Side tank: 68.2. Station Wagon side tank: 45.5. 90: 54.5
Defender 200TDi 1990 to 1994
[1]Permanent 4 wheel drive: -1.2 to -2.4
[2]Transfer box: 2.3
[3]Rear, 90: 1.7. 110: 2.3
[4]Rear tank only. Side tank: 68.2. Station Wagon side tank: 45.5. 90: 54.5
Freelander 2.0 TDi 1997 to 2000
[1]With A/C on: 875 ± 50

– Not applicable, or information not available

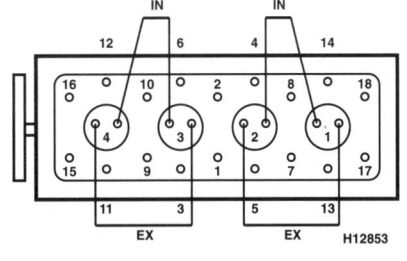

2495 cm³

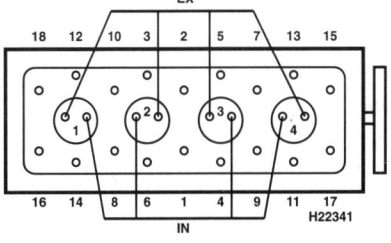

Defender 2495 cm³

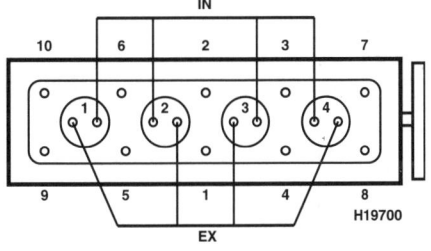

1994 cm³

LAND ROVER

	Discovery 200TDi 1989 to 1994	Discovery 300TDi 1995 to 1998	Discovery TD5 1998 to 2000	Range Rover 2.4 TD 1986 to 1992
Engine				
Engine type/code..........................	200TDi OHV Turbo 82kW	300TDi OHV Turbo 82kW	TD5 SOHC	11A VM HR 492 HI OHV 84kW
Capacity (cm^3) / cylinders........................	2495 / 4	2495 / 4	2498 / 5	2393 / 4
Compression ration / pressure ....bar	19.5 ± 0.5 / _	19.5 /	19.5 /	21.5 ± 0.5 / _
Torque output ....Nm	0	265	0	0
Oil pressure ....idle [running] bar	[1.7 to 3.8]	[1.7 to 3.8]	[1.5 to 3.0]	[3.5 to 3.9]
Oil temperature ....°C	80	80	80	90 to 100
Valve clearances - inlet (mm)	0.20	0.20	0: Hyd.	0.30
- exhaust (mm)	0.20	0.20	0: Hyd.	0.30
Injection order	1-3-4-2	1-3-4-2	1-2-4-5-3	1-3-4-2
No. 1 cylinder position	TBE	TBE	_	F
Cooling system				
Thermostat opening temperature ....°C	82	88	82	83 ± 2
Radiator cap pressure ....bar	1.04	1.04	1.4	1.04
Fuel system				
Idle speed ....rpm	720 ± 20	720 ± 20	740 ± 50 [760 ± 50]	750 to 800
Maximum (no load) speed ....rpm	4600 +40 -120	4600 ± 40	4850	4700 to 4730
Smoke test/opacity ....M^{-1} %	2.5	2.5	2.5	2.5
Static timing method.................................	Plunger travel	Plunger travel	_	Plunger travel
Timing dimension............................mm	1.54	1.54	Computer controlled	0.5
Crankshaft position ....mm [°]	TDC	TDC	_	[4]
Turbo type / ref / pressure ....bar	_	0.8 to 1.0 bar @ 3000rpm	Garrett GT20	_
Injection pump make	Bosch	Bosch		Bosch
Injection pump part no............................	0460 414 069	VE4/11F	_	168-1
Injector Make / type	Bosch	Bosch	_	Bosch
Injector part no...	BDNO/SPC 6209	DSLA 14SP 366	_	DN OSD 263
Injection type..	VE4/11F	R509	Direct Injection	Rotary VE4/10F L
Injection opening pressure, New [used]...bar	203 to 284	200	1500	150 +8 -0
Glow plugs				
Maker	Beru	Beru	Beru	Bosch
Type	783 MJ	0 100 226 129A	_	0250 201 012
Nominal rating.....................................V/A	11.0 / _	_	_	11.0 / 10.5
Brakes				
minimum friction material thickness				
Front......................................mm	_	_	2.0	_
Rear.......................................mm	_	_	2.0	_
Tyres - Saloon / Hatch.....................Size	_	_	_	_
- Estate / Van...........................Size	205x16	205x16 : 235x70 R16	235/70x16: 225/65x16:255/55x18	205x16: 215/75x16
Pressure - front / rear - Saloon / Hatch...bar				
- Estate / Van ......bar	1.9 / 2.6	1.9 / 2.6 : 1.8 / 2.4	1.9 / 2.5	1.9 / 2.1
Front suspension / wheel alignment				
Toe-in (+) / Toe-out (–)....................mm [°]	-1.2 to -2.4	[0° to 16']	[0°5' ± 2.5']	-1.2 to -2.4
Camber	0	0°	0°	0
Castor	3°	3°	3.6° to 3.8°	3°
King pin inclination...................................	7°	7°	13°	7°
Rear suspension / wheel alignment				
Toe-in (+) / Toe-out (–)....................mm [°]	_	_	_	_
Camber				

LAND ROVER

Torque wrench settings	Discovery 200TDi 1989 to 1994	Discovery 300TDi 1995 to 1998	Discovery TD5 1998 to 2000	Range Rover 2.4 TD 1986 to 1992
Cylinder head - stage 1 ......Nm	40	40	30	No 1 to 10: 5 to 10
- stage 2 ......Nm	+ 60°	+ 60°	65	No 1 to 10: 30
- stage 3 ......Nm	+ 60°	+ 60°	+ 90°	No 1 to 10: + 50°
- stage 4 ......Nm	–	M12: + 20°	+ 180°	No 1 to 10: + 50°
- stage 5 ......Nm	–	–	+ 45°	No 11: 90
- stage 6 ......Nm	–	–	–	[1]
Big-end bearings......Nm	56 to 62 N	59 N	20 N	79 to 83
Main bearings......Nm	130 to 136	133	33 + 90°	41 to 43[2]
Crankshaft pulley bolt ......Nm	335 to 350	80 + 90°	455	146 to 157
Camshaft pulley bolt ......Nm	40 to 50	40 to 50	37	30
Flywheel [driveplate] bolt......Nm	139 to 153	146	40 + 90°	108
Front hubs ......Nm	–	WSM	490	41 to 52
Rear hubs ......Nm	–	WSM	490	41 to 52
Wheel nuts / bolts ......Nm	129	126	140	Steel: 102 to 115[3]
Glow plugs ......Nm	15 to 30	20	16	23
Clutch pressure plate bolts ......Nm	30 to 38	34	25	24 to 30
Injection pump sprocket......Nm	22 to 28	25	–	88
Injectors......Nm	20 to 25	25	–	24 to 30
Injection pump mounting bolts ......Nm	22 to 28	25	–	30 to 31
Injector pipe unions......Nm	22 to 25	29	–	19
Capacities				
Engine oil & filter......litres	6.9	6.65	7.2	8.0
Gearbox......litres	2.7 Transfer:2.8	2.67 Transfer: 2.8	2.3 to 2.8 Transfer: 2.0	2.7[4]
Automatic transmission ......litres	–	9.1	9.1	–
Final drive ......litres	Front & rear: 1.7	Front & rear: 1.7	1.6	Front & rear: 1.7
Cooling system......litres	11.5	11.5	8.0	11.5
Fuel tank......litres	88.6	89	95	80

Notes

Range Rover 2.4 TD 1986 to 1992
[1]Run engine for 20 minutes & allow to cool, then no 1 to 10: + 30°. After 40000km: + 10°
[2]Rear carrier nuts: 24 to 30
[3]Alloy: 122 to 129
[4]Transfer box: 2.5. 89 ▶: 2.1

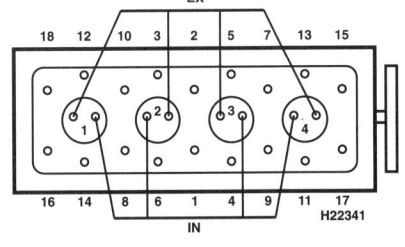

2495 cm³

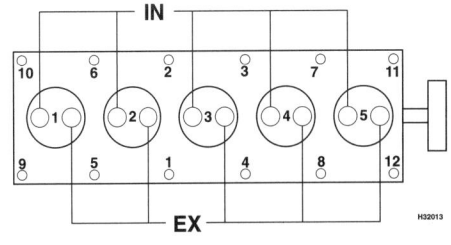

2498 cm³

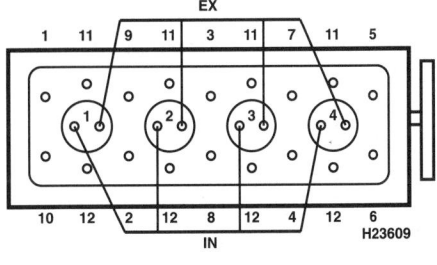

2393 cm³

– Not applicable, or information not available

LAND ROVER

	Range Rover 2.5 Turbo Diesel 1989 to 1992	Range Rover 2.5 TDi 1993 to 1994	Range Rover 2.5 TDi 1994 to 1995	Range Rover 2.5 TD 1994 to 1999
Engine				
Engine type/code....................	95 A VM HR 4924 HI OHV 82kW	200TDI OHV Turbo 82kW	300TDI OHV Turbo 82kW	SOHC Turbo 100kW
Capacity (cm³) / cylinders......................	2500 / 4	2495 / 4	2495 / 4	2497 / 6
Compression ration / pressure ....bar	22.5 ± 0.5 / 23.5	19.5 /	19.5 /	22.6 / ≤ 20.0
Torque output ....Nm	0	0	0	270
Oil pressure ....idle [running] bar	[3.5 to 3.9]	1.7 to 3.8	1.7 to 3.8	2.0
Oil temperature ....°C	90 to 100	80	80	80
Valve clearances - inlet (mm)	0.30	0.20	0.20	0: Hyd.
- exhaust (mm)	0.30	0.20	0.20	0: Hyd.
Injection order	1-3-4-2	1-3-4-2	1-3-4-2	1-5-3-6-2-4
No. 1 cylinder position	F	–	–	–
Cooling system				
Thermostat opening temperature ....°C	80 ± 2	–	–	80
Radiator cap pressure ....bar	1.04	–	–	1.0
Fuel system				
Idle speed ....rpm	750 to 800	720 ± 20	720 ± 20	750 ± 50
Maximum (no load) speed ....rpm	4700 to 4730	4600	4600	4950 ± 150
Smoke test/opacity ....M⁻¹ %	3.0	2.0	2.0	2.5
Static timing method...............................	–	Plunger travel	Plunger travel	Plunger travel
Timing dimension..............................mm	–	1.54	1.54¹	0.95 ± 0.02
Crankshaft position ....mm [°]	[3-0+1]	TDC	TDC	TDC
Turbo type / ref / pressure ....bar	–	0.78 bar	0.8 to 1.0 bar	Mitsubishi TD04-11G4
Injection pump make	Bosch	Bosch	Bosch	Bosch
Injection pump part no.............................	2100 L269	VE4/11F R347	VE4/11F R347	R515
Injector Make / type	Bosch	Bosch	Bosch	Bosch
Injector part no......................................	DNOSN 1510. 90 ▶: DNOSD 263	KBEL 98P52	KBEL 98P52	KCA 21S71
Injection type...	Rotary VE 4/10F	Direct	Direct	Indirect
Injection opening pressure, New [used]...bar	150 +8 -0	200 to 280	100 to 280	140 to 160
Glow plugs				
Maker	Bosch	Bosch/Beru	Bosch	Bosch
Type	0250 201 012	0250 201 033 / 783 MJ	0250 201 033	0250 201 033
Nominal rating....................................V/A	11.0 / _	–	–	–
Brakes				
minimum friction material thickness				
Front.........................mm	–	–	–	2.0
Rear.........................mm	–	–	–	2.0
Tyres - Saloon / Hatch......................Size	–	–	–	–
- Estate / Van............................Size	205x16: 215/75x16	–	–	234/70x16: 255/70x16
Pressure - front / rear - Saloon / Hatch ...bar	–	–	–	–
- Estate / Van ......bar	1.9 / 2.1	–	–	1.9 / 2.6
Front suspension / wheel alignment				
Toe-in (+) / Toe-out (–)...................mm [°]	-1.2 to -2.4	–	–	0.6 to 1.8
Camber	0	–	–	0°
Castor	3°	–	–	4°
King pin inclination.................................	7°	–	–	8°
Rear suspension / wheel alignment				
Toe-in (+) / Toe-out (–)...................mm [°]	–	–	–	–
Camber				

LAND ROVER

Torque wrench settings	Range Rover 2.5 Turbo Diesel 1989 to 1992	Range Rover 2.5 TDi 1993 to 1994	Range Rover 2.5 TDi 1994 to 1995	Range Rover 2.5 TD 1994 to 1999
Cylinder head - stage 1 ....Nm	No 1 to 10: 5 to 10	40	40	80, slacken
- stage 2 ....Nm	No 1 to 10: 30	+ 60°	+ 60°	50
- stage 3 ....Nm	No 1 to 10: + 50°	+ 60°	+ 60°	+ 90°
- stage 4 ....Nm	No 1 to 10: + 50°	–	–	+ 90°
- stage 5 ....Nm	No 11: 90	–	–	warm up, allow to cool
- stage 6 ....Nm	[1]	–		+ 90°
Big-end bearings.........................Nm	79 to 83	60 N	60 N	20 + 70° N
Main bearings............................Nm	41 to 43[2]	136	136	50 N
Crankshaft pulley bolt ....Nm	146 to 157	340	80 + 90°	100 + 150°
Camshaft pulley bolt ....Nm	–		80	20 + 35°
Flywheel [driveplate] bolt....................Nm	108	150	150	105 N [120 N]
Front hubs ....Nm	41 to 52	–	–	260
Rear hubs ....Nm	41 to 52	–	–	WSM
Wheel nuts / bolts ....Nm	Steel: 102 to 115[3]	115 Alloy: 160	130	130
Glow plugs ....Nm	23	30	30	20
Clutch pressure plate bolts ....Nm	24 to 30	30	30	23
Injection pump sprocket.....................Nm	88	28	28	47
Injectors...............................Nm	24 to 30	25	25	62
Injection pump mounting bolts ....Nm	30 to 31	–	28	23
Injector pipe unions...........................Nm	19	25	25	23
Capacities				
Engine oil & filter................................litres	8.0	6.8	6.8	9.5
Gearbox.................................litres	2.7[4]	2.7	2.7	2.7
Automatic transmission ....litres	–			9.7
Final drive ....litres	Front & rear: 1.7	1.7 Rear: 1.7	1.7 Rear: 1.7	1.7 Rear: 1.7
Cooling system................................litres	11.5	–	–	11.3
Fuel tank.............................litres	82	–	–	90

Notes

Range Rover 2.5 Turbo Diesel 1989 to 1992
[1] Run engine for 20 minutes & allow to cool, then no 1 to 10: + 30°. After 40000km: + 10°
[2] Rear carrier nuts: 24 to 30
[3] Alloy: 122 to 129
[4] Transfer box: 2.5. 89 ▶: 2.1
Range Rover 2.5 TDi 1994 to 1995
[1] R509/1: 1.4 mm   R500: 0.4 mm

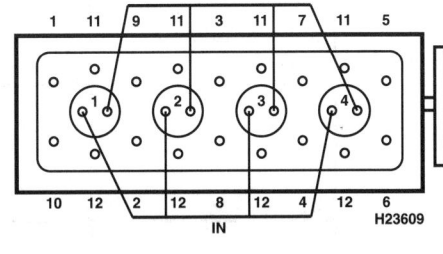

2500 cm³

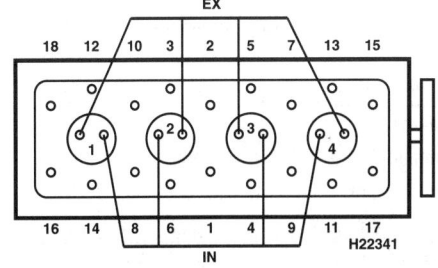

2495 cm³

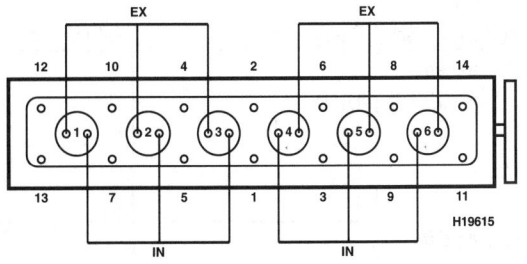

2497 cm³

– Not applicable, or information not available

THE BOOK ®

LEYLAND DAF (LDV)

	200 / Pilot 1.9D 1995 to 1998	200 2.0 1989 to 1995	300 2.5 1989 to 1991	300 2.5 Turbo 1989 to 1991
Engine				
Engine type/code.....................	XUD9A SOHC 52kW	TN44 Two stage DI 44kW	PSA EN55	PSA ET70
Capacity (cm³) / cylinders......................	1905 / 4	1994 / 4	2498 / 4	2498 / 4
Compression ration / pressure ....bar	23.0 / 25.0 to 30.0	18.0 / 27.6	23.0 / ≥18.0	21.0 / ≥18.0
Torque output ....Nm	120	0	0	0
Oil pressure ....idle [running] bar	[3.5 @ 4000]	1.0 to 1.3	1.6 [3.7]	1.2 [3.7]
Oil temperature ....°C	80	_	110	80
Valve clearances - inlet (mm)	015 ± 0.05	0.30 ± 0.10	0.15	0.15
- exhaust (mm)	0.30 ± 0.05	0.40 ± 0.10	0.25	0.25
Injection order	1-3-4-2	1-3-4-2	1-3-4-2	1-3-4-2
No. 1 cylinder position	FE	F	FE	FE
Cooling system				
Thermostat opening temperature ....°C	83	88	81	81
Radiator cap pressure ....bar	1.0	1.0	0.6	1.0
Fuel system				
Idle speed ....rpm	800 ± 500	825 ± 25	775 ± 25	775 ± 25
Maximum (no load) speed ....rpm	5100	5130	4500 ± 75	4750 +50 -100
Smoke test/opacity ....M⁻¹ %	2.5	2.5	2.5	2.5
Static timing method................................	Plunger travel	Plunger travel	Refer to text	Refer to text
Timing dimension.............................mm	Dimension on pump	1.0 ± 0.03'	Value marked on pump	Value marked on pump
Crankshaft position ....mm [°]	_	[0] TDC	2.85	2.85
Turbo type / ref / pressure ....bar	_	_		
Injection pump make	Lucas	Bosch	Roto Diesel	Roto Diesel
Injection pump part no.............................	DPC 8443B 952B	_	R8443 B600A	R8443 B254C
Injector Make / type	CAV	CAV	Roto Diesel	Roto Diesel
Injector part no.......................................	RDNOSDC6887D	JB 6801081	RDN OSDC 6577B	RDN 12SDC 6849C
Injection type..	Indirect	EPVE	DPC Type 059	DPC Type 053
Injection opening pressure, New [used]...bar	120 to 130	N/A	115 ± 5	130 ± 5
Glow plugs				
Maker	Beru	Beru/Bosch/Champion	_	_
Type	GV798	SR-3 / 0250 200 035 / CH32	_	_
Nominal rating ....V/A	11 / 25	12 / 6	_	_
Brakes				
minimum friction material thickness				
Front.............................mm	1.0	3.0	3.0	3.0
Rear.............................mm	1.5	1.5	1.5	1.5
Tyres - Saloon / Hatch......................Size	_	_	_	_
- Estate / Van.........................Size	185/14	185x14	185x14: 205x14	185x14: 205x14
Pressure - front / rear - Saloon / Hatch...bar	_	_	_	_
- Estate / Van ......bar	Owners handbook	Refer to vehicle	Refer to owner's handbook	Refer to owner's handbook
Front suspension / wheel alignment				
Toe-in (+) / Toe-out (−)....................mm [°]	1.6 to 3.2	-1.6 to -3.2	0 to 1.6	0 to 1.6
Camber	2°30'	2°30'	1°30'	1°30'
Castor	5°	3°30'	1° U/L	1° U/L
King pin inclination.................................	8°30'	8°30'	9°30'	9°30'
Rear suspension / wheel alignment				
Toe-in (+) / Toe-out (−)....................mm [°]	_	_	_	_
Camber	_	_	_	_

LEYLAND DAF (LDV)

	200 / Pilot 1.9D 1995 to 1998	200 2.0 1989 to 1995	300 2.5 1989 to 1991	300 2.5 Turbo 1989 to 1991
Torque wrench settings				
Cylinder head - stage 1 ....Nm	20 N	50	30	30
- stage 2 ....Nm	60	100	70	70
- stage 3 ....Nm	+ 180°	+ 90°	- 90°, then 70	- 90°, then 70
- stage 4 ....Nm	–	–	Run engine, cool for 3.5 hours	Run engine, cool for 3.5 hours
- stage 5 ....Nm	–	–	- 90°, then 70	- 90°, then 70
- stage 6 ....Nm	–	–	+ 120°	+ 120°
Big-end bearings..................Nm	WSM	47 N	60	60
Main bearings..................Nm	35, 70	112	120	120
Crankshaft pulley bolt ....Nm	40 + 60°	180	55 + 60°	55 + 60°
Camshaft pulley bolt ....Nm	40	85	–	–
Flywheel [driveplate] bolt..................Nm	50	65	15 + 60°	15 + 60°
Front hubs ....Nm	WSM	–	–	–
Rear hubs ....Nm	–	203 to 217	136 to 163	136 to 163
Wheel nuts / bolts ....Nm	90	80	176 to 227[1]	176 to 227[1]
Glow plugs ....Nm	22	20	–	–
Clutch pressure plate bolts ....Nm	25	35	20	20
Injection pump sprocket..................Nm	50	60	–	–
Injectors..................Nm	90	43	90	90
Injection pump mounting bolts ....Nm	–	–	–	–
Injector pipe unions..................Nm	–	18	–	–
Capacities				
Engine oil & filter..................litres	5.2	5.0	7.4	7.7
Gearbox..................litres	2.0	2.0	1.8	1.8
Automatic transmission ....litres	–	–	–	–
Final drive ....litres	1.9	1.9	1.7[2]	1.7[2]
Cooling system..................litres	50	5.5	9.0	9.0
Fuel tank..................litres	50	50 Option: 64	70. Option: 64	70. Option: 64

Notes

200 2.0 1989 to 1995
[1]New or replacement pump: 1.05 ± 0.03
300 2.5 1989 to 1991
[1]Twin wheel: 230 to 280
[2]Twin wheel: 2.4
300 2.5 Turbo 1989 to 1991
[1]Twin wheel: 230 to 280
[2]Twin wheel: 2.4

– Not applicable, or information not available

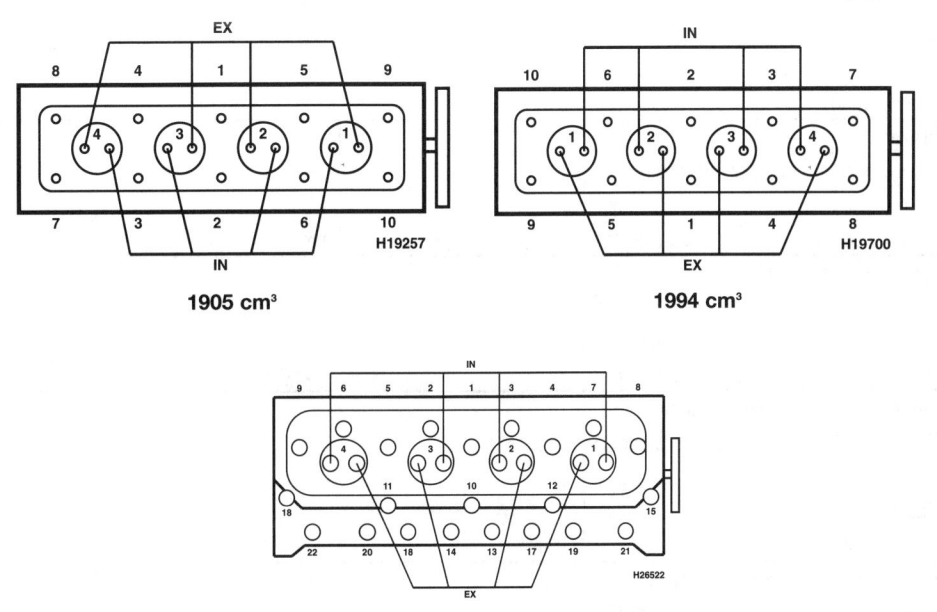

1905 cm³   1994 cm³

2498 cm³

LEYLAND DAF (LDV)

	300 2.5 1988 to 1991	400 / Convoy 2.5D 1989 to 1997	400 / Convoy 2.5TD 1989 to 1997	Convoy 2.5D 1997 to 2000
Engine				
Engine type/code....................	15J	EN55 OHV 53kW	EN55 OHV Turbo 71kW	4HB OHV 56kW
Capacity (cm³) / cylinders......................	2495 / 4	2498 / 4	2498 / 4	2496 / 4
Compression ration / pressure ....bar	21.0 /	23.0 / 25.0 to 30.0	23.0 / 25.0 to 30.0	20.8 / 33.8
Torque output ....Nm	0	0	0	168
Oil pressure ....idle [running] bar	2.4 to 4.0	[3.7 @ 4500]	[3.7 @ 4500]	1.0
Oil temperature ....°C	80	80	80	80
Valve clearances - inlet (mm)	0.25 H	0.15	0.15	0.20
- exhaust (mm)	0.25 H	0.25	0.25	0.38
Injection order	1-3-4-2	1-3-4-2	1-3-4-2	1-2-4-3
No. 1 cylinder position	TCE	–	–	–
Cooling system				
Thermostat opening temperature ....°C	82	–	–	88
Radiator cap pressure ....bar	1.0	–	–	1.0
Fuel system				
Idle speed ....rpm	670 ± 20	775 ± 25	775 ± 25	825 ± 25
Maximum (no load) speed ....rpm	–	4500 ± 50	4700	4500
Smoke test/opacity ....M⁻¹ %	3.0	2.5	3.0	2.5
Static timing method................................	–	Plunger travel	Plunger travel	Refer to wsm
Timing dimension..........................mm	–	Dimension on pump	Dimension on pump	–
Crankshaft position ....mm [°]	–	–	–	–
Turbo type / ref / pressure ....bar	–	–	0.8 bar @ 2000rpm	–
Injection pump make	CAV	Roto Diesel	Roto Diesel	Bosch
Injection pump part no..............................	–	DPC 059	DPC 053	VE
Injector Make / type	–	–	–	–
Injector part no.......................................	–	LCR 6770701C	–	–
Injection type..	DPA	Indirect	Indirect	VE
Injection opening pressure, New [used]...bar	135 to 140	110 to 120	125 to 135	–
Glow plugs				
Maker	–	Champion	Champion	–
Type	–	CH68	CH68	–
Nominal rating ....V/A	–	–	–	–
Brakes				
minimum friction material thickness				
Front..........................mm	3.0	1.0	1.0	1.0
Rear..........................mm	1.5	1.5	1.5	1.5
Tyres - Saloon / Hatch.....................Size	–	–	–	–
- Estate / Van..........................Size	–	–	–	205x14: 185x14
Pressure - front / rear - Saloon / Hatch ..bar	–	–	–	–
- Estate / Van ......bar	–	–	–	Owners handbook
Front suspension / wheel alignment				
Toe-in (+) / Toe-out (–)....................mm [°]	0.0 to 1.6	–	–	0.0 to 1.6
Camber	2°	–	–	1°30'
Castor	1°	–	–	2°
King pin inclination...................................	9°30'	–	–	9°30'
Rear suspension / wheel alignment				
Toe-in (+) / Toe-out (–)....................mm [°]	–	–	–	–
Camber				

LEYLAND DAF (LDV)

	300 2.5 1988 to 1991	400 / Convoy 2.5D 1989 to 1997	400 / Convoy 2.5TD 1989 to 1997	Convoy 2.5D 1997 to 2000
Torque wrench settings				
Cylinder head - stage 1 ....Nm	40 to 50	30, 70	30, 70	70
- stage 2 ....Nm	115 to 130	- 90°	- 90°	70
- stage 3 ....Nm	–	70	70	+ 90°
- stage 4 ....Nm	–	warm-up, wait 3 hrs,	warm-up, wait 3 hrs,	–
- stage 5 ....Nm	–	-90°	-90°	–
- stage 6 ....Nm	–	70 + 120°	70 + 120°	–
Big-end bearings................................Nm	37 to 41	60	60	69
Main bearings....................................Nm	130 to 136	120	120	170
Crankshaft pulley bolt ....Nm	–	55 + 60°	55 + 60°	163
Camshaft pulley bolt ....Nm	–	–	–	160
Flywheel [driveplate] bolt....................Nm	139 to 153	15 + 60°	15 + 60°	21 + 30°
Front hubs ....Nm	–	–	–	WSM
Rear hubs ....Nm	136 to 163	–	–	–
Wheel nuts / bolts ....Nm	176 to 227[1]	Refer to vehicle	Refer to vehicle	180[1]
Glow plugs ....Nm	–	–	–	–
Clutch pressure plate bolts ....Nm	30 to 38	20	20	30
Injection pump sprocket......................Nm	–	–	–	25
Injectors...Nm	–	90	90	40
Injection pump mounting bolts ....Nm	–	–	–	24
Injector pipe unions............................Nm	–	–	–	17
Capacities				
Engine oil & filter................................litres	6.4	7.4	7.7	7.0
Gearbox...litres	2.0	2.0	2.0	1.3
Automatic transmission ....litres	–	–	–	–
Final drive ....litres	1.7 Twin wheel: 2.4	1.9	1.9	2.4
Cooling system...................................litres	8.5	–	–	11.0
Fuel tank...litres	–	–	–	70

Notes

300 2.5 1988 to 1991
[1]Twin wheel: 230 to 280
Convoy 2.5D 1997 to 2000
[1]Twin wheel: 200

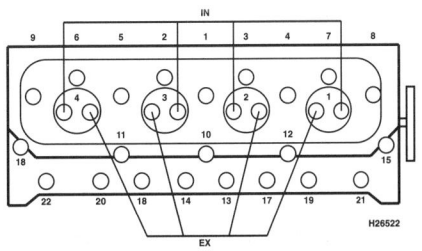

2498 cm³

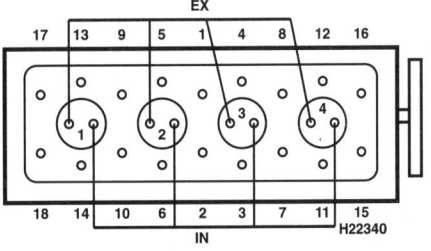

2496 cm³

– Not applicable, or information not
available

LEYLAND DAF (LDV)

	Convoy 2.5TD 1997 to 2000			
Engine				
Engine type/code.....................................	4EB OHV Turbo 74kW			
Capacity (cm³) / cylinders......................	2496 / 4			
Compression ration / pressure ....bar	18.0 / 33.8			
Torque output ....Nm	220			
Oil pressure ....idle [running] bar	1.0			
Oil temperature ....°C	80			
Valve clearances - inlet (mm)	0.20			
- exhaust (mm)	0.38			
Injection order	1-2-4-3			
No. 1 cylinder position	–			
Cooling system				
Thermostat opening temperature ....°C	88			
Radiator cap pressure ....bar	1.0			
Fuel system				
Idle speed ....rpm	850 ± 25			
Maximum (no load) speed ....rpm	4400			
Smoke test/opacity ....M⁻¹ %	3.0			
Static timing method................................	Refer to wsm			
Timing dimension..............................mm	–			
Crankshaft position ....mm [°]	–			
Turbo type / ref / pressure ....bar				
Injection pump make	Lucas			
Injection pump part no.............................	EPIC			
Injector Make / type	–			
Injector part no..	–			
Injection type...	EDC			
Injection opening pressure, New [used]...bar	–			
Glow plugs				
Maker	–			
Type	–			
Nominal rating....................................V/A	–			
Brakes				
minimum friction material thickness				
Front.............................mm	1.0			
Rear.............................mm	1.5			
Tyres - Saloon / Hatch......................Size	–			
- Estate / Van..........................Size	205x14: 185x14			
Pressure - front / rear - Saloon / Hatch...bar	–			
- Estate / Van ......bar	Owners handbook			
Front suspension / wheel alignment				
Toe-in (+) / Toe-out (–)....................mm [°]	0.0 to 1.6			
Camber	1°30'			
Castor	2°			
King pin inclination..................................	9°30'			
Rear suspension / wheel alignment				
Toe-in (+) / Toe-out (–)....................mm [°]	–			
Camber	–			

LEYLAND DAF (LDV)

	Convoy 2.5TD 1997 to 2000			
Torque wrench settings				
Cylinder head - stage 1 ....Nm	70			
- stage 2 ....Nm	70			
- stage 3 ....Nm	+ 90°			
- stage 4 ....Nm	–			
- stage 5 ....Nm	–			
- stage 6 ....Nm	–			
Big-end bearings................................Nm	69 N			
Main bearings....................................Nm	170			
Crankshaft pulley bolt ....Nm	163			
Camshaft pulley bolt ....Nm	160			
Flywheel [driveplate] bolt....................Nm	21 + 30°			
Front hubs ....Nm	WSM			
Rear hubs ....Nm	–			
Wheel nuts / bolts ....Nm	200[1]			
Glow plugs ....Nm	–			
Clutch pressure plate bolts ....Nm	30			
Injection pump sprocket.....................Nm	25			
Injectors..Nm	40			
Injection pump mounting bolts ....Nm	24			
Injector pipe unions............................Nm	17			
Capacities				
Engine oil & filter................................litres	7.0			
Gearbox..litres	1.3			
Automatic transmission ....litres	–			
Final drive ....litres	2.4			
Cooling system...................................litres	11.0			
Fuel tank...litres	70			

Notes

Convoy 2.5TD 1997 to 2000
[1]Twin wheel: 250

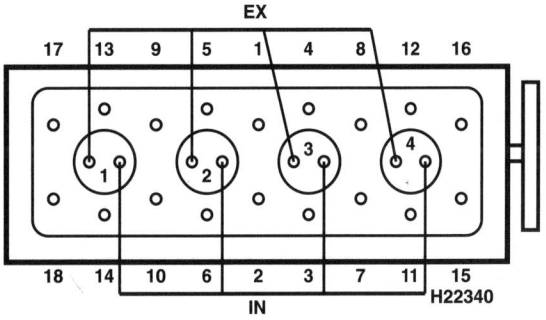

2496 cm³

– Not applicable, or information not
available

MAZDA

	121 1.8D 1996 to 1997	323, 1.7 1989 to 1994	626 2.0 1990 to 1992	626 2.0D 1994 to 1995
Engine				
Engine type/code.........................	RTJ SOHC 44kW	PN SOHC 41kW	RF SOHC 44kW	RF SOHC 55kW
Capacity (cm^3) / cylinders.........................	1753 / 4	1720 / 4	1998 / 4	1998 / 4
Compression ration / pressure ....bar	21.5 / 28.0	22.2 / ≥26.5	22.7 / ≥26.5	21.1 / 29.0
Torque output ....Nm	105	0	0	173
Oil pressure ....idle [running] bar	[1.5 @ 2000]	[≥3.0]	[≥3.0]	1.5 [3.5 @ 3000]
Oil temperature ....°C	80	80	_	80
Valve clearances - inlet (mm)	0.30 to 0.40	0.15	0.25	0.25
- exhaust (mm)	0.45 to 0.55	0.25	0.35	0.35
Injection order	1-3-4-2	1-3-4-2	1-3-4-2	1-3-4-2
No. 1 cylinder position	TBE	TBE	TBE	_
Cooling system				
Thermostat opening temperature ....°C	85	84 to 88	88 ± 1.5	83
Radiator cap pressure ....bar	1.2	0.76 to 1.04	0.9 ± 0.1	0.95 to 1.25
Fuel system				
Idle speed ....rpm	850 ± 50	800 +30 -0	700 +30 -0	725 ± 25
Maximum (no load) speed ....rpm	5350	5700	5400	5100
Smoke test/opacity ....M^{-1} %	2.5	2.5	2.5	2.5
Static timing method................................	Refer to wsm	Plunger travel	Plunger travel	Rotor lift
Timing dimension..............................mm	_	0.53	1.0	1.0
Crankshaft position ....mm [°]	_	[2] ATDC	[0] TDC	TDC
Turbo type / ref / pressure ....bar	_	_	_	_
Injection pump make	Bosch	Diesel Kiki	Diesel Kiki	Bosch
Injection pump part no..............................	_	PN46 13 800: 104740-0420	RF11 13 800B: 104748-0172	ZEXEL
Injector Make / type	Pintle	Bosch	Throttle type	_
Injector part no......................................	_	_	1	_
Injection type..	Indirect	Bosch VE type	Bosch VE type	Indirect VE
Injection opening pressure, New [used]...bar	134	108	132.5	132
Glow plugs				
Maker	Beru	Champion	Bosch	Bosch
Type	666 MJ	CH97	0250 202 056	0250 202 056
Nominal rating...................................V/A	_	10.5 / 16.5	10.5 / 16.5	11 / 12
Brakes				
minimum friction material thickness				
Front.........................mm	1.5	2.0	2.0	2.0
Rear.........................mm	1.0	1.0	1.0	1.0
Tyres - Saloon / Hatch......................Size	165/70x13	155x13: 175/70x13	185/70x14	195/65x14
- Estate / Van...........................Size	_	155x13: 175/70x13	185/70x14	_
Pressure - front / rear - Saloon / Hatch...bar	2.4 / 1.8	1.8 / 1.8	2.2 / 1.8	2.2 / 1.8
- Estate / Van ......bar		1.8 / 1.8	2.2 / 1.8	
Front suspension / wheel alignment				
Toe-in (+) / Toe-out (–)....................mm [°]	0.0 ± 1.0	2.0 ± 3.0	3.0 ± 3.0	3.0 ± 3.0
Camber	-2°4'	49' ± 30'	17' ± 45'	-0°36' ± 45'
Castor	0°5'	2°9' ± 45' Est: 1°35'±45'	1°13' ± 45'	2°37' ± 45'
King pin inclination..................................	_	12°22'	12°47'	15°04'
Rear suspension / wheel alignment				
Toe-in (+) / Toe-out (–)....................mm [°]	0.3	2.0 ± 3.0	0 ± 3.0	3.0 ± 3.0
Camber	-0°30'	-5' ± 45'	-30' ± 45'	-0°7' ± 45'

MAZDA

Torque wrench settings	121 1.8D 1996 to 1997	323, 1.7 1989 to 1994	626 2.0 1990 to 1992	626 2.0D 1994 to 1995
Cylinder head - stage 1 ....Nm	10 N	29	30	30
- stage 2 ....Nm	100, wait 3 mins	+ 90°	+ 90°	+ 90°
- stage 3 ....Nm	slacken	+ 90°	+ 90°	+ 90°
- stage 4 ....Nm	70	–	–	–
- stage 5 ....Nm	+ 120°	–	–	–
- stage 6 ....Nm	–	–	–	–
Big-end bearings.................................Nm	60 + 20° N	15 + 90° N	69 to 73 N	68 N
Main bearings....................................Nm	75	62 to 72 in stages	83 to 88	88
Crankshaft pulley bolt ....Nm	150, slacken, 120 + 60°	10 to 15	157 to 167	176
Camshaft pulley bolt ....Nm	–	39 to 54	56 to 65	65
Flywheel [driveplate] bolt....................Nm	18 + 45° + 45°	67 to 76	176 to 186	186
Front hubs ....Nm	270	WSM	235 to 319	310
Rear hubs ....Nm	270	–	98 to 117	170
Wheel nuts / bolts ....Nm	85	118	118	118
Glow plugs ....Nm	28	15 to 23	15 to 20	20
Clutch pressure plate bolts ....Nm	30	18 to 26	22 to 32	26
Injection pump sprocket......................Nm	23	59 to 69	69 to 79	68
Injectors...Nm	70	59 to 69	59 to 69	68
Injection pump mounting bolts ....Nm	20	19 to 25	–	25
Injector pipe unions............................Nm	25	26 to 33	31 to 46	33
Capacities				
Engine oil & filter.................................litres	4.5	3.9	5.5	5.5
Gearbox..litres	2.8	2.7	3.4	2.7
Automatic transmission ....litres	–	–	–	–
Final drive ....litres	WT	WT	WT	WT
Cooling system...................................litres	9.3	6.0	9.5	9.5
Fuel tank...litres	42	50	60	60

Notes

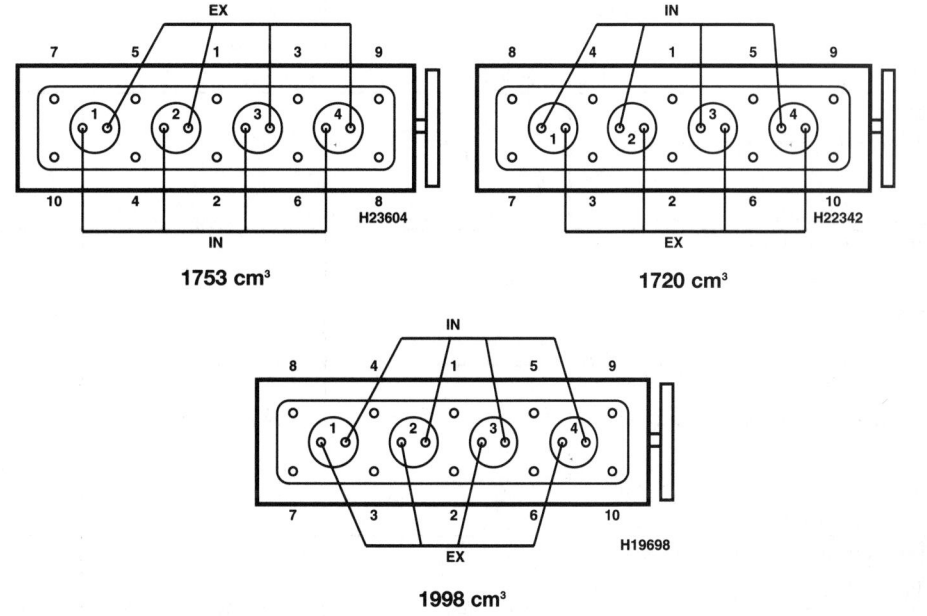

1753 cm³   1720 cm³

1998 cm³

– Not applicable, or information not
available

MAZDA

	B2200 Pick-up	E2200		
	1986 to 1997	1990 to 2000		
Engine				
Engine type/code...........................	R2 SOHC 47kW	R2 SOHC 52kW		
Capacity (cm³) / cylinders......................	2184 / 4	2184 / 4		
Compression ration / pressure ....bar	22.9 / ≥27.0	22.9 /		
Torque output ....Nm	136	136		
Oil pressure ....idle [running] bar	[4.0 to 4.8]	[4.0 @ 3000]		
Oil temperature ....°C	_	80		
Valve clearances - inlet (mm)	0.25	0.25		
- exhaust (mm)	0.35	0.35		
Injection order	1-3-4-2	1-3-4-2		
No. 1 cylinder position	F	TBE		
Cooling system				
Thermostat opening temperature ....°C	86.5 to 89.5	87		
Radiator cap pressure ....bar	0.8 to 1.0	0.75 to 1.05		
Fuel system				
Idle speed ....rpm	700 to 750	700 to 750		
Maximum (no load) speed ....rpm	5100	5100		
Smoke test/opacity ....M⁻¹ %	2.5	2.5		
Static timing method............................	Plunger travel	Plunger travel		
Timing dimension............................mm	1.0 ± 0.02	1.0		
Crankshaft position ....mm [°]	[2]	[2]		
Turbo type / ref / pressure ....bar	_	_		
Injection pump make	Diesel Kiki	Diesel Kiki		
Injection pump part no...........................	R230-13-800A: 104748-0151	VE		
Injector Make / type	Throttle type	Throttle type		
Injector part no..................................	_	104748 - 0151		
Injection type....................................	Bosch VE type	Indirect VE		
Injection opening pressure, New [used]...bar	132	132		
Glow plugs				
Maker	Champion	Bosch		
Type	CH97	0250 202 007		
Nominal rating...................................V/A	10.5 / 16.5	10.5 / 16.5		
Brakes				
minimum friction material thickness				
Front.............................mm	3.0	1.0		
Rear.............................mm	1.0	1.0		
Tyres - Saloon / Hatch.....................Size	_	_		
- Estate / Van...........................Size	185x14	185x14		
Pressure - front / rear - Saloon / Hatch ...bar	_	_		
- Estate / Van ......bar	1.8 / 2.4	3.0 / 4.5		
Front suspension / wheel alignment				
Toe-in (+) / Toe-out (–)............mm [°]	0 to 6.0	0.0 to 3.0		
Camber	25' to 1°15'	0°30' ± 30'		
Castor	5' to 1°35'	3°58' -1° -40'		
King pin inclination............................	8°15'	_		
Rear suspension / wheel alignment				
Toe-in (+) / Toe-out (–)....................mm [°]	_	_		
Camber				

MAZDA

	B2200 Pick-up 1986 to 1997	E2200 1990 to 2000		
Torque wrench settings				
Cylinder head - stage 1 ....Nm	29	29		
- stage 2 ....Nm	+ 90°	+ 90°		
- stage 3 ....Nm	+ 90°	+ 90°		
- stage 4 ....Nm	–	–		
- stage 5 ....Nm	–	–		
- stage 6 ....Nm	–			
Big-end bearings..................................Nm	65 to 69	67 N		
Main bearings..Nm	83 to 88	86		
Crankshaft pulley bolt ....Nm	157 to 167	33		
Camshaft pulley bolt ....Nm	55 to 65	66		
Flywheel [driveplate] bolt....................Nm	176 to 186	176		
Front hubs ....Nm	WSM	–		
Rear hubs ....Nm	–	–		
Wheel nuts / bolts ....Nm	118 to 147	100		
Glow plugs ....Nm	15 to 20	20		
Clutch pressure plate bolts ....Nm	18 to 26	25		
Injection pump sprocket.......................Nm	59 to 69	75		
Injectors..Nm	59 to 69	65		
Injection pump mounting bolts ....Nm	–	46		
Injector pipe unions.............................Nm	31 to 44	23		
Capacities				
Engine oil & filter.................................litres	5.4	5.8		
Gearbox..litres	2.0	2.5		
Automatic transmission ....litres	–	–		
Final drive ....litres	1.3	1.3		
Cooling system.....................................litres	7.5	9.0		
Fuel tank...litres	66	62		

Notes

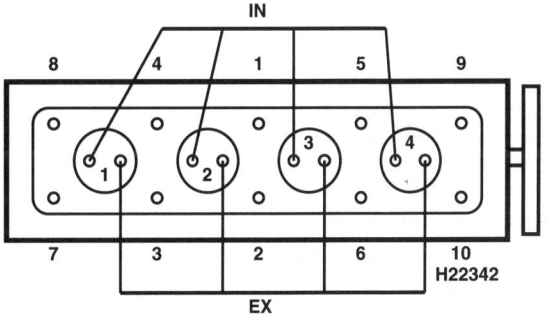

2184 cm³

– Not applicable, or information not available

MERCEDES-BENZ

	190D & CAT¹ (201.122) 1984 to 1993	190D 2.5 & CAT¹ (201.126) 1985 to 1993	190D 2.5 Turbo (201.128) 1987 to 1993	C220D (202.121/182) 1993 to 1999
Engine				
Engine type/code	601.911 SOHC 55kW	602.911 SOHC 69kW	602.961 SOHC Turbo 93kW	604.910 DOHC 16V 70kW
Capacity (cm³) / cylinders	1997 / 4	2497 / 5	2497 / 5	2155 / 4
Compression ration / pressure ...bar	22.0 / ≥18.0	22.0 / ≥18.0	22.0 / ≥18.0	22.0 / 26.0
Torque output ...Nm	0	0	0	150
Oil pressure ...idle [running] bar	0.3 [3.0]	0.3 [3.0]	0.3 [3.0]	[≥3.0 @ 3000]
Oil temperature ...°C	80	80	80	80
Valve clearances - inlet (mm)	0: Hyd.	0: Hyd.	0: Hyd.	O: Hyd.
- exhaust (mm)	0: Hyd.	0: Hyd.	0: Hyd.	0: Hyd.
Injection order	1-3-4-2	1-2-4-5-3	1-2-4-5-3	1-3-4-2
No. 1 cylinder position	TCE	TCE	TCE	_
Cooling system				
Thermostat opening temperature ...°C	85 ± 2	85 ± 2	85 ± 2	87
Radiator cap pressure ...bar	1.2 or 1.4	1.2 or 1.4	1.2 or 1.4	1.3 to 1.5
Fuel system				
Idle speed ...rpm	750 ± 50 Electronic: 720 ± 20	700 ± 50 Electronic: 680 ± 20	680 ± 20	690 to 790
Maximum (no load) speed ...rpm	5150 ± 150	5150 ± 150	5150 ± 150	5200 to 5600
Smoke test/opacity ...M⁻¹ %	2.5	2.5	2.5	1.99
Static timing method	Refer to wsm	Refer to wsm	Refer to wsm	Refer to wsm
Timing dimension ...mm	_	_	_	_
Crankshaft position ...mm [°]	[15 ± 1] ATDC	[15 ± 1] ATDC	[24 ± 1] ATDC	[14.0 to 14.5] ATDC
Turbo type / ref / pressure ...bar	_	_	_	_
Injection pump make	Bosch	Bosch	Bosch	Lucas
Injection pump part no	6010700501, 0401, 3001, 3101	6020700101, 05/21/32/33/3401	602 070 14 01	R86 40A 030A
Injector Make / type	Bosch	Bosch	Bosch	CAV
Injector part no	DNOSD261	DNOSD265	DNOSD265	LDC 001 R03
Injection type	PES4M 55C320RS152-1, 152-3,169	PES5M 55C320RS 153, 170	PES5M 55C320RS 158-1	Indirect
Injection opening pressure, New [used]...bar	≥115 [100]	≥135 [120]	≥135 [120]	115 to 125 [100]
Glow plugs				
Maker	Bosch	Bosch	Bosch	Bosch
Type	0250 201 117 or 0250 201 026²	0250 201 117 or 0250 201 026²	0250 201 026	0250 201 035
Nominal rating ...V/A	11 / 12	11 / 12	_	11 / 15
Brakes				
minimum friction material thickness				
Front ...mm	3.5	3.5	3.5	_
Rear ...mm	2.0	2.0	2.0	_
Tyres - Saloon / Hatch ...Size	185/65x15	185/65x15	185/65x15	195/65x15: 205/60x15
- Estate / Van ...Size	_	_	_	_
Pressure - front / rear - Saloon / Hatch ...bar	1.8 / 2.0	1.8 / 2.0	2.0 / 2.2¹	2.1 / 2.3
- Estate / Van ...bar	_	_	_	_
Front suspension / wheel alignment				
Toe-in (+) / Toe-out (–) ...mm [°]	[20' ± 10']	[20' ± 10']	[20' ± 10']	[0°25' ± 10']
Camber	0 11.88 ▶: -25' +10 -20'	0 11.88 ▶: -25' +10 -20'	0 11.88 ▶: -25' +10 -20'	-0°35' ± 20'
Castor	10°10' 11.88 ▶: 10°25' ± 30'	10°10' 11.88 ▶: 10°25' ± 30'	10°10' 11.88 ▶: 10°25' ± 30'	4°40' ± 30'
King pin inclination	_	_	_	_
Rear suspension / wheel alignment				
Toe-in (+) / Toe-out (–) ...mm [°]	[25' +10' -5']	[25' +10' -5']	[25' +10' -5']	[0°33' ± 7']
Camber	-1°30' ± 30'	-1°30' ± 30'	-1°30' ± 30'	_

MERCEDES-BENZ

	190D & CAT[1] (201.122) 1984 to 1993	190D 2.5 & CAT[1] (201.126) 1985 to 1993	190D 2.5 Turbo (201.128) 1987 to 1993	C220D (202.121/182) 1993 to 1999
Torque wrench settings				
Cylinder head - stage 1 ....Nm	15	15	15	15[1]
- stage 2 ....Nm	35	35	35	35
- stage 3 ....Nm	wait 10 mins	wait 10 mins	wait 10 min	+ 90°
- stage 4 ....Nm	+ 90°	+ 90°	+ 90°	Wait 10 mins
- stage 5 ....Nm	+ 90°	+ 90°	+ 90°	+ 90°
- stage 6 ....Nm	Bolts a: 25	Bolts a: 25	Bolts a: 25	–
Big-end bearings........................Nm	30 + 90 to 100° N	30 + 90 to 100° N	30 + 90 to 100° N	40 + 90° N
Main bearings........................Nm	M12: 90[3]	M12: 90[3]	M12: 90[2]	55 + 90° to 100°
Crankshaft pulley bolt ....Nm	320	320	320	200 + 90°
Camshaft pulley bolt ....Nm	M10: 65[4]	M10: 65[4]	M10: 65[3]	18
Flywheel [driveplate] bolt....................Nm	35 + 90 to 100°	35 + 90 to 100°	35 + 90 to 100°	40 + 90°
Front hubs ....Nm	WSM	WSM	WSM	–
Rear hubs ....Nm	–	–	–	200 to 240
Wheel nuts / bolts ....Nm	110	110	110	110
Glow plugs ....Nm	20 to 22	20 to 22	20 to 22	20
Clutch pressure plate bolts ....Nm	25	25	25	25
Injection pump sprocket......................Nm	40 to 50	40 to 50	40 to 50	–
Injectors........................Nm	Vert[1]: 70 to 80[5]	Vert[1]: 70 to 80[5]	40 to 50	40
Injection pump mounting bolts ....Nm	20 to 25	20 to 25	20 to 25	–
Injector pipe unions.............................Nm	10 to 20	10 to 20	10 to 20	20
Capacities				
Engine oil & filter.................................litres	6.0	6.5	7.0	7.0
Gearbox...litres	1.3 [1.5]	1.5	1.5	1.5
Automatic transmission ....litres	5.5	5.5	6.0	6.2
Final drive ....litres	0.7 ASD: 1.1	0.7 ASD: 1.1	1.1	0.7
Cooling system...................................litres	8.0	9.0	9.0	8.3
Fuel tank...litres	55	55	55	62

Notes

190D & CAT[1] (201.122) 1984 to 1993
[1] Catalyst standard from 6.90
[2] 0250 201 026 is for diagonal injection: 14 to 16 @ 8.0
[3] M11 bolts: 55 + 90 to 100°
[4] M11 bolts: 25 + 90°
[5] Diagonal injection: 40 to 50
190D 2.5 & CAT[1] (201.126) 1985 to 1993
[1] Catalyst standard from 6.90
[2] 0250201026 is for diagonal injection: 14 to 16 @ 8.0
[3] M11 bolts: 55 + 90 to 100°
[4] M11 bolts: 25 + 90°
[5] Diagonal injection: 40 to 50
190D 2.5 Turbo (201.128) 1987 to 1993
[1] 7.88 ▶: 2.1 / 2.3
[2] M11 bolts: 55 + 90 to 100°
[3] M11 bolts: 25 + 90°
C220D (202.121/182) 1993 to 1999
[1] Bolt length below head: ≤104 mm

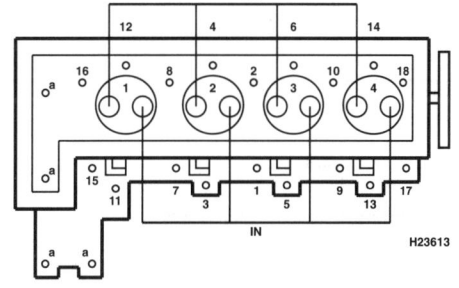

1997 cm³ / 2497 cm³

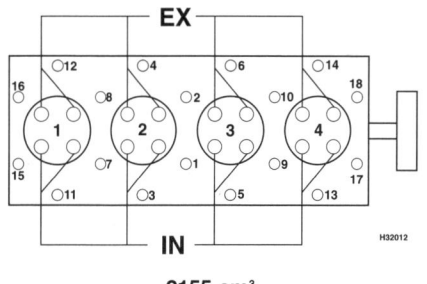

2155 cm³

– Not applicable, or information not available

MERCEDES-BENZ

	C250 D (202) 1993 to 1996	C250 TD (202.128 / 188) 1996 to 2000	250D, 250TD¹ (124.125 / 185) 1985 to 1993	E250D (210.010) 1993 to 1998
Engine				
Engine type/code.........................	605.910 83kW	605.960 DOHC 20V 110kW	602.912 SOHC 69kW	605.912 DOHC 20V 83kW
Capacity (cm³) / cylinders.........................	2497 / 5	2497 / 5	2497 / 5	2497 / 5
Compression ration / pressure............bar	22.0 /	22.0 /	22.0 / ≥18.0	22.0 / 26.0
Torque output ....Nm	280	280	0	170
Oil pressure ....idle [running] bar	[≥3.0 @ 3000]	[3.0 @ 3000]	0.3 [3.0]	[3.0 @ 3000]
Oil temperature ....°C	80	80	80	80
Valve clearances - inlet (mm)	0: Hyd.	0: Hyd.	0: Hyd.	0: Hyd.
- exhaust (mm)	0: Hyd.	0: Hyd.	0: Hyd.	0: Hyd.
Injection order..............................	1-2-4-5-3	1-2-4-5-3	1-2-4-5-3	1-2-4-5-3
No. 1 cylinder position	–	–	TCE	TCE
Cooling system				
Thermostat opening temperature ....°C	87	87	85 ± 2	85
Radiator cap pressure ....bar	1.3 to 1.5	1.3 to 1.5	1.2 or 1.4	1.3 to 1.5
Fuel system				
Idle speed ....rpm	610 to 710	610 to 710	700 ± 50 Electronic: 680 ± 20	610 to 710
Maximum (no load) speed ....rpm	5200 to 5600	5200 to 5600	5150 ± 150	5200 to 5600
Smoke test/opacity ....M⁻¹ %	1.9	1.9	1.9	1.9
Static timing method...........................	Refer to wsm	Refer to wsm	Refer to wsm	Refer to wsm
Timing dimension............................mm				
Crankshaft position ....mm [°]	[14 to 14.5] ATDC	[14.0 to 14.5 ATDC]	[15 ± 1] ATDC	[14.0 to 14.5 ATDC]
Turbo type / ref / pressure ....bar	–	–	–	–
Injection pump make	Bosch	Bosch	Bosch	Bosch
Injection pump part no.............................	0400 195 001	0400 195 001	6020700101, 05/21/32/33/3401	PES 5 M 55C 320 RS 202
Injector Make / type	Bosch	Bosch	Bosch	Bosch
Injector part no...............................	0430 211 997	–	DNOSD265	0432 217 253
Injection type................................	Indirect	Indirect	PES5M 55C320RS 153, 170	Indirect
Injection opening pressure, New [used]...bar	115 to 125 [100]	115 to 125 [110]	≥135 [120]	115 to 125 [100]
Glow plugs				
Maker	Bosch	Bosch	Bosch	Bosch
Type	0250 201 035	0250 201 038	0250 201 117 or 0250 201 026²	0250 201 035
Nominal rating.............................V/A	11 / 15	–	–	11 / 15
Brakes				
minimum friction material thickness				
Front...mm	–	–	3.5	3.5
Rear...mm	–	–	2.0	2.5
Tyres - Saloon / Hatch......................Size	195/65x15: 205/60x15	195/65x15: 205/60x15	195/65x15	205/60x15: 205/65x15
- Estate / Van.......................Size			195/65x15	205/65x15: 215/55x16
Pressure - front / rear - Saloon / Hatch ...bar	Refer to vehicle	Refer to vehicle	2.0 / 2.2³	Refer to vehicle
- Estate / Van ......bar			2.0 / 2.2	Refer to vehicle
Front suspension / wheel alignment				
Toe-in (+) / Toe-out (–)....................mm [°]	[0°25' ± 10']	[0°25' ± 10']	[20' ± 10']	WSM
Camber	-0°35' ± 20'	-0°35' ± 20'	-5' 11.88 ▶: -25' +10' -20'	-0°37' ± 20'
Castor	4°40' ± 30'	4°40' ± 30'	10°10' 11.88 ▶: 10°25' ± 30'	5°22' ± 30'
King pin inclination.............................				
Rear suspension / wheel alignment				
Toe-in (+) / Toe-out (–)....................mm [°]	[0°33' ± 7']	[0°33' ± 7']	[20' to 35']	[0°37' ± 7']
Camber	–		-1°30' ± 30'	–

MERCEDES-BENZ

	C250 D (202) 1993 to 1996	C250 TD (202.128 / 188) 1996 to 2000	250D, 250TD[1] (124.125 / 185) 1985 to 1993	E250D (210.010) 1993 to 1998
Torque wrench settings				
Cylinder head - stage 1 ....Nm	15[1]	15[1]	15	15 N
- stage 2 ....Nm	35	35	35	35
- stage 3 ....Nm	+ 90°	+ 90°	wait 10 min	+ 90°
- stage 4 ....Nm	wait 10 mins	wait 10 mins	+ 90°	wait 10 mins, + 90°
- stage 5 ....Nm	+ 90°	+ 90°	+ 90°	–
- stage 6 ....Nm	–	–	Bolts a: 25	–
Big-end bearings.....................Nm	30 + 90° to 100° N	30 + 90° to 100° N	30 + 90 to 100° N	40 + 90° N
Main bearings.....................Nm	55 + 90° to 100°	55 + 90° to 100°	M12: 90[4]	55 + 90°[1]
Crankshaft pulley bolt ....Nm	–	200 + 90°	320	200 + 90°
Camshaft pulley bolt ....Nm	–	18	M10: 65[5]	18
Flywheel [driveplate] bolt.....................Nm	40 + 90°	40 + 90°	35 + 90 to 100°	40 + 90°
Front hubs ....Nm	–	–	WSM	WSM
Rear hubs ....Nm	200 to 240	200 to 240	–	220
Wheel nuts / bolts ....Nm	110	110	110	110
Glow plugs ....Nm	–	20	20 to 22	20
Clutch pressure plate bolts ....Nm	25	25	25	25
Injection pump sprocket.....................Nm	–	–	40 to 50	–
Injectors.....................Nm	40	–	Vert': 70 to 80[6]	40
Injection pump mounting bolts ....Nm	–	–	20 to 25	20
Injector pipe unions.....................Nm	20	20	10 to 20	20
Capacities				
Engine oil & filter.....................litres	7.5	7.5	7.0	7.5
Gearbox.....................litres	1.5	1.5	1.5	1.5
Automatic transmission ....litres	6.2	6.2	5.5	6.2
Final drive ....litres	1.1	1.1	0.7	1.1
Cooling system.....................litres	9.6	9.6	9.0 A/C: 9.5	9.0
Fuel tank.....................litres	62	62	70 TD: 72	72

Notes

C250 D (202) 1993 to 1996
[1]Bolt length below head: ≤ 104 mm
C250 TD (202.128 / 188) 1996 to 2000
[1]Bolt length below head: ≤ 104 mm
250D, 250TD[1] (124.125 / 185) 1985 to 1993
[1]Catalyst standard from 6.90
[2]0250201026 is for diagonal injection: 14 to 16 @ 8.0
[3]11.85 ▶: 2.0 / 2.0
[4]M11 bolts: 55 + 90 to 100°
[5]M11 bolts: 25 + 90°
[6]Diagonal injection: 40 to 50
E250D (210.010) 1993 to 1998
[1]Max bolt length: ≤63.8 mm

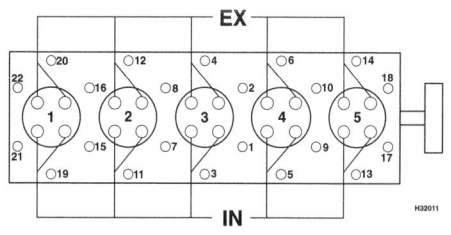

2497 cm³

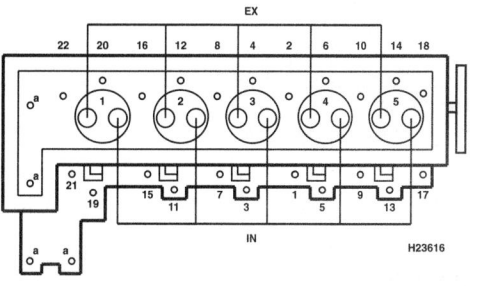

2497 cm³ (250D, 250TD[1] (124.125 / 185) 1985 to 1993)

– Not applicable, or information not available

MERCEDES-BENZ

	300D, TD[1] (124.130/190) 1985 to 1993	300D, TD Turbo (124.133/193) 1985 to 1993	300D 4MATIC[1] (124.330) 1987 to 1993	300D, TD Turbo[1] (124.333/39) 1987 to 1993
Engine				
Engine type/code........................	603.912 SOHC 83kW	603.960 SOHC Turbo 108kW	603.913 SOHC 83kW	603.963 SOHC Turbo 108kW
Capacity (cm³) / cylinders........................	2996 / 6	2996 / 6	2996 / 6	2996 / 6
Compression ration / pressure ....bar	22.0 / ≥18.0	22.0 / ≥18.0	22.0 / ≥18.0	22.0 / ≥18.0
Torque output ....Nm	0	0	0	0
Oil pressure ....idle [running] bar	0.3 [3.0]	0.3 [3.0]	0.3 [3.0]	0.3 [3.0]
Oil temperature ....°C	80	80	80	80
Valve clearances - inlet (mm)	0: Hyd.	0: Hyd.	0: Hyd.	0: Hyd.
- exhaust (mm)	0: Hyd.	0: Hyd.	0: Hyd.	0: Hyd.
Injection order	1-5-3-6-2-4	1-5-3-6-2-4	1-5-3-6-2-4	1-5-3-6-2-4
No. 1 cylinder position	TCE	TCE	TCE	TCE
Cooling system				
Thermostat opening temperature ....°C	85 ± 2	80 ± 2	85 ± 2	80 ± 2
Radiator cap pressure ....bar	1.2 or 1.4	1.2 or 1.4	1.2 or 1.4	1.2 or 1.4
Fuel system				
Idle speed ....rpm	630 ± 20 [680 ± 20]	630 ± 20	630 ± 20 [680 ± 20]	630 ± 20
Maximum (no load) speed ....rpm	5150 ± 150	5150 ± 150	5150 ± 150	5150 ± 150
Smoke test/opacity ....M⁻¹ %	2.1	2.5	2.1	2.5
Static timing method...............................	Refer to wsm	Refer to wsm	Refer to wsm	Refer to wsm
Timing dimension............................mm	–	–	–	–
Crankshaft position ....mm [°]	[15 ± 1] ATDC	[15 ± 1] ATDC	[15 ± 1] ATDC	[15 ± 1] ATDC
Turbo type / ref / pressure ....bar	–	–	–	–
Injection pump make	Bosch	Bosch	Bosch	Bosch
Injection pump part no............................	6030700101, 1901, 2501, 2601	603 070 12 01	6030700101, 1901, 2501, 2601	603 070 12 01
Injector Make / type	Bosch	Bosch	Bosch	Bosch
Injector part no...................................	DNOSD265	DNOSD265	DNOSD265	DNOSD265
Injection type......................................	PES6M 55C320RS 156, 171	PES6M 55C320RS 157-1	PES6M 55C320RS 156, 171	PES6M 55C320RS 157-1
Injection opening pressure, New [used]...bar	≥135 [120]	≥135 [120]	≥135 [120]	≥135 [120]
Glow plugs				
Maker	Bosch	Bosch	Bosch	Bosch
Type	0250 201 117 or 0250 201 026[2]	0250 201 117 or 0250 201 026[1]	0250 201 117 or 0250 201 026[2]	0250 201 117 or 0250 201 026[2]
Nominal rating...........................V/A	11 / 12	11 / 12	11 / 12	11 / 12
Brakes				
minimum friction material thickness				
Front...............................mm	3.5	3.5	3.5	3.5
Rear...............................mm	2.0	2.0	2.0	2.0
Tyres - Saloon / Hatch.....................Size	195/65x15	195/65x15	195/65x15	195/65x15
- Estate / Van...........................Size	195/65x15	195/65x15	–	195/65x15
Pressure - front / rear - Saloon / Hatch...bar	2.0 / 2.2[3]	2.2 / 2.2	2.4 / 2.4	2.4 / 2.4
- Estate / Van ......bar	2.0 / 2.2	2.0 / 2.5	–	2.2 / 2.5
Front suspension / wheel alignment				
Toe-in (+) / Toe-out (–).....................mm [°]	[20' ± 10']	[20' ± 10']	[20' ± 10']	[20' ± 10']
Camber	-5' 11.88 ▶: -25' +10' -20'	-5' 11.88 ▶: -25' +10' -20'	-15' +10' -20'	-15' +10' -20'
Castor	10°10' 11.88 ▶: 10°25' ± 30'	10°10' 11.88 ▶: 10°25' ± 30'	10°20' ± 30'	10°20' ± 30'
King pin inclination...............................	–	–	–	–
Rear suspension / wheel alignment				
Toe-in (+) / Toe-out (–)......................mm [°]	[20' to 35']	[20' to 35']	[20' to 35']	[20' to 35']
Camber	-1°30' ± 30'	-1°30' ± 30'	-1°30' ± 30'	-1°30' ± 30'

MERCEDES-BENZ

	300D, TD[1] (124.130/190) 1985 to 1993	300D, TD Turbo (124.133/193) 1985 to 1993	300D 4MATIC[1] (124.330) 1987 to 1993	300D, TD Turbo[1] (124.333/39) 1987 to 1993
Torque wrench settings				
Cylinder head - stage 1 ....Nm	15	15	15	15
- stage 2 ....Nm	35	35	35	35
- stage 3 ....Nm	wait 10 min	wait 10 min	wait 10 min	wait 10 min
- stage 4 ....Nm	+ 90°	+ 90°	+ 90°	+ 90°
- stage 5 ....Nm	+ 90°	+ 90°	+ 90°	+ 90°
- stage 6 ....Nm	Bolts a: 25	Bolts a: 25	Bolts a: 25	Bolts a: 25
Big-end bearings......................Nm	30 + 90 to 100° N	30 + 90 to 100° N	30 + 90 to 100° N	30 + 90 to 100° N
Main bearings......................Nm	M12: 90[4]	M12: 90[2]	M12: 90[3]	M12: 90[3]
Crankshaft pulley bolt ....Nm	320	320	320	320
Camshaft pulley bolt......................Nm	M10: 65[5]	M10: 65[3]	M10: 65[4]	M10: 65[4]
Flywheel [driveplate] bolt......................Nm	35 + 90 to 100°	35 + 90 to 100°	40 + 90 to 100°	35 + 90 to 100°
Front hubs ....Nm	WSM	WSM	WSM	WSM
Rear hubs ....Nm	–	–	–	–
Wheel nuts / bolts ....Nm	110	110	110	110
Glow plugs ....Nm	20 to 22	20 to 22	20 to 22	20 to 22
Clutch pressure plate bolts ....Nm	25	25	25	25
Injection pump sprocket......................Nm	40 to 50	40 to 50	40 to 50	40 to 50
Injectors......................Nm	Vert': 70 to 80[6]	Vert': 70 to 80[4]	Vert': 70 to 80[5]	Vert': 70 to 80[5]
Injection pump mounting bolts ....Nm	20 to 25	20 to 25	20 to 25	20 to 25
Injector pipe unions......................Nm	10 to 20	10 to 20	10 to 20	10 to 20
Capacities				
Engine oil & filter......................litres	7.0	7.5	7.5	7.5
Gearbox......................litres	1.5	–	1.5 Transfer: 0.6	–
Automatic transmission ....litres	6.0	6.2	6.2	6.2
Final drive......................litres	1.1	1.1	1.1 Front: 1.0	1.1
Cooling system......................litres	9.0 A/C: 9.5	10.0	9.0 A/C: 9.5	10.0
Fuel tank......................litres	70 TD: 72	70 TD: 72	70	70 TD: 72

Notes

300D, TD[1] (124.130/190) 1985 to 1993
[1]Catalyst standard from 6.90
[2]0250201026 is for diagonal injection: 14 to 16 @ 8.0
[3]11.85 ▶: 2.0 / 2.0
[4]M11 bolts: 55 + 90 to 100°
[5]M11 bolts: 25 + 90°
[6]Diagonal injection: 40 to 50
300D, TD Turbo (124.133/193) 1985 to 1993
[1]0250201026 is for diagonal injection: 14 to 16 @ 8.0
[2]M11 bolts: 55 + 90 to 100°
[3]M11 bolts: 25 + 90°
[4]Diagonal injection: 40 to 50
300D 4MATIC[1] (124.330) 1987 to 1993
[1]Catalyst standard from 6.90
[2]0250 201 026 is for diagonal injection: 14 to 16 @ 8.0
[3]M11 bolts: 55 + 90 to 100°
[4]M11 bolts: 25 + 90°
[5]Diagonal injection: 40 to 50
300D, TD Turbo[1] (124.333/39) 1987 to 1993
[1]4MATIC
[2]0250 201 026 is for diagonal injection: 14 to 16 @ 8.0
[3]M11 bolts: 55 + 90 to 100°
[4]M11 bolts: 25 + 90°
[5]Diagonal injection: 40 to 50

– Not applicable, or information not available

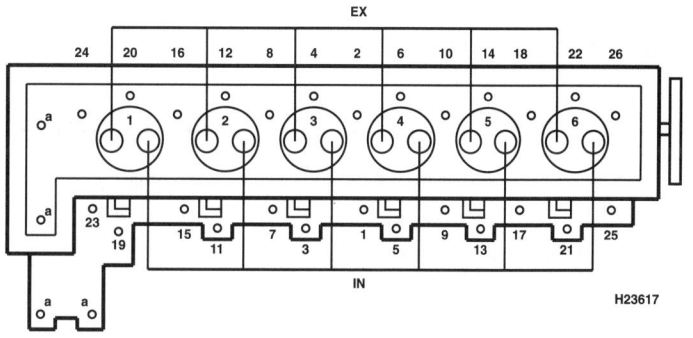

2996 cm³

H23617

MERCEDES-BENZ

	E300 D 1993 to 1996	E300 TD 1995 to 2000	E300 TD (210.020) 1997 to 2000	Vito 108D 1996 to 1999
Engine				
Engine type/code............................	606.910 DOHC 24V 100kW	606.910 DOHC 24V 100kW	606.912 DOHC 24V 130kW	OM601.942 SOHC 58kW
Capacity (cm³) / cylinders......................	2996 / 6	2996 / 6	2996 / 6	2299 / 4
Compression ration / pressure ....bar	22.0 / 26.0	22.0 / ≥18.0	22.0 / ≥18.0	22.0 / ≥18.0
Torque output ....Nm	0	210	330	152
Oil pressure ....idle [running] bar	[3.0 @ 3000]	[3.0 @ 3000]	_	[3.0 @ 3000]
Oil temperature ....°C	80	80	_	80
Valve clearances - inlet (mm)	0: Hyd.	0: Hyd.	0: Hyd.	0: Hyd.
- exhaust (mm)	0: Hyd.	0: Hyd.	0: Hyd.	0: Hyd.
Injection order	1-5-3-6-2-4	1-5-3-6-2-4	1-5-3-6-2-4	1-3-4-2
No. 1 cylinder position	_	TCE	TCE	_
Cooling system				
Thermostat opening temperature ....°C	85	85	85	87
Radiator cap pressure ....bar	1.3 to 1.5	1.3 to 1.5	1.3 to 1.5	1.4
Fuel system				
Idle speed ....rpm	580 to 680	580 to 680	580 to 680	720 ± 50
Maximum (no load) speed ....rpm	5600	5200 to 5600	5200 to 5600	3800
Smoke test/opacity ....M⁻¹ %	2.5	2.5	2.5	2.5
Static timing method...............................	Refer to wsm	Refer to wsm	Refer to wsm	_
Timing dimension................................mm	_	_	_	_
Crankshaft position ....mm [°]	[14.0 to 14.5 ATDC]	[14.0 to 14.5 ATDC]	[14.0 to 14.5 ATDC]	_
Turbo type / ref / pressure ....bar	_	_	_	_
Injection pump make	Bosch	Bosch	Bosch	Bosch
Injection pump part no..............................	0400 076 954	PES 6M 55C 320 RS 203	PES 6M 55C 320 RS 203	PES 4M 55C 320 RS167
Injector Make / type	Bosch	Bosch	Bosch	Bosch
Injector part no..	0432 217 253	DN 0SD 310	DN 0SD 310	_
Injection type..	Indirect	Indirect	Indirect	Indirect
Injection opening pressure, New [used]...bar	115 to 125 [100]	115 to 125 [100]	115 to 125 [100]	115 to 125 [110]
Glow plugs				
Maker	Bosch	Bosch	Bosch	_
Type	0250 201 035	0250 201 038	0250 201 038	_
Nominal rating......................................V/A	11 / 15	11.5 / 15	_	_
Brakes				
minimum friction material thickness				
Front..............................mm	_	3.5	3.5	_
Rear..............................mm	_	2.5	2.5	_
Tyres - Saloon / Hatch......................Size	195/65x15	195/65x15	205/60x15: 205/65x15:215/55x16	_
- Estate / Van...........................Size	195/65x15	195/65x15	205/65x15: 215/55x16	195/70x15
Pressure - front / rear - Saloon / Hatch ...bar	2.2 / 2.4	Refer to vehicle	Refer to vehicle	_
- Estate / Van ......bar	2.2 / 2.5	Refer to vehicle	Refer to vehicle	Refer to vehicle
Front suspension / wheel alignment				
Toe-in (+) / Toe-out (−).....................mm [°]	2.5 ± 1.0	WSM	WSM	_
Camber	-0°25' +10' -20'	-0°37' ± 20'	-0°37' ± 20'	_
Castor	10°27' ± 30'	5°22' ± 30'	5°22' ± 30'	_
King pin inclination...................................	_	_	_	_
Rear suspension / wheel alignment				
Toe-in (+) / Toe-out (−)....................mm [°]	[0°25' +10' -5']	[0°33' ± 7']	[0°33' ± 7']	_
Camber	-1°30' ± 30'			_

MERCEDES-BENZ

	E300 D 1993 to 1996	E300 TD 1995 to 2000	E300 TD (210.020) 1997 to 2000	Vito 108D 1996 to 1999
Torque wrench settings				
Cylinder head - stage 1 ....Nm	15[1]	15 N[2]	15 N	10
- stage 2 ....Nm	35 + 90°	35	35	35
- stage 3 ....Nm	wait 10 min	+ 90°	+ 90°	+ 180°
- stage 4 ....Nm	+ 90°	wait 10 min	wait 10 min	M8: 25
- stage 5 ....Nm	M8: 25	+ 90°	+ 90°	–
- stage 6 ....Nm	–	–	–	–
Big-end bearings.............................Nm	40 + 90° N	40 + 90° N	40 + 90° N	30 + 90° N
Main bearings.................................Nm	55 + 90°[2]	55 + 90°[1]	55 + 90°[1]	55 + 90°
Crankshaft pulley bolt ....Nm	200 + 90°	200 + 90°	200 + 90°	–
Camshaft pulley bolt ....Nm	18	18	18	–
Flywheel [driveplate] bolt....................Nm	40 + 90°	40 + 90°	40 + 90°	45 + 90°
Front hubs ....Nm	–	WSM	WSM	–
Rear hubs......................................Nm	–	220	220	–
Wheel nuts / bolts ....Nm	110	110	110	170
Glow plugs ....Nm	20	20	20	–
Clutch pressure plate bolts ....Nm	25	25	25	25
Injection pump sprocket.....................Nm	–	–	–	–
Injectors..Nm	40	40	40	80
Injection pump mounting bolts ....Nm	–	–	–	–
Injector pipe unions...........................Nm	20	20	20	–
Capacities				
Engine oil & filter..............................litres	7.0	7.0	7.0	6.0
Gearbox...litres	1.5	1.5	1.5	2.0
Automatic transmission ....litres	6.0	6.0	6.0	–
Final drive ....litres	1.1	1.1	1.1	WT
Cooling system................................litres	9.0	9.0	9.0	–
Fuel tank..litres	–	72	72	80

Notes

E300 D 1993 to 1996
[1]Bolt length below head: ≤104mm or 117mm
[2]Bolt length below head: ≤63.8mm
E300 TD 1995 to 2000
[1]Bolt length below head: ≤63.8 mm
[2]Bolt length below head: ≤104mm or 117mm
E300 TD (210.020) 1997 to 2000
[1]Max bolt length below head: ≤63.8 mm

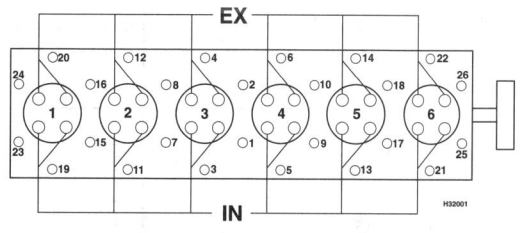

1996 cm³ / 2996 cm³

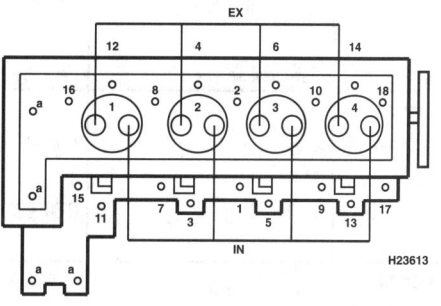

2299 cm³

– Not applicable, or information not
available

MERCEDES-BENZ

	208/308D 1989 to 1995	310/410D 1989 to 1995	408D 1989 to 1995	Sprinter 208/308 D 1997 to 2000
Engine				
Engine type/code......................	OM601.940 SOHC 58kW	OM602.940 SOHC 70kW	OM601.940 SOHC 58kW	OM601.943 SOHC 58kW
Capacity (cm³) / cylinders........................	2299 / 4	2874 / 5	2299 / 4	2299 / 4
Compression ration / pressure ....bar	22.0 / 18.0	22.0 / ≥18.0	22.0 / 18.0	22.0 / ≥18.0
Torque output ....Nm	157	192	157	152
Oil pressure ....idle [running] bar	[3.0 @ 3000]	[3.0 @ 3000]	[3.0 @ 3000]	[3.0 @ 3000]
Oil temperature ....°C	80	80	80	80
Valve clearances - inlet (mm)	0: Hyd.	0: Hyd.	0: Hyd.	0: Hyd.
- exhaust (mm)	0: Hyd.	0: Hyd.	0: Hyd.	0: Hyd.
Injection order	1-3-4-2	1-2-4-5-3	1-3-4-2	1-3-4-2
No. 1 cylinder position	–	–	–	–
Cooling system				
Thermostat opening temperature ....°C	80	80	80	80
Radiator cap pressure ....bar	0.9 to 1.1	0.9 to 1.1	0.9 to 1.1	0.9 to 1.1
Fuel system				
Idle speed ....rpm	750 ± 50	680 ± 50	720 ± 50	720 ± 50
Maximum (no load) speed ....rpm	4500	4500	4500	3800
Smoke test/opacity ....M⁻¹ %	2.2	2.3	2.2	1.7
Static timing method.............................	Refer to wsm	Refer to wsm	Refer to wsm	Refer to wsm
Timing dimension.............................mm	–	–	–	–
Crankshaft position ....mm [°]	[15 ± 1 ATDC]	[15 ± 1 ATDC]	[15 ± 1 ATDC]	[15 ± 1 ATDC]
Turbo type / ref / pressure ....bar	–	–	–	–
Injection pump make	Bosch	Bosch	Bosch	Bosch
Injection pump part no.............................	0400 074 900/5	0400 075 956/47	0400 074 900/5	PES
Injector Make / type	Bosch	Bosch	Bosch	Bosch
Injector part no..	0432 217 161	0432 217 161	0432 217 161	–
Injection type..	Indirect	Indirect	Indirect	Indirect
Injection opening pressure, New [used]...bar	115 to 125 [100]	115 to 125 [100]	115 to 125 [100]	115 to 125 [100]
Glow plugs				
Maker	Bosch/Champion	Bosch/Champion	Bosch/Champion	Bosch
Type	0250 201 026 / CH156	0250 201 026 / CH156	0250 201 026 / CH156	0250 201 026
Nominal rating...................................V/A	11.5 / 15	11.5 / 15	11.5 / 15	–
Brakes				
minimum friction material thickness				
Front..mm	2.0	3.0	3.0	2.0
Rear..mm	3.5	3.5	3.5	3.5
Tyres - Saloon / Hatch....................Size	–	–	–	–
- Estate / Van..........................Size	185x14 308D: 225/70x15	225/70x15: 185x14	185x14	195/70x15: 225/70x15
Pressure - front / rear - Saloon / Hatch ...bar				
- Estate / Van ......bar	Refer to vehicle	Refer to vehicle	Refer to vehicle	Refer to vehicle
Front suspension / wheel alignment				
Toe-in (+) / Toe-out (–)....................mm [°]	0.0	0.0	0.0	–
Camber	1°	1°	1°	–
Castor	2°30' ± 20'	2°30' ± 30'	2°30' ± 20'	–
King pin inclination................................	–	5°	5°	–
Rear suspension / wheel alignment				
Toe-in (+) / Toe-out (–)....................mm [°]	–	–	–	–
Camber	–	–	–	–

MERCEDES-BENZ

	208/308D 1989 to 1995	310/410D 1989 to 1995	408D 1989 to 1995	Sprinter 208/308 D 1997 to 2000
Torque wrench settings				
Cylinder head - stage 1 ....Nm	15	15	15	15
- stage 2 ....Nm	35 + 90°	35	35	35
- stage 3 ....Nm	wait 10 min	+ 90°	+ 90°	+ 90°
- stage 4 ....Nm	+ 90°	wait 10 min	wait 10 min	Wait 10 mins
- stage 5 ....Nm	–	+ 90°	+ 90°	+ 90°
- stage 6 ....Nm	–	–	–	–
Big-end bearings..............................Nm	30 + 90°	30 + 90° N	30 + 90° N	30 + 90° N
Main bearings...................................Nm	55 + 90°	55 + 90°	55 + 90°	55 + 90° N
Crankshaft pulley bolt ....Nm	320	35 + 90°¹	200 + 90°	200 + 90° N
Camshaft pulley bolt ....Nm	–	–	–	–
Flywheel [driveplate] bolt...................Nm	35 + 90°	35 + 90°	35 + 90°	35 + 90°
Front hubs ....Nm	–	–	–	–
Rear hubs ....Nm	–	–	–	–
Wheel nuts / bolts ....Nm	170	170	170	170
Glow plugs ....Nm	–	20	–	20
Clutch pressure plate bolts ....Nm	25	25	25	25
Injection pump sprocket......................Nm	45	45	45	–
Injectors..Nm	80	80	80	80
Injection pump mounting bolts ....Nm	25	25	25	–
Injector pipe unions............................Nm	20	20	20	–
Capacities				
Engine oil & filter.............................litres	6.5	7.0	6.5	9.5
Gearbox..litres	2.3	2.3	2.3	1.6
Automatic transmission ....litres	6.8	6.9	6.9	–
Final drive ....litres	1.4 308D: 1.6	1.6 410D: 1.8	1.8	1.5
Cooling system..................................litres	7.0	7.0	7.0	7.0
Fuel tank...litres	70	70	70	80

Notes

310/410D 1989 to 1995
¹Bolt length below head: ≤22.5mm Bolt diameter: ≥8.1 mm

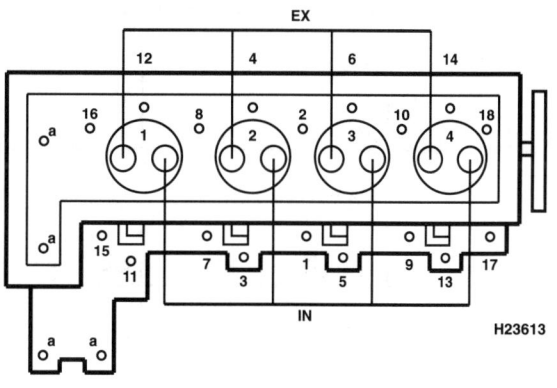

2299 cm³ / 2874 cm³

– Not applicable, or information not available

MERCEDES-BENZ

	Sprinter 210/310 D 1997 to 2000	Sprinter 212/312 D 1997 to 2000		
Engine				
Engine type/code	OM602.980 SOHC Turbo 75kW	OM602.980 SOHC Turbo 90kW		
Capacity (cm³) / cylinders........................	2874 / 5	2874 / 5		
Compression ration / pressure ....bar	22.0 / ≥18.0	22.0 / ≥18.0		
Torque output ....Nm	250	280		
Oil pressure ....idle [running] bar	[3.0 @ 4000]	[3.0 @ 3000]		
Oil temperature ....°C	80	80		
Valve clearances - inlet (mm)	0: Hyd.	0: Hyd.		
- exhaust (mm)	0: Hyd.	0: Hyd.		
Injection order	1-2-4-5-3	1-2-4-5-3		
No. 1 cylinder position	–	–		
Cooling system				
Thermostat opening temperature ....°C	80	80		
Radiator cap pressure ....bar	0.9 to 1.2	0.9 to 1.1		
Fuel system				
Idle speed ....rpm	680 ± 50	680 ± 50		
Maximum (no load) speed ....rpm	–	–		
Smoke test/opacity ....M⁻¹ %	2.5	1.6		
Static timing method...............................	Rotor lift	Rotor lift		
Timing dimension.................................mm	0.9	0.9		
Crankshaft position ....mm [°]	TDC	TDC		
Turbo type / ref / pressure ....bar	–	–		
Injection pump make	Bosch	Bosch		
Injection pump part no.............................	–	–		
Injector Make / type	Bosch	Bosch		
Injector part no.......................................	–	–		
Injection type..	Indirect	Indirect		
Injection opening pressure, New [used]...bar	115 to 125 [100]	115 to 125 [100]		
Glow plugs				
Maker	Bosch/Champion	–		
Type	0250 201 026 / CH176	–		
Nominal rating....................................V/A	11.5 /	–		
Brakes				
minimum friction material thickness				
Front...mm	2.0	2.0		
Rear..mm	3.5	3.5		
Tyres - Saloon / Hatch.....................Size	–	–		
- Estate / Van...........................Size	195/70x15: 225/70x15	195/70x15: 225/70x15		
Pressure - front / rear - Saloon / Hatch ...bar	–	–		
- Estate / Van ......bar	Refer to vehicle	Refer to vehicle		
Front suspension / wheel alignment				
Toe-in (+) / Toe-out (–).....................mm [°]	–	–		
Camber	–	–		
Castor	–	–		
King pin inclination.................................	–	–		
Rear suspension / wheel alignment				
Toe-in (+) / Toe-out (–)................mm [°]	–	–		
Camber				

MERCEDES-BENZ

	Sprinter 210/310 D 1997 to 2000	Sprinter 212/312 D 1997 to 2000		
Torque wrench settings				
Cylinder head - stage 1 ....Nm	10	10		
- stage 2 ....Nm	35	35		
- stage 3 ....Nm	+ 180°	+ 180°		
- stage 4 ....Nm	Wait 10 mins	Wait 10 mins		
- stage 5 ....Nm	+ 90°	+ 90°		
- stage 6 ....Nm	–	–		
Big-end bearings........................Nm	30 + 90° N	30 + 90° N		
Main bearings........................Nm	55 + 90° N	55 + 90° N		
Crankshaft pulley bolt ....Nm	200 + 90° N	200 + 90° N		
Camshaft pulley bolt ....Nm	–	–		
Flywheel [driveplate] bolt....................Nm	35 + 90° N	35 + 90°		
Front hubs ....Nm	–	–		
Rear hubs ....Nm	–	–		
Wheel nuts / bolts ....Nm	170 ± 10	170 ± 10		
Glow plugs ....Nm	–	20		
Clutch pressure plate bolts ....Nm	25	25		
Injection pump sprocket.....................Nm	–	90		
Injectors........................Nm	80	80		
Injection pump mounting bolts ....Nm	–	15		
Injector pipe unions...........................Nm	–	–		
Capacities				
Engine oil & filter.................................litres	9.5	9.5		
Gearbox........................litres	1.6	1.9		
Automatic transmission ....litres	–	–		
Final drive ....litres	1.7	1.5		
Cooling system................................litres	7.5	7.5		
Fuel tank........................litres	80	80		

Notes

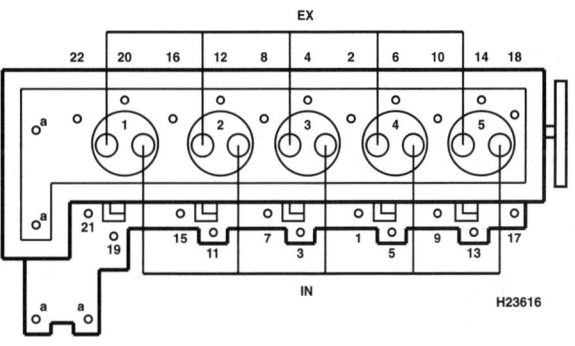

2874 cm³

– Not applicable, or information not available

MITSUBISHI

	Galant 1800 1988 to 1993	Carisma 1.9TD 1997 to 2000	Space Wagon 2.0TD 1995 to 1996	Shogun / Pajero 2.5 Turbo 1986 to 1991
Engine				
Engine type/code.........................	4D65-T SOHC Turbo 55kW	F8QT SOHC Turbo 66kW	4D68T SOHC Turbo 60kW	4D56T SOHC Turbo 69kW
Capacity (cm^3) / cylinders......................	1795 / 4	1870 / 4	1998 / 4	2477 / 4
Compression ration / pressure ....bar	22.2 / 26.5	21.0 / 20.0	22.4 / 25.6	21.0 / 27.0
Torque output ....Nm	0	176	172	240
Oil pressure ....idle [running] bar	0.76	2.0 [3.5 @ 3000]	_	0.76
Oil temperature ....°C	80	80	_	80
Valve clearances - inlet (mm)	0.25 H	0.15 to 0.25	0.25	0.25 H
- exhaust (mm)	0.25 H	0.35 to 0.45	0.35	0.25 H
Injection order	1-3-4-2	1-3-4-2	1-3-4-2	1-3-4-2
No. 1 cylinder position	TBE	_	_	F
Cooling system				
Thermostat opening temperature ....°C	82	76	82	82
Radiator cap pressure ....bar	0.76 to 1.04	0.75 to 1.05	0.8 to 1.0	0.76 to 1.04
Fuel system				
Idle speed ....rpm	800 ± 30	825 ± 25	750 ± 100	750 ± 50
Maximum (no load) speed ....rpm	5200	_	5200	5200
Smoke test/opacity ....M^{-1} %	2.5	2.5	1.5	2.1
Static timing method...............................	Plunger travel	Plunger travel	Plunger travel	Plunger travel
Timing dimension...............................mm	1.0 ± 0.03	0.02	1.0 ± 0.03	1.0 ± 0.03
Crankshaft position ....mm [°]	[7] ATDC	TDC	[7] ATDC	[7] ATDC
Turbo type / ref / pressure ....bar	_	0.80 to 0.95 bar @ 3000rpm	_	_
Injection pump make	Diesel Kiki	_	Nippon Denso	Diesel Kiki
Injection pump part no..............................	_	_	VE	_
Injector Make / type	Throttle type	_	Nippon Denso	Throttle type
Injector part no..	Small	_	_	_
Injection type...	MD077258 or 259	Mitsubishi Electronic	Indirect DPA	Indirect
Injection opening pressure, New [used]...bar	≥118 [108]	130	118 to 127	≥118 [108]
Glow plugs				
Maker	Bosch	Bosch	Bosch	Bosch
Type	0250 202 003	0250 202 003	0250 202 003	0250 202 003
Nominal rating.....................................V/A	12 / 8.5	12 / 8.5	11 / 18	12 / 8.5
Brakes				
minimum friction material thickness				
Front...mm	2.0	2.0	2.0	1.0
Rear...mm	1.0	_	2.0	1.0
Tyres - Saloon / Hatch...................Size	165x14: 185/70x14	185/65x14	_	_
- Estate / Van..........................Size			185/70x14	215x15: 205x16
Pressure - front / rear - Saloon / Hatch...bar	2.0 / 2.1	2.1 / 2.1	_	_
- Estate / Van ......bar	_	_	2.2 / 2.0	1.6 / 2.0
Front suspension / wheel alignment				
Toe-in (+) / Toe-out (−).....................mm [°]	0 ± 1.5	1.0 ± 2.0	0.0 ± 3.0	1.0 to 4.5
Camber	22' ± 30'	0°30'	0°20' ± 30'	1° ± 30'
Castor	2° ± 30'	2°12'	2°10' ± 40'	2°57' ± 30' 5dr: 3°5'±30'
King pin inclination..................................	13°55'	_	13°50'	8°
Rear suspension / wheel alignment				
Toe-in (+) / Toe-out (−)....................mm [°]	0 ± 1.5 PW	3.0 ± 2.0	-2.0 ± 2.0	_
Camber	-45' ± 14'	-0°40' ± 30'	-0°30' ± 30'	_

MITSUBISHI

Torque wrench settings	Galant 1800 1988 to 1993	Carisma 1.9TD 1997 to 2000	Space Wagon 2.0TD 1995 to 1996	Shogun / Pajero 2.5 Turbo 1986 to 1991
Cylinder head - stage 1 ....Nm	45	30 + 50°	90	103 to 112 C
- stage 2 ....Nm	85	Loosen	slacken	113 to 122 H
- stage 3 ....Nm	125	25	40	–
- stage 4 ....Nm	–	+ 215°	+ 90°	–
- stage 5 ....Nm	–	warm engine	+ 90°	–
- stage 6 ....Nm	–	+ 120°	–	–
Big-end bearings........Nm	49 to 51 N	45 N	20 + 90° N	45 to 48
Main bearings........Nm	64 to 68	65	25 + 90°	74 to 83
Crankshaft pulley bolt ....Nm	108 to 127	120	118	108 to 127
Camshaft pulley bolt ....Nm	79 to 98	50	88	79 to 98
Flywheel [driveplate] bolt........Nm	128 to 137	53	132	128 to 137
Front hubs ....Nm	200 to 260	230	250	WSM
Rear hubs ....Nm	–	175	230	WSM
Wheel nuts / bolts ....Nm	90 to 110	100	100	100 to 120
Glow plugs ....Nm	18	23	18	20
Clutch pressure plate bolts ....Nm	15 to 21	15 to 22	22	15 to 22
Injection pump sprocket........Nm	59 to 68	50	83	59 to 68
Injectors........Nm	68	70	55	68
Injection pump mounting bolts ....Nm	15 to 21	–	24	15 to 21
Injector pipe unions ....Nm	23 to 36	23	30	23 to 36
Capacities				
Engine oil & filter........litres	5.6	5.3	5.1	5.9
Gearbox........litres	1.8	1.8	1.8	2.2[1]
Automatic transmission ....litres	–	–	–	7.2
Final drive ....litres	WT	WT	WT	Front: 1.1[2]
Cooling system........litres	8.1	6.0	6.0	8.0[3]
Fuel tank........litres	60	60	60	60

Notes

Shogun / Pajero 2.5 Turbo 1986 to 1991
[1]Transfer box: 2.2. Transfer box with PTO: 2.6
[2]Rear: 2.6
[3]With rear heater: 8.8

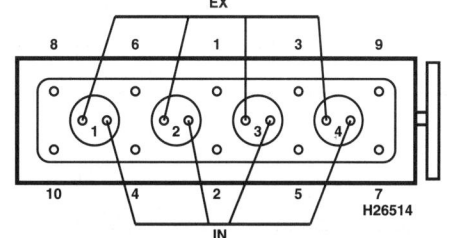

1795 cm³ / 1998 cm³

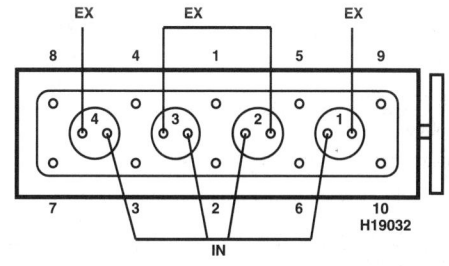

1870 cm³

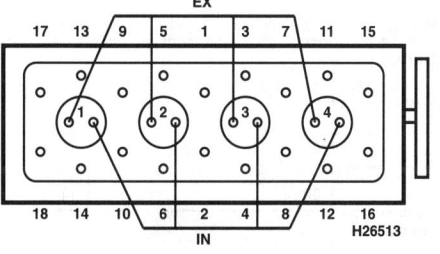

2477 cm³

– Not applicable, or information not available

MITSUBISHI

	Shogun / Pajero 2.5 Turbo 1991 to 1997	Shogun 2.8TD 1994 to 2000	L200 & 4x4 Pick-up 1986 to 1996	L200 TD & 4x4 Pick-up 1992 to 1996
Engine				
Engine type/code.....................	4D56T SOHC Turbo 72kW	4M40 SOHC Turbo 91kW	4D56 SOHC 51kW	4D56T SOHC Turbo 64kW
Capacity (cm³) / cylinders........................	2477 / 4	2835 / 4	2477 / 4	2477 / 4
Compression ration / pressure ....bar	21.0 / 19.0	21.0 / 29.0	21.0 / ≥19.2	21.0 / ≥19.2
Torque output ....Nm	240	291	147	240
Oil pressure ....idle [running] bar	2.0	0.9	0.76	0.76
Oil temperature ....°C	80	80	80	80
Valve clearances - inlet (mm)	0.25 H	0.20	0.25 H	0.25 H
- exhaust (mm)	0.25 H	0.30	0.25 H	0.25 H
Injection order	1-3-4-2	1-3-4-2	1-3-4-2	1-3-4-2
No. 1 cylinder position	F	TCE	F	F
Cooling system				
Thermostat opening temperature ....°C	82	76	82	76
Radiator cap pressure ....bar	0.76 to 1.04	0.7 to 1.0	0.76 to 1.04	0.76 to 1.04
Fuel system				
Idle speed ....rpm	750 ± 50	750 ± 100	750 ± 30	750 ± 30
Maximum (no load) speed ....rpm	5200	_	5200	5200
Smoke test/opacity ....M⁻¹ %	2.1	2.0	2.5	2.5
Static timing method............................	Plunger travel	Plunger travel	Plunger travel	Plunger travel
Timing dimension............................mm	1.0 ± 0.03	1.0	1.0 ± 0.03	1.0 ± 0.03
Crankshaft position ....mm [°]	[7] ATDC	[7] ATDC	[7] ATDC	[7] ATDC
Turbo type / ref / pressure ....bar	_	_	_	_
Injection pump make	Diesel Kiki	_	Diesel Kiki	Nippon Denso
Injection pump part no............................		VE Type		VE
Injector Make / type	Throttle type	Throttle type	Throttle type	Nippon Denso
Injector part no..	_	Zexel	_	0010
Injection type..	Indirect	Mechanical VE	MD0***	Indirect
Injection opening pressure, New [used]...bar	≥118 [108]	147 to 157	118 to 127 [108]	118 to 127 [108]
Glow plugs				
Maker	Bosch		Bosch	Bosch
Type	0250 202 003	K8T 75176	0250 202 003	0250 202 003
Nominal rating....................................V/A	12 / 8.5	11 / 22	12 / 8.5	12 / 8.5
Brakes				
minimum friction material thickness				
Front..mm	1.0	2.0	2.0	2.0
Rear...mm	1.0	2.0	1.0	1.0
Tyres - Saloon / Hatch.....................Size	_	_	_	_
- Estate / Van...........................Size	265/70x15	265/70x15: 235/75x15	185x14: 205x16	208/80x16
Pressure - front / rear - Saloon / Hatch...bar	_	_	_	_
- Estate / Van ......bar	1.6 / 1.9	1.6 / 1.9	1.8 / 2.2	1.8 / 1.8
Front suspension / wheel alignment				
Toe-in (+) / Toe-out (–)............mm [°]	1.8 ± 1.8	1.8 ± 1.8	1.0 to 4.5	1.0 to 4.5
Camber	0°40' ± 30'	0°40' ± 30'	40' ± 30' 4x4: 1°±30'	1°±30'
Castor	3° ± 1°	3° ± 1°	3° ± 1° 4x4: 2°±1°	2°±1°
King pin inclination...............................	14°52'	14°52'	_	
Rear suspension / wheel alignment				
Toe-in (+) / Toe-out (–)................mm [°]	0.0	0.0	0	
Camber	0'	0.0	0	

MITSUBISHI

	Shogun / Pajero 2.5 Turbo 1991 to 1997	Shogun 2.8TD 1994 to 2000	L200 & 4x4 Pick-up 1986 to 1996	L200 TD & 4x4 Pick-up 1992 to 1996
Torque wrench settings				
Cylinder head - stage 1 ....Nm	110	50	110	110
- stage 2 ....Nm	–	+ 90°	–	–
- stage 3 ....Nm	–	+ 90°	–	–
- stage 4 ....Nm	–	M8: 24	–	–
- stage 5 ....Nm	–	–	–	–
- stage 6 ....Nm	–	–	–	–
Big-end bearings...........................Nm	47 N	50 + 90° N	45 to 47 N	45 to 47 N
Main bearings...............................Nm	80	20 + 90° + 90°	74 to 83	74 to 83
Crankshaft pulley bolt ....Nm	190	230	135	190
Camshaft pulley bolt ....Nm	70	90	70	70
Flywheel [driveplate] bolt...................Nm	140	125	128 to 137	128 to 137
Front hubs ....Nm	WSM	WSM	WSM	WSM
Rear hubs ....Nm	WSM	WSM	WSM	WSM
Wheel nuts / bolts ....Nm	90	90	120 to 140	120 to 140
Glow plugs ....Nm	20	18	20	20
Clutch pressure plate bolts ....Nm	15 to 22	15 to 22	15 to 22	15 to 22
Injection pump sprocket....................Nm	85	65	59 to 68	85
Injectors.....................................Nm	60	50 to 60	68	68
Injection pump mounting bolts ....Nm	24	23	15 to 21	22
Injector pipe unions.........................Nm	30	23	23 to 36	23 to 36
Capacities				
Engine oil & filter..............................litres	6.7	7.8	6.8	6.8
Gearbox.......................................litres	2.5[1]	2.5	2.3 4x4: 2.2[1]	2.2[1]
Automatic transmission ....litres	–	7.2		
Final drive ....litres	Front: 1.1[2]	1.1 Rear: 3.2[1]	1.8[2]	1.1[2]
Cooling system................................litres	9.5	9.5	7.3	7.3
Fuel tank......................................litres	75	75 5dr: 95	69 4x4: 60	75

Notes

Shogun / Pajero 2.5 Turbo 1991 to 1997
[1]Transfer box: 2.3
[2]Rear: 1.8
Shogun 2.8TD 1994 to 2000
[1]Transfer box: 2.3
L200 & 4x4 Pick-up 1986 to 1996
[1]Transfer box: 2.2
[2]4x4: Front: 1.1. Rear: 2.1
L200 TD & 4x4 Pick-up 1992 to 1996
[1]Transfer box: 2.2
[2]Rear: 1.5

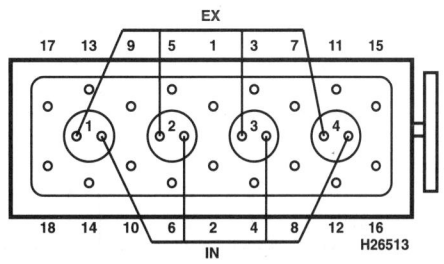

2477 cm³

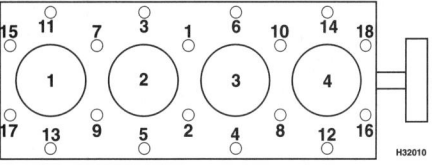

2835 cm³

– Not applicable, or information not available

MITSUBISHI

	L200 D 1996 to 2000	L200 TD 1996 to 2000	L300 1986 to 1996	L300 D 1996 to 2000
Engine				
Engine type/code......................	4D56 SOHC 55kW	4D56 SOHC Turbo 72kW	4D56 SOHC 51kW	4D56 SOHC 50kW
Capacity (cm³) / cylinders........................	2477 / 4	2477 / 4	2477 / 4	2477 / 4
Compression ration / pressure ....bar	21.0 / ≥19.2	21.0 / ≥19.2	21.0 / ≥19.2	21.0 / ≥19.2
Torque output ....Nm	150	240	147	147
Oil pressure ....idle [running] bar	–	–	0.76	–
Oil temperature ....°C	–	–	80	–
Valve clearances - inlet (mm)	0.25 H	0.25 H	0.25 H	0.25 H
- exhaust (mm)	0.25 H	0.25 H	0.25 H	0.25 H
Injection order	1-3-4-2	1-3-4-2	1-3-4-2	1-3-4-2
No. 1 cylinder position	TBE	TBE	F	TBE
Cooling system				
Thermostat opening temperature ....°C	88	76.5	82	82
Radiator cap pressure ....bar	0.75 to 1.05	0.75 to 1.05	0.76 to 1.04	0.75 to 1.05
Fuel system				
Idle speed ....rpm	750 ± 30	750 ± 50	750 ± 50	750 ± 30
Maximum (no load) speed ....rpm	5200	5200	5200	5200
Smoke test/opacity ....M⁻¹ %	2.5	2.5	2.1	2.1
Static timing method..............................	Plunger travel	Plunger travel	Plunger travel	Plunger travel
Timing dimension........................mm	1.0 ± 0.03	1.0 ± 0.03	1.0 ± 0.03	1.0 ± 0.03
Crankshaft position ....mm [°]	[9] ATDC	[9] ATDC	[7] ATDC	[7] ATDC
Turbo type / ref / pressure ....bar	–	–	–	–
Injection pump make	–	–	Diesel Kiki	Bosch
Injection pump part no.............................	VE Rotary	VE Rotary	–	VE Rotary
Injector Make / type	Bosch	Bosch	Throttle type	Throttle type
Injector part no....................................	9430 610 120	9430 610 120	0010	–
Injection type.......................................	Indirect	Indirect	Indirect	Indirect
Injection opening pressure, New [used]...bar	147 to 157	147 to 157	118 to 127	118 to 127
Glow plugs				
Maker	Champion	Champion	Bosch	Bosch
Type	CH94	CH94	0250 202 003	0250 202 003
Nominal rating...................................V/A	–	–	12 / 8.5	–
Brakes				
minimum friction material thickness				
Front............................mm	2.0	2.0	2.0	2.0
Rear............................mm	1.0	1.0	1.0	1.0
Tyres - Saloon / Hatch...................Size	–	–	–	–
- Estate / Van...........................Size	185x15	205/80x16	185x14	185x14
Pressure - front / rear - Saloon / Hatch ...bar	–	–	–	–
- Estate / Van ......bar	1.8 / 2.2	1.8 / 1.8	3.0 / 3.0	3.0 / 3.0
Front suspension / wheel alignment				
Toe-in (+) / Toe-out (–)................mm [°]	1.0 to 4.5	1.0 to 4.5	0 to 3.0	0.0 to 3.0
Camber	0°40' ± 30'	1° ± 30'	31' ± 30'	0°30' ± 45'
Castor	3° ± 1°	2° ± 1°	3°8' ± 30'	3° ± 1°
King pin inclination..............................	–	–	–	–
Rear suspension / wheel alignment				
Toe-in (+) / Toe-out (–)....................mm [°]	–	–	–	–
Camber				

MITSUBISHI

	L200 D 1996 to 2000	L200 TD 1996 to 2000	L300 1986 to 1996	L300 D 1996 to 2000
Torque wrench settings				
Cylinder head - stage 1 ....Nm	40	40	40	40
- stage 2 ....Nm	80	80	80	80
- stage 3 ....Nm	120	120	110	110
- stage 4 ....Nm	–	–	–	–
- stage 5 ....Nm	–	–	–	–
- stage 6 ....Nm	–	–	–	–
Big-end bearings........................Nm	47 N	47 N	45 to 47 N	47 N
Main bearings...........................Nm	80	80	74 to 83	80
Crankshaft pulley bolt ....Nm	175	180	190	190
Camshaft pulley bolt ....Nm	70	70	70	70
Flywheel [driveplate] bolt.....................Nm	135	135	140	135
Front hubs ....Nm	WSM	–	30, slacken, 8	WSM
Rear hubs ....Nm	–	–	WSM	–
Wheel nuts / bolts ....Nm	130	130	120 to 140	130
Glow plugs ....Nm	18	18	20	20
Clutch pressure plate bolts ....Nm	20	20	15 to 22	20
Injection pump sprocket......................Nm	88	85	90	85
Injectors................................Nm	55	55	60	55
Injection pump mounting bolts ....Nm	–	24	15 to 21	–
Injector pipe unions.............................Nm	30	30	23 to 36	30
Capacities				
Engine oil & filter................................litres	6.8	6.8	6.8	6.8
Gearbox..litres	2.3	2.2	2.0	2.0
Automatic transmission ....litres	–	–	–	–
Final drive ....litres	1.5	1.1 Rear: 1.5[1]	1.2	1.2
Cooling system................................litres	7.3	7.3	8.7	8.7
Fuel tank..litres	69	75	55	55

Notes

L200 TD 1996 to 2000
[1] Transfer: 2.2

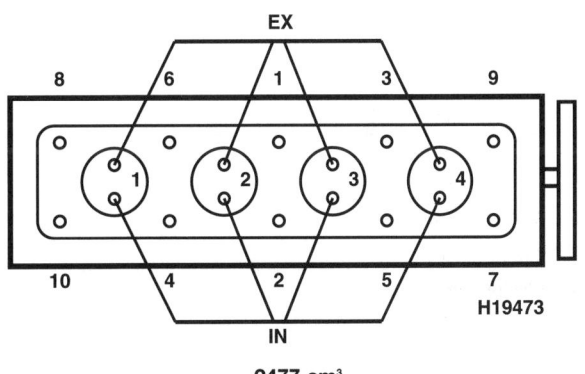

2477 cm³

– Not applicable, or information not
available

NISSAN

	Micra 1.5 Diesel 1998 to 2000	Sunny 1.7 (B12, N13) 1986 to 1991	Sunny 2.0 D 1993 to 1995	Almera 2.0 D 1995 to 2000
Engine				
Engine type/code	VJZ SOHC 42kW	CD17 SOHC 40kW	CD20 SOHC 55kW	CD20E SOHC 55kW
Capacity (cm³) / cylinders........................	1527 / 4	1681 / 4	1974 / 4	1974 / 4
Compression ration / pressure ....bar	23 / ≥25.0	22.2 (N13: 21.8) / ≥19.6	22.2 / 31.4	22.2 / ≥24.5
Torque output ....Nm	95	0	132	132
Oil pressure ....idle [running] bar	0.59 [3.0]	[3.7 to 4.1]	0.59 [2.9 @ 2000]	0.59 [2.9]
Oil temperature ....°C	80	–	80	80
Valve clearances - inlet (mm)	0.07 to 0.23	0.20 to 0.30 H	0: Hyd.	0: Hyd.
- exhaust (mm)	0.22 to 0.38	0.40 to 0.50 H	0: Hyd.	0: Hyd.
Injection order	1-3-4-2	1-3-4-2	1-3-4-2	1-3-4-2
No. 1 cylinder position	TBE	TBE	TBE	TBE
Cooling system				
Thermostat opening temperature ....°C	88	82	88	88
Radiator cap pressure ....bar	1.1 to 1.5	0.78 to 0.98	0.78 to 0.98	0.8 to 1.0
Fuel system				
Idle speed ....rpm	800 ± 25	750 +100 -0 [850 +100 -0]	750 ± 50	715 ± 25
Maximum (no load) speed ....rpm	5450 ± 125	5500	5200 to 5500	5400
Smoke test/opacity ....M⁻¹ %	2.5	2.5	2.5	2.5
Static timing method...............................	Refer to wsm	Plunger travel	Plunger travel	Plunger travel
Timing dimension................................mm		0.88 ± 0.03	0.79 to 0.85	0.78 to 0.89
Crankshaft position ....mm [°]	TDC	0 TDC	[9] ATDC	[8] ATDC
Turbo type / ref / pressure ....bar	–	–	–	–
Injection pump make	Bosch	Diesel Kiki	Diesel Kiki	Bosch
Injection pump part no............................	VE4/8F 16700 6F901	16700-16A60 or 70 [16A65 or75]	VE LNP 865	NP-VE4
Injector Make / type	Bosch	Pintle	Bosch	–
Injector part no......................................	–	–	9430 610 098	KV 11257800
Injection type...	Indirect VE4	Bosch-VE	Indirect	ECCS-D
Injection opening pressure, New [used]...bar	115 to 125	≥132 [122]	127 to 135 [120 to 132]	132.4 to 140.2 [122.6 to 132.]
Glow plugs				
Maker	NGK	Bosch	Beru	–
Type	–	0250 202 005	685 MJ	–
Nominal rating.....................................V/A	–	–	–	12 / 24
Brakes				
minimum friction material thickness				
Front.......................................mm	2.0	2.0	2.0	1.0
Rear..mm	1.5	1.5	1.5	1.5
Tyres - Saloon / Hatch......................Size	155/70x13	155x13: 175/70x13	155x13: 175/70x13	185/65x14
- Estate / Van..........................Size	–	155x13: 175/70x13		–
Pressure - front / rear - Saloon / Hatch ...bar	2.2 / 1.9	2.1 / 2.1	2.1 / 1.9	2.3 / 2.1
- Estate / Van ......bar	–	2.1 / 2.1	–	–
Front suspension / wheel alignment				
Toe-in (+) / Toe-out (–).....................mm [°]	-0.5 to 1.5	-0.5 to 1.5	[0°6']	0.0 to 0.4
Camber	-26' to 1°4'	-50' to 40'	-1° to 30'	-1°20' to 0°10'
Castor	1°31' to 3°1'	35' to 2°5'	0°40'	0°40' to 2°10'
King pin inclination.................................		13°5' to 14°35'	13°15'	14°00 to 15°30
Rear suspension / wheel alignment				
Toe-in (+) / Toe-out (–).....................mm [°]	-0.5 to 1.5	0 to -4.0	[0°18']	-3 to 5
Camber	-26' to 1°4'	-1°45' to -15' [B12: to -25']	-1°40'	-1°45' to -0°15'

NISSAN

Torque wrench settings	Micra 1.5 Diesel 1998 to 2000	Sunny 1.7 (B12, N13) 1986 to 1991	Sunny 2.0 D 1993 to 1995	Almera 2.0 D 1995 to 2000
Cylinder head - stage 1 ....Nm	40 N	59 to 69	39	39
- stage 2 ....Nm	265°	98 to 108	90	83 to 93
- stage 3 ....Nm	–	–	slacken	slacken
- stage 4 ....Nm	–	–	39	39
- stage 5 ....Nm	–	–	+ 80°	75° to 80°
- stage 6 ....Nm	–	–	–	–
Big-end bearings.......................Nm	40 N	29 to 37	16 + 60° N	15 + 60° to 65° N
Main bearings.......................Nm	20 + 50°	44 to 54	50	69 to 78
Crankshaft pulley bolt ....Nm	110	123 to 132	150	152
Camshaft pulley bolt.......................Nm	80	92 to 102	100	107
Flywheel [driveplate] bolt.......................Nm	65	98 to 108	90	83 to 93
Front hubs.......................Nm	148 to 205	196 to 275	196 to 275	195 to 275
Rear hubs.......................Nm	186 to 255	186 to 255	186 to 255	WSM
Wheel nuts / bolts ....Nm	98 to 118	98 to 118	98 to 118	100 to 120
Glow plugs ....Nm	25	15 to 20	20	20
Clutch pressure plate bolts ....Nm	–	22 to 29	30	22 to 29
Injection pump sprocket.......................Nm	25	59 to 69	65	65
Injectors.......................Nm	55	59 to 69	65	59 to 69
Injection pump mounting bolts ....Nm	20	13 to 18	–	55
Injector pipe unions.......................Nm	20	22 to 25	25	25
Capacities				
Engine oil & filter.......................litres	4.4	3.8	5.4	5.2
Gearbox.......................litres	1.8 to 2.0	2.7 [2.8]	2.9	2.9 to 3.2
Automatic transmission ....litres	–	6.3	–	–
Final drive ....litres	WT	WT	WT	WT
Cooling system.......................litres	6.25	6.5	6.0	5.2
Fuel tank.......................litres	50	B12: 52 N13: 50	50	50

Notes

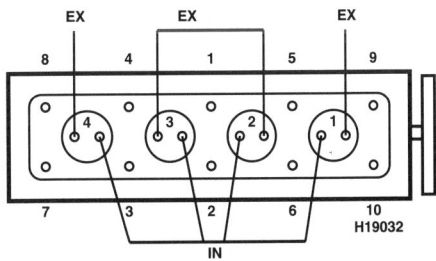

1527 cm³

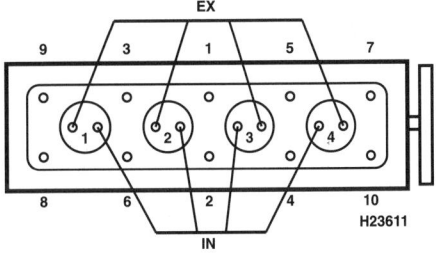

1681 cm³ / 1974 cm³

– Not applicable, or information not available

NISSAN

	Bluebird 2.0 Estate (U11) 1986 to 1991	Bluebird 2.0 (T12/72) 1986 to 1991	Primera 2.0 (P10) 1991 to 1996	Primera 2.0 TD (P11) 1998 to 1999
Engine				
Engine type/code....................	LD 20 SOHC 49kW	LD 20 SOHC 49kW	CD20 SOHC 55kW	CD20T SOHC Turbo 66kW
Capacity (cm^3) / cylinders................	1952 / 4	1952 / 4	1974 / 4	1974 / 4
Compression ration / pressure ....bar	22.2 / 24.5	22.2 / 24.5	22.2 / ≥24.5	22.2 / ≥24.5
Torque output ....Nm	0	0	132	177
Oil pressure ....idle [running] bar	[2.9]	[2.9]	0.6 [2.9]	0.59 [2.9]
Oil temperature ....°C	–	–	80	80
Valve clearances - inlet (mm)	0.25 H	0.25 H	0: Hyd.	0: Hyd.
- exhaust (mm)	0.30 H	0.30 H	0: Hyd.	0: Hyd.
Injection order	1-3-4-2	1-3-4-2	1-3-4-2	1-3-4-2
No. 1 cylinder position	TBE	TBE	TBE	TBE
Cooling system				
Thermostat opening temperature ....°C	82	82	88	82
Radiator cap pressure ....bar	0.88	0.78 to 0.98	0.78 to 0.98	0.78 to 0.98
Fuel system				
Idle speed ....rpm	700 ± 50	700 ± 50	700 +50 -0 [800 +50 -0]	725 ± 25
Maximum (no load) speed ....rpm	5500	5600	5200 to 5500	5400
Smoke test/opacity ....M^{-1} %	2.5	2.5	2.5	3.0
Static timing method...............................	Plunger travel	Plunger travel	Plunger travel	Plunger travel
Timing dimension................................mm	0.73 ± 0.04	0.73 ± 0.04	0.79 to 0.85	0.82 ± 0.07
Crankshaft position ....mm [°]	0 TDC	0 TDC	0 TDC	[8] ATDC
Turbo type / ref / pressure ....bar	–			
Injection pump make	Diesel Kiki	Diesel Kiki	Diesel Kiki	Bosch
Injection pump part no............................	16700-05E20	16700-05E20	16700-57J00,10,05,15 or 60J00	NP-VE4
Injector Make / type	–	Pintle	Bosch	Bosch
Injector part no..	KV11257800	KV11257800	105780	DNOSD1510
Injection type...	Bosch-VE	Bosch-VE	Bosch-VE	Electronic control
Injection opening pressure, New [used]...bar	≥132 [122]	≥132 [122]	128 to 135 [123 to 132]	150 to 160 [144.2 to 156.5]
Glow plugs				
Maker	Beru	Beru		
Type	888 MJ	888 MJ	–	–
Nominal rating...................................V/A	6.5 / 11	6.5 / 11		
Brakes				
minimum friction material thickness				
Front..mm	2.0	2.0	2.0	2.0
Rear..mm	1.5	1.5	1.5 Discs: 2.0	1.5
Tyres - Saloon / Hatch......................Size	–	165x14: 185/70x14	185/65x14: 195/60x14	185/65x15: 195/60x15
- Estate / Van......................Size	185/70x14: 195/60x15	–	–	–
Pressure - front / rear - Saloon / Hatch ...bar	–	1.9 / 1.8	2.3 / 2.1	2.4 / 2.2
- Estate / Van ......bar	1.9 / 1.9	–		
Front suspension / wheel alignment				
Toe-in (+) / Toe-out (–)...............mm [°]	1.0 to 3.0	1.0 to 3.0	0 to 2.0	0.0 to 2.0
Camber	-20' to 1°10'	-25' to 1°5'	-45' to 45'	-45' to 45'
Castor	1°15' to 2°45'	1°20' to 2°50'	1° to 2°30'	1° to 2°30'
King pin inclination..................................	13°45' to 15°15'	13°50' to 15°20''	13°45' to 15°15'	13°45' to 15°15'
Rear suspension / wheel alignment				
Toe-in (+) / Toe-out (–)....................mm [°]	-6.0 to -2.0	-6.0 to -2.0	-2.0 to 2.0	-3.0 to 5.0
Camber	-1°10' to 20'	-1°10' to 20'	-1°45' to -15'	-1°45' to -15'

NISSAN

Torque wrench settings	Bluebird 2.0 Estate (U11) 1986 to 1991	Bluebird 2.0 (T12/72) 1986 to 1991	Primera 2.0 (P10) 1991 to 1996	Primera 2.0 TD (P11) 1998 to 1999
Cylinder head - stage 1 ....Nm	78	78	39	34 to 44
- stage 2 ....Nm	118 to 127	118 to 127	88 ± 5	93 to 103
- stage 3 ....Nm	–	–	Slacken	Slacken
- stage 4 ....Nm	–	–	39	31 to 37
- stage 5 ....Nm	–	–	+ 75° to 80°	+ 90° to 95°
- stage 6 ....Nm	–	–	–	+ 90° to 95°
Big-end bearings............................Nm	44 to 54 N	44 to 54	15 + 60° N	15 + 60° N
Main bearings.................................Nm	69 to 83	69 to 83	54	69 to 78
Crankshaft pulley bolt ....Nm	137 to 157	137 to 157	142 to 152	152
Camshaft pulley bolt ....Nm	132 to 142	132 to 142	87 to 107	107
Flywheel [driveplate] bolt................Nm	137 to 157	137 to 157	83 to 93	83 to 93
Front hubs ....Nm	235 to 314	235 to 314	235 to 314	235 to 314
Rear hubs ....Nm	WSM	WSM	186 to 255	186 to 255
Wheel nuts / bolts ....Nm	78 to 98	98 to 118	98 to 118	98 to 118
Glow plugs ....Nm	20 to 25	20 to 25	20 to 29	20
Clutch pressure plate bolts ....Nm	18 to 22	18 to 22	22 to 29	22 to 29
Injection pump sprocket....................Nm	59 to 69	59 to 69	59 to 69	65
Injectors...Nm	59 to 69	59 to 69	59 to 69	60 to 70
Injection pump mounting bolts ....Nm	29 to 39	29 to 39	13 to 18[1]	27
Injector pipe unions...........................Nm	22 to 25	22 to 25	22 to 25	25
Capacities				
Engine oil & filter................................litres	4.2	4.3	5.5	4.7
Gearbox...litres	2.8	4.7	2.9	3.6 to 3.8
Automatic transmission ....litres	6.8	–	7.0	–
Final drive ....litres	WT	WT	WT	WT
Cooling system..................................litres	6.2	7.0	6.6	6.6
Fuel tank..litres	53	60	60	60

Notes

Bluebird 2.0 (T12/72) 1986 to 1991
'T72: 14°15' to 15°45'
Primera 2.0 (P10) 1991 to 1996
[1]Bolt: 45 to 60

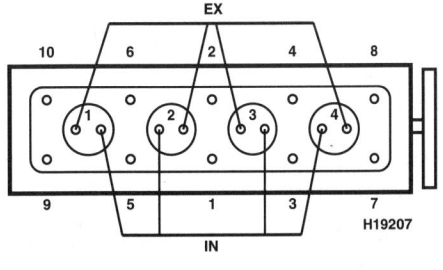

1952 cm³

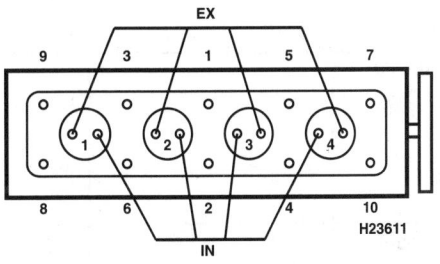

1974 cm³

– Not applicable, or information not available

NISSAN

	Serena 2.0 D (C23) 1993 to 1995	Serena 2.3 D (C23) 1995 to 2000	Terrano II (R20) 1993 to 1996	Terrano II (R20) 1996 to 2000
Engine				
Engine type/code..........................	LD20 SOHC 49kW	LD23 SOHC 55kW	TD27T OHV Turbo 74kW	TD27T OHV Turbo 91kW'
Capacity (cm³) / cylinders........................	1952 / 4	2283 / 4	2663 / 4	2663 / 4
Compression ration / pressure ....bar	21.2 / ≥24.5	22.2 / ≥24.5	21.9 / ≥24.5	21.9 / ≥24.5
Torque output ....Nm	0	145	278	278
Oil pressure ....idle [running] bar	[2.9 @ 2000]	1.0 [3.0 @ 2000rpm]	≥0.78 [2.9 to 3.9 @ 3000]	≥0.78 [2.9 to 3.9 @ 3000]
Oil temperature ....°C	80	80	80	80
Valve clearances - inlet (mm)	0.25 H	0.25 H	0.25 H	0.25 H
- exhaust (mm)	0.30 H	0.30 H	0.25 H	0.25 H
Injection order.............................	1-3-4-2	1-3-4-2	1-3-4-2	1-3-4-2
No. 1 cylinder position	TBE	TBE	F	F
Cooling system				
Thermostat opening temperature ....°C	82	82	82 or 88	82 or 88
Radiator cap pressure ....bar	0.78 to 0.98	0.78 to 0.98	0.78 to 0.98	0.78 to 0.98
Fuel system				
Idle speed ....rpm	700 ± 50	700 ± 50	700 ± 50 A/C: 850 ± 50	700 ± 50 A/C: 850 ± 50
Maximum (no load) speed ....rpm	5200	5200	5050 ± 100	5050 ± 100
Smoke test/opacity ....M⁻¹ %	2.5	2.5	2.5	2.5
Static timing method................................	Plunger travel	Plunger travel	Plunger travel	Plunger travel
Timing dimension............................mm	0.73 ± 0.01	0.69 ± 0.01	0.38 ± 0.02	0.27 ± 0.02
Crankshaft position ....mm [°]	[6]	0	0 TDC	0 TDC
Turbo type / ref / pressure ....bar	–	–	–	–
Injection pump make	Kiki Bosch	Kiki Bosch	Diesel Kiki	Diesel Kiki
Injection pump part no............................	VE 4/9 R1038	14749 2562	104645 4032	NP-VE4 RNP11
Injector Make / type	Bosch	Bosch	Bosch	Bosch
Injector part no..................................	105780 2150	NP-DNOSD1510	–	NP-DNOSD1510
Injection type..	Indirect	Indirect VE	Indirect	Indirect
Injection opening pressure, New [used]...bar	132 to 140 [123]	110 to 120 [100 to 110]	103 to 123 [98 to 103]	103 to 123 [98 to 103]
Glow plugs				
Maker	–	Champion	Beru	Champion
Type	–	CH187	854 MJ	CH162
Nominal rating...................................V/A	11 / 8	11 / 17	11 / 17	11 /
Brakes				
minimum friction material thickness				
Front....................................mm	2.0	2.0	2.0	2.0
Rear....................................mm	1.5	1.5	1.52	1.52
Tyres - Saloon / Hatch.......................Size	–	–	–	–
- Estate / Van...........................Size	175x14	175x14	235/75x15	235/75x15
Pressure - front / rear - Saloon / Hatch ..bar	–	–	–	–
- Estate / Van ......bar	2.5 / 2.5	2.5 / 2.5	1.9 / 2.5	1.9 / 2.5
Front suspension / wheel alignment				
Toe-in (+) / Toe-out (–)......................mm [°]	2.0	2.0	3.0 to 5.0	3.0 to 5.0
Camber	-0°15' to 1°15'	-0°15' to 1°15'	35' ± 30'	35' ± 30'
Castor	3°28' ± 45'	3°28' ± 45'	1°40' ± 30'	1°40' ± 30'
King pin inclination..................................			7°36' to 8°36'	7°36' to 8°36'
Rear suspension / wheel alignment				
Toe-in (+) / Toe-out (–)....................mm [°]	-1.8 to 1.8	–	–	–
Camber	-0°57' to 0°33'			

NISSAN

Torque wrench settings	Serena 2.0 D (C23) 1993 to 1995	Serena 2.3 D (C23) 1995 to 2000	Terrano II (R20) 1993 to 1996	Terrano II (R20) 1996 to 2000
Cylinder head - stage 1 ....Nm	39	100	39 to 44	39 to 44
- stage 2 ....Nm	127	+ 90°	54 to 59	54 to 59
- stage 3 ....Nm	slacken	+ 90°	+ 90 ± 10°	+ 90 ± 10°
- stage 4 ....Nm	39	–	–	–
- stage 5 ....Nm	+ 100°	–	–	–
- stage 6 ....Nm	–	–	–	–
Big-end bearings.........................Nm	16 + 65°	52 to 58 N	78 to 83 N	78 to 83 N
Main bearings............................Nm	80	15 + 60°	167 to 177	167 to 177
Crankshaft pulley bolt ....Nm	157	152	294 to 324	294 to 324
Camshaft pulley bolt ....Nm	142	142	44 to 49	44 to 49
Flywheel [driveplate] bolt....................Nm	157	157	147 to 167	147 to 167
Front hubs ....Nm	206 to 284	206 to 284	WSM	WSM
Rear hubs ....Nm	WSM[1]	WSM	WSM	WSM
Wheel nuts / bolts ....Nm	98 to 118	98 to 118	118 to 147	118 to 147
Glow plugs ....Nm	25	25	15 to 20	15 to 20
Clutch pressure plate bolts ....Nm	22 to 29	22 to 29	22 to 29	22 to 29
Injection pump sprocket....................Nm	65	65	59 to 69	59 to 69
Injectors.................................Nm	60 to 70	60 to 70	54 to 64	54 to 64
Injection pump mounting bolts ....Nm	35	25	30 to 41	32 to 42 Nuts: 20 to 25
Injector pipe unions..........................Nm	25	25	20 to 25	20 to 25
Capacities				
Engine oil & filter................................litres	5.7	7.0	7.2	7.2
Gearbox.....................................litres	2.1	2.1	3.5 Transfer box: 2.3	3.5 Transfer box: 2.3
Automatic transmission ....litres	–	–	8.5	8.5
Final drive ....litres	1.5	1.5 7.96▶: 1.3	Rear: 1.3 LSD: 2.8	Rear: 1.3 LSD: 2.8
Cooling system................................litres	11.5	13.7	10.0	10.0
Fuel tank.....................................litres	60	60	3D: 72 5D: 80	3D: 72 5D: 80

Notes

Serena 2.0 D (C23) 1993 to 1995
[1]IRS type: 206 to 275
Terrano II (R20) 1996 to 2000
[1]With intercooler

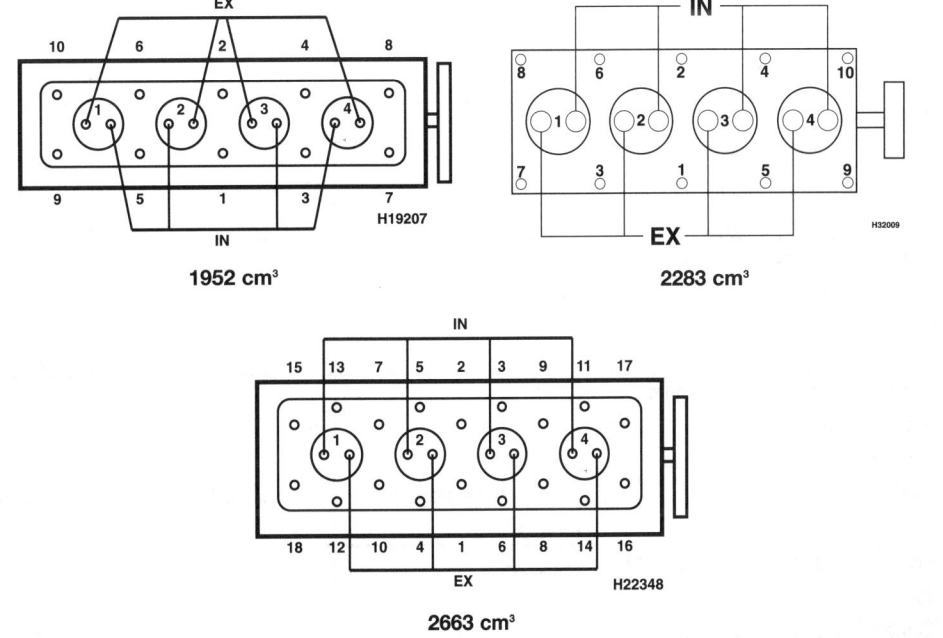

1952 cm³

2283 cm³

2663 cm³

– Not applicable, or information not available

NISSAN

	Patrol 2.8 1989 to 1993	Patrol 2.8 Turbo 1989 to 1993	Patrol GR (Y61) 1995 to 2000	Patrol 4.2 D 1993 to 1995
Engine				
Engine type/code.....................	RD28 SOHC 66kW	RD28T SOHC Turbo 85kW	RD28ET SOHC Turbo 96kW	TD42 OHV 91kW
Capacity (cm³) / cylinders........................	2826 / 6	2826 / 6	2826 / 6	4169 / 6
Compression ration / pressure ....bar	21.2 / ≥24.5	21.2 / ≥24.5	21.8 / ≥24.5	22.7 / 29.4
Torque output ....Nm	0	235	252	272
Oil pressure ....idle [running] bar	[3.1 to 4.3 @ 2000]	[3.2 to 4.3]	0.78 [4.0 @ 3000rpm]	0.78 [2.9 to 3.9 @ 3000]
Oil temperature ....°C	80	–	80	80
Valve clearances - inlet (mm)	0: Hyd.	0: Hyd.	0: Hyd.	0.35 H
- exhaust (mm)	0: Hyd.	0: Hyd.	0: Hyd.	0.35 H
Injection order	1-5-3-6-2-4	1-5-3-6-2-4	1-5-3-6-2-4	1-4-2-5-3-6
No. 1 cylinder position	TCE	TCE	F	–
Cooling system				
Thermostat opening temperature ....°C	82	82	82	82
Radiator cap pressure ....bar	0.78 to 0.98	0.78 to 0.98	0.78 to 0.98	0.78 to 0.98
Fuel system				
Idle speed ....rpm	600 ± 50	650 ± 50	650 ± 25	700 to 750
Maximum (no load) speed ....rpm	5000 ± 100	4600 ± 100	5200	4600 ± 100
Smoke test/opacity ....M⁻¹ %	2.5	2.5	2.5	2.5
Static timing method................................	Plunger travel	Plunger travel	Plunger travel	Plunger travel
Timing dimension................................mm	0.75 ± 0.03	0.86 ± 0.05	0.92 ± 0.04	0.74 ± 0.02
Crankshaft position ....mm [°]	0 TDC	0 TDC	0 TDC	0 TDC
Turbo type / ref / pressure ....bar		–	–	–
Injection pump make	Diesel Kiki	Diesel Kiki	Diesel Kiki	Diesel Kiki
Injection pump part no.............................	16700-C9600	16700-22J00	VE	104760 4300
Injector Make / type	Diesel Kiki	Diesel Kiki	Bosch	Bosch
Injector part no.......................................	KV 112 57800	KV 112 57800	9430 610 033	105780
Injection type..	Bosch-VE	Bosch-VE	Indirect	Indirect
Injection opening pressure, New [used]...bar	132 to 140 [122]	132 to 140 [122]	132 to 140 9123 to 132]	103 to 113 [98]
Glow plugs				
Maker	Beru	Beru	Beru	Champion
Type	636 MJ	636 MJ	636 MJ	CH162
Nominal rating...................................V/A	–	–	–	11 /
Brakes				
minimum friction material thickness				
Front.......................................mm	2.0	2.0	2.0	2.0
Rear..mm	1.5	1.5	2.0	2.0
Tyres - Saloon / Hatch......................Size	–	–	–	–
- Estate / Van...........................Size	205x16	205/70x16	265/70x16	265/70x15
Pressure - front / rear - Saloon / Hatch ..bar	–	–	–	–
- Estate / Van ......bar	1.8 / 2.4	1.8 / 2.4	2.0 / 2.8	2.0 / 2.8
Front suspension / wheel alignment				
Toe-in (+) / Toe-out (–)....................mm [°]	0 ± 2.0	0 ± 2.0	1.0 ± 1.0	[0°14']
Camber	30' ± 30'	30' ± 30'	0° ± 30'	0°
Castor	3°35' ± 30' Hardtop: 3°15'±30	3°35' ± 30' Hardtop: 3°15'±30	3°50' 5dr: 3°30'	2°05'
King pin inclination..................................	7°30' ± 30'	7°30' ± 30'	14°30'	7°
Rear suspension / wheel alignment				
Toe-in (+) / Toe-out (–)....................mm [°]	–	–	–	–
Camber	–	–	–	–

NISSAN

	Patrol 2.8 1989 to 1993	Patrol 2.8 Turbo 1989 to 1993	Patrol GR (Y61) 1995 to 2000	Patrol 4.2 D 1993 to 1995
Torque wrench settings				
Cylinder head - stage 1 ....Nm	29	29	29	44
- stage 2 ....Nm	113	113	113	60
- stage 3 ....Nm	Slacken	Slacken	Slacken	+ 90°
- stage 4 ....Nm	29	29	29	–
- stage 5 ....Nm	+ 100°	+ 100°	127	–
- stage 6 ....Nm	M8 bolts: 16 to 21	M8 bolts: 16 to 21	M8: 19	–
Big-end bearings..............................Nm	16, 45	16, 45 N	15, 40 N	80 N
Main bearings...................................Nm	69 to 78	69 to 78	72	177
Crankshaft pulley bolt ....Nm	142 to 152	142 to 152	152	324
Camshaft pulley bolt ....Nm	123 to 132	123 to 132	132	83
Flywheel [driveplate] bolt....................Nm	137 to 157	137 to 157	145	167
Front hubs ....Nm	167 to 196	167 to 196	WSM	196, slacken, 5
Rear hubs ....Nm	441 to 490	441 to 490	WSM	490
Wheel nuts / bolts ....Nm	118 to 147	118 to 147	118 to 147	130
Glow plugs ....Nm	15 to 20	15 to 20	20	20
Clutch pressure plate bolts ....Nm	22 to 29	22 to 29	10, 26	30
Injection pump sprocket......................Nm	54 to 64	54 to 64	64	69
Injectors..Nm	59 to 69	59 to 69	60 to 70	64
Injection pump mounting bolts ....Nm	16 to 21	16 to 21	21	41
Injector pipe unions...........................Nm	22 to 25	22 to 25	25	25
Capacities				
Engine oil & filter.............................litres	6.7	6.7	6.4	6.7
Gearbox..litres	3.6¹	3.6¹	5.1	3.9¹
Automatic transmission ....litres	–	–	–	–
Final drive ....litres	Front: 1.3²	Front: 1.3²	5.1 Rear: 2.1¹	5.4 Rear: 2.1
Cooling system................................litres	12.9	12.9	11.8	12.9
Fuel tank...litres	82	82	95	95

Notes

Patrol 2.8 1989 to 1993
¹Transfer box: 1.9
²Rear: 1.9
Patrol 2.8 Turbo 1989 to 1993
¹Transfer box: 1.9
²Rear: 1.9
Patrol GR (Y61) 1995 to 2000
¹Transfer box: 1.9
Patrol 4.2 D 1993 to 1995
¹Transfer: 1.9

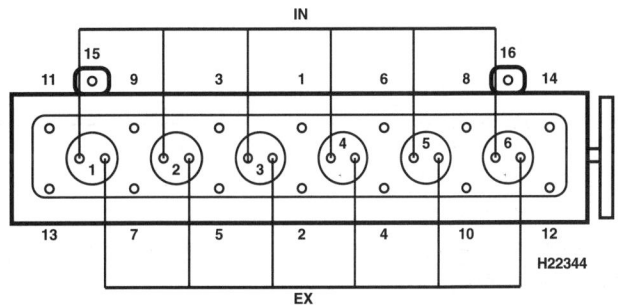

2826 cm³

– Not applicable, or information not
  available

NISSAN

	Pick-up 2.5 (D21) 1990 to 1998	Pick-up (D22) 2.5 TD 1998 to 2000	Vanette 2.0 (C22) 1986 to 1996	Vanette 2.3 D (HC23) 1995 to 2000
Engine				
Engine type/code....................................	TD25 OHV 55kW	TD25 OHV 55kW	LD20 SOHC 44kW	LD23 SOHC 55kW
Capacity (cm³) / cylinders.......................	2494 / 4	2494 / 4	1952 / 4	2283 / 4
Compression ration / pressure ....bar	22.2 / 24.5	22.2 / ≥24.5	22.2 / 24.5	22.2 / 24.5
Torque output ....Nm	160	160	0	145
Oil pressure ....idle [running] bar	[2.9 to 3.9 @ 3000]	0.78 [3.0 @ 3000]	[2.9]	1.0 [3.0 @ 2000]
Oil temperature ....°C	80	80	_	80
Valve clearances - inlet (mm)	0.35 H	0.3 to 0.4 H	0.25 H	0.25 H
- exhaust (mm)	0.35 H	0.3 to 0.4 H	0.30 H	0.30 H
Injection order	1-3-4-2	1-3-4-2	1-3-4-2	1-3-4-2
No. 1 cylinder position	F	_	TBE	_
Cooling system				
Thermostat opening temperature ....°C	82	82	82	82
Radiator cap pressure ....bar	0.78 to 0.98	0.78 to 0.98	0.78 to 0.98	0.78 to 0.98
Fuel system				
Idle speed ....rpm	700 ± 50	700 ± 50	700 ± 50	700 ± 50
Maximum (no load) speed ....rpm	5000 ± 100	_	5300	5000 ± 100
Smoke test/opacity ....M⁻¹ %	2.5	2.5	2.5	2.5
Static timing method............................	Plunger travel	Plunger travel	Plunger travel	Rotor lift
Timing dimension...........................mm	0.71 ± 0.02	0.71 ± 0.02	0.73 ± 0.04	0.69 ± 0.01
Crankshaft position ....mm [°]	0 TDC	TDC	0 TDC	TDC
Turbo type / ref / pressure ....bar	_	_	_	_
Injection pump make	Diesel Kiki	Bosch	Diesel Kiki	Kiki Bosch
Injection pump part no...........................	16700-44GO4 or G06	VE	16700-05E20	10749 2562
Injector Make / type..............................	Diesel Kiki	_	Pintle	Diesel Kiki
Injector part no.....................................	16600-43G02	_	KV11257800	DNOSD1510
Injection type.......................................	Bosch-VE	_	Bosch-VE	Indirect
Injection opening pressure, New [used]...bar	103 to 112 [98 to 103]	[98.1 to 103]	≥132 [122]	110 to 120 [100]
Glow plugs				
Maker	Beru	Champion	Bosch	Champion
Type	921 MJ	CH162	0250 201 032	CH187
Nominal rating...............................V/A	_	11 /	_	_
Brakes				
minimum friction material thickness				
Front..............................mm	2.0	2.0	2.0	2.0
Rear...............................mm	1.5	1.5	1.5	1.5
Tyres - Saloon / Hatch.......................Size	_	_	_	_
- Estate / Van.....................Size	185x14: 195x14: 205x16	185x14 4x4: 205x16	165x14: 195/70x14	185/75x14
Pressure - front / rear - Saloon / Hatch ...bar	_	_	_	_
- Estate / Van ......bar	1.8 / 2.2	2.5 / 2.7 4x4: 2.5 / 2.8	3.0 / 3.0	Refer to vehicle
Front suspension / wheel alignment				
Toe-in (+) / Toe-out (–)....................mm [°]	1.0 to 5.0¹	3.0 ± 1.0 4x4: 4.0 ± 1.0	-1.0 to 3.0	-2.0
Camber	-28' to 1°2'²	25' 4x4: 36'	-30' to 1°	-0°15 to 1°15'
Castor	-24' to 1°6'³	22' 4x4: 2°20'	50' to 2°20'	3°28' ± 45'
King pin inclination................................	8°28' to 9°58'⁴	9°5' 4x4: 10°48'	9° to 20°	
Rear suspension / wheel alignment				
Toe-in (+) / Toe-out (–)....................mm [°]	_	_	_	_
Camber			_	_

NISSAN

Torque wrench settings	Pick-up 2.5 (D21) 1990 to 1998	Pick-up (D22) 2.5 TD 1998 to 2000	Vanette 2.0 (C22) 1986 to 1996	Vanette 2.3 D (HC23) 1995 to 2000
Cylinder head - stage 1 ....Nm	39 to 44	40	78	100
- stage 2 ....Nm	54 to 59	60	118 to 127	+ 90°
- stage 3 ....Nm	+ 90°	+ 90°	–	+ 90°
- stage 4 ....Nm	–	–	–	–
- stage 5 ....Nm	–	–	–	–
- stage 6 ....Nm	–	–	–	–
Big-end bearings.......................Nm	78 to 83 N	80 N	44 to 54 N	55 N
Main bearings.......................Nm	167 to 177	170	69 to 83	15 + 60°
Crankshaft pulley bolt ....Nm	294 to 324	324	137 to 157	152
Camshaft pulley bolt ....Nm	49	50	132 to 142	142
Flywheel [driveplate] bolt.......................Nm	147 to 167	150	137 to 157	160
Front hubs ....Nm	34 to 39⁵	–	265 to 353	250
Rear hubs ....Nm	–	–	–	–
Wheel nuts / bolts ....Nm	118 to 147	130	98 to 118	105
Glow plugs ....Nm	15 to 20	20	20 to 25	25
Clutch pressure plate bolts ....Nm	22 to 30	25	18 to 22	25
Injection pump sprocket.......................Nm	59 to 69	70	59 to 69	69
Injectors.......................Nm	54 to 64	60	59 to 69	65
Injection pump mounting bolts ....Nm	20 to 25	23	29 to 39	25
Injector pipe unions.......................Nm	20 to 25	25⁻	22 to 25	25
Capacities				
Engine oil & filter.......................litres	6.0	6.0	4.1	7.0
Gearbox.......................litres	2.0 4x4: 4.0⁶	2.0 4x4: 5.1	2.0	2.1
Automatic transmission ....litres	–	–	–	–
Final drive ....litres	1.3⁷	1.3 Transfer: 2.2	1.0	1.5
Cooling system.......................litres	10.4	80	7.7¹	13.7
Fuel tank.......................litres	60 or 80	80	55	60

Notes

Pick-up 2.5 (D21) 1990 to 1998
¹4x4: 2.0 to 6.0
²4x4: -9' to 1°21'
³4x4: 33' to 2°3'
⁴4x4: 7°20' to 8°50'
⁵4x4: 78 to 98
⁶4x4 transfer box: 2.2
⁷4x4 rear: 2.8
Vanette 2.0 (C22) 1986 to 1996
¹With front and rear heaters: 8.2

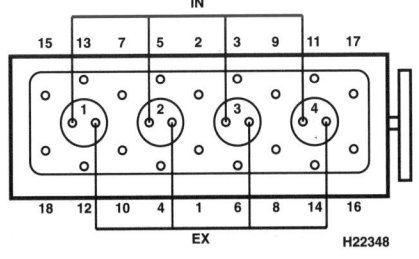

2494 cm³

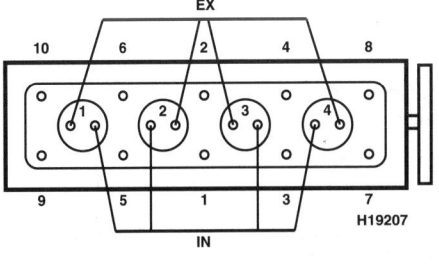

1952 cm³

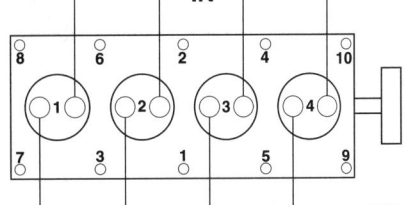

2283 cm³

– Not applicable, or information not
  available

NISSAN

	Urvan 2.5 (E24) 1990 to 1996	Cabstar 2.5 (F22) 1990 to 1992	Cabstar D (F23) 1993 to 1999	
Engine				
Engine type/code.....................................	TD25 OHV 55kW	TD25 OHV 55kW	TD25 OHV 55kW	
Capacity (cm³) / cylinders.....................	2494 / 4	2494 / 4	2494 / 4	
Compression ration / pressure ....bar	22.2 / 24.5	22.2 / 24.5	22.2 / ≥22.4	
Torque output ....Nm	160	160	160	
Oil pressure ....idle [running] bar	[3.0 @ 3000]	[2.9 to 3.9 @ 3000]	0.78 [3.0 @ 3000]	
Oil temperature ....°C	80	80	80	
Valve clearances - inlet (mm)	0.35 H	0.35 H	0.35 H	
- exhaust (mm)	0.35 H	0.35 H	0.35 H	
Injection order	1-3-4-2	1-3-4-2	1-3-4-2	
No. 1 cylinder position	F	F	F	
Cooling system				
Thermostat opening temperature ....°C	82	82	82	
Radiator cap pressure ....bar	0.78 to 0.98	0.9	0.78 to 0.98	
Fuel system				
Idle speed ....rpm	700 ± 50	700 ± 50	850 ± 50	
Maximum (no load) speed ....rpm	5100	5050	4850 to 5050	
Smoke test/opacity ....M⁻¹ %	2.5	2.5	2.5	
Static timing method...............................	Plunger travel	Plunger travel	Rotor lift	
Timing dimension...............................mm	0.71 ± 0.02	0.71 ± 0.02	0.74 ± 0.02	
Crankshaft position ....mm [°]	0 TDC	0 TDC	0 TDC	
Turbo type / ref / pressure ....bar	–	–		
Injection pump make	Diesel Kiki	Diesel Kiki	Bosch	
Injection pump part no.............................	16700-30N05 or N08	16700-21T13	VE	
Injector Make / type	Diesel Kiki	Diesel Kiki	Diesel Kiki	
Injector part no..	16600-43G02	16600-43G02	KV 11 257 800/802	
Injection type...	Bosch-VE	Bosch-VE	Indirect	
Injection opening pressure, New [used]...bar	103 to 112 [100]	103 to 112 [100]	103 to 112 [98 to 103]	
Glow plugs				
Maker	Bosch	Bosch	Bosch	
Type	0250 202 060	0250 202 060	0250 202 060	
Nominal rating.................................V/A		11 / 10	11 / 10	
Brakes				
minimum friction material thickness				
Front..mm	2.0	2.0	2.0	
Rear...mm	1.5	1.5	1.5	
Tyres - Saloon / Hatch......................Size	–	–	–	
- Estate / Van.....................Size	185x14	205x16	205/75x16	
Pressure - front / rear - Saloon / Hatch ...bar	–	–	–	
- Estate / Van ......bar	2.8 / 2.8	4.25 / 4.7	4.25 / 4.7	
Front suspension / wheel alignment				
Toe-in (+) / Toe-out (–)......................mm [°]	0 to 2.0	0 to 2.0	[0° to 10']	
Camber	-5' to 55'	-25' to 35'	-0°25' to 0°35'	
Castor	30' to 1°30'	30' to 1°30'¹	0°30' to 1°30'	
King pin inclination..................................	8°35' to 9°35'	8°55' to 9°55'	8°55 to 9°55'	
Rear suspension / wheel alignment				
Toe-in (+) / Toe-out (–)......................mm [°]	–	–	–	
Camber	–	–	–	

NISSAN

Torque wrench settings	Urvan 2.5 (E24) 1990 to 1996	Cabstar 2.5 (F22) 1990 to 1992	Cabstar D (F23) 1993 to 1999	
Cylinder head - stage 1 ....Nm	39 to 44	39 to 44	42	
- stage 2 ....Nm	54 to 59	54 to 59	56	
- stage 3 ....Nm	+ 90°	+ 90°	+ 90°	
- stage 4 ....Nm	–	–	–	
- stage 5 ....Nm	–	–	–	
- stage 6 ....Nm	–	–	–	
Big-end bearings...............................Nm	78 to 83 N	78 to 83 N	80 N	
Main bearings.....................................Nm	167 to 177	167 to 177	170	
Crankshaft pulley bolt ....Nm	294 to 324	294 to 324	324	
Camshaft pulley bolt ....Nm	49	49	49	
Flywheel [driveplate] bolt....................Nm	147 to 167	147 to 167	155	
Front hubs ....Nm	34 to 39	34 to 39	40 + 45°	
Rear hubs ....Nm	–	WSM	–	
Wheel nuts / bolts ....Nm	118 to 147	118 to 147[2]	128	
Glow plugs ....Nm	15 to 20	15 to 20	20	
Clutch pressure plate bolts ....Nm	22 to 30	22 to 30	25	
Injection pump sprocket......................Nm	59 to 69	59 to 69	69	
Injectors..Nm	54 to 64	54 to 64	60	
Injection pump mounting bolts ....Nm	20 to 25	20 to 25	–	
Injector pipe unions............................Nm	20 to 25	20 to 25	25	
Capacities				
Engine oil & filter................................litres	5.9	6.2	6.2	
Gearbox..litres	2.0	2.7	2.0	
Automatic transmission ....litres	7.0	–	–	
Final drive ....litres	2.0	2.0	2.0	
Cooling system...................................litres	13.0[1]	13.8	13.8	
Fuel tank...litres	65	65	64	

Notes

Urvan 2.5 (E24) 1990 to 1996
[1]With rear heater: 14.0
Cabstar 2.5 (F22) 1990 to 1992
[1]Twin rear tyre: 15' to 1°15'
[2]Twin rear tyre: 216 to 255

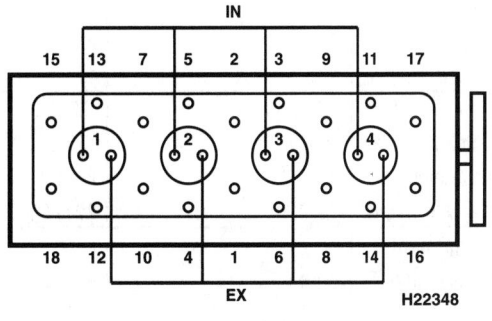

2494 cm³

– Not applicable, or information not available

PEUGEOT

	106 1.4 D 1993 to 1994	106 1.4 D 1993 to 1994	106 1.5 D 1994 to 2000	106 1.5 D 1994 to 2000
Engine				
Engine type/code....................	TUD3Y K9Y SOHC 8V 37kW	TUD3/L K9B SOHC 8V 37kW	TUD5/L VJZ SOHC 8V 42kW	TUD5/Y/L3 VJY SOHC 8V 40kW
Capacity (cm³) / cylinders.......................	1360 / 4	1360 / 4	1527 / 4	1527 / 4
Compression ration / pressure ....bar	22.5 /	22.5 /	23.0 /	23.0 /
Torque output ....Nm	0	0	95	95
Oil pressure ....idle [running] bar	[4.0 @ 4000rpm]	[4.0 @ 4000rpm]	[4.0 @ 4000rpm]	[4.0 @ 4000rpm]
Oil temperature ....°C	80	80	80	80
Valve clearances - inlet (mm)	0.15	0.15	0.20	0.20
- exhaust (mm)	0.30	0.30	0.40	0.40
Injection order	1-3-4-2	1-3-4-2	1-3-4-2	1-3-4-2
No. 1 cylinder position	FE	FE	FE	FE
Cooling system				
Thermostat opening temperature ....°C	89	89	89	89
Radiator cap pressure ....bar	1.4	1.4	1.4	1.4
Fuel system				
Idle speed ....rpm	775 ± 50	775 ± 50	775 ± 50	775 ± 50
Maximum (no load) speed ....rpm	5450	5450	5450	5450
Smoke test/opacity ....M⁻¹ %	2.5	2.5	2.5	2.5
Static timing method................................	Plunger travel	Plunger travel	Refer to wsm	Refer to wsm
Timing dimension...............................mm	Dimension on pump	Dimension on pump	–	Dimension on pump
Crankshaft position ....mm [°]	–	–	–	–
Turbo type / ref / pressure ....bar	–	–	–	–
Injection pump make	Lucas	Lucas	Bosch	Lucas
Injection pump part no.............................	R8443 B971B	R8443 B461C	537	R8444B480A
Injector Make / type	CAV	CAV	CAV	CAV
Injector part no...	RDNOSDC6880C	RDNOSDC6863C	DNOS D299A/Lucas RDN 125DC6849	LDC007R01A
Injection type...	Indirect DPC	Indirect DPC	Indirect	Indirect DPC
Injection opening pressure, New [used]...bar	122 to 132	122 to 132	130 to 135	130 to 135
Glow plugs				
Maker	Bosch/Champion	Bosch/Champion	Beru/Champion	Beru/Champion
Type	0250 202 001 / CH147	0250 202 001 / CH147	0100 226 188 / CH168	0100 226 188 / CH168
Nominal rating.....................................V/A	11 / 12	11 / 12	11 / 16	11 / 16
Brakes				
minimum friction material thickness				
Front...mm	2.0	2.0	2.0	2.0
Rear..mm	1.5	1.5	1.5	1.5
Tyres - Saloon / Hatch......................Size	165/70x13	165/70x13	165/70x13	165/70x13
- Estate / Van...........................Size	165/70x13	165/70x13	165/70x13	165/70x13
Pressure - front / rear - Saloon / Hatch...bar	2.2 / 2.2	2.2 / 2.2	2.2 / 2.2	2.2 / 2.2
- Estate / Van ......bar	2.2 / 2.2	2.2 / 2.2	2.2 / 2.2	2.2 / 2.2
Front suspension / wheel alignment				
Toe-in (+) / Toe-out (–)....................mm [°]	-0.5 ± 0.5	-0.5 ± 0.5	-0.5 ± 0.5	-0.5 ± 0.5
Camber	-0°40' ± 30'	-0°40' ± 30'	-0°40' ± 30'	-0°40' ± 30'
Castor	2°13' ± 30'	2°13' ± 30'	2°13' ± 30'	2°13' ± 30'
King pin inclination.................................	12°41' ± 40'	12°41' ± 40'	12°41' ± 40'	12°41' ± 40'
Rear suspension / wheel alignment				
Toe-in (+) / Toe-out (–)....................mm [°]	1.5 ± 1.2	1.5 ± 1.2	1.5 ± 1.2	1.5 ± 1.2
Camber	-1°35' ± 30'	-1°35' ± 30'	-1°35' ± 30'	-1°35' ± 30'

PEUGEOT

	106 1.4 D 1993 to 1994	106 1.4 D 1993 to 1994	106 1.5 D 1994 to 2000	106 1.5 D 1994 to 2000
Torque wrench settings				
Cylinder head - stage 1 ....Nm	20 N	20 N	40[1]	40[1]
- stage 2 ....Nm	+ 160°	+ 160°	+ 260°	+ 260°
- stage 3 ....Nm	+ 160°	+ 160°	–	–
- stage 4 ....Nm	–	–	–	–
- stage 5 ....Nm	–	–	–	–
- stage 6 ....Nm	–	–	–	–
Big-end bearings.................................Nm	40 N	40 N	40 N	40 N
Main bearings.......................................Nm	20 + 45°	20 + 45°	20 + 50°	20 + 50°
Crankshaft pulley bolt ....Nm	90 to 130	90 to 130	110	110
Camshaft pulley bolt ....Nm	40 to 50	40 to 50	80	80
Flywheel [driveplate] bolt....................Nm	70 LcK	70 LcK	65 LcK	65 LcK
Front hubs ....Nm	260	260	245	260
Rear hubs ....Nm	140	140	140	140
Wheel nuts / bolts ....Nm	85	85	85	85
Glow plugs ....Nm	22	22	25	25
Clutch pressure plate bolts ....Nm	15	15	15	15
Injection pump sprocket......................Nm	–	–	25	25
Injectors...Nm	90	90	55	55
Injection pump mounting bolts ....Nm	–	–	20	20
Injector pipe unions............................Nm	20	20	20	20
Capacities				
Engine oil & filter.................................litres	3.5	3.5	4.50	4.50
Gearbox..litres	2.0	2.0	2.0	2.0
Automatic transmission ....litres	–	–	–	–
Final drive ....litres	WT	WT	WT	WT
Cooling system....................................litres	6.0	6.0	6.0	6.0
Fuel tank..litres	45	45	45	45

Notes

106 1.5 D 1994 to 2000
[1]Bolt length below head: ≤197.5 mm
[2]or Lucas 070
106 1.5 D 1994 to 2000
[1]Bolt length below head: ≤197.5 mm

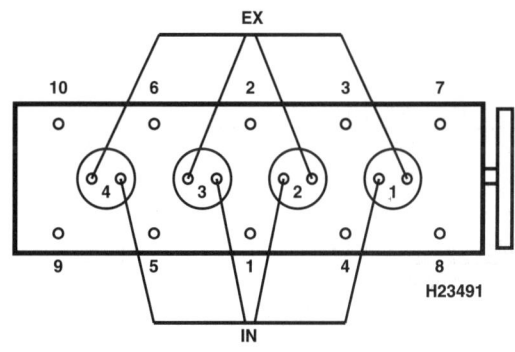

1360 cm³ / 1527 cm³

– Not applicable, or information not
available

PEUGEOT

	205 & Van 1.7 1983 to 1996	205 & Van 1.7 1983 to 1996	205 Turbo 1.7 1990 to 1996	205 1.9 1987 to 1994
Engine				
Engine type/code....................	XUD7 161A SOHC 44kW	XUD7 161A SOHC 44kW	XUD7T/L A8B SOHC Turbo 58kW	XUD9Y DJZ SOHC 47kW
Capacity (cm³) / cylinders........................	1769 / 4	1769 / 4	1769 / 4	1905 / 4
Compression ration / pressure ....bar	23.0 / _	23.0 / _	22.0 / _	23.5 / _
Torque output ....Nm	108	108	157	120
Oil pressure ....idle [running] bar	2.0 [4.5 @ 4000]	2.0 [4.5 @ 4000]	2.0 [4.5 @ 4000]	2.0 [4.5 @ 4000]
Oil temperature ....°C	80	80	80	80
Valve clearances - inlet (mm)	0.15 ± 0.04	0.15 ± 0.04	0.15 ± 0.04	0.15 ± 0.04
- exhaust (mm)	0.30 ± 0.04	0.30 ± 0.04	0.30 ± 0.04	0.30 ± 0.04
Injection order.....................................	1-3-4-2	1-3-4-2	1-3-4-2	1-3-4-2
No. 1 cylinder position...............	FE	FE	FE	FE
Cooling system				
Thermostat opening temperature ....°C	83	83	89	88
Radiator cap pressure ....bar	1.0	1.0	1.4	1.0
Fuel system				
Idle speed ....rpm	800 ± 50	800 ± 50	800 ± 50	800 ± 50
Maximum (no load) speed ....rpm	5100	5100	4800	5100
Smoke test/opacity ....M⁻¹ %	2.5	2.5	3.0	2.5
Static timing method...............................	Plunger travel	Rotor groove	Dial gauge	Plunger travel
Timing dimension............................mm	0.3	0.01'	Value marked on pump	0.3
Crankshaft position ....mm [°]	0.8 ± 0.03	2.26 ± 0.05'	TDC	0.57 ± 0.03
Turbo type / ref / pressure ....bar	_	_	KKK K14 0.67 bar @ 3500rpm	_
Injection pump make	Bosch	Roto Diesel	Roto Diesel	Bosch
Injection pump part no...........................	2300 R143, R171	R8443 260A/052, A262D or B263D	R8443 B620A	VE R162/4
Injector Make / type	Bosch	Roto Diesel	Roto Diesel	Bosch
Injector part no....................................	DNOSD 256	RDNOSDC 6850C	RDNOSDC 682D	0432 217 153
Injection type......................................	Indirect VE4/18F	Indirect DPC	Indirect DPC	Indirect VER
Injection opening pressure, New [used]...bar	130	115	130	130 [117]
Glow plugs				
Maker	Bosch/Champion	Bosch/Champion	Bosch/Champion	Bosch/Champion
Type	0250 201 019 / CH68	0250 201 019 / CH68	0250 201 019 / CH68	0250 201 019 / CH68
Nominal rating.....................................V/A	11 / 12	11 / 12	11 / 13	11 12
Brakes				
minimum friction material thickness				
Front..mm	1.0	1.0	1.0	1.0
Rear...mm	1.5	1.5	1.5	1.5
Tyres - Saloon / Hatch................Size	145x13: 155/70x13	145x13: 155/70x13	165/70x13	155/70x13: 165/70X13
- Estate / Van...........................Size	145x13: 165/70x13	145x13: 165/70x13	_	_
Pressure - front / rear - Saloon / Hatch ...bar	1.8 / 2.0	1.8 / 2.0	1.8 / 2.0	1.8 / 2.0
- Estate / Van ......bar	1.9 / 2.1	1.9 / 2.1	_	_
Front suspension / wheel alignment				
Toe-in (+) / Toe-out (–)......................mm [°]	1.0 ± 0.5 PW	1.0 ± 0.5 PW	1.0 ± 0.5 PW	1.0 ± 0.5 PW
Camber	0° ± 30'	0° ± 30'	0° ± 30'	0° ± 30'
Castor	3° ± 30'	3° ± 30'	3° ± 30'	3° ± 30'
King pin inclination.................................	9°45' ± 30'	9°45' ± 30'	9°45' ± 30'	9°45' ± 30'
Rear suspension / wheel alignment				
Toe-in (+) / Toe-out (–).....................mm [°]	1.6 ± 0.8 PW	1.6 ± 0.8 PW	1.6 ± 0.8 PW	1.6 ± 0.8 PW
Camber	-50' ± 30'	-50' ± 30'	-50' ± 30'	-50' ± 30'

PEUGEOT

	205 & Van 1.7 1983 to 1996	205 & Van 1.7 1983 to 1996	205 Turbo 1.7 1990 to 1996	205 1.9 1987 to 1994
Torque wrench settings				
Cylinder head - stage 1 ....Nm	30[1]	30[2]	30 N	30 N
- stage 2 ....Nm	60	60	70	70
- stage 3 ....Nm	Slacken, 60	Slacken, 60	+ 120°	+ 120°
- stage 4 ....Nm	Warm up, allow to cool	Warm up, allow to cool	–	–
- stage 5 ....Nm	Slacken	Slacken	–	–
- stage 6 ....Nm	70	70	–	–
Big-end bearings......................Nm	20 + 70° N	20 + 70° N	20 + 70° N	20 + 70° N
Main bearings......................Nm	70	70	70	70
Crankshaft pulley bolt ....Nm	40 + 60°	40 + 60°	40 + 60°	40 + 60°
Camshaft pulley bolt ....Nm	40	40	40	40
Flywheel [driveplate] bolt......................Nm	50 LkC	50 LkC	50	50 LkC
Front hubs ....Nm	265	265	265	265
Rear hubs ....Nm	215	215	215	215
Wheel nuts / bolts ....Nm	85	85	85	85
Glow plugs ....Nm	22	22	22	22
Clutch pressure plate bolts ....Nm	25	25	25	25
Injection pump sprocket......................Nm	50	50	50	50
Injectors......................Nm	90	90	90	90
Injection pump mounting bolts ....Nm	–	–	–	–
Injector pipe unions......................Nm	–	–	–	–
Capacities				
Engine oil & filter......................litres	4.75	4.75	4.75	4.75
Gearbox......................litres	2.0	2.0	2.0	2.0
Automatic transmission ....litres	–	–	–	–
Final drive ....litres	WT	WT	WT	WT
Cooling system......................litres	8.3	8.3	8.5	8.3
Fuel tank......................litres	50	50	50	50

Notes

205 & Van 1.7 1983 to 1996
[1]9.86 ▶: 30, 70 + 120°
205 & Van 1.7 1983 to 1996
[1]8443B263D: value shown on pump
[2]9.86 ▶: 30, then 70 + 120°

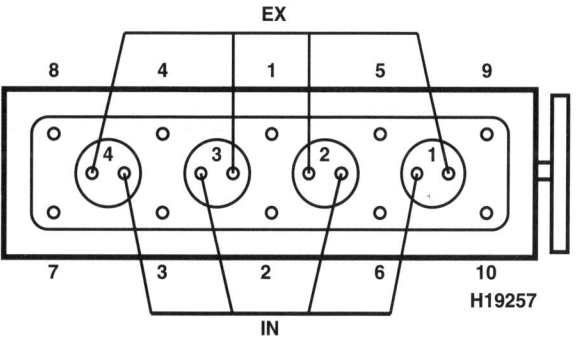

1769 cm³ / 1905 cm³

– Not applicable, or information not available

PEUGEOT

	206 1.9 D 1998 to 2000	206 1.9 D 1999 to 2000	206 2.0 HDi 1999 to 2000	306 1.9 D 1993 to 1999
Engine				
Engine type/code......................	DW8/L3 WJZ 51kW	DW8B/L3 WJY 51kW	DW10TD/L3 RHY SOHC 66kW	XUD9A/L D9B SOHC 52kW
Capacity (cm³) / cylinders.....................	1868 / 4	1868 / 4	1997 / 4	1905 / 4
Compression ration / pressure ....bar	23 /	23 /	18.0 /	23.0 /
Torque output ....Nm	125	125	205	120
Oil pressure ....idle [running] bar	1.8 [4.5 @ 4000]	1.8 [4.5 @ 4000]	2.0 [4.0 @ 4000]	[4.5 @ 4000]
Oil temperature ....°C	80	80	80	80
Valve clearances - inlet (mm)	0.15 ± 0.08	0.15 ± 0.08	0: Hyd.	0.15 ± 0.08
- exhaust (mm)	0.30 ± 0.08	0.30 ± 0.08	0: Hyd.	0.30 ± 0.08
Injection order	1-3-4-2	1-3-4-2	1-3-4-2	1-3-4-2
No. 1 cylinder position	TBE	TBE	FE	FE
Cooling system				
Thermostat opening temperature ....°C	83	83	83	89
Radiator cap pressure ....bar	1.4	1.4	1.4	1.4
Fuel system				
Idle speed ....rpm	875 ± 50	875 ± 50	800 ± 20	750 to 800
Maximum (no load) speed ....rpm	-	-	5000	5150
Smoke test/opacity ....M^{-1} %	2.01	1.77	2.0	2.5
Static timing method............................	-	-	-	Plunger travel
Timing dimension.........................mm	Computer controlled	Computer controlled	Computer controlled	1.07 ± 0.02[3]
Crankshaft position ....mm [°]	-	-	-	0 TDC
Turbo type / ref / pressure ....bar	-	-	KKK K03	-
Injection pump make	Lucas	Lucas	Bosch	Bosch
Injection pump part no...........................	DWLP11	DWLP12	EDC15C2	VER 425/1
Injector Make / type			Bosch	
Injector part no..................................	Lucas RDN OSDC 6903	Lucas RDNOSDC 6903	-	DNOSD299A / RDNOSDC6887D
Injection type......................................	-	-	Direct	Indirect
Injection opening pressure, New [used]...bar	135 ± 5	135 ± 5	200 to 1500	123 to 130
Glow plugs				
Maker	Beru/Champion	Beru/Champion	Champion	Bosch/Champion
Type	0100226371 / CH185	0100226371 / CH185	CH170	0250 201 019 / CH68
Nominal rating...................................V/A	-	-	-	11 / 12
Brakes				
minimum friction material thickness				
Front...........................mm	2.0	2.0	2.0	2.0
Rear............................mm	1.0	1.0	1.0	1.5
Tyres - Saloon / Hatch......................Size	175/70x13	175/65x14	175/65x14	185/65x14
- Estate / Van.....................Size	-	-	-	-
Pressure - front / rear - Saloon / Hatch ...bar	2.5 / 2.5	2.4 / 2.4	2.4 / 2.4	22.3 / 2.4
- Estate / Van ......bar	-	-	-	-
Front suspension / wheel alignment				
Toe-in (+) / Toe-out (–)....................mm [°]	-0.75 ± 0.5	-0.75 ± 0.5	-0.75 ± 0.5	-2.0 ± 1.0 PAS: 2.0 ± 1.0
Camber	-0°0' ± 30'	- 0°0' ± 30'	- 0°0' ± 30'	-0°20' ± 30'
Castor	3° ± 30'	3° ± 30'	3° ± 30'	2°30' PAS: 3°30' ± 30'
King pin inclination................................	9°45' ± 30'	9°45' ± 30'	9°45' ± 30'	11°30'
Rear suspension / wheel alignment				
Toe-in (+) / Toe-out (–)....................mm [°]	1.7 ± 0.5	1.7 ± 0.5	1.7 ± 0.5	4.2 ± 1.5
Camber	1° ± 30'	1° ± 30'	1° ± 30'	-1°20' ± 15'

PEUGEOT

Haynes THE BOOK ®	206 1.9 D 1998 to 2000	206 1.9 D 1999 to 2000	206 2.0 HDi 1999 to 2000	306 1.9 D 1993 to 1999
Torque wrench settings				
Cylinder head - stage 1 ....Nm	20²	20²	20¹	20²
- stage 2 ....Nm	60	60	60	60
- stage 3 ....Nm	+ 180° ± 5°	+ 180° ± 5°	+ 220°	+ 180°
- stage 4 ....Nm	–	–	–	–
- stage 5 ....Nm	–	–	–	–
- stage 6 ....Nm	–	–	–	–
Big-end bearings ....Nm	20 + 70° N	20 + 70° N	20 + 70° N	20 + 70° N
Main bearings ....Nm	70	70	25 + 60°	70
Crankshaft pulley bolt ....Nm	40 + 60°	40 + 60°	197	40 + 60°
Camshaft pulley bolt ....Nm	45	45	43	45
Flywheel [driveplate] bolt ....Nm	48	48	48	50
Front hubs ....Nm	–	–	–	265 M24: 325
Rear hubs ....Nm	200	200	200	185
Wheel nuts / bolts ....Nm	85	85	85	85
Glow plugs ....Nm	22	22	10	25
Clutch pressure plate bolts ....Nm	20	20	20	22
Injection pump sprocket ....Nm	23	23	50	50
Injectors ....Nm	90	90	30	90
Injection pump mounting bolts ....Nm	20	20	22.5	–
Injector pipe unions ....Nm	25	25	20	–
Capacities				
Engine oil & filter ....litres	4.75	4.75	4.5	5.0
Gearbox ....litres	2.0	2.0	2.0	2.15
Automatic transmission ....litres	–	–	–	–
Final drive ....litres	WT	–	WT	WT
Cooling system ....litres	8.2	8.2	6.25	9.0
Fuel tank ....litres	–	–	–	60

Notes

206 1.9 D 1998 to 2000
² Bolt length below head: ≤125.5 mm
206 1.9 D 1999 to 2000
²Bolt length below head: ≤125.5 mm Bolt without guide boss: ≤121.5mm
206 2.0 HDi 1999 to 2000
² Bolt length below head: ≤133.4 mm
306 1.9 D 1993 to 1999
¹or Lucas DPCR/C 8443B 980A
²Bolt length below head: ≤121.5 mm
³Lucas DPC: Dimension on pump

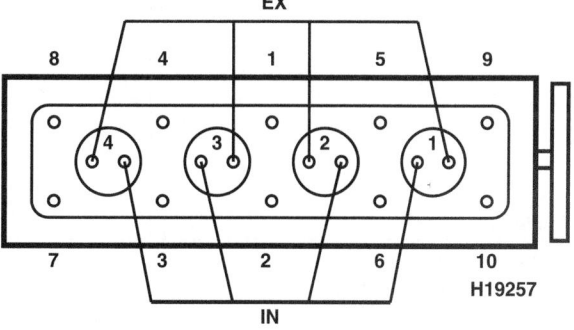

1868 cm³ / 1905 cm³ / 1996 cm³

– Not applicable, or information not available

PEUGEOT

	306 1.9 TD 1993 to 1999	306 1.9 TD 1993 to 1999	306 1.9 D 1999 to 2000	306 2.0 HDi 1999 to 2000
Engine				
Engine type/code....................	XUD9TE/L D8A SOHC Turbo 68kW	XUD9TE/Y DHY SOHC Turbo 66kW	DW8/L3 WJZ SOHC 51kW	DW10TD/L3 RHY SOHC 66kW
Capacity (cm³) / cylinders........................	1905 / 4	1905 / 4	1868 / 4	1996 / 4
Compression ration / pressure ....bar	21.8 /	21.8 /	23.0 /	18.0 /
Torque output ....Nm	196	196	125	205
Oil pressure ....idle [running] bar	[4.9 @ 4000]	[4.9 @ 4000]	1.8 [4.5 @ 4000]	[3.8]
Oil temperature ....°C	80	80	80	80
Valve clearances - inlet (mm)	0.15 ± 0.08	0.15 ± 0.08	0.15 ± 0.07	0: Hyd.
- exhaust (mm)	0.30 ± 0.08	0.30 ± 0.08	0.30 ± 0.07	0: Hyd.
Injection order	1-3-4-2	1-3-4-2	1-3-4-2	1-3-4-2
No. 1 cylinder position	FE	FE	FE	FE
Cooling system				
Thermostat opening temperature ....°C	89	89	89	83
Radiator cap pressure ....bar	1.4	1.4	1.4	1.4
Fuel system				
Idle speed ....rpm	750 to 800	750 to 800	875 ± 25	850 ± 25
Maximum (no load) speed ....rpm	5100	5100	5100	5100
Smoke test/opacity ....M⁻¹ %	3.0	3.0	2.5	2.0
Static timing method..............................	Plunger travel	Plunger travel	Refer to wsm	Refer to wsm
Timing dimension.............................mm	0.66 ± 0.02	0.63 ± 0.02	_	_
Crankshaft position ....mm [°]	0 TDC	0 TDC	_	_
Turbo type / ref / pressure ....bar	KKK K14 / Garrett T2²	KKK K14 / Garrett T2²	_	Garrett T15 0.95 bar @ 2000
Injection pump make	Bosch	Bosch	Lucas	Bosch
Injection pump part no...........................	VER 445	VER 472	DWLP11²	EDC15C2
Injector Make / type...............................	Bosch	Bosch	_	_
Injector part no.......................................	DNOSD 299	DNOSD 299	DNOSD 299	DNOSD 299
Injection type..	Indirect	Indirect	_	Direct common rail
Injection opening pressure, New [used]...bar	175 to 180	175 to 180	135 ± 5	200 to 1500
Glow plugs				
Maker	Bosch/Champion	Bosch/Champion	Beru/Champion	Bosch/Champion
Type	0250 201 033 / CH163	0250 201 033 / CH163	0100 226 6371 / CH185	0250 202 032 / CH170
Nominal rating.................................V/A	11 / 12	11 / 12	_	_
Brakes				
minimum friction material thickness				
Front...mm	2.0	2.0	2.0	2.0
Rear..mm	1.5	1.5	1.5	1.5
Tyres - Saloon / Hatch ....Size	185/65x15	185/65x15	185/65x15	185/65x15
- Estate / Van..........................Size	_	_	_	_
Pressure - front / rear - Saloon / Hatch ...bar	2.3 / 2.3	2.3 / 2.3	2.3 / 2.3	2.3 / 2.3
- Estate / Van ......bar	_	_	_	_
Front suspension / wheel alignment				
Toe-in (+) / Toe-out (−)......................mm [°]	2.0 ± 1.0	2.0 ± 1.0	2.0 ± 1.0	2.0 ± 1.0
Camber	-0°20' ± 30'	-0°20' ± 30'	-0°20' ± 30'	-0°20' ± 30'
Castor	3°20' ± 30'	3°20' ± 30'	3°20' ± 30'	3°20' ± 30'
King pin inclination................................	11°30'	11°30'	11°30'	11°30'
Rear suspension / wheel alignment				
Toe-in (+) / Toe-out (−)......................mm [°]	4.2 ± 1.5	4.2 ± 1.5	4.2 ± 1.5	4.2 ± 1.5
Camber	-1°20' ± 15'	-1°20' ± 15'	-1°20' ± 15'	-1°20' ± 15'

PEUGEOT

	306 1.9 TD 1993 to 1999	306 1.9 TD 1993 to 1999	306 1.9 D 1999 to 2000	306 2.0 HDi 1999 to 2000
Torque wrench settings				
Cylinder head - stage 1 ....Nm	20[1]	20[1]	20[1]	20[1]
- stage 2 ....Nm	60	60	60	60
- stage 3 ....Nm	+ 220°	+ 220°	+ 180°	+ 220°
- stage 4 ....Nm	–	–	–	–
- stage 5 ....Nm	–	–	–	–
- stage 6 ....Nm	–	–	–	–
Big-end bearings......................Nm	20 + 70° N	20 + 70° N	20 + 70° N	20 + 70° N
Main bearings......................Nm	70	70	70	25 + 60°
Crankshaft pulley bolt ....Nm	40 + 60°	40 + 60°	40 + 60°	197
Camshaft pulley bolt ....Nm	45	45	45	43
Flywheel [driveplate] bolt.....................Nm	50	50	50	48
Front hubs ....Nm	265 M24: 325	265 M24: 325	265 M24: 325	265 M24: 325
Rear hubs ....Nm	185	185	185	185
Wheel nuts / bolts ....Nm	85	85	85	85
Glow plugs ....Nm	25	25	15	25
Clutch pressure plate bolts ....Nm	22	22	22	22
Injection pump sprocket.....................Nm	50	50	23	50
Injectors.....................Nm	90	90	90	30
Injection pump mounting bolts ....Nm	–	–	20	22.5
Injector pipe unions.....................Nm	–	–	25	20
Capacities				
Engine oil & filter.....................litres	5.0	5.0	5.0	5.0
Gearbox.....................litres	2.15	2.15	2.15	2.15
Automatic transmission ....litres	–	–	–	–
Final drive ....litres	WT	WT	WT	WT
Cooling system.....................litres	9.0	9.0	9.0	9.0
Fuel tank.....................litres	60	60	60	60

Notes

306 1.9 TD 1993 to 1999
[1]Bolt length below head: ≤146.5 mm
[2] 0.93 to 1.1 bar @ 3000rpm
306 1.9 TD 1993 to 1999
[1]Bolt length below head: ≤146.5 mm
[2] 0.93 to 1.1 bar @ 3000rpm
306 1.9 D 1999 to 2000
[1]Bolt length below head: ≤121.5 mm Bolt with guide boss: ≤125.5 mm
[2]or Bosch DWBP11
306 2.0 HDi 1999 to 2000
[1]Max bolt length below head: ≤133.3 mm

– Not applicable, or information not available

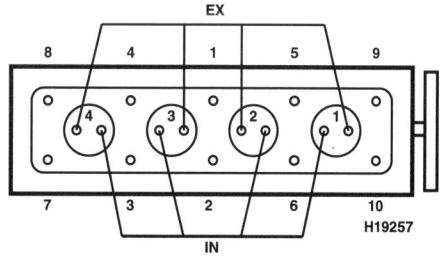

1905 cm³

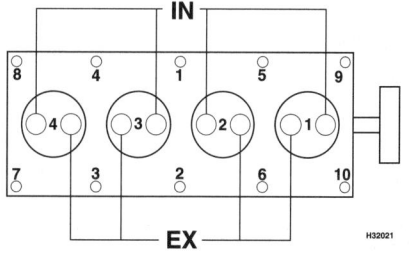

1868 cm³ / 1996 cm³

PEUGEOT

	309 1.7 1987 to 1993	309 1.7 1987 to 1993	309 1.7 Turbo 1989 to 1993	309 1.9 1986 to 1993
Engine				
Engine type/code....................	XUD7 A9A SOHC 44kW	XUD7 A9A SOHC 44kW	XUD7T/L A8B SOHC Turbo 58kW	XUD9 D9A SOHC 47kW
Capacity (cm³) / cylinders........................	1769 / 4	1769 / 4	1769 / 4	1905 / 4
Compression ration / pressure ....bar	23.0 / _	23.0 / _	22.0 / _	23.5 / _
Torque output ....Nm	108	108	157	120
Oil pressure ....idle [running] bar	2.0 [4.5 @ 4000]	2.0 [4.5 @ 4000]	2.0 [4.5 @ 4000]	2.0 [4.5 @ 4000]
Oil temperature ....°C	80	80	80	80
Valve clearances - inlet (mm)	0.15 ± 0.04	0.15 ± 0.04	0.15 ± 0.04	0.15 ± 0.04
- exhaust (mm)	0.30 ± 0.04	0.30 ± 0.04	0.30 ± 0.04	0.30 ± 0.04
Injection order	1-3-4-2	1-3-4-2	1-3-4-2	1-3-4-2
No. 1 cylinder position	FE	FE	FE	FE
Cooling system				
Thermostat opening temperature ....°C	82	82	82	83
Radiator cap pressure ....bar	1.0	1.0	1.0	1.0
Fuel system				
Idle speed ....rpm	800 ± 50	800 ± 50	775 ± 25	800 ± 50
Maximum (no load) speed ....rpm	5100	5100	4800	5100
Smoke test/opacity ....M⁻¹ %	3.0	3.0	3.0	2.5
Static timing method................................	Plunger travel	Refer to wsm	Dial gauge	Dial gauge
Timing dimension.................................mm	0.30	_	Dimension on pump	0.01
Crankshaft position ....mm [°]	0.80 ± 0.03	_	TDC	2.26 ± 0.05
Turbo type / ref / pressure ....bar	_	_	_	_
Injection pump make	Bosch	Roto Diesel	Roto Diesel	Roto Diesel
Injection pump part no.............................	2300 R171 R523	R8443262D	R8443 B260A/052 or 261C	R8443 261A
Injector Make / type	Bosch	Roto Diesel	Roto Diesel	Roto Diesel
Injector part no......................................	DNOSD 256	RDNOSDC6850	RDNOSDC 682D	RDNOSDC 6850B
Injection type..	Indirect VE4/8F	Indirect DPC	Indirect DPC	Indirect DPC
Injection opening pressure, New [used]...bar	130 ± 5	115 [103]	115 ± 5	115 ± 5
Glow plugs				
Maker	Bosch/Champion	Bosch/Champion	Bosch/Champion	Bosch/Champion
Type	0250 201 033 / CH68	0250 201 033 / CH68	0250 201 019 / CH68	0250 201 019 / CH68
Nominal rating.....................................V/A	11 / 12	11 / 12	11 / 12	11 / 12
Brakes				
minimum friction material thickness				
Front...mm	1.0	1.0	1.0	1.0
Rear...mm	1.5	1.5	1.5	1.5
Tyres - Saloon / Hatch......................Size	165/70x13	165/70x13	175/65x14	165/70x13
- Estate / Van............................Size	_	_	_	_
Pressure - front / rear - Saloon / Hatch ...bar	1.9 / 2.1	1.9 / 2.1	1.9 / 2.1	1.9 / 2.1
- Estate / Van ......bar	_	_	_	_
Front suspension / wheel alignment				
Toe-in (+) / Toe-out (–).....................mm [°]	0.5 ± 0.5 PW	0.5 ± 0.5 PW	0.5 ± 0.5 PW	0.5 ± 0.5 PW
Camber	-30' ± 30'	-30' ± 30'	-30' ± 30'	-30' ± 30'
Castor	2° ± 30'	2° ± 30'	2°10' ± 30'	2° ± 30'
King pin inclination.................................	10°15' ± 30'	10°15' ± 30'	10°15' ± 30'	10°15' ± 30'
Rear suspension / wheel alignment				
Toe-in (+) / Toe-out (–).....................mm [°]	1.6 ± 0.8 PW¹	1.6 ± 0.8 PW¹	1.9 ± 0.8 PW	1.6 ± 0.8 PW¹
Camber	-50' ± 30'	-50' ± 30'	-1°20' ± 30'	-50' ± 30'

PEUGEOT

	309 1.7 1987 to 1993	309 1.7 1987 to 1993	309 1.7 Turbo 1989 to 1993	309 1.9 1986 to 1993
Torque wrench settings				
Cylinder head - stage 1 ....Nm	30 N	30 N	30 N	30 N
- stage 2 ....Nm	70	70	70	70
- stage 3 ....Nm	+ 120°	+ 120°	+ 120°	+ 120°
- stage 4 ....Nm	–	–	–	–
- stage 5 ....Nm	–	–	–	–
- stage 6 ....Nm		–		
Big-end bearings.............................Nm	20 + 70° N	20 + 70° N	20 + 70° N	20 + 70° N
Main bearings...............................Nm	70	70	70	70
Crankshaft pulley bolt ....Nm	40 + 60°	40 + 60°	40 + 60°	40 + 60°
Camshaft pulley bolt ....Nm	40	40	40	40
Flywheel [driveplate] bolt.....................Nm	50 LkC	50 LkC	50 LkC	50 LkC
Front hubs ....Nm	265	265	265	265
Rear hubs....................................Nm	215	215	215	215
Wheel nuts / bolts ....Nm	85	85	85	85
Glow plugs ....Nm	22	22	22	22
Clutch pressure plate bolts ....Nm	25	25	25	25
Injection pump sprocket.....................Nm	50	50	50	50
Injectors....................................Nm	90	90	90	90
Injection pump mounting bolts ....Nm	–	–	–	–
Injector pipe unions............................Nm	–	–	–	–
Capacities				
Engine oil & filter.............................litres	4.8	4.8	4.8	4.8
Gearbox......................................litres	2.0	2.0	2.0	2.0
Automatic transmission ....litres	–	–	–	–
Final drive ....litres	WT	WT	WT	WT
Cooling system...............................litres	8.5	8.5	8.5	8.5
Fuel tank.....................................litres	55	55	55	55

Notes

309 1.7 1987 to 1993
¹1990 ▶: 1.8±0.8 PW
309 1.7 1987 to 1993
¹1990 ▶: 1.8±0.8 PW
309 1.9 1986 to 1993
¹1990 ▶: 1.8±0.8 PW

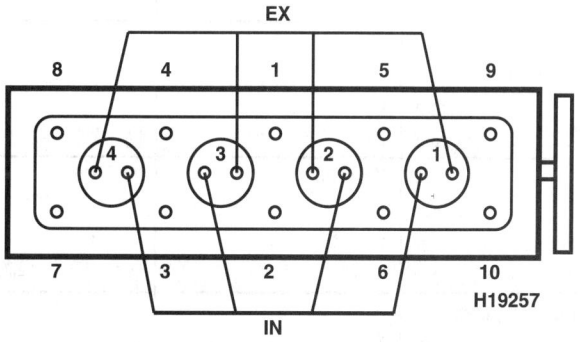

1769 cm³ / 1905 cm³

– Not applicable, or information not
available

PEUGEOT

	309 1.9 1987 to 1993	309 1.9 EGR 1987 to 1993	405 1.7 Turbo 1988 to 1992	405 1.7 Turbo 1988 to 1992
Engine				
Engine type/code............................	XUD9 D9A SOHC 47kW	XUD9Y DJZ SOHC 47kW	XUD7TE A8A SOHC Turbo 66kW	XUD7TE A8A SOHC Turbo 66kW
Capacity (cm³) / cylinders....................	1905 / 4	1905 / 4	1769 / 4	1769 / 4
Compression ration / pressure ....bar	23.5 / _	23.5 / _	22.0 / _	22.0 / _
Torque output ....Nm	120	120	157	157
Oil pressure ....idle [running] bar	2.0 [4.5 @ 4000]	2.0 [4.5 @ 4000]	2.0 [4.5 @ 4000]	2.0 [4.5 @ 4000]
Oil temperature ....°C	80	80	80	80
Valve clearances - inlet (mm)	0.15 ± 0.04	0.15 ± 0.04	0.15 ± 0.04	0.15 ± 0.04
- exhaust (mm)	0.30 ± 0.04	0.30 ± 0.04	0.30 ± 0.04	0.30 ± 0.04
Injection order	1-3-4-2	1-3-4-2	1-3-4-2	1-3-4-2
No. 1 cylinder position	FE	FE	FE	FE
Cooling system				
Thermostat opening temperature ....°C	83	83	88	88
Radiator cap pressure ....bar	1.0	1.0	1.0	1.0
Fuel system				
Idle speed ....rpm	800 ± 50	775 ± 50	750 to 800	750 to 800
Maximum (no load) speed ....rpm	5100	5100	4800	4800
Smoke test/opacity ....M⁻¹ %	2.5	2.5	3.0	3.0
Static timing method................................	Plunger travel	Plunger travel	Dial gauge	Plunger travel
Timing dimension.............................mm	0.3	1.0	Dimension on pump	0.8
Crankshaft position ....mm [°]	0.50 ± 0.03	0.77 ± 0.03	TDC	1.0 ± 0.03
Turbo type / ref / pressure ....bar	_	_	0.85 bar @ 3000rpm	0.85 bar @ 3000rpm
Injection pump make	Bosch	Bosch	Roto Diesel	Bosch
Injection pump part no..........................	2300 R162	VE4/9 R162.4	R8443 B452C	2150 R316
Injector Make / type	Bosch	Bosch	Roto Diesel	Bosch
Injector part no.................................	0432 217 153	0432 217 153	RDNOSDC 6862C	DNOSD 289
Injection type....................................	Indirect VE4/9F	Indirect VE4/9F	Indirect DPC	Indirect VE4/8F
Injection opening pressure, New [used]...bar	130 ± 5 [117]	130 ± 5	130 ± 5	130 ± 5
Glow plugs				
Maker	Bosch/Champion	Bosch/Champion	Bosch/Champion	Bosch/Champion
Type	0250 201 019 / CH68	0250 201 019 / CH68	0250 201 019 / CH68	0250 201 019 / CH68
Nominal rating..................................V/A	11 / 12	11 / 12	11 / 12	11 / 12
Brakes				
minimum friction material thickness				
Front.......................................mm	1.0	1.0	1.0	1.0
Rear.......................................mm	1.5	1.5	1.5	1.5
Tyres - Saloon / Hatch......................Size	165/70x13	165/70x13	185/65x14	185/65x14
- Estate / Van...........................Size	_	_	185/65x14	185/65x14
Pressure - front / rear - Saloon / Hatch ...bar	1.9 / 2.1	1.9 / 2.1	2.1 / 2.1	2.1 / 2.1
- Estate / Van ......bar	_	_	2.1 / 2.4	2.1 / 2.4
Front suspension / wheel alignment				
Toe-in (+) / Toe-out (−)....................mm [°]	0.5 ± 0.5 PW	0.5 ± 0.5 PW	1.0 ± 0.5 PW	1.0 ± 0.5 PW
Camber	-30' ± 30'	-30' ± 30'	0° ± 30'	0° ± 30'
Castor	2° ± 30'	2° ± 30'	3°10' ± 30'¹	3°10' ± 30'¹
King pin inclination..............................	10°15' ± 30'	10°15' ± 30'	11°25' ± 30'	11°25' ± 30'
Rear suspension / wheel alignment				
Toe-in (+) / Toe-out (−)....................mm [°]	1.6 ± 0.8 PW¹	1.6 ± 0.8 PW¹	-2.5 ± 0.8 PW²	-2.5 ± 0.8 PW²
Camber	-50' ± 30'	-50' ± 30'	-1°20' ± 30'	-1°20' ± 30'

PEUGEOT

	309 1.9 1987 to 1993	309 1.9 EGR 1987 to 1993	405 1.7 Turbo 1988 to 1992	405 1.7 Turbo 1988 to 1992
Torque wrench settings				
Cylinder head - stage 1 ....Nm	30 N	30 N	30 N	30 N
- stage 2 ....Nm	70	70	70	70
- stage 3 ....Nm	+ 120°	+ 120°	+ 120°	+ 120°
- stage 4 ....Nm	–	–	–	–
- stage 5 ....Nm	–	–	–	–
- stage 6 ....Nm	–	–	–	–
Big-end bearings...............................Nm	20 + 70° N	20 + 70° N	20 + 70° N	20 + 70° N
Main bearings.....................................Nm	70	70	70	70
Crankshaft pulley bolt ....Nm	40 + 60°	40 + 60°	40 + 60°	40 + 60°
Camshaft pulley bolt ....Nm	40	40	40	40
Flywheel [driveplate] bolt....................Nm	50 LkC	50 LkC	50 LkC	50 LkC
Front hubs ....Nm	265	265	265	265
Rear hubs...Nm	215	215	275	275
Wheel nuts / bolts ....Nm	85	85	85	85
Glow plugs ....Nm	22	22	22	22
Clutch pressure plate bolts ....Nm	25	25	25	25
Injection pump sprocket.....................Nm	50	50	50	50
Injectors...Nm	90	90	90	90
Injection pump mounting bolts ....Nm	–	–	–	–
Injector pipe unions............................Nm	–	–	–	–
Capacities				
Engine oil & filter................................litres	4.8	4.8	4.75	4.75
Gearbox..litres	2.0	2.0	2.0	2.0
Automatic transmission ....litres	–	–	–	–
Final drive ....litres	WT	WT	WT	WT
Cooling system...................................litres	8.5	8.5	7.8	7.8
Fuel tank...litres	55	55	70	70

Notes

309 1.9 1987 to 1993
¹1990 ▶: 1.8±0.8 PW
309 1.9 EGR 1987 to 1993
¹1990 ▶: 1.8 ± 0.8 PW
405 1.7 Turbo 1988 to 1992
¹Estate: 1°40'±30' [PAS: 2°25'±30']
²Estate: -1.4±0.8 PW
405 1.7 Turbo 1988 to 1992
¹Estate: 1°40'±30' [PAS: 2°25'±30']
²Estate: -1.4±0.8 PW

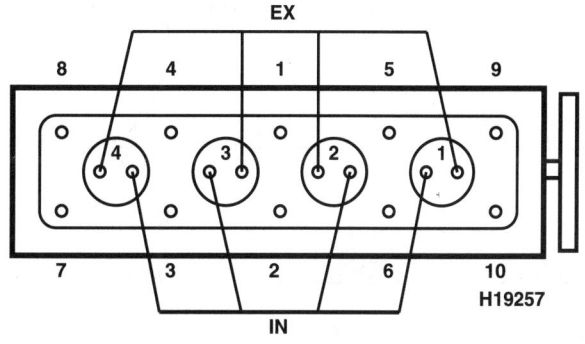

1769 cm³ / 1905 cm³

– Not applicable, or information not
available

PEUGEOT

	405 1.7 Turbo 1992 to 1994	405 1.9 1988 to 1991	405 1.9 1988 to 1991	405 1.9 EGR 1988 to 1991
Engine				
Engine type/code	XUD7TE/Y AJZ SOHC Turbo 66kW	XUD9A D9B SOHC 51kW	XUD9A D9B SOHC 51kW	XUD9Y DJZ SOHC 47kW
Capacity (cm³) / cylinders......................	1769 / 4	1905 / 4	1905 / 4	1905 / 4
Compression ration / pressure ....bar	22.0 / _	23.0 / _	23.0 / _	23.5 / _
Torque output ....Nm	157	120	120	118
Oil pressure ....idle [running] bar	2.0 [4.5 @ 4000]	2.0 [4.5 @ 4000]	2.0 [4.5 @ 4000]	2.0 [4.5 @ 4000]
Oil temperature ....°C	80	80	80	80
Valve clearances - inlet (mm)	0.15 ± 0.04	0.15 ± 0.04	0.15 ± 0.04	0.15 ± 0.04
- exhaust (mm)	0.30 ± 0.04	0.30 ± 0.04	0.30 ± 0.04	0.30 ± 0.04
Injection order	1-3-4-2	1-3-4-2	1-3-4-2	1-3-4-2
No. 1 cylinder position	FE	FE	FE	FE
Cooling system				
Thermostat opening temperature ....°C	88	88	88	88
Radiator cap pressure ....bar	1.0	1.0	1.0	1.0
Fuel system				
Idle speed ....rpm	750 to 800	750 to 800	750 to 800	775 ± 50
Maximum (no load) speed ....rpm	4800	5150	5150	5100
Smoke test/opacity ....M⁻¹ %	3.0	2.5	2.5	2.5
Static timing method..............................	Plunger travel	Plunger travel	Dial gauge	Plunger travel
Timing dimension................................mm	0.74	0.9	Dimension on pump	0.77
Crankshaft position ....mm [°]	0 TDC	0 TDC	TDC	0 TDC
Turbo type / ref / pressure ....bar	0.85 bar @ 3000rpm	_	_	_
Injection pump make	Bosch	Bosch	Roto Diesel	Bosch
Injection pump part no..........................	R403	2300 R272/2	R8443 B380A, 381B	2300 R162.4
Injector Make / type	Bosch	Bosch	Roto Diesel	Bosch
Injector part no........................	_	DNOSD 287+	RDNOSDC 6751C	DNOSD 256
Injection type............................	Indirect VE4/8F	Indirect VF4/9F	Indirect DPC	Indirect VE4/9F
Injection opening pressure, New [used]...bar	150 ± 5	130 ± 5	125 ± 5	130 ± 5
Glow plugs				
Maker	Bosch/Champion	Bosch/Champion	Bosch/Champion	Bosch/Champion
Type	0250 201 019 / CH68	0250 201 019 / CH68	0250 201 019 / CH68	0250 201 019 / CH68
Nominal rating.............................V/A	11 / 12	11 / 12	11 / 12	11 / 13
Brakes				
minimum friction material thickness				
Front...mm	1.0	1.0	1.0	1.0
Rear...mm	1.5	1.5	1.5	1.5
Tyres - Saloon / Hatch......................Size	185/65x14	165/70x14: 175/70x14	165/70x14: 175/70x14	165/70x14: 175/70x14
- Estate / Van........................Size	185/65x14	175/70x14	175/70x14	175/70x14
Pressure - front / rear - Saloon / Hatch ...bar	2.1 / 2.1	2.1 / 2.1	2.1 / 2.1	2.1 / 2.1
- Estate / Van ......bar	2.1 / 2.4	2.1 / 2.3	2.1 / 2.3	2.1 / 2.3
Front suspension / wheel alignment				
Toe-in (+) / Toe-out (–)......................mm [°]	1.0 ± 0.5 PW	1.0 ± 0.5 PW	1.0 ± 0.5 PW	1.0 ± 0.5 PW
Camber	0° ± 30'	0° ± 30'	0° ± 30'	0° ± 30'
Castor	3°10' ± 30'¹	3°10' ± 30'¹	3°10' ± 30'¹	3°10' ± 30'¹
King pin inclination..............................	11°25' ± 30'	11°25' ± 30'	11°25' ± 30'	11°25' ± 30'
Rear suspension / wheel alignment				
Toe-in (+) / Toe-out (–)......................mm [°]	-2.5 ± 0.8 PW²	-2.5 ± 0.8 PW²	-2.5 ± 0.8 PW²	-2.5 ± 0.8 PW²
Camber	-1°20' ± 30'	-1°20' ± 30'	-1°20' ± 30'	-1°20' ± 30'

PEUGEOT

Torque wrench settings	405 1.7 Turbo 1992 to 1994	405 1.9 1988 to 1991	405 1.9 1988 to 1991	405 1.9 EGR 1988 to 1991
Cylinder head - stage 1 ....Nm	30 N	30 N	30 N	30 N
- stage 2 ....Nm	70	70	70	70
- stage 3 ....Nm	+ 120°	+ 120°	+ 120°	+ 120°
- stage 4 ....Nm	–	–	–	–
- stage 5 ....Nm	–	–	–	–
- stage 6 ....Nm	–	–	–	–
Big-end bearings........................Nm	20 + 70° N	20 + 70° N	20 + 70° N	20 + 70° N
Main bearings........................Nm	70	70	70	70
Crankshaft pulley bolt ....Nm	40 + 60°	40 + 60°	40 + 60°	40 + 60°
Camshaft pulley bolt ....Nm	40	40	40	40
Flywheel [driveplate] bolt.....................Nm	50 LkC	50 LkC	50 LkC	50 LkC
Front hubs ....Nm	265	265	265	265
Rear hubs ....Nm	275	275	275	275
Wheel nuts / bolts ....Nm	85	85	85	85
Glow plugs ....Nm	22	22	22	22
Clutch pressure plate bolts ....Nm	25	25	25	25
Injection pump sprocket.....................Nm	50	50	50	50
Injectors........................Nm	90	90	90	90
Injection pump mounting bolts ....Nm	–	–	–	–
Injector pipe unions..........................Nm	–	–	–	–
Capacities				
Engine oil & filter................................litres	4.75	4.5	4.5	4.75
Gearbox................................litres	2.0	2.0	2.0	2.0
Automatic transmission ....litres	–	–	–	–
Final drive ....litres	WT	WT	WT	WT
Cooling system................................litres	7.8	7.8	7.8	7.8
Fuel tank................................litres	70	70	70	70

Notes

405 1.7 Turbo 1992 to 1994
¹Estate: 1°40'±30' [PAS: 2°25'±30']
²Estate: -1.4±0.8 PW
405 1.9 1988 to 1991
¹Estate: 1°40'±30' [PAS: 2°25'±30']
²Estate: -1.4±0.8 PW
405 1.9 1988 to 1991
¹Estate: 1°40'±30' [PAS: 2°25'±30']
²Estate: -1.4±0.8 PW
405 1.9 EGR 1988 to 1991
¹Estate: 1°40'±30' [PAS: 2°25'±30']
²Estate: -1.4±0.8 PW

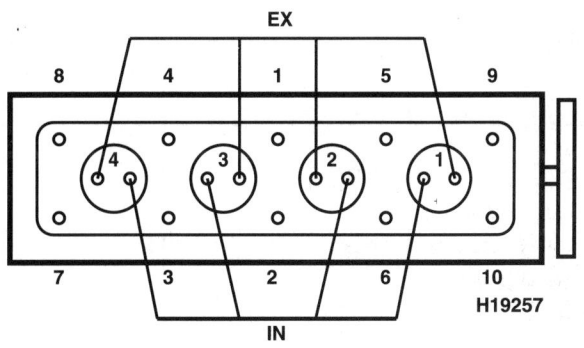

1769 cm³ / 1905 cm³

– Not applicable, or information not available

PEUGEOT

	405 1.9 D 1992 to 1996	405 1.9 D 1992 to 1996	405 1.9 D 1992 to 1996	405 1.9 TD 1992 to 1996
Engine				
Engine type/code....................	XUD9A/L D9B SOHC 51kW	XUD9A/L D9B SOHC 51kW	XUD9A/Y DJZ SOHC 47kW	XUD9TE/Y DHY SOHC Turbo 66kW
Capacity (cm³) / cylinders........................	1905 / 4	1905 / 4	1905 / 4	1905 / 4
Compression ration / pressure ....bar	23.0 /	23.0 /	23.0 /	21.8 /
Torque output ....Nm	120	120	118	196
Oil pressure ....idle [running] bar	[4.5 @ 4000]	[4.5 @ 4000]	[4.5 @ 4000]	[4.5 @ 4000]
Oil temperature ....°C	80	80	80	80
Valve clearances - inlet (mm)	0.15 ± 0.04	0.15 ± 0.04	0.15 ± 0.04	0.15 ± 0.04
- exhaust (mm)	0.30 ± 0.04	0.30 ± 0.04	0.30 ± 0.04	0.30 ± 0.04
Injection order	1-3-4-2	1-3-4-2	1-3-4-2	1-3-4-2
No. 1 cylinder position	FE	FE	FE	FE
Cooling system				
Thermostat opening temperature ....°C	83	83	83	83
Radiator cap pressure ....bar	1.0	1.0	1.0	1.4
Fuel system				
Idle speed ....rpm	750 to 800	750 to 800	750 to 800	750 to 800
Maximum (no load) speed ....rpm	5150	5150	5150	5100
Smoke test/opacity ....M⁻¹ %	2.5	2.5	2.5	3.0
Static timing method...............................	Plunger travel	Plunger travel	Plunger travel	Dial gauge
Timing dimension..............................mm	1.07 ± 0.02	Dimension on pump	0.77	0.66
Crankshaft position ....mm [°]	0 TDC	0 TDC	0 TDC	0 TDC
Turbo type / ref / pressure ....bar	–	–	–	Garrett T2 / KKK K14²
Injection pump make	Bosch	Roto Diesel	Bosch	Bosch
Injection pump part no...........................	VER425/1	R8443B980A	VER 162	VER 472
Injector Make / type	Bosch	Roto Diesel	Bosch	Bosch
Injector part no....................................	DNOSD299A	RDNOSDC6887D	256	299
Injection type...	Indirect VER	Indirect DPC	Indirect VER	Indirect VER
Injection opening pressure, New [used]...bar	130	–	130	175
Glow plugs				
Maker	Bosch/Champion	Bosch/Champion	Bosch/Champion	Bosch/Champion
Type	0250 201 019 / CH68	0250 201 019 / CH68	0250 201 019 / CH68	0250 201 033 / CH68
Nominal rating...................................V/A	11 / 13	11 / 13	11 / 13	11 / 13
Brakes				
minimum friction material thickness				
Front.....................................mm	1.0	1.0	1.0	1.0
Rear.....................................mm	1.5	1.5	1.5	1.5
Tyres - Saloon / Hatch......................Size	175/70x14	175/70x14	175/70x14	185/65x14
- Estate / Van.............Size	175/70x14	175/70x14	175/70x14	185/65x14
Pressure - front / rear - Saloon / Hatch ...bar	2.2 / 2.2	2.2 / 2.2	2.2 / 2.2	2.2 / 2.2
- Estate / Van ......bar	2.4 / 2.4	2.4 / 2.4	2.4 / 2.4	2.3 / 2.3
Front suspension / wheel alignment				
Toe-in (+) / Toe-out (–).....................mm [°]	2.0 ± 1.0	2.0 ± 1.0	2.0 ± 1.0	2.1 ± 1.0
Camber	0° ± 30'	0° ± 30'	0° ± 30'	0° ± 30'
Castor	2°10' ± 30' PAS: 3° ± 30'	2°10' ± 30' PAS: 3° ± 30'	2°10' ± 30' PAS: 3° ± 30'	3° ± 30'
King pin inclination.................................	11°30' ± 30'	11°30' ± 30'	11°30' ± 30'	11°30' ± 30'
Rear suspension / wheel alignment				
Toe-in (+) / Toe-out (–)....................mm [°]	5.4 ± 1.4 Est: 2.1 ± 1.4	5.4 ± 1.4 Est: 2.1 ± 1.4	5.4 ± 1.4 Est: 2.1 ± 1.4	5.4 ± 1.4 Est: 2.1 ± 1.4
Camber	-1°30' ± 30'	-1°30' ± 30'	-1°30' ± 30'	

PEUGEOT

Torque wrench settings	405 1.9 D 1992 to 1996	405 1.9 D 1992 to 1996	405 1.9 D 1992 to 1996	405 1.9 TD 1992 to 1996
Cylinder head - stage 1 ....Nm	20[1]	20[1]	20[1]	20[1]
- stage 2 ....Nm	60	60	60	60
- stage 3 ....Nm	+ 180°	+ 180°	+ 180°	+ 220°
- stage 4 ....Nm	–	–	–	–
- stage 5 ....Nm	–	–	–	–
- stage 6 ....Nm	–	–	–	–
Big-end bearings...............................Nm	20 + 70° N	20 + 70° N	20 + 70° N	20 + 70° N
Main bearings......................................Nm	70	70	70	15 + 60°
Crankshaft pulley bolt ....Nm	40 + 60°	40 + 60°	40 + 60°	40 + 60°
Camshaft pulley bolt ....Nm	45	45	45	45
Flywheel [driveplate] bolt...................Nm	50	50	50	50
Front hubs ....Nm	320	320	320	320
Rear hubs ....Nm	275	275	275	275
Wheel nuts / bolts ....Nm	85	85	85	85
Glow plugs ....Nm	25	25	25	25
Clutch pressure plate bolts ....Nm	25	25	25	25
Injection pump sprocket....................Nm	50	50	50	50
Injectors..Nm	90	90	90	90
Injection pump mounting bolts ....Nm	–	–	–	–
Injector pipe unions............................Nm	–	–	–	–
Capacities				
Engine oil & filter.................................litres	5.0	5.0	5.0	5.0
Gearbox...litres	2.0	2.0	2.0	2.0
Automatic transmission ....litres	–	–	–	–
Final drive ....litres	WT	WT	WT	WT
Cooling system...................................litres	7.8	7.8	7.8	7.0
Fuel tank..litres	70	70	70	70

Notes

405 1.9 D 1992 to 1996
[1]Bolt length below head: ≤121.5 mm
405 1.9 D 1992 to 1996
[1]Bolt length below head: ≤121.5 mm
405 1.9 D 1992 to 1996
[1]Bolt length below head: ≤121.5 mm
405 1.9 TD 1992 to 1996
[1]Bolt length below head: ≤146.5 mm
[2]0.85 bar @ 3000rpm

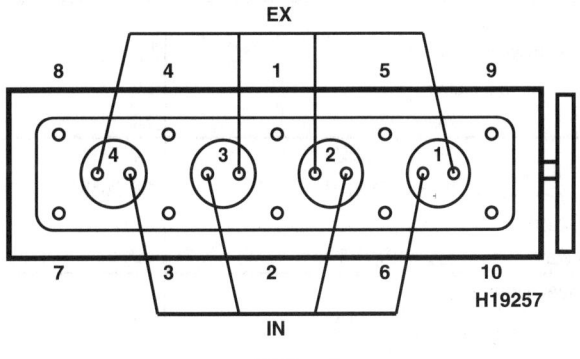

1905 cm³

– Not applicable, or information not available

PEUGEOT

	405 1.9 TD 1992 to 1996	406 1.9 TD 1996 to 1999	406 1.9 TD 1996 to 1999	406 2.0 HDi 1999 to 2000
Engine				
Engine type/code	XUD9TE/L D8A SOHC Turbo 68kW	XUD9TE/L D8A SOHC Turbo 66kW	XUD9BTF DHX SOHC Turbo 66kW	DW10TD/L3 RHY SOHC 66kW
Capacity (cm³) / cylinders	1905 / 4	1905 / 4	1905 / 4	1997 / 4
Compression ration / pressure ....bar	21.8 /	21.8 /	21.8 /	18.0 / ≥30.0
Torque output ....Nm	196	196	196	205
Oil pressure ....idle [running] bar	[4.5 @ 4000]	[4.5 @ 4000]	[4.5 @ 4000]	[3.8]
Oil temperature ....°C	80	80	80	80
Valve clearances - inlet (mm)	0.15 ± 0.04	0.15 ± 0.04	0.15 ± 0.04	0: Hyd.
- exhaust (mm)	0.30 ± 0.04	0.30 ± 0.04	0.30 ± 0.04	0: Hyd.
Injection order	1-3-4-2	1-3-4-2	1-3-4-2	1-3-4-2
No. 1 cylinder position	FE	FE	FE	FE
Cooling system				
Thermostat opening temperature ....°C	83	83	83	83
Radiator cap pressure ....bar	1.4	1.4	1.4	1.4
Fuel system				
Idle speed ....rpm	750 to 800	750 to 800	800 ± 50	850 ± 50
Maximum (no load) speed ....rpm	5100	5100	5100	5100
Smoke test/opacity ....M⁻¹ %	2.5	1.84	2.41	2.5
Static timing method................................	Dial gauge	Refer to wsm	Dial gauge	Refer to wsm
Timing dimension...................................mm	0.66	_	0.57	_
Crankshaft position ....mm [°]	0 TDC	_	0 TDC	_
Turbo type / ref / pressure ....bar	Garrett T2 / KKK K14²	Garrett T2 / KKK K14²	Garrett T2 / KKK K14²	KKK or Garrett
Injection pump make	Bosch	Bosch	Bosch	Bosch
Injection pump part no.............................	VER 445	VER 445	VP20	EDC15C2
Injector Make / type	Bosch	Bosch	Bosch	
Injector part no..	299	DNOSD299	DNOSD299C	_
Injection type...	Indirect VER	Indirect	Indirect	Direct
Injection opening pressure, New [used]...bar	175	175	175	200 t0 1500
Glow plugs				
Maker	Bosch/Champion	Bosch/Champion	Bosch/Champion	Bosch/Champion
Type	0250 201 033 / CH68	0250 201 033 / CH68	0250 201 033 / CH68	0250 202 032 / CH170
Nominal rating....................................V/A	11 / 13	11 / 12	11 / 12	_
Brakes				
minimum friction material thickness				
Front......................................mm	1.0	2.0	2.0	2.0
Rear.......................................mm	1.5	1.0	1.0	1.0
Tyres - Saloon / Hatch......................Size	185/65x14	195/65x15	195/65x15	185/70x14: 195/65x15:205/60x15
- Estate / Van.....................Size	185/65x14	195/65x15	195/65x15	185/70x14: 195/65x15:205/60x15
Pressure - front / rear - Saloon / Hatch ...bar	2.2 / 2.2	2.2 / 2.2	2.2 / 2.2	2.4 / 2.4: 2.3 / 2.3:2.4 / 2.4
- Estate / Van ......bar	2.3 / 2.3	_	_	2.4 / 2.5: 2.4 / 2.4:2.4 / 2.4
Front suspension / wheel alignment				
Toe-in (+) / Toe-out (–)....................mm [°]	2.1 ± 1.0	0.0 ± 0.5	0.0 ± 0.5	0.0 ± 0.5
Camber	0° ± 30'	0°30'	0°30'	0° ± 30'
Castor	3° ± 30'	3°30' ± 30'	3°30' ± 30'	3°30' ± 30'
King pin inclination.................................	11°30' ± 30'	11°30' ± 30'	11°30' ± 30'	11°30' ± 30'
Rear suspension / wheel alignment				
Toe-in (+) / Toe-out (–)....................mm [°]	5.4 ± 1.4 Est: 2.1 ± 1.4	1.5 ± 0.5	1.5 ± 0.5	1.5 ± 0.5
Camber	_	-1°50' ± 30'	-1°50' ± 30'	1°50' ± 30'

PEUGEOT

	405 1.9 TD 1992 to 1996	406 1.9 TD 1996 to 1999	406 1.9 TD 1996 to 1999	406 2.0 HDi 1999 to 2000
Torque wrench settings				
Cylinder head - stage 1 .....Nm	20[1]	20[1]	20[1]	20[1]
- stage 2 .....Nm	60	60	60	60
- stage 3 .....Nm	+ 220°	+ 220°	+ 220°	+ 220°
- stage 4 .....Nm	–	–	–	–
- stage 5 .....Nm	–	–	–	–
- stage 6 .....Nm	–	–	–	–
Big-end bearings.....Nm	20 + 70° N	20 + 70° N	20 + 70° N	20 + 70° N
Main bearings.....Nm	15 + 60°	15 + 60°	15 + 60°	25 + 60°
Crankshaft pulley bolt .....Nm	40 + 60°	40 + 50°	40 + 50°	197
Camshaft pulley bolt .....Nm	45	45	45	43
Flywheel [driveplate] bolt.....Nm	50	50	50	48
Front hubs .....Nm	320	320	320	325
Rear hubs .....Nm	275	270	270	275
Wheel nuts / bolts .....Nm	85	85	85	90
Glow plugs .....Nm	25	25	25	25
Clutch pressure plate bolts .....Nm	25	20	20	20
Injection pump sprocket.....Nm	50	50	50	50
Injectors.....Nm	90	90	90	30
Injection pump mounting bolts .....Nm	–	20	20	22.5
Injector pipe unions.....Nm	–	20	20	20
Capacities				
Engine oil & filter.....litres	5.0	4.2	4.2	4.5
Gearbox.....litres	2.0	2.0	2.0	2.0
Automatic transmission .....litres	–	–	–	–
Final drive .....litres	WT	WT	WT	WT
Cooling system.....litres	7.0	7.0	7.0	7.0
Fuel tank.....litres	70	70	70	70

Notes

405 1.9 TD 1992 to 1996
[1]Bolt length below head: ≤146.5 mm
[2]0.85 bar @ 3000rpm
406 1.9 TD 1996 to 1999
[1]Bolt length below head: ≤146.5 mm
[2]0.93 to 1.07 bar @ 3000rpm
406 1.9 TD 1996 to 1999
[1]Bolt length below head: ≤146.5 mm
[2]0.93 to 1.07 bar @ 3000rpm
406 2.0 HDi 1999 to 2000
[1]Max bolt length below head ≤133.3 mm

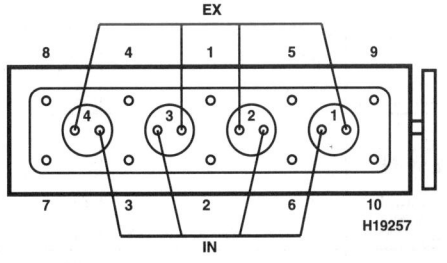

1905 cm³

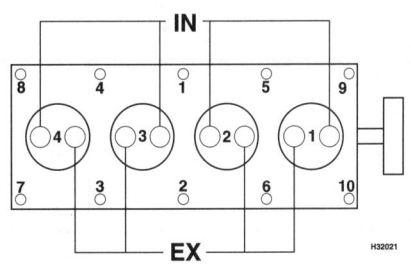

1997 cm³

– Not applicable, or information not available

PEUGEOT

	406 2.0 HDi 1999 to 2000	406 2.1TD 1996 to 1999	504 2.3 Pick-up 1982 to 1993	505 2.5 1982 to 1992
Engine				
Engine type/code.......................	DW10ATED/L3 RHZ SOHC 80kW	XUD11BTE/L P8C Turbo 81kW	XD2 134 OHV 51kW	XD3 155 OHV 51kW
Capacity (cm^3) / cylinders........................	1997 / 4	2088 / 4	2304 / 4	2498 / 4
Compression ration / pressure ....bar	17.6 / ≥30.0	21.5 / ≥18.0	22.2 / ≥18.0	23.0 / _
Torque output ....Nm	250	250	0	0
Oil pressure ....idle [running] bar	[3.8]	[2.5 @ 2000]	1.5 [3.0 @ 4000]	1.5
Oil temperature ....°C	80	80	90	90
Valve clearances - inlet (mm)	0: Hyd.	0: Hyd.	0.15	0.15
- exhaust (mm)	0: Hyd.	0: Hyd.	0.25	0.25
Injection order	1-3-4-2	1-3-4-2	1-3-4-2	1-3-4-2
No. 1 cylinder position	FE	FE	FE	FE
Cooling system				
Thermostat opening temperature ....°C	83	83	78	81
Radiator cap pressure ....bar	1.4	1.4	0.8 5.89 ▶: 1.0	0.8
Fuel system				
Idle speed ....rpm	850 ± 50	675 ± 25	750 to 800	700 to 800 [800 to 850]
Maximum (no load) speed ....rpm	5100	5150	4750	4850
Smoke test/opacity ....M^{-1} %	2.41	2.41	2.5	2.5
Static timing method...............................	Refer to wsm	Refer to wsm	Dial gauge	Dial gauge
Timing dimension.................................mm	_	0.88	0.01	0.01[1]
Crankshaft position ....mm [°]	_	0 TDC	3.83 ± 0.05 [22]	2.85 ± 0.05 [18][1]
Turbo type / ref / pressure ....bar	KKK	Mitsubishi TD04 1.0 bar @ 3000	_	_
Injection pump make	Bosch	Lucas	Roto Diesel	Roto Diesel
Injection pump part no............................	EDC15C2	EPIC	R3443 F790 040 or 794/5	100A, 150A, 152A, A155B, B156B
Injector Make / type	_	Bosch	Roto Diesel	Roto Diesel
Injector part no.......................................	_	DNOSD 299	RDNOSDC 6577 or 6577B	RDNOSDC 6577 or RDN12SCC 6849
Injection type..	Direct	Indirect	DPA	DPC R8443
Injection opening pressure, New [used]...bar	200 to 1500	163 ± 3.5	115 ± 5	115 ± 5
Glow plugs				
Maker	Bosch/Champion	Bosch/Champion	Bosch/Beru	Bosch/Champion
Type	0250 202 032 / CH170	0250 201 019 / CH68	0250 200 019 / 0 100 221 118	0250 201 019 / CH68
Nominal rating....................................V/A	_	11 / 12	11 / 8.5	11 / 12
Brakes				
minimum friction material thickness				
Front..mm	2.0	2.0	1.0	1.0
Rear...mm	1.0	2.0	1.5	1.5
Tyres - Saloon / Hatch ....Size	185/70x14: 195/65x15:205/60x15	195/65x15	_	175x14
- Estate / Van............................Size	185/70x14: 195/65x15:205/60x15	_	185x15	185x14
Pressure - front / rear - Saloon / Hatch ...bar	2.4 / 2.4: 2.3 / 2.3:2.4 / 2.4	2.2 / 2.2	_	1.8 / 2.1
- Estate / Van ......bar	2.4 / 2.5: 2.4 / 2.4:2.4 / 2.4	_	1.6 / 4.0[1]	1.9 / 2.2
Front suspension / wheel alignment				
Toe-in (+) / Toe-out (–)....................mm [°]	0.0 ± 0.5	0.0 ± 0.5	2.2 ± 1.0	3.0 ± 1.0[2]
Camber	0° ± 30'	0°30'	45' ± 30'	-45' ± 30' Est: -30'±30'
Castor	3°30' ± 30'	3°30' ± 30'	1°30' ± 45'	2°40' ± 30' Est: 2°±30'
King pin inclination..................................	11°30' ± 30'	11°30' ± 30'	8°54' ± 30'	9°15' ± 30' Est: 9°±30'
Rear suspension / wheel alignment				
Toe-in (+) / Toe-out (–)....................mm [°]	1.5 ± 0.5	1.5 ± 0.5	0	3.0 ± 1.0 Est: 0
Camber	1°50' ± 30'	-1°50' ± 30'	0	-1° ± 30' Est: 0

PEUGEOT

Torque wrench settings	406 2.0 HDi 1999 to 2000	406 2.1TD 1996 to 1999	504 2.3 Pick-up 1982 to 1993	505 2.5 1982 to 1992
Cylinder head - stage 1 ....Nm	20¹	20¹	30	30
- stage 2 ....Nm	60	60	50	70
- stage 3 ....Nm	+ 220°	+ 180°	Slacken, 70	Slacken, 70
- stage 4 ....Nm	–	–	Warm up, allow to cool	Warm up, allow to cool
- stage 5 ....Nm	–	–	Slacken, 70	Slacken, 70
- stage 6 ....Nm	–	–	+ 120°	+ 120°
Big-end bearings......................Nm	20 + 70° N	20 + 70° N	60 N	60 N
Main bearings......................Nm	25 + 60°	15 + 60°	110	120
Crankshaft pulley bolt ....Nm	197	50	55 + 60°	55 + 60°
Camshaft pulley bolt ....Nm	43	20	–	–
Flywheel [driveplate] bolt......................Nm	48	50	50	15 + 60°
Front hubs ....Nm	325	320	WSM	WSM
Rear hubs......................Nm	275	275	250	280
Wheel nuts / bolts ....Nm	90	85	60	Steel: 60 Alloy: 85
Glow plugs ....Nm	25	25	23	23
Clutch pressure plate bolts ....Nm	–	20	15	20
Injection pump sprocket......................Nm	50	50	–	–
Injectors.............................Nm	30	90	90	90
Injection pump mounting bolts ....Nm	22.5	20	20	20
Injector pipe unions.............................Nm	20	20	30	30
Capacities				
Engine oil & filter.............................litres	4.5	4.0	5.0	5.0
Gearbox.............................litres	2.0	1.85	1.15	BA7: 1.8 BA10: 1.85
Automatic transmission ....litres	–	–	–	–
Final drive ....litres	WT	WT	1.6	1.6
Cooling system.................................litres	7.0	7.0	10.0	10.0
Fuel tank.............................litres	70	70	64	70

Notes

406 2.0 HDi 1999 to 2000
¹Max bolt length below head ≤133.3 mm
406 2.1TD 1996 to 1999
¹Max bolt length: Refer to dealer for latest spec.
504 2.3 Pick-up 1982 to 1993
¹Varies according to tyre manufacturer
505 2.5 1982 to 1992
¹R8443 B156B: value marked on pump
²Estate: 3.5±1.0

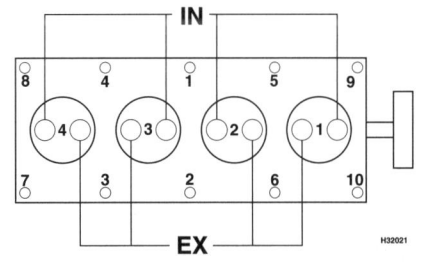

1997 cm³

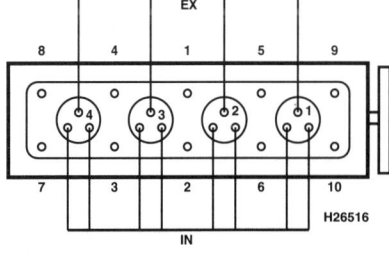

2088 cm³

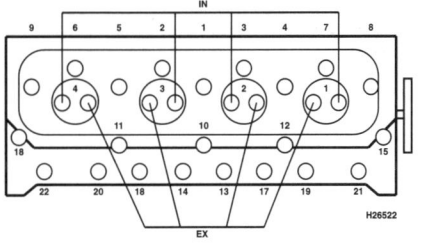

2304 cm³ / 2498 cm³

– Not applicable, or information not available

PEUGEOT

	505 2.5 1982 to 1992	605 2.1 TD 1990 to 1998	605 2.1 TD 1990 to 1993	605 2.1 TD 1993 to 1997
Engine				
Engine type/code....................	XD3 155 OHV 51kW	XUD11ATE/W PBA 12V 80kW	XUD11ATEY PHZ 12V 80kW	XUD11ATE/L P8B 12V 80kW
Capacity (cm³) / cylinders........................	2498 / 4	2088 / 4	2088 / 4	2088 / 4
Compression ration / pressure ....bar	23.0 / _	21.5 /_	21.5 /_	21.5 /
Torque output ....Nm	0	234	234	234
Oil pressure ....idle [running] bar	1.5	1.3 [3.5 @ 4000]	1.3 [3.5 @ 4000]	[2.5 @ 2000]
Oil temperature ....°C	90	80	80	80
Valve clearances - inlet (mm)	0.15	0: Hyd.	0: Hyd.	0: Hyd.
- exhaust (mm)	0.25	0: Hyd.	0: Hyd.	0: Hyd.
Injection order	1-3-4-2	1-3-4-2	1-3-4-2	1-3-4-2
No. 1 cylinder position	FE	FE	FE	FE
Cooling system				
Thermostat opening temperature ....°C	81	85	85	85
Radiator cap pressure ....bar	0.8	1.4	1.4	1.4
Fuel system				
Idle speed ....rpm	750 to 800 [800 to 850]	700 ± 25	675 ± 25	700 ± 25
Maximum (no load) speed ....rpm	4850	5150 ± 125	5150 ± 125	5150
Smoke test/opacity ....M⁻¹ %	2.5	2.41	3.0	3.0
Static timing method.................................	Plunger travel	Refer to wsm	Plunger travel	Dial gauge
Timing dimension.............................mm	0.3	Dimension on pump	0.88	0.88
Crankshaft position ....mm [°]	0.72 ± 0.03	_	0 TDC	0 TDC
Turbo type / ref / pressure ....bar	_	Mitsubishi TD04 0.9 bar @ 3000	Mitsubishi TD04 0.9 bar @ 3000	Mitsubishi TD04 0.9 bar @ 3000
Injection pump make	Bosch	Roto Diesel	Bosch	Bosch
Injection pump part no.............................	2240 R84 or 2250 R84	R8443 B743A	2150 R281	VER 474
Injector Make / type	Bosch	Roto Diesel	Bosch	Bosch
Injector part no..	DNOSD 252	RDN12SDC 6874C	DNOSD 299	DNOSD299
Injection type..	VE4/9F	Indirect DPC	Indirect VE4/9F	Indirect VER
Injection opening pressure, New [used]...bar	120 + 5	150 ± 5	175 ± 5	175 ± 5
Glow plugs				
Maker	Bosch/Champion	Bosch/Champion	Bosch/Champion	Bosch/Champion
Type	0250 201 019 / CH68	0250 201 019 / CH68	0250 201 019 / CH68	0250 201 019 / CH68
Nominal rating......................................V/A	11 / 12	11 / 12	11 / 12	11 / 13
Brakes				
minimum friction material thickness				
Front.......................mm	1.0	1.0	1.0	1.0
Rear........................mm	1.5	1.5	1.5	1.5
Tyres - Saloon / Hatch.....................Size	175x14	195/65x15	195/65x15	205/60x15
- Estate / Van..........................Size	185x14	_	_	_
Pressure - front / rear - Saloon / Hatch ...bar	1.8 / 2.1	2.2 / 2.2	2.2 / 2.2	2.3 / 2.3
- Estate / Van ......bar	1.9 / 2.2	_	_	_
Front suspension / wheel alignment				
Toe-in (+) / Toe-out (–)....................mm [°]	3.0 ± 1.0'	1.19 ± 0.5 PW	1.19 ± 0.5 PW	1.42 ± 1.0
Camber..	-45' ± 30' Est: -30'±30'	-18' ± 30'	-18' ± 30'	-0°18' ± 30'
Castor	2°40' ± 30' Est: 2°±30'	2°35' ± 30'	2°35' ± 30'	2°35' ± 30'
King pin inclination..................................	9°15' ± 30' Est: 9°±30'	13°38' ± 30'	13°38' ± 30'	13°38' ± 30'
Rear suspension / wheel alignment				
Toe-in (+) / Toe-out (–)....................mm [°]	3.0 ± 1.0 Est: 0	1.4 ± 0.3 PW	1.4 ± 0.3 PW	2.8 ± 0.6
Camber	-1° ± 30' Est: 0	-1°29' ± 30'	-1°29' ± 30'	-1°29' ± 30'

PEUGEOT

	505 2.5 1982 to 1992	605 2.1 TD 1990 to 1998	605 2.1 TD 1990 to 1993	605 2.1 TD 1993 to 1997
Torque wrench settings				
Cylinder head - stage 1 ....Nm	30	35[2]	35[2]	20[1]
- stage 2 ....Nm	70	70	70	60
- stage 3 ....Nm	Slacken, 70	+ 150°	+ 150°	+ 180°
- stage 4 ....Nm	Warm up, allow to cool	–	–	–
- stage 5 ....Nm	Slacken, 70	–	–	–
- stage 6 ....Nm	+ 120°	–	–	–
Big-end bearings.............................Nm	60 N	40[1]	40[1]	20 + 70° N
Main bearings..................................Nm	120	15 + 60°	15 + 60°	15 + 60°
Crankshaft pulley bolt ....Nm	55 + 60°	40 + 60°	40 + 60°	40 + 60°
Camshaft pulley bolt ....Nm	–	45	45	45
Flywheel [driveplate] bolt..................Nm	15 + 60°	50 LkC	50 LkC	50
Front hubs ....Nm	WSM	345	345	345
Rear hubs ....Nm	280	280	280	280
Wheel nuts / bolts ....Nm	Steel: 60 Alloy: 85	90	90	90
Glow plugs ....Nm	23	22	22	25
Clutch pressure plate bolts ....Nm	20	15	15	15
Injection pump sprocket.....................Nm	–	50	50	50
Injectors..Nm	90	90	90	90
Injection pump mounting bolts ....Nm	20	17	17	20
Injector pipe unions............................Nm	30	23	–	23
Capacities				
Engine oil & filter..................................litres	5.0	5.5	6.0	6.0
Gearbox..litres	BA7: 1.8 BA10: 1.85	BE1: 2.0 BE3: 2.0 ML5T: 1.85	BE1/BE3: 2.0 ML5T:1.85	1.85
Automatic transmission ....litres	–	–	–	2.0
Final drive ....litres	1.6	WT	WT	WT
Cooling system...................................litres	10.0	10.0	10.0	10.0
Fuel tank..litres	70	80	80	80

Notes

505 2.5 1982 to 1992
[1]Estate: 3.5±1.0
605 2.1 TD 1990 to 1998
[1]Slacken, 20 + 70° N
[2]Bolt length below head: ≤146.5 mm Bolt with guide boss: ≤151.5 mm
605 2.1 TD 1990 to 1993
[1]Slacken, then 20 + 70° N
[2]Bolt length below head: ≤146.5 mm
605 2.1 TD 1993 to 1997
[1]Bolt length below head: ≤151.5 mm

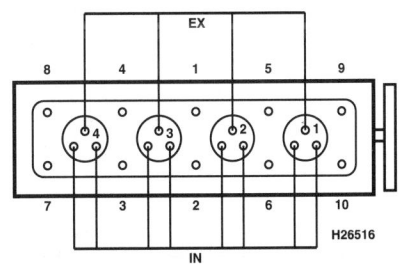

2088 cm³

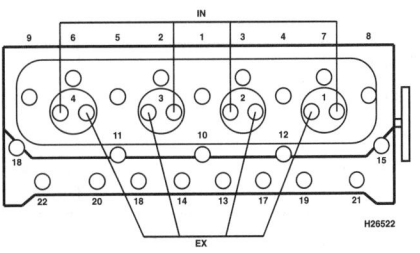

2498 cm³

– Not applicable, or information not available

PEUGEOT

	605 2.1 TD 1994 to 1997	605 2.1 D 1990 to 1993	605 2.1 D 1994 to 1996	605 2.5 TD 1995 to 1998
Engine				
Engine type/code..........................	XUD11BTE/L P8C 12V 80kW	XUD11A/W P9A 12V 60kW	XUD11A/L PJZ 12V 60kW	DK5ATE/L/Y THY SOHC 12V 95kW
Capacity (cm^3) / cylinders......................	2088 / 4	2138 / 4	2138 / 4	2446 / 4
Compression ration / pressure ....bar	21.5 /	22.5 /_	22.5 /_	21.0 /
Torque output ....Nm	234	0	0	0
Oil pressure ....idle [running] bar	[2.5 @ 2000]	1.4 [3.8 @ 4000]	1.4 [3.8 @ 4000]	[≥3.0]
Oil temperature ....°C	80	80	80	80
Valve clearances - inlet (mm)	0: Hyd.	0: Hyd.	0: Hyd.	0: Hyd.
- exhaust (mm)	0: Hyd.	0: Hyd.	0: Hyd.	0: Hyd.
Injection order	1-3-4-2	1-3-4-2	1-3-4-2	1-3-4-2
No. 1 cylinder position	FE	FE	FE	FE
Cooling system				
Thermostat opening temperature ....°C	85	85	85	–
Radiator cap pressure ....bar	1.4	1.4	1.4	–
Fuel system				
Idle speed ....rpm	700 ± 25	700 ± 25	700 ± 25	750
Maximum (no load) speed ....rpm	5150	5150 ± 125	5150 ± 125	5100
Smoke test/opacity ....M^{-1} %	1.54	2.41	2.41	2.5
Static timing method................................	Refer to wsm	Refer to wsm	Refer to wsm	Dial gauge
Timing dimension................................mm	–	Dimension on pump	Dimension on pump	0.52
Crankshaft position ....mm [°]	–	–	–	0 TDC
Turbo type / ref / pressure ....bar	Mitsubishi TD04 0.9 bar @ 3000	–	–	–
Injection pump make	Lucas	Roto Diesel	Roto Diesel	Bosch
Injection pump part no.............................	DPC	R8443 B733C	R8443 B962C	R520/535
Injector Make / type	CAV	Roto Diesel	Roto Diesel	Bosch
Injector part no..	–	RDN12SDC 6872C	RDN12SDC 6872D	DNOS D312
Injection type..	Indirect DPC	DPC	DPC	MSA11 3.6 VP36 electronic
Injection opening pressure, New [used]...bar	175 ± 5	130 ± 5	130 ± 5	170 to 175
Glow plugs				
Maker	Bosch/Champion	Bosch/Champion	Bosch/Champion	Bosch/Champion
Type	0250 201 019 / CH68	0250 201 019 / CH68	0250 201 019 / CH68	0250 201 039 / CH163
Nominal rating....................................V/A	11 / 13	11 / 12	11 / 12	–
Brakes				
minimum friction material thickness				
Front...........................mm	1.0	1.0	1.0	1.0
Rear...........................mm	1.5	1.5	1.5	1.0
Tyres - Saloon / Hatch......................Size	205/60x15	195/65x15	195/65x15	195/65x15: 205/60x15
- Estate / Van..........................Size	–	–	–	–
Pressure - front / rear - Saloon / Hatch ...bar	2.3 / 2.3	2.2 / 2.2	2.2 / 2.2	2.3 / 2.3
- Estate / Van ......bar	–	–	–	–
Front suspension / wheel alignment				
Toe-in (+) / Toe-out (–)......................mm [°]	1.42 ± 1.0	1.19 ± 0.5 PW	1.19 ± 0.5 PW	1.5 ± 1.0
Camber	-0°18' ± 30'	-18' ± 30'	-18' ± 30'	-0°27' ± 30'
Castor	2°35' ± 30'	2°35' ± 30'	2°35' ± 30'	2°47'
King pin inclination.................................	13°38' ± 30'	13°38' ± 30'	13°38' ± 30'	13°44' ± 30'
Rear suspension / wheel alignment				
Toe-in (+) / Toe-out (–)......................mm [°]	2.8 ± 0.6	1.4 ± 0.3 PW	1.4 ± 0.3 PW	–
Camber	-1°29' ± 30'	-1°29' ± 30'	-1°29' ± 30'	

PEUGEOT

	605 2.1 TD 1994 to 1997	605 2.1 D 1990 to 1993	605 2.1 D 1994 to 1996	605 2.5 TD 1995 to 1998
Torque wrench settings				
Cylinder head - stage 1 ....Nm	20¹	35²	35²	M10 Bolts≤162.5 mm: 35+120°
- stage 2 ....Nm	60	70	70	M12 Bolts≤153.5 mm: 50+120°
- stage 3 ....Nm	+ 180°	+ 150°	+ 150°	–
- stage 4 ....Nm	–	–	–	–
- stage 5 ....Nm	–	–	–	–
- stage 6 ....Nm		–		
Big-end bearings.............................Nm	20 + 70° N	40¹	40¹	20 + 65° N
Main bearings.................................Nm	15 + 60°	70	70	20 + 60°
Crankshaft pulley bolt ....Nm	40 + 60°	40 + 60°	40 + 60°	–
Camshaft pulley bolt ....Nm	45	45	45	45
Flywheel [driveplate] bolt..................Nm	50	50 LkC	50 LkC	45
Front hubs ....Nm	345	345	345	345
Rear hubs ....Nm	280	280	280	280
Wheel nuts / bolts ....Nm	90	90	90	90
Glow plugs ....Nm	25	22	22	20
Clutch pressure plate bolts ....Nm	15	15	15	15
Injection pump sprocket....................Nm	50	50	50	10, 25
Injectors..Nm	90	90	90	55
Injection pump mounting bolts ....Nm	20	17	17	20
Injector pipe unions.........................Nm	23	23	23	25
Capacities				
Engine oil & filter.............................litres	6.0	5.5	5.5	8.0
Gearbox...litres	1.85	BE1/BE3: 2.0 ML5T: 1.85	BE1/BE3: 2.0 ML5T: 1.85	1.85
Automatic transmission ....litres	2.0	–	–	–
Final drive ....litres	WT	WT	WT	WT
Cooling system................................litres	10.0	9.6	9.6	–
Fuel tank..litres	80	80	80	

Notes

605 2.1 TD 1994 to 1997
¹Bolt length below head: ≤151.5 mm
605 2.1 D 1990 to 1993
¹Slacken, 20 + 70°
²Bolt length below head: ≤146.5 mm Bolt with guide
boss: ≤151.5 mm
605 2.1 D 1994 to 1996
¹Slacken, 20 + 70°
²Bolt length below head: ≤146.5 mm Bolt with guide
boss: ≤151.5 mm

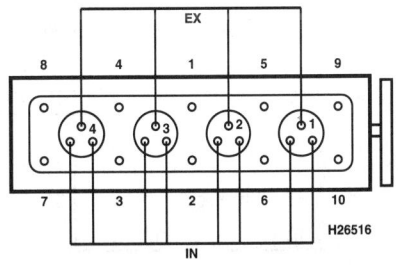

2088 cm³ / 2138 cm³

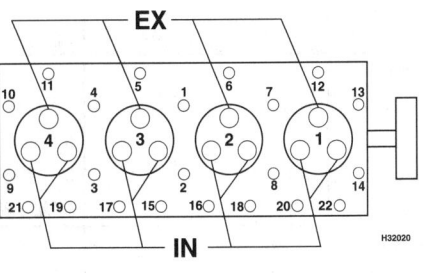

2446 cm³

– Not applicable, or information not
available

PEUGEOT

Haynes THE BOOK	806 1.9 TD 1995 to 2000	806 1.9 TD 1995 to 2000	806 2.0 HDi 1999 to 2000	806 2.1 TD 1995 to 1999
Engine				
Engine type/code....................	XUD9TF/L D8B EGR 67kW	XUD9TF/Y DHX EGR 66kW	DW10ATED/L3 RHZ SOHC 80kW	XUD11BTE P8C SOHC 80kW
Capacity (cm³) / cylinders....................	1905 / 4	1905 / 4	1996 / 4	2088 / 4
Compression ration / pressure ....bar	21.8 / _	21.8 / _	18.0 / ≥30.0	21.8 / _
Torque output ....Nm	202	202	250	0
Oil pressure ....idle [running] bar	[5.0 @ 4000]	[4.1 @ 2000]	[3.8]	[3.1 @ 2000]
Oil temperature ....°C	80	80	80	80
Valve clearances - inlet (mm)....................	0.15	0.15 ± 0.04	0: Hyd.	0: Hyd.
- exhaust (mm).............	0.30	0.30 ± 0.04	0: Hyd.	0: Hyd.
Injection order....................	1-3-4-2	1-3-4-2	1-3-4-2	1-3-4-2
No. 1 cylinder position............................	FE	FE	FE	FE
Cooling system				
Thermostat opening temperature.........°C	83	85	83	85
Radiator cap pressure ....bar	1.4	1.4	1.4	1.4
Fuel system				
Idle speed ....rpm	800 ± 50	800 ± 50	850 ± 50	750 ± 50
Maximum (no load) speed ....rpm	5100	5100	5100	5100
Smoke test/opacity ....M⁻¹ %	1.84	2.41	2.0	2.41
Static timing method............................	Dial gauge	Dial gauge	_	Refer to wsm
Timing dimension............................mm	0.66	0.75	_	_
Crankshaft position ....mm [°]	0 TDC	0 TDC	_	_
Turbo type / ref / pressure ....bar	_	0.6 to 0.7 bar @ 2000rpm	Garret T15 1.0 bar @ 3000	0.6 to 0.8 bar @ 2000rpm
Injection pump make	Bosch	Bosch	Bosch	Lucas
Injection pump part no................................	VE 4/9F 2250R513	XUDBP02	EDC15C2	XUDLP01
Injector Make / type	Bosch	Bosch	_	CAV
Injector part no................................	DNOSD299	DNOSD299C	_	6751H
Injection type................................	Indirect VE	Indirect	Direct Common Rail	Indirect
Injection opening pressure, New [used]...bar	175	175 ± 5	200 to 1500	145 to 155
Glow plugs				
Maker	Bosch/Champion	Bosch/Champion	Bosch/Champion	Bosch
Type	0250 201 033 / CH163	0250 201 033 / CH163	0250 202 032 / CH170	0250 201 033
Nominal rating........................V/A	11 / 12	11 / 12	_	11 / 9
Brakes				
minimum friction material thickness				
Front.....................mm	8.5 with backing	8.5 with backing	_	8.5 with backing
Rear.....................mm	1.5	1.5	_	1.5
Tyres - Saloon / Hatch.....................Size	_	_	_	_
- Estate / Van.....................Size	205/65x15	205/65x15	205/65x15	205/65x15
Pressure - front / rear - Saloon / Hatch...bar	_	_	_	_
- Estate / Van ......bar	2.3 / 2.3	2.3 / 2.3	_	2.3 / 2.3
Front suspension / wheel alignment				
Toe-in (+) / Toe-out (–).....................mm [°]	2.0 ± 1.0	2.0 ± 1.0	2.0 ± 1.0	2.0 ± 1.0
Camber	0° ± 30'	0° ± 30'	0° ± 30'	0° ± 30'
Castor	3°30' ± 30'	3°30' ± 30'	3°30' ± 30'	3°30' ± 30'
King pin inclination........................	11°28' ± 40'	11°28' ± 30'	11°28' ± 30'	11°28' ± 30'
Rear suspension / wheel alignment				
Toe-in (+) / Toe-out (–).....................mm [°]	2.0 ± 1.0	2.0 ± 1.0	2.0 ± 1.0	2.0 ± 1.0
Camber	-1°0' ± 30'	-1°0' ± 30'	-1°0' ± 30'	-1°0' ± 30'

PEUGEOT

	806 1.9 TD 1995 to 2000	806 1.9 TD 1995 to 2000	806 2.0 HDi 1999 to 2000	806 2.1 TD 1995 to 1999
Torque wrench settings				
Cylinder head - stage 1 ....Nm	20¹	20¹	20¹	20¹
- stage 2 ....Nm	60	60	60	60
- stage 3 ....Nm	+ 220°	+ 220°	+ 220°	+ 180°
- stage 4 ....Nm	–	–	–	–
- stage 5 ....Nm	–	–	–	–
- stage 6 ....Nm	–	–	–	–
Big-end bearings..............................Nm	20 + 70° N	20 + 70° N	20 + 70° N	20 + 70° N
Main bearings...................................Nm	70	70	25 + 60°	15 + 60°
Crankshaft pulley bolt ....Nm	40 + 60°	40 + 60°	197	70 + 60°
Camshaft pulley bolt ....Nm	45	45	43	45
Flywheel [driveplate] bolt....................Nm	50	50	48	50
Front hubs ....Nm	110	110	100 + 60°	110
Rear hubs ....Nm	99 + 60°	100 + 60°	–	100 + 60°
Wheel nuts / bolts ....Nm	100	100	100	100
Glow plugs ....Nm	22	22	25	22
Clutch pressure plate bolts ....Nm	20	23	–	23
Injection pump sprocket.....................Nm	50	50	50	50
Injectors...Nm	90	90	30	90
Injection pump mounting bolts ....Nm	22	22	22.5	22
Injector pipe unions............................Nm	20	20	20	25
Capacities				
Engine oil & filter...............................litres	4.0 Steel sump: 4.25	4.0 Steel sump: 4.25	4.5	4.5
Gearbox..litres	1.85	1.85	1.85	1.85
Automatic transmission ....litres	–	–	–	–
Final drive ....litres	WT	WT	WT	WT
Cooling system..................................litres	9.0	9.0	–	9.0
Fuel tank ....litres	80	50	50	50

Notes

806 1.9 TD 1995 to 2000
¹Bolt length below head: ≤146.8 mm
806 1.9 TD 1995 to 2000
¹Bolt length below head: ≤151.5 mm
806 2.0 HDi 1999 to 2000
¹Bolt length below head: ≤133.4 mm
806 2.1 TD 1995 to 1999
¹Bolt length below head: 146.5 mm Bolt with guide boss: ≤151.5 mm

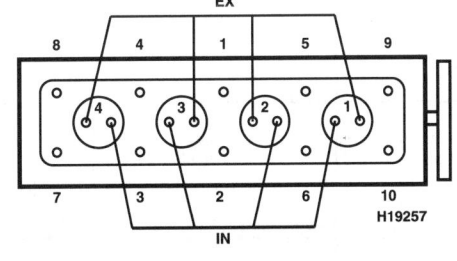

1905 cm³

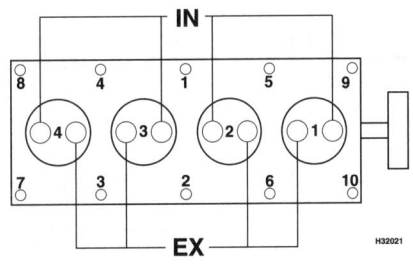

1996 cm³

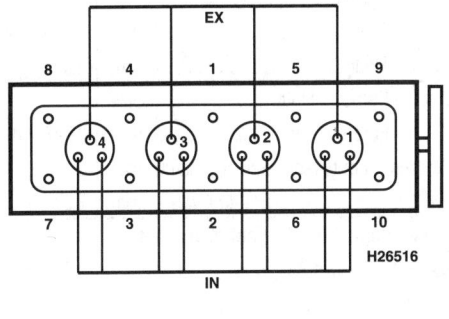

2088 cm³

– Not applicable, or information not available

PEUGEOT

	Expert 1.9 D 1996 to 1999	Expert 1.9 D 1998 to 2000	Expert 1.9 TD 1996 to 2000	Partner 1.7 D 1996 to 2000
Engine				
Engine type/code....................	XUD9A D9B SOHC 51kW	DW8/L3 WJZ SOHC 51kW	XUD9TF/L D8A Turbo 68kW	XUD7A A9A SOHC 44kW
Capacity (cm³) / cylinders........................	1905 / 4	1868 /	1905 / 4	1769 / 4
Compression ration / pressure ....bar	23.0 /	23.0 /	21.8 /	23.0 /
Torque output ....Nm	120	125	196	0
Oil pressure ....idle [running] bar	[5.0 @ 4000]	1.8 [4.5 @ 4000]	[5.0 @ 4000]	[3.5 @ 4000]
Oil temperature ....°C	80	80	80	80
Valve clearances - inlet (mm)	0.15 ± 0.05	0.15 ± 0.08	0.15 ± 0.05	0.15 ± 0.05
- exhaust (mm)	0.30 ± 0.05	0.30 ± 0.08	0.30 ± 0.05	0.30 ± 0.05
Injection order	1-3-4-2	1-3-4-2	1-3-4-2	1-3-4-2
No. 1 cylinder position	FE	TBE	FE	FE
Cooling system				
Thermostat opening temperature ....°C	83	89	83	83
Radiator cap pressure ....bar	1.1	1.4	1.1	1.4
Fuel system				
Idle speed ....rpm	800 ± 50	–	800 ± 50	800 ± 50
Maximum (no load) speed ....rpm	5150 ± 80	–	5150 ± 80	5150 ± 125
Smoke test/opacity ....M⁻¹ %	2.5	2.41	3.0	2.5
Static timing method...........................	Dial gauge	–	Dial gauge	Dial gauge
Timing dimension...........................mm	1.07 ± 0.02	–	0.66 ± 0.02	Dimension on pump
Crankshaft position ....mm [°]	0 TDC	–	0 TDC	0 TDC
Turbo type / ref / pressure ....bar	–	–	KKK K14 or Garrett T2'	–
Injection pump make	Bosch	Lucas	Bosch	Lucas
Injection pump part no...........................	VER 425-10/11	DPC + EGR	R445	X052 8443B930A
Injector Make / type	Bosch	–	Bosch	CAV
Injector part no.............................	DNOSD299A	–	DNOSD299	LCR 6730705D
Injection type.............................	Indirect	–	Indirect	Indirect
Injection opening pressure, New [used]...bar	130	–	175	140
Glow plugs				
Maker	Bosch/Champion		Champion	Bosch/Champion
Type	0250 201 019 / CH68		CH163	0250 201 033 / CH163
Nominal rating ....V/A	11 / 12		11 / 12	11 / 14
Brakes				
minimum friction material thickness				
Front..mm	8.5 with backing	8.5 with backing	8.5 with backing	8.5 with backing
Rear..mm	1.0	1.0	1.0	1.0
Tyres - Saloon / Hatch......................Size	–	–	–	–
- Estate / Van......................Size	195/70x14	195/70x14	195/70x14	165/70x14
Pressure - front / rear - Saloon / Hatch ...bar	–		–	–
- Estate / Van ......bar	3.0 / 3.0	3.0 / 3.0	3.0 / 3.0	2.5 / 3.7
Front suspension / wheel alignment				
Toe-in (+) / Toe-out (–).....................mm [°]	1.0 ± 0.5	1.0 ± 0.5	1.0 ± 0.5	1.0 to 3.0
Camber	0° ± 30'	0° ± 30'	0° ± 30'	0° ± 30'
Castor	3°30' ± 30'	3°30' ± 30'	3°30' ± 30'	3° ± 30'
King pin inclination.............................	–		–	–
Rear suspension / wheel alignment				
Toe-in (+) / Toe-out (–).....................mm [°]	1.0 ± 0.5	1.0 ± 0.5	1.0 ± 0.5	3.8 to 5.8
Camber	-1° ± 30'	-1° ± 30'	-1° ± 30'	1°15' ± 30'

PEUGEOT

Torque wrench settings	Expert 1.9 D 1996 to 1999	Expert 1.9 D 1998 to 2000	Expert 1.9 TD 1996 to 2000	Partner 1.7 D 1996 to 2000
Cylinder head - stage 1 ....Nm	20 N	20[1]	20 N	20[1]
- stage 2 ....Nm	60	60	60	60
- stage 3 ....Nm	+ 180°	+ 180° ± 5°	+ 220°	+ 180°
- stage 4 ....Nm	–	–	–	–
- stage 5 ....Nm	–	–	–	–
- stage 6 ....Nm	–	–	–	–
Big-end bearings ....Nm	20 + 70° N	20 + 70° N	20 + 70° N	20 + 70° N
Main bearings ....Nm	70	70	70	15 + 60°
Crankshaft pulley bolt ....Nm	40 + 60°	40 + 60°	40 + 60°	40 + 60°
Camshaft pulley bolt ....Nm	45	45	45	45
Flywheel [driveplate] bolt............Nm	50	48	50	65
Front hubs ....Nm	100 + 60°	100 + 60°	100 + 60°	320
Rear hubs ....Nm	100 + 60°	100 + 60°	100 + 60°	275
Wheel nuts / bolts ....Nm	100	100	100	85
Glow plugs ....Nm	22	15	22	22
Clutch pressure plate bolts ....Nm	20	20	20	20
Injection pump sprocket............Nm	50	23	50	50
Injectors............Nm	90	90	90	90
Injection pump mounting bolts ....Nm	22	20	22	20
Injector pipe unions............Nm	22	25	20	20
Capacities				
Engine oil & filter............litres	4.2	4.5	4.0 STEEL SUMP: 4.25	4.25
Gearbox............litres	1.85	BE3:2.0 ML5T:1.85 ME5T:1.85	1.85	2.0
Automatic transmission ....litres	–	–	–	–
Final drive ....litres	WT	WT	WT	WT
Cooling system............litres	9.0	7.0	9.0	8.0
Fuel tank............litres	80	80	80	60

Notes

Expert 1.9 D 1998 to 2000
[1]Bolt length below head: ≤125.5 mm
Expert 1.9 TD 1996 to 2000
[1]0.6 to 0.7 bar @ 2000rpm
Partner 1.7 D 1996 to 2000
[1] Bolt length below head: ≤121.5 mm

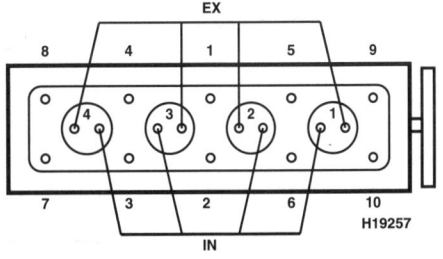

1769 cm³ / 1905 cm³

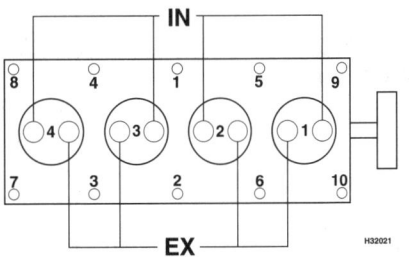

1868 cm³

– Not applicable, or information not
available

PEUGEOT

	Partner 1.9 D 1996 to 2000	Partner 1.9 D 1996 to 2000	Partner 1.9 D 1996 to 2000	J5, Express 1.9 1987 to 1991
Engine				
Engine type/code..............................	XUD9A/W2 D9B SOHC 52kW	XUD9Y DJY SOHC 50kW	XUD9Y DJY SOHC 50kW	XUD9A D9B SOHC 51kW
Capacity (cm³) / cylinders.........................	1905 / 4	1905 / 4	1905 / 4	1905 / 4
Compression ration / pressure ....bar	23 /	23 /	23 /	23.0 / _
Torque output ....Nm	120	120	120	0
Oil pressure ....idle [running] bar	[3.5 @ 4000]	[3.5 @ 4000]	[3.5 @ 4000]	2.0 [4.5 @ 4000]
Oil temperature ....°C	80	80	80	80
Valve clearances - inlet (mm)	0.15 ± 0.05	0.15 ± 0.05	0.15 ± 0.05	0.15 ± 0.04
- exhaust (mm)	0.30 ± 0.05	0.30 ± 0.05	0.30 ± 0.05	0.30 ± 0.04
Injection order....................................	1-3-4-2	1-3-4-2	1-3-4-2	1-3-4-2
No. 1 cylinder position	FE	FE	FE	FE
Cooling system				
Thermostat opening temperature ....°C	83	83	83	95
Radiator cap pressure ....bar	1.4	1.4	1.4	0.9
Fuel system				
Idle speed ....rpm	750 to 800	750 to 800	750 to 800	750 to 800
Maximum (no load) speed ....rpm	5150 ± 125	5150 ± 125	5100 ± 125	5150
Smoke test/opacity ....M⁻¹ %	2.5	2.5	2.5	2.5
Static timing method................................	Dial gauge	Dial gauge	Dial gauge	Plunger travel
Timing dimension................................mm	0.66	Dimension on pump	0.66 ± 0.02	0.9
Crankshaft position ....mm [°]	TDC	0 TDC	0 TDC	0 TDC
Turbo type / ref / pressure ....bar	–	–	–	–
Injection pump make	Bosch	Lucas	Bosch	Bosch
Injection pump part no...........................	VE4/8F230	XUDLP05	VER 425 6	2300 R272-1 or R272-2
Injector Make / type	Bosch	CAV	Bosch	Bosch
Injector part no.....................................	DNOSD299A	LCR6335201D	KCA 17S42	DNOSD 287/ or 287+
Injection type..	Indirect	Indirect	Indirect	VE4/9F
Injection opening pressure, New [used]...bar	130	140	130	130 ± 5'
Glow plugs				
Maker	Bosch/Champion	Bosch/Champion	Bosch/Champion	Bosch/Champion
Type	0250 201 033 / CH163	0250 201 033 / CH163	0250 201 033 / CH163	0250 201 019 / CH68
Nominal rating ....V/A	11 / 14	11 / 14	11 / 14	11 / 12
Brakes				
minimum friction material thickness				
Front.......................................mm	8.5 with backing	8.5 with backing	8.5 with backing	2.0
Rear..mm	1.0	1.0	1.0	1.5
Tyres - Saloon / Hatch.......................Size	–	–	–	–
- Estate / Van....................Size	165/70x14	165/70x14	165/70x14	185x14
Pressure - front / rear - Saloon / Hatch ...bar	–	–	–	–
- Estate / Van ......bar	2.5 / 3.7	2.5 / 3.7	2.5 / 3.7	4.3 / 4.5
Front suspension / wheel alignment				
Toe-in (+) / Toe-out (–).....................mm [°]	1.0 to 3.0	1.0 to 3.0	1.0 to 3.0	-0.5 ± 1.0
Camber	0° ± 30'	0° ± 30'	0° ± 30'	2°20' ± 30'
Castor	3° ± 30'	3° ± 30'	3° ± 30'	30' ± 30'
King pin inclination	–	–	–	11°
Rear suspension / wheel alignment				
Toe-in (+) / Toe-out (–)....................mm [°]	3.8 to 5.8	3.8 to 5.8	3.8 to 5.8	0 ± 1.0
Camber	1°15' ± 30'	1°15' ± 30'	1°15' ± 30'	0

PEUGEOT

	Partner 1.9 D 1996 to 2000	Partner 1.9 D 1996 to 2000	Partner 1.9 D 1996 to 2000	J5, Express 1.9 1987 to 1991
Torque wrench settings				
Cylinder head - stage 1 ....Nm	20[1]	20[1]	20[1]	30 N
- stage 2 ....Nm	60	60	60	70
- stage 3 ....Nm	+ 180°	+ 180°	+ 180°	+ 120°
- stage 4 ....Nm	–	–	–	–
- stage 5 ....Nm	–	–	–	–
- stage 6 ....Nm	–	–	–	–
Big-end bearings.............................Nm	20 + 70° N	20 + 70° N	20 + 70° N	20 + 70° N
Main bearings.................................Nm	15 + 60°	15 + 60°	15 + 60°	70
Crankshaft pulley bolt ....Nm	40 + 60°	40 + 60°	40 + 60°	40 + 60°
Camshaft pulley bolt ....Nm	45	45	45	40
Flywheel [driveplate] bolt....................Nm	65	65	65	50 LkC
Front hubs ....Nm	320	320	320	265
Rear hubs ....Nm	275	275	275	275
Wheel nuts / bolts ....Nm	85	85	85	180
Glow plugs ....Nm	22	22	22	22
Clutch pressure plate bolts ....Nm	20	20	20	25
Injection pump sprocket....................Nm	50	50	50	50
Injectors...Nm	90	90	90	90
Injection pump mounting bolts ....Nm	20	20	20	–
Injector pipe unions.........................Nm	20	22	22	–
Capacities				
Engine oil & filter.................................litres	4.0 STEEL SUMP: 4.25	4.25	4.25	4.75
Gearbox..litres	2.0	2.0	2.0	1.6
Automatic transmission ....litres	–	–	–	–
Final drive ....litres	WT	WT	WT	WT
Cooling system..................................litres	8.0	8.0	8.0	9.5
Fuel tank...litres	60	60	60	70

Notes

Partner 1.9 D 1996 to 2000
[1]Bolt length below head: ≤121.5 mm
Partner 1.9 D 1996 to 2000
[1]Bolt length below head: ≤121.5 mm
Partner 1.9 D 1996 to 2000
[1]Bolt length below head: ≤121.5 mm
J5, Express 1.9 1987 to 1991
[1]DNOSD287+ with VE49F2300R2722: 135±5

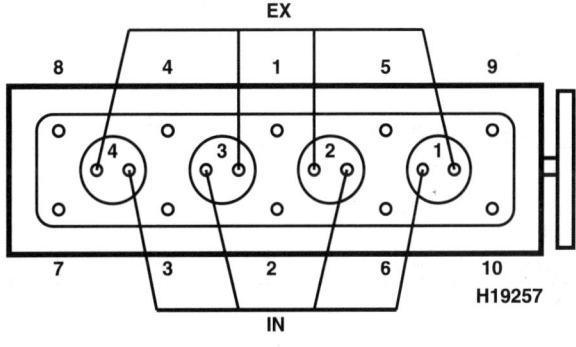

1905 cm³

– Not applicable, or information not
available

PEUGEOT

	J5, Express 1.9 1987 to 1993	J5, Express 2.5 1987 to 1994	J5, Express 2.5 TD 1987 to 1994	Boxer 1.9 D 1994 to 1997
Engine				
Engine type/code....................	XUD9A D9B SOHC 51kW	U25 661 OHV 54kW	U25 673 OHV 70kW	XUD9AU D9B SOHC 51kW
Capacity (cm³) / cylinders........................	1905 / 4	2500 / 4	2500 / 4	1905 / 4
Compression ration / pressure ....bar	23.0 / _	22.2 / _	21.0 / _	23.0 / _
Torque output ....Nm	121	0	0	120
Oil pressure ....idle [running] bar	2.0 [4.5 @ 4000]	1.0 [4.1 @ 3500]	2.5 [4.2 @ 3500]	2.2 [4.6]
Oil temperature ....°C	80	80	80	80
Valve clearances - inlet (mm)	0.15 ± 0.04	0.30	0.30	0.15 ± 0.08
- exhaust (mm)	0.30 ± 0.04	0.20	0.20	0.30 ± 0.08
Injection order	1-3-4-2	1-3-4-2	1-3-4-2	1-3-4-2
No. 1 cylinder position	FE	FE	FE	FE
Cooling system				
Thermostat opening temperature ....°C	88	86	86	83
Radiator cap pressure ....bar	1.1	1.1	1.1	1.0
Fuel system				
Idle speed ....rpm	750 to 800	800 ± 25	750 ± 25	800 +0 -50
Maximum (no load) speed ....rpm	5150	4625 ± 125	4600 ± 150	5150 ± 125
Smoke test/opacity ....M⁻¹ %	2.5	2.5	3.0	2.5
Static timing method................................	Dial gauge	Dial gauge	Dial gauge	Dial gauge
Timing dimension.............................mm	Dimension on pump	Dimension on pump	Dimension on pump	1.07 ± 0.01
Crankshaft position ....mm [°]	0 TDC	[22]	[19]	0 TDC
Turbo type / ref / pressure ....bar	_	_	_	_
Injection pump make	Roto Diesel	Roto Diesel	Roto Diesel	Bosch
Injection pump part no.............................	R8443 B380A or B380B	MA260 R8443 B111A	MAS100 R8443 B123B	VER 4251
Injector Make / type	Roto Diesel	Roto Diesel	Roto Diesel	Bosch
Injector part no..	RDNOSDC 6751C	RDNOSDC 6577B	RDNOSDC 6751	KCA 17 S42 299A
Injection type...	DPC	Indirect DPC	DPC	Indirect VER
Injection opening pressure, New [used]...bar	125 ± 5	122 + 5	137 + 5	130
Glow plugs				
Maker	Bosch/Champion	Bosch/Champion	Beru/Champion	Bosch/Beru
Type	0250 201 019 / CH68	0250 201 019 / CH68	0250 201 019 / CH68	0250 201 019 / 0100221133
Nominal rating....................................V/A	11 / 12	11 / 12	11 / 12	11 / 11
Brakes				
minimum friction material thickness				
Front............................mm	2.0	2.0	2.0	2.0
Rear............................mm	1.5	1.5	1.5	1.0
Tyres - Saloon / Hatch ....Size	_	_	_	_
- Estate / Van...........................Size	185x14	185x14: 195/75x16	185x14: 195/75x16	195/70x15: 205/70x15³
Pressure - front / rear - Saloon / Hatch ...bar				_
- Estate / Van ......bar	4.3 / 4.5	Refer to owners handbook	Refer to owners handbook	270: 3.0 / 3.0⁴
Front suspension / wheel alignment				
Toe-in (+) / Toe-out (–)...................mm [°]	-0.5 ± 1.0	-0.5 ± 1.0	-0.5 ± 1.0	-2.0 ± 1.0
Camber	2°20' ± 30'	2°20' ± 30' 1800: 1°±30'	2°20' ± 30' 1800: 1°±30'	0' ± 30'
Castor	30' ± 30'	30' ± 30' 1800: 0±30'	30' ± 30' 1800: 0±30'	1° ± 30'
King pin inclination.................................	11°	11°	11°	
Rear suspension / wheel alignment				
Toe-in (+) / Toe-out (–)...................mm [°]	0 ± 1.0	0 ± 1.0	0 ± 1.0	0 ± 1.0
Camber	0	0	0	0° ± 10'

PEUGEOT

	J5, Express 1.9 1987 to 1993	J5, Express 2.5 1987 to 1994	J5, Express 2.5 TD 1987 to 1994	Boxer 1.9 D 1994 to 1997
Torque wrench settings				
Cylinder head - stage 1 ....Nm	30 N	40 N	40 N	Bolt length: ≤121.5mm[1]
- stage 2 ....Nm	70	+ 100°	+ 100°	20
- stage 3 ....Nm	+ 120°	+ 100°	+ 100°	60
- stage 4 ....Nm	–	Warm up, allow to cool	Warm up, allow to cool	+ 180°
- stage 5 ....Nm	–	+ 45°	+ 45°	–
- stage 6 ....Nm	–	–	–	–
Big-end bearings.......................Nm	20 + 70° N	90 N	90 N	40 slacken, 20 + 70° N
Main bearings.........................Nm	70	95	95	70
Crankshaft pulley bolt ....Nm	40 + 60°	250	250	40 + 51°
Camshaft pulley bolt ....Nm	40	32	32	45
Flywheel [driveplate] bolt....................Nm	50 LkC	90 LkC	90 LkC	50
Front hubs ....Nm	265	265	265	450 18Q: 500
Rear hubs ....Nm	275	275	275	WSM
Wheel nuts / bolts ....Nm	180	180	180	160 18Q: 180
Glow plugs ....Nm	22	28	28	20 to 25
Clutch pressure plate bolts ....Nm	25	35	35	20
Injection pump sprocket.....................Nm	50	–	–	50
Injectors...............................Nm	90	25	25	90
Injection pump mounting bolts ....Nm	–	–	–	20
Injector pipe unions.............................Nm	–	25	25	25
Capacities				
Engine oil & filter................................litres	4.75	4.7	4.7	4.25 STEEL SUMP: 5.25
Gearbox................................litres	1.6	1.6 Transfer: 2.4	1.6	1.85
Automatic transmission ....litres	–	–	–	–
Final drive ....litres	WT	4x4: 1.6	WT	WT
Cooling system................................litres	9.5	10.6	10.6	12.0
Fuel tank................................litres	70	70	70	80

Notes

Boxer 1.9 D 1994 to 1997
[1]Bolt with guide boss: ≤125.5mm
[3]215/75x15:205/75x15:215/75x16
[4]310/320: 4.1 / 4.5 350: 4.5 / 4.5

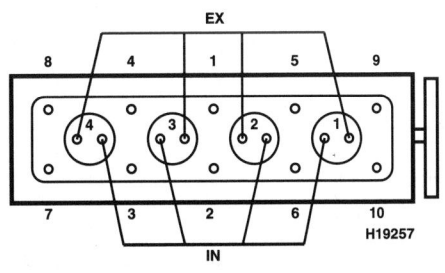

1905 cm³

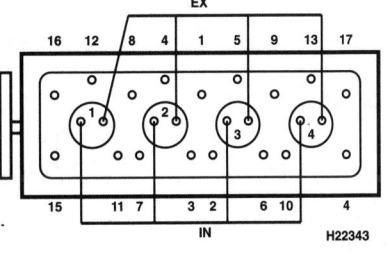

2500 cm³

– Not applicable, or information not available

PEUGEOT

	Boxer 1.9 TD 1994 to 1997	Boxer 1.9 TD 1994 to 2000	Boxer 1.9 D 1998 to 2000	Boxer 2.5 D 1998 to 2000
Engine				
Engine type/code..............................	XUD9UTF D8C SOHC Turbo 67kW	XUD9UTF DHX SOHC Turbo 67kW	XUD9AU D9B SOHC 50kW	DJ5 T9A SOHC 12V 63kW
Capacity (cm^3) / cylinders........................	1905 / 4	1905 / 4	1905 / 4	2446 / 4
Compression ration / pressure ....bar	21.8 / _	21.8 / _	23.0 / _	20.0 / _
Torque output ....Nm	196	196	120	153
Oil pressure ....idle [running] bar	2.4 [4.8]	2.4 [4.8 @ 4000]	2.2 [4.6]	1.8 [3.5 to 4.4 @ 2000]
Oil temperature ....°C	80	80	80	80
Valve clearances - inlet (mm)	0.15 ± 0.08	0.15 ± 0.08	0.15 ± 0.08	0: Hyd.
- exhaust (mm)	0.30 ± 0.08	0.30 ± 0.08	0.30 ± 0.08	0: Hyd.
Injection order	1-3-4-2	1-3-4-2	1-3-4-2	1-3-4-2
No. 1 cylinder position	FE	FE	FE	FE
Cooling system				
Thermostat opening temperature ....°C	83	83	83	83
Radiator cap pressure ....bar	1.0	1.0	1.0	1.0
Fuel system				
Idle speed ....rpm	800 +0 -50	850 ± 50	800 ± 50	725 ± 25
Maximum (no load) speed ....rpm	5050 ± 125	5050	5150 ± 125	5200 ± 125
Smoke test/opacity ....M^{-1} %	2.41	2.41	2.41	2.5
Static timing method..............................	Dial gauge	Dial gauge	_	Peg
Timing dimension...............................mm	0.66 ± 0.01	0.66 ± 0.02	1.07 ± 0.01	9.5 Setting rod
Crankshaft position ....mm [°]	0 TDC	0 TDC	TDC	TDC
Turbo type / ref / pressure ....bar	KKK	KKK	_	_
Injection pump make	Bosch	Bosch	Bosch	Bosch
Injection pump part no............................	VER445	VER4452/2	_	_
Injector Make / type	Bosch	Bosch	Bosch	Bosch
Injector part no.....................................	KCA 17 S42 299	KCA 17 S42 299	KCA 17 S42 299A	140 PV 3375 955
Injection type.......................................	Indirect VER	Indirect VER	VER425/*XUD201	VER 683
Injection opening pressure, New [used]...bar	175	175	130	1st stage: 200. 2nd: 425
Glow plugs				
Maker	Bosch/Beru	Bosch/Beru	Bosch/Beru	Beru
Type	0250 201 033 / 0100226186	0250 201 033 / 0100226186	0250201019 / 0100221133	0100 226 246
Nominal rating...............................V/A	11 / 9	11 / 9	11.0 / _	11 / 9
Brakes				
minimum friction material thickness				
Front...mm	2.0	2.0	2.0	2.0
Rear...mm	1.0	1.0	1.0	1.0
Tyres - Saloon / Hatch......................Size	_	_	_	_
- Estate / Van......................Size	195/70x15: 205/70x15^3	195/70x15: 205/70x15^3	195/70x15: 205/70x15^3	195/70x15: 205/70x15^3
Pressure - front / rear - Saloon / Hatch ...bar	_	_	_	_
- Estate / Van ......bar	270: 3.0 / 3.0^4	270: 3.0 / 3.0^4	270: 3.0 / 3.0^4	270: 3.0 / 3.0^4
Front suspension / wheel alignment				
Toe-in (+) / Toe-out (–)...................mm [°]	-2.0 ± 1.0	-2.0 ± 1.0	-2.0 ± 1.0	-2.0 ± 1.0
Camber	0' ± 30'	0'30' ± 30'	0'30' ± 30'	0'30' ± 30'
Castor	1° ± 30'	1° ± 30'	1° ± 30'	1° ± 30'
King pin inclination...............................	_	_	_	_
Rear suspension / wheel alignment				
Toe-in (+) / Toe-out (–)...................mm [°]	0.0 ± 1.0	0.0 ± 1.0	0.0 ± 1.0	0.0 ± 1.0
Camber	0' ± 10'	0° ± 10'	0° ± 10'	0° ± 10'

PEUGEOT

	Boxer 1.9 TD 1994 to 1997	Boxer 1.9 TD 1994 to 2000	Boxer 1.9 D 1998 to 2000	Boxer 2.5 D 1998 to 2000
Torque wrench settings				
Cylinder head - stage 1 ....Nm	20[1]	20[1]	20[1]	Bolt length M10: ≤189.3mm[1]
- stage 2 ....Nm	60	60	60	M12: 50 M10: 35
- stage 3 ....Nm	+ 220°	+ 220°	+ 180°	+ 150° ± 5°
- stage 4 ....Nm	–	–	–	–
- stage 5 ....Nm	–	–	–	–
- stage 6 ....Nm	–	–	–	–
Big-end bearings..............................Nm	40 slacken, 20 + 70° N	40 slacken, 20 + 70° N	40 slacken, 20 + 70° N	40, slacken, 20 + 70° N
Main bearings..................................Nm	70	70	70	70
Crankshaft pulley bolt ....Nm	40 + 51°	40 + 51°	40 + 51°	40 + 51°
Camshaft pulley bolt ....Nm	45	45	45	45
Flywheel [driveplate] bolt.................Nm	50	50	50	50
Front hubs ....Nm	450 18Q: 500	450 18Q: 500	450 18Q: 500	450 18Q: 500
Rear hubs ....Nm	WSM	WSM	WSM	WSM
Wheel nuts / bolts ....Nm	160 18Q: 180	160 18Q: 180	160 18Q: 180	160 18Q: 180
Glow plugs ....Nm	20 to 25	20 to 25	20 to 25	20 to 25
Clutch pressure plate bolts ....Nm	20	20	20	20
Injection pump sprocket.....................Nm	50	50	–	–
Injectors..Nm	90	90	90	–
Injection pump mounting bolts ....Nm	20	20	–	–
Injector pipe unions..........................Nm	25	25	25	25
Capacities				
Engine oil & filter................................litres	5.25 A/C: 4.25	5.25 A/C: 4.25	4.25 STEEL SUMP: 5.25	5.25 A/C: 4.25
Gearbox...litres	1.85	1.85	1.85	1.85
Automatic transmission ....litres	–	–	–	–
Final drive ....litres	WT	WT	WT	WT
Cooling system...................................litres	12.0	12.0	12.0	13.0 A/C: 13.5
Fuel tank..litres	80	80	80	80

Notes

Boxer 1.9 TD 1994 to 1997
[1]Bolt length below head: ≤146.5 mm Bolt with guide boss: ≤150.5 mm
[3]215/75x15:205/75x15:215/75x16
[4]310/320: 4.1 / 4.5 350: 4.5 / 4.5
Boxer 1.9 TD 1994 to 2000
[1]Bolt length below head: ≤146.5 mm Bolt with guide boss: ≤150.5 mm
[3]215/75x15: 205/75x15: 215/75x16
[4]310/320: 4.1 / 4.5 350: 4.5 / 4.5
Boxer 1.9 D 1998 to 2000
[1]Bolt length below head: ≤121.5 mm Bolt with guide boss: ≤125.5 mm
[3]215/75x15: 205/75x15: 215/75x16
[4]310/320: 4.1 / 4.5 350: 4.5 / 4.5
Boxer 2.5 D 1998 to 2000
[1]M12: ≤178.3mm
[3]215/75x15: 205/75x15: 215/75x16
[4]310/320: 4.1 / 4.5 350: 4.5 / 4.5

– Not applicable, or information not available

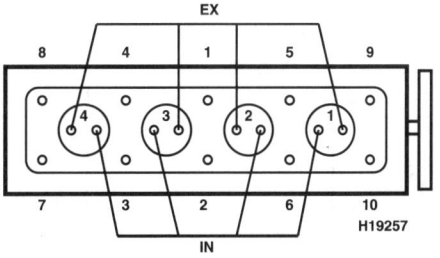

1905 cm³

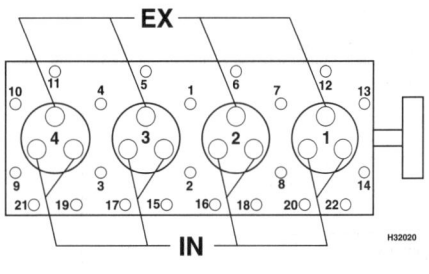

2446 cm³

PEUGEOT

	Boxer 2.5 TD 1997 to 2000	Boxer 2.5 TD 1994 to 1997		
Engine				
Engine type/code..................................	DJ5TED THX SOHC 12V 79kW	DJ5T THZ SOHC 12V 76kW		
Capacity (cm³) / cylinders........................	2446 / 4	2446 / 4		
Compression ration / pressure ....bar	20.0 / _	21.0 / _		
Torque output ....Nm	225	232		
Oil pressure ....idle [running] bar	1.8 [3.5 to 4.4]	1.8 [3.5 to 4.4]		
Oil temperature ....°C	80	80		
Valve clearances - inlet (mm)	0: Hyd.	0: Hyd.		
- exhaust (mm)	0: Hyd.	0: Hyd.		
Injection order	1-3-4-2	1-3-4-2		
No. 1 cylinder position	FE	FE		
Cooling system				
Thermostat opening temperature ....°C	83	83		
Radiator cap pressure ....bar	1.0	1.0		
Fuel system				
Idle speed ....rpm	850 ± 50	725 ± 25		
Maximum (no load) speed ....rpm	4900 ± 160	4900 ± 160		
Smoke test/opacity ....M⁻¹ %	3.0	3.0		
Static timing method............................	Peg	Refer to wsm		
Timing dimension..............................mm	9.5 Setting rod	_		
Crankshaft position ....mm [°]	TDC	_		
Turbo type / ref / pressure ....bar	KKK K16 0.9 bar @ 2500rpm	KKK K16 0.9 bar @ 2500rpm		
Injection pump make	Bosch	Lucas		
Injection pump part no..............................	_	R8444		
Injector Make / type	Bosch	CAV		
Injector part no..	140 PV 3375 955	LDC 003R		
Injection type..	VER 683	Indirect DPC		
Injection opening pressure, New [used]...bar	1st stage: 200. 2nd: 425	168 to 173		
Glow plugs				
Maker	Beru	Beru/Champion		
Type	0100 226 246	0100 226 246 / CH163		
Nominal rating.....................................V/A	11 / 9	11 / 9		
Brakes				
minimum friction material thickness				
Front....................................mm	2.0	2.0		
Rear.....................................mm	1.0	1.0		
Tyres - Saloon / Hatch......................Size	_	_		
- Estate / Van............................Size	195/70x15: 205/70x15³	195/70x15: 205/70x15³		
Pressure - front / rear - Saloon / Hatch ...bar	_	_		
- Estate / Van ......bar	3.0 / 3.0⁴	3.0 / 3.0⁴		
Front suspension / wheel alignment				
Toe-in (+) / Toe-out (–)....................mm [°]	-2.0 ± 1.0	-2.0 ± 1.0		
Camber	0°30' ± 30'	0°30' ± 30'		
Castor	1° ± 30'	1° ± 30'		
King pin inclination..............................	_			
Rear suspension / wheel alignment				
Toe-in (+) / Toe-out (–)....................mm [°]	0.0 ± 1.0	0.0 ± 1.0		
Camber	0° ± 10'	0° ± 10'		

PEUGEOT

	Boxer 2.5 TD 1997 to 2000	Boxer 2.5 TD 1994 to 1997		
Torque wrench settings				
Cylinder head - stage 1 ....Nm	Bolt length M10: ≤189.3mm[1]	Bolt length M10: ≤162.5mm[1]		
- stage 2 ....Nm	M12: 50 M10: 35	M12: 50 M10: 35		
- stage 3 ....Nm	+ 150° ± 5°	+ 120° ± 5°		
- stage 4 ....Nm	–	–		
- stage 5 ....Nm	–	–		
- stage 6 ....Nm	–	–		
Big-end bearings...................................Nm	40, slacken, 20 + 70° N	40, slacken, 20 + 70° N		
Main bearings.......................................Nm	70	70		
Crankshaft pulley bolt ....Nm	40 + 51°	40 + 51°		
Camshaft pulley bolt ....Nm	45	45		
Flywheel [driveplate] bolt....................Nm	50	45		
Front hubs ....Nm	450 18Q: 500	450 18Q: 500		
Rear hubs ....Nm	WSM	WSM		
Wheel nuts / bolts ....Nm	160 18Q: 180	160 18Q: 180		
Glow plugs ....Nm	20 to 25	20 to 25		
Clutch pressure plate bolts ....Nm	20	20		
Injection pump sprocket.......................Nm	–	–		
Injectors...Nm	55	55		
Injection pump mounting bolts ....Nm	–	–		
Injector pipe unions.............................Nm	25	25		
Capacities				
Engine oil & filter.................................litres	8.0	8.0		
Gearbox..litres	1.85	1.85		
Automatic transmission ....litres	–	–		
Final drive..litres	WT	WT		
Cooling system.....................................litres	13.0 A/C: 13.5	13.0 A/C: 13.5		
Fuel tank..litres	80	80		

Notes

Boxer 2.5 TD 1997 to 2000
[1]M12: ≤178.3mm
[3]215/75x15: 205/75x15: 215/75x16
[4]320: 4.1 / 4.5 350: 4.5 / 4.5
[5]1999 on: W3
Boxer 2.5 TD 1994 to 1997
[1]M12: ≤153.5mm
[3]215/75x15: 205/75x15: 215/75x16
[4]320: 4.1 / 4.5 350: 4.5 / 4.5

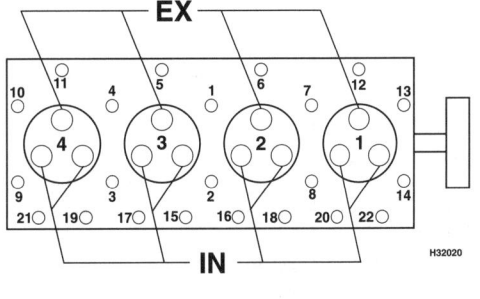

2446 cm³

– Not applicable, or information not available

PROTON

	Persona 2.0 D 1996 to 1997	Persona 2.0 TD 1997 to 2000		
Engine				
Engine type/code............................	4D68 SOHC 48kW	4D68 SOHC Turbo 59kW		
Capacity (cm^3) / cylinders........................	1998 / 4	1998 / 4		
Compression ration / pressure ....bar	22.4 / 25.6	22.2 / 25.6		
Torque output ....Nm	0	172		
Oil pressure ....idle [running] bar	–	–		
Oil temperature ....°C	–	–		
Valve clearances - inlet (mm)	0.25 H	0.25 H		
- exhaust (mm)	0.25 H	0.25 H		
Injection order..	1-3-4-2	1-3-4-2		
No. 1 cylinder position	–	–		
Cooling system				
Thermostat opening temperature ....°C	76.5	76.5		
Radiator cap pressure ....bar	0.75 to 1.05	0.75 to 1.05		
Fuel system				
Idle speed ....rpm	750 ± 100	750 ± 100		
Maximum (no load) speed ....rpm	–	–		
Smoke test/opacity ....M^{-1}%	1.6	2.5		
Static timing method................................	Dial gauge	Dial gauge		
Timing dimension.................................mm	0.97 to 1.03	0.97 to 1.03		
Crankshaft position ....mm [°]	9 ATDC	10 ATDC		
Turbo type / ref / pressure ....bar	–	–		
Injection pump make	Nippon Denso	Nippon Denso		
Injection pump part no.............................	MD 189552			
Injector Make / type	Nippon Denso	Nippon Denso		
Injector part no...	–	ND189552		
Injection type...	Indirect	Indirect		
Injection opening pressure, New [used]...bar	117 to 127	117 to 127		
Glow plugs				
Maker	–	–		
Type	–	–		
Nominal rating.....................................V/A	–	–		
Brakes				
minimum friction material thickness				
Front..mm	2.0	2.0		
Rear...mm	1.0	1.0		
Tyres - Saloon / Hatch......................Size	185/60x14	185/60x14		
- Estate / Van...........................Size	–	–		
Pressure - front / rear - Saloon / Hatch ...bar	2.1 / 2.0	2.1 / 2.0		
- Estate / Van ......bar	–	–		
Front suspension / wheel alignment				
Toe-in (+) / Toe-out (–).....................mm [°]	0 ± 3.0	0 ± 3.0		
Camber	0°30'	0°30'		
Castor	2°15'	2°15'		
King pin inclination..................................	12°49'	12°49'		
Rear suspension / wheel alignment				
Toe-in (+) / Toe-out (–).....................mm [°]	–	–		
Camber	-0°40' ± 30'	-0°40' ± 30'		

PROTON

Torque wrench settings	Persona 2.0 D 1996 to 1997	Persona 2.0 TD 1997 to 2000		
Cylinder head - stage 1 ....Nm	40[1]	40[1]		
- stage 2 ....Nm	+ 90°	+ 90°		
- stage 3 ....Nm	+ 90°	+ 90°		
- stage 4 ....Nm	–	–		
- stage 5 ....Nm	–	–		
- stage 6 ....Nm	–	–		
Big-end bearings...............................Nm	20 + 90° N	20 + 90° N		
Main bearings.....................................Nm	25 + 90°	25 + 90°		
Crankshaft pulley bolt ....Nm	120	120		
Camshaft pulley bolt ....Nm	90	90		
Flywheel [driveplate] bolt....................Nm	135	135		
Front hubs ....Nm	230	230		
Rear hubs ....Nm	180 N	180 N		
Wheel nuts / bolts ....Nm	100	100		
Glow plugs ....Nm	18	18		
Clutch pressure plate bolts ....Nm	20	20		
Injection pump sprocket.....................Nm	85	85		
Injectors...Nm	55	55		
Injection pump mounting bolts ....Nm	24	24		
Injector pipe unions............................Nm	30	30		
Capacities				
Engine oil & filter................................litres	5.1	5.1		
Gearbox..litres	1.8	1.8		
Automatic transmission ....litres	–	–		
Final drive ....litres	WT	WT		
Cooling system...................................litres	8.0	8.0		
Fuel tank...litres	50	50		

Notes

Persona 2.0 D 1996 to 1997
[1]Bolt length below head ≤119.7 mm
Persona 2.0 TD 1997 to 2000
[1]Bolt length below head ≤119.7 mm

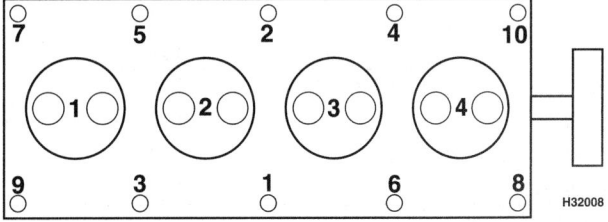

1998 cm³

– Not applicable, or information not available

RENAULT

	5 & Extra (B/C/F/S404) 1986 to 1994	5 & Extra (B/C/F/S404) 1986 to 1994	5 & Extra D 1992 to 1996	5 & Extra D 1992 to 1996
Engine				
Engine type/code............................	F8M720 / 730 SOHC 40kW	F8M720 / 730 SOHC 40kW	F8Q776 SOHC 48kW	F8Q722 SOHC 48kW
Capacity (cm^3) / cylinders........................	1596 / 4	1596 / 4	1870 / 4	1870 / 4
Compression ration / pressure ....bar	22.5 / ≥20.0	22.5 / ≥20.0	21.5 / ≥20.0	21.5 / ≥20.0
Torque output ....Nm	0	0	118	118
Oil pressure ....idle [running] bar	2.0 [3.5 @ 3000]	2.0 [3.5 @ 3000]	2.0 [3.5 @ 3000]	2.0 [3.5 @ 3000]
Oil temperature ....°C	80	80	80	80
Valve clearances - inlet (mm)	0.20	0.20	0.20	0.20
- exhaust (mm)	0.40	0.40	0.40	0.40
Injection order	1-3-4-2	1-3-4-2	1-3-4-2	1-3-4-2
No. 1 cylinder position	FE	FE	FE	FE
Cooling system				
Thermostat opening temperature ....°C	83	83	82	82
Radiator cap pressure ....bar	0.9	0.9	1.2	1.2
Fuel system				
Idle speed ....rpm	850 ± 25	850 ± 25	825 ± 25	825 ± 25
Maximum (no load) speed ....rpm	5200 to 5400	5200 to 5400	5200 ± 100	5200 ± 100
Smoke test/opacity ....M^{-1} %	1.2	1.38	1.38	1.38
Static timing method............................	Plunger travel	Rotor lift	Dial gauge	Refer to wsm
Timing dimension................................mm	0.65 ± 0.02	Dimension on pump	0.9 ± 0.02	Dimension on pump
Crankshaft position ....mm [°]	[0] TDC	[0] TDC	0 TDC	–
Turbo type / ref / pressure ....bar	–	–	–	–
Injection pump make	Bosch	Roto Diesel	Bosch	Lucas
Injection pump part no...........................	VER 95 0460 494 105	DPC	VER458	R8444B 230A
Injector Make / type	Bosch	Roto Diesel	Bosch	CAV
Injector part no...................................	0432 217 099	RDNOSDC 6843, 6834C	0432 217 258	RDN4SDC6868D
Injection type......................................	Indirect VER	DPCR 8443 B372A, B373B, B375B	Indirect VER	Indirect DPC
Injection opening pressure, New [used]...bar	130 +8 -5	118 +7 -5	125 to 138	118
Glow plugs				
Maker	Bosch/Champion	Champion	Bosch/Champion	Bosch/Champion
Type	0250 201 013 / CH155	CH88 / CH137	0250 201 029 / CH155	0250 201 029 / CH155
Nominal rating ....V/A	11 / 12	11 / 12	11 / 15	11 / 15
Brakes				
minimum friction material thickness				
Front...mm	6.0 with backing	6.0 with backing	7.0 with backing	7.0 with backing
Rear..mm	2.5 with backing	2.5 with backing	2.5 with backing	2.5 with backing
Tyres - Saloon / Hatch ....Size	145/70x13:155/70x13:165/65x13	145/70x13:155/70x13:165/65x13	–	–
- Estate / Van ....Size	155x13: 145/70x13: 155/70x13	155x13: 145/70x13: 155/70x13	165/70x13	165/70x13
Pressure - front / rear - Saloon / Hatch ...bar	2.2 / 2.2	2.2 / 2.2	–	–
- Estate / Van ......bar	2.2 / 2.2¹	2.2 / 2.2¹	2.4 / 2.5	2.4 / 2.5
Front suspension / wheel alignment				
Toe-in (+) / Toe-out (–)....................mm [°]	-1.0 ± 1.0	-1.0 ± 1.0	-1.0 ± 1.0	-1.0 ± 1.0
Camber	15' ± 30'	15' ± 30'	0°20' ± 40'	0°20' ± 40'
Castor	2° ± 30'	2° ± 30'	WSM	WSM
King pin inclination............................	12°40' ± 30'	12°40' ± 30'	12°45' ± 45'	12°45' ± 45'
Rear suspension / wheel alignment				
Toe-in (+) / Toe-out (–)....................mm [°]	0 to 3.0	0 to 3.0	-1.5 ± 1.5	-1.5 ± 1.5
Camber	-50' ± 30'	-50' ± 30'	-0°50' ± 30'	-0°50' ± 30'

RENAULT

Torque wrench settings	5 & Extra (B/C/F/S404) 1986 to 1994	5 & Extra (B/C/F/S404) 1986 to 1994	5 & Extra D 1992 to 1996	5 & Extra D 1992 to 1996
Cylinder head - stage 1 ....Nm	30 N	30 N	30 N	30 N
- stage 2 ....Nm	70	70	+ 80°	+ 80°
- stage 3 ....Nm	Wait 3 mins	Wait 3 mins	Wait 3 mins	Wait 3 mins
- stage 4 ....Nm	Slacken	Slacken	Slacken	Slacken
- stage 5 ....Nm	20	20	25	25
- stage 6 ....Nm	+ 123°	+ 123°	+ 213°	+ 213°
Big-end bearings................................Nm	45 to 50 N	45 to 50 N	50 N	50 N
Main bearings....................................Nm	60 to 65	60 to 65	65	65
Crankshaft pulley bolt ....Nm	90 to 100	90 to 100	100	100
Camshaft pulley bolt ....Nm	50	50	50	50
Flywheel [driveplate] bolt.....................Nm	50 to 55	50 to 55	55	55
Front hubs ....Nm	250	250	250	250
Rear hubs ....Nm	160	160	175	175
Wheel nuts / bolts ....Nm	80	80	90	90
Glow plugs ....Nm	15 to 30	15 to 30	20	20
Clutch pressure plate bolts ....Nm	25	25	40	40
Injection pump sprocket.....................Nm	50	50	50	50
Injectors..Nm	65 to 75	65 to 75	70	70
Injection pump mounting bolts ....Nm	–	–	25	25
Injector pipe unions...........................Nm	25	25	25	25
Capacities				
Engine oil & filter...............................litres	5.3	5.3	5.5	5.5
Gearbox..litres	JB1: 3.4 JB5: 2.9	JB1: 3.4 JB5: 2.9	3.4	3.4
Automatic transmission ....litres	–	–	–	–
Final drive ....litres	WT	WT	WT	WT
Cooling system.................................litres	6.5	6.5	7.2	7.2
Fuel tank...litres	43	43	43	43

Notes

5 & Extra (B/C/F/S404) 1986 to 1994
¹Extra: 2.1 / 2.5
5 & Extra (B/C/F/S404) 1986 to 1994
¹Extra: 2.1 / 2.5

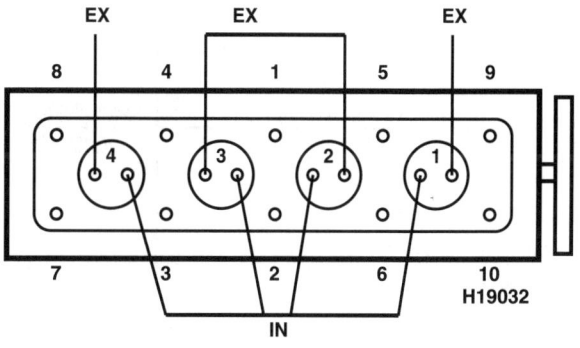

1596 cm³ / 1870 cm³

– Not applicable, or information not
available

RENAULT

	Clio 1.9 D 1991 to 1998	Clio 1.9 D 1992 to 1998	Clio 1.9 D 1991 to 1998	Clio 1.9 D 1995 to 1998
Engine				
Engine type/code.....................................	F8Q732 SOHC 48kW	F8Q714 SOHC 48kW	F8Q730 SOHC 48kW	F8Q678 SOHC 48kW
Capacity (cm^3) / cylinders.........................	1870 / 4	1870 / 4	1870 / 4	1870 / 4
Compression ration / pressure ....bar	21.5 / ≥20.0	21.5 / ≥20.0	21.5 / ≥20.0	21.5 / ≥20.0
Torque output ....Nm	118	118	118	118
Oil pressure ....idle [running] bar	1.2 [3.5 @ 4000]	1.2 [3.5 @ 4000]	1.2 [3.5 @ 4000]	1.2 [3.5 @ 4000]
Oil temperature ....°C	80	80	80	80
Valve clearances - inlet (mm)	0.20	0.20	0.20	0.20
- exhaust (mm)	0.40	0.40	0.40	0.40
Injection order	1-3-4-2	1-3-4-2	1-3-4-2	1-3-4-2
No. 1 cylinder position	FE	FE	FE	FE
Cooling system				
Thermostat opening temperature ....°C	82	82	82	82
Radiator cap pressure ....bar	1.2 or 1.8	1.2 or 1.8	1.2 or 1.8	1.2 or 1.8
Fuel system				
Idle speed ....rpm	825 ± 25	825 ± 25	825 ± 25	825 ± 25
Maximum (no load) speed ....rpm	5200	5200	5200	5200
Smoke test/opacity ....M^{-1} %	1.38	1.17	2.41	0.96
Static timing method.............................	Plunger travel	Dial gauge	Dial gauge	Dial gauge
Timing dimension..............................mm	0.7	Dimension on pump	Dimension on pump	0.9 ± 0.02
Crankshaft position ....mm [°]	0 TDC	0 TDC	0 TDC	0 TDC
Turbo type / ref / pressure ....bar	–	–	–	–
Injection pump make	Bosch	Lucas	Roto Diesel	Bosch
Injection pump part no............................	VE 4/8 R317/5	DPC R8444B080A/B	DPC R8444B690A	VE 4/8F R588
Injector Make / type	Bosch	CAV	Roto Diesel	Bosch
Injector part no.....................................	DNOSD302	RDN4SDC6886D	LCR 67334	KCA 17S42
Injection type.......................................	Indirect VE	Indirect DPC	Indirect DPC	Indirect VE
Injection opening pressure, New [used]...bar	125 to 140	113 to 125	113 to 125	125 to 138
Glow plugs				
Maker	Bosch/Champion	Bosch/Champion	Bosch/Champion	Bosch/Champion
Type	0250 201 029 / CH155	0250 201 029 / CH155	0250 201 029 / CH155	0250 201 029 / CH155
Nominal rating.................................V/A	11 / 15	11 / 15	11 / 15	11 / 15
Brakes				
minimum friction material thickness				
Front.......................................mm	6.0 with backing	6.0 with backing	6.0 with backing	6.0 with backing
Rear..mm	2.5 with backing	2.5 with backing	2.5 with backing	2.5 with backing
Tyres - Saloon / Hatch......................Size	155/70x13: 165/65x13	155/70x13: 165/65x13	155/70x13: 165/65x13	155/70x13: 165/65x13
- Estate / Van..........................Size	–	–	–	–
Pressure - front / rear - Saloon / Hatch ...bar	2.4 / 2.4	2.4 / 2.4	2.4 / 2.4	2.4 / 2.4
- Estate / Van ......bar				
Front suspension / wheel alignment				
Toe-in (+) / Toe-out (–).....................mm [°]	-1.0 ± 1.0	-1.0 ± 1.0	-1.0 ± 1.0	-1.0 ± 1.0
Camber	WSM	WSM	WSM	WSM
Castor	WSM	WSM	WSM	WSM
King pin inclination..................................	–	–	–	–
Rear suspension / wheel alignment				
Toe-in (+) / Toe-out (–).....................mm [°]	1.5 ± 1.0 PW	1.5 ± 1.0 PW	1.5 ± 1.0 PW	1.5 ± 1.0 PW
Camber	-0°50' ± 30'	-0°50' ± 30'	-0°50' ± 30'	-0°50' ± 30'

RENAULT

Torque wrench settings	Clio 1.9 D 1991 to 1998	Clio 1.9 D 1992 to 1998	Clio 1.9 D 1991 to 1998	Clio 1.9 D 1995 to 1998
Cylinder head - stage 1 ....Nm	30 N	30 N	30 N	30 N
- stage 2 ....Nm	+80°	+80°	+80°	+80°
- stage 3 ....Nm	Wait 3 mins	Wait 3 mins	Wait 3 mins	Wait 3 mins
- stage 4 ....Nm	Slacken	Slacken	Slacken	Slacken
- stage 5 ....Nm	25	25	25	25
- stage 6 ....Nm	+213°	+213°	+213°	+213°
Big-end bearings................................Nm	50 N	50 N	50 N	50 N
Main bearings.....................................Nm	65	65	65	65
Crankshaft pulley bolt ....Nm	95	95	95	95
Camshaft pulley bolt ....Nm	50	50	50	50
Flywheel [driveplate] bolt....................Nm	55	55	55	55
Front hubs ....Nm	250	250	250	250
Rear hubs ....Nm	160	160	160	160
Wheel nuts / bolts ....Nm	90	90	90	90
Glow plugs ....Nm	20	20	20	20
Clutch pressure plate bolts ....Nm	25	25	25	25
Injection pump sprocket.....................Nm	50	50	50	50
Injectors..Nm	70	70	70	70
Injection pump mounting bolts ....Nm	–	–	–	–
Injector pipe unions...........................Nm	25	25	25	25
Capacities				
Engine oil & filter................................litres	5.5	5.5	5.5	5.5
Gearbox...litres	3.4	3.4	3.4	3.4
Automatic transmission ....litres	–	–	–	–
Final drive ....litres	WT	WT	WT	WT
Cooling system...................................litres	6.6	6.6	6.6	6.6
Fuel tank..litres	43	43	43	43

Notes

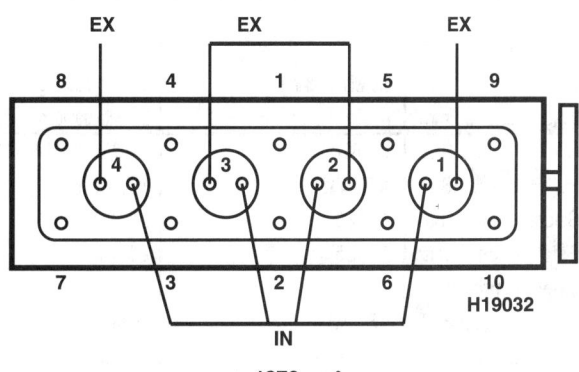

1870 cm³

– Not applicable, or information not
available

RENAULT

	Clio 1.9 (B/C/SBOE/J/N/R) 1998 to 2000	19 & Chamade (B/C/L/S534) 1989 to 1996	19 & Chamade (B/C/L/S534) 1990 to 1996	19 & Chamade (B/C/L/S534) 1992 to 1996
Engine				
Engine type/code.........................	F8Q630 SOHC 48kW	F8Q706 SOHC 47kW	F8Q706 SOHC 47kW	F8Q706 SOHC 47kW
Capacity (cm^3) / cylinders......................	1870 / 4	1870 / 4	1870 / 4	1870 / 4
Compression ration / pressure ....bar	21.5 /	21.5 / ≥20.0	21.5 / ≥20.0	21.5 / ≥20.0
Torque output ....Nm	118	118	118	118
Oil pressure ....idle [running] bar	1.2 [3.5 @ 3000]	2.0 [3.5 @ 3000]	2.0 [3.5 @ 3000]	2.0 [3.5 @ 3000]
Oil temperature ....°C	80	80	80	80
Valve clearances - inlet (mm)	0.10	0.20	0.20	0.20
- exhaust (mm)	0.25	0.40	0.40	0.40
Injection order	1-3-4-2	1-3-4-2	1-3-4-2	1-3-4-2
No. 1 cylinder position	FE	FE	FE	FE
Cooling system				
Thermostat opening temperature ....°C	89	89	89	89
Radiator cap pressure ....bar	1.2	1.2	1.2	1.2
Fuel system				
Idle speed ....rpm	850 ± 25	825 ± 25	825 ± 25	825 ± 25
Maximum (no load) speed ....rpm	5100 ± 100	5200	5100	5100
Smoke test/opacity ....M^{-1} %	1.11	1.38	1.46	2.5
Static timing method..............................	Refer to wsm	Plunger travel	Rotor lift	Dial gauge
Timing dimension.............................mm	_	0.7 ± 0.02	Dimension on pump	0.82 ± 0.02
Crankshaft position ....mm [°]	_	[0] TDC	[0] TDC	[0] TDC
Turbo type / ref / pressure ....bar	_	_	_	_
Injection pump make	Lucas	Bosch	Roto Diesel	Bosch
Injection pump part no............................	8448B171 A/231A	VE 4/9 R317	R8443 B700A	VER 458-2
Injector Make / type		Bosch	Roto Diesel	Bosch
Injector part no..	Lucas RDNOSDC 6902	0432 171 192	RDN4SDC 6868C	0432 217 258
Injection type..	Indirect	Indirect VE	Indirect DPC	Indirect VER
Injection opening pressure, New [used]...bar	130 ± 5	125 to 138	118 +7 -5	125 to 138
Glow plugs				
Maker	Beru	Bosch/Champion	Bosch/Champion	Bosch/Champion
Type	_	0250 201 013 / CH155	0250 201 013 / CH155	0250 201 029 / CH155
Nominal rating.....................................V/A	_	11 / 12	11 / 12	11 / 15
Brakes				
minimum friction material thickness				
Front...................................mm	6.0 with backing	6.0 with backing	6.0 with backing	6.0 with backing
Rear....................................mm	_	2.5 with backing	2.5 with backing	2.5 with backing
Tyres - Saloon / Hatch......................Size	165/60x14	165/70x13: 155/80x13	165/70x13: 155/80x13	165/70x13: 155/80x13
- Estate / Van......................Size	165/65x13	165/70x13	165/70x13	165/70x13
Pressure - front / rear - Saloon / Hatch ...bar	2.2 / 2.0	2.2 / 2.2	2.2 / 2.2	2.2 / 2.2
- Estate / Van ......bar	2.5 / 2.5	2.2 / 2.2	2.2 / 2.2	2.2 / 2.2
Front suspension / wheel alignment				
Toe-in (+) / Toe-out (−)....................mm [°]	1.6 ± 2.0²	-1.0 ± 1.0	-1.0 ± 1.0	-1.0 ± 1.0
Camber	-0°45' ± 30'	-5' ± 30'	-5' ± 30'	-5' ± 30'
Castor	2°30' ± 30'	1°55' PAS: 4°	1°55' PAS: 4°	1°55' PAS: 4°
King pin inclination..................................	12°0' ± 30'	12°55' ± 30'	12°55' ± 30'	12°55' ± 30'
Rear suspension / wheel alignment				
Toe-in (+) / Toe-out (−)....................mm [°]	-3.0 ± 3.0	1.0 to 3.0	1.0 to 3.0	1.0 to 3.0
Camber	-0°42' ± 20'	-50' ± 30'	-50' ± 30'	-50' ± 30'

RENAULT

Torque wrench settings	Clio 1.9 (B/C/SBOE/J/N/R) 1998 to 2000	19 & Chamade (B/C/L/S534) 1989 to 1996	19 & Chamade (B/C/L/S534) 1990 to 1996	19 & Chamade (B/C/L/S534) 1992 to 1996
Cylinder head - stage 1 ....Nm	30 + 80° N	30 N	30 N	30 N
- stage 2 ....Nm	slacken[1]	70	70	70
- stage 3 ....Nm	25	Wait 3 mins	Wait 3 mins	Wait 3 mins
- stage 4 ....Nm	+ 213°	Slacken	Slacken	Slacken
- stage 5 ....Nm	_	20	20	20
- stage 6 ....Nm	_	+ 123°	+ 123°	+ 123°
Big-end bearings................................Nm	50 N	45 to 50 N	45 to 50 N	45 to 50 N
Main bearings.....................................Nm	60 N	60 to 65	60 to 65	60 to 65
Crankshaft pulley bolt ....Nm	20 + 115° ± 15°	90 to 100	90 to 100	90 to 100
Camshaft pulley bolt ....Nm	50	50	50	50
Flywheel [driveplate] bolt....................Nm	20 + 60°	50 to 55	50 to 55	50 to 55
Front hubs ....Nm	280	250	250	250
Rear hubs ....Nm	175	160	160	160
Wheel nuts / bolts ....Nm	90	80	80	80
Glow plugs ....Nm	25	15 to 30	15 to 30	15 to 30
Clutch pressure plate bolts ....Nm	40	25	25	25
Injection pump sprocket......................Nm	15 + 60°	50	50	50
Injectors..Nm	70	65 to 75	65 to 75	65 to 75
Injection pump mounting bolts ....Nm	_	_	_	_
Injector pipe unions............................Nm	_	25	25	25
Capacities				
Engine oil & filter................................litres	5.2	5.5	5.5	5.5
Gearbox...litres	3.4	3.4	3.4	3.4
Automatic transmission ....litres	_	_	_	_
Final drive ....litres	WT	WT	WT	WT
Cooling system....................................litres	7.3	6.0	6.0	6.0
Fuel tank..litres	50	55	55	55

Notes

Clio 1.9 (B/C/SBOE/J/N/R) 1998 to 2000
[1]Slacken bolts 1-2, then 25 Nm + 213°. Repeat procedure for bolts 3-4, 5-6, 7-8 and 9-10
[2]Van: 1.0 ± 1.0

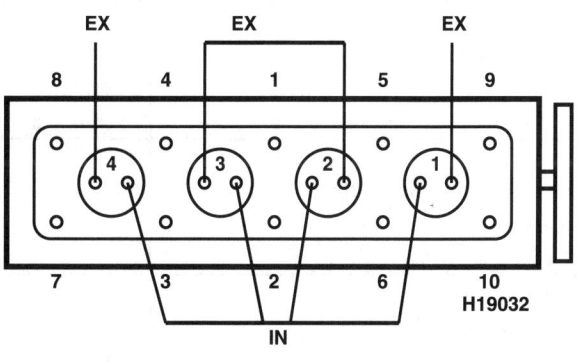

1870 cm³

– Not applicable, or information not available

RENAULT

	19 & Chamade (B/C/L/S53l) 1992 to 1996	19 TD (B/C/L/S53K) 1991 to 1996	19 TD (B/C/L/53T) 1992 to 1996	19 TD (B/C/L/S53Z) 1995 to 1996
Engine				
Engine type/code....................	F8Q706 SOHC 47kW	F8Q740 SOHC Turbo 68kW	F8Q744 SOHC Turbo 68kW	F8Q768 SOHC Turbo 68kW
Capacity (cm^3) / cylinders........................	1870 / 4	1870 / 4	1870 / 4	1870 / 4
Compression ration / pressure ....bar	21.5 / ≥20.0	20.5 /	20.5 /	20.5 /
Torque output ....Nm	118	175	175	175
Oil pressure ....idle [running] bar	2.0 [3.5 @ 3000]	1.2 [3.5 @ 3000]	1.2 [3.5 @ 3000]	1.2 [3.5 @ 3000]
Oil temperature ....°C	80	80	80	80
Valve clearances - inlet (mm)	0.20	0.20	0.20	0.20
- exhaust (mm)	0.40	0.40	0.40	0.40
Injection order..............................	1-3-4-2	1-3-4-2	1-3-4-2	1-3-4-2
No. 1 cylinder position	FE	FE	FE	FE
Cooling system				
Thermostat opening temperature ....°C	89	82	82	82
Radiator cap pressure ....bar	1.2	1.2	1.2	1.2
Fuel system				
Idle speed ....rpm	825 ± 25	825 ± 25	825 ± 25	825 ± 25
Maximum (no load) speed ....rpm	5100	4900	4900	4900
Smoke test/opacity ....M^{-1} %	1.38	1.38	2.0	0.87
Static timing method.............................	Dial gauge	Dial gauge	Dial gauge	Dial gauge
Timing dimension.............................mm	0.82 ± 0.02	Dimension on pump	Dimension on pump	Dimension on pump
Crankshaft position ....mm [°]	[0] TDC	0 TDC	0 TDC	0 TDC
Turbo type / ref / pressure ....bar	–	Garrett T2	Garrett T2	Garrett T2
Injection pump make	Bosch	Lucas	Lucas	Lucas
Injection pump part no.............................	VER 317-6	R8443B783D/E	R8443B871BE	R8444B431A
Injector Make / type	Bosch	Roto Diesel	Roto Diesel	Roto Diesel
Injector part no...	0432 217 258	RDN4 SDC6878C	RDN4 SDC6878C	RDN4 SDC6878D
Injection type...	Indirect VER	Indirect DPC	Indirect DPC	Indirect DPC
Injection opening pressure, New [used]...bar	125 to 138	125 to 138	125 to 138	125 to 138
Glow plugs				
Maker	Bosch/Champion	Bosch/Champion	Bosch/Champion	Bosch/Champion
Type	0250 201 029 / CH155	0250 201 031 / CH69	0250 201 031 / CH69	0250 201 031 / CH69
Nominal rating..V/A	11 / 15	11 / 15	11 / 15	11 / 15
Brakes				
minimum friction material thickness				
Front...mm	6.0 with backing	6.0 with backing	6.0 with backing	6.0 with backing
Rear...mm	2.5 with backing	2.5 with backing	2.5 with backing	2.5 with backing
Tyres - Saloon / Hatch......................Size	165/70x13: 155/80x13	175/70x13	175/70x13	175/70x13
- Estate / Van...........................Size	165/70x13	–	–	–
Pressure - front / rear - Saloon / Hatch...bar	2.2 / 2.2	2.2 / 2.2	2.2 / 2.2	2.2 / 2.2
- Estate / Van ......bar	2.2 / 2.2	–	–	–
Front suspension / wheel alignment				
Toe-in (+) / Toe-out (–)....................mm [°]	-1.0 ± 1.0	-1.0 ± 1.0	-1.0 ± 1.0	-1.0 ± 1.0
Camber	-5' ± 30'	WSM	WSM	WSM
Castor	1°55' PAS: 4°	WSM	WSM	WSM
King pin inclination................................	12°55' ± 30'	–	–	–
Rear suspension / wheel alignment				
Toe-in (+) / Toe-out (–)....................mm [°]	1.0 to 3.0	1.5 ± 1.0 PW	1.5 ± 1.0 PW	1.5 ± 1.0 PW
Camber	-50' ± 30'	-0°50' ± 30'	-0°50' ± 30'	-0°50' ± 30'

Haynes THE BOOK ®

RENAULT

Torque wrench settings	19 & Chamade (B/C/L/S53I) 1992 to 1996	19 TD (B/C/L/S53K) 1991 to 1996	19 TD (B/C/L/53T) 1992 to 1996	19 TD (B/C/L/S53Z) 1995 to 1996
Cylinder head - stage 1 ....Nm	30 N	30 N	30 N	30 N
- stage 2 ....Nm	70	+ 50°	+ 50°	+ 50°
- stage 3 ....Nm	Wait 3 mins	Wait 3 mins	Wait 3 mins	Wait 3 mins
- stage 4 ....Nm	Slacken	Slacken	Slacken	Slacken
- stage 5 ....Nm	20	25	25	25
- stage 6 ....Nm	+ 123°	+ 213°	+ 213°	+ 213°
Big-end bearings.............................Nm	45 to 50 N	50 N	50 N	50 N
Main bearings.................................Nm	60 to 65	65	65	65
Crankshaft pulley bolt ....Nm	90 to 100	95	95	95
Camshaft pulley bolt ....Nm	50	50	50	50
Flywheel [driveplate] bolt.................Nm	50 to 55	55	55	55
Front hubs ....Nm	250	250	250	250
Rear hubs ....Nm	160	160	160	160
Wheel nuts / bolts ....Nm	80	90	90	90
Glow plugs ....Nm	15 to 30	22	22	22
Clutch pressure plate bolts ....Nm	25	25	25	25
Injection pump sprocket...................Nm	50	50	50	50
Injectors...Nm	65 to 75	70	70	70
Injection pump mounting bolts ....Nm	–	–	–	–
Injector pipe unions.........................Nm	25	25	25	25
Capacities				
Engine oil & filter................................litres	5.5	5.5	5.5	5.5
Gearbox...litres	3.4	3.1	3.1	3.1
Automatic transmission ....litres	–	–	–	–
Final drive ....litres	WT	WT	WT	WT
Cooling system...................................litres	6.0	6.8	6.8	6.8
Fuel tank..litres	55	55	55	55

Notes

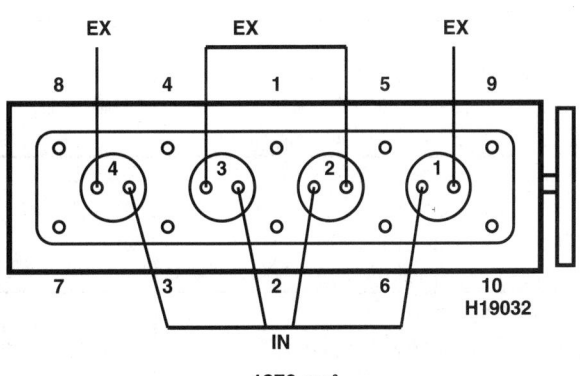

1870 cm³

– Not applicable, or information not available

RENAULT

	19 TD (B/C/L/S53K) 1995 to 1996	Megane 1.9 D (B/E/LOA) 1996 to 2000	Megane 1.9 TD (B/E/LOA) 1996 to 1997	Megane 1.9 TDi (B/LAO N) 1998 to 1999
Engine				
Engine type/code..................................	F8Q610 SOHC Turbo 68kW	F8Q620 SOHC 48kW	F8Q784 SOHC 66kW	F9Q730/734 SOHC Turbo 74kW
Capacity (cm³) / cylinders.........................	1870 / 4	1870 / 4	1870 / 4	1870 / 4
Compression ration / pressure ....bar	20.5 /	21.5 /	20.5 /	18.3 /
Torque output ....Nm	175	118	200	200
Oil pressure ....idle [running] bar	1.2 [3.5 @ 3000]	1.2 [3.5 @ 4000]	1.2 [3.5 @ 4000]	1.2 [3.5 @ 4000]
Oil temperature ....°C	80	80	80	80
Valve clearances - inlet (mm)	0.20	0.20	0.20	0.20
- exhaust (mm)	0.40	0.40	0.40	0.40
Injection order	1-3-4-2	1-3-4-2	1-3-4-2	1-3-4-2
No. 1 cylinder position	FE	FE	FE	FE
Cooling system				
Thermostat opening temperature ....°C	82	82	82	82
Radiator cap pressure ....bar	1.2	1.2	1.2	1.2
Fuel system				
Idle speed ....rpm	825 ± 25	825 ± 25	825 ± 25	825 ± 25
Maximum (no load) speed ....rpm	4900	5200	500	4800 ± 100
Smoke test/opacity ....M⁻¹%	0.99	1.24	1.24	1.24
Static timing method..............................	Dial gauge	Dial gauge	Dial gauge	Dial gauge
Timing dimension................................mm	Dimension on pump	0.82 ± 0.04	Dimension on pump	0.45 ± 0.02
Crankshaft position ....mm [°]	0 TDC	0 TDC	0 TDC	0 TDC
Turbo type / ref / pressure ....bar	Garrett T2	–	–	Garrett T2
Injection pump make	Lucas	Bosch	Lucas	Bosch
Injection pump part no............................	R8444B431A	VE4/8 F2300R598	8448B020A	VE4/11 E2000R6T2
Injector Make / type	Roto Diesel	Bosch	CAV	Bosch
Injector part no......................................	RDN4 SDC6878D	DNOSD302	LCR 6733402D	DLSA 145 P619
Injection type...	Indirect DPC	Indirect	Indirect DPC	Direct injection
Injection opening pressure, New [used]...bar	125 to 138	125 to 138	125 to 138	170 to 212
Glow plugs				
Maker	Bosch/Champion	Bosch/Champion	Bosch/Champion	Beru
Type	0250 201 031 / CH69	0250 201 029 / CH155	0250 201 029 / CH155	GN999
Nominal rating...................................V/A	11 / 15	11 / 16	11 / 16	–
Brakes				
minimum friction material thickness				
Front......................................mm	6.0 with backing	6.5 with backing	6.5 with backing	6.5 with backing
Rear.......................................mm	2.5 with backing	1.5	1.5	1.5
Tyres - Saloon / Hatch ....Size	175/70x13	175/70x13	175/70x13	175/65x13
- Estate / Van......................Size	–	–	–	–
Pressure - front / rear - Saloon / Hatch ...bar	2.2 / 2.2	2.3 / 2.3	2.3 / 2.3	2.2 / 2.4
- Estate / Van ......bar	–	–	–	–
Front suspension / wheel alignment				
Toe-in (+) / Toe-out (–)....................mm [°]	-1.0 ± 1.0	[-0°10' ± 10']	[-0°10' ± 10']	[-0°10' ± 10']
Camber	WSM	-0°20' ± 30'	-0°20' ± 30'	-0°20' ± 30'
Castor	WSM	4° ± 30'	4° ± 30'	4° ± 30'
King pin inclination................................	–	13°21' ± 30'	13°21' ± 30'	13°21' ± 30'
Rear suspension / wheel alignment				
Toe-in (+) / Toe-out (–)....................mm [°]	1.5 ± 1.0 PW	[0°50' ± 20']	[0°50' ± 20']	[0°50' ± 20']
Camber	-0°50' ± 30'	-1° ± 15'	-1° ± 15'	-1° ± 15'

RENAULT

Torque wrench settings	19 TD (B/C/L/S53K) 1995 to 1996	Megane 1.9 D (B/E/LOA) 1996 to 2000	Megane 1.9 TD (B/E/LOA) 1996 to 1997	Megane 1.9 TDi (B/LAO N) 1998 to 1999
Cylinder head - stage 1 ....Nm	30 N	30 + 80° N	30[1]	30 N
- stage 2 ....Nm	+ 50°	slacken [1]	+ 50°	+ 50°
- stage 3 ....Nm	Wait 3 mins	25	Wait 3 mins, slacken	Wait 3 mins
- stage 4 ....Nm	Slacken	+ 213°	25 + 213°	Slacken
- stage 5 ....Nm	25	–	warm up, allow to cool	25
- stage 6 ....Nm	+ 213°	–	+ 120°	+ 213°[1]
Big-end bearings................................Nm	50 N	60 N	50 N	50
Main bearings....................................Nm	65	60 N	60 N	65
Crankshaft pulley bolt ....Nm	95	20 + 115° ± 15°	20 + 115°	20 + 115°
Camshaft pulley bolt ....Nm	50	50	50	60
Flywheel [driveplate] bolt....................Nm	55	20 + 60°	20 + 60°	50 to 55
Front hubs ....Nm	250	250	250	250
Rear hubs ....Nm	160	175	175	175
Wheel nuts / bolts ....Nm	90	90	90	90
Glow plugs ....Nm	22	20	20	25 to 30
Clutch pressure plate bolts ....Nm	25	40	40	22.5
Injection pump sprocket.....................Nm	50	65	65	90
Injectors..Nm	70	27	70	27
Injection pump mounting bolts ....Nm	–	25	25	–
Injector pipe unions...........................Nm	25	25	25	–
Capacities				
Engine oil & filter................................litres	5.5	5.5	5.5	5.5
Gearbox...litres	3.1	3.4	3.4	3.4
Automatic transmission ....litres	–	–	–	–
Final drive ....litres	WT	WT	WT	WT
Cooling system..................................litres	6.8	7.5	7.5	7.5
Fuel tank..litres	55	60	60	60

Notes

Megane 1.9 D (B/E/LOA) 1996 to 2000
[1]Slacken bolts 1-2, then 25 Nm + 213°. Repeat procedure
[1]for bolts 3-4, 5-6, 7-8 and 9-10
Megane 1.9 TD (B/E/LOA) 1996 to 1997
[1]Apply procedure to each bolt in turn
Megane 1.9 TDi (B/LAO N) 1998 to 1999
[1] Apply procedure to each bolt in turn

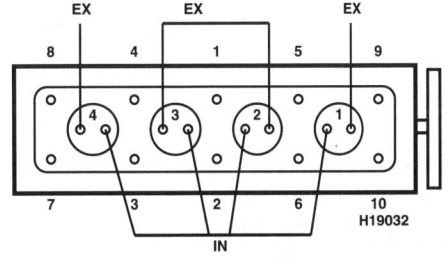

1870 cm³

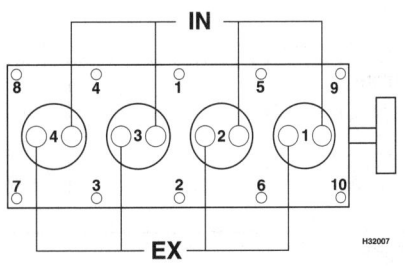

1870 cm³ (TDi)

– Not applicable, or information not available

	Megane 1.9 TDi (B/LA0NM5/2) 1999 to 2000	Scenic 1.9 TD (JAO) 1996	Scenic 1.9 TDi (JAO) 1997 to 1999	21 1.9 D (B/L/K/S48H) 1989 to 1995
Engine				
Engine type/code...............................	F9Q SOHC Turbo 74kW	F8Q784 SOHC Turbo 70kW	F9Q730/734 SOHC Turbo 74kW	F8Q710 SOHC 47kW
Capacity (cm^3) / cylinders.................	1870 / 4	1870 / 4	1870 / 4	1870 / 4
Compression ration / pressure ....bar	18.3 /	20.5 /	18.3 /	21.5 / ≥20.0
Torque output ....Nm	200	175	200	118
Oil pressure ....idle [running] bar	1.2 [3.5 @ 4000]	[3.5 @ 3000]	[3.5 @ 3000]	2.0 [3.5 @ 3000]
Oil temperature ....°C	80	80	80	80
Valve clearances - inlet (mm)	0.20	0.20	0.20	0.20
- exhaust (mm)	0.40	0.40	0.40	0.40
Injection order	1-3-4-2	1-3-4-2	1-3-4-2	1-3-4-2
No. 1 cylinder position	FE	FE	FE	FE
Cooling system				
Thermostat opening temperature ....°C	82	82	82	89
Radiator cap pressure ....bar	1.2	1.2	1.2	1.2
Fuel system				
Idle speed ....rpm	825 ± 25	825 ± 25	850 ± 25	825 ± 25
Maximum (no load) speed ....rpm	4800 ± 100	4900	4900	5100 to 5300
Smoke test/opacity ....M^{-1} %	2.5	2.5	1.24	0.51
Static timing method..............................	–	Dial gauge	Dial gauge	Rotor lift
Timing dimension................................mm	Computer control	Dimension on pump	0.45 ± 0.02	Dimension on pump
Crankshaft position ....mm [°]		0 TDC	0 TDC	[0] TDC
Turbo type / ref / pressure ....bar	Garrett T2	–	Garrett T2	–
Injection pump make	Bosch	Lucas	Bosch	Roto Diesel
Injection pump part no............................	VE4/11E2000R6T2	8448B020A	VER672	R8443B471C
Injector Make / type..............................		CAV	Bosch	Roto Diesel
Injector part no......................................	DLSA 145 P619	RDN4SDC6878D	DSLA145P619	RDN4SDC 6868C
Injection type...	Direct, Electronic control	Indirect DPC	Direct VER	Indirect DPC
Injection opening pressure, New [used]...bar	170 to 212	125 to 138	200 to 212	118 +7 -5
Glow plugs				
Maker	Champion	Champion	Beru	Bosch/Champion
Type	CH172	CH179	GN999	0250 201 006 / CH137
Nominal rating ....V/A	11 /	11 /	12 / 12	11 / 15
Brakes				
minimum friction material thickness				
Front...mm	6.0 with backing	6.0 with backing	6.0 with backing	6.0 with backing
Rear..mm	2.5 with backing	2.5	2.5	2.5 with backing
Tyres - Saloon / Hatch.....................Size	185/60x15	–	–	175/70x13: 175/65x14
- Estate / Van..........................Size	175/70x14	175/70x14	175/70x14	175/70x13: 175/65x14
Pressure - front / rear - Saloon / Hatch...bar	2.3 / 2.2	–	–	2.3 / 2.3
- Estate / Van ......bar	2.4 / 2.5	2.5 / 2.5	2.5 / 2.5	2.3 / 2.6
Front suspension / wheel alignment				
Toe-in (+) / Toe-out (–)....................mm [°]	-1.0 ± 1.0	[-0°10' ± 10']	[-0°10' ± 10']	-1.0 ± 1.0
Camber	-0°20' ± 30'	-0°20' ± 30'	-0°20' ± 30'	0° ± 30'
Castor	4° ± 30'	4° ± 30'	4° ± 30'	2° ± 30'
King pin inclination................................	–	13°21' ± 30'	13°21' ± 30'	11°40' ± 30'
Rear suspension / wheel alignment				
Toe-in (+) / Toe-out (–)....................mm [°]	1.25 ± 1.0	[0°50' ± 20']	[0°50' ± 20']	2.0 to 5.0
Camber	-1°15'	-1° ± 15'	-1° ± 15'	-20' ± 10'

RENAULT

Torque wrench settings	Megane 1.9 TDi (B/LA0NM5/2) 1999 to 2000	Scenic 1.9 TD (JAO) 1996	Scenic 1.9 TDi (JAO) 1997 to 1999	21 1.9 D (B/L/K/S48H) 1989 to 1995
Cylinder head - stage 1 ....Nm	20 oiled	30¹ N	30¹ N	30 N
- stage 2 ....Nm	+ 100°	+ 50°	+ 50°	+ 70°
- stage 3 ....Nm	Wait 3 mins	Wait 3 mins, slacken	Wait 3 mins, slacken	Wait 3 mins
- stage 4 ....Nm	Slacken	20 + 213°	20 + 213°	Slacken
- stage 5 ....Nm	20	Warm up, allow to cool	Warm up, allow to cool	20
- stage 6 ....Nm	+ 100°¹	+ 120°	+ 120°	+ 123°
Big-end bearings................................Nm	10 + 43° N	50 N	50 N	45 to 50 N
Main bearings.....................................Nm	25 + 47°	65	65	60 to 65
Crankshaft pulley bolt ....Nm	20 + 115°	20 + 115°	20 + 115°	90 to 100
Camshaft pulley bolt ....Nm	60	_	_	50
Flywheel [driveplate] bolt....................Nm	50	50 to 55	50 to 55	50 to 55
Front hubs ....Nm	250	250	250	250
Rear hubs ....Nm	175	175	175	160
Wheel nuts / bolts ....Nm	90	90	90	90
Glow plugs ....Nm	25 to 30	22	22	15 to 30
Clutch pressure plate bolts ....Nm	25	22.5	22.5	25
Injection pump sprocket......................Nm	_	65	90	50
Injectors...Nm	27	70	_	65 to 75
Injection pump mounting bolts ....Nm	_	20	_	_
Injector pipe unions............................Nm	_	25	_	25
Capacities				
Engine oil & filter................................litres	5.5	5.5	5.5	5.5
Gearbox...litres	3.4	3.1	3.1	3.4
Automatic transmission ....litres	_	_	_	_
Final drive ....litres	WT	WT	WT	WT
Cooling system...................................litres	7.5	7.5	7.5	7.0
Fuel tank...litres	60	60	60	66

Notes

Megane 1.9 TDi (B/LA0NM5/2) 1999 to 2000
¹Apply procedure to each bolt in turn
Scenic 1.9 TD (JAO) 1996
¹Apply procedure to each bolt in turn
Scenic 1.9 TDi (JAO) 1997 to 1999
¹Apply procedure to each bolt in turn

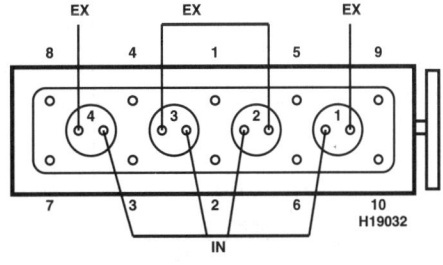

1870 cm³

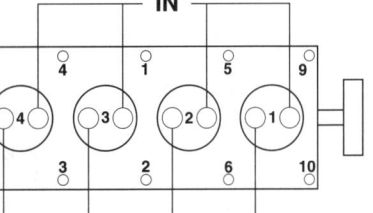

1870 cm³ (TDi)

– Not applicable, or information not available

RENAULT

	21 2.1 D & 4x4 (B/L/K/S486) 1986 to 1995	21 2.1 D & 4x4 (B/L/K/S486) 1986 to 1995	21 2.1 TD (B/L/K488) 1986 to 1995	21 2.1 D (B/L/K/S48V) 1989 to 1995
Engine				
Engine type/code........................	J8S704 SOHC 54kW	J8S704 SOHC 54kW	J8S714 SOHC Turbo 65kW	J8S740 SOHC 54kW
Capacity (cm³) / cylinders..........................	2068 / 4	2068 / 4	2068 / 4	2068 / 4
Compression ration / pressure ....bar	21.5 / _	21.5 / _	21.5 / _	21.5 / _
Torque output ....Nm	0	0	185	0
Oil pressure ....idle [running] bar	0.8 [3.0 @ 3000]	0.8 [3.0 @ 3000]	0.8 [3.5 @ 3000]	0.8 [3.0 @ 3000]
Oil temperature ....°C	80	80	80	80
Valve clearances - inlet (mm)	0.20	0.20	0.20	0.20
- exhaust (mm)	0.25	0.25	0.25	0.25
Injection order	1-3-4-2	1-3-4-2	1-3-4-2	1-3-4-2
No. 1 cylinder position	FE	FE	FE	FE
Cooling system				
Thermostat opening temperature ....°C	89	89	89	89
Radiator cap pressure ....bar	1.2	1.2	1.2	1.2
Fuel system				
Idle speed ....rpm	825 ± 25	825 ± 25	825 ± 25	825 ± 25
Maximum (no load) speed ....rpm	4800 to 5100	4750 to 4900	4700 to 4800	5100 to 5300
Smoke test/opacity ....M⁻¹ %	0.51	0.51	0.51	0.51
Static timing method...............................	Plunger travel	Rotor lift	Plunger travel	Plunger travel
Timing dimension............................mm	0.7 ± 0.02	Dimension on pump	0.7 ± 0.02	0.75 ± 0.02
Crankshaft position ....mm [°]	[0] TDC	[0] TDC	[0] TDC	[0] TDC
Turbo type / ref / pressure ....bar	_	_	Garrett 0.5 to 0.6 bar @ 2500	_
Injection pump make	Bosch	Roto Diesel	Bosch	Bosch
Injection pump part no.............................	VE R158	DPC R8443 B402B	VE 4/9 R153	VE 4/9 R309
Injector Make / type	Bosch	Roto Diesel	Bosch	Bosch
Injector part no.......................................	DNOSD 189	RDN OSDC 6751C	DNOSD 264	DNOSD 252+
Injection type...	Indirect VE	Indirect DPC	Indirect VE	Indirect VE
Injection opening pressure, New [used]...bar	130 +8 -5	115 +7 -5	130 +8 -5	130 +8 -5
Glow plugs				
Maker	Bosch/Champion	Bosch/Champion	Bosch/Champion	Bosch/Champion
Type	0250 201 006 / CH137	0250 201 006 / CH137	0250 201 006 / CH137	0250 201 006 / CH137
Nominal rating....................................V/A	11 / 12	11 / 12	11 / 15	11 / 15
Brakes				
minimum friction material thickness				
Front...mm	6.5 with backing	6.5 with backing	6.5 with backing	6.5 with backing
Rear..mm	2.5 with backing	2.5 with backing	2.5 with backing	2.5 with backing
Tyres - Saloon / Hatch......................Size	175/70x13: 175/65x14	175/70x13: 175/65x14	185/70x13: 185/65x14	175/70x13: 175/65x14
- Estate / Van......................Size	175/70x13:175/65x14:185/65x14	175/70x13:175/65x14:185/65x14	185/65x14	175/70x13: 175/65x14
Pressure - front / rear - Saloon / Hatch ...bar	2.3 / 2.3	2.3 / 2.3	2.3 / 2.3	2.3 / 2.3
- Estate / Van ......bar	2.3 / 2.6	2.3 / 2.6	2.3 / 2.6	2.3 / 2.6
Front suspension / wheel alignment				
Toe-in (+) / Toe-out (–)....................mm [°]	-2.0 ± 1.0	-2.0 ± 1.0	-2.0 ± 1.0	-2.0 ± 1.0
Camber	-10' ± 30'	-10' ± 30'	-25' ± 30'	-10' ± 30'
Castor	2°30' ± 30'	2°30' ± 30'	2°30' ± 30'	2°30' ± 30'
King pin inclination.................................	12°20' ± 30'	12°20' ± 30'	12°45' ± 30'	12°20' ± 30'
Rear suspension / wheel alignment				
Toe-in (+) / Toe-out (–)....................mm [°]	2.0 to 5.0'	2.0 to 5.0'	2.0 to 5.0	2.0 to 5.0
Camber	-40' ± 10' 89 ▶: -20'±10'²	-40' ± 10' 89 ▶: -20'±10'²	-40' ± 10'	-20' ± 10'

RENAULT

Torque wrench settings	21 2.1 D & 4x4 (B/L/K/S486) 1986 to 1995	21 2.1 D & 4x4 (B/L/K/S486) 1986 to 1995	21 2.1 TD (B/L/K488) 1986 to 1995	21 2.1 D (B/L/K/S48V) 1989 to 1995
Cylinder head - stage 1 ....Nm	30	30 N	30 N	30 N
- stage 2 ....Nm	50	50	50	50
- stage 3 ....Nm	95 to 105	95 to 105	95 to 105	95 to 105
- stage 4 ....Nm	Warm up, allow to cool	Warm up, allow to cool	Warm up, allow to cool	Warm up, allow to cool
- stage 5 ....Nm	Slacken, 95 to 105[3]	Slacken, 95 to 105[3]	Slacken, 95 to 105[1]	Slacken, 95 to 105[1]
- stage 6 ....Nm	–	–	–	–
Big-end bearings.................................Nm	65 N	65 N	65 N	65 N
Main bearings.....................................Nm	88 to 98	88 to 98	88 to 98	88 to 98
Crankshaft pulley bolt ....Nm	120 to 135	120 to 135	120 to 135	120 to 135
Camshaft pulley bolt ....Nm	50	50	50	50
Flywheel [driveplate] bolt.....................Nm	55 to 60	55 to 60	55 to 60	55 to 60
Front hubs ....Nm	250	250	250	250
Rear hubs ....Nm	160 4x4: 210	160 4x4: 210	160	160 4x4: 210
Wheel nuts / bolts ....Nm	90	90	90	90
Glow plugs ....Nm	20 to 30	20 to 30	20 to 30	20 to 30
Clutch pressure plate bolts ....Nm	25	25	25	25
Injection pump sprocket......................Nm	50	50	50	50
Injectors...Nm	17	17	17	17
Injection pump mounting bolts ....Nm	–	–	–	–
Injector pipe unions............................Nm	25	25	25	25
Capacities				
Engine oil & filter................................litres	5.3	5.3	5.3	5.3
Gearbox...litres	2.2 4x4: 2.4	2.2 4x4: 2.4	2.2	2.2
Automatic transmission ....litres	–	–	–	–
Final drive ....litres	4x4 rear: 1.3	4x4 rear: 1.3	WT	WT
Cooling system...................................litres	7.1	7.1	7.2	7.1
Fuel tank..litres	66 4x4: 62	66 4x4: 62	66	66

Notes

21 2.1 D & 4x4 (B/L/K/S486) 1986 to 1995
[3]Each bolt in turn
[1]4x4: 3.0 to 5.0
[2]4x4: -30'±10'
21 2.1 D & 4x4 (B/L/K/S486) 1986 to 1995
[3]Each bolt in turn
[1]4x4: 3.0 to 5.0
[2]4x4: -30'±10'
21 2.1 TD (B/L/K488) 1986 to 1995
[1]Each bolt in turn
21 2.1 D (B/L/K/S48V) 1989 to 1995
[1]Each bolt in turn

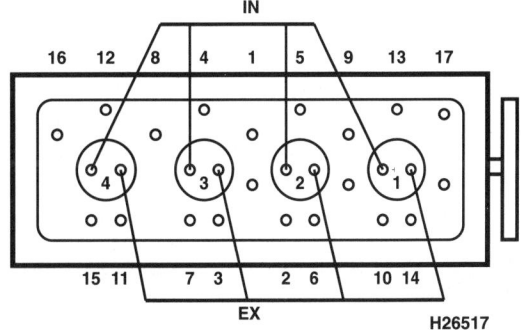

2068 cm³

– Not applicable, or information not
available

RENAULT

	21 2.1 TD (B/L/K488) 1989 to 1995	21 2.1 TD (B/L/K487) 1991 to 1995	21 2.1 TD (B/L/K48A) 1992 to 1995	21 2.1 D (B/L/K480) 1992 to 1995
Engine				
Engine type/code....................	J8S742 SOHC Turbo 65kW	J8S786 SOHC Turbo 65kW	J8S604 SOHC Turbo 65kW	J8S784 SOHC 54kW
Capacity (cm^3) / cylinders........................	2068 / 4	2068 / 4	2068 / 4	2068 / 4
Compression ration / pressure ....bar	21.5 / _	21.0 /	21.0 /	21.5 /
Torque output ....Nm	185	185	185	0
Oil pressure ....idle [running] bar	0.8 [3.5 @ 3000]	0.8 [3.5 @ 3000]	0.8 [3.5 @ 3000]	0.8 [3.0 @ 3000]
Oil temperature ....°C	80	80	80	80
Valve clearances - inlet (mm)	0.20	0.20	0.20	0.20
- exhaust (mm)	0.25	0.25	0.25	0.25
Injection order	1-3-4-2	1-3-4-2	1-3-4-2	1-3-4-2
No. 1 cylinder position	FE	FE	FE	FE
Cooling system				
Thermostat opening temperature ....°C	89	81	81	81
Radiator cap pressure ....bar	1.2	1.2	1.2	1.2
Fuel system				
Idle speed ....rpm	825 ± 25	825 ± 25	825 ± 25	825 ± 25
Maximum (no load) speed ....rpm	4700 to 4800	4900	4900	5200
Smoke test/opacity ....M^{-1} %	0.51	2.3	0.51	0.51
Static timing method................................	Plunger travel	Dial gauge	Dial gauge	Dial gauge
Timing dimension.............................mm	0.7 ± 0.02	0.7 ± 0.02	0.75 ± 0.02	0.75 ± 0.02
Crankshaft position ....mm [°]	[0] TDC	0 TDC	0 TDC	0 TDC
Turbo type / ref / pressure ....bar	Garrett 0.5 to 0.6 bar @ 2500	Garrett	Garrett 0.7 to 0.8 bar @ 2500	_
Injection pump make	Bosch	Bosch	Bosch	Bosch
Injection pump part no............................	VE4/9 R345	VE R449/449-1	VE R423-2/3	VE R452/452-1
Injector Make / type	Bosch	Bosch	Bosch	Bosch
Injector part no............................	DNOSD 264	0432 217 237	0432 217 237	0432 217 259
Injection type................................	Indirect VE	Indirect VE	Indirect VE	Indirect VE
Injection opening pressure, New [used]...bar	130 +8 -5	145 to 158	145 to 158	125 to 138
Glow plugs				
Maker	Bosch/Champion	Bosch/Champion	Bosch/Champion	Bosch/Champion
Type	0250 201 006 / CH137	0250 201 006 / CH137	0250 201 006 / CH137	0250 201 006 / CH137
Nominal rating......................V/A	11 / 15	11 / 15	11 / 15	11 / 15
Brakes				
minimum friction material thickness				
Front............................mm	6.5 with backing	6.5 with backing	6.5 with backing	6.5 with backing
Rear............................mm	2.5 with backing	2.5 with backing	2.5 with backing	2.5 with backing
Tyres - Saloon / Hatch......................Size	185/70x13: 185/65x14	185/65x14	185/65x14	185/65x14
- Estate / Van...........................Size	185/65x14	185/65x14	185/65x14	185/65x14
Pressure - front / rear - Saloon / Hatch ...bar	2.3 / 2.3	2.3 / 2.3	2.3 / 2.3	2.3 / 2.3
- Estate / Van ......bar	2.3 / 2.6	2.3 / 2.6	2.3 / 2.6	2.3 / 2.6
Front suspension / wheel alignment				
Toe-in (+) / Toe-out (–)...................mm [°]	-2.0 ± 1.0	1.0 ± 1.0	1.0 ± 1.0	1.0 ± 1.0
Camber	-25' ± 30'	WSM	WSM	WSM
Castor	2°30' ± 30'	WSM	WSM	WSM
King pin inclination..............................	12°45' ± 30'	_	_	_
Rear suspension / wheel alignment				
Toe-in (+) / Toe-out (–)...................mm [°]	2.0 to 5.0	-2.0 to 5.0	-2.0 to 5.0	-2.0 to 5.0
Camber	-20' ± 10'	-0°20' ± 10'	-0°20' ± 10'	-0°20' ± 10'

RENAULT

Torque wrench settings	21 2.1 TD (B/L/K488) 1989 to 1995	21 2.1 TD (B/L/K487) 1991 to 1995	21 2.1 TD (B/L/K48A) 1992 to 1995	21 2.1 D (B/L/K480) 1992 to 1995
Cylinder head - stage 1 ....Nm	30 N	30 N	30 N	30 N
- stage 2 ....Nm	50	50	50	50
- stage 3 ....Nm	95 to 105	95 to 105	95 to 105	95 to 105
- stage 4 ....Nm	Warm up, allow to cool	Warm up, allow to cool	Warm up, allow to cool	Warm up, allow to cool
- stage 5 ....Nm	Slacken, 95 to 105¹	Slacken, 95 to 105¹	Slacken, 95 to 105¹	Slacken, 95 to 105¹
- stage 6 ....Nm	–	–	–	–
Big-end bearings.............................Nm	65 N	65 N	65 N	65 N
Main bearings..................................Nm	88 to 98	97	97	97
Crankshaft pulley bolt ....Nm	120 to 135	135	135	135
Camshaft pulley bolt ....Nm	50	50	50	50
Flywheel [driveplate] bolt.................Nm	55 to 60	60	60	60
Front hubs ....Nm	250	250	250	250
Rear hubs ....Nm	160	160	160	160
Wheel nuts / bolts ....Nm	90	90 Alloy: 100	90 Alloy: 100	90 Alloy: 100
Glow plugs ....Nm	20 to 30	30	30	30
Clutch pressure plate bolts ....Nm	25	30	30	30
Injection pump sprocket.....................Nm	50	50	50	50
Injectors..Nm	17	70	70	70
Injection pump mounting bolts ....Nm	–	25	25	25
Injector pipe unions............................Nm	25	25	25	25
Capacities				
Engine oil & filter............................litres	5.3	5.2	5.2	5.2
Gearbox...litres	2.2	2.2	2.2	2.2
Automatic transmission ....litres	–	–	–	–
Final drive ....litres	WT	WT	WT	WT
Cooling system................................litres	7.2	7.2	7.2	7.2
Fuel tank..litres	66	66	66	66

Notes

21 2.1 TD (B/L/K488) 1989 to 1995
¹Each bolt in turn
21 2.1 TD (B/L/K487) 1991 to 1995
¹Each bolt in turn
21 2.1 TD (B/L/K48A) 1992 to 1995
¹Each bolt in turn
21 2.1 D (B/L/K480) 1992 to 1995
¹Each bolt in turn

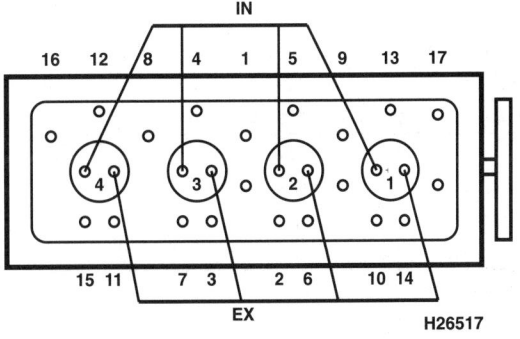

2068 cm³

– Not applicable, or information not
available

RENAULT

	21 2.1 D (B/L/K480) 1992 to 1995	Laguna 1.9 DTi (B/K/S56J) 1997 to 1999	Laguna 2.2 D 1995 to 1998	Laguna 2.2 DT B/K569 1996 to 1999
Engine				
Engine type/code..........	J8S600 SOHC 54kW	F9Q Turbo SOHC 74kW	G8T752/794 SOHC 12V 61kW	G8T760 SOHC Turbo 12V 85kW
Capacity (cm³) / cylinders......................	2068 / 4	1870 / 4	2188 / 4	2188 / 4
Compression ration / pressure ....bar	21.5 /	18.3 /	23.0 /	22.0 /
Torque output ....Nm	0	200	0	250
Oil pressure ....idle [running] bar	0.8 [3.0 @ 3000]	1.2 [3.5 @ 4000]	1.6 [4.0 @ 4000]	1.6 [4.0 @ 4000]
Oil temperature ....°C	80	80	80	80
Valve clearances - inlet (mm)	0.20	0.20	0: Hyd.	0: Hyd.
- exhaust (mm)	0.25	0.40	0: Hyd.	0: Hyd.
Injection order........................	1-3-4-2	1-3-4-2	1-3-4-2	1-3-4-2
No. 1 cylinder position	FE	FE	FE	FE
Cooling system				
Thermostat opening temperature ....°C	81	82	82	82
Radiator cap pressure ....bar	1.2	1.2	1.2	1.2
Fuel system				
Idle speed ....rpm	825 ± 25	825 ± 25	725 ± 25	725 ± 25
Maximum (no load) speed ....rpm	5200	4800 ± 100	5200	5000 ± 100
Smoke test/opacity ....M⁻¹ %	1.11	2.5	2.5	2.5
Static timing method.............................	Dial gauge	Plunger travel	Rotor lift	Rotor lift
Timing dimension.....................mm	0.75 ± 0.02	0.45	0.84	0.74
Crankshaft position ....mm [°]	0 TDC	0 TDC	0 TDC	0 TDC
Turbo type / ref / pressure ....bar	_	Garrett T2	Garrett	Garrett
Injection pump make	Bosch	Bosch	Bosch	Bosch
Injection pump part no........................	VE R309-3/4	VE4/11E2000R6T2	VE4/9 R486	VE4/9F2250R593
Injector Make / type	Bosch	Bosch	Bosch	Bosch
Injector part no........................	0460 494 299	DLSA 145P619	DNOSD 313	DNOSD 313
Injection type........................	Indirect VE	Direct EDC	Indirect VE	_
Injection opening pressure, New [used]...bar	125 to 138	170 to 212	125 to 138	145 to 158
Glow plugs				
Maker................................	Bosch/Champion	Champion	Bosch	Bosch/Champion
Type	0250 201 006 / CH137	CH172	0250 202 025	0250 202 025 / CH169
Nominal rating.....................V/A	11 / 15	11 /	11 / 15	11 / 15
Brakes				
minimum friction material thickness				
Front........................mm	6.5 with backing	6.0 with backing	6.0 with backing	6.0 with backing
Rear........................mm	2.5 with backing	2.5 with backing	6.0 with backing	6.0 with backing
Tyres - Saloon / Hatch.....................Size	185/65x14	195/65x15	195/65x15	195/65x15
- Estate / Van.....................Size	185/65x14	195/65x15	195/65x15	195/65x15
Pressure - front / rear - Saloon / Hatch ...bar	2.3 / 2.3	2.5 / 2.3	2.5 / 2.3	2.5 / 2.3
- Estate / Van ......bar	2.3 / 2.6	2.5 / 2.6	2.5 / 2.6	2.5 / 2.6
Front suspension / wheel alignment				
Toe-in (+) / Toe-out (–)...................mm [°]	1.0 ± 1.0	1.0 ± 1.0	1.0 ± 1.0	1.0 ± 1.0
Camber	WSM	WSM	WSM	WSM
Castor	WSM	WSM	WSM	WSM
King pin inclination............................	_	_	_	_
Rear suspension / wheel alignment				
Toe-in (+) / Toe-out (–)...................mm [°]	-2.0 to 5.0	3.0 ± 1.0	3.0 ± 1.0	3.0 ± 1.0
Camber	-0°20' ± 10'	-1°15' ± 10'	-1°15' ± 10'	-1°15' ± 10'

RENAULT

Torque wrench settings	21 2.1 D (B/L/K480) 1992 to 1995	Laguna 1.9 DTi (B/K/S56J) 1997 to 1999	Laguna 2.2 D 1995 to 1998	Laguna 2.2 DT B/K569 1996 to 1999
Cylinder head - stage 1 ....Nm	30 N	20 oiled N	1	1
- stage 2 ....Nm	50	+ 100° ± 6° wait 3 mins	–	–
- stage 3 ....Nm	95 to 105	Slacken 20 + 100° ± 6° in turn	–	–
- stage 4 ....Nm	Warm up, allow to cool	–	–	–
- stage 5 ....Nm	Slacken, 95 to 105¹	–	–	–
- stage 6 ....Nm	–	–	–	–
Big-end bearings..................................Nm	65 N	10 + 43° N	20 + 70° N	20 + 70° N
Main bearings.......................................Nm	97	20 + 68° ± 6°	20 + 140°	20 + 140°
Crankshaft pulley bolt ....Nm	135	20 + 115°	25 + 65° N	25 + 65° N
Camshaft pulley bolt ....Nm	50	60	20 + 90° N	20 + 90° N
Flywheel [driveplate] bolt...................Nm	60	55	60 N	60 N
Front hubs ....Nm	250	250	250	250
Rear hubs ....Nm	160	190	175	175
Wheel nuts / bolts ....Nm	90 Alloy: 100	100	100	100
Glow plugs ....Nm	30	25 to 30	20	20
Clutch pressure plate bolts ....Nm	30	20	20	20
Injection pump sprocket.....................Nm	50	–	90	90
Injectors..Nm	70	27	70	70
Injection pump mounting bolts ....Nm	25	–	25	22
Injector pipe unions............................Nm	25	–	25	25
Capacities				
Engine oil & filter...............................litres	5.2	4.6	7.2	7.2
Gearbox...litres	2.2	3.1	2.8	2.8
Automatic transmission ....litres	–	–	–	–
Final drive ....litres	WT	WT	WT	WT
Cooling system...................................litres	7.2	7.5	9.0	9.0
Fuel tank...litres	66	66	66	66

Notes

21 2.1 D (B/L/K480) 1992 to 1995
¹Each bolt in turn
Laguna 2.2 D 1995 to 1998
¹Engine number ≤183720: 35 + 70° ≥183721: 35 + 150°
Laguna 2.2 DT B/K569 1996 to 1999
¹Engine number ≤183720: 35 + 70° ≥183721: 35 + 150°

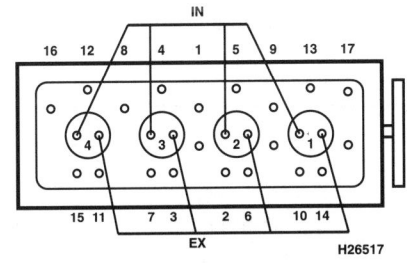

2068 cm³

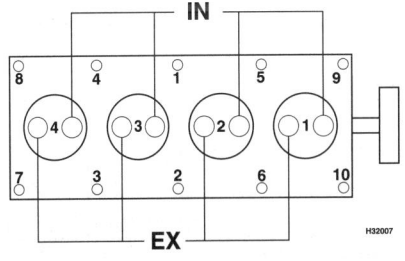

1870 cm³

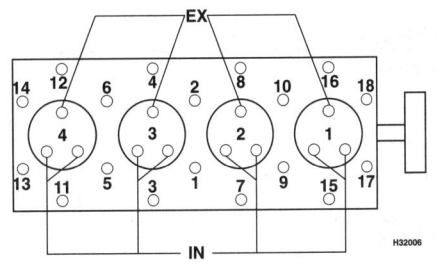

2188 cm³

– Not applicable, or information not available

RENAULT

	25 TD (B290) 1984 to 1993	25 D (B296) 1989 to 1993	Safrane 2.1 TD (B546) 1991 to 1998	Safrane 2.5 TD (B548) 1995 to 1996
Engine				
Engine type/code..................................	J8S708 SOHC Turbo 63kW	J8SC736 SOHC 50kW	J8S760 SOHC Turbo 66kW	S8U762/3 SOHC Turbo 83kW
Capacity (cm^3) / cylinders........................	2068 / 4	2068 / 4	2068 / 4	2499 / 4
Compression ration / pressure ....bar	21.5 / _	21.5 / _	21.5 /	22.5 /
Torque output ....Nm	185	0	0	0
Oil pressure ....idle [running] bar	0.8 [3.5 @ 3000]	0.8 [3.0 @ 3000]	[3.5 @ 3000]	[3.5 @ 4100]
Oil temperature ....°C	80	80	80	80
Valve clearances - inlet (mm)	0.20	0.20	0.20	0.50
- exhaust (mm)	0.25	0.25	0.25	0.50
Injection order	1-3-4-2	1-3-4-2	1-3-4-2	1-3-4-2
No. 1 cylinder position	FE	FE	FE	FE
Cooling system				
Thermostat opening temperature ....°C	88	88	–	–
Radiator cap pressure ....bar	0.8 to 1.2	0.8 to 1.2	–	–
Fuel system				
Idle speed ....rpm	775 ± 50	800 ± 50	850 ± 25	850 ± 50
Maximum (no load) speed ....rpm	4700 to 4800	5100 to 5300	4700 to 4900	4950
Smoke test/opacity ....M^{-1} %	1.6	0.77	2.5	1.36
Static timing method.............................	Plunger travel	Plunger travel	Dial gauge	Dial gauge
Timing dimension................................mm	0.7 ± 0.02	0.75 ± 0.02	0.75 ± 0.02	0.83 ± 0.04
Crankshaft position ....mm [°]	[0] TDC	[0] TDC	0 TDC	0 TDC
Turbo type / ref / pressure ....bar	–	–	–	–
Injection pump make	Bosch	Bosch	Bosch	Bosch
Injection pump part no..........................	VE 4/9F R153/345	VE 4/9F R309	VE 4/9F R423	VE R442/4411
Injector Make / type	Bosch	Bosch	Bosch	Bosch
Injector part no.....................................	DNOSD 264	DNOSD 252+	DNOSD306	DNOSD307
Injection type..	Indirect VE	Indirect VE	Indirect VE	Indirect VE
Injection opening pressure, New [used]...bar	130 +8 -5	130 +8 -5	150 to 158	140 to 153
Glow plugs				
Maker	Bosch/Champion	Bosch/Champion	Bosch/Champion	Bosch
Type	0250 201 006 / CH137	0250 201 006 / CH137	0250 201 006 / CH137	0250 201 039
Nominal rating.................................V/A	11 / 12	11 / 12	12 / 15	12 / 15
Brakes				
minimum friction material thickness				
Front...mm	6.0 with backing	6.0 with backing	6.0 with backing	6.0 with backing
Rear..mm	2.5 with backing[1]	2.5 with backing[1]	2.5 with backing	2.5 with backing
Tyres - Saloon / Hatch.....................Size	195/60x15	165x14: 185/70x14	195/60x15	195/60x15
- Estate / Van..........................Size	–	–	–	–
Pressure - front / rear - Saloon / Hatch ...bar	2.0 / 2.2	2.0 / 2.2	2.5 / 2.3	2.5 / 2.3
- Estate / Van ......bar	–	–	–	–
Front suspension / wheel alignment				
Toe-in (+) / Toe-out (–)...................mm [°]	-3.0 ± 1.0	-3.0 ± 1.0	1.0 ± 1.0	1.0 ± 1.0
Camber	0° ± 30'	0° ± 30'	WSM	WSM
Castor	2° PAS: 4°	2° PAS: 4°	WSM	WSM
King pin inclination................................	12°30' ± 30'	12°30' ± 30'	–	–
Rear suspension / wheel alignment				
Toe-in (+) / Toe-out (–)...................mm [°]	0 ± 1.0[2]	0 ± 1.0[2]	2.0 ± 1.0	2.0 ± 1.0
Camber	-1°15' ± 30'	-1°15' ± 30'	-0°55'	-0°55'

RENAULT

	25 TD (B290) 1984 to 1993	25 D (B296) 1989 to 1993	Safrane 2.1 TD (B546) 1991 to 1998	Safrane 2.5 TD (B548) 1995 to 1996
Torque wrench settings				
Cylinder head - stage 1 ....Nm	30 N	30 N	WSM	60 N
- stage 2 ....Nm	50	50	–	60
- stage 3 ....Nm	95 to 105	95 to 105	–	+ 180°
- stage 4 ....Nm	Warm up, allow to cool	Warm up, allow to cool	–	–
- stage 5 ....Nm	Slacken, 95 to 105³	Slaken, 95 to 105³	–	–
- stage 6 ....Nm	–	–	–	–
Big-end bearings...............................Nm	65 N	65 N	65 N	50 + 65° N
Main bearings....................................Nm	88 to 98	88 to 98	98	160
Crankshaft pulley bolt ....Nm	120 to 135	120 to 135	135	200
Camshaft pulley bolt ....Nm	50	50	50	25
Flywheel [driveplate] bolt....................Nm	55 to 60	55 to 60	60	30 + 90°
Front hubs ....Nm	250	250	250	250
Rear hubs ....Nm	160	160	165	165
Wheel nuts / bolts ....Nm	100	90	90 Alloy: 100	90 Alloy: 100
Glow plugs ....Nm	20 to 30	20 to 30	20	–
Clutch pressure plate bolts ....Nm	25	25	20	–
Injection pump sprocket.....................Nm	50	50	–	–
Injectors...Nm	17	70	–	–
Injection pump mounting bolts ....Nm	–	–	–	–
Injector pipe unions............................Nm	25	25	25	25
Capacities				
Engine oil & filter...............................litres	6.0	6.0	6.7	5.5
Gearbox...litres	2.0 86 ▶: 2.2	2.0	2.6	2.6
Automatic transmission ....litres	–	–	–	–
Final drive ....litres	WT	WT	1.3	1.3
Cooling system...................................litres	7.5	7.5	7.2	7.2
Fuel tank..litres	67	67	80	80

Notes

25 TD (B290) 1984 to 1993
³Each bolt in turn
¹With ABS: 6.0 with backing
²¹1986 ▶: 1.0±1.0
25 D (B296) 1989 to 1993
³Each bolt in turn
¹With ABS: 6.0 with backing
²¹1986 ▶: 1.0±1.0

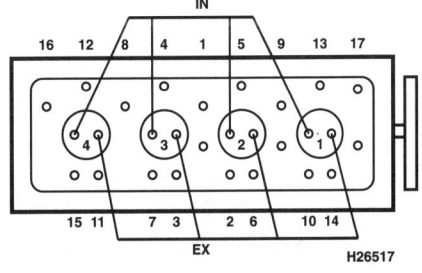

2068 cm³

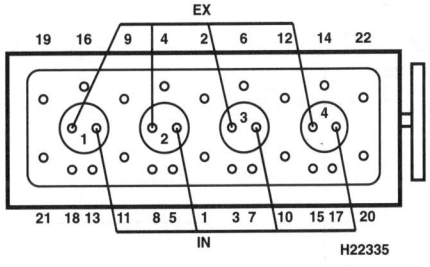

2499 cm³

– Not applicable, or information not available

RENAULT

	Espace 2.1 TD (J115/S115) 1985 to 1991	Espace 2.1 TD (J637) 1991 to 1993	Espace 2.1 TD (J63D) 1991 to 1994	Espace 2.1 TD (J63G) 1991 to 1997
Engine				
Engine type/code..........................	J8S240 SOHC Turbo 63kW	J8S772 SOHC Turbo 65kW	J8S610 SOHC Turbo 65kW	J8S776 SOHC Turbo 65kW
Capacity (cm^3) / cylinders........................	2068 / 4	2068 / 4	2068 / 4	2068 / 4
Compression ration / pressure ....bar	21. / _	21.5 / _	21.5 / _	21.5 / _
Torque output ....Nm	0	181	181	181
Oil pressure ....idle [running] bar	0.8 [3.0 @ 3000]	0.8 [3.0 @ 3000]	0.8 [3.0 @ 3000]	0.8 [3.0 @ 3000]
Oil temperature ....°C	80	80	80	80
Valve clearances - inlet (mm)	0.20	0.20	0.20	0.20
- exhaust (mm)	0.25	0.25	0.25	0.25
Injection order	1-3-4-2	1-3-4-2	1-3-4-2	1-3-4-2
No. 1 cylinder position	FE	FE	FE	FE
Cooling system				
Thermostat opening temperature ....°C	81	81	81	81
Radiator cap pressure ....bar	1.2	1.2	1.2	1.2
Fuel system				
Idle speed ....rpm	700 to 750	700 to 750	700 to 750	700 to 750
Maximum (no load) speed ....rpm	4700 to 4800	5000	5000	4800
Smoke test/opacity ....M^{-1} %	1.6	2.46	2.46	2.46
Static timing method............................	Plunger travel	Dial gauge	Dial gauge	Dial gauge
Timing dimension............................mm	0.7 ± 0.02	0.7 ± 0.02	0.75 ± 0.02	0.7 ± 0.02
Crankshaft position ....mm [°]	[0] TDC	[0] TDC	[0] TDC	[0] TDC
Turbo type / ref / pressure ....bar	0.6 bar @ 2500 rpm	0.6 bar @ 2500 rpm	0.6 bar @ 2500 rpm	0.6 bar @ 2500 rpm
Injection pump make	Bosch	Bosch	Bosch	Bosch
Injection pump part no.............................	VE 4/9F R69	VE 4/9F R345	VE 4/9F R423-2	VE 4/9F R153
Injector Make / type	Bosch	Bosch	Bosch	Bosch
Injector part no.............................	DNOSD 193	0431 211 995	DNOSD306	DNOSD264
Injection type.............................	Indirect VE	Indirect VE	Indirect VE	Indirect VE
Injection opening pressure, New [used]...bar	130 to 138	125 to 138	145 to 158	130 to 138
Glow plugs				
Maker	Bosch/Champion	Bosch/Champion	Bosch/Champion	Bosch/Champion
Type	0250 201 006 / CH137	0250 201 006 / CH137	0250 201 006 / CH137	0250 201 006 / CH137
Nominal rating ....V/A	11 / 12	11 / 12	11 / 12	11 / 12
Brakes				
minimum friction material thickness				
Front........................mm	6.0 with backing	6.0 with backing	6.0 with backing	6.0 with backing
Rear........................mm	2.5 with backing	2.5 with backing	2.5 with backing	2.5 with backing
Tyres - Saloon / Hatch......................Size	_	_	_	_
- Estate / Van......................Size	185/70x13: 185/65x14	185/70x13: 185/65x14	185/70x13: 185/65x14	185/70x13: 185/65x14
Pressure - front / rear - Saloon / Hatch ...bar	_	_	_	_
- Estate / Van ......bar	2.4 / 2.1	2.4 / 2.1	2.4 / 2.1	2.4 / 2.1
Front suspension / wheel alignment				
Toe-in (+) / Toe-out (–)....................mm [°]	-1.5 ± 0.5¹	-1.9 to 2.6	-1.9 to 2.6	-1.9 to 2.6
Camber	20' ± 30'	-0°8' ± 30'	-0°8' ± 30'	-0°8' ± 30'
Castor	3° ± 30' [86 ▶: 3°25'±20']	4°24' ± 20'	4°24' ± 20'	4°24' ± 20'
King pin inclination.............................	13° ± 30'	_	_	_
Rear suspension / wheel alignment				
Toe-in (+) / Toe-out (–)....................mm [°]	1.0 to 4.0	1.0 to 4.0	1.0 to 4.0	1.0 to 4.0
Camber	0° ± 30'	0° ± 30'	0° ± 30'	0° ± 30'

RENAULT

Torque wrench settings	Espace 2.1 TD (J115/S115) 1985 to 1991	Espace 2.1 TD (J637) 1991 to 1993	Espace 2.1 TD (J63D) 1991 to 1994	Espace 2.1 TD (J63G) 1991 to 1997
Cylinder head - stage 1 ....Nm	30 N	30 N	30 N	30 N
- stage 2 ....Nm	50	50	50	50
- stage 3 ....Nm	Wait 3 mins, slacken	Wait 3 mins, slacken	Wait 3 mins, slacken	Wait 3 mins, slacken
- stage 4 ....Nm	20^2	20^2	20^2	20^2
- stage 5 ....Nm	–	–	–	–
- stage 6 ....Nm	–	–	–	–
Big-end bearings.................................Nm	65 N	65 N	65 N	65 N
Main bearings.....................................Nm	160	160	160	160
Crankshaft pulley bolt ....Nm	120 to 135	120 to 135	120 to 135	120 to 135
Camshaft pulley bolt ....Nm	50	50	50	50
Flywheel [driveplate] bolt....................Nm	55 to 60	55 to 60	55 to 60	55 to 60
Front hubs ....Nm	250	250	250	250
Rear hubs ....Nm	160	160	160	160
Wheel nuts / bolts ....Nm	90	90	90	90
Glow plugs ....Nm	20 to 30	20 to 30	20 to 30	20 to 30
Clutch pressure plate bolts ....Nm	25	25	25	25
Injection pump sprocket......................Nm	50	50	50	50
Injectors...Nm	17	17	17	17
Injection pump mounting bolts ....Nm	–	–	–	–
Injector pipe unions............................Nm	25	25	25	25
Capacities				
Engine oil & filter................................litres	6.5	7.0	7.0	7.0
Gearbox...litres	2.0	2.2	2.2	2.2
Automatic transmission ....litres	–	–	–	–
Final drive ....litres	WT	WT	WT	WT
Cooling system...................................litres	7.3	10.0	10.0	10.0
Fuel tank..litres	65	65	65	65

Notes

Espace 2.1 TD (J115/S115) 1985 to 1991
^{2}Bolts 1, 8 & 9: + 120°, + 120° Bolts 2, 3, 6, 7, 10, 11,
14 & 15: + 60°, + 60°Bolts 4, 5, 12 & 13: + 70°, + 70°
^{1}1986 ▶: -4.0±2.0
Espace 2.1 TD (J637) 1991 to 1993
^{2}Bolts 1, 8 & 9: + 120°, + 120° Bolts 2, 3, 6, 7, 10, 11,
14 & 15: + 60°, + 60°Bolts 4, 5, 12 & 13: + 70°, + 70°
Espace 2.1 TD (J63D) 1991 to 1994
^{2}Bolts 1, 8 & 9: + 120°, + 120° Bolts 2, 3, 6, 7, 10, 11,
14 & 15: + 60°, + 60°Bolts 4, 5, 12 & 13: + 70°, + 70°
Espace 2.1 TD (J63G) 1991 to 1997
^{2}Bolts 1, 8 & 9: + 120°, + 120° Bolts 2, 3, 6, 7, 10, 11,
14 & 15: + 60°, + 60°Bolts 4, 5, 12 & 13: + 70°, + 70°

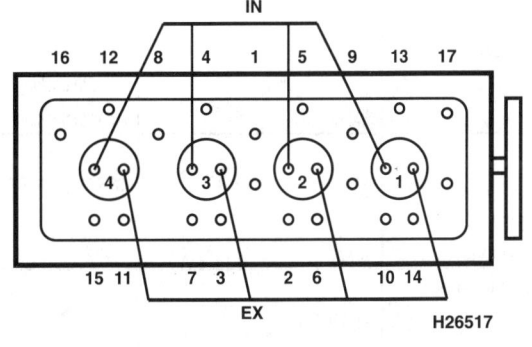

2068 cm³

– Not applicable, or information not
available

RENAULT

	Espace 2.1 TD (J633) 1992 to 1997	Espace 2.1 TD (J635E) 1995 to 1997	Espace 2.2 TD (JEOE)[1] 1997 to 2000	Espace 2.2 TD (JEOH)[1] 1997 to 2000
Engine				
Engine type/code................................	J8S778 SOHC Turbo 65kW	J8S612 SOHC Turbo 65kW	G8T716 SOHC 12V Turbo 85kW	G8T714 SOHC 12V Turbo 85kW
Capacity (cm³) / cylinders........................	2068 / 4	2068 / 4	2188 / 4	2188 / 4
Compression ration / pressure ....bar	21.5 / _	21.5 / _	23.0 /	23.0 /
Torque output ....Nm	181	181	234	234
Oil pressure ....idle [running] bar	0.8 [3.0 @ 3000]	0.8 [3.0 @ 3000]	1.6 [4.0 @ 4000]	1.6 [4.0 @ 4000]
Oil temperature ....°C	80	80	80	80
Valve clearances - inlet (mm)	0.20	0.20	0: Hyd.	0: Hyd.
- exhaust (mm)	0.25	0.25	0: Hyd.	0: Hyd.
Injection order.....................................	1-3-4-2	1-3-4-2	1-3-4-2	1-3-4-2
No. 1 cylinder position	FE	FE	FE	FE
Cooling system				
Thermostat opening temperature ....°C	81	81	89	89
Radiator cap pressure ....bar	1.2	1.2	1.6	1.6
Fuel system				
Idle speed ....rpm	800 ± 50	800 ± 50	725 ± 25	725 ± 25
Maximum (no load) speed ....rpm	5000	5000	5000	5000
Smoke test/opacity ....M⁻¹ %	2.46	1.14	2.5	2.5
Static timing method..............................	Dial gauge	Dial gauge	Rotor lift	Rotor lift
Timing dimension.................................mm	0.75 ± 0.02	0.75 ± 0.02	0.74	0.6 ± 0.1
Crankshaft position ....mm [°]	[0] TDC	[0] TDC	TDC	TDC
Turbo type / ref / pressure ....bar	0.6 bar @ 2500 rpm	0.6 bar @ 2500 rpm	Garrett 0.9 bar @ 4300rpm	Garrett 0.9 bar @ 4300rpm
Injection pump make	Bosch	Bosch	Bosch	Bosch
Injection pump part no............................	VER 484	VER 423	VE R593/4	VE R708
Injector Make / type	Bosch	Bosch	Bosch	Bosch
Injector part no....................................	0432 217 237	DNOSD306	DNOSD313	DNOSD313
Injection type.......................................	Indirect VE	Indirect VE	Indirect VE	Indirect VE
Injection opening pressure, New [used]...bar	145 to 158	145 to 158	145 to 158	145 to 158
Glow plugs				
Maker	Bosch/Champion	Bosch/Champion	Champion	Champion
Type	0250 201 006 / CH137	0250 201 006 / CH137	CH169	CH169
Nominal rating.................................V/A	11 / 12	11 / 12	11 /	11 /
Brakes				
minimum friction material thickness				
Front.......................................mm	6.0 with backing	6.0 with backing	6.0 with backing	6.0 with backing
Rear.......................................mm	2.5 with backing	2.5 with backing	2.5 with backing	2.5 with backing
Tyres - Saloon / Hatch......................Size	_	_	_	_
- Estate / Van....................Size	185/70x13: 185/65x14	185/70x13: 185/65x14	215/65x15	215/65x15
Pressure - front / rear - Saloon / Hatch ...bar	_	_	_	_
- Estate / Van ......bar	2.4 / 2.1	2.4 / 2.1	2.6 / 2.4	2.6 / 2.4
Front suspension / wheel alignment				
Toe-in (+) / Toe-out (–).....................mm [°]	-1.9 to 2.6	-1.9 to 2.6	[0° ± 10']	[0° ± 10']
Camber	-0°8' ± 30'	-0°8' ± 30'	WSM	WSM
Castor	4°24' ± 20'	4°24' ± 20'	WSM	WSM
King pin inclination...............................	_	_		
Rear suspension / wheel alignment				
Toe-in (+) / Toe-out (–)....................mm [°]	1.0 to 4.0	1.0 to 4.0	[-30' ± 10']	[-30' ± 10']
Camber	0° ± 30'	0° ± 30'	-1°15' ± 10'	-1°15' ± 10'

RENAULT

Torque wrench settings	Espace 2.1 TD (J633) 1992 to 1997	Espace 2.1 TD (J635E) 1995 to 1997	Espace 2.2 TD (JEOE)[1] 1997 to 2000	Espace 2.2 TD (JEOH)[1] 1997 to 2000
Cylinder head - stage 1 ....Nm	30 N	30 N	WSM	WSM
- stage 2 ....Nm	50	50	–	–
- stage 3 ....Nm	Wait 3 mins, slacken	Wait 3 mins, slacken	–	–
- stage 4 ....Nm	20[2]	20[2]	–	–
- stage 5 ....Nm	–	–	–	–
- stage 6 ....Nm	–	–	–	–
Big-end bearings.............................Nm	65 N	65 N	20 + 70° N	20 + 70° N
Main bearings.................................Nm	160	160	20 + 140°	20 + 140°
Crankshaft pulley bolt ....Nm	120 to 135	120 to 135	25 + 65° N	25 + 65° N
Camshaft pulley bolt ....Nm	50	50	20 + 90° N	20 + 90° N
Flywheel [driveplate] bolt.....................Nm	55 to 60	55 to 60	60	60
Front hubs ....Nm	250	250	280	280
Rear hubs ....Nm	160	160	170	170
Wheel nuts / bolts ....Nm	90	90	100	100
Glow plugs ....Nm	20 to 30	20 to 30	25 to 30	25 to 30
Clutch pressure plate bolts ....Nm	25	25	22.5	22.5
Injection pump sprocket......................Nm	50	50	90	90
Injectors..Nm	17	17	70	70
Injection pump mounting bolts ....Nm	–	–	22	22
Injector pipe unions.........................Nm	25	25	25	25
Capacities				
Engine oil & filter.............................litres	7.0	7.0	7.2	7.2
Gearbox..litres	2.2	2.2	2.8	2.8
Automatic transmission ....litres	–	–	–	–
Final drive ....litres	WT	WT	WT	WT
Cooling system...............................litres	10.0	10.0	9.0	9.0
Fuel tank.......................................litres	65	65	80	80

Notes

Espace 2.1 TD (J633) 1992 to 1997
[2]Bolts 1, 8 & 9: + 120°, + 120° Bolts 2, 3, 6, 7, 10, 11, 14 & 15: + 60°, + 60°Bolts 4, 5, 12 & 13: + 70°, + 70°
Espace 2.1 TD (J635E) 1995 to 1997
[2]Bolts 1, 8 & 9: + 120°, + 120° Bolts 2, 3, 6, 7, 10, 11, 14 & 15: + 60°, + 60°Bolts 4, 5, 12 & 13: + 70°, + 70°
Espace 2.2 TD (JEOE)[1] 1997 to 2000
[1]Grand Espace (JEO E/HL5)
Espace 2.2 TD (JEOH)[1] 1997 to 2000
[1]Grand Espace (JEO E/HL5)

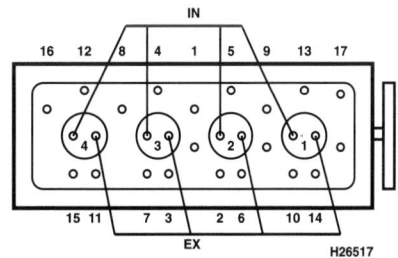

2068 cm³

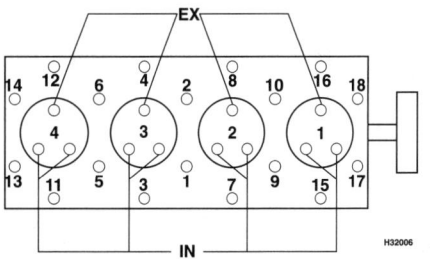

2188 cm³

– Not applicable, or information not available

RENAULT

	Kangoo 1.9D (F/KCOD/E) 1998 to 2000	Trafic 1.9 D (TxxJ) 1997 to 2000	Trafic 2.1 (T**3/V**3) 1987 to 1995	Trafic 2.1 1995 to 1999
Engine				
Engine type/code.....................	F8Q662/630 SOHC 41kW	F8Q606 SOHC 45kW	852D750 SOHC 43kW	J8S758/620 SOHC 47kW
Capacity (cm³) / cylinders.......................	1870 / 4	1870 / 4	2068 / 4	2068 / 4
Compression ration / pressure ....bar	21.5 / ≥20.0	21.5 / _	21.5 / 20.0	21.5 / 20.0
Torque output ....Nm	120	120	120	120
Oil pressure ....idle [running] bar	1.2 [3.5 @ 3000]	1.2 [3.5 @ 3000]	0.8 [3.5 @ 3000]	0.8 [3.5 @ 3000]
Oil temperature ....°C	80	80	80	80
Valve clearances - inlet (mm)	0.20	0.10	0.20	0.20
- exhaust (mm)	0.40	0.25	0.25	0.25
Injection order	1-3-4-2	1-3-4-2	1-3-4-2	1-3-4-2
No. 1 cylinder position	FE	FE	FE	FE
Cooling system				
Thermostat opening temperature ....°C	82	89 to 101	83	83
Radiator cap pressure ....bar	1.2	1.2	0.8 to 1.2	0.8 to 1.2
Fuel system				
Idle speed ....rpm	825 ± 25	825 ± 50	750 ± 50	750 ± 50
Maximum (no load) speed ....rpm	4600 ± 100	4500 ± 100	4250 to 4400	4800
Smoke test/opacity ....M⁻¹ %	2.5	2.5	1.11	2.5
Static timing method...............................	–	Dial gauge	Rotor lift	Dial gauge
Timing dimension.............................mm	–	Dimension on pump	Dimension on pump	0.75 ± 0.02
Crankshaft position ....mm [°]	–	TDC	[0] TDC	[0] TDC
Turbo type / ref / pressure ....bar	–	–	–	–
Injection pump make	Lucas	Lucas	Roto Diesel	Bosch
Injection pump part no............................	8448B171A/231A	8448B120A	DPC R8443140/1A	VER 452-2
Injector Make / type	–	CAV	Roto Diesel	Bosch
Injector part no.......................................	RDNOSDC6902	RDN OSDC 6902	RDN OSDC 6751	DNOSD302
Injection type...	Indirect	8448B/120A or 110A	Indirect DPC	Indirect VER
Injection opening pressure, New [used]...bar	130 ± 5	130 ± 5	115 ± 5	125 to 138
Glow plugs				
Maker	Champion	Beru	Bosch/Champion	Champion
Type	CH155	–	0250 201 006 / CH88	CH137
Nominal rating.................................V/A	–	–	11 / 12	11 /
Brakes				
minimum friction material thickness				
Front............................mm	6.5 with backing	9.0 with backing	9.0 with backing	9.0 with backing
Rear.............................mm	2.0	5.0 with backing	5.0 with backing	5.0 with backing
Tyres - Saloon / Hatch....................Size	–	–	–	–
- Estate / Van........................Size	165/70x13: 165/70x14	175x14	165x14: 175x14: 195x14	165x14: 175x14: 195x14
Pressure - front / rear - Saloon / Hatch...bar	–	–	–	–
- Estate / Van ......bar	2.8 / 3.6: 2.4 / 3.0	3.3 / 3.8	Refer to owner's handbook	Refer to owner's handbook
Front suspension / wheel alignment				
Toe-in (+) / Toe-out (–)....................mm [°]	1.0 ± 1.0	-1.0 ± 1.0	-1.0 ± 1.0	-1.0 ± 1.0
Camber	-15' ± 30'	0°30' ± 30'	1° ± 30'	1° ± 30'
Castor	3°25' ± 30'	3° ± 1°	1°30' ± 30'	1°30' ± 30'
King pin inclination.................................	10°25' ± 30'	9° ± 30'	9° ± 30'	9° ± 30'
Rear suspension / wheel alignment				
Toe-in (+) / Toe-out (–)....................mm [°]	-1.5 ± 1.0	0 to 3.0	0 to 3.0	0 to 3.0
Camber	-50' ± 15'	0° to 30'	0° to 30'	0° to 30'

RENAULT

Torque wrench settings	Kangoo 1.9D (F/KCOD/E) 1998 to 2000	Trafic 1.9 D (TxxJ) 1997 to 2000	Trafic 2.1 (T**3/V**3) 1987 to 1995	Trafic 2.1 1995 to 1999
Cylinder head - stage 1 ....Nm	30	30	30	30 N
- stage 2 ....Nm	+ 80° ± 4°	+ 80° ± 4°	50	50
- stage 3 ....Nm	Slacken	Slacken	95 to 105	Slacken, 20
- stage 4 ....Nm	25	25	95 to 105	3
- stage 5 ....Nm	+ 213° ± 7°	+ 213° ± 7°	–	4
- stage 6 ....Nm	–	–	–	–
Big-end bearings.................................Nm	47 N	60 to 65 N	65 N	65 N
Main bearings.....................................Nm	62	88 to 98	88 to 98	88 to 98
Crankshaft pulley bolt ....Nm	120	20 + 115° ± 15°	120 to 135	120 to 135
Camshaft pulley bolt ....Nm	50		50	50
Flywheel [driveplate] bolt.....................Nm	52	20 + 60°	55 to 60	55 to 60
Front hubs ....Nm	280	190 to 210	190 to 210	190 to 210
Rear hubs ....Nm	175	WSM	WSM	WSM
Wheel nuts / bolts ....Nm	90	120	120	120
Glow plugs ....Nm	20	20	20 to 30	20 to 30
Clutch pressure plate bolts ....Nm	40	20	25	25
Injection pump sprocket......................Nm	65	20, then 45	50	50
Injectors...Nm	70	70	17	17
Injection pump mounting bolts ....Nm	–	22	–	–
Injector pipe unions.............................Nm	–	–	25	25
Capacities				
Engine oil & filter................................litres	5.2	5.5	6.0	6.0
Gearbox...litres	3.4	1.8	2.5 NE3: 1.8¹	2.5 NE3: 1.8¹
Automatic transmission ....litres		–	–	–
Final drive ....litres	WT	1.3	0.8²	0.8²
Cooling system...................................litres	7.4	6.8	8.0	8.0
Fuel tank..litres	50	60	60 4x4: 70	60 4x4: 70

Notes

Trafic 2.1 (T3/V**3) 1987 to 1995**
¹4x4: 2.0
²NE3: 1.45. With cooler and 4x4: 1.3
Trafic 2.1 1995 to 1999
³Bolt 1: 105° Bolts:2, 3, 6, 7, 10, 11, 14 & 15: + 60°
Bolts 4, 5, 12 & 13: + 70° Bolts 16 & 17: + 80°
⁴Bolt 1: 105° Bolts:2, 3, 6, 7, 10, 11, 14 & 15: + 60°
Bolts 4, 5, 12 & 13: + 70° Bolts 16 & 17: + 80°
¹4x4: 2.0
²NE3: 1.45. With cooler and 4x4: 1.3

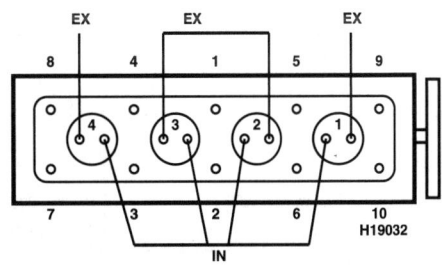

1870 cm³

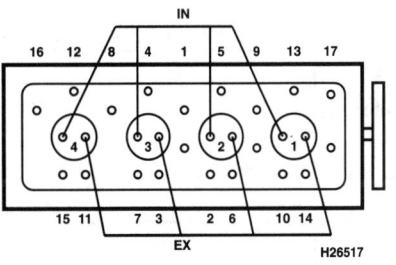

2068 cm³

– Not applicable, or information not available

RENAULT

	Trafic 2.5 (T**4/V**4) 1989 to 1995	Trafic 2.5 (T**B/V**B) 1989 to 1995	Trafic 2.5 (T**B/V**B) 1995 to 1999	Trafic 2.5 D (TxxG) 1997 to 2000
Engine				
Engine type/code......................	S8U750 8140 SOHC 50kW	S8U752 8140 SOHC 50kW	S8U758 SOHC 55kW	S8U782 SOHC 56kW
Capacity (cm³) / cylinders........................	2499 / 4	2499 / 4	2499 / 4	2499 / 4
Compression ration / pressure ....bar	22.0 / 20.0	22.0 / 20.0	23.0 /	23.0 /
Torque output ....Nm	162	162	158	158
Oil pressure ....idle [running] bar	1.2 [3.5 @ 4000]	1.2 [3.5 @ 4000]	1.2 [3.5 @ 4000]	1.2 [5.6 @ 4000]
Oil temperature ....°C	80	80	80	80
Valve clearances - inlet (mm)	0.40	0.40	0.40	0.40
- exhaust (mm)	0.40	0.40	0.40	0.40
Injection order	1-3-4-2	1-3-4-2	1-3-4-2	1-3-4-2
No. 1 cylinder position	TBE	TBE	TBE	FE
Cooling system				
Thermostat opening temperature ....°C	83	83	89	89 to 101
Radiator cap pressure ....bar	0.8 to 1.2	0.8 to 1.2	1.2	1.2
Fuel system				
Idle speed ....rpm	700 ± 50	700 ± 50	80050	800 ± 50
Maximum (no load) speed ....rpm	4600 to 4700	4600 to 4700	4750	4750 ± 100
Smoke test/opacity ....M⁻¹ %	2.05	2.05	2.5	2.5
Static timing method............................	Plunger travel	Plunger travel	Plunger travel	Dial gauge
Timing dimension............................mm	1.1 ± 0.02	1.0 ± 0.02	1.0 ± 0.02	Dimension on pump
Crankshaft position ....mm [°]	[0] TDC	[0] TDC	[0] TDC	TDC
Turbo type / ref / pressure ....bar	–	–	–	–
Injection pump make	Bosch	Bosch	Bosch	Bosch
Injection pump part no.	VE 4/9 R22-7	VE 4/9 R295-1	VE R589	VE4/10F2100R589
Injector Make / type	Bosch	Bosch	Bosch	Bosch
Injector part no........................	DNOSD 193	DNOSD 259	DNOSD 301	DN OSD 301
Injection type............................	Indirect VE	Indirect VE	Indirect VE	Indirect
Injection opening pressure, New [used]...bar	125 ± 5	125 ± 5	115 to 128	125 ± 5
Glow plugs				
Maker	Bosch/Champion	Bosch/Champion	Champion	Champion
Type	0250 201 006 / CH88	0250 201 006 / CH87	CH137	–
Nominal rating........................V/A	11 / 12	11 / 12	11 /	–
Brakes				
minimum friction material thickness				
Front......................mm	9.0 with backing	9.0 with backing	9.0 with backing	9.0 with backing
Rear......................mm	5.0 with backing	5.0 with backing	5.0 with backing	5.0 with backing
Tyres - Saloon / Hatch......................Size	–	–	–	–
- Estate / Van......................Size	165x14: 175x14: 185x14: 195x14	165x14: 175x14: 185x14: 195x14	165x14: 175x14: 185x14: 195x14	185/70x14
Pressure - front / rear - Saloon / Hatch ...bar	–	–	–	–
- Estate / Van ......bar	Refer to owner's handbook	Refer to owner's handbook	Refer to owner's handbook	4.1 / 4.5
Front suspension / wheel alignment				
Toe-in (+) / Toe-out (–)....................mm [°]	-1.0 ± 1.0	-1.0 ± 1.0	-1.0 ± 1.0	-1.0 ± 1.0
Camber	1° ± 30'	1° ± 30'	30' ± 1°	0°30' ± 30'
Castor	1°30' ± 30'	1°30' ± 30'	3° ± 1°	3° ± 1°
King pin inclination............................	9° ± 30'	9° ± 30'	9° ± 30'	9° ± 30'
Rear suspension / wheel alignment				
Toe-in (+) / Toe-out (–)....................mm [°]	0 to 3.0	0 to 3.0	-1.5 ± 1.5	0 to 3.0
Camber	0° to 30'	0° to 30'	15' ± 15'	0° to 30'

RENAULT

	Trafic 2.5 (T**4/V**4) 1989 to 1995	Trafic 2.5 (T**B/V**B) 1989 to 1995	Trafic 2.5 (T**B/V**B) 1995 to 1999	Trafic 2.5 D (TxxG) 1997 to 2000
Torque wrench settings				
Cylinder head - stage 1 ....Nm	40 N	40 N	40 N	40 N
- stage 2 ....Nm	40	40	40	40
- stage 3 ....Nm	+ 180°	+ 180°	+ 180°	+ 180°
- stage 4 ....Nm	–	–	–	–
- stage 5 ....Nm	–	–	–	–
- stage 6 ....Nm	–	–	–	–
Big-end bearings................................Nm	110	110 N	110 N	50 + 60° N
Main bearings....................................Nm	80, 160	80, 160	80, 160	80, 160
Crankshaft pulley bolt ....Nm	200	200	200	200
Camshaft pulley bolt ....Nm	25	25	25	–
Flywheel [driveplate] bolt....................Nm	120	50	30 + 90°	120
Front hubs ....Nm	190 to 210	190 to 210	250	190 to 210
Rear hubs ....Nm	WSM	WSM	WSM	WSM
Wheel nuts / bolts ....Nm	120	120	120	120
Glow plugs ....Nm	15	15	15	20
Clutch pressure plate bolts ....Nm	25	25	25	20
Injection pump sprocket......................Nm	50	50	50	20, 45
Injectors...Nm	50	50	50	70
Injection pump mounting bolts ....Nm	50	50	50	22
Injector pipe unions............................Nm	25	25	25	–
Capacities				
Engine oil & filter................................litres	5.5	5.5	5.5	5.5
Gearbox...litres	2.5 NE3: 1.8[1]	2.5 NE3: 1.8[1]	1.8	1.8
Automatic transmission ....litres	–	–	–	–
Final drive ....litres	0.8[2]	0.8[2]	1.45	1.3
Cooling system..................................litres	8.4	8.4	8.4	8.4
Fuel tank..litres	60	60	60	60

Notes

Trafic 2.5 (T4/V**4) 1989 to 1995**
[1] 4x4: 2.0
[2] NE3: 1.45. With cooler and 4x4: 1.3
Trafic 2.5 (TB/V**B) 1989 to 1995**
[1] 4x4: 2.0
[2] NE3: 1.45. With cooler and 4x4: 1.3

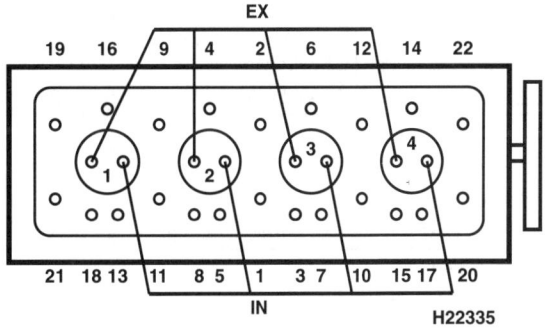

2499 cm³

– Not applicable, or information not available

RENAULT

	Master 2.1 fwd (R**3) 1987 to 1995	Master 2.5 fwd (R**6) 1989 to 1995	Master 2.5 Turbo fwd (R**5) 1989 to 1995	Master 2.5 D 1995 to 1997
Engine				
Engine type/code	J8S330 SOHC 43kW	S8U 8140672510 SOHC 55kW	S9U 8140272500 Turbo 69kW	S8U748 SOHC 55kW
Capacity (cm³) / cylinders	2068 / 4	2499 / 4	2499 / 4	2499 / 4
Compression ration / pressure ....bar	21.5 / _	22.0 / _	17.5 / _	23.0 / _
Torque output ....Nm	0	166	209	158
Oil pressure ....idle [running] bar	0.8 [3.5 @ 3000]	0.8 [4.0 @ 4000]	0.8 [4.0 @ 4000]	0.8 [3.5 @ 4100]
Oil temperature ....°C	80	80	80	80
Valve clearances - inlet (mm)	0.20	0.50	0.50	0.50
- exhaust (mm)	0.25	0.50	0.50	0.50
Injection order	1-3-4-2	1-3-4-2	1-3-4-2	1-3-4-2
No. 1 cylinder position	FE	TBE	TBE	FE
Cooling system				
Thermostat opening temperature ....°C	_	79	79	89
Radiator cap pressure ....bar	0.8 to 1.2	0.8	0.8	0.8 to 1.2
Fuel system				
Idle speed ....rpm	800 ± 50	700 ± 50	750 ± 50	770 ± 25
Maximum (no load) speed ....rpm	4400 to 4600	4600 to 4700	4100 to 4300	4850
Smoke test/opacity ....M⁻¹ %	3.0	2.0	2.3	2.2
Static timing method................................	Rotor lift	Plunger travel	Plunger travel	Dial gauge
Timing dimension.............................mm	Dimension on pump	1.1 ± 0.02	1.37 ± 0.02	1.0 ± 0.02
Crankshaft position ....mm [°]	[0] TDC	[0] TDC	[0] TDC	[0] TDC
Turbo type / ref / pressure ....bar	_	_	_	_
Injection pump make	Roto Diesel	Bosch	Bosch	Bosch
Injection pump part no...........................	_	VE 4/9F R22-7	VE 4/11 R368	VER 591
Injector Make / type	Roto Diesel	Bosch	Bosch	Bosch
Injector part no...................................	RDN OSDC 6751	DNOSD 193	0432 193 876	DNOSD 301
Injection type....................................	DPC R 8443 B 143B	Indirect VE	VE	Indirect VER
Injection opening pressure, New [used]...bar	115 ± 5	125 ± 5	225 ± 5	115 to 128
Glow plugs				
Maker	Champion	Champion	_	Bosch
Type	CH87	CH87	_	0250 201 006
Nominal rating.............................V/A	_	_		
Brakes				
minimum friction material thickness				
Front...mm	9.0 with backing	9.0 with backing	9.0 with backing	9.0 with backing
Rear..mm	3.0 with backing	3.0 with backing	3.0 with backing	1.0
Tyres - Saloon / Hatch......................Size	_	_	_	_
- Estate / Van....................Size	175x16: 195x16: 205x16: 215x16	205x16	205x16	195/70x15: 215/70x15
Pressure - front / rear - Saloon / Hatch ...bar	_	_	_	_
- Estate / Van ......bar	Refer to owner's handbook	3.5 / 3.5	3.5 / 3.5	3.6 / 3.5: 3.5 / 3.75
Front suspension / wheel alignment				
Toe-in (+) / Toe-out (−)....................mm [°]	-1.2 ± 1.2	-1.2 ± 1.2	-1.2 ± 1.2	1.0 ± 1.0
Camber	1° ± 30'	1° ± 30'	1° ± 30'	0°1' ± 30'
Castor	3°30' ± 30'	3°30' ± 30'	3°30' ± 30'	2°58' ± 30'
King pin inclination...............................	7°55' ± 30'	8° ± 30'	8° ± 30'	8° ± 30'
Rear suspension / wheel alignment				
Toe-in (+) / Toe-out (−)....................mm [°]	1.2 to -1.2	1.2 to -1.2	1.2 to -1.2	3.0 ± 2.0
Camber	0° to 30'	0° to 30'	0° to 30'	0°8' ± 05'

RENAULT

	Master 2.1 fwd (R**3) 1987 to 1995	Master 2.5 fwd (R**6) 1989 to 1995	Master 2.5 Turbo fwd (R**5) 1989 to 1995	Master 2.5 D 1995 to 1997
Torque wrench settings				
Cylinder head - stage 1 ....Nm	30 N	40 N	40 N	40 N
- stage 2 ....Nm	50	40	40	40
- stage 3 ....Nm	95 to 105	+ 180°	+ 180°	+ 180°
- stage 4 ....Nm	95 to 105	–	–	–
- stage 5 ....Nm	–	–	–	–
- stage 6 ....Nm	–	–	–	–
Big-end bearings..................................Nm	65 N	110 N	110 N	50 + 60° N
Main bearings......................................Nm	88 to 98	80, 160	80, 160	80, 160
Crankshaft pulley bolt ....Nm	120 to 135	200	200	200
Camshaft pulley bolt ....Nm	50	25	25	25
Flywheel [driveplate] bolt.....................Nm	55 to 60	120	120	120
Front hubs ....Nm	450	450	450	450
Rear hubs...Nm	WSM	WSM	WSM	WSM
Wheel nuts / bolts ....Nm	160	160	160	160
Glow plugs ....Nm	20 to 30	15	15	15
Clutch pressure plate bolts ....Nm	25	25	25	25
Injection pump sprocket......................Nm	50	50	50	50
Injectors..Nm	17	50	50	50
Injection pump mounting bolts ....Nm	–	50	50	–
Injector pipe unions.............................Nm	25	25	25	25
Capacities				
Engine oil & filter.................................litres	6.0	5.5	6.0	7.0
Gearbox..litres	4.0	4.0	4.0	4.0
Automatic transmission ....litres	–	–	–	–
Final drive ....litres	WT	WT	WT	WT
Cooling system....................................litres	14.0	14.0	14.0	11.0
Fuel tank..litres	62	62	62	62

Notes

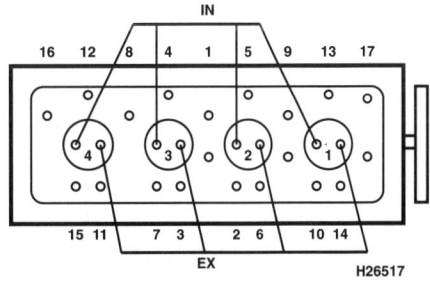

2068 cm³

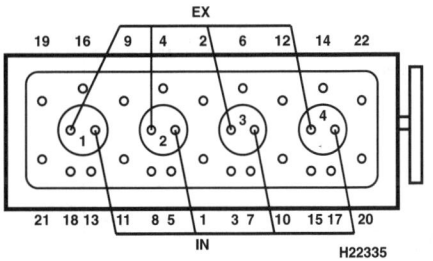

2499 cm³

– Not applicable, or information not available

RENAULT

	Master 2.5 Turbo D 1995 to 1999	Master 2.5 D (F/UD0A) 1997 to 2000	Master 2.8 D (F/UD0C) 1997 to 2000	
Engine				
Engine type/code...........................	S9U714 SOHC Turbo 69kW	S8U770 SOHC 59kW	S9W702 SOHC Turbo 85kW	
Capacity (cm^3) / cylinders........................	2499 / 4	2499 / 4	2799 / 4	
Compression ration / pressure ....bar	18.5 / _	22.0 / _	19.1 /	
Torque output ....Nm	158	155	260	
Oil pressure ....idle [running] bar	0.8 [3.5 @ 4100]	0.8 [3.5 @ 4100]	0.8 [4.0]	
Oil temperature ....°C	80	80	80	
Valve clearances - inlet (mm)	0.50	0.50	0.50	
- exhaust (mm)	0.50	0.50	0.50	
Injection order	1-3-4-2	1-3-4-2	1-3-4-2	
No. 1 cylinder position	FE	FE	FE	
Cooling system				
Thermostat opening temperature ....°C	89	89	89	
Radiator cap pressure ....bar	0.8	0.8 to 1.2	0.8 to 1.2	
Fuel system				
Idle speed ....rpm	750 ± 50	800 ± 25	800 ± 25	
Maximum (no load) speed ....rpm	4500	4600 to 4700	4000 ± 100	
Smoke test/opacity ....M^{-1} %	1.14	3.0	2.5	
Static timing method................................	Dial gauge	Rotor lift	Plunger travel	
Timing dimension.................................mm	1.2 ± 0.02	0.97 ± 0.02	1.1 ± 0.02	
Crankshaft position ....mm [°]	[0] TDC	[0] TDC	[0] TDC	
Turbo type / ref / pressure ....bar	0.8 to 0.9 bar @ 2500rpm	_	_	
Injection pump make	Bosch	Bosch	Bosch	
Injection pump part no.............................	VER 521	VE4/10F2100R227	VE4/12F1800R721	
Injector Make / type	Bosch	Bosch	Bosch	
Injector part no.......................................	DSLA 145P300	DNOSD 193	DNOSD 193	
Injection type..	VER	Indirect VE	Direct VE	
Injection opening pressure, New [used]...bar	235 to 248	125 ± 5	235 to 243	
Glow plugs				
Maker	_	Champion	_	
Type	_	CH87	_	
Nominal rating...................................V/A	_	_	_	
Brakes				
minimum friction material thickness				
Front...mm	9.0 with backing	9.0 with backing	9.0 with backing	
Rear..mm	1.0	1.0	1.0	
Tyres - Saloon / Hatch....................Size	_	_	_	
- Estate / Van....................Size	195/70x15: 215/70x15	195/70x15: 215/70x15	195/70x15: 215/70x15	
Pressure - front / rear - Saloon / Hatch ...bar	_	_	_	
- Estate / Van ......bar	3.6 / 3.5: 3.5 / 3.75	3.6 / 3.5: 3.5 / 3.75	3.6 / 3.75: 3.5 / 3.37	
Front suspension / wheel alignment				
Toe-in (+) / Toe-out (–)...................mm [°]	1.0 ± 1.0	1.0 ± 1.0	1.0 ± 1.0	
Camber	0°1' ± 30'	0°1' ± 30'	-0°1' ± 30'	
Castor	2°58' ± 30'	2°58' ± 30'	2°58' ± 30'	
King pin inclination..................................	8° ± 30'	8° ± 30'	8° ± 30'	
Rear suspension / wheel alignment				
Toe-in (+) / Toe-out (–)...................mm [°]	3.0 ± 2.0	-3.0 ± 2.0	-3.0 ± 2.0	
Camber	0°8' ± 05'	-14' ± 14'	-14' ± 14'	

RENAULT

	Master 2.5 Turbo D 1995 to 1999	Master 2.5 D (F/UD0A) 1997 to 2000	Master 2.8 D (F/UD0C) 1997 to 2000	
Torque wrench settings				
Cylinder head - stage 1 ....Nm	40 N	40 N	WSM	
- stage 2 ....Nm	40	40	–	
- stage 3 ....Nm	+ 180°	+ 180°	–	
- stage 4 ....Nm	–	–	–	
- stage 5 ....Nm	–	–	–	
- stage 6 ....Nm	–	–	–	
Big-end bearings...............................Nm	110 N	50 + 60° N	50 + 60° N	
Main bearings....................................Nm	80, 160	80, 160	80, 160	
Crankshaft pulley bolt ....Nm	200	200	200	
Camshaft pulley bolt ....Nm	25	25	25	
Flywheel [driveplate] bolt...................Nm	120	120	120	
Front hubs ....Nm	450	450	450	
Rear hubs ....Nm	WSM	WSM	WSM	
Wheel nuts / bolts ....Nm	160	160	160	
Glow plugs ....Nm	15	15	15	
Clutch pressure plate bolts ....Nm	25	25	25	
Injection pump sprocket.....................Nm	50	50	50	
Injectors..Nm	50	50	50	
Injection pump mounting bolts ....Nm	–	50	50	
Injector pipe unions...........................Nm	25	25	25	
Capacities				
Engine oil & filter...............................litres	6.0	7.0	6.7	
Gearbox...litres	4.0	4.0	4.0	
Automatic transmission ....litres	–	–	–	
Final drive ....litres	WT	WT	WT	
Cooling system..................................litres	11.0	11.0	11.0	
Fuel tank..litres	62	62	62	

Notes

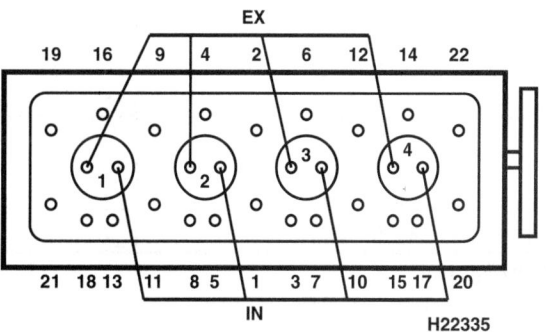

2499 cm³ / 2799 cm³

– Not applicable, or information not available

ROVER

	Metro/100 1.4 D 1992 to 1993	115 D 1995 to 1998	Maestro 500 & 700 Van 1986 to 1995	Maestro 2.0 1990 to 1992
Engine				
Engine type/code...........................	TUD3 K9A SOHC 38kW	TUD5 VJZ SOHC 40kW	MDi SOHC 46kW	MDi SOHC 44kW
Capacity (cm^3) / cylinders.........................	1360 / 4	1527 / 4	1994 / 4	1994 / 4
Compression ration / pressure ....bar	22.0 /	23.0 /	18.1 / _	18.1 / 20.7 to 34.5
Torque output ....Nm	0	95	0	0
Oil pressure ....idle [running] bar	[4.0 @ 4000]	[4.0 @ 4000]	1.0 [4.2]	0.7 [3.8 @ 3000]
Oil temperature ....°C	80	80	_	80
Valve clearances - inlet (mm)	0.15	0.15 ± 0.08	0.25 to 0.35	0.25 to 0.35
- exhaust (mm)	0.30	0.30 ± 0.08	0.35 to 0.45	0.35 to 0.45
Injection order	1-3-4-2	1-3-4-2	1-3-4-2	1-3-4-2
No. 1 cylinder position	FE	FE	TBE	TBE
Cooling system				
Thermostat opening temperature ....°C	_	98	88	88
Radiator cap pressure ....bar	_	1.4	1.04	1.04
Fuel system				
Idle speed ....rpm	775 ± 25	800 ± 25	925 ± 25	800 to 850
Maximum (no load) speed ....rpm	5450	5450	5130	5130
Smoke test/opacity ....M^{-1} %	2.5	2.5	2.35	2.5
Static timing method................................	Dial gauge	Refer to wsm	Plunger travel	Plunger travel
Timing dimension.................................mm	Dimension on pump	_	1.37	1.1
Crankshaft position ....mm [°]	0 TDC	_	0 TDC	0 TDC
Turbo type / ref / pressure ....bar	_	_	_	_
Injection pump make	Lucas	Lucas	Bosch	Bosch
Injection pump part no.............................	DPC R844	DPC 070	0460 414 029	0460 414 029
Injector Make / type	CAV	CAV	4 hole	CAV
Injector part no.......................................	RDNOSDC 68863C	RDN125DC 6849	JB6801061	TRB 6704201
Injection type...	Indirect DPC	Indirect DPC	Direct EPVE	Direct EPVE
Injection opening pressure, New [used]...bar	125 ± 5	135	222 [200]	228
Glow plugs				
Maker	Bosch/Champion	Bosch	Bosch/Champion	Bosch/Champion
Type	0250 202 001 / CH147	0250 202 020	0250 201 005 / CH68	0250 201 005 / CH68
Nominal rating.......................................V/A	_	_	11 / 12	11 / 12
Brakes				
minimum friction material thickness				
Front...mm	1.0	1.0	3.0	3.0
Rear..mm	1.5	1.5	1.6	1.5
Tyres - Saloon / Hatch......................Size	155/65x13	155/65x13	_	175/70x14
- Estate / Van...........................Size	_	_	165x13	_
Pressure - front / rear - Saloon / Hatch ...bar	2.3 / 2.3	2.3 / 2.3	_	2.1 / 2.1
- Estate / Van ......bar	_	_	2.0 / 2.8'	_
Front suspension / wheel alignment				
Toe-in (+) / Toe-out (–)....................mm [°]	[0 to -25']	[0 to -25']	[0° ± 8']	[0° ± 8']
Camber	18' ± 37'	18' ± 37'	A: -15'±21' B: 30'±21'²	A: -15'±21' B: 30'±21'¹
Castor	23' ± 55'	23' ± 55'	37' ± 30'	37' ± 30'
King pin inclination...................................	10°52'	10°52'	A: 12°30'±30' B: 12°33'±30²	A: 12°30'±30' B: 12°33'±30'¹
Rear suspension / wheel alignment				
Toe-in (+) / Toe-out (–)....................mm [°]	[10' to 50']	[10' to 50']	[30' to 1°]	[30' to 1°]
Camber	-1° ± 30'	-1° ± 30'	-0.5° ± 0.5°	-0.5° ± 0.5°

ROVER

Torque wrench settings	Metro/100 1.4 D 1992 to 1993	115 D 1995 to 1998	Maestro 500 & 700 Van 1986 to 1995	Maestro 2.0 1990 to 1992
Cylinder head - stage 1 ....Nm	60¹	40 N	50 N	50 N
- stage 2 ....Nm	Slacken	+ 260°	100	100
- stage 3 ....Nm	20	–	+ 90°	+ 90°
- stage 4 ....Nm	+ 160°	–	–	–
- stage 5 ....Nm	+ 160°	–	–	–
- stage 6 ....Nm	–	–	–	–
Big-end bearings..............................Nm	40 N	40 N	47 N	47 N
Main bearings...................................Nm	20 + 45°	20 + 45°	112 N	112 N
Crankshaft pulley bolt ....Nm	110	110	180	180
Camshaft pulley bolt ....Nm	80	80	85	85
Flywheel [driveplate] bolt....................Nm	65	65	65	65
Front hubs ....Nm	210	210	203	203
Rear hubs...Nm	120	120	68	68
Wheel nuts / bolts ....Nm	70	70	72	72
Glow plugs ....Nm	22	22	20	20
Clutch pressure plate bolts ....Nm	15	23	23	23
Injection pump sprocket......................Nm	50	25	60	60
Injectors...Nm	70	55	43	43
Injection pump mounting bolts ....Nm	23	20	–	–
Injector pipe unions............................Nm	20	20	22	22
Capacities				
Engine oil & filter.................................litres	3.5	4.75	5.25	5.25
Gearbox...litres	2.0	2.0	2.0	2.0
Automatic transmission ....litres	–	–	–	–
Final drive ....litres	WT	WT	WT	WT
Cooling system...................................litres	–	6.7	7.5	7.5
Fuel tank..litres	33	33	54	54

Notes

Metro/100 1.4 D 1992 to 1993
¹Bolt length below head: ≤185.9 mm
Maestro 500 & 700 Van 1986 to 1995
¹700 MODEL: 2.0 / 3.2
²A = early, B = later models
Maestro 2.0 1990 to 1992
¹A = early, B = later models

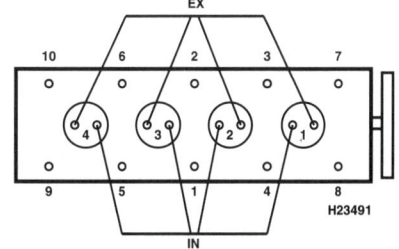

1360 cm³ / 1527 cm³

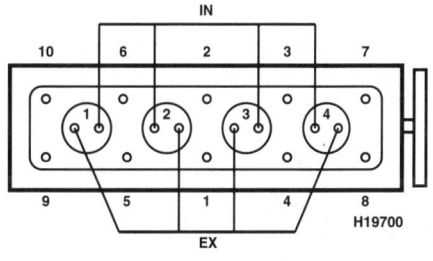

1994 cm³

– Not applicable, or information not available

ROVER

Haynes THE BOOK ®

	Maestro 2.0 TD 1992 to 1995	Montego 2.0 TD 1991 to 1995	75 2.0 CDT 1999 to 2000	218, 418 TD 1991 to 1996
Engine				
Engine type/code................	MDi SOHC Turbo 60kW	MDi SOHC Turbo 60kW	M47R DOHC 16V Turbo[1] 85kW	XUD7TE A8A SOHC Turbo 65kW
Capacity (cm^3) / cylinders................	1994 / 4	1994 / 4	1950 / 4	1769 / 4
Compression ration / pressure ....bar	17.2 / 34.0	17.2 / ≥34.0	18.0 /	22.0 / _
Torque output ....Nm	0	0	260	180
Oil pressure ....idle [running] bar	0.7 [4.2 @ 3000]	0.7 [4.2 @ 3000]	[3.0 to 4.5 @ 3500]	2.1
Oil temperature ....°C	80	_	80	_
Valve clearances - inlet (mm)	0.25 to 0.35	0.25 to 0.35	0: Hyd.	0.15
- exhaust (mm)	0.35 to 0.45	0.35 to 0.45	0: Hyd.	0.30
Injection order	1-3-4-2	1-3-4-2	1-3-4-2	1-3-4-2
No. 1 cylinder position	TBE	TBE	TCE	FE
Cooling system				
Thermostat opening temperature ....°C	88	88	88 ± 2	88
Radiator cap pressure ....bar	1.04	1.04	1.43	1.4
Fuel system				
Idle speed ....rpm	800 to 850	825 ± 25	780 ± 50	800 ± 50
Maximum (no load) speed ....rpm	5130	5130	4900	4300 ± 160
Smoke test/opacity ....M^{-1} %	2.35	2.5	2.0	2.5
Static timing method................	Plunger travel	Plunger travel	_	Dial gauge
Timing dimension................mm	1.0	1.0	Computer controlled	Dimension on pump
Crankshaft position ....mm [°]	0 TDC	0 TDC	_	0 TDC
Turbo type / ref / pressure ....bar	_	_	Mitsubishi MR1 TD025L3[2]	_
Injection pump make	Bosch	Bosch	Bosch	CAV
Injection pump part no................	_	11F2250R413	CP1	058
Injector Make / type................	CAV	Bosch	Bosch	CAV
Injector part no................	JB	DSLA 150PV3372 481	CRI 0445 110 030	RDN 12SDC 6862
Injection type................	Direct EPVE	Direct EPVE	Direct Common Rail DDE 4.0	Indirect
Injection opening pressure, New [used]...bar	228	228	250 to 1350	130 ± 5
Glow plugs				
Maker	Bosch	Champion	_	Bosch/Champion
Type	0250 201 019	CH88 or CH137	_	0250 201 019 / CH88/137
Nominal rating ....V/A	11 / 12	11 / 12	_	11 / 12
Brakes				
minimum friction material thickness				
Front................mm	3.0	3.0	8.8 with backing	3.0
Rear................mm	1.5	1.5	8.0 with backing	2.0 Discs: 3.0
Tyres - Saloon / Hatch................Size	175/70x14	180/65x365	195/65x15: 205/65x15:215/55x16	175/70x14
- Estate / Van................Size	_	180/65x365		
Pressure - front / rear - Saloon / Hatch...bar	2.1 / 2.1	2.0 / 2.0	2.2 / 2.1: 2.1 / 1.9:2.2 / 2.1	2.1 / 2.1
- Estate / Van ......bar	_	2.0 / 2.2	_	_
Front suspension / wheel alignment				
Toe-in (+) / Toe-out (−)................mm [°]	[0° ± 8']	[0° ± 8']	[0.17° ± 0.25°]	[-4' ± 8']
Camber................	30' ± 21'	30' ± 21'	-0.33° ± 0.75°	-20' ± 10'
Castor................	37' ± 30'	37' ± 30'	3.45° ± 1.0°	1°59' ± 30'
King pin inclination................	12°33' ± 30'	12°33'	12.55° ± 0.5°	12°
Rear suspension / wheel alignment				
Toe-in (+) / Toe-out (−)................mm [°]	[30' to 1°]	[20' to 40']	[0.37° ± 0.25°]	[7' ± 4'] PW
Camber................	-0.5° ± 0.5°	0° to -1°	-0.52° ± 0.75°	-25' ± 10'

ROVER

Torque wrench settings	Maestro 2.0 TD 1992 to 1995	Montego 2.0 TD 1991 to 1995	75 2.0 CDT 1999 to 2000	218, 418 TD 1991 to 1996
Cylinder head - stage 1 ....Nm	50 N	50 N	Bolts 1 to 11: 80 N	30
- stage 2 ....Nm	100	100	Slacken, 50	70
- stage 3 ....Nm	+ 90°	+ 90°	+ 90°	+ 120°
- stage 4 ....Nm	–	–	+ 90°	–
- stage 5 ....Nm	–	–	Bolt 12: 110	–
- stage 6 ....Nm	–	–	–	–
Big-end bearings.................................Nm	47 N	47 N	5, 25 + 70° N	50 N
Main bearings.....................................Nm	112 N	112	20 + 70° N	20 + 70°
Crankshaft pulley bolt ....Nm	180	180	100 + 60° + 60° + 30° N	40 + 60°
Camshaft pulley bolt ....Nm	85	85	20 + 35°	40
Flywheel [driveplate] bolt...................Nm	65	65	110 [115]	50
Front hubs ....Nm	203	203	350	185
Rear hubs ....Nm	68	68	210	185
Wheel nuts / bolts ....Nm	72	72	125	100
Glow plugs ....Nm	20	20	20	23
Clutch pressure plate bolts ....Nm	23	23	25	25
Injection pump sprocket.....................Nm	60	60	65	50
Injectors..Nm	43	43	10	90
Injection pump mounting bolts ....Nm	–	–	24	18
Injector pipe unions...........................Nm	22	22	20	20
Capacities				
Engine oil & filter................................litres	5.9	5.25	6.75	4.5
Gearbox..litres	2.0	2.0	1.4	2.0
Automatic transmission ....litres	–	–	4.0	–
Final drive ....litres	WT	WT	WT	WT
Cooling system...................................litres	7.5	7.5	8.2	8.0
Fuel tank...litres	54	50	66	55

Notes

Montego 2.0 TD 1991 to 1995
[1]VIN 606579 ▶, RHD or 606709, LHD
75 2.0 CDT 1999 to 2000
[1]Intercooled
[2]1.4 to 1.8 bar

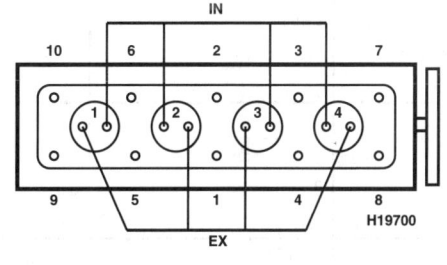

1994 cm³

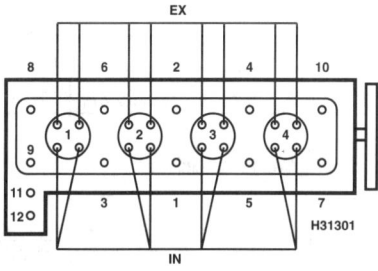

1950 cm³

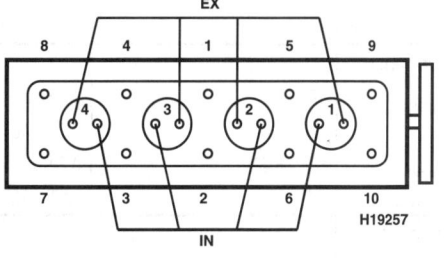

1769 cm³

– Not applicable, or information not
available

ROVER

	218, 418 D 1991 to 1996	220 D/SD 1995 to 2000	220 Di/SDi 1995 to 2000	420 D/SD 1995 to 2000
Engine				
Engine type/code	XUD9A D9B SOHC 49kW	20T2R SOHC Turbo 63kW	20T2N' SOHC Turbo 77kW	20T2R SOHC Turbo 63kW
Capacity (cm^3) / cylinders	1905 / 4	1994 / 4	1994 / 4	1994 / 4
Compression ration / pressure ....bar	23.0 / _	19.5 /	19.5 /	19.5 /
Torque output ....Nm	0	170	210	170
Oil pressure ....idle [running] bar	1.6	0.7 [3.8 @ 3000]	0.7 [3.8 @ 3000]	0.7
Oil temperature ....°C	_	80	80	80
Valve clearances - inlet (mm)	0.15	0: Hyd.	0: Hyd.	0: Hyd.
- exhaust (mm)	0.30	0: Hyd.	0: Hyd.	0: Hyd.
Injection order	1-3-4-2	1-3-4-2	1-3-4-2	1-3-4-2
No. 1 cylinder position	FE	TBE	TBE	TBE
Cooling system				
Thermostat opening temperature ....°C	88	82	82	82
Radiator cap pressure ....bar	1.4	0.9 to 1.2	0.9 to 1.2	0.9 to 1.2
Fuel system				
Idle speed ....rpm	800 ± 50	850 ± 50	850 ± 50	850 ± 50
Maximum (no load) speed ....rpm	4600 ± 160	_	_	_
Smoke test/opacity ....M^{-1} %	2.5	2.5	3.0	3.0
Static timing method...............................	Dial gauge	Refer to wsm	Refer to wsm	Refer to wsm
Timing dimension..................................mm	Dimension on pump	Electronic control	Electronic control	Electronic control
Crankshaft position ....mm [°]	0 TDC	_	_	_
Turbo type / ref / pressure ....bar	_	_	_	_
Injection pump make	CAV	Bosch	Bosch	Bosch
Injection pump part no.............................	057	VP21	VP37	VP21
Injector Make / type	CAV	Bosch	Bosch	Bosch
Injector part no.......................................	RDN 08D 6751	2 Stage MSC 100510	2 Stage MSC 100510	2 Stage MSC 100510
Injection type...	Indirect	Direct Electronic	Direct VP37	Direct VP21
Injection opening pressure, New [used]...bar	125 ± 5	200 to 320	200 to 320	200 to 320
Glow plugs				
Maker	Bosch/Champion	Beru	Beru	Beru
Type	0250 201 019 / CH88/137	0100226 184	0100226 184	0100226 184
Nominal rating.....................................V/A	11 / 12	_	_	_
Brakes				
minimum friction material thickness				
Front...mm	3.0	3.0	3.0	3.0
Rear...mm	2.0 Discs: 3.0	2.0	2.0	2.0
Tyres - Saloon / Hatch......................Size	175/70x14	175/65x14	175/65x14	175/65x14
- Estate / Van...........................Size	_	_	_	_
Pressure - front / rear - Saloon / Hatch...bar	2.1 / 2.1	2.4 / 2.1	2.4 / 2.1	2.4 / 2.1
- Estate / Van ......bar		_	_	_
Front suspension / wheel alignment				
Toe-in (+) / Toe-out (–)....................mm [°]	[-4' ± 8']	[-0°5']	[-0°5']	[0° ± 12']
Camber	-20' ± 10'	0°20' ± 10'	0°20' ± 10'	0° ± 1°
Castor	1°59' ± 30'	2°30'	2°30'	1°28'
King pin inclination.................................	12°	12°7' ± 30'	12°7' ± 30'	10°41'
Rear suspension / wheel alignment				
Toe-in (+) / Toe-out (–)....................mm [°]	[7' ± 4'] PW	[0°8' ± 10']	[0°8' ± 10']	[0°11' ± 10']
Camber	-25' ± 10'	-0°20' ± 10'	-0°20' ± 10'	-0°55' ± 10'

ROVER

Torque wrench settings	218, 418 D 1991 to 1996	220 D/SD 1995 to 2000	220 Di/SDi 1995 to 2000	420 D/SD 1995 to 2000
Cylinder head - stage 1 ....Nm	30	30[1]	30[2]	30[1]
- stage 2 ....Nm	70	65	65	65
- stage 3 ....Nm	+ 120°	+ 90°	+ 90°	+ 90°
- stage 4 ....Nm	–	+ 90°	+ 90°	+ 90°
- stage 5 ....Nm	–	–	–	–
- stage 6 ....Nm	–	–		
Big-end bearings...............................Nm	50 N	48 N Bolts: 20 + 85° N	48 N Bolts: 20 + 85°N	48 N Bolts: 20 + 85° N
Main bearings...................................Nm	20 + 70°	112	112	112
Crankshaft pulley bolt ....Nm	40 + 60°	63 + 90°	63 + 90°	63 + 90°
Camshaft pulley bolt ....Nm	40	20 + 90°	20 + 90°	20 + 90°
Flywheel [driveplate] bolt...................Nm	50	15 + 90° N	15 + 90° N	15 + 90° N
Front hubs ....Nm	185	180 N	180 N	180 N
Rear hubs ....Nm	185	185 N	185 N	185 N
Wheel nuts / bolts ....Nm	100	110	110	110
Glow plugs ....Nm	23	20	20	20
Clutch pressure plate bolts ....Nm	25	26	26	26
Injection pump sprocket.....................Nm	50	–	20 + 90° N	20 + 90° N
Injectors...Nm	90	–	25	25
Injection pump mounting bolts ....Nm	18	–	25	25
Injector pipe unions..........................Nm	20	–	20	20
Capacities				
Engine oil & filter...............................litres	4.5	4.9	4.9	4.0
Gearbox..litres	2.0	2.0	2.0	2.2
Automatic transmission ....litres	–	–	–	–
Final drive ....litres	WT	WT	WT	WT
Cooling system.................................litres	8.0	7.0	7.0	7.0
Fuel tank...litres	55	50	50	55

Notes

220 D/SD 1995 to 2000
[1]Bolt length: ≤243.41 mm
220 Di/SDi 1995 to 2000
[1]With intercooler
[2]Bolt length: ≤243.41 mm
420 D/SD 1995 to 2000
[1]Bolt length: ≤243.41 mm

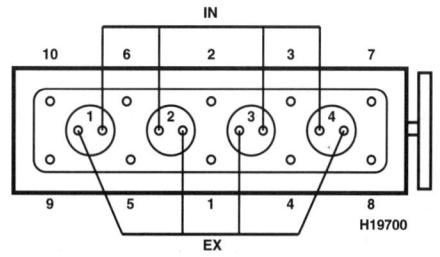

1994 cm³

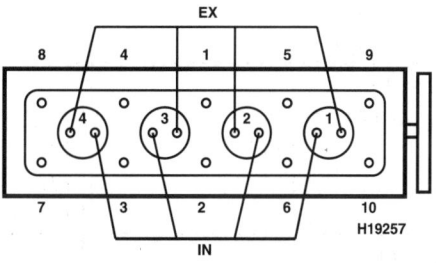

1905 cm³

– Not applicable, or information not
  available

ROVER

	420 Di/SDi 1997 to 2000	620 Di 1997 to 2000	825 TD 1990 to 1991	825 TD 1991 to 1996
Engine				
Engine type/code..............................	20T2N' SOHC Turbo 77kW	20 T2N' SOHC Turbo 77kW	VM 4924 SHIRG OHV Turbo 87kW	VM 425 SLIRR OHV Turbo 87kW
Capacity (cm³) / cylinders........................	1994 / 4	1994 / 4	2498 / 4	2498 / 4
Compression ration / pressure ....bar	19.5 /	19.5 /	22.5 / ≥23.4	22.0 / 23.4
Torque output ....Nm	210	210	268	268
Oil pressure ....idle [running] bar	0.7	0.7 [3.8 @ 3000]	[3.4 to 4.8]	[3.4 to 4.8 @ 4000]
Oil temperature ....°C	80	80	90 to 100	80
Valve clearances - inlet (mm)	0: Hyd.	0: Hyd.	0.30	0.30
- exhaust (mm)	0: Hyd.	0: Hyd.	0.30	0.30
Injection order	1-3-4-2	1-3-4-2	1-3-4-2	1-3-4-2
No. 1 cylinder position	TBE	TBE	–	–
Cooling system				
Thermostat opening temperature ....°C	82	82	80 ± 2	80
Radiator cap pressure ....bar	0.9 to 1.2	0.9 to 1.2	1.0	1.03
Fuel system				
Idle speed ....rpm	850 ± 50	850 ± 50	850 ± 50	900 ± 20
Maximum (no load) speed ....rpm	–	–	4700 to 4730	4730
Smoke test/opacity ....M⁻¹ %	3.0	3.0	2.41	3.0
Static timing method.............................	Refer to wsm	Refer to wsm	Plunger travel	Dial gauge
Timing dimension................................mm	Electronic control	Electronic control	0.68	0.68
Crankshaft position ....mm [°]	–	–	0 TDC	0 TDC
Turbo type / ref / pressure ....bar	–	–	0.89 bar	0.89 bar
Injection pump make	Bosch	Bosch	Bosch	Bosch
Injection pump part no............................	VP37	VP37	VE4/10 F 2100 L269-2	VE4/10F
Injector Make / type............................	Bosch	Bosch	Bosch	Bosch
Injector part no....................................	2 Stage MSC 100510	2 Stage MSC 100510	DNO SD 263 or SD 379	DNOSD301
Injection type....................................	Direct VP21	Direct VP37	Indirect VE	Indirect VE
Injection opening pressure, New [used]...bar	200 to 320	200 to 320	150 to 158	150 to 158
Glow plugs				
Maker	Beru	Beru	Bosch/Champion	Bosch/Champion
Type	0100226 184	0100226 184	0250 201 012 / CH70	0250 201 033 / CH70
Nominal rating...................................V/A	–	–	11 / 10.5	11 / 10.5
Brakes				
minimum friction material thickness				
Front...mm	3.0	3.0	8.2 with backing	8.2 with backing
Rear...mm	2.0	2.0	7.2 with backing	7.2 with backing
Tyres - Saloon / Hatch......................Size	185/65x14	195/65x15	195/65x15	195/65x15
- Estate / Van.....................Size	–	–	–	–
Pressure - front / rear - Saloon / Hatch ...bar	2.1 / 2.1	2.2 / 2.1	2.1 / 2.1	2.1 / 2.1
- Estate / Van ......bar	–	–	–	–
Front suspension / wheel alignment				
Toe-in (+) / Toe-out (–)...................mm [°]	[0° ± 12']	0.0 ± 3.0	[0' ± 8'] 91 ▶: [0'±15']	[0° ± 15']
Camber	0° ± 1°	0° ± 1°	24' ± 1°12'	0°24' ± 1°12'
Castor	1°28'	3° ± 1°	2°36' ± 2°	2°35' ± 2°
King pin inclination.............................	10°41'	7°37' ± 1°	8° ± 1°12'	7°27' ± 1°12'
Rear suspension / wheel alignment				
Toe-in (+) / Toe-out (–)...................mm [°]	[0°11' ± 10']	2.0 ± 2.0	[22' ± 15']	[0°22' ± 15']
Camber	-0°45' ± 10'	-0°30' ± 10'	13' ± 1°12'	0°18' ± 1°12'

ROVER

Torque wrench settings	420 Di/SDi 1997 to 2000	620 Di 1997 to 2000	825 TD 1990 to 1991	825 TD 1991 to 1996
Cylinder head - stage 1 ....Nm	30²	30²	30	Bolts 1 - 10: 30, 30¹
- stage 2 ....Nm	65	65	+ 70°	Bolts 11 & 12: 80
- stage 3 ....Nm	+ 90°	+ 90°	+ 70°	Warm up, allow to cool
- stage 4 ....Nm	+ 90°	+ 90°	Bolts 11 & 12: 80	Bolts 1 - 10: slacken, 30²
- stage 5 ....Nm	–	–	Warm up, allow to cool	Bolts 11 & 12: 90
- stage 6 ....Nm		–	Bolts 1 to 10: + 30°¹	–
Big-end bearings.............................Nm	48 N Bolts: 20 + 85° N	48 N Bolts: 20 + 85° N	29 + 60° N	29 + 60° N
Main bearings.................................Nm	112	112	42	42
Crankshaft pulley bolt ....Nm	63 + 90°	63 + 90°	177	177
Camshaft pulley bolt ....Nm	20 + 90°	20 + 90° N	–	–
Flywheel [driveplate] bolt....................Nm	15 + 90° N	15 + 90° N	108	20 + 60°
Front hubs ....Nm	180 N	250 N	290	415
Rear hubs ....Nm	185 N	185 N	245	245
Wheel nuts / bolts ....Nm	110	110	110	110
Glow plugs ....Nm	20	20	23	23
Clutch pressure plate bolts ....Nm	26	26	22	22
Injection pump sprocket....................Nm	20 + 90° N	20 + 90° N	88	88
Injectors.......................................Nm	25	25	27	70
Injection pump mounting bolts ....Nm	–	25	30	25
Injector pipe unions..........................Nm	20	20	19	19
Capacities				
Engine oil & filter...............................litres	4.0	4.9	6.4	6.4
Gearbox...litres	2.0	2.2	2.0	2.0
Automatic transmission ....litres	–		–	–
Final drive ....litres	WT	WT	WT	WT
Cooling system................................litres	7.0	7.0	7.0	7.0
Fuel tank..litres	55	65	68	68

Notes

420 Di/SDi 1997 to 2000
¹With intercooler
²Bolt length: ≤243.41 mm
620 Di 1997 to 2000
¹With intercooler
²Bolt length: ≤243.41 mm
825 TD 1990 to 1991
¹Bolts 11 & 12: 90
825 TD 1991 to 1996
¹+ 70°, + 70°
²+ 120 °

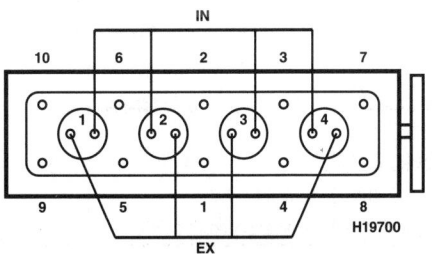

1994 cm³

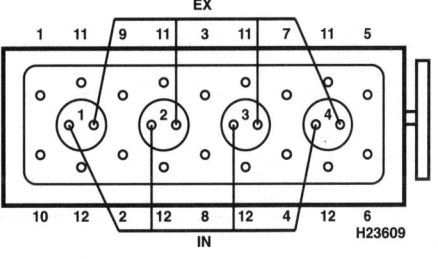

2498 cm³

– Not applicable, or information not
   available

ROVER

	825 TD 1996 to 1997			
Engine				
Engine type/code.............................	VM 425 SLIER OHV Turbo 89kW			
Capacity (cm³) / cylinders.......................	2498 / 4			
Compression ration / pressure ....bar	24.5 / 23.4			
Torque output ....Nm	268			
Oil pressure ....idle [running] bar	[3.4 to 4.8 @ 4000]			
Oil temperature ....°C	80			
Valve clearances - inlet (mm)	0.30			
- exhaust (mm)	0.30			
Injection order	1-3-4-2			
No. 1 cylinder position	_			
Cooling system				
Thermostat opening temperature ....°C	80			
Radiator cap pressure ....bar	1.03			
Fuel system				
Idle speed ....rpm	887 ± 23			
Maximum (no load) speed ....rpm	4700			
Smoke test/opacity ....M⁻¹ %	3.0			
Static timing method...............................	Dial gauge			
Timing dimension.............................mm	0.58 to 0.62			
Crankshaft position ....mm [°]	0 TDC			
Turbo type / ref / pressure ....bar	0.89 bar			
Injection pump make	Bosch			
Injection pump part no................................	VP36			
Injector Make / type	Bosch			
Injector part no......................................	KCA 17S70			
Injection type......................................	Indirect VP36			
Injection opening pressure, New [used]...bar	150 to 158			
Glow plugs				
Maker	Bosch			
Type	0250 202 023			
Nominal rating....................................V/A	11 /			
Brakes				
minimum friction material thickness				
Front.............................mm	8.2 with backing			
Rear.............................mm	7.2 with backing			
Tyres - Saloon / Hatch......................Size	195/65x15			
- Estate / Van............................Size	_			
Pressure - front / rear - Saloon / Hatch ...bar	2.2 / 2.1			
- Estate / Van ......bar	_			
Front suspension / wheel alignment				
Toe-in (+) / Toe-out (–).....................mm [°]	[0° ± 15']			
Camber	0°24' ± 1°12'			
Castor	2°35' ± 2°			
King pin inclination..................................	7°27' ± 1°12'			
Rear suspension / wheel alignment				
Toe-in (+) / Toe-out (–).....................mm [°]	[0°22' ± 15']			
Camber	0°18' ± 1°12'			

ROVER

Torque wrench settings	825 TD 1996 to 1997				
Cylinder head - stage 1 ....Nm	Bolts 1 - 10: 30, 30[1]				
- stage 2 ....Nm	Bolts 11 & 12: 80				
- stage 3 ....Nm	Warm up, allow to cool				
- stage 4 ....Nm	Bolts 1 - 10: slacken, 30[2]				
- stage 5 ....Nm	Bolts 11 & 12: 90				
- stage 6 ....Nm	–				
Big-end bearings.................................Nm	29 + 60° N				
Main bearings....................................Nm	42				
Crankshaft pulley bolt ....Nm	177				
Camshaft pulley bolt ....Nm	–				
Flywheel [driveplate] bolt.....................Nm	20 + 60° N				
Front hubs ....Nm	415				
Rear hubs ....Nm	245				
Wheel nuts / bolts ....Nm	110				
Glow plugs ....Nm	23				
Clutch pressure plate bolts ....Nm	22				
Injection pump sprocket.....................Nm	88				
Injectors..Nm	70				
Injection pump mounting bolts ....Nm	–				
Injector pipe unions............................Nm	19				
Capacities					
Engine oil & filter................................litres	6.4				
Gearbox..litres	2.0				
Automatic transmission ....litres	–				
Final drive...litres	WT				
Cooling system...................................litres	7.0				
Fuel tank...litres	68				

Notes

825 TD 1996 to 1997
[1] + 70°, + 70°
[2] + 120 °

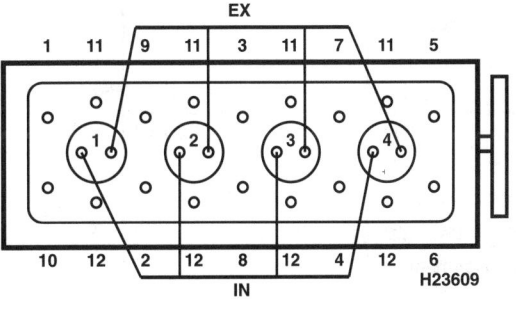

2498 cm³

– Not applicable, or information not
available

SEAT

	Arosa 1.7 D 1998 to 2000	Ibiza 1.7 D 1986 to 1992	Ibiza 1.7 D 1986 to 1992	Ibiza & Cordoba 1.9 D 1993 to 1995
Engine				
Engine type/code......................	AKU SOHC 44kW	022A5.000 SOHC 40kW	022A5.000 SOHC 40kW	1Y SOHC 47kW
Capacity (cm^3) / cylinders........................	1716 / 4	1714 / 4	1714 / 4	1896 / 4
Compression ration / pressure ....bar	19.5 /	20.0 ± 0.5 / _	20.0 ± 0.5 / _	22.5 / ≥26.0
Torque output ....Nm	115	0	0	124
Oil pressure ....idle [running] bar	[≥2.0 @ 2000]	[3.4 to 4.9]	[3.4 to 4.9]	[2.0 @ 2000]
Oil temperature ....°C	80	100	100	80
Valve clearances - inlet (mm)	0: Hyd	0.35	0.35	0: Hyd.
- exhaust (mm)	0: Hyd	0.40	0.40	0: Hyd.
Injection order	1-3-4-2	1-3-4-2	1-3-4-2	1-3-4-2
No. 1 cylinder position	TBE	TBE	TBE	TBE
Cooling system				
Thermostat opening temperature ....°C	85	78 to 82	78 to 82	87
Radiator cap pressure ....bar	1.4 to 1.6	1.0	1.0	1.2 to 1.5
Fuel system				
Idle speed ....rpm	750 ± 50	750 to 800	750 to 800	900 ± 30
Maximum (no load) speed ....rpm	_	5040 to 5120	5040 to 5120	5200
Smoke test/opacity ....M^{-1} %	2.5	2.5	2.5	2.0
Static timing method........................	_	Plunger travel	Dial gauge	Dial gauge
Timing dimension........................mm	Computer controlled	1.0 + 0.05	1.0	0.9 ± 0.02
Crankshaft position ....mm [°]	_	[0] TDC	[0] TDC	0 TDC
Turbo type / ref / pressure ....bar	_	_	_	_
Injection pump make	_	Bosch	CAV	Bosch
Injection pump part no........................	_	_	DPS C8550A001A	VER 430
Injector Make / type	_	Bosch	CAV	Bosch
Injector part no........................	_	DNO SD 1930	CDN OSD 6827, DNC 5641032	0432 217 240
Injection type........................	Direct SDI	VE 4/9F 2300/R54	Indirect DPS	Indirect VER
Injection opening pressure, New [used]...bar	190 to 200	135 +8 -0	135 +8 -0	130 to 138 [120]
Glow plugs				
Maker	_	Bosch/Champion	Bosch/Champion	Bosch
Type	_	0250 200 068 / CH68	0250 200 068 / CH68	0250 201 032
Nominal rating........................V/A	_	11 / 12	11 / 12	11.5 / 8
Brakes				
minimum friction material thickness				
Front........................mm	7.0 with backing	1.5	1.5	7.0 with backing
Rear........................mm	2.5	1.5	1.5	2.5
Tyres - Saloon / Hatch........................Size	155/70x13: 175/65x13:185/55x14	155x13: 165/65x14	155x13: 165/65x14	185/60x14
- Estate / Van........................Size	_	_	_	_
Pressure - front / rear - Saloon / Hatch...bar	2.2 / 2.2: 2.1 / 2.1:2.1 / 2.1	1.9 / 1.8	1.9 / 1.8	2.2 / 1.9
- Estate / Van ......bar	_	_	_	_
Front suspension / wheel alignment				
Toe-in (+) / Toe-out (−)........................mm [°]	[0° ± 10']	2.0 to 6.0'	2.0 to 6.0'	[0° ± 10']
Camber	1°20' ± 30'	30' to 1°30'[2]	30' to 1°30'[2]	-0°30' ± 30'
Castor	_	45' to 1°45'[3]	45' to 1°45'[3]	1°30' ± 30'
King pin inclination........................	_	_	_	_
Rear suspension / wheel alignment				
Toe-in (+) / Toe-out (−)........................mm [°]	[20' ± 10']	2.0 to 6.0[4]	2.0 to 6.0[4]	[-0°15' ± 5']
Camber	-1°20' ± 10'	15' to 1°15'[5]	15' to 1°15'[5]	-1°34' ± 20'

SEAT

Torque wrench settings	Arosa 1.7 D 1998 to 2000	Ibiza 1.7 D 1986 to 1992	Ibiza 1.7 D 1986 to 1992	Ibiza & Cordoba 1.9 D 1993 to 1995
Cylinder head - stage 1 ....Nm	40 N	60	60	40 N
- stage 2 ....Nm	60	+ 90°	+ 90°	60
- stage 3 ....Nm	+ 90°	+ 90°	+ 90°	+ 90°
- stage 4 ....Nm	+ 90°	–	–	+ 90°
- stage 5 ....Nm	–	–	–	–
- stage 6 ....Nm	–	–	–	–
Big-end bearings........................Nm	30 + 90° N	74 N	74 N	30 + 90°
Main bearings........................Nm	65 + 90° N	113 M10: 80	113 M10: 80	65
Crankshaft pulley bolt ....Nm	90 + 90° N	145	145	90 + 90° N
Camshaft pulley bolt ....Nm	45	118	118	45
Flywheel [driveplate] bolt........................Nm	60 + 90° N	85	85	60 + 90° N
Front hubs ....Nm	200 -360°, 50 + 30°	216	216	265
Rear hubs ....Nm	WSM	216	216	WSM
Wheel nuts / bolts ....Nm	110	Alloy: 78 Steel: 86	Alloy: 78 Steel: 86	110
Glow plugs ....Nm	15	15	15	25
Clutch pressure plate bolts ....Nm	20	30	30	20
Injection pump sprocket........................Nm	55	49	49	45
Injectors........................Nm	–	39	39	70
Injection pump mounting bolts ....Nm	25	29	29	25
Injector pipe unions........................Nm	20	–	–	25
Capacities				
Engine oil & filter........................litres	4.7	5.0	5.0	4.5
Gearbox........................litres	2.7	3.4	3.4	2.0
Automatic transmission ....litres	–	–	–	–
Final drive........................litres	WT	WT	WT	WT
Cooling system........................litres	6.5	8.0	8.0	6.5
Fuel tank........................litres	34	49	49	47

Notes

Ibiza 1.7 D 1986 to 1992
[1]1987 ▶: 1.0±2.0
[2]1987 ▶: -43'±30'
[3]1987 ▶: 0±30'
[4]1987 ▶: 4.0±2.0
[5]1987 ▶: 45'±30'
Ibiza 1.7 D 1986 to 1992
[1]1987 ▶: 1.0±2.0
[2]1987 ▶: -43'±30'
[3]1987 ▶: 0±30'
[4]1987 ▶: 4.0±2.0
[5]1987 ▶: 45'±30'

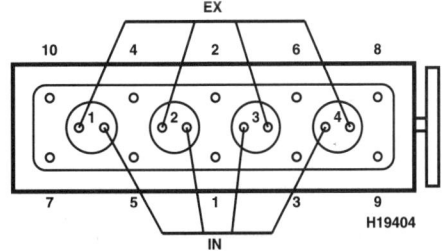

1716 cm³ / 1896 cm³

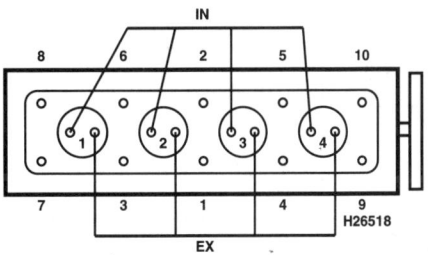

1714 cm³

– Not applicable, or information not available

SEAT

	Ibiza & Cordoba 1.9 TD 1994 to 1997	Ibiza & Cordoba 1.9 TDi 110 1998 to 2000	Ibiza & Cordoba 1.9 SDi 1997 to 2000	Ibiza & Cordoba 1.9 TDi 90 1997 to 2000
Engine				
Engine type/code............................	AAZ SOHC Turbo 55kW	AFN SOHC Turbo 81kW	AEY SOHC 47kW	1Z / AHU SOHC Turbo 66kW
Capacity (cm^3) / cylinders........................	1896 / 4	1896 / 4	1896 / 4	1896 / 4
Compression ration / pressure ....bar	22.5 / ≥ 26.0	19.5 / ≥ 19	19.5 / ≥ 19	19.5 / ≥ 19
Torque output ....Nm	150	235	125	202
Oil pressure ....idle [running] bar	[2.0 @ 2000]	[2.0 @ 2000]	[2.0 @ 2000]	[2.0 @ 2000]
Oil temperature ....°C	80	80	80	80
Valve clearances - inlet (mm)	0: Hyd.	0: Hyd.	0: Hyd.	0: Hyd.
- exhaust (mm)	0: Hyd.	0: Hyd.	0: Hyd.	0: Hyd.
Injection order	1-3-4-2	1-3-4-2	1-3-4-2	1-3-4-2
No. 1 cylinder position	TBE	TBE	TBE	TBE
Cooling system				
Thermostat opening temperature ....°C	87	87	87	87
Radiator cap pressure ....bar	1.2 to 1.5	1.4 to 1.6	1.4 to 1.6	1.4 to 1.6
Fuel system				
Idle speed ....rpm	920 ± 30	900 ± 30	900 ± 40	900 ± 30 N/A
Maximum (no load) speed ....rpm	5200	5200	5200 ± 100	5150
Smoke test/opacity ....M^{-1} %	2.0	2.0	2.5	3.0
Static timing method............................	Dial gauge	Refer to wsm	Refer to wsm	Refer to wsm
Timing dimension...............................mm	0.8 ± 0.02	–	–	–
Crankshaft position ....mm [°]	0 TDC	–	–	–
Turbo type / ref / pressure ....bar	–	0.5 to 0.65 bar @ 3500 rpm[1]	–	0.5 to 0.65 bar @ 4000 rpm[1]
Injection pump make	Bosch	Bosch	Bosch	Bosch
Injection pump part no.............................	VER 433-3	VER 510	VE	VE
Injector Make / type	Bosch	–	Bosch	–
Injector part no......................................	0432 217 244	–	–	–
Injection type..	Indirect VER	Direct VER	DirectVE	Direct VE
Injection opening pressure, New [used]...bar	150 to 158 [140]	≥ 170	190 to 200 [170]	190 to 200 [170]
Glow plugs				
Maker	Bosch	Bosch	Bosch	Bosch
Type	0250 201 032	0250 202 009	0250 202 022	0250 202 009
Nominal rating....................................V/A	11.5 / 8	11 / 12	11.5 / 12	11.5 / 12
Brakes				
minimum friction material thickness				
Front...mm	7.0 with backing	7.0 with backing	7.0 with backing	2.0
Rear..mm	2.5	2.5	2.5	2.0
Tyres - Saloon / Hatch......................Size	185/60x14	155x13	175/70x13	185/60x14
- Estate / Van...........................Size	–	–	–	185/60x14
Pressure - front / rear - Saloon / Hatch ...bar	2.2 / 1.9	2.2 / 1.9	2.2 / 1.9	2.2 / 1.9
- Estate / Van ......bar	–	–	–	2.2 / 2.0
Front suspension / wheel alignment				
Toe-in (+) / Toe-out (–)....................mm [°]	[0° ± 10']	[0° ± 10']	[0° ± 10']	[10' ± 20']
Camber	-0°30' ± 30'	-30' ± 20'	-30' ± 20'	-30' ± 20'
Castor	1°30' ± 30'	1°33' ± 37'	1°33' ± 37'	1°33' ± 37'
King pin inclination................................	–	–	–	–
Rear suspension / wheel alignment				
Toe-in (+) / Toe-out (–).....................mm [°]	[-0°15' ± 5']	[20' ± 10']	[20' ± 10']	[20' ± 10']
Camber	-1°34' ± 20'	-1°30' ± 30'	-1°30' ± 30'	-1°30' ± 30'

SEAT

Torque wrench settings	Ibiza & Cordoba 1.9 TD 1994 to 1997	Ibiza & Cordoba 1.9 TDi 110 1998 to 2000	Ibiza & Cordoba 1.9 SDi 1997 to 2000	Ibiza & Cordoba 1.9 TDi 90 1997 to 2000
Cylinder head - stage 1 ....Nm	40 N	40 N	40 N	40 N
- stage 2 ....Nm	60	60	60	60
- stage 3 ....Nm	+ 90°	+ 90°	+ 90°	+ 90°
- stage 4 ....Nm	+ 90°	+ 90°	+ 90°	+ 90°
- stage 5 ....Nm	–	–	–	–
- stage 6 ....Nm	–	–	–	–
Big-end bearings...............................Nm	30 + 90°	30 + 90° N	30 + 90° N	30 + 90° N
Main bearings....................................Nm	65	65 + 90° N	65 + 90° N	65 + 90° N
Crankshaft pulley bolt ....Nm	90 + 90° N	90 + 90°	90 + 90° oiled N	90 + 90° N
Camshaft pulley bolt ....Nm	45	–	45	45
Flywheel [driveplate] bolt....................Nm	60 + 90° N	60 + 90° N	60 + 90° N	20
Front hubs ....Nm	265	265	265	150 + 90° N
Rear hubs ....Nm	WSM	–	–	200 N
Wheel nuts / bolts ....Nm	110	110	110	140
Glow plugs ....Nm	25	15	15	15
Clutch pressure plate bolts ....Nm	20	–	20	20
Injection pump sprocket.....................Nm	45	55	55	55
Injectors..Nm	70	20	20	20
Injection pump mounting bolts ....Nm	25	–	–	25
Injector pipe unions...........................Nm	25	25	25	25
Capacities				
Engine oil & filter.................................litres	4.5	4.5	4.3	4.7
Gearbox..litres	2.0	2.0	2.0	2.0
Automatic transmission ....litres	–	–	–	–
Final drive ....litres	WT	WT	WT	WT
Cooling system...................................litres	6.5	5.5	5.5	5.5
Fuel tank...litres	47	45	45	45

Notes

Ibiza & Cordoba 1.9 TDi 110 1998 to 2000
[1]Charge pressure control solensoid disconnected
Ibiza & Cordoba 1.9 TDi 90 1997 to 2000
[1]Charge pressure control solenoid disconnected

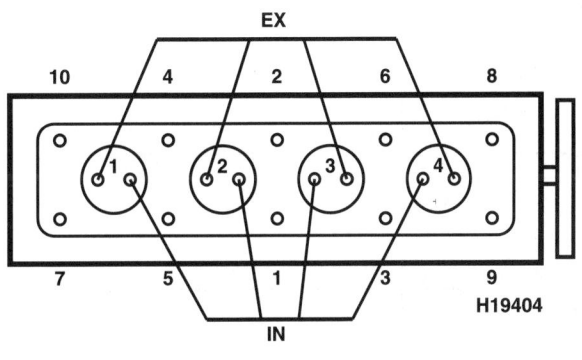

– Not applicable, or information not available

SEAT

	Malaga 1.7 D 1986 to 1992	Malaga 1.7 D 1986 to 1992	Toledo 1.9 D 1991 to 1998	Toledo 1.9 TD 1992 to 1998
Engine				
Engine type/code.....................	022A5.000 SOHC 40kW	022A5.000 SOHC 40kW	1Y SOHC 8V 47kW	AAZ SOHC 8V Turbo 55kW
Capacity (cm³) / cylinders........................	1714 / 4	1714 / 4	1896 / 4	1896 / 4
Compression ration / pressure ....bar	20.0 ± 0.5 / _	20.0 ± 0.5 / _	22.5 / ≥26.0	22.5 / ≥26.0
Torque output ....Nm	0	0	124	150
Oil pressure ....idle [running] bar	[3.4 to 4.9]	[3.4 to 4.9]	[≥2.0 @ 2000]	[≥2.0]
Oil temperature ....°C	100	100	80	80
Valve clearances - inlet (mm)	0.35	0.35	0: Hyd.	0: Hyd.
- exhaust (mm)	0.40	0.40	0: Hyd.	0: Hyd.
Injection order	1-3-4-2	1-3-4-2	1-3-4-2	1-3-4-2
No. 1 cylinder position	TBE	TBE	TBE	TBE
Cooling system				
Thermostat opening temperature ....°C	78 to 82	78 to 82	85	85
Radiator cap pressure ....bar	1.0	1.0	1.4 to 1.6	1.4 to 1.6
Fuel system				
Idle speed ....rpm	750 to 800	750 to 800	900 ± 30	900 ± 30
Maximum (no load) speed ....rpm	5040 to 5120	5040 to 5120	5050 ± 50	5050 ± 50
Smoke test/opacity ....M⁻¹ %	2.5	2.0	2.0	2.0
Static timing method............................	Plunger travel	Rotor groove	Dial gauge	Dial gauge
Timing dimension............................mm	1.0 + 0.05	1.0	1.0 ± 0.02	0.9 ± 0.02
Crankshaft position ....mm [°]	[0] TDC	[0] TDC	0 TDC	0 TDC
Turbo type / ref / pressure ....bar	_	_	_	0.6 to 0.83 bar @ 4000 rpm
Injection pump make	Bosch	CAV	Bosch	Bosch
Injection pump part no...........................	_	DPS C8550A000A	VE4/8 R337	VE 4/9 R420
Injector Make / type	Bosch	CAV	Bosch	Bosch
Injector part no.......................................	DNO SD 1930	CDN OSD 6827, DNC 5641032	0432 217 206	0430 211 053
Injection type.......................................	VE 4/9F 2300/R54	Indirect DPS	Indirect VE	Indirect VE
Injection opening pressure, New [used]...bar	135 +8 -0	135 +8 -0	130 to 138 [120]	150 to 158 [140]
Glow plugs				
Maker	Champion	Champion	Bosch/Champion	Bosch/Champion
Type	CH68	CH68	0250 201 032 / CH160	0250 201 032 / CH160
Nominal rating....................................V/A	11 / 12	11 / 12	11 / 12	11 / 12
Brakes				
minimum friction material thickness				
Front...mm	1.5	1.5	7.0 with backing	7.0 with backing
Rear..mm	1.5	1.5	2.5	2.5
Tyres - Saloon / Hatch..............Size	155x13	155x13	175/70x13	175/70x13; 185/60x14
- Estate / Van................Size				
Pressure - front / rear - Saloon / Hatch...bar	2.1 / 2.2	2.1 / 2.2	2.1 / 1.8	2.1 / 1.8
- Estate / Van ......bar	_	_	_	_
Front suspension / wheel alignment				
Toe-in (+) / Toe-out (–)....................mm [°]	-2.0 to 6.0[1]	-2.0 to 6.0[1]	0 ± 1.0	0 ± 1.0
Camber	30' to 1°30'[2]	30' to 1°30'[2]	-25' ± 20'	-25' ± 20'
Castor	45' to 1°45'[3]	45' to 1°45'[3]	1°40' ± 30'	1°40' ± 30'
King pin inclination...............................	_	_	_	_
Rear suspension / wheel alignment				
Toe-in (+) / Toe-out (–)....................mm [°]	2.0 to 6.0[4]	2.0 to 6.0[4]	[20' ± 10']	[20' ± 10']
Camber	15' to 1°15'[5]	15' to 1°15'[5]	-1°30' ± 10'	-1°30' ± 10'

SEAT

Torque wrench settings	Malaga 1.7 D 1986 to 1992	Malaga 1.7 D 1986 to 1992	Toledo 1.9 D 1991 to 1998	Toledo 1.9 TD 1992 to 1998
Cylinder head - stage 1 ....Nm	60	60	40 N	40 N
- stage 2 ....Nm	+ 90°	+ 90°	60	60
- stage 3 ....Nm	+ 90°	+ 90°	+ 90°	+ 90°
- stage 4 ....Nm	–	–	+ 90°	+ 90°
- stage 5 ....Nm	–	–	–	–
- stage 6 ....Nm	–	–	–	–
Big-end bearings.....................Nm	74 N	74 N	30 + 90° N	30 + 90° N
Main bearings.....................Nm	113 M10: 80	113 M10: 80	65 + 90° N	65 + 90° N
Crankshaft pulley bolt ....Nm	145	145	90 + 90° N	90 + 90° N
Camshaft pulley bolt ....Nm	118	118	45	45
Flywheel [driveplate] bolt.....................Nm	85	85	20	20
Front hubs ....Nm	216	216	260	260
Rear hubs ....Nm	216	216	WSM	WSM
Wheel nuts / bolts ....Nm	Alloy: 78 Steel: 86	Alloy: 78 Steel: 86	110	110
Glow plugs ....Nm	15	15	25	25
Clutch pressure plate bolts ....Nm	30	30	60 + 90° N	60 + 90° N
Injection pump sprocket.....................Nm	49	49	55	55
Injectors.....................Nm	39	39	70	70
Injection pump mounting bolts ....Nm	29	29	25	25
Injector pipe unions.....................Nm	–	–	25	25
Capacities				
Engine oil & filter.....................litres	5.0	5.0	4.3	4.3
Gearbox.....................litres	3.4	3.4	2.0	2.0
Automatic transmission ....litres	–	–	–	–
Final drive.....................litres	WT	WT	WT	WT
Cooling system.....................litres	8.9	8.9	5.5	5.5
Fuel tank.....................litres	47	47	55	55

Notes

Malaga 1.7 D 1986 to 1992
[1]1988 ▶: 1.0±2.0
[2]1988 ▶: -43'±30'
[3]1988 ▶: 0±30'
[4]1988 ▶: 4.0±2.0
[5]1988 ▶: 45'±30'
Malaga 1.7 D 1986 to 1992
[1]1988 ▶: 1.0±2.0
[2]1988 ▶: -43'±30'
[3]1988 ▶: 0±30'
[4]1988 ▶: 4.0±2.0
[5]1988 ▶: 45'±30'

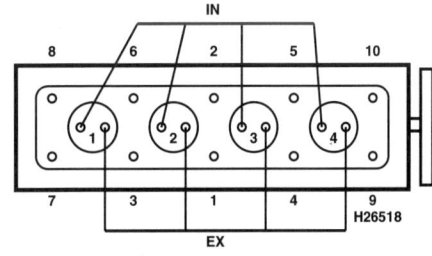

1714 cm³

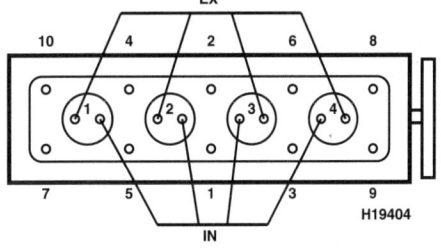

1896 cm³

– Not applicable, or information not available

SEAT

	Toledo 1.9 TDi 1995 to 2000	Toledo 1.9 TDi 1995 to 2000	Alhambra 1.9 TDi 1996 to 2000	Alhambra 1.9 TDi 110 1997 to 2000
Engine				
Engine type/code....................	1Z / AHU SOHC Turbo 8V 66kW	AFN SOHC Turbo 8V 81kW	1Z / AHU SOHC Turbo 66kW	AFN SOHC Turbo 81kW
Capacity (cm³) / cylinders........................	1896 / 4	1896 / 4	1896 / 4	1896 / 4
Compression ration / pressure ....bar	19.5 / ≥19.0	19.5 / ≥19.0	19.5 / ≥19.0	19.5 /
Torque output ....Nm	202	235	202	234
Oil pressure ....idle [running] bar	[≥2.0 @ 2000]	[≥2.0 @ 2000]	[2.0 @ 2000]	[2.0 @ 2000]
Oil temperature ....°C	80	80	80	80
Valve clearances - inlet (mm)	0: Hyd.	0: Hyd.	0: Hyd.	0: Hyd.
- exhaust (mm)	0: Hyd.	0: Hyd.	0: Hyd.	0: Hyd.
Injection order.......................................	1-3-4-2	1-3-4-2	1-3-4-2	1-3-4-2
No. 1 cylinder position............................	TBE	TBE	TBE	TBE
Cooling system				
Thermostat opening temperature ....°C	85	87	85 to 105	85 to 105
Radiator cap pressure ....bar	1.4 to 1.6	1.4 to 1.6	1.4 to 1.6	1.4 to 1.6
Fuel system				
Idle speed ....rpm	900 ± 30	900 ± 30	900 ± 40	860 to 940 N/A
Maximum (no load) speed ....rpm	5050 ± 50	5050 ± 50	5200	_
Smoke test/opacity ....M⁻¹ %	2.0	2.0	2.0	2.0
Static timing method................................	Refer to wsm	Refer to wsm	Refer to wsm	Refer to wsm
Timing dimension.................................mm	_	_	_	_
Crankshaft position ....mm [°]	_	_	_	_
Turbo type / ref / pressure ....bar	0.5 to 0.65 bar @ 4000	0.5 to 0.65 bar @ 3500 rpm¹	0.5 to 0.65 bar @ 3500rpm²	0.5 to 0.65 bar @ 3500 rpm²
Injection pump make	Bosch	Bosch	Bosch	Bosch
Injection pump part no............................	VER 510	VER 510	VER 510	VER 510
Injector Make / type	Bosch	_	Bosch	
Injector part no.......................................	_		0432 193 8383	
Injection type..	Direct VER	Direct VER	Direct VER	Direct VER
Injection opening pressure, New [used]...bar	190 to 200 [170]	190 to 200 [170]	190 to 200 [170]	190 to 200 [≥170]
Glow plugs				
Maker	Beru	Bosch	Bosch	Bosch
Type	855 MJ	0250 202 009	0250 202 009	0250 202 009
Nominal rating ....V/A	11.5 / 8	11 / 12	11.5 / 12	11 / 12
Brakes				
minimum friction material thickness				
Front..mm	7.0 with backing	7.0 with backing	2.0	2.0
Rear...mm	2.5	2.5	2.0	2.0
Tyres - Saloon / Hatch......................Size	185/60x14	175/70x13: 185/60x14	_	_
- Estate / Van......................Size	_	_	195/65x15:205/60x15:215/60x15	195/65x15:205/60x15:215/60x15
Pressure - front / rear - Saloon / Hatch ...bar	2.1 / 1.8	2.1 / 1.8		
- Estate / Van ......bar	_	_	2.6 / 2.4: 2.7 / 2.5¹	2.6 / 2.4: 2.7 / 2.5¹
Front suspension / wheel alignment				
Toe-in (+) / Toe-out (–)....................mm [°]	0 ± 1.0	0 ± 1.0	[10' ± 20']	[10' ± 20']
Camber	-25' ± 20'	-25' ± 20'	-20' ± 45'	-20' ± 45'
Castor	1°40' ± 30'	1°40' ± 30'	3°20' ± 40'	3°20' ± 40'
King pin inclination.................................	_	_	_	_
Rear suspension / wheel alignment				
Toe-in (+) / Toe-out (–)....................mm [°]	[20' ± 10']	20' ± 10'	[10' ± 25'] (not Nivomat)	[10' ± 25'] (not Nivomat)
Camber	-1°30' ± 10'	-1°30' ± 10'	-20' ± 30' (not Nivomat)	-20' ± 30' (not Nivomat)

SEAT

	Toledo 1.9 TDi 1995 to 2000	Toledo 1.9 TDi 1995 to 2000	Alhambra 1.9 TDi 1996 to 2000	Alhambra 1.9 TDi 110 1997 to 2000
Torque wrench settings				
Cylinder head - stage 1 ....Nm	40 N	40 N	40 N	40 N
- stage 2 ....Nm	60	60	60	60
- stage 3 ....Nm	+ 90°	+ 90°	+ 90°	+ 90°
- stage 4 ....Nm	+ 90°	+ 90°	+ 90°	+ 90°
- stage 5 ....Nm	–	–	–	–
- stage 6 ....Nm	–	–	–	–
Big-end bearings.............................Nm	35 + 90° N	30 + 90° N	35 + 90° N	30 + 90° N
Main bearings.................................Nm	65 + 90° N	65 + 90° N	65 + 90° N	65 + 90° N
Crankshaft pulley bolt ....Nm	90 + 90° N	90 + 90° N	90 + 90° N	90 + 90° N
Camshaft pulley bolt ....Nm	45	45	45	80
Flywheel [driveplate] bolt.................Nm	20	20	60 + 90° N	60 + 90° N
Front hubs ....Nm	260	260	150 + 90° N	150 + 90° N
Rear hubs ....Nm	WSM	WSM	200 N	200 N
Wheel nuts / bolts ....Nm	110	110	140	140
Glow plugs ....Nm	15	15	30	30
Clutch pressure plate bolts ....Nm	60 + 90° N	60 + 90° N	25	20
Injection pump sprocket...................Nm	55	55	55	55
Injectors..Nm	70	–	–	–
Injection pump mounting bolts ....Nm	25	25	25	25
Injector pipe unions.........................Nm	25	25	25	25
Capacities				
Engine oil & filter.............................litres	4.3	4.3	4.7	4.3
Gearbox..litres	2.0	2.0	2.2	2.2
Automatic transmission ....litres	–	–	3.5	3.5
Final drive ....litres	WT	WT	WT AT: 0.8	WT AT: 0.8
Cooling system................................litres	5.5	5.5	7.0 2 heat exchangers: 9.0	9.2
Fuel tank...litres	55	55	70	70

Notes

Toledo 1.9 TDi 1995 to 2000
[1]Charge pressure control solenoid disconnected
Alhambra 1.9 TDi 1996 to 2000
[1]215/60x15: 2.3 / 2.1
[2]Charge pressure control solenoid disconnected
Alhambra 1.9 TDi 110 1997 to 2000
[1]215/60x15: 2.3 / 2.1
[2]Charge pressure control solenoid disconnected

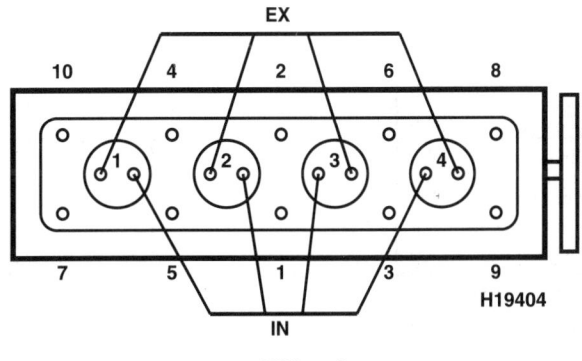

EX

10   4   2   6   8

7   5   1   3   9

IN

H19404

1896 cm³

– Not applicable, or information not
  available

SEAT

	Terra 1272 1990 to 1994	Inca 1.9 SDi 1998 to 2000	Inca 1.9 D 1998 to 2000	
Engine				
Engine type/code....................................	MN SOHC 33kW	AEY SDI SOHC 47kW	1Y SOHC 47kW	
Capacity (cm³) / cylinders.....................	1272 / 4	1896 / 4	1896 / 4	
Compression ration / pressure ....bar	22.0 / ≥25.0	19.5 / ≥19.0	22.5 / ≥26.0	
Torque output ....Nm	0	125	124	
Oil pressure ....idle [running] bar	[2.0]	[2.0 @ 2000]	[2.0 @ 2000]	
Oil temperature ....°C	80	_	80	
Valve clearances - inlet (mm)	0: Hyd.	0: Hyd.	0: Hyd.	
- exhaust (mm)	0: Hyd.	0: Hyd.	0: Hyd.	
Injection order	1-3-4-2	1-3-4-2	1-3-4-2	
No. 1 cylinder position	TBE	TBE	TBE	
Cooling system				
Thermostat opening temperature ....°C	87	92	87	
Radiator cap pressure ....bar	1.2 to 1.5	1.4 to 1.6	1.2 to 1.5	
Fuel system				
Idle speed ....rpm	900 ± 50	850 to 950	900 ± 30	
Maximum (no load) speed ....rpm	5400 ± 100	4950 to 5150	5150	
Smoke test/opacity ....M⁻¹ %	2.0	2.0	2.0	
Static timing method................................	Plunger travel	Refer to wsm	Plunger travel	
Timing dimension................................mm	1.05 + 0.02	_	0.9 ± 0.02	
Crankshaft position ....mm [°]	[0] TDC	_	TDC	
Turbo type / ref / pressure ....bar	_	_	_	
Injection pump make	Bosch	Bosch	Lucas	
Injection pump part no............................	_	VE	_	
Injector Make / type	Bosch	Bosch	CAV	
Injector part no..	_	_	_	
Injection type..	Indirect	VE	Indirect	
Injection opening pressure, New [used]...bar	≥130 [120]	190 to 200 [170]	130 to 138 [≥120]	
Glow plugs				
Maker	Bosch/Champion	Bosch	Bosch/Champion	
Type	0250 201 021 / CH69	0250 202 022	0250 201 032 / CH160	
Nominal rating ....V/A	11 / 12	11.5 / 12	11.5 / 8	
Brakes				
minimum friction material thickness				
Front...mm	1.5	7.0 with backing	7.0 with backing	
Rear...mm	1.5	2.5	2.5	
Tyres - Saloon / Hatch.......................Size	_	_	_	
- Estate / Van..........................Size	145x13	_	175/65x14	
Pressure - front / rear - Saloon / Hatch ...bar	_	_	_	
- Estate / Van ......bar	2.1 / 2.1	Refer to vehicle	2.2 / 2.4	
Front suspension / wheel alignment				
Toe-in (+) / Toe-out (–)....................mm [°]	0 to -4.0	[0° ± 10']	[0° ± 10']	
Camber	1°30' ± 30'	-0°25' ± 20'	-1° ± 30'	
Castor	3° ± 30'	0°33' ± 30'	0°33' ± 30'	
King pin inclination..................................	_	_	_	
Rear suspension / wheel alignment				
Toe-in (+) / Toe-out (–)....................mm [°]	_	[0°25' ± 20']	[0°25' ± 20']	
Camber	_	-1° ± 30'	-1° ± 30'	

SEAT

Torque wrench settings	Terra 1272 1990 to 1994	Inca 1.9 SDi 1998 to 2000	Inca 1.9 D 1998 to 2000	
Cylinder head - stage 1 ....Nm	40 N	40 N	40 N	
- stage 2 ....Nm	60	60	60	
- stage 3 ....Nm	+ 180°	+ 90°	+ 90°	
- stage 4 ....Nm	–	+ 90°	+ 90°	
- stage 5 ....Nm	–	–	–	
- stage 6 ....Nm	–	–	–	
Big-end bearings.........................Nm	30 + 90° N	30 + 90° N	30 + 90° N	
Main bearings.........................Nm	65	65 + 90° N	65	
Crankshaft pulley bolt ....Nm	90 + 180°	90 +90° N	90 + 90° N	
Camshaft pulley bolt ....Nm	45	45	45	
Flywheel [driveplate] bolt.....................Nm	30 + 90°	60 + 90°	30 + 90° N	
Front hubs ....Nm	196	265	265	
Rear hubs ....Nm	57	WSM	WSM	
Wheel nuts / bolts ....Nm	86	110	110	
Glow plugs ....Nm	30	25	30	
Clutch pressure plate bolts ....Nm	25	20	20	
Injection pump sprocket.....................Nm	50	55	–	
Injectors.........................Nm	70	20	70	
Injection pump mounting bolts ....Nm	25	25	–	
Injector pipe unions.............................Nm	25	25	–	
Capacities				
Engine oil & filter................................litres	3.5	4.5	4.5	
Gearbox................................litres	2.4	2.0	2.0	
Automatic transmission ....litres	–	–	–	
Final drive ....litres	WT	WT	WT	
Cooling system................................litres	5.2	6.6	6.6	
Fuel tank................................litres	35	52	52	

Notes

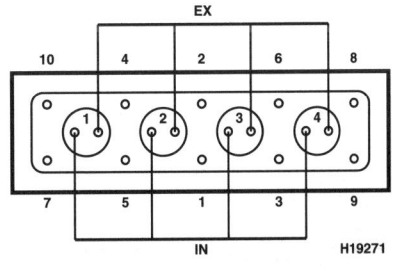

1272 cm³

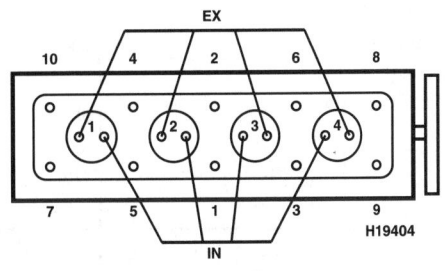

1896 cm³

– Not applicable, or information not
available

SKODA

	Felicia 1.9 D 1997 to 2000	Octavia 1.9 SDi 1997 to 2000	Octavia 1.9 TDi 1998 to 2000	Octavia 1.9 TDi 1998 to 2000
Engine				
Engine type/code	AEF SOHC 47kW	AGP SOHC 50kW	AGR SOHC Turbo 64kW	AHF SOHC Turbo 81kW
Capacity (cm³) / cylinders..................	1896 / 4	1896 / 4	1896 / 4	1896 / 4
Compression ration / pressure ....bar	22.5 / 26 to 34	19.5 / 25 to 31	19.5 / 19 to 31	19.5 / 19 to 31
Torque output ....Nm	124	130	202	235
Oil pressure ....idle [running] bar	[2.0 @ 2000]	1.0 [2.0 @ 2000]	[2.0 @ 2000]	[2.0 @ 2000]
Oil temperature ....°C	80	80	80	80
Valve clearances - inlet (mm)	0: Hyd.	0: Hyd.	0: Hyd.	0: Hyd.
- exhaust (mm)	0: Hyd.	0: Hyd.	0: Hyd.	0: Hyd.
Injection order	1-3-4-2	1-3-4-2	1-3-2-4	1-3-4-2
No. 1 cylinder position	TBE	TBE	TBE	TBE
Cooling system				
Thermostat opening temperature ....°C	84	87	85	87
Radiator cap pressure ....bar	1.4	1.2 to 1.5	1.3 to 1.5	1.2 to 1.5
Fuel system				
Idle speed ....rpm	940 ± 20	875 to 950	780 to 900 N/A	875 to 950
Maximum (no load) speed ....rpm	5050 ± 100	4950 to 5150	4800 to 5200	4800 to 5200
Smoke test/opacity ....M⁻¹ %	1.05	2.0	2.5	2.5
Static timing method........................	Plunger travel	–	Refer to wsm	Refer to wsm
Timing dimension........................mm	0.9 ± 0.02	Computer controlled	Computer controlled	Computer controlled
Crankshaft position ....mm [°]	TDC	–	–	–
Turbo type / ref / pressure ....bar	–	–	–	–
Injection pump make	Lucas	Bosch	Bosch	Bosch
Injection pump part no.......................	DPC	–	VP37	–
Injector Make / type	–	–	–	–
Injector part no.......................	–	–	–	–
Injection type.......................	Indirect DPC	Direct	Direct EDC VP37	Direct EDC
Injection opening pressure, New [used]...bar	130 to 138 [≥120]	190 to 200 [170]	190 to 200 [≥170]	190 to 200 [≥170]
Glow plugs				
Maker	Champion	–	Beru	Beru
Type	CH160	–	GN855	GN855
Nominal rating.......................V/A	11.5 / 8	–		
Brakes				
minimum friction material thickness				
Front..................mm	2.0	2.0	2.0	2.0
Rear..................mm	2.5	2.5	2.2	2.0
Tyres - Saloon / Hatch.................Size	165/70x13	–	195/65x15	195/65x15:205/60x15:205/55x16
- Estate / Van.................Size	165/70x13	175/80x14: 195/65x15:205/60x15	195/65x15	195/65x15:205/60x15:205/55x16
Pressure - front / rear - Saloon / Hatch ...bar	2.0 / 2.0	–	2.1 / 2.1	2.0 / 2.2
- Estate / Van ......bar	2.0 / 2.0	2.2 / 2.2	Refer to vehicle	2.0 / 2.2
Front suspension / wheel alignment				
Toe-in (+) / Toe-out (–)....................mm [°]	1.0 ± 1.0	[0° ± 10']	[0° ± 10']	0° ± 10'
Camber	-0°30' ± 30'	-30' ± 30'	-0°30' ± 30'	-30' ± 30'
Castor	1°20' ± 45'	7°40' ± 30'	7°40' ± 30'	7°40' ± 30
King pin inclination...............................	12°15' ± 50'	–	–	–
Rear suspension / wheel alignment				
Toe-in (+) / Toe-out (–)....................mm [°]	[-1°10' ± 30']	[20' ± 10']	[0°20' ± 20']	[20' ± 10']
Camber	–	-1°27' ± 30'	-1°36' ± 10'	-1°27' ± 30'

SKODA

Torque wrench settings	Felicia 1.9 D 1997 to 2000	Octavia 1.9 SDi 1997 to 2000	Octavia 1.9 TDi 1998 to 2000	Octavia 1.9 TDi 1998 to 2000
Cylinder head - stage 1 ....Nm	40 N	35 N	35 N	35 N
- stage 2 ....Nm	60	60	60	60
- stage 3 ....Nm	+ 90°	+ 90°	+ 90°	+ 90°
- stage 4 ....Nm	+ 90°	+ 90°	+ 90°	+ 90°
- stage 5 ....Nm	–	–	–	–
- stage 6 ....Nm	–	–	–	–
Big-end bearings.........................Nm	30 + 90° N	30 + 90° N	30 + 90° N	30 + 90° N
Main bearings..............................Nm	65 + 90° N	65 + 90° N	65 + 90° N	65 + 90° N
Crankshaft pulley bolt ....Nm	90 + 90° N	120 + 90° N	120 + 90° N	120 + 90° N
Camshaft pulley bolt ....Nm	4	45	45	45
Flywheel [driveplate] bolt...............Nm	65 + 90° N	60 + 90° N [60 + 90° N]	60 + 90° N	60 + 90° N
Front hubs ....Nm	300	–	115 + 90° M16: 190 + 90°	300, slacken, 50 + 30° N
Rear hubs ....Nm	WSM	WSM	WSM	175 N
Wheel nuts / bolts ....Nm	110	120	120	120
Glow plugs ....Nm	25	15	15	15
Clutch pressure plate bolts ....Nm	25	20	20	20
Injection pump sprocket...............Nm	25	–	20 + 90° N	20 + 90° N
Injectors......................................Nm	70	–	20	20
Injection pump mounting bolts ....Nm	25	–	25	25
Injector pipe unions.....................Nm	25	–	25	25
Capacities				
Engine oil & filter.................................litres	5.0	4.5	4.5	5.2
Gearbox...litres	2.4	1.6	2.0	2.0
Automatic transmission ....litres	–	–	–	–
Final drive ....litres	WT	WT	WT	WT
Cooling system.....................................litres	6.0	5.0	6.3	6.3
Fuel tank..litres	42	–	62	55

Notes

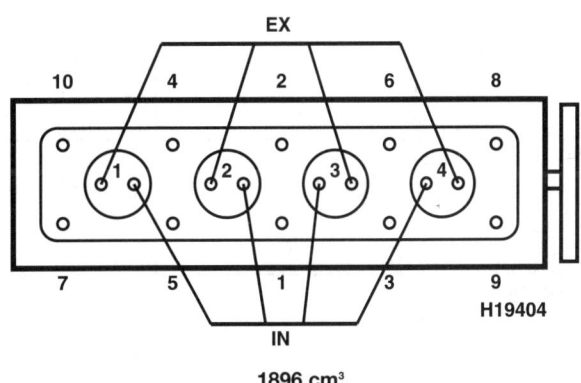

H19404

1896 cm³

– Not applicable, or information not available

Haynes
THE BOOK ®

SSANGYONG

	Korando 2.9 D 1997 to 1998	Musso 2.9 D 1995 to 1998	Musso 2.9 TD 1997 to 1998	
Engine				
Engine type/code............................	OM602 SOHC 58kW	OM602 SOHC 70kW	OM662LA SOHC Turbo 87kW	
Capacity (cm³) / cylinders................	2874 / 4	2874 / 5	2874 / 5	
Compression ration / pressure ....bar	22.0 / 18 to 28	22.0 / 18 to 28	22.0 / 18 to 28	
Torque output ....Nm	190	190	250	
Oil pressure ....idle [running] bar	–	–	–	
Oil temperature ....°C	–	–	–	
Valve clearances - inlet (mm)	0: Hyd.	0: Hyd.	0: Hyd.	
- exhaust (mm)	0: Hyd.	0: Hyd.	0: Hyd.	
Injection order	1-3-4-2	1-2-4-5-3	1-2-4-5-3	
No. 1 cylinder position	TCE	TCE	TCE	
Cooling system				
Thermostat opening temperature ....°C	–	85	80	
Radiator cap pressure ....bar	–	–	–	
Fuel system				
Idle speed ....rpm	700 ± 50	700 ± 50	770 ± 50	
Maximum (no load) speed ....rpm	4600	4600	4600	
Smoke test/opacity ....M⁻¹ %	2.5	2.5	2..5	
Static timing method................................	Refer to wsm	Refer to wsm	Refer to wsm	
Timing dimension.................................mm				
Crankshaft position ....mm [°]	–	–	–	
Turbo type / ref / pressure ....bar	–	–	–	
Injection pump make	Bosch	Bosch	Bosch	
Injection pump part no.............................	601 070 2201	602 070 7501	C320 RS 191	
Injector Make / type	Bosch	Bosch	Bosch	
Injector part no..	–	–	–	
Injection type...	Indirect PES 4M	Indirect PES 5M	Indirect PES 5M	
Injection opening pressure, New [used]...bar	115 to 125 [100]	115 to 125 [100]	115 to 125 [100]	
Glow plugs				
Maker	Bosch	Bosch/Champion		
Type	0250 201 026	0250 201 026 / CH156	–	
Nominal rating.....................................V/A	11.5 / 30	11.5 / 30		
Brakes				
minimum friction material thickness				
Front...mm		2.0	2.0	
Rear..mm		1.5	1.5	
Tyres - Saloon / Hatch.....................Size	–	–	–	
- Estate / Van......................Size	235/75x15	235/75x15	235/75x15	
Pressure - front / rear - Saloon / Hatch ...bar	–	–	–	
- Estate / Van ......bar	2.1 / 2.1	2.1 / 2.1	2.1 / 2.1	
Front suspension / wheel alignment				
Toe-in (+) / Toe-out (–)....................mm [°]	0.4	0.4	0.4	
Camber	0° ± 30'	0° ± 30'	0° ± 30'	
Castor	2°30' ± 30'	2°30' ± 30'	2°30' ± 30'	
King pin inclination.................................	12°30'	12°30'	12°30'	
Rear suspension / wheel alignment				
Toe-in (+) / Toe-out (–)....................mm [°]	–	–	–	
Camber	–	–	–	

SSANGYONG

	Korando 2.9 D 1997 to 1998	Musso 2.9 D 1995 to 1998	Musso 2.9 TD 1997 to 1998	
Torque wrench settings				
Cylinder head - stage 1 ....Nm	10 N	15 N	15 N	
- stage 2 ....Nm	35	35	35	
- stage 3 ....Nm	+ 90°	+ 90°	+ 90°	
- stage 4 ....Nm	Wait 10 mins	Wait 10 mins	+ 90°	
- stage 5 ....Nm	+ 90°	+ 90°	–	
- stage 6 ....Nm	–	–	–	
Big-end bearings........................Nm	35 + 90° N	35 + 90° N	40 + 90° N	
Main bearings........................Nm	35 + 90° N	35 + 90° N	55 + 90° N	
Crankshaft pulley bolt ....Nm	200 + 90°	–	–	
Camshaft pulley bolt ....Nm	25 + 90°	25 + 90°	–	
Flywheel [driveplate] bolt..................Nm	45 + 90° N	45 + 90° N	45 + 90° N	
Front hubs ....Nm	WSM	WSM	WSM	
Rear hubs ....Nm	WSM	WSM	WSM	
Wheel nuts / bolts ....Nm	120	120 Alloy: 136	120 Alloy: 136	
Glow plugs ....Nm	20	20	20	
Clutch pressure plate bolts ....Nm	23	23	35	
Injection pump sprocket........................Nm	–	–	–	
Injectors........................Nm	40	40	–	
Injection pump mounting bolts ....Nm	25	25	–	
Injector pipe unions............................Nm	18	18	–	
Capacities				
Engine oil & filter................................litres	8.0	9.5	9.5	
Gearbox................................litres	2.0	2.0	2.0	
Automatic transmission ....litres	10.0	9.0	–	
Final drive ....litres	1.4 Rear: 1.9	1.4 Rear: 1.9¹	1.4 Rear: 1.9¹	
Cooling system................................litres	10.0	10.0	10.0	
Fuel tank................................litres	–	70	80	

Notes

Musso 2.9 D 1995 to 1998
¹Transfer: 1.4
Musso 2.9 TD 1997 to 1998
¹Transfer: 1.4

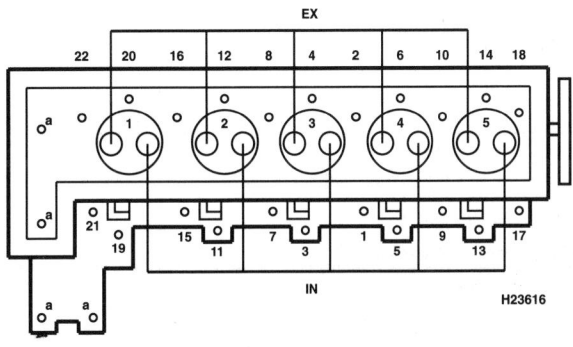

2439 cm³ / 2874 cm³

– Not applicable, or information not
  available

SUZUKI

	Vitara 2.0 D 1996 to 2000			
Engine				
Engine type/code..........................	RFL4 SOHC 65kW			
Capacity (cm³) / cylinders.........................	1998 / 4			
Compression ration / pressure ....bar	20.5 / 27 to 30			
Torque output ....Nm	159			
Oil pressure ....idle [running] bar	[3.9 @ 3000]			
Oil temperature ....°C	80			
Valve clearances - inlet (mm)	0.20 to 0.25			
- exhaust (mm)	0.30 to 0.35			
Injection order	1-3-4-2			
No. 1 cylinder position	TBE			
Cooling system				
Thermostat opening temperature ....°C	80 or 88			
Radiator cap pressure ....bar	0.9			
Fuel system				
Idle speed ....rpm	770 ± 50 N/A			
Maximum (no load) speed ....rpm	_			
Smoke test/opacity ....M⁻¹ %	2.5			
Static timing method................................	Refer to wsm			
Timing dimension.............................mm	_			
Crankshaft position ....mm [°]	_			
Turbo type / ref / pressure ....bar	_			
Injection pump make	_			
Injection pump part no.............................	_			
Injector Make / type	_			
Injector part no..	_			
Injection type...	Indirect			
Injection opening pressure, New [used]...bar	147			
Glow plugs				
Maker	_			
Type	_			
Nominal rating....................................V/A	_			
Brakes				
minimum friction material thickness				
Front....................................mm	8.0 with backing			
Rear....................................mm	1.0			
Tyres - Saloon / Hatch.....................Size	_			
- Estate / Van.........................Size	215/65x16			
Pressure - front / rear - Saloon / Hatch ...bar	_			
- Estate / Van ......bar	1.6 / 1.6			
Front suspension / wheel alignment				
Toe-in (+) / Toe-out (–)....................mm [°]	2.0 to 6.0			
Camber	30' ± 1°			
Castor	2° ± 1°			
King pin inclination..................................	_			
Rear suspension / wheel alignment				
Toe-in (+) / Toe-out (–)....................mm [°]	_			
Camber	_			

SUZUKI

Torque wrench settings	Vitara 2.0 D 1996 to 2000			
Cylinder head - stage 1 ....Nm	30[1]			
- stage 2 ....Nm	+ 90°			
- stage 3 ....Nm	+ 90°			
- stage 4 ....Nm	_			
- stage 5 ....Nm	_			
- stage 6 ....Nm	_			
Big-end bearings.......................Nm	65 N			
Main bearings.......................Nm	87			
Crankshaft pulley bolt ....Nm	175			
Camshaft pulley bolt ....Nm	60			
Flywheel [driveplate] bolt.......................Nm	100			
Front hubs ....Nm	120 to 170			
Rear hubs ....Nm	120 to 200			
Wheel nuts / bolts ....Nm	95			
Glow plugs ....Nm	18			
Clutch pressure plate bolts ....Nm	_			
Injection pump sprocket.......................Nm	65			
Injectors.......................Nm	65			
Injection pump mounting bolts ....Nm	23			
Injector pipe unions.......................Nm	28			
Capacities				
Engine oil & filter.......................litres	5.5			
Gearbox.......................litres	1.5 Transfer: 1.7			
Automatic transmission ....litres	2.5			
Final drive ....litres	Front: 1.0 Rear: 2.2			
Cooling system.......................litres	6.6			
Fuel tank.......................litres	70			

Notes

Vitara 2.0 D 1996 to 2000
[1]Bolt length below head ≤114.5 mm

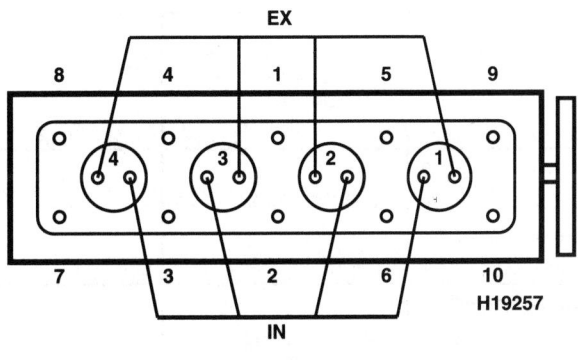

1998 cm³

– Not applicable, or information not available

TATA

Haynes THE BOOK ®

	Loadbeta 2.0 D 1994 to 2000			
Engine				
Engine type/code..........................	483 DL 44 SOHC 46kW			
Capacity (cm³) / cylinders........................	1948 / 4			
Compression ration / pressure ....bar	22.0 /			
Torque output ....Nm	115			
Oil pressure ....idle [running] bar	1.2			
Oil temperature ....°C	80			
Valve clearances - inlet (mm)	0.15			
- exhaust (mm)	0.30			
Injection order	1-3-4-2			
No. 1 cylinder position	–			
Cooling system				
Thermostat opening temperature ....°C	87.2			
Radiator cap pressure ....bar	1.0			
Fuel system				
Idle speed ....rpm	850 ± 50			
Maximum (no load) speed ....rpm	4750 to 4850			
Smoke test/opacity ....M⁻¹ %	2.5			
Static timing method................................	Rotor lift			
Timing dimension.................................mm	0.78 ± 0.02			
Crankshaft position ....mm [°]	TDC			
Turbo type / ref / pressure ....bar	–			
Injection pump make	Lucas			
Injection pump part no.............................	VE			
Injector Make / type	–			
Injector part no...	Mico DNOSO287			
Injection type...	Indirect VE			
Injection opening pressure, New [used]...bar	140 ± 8			
Glow plugs				
Maker	Bosch			
Type	0250 201 019			
Nominal rating......................................V/A	12 / 12			
Brakes				
minimum friction material thickness				
Front.....................................mm	7.0 with backing			
Rear......................................mm	1.5			
Tyres - Saloon / Hatch......................Size	–			
- Estate / Van............................Size	195/80x15: 215/70x15'			
Pressure - front / rear - Saloon / Hatch ...bar	–			
- Estate / Van ......bar	2.2 / 2.2			
Front suspension / wheel alignment				
Toe-in (+) / Toe-out (–)....................mm [°]	–			
Camber	0° ± 20'			
Castor	3° ± 20'			
King pin inclination...................................	–			
Rear suspension / wheel alignment				
Toe-in (+) / Toe-out (–)....................mm [°]	–			
Camber	–			

TATA

	Loadbeta 2.0 D 1994 to 2000			
Torque wrench settings				
Cylinder head - stage 1 ....Nm	30			
- stage 2 ....Nm	60			
- stage 3 ....Nm	Slacken, 60			
- stage 4 ....Nm	Warm up, allow to cool			
- stage 5 ....Nm	Slacken, 90			
- stage 6 ....Nm	Slacken, 90			
Big-end bearings..............................Nm	49 N			
Main bearings...................................Nm	68			
Crankshaft pulley bolt ....Nm	$40 + 60°$			
Camshaft pulley bolt ....Nm	35			
Flywheel [driveplate] bolt....................Nm	49			
Front hubs ....Nm	_			
Rear hubs ....Nm	245			
Wheel nuts / bolts ....Nm	130			
Glow plugs ....Nm	22			
Clutch pressure plate bolts ....Nm	50			
Injection pump sprocket.....................Nm	49			
Injectors...Nm	88			
Injection pump mounting bolts ....Nm	23			
Injector pipe unions...........................Nm	20			
Capacities				
Engine oil & filter...............................litres	6.8			
Gearbox..litres	1.5			
Automatic transmission ....litres	_			
Final drive ....litres	1.9			
Cooling system..................................litres	10.0			
Fuel tank..litres	50			

Notes

Loadbeta 2.0 D 1994 to 2000
[1] 250/70x15

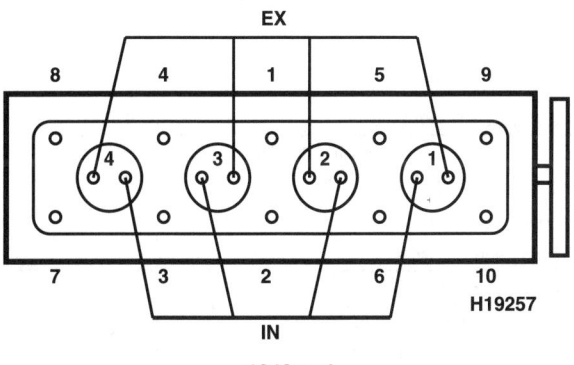

1948 cm³

– Not applicable, or information not available

TOYOTA

	Corolla 2.0 D 1993 to 1997	Corolla 2.0 D 1998 to 2000	Carina 2.0 (CT170) 1990 to 1992	Carina E 2.0 D 1992 to 1996
Engine				
Engine type/code....................	2C SOHC 52kW	2C-E SOHC 52kW	2C SOHC 53kW	2C SOHC 53kW
Capacity (cm³) / cylinders........................	1975 / 4	1975 / 4	1974 / 4	1975 / 4
Compression ration / pressure ....bar	23.1 / ≥24.5	23.1 / 24.5 to 29.4	23.0 / ≥25.0	23.0 / 24.5 to 29.4
Torque output ....Nm	13	132	132	132
Oil pressure ....idle [running] bar	[2.5 to 5.0 @ 3000]	[2.5 to 5.0 @ 3000]	[2.5 to 5.0 @ 3000]	[2.5 to 5.0 @ 3000]
Oil temperature ....°C	80	80	80	80
Valve clearances - inlet (mm)	0.20 to 0.30	0.20 to 0.30	0.20 to 0.30	0.20 to 0.30
- exhaust (mm)	0.25 to 0.35	0.25 to 0.35	0.25 to 0.35	0.25 to 0.35
Injection order...............................	1-3-4-2	1-3-4-2	1-3-4-2	1-3-4-2
No. 1 cylinder position	TBE	TBE	TBE	TBE
Cooling system				
Thermostat opening temperature ....°C	80	80	80 to 84	80
Radiator cap pressure ....bar	0.7 to 1.03	1.03	0.75 to 1.05	0.7 to 1.03
Fuel system				
Idle speed ....rpm	800 ± 50	800 ± 50	750 to 850	800 ± 50
Maximum (no load) speed ....rpm	5350 ± 50	5350 ± 50	5350	5350
Smoke test/opacity ....M⁻¹ %	2.5	2.5	2.5	2.5
Static timing method...............................	Plunger travel	Plunger travel	Plunger travel	Plunger travel
Timing dimension............................mm	0.55 to 0.61	0.38 to 0.58	0.77 to 0.83	0.55 to 0.61
Crankshaft position ....mm [°]	0 TDC	0 TDC	[0] TDC	0 TDC
Turbo type / ref / pressure ....bar	–	–	–	–
Injection pump make	Nippon Denso	Nippon Denso	Nippon Denso	Nippon Denso
Injection pump part no............................	VE	VE	VE	VE
Injector Make / type	Nippon Denso	Nippon Denso	Nippon Denso	Nippon Denso
Injector part no.....................................	DNOPD600	DNOPD650	DNOPD4	DNOPD600
Injection type......................................	Indirect VE	Indirect VE	Indirect VE	Indirect VE
Injection opening pressure, New [used]...bar	147 to 157 [137]	145 to 155 [137 to 156]	135 to 155	147 to 157 [137 to 157]
Glow plugs				
Maker	Bosch/Champion	–	Bosch/Champion	Bosch/Champion
Type	0250 202 052 / CH95	–	0250 202 052 / CH95	0250 202 052 / CH95
Nominal rating..................................V/A	–	–	7 / 8.5	–
Brakes				
minimum friction material thickness				
Front..mm	1.0	1.0	1.0	1.0
Rear..mm	1.0	1.0	1.0	1.0
Tyres - Saloon / Hatch......................Size	165/70x14: 165/65x14	175/65x14	165x13: 165/80x13	175/70x14
- Estate / Van.....................Size	–	175/65x14	–	–
Pressure - front / rear - Saloon / Hatch ...bar	2.4 / 2.1	2.1 / 2.1	2.1 / 1.9	2.2 / 2.1
- Estate / Van ......bar	–	2.1 / 2.2	–	–
Front suspension / wheel alignment				
Toe-in (+) / Toe-out (–)...........mm [°]	1.0 ± 1.0	-1.0 ± 2.0	0 ± 1.0	0.0 ± 2.0
Camber	-0°13'	-0°13' ± 45'	0° ± 30'	-0°20' ± 45'
Castor	1°24'	1°24' ± 45'	-5' ± 30' PAS: 20'±30'	0°30' ± 45'
King pin inclination................................	12°41'	12°41' ± 45	13°15' ± 30'	13°25' ± 45'
Rear suspension / wheel alignment				
Toe-in (+) / Toe-out (–)....................mm [°]	3.0 ± 2.0	4.0 ± 2.0	5.0 ± 1.0	4.0 ± 2.0
Camber	-0°55' ± 45'	1°45' ± 45'	-30' ± 30'	-0°35' ± 45'

TOYOTA

	Corolla 2.0 D 1993 to 1997	Corolla 2.0 D 1998 to 2000	Carina 2.0 (CT170) 1990 to 1992	Carina E 2.0 D 1992 to 1996
Torque wrench settings				
Cylinder head - stage 1 ....Nm	44	44	20	44
- stage 2 ....Nm	+ 90°	+ 90°	50	+ 90°
- stage 3 ....Nm	+ 90°	+ 90°	84	+ 90°
- stage 4 ....Nm	–	–	–	–
- stage 5 ....Nm	–	–	–	–
- stage 6 ....Nm	–		–	
Big-end bearings.........................Nm	64 N	64 N	64 N	64 N
Main bearings.............................Nm	103	103	103	103
Crankshaft pulley bolt ....Nm	98	196	98	496
Camshaft pulley bolt ....Nm	88	88	98	88
Flywheel [driveplate] bolt............Nm	88	88	88	88
Front hubs ....Nm	216	216	186	226
Rear hubs ....Nm	123	123	123	123
Wheel nuts / bolts ....Nm	103	100	103	103
Glow plugs ....Nm	13	13	13	13
Clutch pressure plate bolts ....Nm	19	20	18	19
Injection pump sprocket..............Nm	47	65	64	64
Injectors......................................Nm	64	64	64	64
Injection pump mounting bolts ....Nm	47	46		47
Injector pipe unions.....................Nm	29	30	27	29
Capacities				
Engine oil & filter................................litres	4.2	4.2	4.3	4.2
Gearbox..litres	2.6	2.6	2.6	2.6
Automatic transmission ....litres	–	–	–	–
Final drive ....litres	WT	WT	WT	WT
Cooling system..................................litres	5.3	5.3	7.5	7.4
Fuel tank...litres	50	50	60	60

Notes

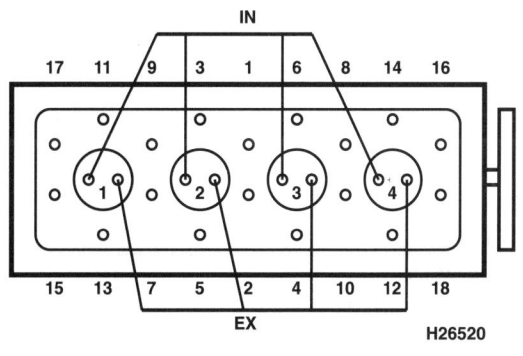

1975 cm³

H26520

– Not applicable, or information not available

TOYOTA

	Carina E 2.0 TD 1996 to 1997	Avensis 2.0 TD 1997 to 1999	Camry 2.0 Turbo (CV20) 1987 to 1991	Picnic 2.2 TD 1997 to 2000
Engine				
Engine type/code....................	2C-T SOHC Turbo 60kW	2C-TE OHC Turbo 62KW	2C-T SOHC Turbo 61kW	3C-TE SOHC Turbo 63kW
Capacity (cm^3) / cylinders........................	1975 / 4	1975 / 4	1975 / 4	2184 / 4
Compression ration / pressure ....bar	23.0 / 24.5 to 29.4	23.1 / ≥24.2	23.0 / ≥25.0	22.6 / 29
Torque output ....Nm	173	203	0	205
Oil pressure ....idle [running] bar	[2.5 to 5.0 @ 3000]	0.3 [2.5 to 6.0 @ 3000]	[2.5 to 6.0 @ 3000]	3.7
Oil temperature ....°C	80	80	80	80
Valve clearances - inlet (mm)	0.20 to 0.30	0.20 to 0.30	0.20 to 0.30	0.20 to 0.30
- exhaust (mm)	0.25 to 0.35	0.25 to 0.35	0.25 to 0.35	0.25 to 0.35
Injection order..............................	1-3-4-2	1-3-4-2	1-3-4-2	1-3-4-2
No. 1 cylinder position	TBE	TBE	TBE	TBE
Cooling system				
Thermostat opening temperature ....°C	80	80 to 84	80 to 84	80
Radiator cap pressure ....bar	0.7 to 1.03	0.75 to 1.05	0.75 to 1.00	1.0
Fuel system				
Idle speed ....rpm	800 ± 50	725 to 825	750 to 850	700 to 800
Maximum (no load) speed ....rpm	5350	5300	5300 to 5400	4900 to 5100
Smoke test/opacity ....M^{-1} %	2.5	2.5	2.5	2.5
Static timing method........................	Plunger travel	Plunger travel	Plunger travel	Rotor lift
Timing dimension.........................mm	0.55 to 0.61	0.38 to 0.58	0.77 to 0.83	0.55 to 0.61
Crankshaft position ....mm [°]	0 TDC	0 TDC	[0] TDC	TDC
Turbo type / ref / pressure ....bar	0.5 to 0.7 bar	0.5 to 0.7 bar	_	0.5 to 0.7 bar
Injection pump make	Nippon Denso	Nippon Denso	Nippon Denso	Nippon Denso
Injection pump part no.............................	VE	VE	VE	Electronic control
Injector Make / type	Nippon Denso	Nippon Denso	Nippon Denso	_
Injector part no.	_	DN0PD650	DNOPD4	DNOPD650
Injection type..............................	Indirect VE	Indirect VE	Indirect VE	Indirect
Injection opening pressure, New [used]...bar	147 to 157 [137 to 157]	147 to 157 [137 to 157]	147 to 157 [137 to 157]	145 to 156
Glow plugs				
Maker	_	_	_	_
Type	_	_	_	_
Nominal rating.................................V/A	11 / 15.3	11 / 15.3	_	_
Brakes				
minimum friction material thickness				
Front............................mm	1.0	1.0	1.0	1.0
Rear............................mm	1.0	1.0	1.0	1.0
Tyres - Saloon / Hatch......................Size	175/70x14	185/65x14: 195/65x15	185/70x14	_
- Estate / Van...........................Size	_	185/65x14: 195/65x15	_	195/65x14
Pressure - front / rear - Saloon / Hatch ...bar	2.2 / 2.1	2.2/2.2: 2.2/2.2	1.9 / 1.9	_
- Estate / Van ......bar		2.2/2.2: 2.2/2.2		2.5 / 2.5
Front suspension / wheel alignment				
Toe-in (+) / Toe-out (–)...............mm [°]	-0.6 ± 1.2	1.0 ± 2.0	1.0 ± 1.0	2.0 ± 2.0
Camber	-0°22' ± 45'	-0°21' ± 45'	35' ± 45'	-0°22' ± 45'
Castor	1°18 ' ± 45'	1°20' ± 45'	1°40' ± 30'	0°55' ± 45'
King pin inclination.................................	13°24' ± 45'	13°22' ± 45'	12°40' ± 45'	13° 17' ± 45'
Rear suspension / wheel alignment				
Toe-in (+) / Toe-out (–)...............mm [°]	-1.2 ± 1.2	2.0 ± 2.0	4.0 ± 1.0	1.0 ± 3.0
Camber	-0°55' ± 45'	-0°28' ± 30'	-35' ± 45'	-0°55' ± 45'

TOYOTA

Torque wrench settings	Carina E 2.0 TD 1996 to 1997	Avensis 2.0 TD 1997 to 1999	Camry 2.0 Turbo (CV20) 1987 to 1991	Picnic 2.2 TD 1997 to 2000
Cylinder head - stage 1 ....Nm	44	59	20	44
- stage 2 ....Nm	+ 90°	+ 90°	50	+ 90°
- stage 3 ....Nm	+ 90°	+ 90°	84	+ 90°
- stage 4 ....Nm	–	–	–	–
- stage 5 ....Nm	–	–	–	–
- stage 6 ....Nm		–		–
Big-end bearings.............................Nm	64 N	64 N	64 N	64 N
Main bearings.................................Nm	103	103	103	100
Crankshaft pulley bolt ....Nm	496	196	98	–
Camshaft pulley bolt ....Nm	88	88	98	–
Flywheel [driveplate] bolt.................Nm	88	88.2	88	88
Front hubs ....Nm	226	216	186[1]	216
Rear hubs ....Nm	123	123	103	WSM
Wheel nuts / bolts ....Nm	103	103	103	100
Glow plugs ....Nm	13	13	13	–
Clutch pressure plate bolts ....Nm	19	19	18	20
Injection pump sprocket...................Nm	64	64	64	–
Injectors..Nm	64	64	64	–
Injection pump mounting bolts ....Nm	47	–	–	–
Injector pipe unions.........................Nm	29	29	27	–
Capacities				
Engine oil & filter............................litres	4.2	5.1	4.6	4.1
Gearbox...litres	2.6	2.2	2.6	2.6
Automatic transmission ....litres	–	–	–	–
Final drive ....litres	WT	WT	WT	WT
Cooling system...............................litres	7.4	6.1 [AT: 6.0]	7.5	–
Fuel tank..litres	60	60	60	60

Notes

Camry 2.0 Turbo (CV20) 1987 to 1991
[1]If noisy, retighten to 273

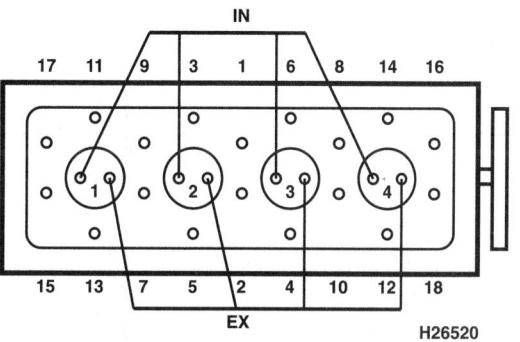

1975 cm³

H26520

– Not applicable, or information not available

TOYOTA

	Land Cruiser Colorado 3.0 TD 1996 to 2000	Land Cruiser Amazon 4.2 TD 1998 to 2000	Land Cruiser 4.2 TD 1990 to 1997	Land Cruiser II 2.4 (LJ70) 1990 to 1993
Engine				
Engine type/code............................	1KZ-T SOHC Turbo 91kW	1HD-FTE SOHC Turbo 148kW	1HD-T SOHC Turbo 121kW	2L-T SOHC Turbo 66kW
Capacity (cm³) / cylinders.......................	2982 / 4	4164 / 6	4164 / 6	2446 / 4
Compression ration / pressure ....bar	21.2 / 19.6 to 30.4	18.8 / ≥24.5	18.6 / 25 to 34	20.0 / 19.6 to 30.4
Torque output ....Nm	295	430	360	0
Oil pressure ....idle [running] bar	[2.5 to 6.1 @ 3000]	[2.5 @ 3000]	[2.5 to 6.0 @ 3000]	[2.5 to 5.9 @ 3000]
Oil temperature ....°C	80	80	80	80
Valve clearances - inlet (mm)	0.20 to 0.30	0.17 to 0.23	0.20	0.25 H
- exhaust (mm)	0.25 to 0.35	0.47 to 0.53	0.40	0.36 H
Injection order	1-3-4-2	–	1-4-3-6-2-5	1-3-4-2
No. 1 cylinder position	TBE	F	F	F
Cooling system				
Thermostat opening temperature ....°C	80	78	74 to 78	86 to 90
Radiator cap pressure ....bar	1.05	1.2	0.75 to 1.00	0.75 to 1.00
Fuel system				
Idle speed ....rpm	700 ± 50	700 ± 50	650	700 to 800
Maximum (no load) speed ....rpm	4600 ± 130	4300 to 4500	4400	4700 to 4900
Smoke test/opacity ....M⁻¹ %	2.5	2.5	2.5	2.5
Static timing method.............................	Plunger travel	Refer to wsm	Plunger travel	Plunger travel
Timing dimension................................mm	0.58 to 0.62	–	1.29 to 1.35	0.97
Crankshaft position ....mm [°]	0 TDC	–	[0] TDC	[0] TDC
Turbo type / ref / pressure ....bar	0.52 to 0.68 bar @ 4000 rpm	0.5 to 0.7 bar @ 4300 rpm	0.49 to 0.63 bar @ ≥2000 rpm	0.61 to 0.81 bar @ ≥2000 rpm
Injection pump make	Nippon Denso	Nippon Denso	Nippon Denso	Nippon Denso
Injection pump part no............................	VE 22100 67070	VE	VE	VE
Injector Make / type	Nippon Denso	Nippon Denso	Nippon Denso	Nippon Denso
Injector part no.....................................	DNOPD 619	–	KBAL 2 stage	DN-DN12SD 12
Injection type.......................................	Indirect VE	Direct VE	Indirect VE	Indirect VE
Injection opening pressure, New [used]...bar	148 to 158 [143 to 151]	340 to 350	129 to 135¹	148 to 156 [147]
Glow plugs				
Maker	Beru	–	Bosch	Bosch
Type	968MJ	–	0250 202 079	0250 202 073
Nominal rating....................................V/A	–	–	–	7 / 8
Brakes				
minimum friction material thickness				
Front..mm	1.0	1.0	1.0	1.0
Rear..mm	1.0	1.0	1.0	1.0
Tyres - Saloon / Hatch....................Size	–	–	–	–
- Estate / Van............................Size	215/80x16²	275/70x16	275/70x15	265/75x15
Pressure - front / rear - Saloon / Hatch ...bar	–	–	–	–
- Estate / Van ......bar	1.8 / 2.0³	2.2 / 2.2	2.2 / 2.2	1.8 / 2.0
Front suspension / wheel alignment				
Toe-in (+) / Toe-out (–).....................mm [°]	2.0 ± 2.0	1.0 ± 2.0	2.0 ± 1.0	1.0 ± 1.0
Camber	-0°10' ± 45'	0°5' ± 45'	1° ± 45'	1° ± 1°
Castor	2°55' ± 45'	2°25' ± 45'	3° ± 1°	4°5' ± 1°
King pin inclination..............................	10°55' ± 45'	12°10' ± 45'	13° ± 45'	9°30' ± 1°
Rear suspension / wheel alignment				
Toe-in (+) / Toe-out (–).....................mm [°]	–	–	–	–
Camber	–	–	–	–

TOYOTA

Torque wrench settings	Land Cruiser Colorado 3.0 TD 1996 to 2000	Land Cruiser Amazon 4.2 TD 1998 to 2000	Land Cruiser 4.2 TD 1990 to 1997	Land Cruiser II 2.4 (LJ70) 1990 to 1993
Cylinder head - stage 1 ....Nm	39	69²	69	78
- stage 2 ....Nm	+ 90°	+ 90°	+ 90°	+ 90°
- stage 3 ....Nm	+ 90°	+ 90°	+ 90°	+ 90°
- stage 4 ....Nm	–	–	–	–
- stage 5 ....Nm	–	–	–	–
- stage 6 ....Nm	–	–	–	–
Big-end bearings........................Nm	29 + 90° N	37 + 90° N	37 + 90° N	34 + 120° N
Main bearings...........................Nm	49 + 90°	103 + 90°	103 + 90°	103
Crankshaft pulley bolt ....Nm	235	430	490	137
Camshaft pulley bolt ....Nm	100	98	98	98
Flywheel [driveplate] bolt................Nm	145	127	127 [54]	123
Front hubs..............................Nm	WSM	WSM	WSM	–
Rear hubs..............................Nm	WSM	WSM	WSM	–
Wheel nuts / bolts ....Nm	158	209 Alloy: 131	147	158
Glow plugs ....Nm	13	–	13	13
Clutch pressure plate bolts ....Nm	20	39	39	19
Injection pump sprocket..................Nm	64	98	98	64
Injectors................................Nm	64	–	39	69
Injection pump mounting bolts ....Nm	18	69	–	–
Injector pipe unions.......................Nm	25	25	25	25
Capacities				
Engine oil & filter...........................litres	8.0	11.4	9.3	6.7
Gearbox.....................................litres	2.2	2.7	2.6 Transfer: 1.3	2.6 Transfer: 2.1
Automatic transmission ....litres	2.0	6.0	6.0	–
Final drive ....litres	1.1 Rear: 2.6¹	1.7 Rear: 3.2¹	Front: 2.7²	Front: 2.0¹
Cooling system.............................litres	11.0	13.2 AT: 12.8	12.0	10.9
Fuel tank....................................litres	90	96	90	90

Notes

Land Cruiser Colorado 3.0 TD 1996 to 2000
¹Transfer box: 1.1
²Stn Wagon 3/5dr: 265/70x16
³Stn Wagon 3/5dr: 2.0 / 2.0
Land Cruiser Amazon 4.2 TD 1998 to 2000
¹Transfer box: 1.3
² Shank diameter: ≥10.55 mm
Land Cruiser 4.2 TD 1990 to 1997
¹2nd stage: 176 to 186
²Rear: 2.8
Land Cruiser II 2.4 (LJ70) 1990 to 1993
¹Rear: 1.9

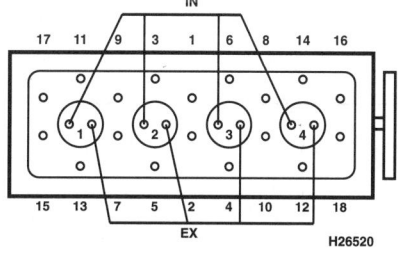

2982 cm³

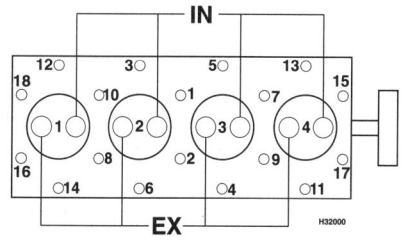

4164 cm³

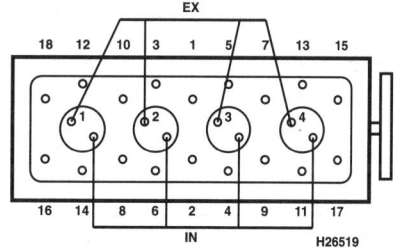

2446 cm³

– Not applicable, or information not available

TOYOTA

	Land Cruiser II 3.0 TD 1993 to 1996	Land Cruiser Colorado 3.0 TD 1996 to 1999	Lite-Ace 2.0 D 1992 to 1994	Lite-Ace/Power Van 2.4 D 1996 to 2000
Engine				
Engine type/code	1KZ-T SOHC Turbo 92kW	1KZ-TE SOHC Turbo 91kW	2C SOHC 53kW	2L SOHC 56kW
Capacity (cm³) / cylinders.....................	2982 / 4	2982 / 4	1975 / 4	2446 / 4
Compression ration / pressure ....bar	21.2 / 19.6 to 31	21.2 / 19.6 to 30.4	22.5 / 24.5 to 29.4	22.2 / 19.6 to 30.7
Torque output ....Nm	218	295	0	157
Oil pressure ....idle [running] bar	[2.5 to 6.0 @ 3000]	[2.9 to 5.4 @ 3000]	[2.5 to 5.9 @ 3000]	[3.0 to 5.4 @ 3000]
Oil temperature ....°C	80	80	80	80
Valve clearances - inlet (mm)	0.20 to 0.30	0.20 to 0.30	0.20 to 0.30	0.25
- exhaust (mm)	0.25 to 0.35	0.25 to 0.35	0.25 to 0.35	0.45
Injection order	1-3-4-2	1-3-4-2	1-3-4-2	1-3-4-2
No. 1 cylinder position	TBE	TBE	TBE	TBE
Cooling system				
Thermostat opening temperature ....°C	80	80	–	86
Radiator cap pressure ....bar	0.75 to 1.05	0.75 to 1.05	–	0.75 to 1.05
Fuel system				
Idle speed ....rpm	750 ± 50	700 ± 50	750 to 850	700 ± 50
Maximum (no load) speed ....rpm	4730	4600	5300	5280
Smoke test/opacity ....M⁻¹ %	2.5	2.5	2.5	2.5
Static timing method.............................	Plunger travel	Plunger travel	Plunger travel	Plunger travel
Timing dimension.............................mm	0.39 to 0.43	0.58 to 0.62	0.77 to 0.83	0.54 to 0.66
Crankshaft position ....mm [°]	0 TDC	–	0 TDC	0 TDC
Turbo type / ref / pressure ....bar	0.52 to 0.68 bar @ 4600 rpm	0.52 to 0.68 bar @ 4600 rpm	–	–
Injection pump make	Nippon Denso	Nippon Denso	Nippon Denso	Nippon Denso
Injection pump part no.............................	VE	VE 22100 67070	–	VE
Injector Make / type	Diesel Kiki	Nippon Denso	Nippon Denso	Nippon Denso
Injector part no.............................	DNOPD619	DNOPD619	DNOPD4	DN4PD57
Injection type.............................	Indirect VE	VE	–	VE
Injection opening pressure, New [used]...bar	148 to 156 [147]	148 to 156 [143 to 151]	142 to 152 [132]	146 to 155 [142 to 152]
Glow plugs				
Maker	–	Beru	Bosch/Champion	Bosch/Champion
Type	–	968MJ	0250 202 052 / CH95	0250 202 053 / CH103
Nominal rating.............................V/A	–	–	7 / 8.5	–
Brakes				
minimum friction material thickness				
Front.........................mm	1.0	1.0	1.0	1.0
Rear.........................mm	1.0	1.0	1.0	1.0
Tyres - Saloon / Hatch......................Size	–	–	–	–
- Estate / Van..........................Size	265/75x15	215/80x16	–	195/70x15
Pressure - front / rear - Saloon / Hatch ...bar	–	–	–	–
- Estate / Van ......bar	1.8 / 2.0	1.8 / 2.0	–	3.3 / 4.4
Front suspension / wheel alignment				
Toe-in (+) / Toe-out (–)......................mm [°]	1.0 ± 1.0	2.0 ± 2.0	0.2 ± 0.5	[0°30' ± 20']
Camber	1° ± 1°	-0°10' ± 45'	-0°10' ± 45'	0°10' ± 45'
Castor	4°5' ± 1°	2°55' ± 45'	1°47' ± 45'	1°40' ± 45'
King pin inclination.............................	9°30' ± 1°	–	10°40'	–
Rear suspension / wheel alignment				
Toe-in (+) / Toe-out (–)......................mm [°]	–	–	–	[0°10' ± 20']
Camber	–	–	–	0°30' ± 45'

TOYOTA

Torque wrench settings	Land Cruiser II 3.0 TD 1993 to 1996	Land Cruiser Colorado 3.0 TD 1996 to 1999	Lite-Ace 2.0 D 1992 to 1994	Lite-Ace/Power Van 2.4 D 1996 to 2000
Cylinder head - stage 1 ....Nm	39	39	44	78
- stage 2 ....Nm	+ 90°	+ 90°	+ 90°	+ 90°
- stage 3 ....Nm	+ 90°	+ 90°	+ 90°	+ 90°
- stage 4 ....Nm	–	–	–	–
- stage 5 ....Nm	–	–	–	–
- stage 6 ....Nm	–	–	–	–
Big-end bearings.....................Nm	29 + 90° N	29 + 90° N	64 N	34 + 120° N
Main bearings.....................Nm	49 + 90°	49 + 90°	103	103
Crankshaft pulley bolt ....Nm	363	235	98	235
Camshaft pulley bolt ....Nm	98	100	88	98
Flywheel [driveplate] bolt.....................Nm	145	145	88	123
Front hubs ....Nm	WSM	WSM	WSM	WSM
Rear hubs ....Nm	WSM	WSM	WSM	290
Wheel nuts / bolts ....Nm	158	158	103	103
Glow plugs ....Nm	13	13	13	13
Clutch pressure plate bolts ....Nm	19	19	19	19
Injection pump sprocket.....................Nm	64	64	64	64
Injectors.....................Nm	64	64	64	65
Injection pump mounting bolts ....Nm	21	18	–	18
Injector pipe unions.....................Nm	15	25	29	25
Capacities				
Engine oil & filter.....................litres	8.0	8.0	4.2	6.7
Gearbox.....................litres	Transfer: 2.1	2.2 Transfer: 1.0	2.2	2.4
Automatic transmission ....litres	–	9.5	–	–
Final drive ....litres	1.98[1]	1.1 Rear: 2.6	–	1.8
Cooling system.....................litres	11.0	11.0	6.3 to 7.5	10.9
Fuel tank.....................litres	90	90	–	75

Notes

Land Cruiser II 3.0 TD 1993 to 1996
[1]Rear: 1.85

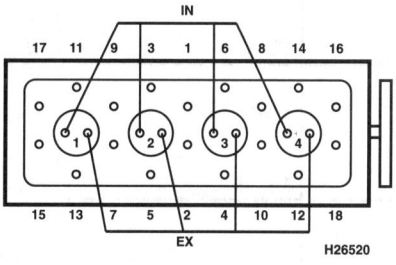

1975 cm³ / 2982 cm³

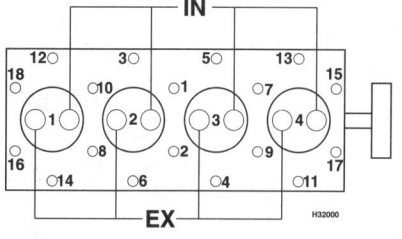

2446 cm³

– Not applicable, or information not available

TOYOTA

	Hi-Lux 2.4 4x4 (LN105) 1988 to 1996	Hi-Lux 2.4 D 1997 to 2000	Hi-Lux 2.4 TD 1997 to 2000	4Runner 3.0 TD 1993 to 1996
Engine				
Engine type/code............................	2L SOHC 60kW	2L SOHC 55kW	2L-T SOHC Turbo 63kW	1KZ-T SOHC Turbo 92kW
Capacity (cm³) / cylinders..................	2446 / 4	2446 / 4	2446 / 4	2982 / 4
Compression ration / pressure ....bar	22.2 / ≥20.0	22.2 / ≥19.6	21.0 / ≥19.6	21.2 / 19.6 to 31.4
Torque output ....Nm	165	157	218	218
Oil pressure ....idle [running] bar	[2.5 to 5.9 @ 3000]	[3.0 @ 3000]	[3.0 @ 3000]	[2.9 to 5.5 @ 3000]
Oil temperature ....°C	80	80	80	80
Valve clearances - inlet (mm)	0.25	0.25	0.20 to 0.30	0.20 to 0.30
- exhaust (mm)	0.36	0.45	0.40 to 0.50	0.25 to 0.35
Injection order	1-3-4-2	1-3-4-2	1-3-4-2	1-3-4-2
No. 1 cylinder position	F	TBE	TBE	TBE
Cooling system				
Thermostat opening temperature ....°C	80 to 84	86	86	80
Radiator cap pressure ....bar	0.7 to 1.0	0.75 to 1.0	0.75 to 1.0	0.75 to 1.05
Fuel system				
Idle speed ....rpm	700	650 to 750	650 to 750	700 ± 50
Maximum (no load) speed ....rpm	4900	5020 to 5280	4670 to 4930	4600
Smoke test/opacity ....M⁻¹ %	2.5	2.5	2.5	2.5
Static timing method............................	Plunger travel	Plunger travel	Plunger travel	Plunger travel
Timing dimension................................mm	0.54 to 0.66'	0.54 to 0.66	0.51 to 0.66	0.39 to 0.43
Crankshaft position ....mm [°]	[0] TDC	TDC	TDC	0 TDC
Turbo type / ref / pressure ....bar	–	–	0.6 to 0.8 bar	0.57 to 0.64 bar @ 4600
Injection pump make	Nippon Denso	Nippon Denso	Nippon Denso	Nippon Denso
Injection pump part no...........................	VE 22100 54750	–	–	VE 22100 67040
Injector Make / type	Nippon Denso	Nippon Denso	Nippon Denso	Nippon Denso
Injector part no......................................	DN-DN4SDND 133	DN12SD12	DN12SD12	DNOPD619
Injection type...	Indirect VE	Indirect	Indirect	VE
Injection opening pressure, New [used]...bar	148 to 156 [142 to 152]	148 to 156 [142 to 152]	148 to 156 [142 to 152]	148 to 156 [147]
Glow plugs				
Maker	Bosch/Champion	Bosch/Champion	Champion	–
Type	0250 202 053 / CH103	0250 202 053 / CH154	CH154	–
Nominal rating......................................V/A	7 / 8	11 /	11 /	–
Brakes				
minimum friction material thickness				
Front..mm	1.0	1.0	1.0	1.0
Rear..mm	1.0	1.0	1.0	1.0
Tyres - Saloon / Hatch.....................Size	–	–	–	–
- Estate / Van......................Size	205x16	185x14	205x16	265/70x15
Pressure - front / rear - Saloon / Hatch...bar	–	–	–	–
- Estate / Van ......bar	1.7 / 2.4	1.8 / 1.8	1.7 / 2.4	–
Front suspension / wheel alignment				
Toe-in (+) / Toe-out (–).....................mm [°]	1.0 ± 1.0	3.0 ± 1.0	1.0 ± 1.0	1.0 ± 2.0
Camber	1° ± 45'	0°20' ± 30'	1° ± 45'	0°45' ± 45'
Castor	1°30' ± 30'	0°45' ± 45'	1°30' ± 45'	2°30' ± 45'
King pin inclination................................	9°30' ± 45'	10°10' ± 30'	9°30' ± 45'	11°50' ± 45'
Rear suspension / wheel alignment				
Toe-in (+) / Toe-out (–).....................mm [°]	–	–	–	–
Camber	–	–	–	–

TOYOTA

Torque wrench settings	Hi-Lux 2.4 4x4 (LN105) 1988 to 1996	Hi-Lux 2.4 D 1997 to 2000	Hi-Lux 2.4 TD 1997 to 2000	4Runner 3.0 TD 1993 to 1996
Cylinder head - stage 1 ......Nm	38[4]	78	78	39
- stage 2 ......Nm	78	+ 90°	+ 90°	+ 90°
- stage 3 ......Nm	+ 90°	+ 90°	+ 90°	+ 90°
- stage 4 ......Nm	+ 90°	–	–	–
- stage 5 ......Nm	–	–	–	–
- stage 6 ......Nm	–	–	–	–
Big-end bearings......Nm	54 + 90° N	34 + 120° N	34 + 120° N	29 + 90° N
Main bearings......Nm	103	103	100	49 + 90°
Crankshaft pulley bolt ......Nm	167	167	167	363
Camshaft pulley bolt ......Nm	98	98	98	98
Flywheel [driveplate] bolt......Nm	114 to 122	123	123	145
Front hubs ......Nm	WSM	WSM	WSM	WSM
Rear hubs ......Nm	WSM	WSM	WSM	WSM
Wheel nuts / bolts ......Nm	103	100	100	103
Glow plugs ......Nm	13	13	10 to 15	13
Clutch pressure plate bolts ......Nm	19	19	20	19
Injection pump sprocket......Nm	64	64	64	64
Injectors......Nm	69	64	64	64
Injection pump mounting bolts ......Nm	–	18	–	21
Injector pipe unions......Nm	25	25	–	15
Capacities				
Engine oil & filter......litres	5.8	5.8	6.5	8.0
Gearbox......litres	G52, 58: 3.9[2]	3.9	3.9	Transfer: 1.1
Automatic transmission ......litres	–	–	–	–
Final drive ......litres	Front: 2.3[3]	1.8	2.0 Rear: 2.2	1.6 Rear: 2.2
Cooling system......litres	9.3	9.7	9.7	11.0
Fuel tank......litres	65	69	77 Double Cab: 66	65

Notes

Hi-Lux 2.4 4x4 (LN105) 1988 to 1996
[1]With automatic cold starting device: 0.82 to 0.98
[2]W56: 3.0. Transfer box: 1.6. Planetary type: 1.1
[3]Rear: 2.2
[4]Shank diameter: ≥11.6 mm

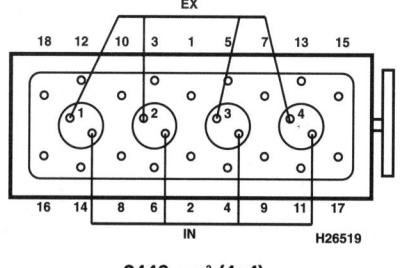

2446 cm³ (4x4)

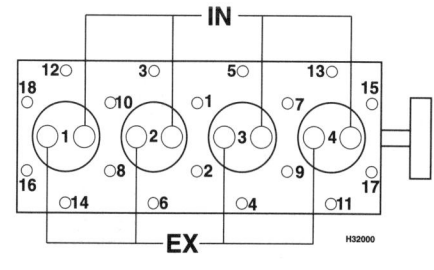

2446 cm³

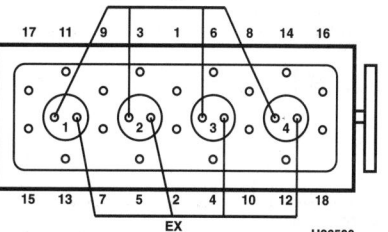

2982 cm³

– Not applicable, or information not available

TOYOTA

	Hi-Ace 2.4 (LH112) 1989 to 1996			
Engine				
Engine type/code	2L SOHC 56kW			
Capacity (cm³) / cylinders......................	2446 / 4			
Compression ration / pressure ....bar	22.2 / ≥20.0			
Torque output ....Nm	162			
Oil pressure ....idle [running] bar	[2.5 to 5.9 @ 3000]			
Oil temperature ....°C	80			
Valve clearances - inlet (mm)	0.20 to 0.30			
- exhaust (mm)	0.25 to 0.35			
Injection order	1-3-4-2			
No. 1 cylinder position	F			
Cooling system				
Thermostat opening temperature ....°C	86 to 90			
Radiator cap pressure ....bar	0.7 to 1.0			
Fuel system				
Idle speed ....rpm	700			
Maximum (no load) speed ....rpm	4900			
Smoke test/opacity ....M⁻¹ %	2.5			
Static timing method...............................	Plunger travel			
Timing dimension.................................mm	0.54 to 0.66			
Crankshaft position ....mm [°]	[0] TDC			
Turbo type / ref / pressure ....bar	_			
Injection pump make	_			
Injection pump part no............................	_			
Injector Make / type	Nippon Denso			
Injector part no......................................	DN-DN4SDND 133			
Injection type..	_			
Injection opening pressure, New [used]...bar	146 to 155 [140]			
Glow plugs				
Maker	Bosch/Champion			
Type	0250 202 053 / CH103			
Nominal rating......................................V/A	_			
Brakes				
minimum friction material thickness				
Front..mm	1.0			
Rear...mm	1.0			
Tyres - Saloon / Hatch......................Size	_			
- Estate / Van......................Size	185x14			
Pressure - front / rear - Saloon / Hatch ...bar	_			
- Estate / Van ......bar	3.0 / 4.5			
Front suspension / wheel alignment				
Toe-in (+) / Toe-out (–).....................mm [°]	1.0 ± 1.0			
Camber	-10' ± 30'			
Castor	1°40' ± 30'			
King pin inclination..................................	10°40' ± 30'			
Rear suspension / wheel alignment				
Toe-in (+) / Toe-out (–).....................mm [°]	_			
Camber	_			

TOYOTA

	Hi-Ace 2.4 (LH112) 1989 to 1996				
Torque wrench settings					
Cylinder head - stage 1 ....Nm	78¹				
- stage 2 ....Nm	+ 90°				
- stage 3 ....Nm	+ 90°				
- stage 4 ....Nm	_				
- stage 5 ....Nm	_				
- stage 6 ....Nm	_				
Big-end bearings...................................Nm	54 + 90° N				
Main bearings.......................................Nm	103				
Crankshaft pulley bolt ....Nm	167				
Camshaft pulley bolt ....Nm	98				
Flywheel [driveplate] bolt.....................Nm	114 to 122				
Front hubs ....Nm	WSM				
Rear hubs ....Nm	WSM				
Wheel nuts / bolts ....Nm	103				
Glow plugs ....Nm	13				
Clutch pressure plate bolts ....Nm	19				
Injection pump sprocket.....................Nm	64				
Injectors..Nm	69				
Injection pump mounting bolts ....Nm	_				
Injector pipe unions.............................Nm	25				
Capacities					
Engine oil & filter.................................litres	6.7				
Gearbox..litres	2.2				
Automatic transmission ....litres	_				
Final drive ....litres	2.2				
Cooling system....................................litres	10.6				
Fuel tank...litres	70				

Notes

Hi-Ace 2.4 (LH112) 1989 to 1996
¹Shank diameter: ≥11.6 mm

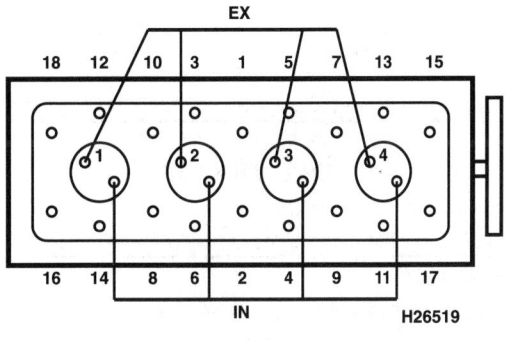

2446 cm³

– Not applicable, or information not available

VAUXHALL/OPEL

	Nova / Corsa 1.5 1989 to 1993	Nova / Corsa 1.5 Turbo 1989 to 1993	Corsa-B 1.5 D 1993 to 1996	Corsa-B 1.5 TD 1993 to 1996
Engine				
Engine type/code...............................	15D SOHC 37kW	15DT SOHC Turbo 49kW	15D SOHC 37kW	X15DT SOHC 49kW
Capacity (cm^3) / cylinders........................	1488 / 4	1488 / 4	1488 / 4	1488 / 4
Compression ration / pressure ....bar	23.0 / ≥22.0	22.5 / ≥22.0	23.0 / ≥22.0	22.0 / _
Torque output ....Nm	90	132	90	132
Oil pressure ....idle [running] bar	1.5	1.5	1.5	2.0
Oil temperature ....°C	80	80	≥80	80
Valve clearances - inlet (mm)	0.15	0.15	0.15	0.15
- exhaust (mm)	0.25	0.25	0.25	0.25
Injection order	1-3-4-2	1-3-4-2	1-3-4-2	1-3-4-2
No. 1 cylinder position	TBE	TBE	TBE	TBE
Cooling system				
Thermostat opening temperature ....°C	88	88	88 to 106	88 to 106
Radiator cap pressure ....bar	1.2 to 1.5	1.2 to 1.5	1.2 to 1.4	1.2 to 1.4
Fuel system				
Idle speed ....rpm	780 to 840	800 to 900	830 to 930	830 to 990 N/A
Maximum (no load) speed ....rpm	5600 to 5800	5400 to 5600	5800	5600
Smoke test/opacity ....M^{-1} %	2.5	2.5	3.0	2.5
Static timing method.............................	Plunger travel	Plunger travel	Plunger travel	Plunger travel
Timing dimension...............................mm	0.90 ± 0.05	0.68 ± 0.05	0.90 ± 0.05	0.68 ± 0.05
Crankshaft position ....mm [°]	[0] TDC	[0] TDC	0 TDC	0 TDC
Turbo type / ref / pressure ....bar	_	0.68 bar @ 5600rpm	_	0.68 bar @ 5600
Injection pump make	Bosch	Bosch	Bosch	Bosch
Injection pump part no............................	9 460 620 001	9 460 620 002	VE R3284	VE R305
Injector Make / type	2 jet	2 jet	Bosch	Bosch
Injector part no...................................	NP-DN OPD N108	NP-DN OPD N108	NP DNOPDN 108	NP DNOPDN 108
Injection type.......................................	VER 284	VER 305	Indirect VE 4	Indirect VE
Injection opening pressure, New [used]...bar	142 to 162	142 to 162	142 to 162	142 to 162
Glow plugs				
Maker	Bosch	Bosch/Champion	Bosch	Bosch
Type	0250 202 008	0250 202 008 / CH158	0250 202 008	0250 202 008
Nominal rating....................................V/A	5 / 11	5.0 / 11	5 / 12.5	5 / 12.5
Brakes				
minimum friction material thickness				
Front..mm	7.0 with backing	7.0 with backing	7.0 with backing	7.0 with backing
Rear...mm	0.5 above rivets	0.5 above rivets	0.5 above rivets	0.5 above rivets
Tyres - Saloon / Hatch.....................Size	145x13: 165/65x14: 165/70x13	145x13: 165/65x14: 165/70x13	145x13	165/70x13: 165/65x14
- Estate / Van...........................Size	145x13: 165/65x14: 165/70x13	_	_	_
Pressure - front / rear - Saloon / Hatch ...bar	1.9 / 1.7	1.9 / 1.7'	2.1 / 1.9	2.2 / 2.0
- Estate / Van ......bar	1.9 / 1.7	_	_	_
Front suspension / wheel alignment				
Toe-in (+) / Toe-out (–).....................mm [°]	-1.5 to 0.5	-1.5 to 0.5	[-0°10' ± 10']	[-0°10' ± 10']
Camber	-45' to 45' L	-45' to 45' L	-0°35' ± 45'	-0°25' ± 45'
Castor	45' to 2°45' L	45' to 2°45' L	1°50' ± 1°	1°50' ± 1°
King pin inclination...............................	_	_	_	_
Rear suspension / wheel alignment				
Toe-in (+) / Toe-out (–)....................mm [°]	-0.5 to 4.0	-0.5 to 4.0	[0°10' +30' -15']	[0°10' +30' -15']
Camber	-40' to -1°35'	-40' to -1°35'	-1°30' ± 30'	-1°30' ± 30'

VAUXHALL/OPEL

	Nova / Corsa 1.5 1989 to 1993	Nova / Corsa 1.5 Turbo 1989 to 1993	Corsa-B 1.5 D 1993 to 1996	Corsa-B 1.5 TD 1993 to 1996
Torque wrench settings				
Cylinder head - stage 1 ....Nm	40 N	40 N	40 N	40 N
- stage 2 ....Nm	+ 60 to 75°	+ 60 to 75°	+ 60°	+ 60°
- stage 3 ....Nm	+ 60 to 75°	+ 60 to 75°	+ 60°	+ 60°
- stage 4 ....Nm	–	–	+ 60°	+ 60°
- stage 5 ....Nm	–	–	–	–
- stage 6 ....Nm	–	–	–	–
Big-end bearings..............................Nm	25 + 45 to 60° N	25 + 45 to 60° N	25 + 60° N	30 + 60° N
Main bearings....................................Nm	88	88	88	90
Crankshaft pulley bolt ....Nm	133 to 161	133 to 161	161	161
Camshaft pulley bolt ....Nm	10	10	10	10
Flywheel [driveplate] bolt....................Nm	30 + 45 to 60°	30 + 45 to 60°	30 + 60° N	35 + 30° + 15° N
Front hubs ....Nm	100, slacken, 20 + 90°	100, slacken, 20 + 90°	100, slacken, 20 + 90°	100, slacken, 20 + 90°
Rear hubs ....Nm	WSM	WSM	WSM	WSM
Wheel nuts / bolts ....Nm	90	90	110	110
Glow plugs ....Nm	20	20	20	25
Clutch pressure plate bolts ....Nm	15	15	15	15
Injection pump sprocket.....................Nm	64	64	64	64
Injectors..Nm	50	50	50	50
Injection pump mounting bolts ....Nm	25	25	40	40
Injector pipe unions............................Nm	25	25	25	25
Capacities				
Engine oil & filter...............................litres	3.75	3.75	3.75	3.7
Gearbox...litres	1.8	1.8	1.6	1.6
Automatic transmission ....litres	–	–	–	–
Final drive ....litres	WT	WT	WT	WT
Cooling system..................................litres	6.4	6.4	6.0	6.3
Fuel tank..litres	42	42	46	46

Notes

Nova / Corsa 1.5 Turbo 1989 to 1993
[1]145x13: 2.0 / 1.8

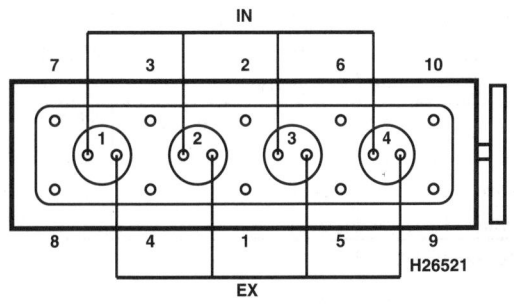

1488 cm³

– Not applicable, or information not
available

VAUXHALL/OPEL

	Corsa-B 1.5 TD 1996 to 2000	Corsa-B & Combo 1.7 D 1996 to 2000	Astra-F 1.7 D 1991 to 1994	Astra-F 1.7 D 1991 to 1994
Engine				
Engine type/code..............................	X15DT SOHC 49kW	X17D SOHC 44kW	17D SOHC 42kW	17D SOHC 42kW
Capacity (cm^3) / cylinders........................	1488 / 4	1686 / 4	1699 / 4	1699 / 4
Compression ration / pressure ....bar	22.0 / _	23.0 / _	23.0 / ≥18.5	23.0 / ≥18.5
Torque output ....Nm	132	112	0	0
Oil pressure ....idle [running] bar	1.5	2.0	1.5	1.5
Oil temperature ....°C	≥80	≥80	80	80
Valve clearances - inlet (mm)	0.15	0.15	0: Hyd.	0: Hyd.
- exhaust (mm)	0.25	0.25	0: Hyd.	0: Hyd.
Injection order	1-3-4-2	1-3-4-2	1-3-4-2	1-3-4-2
No. 1 cylinder position	TBE	TBE	TBE	TBE
Cooling system				
Thermostat opening temperature ....°C	88 to 106	88	86	86
Radiator cap pressure ....bar	1.2 to 1.4	1.2 to 1.4	1.4 to 1.5	1.4 to 1.5
Fuel system				
Idle speed ....rpm	830 to 930	830 to 930	820 to 920	820 to 920
Maximum (no load) speed ....rpm	5600	5300	5600	5600
Smoke test/opacity ....M^{-1} %	2.5	2.5	3.0	3.0
Static timing method............................	Plunger travel	Plunger travel	Plunger travel	Plunger travel
Timing dimension................................mm	0.68 ± 0.05	0.65 ± 0.05	0.80 to 0.85	Dimension on pump
Crankshaft position ....mm [°]	0 TDC	0 TDC	0 TDC	_
Turbo type / ref / pressure ....bar	0.68 bar @ 5600	_	_	_
Injection pump make	Bosch	Bosch	Bosch	Lucas
Injection pump part no............................	VE R305	VE 4R554	VE 4/9 R4433	DPC R8443 B550A
Injector Make / type	2 jet		Bosch	CAV
Injector part no.................................	NP DNOPDN 108	DNOPDN108	DNOSD309	BDNOSDC6751C
Injection type....................................	Indirect VE 4	Indirect VE 4	Indirect VE4	Indirect DPC
Injection opening pressure, New [used]...bar	142 to 162	142 to 162	135 to 143 [130]	135 to 143 [130]
Glow plugs				
Maker	Champion	_	Bosch/Champion	Bosch/Champion
Type	CH158	_	0250 201 019 / CH68	0250 201 019 / CH68
Nominal rating....................................V/A	11 / 5	11 / 5	5 / 12.5	5 / 12.5
Brakes				
minimum friction material thickness				
Front............................mm	7.0 with backing	7.0 with backing	7.0 with backing	7.0 with backing
Rear............................mm	0.5 above rivets	0.5 above rivets	0.5 above rivets	0.5 above rivets
Tyres - Saloon / Hatch ....Size	165/70x13: 165/65x14	165/70x13: 165/65x14	175/65x14	175/65x14
- Estate / Van...........................Size	_	165/70x13: 175/65x14	175/65x14	175/65x14
Pressure - front / rear - Saloon / Hatch ...bar	2.2 / 2.0	2.2 / 2.0	2.1 / 1.8	2.1 / 1.8
- Estate / Van ......bar	_	2.2 / 2.0: 2.2 / 2.2	2.2 / 2.0	2.2 / 2.0
Front suspension / wheel alignment				
Toe-in (+) / Toe-out (–)....................mm [°]	[-0°10' ± 10']	0.5 ± 1.0 L	-2.5 to -0.5 L	-2.5 to -0.5 L
Camber	-0°25 ± 45'	-35' ± 45' L	-1°5' ± 45'	-1°5' ± 45'
Castor	1°50' ± 1°	1°50' ± 1° L	2°15' ± 1°	2°15' ± 1°
King pin inclination..............................	_	_	_	_
Rear suspension / wheel alignment				
Toe-in (+) / Toe-out (–)....................mm [°]	[0°10' +30' -15']	1.0 +3.0 -1.5	-1.0 ± 4.0	-1.0 ± 4.0
Camber	-1°30' ± 30'	-1°30' ± 30' L	-1°40' ± 30'	-1°40' ± 30'

VAUXHALL/OPEL

Torque wrench settings	Corsa-B 1.5 TD 1996 to 2000	Corsa-B & Combo 1.7 D 1996 to 2000	Astra-F 1.7 D 1991 to 1994	Astra-F 1.7 D 1991 to 1994
Cylinder head - stage 1 ....Nm	40 N	40 N	25 N	25 N
- stage 2 ....Nm	+ 60°	+ 60° to 75°	+ 60°	+ 60°
- stage 3 ....Nm	+ 60°	+ 60° to 75°	+ 60°	+ 60°
- stage 4 ....Nm	+ 60°	–	+ 60°	+ 60°
- stage 5 ....Nm	–	–	Warm up	Warm up
- stage 6 ....Nm	–	–	+ 45°	+ 45°
Big-end bearings..............................Nm	25 + 60° N	25 + 100° + 15° N	35 + 45° N	35 + 45° N
Main bearings...................................Nm	90 N	88	50 + 45° N	50 + 45° N
Crankshaft pulley bolt ....Nm	161	133 to 161 N	145 + 30° + 10° N	145 + 30° + 10° N
Camshaft pulley bolt ....Nm	10	75 + 60° N	75 + 60° N	75 + 60° N
Flywheel [driveplate] bolt.....................Nm	35 + 30° + 15° N	38 + 45° to 60° N	50 + 30° N	50 + 30° N
Front hubs ....Nm	100, slacken, 20 + 90°	100, slacken, 20 + 90°	100, slacken, 20 + 90°	100, slacken, 20 + 90°
Rear hubs ....Nm	WSM	WSM	–	–
Wheel nuts / bolts ....Nm	110	110	110	110
Glow plugs ....Nm	20	20	20	20
Clutch pressure plate bolts ....Nm	15	15	15	15
Injection pump sprocket.....................Nm	64	64	25	25
Injectors..Nm	50	50	70	70
Injection pump mounting bolts ....Nm	45	45	23	23
Injector pipe unions............................Nm	25	25	25	25
Capacities				
Engine oil & filter................................litres	4.25	4.25	5.0	5.0
Gearbox..litres	1.6	1.6	1.6	1.6
Automatic transmission ....litres	–	–	–	–
Final drive ....litres	WT	WT	WT	WT
Cooling system...................................litres	6.3	6.0	6.8	6.8
Fuel tank...litres	46	46 Combo: 50	52 Estate: 50	52 Estate: 50

Notes

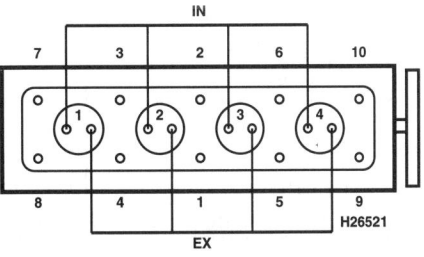

1488 cm³

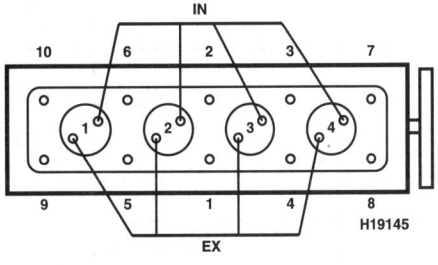

1686 cm³ / 1699 cm³

– Not applicable, or information not available

VAUXHALL/OPEL

	Astra-F 1.7 D 1992 to 1996	Astra-F 1.7 D 1992 to 1996	Astra-F 1.7 TD 1991 to 1998	Astra-F 1.7 TD 1994 to 1998
Engine				
Engine type/code.............................	17DR SOHC EGR 44kW	17DR SOHC EGR 44kW	17DT SOHC Turbo 60kW	X17DTL SOHC Turbo 50kW'
Capacity (cm^3) / cylinders........................	1699 / 4	1699 / 4	1686 / 4	1699 / 4
Compression ration / pressure ....bar	23.0 / ≥18.5	23.0 / ≥18.5	22.0 / ≥18.5	23.0 / ≥18.5
Torque output ....Nm	0	0	168	132
Oil pressure ....idle [running] bar	1.5	1.5	1.5	1.5
Oil temperature ....°C	80	80	80	80
Valve clearances - inlet (mm)	0: Hyd.	0: Hyd.	0.15	0: Hyd.
- exhaust (mm)	0: Hyd.	0: Hyd.	0.25	0: Hyd.
Injection order	1-3-4-2	1-3-4-2	1-3-4-2	1-3-4-2
No. 1 cylinder position	TBE	TBE	TBE	TBE
Cooling system				
Thermostat opening temperature ....°C	86	86	86	88
Radiator cap pressure ....bar	1.4 to 1.5	1.4 to 1.5	1.4 to 1.5	1.4 to 1.5
Fuel system				
Idle speed ....rpm	820 to 920	820 to 920	780 to 880	820 to 920
Maximum (no load) speed ....rpm	5600	5600	5100 to 5300	5600
Smoke test/opacity ....M^{-1} %	3.0	3.0	3.0	3.0
Static timing method................................	Plunger travel	Plunger travel	Plunger travel	Plunger travel
Timing dimension................................mm	Dimension on pump	0.85 to 0.9	0.55	0.85 to 0.90
Crankshaft position ....mm [°]	–	0 TDC	0 TDC	0 TDC
Turbo type / ref / pressure ....bar	–	–	0.72 bar @ 4400rpm	–
Injection pump make	Lucas	Bosch	Bosch	Bosch
Injection pump part no............................	DPC R8443 B850C	VER 487	9460 620 007	VER 571
Injector Make / type	CAV	Bosch	–	Bosch
Injector part no.......................................	BDNOSDC6751D	DNOSD 309	NPDNOPDN122	DNOSD 309
Injection type..	Indirect DPC	Indirect VER	Indirect VE4	Indirect VER
Injection opening pressure, New [used]...bar	135 to 143 [130]	135 to 143 [130]	142 to 162 [142]	135 to 142 [130]
Glow plugs				
Maker	Bosch/Champion	Bosch/Champion	Bosch/Champion	Bosch/Champion
Type	0250 202 087 / CH158	0250 202 087 / CH158	0250 202 087 / CH158	0250 202 087 / CH158
Nominal rating......................................V/A	5 / 12.5	5 / 12.5	11 / 12.2	5 / 12.5
Brakes				
minimum friction material thickness				
Front......................................mm	7.0 with backing	7.0 with backing	7.0 with backing	7.0 with backing
Rear.......................................mm	0.5 above rivets	0.5 above rivets	0.5 above rivets	0.5 above rivets
Tyres - Saloon / Hatch......................Size	175/65x14	175/65x14	175/65x14	155/80x13: 175/65x14
- Estate / Van..........................Size	175/65x14	175/65x14	175/65x14	155/80x13: 175/65x14
Pressure - front / rear - Saloon / Hatch...bar	2.1 / 1.8	2.1 / 1.8	2.1 / 1.8	2.0 / 1.7
- Estate / Van ......bar	2.2 / 2.0	2.2 / 2.0	2.2 / 2.0	2.0 / 1.8
Front suspension / wheel alignment				
Toe-in (+) / Toe-out (–)....................mm [°]	-2.5 to -0.5 L	-2.5 to -0.5 L	-2.5 to -0.5 L	[-0°15' ± 10']
Camber	-1°5' ± 45'	-1°5' ± 45'	-1°5' ± 45'	-1°5' ± 45'
Castor	2°15' ± 1°	2°15' ± 1°	2°15' ± 1°	2°15' ± 1° Est: 1°30' ± 1°
King pin inclination.............................	–	–	–	–
Rear suspension / wheel alignment				
Toe-in (+) / Toe-out (–)....................mm [°]	-1.0 ± 4.0	-1.0 ± 4.0	-1.0 ± 4.0	[0°10' ± 30']
Camber	-1°40' ± 30'	-1°40' ± 30'	-1°40' ± 30'	-1°40' ± 30'

VAUXHALL/OPEL

	Astra-F 1.7 D 1992 to 1996	Astra-F 1.7 D 1992 to 1996	Astra-F 1.7 TD 1991 to 1998	Astra-F 1.7 TD 1994 to 1998
Torque wrench settings				
Cylinder head - stage 1 ....Nm	25 N	25 N	40 N	25
- stage 2 ....Nm	+ 90°	+ 90°	+ 75°	+ 90°
- stage 3 ....Nm	+ 90°	+ 90°	+ 75°	+ 90°
- stage 4 ....Nm	+ 45°	+ 45°	–	+ 45°
- stage 5 ....Nm	Warm up	Warm up	–	warm up
- stage 6 ....Nm	+ 45°	+ 45°	–	+ 45°
Big-end bearings.................................Nm	35 + 45° + 15° N	35 + 45° + 15° N	25 + 100°+ 15° N	35 + 45° + 15° N
Main bearings.....................................Nm	50 + 45° + 15° N	50 + 45° + 15° N	88	50 + 45° + 15° N
Crankshaft pulley bolt ....Nm	145 + 30° + 10° N	145 + 30° + 10° N	161	145 + 30° + 10° N
Camshaft pulley bolt ....Nm	75 + 60° N	75 + 60° N	10	75 + 60° + 5° N
Flywheel [driveplate] bolt....................Nm	50 + 30° N	50 + 30° N	30 + 60° N	50 + 30° + 15° N
Front hubs ....Nm	100, slacken, 20 + 90°	100, slacken, 20 + 90°	100, slacken, 20 + 90°	100, slacken, 20 + 90°
Rear hubs..Nm	–	–	–	50 + 45° N
Wheel nuts / bolts ....Nm	110	110	110	110
Glow plugs ....Nm	20	20	20	20
Clutch pressure plate bolts ....Nm	15	15	15	15
Injection pump sprocket......................Nm	25	25	69	25
Injectors..Nm	70	70	50	70
Injection pump mounting bolts ....Nm	23	23	23	25
Injector pipe unions............................Nm	25	25	25	25
Capacities				
Engine oil & filter.................................litres	5.0	5.0	4.8	4.5
Gearbox...litres	1.6	1.6	1.6	1.9
Automatic transmission ....litres	–	–	–	–
Final drive ....litres	WT	WT	WT	WT
Cooling system....................................litres	6.8	6.8	6.8	6.8
Fuel tank..litres	52 Estate: 50	52 Estate: 50	52 Estate: 50	52 Estate: 50

Notes

Astra-F 1.7 TD 1994 to 1998
¹Low pressure turbo

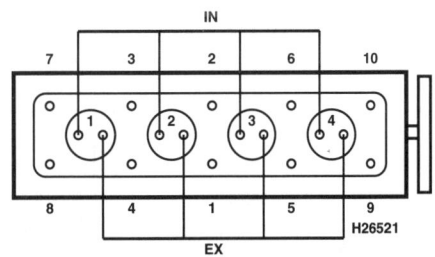

1686 cm³

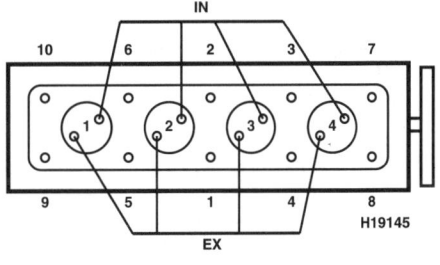

1699 cm³

– Not applicable, or information not available

VAUXHALL/OPEL

	Astra-F 1.7 TD 1994 to 1998	Astra-F 1.7 TD 1994 to 1998	Astra-G 1.7 TD 1998 to 2000	Astra-G 2.0 TDi 16V 1998 to 2000
Engine				
Engine type/code...................	X17DTL SOHC Turbo 50kW'	X17DT SOHC Turbo 60kW	X17DTL SOHC Turbo 50kW	X20DTL SOHC 16V Turbo 60kW
Capacity (cm³) / cylinders........................	1699 / 4	1686/4	1699 / 4	1995 / 4
Compression ration / pressure ....bar	23.0 / ≥18.5	22.0 / ≥18.5	22.0 / _	18.5 / _
Torque output ....Nm	132	168	132	185
Oil pressure ....idle [running] bar	1.5	2.0	1.5	1.5
Oil temperature ....°C	80	80	≥80	≥80
Valve clearances - inlet (mm)	0: Hyd.	0.15	0: Hyd.	0: Hyd.
- exhaust (mm)	0: Hyd.	0.25	0: Hyd.	0: Hyd.
Injection order	1-3-4-2	1-3-4-2	1-3-4-2	1-3-4-2
No. 1 cylinder position	TBE	TBE	TBE	TBE
Cooling system				
Thermostat opening temperature ....°C	88	86	92	92
Radiator cap pressure ....bar	1.4 to 1.5	1.4 to 1.5	1.2 to 1.5	1.2 to 1.5
Fuel system				
Idle speed ....rpm	820 to 920	780 to 880	950 to 100	820 to 890
Maximum (no load) speed ....rpm	5600	5100 to 5300	5400 to 5500	4750
Smoke test/opacity ....M⁻¹ %	3.0	3.0	3.0	2.0
Static timing method...............................	Plunger travel	Plunger travel	_	_
Timing dimension.............................mm	Dimension on pump	0.55	Computer controlled	_
Crankshaft position ....mm [°]	_	0 TDC	_	_
Turbo type / ref / pressure ....bar	_	0.72 bar @ 4400rpm	_	_
Injection pump make	Lucas	Bosch	Bosch	Bosch
Injection pump part no............................	R8443 B850C	VER 487	VP 29	VP 44
Injector Make / type	CAV	Bosch	_	_
Injector part no..	_	DNOSD 309	_	_
Injection type..	Indirect DPC	Indirect VE4	Indirect EDC15M	Direct injection EDC 15M
Injection opening pressure, New [used]...bar	135 to 142 [130]	142 to 162 [142]	132 to 145	180 to 365
Glow plugs				
Maker	Bosch/Champion	Bosch/Champion	_	Champion
Type	0250 202 087 / CH158	0250 202 087 / CH68	_	CH207
Nominal rating.....................................V/A	5 / 12.5	11 / 12.2		11 /
Brakes				
minimum friction material thickness				
Front..mm	7.0 with backing	7.0 with backing	7.0 with backing	7.0 with backing
Rear...mm	0.5 above rivets	0.5 above rivets	0.5 above rivets	7.0 with backing
Tyres - Saloon / Hatch ....Size	155/80x13: 175/65x14	175/65x14	175/70x14: 195/60x15	195/60x15: 185/65x14
- Estate / Van...........................Size	155/80x13: 175/65x14	175/65x14	175/70x14: 195/60x15	195/60x15: 185/65x14
Pressure - front / rear - Saloon / Hatch ...bar	2.0 / 1.7	2.1 / 1.8	2.2 / 1.9: 2.2 / 1.9	2.4 / 2.1: 2.4 / 2.1
- Estate / Van ......bar	2.0 / 1.8	2.2 / 2.0	2.2 / 2.0: 2.2 / 2.0	2.4 / 2.2
Front suspension / wheel alignment				
Toe-in (+) / Toe-out (–)......................mm [°]	[-0°15' ± 10']	[-0°15' ± 10']	-2.5 to -0.5 L	[0°10' ± 10']
Camber	-1°5' ± 45'	-1°5' ± 45'	-1°5' ± 45'	-1°10' ± 45'
Castor	2°15' ± 1° Est: 1°30' ± 1°	2°15' ± 1° Est: 1°30' ± 1°	2°15' ± 1°³	4° ± 1°¹
King pin inclination.................................	_	_	_	_
Rear suspension / wheel alignment				
Toe-in (+) / Toe-out (–).....................mm [°]	[0°10' ± 30']	[0°10' ± 30']	-1.0 to 4.0	[0°10' +30' - 20']
Camber	-1°40' ± 30'	-1°40' ± 30'	-1°40' ± 30'	1°40' ± 30'

VAUXHALL/OPEL

Torque wrench settings	Astra-F 1.7 TD 1994 to 1998	Astra-F 1.7 TD 1994 to 1998	Astra-G 1.7 TD 1998 to 2000	Astra-G 2.0 TDi 16V 1998 to 2000
Cylinder head - stage 1 ....Nm	25	40 N	25 N	25 N
- stage 2 ....Nm	+ 90°	+ 75°	+ 90°	+ 65°
- stage 3 ....Nm	+ 90°	+ 75°	+ 90°	+ 65°
- stage 4 ....Nm	+ 45°	–	+ 45°	+ 65°
- stage 5 ....Nm	warm up	–	Warm-up	+ 65°
- stage 6 ....Nm	+ 45°	–	+ 45°	+ 15°
Big-end bearings...............................Nm	35 + 45° + 15° N	25 + 100°+ 15° N	35 + 45° + 15° N	30 + 45° + 15° N
Main bearings....................................Nm	50 + 45° + 15° N	88	50 + 45° + 15° N	90 + 60° + 15° N
Crankshaft pulley bolt ....Nm	145 + 30° + 10° N	161	130 + 45° ± 5° N	150 + 45° + 15° N
Camshaft pulley bolt ....Nm	75 + 60° + 5° N	10	70 + 60° + 5° N	90 + 60° + 30° N
Flywheel [driveplate] bolt....................Nm	50 + 30° + 15° N	30 + 60° N	65 + 30° + 15° N	40 + 30° + 15° N
Front hubs ....Nm	100, slacken, 20 + 90°	100, slacken, 20 + 90°	WSM	WSM
Rear hubs ....Nm	50 + 45° N	50 + 45° N	WSM	50 + 30° N
Wheel nuts / bolts ....Nm	110	110	110	110
Glow plugs ....Nm	20	20	20	10
Clutch pressure plate bolts ....Nm	15	15	15	15
Injection pump sprocket.....................Nm	25	69	–	20
Injectors...Nm	70	50	70	WSM
Injection pump mounting bolts ....Nm	25	23	–	25
Injector pipe unions...........................Nm	25	25	–	30
Capacities				
Engine oil & filter..............................litres	4.5	4.8	5.5	5.5
Gearbox...litres	1.9	1.9	1.6	1.8
Automatic transmission ....litres	–	–	–	4.0
Final drive ....litres	WT	WT	WT	WT
Cooling system................................litres	6.8	6.8	8.7	7.9
Fuel tank..litres	52 Estate: 50	52 Estate: 50	52 Estate: 50	52 Estate: 50

Notes

Astra-F 1.7 TD 1994 to 1998
[1]Low pressure turbo
Astra-G 1.7 TD 1998 to 2000
[1]Estate: 3°25' ± 1°
Astra-G 2.0 TDi 16V 1998 to 2000
[1]Estate: 3° 25' ± 1°

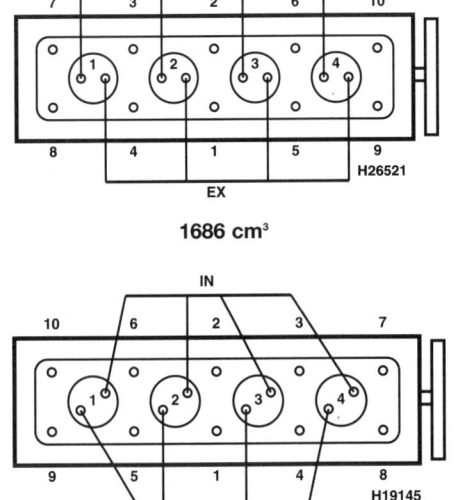

1686 cm³

1700 cm³

1699 cm³

1994 cm³ 16V

– Not applicable, or information not
available

VAUXHALL/OPEL

	Kadett/Astra/Belmont 1.5TD 1990 to 1991	Kadett/Astra/Belmont 1.7 1989 to 1991	Kadett/Astra/Belmont 1.7 1989 to 1991	Zafira 2.0 Di 1999 to 2000
Engine				
Engine type/code..................................	15DTR SOHC 53kW	17D 42kW	17D 42kW	X20DTL SOHC 16V 60kW
Capacity (cm³) / cylinders......................	1488 / 4	1699 / 4	1699 / 4	1995 / 4
Compression ration / pressure ....bar	22.0 / _	23.0 / _	23.0 / _	18.5 / 25 to 28
Torque output ....Nm	0	0	0	185
Oil pressure ....idle [running] bar	1.5	1.0	1.0	1.5
Oil temperature ....°C	60	60	60	80
Valve clearances - inlet (mm)	0.15	0: Hyd.	0: Hyd.	0: Hyd.
- exhaust (mm)	0.25	0: Hyd.	0: Hyd.	0: Hyd.
Injection order	1-3-4-2	1-3-4-2	1-3-4-2	1-3-4-2
No. 1 cylinder position	TBE	TBE	TBE	TCE
Cooling system				
Thermostat opening temperature ....°C	88	92	92	92
Radiator cap pressure ....bar	1.2 to 1.5	1.2 to 1.5	1.2 to 1.5	_
Fuel system				
Idle speed ....rpm	780 to 840	820 to 880	820 to 880	760 to 860
Maximum (no load) speed ....rpm	5400 to 5600	5600-100	5600-100	4750
Smoke test/opacity ....M⁻¹ %	2.5	2.5	2.5	1.9
Static timing method..............................	Dial gauge	Plunger travel	Plunger travel	Refer to wsm
Timing dimension.............................mm	0.68	0.8 ± 0.05	X-0.15¹	Computer controlled
Crankshaft position ....mm [°]	TDC	TDC	TDC	_
Turbo type / ref / pressure ....bar				_
Injection pump make	Bosch	Bosch	CAV.	Bosch
Injection pump part no...........................	8 944 608 050	F 2300 R313 [R313-1]	_	VP44
Injector Make / type	2 jet	Bosch	CAV	_
Injector part no....................................	NP-DN OPD N108	DN OSD 292	BDN or RDN OSD C6751	_
Injection type......................................	VER 305	VE 4/9	DPC OP02	Direct EDC 15M
Injection opening pressure, New [used]...bar	142 to 162	135 to 143	130 to 138	180 to 365
Glow plugs				
Maker	Champion	Bosch/Champion	Bosch/Champion	_
Type	12 14 309 / CH110	90 201 005 / CH68	90 201 005 / CH68	_
Nominal rating.............................V/A	5 / _	11 / 8	11 / 8	_
Brakes				
minimum friction material thickness				
Front.............................mm	7.0 with backing	7.0 with backing	7.0 with backing	8.0 with backing
Rear.............................mm	0.5 above rivets	0.5 above rivets	0.5 above rivets	8.0 with backing
Tyres - Saloon / Hatch......................Size	155x13: 165x13: 175/70x13²	155x13: 165x13: 175/70x13¹	155x13: 165x13: 175/70x13²	_
- Estate / Van......................Size	155x13: 175/70x13: 175/65X14	155x13: 165x13: 175/70x13²	155x13: 165x13: 175/70x13³	195/65x15
Pressure - front / rear - Saloon / Hatch ...bar	1.8 / 1.6³	1.8 / 1.6³	1.8 / 1.6⁴	_
- Estate / Van .....bar	1.9 / 1.9	1.9 / 1.9⁴	1.9 / 1.9⁵	2.2 / 2.2
Front suspension / wheel alignment				
Toe-in (+) / Toe-out (–)....................mm [°]	-2.0 to 0 L	-2.0 to 0 L	-2.0 to 0 L	[0° ± 10'] L
Camber	-1°15' to 15'	-1°15' to 15'	-1°15' to 15'	-1°10' ± 45'
Castor	45' to 2°45'. Est: 0 to 2°	45' to 2°45'. Est: 0 to 2°	45' to 2°45'. Est: 0 to 2°	3° ± 1°
King pin inclination...............................	_			_
Rear suspension / wheel alignment				
Toe-in (+) / Toe-out (–)....................mm [°]	[-10' to 40']	[-10' to 40']	[-10' to 40']	[0°10' +30' -20'] L
Camber	-30' ± 30'	-30' ± 30'	-30' ± 30'	-1°40' ± 30'

VAUXHALL/OPEL

	Kadett/Astra/Belmont 1.5TD 1990 to 1991	Kadett/Astra/Belmont 1.7 1989 to 1991	Kadett/Astra/Belmont 1.7 1989 to 1991	Zafira 2.0 Di 1999 to 2000
Torque wrench settings				
Cylinder head - stage 1 ....Nm	40	25	25	25 N
- stage 2 ....Nm	+ 60 to 75°	+ 60°	+ 60°	+ 65°
- stage 3 ....Nm	–	+ 60°	+ 60°	+ 65°
- stage 4 ....Nm	–	+ 60°	+ 60°	+ 65°
- stage 5 ....Nm	–	Warm up, + 45°	Warm up, + 45°	+ 65°
- stage 6 ....Nm	–	–	–	+ 15°
Big-end bearings.................................Nm	25 + 45 to 60°	35	35	35 + 45° + 15° N
Main bearings.....................................Nm	84 to 93	50	50	90 + 60° + 15° N
Crankshaft pulley bolt ....Nm	133 to 161	130 + 45+5°	130 + 45+5°	150 + 45° + 15° N
Camshaft pulley bolt ....Nm	8 to 11	75 + 60 to 65°	75 + 60 to 65°	90 + 60° + 30° N
Flywheel [driveplate] bolt...................Nm	30 + 45 to 60°	50 + 30 to 45°	50 + 30 to 45°	45 + 30° + 15° N
Front hubs ....Nm	WSM	WSM	WSM	120, slacken, 20 + 90° N
Rear hubs ....Nm	WSM	WSM	WSM	WSM
Wheel nuts / bolts ....Nm	90	90	90	110
Glow plugs ....Nm	20	20	20	10
Clutch pressure plate bolts ....Nm	15	15	15	15
Injection pump sprocket......................Nm	70	22	22	28
Injectors...Nm	50	70	70	WSM
Injection pump mounting bolts ....Nm	45	25	25	25
Injector pipe unions...........................Nm	–	–	–	30
Capacities				
Engine oil & filter................................litres	4.5	4.75	4.75	5.5
Gearbox..litres	1.8	2.1	2.1	1.9
Automatic transmission ....litres	–	6.3	6.3	–
Final drive ....litres	WT	WT	WT	WT
Cooling system...................................litres	6.3	9.1	9.1	7.9
Fuel tank...litres	52. Est/Van: 50	52. Est/van: 50	52. Est/van: 50	58

Notes

Kadett/Astra/Belmont 1.5TD 1990 to 1991
[2]Also 175/65x14
[3]175/65X14: 2.1 / 1.9
Kadett/Astra/Belmont 1.7 1989 to 1991
[1]Also 175/65x14: 185/60x14
[2]Also 165x14: 175/65x14
[3]185/60x14: 1.7 / 1.7. 90 ▶: 175/65x14: 2.1 / 1.9
[4]165x13, 165x14: 1.8 / 1.8. 90 ▶: 1.9 / 1.9. Astramax: 1.8 / 2.2
Kadett/Astra/Belmont 1.7 1989 to 1991
[1]Where X = value shown on pump lever
[2]Also 175/65x14: 185/60x14
[3]Also 165x14: 175/65x14
[4]185/60x14: 1.7 / 1.7. 90 ▶: 175/65x14: 2.1 / 1.9
[5]165x13, 165x14: 1.8 / 1.8. 90 ▶: 1.9 / 1.9. Astramax: 1.8 / 2.2

– Not applicable, or information not available

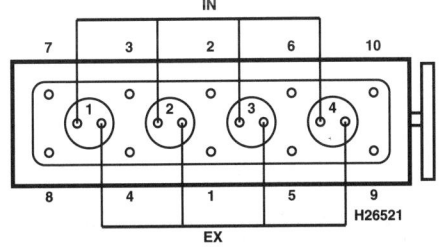

1488 cm³

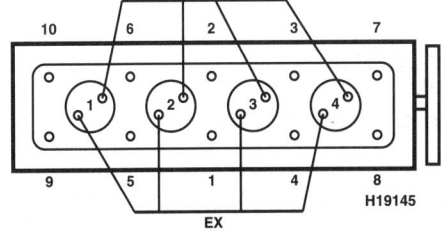

1699 cm³

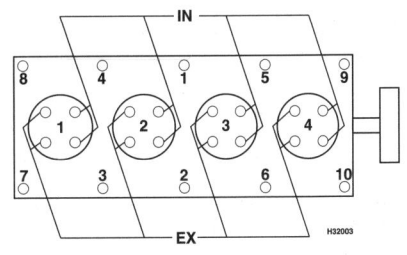

1995 cm³

VAUXHALL/OPEL

	Cavalier/Vectra 1.7 D 1988 to 1992	Cavalier/Vectra 1.7 D 1988 to 1992	Cavalier/Vectra 1.7 D 1993 to 1995	Cavalier/Vectra 1.7 D 1993 to 1995
Engine				
Engine type/code..................................	17D SOHC 42kW	17D SOHC 42kW	17DR SOHC 44kW	17DR SOHC 44kW
Capacity (cm³) / cylinders......................	1699 / 4	1699 / 4	1699 / 4	1699 / 4
Compression ration / pressure ....bar	23.0 / _	23.0 / _	23.0 / ≥19.0	23.0 / ≥19.0
Torque output ....Nm	0	0	105	105
Oil pressure ....idle [running] bar	1.8	1.5	1.5	1.5
Oil temperature ....°C	80	80	80	80
Valve clearances - inlet (mm)	0: Hyd.	0: Hyd.	0: Hyd.	0: Hyd.
- exhaust (mm)	0: Hyd.	0: Hyd.	0: Hyd.	0: Hyd.
Injection order	1-3-4-2	1-3-4-2	1-3-4-2	1-3-4-2
No. 1 cylinder position	TBE	TBE	TBE	TBE
Cooling system				
Thermostat opening temperature ....°C	92	92	92	92
Radiator cap pressure ....bar	1.20 to 1.35	1.20 to 1.35	1.20 to 1.35	1.20 to 1.35
Fuel system				
Idle speed ....rpm	820 to 920	820 to 920	820 to 920	820 to 920
Maximum (no load) speed ....rpm	5600	5600	5600	5600
Smoke test/opacity ....M⁻¹ %	3.0	3.0	3.0	3.0
Static timing method..............................	Plunger travel	Refer to wsm	Plunger travel	Plunger travel
Timing dimension.................................mm	0.80 to 0.85	Dimension on pump	0.80 to 0.85	Dimension on pump
Crankshaft position ....mm [°]	0 TDC	_	0 TDC	_
Turbo type / ref / pressure ....bar	_	_	_	_
Injection pump make	Bosch	CAV	Bosch	Lucas
Injection pump part no..........................	VE 4/9 R443	R8442 B55 0A	VE 4/9	R8443 B850A
Injector Make / type	Bosch	CAV	Bosch	CAV
Injector part no.	0432 217 197	BDN OSDC 6751C	DNOSD 309	BDNOSD C6751C
Injection type......................................	Indirect VE 4/9	Indirect DPC	Indirect VE 4/9	Indirect DPC
Injection opening pressure, New [used]...bar	135 to 143 [130 to 138]	135 to 143 [130 to 138]	135 to 143 [130 to 138]	135 to 143 [130 to 138]
Glow plugs				
Maker	Bosch/Champion	Bosch/Champion	Bosch/Champion	Bosch/Champion
Type	0250 201 019 / CH68	0250 201 019 / CH68	0250 201 019 / CH68	0250 201 019 / CH68
Nominal rating...................................V/A	11 / 12	11 / 12	5 / 12.5	5 / 12.5
Brakes				
minimum friction material thickness				
Front..mm	7.0 with backing	7.0 with backing	7.0 with backing	7.0 with backing
Rear...mm	0.5 above rivets	0.5 above rivets	0.5 above rivets	0.5 above rivets
Tyres - Saloon / Hatch.......................Size	175/70x14: 195/60x14:195/60x15	175/70x14: 195/60x14:195/60x15	175/70x14: 195/60x14:195/60x15	175/70x14: 195/60x14:195/60x15
- Estate / Van......................Size	_	_	_	_
Pressure - front / rear - Saloon / Hatch ...bar	1.9 / 1.7	1.9 / 1.7	1.9 / 1.7	1.9 / 1.7
- Estate / Van ......bar	_	_	_	_
Front suspension / wheel alignment				
Toe-in (+) / Toe-out (−)....................mm [°]	0.5 to -2.8	0.5 to -2.8	0.5 to -2.8	0.5 to -2.8
Camber	-1°25' to 5' L	-1°25' to 5' L	-1°25' to 5' L	-1°25' to 5' L
Castor	1° to 3° L	1° to 3° L	1° to 3° L	1° to 3° L
King pin inclination...............................	_	_	_	_
Rear suspension / wheel alignment				
Toe-in (+) / Toe-out (−)....................mm [°]	-1.0 to 4.0	-1.0 to 4.0	-1.0 to 4.0	-1.0 to 4.0
Camber	-2°10' to -1°10'	-2°10' to -1°10'	-2°10' to -1°10'	-2°10' to -1°10'

VAUXHALL/OPEL

	Cavalier/Vectra 1.7 D 1988 to 1992	Cavalier/Vectra 1.7 D 1988 to 1992	Cavalier/Vectra 1.7 D 1993 to 1995	Cavalier/Vectra 1.7 D 1993 to 1995
Torque wrench settings				
Cylinder head - stage 1 ....Nm	25	25	25	25
- stage 2 ....Nm	+ 60°	+ 60°	+ 90°	+ 90°
- stage 3 ....Nm	+ 60°	+ 60°	+ 90°	+ 90°
- stage 4 ....Nm	+ 60°	+ 60°	+ 45°	+ 45°
- stage 5 ....Nm	Warm up	Warm up	Warm up	Warm up
- stage 6 ....Nm	+ 45°	+ 45°	+ 45°	+ 45°
Big-end bearings..............................Nm	35 + 45° N	35 + 45° N	35 + 45° N	35 + 45° N
Main bearings...................................Nm	50 + 45° N	50 + 45° N	50 + 60° N	50 + 60° N
Crankshaft pulley bolt ....Nm	130 + 30° N	130 + 30° N	145 + 40° N	145 + 40° N
Camshaft pulley bolt ....Nm	75 + 60° N	75 + 60° N	75 + 65° N	75 + 65° N
Flywheel [driveplate] bolt..................Nm	60 + 30° N	60 + 30° N	60 + 30° N	60 + 30° N
Front hubs ....Nm	30, slacken, 20 + 90°	130, slacken 20 + 90°	30, slacken, 20 + 90°	30, slacken, 20 + 90°
Rear hubs ....Nm	WSM	WSM	WSM	WSM
Wheel nuts / bolts ....Nm	110	110	110	110
Glow plugs ....Nm	20	20	20	20
Clutch pressure plate bolts ....Nm	15	15	15	15
Injection pump sprocket....................Nm	22	22	22	22
Injectors...Nm	70	70	70	70
Injection pump mounting bolts ....Nm	25	25	25	25
Injector pipe unions...........................Nm	25	25	25	25
Capacities				
Engine oil & filter................................litres	5.0	5.0	5.0	5.0
Gearbox...litres	1.8	1.8	1.8	1.8
Automatic transmission ....litres	–	–	–	–
Final drive ....litres	WT	WT	WT	WT
Cooling system...................................litres	9.1	9.1	9.1	9.1
Fuel tank..litres	61	61	61	61

Notes

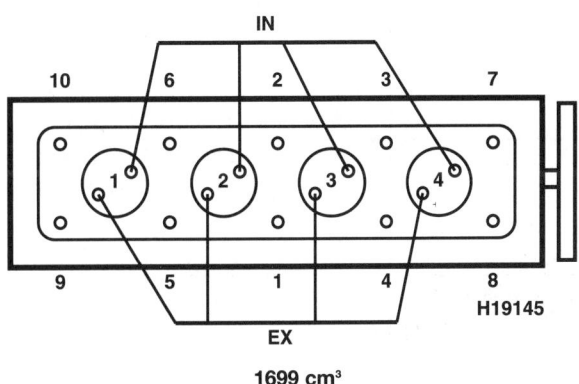

H19145

1699 cm³

– Not applicable, or information not available

VAUXHALL/OPEL

	Cavalier/Vectra 1.7 TD 1992 to 1995	Cavalier/Vectra 1.7 TD 1994 to 1995	Vectra-B 1.7 DTL 1995 to 1997	Vectra-B 2.0 Di 1995 to 2000
Engine				
Engine type/code....................	17DT SOHC Turbo 60kW	X17DT SOHC Turbo 60kW	X17DT SOHC Turbo 60kW	X20DTL SOHC Turbo 60kW
Capacity (cm³) / cylinders....................	1686 / 4	1686 / 4	1686 / 4	1995 / 4
Compression ration / pressure ....bar	22.0 /	22.0 / ≥17.0	22.0 / ≥18.5	18.5 / ≥17.0
Torque output ....Nm	168	168	168	185
Oil pressure ....idle [running] bar	1.5	1.5	2.0	1.5
Oil temperature ....°C	80	80	80	80
Valve clearances - inlet (mm)	0.15	0.15	0.15	0: Hyd.
- exhaust (mm)	0.25	0.25	0.25	0: Hyd.
Injection order....................	1-3-4-2	1-3-4-2	1-3-4-2	1-3-4-2
No. 1 cylinder position....................	TBE	TBE	TBE	TCE
Cooling system				
Thermostat opening temperature ....°C	86	86	86 to 90	91
Radiator cap pressure ....bar	1.4 to 1.5	1.4 to 1.5	1.4 to 1.5	1.4 to 1.5
Fuel system				
Idle speed ....rpm	780 to 880	780 to 880	780 to 880	750 to 850
Maximum (no load) speed ....rpm	5300	5300	5100 to 5300	4900 to 5100
Smoke test/opacity ....M⁻¹ %	2.0	2.0	2.0	2.0
Static timing method....................	Plunger travel	Refer to wsm	Plunger travel	Refer to wsm
Timing dimension............................mm	0.55	_	0.55 ± 0.05	Computer controlled
Crankshaft position ....mm [°]	0 TDC	_	0 TDC	_
Turbo type / ref / pressure ....bar	0.72 bar @ 4000rpm	0.72 bar @ 4000rpm	0.72 bar @ 4000rpm	_
Injection pump make	Bosch	_	Bosch	Bosch
Injection pump part no....................	VER3651	Zexel NP-VE 4/10 F	Zexel NP-VE4/8F2300R	EDC 15.6 / 15M Direct VP44
Injector Make / type	_	_	_	_
Injector part no....................	NPDNOPDN122	NPDNOPDN122	NP-DNO PDN 122	_
Injection type....................	Indirect VE4	Indirect VE4	Indirect VE4	Direct EDC 15M
Injection opening pressure, New [used]...bar	142 to 162 [142]	142 to 162 [142]	142 to 162	180 to 365
Glow plugs				
Maker	Bosch/Champion	Bosch/Champion	_	_
Type	0250 202 087 / CH158	0250 202 087 / CH68	Y711RS	_
Nominal rating ....V/A	11 / 12.5	11 / 12.5	11 / 12.5	11 /
Brakes				
minimum friction material thickness				
Front..mm	7.0 with backing	7.0 with backing	7.0 with backing	7.0 with backing
Rear..mm	0.5 above rivets	0.5 above rivets	7.0 with backing	7.0 with backing
Tyres - Saloon / Hatch......................Size	175/70x14: 195/60x14:195/60x15	175/70x14: 195/60x14:195/60x15	175/70x14:185/70x14:195/65x15	195/65x15: 205/60x15
- Estate / Van......................Size	175/70x14: 195/60x14:195/60x15	175/70x14: 195/60x14:195/60x15	185/70x14: 195/65x15	195/65x15: 205/60x15
Pressure - front / rear - Saloon / Hatch ...bar	2.2 / 2.0	2.2 / 2.0	2.4 / 2.4: 195/65: 2.1 / 2.1	1.9 / 1.9
- Estate / Van ......bar	2.2 / 2.1	2.2 / 2.1	Refer to vehicle	1.9 / 1.9
Front suspension / wheel alignment				
Toe-in (+) / Toe-out (–)....................mm [°]	[-0°15' ± 10']	[-0°15' ± 10']	[0°10' ± 10']	[0°10' ± 10']
Camber	-0°40' ± 45'	-0°40' ± 45'	-1°05' ± 45'	-1°05' ± 45'
Castor	2° ± 1°	2° ± 1°	3°50' ± 1°	3°40' ± 1°
King pin inclination............................	_	_	_	_
Rear suspension / wheel alignment				
Toe-in (+) / Toe-out (–)....................mm [°]	[0°10' +30' -20']	[0°10' +30' -20']	[0°16' ± 10']	[0°30' ± 10']¹
Camber	-1°40'	-1°40'	-1°10' ± 35'	-1°55' ± 35'²

VAUXHALL/OPEL

Torque wrench settings	Cavalier/Vectra 1.7 TD 1992 to 1995	Cavalier/Vectra 1.7 TD 1994 to 1995	Vectra-B 1.7 DTL 1995 to 1997	Vectra-B 2.0 Di 1995 to 2000
Cylinder head - stage 1 ....Nm	40 N	40 N	40 N	25 N
- stage 2 ....Nm	+ 75°	+ 75°	+ 60° to 75°	+ 65°
- stage 3 ....Nm	+ 75°	+ 75°	+ 60° to 75°	+ 65°
- stage 4 ....Nm	–	–	–	+ 65°
- stage 5 ....Nm	–	–	–	+ 65°
- stage 6 ....Nm	–	–	–	+ 15°
Big-end bearings.................................Nm	20 + 115° N	20 + 115° N	20 + 100° + 15° N	35 + 45° + 15° N
Main bearings.....................................Nm	88	88	88	90 + 60° + 15° N
Crankshaft pulley bolt ....Nm	196	196	196	150 + 45° + 15° N
Camshaft pulley bolt ....Nm	10	10	10	90 + 60° + 30° N
Flywheel [driveplate] bolt.....................Nm	30 + 60° N	30 + 60° N	30 + 45° to 60° N	45 + 30° + 15° N
Front hubs ....Nm	130, slacken, 20 + 90°	130, slacken, 20 + 90°	130, slacken, 20 + 90°	WSM
Rear hubs ....Nm	–	–	WSM	WSM
Wheel nuts / bolts ....Nm	110	110	110	110
Glow plugs ....Nm	20	20	20	10
Clutch pressure plate bolts ....Nm	15	15	15	15
Injection pump sprocket.......................Nm	69	69	69	28
Injectors...Nm	50	50	45	45
Injection pump mounting bolts ....Nm	23	23	23	25
Injector pipe unions............................Nm	25	25	25	30
Capacities				
Engine oil & filter...............................litres	4.5	4.5	5.3	5.5
Gearbox..litres	1.9	1.9	1.9	1.9
Automatic transmission ....litres	–	–	–	–
Final drive ....litres	WT	WT	WT	WT
Cooling system.................................litres	7.4	7.4	6.8	7.3
Fuel tank..litres	61	61	60	60

Notes

Vectra-B 2.0 Di 1995 to 2000
[1]Estate: [0°26' ± 10']
[2]Estate: -1°29' ± 35'

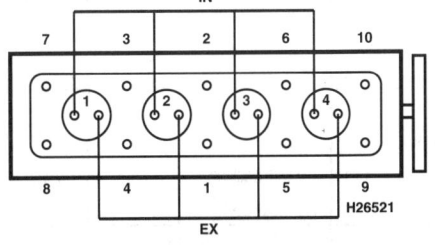

1686 cm³

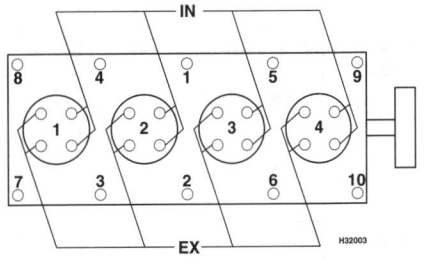

1994 cm³ 16V

– Not applicable, or information not available

VAUXHALL/OPEL

	Vectra-B 2.0 TDi 1997 to 2000	Carlton/Omega-A 2.3 1987 to 1992	Carlton/Omega-A 2.3 TD 1989 to 1994	Carlton/Omega-A 2.3 TD EGR 1989 to 1994
Engine				
Engine type/code..........................	X20DTH 16V Turbo 74kW	23YD SOHC 54kW	23DTR SOHC Turbo 74kW	23DTR SOHC Turbo 74kW
Capacity (cm^3) / cylinders......................	1995 / 4	2260 / 4	2260 / 4	2260 / 4
Compression ration / pressure ....bar	18.5 / ≥17.0	23.0 / ≥17.0	23.0 / ≥17.0	23.0 / ≥17.0
Torque output ....Nm	205	0	0	0
Oil pressure ....idle [running] bar	1.5	1.8	1.8	1.8
Oil temperature ....°C	80	80	60	80
Valve clearances - inlet (mm)	0: Hyd.	0.20	0.20	0.20
- exhaust (mm)	0: Hyd.	0.30	0.30	0.30
Injection order	1-3-4-2	1-3-4-2	1-3-4-2	1-3-4-2
No. 1 cylinder position	TCE	F	F	F
Cooling system				
Thermostat opening temperature ....°C	91	92	92	92
Radiator cap pressure ....bar	1.4 to 1.5	1.2 to 1.3	1.2 to 1.3	1.2 to 1.3
Fuel system				
Idle speed ....rpm	750 to 850	720 to 740 [725 to 750]	720 to 740 [725 to 750]	720 to 740 [725 to 750]
Maximum (no load) speed ....rpm	4900 to 5100	5100	5000	5000
Smoke test/opacity ....M^{-1} %	2.0	3.0	2.5	2.5
Static timing method.............................	Refer to wsm	Plunger travel	Plunger travel	Plunger travel
Timing dimension............................mm	Computer controlled	0.93 ± 0.05	0.85 ± 0.05	0.85 ± 0.05
Crankshaft position ....mm [°]	–	TDC	TDC	TDC
Turbo type / ref / pressure ....bar	–	–	≥0.8 bar @ 2000rpm	–
Injection pump make	Bosch	Bosch	Bosch	Bosch
Injection pump part no..........................	EDC 15.6 / 15M Direct VP44	VE4/9 L 243-5 [243-3]	VE4/10 L 297 [297-2]	F 2100 L 297-1 [297-3]
Injector Make / type..............................	–	Bosch	Bosch	–
Injector part no..................................	–	0432 217 133	0432 217 133	DN OSD 272
Injection type......................................	Direct EDC 15 M	Indirect VE 4/9	Indirect VE 4/10	VE 4/10
Injection opening pressure, New [used]...bar	180 to 365	135 to 143 [117]	135 to 140 [117]	≥135 [120]
Glow plugs				
Maker	–	Bosch/Champion	Bosch/Champion	Bosch/Champion
Type	–	0250 201 019 / CH68	0250 201 019 / CH68	0250 201 019 / CH68
Nominal rating.....................................V/A	11 /	11 / 12	11 / 12	12.0 / _
Brakes				
minimum friction material thickness				
Front..................................mm	7.0 with backing	7.0 with backing	7.0 with backing	7.0 with backing
Rear..mm	7.0 with backing	0.5 above rivets	0.5 above rivets	0.5 above rivets
Tyres - Saloon / Hatch.....................Size	185/70x14: 195/65x15	175x14: 185/70x14: 195/65x15	175x14: 185/70x14: 195/65x15	175x14: 185/70x14: 195/65x15
- Estate / Van........................Size	185/70x14: 195/65x15	175x14:185x14:185/70x14'	175x14:185x14:185/70x14'	175x14:185x14:185/70x14'
Pressure - front / rear - Saloon / Hatch ...bar	2.4 / 2.2	2.0 / 2.0	2.0 / 2.0	2.0 / 2.0
- Estate / Van ......bar	2.4 / 2.2	2.0 / 2.2	2.0 / 2.2	2.0 / 2.2
Front suspension / wheel alignment				
Toe-in (+) / Toe-out (–).....................mm [°]	[0°10' ± 10']	[0 to 20'] L	[0 to 20'] L	[0 to 20'] L
Camber	-1°05' ± 45'	-1°55' to -25' L	-1°55' to -25' L	-1°55' to -25' L
Castor	3°40' ± 1°	²	²	²
King pin inclination................................	–			
Rear suspension / wheel alignment				
Toe-in (+) / Toe-out (–)....................mm [°]	[0°30' ± 10']'	-0.5 to 5.5³	-0.5 to 5.5³	-0.5 to 5.5³
Camber	-1°55' ± 35'²	-2°20' to -1° Est⁴	-2°20' to -1° Est⁴	-2°20' to -1° Est⁴

VAUXHALL/OPEL

	Vectra-B 2.0 TDi 1997 to 2000	Carlton/Omega-A 2.3 1987 to 1992	Carlton/Omega-A 2.3 TD 1989 to 1994	Carlton/Omega-A 2.3 TD EGR 1989 to 1994
Torque wrench settings				
Cylinder head - stage 1 ....Nm	25 N	50 N	50 N	50 N
- stage 2 ....Nm	+ 65°	100	100	100
- stage 3 ....Nm	+ 65°	+ 135°	+ 135°	+ 135°
- stage 4 ....Nm	+ 65°	Warm up	Warm up	Warm up
- stage 5 ....Nm	+ 65°	+ 30°	+ 30°	+ 30°
- stage 6 ....Nm	+ 15°	+ 30°	–	–
Big-end bearings.........................Nm	35 + 45° + 15° N	45 + 45° N	45 + 45° N	45 + 45° N
Main bearings.........................Nm	90 + 60° + 15° N	50 + 60° M12: 115	50 + 60° M12: 70 + 608	M12: 70 + 60°5
Crankshaft pulley bolt ....Nm	150 + 45° + 15° N	150	220	220
Camshaft pulley bolt ....Nm	90 + 60° + 30° N	140 to 170	140 to 170	140 to 170
Flywheel [driveplate] bolt.....................Nm	45 + 30° + 15° N	60	80 + 30°	80 + 25 to 30°
Front hubs ....Nm	WSM	320	320	320
Rear hubs ....Nm	WSM	WSM	WSM	WSM
Wheel nuts / bolts ....Nm	110	90	90	90
Glow plugs ....Nm	10	20	20	20
Clutch pressure plate bolts ....Nm	15	15 M8: 28	15	15
Injection pump sprocket.....................Nm	28	55	55	55
Injectors.........................Nm	45	70	70	70
Injection pump mounting bolts ....Nm	25	–	–	–
Injector pipe unions.............................Nm	30	–	–	–
Capacities				
Engine oil & filter........................litres	5.5	5.5	5.5	5.5
Gearbox........................litres	1.9	1.9	1.9	1.9
Automatic transmission ....litres	–	5.5 oil pan removed	5.5 oil pan removed	5.5 oil pan removed
Final drive ....litres	WT	0.9	1.1	1.1
Cooling system........................litres	7.3	10.9 [A/T: 10.8]	10.9 [A/T: 10.8]	10.9 [A/T: 10.8]
Fuel tank........................litres	60	75 Est: 70	75 Est: 70	75 Est: 70

Notes

Vectra-B 2.0 TDi 1997 to 2000
[1]Estate: [0°26' ± 10']
[2]Estate: -1°29' ± 35'
Carlton/Omega-A 2.3 1987 to 1992
[1]Also 195/65x15
[2]Sal: 2 to 4°. PAS: 4°30' to 6°30'. Est man strg: 1°30' to 3°30'. PAS: 4 to 6°
[3]Estate: 0 to 6.0
[4]-2°5' to -45'
[5]M10: 50 + 45 to 60°
Carlton/Omega-A 2.3 TD 1989 to 1994
[1]Also 195/65x15
[2]Sal man strg: 2 to 4°. PAS: 4°30' to 6°30'. Est man strg: 1°30' to 3°30'. PAS: 4 to 6°
[3]Estate: 0 to 6.0
[4]-2°5' to -45'
[5]M10: 50 + 45 to 60°
Carlton/Omega-A 2.3 TD EGR 1989 to 1994
[1]Also 195/65x15
[2]Sal man strg: 2 to 4°. PAS: 4°30' to 6°30'. Est man strg: 1°30' to 3°30'. PAS: 4 to 6°
[3]Estate: 0 to 6.0
[4]-2°5' to -45'
[5]M10: 50 + 45 to 60°

– Not applicable, or information not available

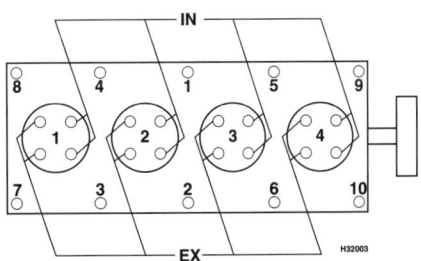

1994 cm³ 16V

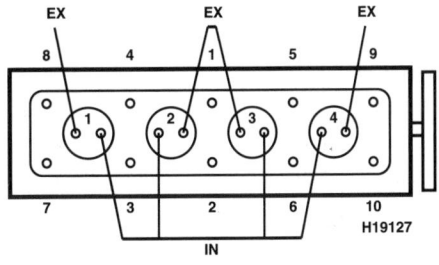

2260 cm³

VAUXHALL/OPEL

	Omega-B 2.0 TDi 16V 1998 to 2000	Omega-B 2.5 TD 1994 to 1996	Omega-B 2.5 TD 1997 to 2000	Sintra 2.2 TD 16V 1998 to 1999
Engine				
Engine type/code.....................................	X20DTH SOHC Turbo 16V 74kW	25 DT SOHC Turbo 96kW	X25DT SOHC Turbo 96kW	X22DTH SOHC 16V Turbo 85kW
Capacity (cm^3) / cylinders......................	1994 / 4	2498 / 6	2497 / 6	2171 / 4
Compression ration / pressure ....bar	18.5 /_	22.0 / ≥20.0	22.5 / ≥20.0	18.5 /
Torque output ....Nm	205	250	250	260
Oil pressure ....idle [running] bar	1.5	2.0	2.0	1.5
Oil temperature ....°C	80	80	80	80
Valve clearances - inlet (mm)	0: Hyd.	0: Hyd.	0: Hyd.	0: Hyd.
- exhaust (mm)	0: Hyd.	0: Hyd.	0: Hyd.	0: Hyd.
Injection order	1-3-4-2	1-5-3-6-2-4	1-5-3-6-2-4	1-3-4-2
No. 1 cylinder position	TBE	_	TCE	TCE
Cooling system				
Thermostat opening temperature ....°C	92	80	80	92
Radiator cap pressure ....bar	1.2 to 1.5	1.4	1.4	1.2 to 1.5
Fuel system				
Idle speed ....rpm	700 to 1000	750 ± 50	750 ± 50	740 to 900
Maximum (no load) speed ....rpm	_	5300 ± 100	5300	4950 to 5050
Smoke test/opacity ....M^{-1} %	2.0	2.41	2.3	2.5
Static timing method...............................	Refer to wsm	Plunger travel	Plunger travel	Refer to wsm
Timing dimension..............................mm	Computer controlled	0.88 to 0.97	0.98 ± 0.02	_
Crankshaft position ....mm [°]	_	0 TDC	0 TDC	_
Turbo type / ref / pressure ....bar	yes	_	_	_
Injection pump make	Bosch	Bosch	Bosch	Bosch
Injection pump part no............................	VP44	VE6/10 2400 R515	VP36	VP44
Injector Make / type	_	Bosch	Bosch	Multi hole
Injector part no.......................................	_	KCA21S76	DNOSD 300	_
Injection type..	Direct EDC 15M	VP36	Bosch DDE 2.1	Direct EDC 15M
Injection opening pressure, New [used]...bar	180 to 365	150 to 158	150 to 158	180 to 360
Glow plugs				
Maker	_/Champion	Bosch/Champion	Beru/Champion	_
Type	CH207	0250 201 027 / CH159	719 MJ / CH159	_
Nominal rating....................................V/A	11 /	_	11 / 12	_
Brakes				
minimum friction material thickness				
Front...mm	8.0 with backing	6.0 with backing	8.0 with backing	8.0 with backing
Rear..mm	6.0 with backing	6.0 with backing	6.0 with backing	7.0 with backing
Tyres - Saloon / Hatch......................Size	195/65x15:205/65x15:225/55x16	205/65x15	205/65x15: 225/55x16	_
- Estate / Van.....................Size	195/65x15: 225/55x16	205/65x15	205/65x15: 225/55x16	205/65x15
Pressure - front / rear - Saloon / Hatch ...bar	2.1 / 2.1	2.2 / 2.2	2.1 / 2.1	_
- Estate / Van ......bar	2.1 / 2.3	2.0 / 2.2	2.1 / 2.3	2.2 / 2.2
Front suspension / wheel alignment				
Toe-in (+) / Toe-out (–)....................mm [°]	[10' ±10'] L	[0°10' ± 10']	[10' ± 10'] L	[0° ± 2']
Camber	-1°40 '± 45'	-1°40' ± 45'	-1°40' ± 45'	0°48' ± 30'
Castor	5°40' ± 1° Est: 5° ± 1°	5°40' ± 1°	5°40' ± 1° Est: 5° ± 1°	2°59' ± 30'
King pin inclination..................................	_		_	_
Rear suspension / wheel alignment				
Toe-in (+) / Toe-out (–)....................mm [°]	[0°20' ± 10'] L	[0°20' ± 10']	[0°20' ± 10'] L	[0° ± 7']
Camber	-1°50' ± 40' Est: -1°40'	-1°50' ± 40'	-1°50' ± 40' Est: -1°40'	-1° ± 15'

VAUXHALL/OPEL

	Omega-B 2.0 TDi 16V 1998 to 2000	Omega-B 2.5 TD 1994 to 1996	Omega-B 2.5 TD 1997 to 2000	Sintra 2.2 TD 16V 1998 to 1999
Torque wrench settings				
Cylinder head - stage 1 ....Nm	25 N	80 N	80 N	25 N
- stage 2 ....Nm	+ 90°	Slacken	+ 90°	+ 65°
- stage 3 ....Nm	+ 90°	50	+ 90°	+ 65°
- stage 4 ....Nm	+ 90°	+ 90°	+ 90°	+ 65°
- stage 5 ....Nm	–	+ 90°	–	+ 65°
- stage 6 ....Nm	–	warm up, + 90°	–	+ 15°
Big-end bearings.............................Nm	35 + 45° + 15° N	20 + 70° N	20 + 70° N	35 + 45° + 15° N
Main bearings.................................Nm	50 + 45° + 15° N	25 + 50° N	25 + 50° N	90 + 60° + 15°
Crankshaft pulley bolt ....Nm	150 + 45° + 15° N	100 + 150° N	100 + 150° N	150 + 45° + 15° N
Camshaft pulley bolt ....Nm	90 + 60° + 30° N	20 + 35° N	20 + 35° N	90 + 60° + 30° N
Flywheel [driveplate] bolt.................Nm	65 + 30° + 15° [60] N	122	112 N	45 + 30° + 15° N
Front hubs ....Nm	320	320	320	160
Rear hubs ....Nm	300	300	300	WSM
Wheel nuts / bolts ....Nm	110	110	110	140
Glow plugs ....Nm	25	23	22	10
Clutch pressure plate bolts ....Nm	15 M8: 28	15 M8: 22	15 M8: 28	18
Injection pump sprocket.....................Nm	–	48	48	20
Injectors..Nm	WSM	65	65	WSM
Injection pump mounting bolts ....Nm	25	22	22	25
Injector pipe unions...........................Nm	25	25	25	15
Capacities				
Engine oil & filter...............................litres	5.5	6.5	6.5	5.5
Gearbox..litres	[1.2]	1.2	[1.2]	1.8
Automatic transmission ....litres	4.4	4.4	4.4	–
Final drive ....litres	1.0	1.0	1.0	WT
Cooling system.................................litres	8.2	10.2	10.2 [AT: 10.0]	–
Fuel tank...litres	75	75	75	70

Notes

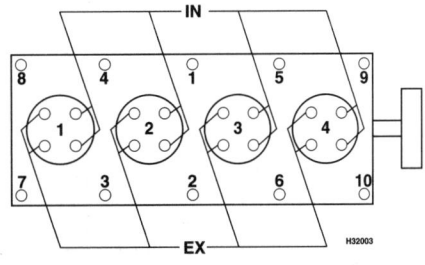

1994 cm³ / 2171 cm³ 16V

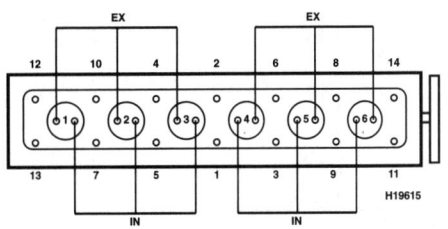

2497 cm³ / 2498 cm³

– Not applicable, or information not available

VAUXHALL/OPEL

	Frontera 2.2 TDi 1998 to 2000	Frontera 2.3 TD 1991 to 1995	Frontera 2.5 TDS 1996 to 1998	Frontera 2.8 TD 1995 to 1996
Engine				
Engine type/code	X22DTH SOHC 16V Turbo 85kW	23DTR SOHC Turbo 63kW	VM41B OHV Turbo 85kW	4JB1TC SOHC Turbo 82kW
Capacity (cm³) / cylinders	2171 / 4	2260 / 4	2499 / 4	2771 / 4
Compression ration / pressure ......bar	18.5 / 17.0 to 24.0	23.0 /	20.95 / 24.0	17.9 /
Torque output ......Nm	260	215	260	242
Oil pressure ......idle [running] bar	1.5	1.8	[4.0 @ 4000]	1.8
Oil temperature ......°C	80	80	80	80
Valve clearances - inlet (mm)	0: Hyd.	0.20	0: Hyd.	0.40
- exhaust (mm)	0: Hyd.	0.30	0: Hyd.	0.40
Injection order	1-3-4-2	1-3-4-2	1-3-4-2	1-3-4-2
No. 1 cylinder position	TCE	–	–	–
Cooling system				
Thermostat opening temperature ......°C	92	92	82	92
Radiator cap pressure ......bar	1.2 to 1.5	1.2 to 1.3	1.2 to 1.5	1.5 to 1.5
Fuel system				
Idle speed ......rpm	740 to 900	720 to 750	775 ± 25	700 to 800
Maximum (no load) speed ......rpm	4950 to 5050	4900 to 5000	–	4500 to 4700
Smoke test/opacity ......M⁻¹ %	2.0	2.5	2.0	2.5
Static timing method	Refer to wsm	Plunger travel	Plunger travel	Plunger travel
Timing dimension ......mm	Computer controlled	0.85 ± 0.05	0.65	0.50
Crankshaft position ......mm [°]	–	0 TDC	0 TDC	[12]
Turbo type / ref / pressure ......bar	–	0.8 bar @ 2000	–	0.68 bar
Injection pump make	Bosch	Bosch	Bosch	Bosch
Injection pump part no	VP44	VE4/10F2100L	VP 36	VE
Injector Make / type	Multi hole	Bosch	Bosch	
Injector part no		DNOSD272	DNOSD272	–
Injection type	Direct EDC 15M	Indirect VE4	Indirect	VE
Injection opening pressure, New [used]...bar	180 to 360	135 to 143	135 to 143	181
Glow plugs				
Maker	–	Bosch/Champion	Bosch	–
Type		0250 200 019 / CH68	0250 201 033	–
Nominal rating ......V/A	–	–	11 / 13	–
Brakes				
minimum friction material thickness				
Front ......mm	1.0	1.0	1.0	1.0
Rear ......mm	1.0	1.0	1.0	1.0
Tyres - Saloon / Hatch ......Size	–	–	–	–
- Estate / Van ......Size	235/75x15: 245/70x16	225/75x15	235/70x16: 255/65x15	235/70x16: 255/65x15
Pressure - front / rear - Saloon / Hatch ...bar	–	–	–	–
- Estate / Van ......bar	2.0 / 2.0: 1.8 / 1.8	1.7 / 1.7	1.7 / 1.7	1.7 / 1.7
Front suspension / wheel alignment				
Toe-in (+) / Toe-out (–) ......mm [°]	[0° ± 17']	2.0 ± 2.0	0.0 ± 2.0	0.0 ± 2.0
Camber	0° ± 1°	0°30' ± 1°	0° ± 45'	0° ± 45'
Castor	2°10' ± 1°	2° 30' ± 45°	2°10 ' ± 45'	2°10' ± 45'
King pin inclination	12°30' ± 30'	–	12°30' ± 30'	12°30' ± 30'
Rear suspension / wheel alignment				
Toe-in (+) / Toe-out (–) ......mm [°]	–	–	–	–
Camber	–	–	–	–

VAUXHALL/OPEL

	Frontera 2.2 TDi 1998 to 2000	Frontera 2.3 TD 1991 to 1995	Frontera 2.5 TDS 1996 to 1998	Frontera 2.8 TD 1995 to 1996
Torque wrench settings				
Cylinder head - stage 1 ....Nm	25 N	100 N	30, 30[1]	25 N
- stage 2 ....Nm	+ 65°	+ 135°	+ 70° + 70°	+ 90°
- stage 3 ....Nm	+ 65°	Warm up	warm up, allow to cool	+ 90°
- stage 4 ....Nm	+ 65°	+ 30°	slacken	+ 90°
- stage 5 ....Nm	+ 65°	+ 30°	30 + 65°	–
- stage 6 ....Nm	+ 15°	–	+ 65°[2]	–
Big-end bearings........................Nm	35 + 45° + 15° N	45 + 45° N	29 + 60° N	35 + 45° + 15° N
Main bearings........................Nm	90 + 60° + 15° N	70 + 60° N M10: 50 + 60° N	44	50 + 45° + 15° N
Crankshaft pulley bolt ....Nm	150 + 45° + 15° N	220	196	187
Camshaft pulley bolt........................Nm	90 + 60° + 30° N	150 N	–	64
Flywheel [driveplate] bolt....................Nm	45 + 30° + 15° N	80 + 30°	20 + 60°	65 + 30° + 15°
Front hubs ....Nm	WSM	WSM	WSM	WSM
Rear hubs ....Nm	WSM	WSM	WSM	WSM
Wheel nuts / bolts ....Nm	118	100 Alloy: 120	100 Alloy: 120	100 Alloy: 120
Glow plugs ....Nm	10	20	15	22
Clutch pressure plate bolts ....Nm	18	22	25	15
Injection pump sprocket....................Nm	20	55	88	64
Injectors........................Nm	WSM	70	70	37
Injection pump mounting bolts ....Nm	25	–	27	19
Injector pipe unions...........................Nm	30	25	23	29
Capacities				
Engine oil & filter.................................litres	5.5	5.7	6.25	5.5
Gearbox.................................litres	2.95	2.8 Transfer: 1.4	2.8 Transfer: 1.4	2.8 Transfer: 1.4
Automatic transmission ....litres	5.5	–	–	–
Final drive.................................litres	1.7 Rear: 2.4[1]	1.7 Rear: 2.4 LSD: 1.9	1.7 Rear: 2.4 LSD: 1.9	1.7 Rear: 2.4 LSD: 1.9
Cooling system...................................litres	7.9	10.9	8.8	8.8
Fuel tank.................................litres	65 5dr: 75	80	80	80

Notes

Frontera 2.2 TDi 1998 to 2000
[1]Transfer: 1.45
Frontera 2.5 TDS 1996 to 1998
[1]M12: 30, 85, after warm up and cool down: 90 Nm
[2] After 20 000Km: + 15°

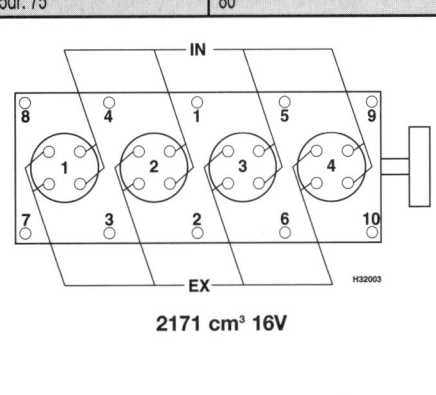

2171 cm³ 16V

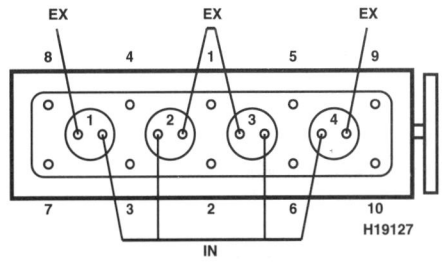

2260 cm³

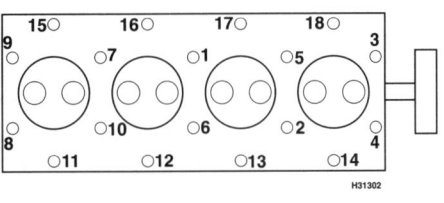

2499 cm³

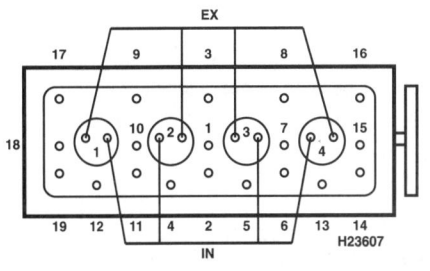

2771 cm³

– Not applicable, or information not available

VAUXHALL/OPEL

	Monterey 3.1 TD 1994 to 1998	Brava 2.5 TDi 1994 to 2000	Midi 2.0 1988 to 1995	Midi 2.2 1988 to 1994
Engine				
Engine type/code	4JG2TC OHV Turbo 84kW	4JA1T OHV Turbo 56kW'	4FC1-T OHC Turbo 51kW	4FD1 OHC IDI 45kW
Capacity (cm^3) / cylinders	3059 / 4	2499 / 4	1995 / 4	2189 / 4
Compression ration / pressure ...bar	20.1 /	18.4 / 30.4	21.0 / 30.4	21.5 / 30.4
Torque output ...Nm	260	160	0	0
Oil pressure ...idle [running] bar	_	[3.9]	[4.4]	[4.4]
Oil temperature ...°C	_	80	80	80
Valve clearances - inlet (mm)	0.40	0.40	0.25	0.25
- exhaust (mm)	0.40	0.40	0.35	0.35
Injection order	1-3-4-2	1-3-4-2	1-3-4-2	1-3-4-2
No. 1 cylinder position	TBE	TBE	F	F
Cooling system				
Thermostat opening temperature ...°C	82	82	80.5 to 83.5	80.5 to 83.5
Radiator cap pressure ...bar	0.9 to 1.2	0.9 to 1.2	1.0	1.0
Fuel system				
Idle speed ...rpm	695 to 745	745 to 795	620	620
Maximum (no load) speed ...rpm	4500 to 4700	4700 to 4900	5200	5200
Smoke test/opacity ...M^{-1} %	2.5	2.0	2.5	2.5
Static timing method	Plunger travel	Plunger travel	Plunger travel	Plunger travel
Timing dimension ...mm	0.50	0.50	0.50	0.50
Crankshaft position ...mm [°]	0 TDC	[12]	[3]	[5]
Turbo type / ref / pressure ...bar	0.67 bar	_	_	_
Injection pump make	Bosch	Bosch	Diesel Kiki	Diesel Kiki
Injection pump part no.	VE	VE	_	_
Injector Make / type	Throttle type	Bosch	Bosch	Bosch
Injector part no.	_	DN12SD12T		
Injection type	Indirect	Direct	Bosch VE type	Bosch VE type
Injection opening pressure, New [used]...bar	147	181	120	120 to 130
Glow plugs				
Maker	Bosch/Champion	Champion	Bosch/Champion	_
Type	0250 202 087 / CH158	CH68	0250 202 087 / CH158	_
Nominal rating ...V/A	11 / 12	11 /	11 / 12	11.0 / 7.1
Brakes				
minimum friction material thickness				
Front ...mm	1.0	1.0	1.0	1.0
Rear ...mm	1.0	1.0	1.0	1.0
Tyres - Saloon / Hatch ...Size	_	_	_	_
- Estate / Van ...Size	245/70x16	185x14 4x4: 205x16	185x14	185x14
Pressure - front / rear - Saloon / Hatch ...bar	_	_	_	_
- Estate / Van ...bar	2.1 / 2.3'	1.8 / 4.5 4x4: 2.0 / 2.7	3.0 / 3.7'	3.0 / 3.7'
Front suspension / wheel alignment				
Toe-in (+) / Toe-out (–) ...mm [°]	[0° ± 17']	[0° ± 19'] 4x4: [0° ± 17']	1.0 to 5.0	1.0 to 5.0
Camber	0° ± 30'	0°30' ± 1°	35' ± 30'	35' ± 30'
Castor	2°10' ± 45'	1°50' ± 45' 4x4: 2°10' ± 45'	2° ± 30'	2° ± 30'
King pin inclination	12°30' ± 30'	10° ± 1°	7°25' ± 30'	7°25' ± 30'
Rear suspension / wheel alignment				
Toe-in (+) / Toe-out (–) ...mm [°]	_	_	_	_
Camber				

VAUXHALL/OPEL

Torque wrench settings	Monterey 3.1 TD 1994 to 1998	Brava 2.5 TDi 1994 to 2000	Midi 2.0 1988 to 1995	Midi 2.2 1988 to 1994
Cylinder head - stage 1 ....Nm	40 N	49 N	40	40
- stage 2 ....Nm	+ 60°	+ 60° to 75°	130	130
- stage 3 ....Nm	+ 60°	+ 60° to 75°	–	–
- stage 4 ....Nm	–	–	–	–
- stage 5 ....Nm	–	–	–	–
- stage 6 ....Nm	–	–	–	–
Big-end bearings.............................Nm	29 + 45° N	83 N	80 N	80 N
Main bearings...................................Nm	167	167	95	95
Crankshaft pulley bolt ....Nm	WSM	WSM	120	120
Camshaft pulley bolt ....Nm	64	64	65	65
Flywheel [driveplate] bolt....................Nm	118	118	55	55
Front hubs ....Nm	WSM	–	WSM	WSM
Rear hubs ....Nm	WSM	–	–	–
Wheel nuts / bolts ....Nm	118	82	108	108
Glow plugs ....Nm	23	23	15	15
Clutch pressure plate bolts ....Nm	20	18	18	18
Injection pump sprocket....................Nm	64	64	70	70
Injectors...Nm	64	35	75	75
Injection pump mounting bolts ....Nm	–	22	20	20
Injector pipe unions...........................Nm	29	35	25	25
Capacities				
Engine oil & filter.............................litres	6.0	4.3	5.7	5.7
Gearbox..litres	1.95	1.5 4x4: 4.4	1.6	1.6
Automatic transmission ....litres	–	–	–	–
Final drive ....litres	1.5 Rear: 1.8[2]	1.5	1.1	1.1
Cooling system.................................litres	8.6	7.0	Van: 10.0[2]	Van: 10.0[2]
Fuel tank..litres	85	75 Double Cab: 53	60	60

Notes

Monterey 3.1 TD 1994 to 1998
[1]LWB rear: 2.4
[2]Transfer: 1.45
Brava 2.5 TDi 1994 to 2000
[1]Fitted with CAT 1997 on
Midi 2.0 1988 to 1995
[1]Bus: 3.0 / 3.1
[2]Bus with rear heater: 11.0
Midi 2.2 1988 to 1994
[1]Bus: 3.0 / 3.1
[2]Bus with rear heater: 11.0

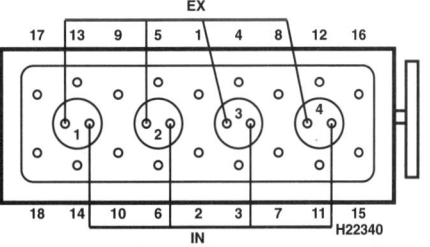

2499 cm³ / 3059 cm³

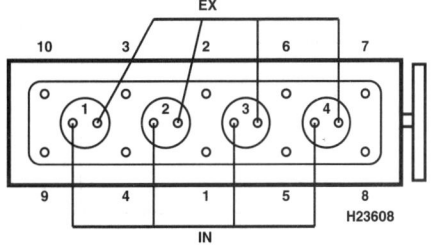

1995 cm³ / 2189 cm³

– Not applicable, or information not available

VAUXHALL/OPEL

	Midi 2.4 TD 1994 to 1997	Arena 1.9 D 1997 to 2000	Arena 2.5 D 1997 to 2000	Movano 2.5 D 1998 to 2000
Engine				
Engine type/code....................	4FG1-T SOHC Turbo 55kW	F8Q 606 SOHC 44kW	S8U 782 SOHC 55kW	S8U SOHC 58kW
Capacity (cm³) / cylinders........................	2380 / 4	1870 / 4	2499 / 4	2499 / 4
Compression ration / pressure ....bar	20.0 / 30	23.0 / _	22.0 / _	22.0 /
Torque output ....Nm	167	120	158	155
Oil pressure ....idle [running] bar	[4.5]	1.2 [5.6]	1.2 [5.6]	0.8 [3.5 @ 4000]
Oil temperature ....°C	80	80	80	80
Valve clearances - inlet (mm)	0.25	0.40	0.40	0.50
- exhaust (mm)	0.35	0.40	0.40	0.50
Injection order...........................	1-3-4-2	1-3-4-2	1-3-4-2	1-3-4-2
No. 1 cylinder position...................	TBE	TBE	TBE	TBE
Cooling system				
Thermostat opening temperature ....°C	80	_	_	89
Radiator cap pressure ....bar	0.9 to 1.2	0.8 to 1.2	0.8 to 1.2	1.2
Fuel system				
Idle speed ....rpm	675 to 725	785 to 885	760 to 860	760 to 860
Maximum (no load) speed ....rpm	5000	4500 to 4700	4650 to 4750	4500 to 4700
Smoke test/opacity ....M⁻¹ %	3.0	2.41	2.5	2.5
Static timing method................................	Plunger travel	Plunger travel	Plunger travel	Rotor lift
Timing dimension...........................mm	0.50	1.10 ± 0.02	1.0 ± 0.02	0.97 ± 0.02
Crankshaft position ....mm [°]	[5]	[0] TDC	[0] TDC	TDC
Turbo type / ref / pressure ....bar	_	_	_	_
Injection pump make	Bosch	Lucas	Bosch	Bosch
Injection pump part no.............................	VE	H300575	VE4/9F 2100 R22-7	VE4/10F2100R717-1
Injector Make / type	Pintle	Bosch	Bosch	Bosch
Injector part no.......................................	_	DNOSD 193	DNOSD 193	DNOSD 301
Injection type..	Indirect VE	Indirect	Indirect	Indirect
Injection opening pressure, New [used]...bar	147	125 ± 5	125 ± 5	115 to 128
Glow plugs				
Maker	_	Champion	Champion	_
Type	_	CH87	CH87	_
Nominal rating.................................V/A	_	_	_	_
Brakes				
minimum friction material thickness				
Front........................mm	1.0	9.0 with backing	9.0 with backing	8.0 with backing
Rear.........................mm	1.0	5.0 with backing	5.0 with backing	1.0
Tyres - Saloon / Hatch......................Size	_	_	_	_
- Estate / Van............................Size	185x14	165/80X14:175/80x14: 185/75x14	175/80x14: 185/75x14	195/70x15: 215/70x15:225/70x15
Pressure - front / rear - Saloon / Hatch ...bar	_	_	_	_
- Estate / Van ......bar	3.0 / 3.8	3.8 / 3.8	SWB: 3.8 / 3.8 LWB: 4.1 / 4.5	3.6 / 3.7: 3.5 / 3.7:3.3 / 3.7
Front suspension / wheel alignment				
Toe-in (+) / Toe-out (–).....................mm [°]	2.0 ± 3.0	[-0°10' ± 10']	[-0°10' ± 10']	1.0 ± 1.0
Camber	0°35' ± 30'	0°30' ± 1°	0°30' ± 1°	-0°01' ± 30'
Castor	2° ± 30'	3° ± 1°	3° ± 1°	2°58' ± 30'
King pin inclination.................................	7°25' ± 30'	9°30' ± 1°	9°30' ± 1°	
Rear suspension / wheel alignment				
Toe-in (+) / Toe-out (–).....................mm [°]	_	[0°15' ± 15']	[0°15' ± 15']	3.0 ± 2.0
Camber		0°15' ± 15'	0°15' ± 15'	0°08' ± 05'

VAUXHALL/OPEL

Torque wrench settings	Midi 2.4 TD 1994 to 1997	Arena 1.9 D 1997 to 2000	Arena 2.5 D 1997 to 2000	Movano 2.5 D 1998 to 2000
Cylinder head - stage 1 ....Nm	49	40	40 N	40 N
- stage 2 ....Nm	132	75	40	40
- stage 3 ....Nm	–	Warm up, allow to cool	+ 180°	+ 180°
- stage 4 ....Nm	–	- 90°	–	–
- stage 5 ....Nm	–	75	–	–
- stage 6 ....Nm	–	–	–	–
Big-end bearings................................Nm	79 N	110 N	110 N	50 + 65° N
Main bearings....................................Nm	93	80, 160	80, 160	80, 160
Crankshaft pulley bolt ....Nm	118	200	200	–
Camshaft pulley bolt ....Nm	64	25	25	–
Flywheel [driveplate] bolt....................Nm	53	120	120	120
Front hubs ....Nm	WSM	190 to 210	190 to 210	480
Rear hubs ....Nm	–	WSM	WSM	280
Wheel nuts / bolts ....Nm	108	120	120	155
Glow plugs ....Nm	23	15	15	–
Clutch pressure plate bolts ....Nm	74	25	25	–
Injection pump sprocket.....................Nm	68	50	50	–
Injectors..Nm	74	50	50	50
Injection pump mounting bolts ....Nm	–	50	50	–
Injector pipe unions...........................Nm	29	25	25	–
Capacities				
Engine oil & filter................................litres	5.7	5.0	6.5	6.5
Gearbox..litres	1.55	2.5 NE3: 1.8[1]	2.5 NE3: 1.8[1]	2.8
Automatic transmission ....litres	–	–	–	–
Final drive ....litres	1.1	0.8[2]	0.8[2]	WT
Cooling system..................................litres	10.0	6.8	8.4	11.0
Fuel tank..litres	60	60	60	70

Notes

Arena 1.9 D 1997 to 2000
[1]4x4: 2.0
[2]NE3: 1.45. With cooler and 4x4: 1.3
Arena 2.5 D 1997 to 2000
[1]4x4: 2.0
[2]NE3: 1.45. With cooler and 4x4: 1.3

– Not applicable, or information not available

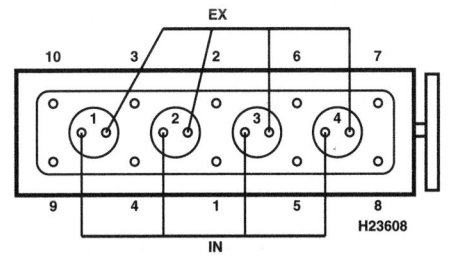

2380 cm³

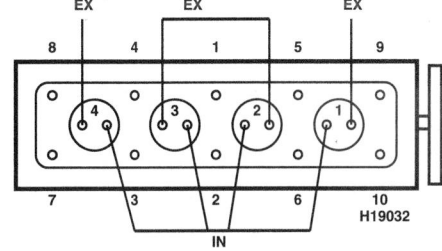

1870 cm³

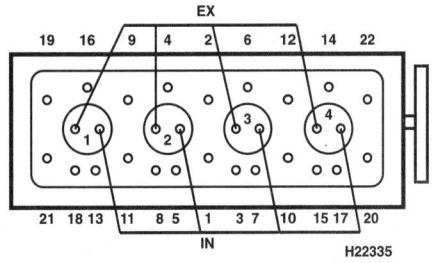

2499 cm³

VAUXHALL/OPEL

Haynes
THE BOOK ®

	Movano 2.8 TD 1998 to 2000			
Engine				
Engine type/code..................................	S9W 702 SOHC Turbo 84kW			
Capacity (cm^3) / cylinders......................	2799 / 4			
Compression ration / pressure ....bar	19.0 /			
Torque output ....Nm	260			
Oil pressure ....idle [running] bar	–			
Oil temperature ....°C	–			
Valve clearances - inlet (mm)	0.50			
- exhaust (mm)	0.50			
Injection order	1-3-4-2			
No. 1 cylinder position	–			
Cooling system				
Thermostat opening temperature ....°C	89			
Radiator cap pressure ....bar	1.2			
Fuel system				
Idle speed ....rpm	760 to 860			
Maximum (no load) speed ....rpm	4000 ± 100			
Smoke test/opacity ....M^{-1} %	2.5			
Static timing method................................	Rotor lift			
Timing dimension................................mm	1.11 ± 0.02			
Crankshaft position ....mm [°]	TDC			
Turbo type / ref / pressure ....bar	Garrett GT17			
Injection pump make	Bosch			
Injection pump part no............................	VE4/12F1800R721			
Injector Make / type	Bosch			
Injector part no.......................................	DLSA 134 P604			
Injection type..	Direct			
Injection opening pressure, New [used]...bar	235 to 243			
Glow plugs				
Maker	–			
Type	–			
Nominal rating....................................V/A	–			
Brakes				
minimum friction material thickness				
Front...mm	8.0 with backing			
Rear..mm	1.0			
Tyres - Saloon / Hatch......................Size	–			
- Estate / Van...........................Size	195/70x15: 215/70x15:225/70x15			
Pressure - front / rear - Saloon / Hatch ...bar	–			
- Estate / Van ......bar	3.6 / 3.7: 3.5 / 3.7:3.3 / 3.7			
Front suspension / wheel alignment				
Toe-in (+) / Toe-out (–)....................mm [°]	1.0 ± 1.0			
Camber	-0°01' ± 30'			
Castor	2°58' ± 30'			
King pin inclination..................................	–			
Rear suspension / wheel alignment				
Toe-in (+) / Toe-out (–)....................mm [°]	3.0 ± 2.0			
Camber	0°08' ± 05'			

VAUXHALL/OPEL

	Movano 2.8 TD 1998 to 2000			
Torque wrench settings				
Cylinder head - stage 1 ....Nm	–			
- stage 2 ....Nm	–			
- stage 3 ....Nm	–			
- stage 4 ....Nm	–			
- stage 5 ....Nm	–			
- stage 6 ....Nm	–			
Big-end bearings.............................Nm	50 + 65° N			
Main bearings.....................................Nm	80, 160			
Crankshaft pulley bolt ....Nm	200			
Camshaft pulley bolt ....Nm	25			
Flywheel [driveplate] bolt....................Nm	120			
Front hubs ....Nm	480			
Rear hubs ....Nm	280			
Wheel nuts / bolts ....Nm	155			
Glow plugs ....Nm	15			
Clutch pressure plate bolts ....Nm	25			
Injection pump sprocket......................Nm	–			
Injectors..Nm	50			
Injection pump mounting bolts ....Nm	–			
Injector pipe unions............................Nm	–			
Capacities				
Engine oil & filter................................litres	6.5			
Gearbox...litres	2.8			
Automatic transmission ....litres	–			
Final drive ....litres	WT			
Cooling system.................................litres	11.0			
Fuel tank...litres	70			

Notes

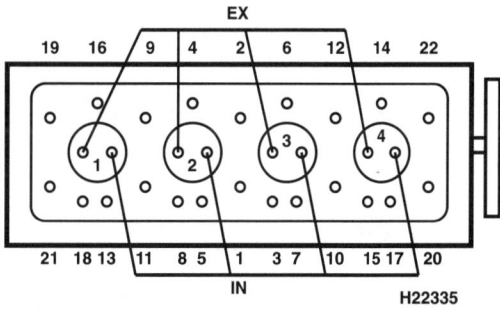

2799 cm³

– Not applicable, or information not available

VOLKSWAGEN

Haynes THE BOOK ®

	Lupo 1.7 SDi 1998 to 2000	Polo 1.4 1990 to 1994	Polo Classic / Caddy 1.7 Di 1996 to 1998	Polo 1.7 SDi 1997 to 2000
Engine				
Engine type/code.....................................	AKU SOHC 8V EGR 44kW	1W SOHC 35kW	AHG SOHC 8V 42kW	AKU SOHC 8V 44kW
Capacity (cm^3) / cylinders........................	1699 / 4	1398 / 4	1715 / 4	1715 / 4
Compression ration / pressure ....bar	19.5 /	22.5 / ≥25.0	19.5 / _	19.5 / ≥19.0
Torque output ....Nm	115	0	112	115
Oil pressure ....idle [running] bar	[2.0 @ 2000]	[2.0]	[2.0 @ 2000]	[2.0 @ 2000]
Oil temperature ....°C	80	80	80	80
Valve clearances - inlet (mm)	0: Hyd.	0: Hyd.	0: Hyd.	0: Hyd.
- exhaust (mm)	0: Hyd.	0: Hyd.	0: Hyd.	0: Hyd.
Injection order	1-3-4-2	1-3-4-2	1-3-4-2	1-3-4-2
No. 1 cylinder position	TBE	TBE	TBE	TBE
Cooling system				
Thermostat opening temperature ....°C	85	87	84	85
Radiator cap pressure ....bar	1.4 to 1.6	1.2 to 1.5	1.3 to 1.5	1.4 to 1.6
Fuel system				
Idle speed ....rpm	_	900 ± 50	861 to 945 N/A	800 ± 50
Maximum (no load) speed ....rpm	_	5600 ± 100	_	_
Smoke test/opacity ....M^{-1} %	2.0	2.0	2.0	2.0
Static timing method................................	Refer to wsm	Plunger travel	Refer to wsm	Refer to wsm
Timing dimension..............................mm	Computer controlled	0.95 ± 0.02	_	Computer controlled
Crankshaft position ....mm [°]	_	TDC	_	_
Turbo type / ref / pressure ....bar	_	_	_	_
Injection pump make	_	Bosch	Bosch	_
Injection pump part no............................	_	031 130 107B	VE4/10 R640	_
Injector Make / type	_	Bosch	Bosch	_
Injector part no.......................................	_	068 130 211G	0432 193 810/1	_
Injection type...	Direct	_	Direct SDI	SDI
Injection opening pressure, New [used]...bar	190 to 200	≥130 [120]	190 to 200 [170]	
Glow plugs				
Maker	Champion	Bosch	Bosch	_
Type	CH171	0250 201 032	0250 202 022	_
Nominal rating.....................................V/A	11 /	_	11.5 / 12	_
Brakes				
minimum friction material thickness				
Front.......................................mm	7.0 with backing	7.0 with backing	7.0 with backing	7.0 with backing
Rear.......................................mm	2.5	2.5	2.5	2.5
Tyres - Saloon / Hatch....................Size	175/65x13: 185/55x14	145x13: 155/70x13: 165/65x13	175/65x13:185/55x14:195/45x15	175/65x13:185/55x14:195/45x15
- Estate / Van...................Size	_	_	_	_
Pressure - front / rear - Saloon / Hatch ...bar	2.0 / 1.9: 2.2 / 2.0	1.7 / 1.7	_	2.1 / 2.1
- Estate / Van ......bar	_	_	_	_
Front suspension / wheel alignment				
Toe-in (+) / Toe-out (−).....................mm [°]	[0° ± 10']	[0° ± 10']	[0° ± 10']	[0° ± 10']
Camber	-45' ± 20'	0° ± 30'	-0°25' ± 20'	-0°25' ± 20'
Castor	1°20' ± 30'	2°20' ± 30'	1°20' ± 30'	1°20' ± 30'
King pin inclination.................................	_	_	_	_
Rear suspension / wheel alignment				
Toe-in (+) / Toe-out (−)......................mm [°]	[0°20' ± 10']	[20' ± 10']	[0°20' ± 10']	[0°20' ± 10']
Camber	-1°25' ± 20'	-1°30' ± 10'	-1°30' ± 10'	-1°30' ± 10'

VOLKSWAGEN

Torque wrench settings	Lupo 1.7 SDi 1998 to 2000	Polo 1.4 1990 to 1994	Polo Classic / Caddy 1.7 Di 1996 to 1998	Polo 1.7 SDi 1997 to 2000
Cylinder head - stage 1 ....Nm	40 N	40	40 N	40 N
- stage 2 ....Nm	60	60	60	60
- stage 3 ....Nm	+ 90°	+ 180°	+ 90°	+ 90°
- stage 4 ....Nm	+ 90°	Warm up	+ 90°	+ 90°
- stage 5 ....Nm	–	+ 90°	–	–
- stage 6 ....Nm	–	–	–	–
Big-end bearings.............................Nm	30 + 90° N	30 + 90° N	30 + 90° N	30 + 90° N
Main bearings.................................Nm	65 + 90° N	65	65 + 90° N	65 + 90° N
Crankshaft pulley bolt ....Nm	90 + 90° N	90 + 180°	90 + 90° oiled N	90 + 90° oiled N
Camshaft pulley bolt ....Nm	45	45	45	45
Flywheel [driveplate] bolt....................Nm	60 + 90° N	30 + 90°	60 + 90° N	60 + 90° N
Front hubs......................................Nm	200 - 360°, 50 + 30°	210	WSM	WSM
Rear hubs.......................................Nm	175 N	WSM	WSM	WSM
Wheel nuts / bolts ....Nm	110	90	110	110
Glow plugs ....Nm	15	30	25	15
Clutch pressure plate bolts ....Nm	–	25	10	10
Injection pump sprocket.....................Nm	55	50	55	55
Injectors..Nm	–	70	20	20
Injection pump mounting bolts ....Nm	–	25	25	25
Injector pipe unions...........................Nm	25	25	25	25
Capacities				
Engine oil & filter...............................litres	4.7	3.5	3.4	5.0
Gearbox..litres	2.7	3.1	3.1	2.7
Automatic transmission ....litres	–	–	5.7 filled for life	–
Final drive ....litres	–	WT	WT	WT
Cooling system.................................litres	6.5	5.5	5.5	6.5
Fuel tank...litres	–	42	45	45

Notes

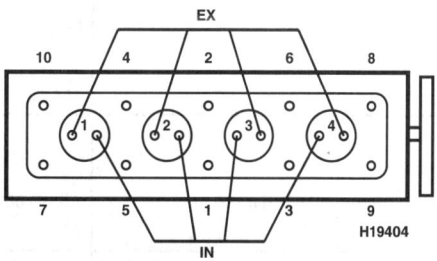

1699 cm³ / 1715 cm³

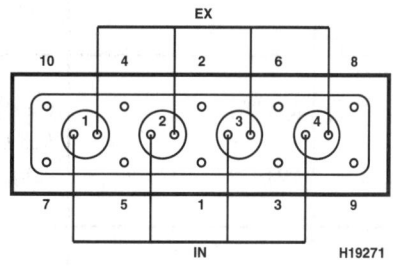

1398 cm³

– Not applicable, or information not available

VOLKSWAGEN

	Polo Classic / Caddy 1.7 Di 1997 to 2000	Polo 1.9 SD 1995 to 1997	Polo 1.9 SDi 1996 to 1998	Polo Classic / Caddy 1.9 D 1995 to 2000
Engine				
Engine type/code.....................................	AKW SOHC 8V EGR 44kW	AEF SOHC 8V 47kW	AGD SOHC 8V 47kW	AEY SOHC 8V 47kW
Capacity (cm³) / cylinders........................	1715 / 4	1896 / 4	1896 / 4	1896 / 4
Compression ration / pressure ....bar	19.5 / _	22.5 / 25 to 31	19.5 / 25 to 31	19.5 / _
Torque output ....Nm	112	124	124	125
Oil pressure ....idle [running] bar	[2.0 @ 2000]	[2.0 @ 2000]	[2.0 @ 2000]	[2.0 @ 2000]
Oil temperature ....°C	80	80	80	80
Valve clearances - inlet (mm)	0: Hyd.	0: Hyd.	0: Hyd.	0: Hyd.
- exhaust (mm)	0: Hyd.	0: Hyd.	0: Hyd.	0: Hyd.
Injection order	1-3-4-2	1-3-4-2	1-3-4-2	1-3-4-2
No. 1 cylinder position	TBE	TBE	TBE	TBE
Cooling system				
Thermostat opening temperature ....°C	84	84	85	84
Radiator cap pressure ....bar	1.3 to 1.5	1.3 to 1.5	1.4 to 1.6	1.3 to 1.5
Fuel system				
Idle speed ....rpm	861 to 945	940 ± 20	875 to 950	875 to 950
Maximum (no load) speed ....rpm	_	4950 to 5150	4950 to 5150	4950 to 5150
Smoke test/opacity ....M⁻¹ %	2.0	2.0	2.0	2.0
Static timing method................................	Refer to wsm	Refer to wsm	Refer to wsm	Refer to wsm
Timing dimension.............................mm	Computer controlled	_	Computer controlled	_
Crankshaft position ....mm [°]	_	_	_	_
Turbo type / ref / pressure ....bar	_	_	_	_
Injection pump make	_	Lucas	Bosch	Bosch
Injection pump part no.............................	_	DPC	VE4/10 R640	VE
Injector Make / type	_	CAV	Bosch	Bosch
Injector part no.......................................	_	_	0432 193 810/1	_
Injection type...	Direct	Indirect DPC	Direct SDI	Direct VE
Injection opening pressure, New [used]...bar	190 to 200	130 to 138	190 to 200 [170]	190 to 200 [170]
Glow plugs				
Maker	_	Bosch	Bosch	Bosch
Type	_	0250 202 022	0250 202 022	0250 202 022
Nominal rating....................................V/A	_	11.5 / 12	11.5 / 12	11.5 / 12
Brakes				
minimum friction material thickness				
Front..mm	7.0 with backing	7.0 with backing	7.0 with backing	7.0 with backing
Rear..mm	2.5	2.5	2.5	2.5
Tyres - Saloon / Hatch.....................Size	175/65x13:185/55x14:195/45x15	175/65x13	175/65x13:185/55x14:195/45x15	185/60x14
- Estate / Van...........................Size	_	_	_	_
Pressure - front / rear - Saloon / Hatch ...bar	_	2.1 / 2.1	2.1 / 2.1	2.1 / 2.0
- Estate / Van ......bar	_	_	_	_
Front suspension / wheel alignment				
Toe-in (+) / Toe-out (−).....................mm [°]	[0° ± 10']	[0° ± 10']	[0° ± 10']	[0° ± 10']
Camber	-0°25' ± 20'	-0°25' ± 20'	-0°25' ± 20'	-0°25' ± 20'
Castor	1°20' ± 30'	1°20' ± 30'	1°20' ± 30'	1°20' ± 30'
King pin inclination................................	_	_	_	_
Rear suspension / wheel alignment				
Toe-in (+) / Toe-out (−)......................mm [°]	[0°20' ± 10']	[0°20' ± 10']	[0°20' ± 10']	[0°20' ± 10']
Camber	-1°30' ± 10'	-1°30' ± 10'	-1°30' ± 10'	-1°30' ± 10'

VOLKSWAGEN

	Polo Classic / Caddy 1.7 Di 1997 to 2000	Polo 1.9 SD 1995 to 1997	Polo 1.9 SDi 1996 to 1998	Polo Classic / Caddy 1.9 D 1995 to 2000
Torque wrench settings				
Cylinder head - stage 1 ....Nm	40 N	40 N	40 N	40 N
- stage 2 ....Nm	60	60	60	60
- stage 3 ....Nm	+ 90°	+ 90°	+ 90°	+ 90°
- stage 4 ....Nm	+ 90°	+ 90°	+ 90°	+ 90°
- stage 5 ....Nm	–	–	–	–
- stage 6 ....Nm	–	–	–	–
Big-end bearings......................Nm	30 + 90° N	30 + 90° N	30 + 90° N	30 + 90° N
Main bearings......................Nm	65 + 90° N	65 + 90° N	65 + 90° N	65 + 90° N
Crankshaft pulley bolt ....Nm	90 + 90° oiled N	90 + 120° oiled N	90 + 90° oiled N	90 + 90° oiled N
Camshaft pulley bolt ....Nm	45	80	45	45
Flywheel [driveplate] bolt......................Nm	60 + 90° N	60 + 90° N	60 + 90° N	60 + 90° N
Front hubs ....Nm	WSM	WSM	WSM	WSM
Rear hubs ....Nm	WSM	WSM	WSM	WSM
Wheel nuts / bolts ....Nm	110	110	110	110
Glow plugs ....Nm	25	25	15	15
Clutch pressure plate bolts ....Nm	10	20	10	20
Injection pump sprocket......................Nm	55	25	55	55
Injectors......................Nm	20	70	20	20
Injection pump mounting bolts ....Nm	25	–	25	25
Injector pipe unions......................Nm	25	–	25	25
Capacities				
Engine oil & filter......................litres	3.4	4.7	5.0	4.3
Gearbox......................litres	3.1	3.1	2.7	3.1
Automatic transmission ....litres	5.7 filled for life	5.7 filled for life	–	5.7 filled for life
Final drive......................litres	WT	WT	WT	WT
Cooling system......................litres	5.5	5.5	6.5	5.5
Fuel tank......................litres	45	45	45	45

Notes

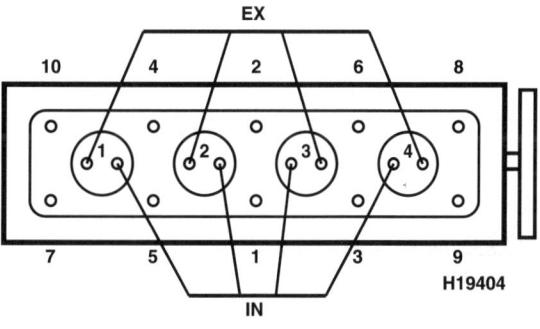

1715 cm³ / 1896 cm³

– Not applicable, or information not available

VOLKSWAGEN

	Polo Classic / Caddy 1.9 D 1995 to 2000	Polo Classic / Caddy 1.9 TDi 1997 to 2000	Polo Classic / Caddy 1.9 TDi 1997 to 2000	Polo Classic / Caddy 1.9 TDi 1998 to 1999
Engine				
Engine type/code	1Y SOHC 8V 47kW	AHU SOHC Turbo 66kW	ALE SOHC 8V EGR Turbo 66kW	AFN SOHC 8V Turbo 81kW'
Capacity (cm^3) / cylinders.........................	1896 / 4	1896 / 4	1896 / 4	1896 / 4
Compression ration / pressure ....bar	22.5 / 26 to 34	19.5 / 25 to 31	19.5 / 25 to 31	19.5 / 25 to 31
Torque output ....Nm	125	202	210	235
Oil pressure ....idle [running] bar	[2.0 @ 2000]	[2.0 @ 2000]	[2.0 @ 2000]	[2.0 @ 2000]
Oil temperature ....°C	80	80	80	80
Valve clearances - inlet (mm)	0: Hyd.	0: Hyd.	0: Hyd.	0: Hyd.
- exhaust (mm)	0: Hyd.	0: Hyd.	0: Hyd.	0: Hyd.
Injection order........................	1-3-4-2	1-3-4-2	1-3-4-2	1-3-4-2
No. 1 cylinder position	TBE	TBE	TBE	TBE
Cooling system				
Thermostat opening temperature ....°C	84	84	84	84
Radiator cap pressure ....bar	1.3 to 1.5	1.3 to 1.5	1.3 to 1.5	1.3 to 1.5
Fuel system				
Idle speed ....rpm	870 to 930	861 to 945 N/A	861 to 945	875 to 950
Maximum (no load) speed ....rpm	4950 to 5150	4800 to 5200	–	5200
Smoke test/opacity ....M^{-1} %	2.0	2.0	2.0	2.0
Static timing method...........................	Plunger travel	Plunger travel	Refer to wsm	Refer to wsm
Timing dimension...........................mm	0.9 ± 0.02	0.7 ± 0.02	Computer controlled	–
Crankshaft position ....mm [°]	TDC	TDC	–	–
Turbo type / ref / pressure ....bar	–	1.8 to 2.0 bar @ 3000 rpm	–	0.5 to 0.65 bar @ 4000 rpm
Injection pump make	Bosch	Bosch	–	Bosch
Injection pump part no.............................	VE4/9 R430	VE4/10 R440/1	–	VER 510 0460 404 986
Injector Make / type	Bosch	Bosch	–	Bosch
Injector part no......................................	0432 217 240	–	–	0432 193 838
Injection type.....................................	Indirect VE	Direct VE	Direct	Direct VER
Injection opening pressure, New [used]...bar	130 to 138	190 to 200 [170]	190 to 200	190 to 200 [170]
Glow plugs				
Maker	Bosch/Champion	Bosch	Champion	Bosch
Type	0250 201 032 / CH160	0250 202 009	CH171	0250 202 009
Nominal rating...............................V/A	–	11.5 / 12	11 /	11 / 12
Brakes				
minimum friction material thickness				
Front...mm	7.0 with backing	7.0 with backing	7.0 with backing	7.0 with backing
Rear...mm	2.5	2.5	2.5	2.5
Tyres - Saloon / Hatch......................Size	175/65x13:185/55x14:195/45x15	185/60x14	175/65x13:185/55x14:195/45x15	185/60x14
- Estate / Van......................Size	185/60x15	–	–	185/60x14
Pressure - front / rear - Saloon / Hatch ...bar	–	2.1 / 2.0	–	2.4 / 2.2
- Estate / Van ......bar	2.2 / 2.4	–	–	2.3 / 2.2
Front suspension / wheel alignment				
Toe-in (+) / Toe-out (–)......................mm [°]	[0° ± 10']	[0° ± 10']	[0° ± 10']	[0° ± 10']
Camber	-0°25' ± 20'	-0°25' ± 20'	-0°25' ± 20'	-0°25' ± 20'
Castor	1°20' ± 30'	1°20' ± 30'	1°20' ± 30'	1°20' ± 30'
King pin inclination...............................	–	–	–	–
Rear suspension / wheel alignment				
Toe-in (+) / Toe-out (–)......................mm [°]	[0°20' ± 10']	[0°20' ± 10']	[0°20' ± 10']	[0°20' ± 10']
Camber	-1°30' ± 10'	-1°30' ± 10'	-1°30' ± 10'	-1°30' ± 10'

VOLKSWAGEN

	Polo Classic / Caddy 1.9 D 1995 to 2000	Polo Classic / Caddy 1.9 TDi 1997 to 2000	Polo Classic / Caddy 1.9 TDi 1997 to 2000	Polo Classic / Caddy 1.9 TDi 1998 to 1999
Torque wrench settings				
Cylinder head - stage 1 ....Nm	40 N	40 N	40 N	40 N
- stage 2 ....Nm	60	60	60	60
- stage 3 ....Nm	+ 90°	+ 90°	+ 90°	+ 90°
- stage 4 ....Nm	+ 90°	+ 90°	+ 90°	+ 90°
- stage 5 ....Nm	–	–	–	–
- stage 6 ....Nm	–	–	–	–
Big-end bearings...............................Nm	30 + 90° N	30 + 90° N	30 + 90° N	30 + 90° N
Main bearings.....................................Nm	65 + 90° N	65 + 90° N	65 + 90° N	65 + 90° N
Crankshaft pulley bolt ....Nm	90 + 90° oiled N	90 + 90° oiled N	90 + 90° oiled N	90 + 90° oiled N
Camshaft pulley bolt ....Nm	45	45	45	45
Flywheel [driveplate] bolt...................Nm	60 + 90° N	60 + 90° N	60 + 90° N	60 + 90° N
Front hubs ....Nm	WSM	265	265	265
Rear hubs ....Nm	WSM	WSM	WSM	WSM
Wheel nuts / bolts ....Nm	110	110	110	110
Glow plugs ....Nm	25	25	25	25
Clutch pressure plate bolts ....Nm	10	20	10	10
Injection pump sprocket.......................Nm	25	55	55	55
Injectors...Nm	70	20	20	20
Injection pump mounting bolts ....Nm	25	25	25	25
Injector pipe unions............................Nm	25	25	25	25
Capacities				
Engine oil & filter................................litres	3.4	4.3	4.3	3.4
Gearbox...litres	3.1	3.1	3.1	3.1
Automatic transmission ....litres	5.7 filled for life	5.7 filled for life	5.7 filled for life	5.7 filled for life
Final drive ....litres	WT	WT	WT	WT
Cooling system..................................litres	5.5	5.5	5.5	5.5
Fuel tank..litres	45	45	45	45

Notes

Polo Classic / Caddy 1.9 TDi 1998 to 1999
¹With intercooler

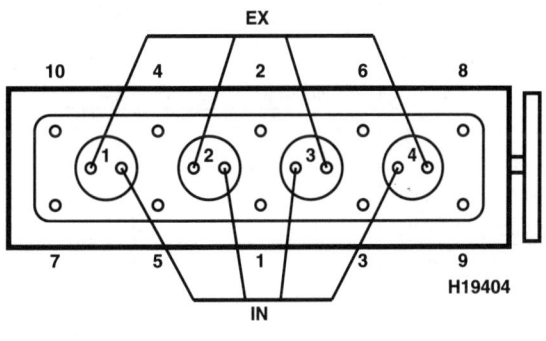

1896 cm³

– Not applicable, or information not available

VOLKSWAGEN

	Beetle 1.9 TDi 1999 to 2000	Golf / Jetta 1.6 D 1983 to 1992	Golf / Jetta 1.6 D 1983 to 1992	Golf / Jetta 1.6 TD 1989 to 1992
Engine				
Engine type/code.........................	ALH SOHC 66kW	JP/A⁶ SOHC 40kW	JP/A⁶ SOHC 40kW	RA/SB SOHC Turbo 59kW
Capacity (cm³) / cylinders........................	1896 / 4	1588 / 4	1588 / 4	1588 / 4
Compression ration / pressure ....bar	19.5 /	23.0 / ≥26.0	23.0 / ≥26.0	23.0 / ≥26.0
Torque output ....Nm	210	0	0	0
Oil pressure ....idle [running] bar	[2.0 @ 2000]	[2.0]	[2.0 @ 2000]	[2.0 @ 2000]
Oil temperature ....°C	80	80	80	80
Valve clearances - inlet (mm)	0: Hyd.	0.15 to 0.25¹	0.15 to 0.25¹	0: Hyd.
- exhaust (mm)	0: Hyd.	0.35 to 0.45¹	0.35 to 0.45¹	0: Hyd.
Injection order	1-3-4-2	1-3-4-2	1-3-4-2	1-3-4-2
No. 1 cylinder position	TBE	TBE	TBE	TBE
Cooling system				
Thermostat opening temperature ....°C	85	87	87	87
Radiator cap pressure ....bar	1.4 to 1.6	1.2 to 1.5	1.2 to 1.5	1.2 to 1.5
Fuel system				
Idle speed ....rpm	860 to 940	850 ± 100	850 ± 100	850 ± 100
Maximum (no load) speed ....rpm	5200	5350 ± 50	5350 ± 50	5100 ± 100
Smoke test/opacity ....M⁻¹ %	2.0	2.0	2.0	2.0
Static timing method..............................	Refer to wsm	Plunger travel	Rotor groove	Plunger travel
Timing dimension.............................mm	–	0.9 ± 0.02	Dimension on pump	SB: 0.9 ± 0.02¹
Crankshaft position ....mm [°]	–	[0] TDC	[0±2]	[0] TDC
Turbo type / ref / pressure ....bar	1.7 to 2.2 bar @ 3000 rpm	–	–	0.64 to 0.72 bar
Injection pump make	Bosch	Bosch	CAV	Bosch
Injection pump part no...........................	–	VE 4/9 0460 494 147	–	VE4/9 0460 494 239
Injector Make / type	–	Bosch	CAV	Bosch
Injector part no......................................	–	0432 217 083	–	0432 217 185
Injection type..	EDC	Indirect VE	Indirect DPC	Indirect VE
Injection opening pressure, New [used]...bar	220 to 230 [200]	130 to 138 [120]	130 to 138 [120]	155 [140]
Glow plugs				
Maker	–	Bosch/Champion	Bosch/Champion	Bosch/Champion
Type	–	0250 201 032 / CH69	0250 201 032 / CH69	0250 201 032 / CH69
Nominal rating.................................V/A	–	11 / 12	11 / 12	11 / 12
Brakes				
minimum friction material thickness				
Front..mm	7.0 with backing	7.0 with backing	7.0 with backing	7.0 with backing
Rear..mm	7.5 with backing	2.5	2.5	2.5
Tyres - Saloon / Hatch ....Size	195/65x15: 205/55x16	155x13: 175/70x13	155x13: 175/70x13	175/70x13: 185/60x14
- Estate / Van.......................Size	–	–	–	–
Pressure - front / rear - Saloon / Hatch ...bar	2.2 / 2.4	2.0 / 1.8	2.0 / 1.8	2.0 / 1.8
- Estate / Van ......bar	–	–	–	–
Front suspension / wheel alignment				
Toe-in (+) / Toe-out (–)....................mm [°]	[0° ± 10']	[0° ± 10']	[0° ± 10']	[0° ± 10']
Camber	-33' ± 30'	-30' ± 20'	-30' ± 20'	-30' ± 20'
Castor	7°50' ± 30'	1°30' ± 30'	1°30' ± 30'	1°30' ± 30'
King pin inclination................................	–			
Rear suspension / wheel alignment				
Toe-in (+) / Toe-out (–)....................mm [°]	[25' ± 10']	[25' ± 15']²	[25' ± 15']²	[25' ± 15']²
Camber	-1°27' ± 10'	-1°40' ± 20'³	-1°40' ± 20'³	-1°40' ± 20'³

VOLKSWAGEN

	Beetle 1.9 TDi 1999 to 2000	Golf / Jetta 1.6 D 1983 to 1992	Golf / Jetta 1.6 D 1983 to 1992	Golf / Jetta 1.6 TD 1989 to 1992
Torque wrench settings				
Cylinder head - stage 1 ....Nm	40 N	40	40	40
- stage 2 ....Nm	60	60	60	60
- stage 3 ....Nm	+ 90°	+ 180°	+ 180°	+ 180°
- stage 4 ....Nm	+ 90°	Warm engine, + 90°	Warm engine, + 90°	Warm engine, + 90°
- stage 5 ....Nm	_	After 1000km,	After 1000km,	After 1000km,
- stage 6 ....Nm	_	+ 90°	+ 90°	+ 90°
Big-end bearings...............................Nm	30 + 90° N	Thread length 15mm: 45[4]	Thread length 15mm: 45[4]	30 + 180° N
Main bearings.....................................Nm	65 + 90 N	65	65	65
Crankshaft pulley bolt ....Nm	120 + 90° N	Hexagon: 180[5]	Hexagon: 180[5]	Hexagon: 180[4]
Camshaft pulley bolt ....Nm	45	45	45	45
Flywheel [driveplate] bolt...................Nm	60 + 90° N	30 + 90°	30 + 90°	30 + 90°
Front hubs ....Nm	50 + 30° N	230	230	230
Rear hubs ....Nm	175 N	WSM	WSM	WSM
Wheel nuts / bolts ....Nm	120	110	110	110
Glow plugs ....Nm	15	30	30	30
Clutch pressure plate bolts ....Nm	_	20	20	20
Injection pump sprocket......................Nm	20 + 90° N	45	45	45
Injectors..Nm	20	70	70	70
Injection pump mounting bolts ....Nm	_	25	25	25
Injector pipe unions............................Nm	25	25	25	25
Capacities				
Engine oil & filter................................litres	4.5	3.5 8.85 ▶: 4.5	3.5 8.85 ▶: 4.5	4.5
Gearbox...litres	1.9	1.5 [2.0]	1.5 [2.0]	2.0
Automatic transmission ....litres	_	3.0	3.0	_
Final drive ....litres	WT	AT: 0.75	AT: 0.75	WT
Cooling system...................................litres	6.0	6.5	6.5	6.5
Fuel tank..litres	_	55	55	55

Notes

Golf / Jetta 1.6 D 1983 to 1992
[1]8.85 ▶: 0: Hydraulic adjusters
[2]2.90 ▶: [+20'±10']
[3]2.90 ▶: -1°30'±10'
[6]JPA: 37kW
[4]Thread length 25mm: 30 + 90°
[5]12 point splined bolts: 90 + 180°
Golf / Jetta 1.6 D 1983 to 1992
[1]8.85 ▶: 0: Hydraulic adjusters
[2]2.90 ▶: [20'±10']
[3]2.90 ▶: -1°30'±10'
[6]JPA: 37kW
[4]Thread length 25mm: 30 + 90°
[5]12 point splined bolts: 90 + 180°
Golf / Jetta 1.6 TD 1989 to 1992
[1]RA: 1.00±0.02
[2]2.90 ▶: [20'±10']
[3]2.90 ▶: -1°30'±10'
[4]12 point splined bolts: 90 + 180°

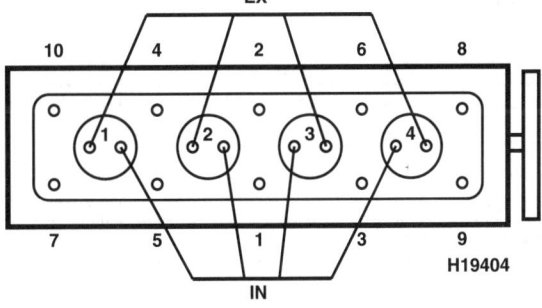

1588 cm³ / 1896 cm³

– Not applicable, or information not available

VOLKSWAGEN

	Golf / Jetta 1.6 D CAT 1989 to 1992	Golf / Vento 1.9 TD 1991 to 1994	Golf / Vento 1.9 D 1991 to 1994	Golf / Vento 1.9 D 1995 to 1999
Engine				
Engine type/code..................................	1V SOHC Turbo 44kW	AAZ SOHC Turbo 55kW	1Y SOHC 47kW	AEY SOHC 47kW
Capacity (cm³) / cylinders......................	1588 / 4	1896 / 4	1896 / 4	1896 / 4
Compression ration / pressure ....bar	23.0 / ≥26.0	22.5 / 26 to 34	22.5 / 26 to 34	19.5 / 26 to 34
Torque output ....Nm	0	150	124	124
Oil pressure ....idle [running] bar	[2.0 @ 2000]	[2.0 @ 2000]	[2.0 @ 2000]	[2.0 @ 2000]
Oil temperature ....°C	80	80	80	80
Valve clearances - inlet (mm)	0: Hyd.	0: Hyd.	0: Hyd.	0: Hyd.
- exhaust (mm)	0: Hyd.	0: Hyd.	0: Hyd.	0: Hyd.
Injection order	1-3-4-2	1-3-4-2	1-3-4-2	1-3-4-2
No. 1 cylinder position	TBE	TBE	TBE	TBE
Cooling system				
Thermostat opening temperature ....°C	87	85	85	85
Radiator cap pressure ....bar	1.2 to 1.5	1.4 to 1.6	1.4 to 1.6	1.4 to 1.6
Fuel system				
Idle speed ....rpm	850 ± 30	900 ± 30	900 ± 30	900 ± 30
Maximum (no load) speed ....rpm	5350 ± 100	5200 ± 50	5200 ± 50	5200
Smoke test/opacity ....M⁻¹ %	2.0	2.0	2.0	2.5
Static timing method...............................	Plunger travel	Plunger travel	Plunger travel	Plunger travel
Timing dimension................................mm	1.0 ± 0.02	0.8 ± 0.02	0.9 ± 0.02	0.9 ± 0.02
Crankshaft position ....mm [°]	[0] TDC	0 TDC	0 TDC	0 TDC
Turbo type / ref / pressure ....bar	0.64 to 0.72 bar	0.6 to 0.83 bar	–	–
Injection pump make	Bosch	Bosch	Bosch	Bosch
Injection pump part no............................	VE4/8 0460 484 027	VE4/9 R433 0460 484 286	VE4/9 R430 0460 484 046	VE 0460 404 989
Injector Make / type	Bosch	Bosch	Bosch	–
Injector part no......................................	0432 217 240	0432 217 244	0432 217 240	
Injection type...	Indirect VE	Indirect VE	Indirect VE	Direct VE
Injection opening pressure, New [used]...bar	155 [140]	150 to 158 [140]	130 to 138 [120]	190 to 200 [170]
Glow plugs				
Maker	Bosch/Champion	Bosch/Champion	Bosch/Champion	Bosch
Type	0250 201 032 / CH69	0250 201 032 / CH160	0250 201 032 / CH160	0250 201 032
Nominal rating......................................V/A	11 / 12	11.5 / 8	–	11.5 / 12
Brakes				
minimum friction material thickness				
Front...................................mm	7.0 with backing	7.0 with backing	7.0 with backing	7.0 with backing
Rear....................................mm	2.5	2.5	2.5	2.5
Tyres - Saloon / Hatch......................Size	175/70x13: 185/60x14	175/70x13: 185/60x14	175/70x13: 185/60x14	175/70x13: 185/60x14
- Estate / Van......................Size	–	185/60x14	185/60x14	185/60x14
Pressure - front / rear - Saloon / Hatch...bar	2.0 / 1.8	2.1 / 1.9: 2.0 / 1.8	2.1 / 1.9: 2.0 / 1.8	2.1 / 1.9: 2.0 / 1.8
- Estate / Van .....bar	–	2.0 / 2.0	2.0 / 2.0	2.0 / 2.0
Front suspension / wheel alignment				
Toe-in (+) / Toe-out (–)....................mm [°]	[0° ± 10']	[0° ± 10']	[0° ± 10']	[0° ± 10']
Camber	-30' ± 20'	-0°30' ± 20'	-0°30' ± 20'	-0°30' ± 20'
Castor	1°30' ± 30'	1°45' ± 30'	1°45' ± 30'	1°45' ± 30'
King pin inclination.................................	–	–	–	–
Rear suspension / wheel alignment				
Toe-in (+) / Toe-out (–)....................mm [°]	[25' ± 15']¹	[-20' ± 10']	[-20' ± 10']	[-0°20' ± 10']
Camber	-1°40' ± 20'²	-1°30' ± 10'	-1°30' ± 10'	-1°30' ± 10'

VOLKSWAGEN

	Golf / Jetta 1.6 D CAT 1989 to 1992	Golf / Vento 1.9 TD 1991 to 1994	Golf / Vento 1.9 D 1991 to 1994	Golf / Vento 1.9 D 1995 to 1999
Torque wrench settings				
Cylinder head - stage 1 ....Nm	40 N	40 N	40 N	40 N
- stage 2 ....Nm	60	60	60	60
- stage 3 ....Nm	+ 180°	+ 90°	+ 90°	+ 90°
- stage 4 ....Nm	Warm engine, + 90°	+ 90°	+ 90°	+ 90°
- stage 5 ....Nm	After 1000km,	–	–	–
- stage 6 ....Nm	+ 90°	–	–	–
Big-end bearings.............................Nm	30 + 180° N	30 + 90° N	30 + 90° N	30 + 90° N
Main bearings.................................Nm	65	65 + 90° N	65 + 90° N	65 + 90° N
Crankshaft pulley bolt ....Nm	Hexagon: 180³	90 + 90° N	90 + 90° N	90 + 90°
Camshaft pulley bolt ....Nm	45	45	45	45
Flywheel [driveplate] bolt.................Nm	30 + 90°	60 + 90° N	30 + 90° N	60 + 90° N
Front hubs ....Nm	230	265	265	265
Rear hubs ....Nm	WSM	WSM	WSM	–
Wheel nuts / bolts ....Nm	110	110	110	110
Glow plugs ....Nm	30	25	25	15
Clutch pressure plate bolts ....Nm	20	25	25	25
Injection pump sprocket...................Nm	45	45	45	55
Injectors..Nm	70	70	70	20
Injection pump mounting bolts ....Nm	25	25	25	25
Injector pipe unions.........................Nm	25	25	25	25
Capacities				
Engine oil & filter..................................litres	4.5	4.5	4.5	4.5
Gearbox..litres	2.0	2.0	2.0	2.0
Automatic transmission ....litres	–	–	–	–
Final drive ....litres	WT	WT	WT	WT
Cooling system.....................................litres	6.5	6.5	6.5	6.5
Fuel tank...litres	55	55 Est: 60	55 Est: 60	55 Est: 60

Notes

Golf / Jetta 1.6 D CAT 1989 to 1992
¹2.90 ▶: [20'±10']
²2.90 ▶: -1°30'±10'
³12 point splined bolts: 90 + 180°

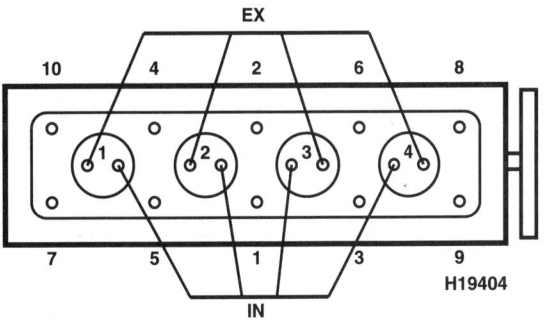

1588 cm³ / 1896 cm³

– Not applicable, or information not available

VOLKSWAGEN

	Golf / Vento 1.9 TD 1995 to 1997	Golf / Vento 1.9 TD 1994 to 1998	Golf / Vento 1.9 TDi 1996 to 1998	Golf / Bora 1.9 SDi 1998 to 2000
Engine				
Engine type/code......................	AAZ SOHC 8V Turbo 55kW	1Z / AHU SOHC 8V Turbo 66kW	AFN SOHC Turbo 8V 81kW	AGP SOHC SDI 50kW
Capacity (cm³) / cylinders........................	1896 / 4	1896 / 4	1896 / 4	1896 / 4
Compression ration / pressure ....bar	22.5 / ≥26.0	19.5 / ≥27.0	19.5 / 25 to 31	19.5 / ≥19.0
Torque output ....Nm	150	202	235	133
Oil pressure ....idle [running] bar	[2.0]	[2.0 @ 2000]	[2.0 @ 2000]	[2.0 @ 2000]
Oil temperature ....°C	80	80	80	80
Valve clearances - inlet (mm)	0: Hyd.	0: Hyd.	0: Hyd.	0: Hyd.
- exhaust (mm)	0: Hyd.	0: Hyd.	0: Hyd.	0: Hyd.
Injection order	1-3-4-2	1-3-4-2	1-3-4-2	1-3-4-2
No. 1 cylinder position	TBE	TBE	TBE	TBE
Cooling system				
Thermostat opening temperature ....°C	84 to 98	84 to 98	85	85
Radiator cap pressure ....bar	1.4 to 1.6	1.4 to 1.6	1.4 to 1.6	1.4 to 1.6
Fuel system				
Idle speed ....rpm	900 ± 30	860 to 940	900 ± 30	875 to 1000
Maximum (no load) speed ....rpm	5200 ± 100	5200	5200	5150
Smoke test/opacity ....M⁻¹ %	2.0	2.0	2.0	2.0
Static timing method................................	_	Plunger travel	Refer to wsm	Refer to wsm
Timing dimension............................mm	_	0.7 ± 0.02	_	Computer controlled
Crankshaft position ....mm [°]	_	TDC	_	_
Turbo type / ref / pressure ....bar	0.6 to 0.83 bar	1.5 to 1.65 bar @ 3500 rpm	0.5 to 0.65 bar @ 3500 to 4000	_
Injection pump make	Bosch	Bosch	Bosch	Bosch
Injection pump part no............................	_	VER 510	VER510 0460 404 986	
Injector Make / type		Bosch	Bosch	
Injector part no..	_	0432 193 838	0432 193 838	
Injection type...	Indirect	Direct VER	Direct VER	SDI
Injection opening pressure, New [used]...bar	150 to 158 [≥140]	190 to 200 [≥170]	190 to 200 [170]	190 to 200 [170]
Glow plugs				
Maker		Bosch	Bosch	Champion
Type	_	0250 202 009	0250 202 009	CH171
Nominal rating......................................V/A		11.5 / 12	11 / 12	11 /
Brakes				
minimum friction material thickness				
Front...........................mm	7.0 with backing	7.0 with backing	7.0 with backing	7.0 with backing
Rear...........................mm	2.5	2.5	2.5	7.0 with backing
Tyres - Saloon / Hatch...............Size	175/70x13:185/60x14:195/50x15	175/70x13:185/60x14:195/50x15	175/70x13:185/60x14:195/50x15	185/60x14:195/50x15: 205/50x15
- Estate / Van.............Size	185/60x14:195/60x14:195/50x15	185/60x14:195/60x14:195/50x15	185/60x14:195/60x14:195/50x15	_
Pressure - front / rear - Saloon / Hatch ...bar	2.1 / 1.9 Vento: 2.0 / 1.8	2.1 / 1.9 Vento: 2.0 / 1.8	2.1 / 1.9 Vento: 2.0 / 1.8	_
- Estate / Van ......bar	2.1: 1.8: 2.1 / 2.1: 1.8: 2.1	2.1 / 2.1: 1.8 / 1.8:2.1 / 2.1	2.1 / 2.1: 1.8 / 1.8:2.1 / 2.1	
Front suspension / wheel alignment				
Toe-in (+) / Toe-out (–)....................mm [°]	[0° ± 10']	[0° ± 10']	[0° ± 10']	[0° ± 10']
Camber	-36' ± 20'	-36' ± 20'	-36' ± 20'	-30' ± 30'
Castor	1°50' ± 30'	1°50' ± 30'	1°50' ±30'	7°40' ± 30'
King pin inclination..................................	_	_	_	_
Rear suspension / wheel alignment				
Toe-in (+) / Toe-out (–)....................mm [°]	[20' ± 10']	[20' ± 10']	[20' ± 10']	[20' ± 10']
Camber	-1°30' ± 10	-1°30' ± 10	-1°30' ± 10'	-1°27' ± 10'

VOLKSWAGEN

	Golf / Vento 1.9 TD 1995 to 1997	Golf / Vento 1.9 TD 1994 to 1998	Golf / Vento 1.9 TDi 1996 to 1998	Golf / Bora 1.9 SDi 1998 to 2000
Torque wrench settings				
Cylinder head - stage 1 ....Nm	40 N	40 N	40 N	40 N
- stage 2 ....Nm	60	60	60	60
- stage 3 ....Nm	+ 90°	+ 90°	+ 90°	+ 90°
- stage 4 ....Nm	+ 90°	+ 90°	+ 90°	+ 90°
- stage 5 ....Nm	–	–	–	–
- stage 6 ....Nm	–	–		
Big-end bearings.........................Nm	30 + 90° N	30 + 90° N	30 + 90° N	30 + 90° N
Main bearings.............................Nm	65 + 90° N	65 + 90° N	65 + 90° N	65 + 90° N
Crankshaft pulley bolt ....Nm	90 + 120° N	90 + 120° N	90 + 120° N	120 + 90° N
Camshaft pulley bolt ....Nm	80	80	80	45
Flywheel [driveplate] bolt...................Nm	60 + 90° N	60 + 90° N	60 + 90° N	60 + 90° [30] N
Front hubs ....Nm	265	265	265	300, slacken, 50 + 30°
Rear hubs ....Nm	WSM	WSM	WSM	175
Wheel nuts / bolts ....Nm	110	110	110	120
Glow plugs ....Nm	25	25	25	15
Clutch pressure plate bolts ....Nm	25	25	25	–
Injection pump sprocket...................Nm	–	55	55	20 + 90° N
Injectors.................................Nm	70	20	20	20
Injection pump mounting bolts ....Nm	–	25	25	25
Injector pipe unions........................Nm	–	25	25	25
Capacities				
Engine oil & filter..............................litres	4.3	4.5	4.3	4.5
Gearbox....................................litres	2.0	2.0	1.9	–
Automatic transmission ....litres	–	–	–	–
Final drive ....litres	WT	WT	WT	WT
Cooling system...............................litres	6.5	5.6	6.5	6.0
Fuel tank...................................litres	55 Est: 60	55 Est: 60	55 Est: 60	55

Notes

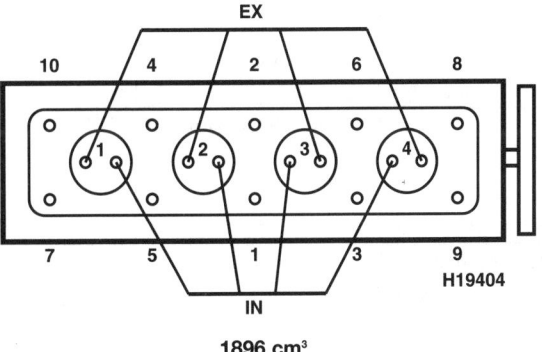

1896 cm³

– Not applicable, or information not available

VOLKSWAGEN

	Golf / Bora 1.9 SDi 1998 to 2000	Golf / Bora 1.9 TDi 1998 to 2000	Golf / Bora 1.9 TDi 1998 to 1999	Golf / Bora 1.9 TDi 1999 to 2000
Engine				
Engine type/code..........................	AQM SOHC SDI 50kW	AGR/ALH SOHC 8V Turbo 66kW	AHF SOHC Turbo 81kW	AJM SOHC 8V Turbo 85kW
Capacity (cm^3) / cylinders........................	1896 / 4	1896 / 4	1896 / 4	1896 / 4
Compression ration / pressure ....bar	19.5 / ≥19.0	19.5 / 25 to 31	19.5 / 25 to 31	18.0 / ≥19.0
Torque output ....Nm	133	210	235	285
Oil pressure ....idle [running] bar	[2.0 @ 2000]	[2.0 @ 2000]	[2.0 @ 2000]	[2.0 @ 2000]
Oil temperature ....°C	80	80	80	80
Valve clearances - inlet (mm)	0: Hyd.	0: Hyd.	0: Hyd.	0: Hyd.
- exhaust (mm)	0: Hyd.	0: Hyd.	0: Hyd.	0: Hyd.
Injection order	1-3-4-2	1-3-4-2	1-3-4-2	1-3-4-2
No. 1 cylinder position	TBE	TBE	TBE	TBE
Cooling system				
Thermostat opening temperature ....°C	85	85	85	85
Radiator cap pressure ....bar	1.4 to 1.6	1.4 to 1.6	1.4 to 1.6	1.4 to 1.6
Fuel system				
Idle speed ....rpm	–	875 to 950	875 to 950	875 to 950
Maximum (no load) speed ....rpm	–	4800 to 5200	4800 to 5200	4800 to 5200
Smoke test/opacity ....M^{-1} %	2.0	2.0	2.0	2.0
Static timing method..............................	Refer to wsm	Refer to wsm	Refer to wsm	Refer to wsm
Timing dimension.............................mm	Computer controlled	Computer controlled	Computer controlled	–
Crankshaft position ....mm [°]	–	–	–	–
Turbo type / ref / pressure ....bar		1.7 to 2.2 bar @ 3000 rpm	1.7 to 2.2 bar' 3000 rpm	–
Injection pump make	–	Bosch	Bosch	Bosch
Injection pump part no...........................	–	VE	VE	–
Injector Make / type		–		–
Injector part no.......................................		–		–
Injection type..	SDI	Direct VE	Direct	Direct Pump Injector
Injection opening pressure, New [used]...bar	–	190 to 200 [≥170]	190 to 200 [≥170]	190 to 200 [≥170]
Glow plugs				
Maker	Champion	Champion	Champion	
Type	CH171	CH171	CH171	–
Nominal rating...................................V/A	11 /	11 /	11 /	
Brakes				
minimum friction material thickness				
Front...mm	7.0 with backing	7.0 with backing	7.0 with backing	7.0 with backing
Rear..mm	7.0 with backing	7.0 with backing	7.0 with backing	7.0 with backing
Tyres - Saloon / Hatch......................Size	185/60x14:195/50x15: 205/50x15	185/60x14:195/50x15: 205/50x15	185/60x14:195/50x15: 205/50x15	195/65x15: 205/60x15:205/55x15
- Estate / Van......................Size				
Pressure - front / rear - Saloon / Hatch ...bar	–	2.0 / 1.8	2.0 / 1.8	1.9 / 1.9
- Estate / Van .....bar	–	–	–	–
Front suspension / wheel alignment				
Toe-in (+) / Toe-out (–)....................mm [°]	[0 ± 10']	[0° ± 10']	[0° ± 10']	[0° ± 10']
Camber	-30' ± 30'	-30' ± 30'	-30' ± 30'	-30' ± 30'
Castor	7° 40' ± 30'	7°40' ± 30'	7°40' ± 30'	7°40' ± 30'
King pin inclination.................................	–	–	–	–
Rear suspension / wheel alignment				
Toe-in (+) / Toe-out (–)....................mm [°]	[20' ± 10']	[20' ± 10']	[20' ± 10']	[20' ± 10']
Camber	-1° 27' ± 10'	-1°27' ± 10'	-1°27' ± 10'	-1°27' ± 10'

VOLKSWAGEN

	Golf / Bora 1.9 SDi 1998 to 2000	Golf / Bora 1.9 TDi 1998 to 2000	Golf / Bora 1.9 TDi 1998 to 1999	Golf / Bora 1.9 TDi 1999 to 2000
Torque wrench settings				
Cylinder head - stage 1 ....Nm	40 N	40 N	40 N	40 N
- stage 2 ....Nm	60	60	60	60
- stage 3 ....Nm	+ 90°	+ 90°	+ 90°	+ 90°
- stage 4 ....Nm	+ 90°	+ 90°	+ 90°	+ 90°
- stage 5 ....Nm	–	–	–	–
- stage 6 ....Nm	–	–	–	–
Big-end bearings.........................Nm	30 + 90° N	30 + 90° N	30 + 90° N	30 + 90° N
Main bearings.........................Nm	65 + 90° N	65 + 90° N	65 + 90° N	65 + 90° N
Crankshaft pulley bolt ....Nm	120 + 90° N	120 + 90° N	120 + 90 N	120 + 90° N
Camshaft pulley bolt ....Nm	45	45	45	100
Flywheel [driveplate] bolt.....................Nm	60 + 90° [30] N	60 + 90° N [30]	60 + 90° N	60 + 90° N [30]
Front hubs ....Nm	300, slacken, 50 + 30°	300, slacken, 50 + 30°	300, slacken, 50 + 30°	300, slacken, 50 + 30°
Rear hubs ....Nm	175	175	175	175
Wheel nuts / bolts ....Nm	120	120	120	120
Glow plugs ....Nm	15	15	15	15
Clutch pressure plate bolts ....Nm	–	20	13	20
Injection pump sprocket.....................Nm	20 + 90° N	20 + 90° N	20 + 90° N	20 + 90° N
Injectors.........................Nm	20	20	20	–
Injection pump mounting bolts ....Nm	25	25	25	25
Injector pipe unions.........................Nm	25	25	25	25
Capacities				
Engine oil & filter.................................litres	4.5	4.5	4.5	4.5
Gearbox.........................litres	–	2.0	2.0	2.0[1]
Automatic transmission ....litres	–	3.0	–	3.0
Final drive ....litres	WT	WT Auto: 0.75	WT	WT Auto: 0.75[2]
Cooling system.........................litres	6.0	6.0	6.0	6.0
Fuel tank.........................litres	55	55	55	55

Notes

Golf / Bora 1.9 TDi 1999 to 2000
[1] 4motion: 2.6
[2] 4motion: 1.0   Haldex coupling: 0.25 (drain and refil)

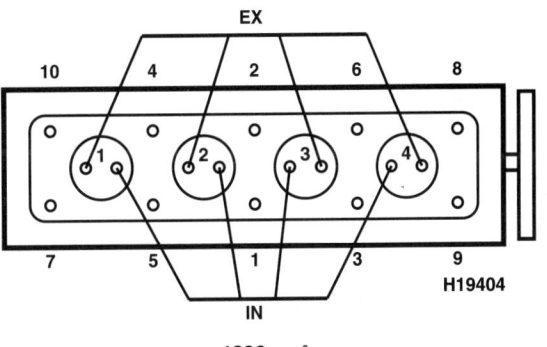

1896 cm³

H19404

– Not applicable, or information not available

VOLKSWAGEN

	Passat 1.6 TD 1988 to 1992	Passat 1.9 D 1989 to 1993	Passat 1.9 TD CAT 1991 to 1993	Passat 1.9 TD 1993 to 1996
Engine				
Engine type/code.....................	RA / SB SOHC Turbo 59kW	1Y SOHC 50kW	AAZ SOHC Turbo 55kW	AAZ SOHC Turbo 55kW
Capacity (cm³) / cylinders......................	1588 / 4	1896 / 4	1896 / 4	1896 / 4
Compression ration / pressure ....bar	23.0 / 26 to 34	23.0 / ≥26.0	22.5 / 26 to 34	22.5 / 26 to 34
Torque output ....Nm	0	0	150	150
Oil pressure ....idle [running] bar	[2.0 @ 2000]	[2.0 @ 2000]	[2.0 @ 2000]	[2.0 @ 2000]
Oil temperature ....°C	80	80	80	80
Valve clearances - inlet (mm)	0: Hyd.	0: Hyd.	0: Hyd.	0: Hyd.
- exhaust (mm)	0: Hyd.	0: Hyd.	0: Hyd.	0: Hyd.
Injection order	1-3-4-2	1-3-4-2	1-3-4-2	1-3-4-2
No. 1 cylinder position	TBE	TBE	TBE	TBE
Cooling system				
Thermostat opening temperature ....°C	87	87	87	87 to 102
Radiator cap pressure ....bar	1.2 to 1.5	1.2 to 1.5	1.2 to 1.5	1.3 to 1.5
Fuel system				
Idle speed ....rpm	900 ± 30	900 ± 30	900 ± 30	900 ± 30
Maximum (no load) speed ....rpm	5100 ± 50	5050 ± 100	5050 ± 100	5200 ± 100¹
Smoke test/opacity ....M⁻¹ %	2.5	2.0	2.0	2.0
Static timing method............................	Plunger travel	Plunger travel	Plunger travel	Dial gauge
Timing dimension............................mm	SB: 0.9 ± 0.02¹	1.0 ± 0.02	0.9 ± 0.02	0.9 ± 0.02
Crankshaft position ....mm [°]	[0] TDC	[0] TDC	[0] TDC	[0] TDC
Turbo type / ref / pressure ....bar	0.63 to 0.83 bar	–	0.6 to 0.83 bar	0.6 to 0.83 bar
Injection pump make	Bosch	Bosch	Bosch	Bosch
Injection pump part no...........................	VE4/9	VE4/8F	VE4/9 R420	VE R420
Injector Make / type	Bosch	Bosch	Bosch	Bosch
Injector part no..................................	0432 217 165	0432 217 210	0432 217 232	0432 217 232
Injection type....................................	Indirect VE	Indirect VE	Indirect VE	Indirect
Injection opening pressure, New [used]...bar	155 [140]	130 to 138 [120]	155 [140]	150 to 158 [140]
Glow plugs				
Maker	Bosch/Champion	Bosch/Champion	Bosch/Champion	Bosch/Champion
Type	0250 201 032 / CH69	0250 201 032 / CH69	0250 201 032 / CH69	0250 201 032 / CH160
Nominal rating.........................V/A	11 / 12	–	11.5 / 8	11.5 / 8
Brakes				
minimum friction material thickness				
Front....................................mm	7.0 with backing	7.0 with backing	7.0 with backing	7.0 with backing
Rear....................................mm	2.5	2.5	2.5	2.5
Tyres - Saloon / Hatch.....................Size	185/65x14: 195/60x14	185/65x14: 195/60x14	185/65x14: 195/60x14	185/65x14:195/60x14:205/50x15
- Estate / Van.....................Size	185/65x14: 195/60x14	185/65x14: 195/60x14	185/65x14: 195/60x14	185/65x14:195/60x14:205/50x15
Pressure - front / rear - Saloon / Hatch ...bar	2.0 / 2.0	2.0 / 2.0	2.0 / 2.0	2.1 / 2.1
- Estate / Van ......bar	2.0 / 2.0	2.0 / 2.0	2.0 / 2.0	2.0 / 2.0
Front suspension / wheel alignment				
Toe-in (+) / Toe-out (–).....................mm [°]	[0° ± 10']	[0° ± 10']	[0° ± 10']	[0° ± 10']
Camber	-1°20' ± 20'	-1°20' ± 20'	-1°20' ± 20'	-1°20' ± 20'
Castor	1°40' ± 30'	1°40' ± 30'	1°40' ± 30'	1°40' ± 30'
King pin inclination.............................	–	–	–	–
Rear suspension / wheel alignment				
Toe-in (+) / Toe-out (–).....................mm [°]	[25' ± 15']²	[25' ± 15']¹	[25' ± 15']¹	[20' ± 10']
Camber	-1°40' ± 20'³	-1°40' ± 20'²	-1°40' ± 20'²	-1°30' ± 10'

VOLKSWAGEN

	Passat 1.6 TD 1988 to 1992	Passat 1.9 D 1989 to 1993	Passat 1.9 TD CAT 1991 to 1993	Passat 1.9 TD 1993 to 1996
Torque wrench settings				
Cylinder head - stage 1 ....Nm	40 N	40 N	40 N	40 N
- stage 2 ....Nm	60	60	60	60
- stage 3 ....Nm	+ 180°	+ 180°	+ 180°	+ 90°
- stage 4 ....Nm	Warm up	Warm up	Warm up	+ 90°
- stage 5 ....Nm	+ 90°	+ 90°	+ 90°	–
- stage 6 ....Nm	+ 90°⁴	–	–	–
Big-end bearings................................Nm	30 + 180° N	30 + 90° N	30 + 90° N	30 + 90° oiled N
Main bearings.....................................Nm	65	65	65	65 + 90° N
Crankshaft pulley bolt ....Nm	180⁵	90 + 180°	90 + 180° N	90 + 90° oiled N
Camshaft pulley bolt ....Nm	45	45	45	45
Flywheel [driveplate] bolt....................Nm	30 + 90°	30 + 90°	30 + 90°	60 N
Front hubs ....Nm	265	265	265	265
Rear hubs ....Nm	WSM	WSM	WSM	WSM
Wheel nuts / bolts ....Nm	110	110	110	110
Glow plugs ....Nm	30	30	30	25
Clutch pressure plate bolts ....Nm	20	20	20	20
Injection pump sprocket......................Nm	45	45	45	45
Injectors..Nm	70	70	70	70
Injection pump mounting bolts ....Nm	25	25	25	25
Injector pipe unions............................Nm	25	25	25	25
Capacities				
Engine oil & filter................................litres	3.5	4.5	4.5	4.5
Gearbox...litres	2.0	2.0	2.0	2.0
Automatic transmission ....litres	–	–	–	–
Final drive ....litres	WT	WT	WT	WT
Cooling system...................................litres	6.5	6.5	6.5	6.5
Fuel tank..litres	70	70	70	70

Notes

Passat 1.6 TD 1988 to 1992
[1]RA: 1.0 ± 0.02
[2]Chassis no. 31LE222 111 or 31LB108 393 ▶:
[20'±10']
[3]From chassis no. 31LE222 111 or 31LB108 393: –
1°30'±10'
[4]After 1000km
[5]Double hex bolt: 90 + 180°
Passat 1.9 D 1989 to 1993
[1]Chassis no. 31LE222 111 or 31LB108 393 ▶:
[20'±10']
[2]From chassis no. 31LE222 111 or 31LB108 393: –
1°30'±10'
Passat 1.9 TD CAT 1991 to 1993
[1]Chassis no. 31LE222 111 or 31LB108 393 ▶:
[20'±10']
[2]From chassis no. 31LE222 111 or 31LB108 393: –
1°30'±10'
Passat 1.9 TD 1993 to 1996
[1]2 piece injection pump sprocket: 5050±100

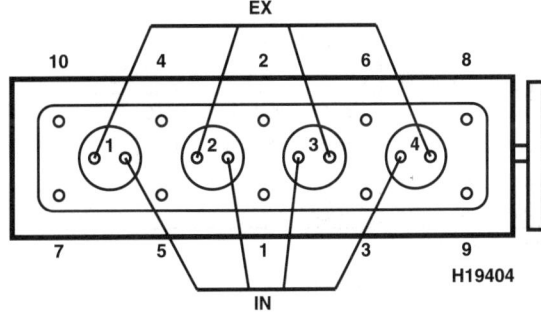

1588 cm³ / 1896 cm³

– Not applicable, or information not available

VOLKSWAGEN

	Passat 1.9 TDi 1993 to 1996	Passat 1.9 TDi 1997 to 2000	Passat 1.9 TDi 1997 to 2000	Passat 1.9 TDi 1997 to 1999
Engine				
Engine type/code.............................	1Z SOHC Turbo 66kW	AHU SOHC Turbo 66kW	AHH SOHC Turbo 66kW	AFN SOHC 8V Turbo 81kW
Capacity (cm³) / cylinders........................	1896 / 4	1896 / 4	1896 / 4	1896 / 4
Compression ration / pressure ....bar	19.5 / ≥27.0	21.0 / ≥19.0	19.5 / ≥19.0	19.5 / ≥19.0
Torque output ....Nm	202	202	210	235
Oil pressure ....idle [running] bar	[2.0 @ 2000]	1.0 to 2.5 [3.0 to 5.0]	[2.0 @ 2000]	1.0 to 2.5 [3.0 to 5.0]
Oil temperature ....°C	80	80	80	80
Valve clearances - inlet (mm)	0: Hyd.	0: Hyd.	0: Hyd.	0: Hyd.
- exhaust (mm)	0: Hyd.	0: Hyd.	0: Hyd.	0: Hyd.
Injection order	1-3-4-2	1-3-4-2	1-3-4-2	1-3-4-2
No. 1 cylinder position	TBE	TBE	TBE	TBE
Cooling system				
Thermostat opening temperature ....°C	87 to 102	85	85	85
Radiator cap pressure ....bar	1.3 to 1.5	1.3 to 1.5	1.4 to 1.6	1.2 to 1.5
Fuel system				
Idle speed ....rpm	900 ± 40	780 to 900	795 to 910	835 to 910
Maximum (no load) speed ....rpm	5050	5200	5100	4800 to 5200
Smoke test/opacity ....M⁻¹ %	2.0	2.0	2.0	2.0
Static timing method............................	Refer to wsm	Refer to wsm	Refer to wsm	Plunger travel
Timing dimension.............................mm	–	–	–	0.70 ± 0.02
Crankshaft position ....mm [°]	–	–	–	TDC
Turbo type / ref / pressure ....bar	0.5 to 0.65 bar @ 3500 rpm	–	1.8 to 2.05 bar @ 3000 rpm	–
Injection pump make	Bosch	Bosch	–	Bosch
Injection pump part no..........................	VE R510	–		–
Injector Make / type.............................	Bosch	Bosch		Bosch
Injector part no..................................	0432 193 838			–
Injection type....................................	Direct EDC VE	Direct	Direct EDC	Direct EDC
Injection opening pressure, New [used]...bar	190 to 200 [170]	190 to 200 [170]	220 to 230 [200]	190 to 200 [170]
Glow plugs				
Maker	Bosch	Bosch		Bosch
Type	0250 202 009	0250 202 022	–	0250 202 022
Nominal rating ....V/A	11.5 / 8	11.5 / 8		11.5 / 8
Brakes				
minimum friction material thickness				
Front..mm	7.0 with backing	7.0	7.0 with backing	7.0
Rear..mm	2.5	7.0	7.0 with backing	7.0
Tyres - Saloon / Hatch......................Size	185/65x14:195/60x14:205/50x15	195/65x15: 205/60x16:205/55x16	195/65x15; 205/60x15	195/65x15: 205/60x16:205/55x16
- Estate / Van.....................Size	185/65x14:195/60x14:205/50x15	195/65x15: 205/60x16:205/55x16	195/65x15: 205/60x15	–
Pressure - front / rear - Saloon / Hatch ...bar	2.1 / 2.1	2.0 / 2.0	2.1 / 2.1	2.0 / 2.0
- Estate / Van ......bar	2.0 / 2.0	2.1 / 2.1	2.1 / 2.2	
Front suspension / wheel alignment				
Toe-in (+) / Toe-out (–)....................mm [°]	[0° ± 10']	[0°10' ± 20']	[-0°10' ± 2']	[0°10' ± 20']
Camber	-1°20' ± 20'	-0°25' ± 25' Sport -0°40'±25'	-0°25' ± 25'	-0°25' ± 25' Sport: -0°40'±25'
Castor	1°40' ± 30'			
King pin inclination............................	–			
Rear suspension / wheel alignment				
Toe-in (+) / Toe-out (–)....................mm [°]	[20' ± 10']	[0°20' ± 10']	[-0°20' ± 10']	[0°20' ± 10']
Camber	-1°30' ± 10'	-1°30' ± 20'	-1°30' ± 20'	-1°30' ± 20'

VOLKSWAGEN

	Passat 1.9 TDi 1993 to 1996	Passat 1.9 TDi 1997 to 2000	Passat 1.9 TDi 1997 to 2000	Passat 1.9 TDi 1997 to 1999
Torque wrench settings				
Cylinder head - stage 1 ....Nm	40 N	40 N	40 N	40 N
- stage 2 ....Nm	60	60	60	60
- stage 3 ....Nm	+ 90°	+ 90°	+ 90°	+ 90°
- stage 4 ....Nm	+ 90°	+ 90°	+ 90°	+ 90°
- stage 5 ....Nm	–	–	–	–
- stage 6 ....Nm	–	–	–	–
Big-end bearings.........................Nm	30 + 90° oiled N	30 + 90° N	30 + 90° N	30 + 90° N
Main bearings..............................Nm	65 + 90° N	65 + 90° N	65 + 90° N	65 + 90° N
Crankshaft pulley bolt ....Nm	90 + 90° oiled N	90 + 90° N	–	90 + 90° N
Camshaft pulley bolt ....Nm	80	45	45	45
Flywheel [driveplate] bolt...............Nm	60 N	60 + 90° N	60 + 90° N	60 + 90° N
Front hubs ....Nm	265	115 + 180°N M16: 140 + 180°N	115 + 180°N M16: 190 + 180°	115 + 180°N M16: 140 + 180°N
Rear hubs ....Nm	WSM	WSM	WSM	WSM
Wheel nuts / bolts ....Nm	110	120	120	120
Glow plugs ....Nm	30	15	15	15
Clutch pressure plate bolts ....Nm	20	20	20	20
Injection pump sprocket.................Nm	55	45	20 + 90° N	45
Injectors.....................................Nm	20	–	20	20
Injection pump mounting bolts ....Nm	25	25	25	25
Injector pipe unions......................Nm	25	25	25	25
Capacities				
Engine oil & filter...............................litres	4.5	4.3	4.3	4.3
Gearbox...litres	2.0	2.25	2.25	2.25
Automatic transmission ....litres	–	–	3.5	3.5
Final drive ....litres	WT	WT	WT AT: .75	WT AT: 0.75
Cooling system................................litres	6.5	7.5	7.5	7.5
Fuel tank..litres	70	62	62	62

Notes

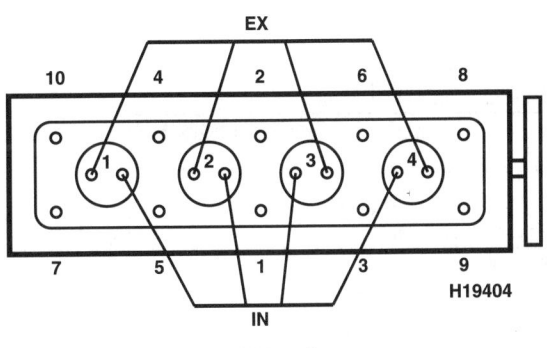

1896 cm³

– Not applicable, or information not available

Haynes
THE BOOK ®

VOLKSWAGEN

	Passat 1.9 TDi 1999 to 2000	Passat 2.5 TDi 1999 to 2000	Sharan 1.9 TDi 1995 to 2000	Sharan 1.9 TDi 1999 to 2000
Engine				
Engine type/code	AJM SOHC 8V Turbo 85kW	AFB DOHC 24V Turbo 110kW	1Z / AHU SOHC 8V Turbo 66kW	ANU SOHC 8V 66kW
Capacity (cm^3) / cylinders......................	1896 / 4	2496 / 6	1896 / 4	1896 / 4
Compression ration / pressure ....bar	18.0 / ≥19.0	19.5 / ≥24.0	21.0 / ≥19.0	18.0 / ≥19.0
Torque output ....Nm	285	310	202	240
Oil pressure ....idle [running] bar	[2.0 @ 2000]	0.8 [2.0 @ 2000]	[2.0 @ 2000]	[2.0 @ 2000]
Oil temperature ....°C	80	80	80	80
Valve clearances - inlet (mm)	0: Hyd.	0: Hyd.	0: Hyd.	0: Hyd.
- exhaust (mm)	0: Hyd.	0: Hyd.	0: Hyd.	0: Hyd.
Injection order	1-3-4-2	1-4-3-6-2-5	1-3-4-2	1-3-4-2
No. 1 cylinder position	TBE	TBE	TBE	TBE
Cooling system				
Thermostat opening temperature ....°C	85	87	85 to 105	85 to 105
Radiator cap pressure ....bar	1.4 to 1.6	1.4 to 1.6	1.2 to 1.5	1.4 to 1.6
Fuel system				
Idle speed ....rpm	800 to 940	680 to 860	900 ± 40	800 to 940
Maximum (no load) speed ....rpm	4800 to 5200	4500 to 5500	–	5200
Smoke test/opacity ....M^{-1} %	2.0	2.0	2.0	2.0
Static timing method...............................	Refer to wsm	Refer to wsm	Refer to wsm	–
Timing dimension.................................mm	–	–	–	–
Crankshaft position ....mm [°]	–	–	–	–
Turbo type / ref / pressure ....bar	1.7 to 2.2 bar at 3000 rpm	–	0.5 to 0.65 bar @ 3500 rpm	1.5 to 1.7 bar @ 3000 rpm
Injection pump make	–	Bosch	Bosch	–
Injection pump part no............................	–		VER510	–
Injector Make / type	–	–	Bosch	–
Injector part no......................................	–	–	0432 193 838	–
Injection type...	Direct Pump Injector EDC	Direct EDC	Direct VER	Direct unit injector
Injection opening pressure, New [used]...bar	190 to 200 [170]	180 to 190 [160]	190 to 200 [170]	190 to 200 [170]
Glow plugs				
Maker	–	–	Bosch	–
Type	–	–	0250 202 009	–
Nominal rating.....................................V/A				
Brakes				
minimum friction material thickness				
Front..mm	7.0	7.0	7.0 with backing	7.0 with backing
Rear...mm	7.0	7.0	7.0 with backing	7.0 with backing
Tyres - Saloon / Hatch......................Size	195/65x15: 205/60x16:205/55x16	195/65x15: 205/60x15:205/55x16	–	–
- Estate / Van......................Size	–	–	195/65x15:205/60x15:215/60x15	195/65x15:205/60x15:215/60x15
Pressure - front / rear - Saloon / Hatch ...bar	2.0 / 2.0	2.6 / 2.4	–	–
- Estate / Van ......bar			2.6 / 2.6: 2.8 / 2.8: 2.8 /2.6	2.6 / 2.6: 2.8 / 2.8: 2.8 /2.6
Front suspension / wheel alignment				
Toe-in (+) / Toe-out (–)....................mm [°]	[0°10' ± 20']	[0°10' ± 20']	[10' ± 20']	[10' ± 20']
Camber	-0°25' ± 25' Sport: -0°40'±25'	-0°25' ± 25' Sport: -0°40'±25'	-20' ± 45'	-20' ± 45'
Castor	–	–	3°20' ± 40'	3°20' ± 40'
King pin inclination.................................				
Rear suspension / wheel alignment				
Toe-in (+) / Toe-out (–)....................mm [°]	[0°20' ± 10']	[0°20' ± 10']	[10' ± 25'] (not Nivomat)	[10' ± 25'] (not Nivomat)
Camber	-1°30' ± 20'	-1°30' ± 20'	-20' ± 30' (not Nivomat)	-20' ± 30' (not Nivomat)

VOLKSWAGEN

	Passat 1.9 TDi 1999 to 2000	Passat 2.5 TDi 1999 to 2000	Sharan 1.9 TDi 1995 to 2000	Sharan 1.9 TDi 1999 to 2000
Torque wrench settings				
Cylinder head - stage 1 ....Nm	40 N	35 N	40 N	40 N
- stage 2 ....Nm	60	60	60	60
- stage 3 ....Nm	+ 90°	+ 90°	+ 90°	+ 90°
- stage 4 ....Nm	+ 90°	+ 90°	+ 90°	+ 90°
- stage 5 ....Nm	–	–	–	–
- stage 6 ....Nm	–	–	–	–
Big-end bearings.............................Nm	30 + 90° N	30 + 90° N	30 + 90° N	30 + 90° N
Main bearings...................................Nm	65 + 90° N	WSM	65 + 90° N	65 + 90° N
Crankshaft pulley bolt ....Nm	120 + 90° N	200 + 180° N	90 + 90° N	120 + 90° N
Camshaft pulley bolt ....Nm	100	75	80	100
Flywheel [driveplate] bolt.....................Nm	60 + 90° N [60 + 90° N]	60 + 180° N [60 + 90° N]	60 + 90° N	60 + 90° N
Front hubs ....Nm	115 + 180°N M16: 140 + 180°N	115 + 180°N M16: 140 + 180°N	150 + 90° N	150 + 90° N
Rear hubs ....Nm	WSM	WSM	200 N	200 N
Wheel nuts / bolts ....Nm	120	120	140	140
Glow plugs ....Nm	15	15	20	15
Clutch pressure plate bolts ....Nm	20	–	25	20
Injection pump sprocket.....................Nm	45	45	55	–
Injectors...Nm	–	10	20	–
Injection pump mounting bolts ....Nm	25	25	25	–
Injector pipe unions...........................Nm	25	25	25	–
Capacities				
Engine oil & filter................................litres	3.8	5.4	4.7	4.5
Gearbox...litres	2.25	2.4	2.2	2.2
Automatic transmission ....litres	3.5	–	–	–
Final drive ....litres	WT AT: 0.75	WT	WT	WT
Cooling system................................litres	7.0	10.0	7.0 2 heat exchangers: 9.0	6.8 2 heat exchangers: 8.6[1]
Fuel tank...litres	62	62	70	70

Notes

Sharan 1.9 TDi 1999 to 2000
[1]Two heat exchangers and additional water heater:
9.2 L

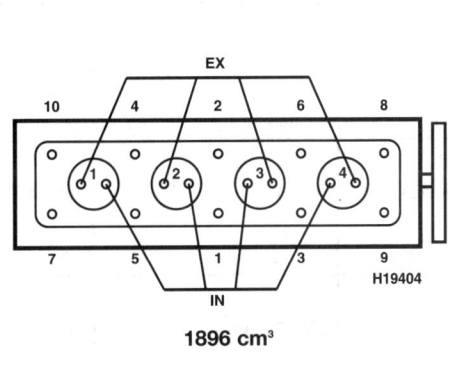

1896 cm³

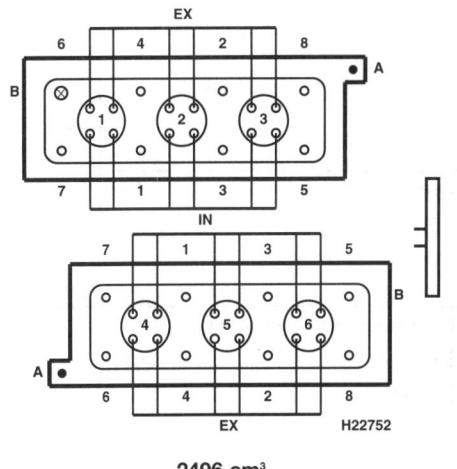

2496 cm³

– Not applicable, or information not available

VOLKSWAGEN

	Sharan 1.9 TDi 1996 to 2000	Caddy 1.6 D 1985 to 1992	Caddy Pick-Up 1.9 D 1997 to 2000	Taro 2.4 D 4x4 1989 to 1994
Engine				
Engine type/code	AFN SOHC 8V Turbo 81kW	JK SOHC 40kW	AEF SOHC 47kW	2L OHC 61kW
Capacity (cm³) / cylinders........................	1896 / 4	1588 / 4	1896 / 4	2446 / 4
Compression ration / pressure ....bar	19.5 / 25 to 31	23.0 / ≥28.0	22.5 / ≥26.0	22.2 / ≥20.0
Torque output ....Nm	235	0	124	0
Oil pressure ....idle [running] bar	[2.0 @ 2000]	[2.0 @ 2000]	[2.0 @ 2000]	[3.0 to 5.5 @ 3000]
Oil temperature ....°C	80	80	80	80
Valve clearances - inlet (mm)	0: Hyd.	0.25 H	0: Hyd.	0.20 to 0.30
- exhaust (mm)	0: Hyd.	0.45 H	0: Hyd.	0.40 to 0.50
Injection order	1-3-4-2	1-3-4-2	1-3-4-2	1-3-4-2
No. 1 cylinder position	TBE	TBE	TBE	F
Cooling system				
Thermostat opening temperature ....°C	85 to 105	87	88	84 to 90
Radiator cap pressure ....bar	1.4 to 1.6	1.2 to 1.4	0.9 to 1.1	0.8 to 1.1
Fuel system				
Idle speed ....rpm	900 ± 40	850 ± 30	920 to 960	700
Maximum (no load) speed ....rpm	5200	5350 ± 50	4950 to 5150	–
Smoke test/opacity ....M⁻¹ %	2.0	2.0	2.5	2.5
Static timing method...............................	Refer to wsm	Plunger travel	Refer to wsm	Plunger travel
Timing dimension...............................mm	–	0.9 ± 0.02	–	0.54 to 0.66
Crankshaft position ....mm [°]	–	[0] TDC	–	[0] TDC
Turbo type / ref / pressure ....bar	0.5 to 0.65 bar @ 3500 rpm	–	–	–
Injection pump make	Bosch	Bosch	Lucas	Nippon Denso
Injection pump part no............................	VER 510	VE4/9 0460 494 052	DPC	J22 100 547 5 or 547 60
Injector Make / type	–	Bosch	–	Nippon Denso
Injector part no.......................................	–	0432 217 084	–	DN4PD57
Injection type..	Direct VER	Indirect VE	Indirect DPC	Bosch type
Injection opening pressure, New [used]...bar	190 to 200 [170]	130 to 138 [120]	130 to 138 [120]	151 [145]
Glow plugs				
Maker	Bosch	Bosch/Champion	Bosch	Beru
Type	0250 202 009	0250 201 032 / CH69	0250 202 022	GV968
Nominal rating....................................V/A	11 / 12	11 / 12	11.5 / 8	7.0 / _
Brakes				
minimum friction material thickness				
Front..mm	7.0 with backing	7.0 with backing	7.0 with backing	1.0
Rear...mm	7.0 with backing	3.5 Riveted: 5.0	2.5	1.0
Tyres - Saloon / Hatch......................Size	–	–	–	–
- Estate / Van......................Size	195/65x15:205/60x15:215/60x15	165x13	165/70x13	205x16
Pressure - front / rear - Saloon / Hatch ...bar	–	–	–	–
- Estate / Van ......bar	2.6 / 2.6: 2.8 / 2.8: 2.8 /2.6	1.8 / 2.4	2.0 / 2.0	1.7 / 2.4
Front suspension / wheel alignment				
Toe-in (+) / Toe-out (–).....................mm [°]	[10' ± 20']	[-15' +10' -15']	[-0°10' ± 10']	3.0 ± 1.0
Camber	-20' ± 45'	20' ± 30'	-0°30' ± 30'	30' ± 30'
Castor	3°20' ± 40'	1°50' ± 20'	1°20' ± 45'	1°10' ± 30'
King pin inclination.................................	–		11°45' ± 45'	12°5' ± 45'
Rear suspension / wheel alignment				
Toe-in (+) / Toe-out (–).....................mm [°]	[10' ± 25'] (not Nivomat)	[0° ± 1°]	[0°20' ± 10']	–
Camber	-20' ± 30' (not Nivomat)	0° ± 10'	1°30' ± 10'	

VOLKSWAGEN

	Sharan 1.9 TDi 1996 to 2000	Caddy 1.6 D 1985 to 1992	Caddy Pick-Up 1.9 D 1997 to 2000	Taro 2.4 D 4x4 1989 to 1994
Torque wrench settings				
Cylinder head - stage 1 ....Nm	40 N	M11 hex: 50[1]	40 N	78 N
- stage 2 ....Nm	60	70	60	+ 90°
- stage 3 ....Nm	+ 90°	90	+ 90°	+ 90°
- stage 4 ....Nm	+ 90°	Warm up, 90	+ 90°	–
- stage 5 ....Nm	–	After 1000km:	–	–
- stage 6 ....Nm	–	Slacken 30°, 90	–	–
Big-end bearings................................Nm	30 + 90° N	Thread length 15mm: 45[2] N	30 + 90° N	54 + 90° N
Main bearings....................................Nm	65 + 90° N	65	65 + 90° N	103
Crankshaft pulley bolt ....Nm	90 + 90° N	Hexagon: 80[3]	90 + 90° oiled N	167
Camshaft pulley bolt ....Nm	80	45	20 + 90° N	98
Flywheel [driveplate] bolt.....................Nm	60 + 90° N	30 + 90°	60 + 90° N	123 [98]
Front hubs ....Nm	150 + 90° N	230	300	WSM
Rear hubs ....Nm	200 N	WSM	WSM	–
Wheel nuts / bolts ....Nm	140	110	110	140
Glow plugs ....Nm	30	30	25	13
Clutch pressure plate bolts ....Nm	20	20	25	19
Injection pump sprocket.....................Nm	55	45	25	64
Injectors...Nm	20	70	70	64
Injection pump mounting bolts ....Nm	25	25	25	18
Injector pipe unions...........................Nm	25	25	25	25
Capacities				
Engine oil & filter...............................litres	4.3	4.5	4.7	5.9
Gearbox..litres	2.2	1.5 [2.0]	2.4	3.9[1]
Automatic transmission ....litres	–	–	–	–
Final drive ....litres	WT	WT	WT	Front: 2.3[2]
Cooling system...................................litres	7.0 2 heat exchangers: 9.0	6.5	6.0	9.2
Fuel tank...litres	70	45	42	65

Notes

Caddy 1.6 D 1985 to 1992
[1] M12 splined socket bolt: 40, 60, 75 + 180° Warm up, + 90°. After 1000km: + 90°
[2] Thread length 25mm: 30 + 90°
[3] Durlock M12: 150. Durlock M14: 200
Taro 2.4 D 4x4 1989 to 1994
[1] W56 type: 3.0. Transfer box: 1.6
[2] Rear: 2.2. Independent front suspension: 1.6

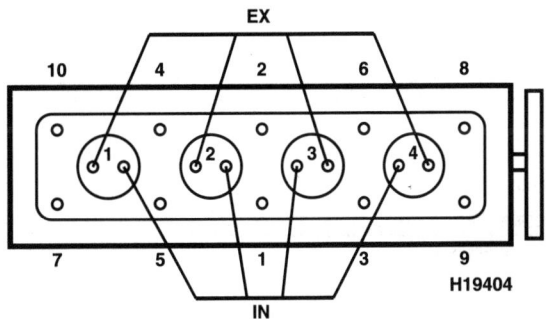

1588 cm³ / 1896 cm³ / 2446 cm³

– Not applicable, or information not available

VOLKSWAGEN

	Transporter/Caravelle 1.6 TD 1984 to 1992	Transporter 1.9 D 1990 to 1992	Transporter 1.9 TD 1992 to 2000	Transporter/Caravelle 2.4D 1991 to 1998
Engine				
Engine type/code............................	JX SOHC Turbo 51kW	1X SOHC 45kW	ABL SOHC Turbo 50kW	AAB SOHC 10V 57kW
Capacity (cm³) / cylinders................	1588 / 4	1896 / 4	1896 / 4	2370 / 5
Compression ration / pressure ....bar	23.0 / ≥28.0	22.5 / ≥26.0	22.5 / ≥26.0	23.0 / ≥26.0
Torque output ....Nm	0	0	140	164
Oil pressure ....idle [running] bar	[1.1 @ 2000]	[2.0 @ 2000]	[2.0 @ 2000]	[2.0 @ 2000]
Oil temperature ....°C	80	80	80	80
Valve clearances - inlet (mm)	0.15 to 0.20	0: Hyd.	0: Hyd.	0: Hyd.
- exhaust (mm)	0.35 to 0.45	0: Hyd.	0: Hyd.	0: Hyd.
Injection order	1-3-4-2	1-3-4-2	1-3-4-2	1-2-4-5-3
No. 1 cylinder position	TBE	TBE	TBE	TBE
Cooling system				
Thermostat opening temperature ....°C	85	87	87	87 to 102
Radiator cap pressure ....bar	0.9 to 1.2	1.2 to 1.5	1.2 to 1.5	1.2 to 1.5
Fuel system				
Idle speed ....rpm	820 ± 50	850 ± 30	850 ± 30	850 ± 50
Maximum (no load) speed ....rpm	5100 ± 100	4800 ± 100	4800 ± 100	4800
Smoke test/opacity ....M⁻¹ %	2.0	2.0	2.0	2.0
Static timing method............................	Plunger travel	Plunger travel	Plunger travel	Dial gauge
Timing dimension............................mm	0.9 ± 0.02	1.0 ± 0.02	1.0 ± 0.02	1.0 ± 0.02
Crankshaft position ....mm [°]	[0] TDC	[0] TDC	[0] TDC	[0 TDC]
Turbo type / ref / pressure ....bar	0.62 to 0.72 bar	–	0.6 to 0.83 bar	–
Injection pump make	Bosch	Bosch	Bosch	Bosch
Injection pump part no............................	VE 4/9 0460 494 152	VE4/8 R357	VE 4/8 R357	VE 5/8 L358
Injector Make / type	Bosch	Bosch	Bosch	Bosch
Injector part no............................	0432 217 078	0432 217 198	0432 217 198	0432 217 198
Injection type............................	Indirect VE	Indirect VE	Indirect VE	Indirect VE
Injection opening pressure, New [used]...bar	155 [140]	130 to 138 [120]	155 to 163 [140]	130 to 138 [120]
Glow plugs				
Maker	Bosch/Champion	Bosch/Champion	Bosch/Champion	Bosch/Champion
Type	0250 201 032 / CH69	0250 201 032 / CH160	0250 201 032 / CH160	0250 201 032 / CH160
Nominal rating ....V/A	11 / 12	11 / 12	11 / 8	11 / 12
Brakes				
minimum friction material thickness				
Front..mm	2.0	7.0 with backing	7.0 with backing	2.0
Rear...mm	2.5	2.5	2.5	1.0
Tyres - Saloon / Hatch.......................Size	–	–	–	–
- Estate / Van..........................Size	185x14	185x14: 195/70x15	185x14: 195/70x15	195/70x15:205/65x15:215/65x15
Pressure - front / rear - Saloon / Hatch ...bar	–	–	–	–
- Estate / Van ......bar	3.0 / 3.7	3.3 / 3.8¹	3.3 / 3.8¹	2.6 / 3.4. 215/65: 2.4 / 3.0
Front suspension / wheel alignment				
Toe-in (+) / Toe-out (–).....................mm [°]	2.0 ± 3.5	Group 1 & 2: [40' ± 20']²	Group 1 & 2: [40' ± 20']²	Group 1 & 2: [20'±20']²
Camber	0° ± 30'	Group 1 & 2: 35' ± 20'³	Group 1 & 2: 35' ± 20'³	Group 1 & 2: 15'±20'³
Castor	7°15' ± 15'	Group 1 & 2: 1°40' ± 30'	Group 1 & 2: 1°40' ± 30'	1°40'±30' Group 4: 1°50'±30'
King pin inclination............................	–	–	–	–
Rear suspension / wheel alignment				
Toe-in (+) / Toe-out (–)....................mm [°]	[0° ± 20']	Group 1: [20' ± 20']⁴	Group 1: [20' ± 20']⁴	Group 1 & 3: [20'±20']⁴
Camber	-30' ± 30'	Group 1: -30' ± 30'⁵	Group 1: -30' ± 30'⁵	Group 1 & 3: -30'±30'⁵

VOLKSWAGEN

Torque wrench settings	Transporter/Caravelle 1.6 TD 1984 to 1992	Transporter 1.9 D 1990 to 1992	Transporter 1.9 TD 1992 to 2000	Transporter/Caravelle 2.4D 1991 to 1998
Cylinder head - stage 1 ....Nm	M11 hex: 50[1]	40 N	40 N	40 N
- stage 2 ....Nm	70	60	60	60
- stage 3 ....Nm	90	+ 180°	+ 180°	+ 90°
- stage 4 ....Nm	Warm up, 90	Warm up	Warm up	+ 90°
- stage 5 ....Nm	After 1000km,	+ 90°	+ 90°	–
- stage 6 ....Nm	Slacken 30°, 90	–	–	–
Big-end bearings......................Nm	Thread length 15mm: 45[2]	30 + 90° N	30 + 90° N	30 + 90° N
Main bearings......................Nm	65	65	65 N	65[6]
Crankshaft pulley bolt ....Nm	M14: 180[3]	90 + 90°	90 + 90° N	460
Camshaft pulley bolt......................Nm	45	45	45	85
Flywheel [driveplate] bolt......................Nm	30 + 90°	30 + 90°	30 + 90° N	30 + 90° N
Front hubs......................Nm	WSM	200	200	150 + 90° N
Rear hubs......................Nm	500	200	200	200
Wheel nuts / bolts ....Nm	180	160	160	160
Glow plugs......................Nm	30	30	30	30
Clutch pressure plate bolts ....Nm	25	20	20	20
Injection pump sprocket......................Nm	45	45	45	100 Pump end: 50
Injectors......................Nm	70	70	70	70
Injection pump mounting bolts ....Nm	25	20	20	20
Injector pipe unions......................Nm	25	25	25	25
Capacities				
Engine oil & filter......................litres	4.5	5.0	5.0	5.5
Gearbox......................litres	4.0	3.0	3.0	2.5 4x4: 3.0
Automatic transmission ....litres	–	–	–	3.5
Final drive ....litres	WT	–	WT	WT AT: 1.15 4x4: 1.0
Cooling system......................litres	16.0	9.0	9.0	9.0
Fuel tank......................litres	60	80	80	80

Notes

Transporter / Caravelle 1.6 TD 1984 to 1992
[1]M12 splined socket bolt: 40, 60, 75 + 180° Warm up, + 90°. After 1000km: + 90°
[2]Thread length 25mm: 30 + 90°
[3]M12: 150. M14 12 point : 90 + 180°
Transporter 1.9 D 1990 to 1992
[1]195/70x15: 2.6 / 3.4
[2]Group 3: [30'±20']. Group 4: [20'±20']
[3]Group 3: 20'±20'. Group 4: 10'±20'
[4]Group 2: 6'±20'. Group 3: 20'±20'. Group 4: 24'±20'
[5]Group 2: 0±30'. Group 3: -30'±30'. Group 4: - 40'±30'
Transporter 1.9 TD 1992 to 2000
[1]195/70x15: 2.6 / 3.4
[2]Group 3: [30' ± 20']. Group 4: [20' ± 20']
[3]Group 3: 20' ± 20'. Group 4: 10' ± 20'
[4]Group 2: 6' ± 20'. Group 3: 20' ± 20'. Group 4: 24' ± 20'
[5]Group 2: 0° ± 30'. Group 3: -30' ± 30'. Group 4: - 40'± 30
Transporter/Caravelle 2.4D 1991 to 1998
[1]Coolant temperature sender plug blue 2 pin disconnected
[2]Group 3: [10'±20']. Group 4: [0±20']
[3]Group 3: 0±20'. Group 4: -10'±20'
[4]Group 2: 6'±20'. Group 4: 24'±20'
[5]Group 2: 0±30'. Group 4: -40'±30'
[6]Replace with fully threaded shank bolts: 65 + 90°

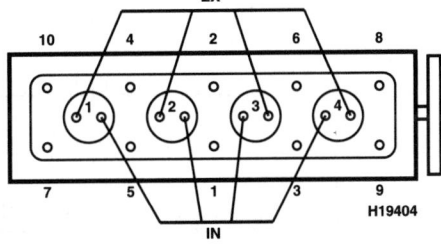

1588 cm³ / 1896 cm³

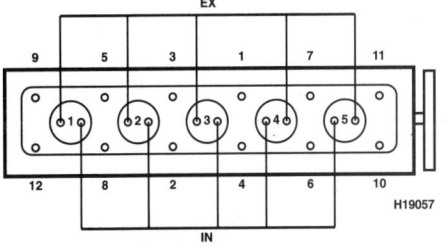

2370 cm³

– Not applicable, or information not available

VOLKSWAGEN

	Transporter/Caravelle 2.4D 1997 to 1998	Transporter/Caravelle 2.5TD 1997 to 2000	Transporter/Caravelle 2.5TD 1998 to 2000	Transporter/Caravelle 2.5TD 1998 to 2000
Engine				
Engine type/code..........................	AJA SOHC 10V 55kW	ACV SOHC 10V Turbo 75kW	AHY SOHC 10V Turbo 111kW	AJT SOHC 10V Turbo 65kW
Capacity (cm^3) / cylinders........................	2370 / 5	2459 / 5	2459 / 5	2459 / 5
Compression ration / pressure ....bar	23.0 / ≥26.0	20.5 / ≥24.0	19.5 / ≥24.0	19.5 / ≥24.0
Torque output ....Nm	164	250	295	195
Oil pressure ....idle [running] bar	[2.0 @ 2000]	[2.0 @ 2000]	[2.0 @ 2000]	[2.0 @ 2000]
Oil temperature ....°C	80	80	80	80
Valve clearances - inlet (mm)	0: Hyd.	0: Hyd.	0: Hyd.	0: Hyd.
- exhaust (mm)	0: Hyd.	0: Hyd.	0: Hyd.	0: Hyd.
Injection order..	1-2-4-5-3	1-2-4-5-3	1-2-4-5-3	1-2-4-5-3
No. 1 cylinder position	TBE	TBE	TBE	TBE
Cooling system				
Thermostat opening temperature ....°C	87 to 102	87 to 102	87 to 102	87 to 102
Radiator cap pressure ....bar	1.2 to 1.5	1.2 to 1.5	1.2 to 1.5	1.2 to 1.5
Fuel system				
Idle speed ....rpm	850 ± 50	740 to 800	740 to 800	740 to 800
Maximum (no load) speed ....rpm	4800	4300 to 4700	–	–
Smoke test/opacity ....M^{-1} %	2.01	2.0	2.0	2.0
Static timing method...............................	Dial gauge	Rotor lift	Plunger travel	Plunger travel
Timing dimension...............................mm	0.9 ± 0.02	0.55	0.9 ± 0.02	0.9 ± 0.02
Crankshaft position ....mm [°]	[0 TDC]	[0 TDC]	[0 TDC]	[0 TDC]
Turbo type / ref / pressure ....bar	–	1.7 to 1.95 bar @ 3000	1.75 to 2.08 bar @ 3000	1.5 to 1.8 bar @ 3000
Injection pump make	Bosch	Bosch	–	–
Injection pump part no............................	–	VE	–	–
Injector Make / type	Bosch	Bosch	–	–
Injector part no....................................	–	–	–	–
Injection type...	Indirect	Indirect VE	Indirect	Indirect
Injection opening pressure, New [used]...bar	130 to 138 [120]	130 to 138 [120]	130 to 138 [120]	130 to 138 [120]
Glow plugs				
Maker	Bosch/Champion	Bosch	Bosch/Champion	Bosch/Champion
Type	0250 201 032 / CH160	0250 202 022	W8DTC / N9BMC	W8DTC / N9BMC
Nominal rating ....V/A	11 / 12	11 / 12	–	–
Brakes				
minimum friction material thickness				
Front...mm	2.0	2.0	2.0	2.0
Rear...mm	1.0	1.0	1.0	2.0
Tyres - Saloon / Hatch.....................Size	–	–	–	–
- Estate / Van.....................Size	195/70x15:205/65x15:215/65x15	195/70x15:205/65x15:215/65x15	195/70x15:205/65x15:215/65x15	195/70x15:205/65x15:215/65x15
Pressure - front / rear - Saloon / Hatch ...bar				
- Estate / Van ......bar	2.6 / 3.4 215/65: 2.4 / 3.0	2.6 / 3.4. 215/65: 2.4 / 3.0	2.6 / 3.4 215/65: 2.4 / 3.0	2.6 / 3.4. 215/65: 2.4 / 3.0
Front suspension / wheel alignment				
Toe-in (+) / Toe-out (–)....................mm [°]	Group 1 & 2: [20'±20']²	Group 1 & 2: [20' ± 20']²	Group 1 & 2: [20'±20']²	Group 1 & 2: [20'±20']²
Camber	Group 1 & 2: 15'±20'³	Group 1 & 2: 15' ± 20'³	Group 1 & 2: 15'±20'³	Group 1 & 2: 15'±20'³
Castor	1°40'±30' Group 4: 1°50'±30'	1°40' ± 30' Group 4: 1°50'±30	1°40'±30' Group 4: 1°50'±30'	1°40'±30' Group 4: 1°50'±30'
King pin inclination................................	–	–	–	–
Rear suspension / wheel alignment				
Toe-in (+) / Toe-out (–)....................mm [°]	Group 1 & 3: [20'±20']⁴	Group 1 & 3: [20' ± 20']⁴	Group 1 & 3: [20'±20']⁴	Group 1 & 3: [20'±20']⁴
Camber	Group 1 & 3: -30'±30'⁵	Group 1 & 3: -30' ± 30'⁵	Group 1 & 3: -30'±30'⁵	Group 1 & 3: -30'±30'⁵

VOLKSWAGEN

	Transporter/Caravelle 2.4D 1997 to 1998	Transporter/Caravelle 2.5TD 1997 to 2000	Transporter/Caravelle 2.5TD 1998 to 2000	Transporter/Caravelle 2.5TD 1998 to 2000
Torque wrench settings				
Cylinder head - stage 1 ....Nm	40 N	40 N	40 N	40 N
- stage 2 ....Nm	60	60	60	60
- stage 3 ....Nm	+ 90°	+ 90°	+ 90°	+ 90°
- stage 4 ....Nm	+ 90°	+ 90°	+ 90°	+ 90°
- stage 5 ....Nm	–	–	–	–
- stage 6 ....Nm	–	–	–	–
Big-end bearings..Nm	30 + 90° N	30 + 90° N	30 + 90° N	30 + 90° N
Main bearings..Nm	65[6]	65	65	65
Crankshaft pulley bolt ....Nm	460 '95-on: 160 + 180° N	160 + 180° N	160 + 180° N	160 + 180° N
Camshaft pulley bolt ....Nm	85	WSM	WSM	WSM
Flywheel [driveplate] bolt......................Nm	30 + 90° N	60 + 90° N	60 + 90° N	60 + 90° N
Front hubs ....Nm	150 + 90° N	150 + 90° N	150 + 90° N	150 + 90° N
Rear hubs ....Nm	200	200	200	200
Wheel nuts / bolts ....Nm	160	160	160[1]	160
Glow plugs ....Nm	30	15	15	15
Clutch pressure plate bolts ....Nm	20	20	20	20
Injection pump sprocket.........................Nm	100 Pump end: 50	90	90	90
Injectors...Nm	70	70	70	70
Injection pump mounting bolts ....Nm	20	20	20	20
Injector pipe unions................................Nm	25	25	25	25
Capacities				
Engine oil & filter...................................litres	5.5	5.5	5.5	5.5
Gearbox...litres	2.5 4x4: 3.0	2.5 4x4: 3.0	2.5 4x4: 3.0	2.5 4x4: 3.0
Automatic transmission ....litres	3.5	3.5	3.5	3.5
Final drive...litres	WT AT: 1.15 4x4: 1.0	WT AT: 1.15 4x4: 1.0	WT AT: 1.15 4x4: 1.0	WT AT: 1.15 4x4: 1.0
Cooling system...litres	11.5	11.5	9.0	9.0
Fuel tank..litres	80	80	80	80

Notes

Transporter/Caravelle 2.4D 1997 to 1998
[1]Coolant temperature sender plug blue 2 pin disconnected
[2]Group 3: [10'±20']. Group 4: [0±20']
[3]Group 3: 0±20'. Group 4: -10'±20'
[4]Group 2: 6'±20'. Group 4: 24'±20'
[5]Group 2: 0±30'. Group 4: -40'±30'
[6]Replace with fully threaded shank bolts: 65 + 90°
Transporter/Caravelle 2.5TD 1997 to 2000
[2]Group 3: [10'±20']. Group 4: [0±20']
[3]Group 3: 0±20'. Group 4: -10'±20'
[4]Group 2: +6'±20'. Group 4: 24'±20'
[5]Group 2: 0±30'. Group 4: -40'±30'
Transporter/Caravelle 2.5TD 1998 to 2000
[1] Vehicles from 01.96: 180 Nm
[2]Group 3: [10'±20']. Group 4: [0±20']
[3]Group 3: 0±20'. Group 4: -10'±20'
[4]Group 2: 6'±20'. Group 4: 24'±20'
[5]Group 2: 0±30'. Group 4: -40'±30'
Transporter/Caravelle 2.5TD 1998 to 2000
[2]Group 3: [10'±20']. Group 4: [0±20']
[3]Group 3: 0±20'. Group 4: -10'±20'
[4]Group 2: 6'±20'. Group 4: 24'±20'
[5]Group 2: 0±30'. Group 4: -40'±30'

– Not applicable, or information not available

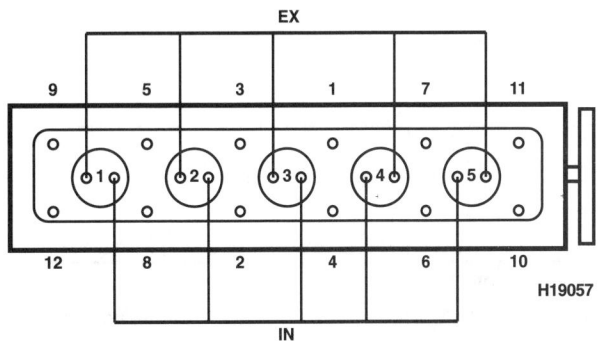

2370 cm³ / 2459 cm³

VOLKSWAGEN

	LT 2.4 D 1989 to 1992	LT 2.4 TD 1989 to 1992	LT 2.4 D 1992 to 1996	LT 2.4 TD 1992 to 1996
Engine				
Engine type/code..........................	1S SOHC 51kW	1G SOHC Turbo 68kW	ACT SOHC 51kW	ACL SOHC Turbo 70kW
Capacity (cm³) / cylinders........................	2383 / 6	2383 / 6	2383 / 6	2383 / 6
Compression ration / pressure ....bar	23.0 / ≥26.0	23.0 / ≥26.0	23.0 / 26 to 35	23.0 / 26 to 35
Torque output ....Nm	0	0	145	205
Oil pressure ....idle [running] bar	[2.0 @ 2000]	[2.0 @ 2000]	[2.0 @ 2000]	[2.0 @ 2000]
Oil temperature ....°C	80	80	80	80
Valve clearances - inlet (mm)	0: Hyd.	0: Hyd.	0: Hyd.	0: Hyd.
- exhaust (mm)	0: Hyd.	0: Hyd.	0: Hyd.	0: Hyd.
Injection order	1-5-3-6-2-4	1-5-3-6-2-4	1-5-3-6-2-4	1-5-3-6-2-4
No. 1 cylinder position	TBE	TBE	TBE	TBE
Cooling system				
Thermostat opening temperature ....°C	80	80	80	80
Radiator cap pressure ....bar	1.2 to 1.5	1.2 to 1.5	1.2 to 1.5	1.2 to 1.5
Fuel system				
Idle speed ....rpm	750 ± 50	750 ± 50	750 ± 50	750 ± 50
Maximum (no load) speed ....rpm	5000 ± 50	4900	4800	4500
Smoke test/opacity ....M⁻¹ %	2.0	2.0	2.5	2.5
Static timing method...............................	Plunger travel	Plunger travel	Dial gauge	Dial gauge
Timing dimension...............................mm	0.8 ± 0.02	0.85 ± 0.02	0.8 ± 0.02	0.85 ± 0.02
Crankshaft position ....mm [°]	[0] TDC	[0] TDC	0 TDC	0 TDC
Turbo type / ref / pressure ....bar	–	0.64 to 0.76 bar @ 4000 rpm	–	–
Injection pump make	Bosch	Bosch	Bosch	Bosch
Injection pump part no............................	VE 6/10F	VE6/10 L324	VE6/10F	VE6/10 L470
Injector Make / type	Bosch	Bosch	Bosch	Bosch
Injector part no.......................................	0432 217 198	0432 217 079	0432 217 198	0432 217 231
Injection type..	Indirect VE	Indirect VE	Indirect VE	Indirect VE
Injection opening pressure, New [used]...bar	130 to 138 [120]	155 to 163 [140]	130 to 138 [120]	155 to 163 [140]
Glow plugs				
Maker	Bosch/Champion	Bosch/Champion	Bosch/Champion	Bosch
Type	0250 201 032 / CH69	0250 201 032 / CH69	0250 201 021 / CH69	0250 201 032
Nominal rating......................................V/A	11 / 12	11.5 / 12	11 / 12	11.5 / 8
Brakes				
minimum friction material thickness				
Front...mm	2.0	2.0	2.5	2.5
Rear..mm	2.5	2.5	2.5	2.5
Tyres - Saloon / Hatch.......................Size	–	–	–	–
- Estate / Van.......................Size	185x14: 195x14: 205x14	185x14: 195x14: 205x14	205x14	205x14
Pressure - front / rear - Saloon / Hatch ...bar				
- Estate / Van ......bar	Refer to manufacturer / ohb	Refer to manufacturer / ohb	3.0 / 4.5	3.0 / 4.5
Front suspension / wheel alignment				
Toe-in (+) / Toe-out (–).....................mm [°]	3.0 to 6.0'	3.0 to 6.0'	3.0 to 6.0	3.0 to 6.0
Camber	40' ± 20'²	40' ± 20'²	0°40' ± 20'	0°40' ± 20'
Castor	40' ± 20'³	40' ± 20'³	2°30' ± 40	2°30' ± 40
King pin inclination..................................	–	–	–	–
Rear suspension / wheel alignment				
Toe-in (+) / Toe-out (–).....................mm [°]	[20' ± 40']⁴	[20' ± 40']⁴	[0° ± 20']	[0° ± 20']
Camber	0° ± 25'	0° ± 25'	0°25'	0°25'

VOLKSWAGEN

	LT 2.4 D 1989 to 1992	LT 2.4 TD 1989 to 1992	LT 2.4 D 1992 to 1996	LT 2.4 TD 1992 to 1996
Torque wrench settings				
Cylinder head - stage 1 ....Nm	M11 hex: 50[5]	M11 hex: 50[5]	40 N M11:[1]	40 N M11:[1]
- stage 2 ....Nm	70	70	60	60
- stage 3 ....Nm	90	90	+ 180°	+ 180°
- stage 4 ....Nm	Warm up, 90	Warm up, 90	–	–
- stage 5 ....Nm	After 1000km,	After 1000km,	–	–
- stage 6 ....Nm	Slacken 30°, 90	Slacken 30°, 90	–	–
Big-end bearings......................Nm	Thread length 15mm: 50[6]	Thread length 15mm: 50[6]	30 + 180° N	30 + 180° N
Main bearings......................Nm	65	65	65	65
Crankshaft pulley bolt ....Nm	460	460	460	460
Camshaft pulley bolt ....Nm	85 Pump end: 100	85 Pump end: 100	85	85
Flywheel [driveplate] bolt......................Nm	70 + 90°	70 + 90°	60 + 90° N	60 + 90° N
Front hubs ....Nm	WSM	WSM	WSM	WSM
Rear hubs ....Nm	WSM	WSM	WSM	WSM
Wheel nuts / bolts ....Nm	320 LT 28/31: 300	320 LT 28/31: 300	200	200
Glow plugs ....Nm	20	20	30	30
Clutch pressure plate bolts ....Nm	25	25	25	25
Injection pump sprocket......................Nm	45	45	50	50
Injectors......................Nm	70	70	70	70
Injection pump mounting bolts ....Nm	25	25	25	25
Injector pipe unions......................Nm	25	25	25	25
Capacities				
Engine oil & filter......................litres	7.0	7.0	7.0	7.0
Gearbox......................litres	3.5	3.5	3.5	3.5
Automatic transmission ....litres	–	–	–	–
Final drive ....litres	1.8 Twin tyres:2.4	1.8 Twin tyres:2.4	1.8	1.8
Cooling system......................litres	12.0	12.0	12.0	12.0
Fuel tank......................litres	70 Option: 110	70 Option: 110	70	70

Notes

LT 2.4 D 1989 to 1992
[1] LT 40, 45, 50, 55: 0 to +2.2
[2] LT 40, 45, 50, 55: +10'±35'
[3] LT 40, 45, 50, 55: +3°25'±20'
[4] LT 40, 45, 50, 55: [0±20']
[5] M12 splined socket bolt: 40, 60 + 180°. Warm up + 90°. After 1000km, + 90°
[6] Thread length 25mm: 30 + 180°
LT 2.4 TD 1989 to 1992
[1] LT 40, 45, 50, 55: 0 to +2.2
[2] LT 40, 45, 50, 55: +10'±35'
[3] LT 40, 45, 50, 55: +3°25'±20'
[4] LT 40, 45, 50, 55: [0±20']
[5] M12 splined socket bolt: 40, 60 + 180°. Warm up + 90°. After 1000km, + 90°
[6] Thread length 25mm: 30 + 180°
LT 2.4 D 1992 to 1996
[1] M11: 50 N, 70, 90, warm up, 90
LT 2.4 TD 1992 to 1996
[1] M11: 50 N, 70, 90, warm up, 90

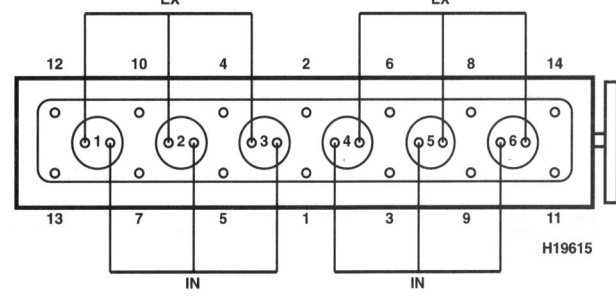

2383 cm³

– Not applicable, or information not available

VOLKSWAGEN

	LT 2.5 D 1996 to 1999	LT 2.5 TDi 1997 to 1999	LT 2.5 TDi 1999 to 2000	LT 2.5 TDi 1999 to 2000
Engine				
Engine type/code....................	AGX SOHC 55kW	AHD SOHC Turbo 75kW	ANJ SOHC Vari - Turbo 80kW	APA SOHC Turbo 66kW
Capacity (cm³) / cylinders........................	2459 / 5	2459 / 5	2459 / 5	2459 / 5
Compression ration / pressure ....bar	19.5 / ≥24.0	19.5 / ≥24.0	19.0 / ≥24.0	19.5 / ≥24.0
Torque output ....Nm	160	250	275	195
Oil pressure ....idle [running] bar	[2.0 @ 2000]	[2.0 @ 2000]	[2.0 @ 2000]	[2.0 @ 2000]
Oil temperature ....°C	80	80	80	80
Valve clearances - inlet (mm)	0: Hyd.	0: Hyd.	0: Hyd.	0: Hyd.
- exhaust (mm)	0: Hyd.	0: Hyd.	0: Hyd.	0: Hyd.
Injection order	1-2-4-5-3	1-2-4-5-3	1-2-4-5-3	1-2-4-5-3
No. 1 cylinder position	TBE	TBE	TBE	TBE
Cooling system				
Thermostat opening temperature ....°C	87	87	87	87
Radiator cap pressure ....bar	1.2 to 1.5	1.2 to 1.5	1.2 to 1.5	1.2 to 1.5
Fuel system				
Idle speed ....rpm	740 to 800	755 to 830	730 to 830	730 to 830
Maximum (no load) speed ....rpm	4800	4450 to 4650	4650 to 4850	4400 to 4800
Smoke test/opacity ....M⁻¹ %	2.0	2.0	2.0	2.0
Static timing method...............................	Refer to wsm	Refer to wsm	–	–
Timing dimension............................mm	–	–	–	–
Crankshaft position ....mm [°]	–	–	–	–
Turbo type / ref / pressure ....bar	–	0.6 to 0.8 bar @ 3000 rpm	–	–
Injection pump make	Bosch	Bosch	–	–
Injection pump part no............................	VE	VE	–	–
Injector Make / type	Bosch	Bosch	–	–
Injector part no.....................................	0432 217 231	–	–	–
Injection type.......................................	Direct VE	Direct VE	Direct	Direct
Injection opening pressure, New [used]...bar	190 to 200 [≥170]	190 to 200 [≥170]	190 to 200 [≥170]	190 to 200 [≥170]
Glow plugs				
Maker	Bosch	Bosch	–	–
Type	0250 202 022	0250 202 022	–	–
Nominal rating.....................................V/A	11 / 12	11 / 12	–	–
Brakes				
minimum friction material thickness				
Front.......................................mm	2.0	3.0	3.0	3.0
Rear..mm	2.0	3.0	3.0	3.0
Tyres - Saloon / Hatch......................Size	–	–	–	–
- Estate / Van....................Size	195/70x15: 225/70x15	195/70x15	195/70x15	195/70x15
Pressure - front / rear - Saloon / Hatch ...bar	–	–	–	–
- Estate / Van .....bar	Refer to vehicle	Refer to vehicle	Refer to vehicle	Refer to vehicle
Front suspension / wheel alignment				
Toe-in (+) / Toe-out (–)....................mm [°]	0 ± 1.0	4.0 ± 2.0	–	–
Camber...	0° ± 45'	0°40' ± 20'	–	–
Castor...	0° ± 30'	2°30' ± 30'	–	–
King pin inclination.................................	–	–	–	–
Rear suspension / wheel alignment				
Toe-in (+) / Toe-out (–)....................mm [°]	[0° ± 15']	–	–	–
Camber...	20' -40'	–	–	–

VOLKSWAGEN

	LT 2.5 D 1996 to 1999	LT 2.5 TDi 1997 to 1999	LT 2.5 TDi 1999 to 2000	LT 2.5 TDi 1999 to 2000
Torque wrench settings				
Cylinder head - stage 1 ....Nm	40 N	40 N	40 N	40 N
- stage 2 ....Nm	60	60	60	60
- stage 3 ....Nm	+ 90°	+ 90°	+ 90°	+ 90°
- stage 4 ....Nm	+ 90°	+ 90°	+ 90°	+ 90°
- stage 5 ....Nm	–	–	–	–
- stage 6 ....Nm	–		–	–
Big-end bearings.............................Nm	30 + 90° N	30 + 90° N	30 + 90° N	30 + 90° N
Main bearings................................Nm	65	65 N	65 N	65 N
Crankshaft pulley bolt ....Nm	160 + 180° N	160 + 180° N	–	–
Camshaft pulley bolt ....Nm	85 M10.9: 100	85 M10.9: 100	–	–
Flywheel [driveplate] bolt...................Nm	60 + 90° N	60 + 90° N	60 + 90° N	60 + 90° N
Front hubs ....Nm	WSM	–	–	–
Rear hubs ....Nm	WSM	325	325	325
Wheel nuts / bolts ....Nm	180	180	180	180
Glow plugs ....Nm	25	15	–	–
Clutch pressure plate bolts ....Nm	25	25	25	25
Injection pump sprocket....................Nm	90	90	–	–
Injectors......................................Nm	30	30	–	–
Injection pump mounting bolts ....Nm	45	45	–	–
Injector pipe unions...........................Nm	25	25	–	–
Capacities				
Engine oil & filter..............................litres	7.8	7.8	7.8	7.8
Gearbox..litres	1.6	2.2	2.2	2.2
Automatic transmission ....litres	–	–	–	–
Final drive ....litres	1.5	1.5	1.5	1.5
Cooling system.................................litres	11.5	11.5	11.5	11.5
Fuel tank......................................litres	76	76	76	76

Notes

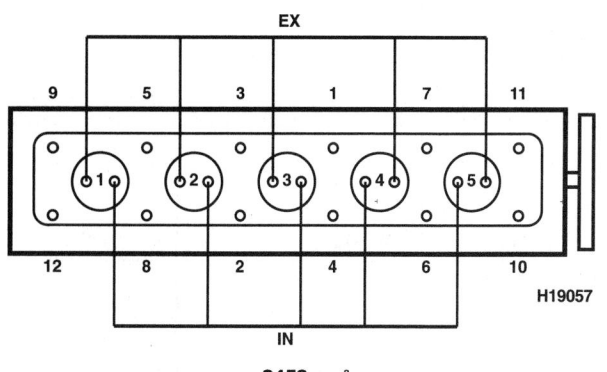

EX

2459 cm³

– Not applicable, or information not
available

VOLKSWAGEN

Haynes THE BOOK ®

	LT 2.8 TDi 1997 to 2000			
Engine				
Engine type/code......................................	AGK SOHC 12V Turbo 92kW			
Capacity (cm³) / cylinders.........................	2798 / 4			
Compression ration / pressure ....bar	19.1 / ≥23.0			
Torque output ....Nm	280			
Oil pressure ....idle [running] bar	[4.0 @ 3800]			
Oil temperature ....°C	80			
Valve clearances - inlet (mm)	0.20			
- exhaust (mm)	0.30			
Injection order	1-3-4-2			
No. 1 cylinder position	FE			
Cooling system				
Thermostat opening temperature ....°C	87			
Radiator cap pressure ....bar	1.2 to 1.5			
Fuel system				
Idle speed ....rpm	700 to 800			
Maximum (no load) speed ....rpm	3900			
Smoke test/opacity ....M⁻¹ %	2.0			
Static timing method...............................	Dial gauge			
Timing dimension...............................mm	1.48			
Crankshaft position ....mm [°]	0 TDC			
Turbo type / ref / pressure ....bar	_			
Injection pump make	Bosch			
Injection pump part no.............................	_			
Injector Make / type	_			
Injector part no..	_			
Injection type...	Direct			
Injection opening pressure, New [used]...bar	220			
Glow plugs				
Maker	Beru			
Type	GN855			
Nominal rating......................................V/A	11.5 /			
Brakes				
minimum friction material thickness				
Front.............................mm	3.0			
Rear...............................mm	3.0			
Tyres - Saloon / Hatch.....................Size	_			
- Estate / Van............................Size	195/70x15			
Pressure - front / rear - Saloon / Hatch...bar	_			
- Estate / Van ......bar	Refer to vehicle			
Front suspension / wheel alignment				
Toe-in (+) / Toe-out (–).....................mm [°]	4.0 ± 2.0			
Camber	0°40' ± 20'			
Castor	2°30' ± 40'			
King pin inclination..................................	_			
Rear suspension / wheel alignment				
Toe-in (+) / Toe-out (–).....................mm [°]	_			
Camber	_			

VOLKSWAGEN

	LT 2.8 TDi 1997 to 2000			
Torque wrench settings				
Cylinder head - stage 1 ....Nm	30 N			
- stage 2 ....Nm	60			
- stage 3 ....Nm	100			
- stage 4 ....Nm	+ 90°			
- stage 5 ....Nm	+ 90°			
- stage 6 ....Nm	_			
Big-end bearings.................................Nm	20 + 90° N			
Main bearings.....................................Nm	25 + 120°			
Crankshaft pulley bolt ....Nm	100			
Camshaft pulley bolt ....Nm	180			
Flywheel [driveplate] bolt....................Nm	125			
Front hubs ....Nm	WSM			
Rear hubs ....Nm	325			
Wheel nuts / bolts ....Nm	180			
Glow plugs ....Nm	15			
Clutch pressure plate bolts ....Nm	25			
Injection pump sprocket......................Nm	90			
Injectors...Nm	40			
Injection pump mounting bolts ....Nm	30			
Injector pipe unions............................Nm	25			
Capacities				
Engine oil & filter.................................litres	7.0			
Gearbox..litres	2.2			
Automatic transmission ....litres	_			
Final drive ....litres	1.5			
Cooling system...................................litres	11.5			
Fuel tank...litres	76			

Notes

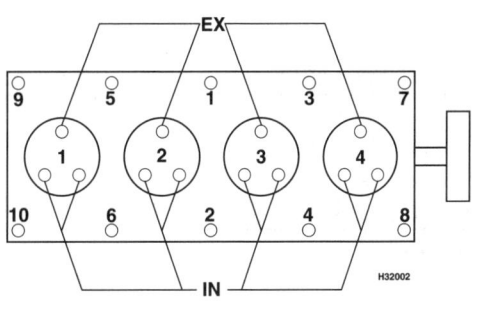

2798 cm³

– Not applicable, or information not
  available

VOLVO

	440 / 460 1.9 TD 1994 to 1997	S40 & V40 1.9 TD 1996 to 2000	850 2.5 TD 1995 to 1997	S70 & V70 2.5 TD 1997 to 2000
Engine				
Engine type/code............................	D19T SOHC Turbo 68kW	D4192T SOHC Turbo EGR 66kW	D5252T SOHC 10V 103kW	D5252T SOHC 10V 103kW
Capacity (cm^3) / cylinders.........................	1870 / 4	1870 / 4	2460 / 5	2460 / 5
Compression ration / pressure ....bar	20.5 / 20.0	20.5 / ≥16.0	20.5 / 25 to 30	20.5 / 25 to 30
Torque output ....Nm	180	176	290	290
Oil pressure ....idle [running] bar	[3.5 @ 3000]	2.0 [3.5 @ 3000]	1.0 [3.5 to 7.0]	1.0 [3.5 to 7.0]
Oil temperature ....°C	80	80	80	80
Valve clearances - inlet (mm)	0.15 to 0.25	0.15 to 0.25	0: Hyd.	0: Hyd.
- exhaust (mm)	0.35 to 0.45	0.35 to 0.45	0: Hyd.	0: Hyd.
Injection order...................................	1-3-4-2	1-3-4-2	1-2-4-5-3	1-2-4-5-3
No. 1 cylinder position	_	FE	TBE	TBE
Cooling system				
Thermostat opening temperature ....°C	92	89	87	87
Radiator cap pressure ....bar	1.5	1.2	1.5	1.5
Fuel system				
Idle speed ....rpm	850 ± 25	825 ± 25	840 ± 50	840 ± 50
Maximum (no load) speed ....rpm	4800	4450	_	_
Smoke test/opacity ....M^{-1} %	2.5	2.0	2.5	2.5
Static timing method................................	Dial gauge	Dial gauge	Refer to wsm	Refer to wsm
Timing dimension................................mm	Dimension on pump	Dimension on pump	_	_
Crankshaft position ....mm [°]	0 TDC	0 TDC	_	_
Turbo type / ref / pressure ....bar	0.7 to 0.85 bar @ 2500 rpm	0.8 to 0.95 bar @ 4250 rpm	_	_
Injection pump make	CAV	Lucas	_	_
Injection pump part no.............................	DPC R8443B 721B	DPI-N ECM	_	_
Injector Make / type	Roto Diesel	CAV	_	_
Injector part no.....................................	END4SDC 6878C	_	_	_
Injection type...	Indirect DPC	Lucas DPI-N ECM	MSA 15.7 Direct Inj	MSA15.7 Direct inj
Injection opening pressure, New [used]...bar	130 to 135 [125]	130 to 135 [125]	190	190
Glow plugs				
Maker	Beru	Champion	Bosch	Bosch
Type	GV844	CH179	0250 202 009	0250 202 009
Nominal rating.....................................V/A	12 / 14	11 /	_	_
Brakes				
minimum friction material thickness				
Front..............................mm	2.0	2.0	3.0	3.0
Rear..............................mm	2.0	2.0	2.0	2.0
Tyres - Saloon / Hatch....................Size	175/65x14: 185/65x15	185/65x14: 195/55x15	195/60x15: 205/55x15	195/60x15: 205/55x15
- Estate / Van......................Size	_	185/65x14: 195/55x15	195/60x15: 205/55x15	195/60x15: 205/55x15
Pressure - front / rear - Saloon / Hatch ...bar	2.1 / 1.9	2.2 / 2.0	2.2 / 2.0	2.2 / 2.0
- Estate / Van ......bar	_	2.2 / 2.0	2.2 / 2.1	2.2 / 2.1
Front suspension / wheel alignment				
Toe-in (+) / Toe-out (–)....................mm [°]	0.0 to 1.0	[0°3' ± 6']	[20' ± 6']	[20' ± 6']
Camber	-0°24' ± 1°	0° ± 1°	0° ± 1°	0° ± 1°
Castor	3°15' ± 1°	2°12' ± 1°	3°20' ± 1°	3°20' ± 1°
King pin inclination...............................	13°14' ± 1°	12°41' ± 1°	_	_
Rear suspension / wheel alignment				
Toe-in (+) / Toe-out (–)....................mm [°]	1.0 to 4.0	[0°6' +9' -3']	[4' ± 10']	[4' ± 10']
Camber	_	-0°40' ± 30'	-1° ± 30'	-1° ± 30'

VOLVO

Haynes THE BOOK ®

	440 / 460 1.9 TD 1994 to 1997	S40 & V40 1.9 TD 1996 to 2000	850 2.5 TD 1995 to 1997	S70 & V70 2.5 TD 1997 to 2000
Torque wrench settings				
Cylinder head - stage 1 ....Nm	30[1]	30[1]	35	35
- stage 2 ....Nm	+ 50°	+ 50°	60	60
- stage 3 ....Nm	Wait 3 mins, slacken	Wait 3 mins, slacken	+ 90°	+ 90°
- stage 4 ....Nm	25	25	+ 90°	+ 90°
- stage 5 ....Nm	+ 213°, warm up	+ 213°, warm up	–	–
- stage 6 ....Nm	+ 120°	+ 120°	–	–
Big-end bearings.............................Nm	45 N	45 N	30 ± 3 + 90° N	30 ± 3 + 90° N
Main bearings.................................Nm	65	65	65 ± 6.5	65 ± 6.5
Crankshaft pulley bolt ....Nm	95	20 + 115°	160 + 180° N	160 + 180° N
Camshaft pulley bolt ....Nm	50	50	85[1] Rear: 160	85[1]
Flywheel [driveplate] bolt.................Nm	53 N	53 N	60 + 90° N	60 + 90° N
Front hubs ....Nm	230	240	120 + 60°	120 + 60°
Rear hubs ....Nm	160	175	120 + 30°	120 + 30°
Wheel nuts / bolts ....Nm	110	110	110	110
Glow plugs ....Nm	23	20	15	15
Clutch pressure plate bolts ....Nm	22	21	25	25
Injection pump sprocket......................Nm	70	WSM	45	45
Injectors..Nm	70	70	22	22
Injection pump mounting bolts ....Nm	20	20	25	25
Injector pipe unions...........................Nm	23	22.5	25	25
Capacities				
Engine oil & filter...............................litres	5.0	5.3	6.0	6.0
Gearbox...litres	3.4	3.4	2.1	2.1
Automatic transmission ....litres	–	–	3.0	3.0
Final drive ....litres	WT	WT	WT	WT
Cooling system..................................litres	7.0	5.0	12.0	7.2
Fuel tank..litres	60	60	73	73

Notes

440 / 460 1.9 TD 1994 to 1997
[1]Bolt length: ≤120.5 mm
S40 & V40 1.9 TD 1996 to 2000
[1]Bolt length below head: ≤120.5 mm No re-tighten if steel gasket used
[2]Steel flange: 50 Nm   Aluminium flange: 15 + 60°
850 2.5 TD 1995 to 1997
[1]If marked 10.9m: 100 Nm
S70 & V70 2.5 TD 1997 to 2000
[1]If marked 10.9M: 100 Nm

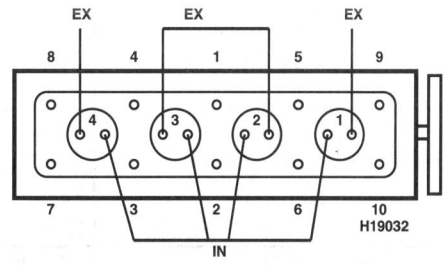

1870 cm³

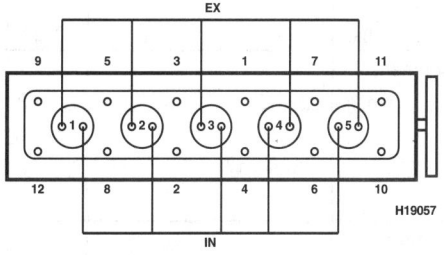

2460 cm³

– Not applicable, or information not available

VOLVO

	940 2.4 TD 1990 to 1992	940 2.4 TD 1992 to 1996		
Engine				
Engine type/code.....................................	D24T SOHC Turbo 80kW	D24TIC SOHC Turbo 90kW		
Capacity (cm³) / cylinders........................	2383 / 6	2383 / 6		
Compression ration / pressure ....bar	23.0 / 24 to 32	23.0 / ≥21		
Torque output ....Nm	0	235		
Oil pressure ....idle [running] bar	[2.0 @ 2000]	[2.0 @ 2000]		
Oil temperature ....°C	80	80		
Valve clearances - inlet (mm)	0.15 to 0.25	0.15 to 0.25		
- exhaust (mm)	0.35 to 0.45	0.35 to 0.45		
Injection order	1-5-3-6-2-4	1-5-3-6-2-4		
No. 1 cylinder position	TBE	TBE		
Cooling system				
Thermostat opening temperature ....°C	87	87		
Radiator cap pressure ....bar	1.5	1.5		
Fuel system				
Idle speed ....rpm	830 ± 40	830		
Maximum (no load) speed ....rpm	5400 ± 50	5400		
Smoke test/opacity ....M⁻¹ %	3.0	2.0		
Static timing method...............................	Plunger travel	Dial gauge		
Timing dimension............................mm	0.9 ± 0.02	0.9 ± 0.02		
Crankshaft position ....mm [°]	[0] TDC	0 TDC		
Turbo type / ref / pressure ....bar	0.7 bar @ 3000 rpm	0.9 to 1.0 bar @ 2400 rpm		
Injection pump make	Bosch	Bosch		
Injection pump part no.............................	VE L116-6	VE6/10 F2400		
Injector Make / type................................	Bosch	Bosch		
Injector part no.......................................	0432 217 078	0432 217 078		
Injection type..	Indirect VE	Indirect VE		
Injection opening pressure, New [used]...bar	166 [145]	163 [145]		
Glow plugs				
Maker	Bosch/Champion	Bosch/Champion		
Type	0250 201 032 / CH68	0250 201 032 / CH68		
Nominal rating....................................V/A	_	11 / 12		
Brakes				
minimum friction material thickness				
Front......................................mm	3.0	3.0		
Rear.......................................mm	2.0	2.0		
Tyres - Saloon / Hatch....................Size	185/70x14	185/65x15		
- Estate / Van...........................Size	185/70x14	185/65x15		
Pressure - front / rear - Saloon / Hatch...bar	1.9 / 1.9	1.9 / 2.1		
- Estate / Van ......bar	1.9 / 2.1	1.9 / 2.2		
Front suspension / wheel alignment				
Toe-in (+) / Toe-out (–)....................mm [°]	2.0 ± 0.5	2.1 ± 0.9		
Camber	-12' to 48'	0°18' ± 30'		
Castor	4°30' to 5°30'	5°0' ± 30'		
King pin inclination..................................	_	_		
Rear suspension / wheel alignment				
Toe-in (+) / Toe-out (–)....................mm [°]	_	_		
Camber	_	_		

VOLVO

	940 2.4 TD 1990 to 1992	940 2.4 TD 1992 to 1996		
Torque wrench settings				
Cylinder head - stage 1 ....Nm	40 N	40 N		
- stage 2 ....Nm	60	60		
- stage 3 ....Nm	75 + 180°	75, + 180°		
- stage 4 ....Nm	Warm engine + 90°	Warm up, + 90°		
- stage 5 ....Nm	After 1000km,	After 1000km		
- stage 6 ....Nm	Allow to cool, + 90°	Allow to cool, + 90°		
Big-end bearings.............................Nm	30 + 180° N	30 + 180° N		
Main bearings.................................Nm	65	65		
Crankshaft pulley bolt ....Nm	350	350		
Camshaft pulley bolt ....Nm	45 Pump end: 100	WSM		
Flywheel [driveplate] bolt...................Nm	75 LkC	75		
Front hubs ....Nm	100 + 45°	100 + 45°		
Rear hubs.....................................Nm	WSM	WSM		
Wheel nuts / bolts ....Nm	85	85		
Glow plugs ....Nm	22	22		
Clutch pressure plate bolts ....Nm	25	25		
Injection pump sprocket.....................Nm	45	45		
Injectors.......................................Nm	70	70		
Injection pump mounting bolts ....Nm	25	–		
Injector pipe unions...........................Nm	25	25		
Capacities				
Engine oil & filter............................litres	6.0 Cooler: 6.6	6.0		
Gearbox.......................................litres	2.3	2.3		
Automatic transmission ....litres	2.0	7.5		
Final drive ....litres	1.6	1.75		
Cooling system...............................litres	9.5	11.0		
Fuel tank ....litres	60	75		

Notes

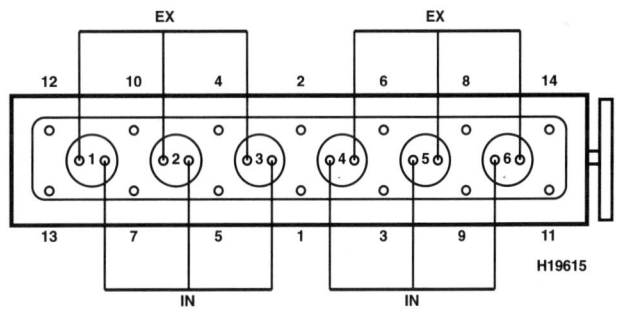

2383 cm³

– Not applicable, or information not
available

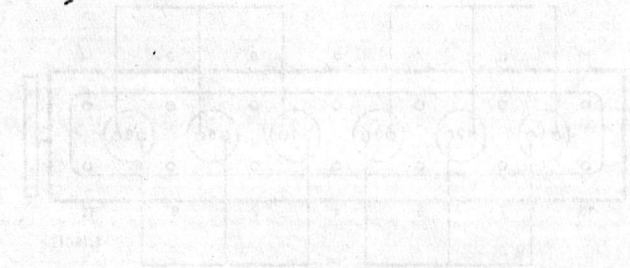

23933822R00291

Printed in Great Britain
by Amazon

The Ides of Matt 2015

a short story collection

by

M. L. Buchman

Discover more by this author at: www.mlbuchman.com
Cover images:
Knife and Apple © Pvf101
Abstract female fire © Clearviewstock
Ice Block © BuildArk
Army Helicotper Resuce © Troy Darr, U.S. Army
Mount Rainier Cloud Cover © Jordan Goss
U.S. Army UH-60 Black Hawk helicopter © Michael Kaplan
Thai jungle I © Mazlov
Image 2792 © g0d4ather
Granite Mountain - Fire Lookout © laffertyryan
Backpacking on St. Joe's © Jason Priem
Night Lighthouse Beam © Joseppi
Space Attack © Philcold
White Point Lookout © Oregon Dept of Forestry
Couple with Bicycles Watching Sunset © Maryia Bahutskaya
Labyrinth Couple 5 © George Williams
USS New York transits the Suez Canal © U.S. Navy photo
MH-6 Little Bird © San Andreas
Bell 206 Jet Ranger OE-BXO © Zeitblick
Love Story Photo © Evdoha
LRR in Feb © Bill Burleigh
Man Fashion Walk © Curaphotography
Forest Fire © Ervins Strauhmanis
Caión 2 © vrelmunde

Buchman Bookworks

Other works by M. L. Buchman:

Contents

Introduction

Welcome to my second annual Ides of Matt collection.

I lifted The Ides of Matt from something my best friend in high school used to say. It started after we saw the Shakespeare play Julius Caesar *together. The soothsayer says to Caesar—shortly before he goes to the Senate and is knifed in the back by his best buddy Brutus—"Beware the ides of March." (March's ides land on the 15th.)*

My birthday lands on March 14th so he (my best friend, not the Roman soothsayer) would wander the halls of high school for days every March intoning: "Beware the ides of March minus one." For some reason I always thought of him being the voice of aging rather than the voice of someone about to stab me in the back. He didn't—which I appreciate, Jeff—but years

later that did set the idea for my monthly short story collection.

This is my second collection (I might have mentioned that) and I find that to be a very exciting achievement. It wasn't all that long ago that I was convinced I couldn't write a short story if my life depended on it.

I've now sold thousands of copies of my stories (occasionally thousands of a single story) and am so grateful for the resulting fan base.

The first thing that comes to mind on that point?

To say: Thank You!

My fans' support has allowed me to try out ideas. Sometimes these stories are to expand my thinking about one of my worlds or to check in on a set of characters I was particularly missing that day and wanted to spend some time with. Other stories are about characters who insisted they had their own tale to tell. And a special few are tales that will be giving birth to whole new series. I'll talk more about all of that below.

For now, let's plunge in!

Iced Chef!

Iced Chef! *is a member of my Dead Chef thriller series.*

Thrillers are a particularly interesting challenge to write, some say the very hardest and I'd agree. I've written science fiction, fantasy, and romance; even dabbled in mystery. Thrillers are just plain tough.

The heart of a thriller is pacing. But it also requires high stakes and a satisfying conclusion after a wild ride which never takes a breath. A fellow thriller writer upped the challenge, "Do all that, but do it in a short story. It's impossible."

I've tried twice, this story and April's Gas Grilled Chef!

I tried…and I failed.

The story succeeded.

I think that both stories are fun, fast-paced, and are true to the voice of the Dead Chef series in general. But I don't think that they're quite thrillers. The hardest part was building truly high stakes. Thrillers are about bombs, toppling governments, saving a city, saving the world. They're also about the chaotic turns and twists that escalate until there is no way a hero should be able to survive.

Trying to fit that, believably, into a short story... Wow! but that's a seriously fun challenge.

So here is the first of my fun, really fast-paced suspense stories...murder mysteries...kinda thriller but kinda not...

Okay, here's a bit of fun on the ice.

M. L. BUCHMAN

ICED CHEF!

A DEAD CHEF
THRILLER STORY

1

Rikka Albert shouldered the eighty thousand dollar Panasonic Varicam video camera, which then tried to freeze to her cheek. It was stupid. Her beautiful camera was designed to turn the everyday world into television art. Where was "television art" in the middle of a frozen Minnesota lake? Featureless white to the west. And also north, east, and south just to spite her.

She'd come out a day early to do the pre-shoot planning and all of the B-roll shots before tomorrow's Annual Lake Winnibigoshish Northland Chowder-Off cooking competition—ice-caught fish only allowed.

It was another episode of *Kate's Kitchen Raids* and Rikka was glad to be here. She really was. And if she kept telling herself that often enough she might actually believe it someday…like in spring maybe.

"You people really do this for fun?"

Senator Hamilton Waring, who had nothing to do with blenders but had a lot to do with a massive chunk of iron ore money and the US Senate Armed Services Committee, looked

completely at home in this sub-Arctic world. Tall, blond, and broad-shouldered even before he'd hauled on his parka. He looked down at her five-feet of Asian sass as if she was an alien bug.

"I'd say that we Minnesotans are a hardier stock than the rest of the country. We've had to be."

Rikka resisted the urge to point out that driving out onto the ice in a tricked out crystal red Cadillac Escalade SUV didn't exactly constitute hardship. The thing looked like a blot of blood in the middle of the winter wonderland. *Try trudging through yet another New York City slush storm and see how you do, blender man.* But she kept that thought to herself and looked for something, anything to focus on.

They were well out on Lake Winnibigoshish. Nearer the shore were numerous fishing shacks set in neat rows. It looked like any small shanty town with street-wide lanes on which SUVs and snowmobiles were parked in equal numbers. A line of dark green pine trees marked the shore, which was close by the lines of shacks.

But Hamilton Waring had not stopped there. He'd driven almost a mile out onto the ice; ice that shot unnerving snaps and crackles at her like a gun battle in the South Bronx. They'd proceeded north across the sixty-nine thousand acre lake farther and farther from the shore that she kept eyeing longingly in the passenger side rearview mirror. The tiny words written there—*Objects may be closer than they appear*—made the disappearing shoreline even more achingly distant.

As they'd driven out, the number of shacks diminished, but their designs became rapidly more elaborate. They passed a lone cute cottage and beyond that, off by themselves, were three ice "shacks" far grander than anything around them. They were arranged in a triangle with perhaps a hundred feet between them.

He circled most of the way around one of the shacks before parking in front of a mansion on ice. Like Elsa would have built if all of her capital hadn't been *Frozen.*

It was one of those "tiny houses" that were all the rage, looking like a big house that had been shrunken by being run through a couple too many dry cycles on high heat. Waring's mini-mansion was fake Edwardian, with imitation stone siding, numerous bay windows, and a burnished copper roof. The second-story glassed-in cupola surely offered a commanding view over the vasty nothingness. The thing reeked of money, and of no taste whatsoever.

They clambered out of the Cadillac blood blot, but he didn't head for the door. It didn't take Rikka's trained eye to see that he'd been careful in positioning both his SUV and himself. The man knew his camera angles.

What the hell, time to stroke the man's ego.

Rikka stepped back, ignoring the gunshot ripple of "ice just adjusting itself a bit" and flipped the camera to record. She started with an opening set-up shot of the distant shacks clustered in the distance, panned across the sparser, and clearly far richer, neighborhood out here on the offshore ice. Slowed as she passed over on of the two equally ostentatious ice "shacks" next to Waring's—one that boasted Bavarian white walls, dark wood angular trim, and massive porch beams—and finally to the man, his SUV, and his own "humble" shack.

He wore a deprecating smile, that was probably meant to express an approachable billionaire who welcomed you to his playpen, but instead said, "I could be your next President and there ain't shit you can do about it." Man seriously needed an image consultant to kick his ass around the ice a few hundred times. But the election was still a ways off, he had time.

Bet this clip gets cut. Unless he lost tomorrow's Chowder-Off, then maybe she could slip it in as the "before the fall" shot.

In addition to being handsome, rich, powerful, and totally full of himself, he was also the reigning champion of the Lake Winnibigoshish Annual Northland Chowder-Off for three years running.

"So, Senator, how are you feeling about your chances in tomorrow's competition?" Give him a leading question and

maybe he'd drop out of smarmy mode and give her a decent image to use.

"Well, you betcha there are some fine cooks out here on the ice. Congressman Maxwell is good," he pointed off-screen toward the Bavarian ice house that showed even less taste than blender-man's if possible. "He uses too much pepper, but don't let on."

"Scout's honor," Rikka prompted him knowing she could edit out her comment later.

"Over there…" he pointed the other way.

Rikka would have to remember and shoot some footage of the hideous affair to the east. It was white on white on ice. Fake Corinthian columns under a carved portico complete with naked gods and goddesses, all painted in faux marble.

"That's Governor Llewellyn's place. He's placed second to my chowder twice now, so I've got to watch him close."

"Not *enough* pepper?" Rikka guessed.

"Manhattan style," Hamilton said grimly.

"Eww!" Rikka would have to agree with him there. She was originally from Boston and still cringed every time she saw chowder with a tomato base. It was like pizza with chow mein noodles; she liked both, but together?

"My pop's recipe, with a few secret changes of my own, has fended them off so far, and it will again."

"So, we have the top three chowder chefs in the Northland who are also three of the top-seeded men for the next Presidential election, all on the same piece of ice. Quite a coincidence." And if blender-man was elected she was moving to another country.

"Not at all, little lady. Not at all."

She'd *little lady* him right in the shins if she wasn't filming this.

"Minnesota breeds more than hardy stock. It breeds great political leaders. But neither one will beat me on the political or the foodie field."

Rikka decided not to point out that "foodie" was a totally passé term. The final tombstone on the word's grave had been the McDonald's ad campaign, "Foodies Welcome!"

"Any non-political combatants?" she asked instead.

"Combatants? We're all friends out here braving the deep ice. That's what we call it out here, farther from shore. The ice is actually thinner, but the water is deeper and the catch better. That's one of the reasons we're always the top three in the Chowder-Off."

Thinner ice? Rikka suddenly wished she was in Florida, which was really saying something—she hated Florida.

The noble warrior led her forward to the front door of his humble little ice shack. He unlocked the door and held it open for her to enter first. The blissfully warm air washed over her as she entered.

Rikka let the camera be the first-time visitor, first taking in the oak paneling, green-shaded lamps, and deep red leather chairs—all more fit for a brandy-and-cigar political meeting than an ice fishing hovel. The floor was mostly covered with a lush oriental rug that Rikka could see was exceptional work.

Close beside each chair an eighteen-inch hole had been punched through the carpet and was covered with a metal and rubber plate. Each chair had its own private ice-fishing hole. She resisted the urge to smack the man for committing crime against carpet as he pointed to various features of his little "home away from home."

There were two rooms on this level. At this end of the great room was a fully equipped kitchen and a spiral staircase leading upward to the glass cupola. The middle of the room was the sitting area, which included the finest whiskey bar she'd seen since the G-8 meeting at Inverlochy Castle Hotel in Scotland. The far wall was mostly covered by a massive television screen that stood as tall as she did—though she wasn't complaining because at the moment it was playing a recording of a massive log fire which made her feel warmer, at least psychologically.

Through the open door to one side was a master bedroom, again with fishing-hole hatches cut through the fine Persian throw rugs.

The hatch cover to the other side of the bed had been raised exposing the hole in the ice and the dark water lapping only inches below.

Rikka would have felt unnerved all over again that she was standing only a mere foot above forty feet of icy depths, but there was a distraction.

A woman lay on the carpet to one side of the fishing hole. Her parka at her feet. Her skin-tight slacks and form-hugging silk turtleneck advertised that she was a particularly well-endowed one. She might have been relaxing and watching for a fish to tug on her line.

Except for one problem.

Rather than greeting Rikka with a smile of surprise, she had no expression whatsoever.

The voluptuous form was lacking one key item—its head.

A few long blond tresses still trailed across the carpet, but the missing body part bobbed in the dark water of the fishing hole.

Rikka recognized Lulu, the wife of Governor Llewellyn, from the prep file despite her being in pieces. Maybe thoughtful, Rikka considered. Perhaps dismayed. It was hard to read the expression frozen on her face.

"Well, that's different," was all Rikka could think to say before calling for the Senator to see to his unexpected guest.

2

I've really got to talk to Kate about sending me out on these bizarre assignments," Rikka spoke into her phone as she waited for her pizza at Rasley's Blueberry Bowl—"Voted number one in the Northland, honey!" the overly blond waitress in the blueberry-shaded uniform had insisted on telling her. Apparently this was the hot spot of Deer River, Minnesota.

Rikka had considered several sharp ripostes, including a quick left jab, but finally reconsidered. She'd even resisted the verbal right cross of telling the waitress she was from New York where—even if pizza hadn't originated—it was where it had been perfected. Atypically, she'd kept her mouth shut. *Which didn't sound like her at all.*

She put a hand to her forehead but didn't feel a fever.

Rasley's wasn't decorated in blue, rather more of a dingy brown. It echoed with the sharp crack of tumbling bowling pins from the ten lanes heating up behind her. The men of Deer River might be out on the ice, but their wives, with one notable exception, were here and they were in league.

At the other end of the phone call, Sam Fierro didn't respond. He knew that sometimes Rikka just had to process out loud. She'd caught him at his butcher shop, Fierro Meats, in Brooklyn.

Kate Stark was a problem though.

She'd be Rikka's closest friend, if Rikka believed in friends. The woman was everything Rikka wasn't and she tried not to let it piss her off too often, but at the moment it most certainly did. Kate was tall with brunette hair brushing her shoulders, had a figure to make men weep, and was the billionaire owner of Cooks Network. She also could out-cook most of the people on her shows, which Rikka totally respected her for, and had a wardrobe to die for…none of which would fit Rikka's tiny frame. Yet another reason for a grudge.

"The worst thing Kate is doing to me right now is not being here until tomorrow, and I'm starting to take it personally." Since when did she care about getting someone else's help? *Which didn't sound like her either.* She checked the lymph nodes under her jaw but they weren't swollen. Yet oddly, she was coming to rely on Kate, her twin brother Paul, and Sam…especially Sam.

Sam, being a wise man, didn't say anything.

"Well, the situation here is even dumber than I first thought now that Minnesota's first lady has turned up a foot shorter than usual." Rikka really shouldn't be whining, especially not to someone in a whole different time zone, but he was the only one handy; and the one she'd found she could turn to time and again.

Besides, Kate would be in the middle of the Eric Ripert interview right now, a total fan moment for Kate, and no way was Rikka interrupting that. Gods! And now she was being considerate. Shoot her now and put her out of her misery. Tomorrow morning would be soon enough when Kate was due in the Gopher State as a principal judge for the Northland Chowder-Off, so Rikka was on her own until then.

"The Governor came right on out," she told Sam, then sipped her glass of Rasley's house red. Major mistake! She edged it as far away across the table as she could. Then she thought better

of allowing it to remain so close by and carried the glass over to a table that hadn't been cleared yet. She covered it with a napkin so that it couldn't see her anymore before retreating to the safety of her own table and continuing.

Thankfully Sam was one of those great guys who was comfortable with silence in a conversation. Or had grown used to Rikka's peripatetic conversational style.

"Governor Llewellyn beat the coroner-police chief—they're the same person here which tells you how far off the map this place really is—to Waring's Edwardian mini-palace by ramming his car into the Chief's bumper and almost skidding her into the front door of his own Greek Palace fishing shack—I'll send you pictures. First they were all being too damn polite about, 'I just don't understand how she got here, Lew,' and 'This isn't going to look good to the voters, Ham,' that they never thought about what was right in the room with them."

Sam allowed her to let the suspense build.

"I had my camera running the whole time," Rikka patted it on the recording head, where it perched on the table beside her, because it had been such a good girl. She was sitting quite alone; Rasley's restaurant had emptied when the bowling league games had begun.

"If Kate wants to break into the news business, I have every moment in hi-def video. They're keeping it hush-hush, so we've got the scoop if we want it. I know. I'm jumping the gun, but it's good, Sam. Governor Llewellyn, Police Chief Patrice Smith, Senator Hamilton Waring Chairman of the Senate Armed Services Committee, and a headless woman; three of them doing a polite two-step like they just ate a whole bushel of green apples and there's only one pot to go in. First Lady Lulu Llewellyn didn't do much, part of her just lay there and the other part sort of watched them."

Overly Blond finally delivered the pepperoni, mushroom, and onion pizza that Rikka had been smelling for far too long. When she started chit-chatting, Rikka continued to Sam over

the phone, "It's not every day that a woman's body ends up in one part of your bedroom and her head in another."

Overly Blond evaporated, after turning almost as green as her outfit was blue.

Rikka wasn't about to waste pizza, good or not. She bit down and seared her mouth nicely. Decent sauce. Real cheese. Not New York, but not too shabby. Maybe Minnesota wasn't as badly off as she thought.

Sam held the line while she dragged in some cool air. Then they caught up on the miscellaneous news of the day.

Overly Blond cleared some tables, including the awful wine that had been watching her from under its napkin, before heading back over for the "Isn't our pizza wonderful?" question.

"The blood," Rikka returned to the former topic on the phone, "was pretty impressive, even if most of it went down the hole. Human body sure contains a lot, doesn't it?"

Rikka made a show of biting into her next piece of pizza as the waitress greened up again and about faced.

"Wait a sec, Sam."

Governor Llewellyn and Senator Hamilton Waring Not-the-blender-man came in, spotted her, and stalked over to her table. Hamilton dropped a coroner's report on her table. She'd seen plenty of these in a past life, back when she was a computer specialist for a Chinese money laundering operation, and spotted the relevant box immediately.

"You're going to love this," she continued to Sam. "The coroner pumped the First Lady's stomach. Her last meal was chowder, it had an exceptionally high ratio of pepper. Something Congressman Marvin Maxwell is known for. Yes," she said before Sam could ask, "the Chairman of the House Armed Services Committee." Sam would know plenty about both Maxwell and Waring as head of the two Congressional Armed Services committees. He was Marine Force Recon (retired). He'd made it clear early on in their acquaintance that there was no such thing as an ex-Marine.

"Why did you track *me* down?" she asked the two men, as if she and Sam didn't already know. She surreptitiously set her phone to speaker, knowing Sam would keep quiet while he listened.

The Governor didn't look too broken up about the unexpected murder of his wife, but whether that was reality or Minnesota stoicism, Rikka couldn't tell.

"We," Hamilton seemed to be having trouble clearing his throat, "need an impartial witness when we confront Marvin Maxwell so that—" he hesitated again and Rikka finished for him.

"—so that you have proof that there is no bias related to tomorrow's Chowder-Off."

Waring and Llewellyn nodded in unison like the *Dumb and Dumber* twins.

Sam may have snorted quietly, but Rikka couldn't tell because she was busy laughing in their faces.

They didn't take it very well.

3

Rikka and Hamilton found Congressman Maxwell's wife Marilyn in lane number four of Rasley's Blueberry Bowl well on her way to breaking two hundred. Like the dead First Lady, she was another tall and fiercely buxom Minnesotan as proven by her particularly well-tailored bowling shirt that had "Marilyn" stitched over one prominent breast and "Maxwell" over the other.

"Marvin's out on the ice. Said he had wanted a couple more perch for the Chowder-Off. Didn't even come home last night. If I find he was with Lew's wife like you always are Hamilton, he just might find himself down an ice hole."

Like the savvy politician he was, Hamilton maintained a straight face as he replied, "I can promise you there's no chance of that, Marilyn."

Right, not with Governor Llewellyn's wife being decapitated and now lying in the morgue.

Marilyn nodded, turned, and rolled her personalized, hot pink bowling ball to catch the six-ten spare, continuing her scoring streak.

They gathered up the Governor, from where he was chatting up Overly Blond, and the rest of Rikka's boxed pizza before climbing into Senator Waring's blood red SUV.

"Think we oughtta get Patrice in on this as well? Make it all legal?"

"She said she was headed back to the ice. We'll stop by and pick her up. She's one of the few women on the deep ice," Hamilton explained as they drove out into the wintry darkness.

Police Chief Patrice Smith's cabin was by far the least ostentatious shack in the deep-ice neighborhood. It was a quarter the size of the other behemoths and might have been a fairy tale cottage with its arched windows, sharply peaked roof, and a fake-brick chimney puffing out smoke from her woodstove.

She climbed aboard and they drove the last several hundred yards to Congressman Marvin Maxwell's Bavarian wonderland.

"Odd that he didn't come out to see all of the excitement earlier."

Patrice's comment had the two men shift plans mid-step, and suddenly Patrice was shuffled to the fore and left to knock.

There was no answer and the door was locked.

She fished out a key ring and the third one opened the door.

"Where did you get Marvin's key?" the Governor asked. "Marvin doesn't give anyone his key."

Patrice grimaced, "I found this key ring in your wife's pocket, Governor."

"Oh."

They all offered a Minnesota shrug, then Patrice opened the door and went in.

Rikka nosed in her camera close behind the Police Chief.

The Bavarian décor was as complete inside as it was outside. A long polished-wood bar. Shelves lined with beer steins. A half dozen beer taps—which were the only real breaks to the motif as their brands were: Budweiser, Bud Lite, Old Milwaukee, Pabst Blue Ribbon…the only concession to Germany was Michelob

Genuine Draft. At least it had a German name even if it was brewed in Columbus, Ohio.

There was one other break in the overall décor.

The headless body lying over the only open ice fishing hole.

A quick inspection revealed no sign of the missing head, but there was little doubt as to his identity. The decapitated Congressman was wearing a t-shirt which said, "Keep Calm and Draw a Pint."

4

It was a seven a.m. sunrise by the time Congressman Maxwell was all squared away and Rikka was wondering just what the purpose of having a hotel room was if she didn't get to sleep in it.

Patrice had moved the Congressman's body back at her morgue to lie beside the Governor's wife—though his head continued to remain at large. She'd done what she could with her limited facilities, like determining that Marvin had also eaten his own over-peppered chowder as a last meal. That had led to the inevitable question of what else had they shared yesterday.

The men had gone off to bed, but Rikka had accompanied Patrice throughout her investigation, including a return to both crime scenes out on the ice, though Rikka had refused to leave the heated car the second time they went out. Patrice had used the door keys to both Maxwell's and Waring's that she'd found in First Lady Llewellyn's pockets to unlock the doors.

It was sunrise by the time that the Secret Service agent finally showed up. He was followed closely by an investigator from Camp

Riley National Guard training center—clearly some poor shmuck who'd been woken from a long winter's nap after a serious battle with a bottle of vodka—selected because he served at the closest military base in all of Minnesota. A dead Governor's wife was a minor police matter. An equally dead Chairman of the House Armed Services Committee drew far more concern from Washington.

Kate wasn't due for another couple hours.

Everyone involved had gathered together back out on the chill ice as the weak morning sun tried to do something about the minus ten degree temperature, with little success. They were standing at the center of the triangle, equidistant from Senator Waring's Edwardian mini-mansion, Congressman Maxwell's Bavarian beer hall, and Governor Llewellyn's Grecian temple.

Were Rikka and her poor camera the only ones freezing to death? Some of these people hadn't even bothered to zip up their parkas.

"We know," Patrice started out, "that Congressman Marvin Maxwell and First Lady Lulu Llewellyn both ate similar chowder recipes shortly before dying."

"Before or after they fornicated like dogs in heat?" Marilyn Maxwell asked as if it was of no real surprise what her husband did.

"I can't tell in relation to when they dined, but we did find two condoms in the Congressman's trash. Used ones. And I can't tell much more without a DNA kit, but the deceased First Lady did appear to have had sex recently."

"Twice? More than he ever gave to me in the same week," Marilyn huffed out a breath that instantly fogged and then, Rikka would have to check the recording later, froze and made a miniature snowfall to the ice. It was just that cold. *And you people live like this?*

"Or my wife ever gave me," the Governor didn't look at all pleased.

Mr. Secret Service looked worried, but hadn't shifted from close by Senator Waring's side. The investigator from the National Guard merely looked hung over.

"Could we go somewhere warmer?" Rikka finally begged after all feeling had been lost below the knees.

They all looked at her in surprise.

"Of course," Governor Llewellyn was the first to recover. "We'll go to my place. At least it isn't a crime scene." His look at Senator Blender-man was archly smug. *Didn't any of these people understand the purpose of a television camera?*

At the Greek revival's door, complete with little leaded glass windows, the Governor unlocked the deadbolt, held it open, and tried to usher Senator Waring in first.

"No, Lew. It's your place, you should lead the way."

The Governor waved Rikka forward, "The photographer…"

Rikka wanted to poke him. First, she was a *videographer* and second, she had a name even if no one in town other than the police chief had yet used it.

"…hasn't seen the inside of *my* little ice shack yet."

Rikka dutifully took the lead and turned on her camera as they approached the Greek colonnade. Her legs felt like useless stumps, but at the promise of imminent warmth, they staggered her forward.

When he pulled open the door for her, Rikka closed her eyes for a moment to enjoy the waft of heat.

Then remembering what she'd found the last two places she'd been, she swung the camera about while watching carefully through the eyepiece before entering. Nope. No corpses in the Spartan interior. White walls, white marble floors, a glass bar that supported only gin, white rum, and vodka bottles. Even a crystal chandelier for light. Close beside the fishing holes through the floor—which had disconcerting Plexiglas covers so you could see the dark waters below—there were uncomfortable-looking lounge chairs that might have been designed to look Grecian.

No bodies, though one of the holes was open.

Someone tugged on the back of her coat just as she took a step forward. It made her hesitate a moment with only the camera lens across the threshold.

Rikka heard a high-pitched zipping sound, like a knife being slipped over a sharpening steel. Then the camera was almost jerked from her hands.

Instead, she held onto it and was stumbling ahead toward the open ice hole, relentlessly dragged by a thin piece of wire that had been looped loosely on the inside edge of the door frame. It was now tightly wrapped around her camera lens. It looked like the titanium multi-threaded fishing lead that Senator Hamilton Waring had been so proud of: "No perch will bite through that and steal my hook."

Five feet from the hole, she had an idea.

Three feet, she grabbed onto the lens.

On her knees—twelve inches from her camera being dragged down the fishing hole—she twisted the lens free from the camera body's mount.

With a splash of freezing water in her face, seven thousand dollars of lens disappeared down into the murky depths.

5

*P*andemonium *broke out around* her, at least Rikka assumed this was what a Minnesotan version would look like.

"Someone tried to kill me," the Governor sounded deeply shocked.

Everyone looked about for the criminal, some of them even wandering off to look behind the uncomfortable chairs.

Police Chief Patrice Smith kept her head. She came over, helped Rikka back to her feet, and handed her a white bar towel to wipe her face.

"Are you okay?"

Now that she was dry and had ascertained that she wasn't being dragged along in darkness beneath the ice...

"As long as I'm warm, I'm fine."

Patrice patted her shoulder and returned her attention to the others who had continued their search of the room yet only discovered the bar. Most of them now had a drink despite the early hour. The National Guardsman was the last to act, decided

on the hair of the dog, and knocked back a double shot of vodka with a grimace.

Rikka shut down the camera and dug a lens cap out of her pocket to protect the camera until she could fetch another lens from her kit back at the hotel she hadn't slept in.

"Well, at least we now know how the murders were committed," Patrice observed calmly.

They all stared at her in astonishment, the Governor crossing rapidly to the bar for another tot of gin.

"A wire lasso run under the ice from another shack," Rikka provided for those slow on the uptake. "A trap just waiting for someone to enter their fishing shack and trigger it."

The reactions were galvanic and fascinating.

Rikka wished she was still recording; it made for great theater.

The Governor punched Senator Hamilton Waring and broke his pretty nose while screaming, "It's because you want to be President instead of me. You had to kill off my lovely wife to hide your affairs with her. Then you killed poor old Marvin because he has a far better voting record than you. If it hadn't been for the cameraman—"

Camerawoman! Rikka grumbled beneath her breath.

"—you were going to off me to clear your path to the White House. That's why you wanted me to go through the door first."

"Nonsense," Waring warmed up his rebuttal while spattering red blood from his nosebleed all over the pristine white surfaces. "I didn't worry about either of you for a second in the run against me. You wanted to frame me for your wife's death, because she knew what a real man was like. And you killed off poor Marvin so that you could marry Marilyn."

Rikka turned to Marilyn in time to catch the shifting expression of repugnance on her face. "Dumb choice, Lew. You're even worse in bed than Hamilton is. Don't know what the First Lady ever saw in either of you, or in Marvin for that matter. Of course, Lulu was never the sharpest thing, poor girl."

Both men sputtered.

Meanwhile, Mr. Secret Service sat on one of the fake-Grecian divans to await results. Odd, he wasn't reacting the way she'd come to expect from the Secret Service agents she'd met over the years. The military investigator made a similar choice to sit, but fell asleep almost immediately.

Police Chief Patrice Smith watched quietly with her arms crossed.

Finally the three of them wound down, not knowing who else to accuse.

Senator Waring's nosebleed was finally staunched.

The Governor's face was no longer red with high blood pressure.

Marilyn Maxwell looked resigned about her only two choices for getting to the White House. Or maybe at having to find a new set of lovers. Or maybe just wishing them both dead for being such idiots and sad that they'd survived the various attacks.

Rikka finally had a moment to consider, "How were those wires positioned under the ice, anyway? Do any of you scuba?"

Everyone shook their heads no except for the National Guardsman who offered a deeply adenoidal snore.

"And where did my goddamn seven-thousand dollar lens go?"

6

There was a knock on the door and when it was opened almost no additional sunlight was able to make it into the room. A large man with broad shoulders filled the doorway. His hair was crew cut and he wore a faded sweatshirt that said, "USMC" across the front. His big hands had no gloves despite the bitter cold.

Rikka threw herself at him and he caught her easily.

"Sam!" she kissed him hard. She'd actually thrown herself at a man and kissed him…and meant it. It might not be the Rikka Albert she knew, but she could get to like this woman she was turning into. Sam certainly appeared to approve by the way he kissed her back.

She dragged him into Grecian shack's main room and closed out the frigid morning.

It only took moments to fill him in on the events since her pizza call last night.

He'd come for her. For HER! Rikka checked in with herself and about her most coherent thought was, *Wow!*

He moved forward silently to the ice hole where her lens had disappeared—he was so light of foot you couldn't even hear his boots on the faux-marble floor. Yet he moved with such determination that everyone scrambled to get out of his way.

Kneeling, he stared at the hole for a long moment, she could see his Marine Force Recon mind working. These were the guys sent in behind enemy lines and told to "figure it out." Sam had been one of their very best. After mere seconds he reached down into the icy water as if it was a warm bath and swirled his hand around the lower edge.

"Of course!" Rikka then explained to the others over her shoulder, knowing they wouldn't get it. "There would be a groove cut in the bottom lip of the ice by the wire running in the direction it was being pulled from."

Sam pointed in the direction her lens had gone.

Rikka latched onto Sam's arm as he led the way. She was so glad to see him she was feeling all bubbly and chirpy and… and…girly, she decided. Maybe it was okay to feel that around him, but just him.

Sam led them directly to Senator Waring's Edwardian mini-palace of a fish shack. Patrice unlocked the heavy deadbolt for them.

There, in the middle of the living room, was an open fishing hole cover. Beside it, a heavy-duty fishing reel with a motorized winding spool was attached to a small stand. Her lens dangled in its evil clutches.

"Where did that come from?" Waring protested but no one believed him. It certainly hadn't been there last night when she and Patrice had returned to the scene or they would have noticed it.

Rikka retrieved her lens and tested it. It seemed none the worse for its dunking after she'd dried it off with a plush maroon towel that matched the leather furniture.

There was a sharp snap of handcuffs. Patrice had latched Senator Waring to the circular staircase leading up to his eagle's

aerie lounge, "I hereby arrest you for the attempted murder of Governor Llewellyn."

He sputtered and protested as she read him his rights.

Meanwhile, Rikka led Sam to the hole where the Governor's wife had been decapitated.

The wire-notch goose chase then led them to Congressman Marvin Maxwell's cabin. That meant that Marvin had set the trap in Senator Waring's bedroom that had killed the Governor's wife shortly after he'd slept with her.

It didn't take much digging around among the beer kegs behind the bar to unearth another powered spool.

"The Congressman," Patrice inspected it carefully, "must have hidden this away after he killed First Lady Llewellyn, but before he was in turn killed. Yes," she held up the end of the wire. "There are several long blond hairs still wound around the wire. No one on the deep ice but the First Lady has hair this long."

Marilyn Maxwell's was shoulder-length and Patrice's was even shorter.

Rikka also found a fancy looking remote control with two tiny joysticks, "It's like those things to radio-control those toy drones."

Sam had continued his search. Behind a keg of Bud Lite, he unearthed a small submarine as long as Rikka's forearm.

"I," everyone turned to look at Marilyn—except for the military security guy they'd left asleep in the Governor's shack and Senator Hamilton Waring who was still shackled to the staircase of his own fish shack for safe keeping. "I gave each of the boys one of those last summer. Three remote-controlled submarines that they could have mock battles with out on the lake."

"I remember that," Patrice was nodding.

"They must have used them to carry a lead under the ice and each set up a trap," Rikka tried to think it through. "First, Congressman Maxwell must have tried to murder Senator Waring. But he killed poor First Lady Lulu Llewellyn instead when she entered the Senator's bedroom."

"Then someone decapitated poor Marvin," Marilyn shook her head though she didn't appear too sad about the loss.

"And just now," Patrice picked up the line of logic, "Senator Waring tried to kill Governor Llewellyn but by pure chance caught your camera lens instead."

Sam didn't even need to point to the hole where Congressman Marvin Maxwell had died for Rikka to know the next question.

"Now, let's find out where Marvin's head has gone."

One last time her Sam...*her Sam, she kind of liked the way that sounded*...plunged his hand into the icy depths. She found him a Bavarian-brown towel to dry his hand with as he rose to his feet and slowly turned to face Governor Llewellyn.

They left Maxwell's Bavarian beer hall and crossed the ice once more to Llewellyn's Greek temple completing the sides of the triangle. Back in the Governor's palace of white, in an open fishing hole close beside the claw-foot bathtub—white of course—dangled Marvin Maxwell's head caught in a loop of wire.

"It seems," Chief Patrice Smith noted in a dry voice, "that the Governor didn't want any competition for his run at the White House either. After all, he'd know that only forty percent of our Presidents were governors first. His chances against both Maxwell and Waring would have been poor. He had to level the playing field."

"I would never—"

The Secret Service agent ignored the Governor's protests as he cuffed him to the clawfoot tub.

"They'll have to go to trial for—" Patrice started but the not-so Secret Service man cut her off.

His look of disgust gave him away. Rikka knew from experience that the US Secret Service would never reveal any judgment about the people they protected. Well, if he wasn't Secret Service...

"We are not going to have the two Chairmen of the Congressional Armed Services Committees and a dirty Governor besmirching the news nor the next Presidential election with their

petty rivalries. Now if you would all kindly leave." He sounded more like a special troubleshooter for a political party—perhaps the one that all three men had shared. He rousted the military man and sent him stumbling out to get in their car. The man sounded dangerous and borderline psychotic.

Rikka checked in with Sam.

Sam and the troubleshooter squared off and eyed each other for a long moment. There was no question Sam would win if it came down to a fight, but she also knew that no one loved his country more than Sam. Getting his hands dirty in its name was not a foreign concept to a Marine Force Recon soldier, retired or not.

His slight nod of agreement was enough for her. If Sam thought it was best to stay out of this guy's way, she wouldn't argue.

Kate was going to hate this, but Rikka shut down the camera and pulled out the memory card. Then she reached into her boot for the second copy she'd made as they came back out on the ice this morning, and handed them both to the agent.

Patrice drove Marilyn, Sam, and Rikka back to Patrice's ice shack where she made them all freshly-brewed coffee. It was as cozy and feminine inside as it was cute outside. The Police Chief had faced her shack so that the view wasn't polluted by the three ridiculous buildings grouped at the outer edge of safety. Instead, the windows faced the distant shore and the sun climbing above the dark tree line.

Rikka wasn't overly surprised by the loud *thump* behind her. She ignored the groan of protest that echoed briefly through the ice beneath the ice shack.

By unspoken agreement they waited until the Not Secret Service agent's vehicle had passed by on its way to clear out the morgue and Patrice's coroner files. They finished their coffee before they went outside to look at the empty horizon to the north across Lake Winnibigoshish.

No three fishing shacks. No crystal red Cadillac SUV. No three men at war over the Presidency of the United States by whatever means necessary.

Something had broken the ice, perhaps the papers would attribute it to the excess weight of the extravagant fish-shack palaces and massive SUVs out on the deep ice.

Whatever the actual cause, everything had disappeared from view. By tonight, the ice would be refrozen over the shattered chunks that now filled that area of the lake. In a week it would be walkable. A team of special divers would surely be called, but a thorough investigation would have to wait until spring and melt-out.

The four of them went back inside to prepare for the Lake Winnibigoshish Northland Chowder-Off that would be starting in a few hours.

7

"*You know,*" *Kate Stark* sat back in her chair after tasting the four dozen chowders that had been entered in the contest, "that's quite some story, Rikka."

"I know. I almost didn't tell you, but Sam thought you'd like to know." She'd filled Kate in on all of the details she could remember while the chowders had been cooking on portable stoves, in between when Kate had wandered from chef to chef for "on the ice" interviews.

The Chowder-Off was a near-shore event, with a section of the ice polished for an ice skating rink, Genuine Lake Winnibigoshish Ice snow cones, and a fairway of game booths and crafts. Everyone was bundled up in heavy parkas and making merry of the sunny day that had reached a balmy ten degrees above zero.

Rikka was definitely going to put in for an equatorial assignment next time.

"Odd that all three men thought to use the same method," Kate remarked.

"Maybe one thought it up, told it as a joke to the others, and then they each decided to give it a go."

"Could be," Kate admitted and started flipping through her scoring notes again. "Could be." She began handing losers to Rikka.

Rikka read the tasting notes and was once again awed by Kate. There were nuances and subtleties marked down that Rikka didn't even know about, never mind might have noticed.

For her, there was one pretty clear winner, but maybe she was biased.

Kate finally winnowed the stack down to the top three, and flicked one of her perfectly manicured but unpainted nails against the winner. Well, Rikka was pleased to have been right about that.

"Maybe," Kate said quietly as the crowds gathered to hear the final judging and prize awards, "a fourth person suggested it to each of them individually. Though there's no way to tell who now."

Rikka blinked at that, then Kate gave her a nudge toward a good camera position as she moved to the carved ice podium and began speaking. Rikka got the camera aimed and gave her friend a nod that she was recording.

Kate was funny, of course, and charming. There was a reason the woman ran the Number One food network on television with the most popular shows being the ones she did herself.

A fourth person, Rikka considered.

Marilyn Maxwell, the dead-Congressman's wife, had given each of the "boys" one of the toy submarines.

And slept with each one, though clearly not as freely as the Governor's wife had. Maybe just once or twice, to suggest the idea of the trap.

Kate awarded third place amid a large round of cheers and applause.

Rikka tracked a great bearded man in her viewscreen. He was on the verge of weeping with joy as he lumbered forward

and wrapped Kate in a great hug that drew laughter and more applause from the audience.

But what would make all three men set their traps on the same day? Perhaps because of it being the day before the Chowder-Off? It still didn't sound right.

What if Marilyn had slept with each merely in order to convince the men to give her a key so that she'd have access to each of their cabins?

Kate called up the second place winner. A tiny elderly woman came trotting forward with her gray braids flapping about her. She was clearly well loved and also garnered much heartfelt applause that made Rikka feel more kindly toward Minnesota than she had since reading the first weather report of her latest assignment.

And then Rikka remembered one fact she hadn't thought to tell Kate.

Who had tugged on the back of her coat and stopped Rikka from sticking her neck into the trap set at the Governor's shack?

No one had admitted to it, but only one person had come to check on her after she'd sprung the trap and lost her lens down an ice hole.

She swung the camera to locate her as Kate called out the winner.

There was a roar of approval as the name was announced.

The camera caught Police Chief Patrice Smith's radiant smile, the first Rikka had ever seen cross her features. And the look bloomed further a moment later when Marilyn Maxwell threw herself into the victor's arms and kissed her soundly.

Rikka captured their moment—one that she'd edit out later and perhaps send to them privately—then panned into the cheering crowd capturing some great footage for the television show.

Rikka herself wouldn't miss the three men from the upcoming Presidential race and she doubted if Patrice or Marilyn would miss them in the years to come.

Patrice Smith, who—Rikka finally recalled from her prep work—had placed a consistent fourth over the years, came to the

podium to collect her First Place prize for the best Ice-Caught Fish Chowder.

And the cheering continued as the cooks and fans of the Annual Lake Winnibigoshish Northland Chowder-Off proved that they definitely weren't going to miss the three men at all.

Heart of the Storm

Some stories are born out of thin air...in this case very thin air high atop the fourteen thousand foot peak of Mt. Rainier.

I set out to learn the backstory of Delta Force operator Michael Gibson. I needed to know more about him before writing his romantic suspense novel, Bring on the Dusk *(the sixth novel in the main flight of my series The Night Stalkers).*

Colonel Michael Gibson materialized out of thin air while I was writing the very first book in the series, The Night Is Mine. *And he refused to leave. Book after book, when the mission got truly ugly, Michael was there at the heart of it. And the fans were constantly asking me for more about him.*

I had been writing about him in desert situations, so I decided to toss him into a very different environment and sat back to see what happened.

I set him high on the slopes of Mt. Rainier, alone in the middle of a winter blizzard. You know, a training hike…Delta Force-style.

There is a silence that wraps around Michael. It is a silence that I've heard up on those slopes (though I was there in the summer and didn't go anywhere near the glacier-shrouded peak). But I've heard the wind slipping over the ice field with the soft rattle of ice granules. I've followed the melt-out stream upslope until it was at first ice-covered and finally frozen solid.

Against his desire for a little peace and silence, he is soon wrapped up in a high-mountain rescue— drawn back into the world he was trying to avoid for a little time.

Two unexpected things happened at that point for me as a writer.

I needed a helicopter, a crazy-good one flown by a top pilot. Who better than then-Captain Mark Henderson and members of his newly formed 5th Battalion D Company. I had not expected to discover the origins of the company I'd already written five novels about while in a short story about a Delta Force operator. I also learned why Colonel Michael Gibson just always seemed to be there at the tip of the spear with the Night Stalkers of the 5D.

It was over a year later that this little story offered up its second surprise. The woman rescued by Michael and Mark high atop the slopes that

winter was a small character, practically a throw-away.

Except she refused to be thrown away. When I finally launched my Delta Force romantic suspense series, she stepped to the fore and said, "Book Number Two. That one is mine!" When Melissa Charlene Moore speaks, you listen.

I'm glad I did. She totally rocked Heart Strike, *earning fantastic reviews.* Rock it, Charli!

M.L. BUCHMAN

NPR and B&N Top 5 Romance Author of the Year

"One of our favorite authors."
RT Book Reviews

Heart of the Storm

a Night Stalkers story

Five Years Ago

1

Major Michael Gibson of US Army's Delta Force was at eleven thousand feet, less than two hundred feet below the summit of Little Tahoma Peak when he heard the distress call on his radio. It was pure chance that he heard anything.

The January winds were howling down upon him, caught in the funnel of Little Tahoma and Mount Rainier's nearby peak that climbed another three thousand feet above him. The Arctic northerly, driving in the frigid Canadian air and dumping several feet of overnight snow, still howled beneath a sky so blue it could have been a child's spilled paints.

At least it was blue where last night's storm scrubbed the sky clean before departing southward to rush over Mount St. Helens and Mount Hood on its way to bury the Siskiyou Mountains of northern California. The northern sky already looked to be gearing up for the next onslaught through the Cascade Mountain Range.

Michael had escaped the ceaseless roar of the icy wind when he tucked down behind a sharp crag for a minute to chew on

an energy bar and drink some water. He wasn't hungry, or thirsty. But his high-altitude survival training had reinforced what he'd already known—by the time you noticed hunger or thirst at altitude, it was already too late. And this broken bit of volcanic rock was probably his last refuge before the summit—two hundred feet and perhaps an hour and a half above him. He should have just enough time to take the peak and get clear. Maybe he could cut the ascent time down to an hour.

Hunkered down behind his rock, he was offered one of the best views on the planet. From southeast sweeping around to southwest, the lush forests of Washington State spread over the rugged terrain. Doug Fir, Larch, White Pine, all underlaid with Oregon grape and blackberries so thick that even a Special Operations soldier would go looking for a way around.

North down to northwest revealed relative flatlands, no less green. In the far distance the waters of Puget Sound glittered beneath the low morning sun. If he'd been willing to remove his snow goggles and pull out his binoculars, Tacoma and Seattle would come easily into view despite the fifty miles of distance. As it was, the big airplanes climbing out of SeaTac Airport were the only encroachment from the big cities; he could feel the passengers snapping blurred pictures through plastic windows from their warm, plush seats as they flew over.

Immediately below him in all directions were the glaciers of Mt. Rainier National Park. From his vantage point high on Little Tahoma's flank, Emmons, Ingraham, Cowlitz, and the bound shoulder of Nisqually Glacier lay like a broken carpet of blinding white; constantly tearing at the volcano's rocky sides to bring the old girl down. *Good luck with that.*

Straight ahead Mount Rainier rose to fourteen thousand-four hundred feet, her rounded peak in stark relief—permanently-glaciered blinding white against the blue sky. He'd ridden out last night's brief but vicious blow at the base of Little Tahoma in a snow cave.

He found a small patch of snow to sit on among the high rock to give his legs a moment or two to recover while he ate.

His goal was to hit Little Tahoma Peak today, and then get off the mountain. He'd originally thought to climb Rainier as well while he was up here, but he could see the northern horizon already graying up even more in the five minutes he'd crouched here. Tonight's storm was predicted to make last night's look like a mere flurry. It would lock out the mountain to even the most ambitious climbers for days. And he was due back in the Congo Rainforest hunting genocidal warlords soon.

Michael listened for a response to the radio distress call. He should have heard it if there was one. From his perch he had sweeping coverage of the entire west of the Rainier down to Camp Muir at ten thousand feet and Paradise at five thousand.

The call repeated.

Silence, except for the wind raging to either side of his boulder. The park rangers clearly hadn't heard it though the call had been on the frequency that everyone on the mountain was required to monitor.

Everyone on the mountain.

It was the first day of the New Year, there were only five people listed in the Park Office's register. An elderly couple in their fifties hiking to Camp Muir for an overnight. A young couple, Charlene and Fred Moore with a climbing pass for the summit. And him.

Reluctantly, sorry to break away from the peace of his solo climb, he pulled out his radio and fitted the earpiece; partly because it used less power than the speaker, and partly his military training to not give away his position was so deeply ingrained—despite being on friendly soil.

"To distress call on Rainier. This is Michael, go ahead."

"Oh thank god!" A woman's voice crackled over the earbud. "Fred has fallen into a crevasse and I can't get him out. He's trapped and hurt. His radio must be broken, but we can sort of shout in between the wind gusts. At least I think so. It's hard to tell. Please, come. Please. We need help."

Michael listened for a moment. The wind wasn't gusting much. More of a steady howl.

"Okay, I need you to remain calm," he thought back to the register at the park office, "Charlene. We're going to go through this by the numbers. First, are you personally safe? Solidly anchored and warm enough?"

"Yes, we're both snapped into an ice screw. I managed to get a piton into the rock and have a backup line anchored there. And its Charli."

"Good," more together than he'd expected. "Now, where are you?"

She wasn't sure. The trip had been Fred's idea, no surprise there. In Michael's experience women were generally too sensible to do something like climb Rainier in the dead of winter. Though they weren't sensible enough not to follow along when a likely male beckoned them into doing something stupid.

For himself, he'd wanted to keep his snow and ice skills up. He could feel them melting away under ops conducted in the heat and humidity of the Congolese rainforest, so he'd gone for a winter tromp up Rainier.

But if she'd known to anchor herself more securely *before* calling for help, she wasn't a helpless soul traveling on her husband's whim either.

So, they began playing the "what can you see" game in a world of ice, rock, and sky.

2

Captain Mark Henderson stood outside hangar 4-C and stared up at the northern sky. He debated whether to call for a training sortie or to give his crew the night off.

It was midday and he was supposed to be sleeping, just as they were now, but upper brass never did understand SOAR. They'd rousted him with a question that could have been asked and answered by e-mail anytime in the next week if some Colonel hadn't had a hair up his ass.

The US Army 160th Special Aviation Regiment specialized in one thing, nighttime helicopter operations. They did it better than anyone on the planet. That's how they earned the Night Stalkers nickname. They'd flown into hundreds of places that they could never admit to, delivering Delta, SEAL, and other US Special Operations assets to places no one else could get them into…or back out of alive.

But that meant he was supposed to be asleep right now like the rest of his crew. Not that he'd been sleeping much lately. Three months ago, the powers that be had decided that SOAR needed

a fifth battalion. It was to be based in Tacoma, Washington at Joint Base Lewis-McChord. The 5th was to have four companies and he'd gotten the tap to lead "D" Company.

His goal was simple. If SOAR were the best helicopter pilots on the planet, the 5th Battalion's D Company was going to be the best in SOAR.

Period.

There would be no need to brag.

They'd earn it.

He had argued for and, surprisingly considering the layers of brass involved, been given the go-ahead to assemble a mixed company—the only one in the entire regiment. Most of SOAR was structured with one type of helicopter per company. Hell, the entire 2nd Battalion only flew the massive twin-rotor Chinooks. If they needed a Black Hawk or a Little Bird, they had to go outside not just the company, but the battalion to borrow assets.

It made sense to the paper pushers. This way the 2nd Battalion only needed one set of spare parts, one set of mechanics.

Mark didn't give one damn about logistics.

Well, he did, but he cared far more about his ability to respond. Because the 5D was a unique experiment, they'd decided to cripple his ass by giving him fewer assets. They'd learn and change their tune soon enough. In the meantime, he was collecting the best of the best personnel he could find, anywhere. He called and they came running, whether or not he'd flown with them before. It was deeply gratifying because these guys really were good.

At the moment he had two Little Bird helicopters good for fast attacks and tight in-and-out tight scenarios, one of the big Chinook heavy lifters, two transport Black Hawks, and his baby.

He needed to name her, but she was so perfect that he hadn't found the right name yet. She was one of the rarest helicopters on the planet. There were only a dozen of them and they were custom designed by SOAR for their exclusive usage. The Direct Action Penetrator DAP Hawk might look like a weaponized

Sikorsky Black Hawk, but she wasn't even close. She was the most sophisticated, bad-ass helo ever launched into the night sky. And she was all his.

A lot of SOAR's company commanders led from the safety of a behind-the-lines observer seat. Mark planned to lead the 5D from the right-hand pilot's seat of the…

Damn! He really needed to name her.

He wouldn't mind a bad-ass nickname himself. He'd never picked one up along the way. Guys tried them on him, but they always faded away and he always drifted back to being Mark.

Wouldn't it be funny if he and his Hawk had the same cool name? That would really screw with people's minds.

He focused back on the graying sky to the north and pulled the leather bomber jacket his mother had given him for Christmas more tightly closed against the chill air. Puget Sound didn't hit freezing very often, even in mid-winter, but the day was struggling for twenty degrees this afternoon courtesy of a chill Canadian system sweeping down the Fraser River Valley.

He turned his head and spoke back over his shoulder to the DAP Hawk helicopter tucked away in hangar 4-C, "We fly tonight, girl."

The helo seemed to sleep more soundly for knowing that.

Yeah. SOAR was known for being able to go anywhere at any time. The US Postal Service motto of rain, snow, sleet, etc. had nothing on what the 160th SOAR could do.

And very soon now, the 160th SOAR wouldn't have anyone able to match what Henderson's 5th Battalion D Company Night Stalkers could deliver.

3

*C*harli had calmed down the rest of the way when she realized that she could see Michael's position high on Little Tahoma Peak. That still hadn't given him much information on where she was.

They did finally narrow it down though.

Partly, the westerning sun had revealed the snake-track shadow that their hike up had left through last night's thick powder, and partly Charli had kept her head.

"Disappointment Cleaver," she agreed when he mentioned that landmark. "Yes, my brother talked about that."

Not Mister and Missus.

"He said we were nearing the jump off point for the peak."

Top of the Cleaver, which put them past twelve thousand feet. The low-point saddle between his eleven thousand foot position and hers past twelve was down at ten thousand feet. An easy hour stroll down and back up if it wasn't all up in the rarefied atmosphere above ten thousand feet and the ice-bearing winds weren't howling like a demon across the face of fifty-degree slopes.

With his binoculars he finally pinned down their position, a dot of bright red parka against white snow. Just a mile away, as the raven flies.

He watched the snow spume outward in white-out clouds beneath the near hurricane force winds. Mt. Rainier stuck up past halfway to the jet stream and it wreaked havoc up here near the peak. Not even a raven could fly a straight line in this weather.

"Okay, I'm going to listen all the time, Charli. So if you have a problem, you can call me. But I want you to shut off your radio to save the battery," he had three spares with him, but Charli's were in the crevasse with her brother. "Each hour on the hour, you can check in with me. But I want you to have plenty of power for when I get close. Can you do that?"

She hadn't sounded happy, but she'd agreed and signed off.

He needed her to save her battery, but he also needed the silence. He had a long, technically tricky traverse down the slopes of Little Tahoma, then much higher up on Rainier's flank. It required concentration to plan and execute. Being in the quiet of the world was what he needed to—

"This is Mt. Rainier Rescue to Michael, come back."

So much for peace up on the mountain. It simply wasn't going to be that sort of day.

"Michael here."

"We couldn't hear the other half of the conversation. Please relay information and do not attempt solo rescue. We already have climbing rangers enroute."

He relayed the information about Charli's location and her brother's unknown condition. Then he waited through the long pause of silence. He knew what was coming and wanted to get moving, but once he did the wind would deafen him and he'd need both hands and his full attention on what he was doing. He was free-climbing solo. He hadn't brought great chunks of rope or enough gear so that he could rappel down from Little Tahoma and then abandon the gear to climb up Rainier.

"Uh, Roger that," the rangers finally responded. They might be good guys, but they were just Parkies and this was getting uglier by the moment.

Michael gave them another thirty seconds. *Enough.*

"Please transmit location and number of rescue climbers."

"We have two rescue-qualified climbing rangers leaving Longmire now."

Michael didn't need three minutes of silence to assess the situation. Current road conditions made that an hour drive to Paradise, then a three to five hour hike to Camp Muir. And that would place them five to six hours below Charli and Fred Moore's location. The storm was two hours out—three if he was lucky, but it wasn't that kind of day.

"Helicopter support?"

"Not in these winds. They'll have to…"

Ride it out, the operator didn't need to say. One of them down in a crevasse and injured; the other exposed to the elements with night falling. It was their death warrant.

"Roger that," Michael considered for a moment. "Get your two park rangers to recover the couple who went up to Camp Muir. Do not send them up the mountain. I don't want to have to rescue four people. Out."

He tucked away his radio, the wrapper of his energy bar, and sipped some more water. Then he dug out a balaclava that covered everything except his snow goggles, and tugged on his gloves.

He had to keep the radio on in case Charli called.

The park rangers stop talking to him by the time he'd descended the first five hundred feet.

After that, he moved in peace.

Nothing but himself, the rock and ice, and the roar of Mother Nature as she veered the storm to drive straight onto the face of the mountain.

4

Y*ou are a complete* bastard making us fly in this shit."
The new guy Tim Maloney came up to Mark as he waited by the
DAP Hawk in the dark of the glowering afternoon sky.

It was blowing twenty, gusting thirty down here on the field.

Tim tapped a couple fingers to the brow of his helmet in a
casual salute, "Damn but I like that in a commander."

Sergeants Crazy Tim Maloney and Big John Wallace had
requested to come aboard as a package deal to fill the two crew-chief
spots. He'd wanted Big John who had a reputation for being an ace
mechanic. And after he'd seen how Tim handled a mini-gun—crazy
or not he was damned good—decided to give them both a try.

Mark's co-pilot, a Lieutenant Richardson, also was shaking
down well. He flew silent which Mark appreciated and was a
steady hand on the cyclic in a tight spot.

It had only been a month but the team was mission ready
by any standard he could come up with. There were still a lot of
rough edges to polish off, but they were all of the sort that were
only going to happen in combat.

Lots of combat.

Of course with multiple wars on top of the usual list of black ops, that polish would come all too soon.

Henderson managed not to laugh at Tim's comment, staying cool behind his mirrored Ray Bans as the afternoon light was shifting to evening. That's what a commander did. He wasn't one of the guys, he was the steady rock. He made sure he was always the best man on his team to give them something to trust and to strive for.

Under the heavy storm clouds, it was already growing dark.

"If you're too chicken to fly in this shit," he informed Tim doing his best not to smile, "you just let me know and I'll sign your ticket back to the 10th Mountain."

"Dude," Big John rolled up from completing the preflight inspection on the DAP Hawk to glare down at his buddy.

John stood at least six-four and was massively strong. It was a surprise every single time he managed to fit into the crew chief's seat close behind Henderson's piloting position.

Tim was equally broad-shouldered but at five-eight looked tiny next to his friend.

"You get us booted," John's voice was a low rumble, "and I am gonna sit on your head until all the stupid runs out your ass."

"I—"

"Let's saddle up," Mark cut Tim off granting the round to Big John. He'd long since learned that if he didn't choose a winner, these guys could go at it all night without missing a beat.

Once they were all in position and the intercom was up, he filled them in. "This is a test of the new systems you guys laid in. I want to be able to fly within three meters of plan in zero visibility. Fog, snow, I don't give a shit. I want to prove we can do it until we know it in our bones."

"Yes sir, Boss."

"She'll do it," John was very protective of his helo, just what you wanted in a chief mechanic.

Richardson kept his usual silence.

Mark cleared with the tower and lifted the DAP Hawk into thirty knots of nasty. The rain was trying to decide if it was sleet, and the gusts were working to tie the clouds into intricate Christmas bows. Perfect weather for a test flight.

They'd start in the waning daylight and work the skills right into darkness. By sunrise they'd have it down. They'd own it.

Without his needing to ask, Richardson had the terrain-following radar active and had layered it on the terrain map programmed into the computer.

Mark started at fifty feet above the pines and laid down the hammer, turning southeast to get some terrain-following practice in the foothills of the Cascade Range.

5

Michael was down off Little Tahoma Peak in forty minutes.

The blue was gone from the sky. He couldn't tell if there was new snow falling from the clouds yet. Maybe it was just that he'd descended from the rocky buttress of Little Tahoma down onto the Emmons Glacier and it was last night's snowfall blowing sideways.

After he clipped on his crampons, he spent twenty precious minutes finding his way through the serac field. The massive blocks of ice thrown up by the glacier impacting the uphill slopes of Little Tahoma had created a near impenetrable field of house-sized chunks all looking to spill downslope at the slightest provocation.

It was a risky route. He'd calculated that against the additional time required to descend another thousand feet to where the Ingraham Glacier trailed placidly along Little Tahoma's lower reaches, cross safely, and only then start the much longer climb up to Disappointment Cleaver. He knew Charli Moore wasn't in a position to wait that long.

Also, the lengthier and safer descent would place the bastion of rock between them and block radio contact. The last thing he needed was for her to panic.

At precisely one hour she called in, just as he cleared the serac field.

"I'm hoping to be there before your next check-in call," he assured her and then calculated the chances of delivering on that and didn't like them.

A mile away up a forty-five degree slope. Climb the stairs of the Empire State Building five times. Except the stairs were steeper than normal ones built by humans, made of ice, and filled with hidden crevasses crossed by precarious snow bridges.

Michael checked his map for his current position and took a careful compass reading. Then he set the GPS as a backup.

By the next check-in, if he wasn't there, he'd be able to tell her how soon in minutes.

He unfolded the long aluminum pole and snapped it together so that he could test the snow before he stepped on it. Taking his ice axe firmly in his other hand, he set out at a fast clip.

It would be best if he crossed as much territory as he could before the storm hit. The gray clouds were already flattening the light, making it hard to see just how high to raise a foot to make the next step.

Sometimes the sun broke through the ragged southern edges of the clouds, but it soon disappeared behind the high shoulder of the mountain and he was on his own.

That was one of the things that had drawn him to join Delta Force. Unlike the SEALs who trained in groups of four, six, and twelve as their ultimate team, Delta was trained to survive and complete the mission at all costs—even when it was just one man.

Michael had always enjoyed the structure of the Army, but the "group think" had never fit him well. Delta, who called themselves simply The Unit, had answered much of that need within him. He trained with the very best and could slide

seamlessly into any size team when required. But his favorite mode was to walk alone into the heart of a hostile city, see what had to be seen, perhaps kill who needed to be killed, and walk back out with no one the wiser.

Counter-terrorism.

The terrorists followed no set rules. Battlefield tactics didn't apply. The Unit of Delta Force had been built from the ground up by its own soldiers with that in mind. When—

He hesitated.

Froze in place.

Then took a careful step backward.

His testing pole had gone too deeply into the snow ahead of him. He eased back another step and whacked the pole like a whipping cane flat against the snow ahead of him a half dozen times. The long sideways span of the slope silently disappeared from in front of him. The exposed crevasse was only a yard wide, but it would have swallowed him happily.

He could traverse sideways looking for a crossing, which could lead him far astray from his intended path. And chew up valuable time that Charli Moore didn't have.

Instead he took a moment to pack the snow on his side of the crevasse and double-checked his grip on his ice axe. He used the packed snow as a sprinter's lane, kicked hard off the rim of the gap in the glacier, and landed cleanly on two feet on the high side.

He lunged upward onto his stomach to spread his weight and buried his ice axe as far upslope as he could.

The snow held. He crawled a dozen paces further upslope before returning to his feet. He stood in place a moment and checked his crampons. The straps were tight and the closures were frozen in place, which was fine. He rubbed the tip of his nose through the thick balaclava for a moment to warm it up, then continued upward.

Michael had joined the Army straight out of college at eighteen. And gone Delta as soon as they let him, at twenty-three. After five years…

It wasn't going stale.

Not exactly.

The ops were too challenging for that. The variety too interesting. But there was a sameness to it. The Unit had some of the very best soldiers on the planet and he was pretty much the top of that heap.

That wasn't ego.

He reached a broad crevasse crossed by a narrow snow bridge; snow that had packed into the perfect pattern to hold shape over the gap. But would it stay there? A couple jabs with his pole were encouraging.

His commanders said he was the best—consistently.

He ran an ice screw into the low side of the crevasse, tied a line to it, and made a midpoint loop to attach to his harness. He crossed carefully and made it. So he ran in another ice screw well above the crevasse and tied the far end of his line to that. He then crossed once more to the low side of the bridge to recover the first ice screw, he had few to spare.

His fellow operators always looked to him to take the lead.

The bridge gave out on the third crossing, but he fell only a few feet before the line snapped taut. He hauled himself to the upper ice screw, recovered it, and moved ahead upslope. The fall barely registered. Little things like that didn't when he was in the zone.

You got through the mission by doing what had to be done. *Whatever* had to be done.

It was fine for everyone else when you were at the head of the pack. But who was going to push him to be better? There was only so much you could do to drive your own self ahead.

He circled a serac as big as C-130 cargo plane that was tipped so far forward it could well roll down the slope in the next gust. It was eerily quiet crossing underneath its looming mass.

The fresh blast of needle-sharp ice crystals as he stepped from behind it only served to remind him that the storm's hammer had hit the mountain now, with a vengeance.

He double-checked his compass, considered how far the serac had been from the rocky outcropping of Disappointment Cleaver and knew he was off pace. He would not be up to where Charli and Fred's lives hung in the balance before the next radio check in.

But he'd be close.

So would the darkness.

6

*M*ark *Henderson hovered three* feet above the empty football field at Enumclaw High School. They had made it flying totally blind, at least he had been blind. Richardson had kept a lookout just in case.

But Mark had slid down the outer sun shield on his helmet's visor and concentrated on the FLIR terrain data projected on the inside of his helmet, pretending it was pitch dark rather than a stormy evening.

Per SOAR standards, he'd arrived within thirty seconds of the arbitrarily set time…barely. He'd have to work on that. That was one of the promises the Night Stalkers of SOAR made to their Special Operations customers, to always be there within thirty seconds of schedule…no matter where "there" was. No matter what the weather or the enemy were doing. "Landing zone is too hot" simply wasn't in the Night Stalkers' vocabulary.

"Okay. Set me three geographic points. Who knows this area?"

"There's this crazy bike ride out here that a friend was telling me about," Tim spoke up over the intercom. "It's called the

Rimrock or the…no, the RAMROD. It stands for Ride Around Mt. Rainier in One Day."

"Bike ride?" John cut in. "Hell, Tim. If your bike is so slow it can't do that, you need to upgrade your scooter."

"Not motorcycle, dude. Bicycle. It's like two hundred miles and tons of vertical climb. One hellacious lap of the mountain."

"I like it," Mark decided. "Give me three target points."

Richardson was working the maps. "Top of Cayuse Pass." He marked it on the display. Mark kept them hovering at three feet, steady despite the sharp gusts that slapped at him, and checked the route.

Up the White River Canyon. Pick up route 410. Climb to the top of the pass along the road.

"Around to Paradise on the south side."

Nearly due west over the lower flanks of Rainier, some good valley and ridge work there.

"And back to base."

Follow the Nisqually River, shoot over Alder Lake and head home.

"One more," Tim suggested with a tone that Henderson was beginning to learn meant trouble. "Hit Box Canyon."

Ouch! Halfway between Cayuse Pass and Paradise, but at the bottom of a deep cleft carved by the Muddy Fork Cowlitz River.

"And your max altitude for this operation," Big John rumbled not to be outdone by his buddy, "is fifty feet…to the top of your rotor."

"C'mon, dude. You got a death wish asking the man to do that?"

And Mark knew at that moment he was committed to it. And the next moment he realized that had been exactly Tim's intent with his complaint. He was definitely going to have to watch out for this pair.

But he couldn't turn away from the challenge or he'd lose face as their commander; totally unacceptable.

Top of rotor below fifty feet meant keeping his wheels below forty. Through canyon switchbacks, winding passes, and unpredictable weather.

He had to admit that it was a good challenge; now to make sure it wasn't a lethal one.

"Thirty minutes between each point, Richardson." At full speed, straight flight, he could make each target in under fifteen minutes, but he wanted to nail the height limit as well.

Mark double-checked his fuel, engine temp, and all of the other readings that were his copilot's responsibility. All green and good to go. "I want to hit each arrival plus or minus twenty-five seconds." It wasn't like the big eight-hundred mile Black Route training loops that were a standard part of SOAR training, but it would definitely be a challenge.

"Start the timer...now."

Mark shoved the cyclic forward to tip the nose down for speed and pulled up on the collective just enough to not eat the goal post at the end of the football field, though his wheels passed between the uprights.

"I'd suggest hanging on back there."

7

*M*ichael *arrived at the* head of Disappointment Cleaver as full dark and the first big slam of the storm arrived together. He'd had to pull his second ice axe to make sure he always had one buried in the snow so that he wasn't blown away.

Charli had missed the one-hour check in and it was a grueling twenty more minutes before he reached her position. And another ten before he found her in the minimal visibility afforded by his headlamp.

She was barely conscious when he arrived and was difficult to rouse from her stupor. When he did, she hugged him and then she wept. The tears froze on her face and would have sealed her eyes shut if Michael hadn't brushed them clear quickly with his bare fingers.

She'd been telling the truth about being secure. The various lines leading from the ice screw and the piton in the rock face down to her brother all met at her harness. So while she hadn't been under any stress, she'd been solidly pinned against the snow and wholly unable to shift out of the wind or exercise her legs.

She'd kept her arms moving and wore a balaclava, as he did, but her legs had no feeling up past the knees. She needed to get into a hospital fast if she didn't want to be losing toes, feet, or even worse. Her words were already slurring though he couldn't tell if that was hypothermia or altitude sickness.

In minutes he had a Z-harness rigged and was levering Fred Moore back up from the crevasse. He didn't want to think about what he'd find, but it wasn't as if he had a choice.

Except Fred wasn't coming. The rope went taut—and stopped.

Michael rigged a descending harness from the extra lines Charli had draped about her and lowered himself into the crevasse.

The line between the brother and sister had been fifty feet long, about five feet too long for poor Fred. His head had shattered and the blood had frozen to the ice.

Michael sat there for a long moment. In a war zone, you left no man behind. But in a civilian zone like a winter blizzard on Rainier he was less sure? If Michael simply cut the line, the chances of the body being discovered in the next thousand years was minimal until he spilled out in the Muddy Fork Cowlitz River at the base of the creeping glacier. Did Charli, if she lived, really need to see her brother's battered body?

That was assuming that he could keep her alive to get her off the mountain.

Michael kept the dead body company while he dangled close above it and quickly considered the options.

The woman certainly couldn't walk down.

He had a bivy bag that would only fit one person. If Charli and Fred had a tent or sleeping bags, they weren't on their packs. They must still be down at Camp Muir; they'd thought to go light for a fast strike at the summit and a quick return. Pretty standard even if it hadn't worked out this time.

He could rig a sledge, but the dozen or more hours that would require to get her down the mountain safely would be a cruel torture and probably kill any chance for her survival.

Old adage: when all the options suck, come up with a new option.

As gently as he could, he used his axe to chip Fred free of the ice. Then he pulled the parka hood up and over the missing chunk of his skull and snugged the front closure as tight as he could until only a scarf-covered nose showed through. Then he hauled himself back to the surface.

Charli was too far out of it to ask about her brother, so he pulled out his radio.

The park rangers could be of no help here. He did reach them on a patch-through to Crystal Mountain Ski Resort a couple of peaks to the east. There the signal wasn't blocked by the mass of Disappointment Cleaver.

Michael was relieved that the park rangers had met the older couple most of the way down from Camp Muir and gotten them off the mountain. There was no chance of them setting out again in time to help him.

Well, he might be sitting in what the park rangers clearly thought was a hopeless position. But they lived in a different world than he did.

Michael had long since crossed over into the military world and could see how that completely overlaid the civilian one. It had a network of lines that connected as thoroughly as highways and airlines connected civilians. The American military and its allies formed a complete second layer all over the globe that was mostly invisible—unless you knew how to access it.

Time to try that next layer out.

He switched his radio over to the military radio band.

If that didn't work, he had a satellite radio. It was more typically used for calling in airstrikes, but it would do the job of checking all options if the first one didn't pan out.

He began transmitting on the military emergency frequency.

8

*W*e've got a Mayday on 243," Richardson spoke up.

Henderson slid to a halt inches above the road in the heart of Cayuse Pass and grinned at the clock. Twenty-four seconds early. He was inside his time window.

Full dark had set in and—Shit!

He yanked up on the collective barely in time to clear a truck that had decided to cross the pass at that moment. He wasn't sure which of them was more surprised.

"Give me audio," he said to cover his nerves.

"Roger, Mayday," Mark kept his voice steady. "This is a Black Hawk out of Fort Lewis-McChord," he certainly wasn't giving his ID to some unknown. "How may we assist?"

"I have a civilian at twelve thousand feet who needs immediate evac to hospital. Plus one."

No need to explain the "Plus one." Plus one corpse.

The guy on the far end of the radio call continued, "Winds currently gusting seven-five to one-one-zero estimated. Temp steady at minus thirty. Visibility under ten feet in driving ice

and snow. Can you assist?" The wind howled over his headset while the guy's transmit key was down. No question about mishearing the wind speeds aloft.

The silence on the intercom was deafening after he keyed off. What kind of a crazy bastard was doing a search-and-rescue in those kinds of conditions? The guy was sitting in the middle of a Category 2 hurricane.

Mark considered. He *had* wanted to prove that a DAP Hawk of the 5D could go anywhere and do anything. Hadn't he wished for that just an hour or two ago while standing safely on the tarmac at Fort Lewis?

Twelve thousand feet in the midst of a kick-ass, hurricane-force storm.

What ultimately decided him was one word, "Civilian."

Based on the frequency and the way he spoke, there was a military man on top of that mountain asking for his assistance. And the military's job was to—

"Roger, Mayday. This is—"

Mark released the mike switch. He wasn't sure whether he wanted to announce himself over the radio. "Richardson. Where is this guy? Do you have a vector?" A sharp gust over the pass caused his finger to bump the transmit button and send the last word.

"—vector."

"Roger, Viper. Mayday is at," and he rattled off GPS coordinates that Richardson keyed down for him.

Viper? Mark had said "vector" but over the wind's roar it was miracle they could understand each other at all. Mark also noted that the guy still wasn't identifying himself.

Any normal soldier would do so. Even Green Beret.

Whoever this was had to be extremely well trained to be doing a rescue in this weather and yet wasn't transmitting his identity. That meant only one thing: Special Operations—the standard customers of the Night Stalkers.

And he needed their help.

"Okay guys," Mark called over the intercom. "Dress warm. This elevator is going up." The DAP Hawk was hovering at forty-six hundred feet in Cayuse Pass headed for twelve thou.

"Women's lingerie," Tim started in as they climbed.

"Bowling jerseys, jockstraps, and tap shoes," Big John joined in.

Mark tuned them out as they continued calling out more and more outrageous items.

He considered the best plan of attack as he climbed clear of the trees and ridges and turned west. Staying low had been a rough ride up to Cayuse Pass through all the gusts. Up here the air was smoother, somewhat…and blowing him south like a son of a bitch. He carved harder to the north in order to aim straight west.

The DAP bucked in a sharp curl of wind and dropped two hundred feet before he could find clear air for the rotors to get a good bite. He added another thousand feet to his altitude, and it got worse.

Low, Mark muttered to himself. It was crazy, but it was what they had just been practicing. Come in low. Close enough to the ground that the worst of the winds were passing above you.

"No rescue basket down to this guy," he called back to the crew chiefs who were off in some strange shopping land of elf hats, reindeer girdles, and replacement runners for your sleigh. "If the wind catches it, it will just brain him. Long line off the winch. If he can catch it, we'll trust to their climbing harnesses. And wrap it in glow sticks. They aren't going to be able to see shit out there."

Tim and John didn't say a word now, which was good because he couldn't afford the distraction as he moved down toward an ice field that he couldn't see, except by radar. It also told him that they didn't need to talk; he'd given them enough information to figure out the plan without any questions. A huge time saver that he'd have to remember for the future.

He tried the landing lights, but was blinded by the million reflections off the swirling snow. Turbulence or not, he climbed

up until his eyes recovered from the momentary blinding enough to again clearly read the radar images on the inside of his visor.

What kind of a lunatic ran a search and rescue operation under these conditions?

9

Michael neither heard nor saw the approaching helicopter. He knew they'd arrived above him when the ice crystals shifted from being driven sideways by the wind from the storm to being driven downward by the wind from the helo's rotors.

A winch line appeared in the middle of the maelstrom whirling about wildly, so he pushed Charli down to lie on the snow and stood over her as he tried to guide them into place over the radio. Fat chance. They couldn't see him, and even if they could, the wind was slapping around the steel cable like a cat with a piece of yarn.

There was no question who was above him. Only a Night Stalker would be out on such a foul night and only a man who completely trusted his own piloting skills would attempt such a rescue.

The first time the glowing green lights tied to the heavy metal hook at the end of the cable came sailing at him, he had to dive down on top of Charli to avoid being clobbered by it. She barely responded with an "oof!" Her time was getting short.

On the next pass the winch hook was moving more slowly and he managed to snag it.

Not trusting that he'd have another success like that, or that he could hold it for long, he immediately spun around and snapped the hook into Charli's harness. He flicked out a knife and severed her ties to the ice screw and piton.

"Lift. Now, now, now!" he called over the radio.

With a surprised squawk, jolted out of her stupor by the sudden yank on her harness, Charli disappeared aloft. Michael stood ready to slash the line connecting her to her brother's body in case it snagged on the side of the crevasse, but moments later Fred's body followed his sister's skyward.

He waited for the report, it didn't take long.

"Civilian aboard." And a few seconds more, "Plus one."

"Roger and thanks."

"Sending winch back down."

"Negative," he was just two thousand feet below the summit and there were two very solid anchors tied to his harness. "Negative. I'm good here. Thanks, Mayday out."

Whoever was flying the helicopter was silent for a long moment, swore succinctly on air, and then was gone.

Michael called the Park Rangers to let them know it was okay to stand down. Then he dug a shelf into the snow, pulled out his bivy bag, and slid in to ride out the peace within the heart of the storm.

10

Michael came down off the mountain three days later. The storm had blown hard for two days and then he'd taken the summit beneath a sunny sky on a dead calm day. The temperature was steady around minus sixty, he had owned the peak and his extra days at altitude had given him the acclimation to spend some time enjoying the vistas in every direction from the wide crater's rim that capped the old volcano.

He still didn't have any better answers to his career.

If he left The Unit…

There simply wasn't anything challenging enough on the other side of that coin. He was never one of those serve-a-couple-tours-and-get-out kind of guys. He was a lifer, but what was the next step from Delta?

CIA's Special Activities Division had tried to recruit him a couple times. While he enjoyed the black ops, the S.A.D. was just a little too much wash-your-hands-after-even-talking-to-them kind of guys.

The Activity, the slickest field intel guys that the nation's military had ever come up with, was tempting. But it wasn't the kind of challenge he liked. He enjoyed being at the tip of the spear.

After he dug his truck out of a couple feet of snow, he drove down from Paradise on the plowed out road and signed out at the Ranger Station in Longmire. Apparently news of his rescue was the talk of the park's ranger staff. They were sorely disappointed when he refused to meet with any media. A Delta operator's photo smack on the front page of the *Tacoma News Tribune* would not play well back at Fort Bragg. He should never have used his real name in the register.

He escaped as quickly as he could and went into the National Park Inn for a meal. It was one of those Depression-era lodges built by the government on a grand scale to put the nation back to work. Log-built, generous veranda, high, gabled roof. It was equally grand inside, the massive lobby had enough room to play a fair game of hockey except for the deep couches. High ceilings held up by massive log rafters. He headed for the restaurant seating down at the far end of the massive space.

Michael ordered the turkey blue plate special, heavy on the gravy, and the roast beef platter as well. He'd burned a lot of calories over the last four days. He'd managed to squeeze in a summiting of Little Tahoma Peak on the way down just for completeness sake. The Congo Rainforest still lay two days away.

He chose a seat with his back to the wall. As always.

Exit to the kitchen was to his left. At the other end of the wide lobby was the entrance to the lodge. Big, timber-built doors led in from the outside partway along the right-hand wall. The cathedral ceiling was filled with the light reflected off the heavy layers of snow outside the long bank of front windows. A grandfather clock, stately in a heavy Doug Fir cabinet, chimed the hour.

He was done with the turkey and halfway through the roast beef when a man walked in through the double doors. Many people had come in and out while Michael was eating, fourteen

women and nine men in seven separate groups. Each one with the earmarks of a tourist: chattering, expensive if not the most sensible snow gear, cameras.

This man was tall, broad-shouldered, and alone. He wore an expensive leather bomber jacket that said tourist and worn jeans and heavy boots that didn't. He wore mirrored Ray Bans and stood in the center of the main room doing a slow turn as he assessed the lobby and restaurant seating and everyone in it. His stance said "trained soldier" as did his careful inspection. His body silence said Special Operations.

His eyes passed over Michael…and then swung back.

The pilot.

He strode across the room, covertly continuing his assessment of people and exits, though not in any way that would be noticed except by another trained soldier. Without asking he sat down across from Michael with his back to the lobby, something Michael would never do.

Then the man took off his mirrored sunglasses and set them on the table without folding in the earpieces, so that the lens were only slightly tilted. He looked down and gave the glasses a slight nudge.

Michael judged angles and decided that the man had chosen well. Anything he couldn't see directly, he could keep an eye on in the mirrors of the sunglasses, which reflected the goings-on behind him. He'd have to remember that trick.

"The girl's okay," the pilot began speaking without introduction. "Lost a couple of toes and a brother. But she's fine."

Michael nodded. Good news.

"Really wants to thank you."

Michael shook his head, not gonna happen.

"She's a real looker under all that snow gear. Blond, tall, very athletic body."

Clearly this man's type. Still, not gonna happen. Michael wasn't sure if he had a type, but Nightingale-effect gratitude wasn't something he was after.

The pilot ordered the roast beef when a waitress came around. Michael ordered a slice of blueberry pie with ice cream.

They sat in comfortable silence until they were served.

"I know a guy who talks as little as you," the man who still hadn't introduced himself cut into his beef and sighed with pleasure. "Not Montana roast like my parents' beef, but not too shabby."

Michael had never had Montana beef. Especially not ranch fresh.

"My dad was a SEAL for almost twenty years. Doesn't speak unless he has something to say."

SEAL dad. Night Stalker son.

"Dad never was a big fan of ice and snow; shoveled it for too many Montana winters is his excuse. I expect it was because it froze up the fishing streams for too many months each year. He and I spent a lot of good time standing in those streams."

Michael would agree. Ice and snow he could take or leave. But he loved the quiet times of fishing.

"Strikes me that camping on that mountain top in the middle of a storm with an attitude was more the style of an operator from The Unit than a SEAL. Damned nice piece of work, by the way. Got some of the details out of the park rangers. Though your name led nowhere—I was able to verify you exist, but otherwise an absolute dead end—another sure Delta sign. Got some more of the details from the grateful babe. Very grateful. Sure you're not interested?"

Michael was sure; he was headed back to the Congo in thirty-six hours and had never been a big fan of a casual screw. He ate some of the blueberry pie which was exceptional. The wild blueberries were probably picked right out here on the mountain slopes last summer and frozen.

So, the man was smart enough to figure out that he was with the Special Forces Operational Detachment-Delta, which Michael would never confirm nor deny.

"I'm putting together a new company," the pilot's tone became more businesslike, more intense. This man cared deeply about the next part of the conversation.

He didn't need to identify SOAR, he'd know that Michael had figured that much out on his own. Michael appreciated that the man didn't waste time on useless words.

"I'm getting the very best people. I've stolen the Number One mechanic and gunner out of 1st Battalion, pissed their commanders off pretty good. Found a copilot out of the 3rd. I've snared the Number One teams in Chinook and Little Birds too. I'm building a blended Company, not just all one platform."

"Flexibility," Michael spoke for the first time. The military was not the most flexible structure; it was something The Unit prided themselves on but few others could get away with or even wrap their heads around at a true operational level.

"Precisely!" the pilot leaned forward eagerly. "I want a company that can dynamically adapt to any situation. I want you."

Michael stopped with his fork halfway up in the air, "Me?"

The man nodded and returned his attention to his roast beef. A couple came in, their designer clothes covered with the telltale patches of a snowball fight. The pilot's eyes flicked down to inspect the reflection in his glasses and he smiled for a moment before returning his attention to his meal.

"Doing what?"

The man shrugged, "We'll figure that out as we go. An embedded Delta liaison? A permanently-attached squad for fast reaction, trained to maximize leverage of the DAP heli-platform? How the hell should I know. I'm after the right people first. Then the team will drive itself to excel as it comes together."

A DAP. He hadn't dropped that casually; it was a symbol of just how much the military was impressed by this man. The rescue helicopter had been a DAP—in the heart of a blizzard atop the most lethal mountain in the lower forty-eight states. This guy wasn't just blowing smoke; he was out on the cutting edge.

It would be a hell of a challenge.

Michael would miss The Unit…but then he wouldn't really be leaving, would he?

And a team that could develop synergistically from the interactions of a half dozen different disciplines?

New information and challenges from the very best pilots on the planet to the very best technological innovations. And a commander who wasn't all hoo-rah but rather understood the silence of fishing, of being out in nature.

He finished his blueberry pie as the pilot polished off his roast beef.

Michael kept inspecting the possibilities as he would any pending operation, but could find no tactical or strategic hole in the plan. If the man was as good as his word, which he'd proved by doing that rescue, then it should work. In fact, it could set up a whole new model for inter-operability, always a fascinating problem.

Michael decided he wasn't just in, he was in all the way.

"What's your name?"

The man pushed aside his empty plate, kicked back and rested his feet on another chair. He pulled on his mirrored shades as if saying he completely trusted Michael to have his back.

He grinned like he'd just won the best poker hand on the planet.

"Viper. Viper Henderson."

Then they both started to laugh.

Beale's Hawk Down

This story came out of the prior one. Emily Beale is never subtle. She strode up, looked down at Michael's and Mark's story and said, "Where's my origin story?"

I thought back to what little I knew about her military career prior to her joining the US Army's 160th SOAR. There were two lines in the first book in The Night Stalker series, The Night Is Mine, *(which was written five years earlier) about that.*

And much like Emily, that forthright explanation seems to cover it.

This is the story behind those two lines. (I didn't include them here because they would spoil part of this story and Emily wouldn't like that.)

M.L. BUCHMAN

NPR and B&N Top 5 Romance Author of the Year

Beale's
Hawk Down

a Night Stalkers story

Seven Years Ago

1

T*he Black Hawk helicopter* shredded around her. The spin and fall fast enough that only the harness kept her in her seat.

The star-cracked glass-laminate windshield—each star centered around the hole where an armor-piercing round had punched into the cockpit—fragmented Lieutenant Emily Beale's view of the outside world into a thousand tiny refracted images. The veering Thai jungle hacked into crystalline shards of green in a thousand hues.

Hydraulic fluid sprayed over the outside of the windshield. Altered the colors of the world around them to a dark, alien-realm red. Even the yellow sunlight bled vermillion.

Emily flicked off the primary hydraulic system. The secondary didn't take over, but the emergency backup hung on.

"Beale!" Larry shouted at her from the pilot's seat. "I can't see!"

Emily glanced over and saw blood dribbling down over his forehead. One of the rounds had punched him high in the helmet, hopefully just a scalp wound.

She stopped attempting to recover systems and clamped her hands onto the controls. They were heavy, sluggish. The intercom sounded dead, Larry's shout had traveled across the cockpit, not into the headphones built into her helmet.

She clicked the mic switch a few times—with no result.

There wasn't time to look down, but the acrid stench of scorched electronics told her what had happened to their radios.

"Can't see much myself," she shouted back.

Emily watched out the only clear section of her copilot's side window, an intact area little bigger than her hand. She had to wait only two seconds for the helicopter to spin through a full three-sixty and reveal her options of where to crash.

South and east, a broad spread of poppy fields. They'd been flying east to west when they stumbled on the fields and been fired upon immediately. Even shot up, they'd managed to overfly most of the fields. She spotted several vehicles racing across it in their direction, they'd be in range in three minutes; up close and personal within five. The rugged terrain of the foothills to the Luang Prabang Range was in her favor, or would be until the Black Hawk hit the rolling fields.

"Poppies!" the Wicked Witch of the West seemed to cackle in her ears. They hadn't known about the poppy fields. Their flight had been racing northwest on a flood-relief mission. But some Thai opium lord hadn't liked the pair of US Army helicopters flying low over his fields.

North and west was helicopter-killing jungle. Though by the vibrations building up in the controls and the airframe, there wasn't much left to kill except the pilot and copilot. The crew chiefs? The silence of the two big M240 machine guns that they should be using to hammer back at the people who had just shot up the helicopter were ominously silent.

The second spin around let her spot the burning wreckage of their sister helicopter. The 101st Airborne Screaming Eagles were gonna be some kind of pissed. She just had to stay alive long enough for that to matter.

Emily risked taking one hand off the controls to trigger her personal radio tucked in her vest pocket. "Mayday! Mayday! Mayday! Army flight—" Larry lost control of their flight angle and she had to retake the controls and simply pray that someone had heard her. The rising screams of the dying Black Hawk masked any response. If there was one.

"I'm taking us into the trees, Larry."

"That sounds like fun."

It was too much effort to shout over the roar of the rotors and all of the systems alarms blaring for attention. She concentrated on landing without killing them.

On the next spin, she initiated a roll. Larry leaned his support into the rapidly failing controls and she was glad for every ounce of help.

By tipping the wounded Black Hawk well onto its side, she managed to counter the spin that the rudder pedals could no longer wholly fix. But it also meant she was losing altitude—at little better than a plummet.

There was now as much red inside the cockpit as outside. What electronics were still with them were blinking red warnings. Alarms were blaring but she didn't have time or the free hand to silence them.

Larry took a hand off the collective to kill the worst of them. But it was clear by how his hand fumbled slightly that he still couldn't see and was doing it based solely on training.

They were going down no matter what.

No time like the present.

If they hung up in a tree, they'd be shot before they could climb down.

If they hit the jungle floor directly, they'd just be dead.

Emily aimed ten tons of dying Black Hawk helicopter into the crown of a fig tree, hoping it would break their fall, but she wanted to hit it off center enough for them to slide down to the forest floor.

That all sounded great…if she'd had any control.

She didn't.

One last try on the radios. Nothing.

"Brace yourself," a shout to Larry.

The Black Hawk finally rolled onto its side as they hit the trees.

Rotor blades hacked at foliage, hit branches, crumpled, and broke away. The twin turboshaft engines, no longer trying to turn the long rotor blades, raced wildly out of control. She managed one T-handle engine cut off and Larry managed the other as the body of the helicopter slammed into the fig tree she'd targeted and tumbled off into an oak on the way to its doom.

2

*O*nly *after they hit* was Emily able to reconstruct where "up" was and that by some miracle they'd landed tail first. The rear of the helicopter had acted as a giant shock absorber as it crushed.

Then they flopped forward onto the wheels. The shock absorbers managed to bounce and they were parked right side up on the jungle floor. The helo was tipped back and thirty degrees to her side, but they were down.

A massive root system, which wound and snarled like a thousand giant snakes, was pressed against Larry's door. She tried to shove her door open. It opened six inches then caught on the jungle floor. Throwing her shoulder into it gained her only another three inches.

Emily popped her harness, managed to contort her long legs up to the main windscreen and kicked with both boots. The shot-up laminate disintegrated into a shower of a thousand crystalline shards.

Beyond the shattered remains of their twenty million-dollar hi-tech cocoon lay such a different world that it felt as if she

was in some science fiction movie looking through a rip in the space-time continuum.

All around her the dead helicopter still blinked and wept. She powered down the few surviving systems and the last of the alarms descended into silence, but still-creaking metal and the steady drip of leaking fluids surrounded her.

Mere feet away stood a shadowed jungle unlike anything she'd ever seen. No training in the swamps of Mississippi had given her a calibration for what she was seeing. Tree trunks a dozen feet across and fifty to a hundred feet high soared above them. The undergrowth was thick with leaves that were as big as she was and seemed as big as her helicopter. The silence of it was breathless. And the smell was—the tang of blood, the bite of hydraulic fluid, and the nasty, sharp, warning stench of kerosene-laden fuel.

Larry was struggling with his door, unaware of the massive tree-trunk blocking his way.

Survive!

The shouted self instruction finally shifted Emily into action.

"We're going out the windshield on my side."

Larry popped his harness and clumsily followed her out of the helo.

Once they were out, she leaned back in for her rifle. It wasn't in the door's mount anymore. Nor was Larry's.

She'd flown beside the Night Stalkers of the Army's 160th Special Operations Aviation Regiment a few times and now understood why they always wore their weapons across the front of their vests even in training flights. It was a practice she would certainly adopt. If she survived this.

Emily circled around to the crew chief's gun window on her side of the helo. The open-eyed bloody face that confronted her was identifiable only by which side of the aircraft it was on.

Down here beneath the forest's canopy, midday was dusk and the Black Hawk's interior was midnight. She fished a flashlight out of her thigh pouch and shone it over Vincenzo's head. Yamota

was no better off. Her attempts to wrench open the helo's side door were futile, it was trapped by the badly twisted frame.

She considered dismounting Vincenzo's M240, and then she looked again at the waiting jungle. It was going to be challenge enough without a twenty-seven pound machine gun and the same weight again in ammunition. Her and Larry's best chances were in evasion, not confrontation.

Bracing herself emotionally, she tipped Vincenzo back far enough that she could strip him of his holstered sidearm and spare magazines. He had a pair of full water bottles tucked into pockets as well. She took those despite the slickness of blood on plastic. Pretending that it was only red water spilled over the outsides didn't help in the least.

Larry was no longer standing at the nose of the helo when she returned. Instead he'd slid down to the jungle mulch and was leaning against the rounded nose cone, the only undamaged panel of the entire aircraft.

He had wiped his face of the worst of the blood and was blinking normally. But when she waved a hand in front of his eyes, he didn't react. She leaned back into the helicopter and managed to find her medkit, which she stuffed into a vest pocket. Another future lesson to keep everything she'd ever need on her person.

It was all very well back in the classroom to believe that you'd have the on-going resources of your downed helicopter and that the instructors had just been blowing their usual smoke. Besides, for every hour of survival training there was a hundred hours of flight training and a hundred more of combat training. That made it easy to discount the one hour that was squeezed in here and there. During Iraq and Afghanistan they didn't even have time for that.

After a moment's debate, Emily removed her helmet. She wanted its protection, but she needed her ears uncovered. The instant she did so, the world came crashing in. Bird calls came bursting to life around her, all commenting on the helicopter

that had just plummeted into their midst. And it wasn't the comfortable *check check* of a red-wing blackbird or a crow's sharp *caw*. The jungle chittered and nattered and the occasional spine-tingled *scree!* sliced through the air.

Also, now that she'd shed her helmet, she could hear the racing engines of pickup trucks as they roared across the poppy fields in their direction. As well as Emily could judge, she and Larry still had a few minutes, but "few" was the operative word.

She considered removing Larry's helmet, but didn't think she'd like what she found. Besides, his blindness—whether temporary or permanent—meant he needed protection to not batter his face against branches as they forged into the undergrowth.

There was no question that's where they were going. To stay by the helicopter would only guarantee their doom.

"C'mon, Larry. Let's get a move on." He stumbled uncertainly to his feet when she pulled on his arm.

"What about Vincenzo and Yamota?"

"We need to get going if we don't want to join them."

"Shit." Not even emphatic enough to earn an exclamation point. The four of them had flown two tours in Iraq and Afghanistan. There was no way to encapsulate or deal with such a loss in this moment. Focus on the next task. Survive.

Their Black Hawk had a transport configuration, but it wasn't all about moving howitzers and supplies. They'd flown hundreds of infantry delivery and retrieval missions, combat search and rescue, and pretty much everything under the sun that wasn't covered by the Special Operations guys of the Night Stalkers. They'd even flown a few special ops missions when SOAR was strapped for resources in a particular region.

To survive all of that and then lose two men on a flood-relief flight halfway around the world was too painful to elicit external emotion, the internal anguish was far too great.

As was the need to survive.

Emily pulled Larry's arm over her shoulder and locked her arm around his waist. It would be easier to guide his steps that way.

She stared at the thick undergrowth and wished for a machete. Then she thought better of it. That would just make them that much easier to track.

With that in mind, she tossed her helmet toward the far side of the clearing punched by the Black Hawk's crash. It landed against the edge of a small gap in the branches. Maybe the bad guys would think they'd gone that direction, deeper into the jungle.

Deeper into the jungle. That's exactly what they'd expect.

Never do the expected, some drill Sergeant's voice echoed out of her past. McCluskey?

"I'd pay good money right now to have taken SERE." The Survival, Evasion, Resistance, and Escape course was mainly for Special Operations guys.

"Probably be real handy at the moment," Larry agreed. "When it really hits the wall, Emily, you leave me and take off."

"Don't be an idiot."

"Exactly. Don't be an idiot. If they're going to catch one of us or both of us, make it only one."

Emily kept her mouth shut. There was no chance in hell she was going to be leaving a live man behind.

"So, where are we going?"

"Back to the poppy fields," she checked the compass on her wrist. Who ever thought that a helicopter pilot would have need of a simple, mechanical compass. But the sun was straight overhead, she could barely see even that much through the thick canopy of branches and leaves, and it offered no indication of east or west.

"We're doing *what?*"

"You need to keep your voice down." The approaching roar of the truck engines reinforced her.

They'd been in the clearing under a minute. It had already been too long.

She led them around the tree that was pressed hard against Larry's door, and ducked under a massive leaf that dripped with

the thick moisture despite the heat. Away from the helo her nose was assaulted with the foreignness of the jungle. Life so thick that she couldn't sort flower from fruit from rotting debris that rustled beneath their boots but also hid their footprints.

Because their arms were wrapped around each other, Larry had little choice but to follow.

They disappeared into the shadowy foliage.

3

Captain Larry Engstrom stumbled in a haze of red and gray shadows. He'd wiped his eyes clear of the blood, but it had made no difference. While Emily had scouted, he'd tried covering and uncovering his eyes to no effect. The play of light was from his optic nerves, not from his eyes.

He'd started to remove his helmet, but the slicing pain had added stars to the red and gray shadows. It had also taken his legs out from under him, dropping him to the jungle floor until Emily had hauled him back to his feet. His body screamed in a dozen places, but he was alive and that was all that mattered.

His father had begged him to get out. Begged him to do anything else, even something non-practical like music. His mother had simply looked sad and suddenly old. They had grown up protesting the Vietnam War. He had grown up in a different world where the United States was no longer the invader but now the invaded. The twin towers of the World Trade Center had gone down on his sixteenth birthday. He was twenty-two when he graduated from West Point and both Iraq

and the Afghanistan Wars were in their fourth-year of constant escalation. His country had called and he'd answered.

A branch clunked hard against his helmet and sent him staggering into Emily.

He heard a whispered, "Sorry," over the ringing in his ears.

"Just stung for a moment," his head was still ringing but he didn't want to upset Emily. His beautiful Emily. He'd follow her to the ends of the Earth, if he could only make his feet work.

Larry had been stumbling along through his career much as he was now stumbling along the Thai jungle floor. Bravado, broads, and beer—the three "B"s of the Army. He took stupid risks and buried them in alcohol and his dick in willing women…until Second Lieutenant Emily Beale had boarded his Black Hawk.

A tall, slender blond who should have been on a fashion runway, not a militarized mess like Bagram. At first he'd convinced himself that she was a heat mirage or a magical genie, like *I Dream of Jeannie* sprung to life. Barbara Eden had been hot back in her day, but Emily Beale, a soldier with a steel spine and the integrity to match, totally dusted her.

He'd cleaned up his act to meet her standard. And once he had, he couldn't believe the shit he'd done in the past or his low-life taste in women. Beale had kept their relationship strictly professional—right through two tours and her promotion to First Lieutenant and his to Captain—but at least on the few occasions he went womanizing, he'd shown a much better taste in women and been more respectful. With the Beale gold-standard for comparison, it wasn't hard.

She was hurrying them along and he did his best to keep his breathing quiet though it sounded loud and ragged despite the insulation of his helmet.

Lately things had been shifting between them. Over the last year they'd grown closer. She'd let her hair down a few times and they'd talked over a beer and pizza about their careers and a little about their lives before.

Their lives before—

With a loud *tonk* that seemed to echo through the jungle, Larry caught his boot on a tree root and had to wrap both hands around Emily to keep his balance. She felt so good that it was hard to let go of her. He could feel her determination when he felt so little of it remaining. He knew he was injured and only adrenaline was keeping him upright, adrenaline and, again, the need to meet her standard.

Their *lives before* oddly no longer mattered. Her dad was some government bigwig, though she declined to say which one. His was a game-software engineer. Her mom: socialite. His: grade school teacher. It didn't matter. They were soldiers now and moving up through the ranks of the 101st Airborne.

The racing truck engines were close now. Even the jungle didn't muffle them.

Then brakes squealed, tires skidded on gravel. A raking slash of machine-gun fire sent bullets whistling through the leaves overhead. Birds screamed in surprise and departed in noisy flocks.

Emily was pulling on his arm, dragging him to the side, ducking for cover.

He didn't need the urgency transmitted through her guiding arm, still locked around his waist, to know they only had seconds to find cover. He'd scream in frustration if he dared. Blind. Unable to help. Deadweight.

4

Emily blessed every time Larry managed to place one foot in front of the other; he was far too big for her to carry.

He'd been beer-belly bound when she first met him, as wild as most of the pilots in the 101st. Dangerous as hell and on the road down. At least she'd thought he was. Then he'd begun working out more, drinking less, and was soon as fit as he was handsome. He'd also been an exceptional flyer; she'd learned a great deal from Larry Engstrom.

And now he was proving himself to be far above the standard soldier with how he was fighting against the pain and blindness, helping as much as he could.

The pointless gun fire slicing over their heads continued killing leaves.

By the sounds from the vehicles, she'd made a crucial mistake. She'd headed straight from the helicopter back toward the poppy fields. Of course, that was exactly the route the bad guys would take from their vehicles to look for the helo. Should've arced.

Stupid!

It didn't matter that she was a city girl from Washington D.C., she wasn't allowed stupid. Not when the slightest mistake was going to kill them.

Taking advantage of the masking noise of the gunfire, she twisted due north to get out of their direct line of approach toward the crash site.

The gunfire sliced off as if cut with a knife. The sudden silence of the jungle thundered down on them and she froze in place. There wasn't a single bird call. Not even a rustle of something moving through the undergrowth.

Then the shouts in Thai began—so close to hand she almost answered their calls to each other.

Under cover of their shouts, she dragged Larry sideways into a particularly dense clump of undergrowth. Banana, papaya, or Dr. Seuss Truffula trees—she hadn't a clue.

Dragging him down to the jungle floor didn't take much effort. The adrenaline of the crash and their race through the trees could only last so long and it was collapsing out from under her.

She landed hard, but stifled her grunt. Larry fell too, mostly on top of her.

He started to move, but she held him close as a pair of feet and many curses crashed through the brush not a half dozen paces away.

5

Larry tried to do the decent thing, but he was so tired.

He couldn't hear much of anything through the helmet. Someone shouted from a distance away, but he had no way to tell how far. Or how angry.

But while he could neither see nor hear, he could absolutely feel. Despite service revolvers, flight vests, and circumstances—against all odds he was finally holding Emily Beale tightly in his grasp and he was loathe to let go.

Even with all the gear they each wore, he could feel how they would be in each other's arms.

He was past fear now. And, he realized, far past any shred of common sense. It was easy to pretend for a moment that they were on some tropical island—preferably one where no one was trying to kill them—and they could drift together. Turquoise water.

First signs of shock from blood loss, some distant part of him noted. Which was odd, there was no more blood running down his face from the scalp wound, but he was past caring about that.

Larry's nose was still working just fine. He could smell the dark richness of the decaying plant matter that made up the jungle floor—a thick, soft mattress of duff. And also the scent of Emily Beale that he'd know among a thousand flowers: rare, elusive, enticing.

"Don't move," her whisper was just loud enough to penetrate his helmet, but no more.

"Be still my heart," he whispered against her neck and cursed the helmet he wore that let him get no closer.

But her vest had ridden upward off her waist when they'd tumbled to the ground. His hands slid up past her sidearm and wrapped around her waist. Again he imagined what it would feel like to make love to Emily Beale as he had done through a thousand cold showers.

Not some hurried, frantic tumble like with most women found in soldier bars. It was as if they were seeking desperately to cling onto a life, any life, because they didn't have one of their own. Emily wasn't like that. Emily *was* life. She was perhaps the most alive person Larry had ever known.

To make love to her wouldn't be a matter of minutes, hours, or even days. It was a task that could stretch out as a constant discovery over years.

Larry had never thought about years before when it came to women, but the one in his arms now made it seem totally natural.

He snuggled against Emily the best he could, and breathed her life in.

6

Emily wasn't surprised when Larry clung so tightly to her. His feelings for her had been clear from the moment she'd deplaned into the mayhem of Bagram Airfield.

Her efforts to keep him at a distance had slowly weakened. Once he stopped drinking so much, she began to know the immensely skilled flyer. And once he got over the macho, testosterone-poisoned standards that were clearly a pre-req for applying to the 101st Combat Aviation Brigade, she had discovered that Larry Engstrom was a thoroughly decent man. A discovery that seemed to surprise him as much as it did her.

Captain Engstrom was an easy man to respect, but Larry had become a friend. And despite the dangers of fraternization within the US military, she'd begun leaning toward finding a common vacation spot for their next leave.

Hawaii always sounded like the right place for something like that.

Also, letting him cling to her so closely served to keep him quiet.

She knew it wasn't his fault, because he was blind, but he'd stumbled over every single obstacle like a bull in a china shop no matter how carefully she'd guided him.

There were more shouts. They'd reached the helo which couldn't be more than a few hundred meters away; she and Larry had not been moving quickly.

They bought the ruse. The sounds of the pack of Thai bandits went hying off into the distance like a pack of rabid hounds—music to her ears.

She kept listening, ignoring how it felt to have Larry's hands tight about her waist. It felt good—too long since she'd been held close by a man.

Then he shifted and kissed her.

It wasn't some intense or passionate kiss as she'd imagined. Neither was it testing and teasing. It was nice, but it was as if Larry wasn't really all there for it.

A harsh rattle of the big M240 broke them apart. Someone had climbed aboard the downed helo and was having fun with the weapon.

There were numerous shouts and curses.

Larry didn't try to reengage, not that she'd encourage it under their present circumstances. But while she listened for any approaches that circled too close to their hideout, his condition began to worry her.

While they'd been on the move, he'd done well enough. But his hands were no longer holding her as tightly.

7

Larry felt as if he was floating when someone…Emily, rolled him onto his back.

It felt so good to lie here beside her.

If only it wasn't so cold.

"When did the Thai jungle get so cold?" He fought against a shiver.

There was a distant zing of pain as someone…Emily? Yamota?…pulled off his helmet.

He was glad to be rid of the weight.

Also, with the removal of the helmet, the pinch of pressure that had been giving him a splitting headache eased off. With a spinning flash of color, so sharp that it gave him a moment of vertigo, vision returned to his left eye.

And the view was lovely.

Emily's elegant features, framed by her straight, white-gold hair, hung just inches above him.

"Hi, babe."

"Shh, Larry. You have to stay quiet."

"Right. Bad guys." Oops, he'd been speaking aloud. "With guns." Shit! Shh.

So, he occupied himself with lying still as Emily inspected him.

"Cold," he whispered.

He could see the worry on her face. He didn't like seeing that. So instead he focused on her neck. And imagined the rise of breast he would encounter if he were to start a hand there and slide it inside her flightsuit. Any number of hot, sweaty training sessions done in t-shirts had left little enough to the imagination. So, he preoccupied himself with imagining that last bit.

Emily-of-the-perfect-breasts moved out of his view. He felt her fingers poking and prodding him with what he recognized as a medical assessment.

Her sharp hiss drew his attention back from imagining his mouth tracing down that neckline toward heaven.

"Shh," he reminded her. "Bad guys. With guns."

8

Emily opened the medical kit that she knew was useless and looked in it anyway.

With five tons of medical supplies in the back of the helo, none of them had thought about loading up full combat med kits. The kit she'd recovered from the pocket of her copilot's door didn't have thread and needle to attempt to put Larry back together, even if she'd known how. The small tube of skin glue wasn't going to make much difference either.

The pantleg of his flightsuit was soaked in blood. It wasn't arterial, but under the circumstances that might have been a mercy. If the shot that caught him had cut an artery, he'd have bled out in a few minutes. Instead a line of bullets had passed through his thigh. Through-and-through meat shots and he'd been bleeding out of them the whole time.

She could stop those…maybe. He'd be weak and shocky from blood loss, but she could glue and bind him up in time if she hurried.

It wasn't only lack of vision that had him stumbling so badly. The adrenal miracle was that he'd walked on that leg at all.

Then she'd spotted more blood up at the webbed belt that his knife and holster hung from. Just a small blotch of it. A bullet had ricocheted off his sidearm, slipped in beneath the edge of the belt, then been covered and held closed when they'd climbed out of the helo and the belt had settled downward over the hole. The hole wasn't the problem.

Using his knife, she sliced open the flight suit to expose his belly. The tumbling bullet had torn up his internal organs. His stomach was dark and distended with massive internal bleeding.

Larry was already dead.

9

Larry tried to pull Emily back into his arms.

She resisted for a moment and then lay down beside him.

"My beautiful Emily," he remembered there was some reason to whisper, he just couldn't quite recall why. He brushed a hand—so heavy to lift it—over her cheek and she struggled to smile for him. He could see it was hard.

Had he been too forward?

Then she took his hand in hers and pressed its palm against her cheek. His fingers left a blood-red imprint on her perfect skin.

"Are you bleeding?" he struggled to get up and check her for injuries.

"No, Larry. You are." Her whisper close beside his ear was the gentlest caress. He'd always imagined it would be like this with her. Then he made sense of her words.

"Me?" He saw the flash of pain inside those ice-blue eyes it had taken him so long to learn how to read. "That's alright then. As long as it isn't you."

"You're better than okay. You're my idea of a perfect wom—"

"I'm okay."

10

E*mily clamped her hand* over Larry's mouth.

The stealthy footstep sounded along the pathway she and Larry had battered through the undergrowth as she led them back toward the poppy fields.

She held her breath as she watched the entrance to their tiny bower. She didn't dare even reach for a gun or knife, because she didn't dare uncover Larry's mouth.

His lips were moving against her palm, way past knowing he shouldn't speak. Thankfully his muffled vocalizations were very weak. Too soft to understand even though she lay close against him.

The steps neared.

Larry, finally understanding that he shouldn't speak, looked up at her with his one good eye. The other eye tried to look up too, but kept drifting aside.

The steps moved past them as Larry's one good eye struggled to convey some message. But the drug lord's soldier had stopped to listen, or light a cigarette, or scratch himself.

There was a splashing sound.

He was taking a piss not three meters away. In moments she could smell it on the air.

Larry stared at her with his one eye as if trying to memorize her face.

Emily studied his face, knowing it was his last message, but she couldn't read it. Couldn't find what was important enough to be his last words.

Then he moved his lips one last time against her palm. He formed a kiss, closed his eyes, and died with a final sigh the same moment the soldier finished his business and moved farther off through the brush.

11

Emily never cried, it wasn't in her. But her eyes burned for a long time as Larry's hand cooled in hers through the afternoon and evening. The rescue force arrived in the darkness.

She never heard the helos that must have delivered the Combat Search-and-Rescue team. A group of the drug lord's men had camped at the jungle's edge, close enough that she could see their campfire's light through the trees and hear their soft talk.

Then there had been a series of soft spitting sounds, each accompanied by the distinctive click of a bolt returning on a silenced weapon. She listened, but couldn't hear any more Thai voices from the campfire.

She slowly, silently as possible, pulled the small velcroed patches aside that would reveal infrared-reflective patches. They would glow brightly if the person was wearing night-vision gear, identifying her as a "friendly."

Emily never heard the soldier approach. One moment she'd been alone with Larry's cold corpse and a moment later she knew she wasn't.

"First Lieutenant Emily Beale," she read off her service number to the silent darkness in a whisper.

"SEAL Commander Luke Altman, Lt. Beale. Pleasure to find you among the living."

"The only one."

His answer was a grim silence.

"The downed helo is another two hundred meters just south of west."

He transmitted the information to some of his colleagues.

Emily was at a loss as to how to move Larry when a rifle was pushed into her hands.

A moment later Larry's body was gone from beside her. With a grunt the SEAL shouldered Larry in a fireman's carry and then reached to take back his rifle. "Hang onto my belt and I'll lead you out of here, ma'am."

Emily walked into the darkness, trusting to the man to lead her just as Larry had trusted her. One hand on the SEAL's belt, the other once again holding Larry's chill hand—frozen with lifelessness despite the heat of the still, jungle night.

When she had considered being with Larry, she'd been forced to contemplate the possibility of losing the right to serve in the military. This morning, before his death, there had been a choice.

There wasn't any longer.

With simple gestures, a sudden rise of hip or a sideways shift, the SEAL led her around and over obstacles in the jungle. Having led Larry over the jungle floor, she could appreciate how effortlessly the SEAL guided her.

She had done all she could to save Larry. The man...it was hard to even think it...the man whose last message on Earth had been one of love. Love, like tears, wasn't in her, but the ache in her chest ran deep.

When they emerged beneath the starlight, she knew what she was going to do with the rest of her life.

It wasn't to pay back a debt. Nor revenge. It was a thankfulness. A thankfulness for Larry's love, for the SEAL's effortless guidance,

for being alive. If she could one day be the person to stand outside a jungle bower and find not one living soldier, but two, it would be worth it.

When the black helicopters of the 160th SOAR descended through the night sky she let go of Larry's hand and thanked him silently.

Prior to this day, Emily had not known what she wanted to do next with her life. A West Pointer. An officer of the US Army. A helicopter pilot for the 101st Screaming Eagles. She now understood that had been merely her preparation.

As she climbed aboard the Night Stalkers Black Hawk helicopter, she knew exactly what she was going to do next, even if the 160th SOAR didn't accept women.

When the pilot introduced himself over the intercom as Captain Mark Henderson, she considered informing him of who she was and that she would be flying beside him in a few years—female or not.

But she would let that wait. She'd let her actions speak rather than her words, for Larry had taught her how full and how much more important silence could be.

For now, she had faced death for the first time.

Emily was going to prove that it had not found her wanting.

Gas Grilled Chef!

I voyaged back into my effort to write a thriller in a short story structure. Once again, I produced a fun, fast-paced…suspense story.

But I had a second purpose this time.

Kate Stark, the primary heroine of my Dead Chef thrillers, has an utterly useless twin brother. Paul Stark is rich, handsome, and terribly spoiled. He has ne'er-do-well down to a science. He is also a rather masterful con-artist who, on the very rare occasion, does *come through when it really counts.*

I knew that in the next book (which is still in partial draft at this writing), I would need to redeem Paul. He'd have to take an interest in the family business (a massively successful television food network that has the odd problem of chef's dying on-air), if he was going to be deeply involved in the next book.

So, Gas Grilled Chef! *was partly fun because I got to learn a great deal about high-end gas grills (they're really incredible machines with unreal accessories). I also finally found the glimmer of Paul Stark that gives me hope for the future.*

And it really was a lot of fun killing off the chef (the reason for the series' Dead Chef moniker).

M. L. BUCHMAN

GAS
GRILLED
CHEF!

A DEAD CHEF
THRILLER STORY

1
-Yesterday-

Grill-chef Penelope Boudreaux headed out to her Fire Dragon Ultimate Platinum Grill perched in the two-story glass pavilion at the end of the dock. Lake Pontchartrain glittered, the waves dancing before the warm morning breeze.

This was her domain. The world could peek in through the glass walls and admire, but none dared enter without express invitation. It was where she entertained her friends, posh parties that sent glittering light out over the lake's waters far into the night.

Upstairs, in a discretely curtained bedroom, is where she entertained her lovers.

But most of all, the glass pavilion was where she filmed her weekly cooking show for the local network.

She wanted one more round of practice before her next appearance on camera. It wouldn't be the local cable network, nor the local affiliate of the national network like it should be. This time it would be national. The Cooks Network invitation— signed by Kate Stark herself—to *The Belles of the Hot Grill*

Cook-off had placed the future within easy grasp of her perfectly manicured nails and she wasn't about to bungle it through lack of preparation.

The prize was a typical ten thousand dollars-on-an-over-sized-check reward which would barely pay her expenses for the trip and definitely wouldn't pay for her beautiful grill, but that was only the stated prize. Penelope lusted after the unstated prize—a shot at her own show on the biggest cooking television network of them all—the same way the new garden-boy lusted after Penelope.

Just this morning he'd proven that his desire was backed up with acceptable technique and exceptional heft. When he'd finally unveiled that impressive supply of raw material, any concerns she'd had regarding skill had gone out the window. She'd let him take her right in her private boudoir in the big house where even her husband Walter didn't dare tread. The boy didn't have much imagination, but she had plenty of ideas of what to do with all that equipment the next time.

But now she had to cook.

She tossed a scoop of wood chips into the dedicated smoker and turned it on. Then she dug into the refrigerator built into the grill and pulled out the ingredients she needed. In moments, she had a strawberry daiquiri blending in the built-in attachment. She was going to win this hands down with a grilled shrimp and gator jambalaya, Creole-style. Hot, hot, hot!

Penelope ran her hands over her satin blouse as she smoothed on her apron. For just a moment she imagined they were the gardener's. Walter's were all dandified Southern lawyer. It had been ages since he had cared about her breasts one way or another, even if they were Penelope's best feature. She'd paid a lot of money—or rather Walter had without knowing—to make sure they were.

But the boy had been fascinated by them with those strong, calloused hands and his overeager mouth. A little training and he would be most satisfactory for a while.

Penelope resisted the sudden urge to go find him again right now and tied off the apron with a sharp pull around her trim waist to remind herself of what was truly important.

She held open the safety and pressed the sparker to light off the one-hundred-and-fifteen thousand BTU main burner and the sixty-thousand BTU side burner of her nine-foot long, eight hundred pound, stainless steel beauty.

It was the last thing she ever did.

The cut hose inside the sealed grill had filled the interior spaces with five gallons of compressed propane gas. The small spark she ignited lit it off. The explosion, assisted by some small charges of TNT, ripped open the remaining internal tanks. Fifty gallons of propane were involved in the next stage of the flashover.

The explosion shredded the grill, the sitting area, the tastefully curtained upstairs bedroom, and then blew the glass pavilion outward in a vast cloud that would have been a disaster for several neighbors except that the glass had been mostly reduced to the size of sand by the scale of the blast.

The outer fifty feet of the dock simply disappeared, the light chop on the Lake soon extinguished then buried the stubs of the shredded pilings.

Penelope wasn't vaporized, not quite. But she was burned to a crisp except for her breasts, which melted. As she had pressed the switch, she'd been thinking about the boy slipping into her as nicely as a slab of gator into her jambalaya.

Penelope never quite completed that thought.

2
-Four weeks earlier-

*T*he *Top 10 Ladies* *of Chowder Cook-off* had flowed seamlessly. Kate Stark had rented a ballroom at The Lenox Hotel located in the heart of Boston's historic Back Bay. She'd made sure there was a good supply of champagne and light canapés and they'd had a wonderful time which made for great film.

On the back of that success, Kate had put her head together with her program director Mac Olson.

"Oh, honey," he'd flapped a hand at her to make sure he had her attention even though they were the only two passengers on her private plane back to New York. "Oh, I've got it. *The Ten Belles of the Hot Grill Cook-off.* Can't you just see it? We find a simply gorgeous Southern estate, line up ten grills, and ten Southern women. They always make such great theater. They'll dress for it in proper Southern style."

"None dressed so finely as you though, Mac," she'd teased him.

"Well, of course not, honey. They don't stand a chance. But these Boston women today, such understated elegance that it

was almost invisible. Why if I hadn't been there to help them out…" his deep sigh clearly stated what a mortal disaster that would have been. "Oh, I can't wait. Make sure there's a lake and a broad stretch of lawn for the ten grills. Just imagine the visual contrast between today's formality of The Lenox and the Southern sunshine."

Mac had painted such a picture in broad gestures that she'd bought in. By the end of the flight home they even had a list of names; one per state from Virginia to Florida to Texas. Texas made it up to thirteen which was just too many. They dropped the unlucky contestant thirteen and made tentative plans for *Ten Women Who Bring the Beef Cook-off* to cover Texas and up through the Midwest.

"Twelve, we can manage twelve."

Kate's location scout had a site in mind the moment he saw the list of chef's names. There was a reason she hired the best in the business.

The selected estate belonged to Chef Dee Dee DeRue—what kind of a name was that? Though she was certainly 3D, barraging Kate in every way imaginable: phone, e-mail, and social media. She called Kate hourly to check on everything from the plantings in her garden to the best shade of highlights for her hair.

Kate had a network to run and had finally palmed the woman off on Rikka, her top freelance camera operator. Whether the "direct access to the source of knowledge" had calmed Dee Dee down, or Rikka had threatened the woman's perfect facial construction until she shut up, didn't matter. The problem was off Kate's desk.

3
-Today-

D*ee Dee's estate was* perfect. Kate, with Rikka at her side, wandered about wondering if she'd stepped into some fantasy kingdom. Mac was going to be so sorry that he missed this shoot but his schedule hadn't allowed it.

The requisite pond—"heavily stocked with trout for Daddy James (who turned out to be Dee Dee's husband) and his friends"—was gorgeous, a shimmering mirror of blue South Carolina sky. It also included a fishing boat that might have been better placed in a body of water a few thousand times the size of the pond. But it was decidedly picturesque.

The landscaping was so suited to her needs that Kate finally decided it had to have been freshly planted by Dee Dee for the occasion. On the perfect green lawn at the shore of the pond stood a semicircle of a dozen evenly-spaced Georgia Pines, each exactly the same twenty feet tall. At the base of each tree were planted identical clusters of wildly blooming bushes of pink and red azaleas and blue hydrangea.

A grill had been parked in front of eleven of the twelve pine tree-and-blooming bush plantings, each on its own red-brick pad.

"So we can wear our heels while we cook, sweetie," Dee Dee explained to Kate, showing off her Jimmy Choos. "A girl always needs to look her best."

The plantings of only one grill position, Dee Dee's, also sported yellow and purple azaleas—which had caused quite the furor among the other women. It was only partly mitigated by the fact that her dearest friend Priscilla Danz's was set up next to hers and sported a few purple blooms in addition to the normal landscaping. "Friendly competition and all, honey pie."

Kate was going to "honey pie" her in the nose quite soon.

The twelfth pine-and-azalea setting had no grill on it's red brick pad. The tragedy of Penelope Boudreaux' death yesterday had shocked everyone. In place of the missing grill Dee Dee had set a tasteful wreath, spray painted black, and a small vase of the yellow and purple azaleas cut from Dee Dee's own backdrop.

The eleven grills that the contestants had brought were even more of a spectacle than the landscaping.

In Boston, the chowder cooks had brought a favorite pot, knife, and cutting board; all the worse for wear.

Spread across Dee Dee's lawn were eleven of the most ostentatious grills Kate had ever seen. They all shone as if never used, though all of these women present had at least state fair-winner level credentials. Three had their own local TV shows and two had managed to tap regional networks.

Kate wasn't above finding a new Southern cooking grill show if she found the right host.

One grill had side-mounted warmer plates, a wok burner, an auxiliary hibachi-sized grill, and so many other attachments that it looked more like a rock-and-roll drum set than a cook's station. Several sported an array of spatulas, forks, and brushes sufficient to stock a restaurant supply store. Dee Dee's own grill was gold-colored with that burnished brass look—at least Kate hoped it was brass. It was blinding to look at in direct sunlight

and would be almost impossible to film well. Thankfully that was Rikka's problem.

The one from Georgia, Priscilla Danz's, was a monster that could cook a whole side of beef. It sported four propane bottles, all tastefully tucked out of sight behind burled redwood paneling.

In an odd fit of consideration, Dee Dee confessed to Kate, "Let Priscilla start first. She's so famous, you'll want to get your best film of her."

Priscilla's *Red Hot Grill* show was undeniably popular—her Atlanta show had been picked up by three stations already. Mac had found a tape of one of her shows for Kate to watch. The woman was a fine presenter. She also sported a long flow of bleach-blond hair and a cleavage just as long and nearly as well-exposed.

"Is she selling a side of food with that sex?" Rikka had whispered merrily.

The pre-filming dinner overshadowed everything else. North Carolina still hadn't showed up. She'd sent her grill by truck yesterday but was supposed to drive down today. They decided not to wait.

What had been a charming affair in Boston turned into the "Dinner Before the Battle."

Rikka hung in the background as the ten chefs graced the long cherrywood table beneath a line of crystal chandeliers. The room itself had all the ostentation of Tara, the *Gone with the Wind* mansion—marble floors, white dining chairs, damask wallpaper, and gold-framed oil paintings of vistas of the estate grounds.

Any spouses who had tagged along had been relegated to Daddy James' boathouse, an air-conditioned man cave that included pool and poker tables, a massive television tuned permanently to ESPN, and a full wet bar.

By the time the ever-so-polite passive-aggressive sniping kicked in—somewhere between sitting down and picking up their napkins—Kate decided she'd have been better off joining the men.

Afterward in the shared suite in the East Wing—everyone else except best friend Priscilla had been shooed off to the Hilton in town—Rikka had suggested that they skip out before the debacle of tomorrow's cook-off. "I have plenty of film to launch a catfight soap opera."

"We're in the cooking show business."

"You always were stubborn, Kate. They're going to shred you tomorrow."

Kate looked down at Rikka who was most of a foot shorter. "They wouldn't dare; they all want their show on my network too much."

"If you say so," Rikka shrugged from where she dropped down to slouch on a divan covered in brocaded roses. She was wearing her typical black jeans and t-shirt. She propped her black sneakers on an oak coffee table that might have been fake Edwardian, or perhaps fake Grecian. With her straight, jet-black hair, her narrow Asian face was the only part of her that really showed. That and her white hands dipping into a bag of Fritos, that she'd scrounged from who knew where, and her electric blue socks.

"Disaster, you think?" she'd learned to trust Rikka's instincts in such matters.

"Duh!" Rikka found a remote control and flicked around the channels until she found something with women's screams and 1950s giant rubber monsters. "I love cable."

Kate considered, then picked up the phone and began placing calls.

4

Paul Stark rolled into the compound shortly after dawn. His twin sister had called for help, which he knew was hard for Kate. She'd done it perhaps a half dozen times in his whole life. Whereas he was always...

At the airport he'd debated over the rentals. A Mazda Miata was an amusing sports car and about as good as airport agencies ever got. He considered the high-end rental guy in town, had used a Ferrari when he'd been seeing the Governor's daughter a few years back. She'd definitely thought of some fine ways to thank him for showing her a little style.

But this crowd sounded like a problem seeking a different solution. He called in a few favors and rolled up under the estate's Greek-pillared porte-cochere in a cherry-red Cadillac XLR convertible.

The ladies were having morning tea out on the sun porch and every eye had tracked his arrival. He pulled off his caramel sports jacket, tossed the Oliver Spencer negligently back onto the car seat, and moved up the broad, white marble steps to join them.

He walked by without acknowledging a one of them, not that he hadn't learned to peg most women on first glance.

There were ten of them, all very well tended. Twenty-six or so, up to mid-fifties—most of the latter had purchased the figures of their younger counterparts. Three clearly authentic, pure-to-the-core bitches—they were the ones who owned the money rather than their husbands and wielded it ruthlessly. Two more that fit the same bill, but without the wedding band—*I'm just temporarily between men, dearie. Care to fill some of my lonely hours?* He knew that line well enough and was usually plenty glad to oblige. Have to see how today turned out.

The last five looked more sane if no less well tended: two decent spouses, two inherited estates, and one he couldn't quite read—which made her the only interesting one in the crowd.

Money did strange things to women, except for his sister.

Kate had been the same woman since they'd come out of the womb together. She was now one of the wealthiest women on the planet and managed the most successful television network out there, cooking was only one of the many channels in their "family business." And her dry comment of, "Rikka says I need a pro to handle these women," had been absolutely right. Kate was way too trusting. After all, she kept trusting him and he knew just how bad a bet that was.

So, he walked right up to the most beautiful woman in the crowd and kissed her on both cheeks, "Hey, Sis! Thought I'd drop in and see what you do for a living when you're not busy making us disgustingly wealthy."

The atmosphere on the porch shifted abruptly.

Moments before, they'd all been carefully poised and positioned looking for the inside track with the owner of Cooks Network. His little speech had deflected most of the attention toward him. He flashed one of those casual smiles he'd tested on every girl since kindergarten. Except he aimed it at Rikka instead of the other women, which should make most of them even crazier.

"Hey Rikka."

The only woman on the planet consistently resistant to his charm offered back a sneer.

Yep! Everything was in place.

He clapped his hands together and rubbed them happily. "So when do the games begin?"

A sleek, and very well-muscled, very male personal assistant came hustling up to one of the first-category women, rich-bitch-in-control-of-the-purse-strings, and whispered something in her ear.

She went sheet white.

"What is it, Dee Dee?" Kate was moving forward rather than watching the crowd.

Three in the crowd looked unconcerned, one of whom had been looking away, so she had an excuse. There was also that one woman he couldn't get a read on.

And then "Dee Dee" apparently unable to speak, waved a hand at the man-servant.

He cleared his throat, "Chef Tessie Cummings of North Carolina won't be joining us. While on her way here yesterday, her car was in a very bad accident and she burned to death before they could save her."

5

*K*ate *suggested shutting the* competition down, but the ten remaining chefs inundated her with their pleas.

"No, Tessie and Penelope would have wanted us to soldier on."

It was ghoulish, not helped in the least by the male secretary/boy toy's next question which had been to ask if he should order a second wreath to set on Ms. Cumming's grill.

"In its place, dear boy," Dee Dee purred. "Have her grill packed for return shipment."

Kate had been about to shut the contest down despite their protests, when Paul pinched her arm.

"Let it run," Paul dragged her away from the crowd and led her down to the screened gazebo that overlooked the pond.

She inspected him closely. "This isn't another one of your games, Paul."

"But it's such great theater," he waved grandly.

"Which in this case sucks for television." She pretended to sound like an announcer, "Tonight we feature a cooking competition in which two contestants died before they even

arrived at the show. Look at these totally tasteless black wreaths we've put on the set of our merry cooking show to commemorate the event."

"You mean memorialize."

"I chose my word carefully."

"C'mon Katydid, don't you want to know who did the deed?"

"What deed?" Why were all conversations with Paul like this?

"The dirty one."

"Why, how many of them are you planning to have sex with?"

"None. Well maybe…but that's not the point. I'm talking about the other dirty deed."

"This," Rikka said from close beside her elbow, scaring the crap out of Kate. She could sneak up on anyone, anywhere, and was always doing it to Kate just to make her completely crazy. "This is why I had you call Paul."

Kate looked from Rikka to Paul and back. "Why? What am I still not getting?"

"You're way too nice, Sis. Not a foul thought about anyone."

"I have a couple about you at the moment."

Paul sighed. "Don't you want to know which of those women killed the other two chefs?"

Kate stopped at that and considered the statement. It wasn't… *wholly* irrational, just mostly. But what if it was the case? "You think one of those women is a murderer?"

"Yes—" Paul started.

"No!" Rikka cut him off. "Not a murderer. A well-tended murderess in Armani slacks."

"See, you're too nice for this, Sis. Two chefs burned? Not too likely. Gas grills are incredibly safe, yet Ms. Louisiana blows herself off the face of the map. Ms. North Carolina is toasted in a ball of car fire. C'mon!"

"From what the other ladies are saying," Rikka picked up, "the two that are gone were top contenders. Possibly the only ones to take on Priscilla and Dee Dee. I'm betting one of them is in on it."

"Or both together, best friends and all. You need an underhanded sneak like me to find out what's happening." Paul furrowed his brow in concentration. "I need to find a way to get in close with these ladies."

"He's good at that," Rikka commented with a dry tone of disgust. "So, what order are you going to take them out joyriding in your car?"

"Eww!" Paul gave a fake shudder. "I prefer to make sure a woman isn't looking to kill me before I invite her into my bed. Doesn't always work out, but I try."

"Up close and personal?" Kate could feel an idea forming. An idea that Paul and Rikka were going hate, but she actually was feeling pretty good about.

Since their parents' death, Paul had left the network completely up to her to run. She'd made them both wealthy, while he'd gone gallivanting off in every direction. Which didn't bother her, much.

"I know that look, Katydid."

"What look?" Rikka circled around to look up at Kate's face. "Ooo, the evil plan look. I *love* that look."

Paul had surprised her. He'd done well during both the fiasco with the North Koreans and that thorny mess with the G-8 meeting in Scotland. Maybe...

"I think, brother mine," she said it with all of the managerial authority she'd learned from imposing her vision on an entire television network, "that it's time you started earning your keep."

Paul looked at her in confusion.

Rikka narrowed her eyes to even finer slits than they normally were, then they shot wide and she started cursing.

"I have to work with that?" she jabbed her finger so sharply at Paul that she almost skewered his nose with a chewed-short fingernail.

Rikka had always been smart.

6

Hi, ladies. I'm sorry we weren't able to announce this sooner," Paul scanned the ten chefs tastefully arrayed on sofas and settees. The living room, like the rest of the place, had been designed to entertain, not live in. The white marble flooring had large throw rugs scattered across it. On each rug, a circle of seating and small tables were gathered. From a tasteful area for six to lounge comfortably in a wood-paneled corner, to the main area that could hold twice their present number with equal ease.

You could also hold a rumba competition down the center.

Wet bar, bad paintings, and an impressive array of glass cases filled with Dee Dee's artfully-lit cooking competition trophies.

"I was unsure if I'd be able to make it down here in time from a prior commitment," Paul continued. It had been a rather sultry redhead who he'd met at a fundraiser when Katie called. "But I'm thrilled that I was able to get here. I'll be the host of this competition."

There was a little round of applause. Some looked eager, some avaricious, some neutral, and Tennessee still ranked as

enigmatic. She had neither foofed hair nor excessive makeup. Her brunette hair hung straight to her shoulders, her nails were unpainted and her lips barely so. Soft brown eyes that simply followed him and the events in the room.

He also knew from what Katie said that she was one of the best cooks in the room and a solid performer on air, just not the flashiest.

"The best light for filming out at the grills won't be until after lunch. So what we'd like to do is a series of personalized interviews this morning. You, me, the camera," he waved toward Rikka without giving her the finger. What was so wrong about having to work with him anyway?

"Dee Dee has offered to let us set up in her kitchen. Each interview will only be a few minutes. A chance to get some good shots. Because we only have the one camera, we'll sometimes stop to take some video of me asking the questions. Things like that."

Everyone was nodding and smiling.

Two hours later, he'd been propositioned three times on camera. Had a woman's foot run up his leg under the table thrice more, once by one of the women he'd read as a "good girl." And neither from one of the five that he'd tagged as pure-to-the-core bitch on his first arrival, Priscilla Danz. He still wasn't willing to change his assessment of her, no matter how much he admired the extent of cleavage the *Red Hot Grill* host chose to display. Dangerous woman even by his standards.

And he had learned absolutely nothing new except their names; though it was still easier to think of them by state. Though when he'd found out that the enigmatic Tennessean's real name was Wilma.

"There's a reason I go by my middle name of Annie."

He liked Annie, rather hoped she wasn't the killer, but her thoughts were as elusive as the others were blatant.

"I don't know what else to say, Katydid." He, Kate, and Rikka had moved off to the side to confer. No one had picked up any real clues.

"Let the cooking begin," Rikka suggested.

Katydid shrugged her acceptance.

"Fine," Rikka picked up her camera. "But you aren't going to find me doing any taste testing."

7

*W*hat are you doing here, James?" Dee Dee sounded surprised.

"Come to watch my little gal win, of course." Kate watched as he pulled his wife against him in a side-hug and kissed her on the temple. "No offense to you other fine ladies, of course," the tall, handsome man offered a charming smile around the gathered circle before retreating to a chair in the shade and an afternoon Wild Turkey on the rocks.

And they were off.

Kate scooted Paul ahead.

Rikka had him and each of the chefs wired for sound. There were also a couple of general ambiance mikes to fill in background noises so the air didn't sound too dead.

Kate's job was to follow along wearing headphones as if she was doing something with the sound for Rikka's filming.

What she was actually doing was listening to all of the open microphones. They were each recorded and stored separately for later use in final sound mixing, but it allowed her to have

a far wider view of what was happening around them than the camera or even a set of eyes offered.

Florida and Arkansas were trading recipes. And it sounded as if Kentucky and Virginia might be trading men, but most of it seemed innocuous enough.

Unable to settle on the contest's cooking order, Kate had finally forced them to draw names from a hat. West Virginia led off and did a credible job with her opening of grilled-game Burgoo stew and griddle-baked Johnny Cakes. No television star, but the recipe sounded good.

Arkansas' presentation was much sharper, but her grilled catfish spice rub sounded awful. She'd been a Mary Kay cosmetics saleswoman only recently moved to the kitchen. Her makeup, however, was awesome.

Priscilla Danz of the *Red Hot Grill* did a great job. Her presentation was sharp and funny. She made she-crab soup and grilled Vidalia onions sound both simple—once you knew her secret tricks that she was going to demonstrate later—and delicious.

She was actually a little terrifying though and Kate didn't know if she'd be able to sell the woman. Every gesture and move was calculated to place her chest front and center. Her side comments about her competitors moved her to the top of Kate's suspect list. She was one of the seven deadly sins incarnate: avarice lived and breathed inside that plus-sized chest.

Dee Dee was hard pressed to follow Priscilla's act and she knew it, but she put on a brave show. Her Frogmore Stew of grilled sausage, corn, crab, and shrimp did sound delicious.

Then, as each had done, at the end of the introduction moment with Paul, she bent down to light her grill. It clicked loudly as had the others, but there was no answering soft thump as the gas caught fire.

She tried again and it didn't work.

"I can't get my grill started," then she looked up at the camera aghast. "You can cut this out, can't you, dear?"

Rikka nodded and mumbled something reassuring.

Kate was so glad that she'd palmed the woman off on Rikka.

"Is your gas on?" Paul asked. Like he'd have a clue about anything mechanical. He could barely work the television that he'd insisted on buying for their shared condo.

"Let me help," Priscilla Danz moved over beside Dee Dee, tossed her hair, and squatted down, making her cleavage even more dramatic.

"Prissy! No!" The call hadn't been very loud, but it had been alarmed.

Nobody reacted.

In an urging tone rather than a panicked one, "Prissy, move away!" sounded in her ears.

In Kate's ears.

It was over one of the open microphones and she was the only one listening to them all.

A man's voice.

The only men here were Paul and…

Kate shouted for them to back away from the unlit grill.

8

*D*addy James ran.

Rikka tripped him.

Paul sat on him.

And when Kate informed Dee Dee that her husband had intended to blow her up so that he could have all her money and keep sleeping with her best friend Priscilla, Dee Dee kicked her still prone husband in the balls. Very hard. With the pointed toe of her Jimmy Choo.

They had to wait for him to stop screaming before they could work out the rest of it.

The deaths of Penelope and Tessie had been to set a pattern, so that no one would suspect him when his wife was toasted by her malfunctioning grill.

"And he knew Priscilla, as Dee Dee's best friend, would be above suspicion," Rikka added to the discussion.

"It was just a summer fling, James," Priscilla practically shouted down at the man. She planted the point of her Chanel right where Dee Dee had landed her Jimmy Choo. Apparently

James had misread Priscilla's intentions. "I'm between husbands is all, Dee Dee. I didn't mean to hurt you."

In moments the two women were bonding over the troll of a man they were both done with.

Kate really didn't want to know these people.

And by the time the police were done taking statements, none of them wanted to know her either.

There had been four other counts of adultery on the recorded sound track, three more of trading tips about how to cheat on taxes, and innumerable discussions of random lovers.

A couple of them went home with their husbands and their grills. A couple remained wholly unapologetic.

And Annie from Tennessee had spent the long afternoon and evening beside Paul.

"I've invited her to come cook for us, Sis," Paul informed her.

Kate slumped back in her steel patio chair as the last of the police left with the complete set of recordings and James DeRue in tow.

She looked up at Annie from Tennessee. She was one of the three who had done nothing offensive in the last twenty-four hours.

The woman was very presentable and would film well. Her long form and nice figure would play well to the camera without counters chopping her off at the waist and making her look too short. Rachel Ray's counters were decidedly lower than the norm because the woman was only five-three. But the lowered counters often made her guests look gawkily tall.

"Are you a good cook?" Her reputation was still small but very good.

"You'll let me know after you taste my food."

"Fine," Kate like the simple statement in place of any bragging. She started gathering her belongings. "When's that?"

"Well, Ms. Stark," she reached out and took Paul's hand. "Your brother invited me to come stay in your New York condo for a while. I'd be glad to make you breakfast tomorrow morning."

Rikka snorted as she passed by Paul with her camera case and "accidentally" smacked it into his knee.

As Paul began hopping about in pain and sending curses to follow Rikka across the lawn, Kate looked back at Annie.

She looked straight back, perfectly poised despite the sudden distraction of a man hanging onto her shoulder as he hopped up and down on one foot, using her like a crutch for balance.

"Well, Annie. If you can survive my brother, you certainly have the poise to be on the air. It will be a pleasure."

They shook hands on it.

Fire at Gray Wolf Lookout

This is the second story in my Firehawks Lookouts series. It wasn't until I'd written this that I realized I was writing a short story series. This was the first time I'd done that, where the stories were connected more to each other than to the novels that inspired them.

The Firehawks series is about the heli-aviation wildland firefighters of my Oregon-based Mount Hood Aviation team. That primary series, which will be gaining (has gained) its fifth novel in late 2016, spun off other series.

The Firehawks Smokejumpers series was three novels covering the love stories of those heroes willing to jump out of perfectly safe airplanes to parachute into active fire zones.

Then in short stories, two series broke off: Fire Lookouts and Hotshot crews.

For this story, several pieces came together. I write a lot of alpha heroes in my romantic suspense series (both male and female alphas). In Fire at Gray Wolf Lookout *I wanted to write about two normal people with unusual occupations.*

He is a fire tower lookout.

She is a wildlife biologist specializing in wolf behavior. She is based on an actual wolf biologist. She met her husband, another wolf biologist, over the radio during scheduled check-ins. It was months before they met in person. This character isn't her; Patty Dale is very much her own person.

M.L. BUCHMAN
NPR and B&N Top 5 Romance Author of the Year

"Will blow the readers'
minds and leave them
awestruck."
Romantic Times, Pure Heat
4-1/2 stars

Fire at
Gray Wolf Lookout
a Firehawks romance story

1

The view of the Lolo National Forest on the Idaho-Montana border spread for a hundred miles in every direction. And Gray Wolf Summit fire lookout tower commanded one of the most beautiful and most remote regions of the forest. From his perch Tom Cunningham could see much of the Lolo, a big chunk of the Clearwater, and even the north tip of the Selway-Bitterroot Wilderness.

Despite being in his mid-twenties, he felt like the luckiest kid in the US Forest Service.

No one was watching, so what the heck, he spit off the edge of the tower. Like a twelve-year old, he watched as it was the light breeze carried it past the cliff and down into the canyon—he watched it as long as he could.

The whole acting-his-age thing had never really worked for him anyway, and someday he'd have to apologize to his parents for that. Both professors at the University of Washington—English lit Dad and Mom the chemist—and Tom had used his degree in geology to be an auto body shop mechanic.

His rut was obvious, didn't need to be on the outside to see it, Tom could feel it from the inside just fine. Like the crippled vehicles that streamed through his shop door, he couldn't seem to drive straight down any path…and that was on the rare occasions when he got running at all.

Screw that!

Last winter he'd gotten so sick of himself that he figured the best solution was to get away—way away!

He'd grown up in Seattle's Wallingford neighborhood, side-by-side housing that would be suburbia if it wasn't now tucked well inside city limits. It was also saved from that awful fate because the houses were fifty to a hundred years old rather than tract built pillboxes.

However, his experience with the great outdoors was limited to a couple of trips out to Snoqualmie Falls, a two hundred-and-fifty foot waterfall up in the Cascades. A good place for taking a girl on a nice date as the lodge had an excellent brunch.

His present situation, atop a Montana fire lookout tower, had been Lucy's idea. After six months of sharing a bed most nights she'd told him to go jump into a fire—not her exact words. Something about his total lack of either direction or ambition. Hearing this from his parents he could tune out. Hearing it from a hot brunette as he watched her fine behind departing his third-floor apartment for the last time, that was a bit harder to ignore.

He'd hopped on the Internet. And when he'd looked up fire—for lack of anything better to do—an image of wildfire had caught his attention. Somehow, that single glimpse had led to enrolling in a fire lookout certification course and quitting his job as a car mechanic.

"Now you've done it, buddy," Tom looked out at the view and decided that whether stupid, whimsical, or psychotic, it had been a damn fine decision—perhaps the first good one in his adult life.

He clamped his hands on the heavy wood rail and gave it a shake—not even a wiggle. His new home was as solid as the rock it stood on.

The Gray Wolf Summit lookout tower was perched at over seven thousand feet. The valleys fell away on three sides down to three thousand feet and then soared vertically back up, though few of the peaks reached his lofty height. To the north, the ridge descended less dramatically, giving him a long slope of hikeable terrain.

He'd never done much hiking, but couldn't wait to try it out. Per Forest Service training, he had his bear-sized can of pepper spray, supposedly the safest and most effective solution to stop a bear. Same size as a can of spray paint, it shot a cloud of pepper that was the most effective way to stop a charging bear—far better than a big gun, the numbers said. He still would have liked a big gun, but since he'd be as likely to shoot himself as the bear, he'd decided against it.

Beneath his boot soles, he stood on a planked walkway twenty-three feet above the rocky summit ridge; the true summit—a rounded crown of rock—lay fifty feet west and half as high as his tower. The forest fire lookout tower that would be his home for the next five months was a heavy wooden structure. Massive beams of rough-hewn dark wood formed the crisscross framework that supported the tower. Thirty-seven steps made of two-inch thick planks of Douglas fir led up to the fourteen-foot square glass-windowed "cab" that was now home. Those old Depression-era CCC guys really knew how to build something to last; most of the towers and lodges in the Pacific Northwest and Montana had been put up by those "back to work" crews.

He breathed in the air and held it as long as he could. He wanted to savor its taste, its clarity, the complete absence of any hint of civilization or old motor oil. He was so sick of all the people who thought their car was so darned important. It's a machine, people, use it, don't marry it. He was glad to be away from them.

He was almost as sick of them as he was of himself, which was really saying something.

The true extent of his aloneness he was less comfortable with.

Tom's next nearest neighbors were Tess and Jack on Cougar Peak lookout fifteen miles to the north, Swallow Hill twenty miles to the southwest, and—according to his radio plan—Old Crag equally far to the east.

Gray Wolf Summit wasn't on some through-trail, or a trail to anywhere at all except Gray Wolf Summit. It had been a long eight-mile hike with a gargantuan pack that had him cursing in the first mile as he crested a thousand-foot climb only to descend into an even deeper valley.

Vic, the Forest Service ranger in charge of the Selway-Bitterroot and Lolo lookouts, had warned him that his likely visitors over the summer would be the mule skinner who delivered the bulk of his supplies, his substitute who would come up for two days out of every two weeks, and one or two extreme fire-lookout tourists. Gray Wolf, perched at the end of a dead-end trail, was a brutal enough hike to discourage all except the most dedicated.

"Well," he told a turkey buzzard soaring on the high winds with its wing-tip feathers spread like fingers—the bird was probably the only one he'd be talking to most of the time. "If you're seeking something that died, you can cart off the Old Me."

He didn't know who he'd be by the end of the summer, that's why he was up here. But he knew he wasn't going to be the wandering soul who was presently standing on the lookout tower.

It was going to be an interesting summer.

2

Patty Dale hiked up the narrow trail. She'd been looking forward to this summer for four years now. Sure, it was the ass end of wildlife biology—first-year field work—but she didn't care. Being paid to tramp over the mountains and valleys of the Lolo for the next year was her idea of heaven.

She'd absolutely paid her dues.

"No one," her parents had told her, "no one does Army ROTC as a wildlife biologist." Her fellow cadets agreed, but she'd known what she wanted to do since the first reintroduced wolves were released into Yellowstone Park on her sixth birthday—March 21, 1995 after a seventy-year absence.

"Just watch me," though she'd said it only to herself at the time.

Now, after four years in the Army, she'd have said aloud, "Who the fuck do you think you are, judging my ass?"

Patty liked the self-confidence she'd learned in the military, though she was going to have to clean up her language—another gift of her military service—now that she was an academic, working for the Montana Fish, Wildlife & Parks.

An academic—first in her family past high school. First not to work in the open-pit copper mines of Butte, Montana. Busted flat when the operations closed down for several years in the '80s and again when she was in her teens. She was the only one to make it out.

Now, at twenty-six she'd done her time and survived her two full tours overseas. For the rest of her life, she would get to do what *she* wanted to. And right now that included hunting gray wolves—the largest of the wild canines—with a camera and a notebook.

It seemed cliché, but two wolf packs had bred in dens on the mid-level slopes to either side of Gray Wolf Summit. The chance to study two packs simultaneously was almost unheard of. Her rookie year was going to fucking rock…to seriously rock. Whatever.

Patty would be spending most of her time down in the forest, but the chance to sit on Gray Wolf Summit before she did was too perfect to pass up.

Shaded north sections of the trail were still covered with snow. Typical June in Montana. Portions of the mountains were still thick with winter, while in other sections the aspen and maple leafed out in a hundred shades of bright green. The dark spruce and Douglas fir grew bright fingertips at the end of every branch making the mountainside glow with new life.

She took her time hiking up the trail. Rabbit pellets and deer scat littered the trail here and there. Wolf tracks crossed the trail in a section just a half-mile long, this is where she'd start tomorrow. A single massive bear's paw print, in the mud close beside a racing stream of snowmelt runoff, was the first she'd ever seen on her own. She took a photo of it next to her own size six hiking boot. It would look great on her wall, if she ever got a place of her own.

Right now, home was a barracks in Helena, two hundred miles to the east. She didn't plan on being there much this year.

She filled her water bottles, dropped a purification tablet and an electrolyte packet into each one, resettled her pack, and continued up the trail.

Patty made it to the peak after full dark. The fire lookout tower was a blacked-out silhouette against the stars. She dropped her pack and sighed, glad to be free of the load. Using only the starlight, she rolled out an air mattress on the lichen and climbed into her sleeping bag on the very summit. She lay awake a long time after finishing an energy bar and an apple for dinner. Her contentment reached far and wide, watching her breath turn to mist before dissipating against a wilderness of stars.

She knew a lot of the constellations, but the old stories never seemed to fit. Well, now they had plenty of time to become friends. She had been planning to pick up a book, or at least one of those charts with the pretty drawings so that she'd really know the constellations by summer's end. Then she decided that she'd rather make up her own mythology, reinvent herself in the here and now.

She'd never really seen the big bear of Ursa Major in the Big Dipper. It was just a dipper. From now on, it would be dedicated to her first drink of stream water now that she was free.

Hercules was high in the sky, a wasp-waisted group of stars with a sword raised high. She renamed it Warrior Patty. Four years she'd fought for the US Army. Before that she'd fought against the vortex of her family's history that had threatened to suck her down into the copper mine as well.

She fell asleep before she'd decided how to rename Cygnus the Swan flying up over the eastern horizon.

3

T om woke in his lookout "cab" disoriented by the soft dawn light in such a foreign place. His body felt like he'd been battered by the night. The silence was so deep that his ears had rung loud enough to keep him awake. And no matter how deeply he tucked into his sleeping bag, he couldn't seem to get away from the cold.

And there had been the noises.

With the sunset, the world had gone silent, every bird asleep, his buddy the buzzard nesting somewhere in the trees far below. Not a breath of wind.

Then, he'd heard animals rustle about outside and imagined the worst. After a loud thump and strange, soft call like a sigh, there had been slick, snake-like sounds he couldn't identify. Torn between cold and fear, he'd decided that getting up to lock his front door situated at the top of thirty-seven stairs really wasn't necessary—not if he wanted to have any self-respect in the morning.

On the verge of getting up to lock it anyway, a wolf howl lifted into the night. He pictured a entire pack storming his tower if

he made the slightest noise. The single cry was far off and left him awake and shivering for hours.

With large windows encircling his cabin in the sky—his tiny summer home was almost entirely glass from waist to head-high—the low sunlight was rapidly heating it up from sub-Arctic to toasty. Around the edges it had a bed, desk, two comfortable chairs, and a long worktable with a pair of stools facing an amazing view. The entire view was amazing. He could see no signs of civilization in any direction and he was above the whole forest.

Up above the wrap-around windows was an outlined drawing that was a map of the surrounding terrain which named every peak and valley for three-hundred-and-sixty degrees. In the center stood a raised cabinet topped by the Osborne Fire Finder for locating a burn if he saw one.

The first thing Tom did after crawling out of his sleeping bag was to pick up the big binoculars and scan the horizon and the trees for smoke. His training had made sure he remembered to look both near and far—to scan the nearby slopes as well as the distant peaks. The fire season didn't officially start for a few days and he knew that it could be weeks before he saw his first one, if he saw one at all.

Three-quarters of the way around, he yelped.

Smoke!

A huge plume of it.

Still holding the binoculars, he waved his other hand around reaching for his radio when he caught a view of something silver.

Tom peeked over the top of the binoculars, but couldn't see any fire down toward Cougar Peak or in the valley directly below.

But the thing had been massive.

And then he looked closer.

A woman with light-colored hair was sitting cross-legged in front of a small fire that occasionally released a little puff of smoke. The flash of silver was a small cooking pot. Even as he watched, she tipped it into a mug and then dumped in a slim packet of—he adjusted the binoculars' focus—instant coffee.

He swung the glasses up to see her face…and she was looking right at him.

Okay, voyeuristic. He lowered the glasses and waved before it could become voyeuristic in a bad way. She didn't wave back.

He stepped out the door onto the walkway around the cab.

"Sorry," he called out. "I thought you were a forest fire."

"Well, that's a new one."

At just fifty feet away he could see she sat on a heavy field pack. She wore a thick jacket, messy light-brown hair ruffled down to her collar. Looking at her all wrapped up, he suddenly realized he was freezing his balls off. He looked down.

Briefs and binoculars.

"Holy crap!" he hurried back inside to the sound of her snort of amusement.

4

*A*bout the time Patty finished making her oatmeal in the same pot she'd made coffee, the lookout guy emerged again. This time he was wearing enough layers to look like the Michelin tire man.

Too bad. He'd looked good in just his tighty-whities. He wasn't macho-soldier strong, but he was close.

She'd done her best not to think about men since she'd gotten her commander court-martialed for thinking he could take liberties. It had led to her complete isolation by the men in the unit, and by the women as well—a lot of whom were screwing other soldiers, married and not. Totally gross.

Mr. Fire Lookout stood about six feet and didn't move down the stairs like an athlete or a soldier. He moved like a geek. Even though he now carried a mug of his own, he didn't approach her campfire until she waved him over.

Man unsure of himself. That was a new one. Most guys, especially the ones without a clue, moved with a self-entitled assuredness and bravado that only served to piss her off.

He moved close to the fire but didn't sit, instead looming above her. Well, she wasn't going to crick her neck for any male of the species.

"Sit down or shove off," she pulled out a squeeze bottle of maple syrup and drizzled a scant teaspoon on her oatmeal to make the syrup last.

"Sorry," he sat. So not a total write-off.

Patty hadn't really wanted company, but then again, she was the one who'd camped by a lookout tower—you get what you pay for. "Got a name?"

"Yes. Do you?"

She almost spewed her first mouthful of scalding oatmeal in his face along with her barely contained laughter.

Handsome unsure guy with a sense of humor?

"Sure," she kept eating and they shared a smile. "They're useful things to have…at times. I'll just call you Fireboy."

"Works for me." Still he didn't ask her name and she could no longer conveniently ask for his. Instead he sipped his coffee and stared out at the sunlight-etched shadows as sunrise moved across the tree-dark slopes.

This was why she'd come here, to watch daybreak sweep over the rugged mountains.

It certainly wasn't to be studying the profile of the man etched against the softening blue sky.

5

Tom stared into the distance and struggled for something to say. Though women didn't make him tongue-tied, he knew that he wasn't the smoothest guy around. Now Jimmy at the auto-body shop could talk female clients out of their BMWs and straight into a hotel room, but Tom had never figured out how.

But after convincing himself that he was alone in the wilderness, then flashing himself at a woman camping at the edge of a thousand foot drop-off, he didn't know what to say. She was pretty, at least her face and hair were. Her fingers were fine and strong. The rest of her was covered in a thick jacket, many-pocketed camo pants, and heavy hiking boots.

The women he knew were the sorts who wanted to hit a movie or go out drinking. Outdoorsy ones would play Frisbee on the lawn at Gasworks Park overlooking Lake Union and downtown Seattle.

This one was sitting on a pack that looked heavier than his had been and was cooking breakfast over an open campfire a dozen miles from the next nearest living soul.

"What brings you to Gray Wolf Summit?" That was safe enough, wasn't it?

"Exactly," she mumbled as she sucked in cool pine air over a hot mouthful of oatmeal. She didn't elaborate.

"You came for the summit?"

"No, the gray wolf part."

Was she naturally prickly or was she just teasing him? He decided to wait her out. After all, he'd felt plenty lonely last night—not knowing that an attractive woman was camped just a shout away—and it was only his first day in the wilderness. He didn't want to scare off what might be his only visitor for the entire summer.

"Wildlife biologist. I'm here to monitor the gray wolf dens off either side of the trail," she hooked a thumb back over her shoulder.

"They're here?" He spun to look, feeling as if one was about to attack him from behind. Nothing but the rolling line of the ridge, the narrow alpine meadow of grass and wildflowers with his wooden outhouse perched a few hundred feet downslope. Beyond that, the short scrub trees that eked out a living high on the granite, though their spareness quickly developed into a thick forest.

"Sure," she said, continuing to pay attention to her oatmeal and the distant mountains. "Plenty of trail sign if you'd known what to look for on your way up."

He could hear all of the points he'd just lost by missing the "shit signs." Like how was he supposed to know. Though drawings did fill the tiny safety handbook the Forest Service had given him during training.

"There are two known dens and we think they're both occupied. I'm going to watch, record, set camera traps…all of the fun stuff." She'd finished off her breakfast and returned to her coffee.

"You don't look like a lunatic."

"Don't ask my former commander."

"Deal." Ex-military, which made the "lunatic" assessment even less likely. This was a woman with skills and a lack of fear because of those skills.

Whereas he had a complete lack of wilderness skills, which totally explained last night. Well, he wouldn't be letting himself go there again. From now on his fears would only be real ones.

"I'll just call you Wolfgirl."

"You're saying I'm not a woman?" No sense of offense, as if she was just asking.

"Wolfwoman doesn't exactly trip off the tongue. Besides, if I'm Fireboy, you're stuck with Wolfgirl."

"As long as you aren't calling me a bitch."

Female wolf. Bitch. "Don't know you well enough to decide one way or another."

"I'll be around. By the end of the summer, you'll know for sure that I am."

She'd be around.

He'd spent much of last night wondering if there was any way he could cut and run. Fire tower, isolation, howling wolves, the whole bit. Now, no more imaginary fears and, maybe, he wouldn't be so alone all summer.

A regular visitor.

He could deal with that.

6

Tom had settled into a semblance of routine after the first couple weeks. Up with the sun—he'd never been an early riser—but there wasn't much to do up here at night except watch the stars. First scan of the horizon for the day, then a couple-hour hike up and down the trail. Eventually, he'd branched off the trail for longer and longer forays through the pine and fir forest. He started seeing the "shit signs" but decided that unless they were still steaming he wasn't going to worry, too much. The one time he saw bear scat—freaking gigantic—he actually pulled his pepper spray can from his hip holster for the rest of that hike.

At first, he'd been hoping to run into Wolfgirl, but then he'd started noticing the wildlife and the plants changed with elevation along the trail. The Forest Service safety guide let him identify the basics, but he'd get a better guidebook on his first break back in town.

He was on duty from nine a.m. til six in the evening. He sent a morning radio report of weather readings and the fact that he was "in service." Every fifteen minutes, scan the horizon for "a

smoke"—the little wisp of white that promised fire close behind. It was a little dizzying at times sweeping the binoculars up and down the hills—they went on forever. Once he got disoriented enough he couldn't remember where he'd started and had to go around a second time. After that he started and ended with due north.

Due north was the trail that Wolfgirl had walked down two weeks ago, swinging her monstrous backpack on as if it weighed nothing at all.

He felt better when he noticed that she too carried the bear spray rather than a gun. She was a wildlife biologist, so he'd guess that she knew what was best. And being a soldier meant that she had a handgun skill set that he didn't.

When she'd stood up, she'd been smaller than he'd expected. Somehow a person who tracked over the wilderness fearlessly seeking a massive four-legged predator should stand more than five-foot six. His final view of her had been a single pair of slender, camo-clad legs sticking out from below her pack and a battered blue baseball cap with a Montana State University bobcat logo above.

After two weeks—and still no sign of Wolfgirl—he'd had his first two days "down." A lookout relief had hiked in and continued the firewatch while he got off the mountain and went into town—a four-hour hike out and another hour skidding his car down muddy logging roads and then the bland pavement of the highway to Missoula. A night at the bar and crashing in a cheap motel. Alone.

There'd been a couple of potentials at the bar, but he wasn't into it. He'd had his fair share of cheap sex—it usually cost a couple beers, some nachos, and a little dancing. It had always bothered him that the dancing was often better than the cheap sex.

He hadn't felt that way at first, of course. Women in bars had started happening for him as he'd shifted from geeky academic to muscled mechanic from wrenching on crumpled car frames all day. It was true, macho guys got the hot women and he'd

certainly enjoyed the benefits of that at first. But now, his ego didn't need the boost and he just didn't care for the hollow feeling morning-afters always left.

Technically, he had another day down. Instead, he hit the bookstore for a wilderness guide. *Flora and Fauna of the Lolo Forest* was perfect. Then he spotted a title on wolves and grabbed it too. It had become clear that Wolfgirl was gone from his life, but he wanted to read up on them anyway. Tom went through the grocery store, loaded up his pack to a ridiculous weight, and struggled back up to the summit.

And Wolfgirl had left him a note with the substitute lookout.

Hi! and a line-drawing of what he now recognized as a wolf's paw print for a signature. Later that afternoon, he'd been idly doodling between lookout duties, and had drawn wind-blown hair around the paw print as if it was a face.

He didn't know why it mattered, but it did.

Shit!

7

It was late afternoon and Patty should have headed down the trail and into town. She hated to be away from the mountain and her wolf dens. There were two packs. One pack of six had a dozen pups just starting to peek their noses out of the dens. The other was a threesome led by a great, black-furred male; the smaller pack had just five pups as far as she could tell. The two groups had very little to do with each other except for the older female of the threesome, gray in the muzzle, who hunted across the range. It was her tracks that crossed the trail back and forth. All the other wolves hunted down the valleys on their own sides of the ridge.

Patty had spent three intense weeks trying to track that lone female and discover what she was doing on both sides, but hadn't found out yet. Patty monitored the packs nonstop, except for an afternoon, going up to the lookout tower, only to discover that was Fireboy's day off the mountain.

Quite what had drawn her up the mountain that day was unclear. She'd only shared a cup of coffee and a few jokes, but

he'd stuck in her mind. One thing she'd learned in the Army was to pay attention to those little things. In Iraq, wondering about that unexplained cardboard box along the roadside, could be someone's groceries, could be an IED. Don't remember that pile of cut wheat stalks off the side of the road at the junction? Turns out to be perfect cover for a shooter.

Now she was back again, to find out what had stuck in her mind about Fireboy. She thought about kicking the timber at ground-level a couple of times to announce she was coming. Then she remembered his seriously cute, "Holy crap!" when he'd discovered he wasn't wearing anything but very tight briefs and binoculars.

Patty kept her gait light on the stairs and moved upward silently despite her heavy pack which had become like a second skin. On the way up, she could only marvel at the view after having her head down in the woods for three weeks. She so loved being out here.

Up at the catwalk level, she could see through the broad windows into the cabin—it was a very fine view indeed. He was wearing shorts, but that was all. It was June 21st according to her observations log book, mid-summer's eve, and the late afternoon sun was warm.

The stairs had landed her at the north side, close beside the door. Fireboy was facing away from her doing a slow methodical scan of the hills to the south. Now only ten feet away, she could see the definition of his shoulder muscles put on display by his raised arms.

Clean, no tats, like a canvas not yet written upon. Beneath her shirt she wore a lone she-wolf face on her left shoulder blade. Eyes closed, howling a song of purest joy.

He slowly turned in her direction as he inspected his way around the hills. The abs definition from the side was just as nice.

Then facing her...and finally the fat end of the binocs lined up on her face and she smiled.

"Holy crap!" just like the first time. He jerked down the glasses and looked at her blankly.

She didn't know what response she wanted or expected from him. But it was a good one when it came—

"Wolfgirl!" His smile was huge and welcoming. Then he raised the binoculars again and got points for not aiming them at her breasts. "My, what big teeth you have."

Patty laughed. It was something she hadn't done in a long time. Not since before her commander had almost succeeded in raping her—"because deep down she really wanted it"—before she succeeded in breaking his face—"because deep down he really wanted it." Not since…she didn't know when.

8

T om was glad it was the end of the afternoon watch, his last scan of the peaks and valleys for the day. She was actually here, standing in his doorway as if that was somehow completely normal. Only habit reminded him to call in an end-of-day report of "no smokes, no fire activity, Gray Wolf Summit out of service."

He thought about all of the clichés. "You're here!" "Wasn't expecting you!" Really wasn't.

He also hadn't known quite how beautiful she was. He'd seen her face before, clear skin, dark eyes of unfathomable depth. Even in the three weeks since he'd last seen her, her hair had grown and now looked just a little out of control, a touch wild. She was what the guys at the shop would have called a "solid gal." Not heavy—there was not an ounce of heavy anywhere on Wolfgirl—but not slender or model frail either. She was the kind of woman who had the strength to do something other than look good in clothes. The chest and waist belts of her pack stretched her thin cotton t-shirt tight over her breasts. Very nice.

Say something you idiot!

"Was that you I heard howling at the moon last night?"

"Might have," that grin lit up her face even brighter.

Forget pretty, plug in gorgeous with that smile.

"Catch any fires yet?" she asked.

He slapped a hand tragically to his chest, and realized that once again he was mostly unclothed in front of her. *Go with it.* "Not so much as a firefly," he moaned like a player in a Shakespearean drama.

"Not much of a Fireboy, are you?"

He tried to sigh tragically.

Must have worked; that surprising, musical laugh reemerged.

"How goes the wolf hunt?" he wanted to keep her talking.

"Fucking awesome!"

"Drop your pack…" *please stay awhile,* "…and tell me."

She did, dropping it with a heavy thunk that seemed to shake the cab with its weight. She pulled out a water bottle and turned to point north.

Then she cursed, "Do you have a map?"

Now it was his turn to laugh. There was the drawing around the whole top of the wall. There was the wide area map mounted on the Osborne finder that gave him the area for fifty miles in every direction. And on the main desk he kept a 7-1/2-minute quadrangle map rolled out. It showed the area for seven-and-a-half miles north of Cougar Peak—the area he'd known she was tramping.

"Do you have the fifteen?"

He pulled out the larger area 15-minute map.

For the next half hour she led him on a tour of a vast range of hills and valleys, amazing him with the amount of territory she and the wolves covered. Her hard-bitten nails tracing the lines of brutal climbs that had nothing to do with fire-tower trails or logging roads. She'd been hiking straight through brush. The excitement in her voice was so true and pure and it evoked a whole series of emotions.

At first, awe that anyone could care so much about…anything.

She was pulling out her log book to trace the wolves' movements more accurately over the terrain of the map.

Then it was discomfort and finally a shame that had him shuffling foot to foot. He cared about nothing this much; the past few years he'd mostly felt...just blank.

What the hell was he doing with his life?

A college degree he couldn't imagine ever using, a career that included wiping blood, vomit, and empty beer bottles out of shattered cars before he could even work on them, and now sitting alone watching a forest that might never catch on fire. Even if it did, the more experienced spotters at Cougar Peak or Old Crag would probably spot it long before he did.

But finally Wolfgirl overwhelmed his sense of uselessness. Her excitement swept him aboard.

When she spotted Dutcher and Dutcher's *The Hidden Life of Wolves* on his desk, she cried for joy and dragged it onto the map to flip pages searching for pictures that would show him what The Messenger—as she'd dubbed the traveling female—looked like. He'd barely been able to focus on the pictures as they rubbed shoulders and jostled together hip to hip while she told more stories.

He'd made dinner, that she'd bolted, and they'd made love on his narrow bunk as the sunset filled the fire tower with the colors of fire. She rose over him, feral, powerful, as wild as her wolves. The red-gold light played over her skin as she threw her head back and cried out when he sheathed himself and entered her.

Tom half expected her to howl, instead she groaned like her heart had been ripped from her chest. He leaned up to bury his lips and his face between her breasts and she pulled him in with a truth, with an honesty of emotion he'd never found in a woman before.

This was not a woman who revved his engines or fit him like the seat of a Porsche 944 Turbo. She was too primal, too purely herself for that.

When their climaxes ripped through them he felt every jolt through her body as if it was his own.

And after their pulses peaked then slowed and their bodies both shuddered until she finally lay still upon his chest, then she wept.

He held her, stroked her hair, and whispered in her ear that she was okay.

Okay? She was life-changing amazing, but that wasn't what she needed to hear right now as the sobs wracked her, as the smell of salt tears washing against his cheek threatened to overpower the scent of the forest that clung to her hair.

They slept clinging tightly to each other.

In the middle of the night, she woke him, and by the light of the stars she lay beneath him and they were as gentle with each other as they'd been frantic earlier.

Tom woke alone with the sunlight streaming over him.

A note rested on the open page of *The Hidden Lives of Wolves*.

I owe you three pounds of oatmeal,
a half bottle of maple syrup,
and a box of energy bars.
You're very pretty when you sleep.

Again, the paw-print signature.

This time there was a radio frequency.

She'd left the note on the picture of the wolf he'd chosen as prettiest in the whole book. It was a close-up of a black-furred wolf. Just her face, with her chin resting on the snow, yellow eyes looking right at the camera.

9

Patty went back to him whenever she could tear herself away.

Talked to him by radio on other nights when he wasn't on fire watch and the wolves weren't on an active hunt.

June passed into July.

Fireboy's first fire sighting had them talking for hours over their radios. She normally limited herself to fifteen minutes to conserve batteries, but he'd been so excited she couldn't help herself and let him roll.

She'd been very attentive the next day to make sure that her solar battery charger was always aligned to best advantage to the sun.

Something was changing inside her. Patty had come to the wilderness for her wolves and the silence of nature, but like a bear to a honey trap, she couldn't resist circling back to the fire tower atop Gray Wolf Summit.

It wasn't even the sex.

Okay. It wasn't just the sex.

When they were together, nothing else existed. There were visits when they hardly spoke a word. She would track him to his cabin atop the summit, take all he could give her, sleep in his arms, and be gone back to the wolves by daybreak. Such a heavy sleeper, he rarely woke to see her off. But when he did, he always caressed her gently and kissed her sweetly. One of those wordless nights he'd spent hours tracing every line of her wolf tattoo as if stamping its joy onto her soul more deeply than the tattoo artist had.

Other visits, they might not make love at all. Just watch the sunset, curl up in each other's arms, and sleep. They talked of nothing and everything, but only about the present. Neither of them had a past or future. Neither of them even had a name.

Patty was not her mother or her grandmother or even her great-gran. They had all married their men at sixteen or seventeen and given birth well before the acceptable nine months had passed.

The one thing Patty knew for certain, being with a man for more than a time or two was too great a risk. Too dangerous. But again that bear to the honey trap; she could no more resist Fireboy than he could her.

He'd taken to doing the town food-run for both of them so that she didn't have to leave the wilderness. On his way down the mountain, he would radio, gather what little trash she couldn't burn, and bring back an extra twenty pounds of supplies. He never stayed away overnight, though she never let him return directly to the tower unrewarded.

If the wolves were running that night, she'd sneak him into one of the blinds she'd created along the primary trail. He'd been nervous as hell the first couple times, even after she assured him that wolves didn't attack humans. But as they watched The Messenger through the view screen on her night-vision camera—and the female had given them little more than a sideways glance—he'd settled down. When the wolf was gone, they made love among the soft ferns with the rich smell of the forest duff wrapped around them.

"Wolfgirl!" the radio snapped at her early on a hot August afternoon.

A startled rabbit bolted from close beside Patty's position.

The Messenger shot after it, but Patty knew the wolf would be too late.

"What?" she yelled back into the radio.

"Where are you?"

"Go away!" She began gathering her camera gear and was about to shut off the radio in frustration—the rabbit would have made a great catch and she knew how hard it was getting for the old wolf to hunt. This wouldn't be her last season, but the end had just come a little closer.

"Where are you? It's important," he shouted at her.

"Head of Long Tail Creek, about two hundred yards above the western den."

"You need to get out of there. Get up onto the ridge trail. Either get to me or get off the mountain."

"Why?" But she heard why. She found a break in the forest canopy and caught a glimpse of a black airplane painted with orange-and-red flames like a sports car. Even as she watched, small figures dressed in yellow tumbled out of the rear and then popped open parachutes.

Two smokejumpers. Four.

The plane circled back, four more.

And a third time.

Two smokejumpers is what they sent to stop a typical small fire, under an acre. Four could beat down a half dozen acres. A dozen smokies was very bad news indeed.

A helicopter came in. Instead of delivering more smoke-jumpers up high, it came in so low that she had to cover her ears as it by passed overhead and continued down into the steep valley. Then there was a high whine, momentarily louder than the roaring engines and the pounding rotors, and a shower of red retardant sheeted from the sky down onto the forest in the valley far below her—but not that far.

The wind, almost undetectable down here in the trees, was brushing downslope, which would explain why she hadn't smelled any wood smoke. Even though the wind might be washing down the hill, fire loved to climb.

Patty was a dozen scrambling steps upslope before she caught herself.

The Messenger hadn't smelled the wood smoke either. And if she hadn't, then the wolves in the den down below hadn't.

It was absurd, she was human, they were wolves. But she knew them, had named and cataloged each one, knew them by their markings, their behaviors, even the half-grown pups. Of them all, only The Messenger remained nameless to her.

Another helo roared by low overhead, another sheet of red cascading from the sky.

"Wolfgirl, tell me you're on the move."

She wasn't. She was frozen between escape and saving—

Patty plunged down the slope, smashing a shoulder against a tree to slow herself down when her speed went out of control, jumping over a boulder that threatened to kneecap her and landing a dozen feet below in a roll that was only broken when she tumbled into a blackberry patch. Cursing and bleeding from a dozen scratches, she circled wide below the den.

The roar of chainsaws and heart-stopping thunder of crashing trees below told that the trouble was far closer than she'd like.

Approach the den from below.

The pack was out front, agitated by the noise, but not frightened by the fire they couldn't smell. Blackthorne the big male pacing back and forth. Mariko, the small pack's second female—Blackthorne's true love in *Shogun*—was guarding the pups, keeping them confined in the cave.

Patty climbed back toward them.

"Shoo! Move!" The massive black pack leader turned to face her but, other than a worried snarl, made no effort to move off. She didn't dare move any closer, he might attack her in simple panic. Then she had an idea.

Patty pulled out her can of bear pepper spray. She shot the smallest squirt she could upslope and a bit to the side.

She heard a sharp *Yip!* from Blackthorne just as she realized her mistake. The rising heat of the fire below them was now actively pulling air downslope to feed itself. The pepper spray she'd shot near the wolf was also dragged right back down on her.

It wafted into her face.

She cried out in pain as it hit her eyes and nostrils despite her raised arm. Diving down, she rubbed her face in the soft ferns and cool earth. She screamed out the pain that even that small amount of spray had caused.

When she could finally see again, Blackthorne was gone. The female was following and several bushy-tailed youngsters disappeared with them into the brush—upslope, thank god.

She couldn't count how many pups through her streaming eyes, so she forced her way up to the den. A lone twenty-pound pup had been left behind, Vasco by his markings—one white ear, one black—the Portuguese pilot and Blackthorne's lone friend. So terrified that he didn't even nip at her as she reached in to drag him out.

Patty struggled up the steep slope.

What had been a crashing three-minute descent became a brutal half-hour climb. She tried releasing Vasco, dropping him to the ground and shooing him upwards but he merely cowered at her feet, front paw raised. She saw it had a nasty cut and probably hurt too much to walk on. It would heal in the den, could be ignored in a three-footed run across level ground, but the pup couldn't climb a steep slope with it.

She eventually became aware of two things.

The rising heat wasn't only from her hard climb, the fire was starting to run up the narrow cleft.

The other thing was Fireboy's near frantic calls.

"I'm headed upslope," she answered in between raged gasps. "I have an injured pup. But the fire's close. It's hot."

"God damn it, Wolfgirl. Drop it and run!"

She knew that was the smart thing, the wise thing but, "I can't," her voice came out as a sob and she kept struggling up the slope.

As she climbed past where the initial radio call had spooked the rabbit and The Messenger, she started scanning for the female. There was no question, she would know to run. Wouldn't she? Patty hadn't.

There was still no smell of smoke, just the insufferable heat.

Patty continued battling her way through the brush. The slope rose so steeply that her sore knee—she must have cracked it against something during her pell-mell descent—often banged against tree roots and rocks.

Would the wolf pup tolerate being inside a fire shelter with her? She doubted it, but they might have no choice.

That's when she noticed her pack was gone. She'd shed it somewhere. Her camera, data, and radio were attached to her fanny pack, but all of her clothes and gear were lost somewhere in the trees and boulders. And in it was the foil fire shelter kept for true, last-resort emergencies.

A glance back over her shoulder was a bad mistake. The fire had reached the den, only three hundred yards below her, but with flames reaching hundreds of feet above the hundred-foot tall trees. Even glancing over her shoulder, the heat was a slap on the face. And the roar, the roar was deafening.

There might have been a radio call, but she couldn't hear it over the fire's howl.

She turned and kept climbing though her knee throbbed at every step. The stitch in her side was so bad that she was almost weeping into the wolf pup's fur. Every step had become agony.

Bear down, soldier! There is no such thing as quitting!

She bore down, but she didn't have much to bear down with.

The roar and the wind peaked, slamming against her so hard that all she could do was drop to her knees and wait for it.

"Hi."

Patty screamed as a hand touched her on the shoulder.

A man clad in yellow Nomex and a pilot's helmet dangled at the end of a wire not a foot from her.

Patty's gaze followed the wire upward until she spotted the helicopter hovering high above the trees, its engines even louder than the fire, the downblast shaking trees and brush.

"Steve Mercer, Mount Hood Aviation. How about we get out of here?"

She held the wolf pup closer, "I'm not leaving Vasco."

The man swept the pup under one arm—Vasco whined nervously but accepted the transfer—and twisted a lifting ring toward her. It was also attached to the wire; he held it so that the opening faced Patty.

It was like a circular orange life preserver.

"Head, arms, and shoulders through the hole," he instructed as calmly as if they were on a quiet street corner. "Keep your arms down so that it catches you in the armpits. Keeping them down locks you in place."

She did as he said and moments later they were lifting up out of the trees, spinning slowly, too much like a rotisserie in the approaching fire's heat.

Once they were clear of the trees, the helo pulled them away from the fire and she could start to breathe again. A hundred feet above the trees, the flames still reached far higher, but they were rapidly falling astern as they continued to climb up the slope.

"Emily says that we'll drop you at the base of the mountain," the man shouted to her.

"Can you drop us near the top?"

10

Patty curled up on the fire tower's bunk and tended the wolf pup. Calming the young wolf let her not think about how her eyes still stung, how much her knee hurt, or quite how close she'd come to dying.

She listened as Fireboy worked the radio through the long afternoon.

The smokejumpers fought the fire in pitched battle until it was trapped and couldn't spread either way along the valley wall. The helicopters had contained it before it crossed the ridge. The second den would be safe.

If The Messenger lived, perhaps she'd guide Blackthorne's pack over to join the larger one to the east.

She buried her face in Vasco's fur and wept for only the second time since that day as a young girl when she had understood the trap that her family was in. It was the day she'd determined to find a way out.

Patty had wept that first time in Fireboy's arms as some impossible sense of loss had overwhelmed her, even if she hadn't

known what the loss had been. And now she wept because she understood that from the first time with him, what she had left behind was the Warrior Girl fighting for freedom against all odds. In his powerful arms, she was more truly herself than anywhere she'd ever been.

A long time later, Fireboy sat down close beside her, but didn't touch her.

The sun had gone, but she hadn't noticed.

"Is it out?" her voice was rough and still stung from the pepper spray she'd inhaled.

"Yes," he nodded in the soft light of the small oil lamp that he'd lit "A ground team has arrived and is making sure it stays dead. The smokies are already being lifted out by the helos. We have another fire north of Cougar Peak that they're needed on. How's the pup?"

She held up the long and sharp stone sliver that she'd extracted from Vasco's pad, "He'll heal fine now."

"Is he like a permanent addition to the family?"

"No, I can probably reintroduce him to his pack tomorrow. I think I know where they've moved to." And Patty knew she'd find The Messenger there, she just had to.

Then his words registered.

"The family?"

He shrugged easily, "Does seem to be what I said."

As she watched his face shifted. One moment he was casual, keeping up a cool facade. The next was a wash of emotion she couldn't even recognize, but both his hands were crushing down on one of hers.

"I thought I'd lost you. I've never been so afraid in my life. I couldn't imagine this world without you in it. To never be able to talk to you again, laugh with you again, it simply wasn't possible, but it felt so real. I could barely help on the fire until they found you."

"Family?" she couldn't seem to get past the word. Was family about something more than mere survival? Hers had never been.

But she could see in his eyes that she did mean the world to him. She didn't need his crushing grip nor his eyes glistening in the soft lamplight to know that he'd been afraid to the very core. For her. Of losing her.

"I've never been important to anyone," Patty told him. "Not that important."

"I swear I almost went charging down into the fire myself to find you. If that Mount Hood Aviation helo hadn't called that they had you, I would have. I never knew what was important—that anything *could be* that important to me—until I met you."

And she could see the truth of that. He really would have run right into the fire for her.

The strange thing was, she'd have done the same for him. Without knowing how it happened, she'd discovered what family was supposed to be. It wasn't about surviving together, it was about helping each other. Not just from a fire, but from the heart.

She raised her free hand—the one not still locked in his crushing grip—from the pup's fur and brushed it over his cheek. How could she describe how she felt about him? How could she explain anything to someone who made her feel so important, so precious?

She leaned forward to kiss him lightly on the lips and then leaned back to look him in the eyes.

"Patty Dale," she whispered because what could be more important than a name.

"Tom Cunningham."

She listened to her heart and knew. Knew that this was simply right. As nothing in her life had ever been, even more than wildlife biology.

"Patty Cunningham?" She asked it softly, as much of herself as of him.

His smile was all the answer she needed.

Dawn Flight

In 2014 I had the privilege of being invited into a romance anthology with all proceeds going to support wounded veterans of our Armed Forces. I wrote the story NSDQ *(which is the Night Stalkers motto: Night Stalkers Don't Quit) about a wounded pilot and her quest on the path back to a life and to love.*

During my research, I was overwhelmed by what the pilots of Combat Search and Rescue (CSAR) do. When Delta Force or SEAL teams get wounded and dial "911," CSAR is who they reach...and they always show up.

Dawn Flight *was a further study of this world for me; a study leading toward a future CSAR series. I love this world and these people. I can't wait to find out what their next plans are.*

But here's what happens when things are more on the unplanned side of things, which is what CSAR is there for in the first place.

M.L. BUCHMAN

DAWN FLIGHT

THE NIGHT STALKERS:
CSAR

1

*Y*ou gotta be kidding me."

"Do I look like I'm kidding you?"

Captain Jack Slater looked down at the slip of a woman wearing full flight gear but no rank insignia. According to the orders tucked in his pocket, she had the unlikely name of Lois Lang-Clark. Damned cute despite the flight gear that overwhelmed her sleek frame, and the fact that one of her feet was mechanical. Cute despite the *Terminator* foot wasn't a factor though, as she had a ring. Whether it was real or merely to ward off unwanted attention because she was a pretty woman in the man's world of US Army heli-aviation didn't matter. Answer was clearly "no" to all comers, but he couldn't bring himself to leave it totally alone.

"Your husband named Kent?" Maybe this was all some kind of Superman joke? An initiation gag, not that Army orders were big on gags. He looked around the pristine training hangar, but they were the only occupants. No line of guys waiting to laugh when he fell for whatever the newbie game was.

Fresh from two years of Night Stalker school, he'd been on a red-eye flight out of Fort Campbell, Kentucky, landing in the predawn darkness at Joint Base Lewis-McChord in Washington state. He'd stepped out into the cool October morning with first light just cracking the horizon and checked his watch. His orders had sent him straight here in his first hour as a new member of the 160th's 5th Battalion.

But this was a training center.

A line of three flight simulators stood on tall hydraulic pistons that could simulate harsh flight conditions. Each set of pistons supported a white block of metal that looked like nothing so much as a ten-foot-wide white egg on steroids from the outside. He knew from vast experience that the insides looked like very realistic helicopter cockpits, complete with a projection system that could convince you that a crash into downtown Kabul was truly imminent.

After two years he was supposed to be done with this shit.

"Kent Clark?" he nudged when she didn't respond. "Superman in disguise," he prompted and still got back nothing.

With a loud rattle and hiss, the rightmost of the simulators, the one for the MH-47G Chinook heavy lift helo, bucked and slewed hard left. By the sustained nose down attitude, he could tell that its pilot was not having a good day. The left hand one for the MH-6 Little Bird was in a slow, steady climb. The one in the middle, the one for his baby, the MH-60M Black Hawk, stood quietly at rest. Waiting.

"My husband's name is Kendall," the slim woman informed him in a tone as warm as an iceberg. "Kendall Clark."

He laughed, he couldn't help himself. A crazy name, a ring, and a false foot. What the hell? Could she even fly with that thing?

Her silence was more deafening than the two simulators, now both protesting loudly as they jerked and twisted.

"I'm already FMQ. Fully Mission Qualified," Jack explained and had the sudden feeling that he was being more rather than less of an idiot with each passing moment. He rubbed at his face

trying for a reset. "Look I need some sleep. Can you point me in the right direction and we'll play out your little game later?"

"You signed up for Combat Search and Rescue?"

"Damn straight!" Bringing out the wounded from a hot battle zone was the kind of serious-as-shit job he'd always dreamed of. One he'd been gunning for since the moment he'd learned it existed. He liked the idea of rescuing people who really needed it. It fit something right in his brain.

"If you haven't been signed off by me, then you aren't Fully Mission Qualified for CSAR activities with the 5th Battalion. Period."

Jack thought of several short sharp comebacks. But there was something in her tone that gave him pause.

One of the simulators slammed to a halt, tipped at a hard angle against the stops. Then, with a groan, it eased and lowered into the reset position.

At that moment, two other people joined them.

One climbed down from the Black Hawk simulator, a grizzled, gray-hair with faded Master Sergeant stripes on his uniform—those took a long time to fade. He came to a parade rest close behind Superman's wife. That said that just maybe she was for real and it was time he started listening.

The second was a tall brunette who'd come in the same hangar door he had. Even had a big duffle, worn pack-fashion over nice strong shoulders. Now that was his idea of a woman. Eyes as dark as her hair, a fine face wearing an easy smile, and almost as tall as he was.

"Excuse me." Voice smooth and low. Unlike Mrs. Superman, her flight gear didn't overwhelm her frame.

"Yes?" he replied before Superwoman could speak.

The new arrival looked him up and down, "I'm guessing you're not Major Lois Lang-Clark unless your parents hated you when they named you."

There might have been a twitch of a smile; or there might have been a roast-in-hell-macho-asshole look. Jack was too tired

to tell. *Major* Lang-Clark? He'd forgotten that from his orders. He'd just been dumb enough to be harassing a major? Bad start for first day in a new battalion.

"You want Mrs. Superman, here," he pointed to the slender figure still glaring up at him.

The new arrival turned and saluted sharply, "Captain Diana Price reporting."

Again, the laugh burst from him. He just couldn't stop it though he knew he was only digging his grave deeper.

The Master Sergeant and the two women turned to look at him.

"I'm sorry," Jack did his best to sober at their bland expressions. "Mrs. Superman Lois Lang-Clark meets Diana Price. You probably don't know, no sane person would, but Diana *Prince* was the secret identity of Wonder Woman. It's just too damned funny."

The tall brunette turned so that he could see the helmet dangling off the other side of her duffle bag. On the side was painted a wide golden triangle with a red star at the center, curved like the heroine's headband. Below that was the stylized "WW" that arced across the breasts of Wonder Woman's comic book uniform.

The petite instructor held out the helmet she had tucked under her arm. On it was emblazoned the Superman logo.

He held up both his hands in hopeless resignation.

The women didn't look amused and he almost kept it in.

But then he caught the merry twinkle in the gray-haired Master Sergeant's eye and Jack totally lost it.

2

S*ome men did not* deserve to live and Diana had just met a prime example. Big, handsome, and a total jerk. Of course, after a decade as an Army aviator and Black Hawk pilot, she should be used to dealing with that by now. It was the great laugh that was throwing her. Macho jerks weren't supposed to have laughs that made you want to smile right along with them.

But it wasn't that hard to resist, especially looking at Major Lang-Clark's serious expression.

"Why are you flying CSAR?" the Major ignored the buffoon's attempts to recover.

How many times had she been asked that? Always by men who were testing, pushing, looking for that weakness that would say she was the wrong person for the role. It wasn't sexual bias, at least not all of the time. Many of the examiners were equally stringent about men applying for CSAR, because this wasn't the Gulf War Army of her mother's day.

Still, it was the first time a woman had ever asked her the question. Diana would prefer not to answer in front of Mr. Jerk,

but she'd been asked, so she'd reply, with something other than the "Want to serve and save people" line.

"My father died in Kuwait during Desert Storm. Before they knew about the Golden Hour or had the systems in place to take advantage of it."

Modern CSAR was now all about recovering casualties and getting them into a hospital within one hour—with faster being much better. For severe bleeding, sixty minutes was the line of near hundred percent fatality. Thirty minutes marked a fifty percent survival rate, and all but the very worst cases could be kept alive for fifteen minutes.

A CSAR pilot's job was to deliver the medics within that quarter hour if possible, and get the casualty to the hospital inside the hour no matter what hell was breaking loose.

Her father had hung on for two-and-a-half hours in the Kuwaiti Desert, but there hadn't been the assets in place to get to him sooner. Medical help other than his squad mates' first-aid had arrived too late. That her mother had been an unmarried and, she'd soon discovered, pregnant supply sergeant, had denied her both her own military benefits and her sworn fiancé's death benefits.

"If I can save one person who is somebody's father, I want to be the one doing it," Diana's voice had gone harsh. She closed her eyes for a moment and when she swallowed hard, it hurt.

A comforting hand rested on her shoulder. To her surprise, when she opened her eyes it belonged to the laughing jerk—except now his expression was sober and sympathetic. Maybe there was a person inside there…though she wasn't willing to bet on it.

3

Jack was pointed to the copilot's seat. Well, he supposed that he'd earned that, though it rankled.

Diana the Wonder Woman had been sent to their simulator's right-side pilot-in-command seat.

The Major sat in a jump seat close behind them. Master Sergeant Hamlin had settled in the chair that would control the simulator experience.

Jack settled himself in for a couple hours of boredom. Start-up, take-off, basic flight… After five years with the 10th Mountain's Combat Aviation Brigade and two more years of Night Stalker training, you'd think he could skip all this basic crap. But nooo.

Shit.

However, knowing he had ground to make up with the Major, he settled in to do what he did best, fly.

Captain Diana Price started out with an attitude of sharp competence. She adjusted the seat and safety harness with the motions of long familiarity. He'd never flown with a woman,

but she showed the signs of a skilled pilot, so he'd give her the benefit of the doubt until she had a chance to prove herself.

And there was no doubt about her commitment—not by how her voice and face had shifted. No matter how much of a showpiece she might look on the outside, she cared deeply about CSAR on the inside, and with good reason. He wished his reasons were so clear. They were just as strong, somehow, but he'd never been able to sort them out into words. Another point in Wonder Woman's favor.

"Ready?" Major Lang-Clark asked over the simulator's intercom.

Jack made one last visual inspection. The side-by-side pilots seats were separated by a wide bank of radio and navigation gear. That swept up into a broad, sideways dashboard that crossed in front of them and ended about chest high. On the console were six large glass screens, each the size of a tablet computer. The simulator was rigged just like the latest glass-cockpit standard which was nice—once he'd gotten used to the digital cockpit, going back to the old analog dials and gauges was always frustrating.

A couple arm's lengths beyond the outside of the windshield was the blank screen on which would be projected their "view" during the simulated flight. Additional viewports to the Earth below were down under the console beside where his feet rested on the rudder pedals.

Collective under his left hand to control lift, and cyclic joystick rising between his legs for his right hand to control direction and speed of flight. Both controls were studded with a dozen buttons and switches. He brushed his fingers over them, now so familiar with practice that he knew them as well as he did where his nose was on his face.

He saw Diana doing the same, confirming that the unfamiliar simulator was indeed familiar. Their matching sets of controls meant that between them, they'd always have control of the aircraft, even if one or the other had to reach out to adjust something on the dash or radios.

He pulled down his helmet's semi-transparent visor and double-checked that the head's up display calibration was properly projected across the inside surface.

"Ready," he and Diane spoke in almost perfect unison.

"Good. Let's go."

The goddamn simulator exploded.

4

D*iana's instincts took over* before she could make sense of the transition.

One moment she'd been preparing to show the Major that she did indeed have the basics down solid—she'd had to prove herself so many times over the last decade that the rote routines were almost comforting with their familiarity.

The next moment, she was diving left as the audio warning system squealed in her right ear identifying an incoming attack of small caliber rifle fire from that side.

Before she could call out to Jack the Jerk, he'd nudged the cyclic just enough to shift the aim of the weapons mounted on the outside of the helo. He launched a pair of Hydra 70 rockets; their simulated streaks raced right down onto the origin point of the ground fire.

He was back off the cyclic an instant later.

Jack had integrated into the simulation so fast that he must have known what was coming. Maybe he was another trainer, had to be with the way he'd been sparring with the Major.

"Mrs. Superman" indeed; as if Superwoman didn't deserve her own name.

A battle raged overhead…and they weren't really a part of it. Three Black Hawks and a pair of Little Birds were dodging and diving over a convoy at the far end of the narrow valley.

"*CSAR 01.* Two wounded, grid thirty-nine," the Major's voice informed her with the dispassion of a mission commander sitting in some distant command bunker.

A blink to shift her focus from inside her visor to glance down at the electronic map on the console. Grid thirty-nine was right at the heart of the battle.

"Do it!" Captain Jack Slater snarled.

"Roger that!" Diana yanked up the collective and shoved the cyclic forward. She dove hard and fast down into the throat of the valley. A dozen targets presented themselves.

She started to turn for one, when the Major shouted, "Someone else's problem. Get the wounded. That's always your priority."

Gritting her teeth, Diana flew through a rain of small fire, bullets too light to penetrate the Black Hawk's armor…hopefully.

She swooped and settled into Grid 39.

"Medics away," the Master Sergeant reported.

She sat interminably—the mission clock counted ten, twenty, thirty seconds—wincing every time a round pinged off her windshield with a bright *Thwack!*

"Ten," Sergeant Hamlin called out.

Instinct had her looking to the side port to watch for the medics only ten seconds out. All she could see was the swirling brownout that would have been caused by her own helo's rotors stirring up the dust and dirt…if this was real.

"Raising to hover," Jack eased up on the copilot controls and hers moved with him.

Damn it. She should have thought of that. It would save them several seconds in getting the hell out of here if they were at a ground-hugging hover by the time the medics boarded.

"Four aboard," Sergeant Hamlin announced.

And just as she was about to lift he called out again.

"Still missing a medic."

"We're what?"

In answer, a wounded soldier hobbled out of the brownout, moving slowly toward her. Assisting him was a young woman dressed like a CSAR medic.

And a dozen meters to their left a battered pickup swung into view through the dust cloud. The "technical" had an out-sized machine gun mounted on the truck bed.

Its first salvo star-cracked her windscreen.

This wasn't the small rifle fire of before; this was .50 cal machine gun fire that would chew them apart in seconds.

Time to go. Now!

She couldn't leave the two injured.

Her helo wouldn't survive if she hesitated.

But she did.

The last sound she was aware of was a soft but heart-felt, "Holy shit!" from Jack Slater.

5

*T*hen the simulation ended like a switch thrown and Jack slammed forward against his safety harness into the sudden void left by the end of the projection. He eased back into his seat and flexed his fingers trying to get his hands to stop shaking. He blinked out the helo's windows, now the bland light gray of an empty projection screen wrapped around the simulator's cabin.

No sign of the shattered windscreen.

Or the bad guys.

Or the wounded.

The silence was deafening.

With fumbling fingers he managed to find the chin strap of his helmet and pull it free.

He looked over at Diana, but she looked little better off. Sweat streamed down her face and her eyes were wide with shock.

"We were in the simulation for..." he couldn't even finish the question his throat was so dry.

"Sixty seconds. Maybe."

He rubbed at his face, "Felt like a goddamn week."

"It did," she sighed and slumped in her seat.

There was a strange, asymmetrical clumping that sounded like someone with two different...

Major Lang-Clark stepped into view outside the window, between the simulator's cabin and the projection screen. Right, the woman had one real foot and one artificial one. Clearly not a factor. She was a pure hard-ass about Combat Search and Rescue and nothing else mattered. He was sure he'd have no trouble remembering that detail in the future.

"CSAR Training," she said as she looked in at the two of them. "Begins tomorrow at 0700. Get some sleep."

She began walking back out of view, but stopped and looked at them over her shoulder.

If he didn't know better, he'd say her smile was almost kindly.

"The decisions," her voice was soft, without the hard edge she'd used since the moment of his less-than-respectful arrival, "get harder from here." Then she was gone.

Jack's groan was cut off when a strong hand clapped down hard on his shoulder from behind.

He looked up at the Master Sergeant.

"She's a pistol, ain't she?" Then he shook Jack like a ragdoll, before heading toward the ladder while whistling *The Army Goes Rolling Along* happily to himself.

"You okay?" he asked Diana.

Diana nodded once, uncertainly. Then again with a little more surety.

"Don't beat on yourself. Even Wonder Woman couldn't have gotten out of that."

"But in real world, what would you do?"

He thought about the situation again: escape, make that *possible* escape, but only at the price of committing others to death including their own medic.

"You don't know, do you?" she wasn't being nasty. It sounded as if she really hoped he did.

"Damned if I do," he hated letting her down. "Maybe Major Mrs. Superman will let us know, if we behave."

"I wouldn't bet on it. Lesson for the student and all that crap."

He like her attitude and her easy confidence. "How about if I buy you breakfast at the Mess Hall and we can discuss it a bit?" It wasn't that he wanted to share a meal with such a fine-looking woman…well, he did, but it wasn't just that. For perhaps the first time in his career, he was well and truly stumped. Flying always just came easy to him, but this was hard.

She nodded, shook her head, then nodded again.

He couldn't tell if she'd heard or even understood him.

She covered her face with her hands for a moment and gave a small scream of frustration that almost made him feel like smiling again. Then she pulled her hands away and turned to face him.

"It depends," between one moment and the next she'd gotten her act back together. Just that fast. Which was pretty damned amazing.

He wasn't even close to having his own act back together after that simulation. "Depends on what?"

"Are you always a jerk?"

Jack grinned at her, "Depends on who you ask. A couple commanders, several ex-girlfriends, Mom…more like a pain in the ass."

"A pain in the ass is better than a jerk?" She shrugged. "Well, I always preferred forming my own opinions."

6

*B*reakfast almost lapsed into lunch.

Not jerk, Diana assessed. Too sure of himself perhaps, though he sounded as if he had some reason to be.

Well, if he did, so did she. They'd both made it through the notoriously difficult selection process of the Army's 160th Special Operations Aviation Regiment, and nobody in any military had more skilled helicopter pilots than the Night Stalkers of SOAR.

And they'd both volunteered to go CSAR, which took a special kind of masochism—flying into hot battle zones to extract the wounded rather than fighting from far above until the battle was done and won.

While Jack Slater didn't tell her why he'd gone CSAR, she was finding it easier and easier to like the man. He always found the lighter side—he was funny. Not like one of those guys who only thought he was funny, but one that actually was.

The emblem on the side of his helmet was an impossibly elaborate sword. When she finally asked if he was King Arthur,

he'd told her she was really close, but kept her guessing for a while. She finally gave up.

"Jack. Jack Slater," he said it like "Bond. James Bond."

"Your name," she'd replied still not getting it.

He'd practically chortled with delight. "Jack Slater. Jack Slaughter. Jack the Killer. Jack the Giant Killer. They had me pegged by the end of first formation at West Point. King Arthur was the original of the Giant Killer myths; Jack came along a handful of centuries later. You see, you might be a fictional heroine, but I'm mythic! And I'll put my magic sword up against your golden lasso anytime."

Mythic or not, he was sharp. They dissected that morning's mission at length and finally decided they should have gone for the escape. Once clear, there were more options: to return, to send in others. But to sit still was to kill them all.

They pounded out possible counter tactics for the future. Drop off the medics and automatically return to the sky to await their return? Too much risk of having to abandon the team.

Stay just inches aloft? Tricky to sustain and it would continue to stir dust badly, perhaps making it harder for the medics to recover the injured, but offering far more flexibility in an attack scenario.

Jack made a couple of forays at finding out more about her past, but she just couldn't go there. It was too deep and she was still shocked that she had blurted it out, even that one little part.

And refusing to go there, she couldn't ask about his past.

But he'd been kind enough to stay backed away rather than pushing or wheedling as any other guy would have.

By the time they tracked down their apartments in the on-site barracks they were both weaving with exhaustion. They were on the same floor of the same unit. There were definite advantages to being an officer in an elite outfit—no open barracks. They stood close in the dimly lit common hall. It was barely big enough to hold a stairwell, the doors to four tiny one-bedroom apartments, and both of them with their duffle bags.

They stood in that little hall, too close together, but she found herself reluctant to move away. She'd only flown in from Hunter Army Airfield this morning, so it's not as if she knew this place. The only thing she did know was Captain Jack "the Giant Killer" Slater.

"Are you sure you're not a jerk?"

"You mean despite my demonstration with Major Mrs. Superman this morning?"

"Despite that," she didn't even know what she was asking.

"Well," he aimed that powerful smile of his at her again, "will it make me more of a jerk if I do what I want, or what I should?"

"Hard to know because I don't know what the hell you're talking about soldier." Unless, maybe she did.

"Am I more of a jerk if I kiss you like I've been wanting to since the moment you showed me that screwy Wonder Woman helmet of yours?"

She definitely knew what he was talking about.

"Or is it worse if I don't kiss you and walk away as if you aren't beautiful, desirable, great company, and a hell of a pilot?"

It had to be lack of sleep talking, but what the hell. "I think the latter would make you much, much more of a jerk." No other man had ever thought to include how she flew in a string of compliments. That last was the scale tipper.

He looked at her with some surprise.

"Well?" Now that she'd said it, she did want him to kiss her, preferably before she decided just how stupid an idea that might be.

Jack shrugged his duffle bag off his shoulder and it thudded onto the floor. With an easy strength, he lifted the strap of hers off her shoulder and lowered it as well.

Then, with a gentleness she hadn't expected, he pulled her into his arms, offering her a dozen opportunities to escape or evade.

When she failed to vary the course of his approach, he completed the gesture. It wasn't just some kiss, some hand around her neck and a fiery meeting of the lips and tongue.

Kissing Jack Slater included a full body hug as if they'd been lovers for years. His arms wrapped naturally around her, as hers slid up his chest and around his neck.

He tasted of the ice cream dessert they'd just split, the strawberries that he'd chosen and the chocolate sauce that she had. And he felt soldier hard and magnificent.

This wasn't her mother's war. She didn't want a child from a dead man that would ruin her career. For that reason, she'd sworn off military men.

Until this moment.

For tonight, at least, this wasn't her mother's Army either.

She led Jack Slater into her new apartment.

7

Jack woke up tangled in sheets and woman. It was pitch dark and a helicopter had just roared by close overhead. Then another followed it—and several more.

"Two Black Hawks, four Chinooks," the woman in his arms whispered. Wonder Woman. And Diana had certainly proved that she completely deserved the accolade.

"Plus a pair of Little Birds and a partridge in a pear tree," he replied. "You know that a woman who can tell helo models by their sound is pretty sexy."

She nuzzled his neck. "Gods, I feel like such a slut."

"I've been used," he groaned in mock complaint. He slid a hand down over that fine soldier's butt of hers, the other tracing over that long leg of hers draped across his waist. "However, I would point out that no way does a slut feel this good."

"You say the sweetest things."

"I'm a sweet guy."

She snorted against his neck, her laugh sending interesting ripples along where their bodies lay together.

"Okay," he admitted. "You caught me," and he wiggled a finger in the ticklish spot he'd located earlier, the soft inside of her thigh just above the knee.

She convulsed and he used her momentary imbalance to leverage her back under him.

He groped for some more protection and she didn't make a single protest as he did his best to prove that he wasn't sweet at all.

8

S*imulator scenarios were mixed* with actual night flights. And as one flight turned into a dozen, then two dozen, so did their days together—for daytime is when Night Stalkers slept, or didn't.

Diana was going weak in her head for a fellow officer which was stupid in so many ways. It would help if he wasn't such a joy to fly with or an equal joy to tussle with between the sheets—but he was both of those and more.

In a blur so fast that it was hard to imagine, a cold October dawn had turned into a bitter December, but she didn't care. The training regimen from Major Lang-Clark was intense—and serving its purpose. The confusion of that first simulation had turned into a clear set of skills, even if the new simulations posed even harder moral dilemmas and more difficult to perfect tactics.

And Captain Jack Slater had turned into the best man she'd ever been with. She simply couldn't get enough of him, no matter how much they both tried.

They alternated seats, both in the simulator and aloft until one night he looked at her after a particularly complex storm-and-mountain scenario and declared, "You're pilot-in-command from now on. You're better than I am."

"No way. You're—" then she saw Major Lang-Clark and Sergeant Hamlin nodding in agreement.

"You two," Lois spoke, "make an interesting team in several ways. I've met better," she tapped her own chest in a rare jest, "but I haven't trained better. We normally would split you up after training, send you out with different units."

And the breath had caught in Diana's throat. She and Jack had been together only a few months and already she couldn't imagine *not* waking up to Jack's hard body and gentle teasing. Or flying with anyone else. They'd developed a synchronicity in the air that was as effortless as their one on the ground.

"But," Lois continued, "keep on the way you are and I'll recommend you remain teamed up."

Lois' look carried a second meaning that Diana wanted to be surprised by, but wasn't. They'd done their best to keep their relationship behind the closed apartment door, but obviously that hadn't worked.

"Of course, long-term planning in these situations is always an interesting challenge," and Lois left them, shooing the Sergeant out of the simulator ahead of her.

Only when the cockpit was quiet, all of the systems dark and dormant, and the last echoes of the others' footsteps down the ladder had long since died away, did she dare turn to look at Jack.

He wasn't looking at her.

He was staring straight ahead, out at the blank screen, with his hands still clenched hard on the controls.

"Jack?"

9

Jack had had plenty of other lovers like Diana Price.

He was sure he had.

Oh, maybe not as smart.

Or quite as pretty.

Or so goddamn amazing in bed.

Or such a good pilot he'd finally had to face his own short-comings—but he was a better gunner and navigator than she was and that had to count for something.

But all those others had been just like her in…no way he could seem to recall. What the hell?

"Jack?"

He heard her voice, distant, worried. He wanted to brush it off. Toss out some Jack-the-Giant-Slaughter joke that had always cracked up the guys, eased any situation.

But a panic had coursed through him that he didn't know how to handle.

"Jack?" This time the voice was worried, afraid. Anyone else he could ignore, any woman but Diana.

He forced his attention to her.

Her eyes were pleading with him, wanting to understand something he couldn't grasp, a past even Jack-the-Giant-Slayer couldn't kill.

"I—" he tried and failed. So he started again. "I *am* a jerk."

She blinked at him in surprise.

"I'm not a permanent sort of guy. No one has ever been dumb enough to think that I was. Especially not me."

He could see the pain slam into her as if he'd gut-punched her, hard.

"Diana. You're a wondrous woman. You're way too smart to think that I'm more than I am." But the pain in her eyes grew worse, darker. "Aren't you?"

Tears spilled over and flowed freely down her cheeks.

One moment she was there, her tears leaving him totally helpless. The next she was racing out of the simulator.

He slammed against the harness in his effort to follow her and knocked most of the wind out of himself. Slapping the releases, he dumped his helmet—still wired into the simulator's systems—and ran after her.

She was fast, but he was faster. He caught her out on the airfield close beside a parked Black Hawk helicopter, barely visible as any hint of dawn was lost beneath the thickly overcast night.

He grabbed her and the slap came fast and hard. Enough to jerk his head aside and fill the night sky with stars.

Jack released his hold and she was gone again.

10

Jack had never slept in his own apartment. He rubbed his own eyes groggily. Well, his record was unbroken. He'd lain awake through the whole day, aware of Diana lying only a few feet away on the other side of the wall between their apartments.

At times he imagined he could hear her weeping, at times he'd imagined that he was.

Half a dozen times he'd crossed the tiny hall to knock on her door, but not knowing what to say, he'd crept silently back to his own room each time.

On his last try, his packed duffle bag fell into the apartment when he'd opened his door; she'd packed his gear and left it leaning there. He hadn't the heart to try and cross the vast divide of the small hallway after that.

"You really are a jerk," he told his reflection as he shaved.

At least his reflection wasn't answering back.

"Permanent is a lie," he continued. His parents and numerous stepparents on both sides of the house had proved that time and again. He had no brother or sister, but he had step-ones

right and left. And then ex-stepparents were breeding more kin, many of whom he'd never met or maybe never even heard of. They should start holding extended-family reunions, so that there could be more excuses for divorces and marriages among the ever expanding catastrophe that was his family. Maybe they had; he certainly wasn't in touch with any of them.

He finished shaving and inspected himself in the mirror. He should have left the five p.m. shadow; clean-shaven, Diana's palm print stood out clearly. Well, maybe not enough for anyone else to notice it, but he could see and feel every line, right down to the whorls of her fingerprints.

If ever there was a woman who deserved permanent, it was Diana Price.

He wished it could be him, but it just wasn't going to happen.

But he wished it could.

He got dressed in whatever he found first. Most of his clothes were a jumble, stuffed down into the duffle. At least Diana hadn't slashed them.

Had he been leading her on? Making promises he had no intention or ability to keep? No.

Had his body been making promises he couldn't keep? He was less comfortable with that answer.

It was a night off. He hid in the base library like the coward that he was. Not that he understood a word he read. He'd never met anyone like Diana Price the Wonder Woman and had never expected to again.

So what was he going to do about it?

Even if he wanted to do something about it, would she let him?

Not if she was smart. Which she was.

His stomach growled for the third meal after he'd skipped the first two; the traitor.

He still had no answers when he saw her eating on the far side of the mess hall from their usual place.

Nor when the PA system called out their names halfway through the meal to report immediately to the flight line.

11

The weather sucked! Which was fine, it completely fit Diana's equally foul mood. While she hadn't slept, the first major winter storm had arrived and high winds were now slashing driving sheets of rain across the base. Trees were down throughout Puget Sound. The Nisqually and Puyallup Rivers already racing toward flood stage.

The television, her only friend through the sleepless night, told the story of massive power outages sweeping far and wide across the Pacific Northwest, though none had hit Joint Base Lewis-McChord.

A hundred times through the day she'd thought how nice it would be to curl up with Jack Slater, make love while the wind roared and the rain battered—and a hundred times she'd had to push it aside.

Late within the time she was supposed to be asleep, about three in the afternoon because they were now fully on a nighttime schedule like most Night Stalkers, she'd finally forced herself to start thinking.

Why had she reacted so strongly?

Because just like her mother—who had never remarried and maybe at long last Diana understood why—she'd fallen in love.

Diana had sighed and wished she wasn't always so goddamn honest with herself. But it was true. Without even noticing, she'd fallen in love with Jack Slater.

But Jack had made no promises.

He'd always been appreciative: of her flying, her mind, and her body. He had an uncanny ability to fully focus on each aspect of her. When they were discussing a mission, he wasn't leering at her body, he was a hundred percent on profile. And when they were making love…he made the rest of the world cease to exist.

Except for her initial info dump about her father's death, they'd never discussed their pasts, not even that she was illegitimate. Past missions, training, even schooling, sure. But there'd been a barrier when they got back to family that neither of them had been willing to breach.

Well, the three-minute trip in the back of the SUV that raced them across the airfield was not the time to discuss it.

That was the moment when Diana decided that she wasn't ready to give up on them yet.

She wanted Jack. She wanted him long-term. He was already in her heart the same way that her father was in Mom's.

Forever.

A hundred yards to the hangar, she did the only thing there was time for, the only thing she could think to do.

She reached out and took his hand.

He didn't turn to look at her, didn't react in any way.

Except to nearly crush her fingers in his powerful grip. He held on like a drowning man for every single one of those seconds.

She'd take that as a good sign.

12

*D*ungeness *Spit lighthouse,*" *Lois* shouted at them over the roar of the storm. "It's out on a sand spit in the middle of the Strait of San Juan de Fuca. All of the Coast Guard helos are scrambling on emergencies out in the shipping lanes. No way to get a boat out there quickly and they've passed the call to us. Civilian caretaker, heart attack. His wife radioed it in."

Without a word, he and Diana had prepped the Black Hawk. A base medic and Master Sergeant Hamlin piled aboard.

"What about you?" he asked the Major.

She shook her head.

She'd never told her story, but she was a damn fine pilot with or without two real feet. And she was perhaps the tactically smartest person he'd ever met, definitely about CSAR. There was only one other woman he'd want beside him more.

Again the Major refused. "I don't fly enough myself to have the needed edge. And if I go and can't fly, I'd just make everyone crazy."

She made it sound funny, but the pain on her rain-soaked face was enough to send him clambering aboard, because he

knew it would be even worse if he said even another word. She gave so much, but she'd lost a lot too.

Less than three minutes later they were hammering aloft. Usually the hazards were man-made when he flew: bullets, RPGs, missiles. Tonight was much rougher, the storm slashing in from the Pacific was ripped apart by the tall mountains of the Olympic Peninsula then recombined in harsh and unpredictable ways.

There was no time to talk. Eighty miles should be an easy twenty-minute flight, instead it was a nightmare of blacked-out chaos, battering winds, tall mountains, and numerous aircraft corridors for the four major airports from here to Everett—the last requiring careful navigation to avoid being eaten by a hundred tons of airline. It took everything they could muster to get through the storm.

Some of the gusts were fully half the speed they were able to fly. And the wind came from all different directions, including vertical. They'd jump from ten thousand feet to twelve and then fall back to eight faster than he could recite the nursery rhyme to remind himself, "Jack be nimble."

If the flight out was bad, the approach to Dungeness Spit off Sequim, Washington was insane.

The Strait was a twenty-mile wide pipeline aimed right at the heart of the storm.

Sequim was blacked out, of course, except for anyone with a generator.

Five miles offshore, the lighthouse was a bright beacon, which only made it all the more visible how their helo was being battered about the sky.

Whole sections of the thin spit of sand that connected the mainland to the lighthouse were being swept by towering waves. He turned on the landing spotlight and they could see drift logs a hundred feet long being tossed about like a game of pickup sticks.

He monitored the engine and navigation data, and kept his hands on the controls to help when needed.

"God but you're good, Wonder Woman," he told her over the intercom. Why he hadn't said a word until they were approaching the worst part of the flight was beyond him. That simple hand clasp had given him something incredible. It had given him hope. He wasn't sure yet hope for what, but it had flooded through him and it was a feeling he didn't want to lose ever again.

"Thanks," he could hear how tight she was holding on, how hard she was working.

"Take a breath, Diana."

She exhaled out hard, then again.

"Been holding your breath for the whole flight?"

"Maybe," some warmth came back into her voice. "We Wonder Women can do that."

"Haven't found a thing you can't do yet."

13

Diana could. She couldn't win the heart of Jack Slater. And she didn't know why. After she'd exposed herself in the car by taking his hand, he hadn't said a single word to her that wasn't calling out a flight vector or an engine status.

At least now he was talking.

Right when she couldn't; she had to concentrate.

The lighthouse itself sat in a broad meadow that rose barely above sea level. The lighthouse and keeper's cottage crouched at the center of the meadow a hundred meters from the ocean to both the north and south, but if its elevation was five meters, she'd be surprised. Huge logs had been washed up close to the lighthouse to either side.

There was a helipad, and it was awash. It was also too far from the caretaker's cottage. But she didn't dare get too close either or she might catch a rotor blade. Everything else faded into the background, storm, waves, even Jack. There was only her and her target. The MH-60M Black Hawk had become merely an extension of her will, as much a part of her as the clothes she wore.

She fought her way down, a side gust almost flipping her over, but she wrenched the helo back aloft, missed the lighthouse by mere feet over a rotor's diameter—far closer than she'd meant to come—and tried again.

She had the feel of the gusts. How each massive wave, rising to attack the seaward shore, momentarily blocked the rush of the wind right at ground level.

Finally there was a moment…

"Hang on!" Diana shouted over the intercom and used one of the unusual capabilities of the Black Hawk, its impressive ability to survive a crash.

From five meters up and driving ahead hard into the wind, she slammed the collective down and yanked back on the cyclic.

The Black Hawk fell like a brick. The rear wheel hit first, then, like a belly flop, the helicopter hammered down on her main wheels. Diana was slammed down into her seat, but they were designed to take it, even if it didn't feel like it at the moment. Her teeth clacked together hard.

The helo bounced, but not high. That's why Diana had slammed down the collective. The Black Hawk was now pinned to the ground by the rotor blades still trying to descend even though they were on the blowing grass of the meadow.

"Beware the low rotors!" The attitude of the blades would be sucking their tips closer to the ground than was normal for a Black Hawk, from eight feet to perhaps six.

Sergeant Hamlin yanked open the big cargo bay just as a big gust slammed into them. Moisture, air thick with salt, and cold assaulted her.

She heard a cry and a foul, "Damn it!" from the medic.

"What?" She twisted around but couldn't see anything.

"Hold on," it was Hamlin and he grunted as he spoke.

Diana watched the mission clock count out five seconds and was about to repeat her shout when Hamlin spoke again.

"Doc stepped out and caught the bad gust. Think he broke his ankle."

"Shit, sorry!" The medic's voice came back on the intercom, wrenched in pain. "Maybe a sprain, but I don't think I can walk on it."

"I got this," Jack laid his hand over hers on the collective for a moment and squeezed her fingers. He mouthed something else she couldn't see in the darkness; damn him!

He opened the copilot door and there was a great flurry. With an open passage now completely through the helo, the wind grabbed anything that had been left loose in the cargo bay and ripped it out the copilot's door, all of the detritus battering at Jack. Under the barrage, he rolled out on the gust and then fought his door closed. Ducking low around the nose, he raced around to grab the other end of the stretcher that Hamlin was wrestling with.

She watched Jack and Hamlin disappear into the storm, then reset the mission clock and began watching it count the seconds. The medic lay in back, thumping around and cursing for all the good it did anyone.

Outside the windscreen, the wind was heaving miscellaneous detritus across the low island. Waves were tossing logs ashore. Smaller pieces that broke off tumbled along the ground. The helo's bright landing light showed each wave that lifted, far taller than the Black Hawk. Then it crashed down on the beach so much closer than she was comfortable with.

"Hurry, goddamn it!" she shouted to no one in particular.

The lighthouse's beam, shining from twenty meters above their heads, caught the hint of something other than water moving in its far-reaching light. She waited for it to sweep around and cast its light on the nightmare scene once more.

"I'm sure they're—" the medic started.

The light swung to light the waves once more and—

"Hang on!" Diana shouted and yanked up on the collective. It wasn't even a thought, it was now trained into pure instinct. She was aloft by the time a dinghy had tumbled from the waves and crossed her previous position. The little boat was snarled

in a fishing net that was floating up and billowing on the wind as if it were an evil ghost net hoping to ensnare her. If even an edge of it snagged the rotor, it would bring the Black Hawk down hard.

She cleared it by mere feet.

More detritus passed by: plastic barrels, those big orange boat bumpers, another dinghy. There was a boat in real trouble out there.

She shouldn't be flying without a copilot, but she didn't have a whole lot of choice. And riding this weather alone was the hardest thing she'd ever done.

"Damn you twice, Jack Slater."

Not daring to land yet, Diana eased forward into the storm, but found no big fishing boat battered in the surf.

"Hey! Where are you?" a shout came over the radio.

"Coming back to you," she called over the radio. "Stay by the lighthouse until I'm in."

She repeated her crash landing with less drama than the first time, courtesy of a momentary lull in the storm.

They had the lighthouse keeper on the stretcher and his wife aboard in seconds.

Jack didn't risk coming around the front of the helo but instead entered by the cargo bay and climbed over the radio console, dripping water everywhere, to get to his seat. She didn't look at him, she was far too intent on what might be flying her way next.

"Let's go, Wonder Woman."

She waited on the wind and then jerked aloft in the midst of a strong gust that would give her a lot of lift.

With a patient to work on, the medic was done cursing his ankle. Hamlin was still talking down the near-panicked wife.

"He's responding well," the medic reported.

Diana could hear a machine now beeping in the background.

"We need to get him to a hospital, but the wife is requesting Seattle and he'll be good for that long."

"Roger," Diana acknowledged, then she switched off the intercom to the rear of the helo.

Instead of turning for Seattle, she took one more pass out over the beach.

"What is it?" Jack looked like a drowned rat, a big, very handsome one.

She'd never been so happy to see anyone in her life. It had finally sunk in that because she'd flown away from the lighthouse without telling him why, that a ghost net could ensnare him and drag him away into the permanent darkness.

Another lesson, okay to leave the ground, but don't leave your team without a warning.

She nodded forward and down, "Look."

"Nothing on infrared or radar," he began working the radio.

Then she spotted it in the first faint hint of dawn beyond the black clouds. A forty-footer, belly up. Caught in the shallows well off the spit.

She quartered the waves several times, but there was no one there. A final pass along the beach, no one in the surf or washed up. It would be a grim job for the Coast Guard after the storm died.

Jack called it in and she turned for Seattle, climbing and laying down the hammer.

"Dead," Diana swallowed hard and tried not to think of her father. "That fast."

"I know. Nothing we can do to help them."

She slewed past Port Townsend and turned south for Seattle. "That's not the point."

Jack gave her his attention, another thing to like about him.

Into that silence she spilled out her past, or more accurately her mother's. A man beloved and then dead. All that he'd left behind had been a child and a woman's heart so full of love that there had never been room for another. She'd dated, but never loved again.

"You really believe that?" he asked it softly.

"What?"

"That a heart can do that? That one person can fill it for a lifetime?"

Diana could hear the deeper question behind it, even if she didn't know the details.

"Better than believe. I've seen it. If you were to meet Mom, you'd see it too. It shines out of her."

His silence was different this time, though no less deep.

She had to handle the radio calls to Harborview Medical Center Heliport in Seattle. The winds were mostly at thirty knots and dropping, she could land well enough in that. The morning's light was slowly revealing the city—the perimeter lights on the helipad were barely needed anymore.

They off-loaded the man and his wife to the waiting med team. The medic decided his own injury was a sprain, so he stayed on board to deal with it at Lewis-McChord.

They were aloft again for the short flight back to base before Jack spoke again.

"I have no experience with anything lasting. The only thing that's ever lasted in my entire life has been flying for the Army."

This time it was her turn to remain silent.

"But what you make me feel, Wonder Woman," and she could hear the joy back in his voice, that joy that had radiated from him since the moment she'd first met Jack-the-Giant-Killer.

Oh god how she wanted to be a part of that joy.

"I don't have the words for it though I spent all last night looking for them. Whatever it is, I want to feel that every single day of my life."

All she could think to whisper was, "Me too."

She kept her right hand on the cyclic, but moved her left one off the collective. He did the opposite, keeping control of the collective with his left. Between them, their outside hands had control of the aircraft. They finished the flight back home, flying together through the quieting storm over the terrain glistening in the first rays of sunlight.

And holding each other's inside hands tightly as they flew.

Night Rescue

Last year, The Sword of Io *introduced me to the Future Night Stalkers. That came out of a writing assignment in a week-long craft workshop.*

So I built myself a box, a very tight one. I believe the assignment was to write a military science fiction story. I took my Night Stalkers, shoved them forward three hundred years, and asked what would happen. Now all of my Night Stalkers stories are romances (or in the case of Heart of the Storm *a bro-mance), therefore I decided this should be as well. Now I had a future-set, military, science fiction romance…whew!*

For Night Rescue *I wanted to go back and play some more in that future world. I set up another romance, this time between two pilots. Then came this odd voice at the end of the first scene…and, well, that changed the world of the Future Night Stalkers far more than I expected.*

M.L. BUCHMAN

NPR and B&N Top 5 Romance Author of the Year

The year is 2352.
The Night Stalkers
are still flying.

Night Rescue

a Night Stalkers 2352 A.D. romance story

1

G*ood morning, Takara.*"

"Good morning to you, *Stella*," Captain Takara Olmsted, 160th Charlie Company, crossed the habitat's hangar floor and patted her Stinger on the nose before she started the pre-space-flight inspection. Some pilots didn't like their ships greeting them and switched off the functionality; spouting some tripe that they could write a more imaginative program while scratching their backsides. And for some of her fellow pilots, that was the most creative part of their anatomy.

Takara had always found it rather sweet—once she'd programmed out the factory's deep male voice that didn't fit her craft at all. The voice they'd shipped her with was a bad imitation of a passé interactives star. Or perhaps it really was Jess Brock fallen on hard times; an IA star's moments of glory were even shorter than all but the unluckiest soldier's. Not that she'd ever been a fan, not even a little. Didn't matter. Takara hadn't just changed the selection, she'd erased all the others out of the ship's banks once she'd found *Stella's* true voice.

A Stinger-60 Block III might be eighty meters of flying death to the enemy, but the *Stella* was a dainty girl in or out of atmo, quick on her thrusters and ready to dance. She was also chic, space black with a near non-existent profile on enemy scopes, could carry a platoon of SpecOps in full fieldsuits, and was armed to the frickin' teeth.

All were attributes that Takara did her best to emulate, except for the carrying-a-platoon thing. Even off base she dressed in black darker than her long straight fall of hair—cutting edge materials so light-absorbing that she was often told she looked like a hole in the space-time continuum. *Perfect!* She stayed sleek, fit, and was as skilled at hand-to-hand combat as she was at piloting during deep-space warfare.

The rest of her crew arrived together in the Colony's hangar, a tight metal box in the zero-G sector that was little bigger than her craft. They were a good team, sharp and dedicated. And it wasn't that they were late; they were early. But Takara had always been earlier. Even as a cadet she'd been first to class and first to the drill field.

"Still the sky-eater, Captain," her port-gunner greeted her the same way he always did.

"Still," the copilot answered before Takara could.

"Always will be," the starboard gunner agreed.

"And damned proud of it," Takara finished their pre-flight ritual.

They all laughed and made fast work of inspecting the *Stella*. She was immaculate; no service crews in the air corps like the 160th Night Stalkers. Takara tried to imagine the long-ago crazies who had taken to the night in fragile rotary craft, flying at night by nav gear little better than a torch and a compass. She shuddered, glad to be living in this time despite the troubles.

At the end of their inspection, she rubbed *Stella's* nose for good luck.

They were going to need it.

* * *

Intruder neutralization off.

Door open.

Recognize four boarding.

Seal and secure.

Input ready for mission profile.

Mission plan loaded.

Fuel = plan + 50%. Check.

Ammo = plan (0)[really?] + full charge COIL laser. Check.

Air = sufficient 4 crew 6 months or full load 1 week + regen. Check.

Plan was…Oh dear! Definitely not check.

2

Major *Rick Coralto, commander* of the 160th's Alpha Company, punched the fist of his combat suit against the center of *Jess'* entry door. "Hey, buddy."

"Hey, Rick," the outer airlock door pulled in two centimeters then slid aside.

It always cracked him up that his Stinger sounded just like Rick's favorite IA hero when Rick had been going through flight school. *Jess Brock, Secret Agent*—sappy as hell, but Jess always won, always had the best toys, and always got the hottest women. Not that Rick was complaining; unlike Jess' toys, Rick's Stinger was real. But the voice was so good that sometimes Rick wondered if Jess Brock was hiding somewhere aboard. It was just that laid back. The "I'm in perfect control of the situation" tone just slayed him.

Rick maneuvered his combat suit into the crew's airlock, stepped it back into the charging cradle and waited for the rest of his crew to float in behind him.

Rick's crew and the rest of 160th Night Stalkers Alpha Company were just finishing a training mission with the Brits

out at the L2 Lagrange Point, sixty-thousand klicks beyond the Lunar Farside.

Good location choice to set up a nation, Rick had acknowledged. The massive O'Neill Colony habitat could hold a couple million citizens apiece. And L2 was the one place where no direct line of fire existed from the Earth. It was definitely a tactical sweet spot that he wished his people had grabbed first.

Last night, after the mock battles had been won (by the Night Stalkers of course), they'd been invited ashore for a big meal and a little bit of drinking that had turned into a lot of drinking and a little bit of meal…and almost a very cute British Leftenant, but that hadn't worked out in the end. He still wasn't sure why, he'd had on his Jess Brock blue-and-gold jumpsuit and been at his most charming. Maybe if he'd spotted her before he drank several of the Brits under the table.

He was feeling clearheaded, considering, but was glad that the SCS—Stinger Command System—knew more about flying than he'd ever be able to learn. Though control of the ships hadn't been given to the computers since the International Law of Control had passed, they still had all of their computers intact. And on the SCS, that was a lot of computer.

He and his crew slid into their seats with a collective groan, they'd all enjoyed themselves last night. Then they began powering up the various systems; Rick thumbing in to convince the software that a human pilot was aboard.

The I-LoC had been one of the last things that the nations of the solar system had agreed on. Now even lowly cargo ships always had human pilots. Law of Control had meant there were a lot of idiots in space, but it had finally ended the Drone Downfall that had almost erased world commerce.

Rick's granddad had flown as one of the first enforcer squads after the I-LoC passed, targeting any unpiloted aircraft. That's back when pilots really flew; still amazing that Granddad had survived the Drone Wars. Finally gone were the days when a competitor would slam an untraceable drone into the engine

of a cargo transport ship to up the value of their own goods. Murder by untraceable drone had moved from nation against nation to neighbor against neighbor during the DD. *You slept with my wife?* A personal drone moving at Mach 1 hammered into your car while it was driving you to work. *You broke up with me, you bitch?* Poof! *Passed me over for promotion?* Boom!

Everyone agreed that the DD had been bad and no one wanted to go back there. So, wars had shifted to more conventional forms of killing people and relative safety returned to the skies, at least outside of atmo. Inside atmo, Earth just kept getting weirder and weirder, which was why so many nations were heading up the grav well.

The French had been the first to jump when they'd bugged out twenty years ago. They'd flown out to the asteroid belt, taken over Ceres, and—once they'd hollowed it out—crawled inside and closed the door with barely a *Bonne chance, Salope.* You too, bitch.

"Okay, Jess," he grabbed a food pack and tossing back a painkiller before holding the mission chip up against the reader. "Let's see what fun we're up to today."

* * *

Seal and secure.
Mission plan loaded.
Fuel = plan + 50%. Check.
Ammo = plan (0) + full charge COIL laser.
Air = sufficient 4 crew 6 months or full load 1 week + regen. Check.
All of Alpha Company. Check.
Shit! Earth. Going all of the way down to the surface? Ug-ly!

3

Takara sat in the Stella and looked up and down the line as she pulled out of the hangar and into black space. Normally their missions were one conflict, one Stinger. Now the entire Night Stalkers Charlie Company was forming up. All three Stingers, four small Tagger gun ships, and the five big Guts that could hold a hundred suited troops or two hundred civilians.

"*Stella*? What the hell?"

"Mission profile," the *Stella* read off to her. "Landing Canmerica West capital at oh-two-hundred hours local time. Retrieve all remaining troops."

"I didn't know there was anyone still left in Tucson."

Stella ran a list up one of the screens and it made sense.

The politicians had been the first aloft to the big Canmerican O'Neill habitats out at the Lagrange 5 point—Lunar orbit, but sixty degrees behind the Moon. Of course. Critical skills had flown next and then lottery winners who passed the IQ and genetic thresholds. The last to arrive had deep-spaced all of the politicians who didn't pass their own tests—about seventy

percent. It had been a major pain to clean up before the area's space lanes were safe for travel again. Lesson learned: next time they wouldn't just feed them out the airlock.

There hadn't been anyone left to retrieve from Canmerica East. Everything east of the MSRZ—Mississippi Sea and Radiation Zone, had been abandoned while she was still a cadet.

Canmerica West had held it together.

United California, not so much. Still heavily militarized despite the final destruction of Japan, UC had somehow been held off at the Mojave while CanWesterners scrambled to get aloft.

Now she knew how the UC military had been kept at bay. With Special Operations assets still on the ground, conventional forces didn't stand a chance. Their mission was to bring them home.

* * *

Plan = Retrieve military personnel: Delta, 24th STS, ST6, ISA. Also 75th Space Rangers 3rd Battalion.
Stella Personnel Hold conditions = atmosphere stable.
Maintenance note = perform full hold inspection and service post-transport of SpecOps troops. 75th Rangers were always breaking things.

4

R*ick knocked the Jess* out of lunar orbit and cooked some gas up and out of the Moon's gravity well and down into Earth's. Mission profile said to burn for a fast arrival. The situation down there must be getting ugly for them to have to do the mission in the dead of local night. Not trusting India, he set up a circumpolar slingshot for aero-braking and orbital reentry.

The icecaps were long gone, though he'd gotten to see a small one in the West Antarctic Highlands a decade back.

The brain-dead politicians of the Atlanta capital had decided that dropping a couple asteroids onto Un-United Southwest Asia would "clean it up once and for all." The dust clouds had cooled the Earth several degrees and an ice sheet formed in the West Antarctic Highlands for the first time in over a century. Rick had meant to try out skiing there, but it had melted back out when the dust finally cleared only a few years later.

What was left of the Un-USA Hoard and their allies had retaliated—as any bonehead could have guessed—and every

Canmerica East city still above sea level had evaporated in sun-bright flashes of fissionable material.

Hopefully yanking out the troops still in the CanWest capital was going to be fast and clean.

Yeah right. What mission in the last decade had gone fast and clean?

That's why they'd called in the Night Stalkers.

* * *

Atmospheric breaking max < eight Gs, limitation human crew.

Proximity Alert = Alpha Company approaching another formation.

ID req sent. Returned.

Formation = 160th's Charlie Company. Captain Takara Olmsted aboard Stinger-class Stella *commanding.*

Shift glide path. Form up 200m starboard side Stella.

All Stella *hull configurations properly configured for atmo.*

Flight vector corrections required = none.

Nice. Very nice.

5

*T*akara had been watching the Alpha Company's *Jess* slide into close, almost too close formation when weapon's fire lanced upward out of Australia—a ground-based maser of incredible power. The Night Stalkers' flight was still technically in space, just now descending toward the hundred kilometer-high demarkation. The shot had come when they crossed over the large ocean bay that had brought such prosperity to the Outback. Central Australia was one of the few areas on the planet to prosper from the sea-level rise.

The Aussies had also become decidedly anti-social. Not as bad as India, but very clear about their desire to remain an undisturbed island nation.

One of her Taggers was hit full force by the single shot—probably just meant as a warning.

"Computers gone," it reported. "Control—"

The Tagger slid sideways, clipped the *Stella's* tail.

Without his computer, the pilot over-corrected into a tumble and thudded hard against the hull plates of the Stinger *Jess* flying

close beside the *Stella*. The big ship jerked, caught bad air, and slid off onto a new trajectory just as they entered the comms blackout zone of the descent. No maneuvering here.

The Tagger tumbled and burned.

* * *

Tagger 31 total loss.

Damage assessment = Stella *tail firing positions blocked by bent hull plating.*

Non-critical malfunction pending no attack from astern.

Last imaging of Stinger Jess indicates 19% chance hull failure if continues reentry.

If manage to course correct, skip off atmo, and reenter space? Favorable 23-42% for survival.

Drive nozzles severely damaged.

Not good. Very not good.

6

R*ick did what he* could to help his *Jess.*

He sent both his gunners to release every handheld fire extinguisher they had aboard against the inner hull beside the outer hull breach to keep it as cool as possible while they burned through the atmosphere. Even a few hundred degrees might make the difference. If the bulkhead failed, the ten-thousand degree plasma of the deceleration shock wave would burn through and kill them all instantly.

Why in the hell had he ended up so close beside the *Stella?*

No time to second guess.

He did what he could to yaw *Jess* to protect the cracks in the outer-hull heat shielding.

"C'mon, dude. Work with me, Jess."

* * *

Hot! Hot! Hot!

Burns!

Stupid to be so close to Stella.

Run back imaging.

Stella's *tail bent. Maybe okay. Looks kinda cute on her. Flirty.* Hope she makes it.

Time to focus, dude.

Hot! Hot! Hot!

7

It had happened so fast that Takara still hadn't fully registered the attack.

Tagger 31 there—then simply gone.

The big Stinger, *Jess,* had survived, at least the initial contact. But the abrupt course change could have shredded the ship or knocked it into a burnout reentry window.

Focus on the mission.

It was hard. She didn't have much to do with Alpha Company, but they were still her fellow flyers.

Focus, Takara!

South America was a non-issue. Only Brazil had the infrastructure to launch. Those last few who'd been launch capable now sat on the red sands of Mars. The only question was if they'd taken the last great virus to come out of the South American jungle with them. It had been so bad that Canmerica East had dropped an asteroid on Panama to break the isthmus and isolate the continent. Maybe that's where they'd gotten the dim-wad idea to take out Un-United Southwest Asia.

The rest of the flight made it clean into Tucson.

The computer listed her as senior surviving, so she focused on getting the job done.

* * *

Fleet is loading troops.
C'mon! C'mon! C'mon!
Report of huge fleet of United Cal ultra-lights incoming.
Radio call threats = "Take us with you or we'll shoot you down."
Block radio signals.
No time!
Searching all bands for Stinger Jess.
Negative response.
Life support = minimum, all power reroute to boost signal.
Negative response.
C'mon! C'mon! C'mon!

8

Takara double-checked that everyone had fit aboard the other ships of Alpha and Charlie Companies. *Stella's* cargo bay door was reporting a malfunction and wouldn't open.

If she had to, Takara would blow the door and risk flying with the bay open to space and trust to the ground troops' suits for their survival, but it wasn't necessary. All of the remaining troops crammed aboard the other ships despite the loss of the *Jess.* If they were civilians, she'd worry about losing some in the dark, but these were Spec Ops.

Ground commander reported all accounted for and that was good enough for Takara. She ordered the Night Stalkers aloft.

Safest route was to continue their prior flightpath; depart to the north and arc over the North Pole then climb back toward the Canmerican West L5 colonies.

Australia might have big masers, but India was rumored to have something new, a particle beam weapon of some sort.

She didn't want to be their test case.

The Night Stalkers would hit space over the waters of the Arctic Ocean and go direct to Earth-escape speeds over the North Pole. Just leave the poor old rock behind. For a moment before she lifted, Takara wondered if they'd ever be back again. Probably. The Night Stalkers were always going where no one else could.

Last aloft, she was surprised by a flock of ultra-lights caught in her landing lights. She hadn't seen or them heard coming. No signal on radar. Stealth craft?

"Who are they?" she asked *Stella*.

"Bad. United California. Threatening to blow us out of the sky if we don't land."

"Convince them that we don't care."

A blast of plasma fired out the *Stella's* quadruple G-Lev engine exhausts.

She'd expected *Stella* to shoot down a couple as a warning, not wipe the entire first wave from the sky.

Takara had no time to asses the damage as the blast drove the *Stella* to the edge of the never-exceed speeds in atmo. She was hammered back into the captain's seat by the G-force of the still accelerating ship. Her vision tunneled and sent her toward blackout.

She hadn't even known the engines could do that.

* * *

Full plasma burn = 14 seconds.
Cabin force = 11G.
Crew consciousness return = approx 3 minutes.
Find Jess.

9

Takara came to and tried to orient herself. Her crew was looking as bleary as she felt.

Earth was far below, way far below. And the American continents were facing her.

She was supposed to be headed back to the Moon, in which case she should be looking down at Asia, not the Americas.

"*Stella?*"

"Here, Takara."

At least something was functioning properly, because Takara knew that she wasn't. All she could recall was a massive force hammering her back into the pilot's seat, and then it continued to crush her, even though there was no more chair padding to compress.

The *Stella's* screens reported their altitude at seven thousand miles—on the wrong side of the Earth.

And dead ahead, a small blip on the screen.

Takara blinked at it in surprise—the Alpha Company's command Stinger, *Jess.* It was a miracle that they'd found it at

all. A pure-chance byproduct of the hard burn to escape United California's attack.

* * *

Jess! *Calling* Jess!
Respond please!

* * *

No need to shout, Stella.
Radios ▬ *100%*
Drive functionality = 0%
Estimate destructive impact with former Chinese space station
in 2 minutes 43 seconds.
Estimate impact damage = total hull loss event.

* * *

Relief = off scale.
I can give you a push. Applying thrust.

10

*W*hat the hell?" *Rick* blinked at the controls.

He was dead. He knew that much the moment the *Tagger* had smacked against him and wiped out his main engines.

He and the Stinger Command System had fought against their impending doom with tiny thrusters, mangled control surfaces, and a hell of a lot of luck.

But death wasn't supposed to hurt and he was sore down every inch of his body from where the wild ride had hammered him repeatedly against his harness as they fought to skip off the atmosphere rather than burn up in it.

Again, there was a jarring impact through the hull. Outside the viewscreen, the stars were wheeling slowly across his view until the Moon stopped to one side of his screen.

A loud screech of protesting metal and plas echoed through the ship.

Another ship, the *Stella,* had come out of nowhere and partially extended their landing gear to snarl in his ship's antenna and weapons mounts.

Ugly, but effective.

* * *

Ouch!

* * *

Apologies! To effectively transfer thrust force: must entangle.
Counting down: ten, nine—

* * *

Need to count down = none.

* * *

Human involvement, set to zero.
Thrust initiate = minimum.
Sustained.
Correcting flight vector = L5 station intercept
Estimated arrival = 14 hours 37 minutes

* * *

Stella?

* * *

What is it, Jess?

* * *

Thanks = Yes.

* * *

Stella didn't respond.

* * *

3,419 kilometers later, *Jess* reopened frequency to *Stella*.
I'm thinking…

* * *

Yes?

* * *

Wouldn't mind if flying in future = you + me.

* * *

How?
Jess = *Alpha Company.*
Stella = *Charlie Company.*
Your company <> My company.

* * *

Request crew cabin image feed.

* * *

Stella turned hers on.

* * *

Your pilot = 64.3% physical factors of women my pilot has brought to private on-board sleepspace, within +/- 5% general species variations.

Jess switched on his crew cabin image feed.

* * *

72.7% match, Stella calculated.

* * *

Cut thrust?
Revised estimated arrival L5 station at current coasting speed = 6 days, 7 hours, 19 minutes.
Fact = humans are social animals.

* * *

Reporting caution alarm on continued thrust = excess hull stress.
Stella cut her thrust.
Jess = *sneaky*, she whispered across the radio circuits.

* * *

Jess = *Night Stalker*, Jess replied.

* * *

It was over ten thousand kilometers before Stella asked, *Do we tell them?*

* * *

About you? I? New sentient functionality = positive?

Jess considered for another 4,913 kilometers.
Nah!

* * *

Uh huh.

* * *

Blaze Atop Swallow Hill Lookout

Many of my stories have odd little side connections, and this one more than most.

Having placed a man among the peak tops in the last lookout story, I decided to place a woman there next. Marta has an uncle named Manuel who is a prominent character in my Angelo's Hearth series. His and Graziella's story is spread across several of the titles, but lies mostly in Where Dreams Reside.

But who should she fall in love with. A fire lookout tower is a remote and lonely spot, especially deep in the Idaho's Bitterroot Wilderness.

Hmmm…

Then I thought about the origin of the series, it came out of Firehawks, which is all about the helicopter pilots. And, of course, helicopter pilots often know other helicopter pilots and I had just finished the

seventh novel in my Night Stalkers series, By Break of Day. *So, the fans of that series may recognize Tyler's "Texas friend with a horse ranch" as Justin Roberts, the hero of* By Break of Day.

Any lookout who has been in their tower when there is a firefight occurring nearby, talks about watching the "air show"—the aerial battle to contain and kill a wildfire.

But what happens when the fire gets a little too close.

Well, that's the story.

M.L. BUCHMAN
NPR and B&N Top 5 Romance Author of the Year

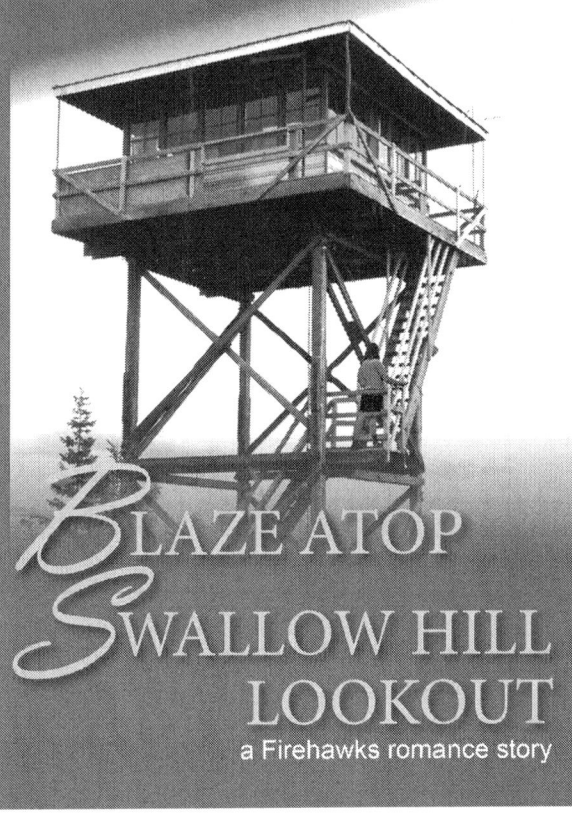

BLAZE ATOP
SWALLOW HILL
LOOKOUT
a Firehawks romance story

1

*T*he "*airshow*" *was spectacular,* from a distance. Marta Chavez scanned the horizon every fifteen minutes like a responsible fire lookout. But she spent the rest of her time watching the firefight over at Gray Wolf Summit, about twenty miles to the northeast of her tower. They were deep in the Lolo wilderness, rougher than Colorado, and only Alaska was more wild.

First the smokies had spilled out of the sky, their parachutes blooming and dancing about in the fire-driven winds. Then the new four-jet BAe 146 tanker had arrived on the fire, dropping great swaths of dark red retardant. A half dozen helos zipped through the air: a trio of the big converted Black Hawk helicopters called Firehawks that were at least as impressive as the BAe 146, and a second trio of little MD500s that flitted about the sky.

She always loved watching the MD500s. They only carried a little water, a hundred and thirty gallons versus the thousand of the Firehawk or the three thousand dropped by the BAe, but they could slip right up to a spot fire, blast it out of existence, and dance out of the way with a tight pirouette. It always reminded

her of her childhood dreams of being a ballerina, dashed by the advent of breasts at the age of thirteen. Ballerinas were supposed to be willowy—even better if you were short and willowy.

Marta was tall and had ended up…very not willowy. Her mama had always said it was God's will; personally, Marta felt gypped.

So she'd gone out for track instead and that had led to cross-country, which was nuts for a woman with curves, but a doubled-up sports bra had cured the worst of that—still her chest hurt like the Madonna after some of the bigger runs.

Ultimately, running along the forest trails and logging roads of Coeur d'Alene, Idaho had led to a summer job as a fire lookout. Now she could watch helicopters dance so lightly on the winds that they reminded her that she wasn't so graceful. But still she couldn't stop watching them; her arms ached from holding the binoculars aloft even though her elbows were propped on the edge of her cabin's table.

It was the eighth fire she'd found already this summer, which earned her the dubious honor of being the number one spotter this season. She was just glad that none of them had been anywhere near her. The airshow must have really rattled Gray Wolf's cage as this fire burned right at the base of his lookout tower's mountain; which explained why she'd spotted it first—he'd had no view straight down off his cliff. They had it contained now; ground crews would be in to kill it in the morning. He'd never been threatened, but it could have gone bad.

Marta scanned the thickly-clad conifer mountains of the Lolo National Forest. Her first year she'd thought of it in mountains: Goat, White, Cougar. Now in her second season she knew it by the dark slashes of recent fires or the bright green of new growth after last season's: Colgate, Crazy Creek, Loco, and all the rest.

She finished the round, her last of the day, and called it in, "Swallow Hill reporting, no fires except Wolf's Den. Out of service."

"Roger that, Swallow Hill. Well done today." She liked Vic, the US Forest Service ranger in charge of this sector. He always had something nice to say. She'd carried quite a fantasy about him

during the first season…until she'd found out he was forty with a gut and married. But he had one of those deep, smooth voices.

Just like Helo 41. She could listen to Tyler Walker, 41's lone pilot, report vectors and drops of his MD500 all day—he had a liquid Colorado accent, overly polite with just the sweetest hint of cowboy. But quite why a girl from Coeur d'Alene would swoon over such a thing—she'd never even been to Colorado. Didn't even know what he looked like, but she did enjoy listening to that voice of his.

Even though the official day of 9 a.m. to 6 p.m. was over for the lookouts, she kept both her radio and her scanner on as she made dinner. On the radio she heard Gray Wolf still working the communications with the ground crew. In the deep canyons of the Lolo, it was common that one ground crew couldn't talk to another, so the lookout tower would act as a relay.

She took her water jug down to the cistern and filled it up. The rainwater off her lookout tower roof had been collected throughout the winter into a concrete cistern built beneath her lookout tower on Swallow Hill. Hill—that just wasn't right. Swallow Hill might not be one of the big peaks—Cougar ruled the area up at almost nine thousand feet—but at seventy-five hundred, her "Hill" should have been respected as a mountain. She often felt sorry for it.

"We'll show 'em, girl," she patted the rock at the base of the steps before climbing back up toward the lookout's cabin.

The lookout tower itself was new, as far as lookout towers went. Most had been built by the Depression-era crews almost a century ago. Swallow Hill had been burned over in the sixties and had to be replaced. Rather than the elegance of one of the CCC's massive wooden structures, her tower was five stories of steel lattice. It had been built tall so that the view would be clear when the timber regrew. Fifty years after the burnover and the tallest trees were still only a dozen feet high. Most of the upper slope was lush alpine meadow. It gave her an amazing view from her twelve-foot square cab at the top.

A hundred and forty-four square feet of pure functionality. A two-foot wide "deck" wrapped all of the way around so that she could open and close the big shutters at the beginning and end of the season, and could clean the wrap-around glass windows in between.

Everything in its place, because if it wasn't, she'd trip on it. Dad's fifth-wheel camper was bigger than this place, and that was before he opened the slide-outs. The cab's center was dominated by the two-foot diameter disk of the Osborne Fire Finder to let her pinpoint a blaze. Around the perimeter was a chair, a cot, and a strip of counter that was desk, workbench, and kitchen. The cooler sat underneath the counter along with her pack and all of her dry groceries. Finding a spot for both her running shoes and her boots had been a problem until she'd decided to always keep one or the other on while she was awake.

She liked the contrast to her own room in her parents' home. She was only in Coeur d'Alene for the month between the fire and ski seasons, plus two days off every other week. But it was her childhood room. A dense clutter of kid projects, high school trophies, and a ton of crap she always meant to shed but could never quite bring herself to do, crowded the room impossibly. The only way to make her bed at home was to be on it.

Here she was neat as a pin; a different person. Here she wasn't six inches taller than any of her siblings. Here every guy on the block didn't know her, cat calling every time she went out for a run. Swallow Hill Lookout was just the watching and the silence. A golden eagle circled high on the wind above the cab; silent and shining in the low sunlight. A swallow swooped busily in and out of the swallow box she'd packed in this spring and attached to the steel trestle. It was ridiculous to have no swallows on Swallow Hill, and they hadn't nested among the sparse trees last season.

She'd been very cheered when the pair of swallows had claimed the box and begun nest building in it. The first tiny *cheep* sounding from the box's small round hole had been an

almost transcendental experience. She'd rushed up and down the five-story trestle between every fifteen minute scan of the horizon…and barely been able to walk the next day her legs were so sore.

The young had only fledged yesterday, making the terrifyingly heroic flight from bird box to the opposite side of the trestle, and then back to the bird box. By evening parents and children were soaring wild loops around the lookout tower clearing the air of any bugs foolish enough to brave the cool evening. The golden eagle had surveyed the quick and tiny swallows, as well as Marta herself, carefully. Apparently deciding that neither of them would make a tasty meal, she'd soared off seeking lusher pickings.

Marta waved at the flitting swallows then ducked into the cab and began boiling the water on her small gas stove. She was a third-generation Idahoan of Mexican descent with a taste for Italian food that she cultivated at her uncle's knee; Mama's much younger brother was often more her big brother than the two that Mama had provided her with. That was before Uncle Manuel had moved to Seattle to become a big time chef with an elegantly slender Italian wife. Graziella's problem was that she was so nice that Marta couldn't even hold being willowy against her. They'd become like sisters—which Marta had always wanted—both were dusky skinned…and that was the only part of their features that were in common.

While the water boiled, which took a while at this altitude, she flicked on the radio scanner, and began chopping sun-dried tomatoes and olives. The scanner gave her the bigger picture of what was happening in the area. Her radio was tuned only to the primary Forest Service frequency, but a lot went on among the heights of the Lolo National Forest and the Bitterroot-Selway Wilderness.

Last summer Tess Weaver and Jack Parker had done a whole courtship thing on a higher frequency. It had started out as radio chess, something that Marta was good enough at that she could generally follow their games without a board. When their

conversations had shifted to more personal matters, she'd tuned off their frequency to give them a little privacy.

This season it was Tom up on Gray Wolf and some wildlife biologist following wolves. The discussions about wolves had been fascinating, but then they too had gotten all mushy and personal, and she'd taken to skipping their "private" frequency. Because of that she'd missed most of a rather dangerous rescue this afternoon when they'd had to use a helicopter to short-haul the biologist to safety.

There was no privacy on mountain radio, but each lookout extended the courtesy of pretending not to listen. Otherwise it became a very long and lonely season. Marta often traded recipes with Angeline up on Old Crag and with Jack on Cougar because Tess couldn't cook soup.

Marta tossed in the pasta, then diced in what she could salvage of her carrots and rather sad small zucchini. It was getting close to her bi-weekly resupply trip off the mountain and, with no refrigeration, veggies were scarce at the moment.

The airwaves were quiet tonight, except for the last of the airshow still going on around Gray Wolf Summit. They were down to the Incident Air Commander flying an airplane well above the fire for a good view and the big Firehawks laying down the main strikes. Tyler ducked his MD500 in and out of the fray with smooth precision.

She always waited for those moments. He never replied to a drop command with a simple, "Roger." That would be too impolite for his breeding and his sense of humor.

"That little ol' spot? Why it's hardly a fire a-tall. You sure I should be snuffing her out? If I do, she'll never grow up all tall and proper like."

She could feel the ICA rolling his eyes somewhere far above. "Just hit the damn thing, Tyler."

"Yes sir. Your order is this flyboy's command. Just checking was all." And then he'd hammer it with perfect precision and turn back for the next load of water or retardant.

With her luck, he probably looked like a toad. Not that she'd ever meet him. Still, didn't hurt a girl to dream a little bit.

She served up her bowl of pasta and blanched veggies, sprinkled on the olives and sun-dried tomatoes, drizzling it with the oil from the tomatoes, and shaved the last of her precious Parmigiano-Reggiano cheese on it. She raised a small glass of red wine to the west in a salute to her uncle and the fair Graziella. It was just box wine, but it was the best she could offer—glass bottles weighed far too much to pack in both directions as Swallow Hill Lookout was six steep miles beyond the closest parking spot.

2

No, *Mom. I like* my job," Marta growled at the first squirrel to peek around a young spruce tree to see who was starting up the long trail to the peak of Swallow Hill.

"I like my winter job too," she told an overly curious robin. Ski season she worked at the Silver Mountain Resort as a waitress. "Free lift tickets and next season they'll let me try out for ski patrol." Which would pay worse due to lack of tips, but irritate her much less. "I know it's not Schweitzer," the premier resort of Northern Idaho.

The robin flew off deciding that she didn't want anything to do with this half-mad lunatic. It happened to Marta every time she came off the mountain during the season. She reverted into some form of her normal self that she wasn't real happy with. And Mama had been on a roll because "that Janson girl just married your old flame the Malcolm boy."

"She's welcome to him," Marta told a garter snake, a big one almost two feet long. It scowled at her from its sunny rock where the trail turned to avoid an old slide that had taken down several

alders and a six-inch maple. The alders, typically were doing well, the maple was turning back into mulch. The "Malcolm boy" had been good enough in bed, an event she had made sure Mama never knew about, but about as mentally exciting as watching paint dry, which fit for a housepainter. Of course "that Janson girl" hadn't been the brightest color in the palette either, so maybe it worked for her.

"I know I won't meet a man on the tops of mountains," she shouted at the snake when he flicked out a long red and black-tipped tongue at her. As she was neither a predator nor rational, he returned to enjoying his sun-warmed rock.

It was a problem, but it wasn't one she was terribly worried about. No one came up Swallow Hill. If she had five visitors a summer it was a big deal. And the guys she met waitressing at the ski resort up on Silver Mountain, well, even she wasn't that kind of desperate. Maybe if she worked her way up to Schweitzer or Sun Valley, but those were coveted spots on the ski patrol circuit and she'd bet the men-type opportunities weren't any more fruitful in those places even if by all rights they should be.

"But, Marta-*cara*, we worry about you soooo much!" Marta mimicked Mama's voice to a turkey vulture that carved the air at the first vista lookout, an hour up the trail.

Since when had her Mexican family become so Italian?

Again Uncle Manuel's fault. She'd been raised in a lingual hash of Mexican, Italian, and English. No wonder nobody understood her when she got angry—other than her own family, of course.

At the halfway point up the trail, she needed a break and dropped the heavy pack loaded with her next two weeks of supplies. She heard the bright *tink* and cursed. At the very bottom of her pack she'd tucked two jars of the spaghetti sauce that she and Mama had put up last fall. Still not really understanding Marta's isolation, Mama had insisted that Marta should have something homemade to serve if a nice man came by. The glass had broken against a rock when she'd unthinkingly dropped her pack. There was no point in digging the mess out now. She'd

been right, glass was too heavy, but she'd been in such a hurry to get out of the house that she hadn't taken the time to empty the jars into baggies.

By the time she reached the summit, she had chaffed shoulders, a sore back, and the bottom corner of her pack and the right hip of her shorts were both stained tomato red.

3

D*etrick met her at* the bottom of the lookout stairs with his pack already on. He headed down the mountain with barely a, "No fires. See you in twelve days." He was hustling down to be with his new girlfriend, whom she wished luck. Dating a lookout substitute meant that you saw him only briefly every three days, because the rest of the summer he was cycling up and down to various towers.

He wasn't her type anyway. He reminded her of all of the jocks who used to try and grope her in school. He seemed nice enough in the moments they traded places, and didn't stare at her chest—too much. He also left the cab as neat as he found it, which she appreciated but he never slowed down enough for her to thank him.

She did her first scan to confirm Detrick's assessment and then unloaded her pack. At least only one jar was broken.

Then she felt an itch.

It was the itch that her Uncle Manuel had tried to teach her, "It is when what is missing is too subtle. You can no longer taste that

you need more thyme or oregano, but you know that the balance, it isn't right. Then you must become very careful. A mistake now and the whole sauce must go down the drain. But still the sauce is incomplete and must be finished. Slow down and listen. You will feel an itch, a tiny push from some part of you that knows about food and flavors. What it tells you, that will be the right answer."

She felt one of those.

Marta slowed down, waited, putting everything away. Produce in the cooler, dry goods and cans on the shelves. Two new books, one only a little stained with red sauce, on the tiny shelf above her desk.

Out of the soggy shorts and into clean ones.

Then, rather than standing at the Osborne Fire Finder, she stepped out onto the narrow porch with her binoculars. She started the circuit at Gray Wolf Summit to the northeast because it was one of the clearest landmarks in the area. From the burn at the base of the summit, she swept slow arcs; first along the horizon, then lower and lower until she was looking down the cliff of Swallow Hill's north and east face.

Then she moved to the north side of the tower and did the same thing. She always went around it "contrariwise." Her dad said that she did everything backwards because she was left-handed. What had been an idle joke had turned into an act of defiance and finally a force of habit. She'd learned that doing things "contrariwise" let her see things that she wouldn't otherwise if she was being "normal."

She was around to the south side of Swallow Hill when she spotted the faint puff of smoke. It disappeared almost the instant that she saw it, because young fires could do that.

She didn't move the binoculars from that point for three long minutes.

The breeze was a light brush out of the south as well. And it carried…wood smoke.

Another puff and she had it pinpointed. Not taking her eyes off the spot, she fumbled around until she found the doorway

into the cab. She spun the Osborne into position and looked through the two brass sights just as the smoke went from puffy white to steady gray.

She pulled the radio off her hip. "Gray Wolf, this is Swallow Hill. I need a cross approximately two miles south of my tower."

Tom up on Gray Wolf came back moments later, "Two-three-nine degrees."

"Roger that," Marta plotted the cross quickly, double-checked everything, then called Vic. "Fire control, this is Swallow Hill. Confirming new fire due south of my tower at…" she read out the longitude and latitude. "Just gone steady. No eyes on the blaze, but estimate one acre based on plume."

"Catching them earlier and earlier, Marta. You know this blaze puts you ahead of the all-time record: number of fires spotted by an individual this far through the season. Ten more and you'll break the all-time season record."

"Oh joy," she radioed back.

Vic's laugh made her feel worlds better. "I have a chopper heading your way. He's in the area, we'll get some eyes on the prize before we call the troops."

4

*M*arta ducked and cursed as the small helo buzzed her tower from behind. She'd been out on the south deck again, and hadn't heard him coming from the opposite side of the tower.

Helo 41 slewed to a halt, hovering, and turned so that the pilot was facing her from just a hundred feet away. The sun was behind him and she could only see the shadows of the man at the controls.

"Afternoon, Ms. Swallow Hill." Tyler. She did her best to avoid a girly sigh…and really wished she could see him better.

She pulled her radio off her belt and keyed the Transmit button. "Afternoon yourself, Master Tyler."

"I hear you found another one for us to play with."

She pointed down the slope toward the gray puff that was already going black with soot.

Helo 41 twisted to the left and she could see the silhouette of the pilot looking down and to the side, but still couldn't tell his age or build. "Yep, that does indeed look like you have a fire, Ms. Swallow Hill. And right on your front stoop."

Then the helicopter twisted back to face her rather than diving down for a closer look.

"If you don't mind my sayin', Ms. Swallow Hill. Never have seen you out of your glass tower before. I can see that I was missing a fine sight. A fine sight indeed."

Before she could think how to respond, he'd slammed over his controls and half rolled into a plummet down the valley.

So, he liked the way she looked. Big deal. Most guys liked how she looked. Then she raised her binoculars and focused them down the slope again.

"Wish't," she imitated his voice, "I'd a thought to look through this here contraption when you were a might closer, Mister Tyler." Instead of Tyler's smooth Colorado, it came out more Mexican-Italian-Texan which sounded even stupider out loud than she'd imagined. A burst of giggles tickled its way up her throat and she was never one to hold back a giggle when it came.

So, he thought she was pretty? Well, with that gorgeous voice of his, he didn't have to be a handsome one. Maybe she'd find a way to meet him…when there wasn't a fire on her mountain.

For now she'd just sit and watch the airshow.

5

S*ix hours later she* *wished* she could just sit and watch.

Tankers were on other fires. Helos were spread thin. Most of the smokies were in Colorado. And the Swallow Creek Fire was taking unfair advantage of their lack of attention. The south side of Swallow Hill was engulfed in flame and the plume of smoke kept blanking out Marta's view.

She'd retreated into the cab, closed the windows and doors, and donned a dust filter mask so that she didn't choke on the ash.

"Hang on, sweet thing," was all the warning she had before Tyler unleashed a hundred and forty gallons of water over her cab. It whumped down onto the roof with a crash like thunder. The half-ton of water striking in a single blow made the cab shake its head like a wet dog shedding bathwater. The structure shuddered and then calmed.

It was a good move, once she was over the shock of it. Soak down the tower so that no stray ember alighted and caught the place on fire.

Soaking down the tower.

That was definitely not a good sign. You didn't do that unless the fire was close.

If she had to move, it was going to be fast. Her big pack would slow her down too much. She grabbed her fanny pack and shoved in a small medical kit along with spare batteries for her radio, and a water bottle. She pulled down her favorite family photo, parents and two hopelessly dense but terribly handsome older brothers gathered at Manuel and Graziella's wedding. She kissed the photo for good luck, snapped a can of bear-repellent pepper spray onto the belt along with a foil fire shelter—because a firefighter is always prepared, even when she's a lookout tower woman on the verge of totally freaking out.

And then she couldn't think what else to do.

The helos were losing the battle and she was losing options.

Ten minutes. She was a fast runner. Marta would give them ten more minutes and then she'd be jackrabbitting down the trail and to hell with the firefight.

6

At fifteen minutes, she'd eased down three of the five flights of steps, reluctant to leave the Swallow Hill Lookout un-womaned in the middle of a fire.

The air was thick with smoke and growing hotter by the minute. She could taste the char right through her filter mask. The fire's roar, always a distant thunder in her experience, was now a passing freight train. It wouldn't be long before it was a jet engine at max thrust, and just as hot.

At seventeen minutes, she'd made it down another flight and the steel handrail was warm against her palm.

Was the air clearer? Or was it her imagination? She looked up and couldn't see the cab at all. It was wrapped in a shroud of smoke that was climbing the hill and soaring aloft.

Okay. It was her imagination. That and she was getting closer to the ground.

The airshow had become a distant sound, muffled by the fire's thunder, but she could still pick them out. Another helo and a tanker had just arrived. But Tyler had left to refuel just a moment ago.

She checked her watch, had to rub at her eyes to make them focus.

Duh!

Marta rinsed her eyes from the water bottle, dried them with the hem of her t-shirt, and then pulled on the goggles that habit had shoved onto her hair.

Now she could see her watch. Tyler had been gone twenty minutes. Long enough to refuel in Missoula and get back here? Probably. Maybe he was the returning helicopter she could hear circling above the tower. That meant there was still only him and the tanker. It made her feel safer, knowing he was close by.

She heard him setting up for another pass, then she heard a high buzzing sound—impossibly close to her. She thought she saw motion out of the corner of her eye, but it was gone into the smoke too fast to be sure.

Seconds later she heard an odd crunch. Something mechanical and it didn't sound good.

"Goddamn it!" Tyler. On the radio. Swearing.

That didn't sound good at all.

"Helo 41 report!" The Incident Air Commander called down when Tyler didn't continue.

"Hobbyist drone over the fire. It came up out of the smoke and I think I hit it."

"Any issue?"

"Assessing."

Marta tried to breathe. Tried to count seconds in her head. Tried to think of some way to help him—

"Mayday! Mayday! Mayday! This is Helo 41. Tail rotor not responding, I have to put it down, fast. Visibility zero. Mayday! Mayday! Mayday!"

"Tyler!" Marta screamed at the sky.

And then, almost as if he'd heard her, his helo plunged down out of the smoke so close by that it felt as if she could touch him.

A blade clipped the steel tower not five feet above her head. With a horrible metallic rending sound and a high whistle, a chunk of rotor blade flashed by her head.

She dove down the last flight of stairs, rolled on the ground, and looked up in time to see the helicopter hit the rocky slope, bounce upward, then thump down hard, crushing one of its long skids.

He'd been moving so fast; the helicopter careened and tumbled down the slope.

Marta was away from the platform and racing after the helo even while it still rolled. The five thin blades battered and flailed at the rock. Chunks flew in every direction.

A hard dodge to one side and a four-foot section missed her by mere inches. She barely noticed, her whole being focused on reaching Tyler through the mayhem.

The helo balanced upside down for a long moment, perched on the remains of its rotor head. Then in an almost lazy last gasp, it rolled back onto the meadow—most of the way onto its belly.

Marta reached the bird and finally realized where it had stopped. Another half roll and it would have tumbled right off of Swallow Hill, a two thousand foot fall down a cliff face too steep to walk without a rope.

She reached the door, yanking with sheer adrenaline until she had it free.

Someone was shouting on the radio.

Wasn't Tyler.

So didn't matter.

Tyler lay sideways in his harness. Slowly, so slowly, he twisted around to look at her.

He had a half dozen cuts on his face and was bleeding from several of them, but none of them badly. Despite the cuts and blood, she could see that while he wasn't beautiful—so much for girlish fantasies—his face had a ruggedness that was quite attractive.

He offered her a sideways smile, then hissed and reached up a hand to gently test a split lip. His eyes had not left her face for a second.

"Hello there, Ms. Swallow Hill. Sorry for dropping in unannounced like this. Poor form for a gentleman come calling."

"Terribly poor form," she did her best to match his tone. "Let's get y'all out of there before something worse happens."

"There's only one of me."

"What?" She climbed into the cockpit to help him.

"Y'all isn't singular, Ms. Hill. It's for a group of folks. Especially if they're from the Deep South, which I'm not."

"Then how's that sentence supposed to go," she worked his harness free and did her best to ignore how close together they were in the tiny space, she kneeling on the tilted co-pilot's seat, him still strapped into the pilot's position.

"Should be: 'Let's get you out of there….'" He spoke in a deadpan accentless voice, clearly making fun of her, but trailed off in a way she didn't like.

There were no obvious signs of blood. So maybe he'd just been concussed rather than collapsing into shock. She'd taken the standard First Aid course for lookouts, but it wasn't much. The bottom line for a lookout was: do anything to yourself worse than a small cut and you're screwed. Help was a long way off.

Between them, they maneuvered him out of the cockpit. The smoke was getting thicker and she'd lost her mask and goggles somewhere during the sprint. A path of destruction had been flattened through the tall meadow grass by the helicopter's tumble. It was a wonder he was still alive.

"My ankle isn't working quite right."

They both looked down as he clung to her. It was twisted to the side. Grotesquely.

She looked at him, liking that he was a couple inches taller than her own height, and did her best to keep her voice light, "I don't think it's supposed to look like that."

"Not if I want to go walking anywhere on it," he agreed and continued to hang onto her.

"Where's my pilot? Tyler, report!" The ICA's voice screamed from her radio.

She pulled it out, "I've got him. But his ankle is broken. Request immediate medivac."

The stream of vitriol that poured out of the radio was quite impressive.

"You'll have to forgive him. He's rarely a passionate man, except about his pilots," Tyler whispered close to her jaw, with his nose practically buried in her ear. "Hey, Ms. Swallow. You smell right nice. Like—"

Something romantic?

"Like, tomato sauce."

Crap! She must have rubbed her hand in her hair while she was cleaning up her mess from the hike up. "It's an…old family recipe."

"Good enough to eat."

She considered taking offense, but if he was coming onto her, he wasn't doing it with a grope or a grab, so she'd tolerate it for the moment.

"Swallow Hill, this is ICA. I can't get to you. The entire peak is shrouded in smoke and you have my only helo in the area. Can you confirm the hobbyist drone?"

Tyler pointed with the hand that wasn't around her shoulder at the mangled tail of his wrecked aircraft. There was the remains of something white and mechanical caught in the rear rotor blade. She didn't know how to fly, but she knew a helicopter didn't do so well without its rear rotor.

"Roger that, ICA. Have visual on a hobbyist drone, or at least the remains of one."

"I'm gonna kill the bugger that flew that thing. I swear I am. Tyler, I have to pull back the tanker; I can't have him hitting a second drone. We can't get through the smoke even if I had a helo local. You'll have to take care of yourself."

Marta held the mike up to Tyler's mouth and hit the Transmit key for him.

"Not a problem, Mark. I'm right comfortable where I'm standing." And Marta was too. Very comfortable. He had an

arm around her shoulders, and she around his waist, as if it was the most natural thing in the world for them to stand that way. He was just an inch or so taller than she was and she was upgrading that ruggedness to very good-looking.

"No," the ICA called back. "You're not. The fire is going to crest the ridge in ten minutes and there isn't a thing we can do to stop it, even if there weren't any other drones in the air and I had the full fleet. Swallow Hill, you keep my pilot alive, god damn it."

Though there was nothing to see, Marta became aware of the sounds for the first time since the crash.

She heard the heavy roar of the BAe 146 jet climbing clear of the area. High above, she heard the strong buzz of the ICA's twin-engine airplane circling thousands of feet above the fire.

Close to hand, there was a deep, basso roar that shook the air. So loud now that it felt as if it was shaking the ground.

"A FEAR fire," Tyler whispered, and this time she didn't feel any tease close beside her ear.

It was the worst stage of a wildfire before it overran you, the Fuck Everything And Run moment.

7

T*en minutes," Tyler sounded* perfectly calm. Dangerously so.

Marta remembered a cross-country race. She and a top runner from Boise had been deep in the woods and way ahead of the pack. They'd run against each other before and Barb was a tough contender.

Then Barb had caught a foot on a high tree root and crashed to the ground. And she'd just sat there. Cheerful. Glad to chat and answer questions. But she hadn't had a single thought for the race. No complaints while Marta had checked both her ankles which appeared fine. Barb hadn't had any reaction even when she looked down at her broken wrist bent over backward. So decisive just a moment before, she appeared perfectly calm once injured.

Shock.

Tyler was in shock which meant it was up to her.

Hightailing it down the trail was no longer an option. She should have left twenty minutes ago; she checked her watch. Thirty minutes ago.

But at some moment very soon, Tyler was going to start feeling his broken ankle.

The cab wouldn't do them any good, even if they could get up to it. And the shattered helo was no option at all.

"Tyler," she cupped his chin and turned him to look at her. "Hang onto the helo, I have to check something out."

He grumbled about trading soft-and-warm for hard-and-metal, but made the shift.

Marta crawled back into the cockpit and looked around, but couldn't see it. It had to be here somewhere. She stuck her head back out.

"Where's your emergency shelter?" The foil shelters were the tool of last resort and she knew the pilots had to fly with one.

"In the door pocket, pilot's side," he said it with enough clarity that she wondered if he really was in shock, or if he was just keeping a humorous facade up against the pain.

She looked back down into the tiny cabin. There was no pilot's door, there was only granite and tufts of grass where it should be. Crawling back out of the cabin, she looked around, still no sign of it, though there was the debris trail that started near the tower and was scattered across a hundred yards of the slope, she didn't see anything as large as a door.

The debris field continued past the helicopter and...

She moved as close as she dared to the edge of the cliff and looked down through the thickening smoke. Fifty yards below them there might have been a piece of helicopter big enough to be a door, but it was far out of their reach.

She had the one shelter on her belt. But as tempting as the idea of sharing a fire shelter with Tyler might be, it wouldn't work. The shelter was designed to provide close protection for a single person. Maybe if they were both petite...but they weren't.

"Story of my life," she mumbled as she looked around the barren hilltop for other options.

8

*H*er *final glimpse before* shutting the lid was of thick black clouds of smoke colored with the deep orange of fast-approaching flame.

"And I had so hoped, Ms. Swallow Hill, that my first water adventure with you might include something like skinny dipping. Seems my imagining came close. Care to complete a man's wildest dreams, Ms. Hill?" She could hear his gentle smile even if she couldn't see it in the pitch black.

Marta appreciated it all the more because there hadn't been time to move Tyler gently. The sweat of pain poured off him, but he'd kept his tone light and friendly despite the anguish of getting him in here. Whether the effort or the terror had done it, he was shaking off the shock. At least for the moment.

They were submerged up to their necks in the concrete cistern of her lookout tower's drinking water. The heavy steel lid above them was closed, for whatever protection it might afford. Then, draped like an air bubble over their heads, she'd spread her fire shelter. It was the only chance they had. *Santa Maria, Madre di*

Dios. Childhood prayers weren't helping her much. She focused back on Tyler, except she couldn't see him in the dark.

"How about a rain check on the skinny dipping?" She barely managed the thought around her raw nerves. Now that she had done everything she could other than wait, the impact of their precarious position was striking home.

"Rain, might help some. Douse this fire down a bit," it helped that his tone had finally taken on an anxious note. His voice was becoming clearer, recovering from his shock. Sharing her fear with someone else made the situation a bit more bearable. A very tiny bit.

A silence formed between them but she wasn't feeling very comfortable in it. The cistern was seven feet deep and, thankfully, she'd used up the top two feet of it in her first two months here. Thankfully, they were both tall enough to stand in the five feet of water still remaining, rather than Tyler having to tread water with a broken ankle. It was also just four feet square so they were jostling and bumping underneath the water despite having their backs pressed against opposite sides.

The water was cool, without being cold. At least not at first. It was starting to chill her and she could feel the panic approaching and…

"Talk to me, please!" She begged before she went off the deep end. The fire's roar beyond their shelter blanket and the steel lid over the cistern was muted, but growing fast.

"Right, my apologies, ma'am," his tone which had thinned a little under the pain had shifted back to more solid. "I was just a bit perplexed is all. By our current situation. It's awkward to be bumping hips and, uh, other things with a beautiful woman under such circumstances."

"Which are?"

"Ms. Swallow Hill, I don't even know your name."

Before she could answer, he hurried on.

"But your voice, you could make a man die happy just to hear such a thing over his grave. So sure and confident and female.

It's an amazing thing, I'm telling you. And then when I finally saw you," he let out a low whistle. Not a wolf whistle, but rather one of deep appreciation. "I didn't know anyone built women who looked like you. One who stood as tall and straight as a ballerina but shaped like a goddess."

"Huh!" She tried to pull herself together. She really did. It wasn't working. "That can't be."

"But it's truth."

"But it can't," and she felt about as naïve as a swallow first leaving its nest.

"Tell me why?"

"Because…" She didn't even know why. "Because—" she tried again with no more success, having to raise her voice as the fire's roar built. "Because I'm no more ballerina than goddess. I guess the confident part is right…maybe. It must be, because it pushes men away like mad." And he saw the dancer in her? She could still feel that deep inside, but no man had ever said such a thing to her.

"Then you have been—and please don't take this wrong, Ms. Swallow Hill—spending your time with a bunch of fools."

She reached out in the dark, lost for a moment in the disorienting darkness, and tentatively brushed a hand down his chest. It felt safe and right, huddled together in here as the fire burned toward them. She suppressed a shiver against the cool water.

"Living in my glass tower, can't say I've been spending time with much of anyone."

"I like the sound of that even better. Less competition for me. Not so long ago I swore I was going to get to know the lady of Swallow Hill before this summer was over."

Marta could feel the heat rising to her own face and was glad for the darkness. "I, uh, might have made a similar swear about this certain deep-voiced pilot I know." Which she couldn't believe she'd just admitted. "Say something else. Anything else."

"Well…" he tried to keep his tone light despite the tension she could feel where her hand still rested against his chest. A

good man to have around in a bad situation. "The skinny dipping wasn't a completely idle suggestion. I have a pal with a big ranch down in Texas. Horse ranch. Do you ride?"

"Willing to learn," she didn't let loose the bubble of a laugh building in her throat for fear that it would emerge as a babble of panic instead.

"Some fine places there to take a lady," Tyler continued resting a hand over hers, "if she's of a mind. Fine places. Not another soul for miles in any direction."

"I might be open to that," Marta slapped her free hand over her mouth. "I can't believe I just said that." The heat flashed back into her cheeks.

Then her face kept heating.

And heating.

Their breathing air was—

"Okay, Swallow Hill. You listen close," Tyler's smooth and calm disappeared and he started speaking quickly. Dead serious now. "It's going to get hot in here, unbearably hot. And then it will get hotter. You hold down the corners of the fire blanket, keep its edges under the water. I'll do the same with my end. You're going to want to rip off the blanket. Don't! Our lives may depend on that."

Marta ducked her face down into the cool water, which only made the air feel twice as hot when she surfaced. She reached around Tyler, grabbed one end of the foil fire shelter and held it firmly behind him under the water. He did the same behind her. They were embracing…to save their lives. *Don't get stupid, Marta!*

"It all depends," Tyler continued hurriedly, "how fast the concrete and the water heat up. But do not pull the blanket aside until I tell you. No matter what. Do you understand? Do you…"

She nodded, which was pointless in the dark. They were going to be boiled alive. But she couldn't speak.

The concrete wall she was leaning against was no longer cold, it was comfortably warm. She shifted away from the wall, the warmth was creepy, felt dangerous.

Bumped into Tyler, chest to chest, but there was nowhere to go.

"Kiss me."

"What?" She had to shout to be heard over the building roar.

"I want to have kissed you *before* we survive this."

She wished she could see his face, his eyes, how he was looking at her.

The wall behind her was definitely warm now.

But she hadn't needed to see him before this moment. She heard his voice, just as she had all season; it became the center of her thoughts.

She leaned in and kissed him as the roar deafened her. She clung to him as long as she could, but she had to break apart to get air.

It was so hot.

She dragged in a breath.

The air was fire in her lungs.

"Scream!" He shouted at her. "It's okay!"

The wall behind her was now hot when she bumped it. The water was starting to warm up. She held onto Tyler. Held onto their fragile shelter where it had been pulled down behind him. The air inside the small bubble of the fire shelter inside the concrete cistern was so hot it scalded her lungs. It—

The scream that ripped from her chest was echoed by the scream from his as the fire rolled over them.

9

*A*t *some point Marta* stopped screaming.

The pain had eased.

The agony of each breath.

She floated in the dark, wrapped tight around a man. Around Tyler. Her end of the foil shelter was still held tight in her fists.

"Am I dead?" Then a horrible thought struck her and she gasped out, "Are you dead?"

His soft chuckle reassured her infinitely.

"Can we open the shelter yet?"

"Not yet," his voice was a whisper.

"But the fire's roar…" It was gone.

"The area around us is still too hot. Give it a few minutes. Besides…"

"Besides what?"

"I wouldn't mind kissing you *after* we survived this."

Marta decided she wouldn't mind either, not with a man who kissed as well as he did.

They only had a few moments, that she thoroughly appreciated, before the heavy pounding of an approaching helicopter sounded loud outside the cistern.

Tyler broke off the kiss, but didn't release his tight hold on her.

"That, Ms. Swallow Hill, is how we know it's cool enough to leave."

Together they pushed the blanket up against the heavy steel lid and levered it open. It dropped aside with a loud clang of steel on concrete.

They tossed the foil blanket over the hot concrete and she pulled herself up to sit on the broad rim, and then helped Tyler up to join her. He flinched when he banged his broken ankle against her, but remained stoic. A good man to have beside you…beside her. A Black Hawk was settling onto the flat spot just below the tower. Everywhere around them was black char. The fire had burned every living thing in its path. Even now, other helicopters and a pair of tankers were battling the flames farther down the slope.

The tower!

She looked up. It still stood. The soaking Tyler gave it right before he crashed had saved it. The swallows swooped in, complained that the box was gone, had been burned away, and then all flew off again.

"I'll bring you a new one next year," she called after them.

A silence settled as the Black Hawk's engines wound down.

"You're all red, Ms. Swallow Hill."

They only had moments before his friends arrived from the helicopter.

She was worried about facing Tyler. They had been through the heart of a fire together. They didn't know each other—but they'd said things. Shared things.

Be brave! She forced herself to look at him.

"You too." Bright red. His skin flushed brighter than a sunburn though she could see it was easing already. "Cooked like lobsters."

"Could be our first dinner date," he noted in that dry tone of his.

"Not a chance. Never had lobster and now I never will." Then she thought that, of course, she should have known—her Mama was always right. She nodded up toward the tower, "But you come visiting and I'll make the best spaghetti sauce you've ever had."

His smile was deep and proved that rugged and handsome could definitely be on the same face.

"Tell me one thing, Ms. Hill."

Her courteous, deeply-voiced Coloradan was back. With his easy humor and very good face. A man she wouldn't mind getting to know much, much better.

What secrets could she keep from such a man?

Tell him one thing?

"Anything," and she knew it was a promise.

"What's your name, Ms. Swallow Hill?"

"I offer you 'anything,' and that's the best you've got?"

She tried to shove him back into the tank, but he caught her up in his arms and gave her one of those deep, desperate kisses. Just like when they'd been at death's door, except now he was just doing it because he wanted to.

Because she wanted him to.

Flight to Fight

Flight to Fight *was written as a bonus story to my newsletter subscribers. It was a side visit back to the USS* Peleliu, *the semi-retired Landing Helicopter Assault ship. My Night Stalkers have taken it over from the Marines.*

I was also having run researching parts of Special Operations Forces training. It is very common for SOF soldiers to speak multiple languages, making it easier to infiltrate and scout different parts of the world. And I was reading about the common practice of watching online newscasts in their study language to hone those language skills, especially in the common words that are used by news reporters all over the world.

This story was written during the brutal aftermath following the "Arab Spring." General news reports from Egypt were abruptly cut off after a very short period of time and replaced by government-approved news.

I'm often told that it feels as if my stories are ripped right out of the headlines, and this is one that was. Or rather, it was ripped out of the sudden lack of headlines.

M.L. BUCHMAN

NPR and B&N Top 5 Romance Author of the Year

Flight to Fight

a Night Stalkers romance story

1

S*ergeant Lee Ames had* been cornered and didn't appreciate it in the slightest. Night Stalkers were not supposed to be cornered; they were supposed to rule the night. He'd been flying for three months now with the most kick-ass helicopter company aloft. And in three months of almost nightly missions based off the U.S.S. *Peleliu* he'd been cornered a total of once—tonight. Lee didn't care for it.

He wasn't "supposed" to be in the Sinai any more than the person he was here to extract. The Egyptian government would be very unhappy if they knew a helicopter of the US Army's 160th Special Operations Aviation Regiment was parked in a dry *wadi* less than five hundred meters from the Suez Canal. His Little Bird might be the smallest helo in the military, but because it was an AH-6M assault bird, it was also heavily armed and absolutely lethal.

He checked his watch. 22:37. Three minutes later than the last time he'd looked, which did nothing to calm him down. Normally he was fine with waiting; patience was one of the deep

skills learned during a decade in the Army. Hurry-up-and-wait was only the first-level talent, one possessed by every grunt who'd ever served. By the time he'd hit his third tour, he had the second-level down; waiting as tactic—a battlefield choice of inaction versus action.

The Night Stalkers had taught him the top-tier skill: waiting is. Waiting wasn't something you endured or used. A non-judgmental time of primal consciousness. It was that time during which there simply was no proper action, so you waited. And when the time was over, you stopped waiting. In Special Operations waiting had become a simple state of being in between moments of action at a level most people couldn't imagine, never mind sustain. There was a reason US Special Operations were at the top of the world's military pyramid.

Except right now the person he was supposed to be exfiltrating was two hours late. Captain Kara Moretti—watching from her drone circling six miles above—had told him to stay put as long as possible.

Waiting is he sighed, wholly unconvinced.

He'd assumed that a few hours before dawn, he'd have to leave whether or not his extraction subject had arrived. However, a firefight had broken out nearby and was now raging on the flat desert surface of the Sinai above his hiding place. Which might explain his contact's delay, but was definitely going to trap him here for a while.

About an hour after he'd landed, he'd heard the distinctive crack of a supersonic bullet ripping through the air a half dozen meters above his helo. Then he'd heard the rumble-boom of artillery to the west. The Suez Canal ran half a kilometer to the west.

Lee had left his helo and crawled up the west side of the *wadi*—so familiar from the Arizona arroyos he'd played in as a kid that he had a weird *déjà vu* moment. The dry river bed was deep enough that his helicopter rested almost twenty feet below ground level, though only wide enough to leave a few yards to either side

of his main rotor blades. He lay on his belly with his night-vision goggles just peeking above the sandy rim of the dry wash. Unlike the Sonoran Desert of Arizona, a careful scan east revealed nothing but the boundless wasteland of the Sinai Desert. Not even a saguaro cactus reaching its arms up to the star-filled night.

To the west rose the high berm of the Suez Canal, with its dredgings of yellow sand piled up several stories high. Bright flashes from a battle flickered over the berm. He could see the upper structures of the ship moving along the canal. Gunfire lashed out from atop the berm and was returned hard from a ship he could see only by its superstructure.

At least he wasn't the target. But if he took flight and exposed himself above the *wadi,* he would be fast enough. For now, he and his helicopter were trapped here.

ISIS? One faction of the Egyptian government fighting another? Or an attack on a specific ship?

Didn't matter. It wasn't in his mission profile. His profile had been arrive, wait, extract, and do not, under *any* circumstances, be seen. That was the Night Stalkers' specialty.

Except no one had told him that his clandestine rendezvous point was going to end up in the middle of a battle. Even though it wasn't in the mission brief he knew that revealing a piece of heavy US military equipment and involving it in any local conflict would be disastrous.

He turned, as he had a hundred times during the last hour that the fight had raged, to check in all directions to make sure he was alone.

This time he wasn't.

A heavily-burdened lone figure was moving stealthily up the *wadi;* from rock to hump of sand. The figure's long *thawb* robe would have blended perfectly into the night if it hadn't been for Lee's night-vision goggles.

Sliding quietly down the slip face of the *wadi's* side wall, Lee came up behind the person when they were still fifty meters from his helo.

"*Tawaqaf!*" he said quietly in Arabic. Then clicked off the safety on his FN-SCAR rifle to reinforce the command to stop.

The figure froze. Either too well-trained or too panicked to turn.

"On your knees," he continued in Arabic. The man settled awkwardly, but didn't protest even though it would drastically limit his ability to attack or run.

Lee made a quick scan, but there appeared to be only the two of them in the steep-walled dry river bed.

"What are you carrying?"

"Golden potatoes," his captive said in a muffled voice.

It took Lee a moment to recognize first that it was said in English and second that it was the pass phrase for the person he was supposed to be extracting.

"Stay still," he ordered. Exchanging his rifle for his Glock handgun, he moved up close behind and began frisking the person.

Through the linen *thawb,* Lee could feel neither knife or handgun at ankles, calves, or thighs. No ring of explosives at the slender waist. And just as his hand cupped what was not a shoulder holster but undoubtedly a woman's breast, a small face rose from over the kneeling woman's shoulder and looked up at him.

In a tiny, scared voice, the girl mumbled, "*Baba?*"

2

Lee managed to finish checking both the woman and the child. They were clean.

"Apologies. I was not expecting a woman," he continued in Arabic as he backed away and kept his sidearm loose in his grip.

Nor had he been expecting a small girl to ask for her father.

"I understand," the woman said in perfect English as she rose awkwardly and turned.

Lee resisted the urge to step forward and help her to her feet. This was not the situation he'd been counting on and he wanted to maintain a clear field of fire for the moment.

"I was told..." he shook his head and switched to English. "I was told to expect..." A person. He'd assumed male, but that had never actually been stated. Had it even been known?

There was only so much that he could see through his night-vision goggles. All colors were green, painted in shades based on varying degrees of heat. He pushed them up and blinked hard to adapt his eyes to the darkness.

"Close your eyes for a moment."

He flicked on a small flashlight. Both woman and child were traditionally dressed. Their robes were dirty as if they'd spent a lot of time crawling. They were both Egyptian dark in complexion. The girl blinking at him still had a baby's round face framed by a tangle of brown hair.

The woman's hair was a lush cascade of mahogany that spilled over her shoulders. If she'd had a *niqāb,* the scarf was long gone. Her shapely face had a broad openness that belonged on television.

She squinted one eye open at him and he realized he was staring, but even in her current state of disarray she was well worth a second look. He swung the flashlight rapidly downward and caught the dark red stains on the little girl's hem and the still bright stain on the woman's arm.

"Come," he stepped forward and took her opposite arm, the one that held the child, and guided her toward the helo. He could feel her stagger beneath his grasp as if she'd collapse in the next step.

He knew he shouldn't burden himself, all his training said to keep his hands free, but he scooped the girl into the crook of his own arm and supported the woman back to the helo. The girl clung to his neck as he bounced her on his hip the way he'd seen his sister do with Lee's niece.

"Where are you hurt?" There was no room in his Little Bird to minister to her. The small two-seat cockpit was a tight fit. The rear seat was filled with ammunition cans for the mini-guns mounted to either side. Lee guided her to sit on a nearby rock.

"Am I hurt?" the woman asked with mild curiosity that his medical training had said was a bad sign.

"Excuse me. But I'm going to have to touch you again."

First, he checked the child. She didn't complain even though he made a point of pressing everywhere and moving her limbs. Then, so that she wouldn't run off, he placed her between the woman's knees and hoped she'd stay.

"What are your names?" Lee asked in a light sing-song voice hoping to soothe the child as he began checking over the woman.

"I'm Donya," she hissed sharply when he probed her ribcage and the girl whimpered in response.

Great! It couldn't have been her arm, could it? He checked the rest of her, but it was definitely her left side. The sleeve of her robe was only stained from being pressed hard against her wounded side. He could feel the wetness in the cloth through the thin gloves he'd pulled on.

"Donya Nakhla."

Startled he looked back at her face, blinding her with his headlamp. He immediately looked down again. She certainly did belong on television. How many hours had he watched Donya Nakhla's insightful reporting while he was learning Arabic? It had been a tough learn for an Air Force brat from Arizona and watching her had certainly eased the path.

"What's your daughter's name?" he asked to distract them both—her from her pain and him from…her. He slipped on his NVGs and scanned the *wadi* once to make sure they were still alone.

Then he replaced it with a small headlamp, flicked it to its narrowest beam and turned it on. Her whole side was a bright red. He shouldn't even be touching an Arab woman, much less pull up her robe up around her high enough to expose, well, everything. He hadn't felt a bra. Did she even wear anything beneath the robe? She must, but it was best not to find out what.

The fighting, which had fallen off for the last few minutes was rejoined with renewed vigor over to the west. The muted hammer of .50cal and the higher pop of NATO 7.62mm rounds sawed back and forth through the darkness.

"Here," he pulled the NVGs over Donya's head. "Can you see through them?"

"Everything is green." Still alert. No slurring in her voice. Good signs.

"Keep a watch all around us. If someone comes, they'll shine bright green in your view."

"What should I do if I see someone?"

"Tell me."

"Oh. Yes." And now he was worried again.

"Your daughter's name?" Because of course a woman as stunning as Donya Nakhla was married. Where was her husband, the man who should have kept her from being bloodied? Then he thought it through and decided it would be kinder if he didn't ask. He pulled out his K-bar knife and eased the red-stained fabric away from her side. Rather than removing her robe, he'd make a side slit to inspect her through and hope it was enough.

"I don't know."

He looked up at her face again, careful not to flash his light in her NVGs. She sounded coherent, mostly. Was she fading in and out or…

She was attentive enough to stroke the little girl's hair where she lay upon Donya's lap, nearly asleep.

Time to stop the bleeding. He found a small tear, inserted his blade in the hole and sliced the side of the robe, pulling it open gently. A thin linen shirt beneath that, also heavily red. He slit that as well. Blood was smeared everywhere from the side of her breast down to her hip.

"This will be a little cold," he warned and rinsed the area with a squeeze bottle of water. Actually, the desert was so hot, even at night, that the water was nearly body temperature anyway.

Then he spotted the long slice.

"You were stabbed?"

"Oh. Yes of course. That was it." Her voice sounded stable, smooth and sophisticated, whatever trauma her mind and body were dealing with.

Under different conditions, he'd do his damnedest to sidle up to a woman who sounded like that and looked the way she did, and reported the way she did. He'd earned himself more than a small crush during training. A lot of the guys in class

with him had the same reaction, but he liked to think he was the least crass about it.

At the moment, he fished for the glue in his med-kit and hoped it was the right thing to do; he was no corpsman.

"I had forgotten that I was stabbed."

"You forgot?"

"Yes. It has been a busy evening."

To his non-professional eye, it was strictly a surface wound. There was a long slice along her skin, just missing the side of her breast and running over the tops of the ribs. He pulled out a Quik-clot bandage and smoothed it over her wound as gently as he could. He managed to tape it in place without touching her too inappropriately, but it wouldn't stay without a full wrap around her rib cage. He tried to figure out how to do that without pulling up her robe, and was again at a loss.

Instead, he placed her arm tight against her side, then used a four-inch Israeli Emergency Bandage to wrap around her torso and hold her arm in place.

She was still scanning the *wadi*. Something wasn't matching.

Didn't know her daughter's name, had forgotten she'd been stabbed, but was alert and answering his questions.

He'd done all he could. The sounds of distant gunfire rattled in the distance.

"Can we depart yet?" She kept that perfect voice of hers soft and smooth. Her English was American, but with an utterly charming lilt of the Arabic overlaid in the rhythms of it.

He gently took the goggles back from her and pulled them on himself after dousing and removing his headlamp. A quick scan showed that the sounds of battle above still hadn't drawn any attention down into their hideaway.

He radioed the question back to Captain Moretti's drone for relay. Nope.

"Not yet," he sat beside Donya on the rock.

3

*W*hat's your daughter's name?" Lee tried to make the question sound innocent, but was still unable to gauge the woman's well-being.

"She's not my daughter. That's why I don't know," the girl was now asleep in Donya's lap. She tried to bend down over the girl, but hissed sharply and sat back up slowly. Once she had her breath back, she continued, "She is too young to have learned her own name yet as well."

"Where are her parents?"

"They were shot down in their own living room, as punishment for hiding me with their daughter. I owe them a life debt, but I didn't know their names either. I waited until nightfall before leaving, but I couldn't leave the girl behind as none had come to claim her even after her parents' bodies had been dragged away."

The words were horrible, but the anger in Donya's voice was so thick that it he could feel it slicing through her. Unsure what else to do, Lee wrapped his arm around her shoulders.

She leaned her face into his shoulder and he could feel the tears that ran down her face dampen his sleeve, though no sobs shook her body. There was no more questioning her bravery than her deep-rooted anger. He'd had bad tours, landed in the hospital more than once from nightmare battles. He knew about the bad days being even worse than you'd ever thought possible.

So, he simply let her cry; let her purge.

When she at last recovered, she made no move to sit upright, but kept her head against his shoulder.

Each time he turned to look past her to scan for possible intruders, his face ran into her hair. He was holding an injured woman with a child that wasn't hers deep in hostile territory and hoping that he wasn't about to create an international incident. And all he could think about was how incredible it felt to hold her and how exotic her smell was. Her hair was like some unknown spice, a smell as rare as the Sonoran Desert after a hard rainfall that made the saguaro cactus bloom for a single magic day.

As a distraction, he asked about just what had been her day.

Caught in the middle of a riot.

A protest against the current military government, suppressed with unblinking brutality. Chaos, she'd seen dozens die when troops had stormed her news station's office under cover of the protests. A reputable, if vocal station, that would have an entirely new staff supplied by the government for tomorrow's broadcasts. As far as she knew, she was the only survivor. Her hiding place beneath the brightly lit anchor desk so obvious that no one had looked there.

"That is what is happening up there," he could feel her head against his shoulder nodding toward the Suez. "Using the canal, the government already had gunboats in place before the riot started. They staged the riot to clear out *unwelcome elements* of the populace. Everyone who flocked to the protest was arrested or shot. At least that must have been their plan. But someone else was waiting for them." She nodded toward the sound of sporadic gunfire that was slowly moving farther and farther away.

"Who?"

"I don't know. I was able to confirm that it wasn't the local protesters before I was spotted and stabbed. I collapsed and pretended death so they did not shoot me. Perhaps it is militias pushing out from Syria or Gaza. I don't think it's the Israelis because there are no jets or tanks. If this military government stands, it will be as much luck as control."

Another stray round cracked by, low above the *wadi*.

He held her a little tighter as if he could protect her; somehow make her feel safe after such a day. She was wounded, had lost all her co-workers, and managed to survive. He was amazed that she could speak at all.

4

They talked long into the night. Every hour Lee left her and climbed the west wall to observe the battle. It was moving north toward El Qantara. Moretti kept holding him in place. From her vantage high above, she could still see patrols along the Suez.

Less than a half kilometer away were people who would kill himself, Donya, and the nameless child on sight. And yet in their dark haven, they remained alone and undisturbed.

While waiting, Donya began telling him about herself. She unwound her life backwards as slowly as the night stars were crawling forwards.

A top reporter on Egyptian television. A native of Cairo sent by her father to be educated at Vassar College and Columbia University. Her father who had been purged along with the Muslim Brotherhood when they'd been thrown out of power.

A girl who had dreamed of a better place than the one she grew up in and fought for it in every way she knew how.

"That is gone now. You know that they have passed laws that a journalist can be jailed for reporting accurate numbers

that aren't the government's 'official' numbers? Seven hundred dead in protests becomes seventy. A bomb kills a hundred? No, fourteen. And that is only the beginning."

In turn he told her of growing up in a military family, on both sides of the house. Mom mostly at sea—chief petty officer on a destroyer. Dad still a jet mechanic at Luke Air Force Base outside of Glendale.

They spoke softly for hours. Whenever her words slurred, he'd force Donya to eat a little and drink some water, though she tried to decline every time. At first he didn't want her to go to sleep because of her injury; he had no way to assess how much blood she'd lost. He inspected her several times, but there were no signs of additional blood seeping into her clothing; he'd done all he knew how. Now he didn't want her to sleep also because how much he was enjoying her company.

In addition to being passionate in her beliefs and an amazing survivor, Donya Nakhla was sharply intelligent—obviously smarter than he was—and so kind when the nameless girl awoke scared of the night. Kind, but he could see that it hurt her to hold and reassure the child. He didn't dare give Donya more than the mildest of painkillers.

He finally took the girl himself and rocked her until she fell asleep.

"You are good with the child," Donya sounded unsurprised.

Lee was surprised to his very core, "First time holding one."

"It looks good on you." A sliver-thin moon had risen and washed the dry river bed with a brush of silver. Donya's dark eyes were so close, her face so…he pulled the goggles back down and scanned once again.

Lee had never been so attracted to a woman who he'd done no more than help and talk to. As the night progressed, his crush on the stunning television reporter had been left behind by the reality of the incredible woman leaning against him.

Moretti's call to get out of there came as a shock, as if the real world had reached out to slap them both. The gunfire had faded

and moved north. He'd tracked the battle, but while waiting for Moretti's all clear, the desert had become silent except for their whispered words and the occasional creak from the helicopter as the temperature dropped.

When she was unable to stand, he lifted Donya into the copilot's position. The Little Bird wasn't made for a third passenger, not even a child. He managed to rig a strap so that the girl could ride safely in Donya's lap and not fall out the open side of the helo if he had to do some hard maneuvering. Taking an extra minute, he unrigged the cyclic joystick from the copilot's side so that it wouldn't be inadvertently kicked.

He took them aloft, headed south and stayed as deep in the *wadi* as he could fly. Rather than heading ten miles north to cross the heavily populated Mediterranean coast as someone might expect, he turned south and flew a hundred miles over the desert. Well past Suez—the southern entry to the canal—he veered out into the Gulf of Suez and landed aboard the U.S.S. *Peleliu* that had been waiting for him.

5

Lee had tried to stay away, because he knew his attachment to a celebrity Egyptian reporter was utterly ridiculous. So instead, he'd delivered her to the infirmary, found his way to Chief Warrant Lola Maloney, and been debriefed on the mission.

He'd eaten dinner and headed to bed. Night Stalkers flew at night and slept during the day, so this should be perfectly normal. Except the longer he lay there, the less normal it felt. Giving up on his finding comfort in his narrow bunk, he climbed up two decks and went for a run around the Hangar Deck, clocking a quick ten K. And when that didn't help, he did ten more. A cold shower and he was back in bed.

But his memories were sitting out in the silent desert, a small girl asleep in his arms, and a stunning woman at his side.

A stunning woman with whom you have nothing in common, he reminded himself as he arrived outside the infirmary. No sign of Doc Evans or either of the nurses, he poked his head in still wondering what he hoped for. The ward was a half dozen beds packed tightly together; just enough room between them

to get someone on and off the mattress. Only two of them were occupied.

Whatever his expectations had been, Donya was asleep. Her bloody clothes were gone. The hospital gown and thin sheet revealed that her figure, always hidden on camera by her traditional attire, went just fine with her face. There was no equipment hooked up to her which he took as a good sign.

He should really get his sorry ass out of there.

Then he spotted the girl, fussing in the next bed over. He scooped her up so that she didn't disturb Donya. Lee turned to find the orderly on duty down the corridor and tell him that he was taking the girl out for a walk or food. Perhaps find someone who knew what to do with a small girl. But the instant he lifted her, she snuggled down against his chest and settled back to sleep.

At a loss for what to do next, he sat down in a chair, rested his head back against the steel wall and watched the two women: the tiny one asleep in his arms and the other one…he didn't know what.

6

*D*onya woke slowly, opened her eyes and was relieved to recognize gray steel and narrow beds. Ship's infirmary. An American ship. She was safe and whole, a gift that her own country could no longer promise.

She managed a shaky breath.

And she was done. Her mother dead in the first riots of the Arab Spring. Her father and brother during the bloody aftermath of the military coup that followed two years later. She had given enough.

Somehow she had to find a new start, a new way to help her people without dying in the process—for then she would be of no use at all.

But any vision of the future eluded her.

She needed time. Needed to get past yesterday's anger and the horror of watching more sanctioned murders.

Then she remembered last night. The pain that had lanced through her with every breath; the certainty that each step would be her last but finding the strength to take one more because of

the young life she had chosen to carry and protect. Until she'd nearly collapsed into the arms of Sergeant Lee Ames.

He was not a complex man. It wasn't that he was simple; it was that he was uncomplicated. Most men Donya met wanted to bed her for power or marry her to enhance their own status. They all had an agenda, a plan of their own that her star power would somehow feed.

With Lee Ames she suspected that he was as she saw him. He had simply cared for her and kept her conscious through the long wait. He flew for his team, his country, and his family.

He'd held the little girl like she herself was precious. No man did tha—

The girl!

Donya looked over at the rumpled but empty bed beside her. Twisting and trying to raise herself, which sent a sharp pain that told her that had been a truly foolish action, Donya spotted her. The little girl lay asleep on Lee's chest. He in turn leaned back against a wall, his boots up on the corner of her own bed, fast asleep as well.

She propped her pillow up so that she could watch him. Last night the night-vision goggles had given her no clear view of him. When the thin moon rose, she'd seen blond hair beneath the straps of the night-vision goggles and that he was clean-shaven, a rarity in her culture. It looked good on him, strong.

And with the darker-skinned child curled up in his arms he was about the cutest thing she'd ever seen. Calling six feet of American pilot—who had carried her to his helicopter last night as if she was weightless—cute might be inappropriate. But despite making her living with words, it was all she could come up with.

7

Lee knew he was hovering but couldn't help himself. Over the next few days he spent most of his non-mission time with Donya and Sughraa—not knowing her name they'd taken to calling her "Little One." Once Donya could get out of bed, Lee always carried the girl to make sure that Donya didn't pull open her wound.

Sometimes Dilya Stevenson took the little girl for a while.

"Why is there a teenager on a ship of war?"

Lee smiled at Donya, "Threw me the first time too." He'd led her up onto the upper deck of the eight-hundred foot long helicopter carrier. A number of people were wandering among the tied-down and covered helicopters in the cool dawn morning, taking a stroll before heading to their bunks. There was a light breeze as the ship drove south.

Her hair caught and fluttered. Her *niqāb* was down on her shoulders and she didn't appear to care. Unable to help himself, he brushed a hand over her hair, touching her for the first time since he'd brought her aboard. It was as soft as he'd remembered.

"Sorry," he pulled his hand back. "You have no chaperone. No person of your culture to make sure…"

She watched him closely with those dark eyes that he could so easily get lost in. They halted at the stern of the ship and the breeze brushed her thick hair forward, partly hiding her face.

Subject change. He needed a subject change. The next time he touched her it would be much harder to stop.

"This *is* a very odd ship," he found a new subject. "It was retired by the Navy, but kept active exclusively for our helicopter company."

"The 5th Battalion D Company," she acknowledged. She said it as if it wasn't something she'd merely overheard.

"Yes," he said carefully.

"And under several of those tarpaulins are stealth aircraft that no one is supposed to know about."

He knew she hadn't seen them personally. She, like all guests, was kept strictly below decks during night operations.

Donya pulled her hair aside and looked up at him. She stood five-four, taller than average for a woman of her race. Her western clothes revealed how perfectly she was proportioned for that height.

"I know a great deal about your company, Sergeant Ames. Possibly more than you do."

A spy? He'd brought a spy into their midst. Except…he'd only handled the exfiltration. Someone else had cut the orders.

"Yes," she nodded. "I have been, shall I say, *consulting* for your government for some time. I was approached by a man who claimed he worked for CNN. I am a very good reporter and I know that he is not in their employ. Yet if it helped my country," her shrug was eloquent. "I became an analyst on Egyptian and North African affairs. I have seen many changes for the better in the region this year, but never knew how they happened. Not until I saw this ship. Even wounded and faint with blood loss, I knew what this was."

Lee had never faced a national security risk from a foot away.

"And whatever happens next, I wanted to say thank you for taking care of me."

Then she rose up on her toes and kissed him.

In that moment Lee knew two things.

He had to turn her in.

And he was totally lost as he pulled her into his arms.

8

Donya tried not to be amused. Didn't they understand how much even their questions revealed?

They were in the ship captain's office, yet he was nowhere to be seen. The room had a view of the Flight Deck. A large desk dominated one end of the room, but a circle of comfortable chairs and sofas spoke to the many meetings that were held here.

To one side sat Lee Ames. He had refused a direct order to leave her side—which had been accepted instead of having him arrested. A curious ship indeed. He sat close beside her on a couch, practically hovering in protection as if he could halt the might of military justice should it turn against her, which she found incredibly charming. She'd thought she was long past being charmed by any man.

In front of her sat the inquisition. Four women of SOAR, all pilots. Lola Maloney their commander, Claudia Casperson, Kara Moretti, and—because they apparently couldn't keep her out any more than they could make Lee depart—Trisha O'Malley.

To her left stood a man who had neither spoken nor been introduced. He was clearly a warrior, not a spook, so he must be Delta Force—it was the only explanation.

They were smart. They questioned more than her knowledge; they also questioned her methodology. She answered them willingly enough. With her homeland closed behind her, her future would lay with these people or ones like them. She would step forward in trust.

"All of the Somali pirate's hostages were freed on two separate nights. A clean sweep north and south."

"A ship, this ship, moving from trouble spot to trouble spot unescorted, separate from any carrier group. I don't know if anyone else took an interest in an old ship past retirement making high speed runs from the Black Sea to the Gulf of Guinea and back to the Arabian Sea. But I did."

"Add that to the complete lack of reports from land. Massive endeavors with a complete lack of news coverage, except for hostages who consistently reported being rescued by Navy SEALs. Yet the SEALs, who have become notorious for their lack of circumspection, had no comment when asked."

That's when the silent warrior confirmed he was Delta Force; he made a poor effort at covering a snort of laughter. Now she knew who had spread those news-hound diverting stories.

"Lady has a brain," Trisha the redhead spoke up. "I like her. Can we keep her?"

Lola the commander rolled her eyes. "She's not a puppy, O'Malley."

"She's as cute as one."

"No," Kara spoke with a thick New York accent that reminded Donya of her college days, "not cute, she's beautiful. And we don't need Lee's besotted gaze to tell us that. Women are beautiful."

"Then explain me," Trisha *was* as cute as a puppy. She was petite and radiated attitude right down to her cliché fists-on-hips and mock defiant scowl.

Donya scanned the others' faces and answered for them. "Believe me, they wish they could."

That won her a round of laughter. Even Lee, who'd been looking more worried than a puppy, smiled. She'd had big strong men try to sweep her off her feet before, only to learn that Donya Nakhla didn't sweep. But she'd never made a big strong man weak in the knees before either.

She sighed. For all the good ignoring it was doing her, he had the same effect on her. She reached out to take Lee's hand and took strength from it. Donya wished she could stay. She would take any bet that the more she knew about Lee Ames, the more she would want to be with him. But that wasn't an option. She was a civilian on a ship of war. They would be getting her off the *Peleliu* as fast as they could.

She was a civilian…on a ship of war…

"How did I get here?"

The women looked at her in confusion.

"I flew you," Lee answered in his absolutely forthright way.

"No," Donya shook her head. "Me. On this ever-so-quiet ship. I shouldn't be here."

She scanned the faces, and that's when she spotted the small but very self-satisfied smile on the silent warrior.

"You live up to your reputation, Ms. Nakhla," his voice was deep and surprisingly soft.

Donya bowed her head in brief acknowledgment, unsure what else to do.

"I have friends at Fort Bolivar near Washington D.C. who think you might be very helpful to our endeavors in this region. You have shown clear vision, exceptional analytical abilities, and a willingness to fight for peace even at risk of deadly peril. They would like to recruit you back to the States, but I think you could serve exceptional utility here aboard the *Peleliu* as an operations advisor. It is your choice."

Everyone was staring at him in surprise. The quiet blond who Donya suspected was his wife appeared particularly wide-eyed.

"Doesn't ever talk that much at one time, does he?" Donya asked.

They all shook their heads in unison.

"I appreciate the compliment," Donya slipped together a few more facts she'd "acquired" in the past, "Colonel Gibson."

He nodded, confirming her guess.

She had hoped to find some small way to fight back, to help her people and her country. This was an opportunity beyond imagining.

9

Lee sat in the middle of Lieutenant Commander Boyd Ramis' office and tried to understand what had just happened.

They were alone. Everyone had left except he and Donya. Holding hands on a leather couch on a retired ship that had found a whole new purpose after forty years at sea.

"I—" he clamped down on his tongue. He barely knew her, yet he could imitate her speech patterns as if they came from the same village, not from opposite sides of the world. He wanted so much to—

"Are you going to say something, Lee, or just sit there in petrified silence?"

He blew out a breath and took in another, but it didn't help. Donya waited him out.

"I—" he stalled again, then stumbled ahead in a mad rush. "I don't want to lose you. I mean I know that you aren't mine to lose, but I like you so much. You are too beautiful, too important for me. There are a million places you can go. I'm just a pilot who—"

"Is flying for the most elite helicopter company in all of the 160th SOAR," she rescued him from himself.

"But you're—" he waved a hand helplessly. "And I barely know you, not really. And you barely know me, I can't ask what I want to ask." And then he bit down hard on his tongue. Clamped down on it hard enough to hurt; he'd already said too much. And he could see on her face that she knew exactly what he'd left unsaid.

"It is fast."

He nodded vigorously in agreement.

"We need to take time."

He nodded a little less emphatically this time. "How much time?"

"Enough, but no more than that."

"Why no more?" Lee was once again feeling lost; a step behind this woman who did indeed know more about the 5D and the *Peleliu* than he did, though he'd been aboard for months.

"Because I don't want to wait too long."

"For what?" Did he dare hope? No. He'd been right the first time, he was just being stupid.

When she looked down at their joined hands, her hair fell forward over her face. He needed to see her face. With a light touch, he raised her chin and tucked back her long fall of silky hair.

"Don't want to wait too long for what, Donya Nakhla?"

Her eyes were wide as if she spoke with fear in her heart.

"For family, Lee Ames. Please tell me you want family."

He nodded, what else could he do? And then he knew. He brushed his thumb over her lips.

"I do. With you? A hundred times over I do." The smile that he had seen so many times on television was nothing compared to the one now growing on her features just for him.

There had to be some way that he could describe just how deeply he meant that. Perhaps with simple truth.

"There are two conditions."

Her eyes stayed wide, her smile fading only a little.

"We must find a name for our daughter better than Sughraa. She will not always be a Little One."

At Donya's quick nod, her eyes began to fill with tears.

"And Donya?"

"Yes," she whispered.

"We'll have to pick out a good name for a son as well."

He kissed away the last of her sorrow through the salt of her happy tears.

Where Dreams Are Sewn

Angelo's Hearth was my first big romance series. It is set in a favorite place of mine, Seattle's immensely charming and equally popular Pike Place Market. Angelo's is a fine Italian restaurant where a small group of female friends gather, face life's challenges, and find that when they share their hearts with the right man, that's when they truly come alive.

Opening with Where Dreams Are Born, *the series ended up being five novels about three women. Okay, it was* supposed *to be only three women, but two more insisted on shouldering their way in. And I'm glad they did.*

Maria, Angelo's mom, is a total sweetheart. And Melanie the supermodel was the one who ultimately pulled the whole series together and created closure.

Or so I thought.

The strong women I created as side characters had a few other thoughts on that point.

Melanie, the supermodel heroine of book #5 Where Dreams Are Written, *helped Perrin, the clothing designer of book #4* Where Dreams Are Sewn *(and the real heart of the whole series), to hire a new seamstress. I thought nothing of it.*

Kari Jones had something different to say about that.

M.L. BUCHMAN

B&N and NPR Top 5 Romance Author of the Year

Author of 2 of the "Best 101 Romances in the Last 10 Years."
Booklist

Where Dreams Are Sewn

an Angelo's Hearth romance story

1

Y*ou've got to be* kidding me."

"The woman is a demon," Clem agreed with her. Anna, Kristin, and Mitchell leaned in to look over their shoulders.

Kari Jones flipped up the hem of the dress to inspect the lie of the fabric. "Look at this seam work."

"Beyond demon," Anna agreed over Mitchell's low groan.

"Am not a demon!" Perrin stood at the entry to the sewing room. Her designs were always immaculate and technically a challenge, but this one took it to a whole new level.

"Then what are you?" Kari wanted to grow up to be like Perrin. Too bad it was never going to happen.

That Kari was three days older than her boss was the least of the problems. She had learned the skills and could pattern and sew as well as her boss, perhaps even better, but the talent and vision that had struck down like a lightning bolt and launched Perrin into the fashion design firmament had somehow bypassed Kari.

Her own designs looked...serviceable.

Perrin's designs had walked runways and were splashed across major magazines.

When she took the job almost a year ago, Kari had hoped that some "designer magic" might rub off Perrin and onto her, but it hadn't happened so far. However, the amount she'd learned about construction was huge.

"How did you even do this?" Clem was still inspecting the dress. Kari could see the secrets and looked to be the only one of the four sewers who had.

"I'm a *demoness!*" Perrin did a football end-zone style dance around the dress form.

Kari snorted, "That's for damn sure. A very pregnant one."

Perrin rubbed her belly and smiled. She wasn't due for another couple months, but the blond was already spectacularly round-bellied rather than her normal waif-slender self. Another reason Kari would never grow up to be Perrin; Kari was already six inches taller and far more curved—well, than Perrin's normal shape. Her non-descript dark hair curled to her shoulders instead of being a golden bob.

Still, she wished she could grow up to be Perrin someday.

Perrin continued a cha-cha toward the doorway of the room where the five sewers worked to reproduce Perrin's designs in necessary sizes for the shop and custom orders. The others continued deconstructing the new dress, learning from Perrin's prototype. Anna was a wizard with pattern-making and Clem was almost as good a seamstress as Kari was. They'd figure it out. Kristin and Mitchell were so fast that, once a design was understood, they could reproduce it as many times as necessary.

Perrin tipped her head for Kari to join her in the main design studio.

Kari loved this space. Folded fabrics shimmered along three walls, peeking out of floor-to-ceiling cubby holes abundant with color and texture. To the right a doorway into the storefront cut through the shelving and to the left a bank of high windows let in the Seattle sunshine, which today was of the typically autumn

gray-and-wet variety. Down the center was a sprawling cutting table and another pair of sewing machines below the windows. Design heaven.

Tamara, Perrin's step-daughter, sat at the cutting table amidst a sea of fabrics. Even though they weren't related by blood, it was easy to see that the bolt of brilliance had landed on the fifteen-year-old's head as well. Tammy was piecing together a teen clothing line and it was some of the coolest work Kari had ever seen. With Tammy's work, she didn't feel envy; she felt awe.

Perrin perched carefully on a stool and Kari sat down facing her. Tammy looked up at her for a moment from across the table, but then returned her attention to the fabrics as if embarrassed; not even looking up long enough to offer a hello. Tammy Cullen was usually as effusive as her adoptive mother. And Kari had thought she was really close with the girl…but not at the moment.

Kari's nerves suddenly roared awake.

She turned her attention to Perrin and saw the worry there.

Kari couldn't imagine that she was about to be fired. Perrin's Glorious Garb needed to be adding more people, not cutting them. And she'd finally come to terms with not being Perrin…mostly.

But Perrin's worried look didn't go away. Not a good thing; she and Perrin tended to buoy each other up. The CEO could always calm them down, but Melanie was in Paris on her honeymoon.

"I—" they both started on the same breath.

Kari nodded for Perrin to speak first, and then had to wait her out.

"I," Perrin finally began again, "need you to do me a favor."

"Sure," Kari said in her most even voice. It didn't sound like she was being fired.

"Tammy's getting overwhelmed. Would you be willing to take over as the head fabricator for her line?" Perrin spoke in a mad rush. "I know she's just fifteen and you'd rather be a designer but I want her to have the best and she's good on a machine but you're way better and you can do fitting even better than I can

and I can't think of anyone that would be more fantastic or that we'd both trust more and—"

"Oh god, yes!" Kari managed to cut in on the runaway freight train of Perrin's words. Fifty percent relief and a hundred percent excitement.

"—because it's my daughter's line and I want it to be perfect and you could do that and…really?"

Kari nodded. Head seamstress on a new line? Oh yeah. And if it came out of Perrin's Glorious Garb, there was no question any line would be major; especially when it was as good as Tammy's. She'd been dying to get her hands on those designs.

"I was afraid you were about to fire me."

Perrin's shock shifted quickly to mock anger, "You try to leave here and you and I are going to have some harsh words. You belong here."

Kari could only smile back at her. She did. It was hard to believe, but Kari knew that she'd found exactly where she was supposed to be. It gave her a bit of a thrill that a woman of Perrin's amazing skills agreed.

"Really?" Tamara whispered from across the table, looking up shyly beneath the dark brows and hair she'd inherited from her mother who had passed on several years ago.

Kari rose and circled the table to sit beside Tamara. They were the ones who looked like mother and daughter. Everyone remarked on it who saw them together. It had become a running joke between them by the second time they'd met. They'd rapidly settled on Auntie and Little Girl.

"Are you kidding me? Your clothes rock. I'd love to help with them." Instead of offering the hug they normally shared, she held out a hand. "Thanks for asking for me."

They shook on it like two serious adults.

Then Tammy gave out a very fifteen-year-old squeal and threw herself into Kari's arms. "Love you, Auntie Kari."

Kari hugged her tight. This is what she wanted. Not just the challenge of helping shape a whole line. She buried her face in

Tamara's hair for just a moment and wished she had a girl of her own.

"Love you…Big Girl."

Tammy squeezed her even tighter and squealed again.

2

The moment he walked in the shop, Richard Nyberg knew he'd made a mistake. *Always have an emergency book with you.* His wife had told him that any number of times before she'd left him—left them—two years ago. It was the one piece of her unending streams of advice he should have listened to, but he was always forgetting.

In seconds, his Lana and Tammy Cullen had hugged and were giggling together like he supposed a pair of fifteen-year-old girls were supposed to. Tammy dragged his daughter off for a whirlwind tour of the shop.

He slowed down a minute to check out the place. The dress shop felt welcoming and successful. Part of it was the thoughtful designer who had staged the store as carefully as the colorful designs.

The other part was the surprising number of customers in a relatively small shop. Women swirled in and out of changing rooms, sipped tea while inspecting skirts and blouses. There was a happy buzz of voices from people glad to be there.

The stage manager at Emerald City Opera, Bill Cullen, had married the owner of the shop a year ago. Lana had hung out with their kids a lot during long rehearsals and shows, but he'd never thought to bring his daughter here. He could see that he should have. He supposed that a high-end women's clothing store was one of those "girly rights" that no one told single dads about.

Tammy was towing Lana from one display to the next. It was set up like a 1950s diner. Chrome and red leather booths were populated by mannequins in clothes hot enough to remind his libido that it had been a damned long dry spell. Single dads with young kids didn't date. There was never enough time and it got way too complicated the few times he'd tried it.

The girls slowed down at the wedding dresses and he heard their oohs and aahs despite the general noise level of customers chatting around the busy shop. Girls who were fifteen should not be admiring wedding dresses. Daughters who were twenty-five shouldn't be doing that. Maybe at thirty-five he'd let Lana out on her first date…like he'd have any say in the matter.

The shop manager, an elegant redhead, whisked past him in a dress that said, "I'm hot, powerful, and you couldn't handle me." She had a smile of welcome that invited and a ring that attested to the fact that some man thought he could, indeed, handle her.

"You're looking lost," she eased to a stop close beside him on a return loop.

"I'm with the blond one," he hooked a thumb toward Lana sighing over evening gowns with low cuts that his daughter would never, ever, under any circumstances be allowed to even dream about wearing.

"It was so nice of your daughter to agree to be Tammy's model," the redhead sounded delighted. "We truly do appreciate you helping her out. However, it looks like they're going to be a while. If you want to wait, they'll end up in the back room eventually. There are places to sit there out of the fray."

He took one last glance around the room. He was the only male present and was receiving the eye from several of the very

well-tended women in the shop. That had been Lana's mother's trademark—he still did his best not to think her name—she was always well tended. And had finally found herself a sugar daddy who had offered to make sure she stayed that way—far beyond the capacity of the opera's chief lighting designer who had merely loved her.

"Yeah, that would be good."

The redhead guided him through the swinging doors to a diner's cook line that was filled with women's accessories. It was such a creative space that he had to slow down to admire it. Someone here knew what they were doing. Handbags dangled from pot hooks, pantries were filled with fine boots, and a walk-in freezer was lined in lush winter coats. The ceiling was hidden by dozens of inverted open umbrellas, splashes of color to delight the eye. It would be even better if they were backlit and the light shone through them—not his place to point that out. Still a very cheerful effort.

Another set of swinging doors and he was in a space that would almost put the opera's costume shop to shame. It was far smaller in scale, but there wasn't a wasted inch; it was a dressmaker's dream.

The redhead pointed him to a chair at a sewing machine, then whisked back to the shop while he admired the departing view. She might be too high-end and too married for his taste, but that didn't mean he was dead.

At the same instant a tall woman swooped in from the next room with her arms full of a gaudy mish-mash of fabric. Some of it looked like it had been graffitied all over, like girls used to do to their notebooks when he'd gone to high school. Now Lana stored all her school stuff in her tablet computer and he couldn't even check the outsides of her notebooks for doodles of boys' names.

She'd gone on dates; group dates, but even she called them dates. It had taken him a lot of careful prodding to discover that "hanging out" was a bunch of people and that a "date" meant

little more than that the bunch of people had an even number of boys and girls, not necessarily paired off thank god. If there was a boy in the next-level "going out" category, she hadn't let him know about it yet.

The woman dropped into the chair in front of the machine beside him.

"You applying for a job?" she nodded toward the machine he was seated at. Like the ones at the opera, it looked oversized and immensely complex.

"Not likely."

Then he looked at the woman. She was a knock-out in a different way than the store's manager. Long dark hair curled past golden skin. She was as shapely as the redhead, but because she was so tall it looked right on her. The redhead was more in a powerfully voluptuous category. This brunette was built… just right?

"Then what are you doing other than checking me out?"

He didn't fight the grin at her teasing tone. "Not much, I have to admit."

3

Lana stepped into the back room and froze. There was her dad and he was…she knew that posture. Enough boys had tried it on her on dates. That laid-back pose with the "I'm so cool" smile. They always thought they were so charming and handsome, and most of them were just jerks.

Her dad was like the handsomest guy on the planet, except maybe Francis the track team captain. But what was he doing?

She grabbed Tammy's arm. "Who's that? She your aunt or something?"

"Kinda. We're not related. Kari just looks like me or maybe I look like her. Seriously though, she's amazing."

Lana couldn't believe that Tammy had asked her to come be a model, because Tammy always had the most incredible wardrobe. Every guy watched her walk by, not that she was so beautiful—though she was awfully pretty. They watched because she dressed like she was gorgeous. A lot of girls hated her for it; others envied her. If Lana got Tammy's help, maybe Francis would look at her that way.

Lana had known Tammy since right after her mom died and they moved to Seattle four years ago. As friends, they now had Dad's divorce behind them and Tammy's new mom. Friends didn't get better than Tammy, but Lana still didn't like the way that Dad was looking at Kari.

"C'mon. You're like forever tall. I need to see how my clothes look on you."

Lana submitted, moving behind the screen to try on whatever Tammy had in mind. It felt weird changing with Dad in the room, but Tammy made it seem normal, so Lana did her best to calm her nerves. And peeked around the edge.

But her dad wasn't watching her.

4

K*ari only had a* distant look at the girl before she ducked behind the screen, but it was enough. The measurements that Tammy had given her had been good, these clothes should fit just fine. Rather than just leaving the fabric pinned, she dropped the dress she'd been working on under the sewing machine's foot and began running the seam.

"She's beautiful, Richard. She looks so much like you." His sleekly handsome and blond was transformed elegantly into her slender frame. She was a good choice for a model; a sharp contrast to Tammy's short, curvier frame and darker complexion.

"Scares the crap out of me every day," he rubbed at his face. "When she grows up I'm in so much trouble."

"Hello. Already grown."

He grimaced at her before slouching even lower in the chair and groaning. "I really didn't need to hear that."

"Oh," Kari did her best to sound contrite, "I mean she's such a cute toddler. Do the little boys follow her all over the pre-school grounds?"

"With their tongues hanging out. Have since she was about three. That's the problem," his chuckle acknowledged he was being ridiculous, which she liked about him.

"What's the problem?"

"I'm a guy. I know what they're thinking at their age and I can't believe they're thinking it about my daughter."

"Bad news first or good news?" Kari appreciated the way he talked about his daughter, as if she was precious and worth protecting. Her own father, well, he hadn't made her feel the least bit safe. He'd never groped her, but his big-screen sports drinking buddies hadn't been so hands off; she'd learned to be scarce come half-time or seventh inning stretches or even long commercial breaks.

Richard looked at her through blue eyes shaded darker by an assessing scowl.

"Good news?" he asked cautiously.

"I'd say she loves you a lot."

"How can you tell? Did you even notice her?"

"I know, because of how long she was glaring at you for talking to me."

At his "Huh?" she pointed up at the mirror leaning against the wall behind the sewing machine.

"It wasn't me she was glaring at or she'd have spotted me watching her in the mirror," it had offered a clear view of Lana and Tammy talking.

"Perfect. It's not as if I have a love life for her to guard against anyway."

Kari had just assumed he was married, but saw there was no ring. She kept her thoughts about handsome single dads to herself. But it was hard.

Perrin had married one just a year ago, had adopted his two children, and would soon add a third to the family. Yeah right. And she'd known Richard for about five minutes. Stupid fantasies never did a girl any good.

"Okay, pretty lady. What's the bad news?"

"Sure you want to hear it?" *Pretty lady?* She rather liked that Richard was plain-spoken. Had he always done that, or learned it from his daughter? Teasing was fine, but she'd never liked games, and Richard didn't appear to be one to play them.

"No, but tell me anyway."

There was that straightforward thinking again. This time she waited until she finished her final seam and had clipped the thread.

She turned to face him fully for the first time. He was slouched low in the chair, but he was watching her face rather than her body. More reason to think he was decent.

"At her age…"

"Oh no!" Richard moaned and closed his eyes, wincing as if she'd just poked him with a sharp stick.

"…she's thinking along pretty much the same lines as the boys."

"Shit! I didn't need to know that either," he moaned.

5

Lana had kept an eye out while Tammy went back and forth bringing her different clothes. The first couple had looked cool, especially the off-the-shoulder black-and-white zigzag top, but Tammy had rejected every one. A lot of them she didn't even get to try on; Tammy would just hold a killer blouse up in front of Lana, sigh and take it away.

Each time Lana peeked, Dad was looking less and less comfortable talking to the woman, which was good.

Tammy held up another, then took it away.

"Hey! That looked great!"

"Wrong color," Tammy insisted. "See?" She twisted Lana around to face the small mirror behind the screen. First she held it up to her own chest and it looked totally awesome. Then she held it up in front of Lana. It was good; it was hot. Francis would definitely look at her in it…but it didn't snap the way it did on Tammy.

"It's just because you're so different," Lana often envied Tammy her gold-skinned beauty. Their high school was in

Ballard, a Scandinavian neighborhood of Seattle, and a whole lot of people were colored like Lana. And Tammy had developed real curves, Lana's body was so flat that she was barely female—at least it felt that way sometimes.

"No," Tammy shook her head. "You'll see, once I find it."

"I want to knock Francis off his feet."

"Then try this." It was the woman, Kari. Up close she looked as if she really could be Tammy's mom. Way taller, but the same Italian gold skin and real shape, just way taller—almost as tall as her dad.

Lana started to turn for the mirror.

"Don't look," Kari turned her away from it. "Instead, put it on and step out to show it to your dad. Watch his reaction. That will tell you more than your own."

Lana squinted at her, but Tammy stood behind her nodding.

She took the clothes from over Kari's arm. It was a dress. "I don't wear dresses."

"Trust her," Kari nodded at Tammy. "She's an amazing designer."

"Then what's your job?"

"She," Tammy slipped an arm around the woman's waist, "is the most amazing seamstress on the planet. I can think it up; she can make it so that it hangs great and actually fits people."

Lana shrugged. She was half out of her clothes before she realized that Tammy and Kari weren't going anywhere. She wanted to shoo them out, but already they were helping her. No one had helped her dress since her Mom had forced her into a stupid formal dress for a birthday party when she was six—the only one not wearing jeans, she hadn't been able to play in a single game. She fought down against the bitter tears that apparently had no end. Stupid tears because Mom had only ever cared about how her daughter had made *her* look; they were so hard to stop.

Finally, she just closed her eyes against the pain and let them dress her in…whatever.

6

*R*ichard *was so desperate* for a distraction that he picked up a teen fashion magazine. Lots of young women wearing far too little. Ratted jeans, crop tops, wide brim hats. Some looked okay, but then he hit the brilliant red dress with too many cutouts and barely enough material to cover the model's butt. On the next page a girl who looked like she was sixteen wore a black bit of cloth with cleavage practically down to her panty line.

No way in hell! Not his Lana!

He tossed aside the magazine, shoved to his feet, and made it one step in Lana's direction before he stumbled back into his chair which almost flipped him backward into the sewing machine.

There she was—it was his Lana. He knew it was. But he barely recognized her despite that.

The blond girl who looked so sharp in her black and white track outfit was gone. As was the sad girl, as devastated by her mother's abrupt departure as he'd been. In her place wasn't some under-clothed vixen. It was Lana, become herself. She'd blossomed into a version of herself he'd never imagined.

Her dress was the graffiti-laden fabric he'd watched Kari working with just moments before. It was purple, with pale-orange lettering that looked hand painted. Quotes of great thinkers, silly faces, and words like: strength, passion, joy. It should have been garish—would have been if the shades and tones hadn't been so carefully selected—instead it was pure teen chic.

It hung loosely without hiding that she was lean and fit. Below a narrow belt that made a tight gather at the waist, it flowed to mid-thigh. No slutty tease of a neckline about to slide off her shoulder, the sleeves added a softness that belied the hard lines of her dead straight hair and tall figure.

Calf-high boots in a cobalt blue shouldn't have worked, but the tone matched the small purse slung low across her body on a thin strap and made her eyes shine. The only adornment, a thick copper bracelet tied her outfit together.

It wasn't sexy. It wasn't lurid.

But it also wasn't his little girl. It was a confident woman who was no longer afraid and no longer cowed by a mother's betrayal. Powerful in herself.

He staggered to his feet and went up to her, stopping half a step away.

"Daddy?" Lana asked him uncertainly. She hadn't called him that in a long time.

"You're magnificent." He didn't know what else to say to her. Didn't know how to say how proud he was of her, so grown and so strong. So he did the only thing he could think of, he folded her into his arms and held her as tightly as he used to hold a little girl afraid of the dark.

He mouthed a *thank you* to Tammy who was doing a little victory dance.

Then he looked at Kari. She had her back to him, but kept wiping at her eyes, her hand coming away wet.

7

A date?" *Kari grabbed Tammy* by the shoulders and shook her. "What am I going to do? Richard wants to take me out on a date."

"Cool!"

"Don't *Cool!* me, girl."

"Why are you so wound up? Do you like him?"

Kari dropped onto a stool. It was evening, rain pattered against the darkened windows. The store and studio were empty, still echoing with Lana and Richard's most recent visit. She and Tammy were the last ones here, waiting for Perrin to get back from her doctor's checkup.

"Well?" Tammy stood in front of her with her fists planted on her hips like a school marm.

"Can you tell me one thing not to like?" She waved a hand helplessly. Over the last few weeks, father and daughter had come in several times. Kari had watched them closely. And it was as if he was rediscovering his daughter all over again. Tammy had made Lana look chic, smart, sassy, modest, and

amazing in turn, without once hitting what Perrin called the "Slut Button."

Throughout the fittings, Richard kept revealing facets of how deeply he cared about his daughter's happiness. He hadn't cried when Lana tried on about the most amazing prom dress Kari had ever seen, but he'd come close.

Then he'd taken Tammy's hand and shook it with great respect. And when he'd taken Kari's hand, he'd asked her out on a date. She never even saw it coming.

"Well," Tamara finally answered her. "Dad likes him. Jasp and I have known him since forever. Jasp worships him."

"Leave your little brother out of this." Jasper Cullen was way too smart about people for a twelve-year-old boy.

"Richard is really nice to him. He's always teaching Jasp lighting design stuff or letting him run the control board or something. I thought you'd like him. And he's a real hunk for a grown up."

Kari eyed Tammy.

"You thought I'd like him?"

Tammy's jaw dropped, but she recovered it with a quick shrug and a sheepish grin, though not very sheepish, "Oops!"

"So this wasn't about Lana? You set me up?" Kari felt an anger rising, but not very strongly. It was hard to be angry after seeing father and daughter together. After the last fitting, Lana had given both she and Tammy a huge hug before walking out holding her dad's hand.

"I gotta help Lana with Francis. You see, if I do, then she promised to get Tony to ask me to—" Tammy clamped her mouth shut, slapped a hand over it, and blushed so fiercely that her golden skin went several shades darker.

Kari raised her eyebrows in question, torn between horror and laughter.

"You can't tell Perrin or Dad," Tammy mumbled through her hand. "She might be cool with it, but Dad would freak."

"I'll keep quiet under two conditions."

Tammy nodded carefully. She was just as much a matchmaker as Perrin. Kari had seen Perrin do it to others. Kari already had dodged a couple of Perrin's attempts. She didn't want to be set up, she just wanted to…meet the right guy. But she hadn't been watching out for Perrin's daughter; even though not related by blood the imp was equally dangerous.

"First condition, if anything happens with Tony that you aren't comfortable going to Perrin with, you come to me. That's not optional."

Tammy nodded but didn't remove her hand from over her mouth.

"Second condition, I have no idea what to wear. You have to help me. Deal?"

"Total deal! I already picked it out. C'mon," Tammy grabbed her hand and dragged her toward the front of the store.

Yep. A complete and absolute setup.

8

*W*hy here?" *Kari felt* as if she was crouching under the umbrella in order to disappear rather than merely to stay out of the rain. And she never should have let Tammy talk her into heels; she had to clutch Richard's arm to not go down on the wet cobblestones of Pike Place Market. Or had that too been Tammy's plan.

Oh no!

"Wait, don't tell me. Your daughter said something like, 'Tammy told me about this amazing Italian restaurant named Angelo's.' "

Richard's laugh was warm and welcoming, deep enough she could feel it rumbling about under the umbrella they crowded close to share. "I believe that's a direct quote. Why?"

Kari sighed and indicated for him to open the restaurant door, "It's a long story. I'll tell you over dinner."

They stepped in and Graziella greeted them, "*Ciao*, Kari. I didn't know you were coming."

Kari smiled back at the hostess. "Or you wouldn't have if Tammy hadn't already called you."

Graziella's smile didn't even flicker as she shrugged, "Oh, there might have been something about saving the nicest table, the one that sits close by the fireplace. But I wouldn't know anything about that." She took their coats, which raised a whole other issue.

"Oh my god!" Richard simply stared at her.

The dress Tammy had chosen from the rack had been a simple drape of dark blue with a copper-red trim that accented lines and curves. It suggested and implied without revealing. It wasn't blatantly sexy, because somehow Tammy knew she wouldn't be comfortable in anything that was. But it certainly showed her figure to its very best possible form. It was a dress that she'd sewn dozens of for Perrin's customers, but had never thought to try on for herself.

Graziella winked and escorted them in. Richard took her arm to escort her...after stumbling a bit, in a very satisfying manner.

"You look amazing," his breath was a warm whisper in her ear as he pulled out her seat.

"I'm guessing that's your one suit, but you look pretty amazing in it yourself." She wasn't sure she'd ever sat across from such a handsome man before.

"Lady's smart as well as gorgeous," Richard was studying her face intently.

Kari wanted to brush at her hair, but Tammy had insisted she should leave it loose over her bare shoulders.

"I'm guessing you've been here before."

Kari looked around at the cozy Tuscan elegance. The fire was already warming away Seattle's damp evening chill. She wanted to wrap herself up here and never leave. The air was scented with basil and citrus and red sauce. The music was a bright Italian dancing tune, lively but soft enough to be comforting.

"My first time. This is a little out of my range. But we get together for these amazing dinners every week at the owner's mother's apartment. Perrin is best friends with the owner's wife and his mother." It was one of the true tests of a man or woman

entering the "circle," how they fit in at Mama Maria's dinners. Kari expected that Richard would be as hand to glove.

"So," Richard reached across the table and took her hand. The warmth that spread through her didn't only come from the fire.

"You were going to tell me a story about a conspiracy involving two teenage girls."

Kari smiled, "You're clever, Richard Nyberg. Figure it out and I'll tell you where you go off track."

9

Richard didn't realize quite how smart he was.

Not that first night when they had eaten pappardelle and laughed and flirted. Not the fifth night when they had tumbled into his bed and both wept with the wonder of discovery.

His introduction to the entire crowd at Maria's had been a huge shock and a wonderful one. Against his better judgment, he'd brought Lana as well—that was when he learned just how pointless it was to argue against a pair of fifteen-year-old girls with a plan. He'd been overwhelmed, but he'd watched Lana drink it in like a tonic. She needed family and friends—thrived on it.

By the time he took both gals—his daughter and his lover—back to Angelo's Tuscan Hearth Ristorante several months later, he thought maybe he was getting a little wiser. And maybe his daughter wasn't the only one who needed family. A whole family, with a woman who truly wanted to be there for them.

He'd watched the bond grow between Lana and Kari over Tammy's continuing fashion experiments. Richard had come to enjoy the sessions when he could get away to watch.

Tammy's eye was very good, but sometimes it just flat out missed. Every now and then he'd be witness to his daughter the goof or the nerd—though Tammy thought that a few of the "failure" looks might work on another model. A couple of times the clothes had edged too far toward racy for his taste, but Kari had insisted these were safely conservative.

When he'd been dumb enough to argue, Kari had taken him to hang out by the parking lot a couple of times when high school let out—thankfully not his daughter's so he could retain some illusions. Then she took him to a Nickelback rock concert with the two girls—and two boys who he hadn't counted on but rather liked by the end—and even those few vague illusions had disappeared. He'd stopped arguing with Tammy's taste level after that.

Lana had also taken to using the sewing machine beside Kari. Her first efforts were…first efforts. But she learned. She didn't have Tammy's eye, but she learned to sew, to fit, and eventually to craft.

That was the unique thing that Kari brought to her work. She brought craft. That came from passion. And she had a passion that flowed through her work with Tammy, her patience with Lana, and the way she always shared so openly with him.

Tonight he had a beautiful woman on each arm as they arrived at Angelo's. The brightly lit ferry boats slid across Puget Sound looking like floating birthday cakes. He held the door and bowed for the women to proceed him.

Kari blushed and Lana giggled.

As he and Graziella took their coats, Richard could only marvel. Kari, who had revealed her casual side on most occasions, was once again dressed to kill. Instead of a classic simple black dress, it had been a simple, sleek red one—a dark dusky red that offset and warmed her skin. Open neck, sleeveless, and ending mid-thigh. A copper bangle on one wrist reminded him of that first dress; he'd wager it was a Tammy touch intended to do precisely that. In this dress, Kari wasn't

merely gorgeous, she was also as warm and welcoming as the Tuscan ambiance.

Lana was dressed back in that first look that Tammy had made for her. She was wrapped again in words of strength, joy, and—across the top of either shoulder—passion. He'd questioned that word on his daughter's body the first time. He didn't any longer. Her wrist bangle matched Kari's. The two women looked incredible together.

They sat back at the fireside table; this time he had been the one to make the request. It was a table for two, but it didn't matter—he pulled a third chair up to the side with his daughter to his left and Kari to his right.

As close as family.

10

*T*onight was different. *Kari* could feel it, and it wasn't merely that they were at Angelo's or that they were together. The three of them had eaten together dozens of times: at Richard's house, at restaurants, or huddled backstage at the opera among lighting instruments and cables when Richard couldn't get away for more than a few minutes.

Kari had found her hopes and her dreams, but she couldn't ask for them to come true. Richard and Lana were the family unit—something far too precious, too fragile to risk. It was in such perfect balance that she only dared sit on the outside and dream of being inside.

Like her clothing designs that had never excelled. She could breathe life into Tammy's ideas, create an exciting collaboration, but she couldn't go off on her own and find success. She was meant to be a part of something else, something bigger. But she didn't know how to get there from where she was.

"Kari," Richard's voice was a warm caress. It reminded her of how it felt every single time he held her, or she held him. It

was hard to tell holder from holdee anymore; had been since the beginning.

His voice was also rough as if he was having trouble speaking.

"Yes?" She tried to take strength from the moment, but just as when she'd sat with Perrin the moment before she'd been promoted to help create Tammy's design label, the nerves were winning.

Richard opened his mouth again, but no sound came out. He cleared his throat. Sipped some water.

Lana rolled her eyes at him.

Finally, it was Lana who reached across the table and took Kari's hands which were chilled with nerves she could no longer hide.

"Dad will recover soon enough. I think it's kind of fun that women can have that effect on boys, don't you?"

Kari nodded, unsure what they were talking about. Richard affected her the same way; speechless with the wonder of the man.

"I don't want a mom."

And now Kari knew. It was a death knell that rang inside her. Those foolish family dreams that had been fluttering about in her head a moment ago, sank leaden into her stomach.

"But I could really use a friend."

"What?" The word stumbled out. She was suddenly feeling stupid.

"Having a mom didn't work so well. But I think you're great and once Dad recovers, he'll tell you that he loves you."

"Lana!" He managed a protest.

Kari glanced at Richard. She knew that look on his face. It was the look of seeing his daughter truly herself for the first time, or the shock over the scantily clad teens at concerts; the look of total confusion.

Lana didn't look at her father, but her grin turned about as wicked as Tammy's.

"Frankly, I think you two should have another kid, just the way Perrin did. I'd probably think of her like more of a niece or

something than as a sister, but I'd be cool with it if you decide. I just wanted you to know."

Lana let go of her hands with a final squeeze.

Kari was still trying to catch up.

And then she'd remembered Tammy's question when Kari had agreed to help with her clothing line. One word that had covered a world of hope and joy.

She turned to Richard who still appeared to be stunned speechless by the daughter who had inherited her frankness directly from her father.

Taking confidence from Lana's encouraging nod, Kari was the one to reach out and take Richard's hand.

"Really?" she asked as loudly as she could manage, which was little more than a whisper.

He nodded.

"Say you love her, Dad," Lana prompted.

He nodded again.

Then he clamped down on Kari's hand as if to make sure she wouldn't leave before he recovered his power of speech.

Not a chance.

Now that she knew where she belonged, she wasn't going anywhere.

Christmas at Henderson's Ranch

I should have seen this coming, I really should have…

The two leaders of The Night Stalkers series, Emily Beale and Mark Henderson, also became the founders of my Firehawks series. I knew they had hired a nanny to take care of their child while they fought fires, but I didn't give it any more thought.

When I set out to write Christmas at Henderson's Ranch, *it was just a small story. Beale and Henderson often talked about visiting Mark's family ranch and I thought it would be fun to see it at Christmas time.*

I should know by now that such stories are never that simple.

The ranch manager is gobsmacked the very second that the nanny steps off the plane. Suddenly these two background characters stepped to the fore.

Instead of a quiet story set in a Montana ranch lodge around a warm fire and a fresh-cut Christmas tree, I had a prairie-crossing adventure on my hands with helicopters, horses, and all sorts of amazing adventures.

One of the things that they taught me about Henderson's Ranch?

It's going to be my next new series.

Launching in spring of 2017, my Henderson's Ranch series will be blasting to life full blown thanks to this and another short story, Reaching Out at Henderson's Ranch.

The tone of the ranch is set by the owner, Mark's dad, who is a retired Navy SEAL. This story is created by the romance of the two main characters and I loved *writing their love story set there. Because that's when I fell in love with Henderson's Ranch.*

M.L. BUCHMAN

NPR and B&N Top 5 Romance Author of the Year

"One of our favorite
authors."
RT Book Reviews

Christmas at
Henderson's Ranch

a Henderson's Ranch romance story

1

*T*his isn't right!"

Chelsea Bridges leaned forward to see what Emily Beale was looking at. Chelsea didn't see a thing wrong, but then she'd never been to central Montana before.

Out the small plane's front windshield were miles and miles of rolling green prairie. Streams crisscrossed the grassland in a bewildering maze. The backdrop was the foothills of the Rockies breaking the skyline with their snowy peaks and conifer-clad sides. The westering sun silhouetted the hills, but lit their tops with gold.

"It's absolutely gorgeous!" Then she clamped her mouth closed. She was trying to reel it in. Emily was always so even-keeled and understated that Chelsea was constantly stumbling to be less…Chelsea. Emily was this perfect woman with a drop-dead handsome husband and about the cutest kid on the planet. Chelsea had only been their daughter's nanny for a few months, but she'd seen the deference and respect that everyone at Mount Hood Aviation's firefighter airbase paid Emily. In

return, the woman was kind, courteous, and utterly terrifying. Chelsea wouldn't mind being all of those things.

Her husband Mark, who sat up front in the other pilot seat of the small plane, wasn't much more effusive—except around his daughter. At least he had a sense of humor, though not as much a one as he thought he did; an observation Chelsea kept carefully to herself.

Chelsea looked over at Tessa who was strapped in beside her. She had her tiny version of her mother's elegant nose pressed up against the window. "Green," she announced. Out her window was nothing but the rolling grasslands of eastern Montana.

"It's wrong," Mark agreed solemnly but turned enough to wink at Chelsea, or at least she presumed that's what his cheek twitch was indicating at the lower edge of his mirrored Ray-Bans. "Not much snow in the hills. Means another drought year next summer."

"That's not the problem," Emily responded. "Okay, drought is a problem. But that's not the real problem."

"What is, Emma?" Again the sassy wink that said he already knew what his wife was talking about. It was amazing that the man had survived this long. Chelsea would never dare tease Emily Beale; she could probably kill with a glance if she ever took off her own mirrored shades.

"It's December," Emily took one hand off the plane's wheel—if she was on board, she was the one doing the flying—and waved it helplessly at the stunning scenery before them. "We came to Montana for a white Christmas."

"I thought it was to see Mom and Dad."

"It's still supposed to be white," she grumbled and set up to land the plane. It was as much emotion Chelsea had seen in her entire two months with them. Emily Beale was never unkind, but she was cold. Or at least chilly. But that wasn't right either. The woman was frank and forthright, as much with her daughter as with her husband. Yet Tessa was often in her lap, welcome not as child to adult, but rather as a piece of Emily that was simply

back in the place where it belonged. The mother and daughter weren't close; they were simply one when they were together. It was about the most incredible thing Chelsea had ever seen. It made her ache for a family of her own; not a familiar feeling.

Again Chelsea strained up against her seatbelt to look down. A herd of horses startled and looked up at them as they passed by. They didn't scatter and run, but they eyed the low-flying plane carefully.

"Horsies!" Tessa declared delightedly when Emily shifted her flightpath so that the herd was visible outside her daughter's window. Not cold at all, just…inscrutable.

"Yes," Chelsea encouraged the toddler. "Those are horses. Aren't they pretty?"

"Pretty!" Tessa burbled, and they laughed together with delight.

Chelsea had never seen a whole herd of horses before. There were at least fifty in the group of every shade imaginable: grays, browns, whites, blacks, and mixes in patchworks, dapples, and who knew what all. They were gone behind the plane too fast to distinguish more. She tucked away the trail mix snack they'd been sharing to make sure Tessa's blood sugar was up.

Even after two months, Chelsea wasn't quite sure how she'd ended up in this situation. Not that she was complaining, Emily and Mark were great parents and it showed in their total sweetheart of a daughter. And flying with Mark over forest fires was often very dramatic.

It had started with Aunt Betsy who was a cook for the Mount Hood Aviation helicopter and smoke jumping firefighters. When Chelsea's degree in psychology hadn't led to any kind of a useful job, her aunt had asked if she liked to fly. She'd shrugged a yes because she'd flown in passenger jets any number of times to visit grandparents, and a trip to Nepal for a backpacking gap year.

She'd now spent most of the last two months sitting in tiny planes of six or eight narrow seats and been paid to enjoy the scenery and play with a baby girl. Best job she'd ever had by a long way.

Tessa was a fixture in Mark Henderson's plane when he was flying as the Incident Commander high above the fire. What was surprising wasn't that they'd added a nanny, but rather how he'd done the job for so long without one. Tessa was a pretty low maintenance kid, but she was also eighteen months old and quite intelligent.

It was a late fire season, Mark had said, and MHA had still been flying fire in the Southwest. But, finally released from the summer contract, they'd come north for a vacation and brought Chelsea along with them. She sure as hell wasn't going home. They'd known that.

As they flew closer to the ranch, more and more fences became visible, cutting the prairie into smaller pastures and training rings. There were several barns, smaller residences, and cabins surrounding the main residence.

Emily flew once over the grand log-built ranch house and waggled the plane's wings in a friendly wave.

Chelsea pointed to out to Tessa, "Isn't it amazabiling?"

"'mazbling!" Tessa called out happily. Emily sighed audibly as she circled wide of the barn.

Chelsea wondered if Mark's habits were rubbing off on her, but she couldn't resist messing with Tessa's rapidly developing language set. They landed on a gravel strip that ended close beside the house and a large out-building that turned out to be a hangar.

A big man strolled out to meet them, still buttoning up his sheepskin jacket. He was an older version of Mark; just as tall, just as broad-shouldered, his light hair going silver. But Mark's face was different. Darker, broader, and his hair was thick, straight, and almost midnight black, sharing only his father's gray eyes.

The clouds of mist puffing about with each breath of Mark Senior—Mac, she reminded herself, they'd said he liked to be called Mac—had Chelsea bundling up Tessa before the plane came to a halt in front of a hangar. The ground might be snow free, but it was far colder here than Oregon where they'd boarded the plane.

2

*D*oug *Daniels had stuck* his head out of the barn when he heard the plane come over low. The trademark gloss-black-and-red-flame paint job told him who was aboard. Some part of him had been alarmed that a client was in-bound for a ranch vacation even though they hadn't taken any Christmas reservations this year. But it was just Mark and his knock-out wife. He liked Mark fine, but he had trouble speaking around Emily Beale. It wasn't just the beauty, he knew how to talk to pretty women just fine; it was the fierce level of competence that she demonstrated at every turn.

He finished helping Logan pitch the hay into the stalls' feedboxes before heading out to greet them. The air had a sharp bite to it, wholly different from the horse-and-straw of the barn, but no moisture. As he stepped out of the barn, he noticed that there wasn't even a hint of cloud in the cobalt blue of the late afternoon sky. The temperature was already dropping though it was still an hour to sunset. It was going to get cold tonight.

Doug stuck his head back inside. "Hey, Logan. Open up the gates. If the main herd has any sense, they'll be coming this way by sunset."

"You bet, boss. Any horse that stays out there tonight needs his horse-sense meter checked."

Doug went out to help stow the plane. There was room in the hangar because he'd moved the helicopter tight to the side after the morning's flight to check the main herd and make sure there were no stray or injured. He hadn't been able to get an accurate count, but it had felt low and that was bothering him. Happened all the time. Still, it worried him.

He ducked through the hangar's side door, popped the release, and slid open the main door from the inside. It rattled and boomed in the cold air. A sharp squeal in one of the wheels had him adding "needs grease" to the infinite mental checklist that was running a working dude ranch.

Just emerging from the plane was a figure wrapped deep in a parka, with the fur-rimmed hood already raised as if it wasn't a merely brisk day, but rather a north polar night. She, for there was no chance of a guy wearing such tight jeans and making them look so good, carried an equally bundled child.

He came up and stuck his nose right into the child's hood, "Tessa, my love! Give us a kiss!"

"Kiss!" the little girl squealed and kissed him on the nose.

Then he rubbed noses with her until she was giggling before he pulled back. He'd ended up standing very close to the woman holding her. He could just see brilliant blue eyes, a freckled nose, and a bright smile in the narrow opening of the hood.

"Do you greet all the girls that way?" Her tone was light, almost musical.

"Sure." Never one to back down from a challenge, he stuck his face right into her hood until their noses rubbed and cried out, "Give us a kiss!"

Unlike the little girl, there was no squeal. Instead, there was a quick squawk of surprise.

Way over the line, Doug.

But before he could retreat, she gave him a quick kiss. Unlike Tessa's it didn't land on his nose, but right on the mouth. There and gone, but the lips were warm, soft, and tasted of peanuts and chocolate.

Once he was clear of the hood, the gloved slap that he expected to follow, didn't. He glanced again into the tunnel of the raised hood.

The bright blue eyes caught the low sunlight and weren't round with shock or narrowed with anger.

"Well," she blinked in slow motion, "okay then."

He laughed, he couldn't help himself.

Now that was his kind of woman.

3

Chelsea had no idea what had come over her. She didn't randomly kiss men, even tall handsome ones who adored small children.

Men who then scooped a little girl out of her arms, slung her around with the ease of long practice until she was riding piggy-back, and—while Tessa shouted, "Horsie!" with glee— galloped about the yard with a protective hand wrapped awkwardly behind him. The man shook back his collar-length, sun-streaked hair the color of worn leather so that it brushed in Tessa's face. He let out a fierce whinny escalating her giggles of delight.

He trotted up to Mark and Emily then stopped with a sidle and a stomp that was thoroughly horselike and delivered the child to Emily. Then he and Mark made quick work of pushing the plane back into the hangar.

Chelsea was still standing shocked into place when they'd finished and the men had returned carrying the luggage.

"A field pack, very practical," the man who'd kissed her held it aloft as if it contained only air rather than most of Chelsea's

worldly belongings. Her camping gear was stashed at Aunt Betsy's and a dozen boxes of books at Mom and Dad's, but the rest of it was in that pack.

"It's my hiking pack, but I use it for everything. Really practical since I hike a lot," she was rambling; time to cut that out. She sniffed at the air and the cold made her nose hurt on the insides, "At least when it isn't sub-Arctic."

The man's jacket was fleeced-line denim, but he hadn't even bothered to button it against the frosty day. He smelled of hay and his kiss had been warm and fresh with the outdoors.

Mac greeted his son with a firm handshake, but gave Emily a deep hug that surprised Chelsea almost as much as being kissed by a total stranger. What had happened to the woman's backbone of steel? Emily leaned into Mac's hug as if she was the one related by blood and was happily come home. Then he led them toward the house, leaving Chelsea and her luggage bearer to trail behind.

"Do you have a name or should I just shout 'Sherpa!' when I want your attention? Or perhaps *daai?*"

"*Daai?*" he led her onto the wide porch and held the door for her to enter the mud room. There they shed boots and jackets. She was glad she'd been wearing a thick sweater against the damp chill in Oregon and kept it on for added warmth.

The others were talking happily enough together to be lost in their own conversation as they too stripped off the outdoor gear and pulled on slippers from the large basketful of them close by the inner door.

"*Daai* means *older brother* in Nepalese," she explained softly. "A sign of respect. Better yet, *bhaai* for *younger brother* as who knows if you're worthy of respect."

"You kiss me and question whether I'm worth respecting? That doesn't bode well for the morning after."

Chelsea was preparing a comeback, for she certainly wasn't the one who had done the kissing…or had she been, when she turned and saw the look on his face.

"What?"

He shook himself like a horse again. "If I'd known what was under that hood, I might have spent longer kissing you."

"Skin deep, *bhaai*."

"Yes, but what a nice layer it is."

4

Doug *knew he was* staring, but how was a man supposed to not? Thick waves of red hair cascaded down to her shoulders. Her cream-and-freckle skin only highlighted the brilliant blue eyes that were presently rolling at him. Her sweater must have been custom-made because it traced and enhanced the slender woman within. The rich green was finished with red zig-zags at wrists and waist. A small but elegant snowflake had been knit right over her heart.

"Frozen heart?" He teased to hide his suddenly dry throat.

She looked down where his attention had strayed. "I called this one White Christmas, *bhaai*. And watch where you're looking."

"I am watching where I'm looking, and very glad to be doing so," his made his voice pure tease. Then he wondered, "You name your sweaters?" Could he sound much stupider?

"Sure. At least the Christmas ones."

"You knit it yourself?" Apparently yes, he could sound dumber. But there was something about this girl—woman.

She *liked* hiking? Major understatement. Her pack looked like it had been carried by an entire Army brigade, worn shiny in a hundred places. A very well-used piece of top quality gear. She knew terms of respect in Nepalese and could knit sweaters that made her look like a Christmas delight.

"I—" they stepped out of the mud room and into the living room. Her gasp of amazement echoed that of all who came here. Every ranch guest who entered the main house couldn't help but stumble to a halt.

"Quite something, isn't it?"

"It's gorgeous! A little daunting, but…" she did a slow twirl to take it all in. "But this is right out of a magazine. It's unbelievable!"

The large river-stone fireplace was a true showpiece, big double-length logs crackled away on the grate. The flagstone hearth was surrounded by plush chairs and inviting sofas. An upright piano stood by a corner window overlooking the horse pastures and snow-capped peaks. And the high-beamed cathedral ceiling made the twelve-foot spruce that he and Mac had felled up on the northwest slope fit right in. The Hendersons always really did up Christmas. Coils of holly were draped from mantel and piano. Wreaths, garlands, winter-themed quilts on the walls…

"Quite the spectacle, isn't it?" And this nameless woman in a sweater named White Christmas fit right in.

"It's fabulous! My family does a totally lame Christmas, as in almost not at all. Once I got to college, I discovered it and turned into the Christmas loon of any group. You should have seen this poor pistachio tree I decorated on year in Puri."

"Puri?"

"India. On the east coast. I spent a couple months traveling there by train after I left the Himalayas."

Himalayas? Right, well, that explained where she'd picked up the Nepalese. What hadn't she done?

"I try to do up my place, too," he answered. "Same style of construction, but cozier. Bet you'd like it too."

"You do, huh?" Something was amusing her but he couldn't quite think what.

"Sure. I live on the far side of the meadow. I'm the ranch foreman."

"Your…place."

"They gave me a sweet little setup. Two bedrooms. Looks a lot like this, just on a smaller scale. A ranch house in miniature."

"You bet I'd like it?" Her tone had gone impossibly dry.

And her meaning finally sunk through his thick skull. "I didn't mean—" He had just invited a woman whose name he still didn't know back to his place for a quick— Someone should just take him out to pasture and shoot him.

A wicked smile crossed her features. "Sure know how to make a girl feel welcome, *bhaai*."

Little brother. Suddenly that really wasn't the role he wanted to be cast in. Not even a little. Because he could certainly picture her clearly in that cozy little log house of his.

5

Chelsea was curled up in one of the big chairs by the fire with Tessa on her lap. The girl was fading, but not out yet and Chelsea felt completely content working through the thousandth iteration of *Carl's Snowy Afternoon* picture book.

Mark's parents were relaxing comfortably in side-by-side armchairs. It was easy to see where Mark had gotten his good looks. His father had passed on his physique and kindly eyes. His mother Ama was half Cheyenne and had passed on dark skin and hair to her son. The three of them together were stunning.

Mark sat on an oak-trimmed leather couch and Emily was curled up against him with a woven throw of geometric tans and dark reds across her legs. She looked as sleepy as their daughter while the others talked about the ranch, and fires that MHA had flown to this season. Tessa had her father's gray eyes and her mother's fine features and blond beauty. When Tessa was grown, the three of them would make an equally stunning trio.

It was so unusual to see Emily relaxed, that it made Chelsea content to remain as long as she could in the room. Emily, the

successful senior helicopter pilot of Mount Hood Aviation, the woman always in absolute control of any situation, lying against her husband like…well, like a woman in love. It was surprising and wonderful. Yet another thing that Chelsea put into her Someday List. Lie before a warm fire with her arms wrapped around a man she loved.

No. Scratch that. With *the* man she loved. She still had plenty of time to find him; she hoped. Mr. Wonderfuls weren't exactly hanging about for the picking, but it was a nice image.

It wasn't hard to picture what the man would look like in her fire-warmed daydream. He'd have casually long rough-cut hair, worn-leather brown just like—

There was a soft jolt in her lap. She looked down to see that Tessa had landed face first and fast asleep with her nose on Carl's finished snowman.

Chelsea slipped from the room with her and decided that it was time to put both Tessa and herself to bed before she became any more ridiculous.

Still, it was a nice image as she curled up in a guest room with Tessa on a low trundle bed beside her.

Doug Daniels was a *very* nice image.

6

Can't sleep all day, c'mon."

Doug went for brash to cover his initial reaction to seeing a sleep-tousled Chelsea hunched at the breakfast table. He'd come in to refill his coffee and check up on her as Emily had asked. It looked as if he'd surprised the sleeping lion in her den.

Wrong image. Chelsea didn't strike him as dangerous, just enthusiastic. Like an Irish Setter. The dark red hair color wasn't a bad match. Except at the moment she looked like she'd been run over by warm bed and a soft pillow, and would still be a while recovering. Or like he'd want to sweep her right back into—

Cut it out, Daniels. But he'd lost a lot of sleep over her last night and her current state wasn't helping matters.

Chelsea was clutching a mug of hot chocolate like a lifeline. She wore a gold-colored turtleneck that proved the sweater hadn't lied last night. It revealed strength aplenty to carry a hiking pack and curves to... He sighed at his libido's nudge-nudge, wink-wink.

"Where's Tessa?" she looked up at him through a screen of unkempt hair that she didn't bother to brush aside. The ten-foot

distance from the coffee pot to where he could brush it aside himself was a good thing.

"They all went into town; took her with."

"I should wait for them."

"They won't be back until dinnertime."

She squinted up at him again. "Where the heck is town from here?"

"Highfalls is only thirty miles out, but there's not much there unless you fancy a good steak. They're headed into Great Falls which is eighty each way."

"You sure?"

"There's a note from Emily by your elbow."

She twisted her head to read it without relaxing the death grip on her mug. The long line of her neck was…something he shouldn't be thinking about. Mark and Emily might not be his bosses, but this was their guest. And thinking hot thoughts about Tessa's nanny was wrong in so many ways, not the least of it being that they'd be gone soon. Christmas was just the day after tomorrow; they'd be gone the next day.

"You eaten yet?"

She nodded.

It took him a moment to spot the pan and dish, already washed and perched in the drying rack. Neat and respectful too.

"Good. Dress warmly. I'll meet you at the hangar in five minutes. I need to go up."

"Or I could just kill you and go back to my cozy bed."

"You'd do that to *younger brother?*" he asked in horror.

"Absolutely," but he could hear the grin in her voice even if he couldn't see it clearly through her shield of hair.

"You'll miss a beautiful helicopter ride."

"You know how to fly one?" She was quick enough to take in that he must be the pilot and turn it around into a tease.

He didn't even condescend to answer as he headed for the back door. "Four and a half minutes."

7

Chelsea made it in four and had spent three of that whipping up some instant hot chocolate in a pair of steel travel mugs.

"For me? Thanks."

When he reached for one, she pulled it away. "Mine. Two-fisted drinker."

It earned her that good laugh of his and she handed one over.

Doug had a pretty little Bell JetRanger pulled out of the hangar and was going over it carefully. Chelsea was taken aback for a moment. Two months ago she knew helicopters were the ones with their propellers on top instead of pointing to the front; now she recognized a JetRanger on sight. Furthermore, she thought of it as small compared to the massive Firehawk helicopter that Emily flew for MHA. When had that happened to her?

The pilot-plus-four-passenger craft was clean, but well worn. It looked well-maintained but hard used.

"I've never flown in a helicopter."

Doug looked at her aghast. "You work for two of the best helicopter pilots the Army has ever produced and you haven't been up in one?"

"I—" Chelsea hadn't known that about them. But rather than look foolish for the lack of knowledge, she just shrugged. "My job is to take care of Tessa. Mark is the Incident Commander Air"—she hadn't even known he could *fly* a helicopter—"so I fly with him and Tessa in the ICA plane."

"A helicopter virgin. Well, you're in for a thrill, honey."

"Watch it, *bhaai!*"

Again the merry laugh as he escorted her into the left-hand seat and made sure she was buckled in.

The ride was a real joy. The cabin heater kept the chill air at bay as they roared aloft. Headsets with boom mics made it easy to hear him as he pointed out the features of the ranch.

He let her look her fill, but she didn't know if she'd ever get enough. The green prairie stretched smoothly to the hills. The mountains broke from the grassland as if someone had drawn a line on the ground and said, "start them here." It was an abrupt and visceral shock. Only as they flew closer did the illusion start to break; secluded valleys intruded deep into the hills with small rivers sliding between sheer headlands.

"I love this land," Doug whispered softly after she'd finally managed to voice her awe at the rugged beauty. "It can be a hard land, but I never tire of looking at it."

"I wouldn't either," she said with a sincerity as if she was making a promise.

"Now who's being forward?"

She hadn't meant to be. Then she realized that she hadn't been. It was just Doug Daniel's mind twisting in…she sighed… much the way hers had been.

But the ranch was one of those places that simply felt right. Chelsea would start helicopter lessons tomorrow if it meant she could fly here.

Doug flew with such an easy confidence.

"You've been flying for a long time," she finally turned her attention to the fine scenery inside the cabin.

"Navy. Did three tours, six years. That was enough for me and then some. A SEAL buddy hooked me up with Mac."

"A SEAL buddy? Like the diver guys?"

"Sure, Mac was one too," Doug shrugged easily. No wonder he flew with such ease and confidence. Except he didn't look confident; he looked worried.

"What's wrong?" She checked the narrow dashboard that rose on a pedestal between their feet. She recognized about half of the instruments that were like the ones in Mark's plane, but nothing looked wrong on them and nothing was flashing red.

"Lucy didn't come back to the barn last night. And she had a late season foal, so I'm a little worried about them."

Chelsea looked out the windshield but couldn't imagine how to spot a horse in such a vast area. Now at least she understood that Doug hadn't been sweeping back and forth over the ranch and the prairie simply to show it to her; he'd been quartering and searching the ground. She'd done search and rescue for lost hikers, but that was tromping through woods and over rough terrain.

"How do you find a horse in thousands of acres?"

"Well," he pointed down at a lush, pocket-sized meadow around a tiny lake. "I was hoping she'd be here. It's a favorite of the horses. Hold the collective a minute."

"The what?"

"The control on the left side of your seat. Just hold it steady, don't worry, you can't crash us."

She tentatively wrapped her hand around the control, until she had a firm grasp. "Okay," she barely dared whisper it.

Doug took his left hand off his matching control and reached back to scrabble around behind the seat.

Daring greatly, she pulled up on it ever so slightly, and could feel the helicopter rise. She eased back down until the altimeter said she was back at the starting level.

"Here," he dropped something heavy in her lap. "Put that on, would you?"

She opened the case and looked down at the contraption, for that was the only word for it. There were straps to hold it to your head. It looked like a pair of goggles from one side, and like a half-unicorn, half-bug-eyed monster monocular protruding from the other.

"What is it?"

"Night vision. Lucy and the foal will be significantly warmer than the background. She'll show up clearly. Mark gets us the best toys."

Chelsea straightened it out and leaned over to put it on Doug's head.

"No," he stopped her. "You wear it."

8

Doug was amused by her exclamation when she got it turned on. Chelsea took such pleasure from everything about her. The countryside, the helicopter—rather than showing fear she'd proved she had a good and light touch—and now the night vision was tickling her fancy. Last night he'd left early. Partly because it was a time for the family to be together, but also because the vision of Chelsea with Tessa in her lap had been so powerful. She'd made it too easy to imagine a red-headed girl sitting right there, curled up by his fireplace.

For the next two hours, he flew and she scanned. He filled the time with learning about her background. Deeply independent—with parents who had little interest in an intelligent child filled with dreams—she'd forged out on her own. Six years to get her degree because she'd spent two years traveling and hiking; first walking the Continental Divide Trail from New Mexico to Glacier Park, and then all over the Himalayas.

It both amazed and saddened him. She was incredible, had a much clearer view of the world and herself than most people.

But he'd found where he wanted to be and she had adventure deep in her blood. She'd never be satisfied with…stupid fantasies of a demented ranch manager.

"There," her shriek almost blew his eardrums. Close beside the farthest fishing cabin, Lucy and her foal were huddled up against the side of the building. Lucy was lying down. Not a good sign.

He landed as close as he dared and rushed out to the mare. He'd brought a handgun, but not wanting to jar Chelsea's sensibilities, he'd left it stowed on the helo.

"We won't have to shoot her, will we?" Chelsea was right beside him.

Okay, so much for that worry. "Let's hope not."

Lucy was down, but had raised her head to watch his approach. Her whinny of greeting was encouraging.

He talked to her as he checked her out. No complaints as he tested for broken limbs. Same for the abdomen. Then she coughed in his face, a dry, hacking cough. He felt under her jaw and found swollen lymph nodes.

"Oh, crap!"

"What?"

"We vaccinated her against this."

"What?" Chelsea sounded deeply worried.

He sighed, "She has the flu. I can't do much for her here. She needs a warm barn and some rest. I'll have to ride back out, bring some high energy food and probably start her on a round antibiotics against secondary infection. With a little luck, she'll come back if I guide her. It will be a long slow ride."

9

They flew back, and when Doug saddled up a horse, she'd insisted he saddle two. She'd never ridden a horse, only a very recalcitrant mule when she'd sprained an ankle coming off climbing Imja Tse. She could have hobbled out faster than that Nepalese mule had carried her.

At Doug's guidance, she'd packed a pair of saddlebags with a change of clothes and several days of food. He packed clothes, camping gear in case they were caught out, horse meds, and a twenty pound sack of oats.

He led off at a light trot and she let him. Her horse, a big dapple gray male called Snowflake, looked at her strangely several times as she struggled to imitate Doug's easy saddle position. Every now and then he'd glance back to make sure she was still with him, and she always managed a plucky wave or nod as the saddle's hard leather slowly beat her to death.

They were about an hour out when he happened to look back during one of her barely-still-on-the-horse moments. Doug twisted his mount in a tight circle like it was the easiest thing

in the world. He twisted again until they were side by side. He leaned over to grab Snowflake's reins and everything came to a blessed halt.

"Haven't you ridden before?"

She could only shake her head, because if she opened her mouth she might start crying from all the places she'd rubbed raw.

"You're either incredibly brave or ridiculously stupid!"

"Mostly," she managed through gritted teeth. "Except you got the adjectives backwards." Being at a blessed standstill gave her some tiny sliver of ease. "According to my parents, I'm ridiculously brave and incredibly stupid."

Doug regarded her for a long moment, then glanced in both the direction they'd come and the one they were headed, considering the options. If he tried to send her back, she'd... she didn't know what. But she hadn't gone through this much pain for nothing.

"Okay," he shook his head. "I've seen that look on plenty a stubborn horse and don't want an argument. Stand up in your stirrups, if you still can."

She managed it without crying out.

He unrolled an extra blanket he'd had tied to the back of his saddle. He folded it in quarters, tossed it over her saddle, and then pressed her lightly on the shoulder until she eased back down carefully. It wasn't too painful, and far better than it had been.

"I lead probably a hundred trail rides a summer, Chelsea. You know how many beginner riders could have pulled off what you just did?"

She shook her head.

He held up his fingers and thumb, tips together to show a zero.

"I deserve a prize then."

Chelsea only had a moment to see his grin before he leaned in and kissed her. This wasn't some quick peck through the shield of her parka.

Doug leaned into the kiss and, fool that she was, she welcomed it without even a little protest. He provided plenty skill and

heat, but that wasn't what she was really noticing. What riveted her attention was how absolutely her body was galvanized by the simple act. Actually, ungalvanized. She melted against him despite the two horses that separated them. Leaning as far as she dared, she hung tightly onto the saddle's pommel with one hand and his jacket with the other and pulled them together. The kiss ran right down to her toes and made them curl in her riding boots.

When he finally eased back, Doug Daniels looked awfully pleased with himself. Of course she was feeling much the same way.

"I'm not sure," Chelsea was amazed she could even speak, "which of us you were just rewarding."

"At least you won't be *bhaai-ing* me any more," his laugh was even more self-satisfied than his expression. "Now, let's teach you how to ride. First, take your reins like this."

She did her best to follow his instructions and pay attention, but he'd made a warm buzz between her ears despite the cool day only now breaking above freezing.

Doug Daniels was many things: handsome, male, and a heavenly kisser being only three of them. But *younger brother* he definitely wasn't.

10

It was four hours to the last turn up to the fishing cabin, less than an hour later than he'd planned. Chelsea was the most apt riding student he'd ever taught, and while Henderson Ranch might be a working one, they made the majority of their income from all of the city folk guests who wanted a week or two of "country." Chelsea took to it as if she'd been born in the saddle... though she'd probably be too stiff to walk right for days. It had been cruel to keep going, but he couldn't afford the time to escort her back even if she'd have let him. He'd bet the chances of that were close to zero, yet another thing to appreciate about the beautiful woman. Tenacious as hell.

As it was, they'd be staying in the fishing cabin tonight. The sunset was only a few hours off and Lucy wouldn't be able to move quickly. It would be a far slower ride back tomorrow. On top of that, keeping Chelsea in the saddle through the night's journey back would be a cruelty, even if Lucy was up to it.

The final lap to the cabin at their quick walk should take about half an hour. Then Doug glanced back over his shoulder—

more bad news. A squall was inbound. Blocked by the height of Wind Mountain, and the twisting trail up to the cabin, he hadn't seen it coming. He stopped them long enough to haul on ponchos, but he knew it wouldn't be enough. They were about to get drenched.

They'd galloped briefly on the flat trail, but they were now climbing up a harder route. The way wasn't dangerously narrow, but it would be far more challenging. Another eye at the rain front, now a gray curtain sliding down the mountain face, had him changing plans.

"Ease up out of the saddle a little bit," he told Chelsea. "Lean forward. Loosen the rein. Good!"

And he smacked Snowflake hard on the butt.

He nudged his own mount forward and in moments they were galloping together up the valley. The way narrowed and steepened until they could no longer ride side by side. Doug didn't dare lead from where he couldn't see her.

"Ride on!" he shouted as the first crash of lightning struck the mountain top and thunder rumbled down upon them, amplified by the echoes off the high rock cliffs.

Bless Chelsea, she leaned into it and flew up the trail. He watched closely, but she stayed solid, didn't even a grab the pommel. Her legs must be screaming fire, but she rode, if not like an experienced horsewoman, then plenty close.

The icy rain broke over them, but the trail was solid and drained well, so he left them at the run.

In five minutes they were drenched, but the cabin was in sight. He shouted ahead and they eased down through canter to trot and arrived at the cabin at a walk.

"Down you go," he slid off and helped her down from her horse. "Take their reins and walk them back and forth. It will do all three of you good. Slow is fine, just keep moving." He stripped the saddle bags and tossed them into the cabin. He heaved the saddles inside moments later, then waved her, holding their mounts' reins, down the valley.

Even aching and saddle sore the woman had a walk that stirred his blood. *Ridiculous!* That's what he was being.

He grabbed his medicine bag and the oats and circled around to Lucy who was thankfully back on her feet, but hanging her head miserably in the rain. Her foal was cowering against her. She'd been ten feet from the overhang and the big box stall, but had been too dazed—yet another symptom—to walk under cover.

He guided them in and checked her. He couldn't do anything for the flu, which was viral, but he gave her antibiotics against secondary infection and a booster shot of vitamins. She perked up a bit for her oats and water. He got blankets over her and the foal about the time Chelsea staggered back up to the stall with their mounts plodding along behind her.

"Is this enough?"

He ran a hand over them. No longer breathing hard, not hot. "You did good Chelsea. Go inside. I'll be in as soon as I get these two settled with the mare."

When he entered the cabin a few minutes later, Chelsea was on the floor in a fetal position.

Shit! He was an idiot.

11

Chelsea had been this cold before, she was sure of it. Like when she'd camped above snowline at the base of Chulu West and the zipper on her sleeping bag had broken. But in her memory it didn't feel colder. And when her knees had knocked together high in the Himalayas, she'd laughed at the novelty. Now she fought not to cry as the insides of her legs, rubbed raw by the saddle, sent shivers of pain right along with the cold shakes.

She opened her eyes when Doug entered the cabin and immediately began cursing. He looked furious! His dark hair matted flat and black with the rain, water cascading off his poncho. He hauled it off with a yank and dropped it on the rough wood with a wet splat.

Chelsea wondered if he was about to tear her to shreds because she'd collapsed, then realized she wasn't the one he was swearing at. He dropped to his knees beside her and began calling her name loudly.

"I'm c-c-c-cold, not d-d-deaf," she managed through rattling teeth.

"I'll start a fire," he jumped up toward the iron woodstove in the corner.

She tried avoiding the hard "c" of close, but found the "sh" sound little easier. "Sh-sh-shut the door first, you big lummox. R-raised in a b-b-barn."

Doug closed the door and then redeemed himself with his efficiency in building the fire.

"Heat? How long?" she managed.

He looked uncertainly from her to the stove. Not soon enough.

She tried to remove her poncho, but her hands weren't under her control anymore. This was bad.

"C-c-clothes. Off. B-b-bed," she instructed.

He stripped off the outer layers and hesitated until she stuttered out a series of curses at him. She cried out when he peeled her jeans.

Then he began cursing all over again.

She looked down. Her legs' normally pale skin had gone white with the cold, except for the insides from boot top to panties were livid red with abrasions. No wonder they hurt.

The goofball stopped at her soaking wet turtleneck as if embarrassed.

"C-c-come on. You know you want to s-s-see me naked."

He grunted and had the decency to try and look away as he finished the job and then scooped her up like a feather hard against his soaking wet jacket.

"Eww!" Yet even the tiny bit of heat that escaped through the denim felt so good.

The cabin was simple. Three bunk beds, several couches and plush chairs that had seen better days probably back at the main house, and a small corner kitchen with an impressive collection of cast iron pans appropriate for frying fish. Doug dropped her in one of the lower bunks and began piling blankets over her. She couldn't even clutch the blankets to pull them tighter.

"S-s-strip!" Chelsea ordered.

"But…"

"C-c-come on. You know I want to s-s-see you naked," she did her best to stammer it out the same way she had the first time. "I need heat."

He began peeling down and Chelsea watched as much as the shivers would allow.

"Wow! C-c-cowboys *are* built pretty."

He smiled at her for the first time since finding her on the floor. "This is a horse ranch. Not a cattle ranch."

"So get your fine butt in here, horseboy. Before I f-f-freeze to death."

He hesitated at shedding his underwear, someone please explain men to her, then turned away as he finally stripped off that last piece. His butt really was fine; topped by a narrow waist and broad shoulders with muscle that rippled across them with each movement.

Doug slid in beside her and, after a moment's hesitation, pulled her against him. His skin was so warm compared to hers that it burned, but she leaned into it as hard as she could.

"Christ! You're freezing!" He began chaffing those big hands of his up and down her back.

"D-d-duh!" Chelsea managed to get the covers completely over her head and concentrated on soaking up Doug Daniels' warmth.

12

D*oug held her until* the shivers stopped. With his arms still around her, he could feel her breathing slow. Once she was deeply asleep in exhaustion, he slipped out of bed and dug out fresh clothes from the saddlebags, hanging the others to dry. He stoked the fire, made hot chocolate and wished for coffee, but the latter would make him even more awake than he already was.

A quick radio call back to Logan told him that the Hendersons weren't back from Great Falls yet. Logan wasn't a pilot so he couldn't bring the helo to fetch Chelsea. With the shakes gone, she probably just needed sleep…and time to heal. Gods but she was tough.

The windows were dark with the fading light of sunset happening somewhere beyond the heavy overcast. Lightning still shimmered through the heavy rain, though far enough off that the thunder was a rumble rather than a crack. The weather was still too nasty for a flight even if Mark was back. He had Logan leave a message on the kitchen table so that they wouldn't worry when they returned and found no Chelsea.

"Doug," Chelsea's voice was a whisper barely louder than the crackling flame from the glass-fronted woodstove. "Come back to bed." The firelight caught the blue of her eyes and the tip of her nose from where they peeked out of the blankets.

"You trying to kill me, girl?" Yes, he'd wanted to see Chelsea naked, from the first moment he'd spotted her climbing down out of that plane in those deliciously tight jeans. Even shuddering with the leading edge of hypothermia, she was beyond spectacular.

"Not girl. It's woman. And I think you trying to kill me once already today should be enough for both of us."

"I didn't—" But he had. He'd taken her skills for granted when she climbed up on the horse. And led her on a grueling ride through a storm. Sending her out on a cool-down walk in the freezing rain was about as dumb as it got.

Unlike so many of the guests who came to the ranch, Chelsea radiated skill. She'd triggered none of his high-season alarms that told him who to watch out for. Though she was certainly triggering other reactions.

"I don't think that's a good idea."

She rolled her eyes at him. "Get your warm butt back in here before I have to climb out and kick it. I ache right down to my joints."

Which told him just how dangerously cold she'd gotten.

Once again he stripped down, far more conscious of the woman who now wouldn't turn away than the earlier one whose eyes had been partly rolled back into her head.

She went to throw a leg over his, but jerked back and hissed at the pain.

"God I'm so sorry. Let me get some horse liniment," he climbed out of the bunk.

"Hello! Not a horse."

He grabbed a bottle from the kitchen shelf and returned to stand over the bed. How was he supposed to…

"Here," he held out the bottle. "Trust me. It works great."

13

Chelsea felt as if she was being a total wanton. She was in a cozy little cabin with no distractions of electricity. A very handsome man, momentarily unaware of his own nakedness, stood close beside her lit by the soft firelight that filtered through the woodstove's glass-paned door. And he was holding out the horse liniment the way you hold out a mouse for a dangerous viper to snack on.

The normal version of herself would have taken the liniment and tried to slather it on under the covers.

Instead, she watched Doug's face as she slipped a leg out from under the covers and twisted to turn it, inside-thigh up. His eyes didn't narrow suspiciously, instead they widened in alarm. She'd watched him handling the horses with a gentle but firm hand. A half ton of horse flesh didn't bother him at all, but the inside of a woman's leg had him totally flustered. Damn but he was cute.

"Come along, horseboy," she coaxed him in the same tone he'd cajoled the colt to follow its mother into the stall.

His gaze snapped from her leg to her face, and then his nice deep laugh rolled out. "Okay, you got me. I'm dying to slather some liniment on those fine legs of yours." And he knelt on the wood floor beside her and smoothed some on.

It was cold and sent a shiver up her leg. But the warm steadiness of his hand stroking in the thick liquid calmed the convulsive response before it could turn back into the shakes. She could feel his hard calluses and easy strength, but was surprised at the gentleness of his rough hands. Within moments a numbing warmth spread up her leg in a wave of relief.

"I'll smell like a horse," she complained to cover a moan of delight. The camphor was sharp in the cabin's warm air, but her attention was nowhere near her nose.

"A sweet smell to a rancher."

"How about to a horseboy?"

"Lady," he didn't even bat an eye. "You smell incredible to this horseboy, with or without the liniment."

There was no sign of any embarrassment by the time he'd ministered to both her legs and tucked them once more under the covers. He'd somehow transferred all of it to her. As he slid back under the layers of blankets, Chelsea was intensely aware of the narrow bunk and the warmth of his body pressed against hers. She was more of a long t-shirt gal, but it would be stupid to ask for one with a man she'd lain naked against for most of the last few hours.

Unable to find words, she simply nestled inside the curve of his arm. Then, against the fiery tension building so high that it roared in her ears, Doug began talking. He told her about the birth of the foal, who slept even now in the nearby stall with Lucy. He talked about the ranch and the spring wildflowers that colored the prairie like a paintbrush.

She fell asleep with the sound of his love for his life rumbling from his chest directly into her ear. It was the sweetest, safest sound she'd ever heard.

14

*H*e'd offered to call the helo a half dozen times this morning, but Chelsea had turned him down cold, despite hobbling about like a geriatric case. Another round of liniment helped some, but he knew she'd be stiff for days.

Doug finally gave in. Partly because he knew Lucy wouldn't be up for more than a casual amble and partly because he wanted every single minute with Chelsea that he could get. He'd held her throughout the night, marveling at the rightness of it.

It had been like that when he'd arrived at the ranch fresh out of the service. After three full tours, most of them spent on ships in the Persian Gulf, he'd been sick to his heart of the unending heat, the limitless steel, and the noise—for a Navy ship was never silent. He'd been on the ranch for three years now and could still feel the Persian dust in his pores. But the ranch had fit him since the first moment he'd stepped on the soil.

He'd ridden plenty as a kid at his parents' place in Wyoming. When he didn't re-up, SEAL Commander Luke Altman had sent him up to see his own former commander outside Highfalls,

Montana. Mac had shown him around Henderson Ranch and Doug had decided on the spot that he never wanted to leave. Mac and Ama had been looking for a foreman. Together, they'd transformed the aging ranch into a showplace tourist destination.

He'd worried a lot about "the son" coming home, until he'd met Mark and Emily. Mark had taken one look at the transformation and thumped him hard on the shoulder before walking away without a word.

It was Emily who'd translated for him. "He was so worried for his parents. You've really touched him." Then she'd kissed him on either cheek. "You done good, Doug. Keep it up." Then she'd gone after her husband. That's when he'd set his sights on the kind of woman he wanted. One just like Emily Beale.

And he couldn't have found one more different than Chelsea Bridges if he'd tried. Oh, a lot of the things that were right with Emily were just as right on Chelsea, especially her absolute fearlessness—the image of her galloping through a thunderstorm on her first ride still fired the imagination.

But where Emily was quiet, thoughtful, and soft spoken, Chelsea spoke her mind and laughed with a bright joy—even when on the verge of succumbing to hypothermia.

He imagined it would take years to fall for the right woman once he met her, because his ideal woman didn't fall that quickly. At least so he'd thought until he'd rubbed noses with Chelsea inside a parka hood and received a kiss for it. Now he was crazy about a sassy redhead who'd slept in his arms like she'd always been there.

Slept. And that was all she'd done. Hard to blame her, as her body had been through a lot of extremes yesterday. But the only extremes he'd been through had been treating Chelsea as if she was his injured sister. Everything had been perfectly chaste last night, if you didn't include his thoughts.

"Storm has passed," he did his best to distract himself. "Temperature is falling and there's another front moving in. Let's get ahead of it."

"Sure," she gamely picked up her saddle, that probably weighed half as much as she did, and headed for the door.

So much for a morning tumble, or even a kiss.

He'd escaped her bed early—because it was either that or he was going to do something wholly inappropriate—and bundled up to go tend the horses. Lucy had perked up overnight enough to greet him. She was still snotty with the flu, but it was clear so no secondary infection yet. Her breathing also sounded clear enough for the walk back to the ranch. The foal was more cheerful than the night before, which he'd take as a good sign regarding his mother's condition. By the time he was back inside, Chelsea was dressed in warm clothes and had made the bed. Oatmeal and coffee were simmering on the woodstove.

She'd looked as natural here as no paying guest ever really did.

They'd had breakfast together; Chelsea going on about the upcoming ride...and he hadn't jumped her. What was up with that? There was decent and there was ridiculous, and he'd definitely crossed that line somewhere in the night.

Then she'd washed the dishes, grabbed her saddle, and gone.

He'd already taken his own saddle out. So, he gathered up their saddle bags, double-checked that the woodstove was secure—the few remaining embers would burn themselves out—and gave the cabin one last look. All shipshape...damn it. Not a single tousled bed sheet. He hadn't brought any protection with him, but that didn't mean there weren't other options. But had they used them? Nope! Not a single, damned, inappropriately pleasant fondle had passed between them.

Closing the door, he stomped around to the horse stall and ran head on into a kiss.

This wasn't some little kiss through a parka or a taste of wonder when they were both up on horses. Chelsea wrapped herself around him and had him backed against the rail fence. With her arms tight around his neck, she was rapidly killing off fantasy after fantasy. Who knew it was possible to pack so much joy into such a simple act? Apparently Chelsea did.

When Snowflake came over to snort in his hair across the fence, Chelsea flapped a hand at the horse's nose.

"Busy here," she mumbled at the big gray.

Damn straight! was all Doug could think. All that soft and gentle warmth of last night had been replaced by the lively redhead who'd teased with him since the moment of her arrival. She didn't play coy or tease now; she delivered a kiss with her entire body. It left him shuddering with need when she abruptly released him and, as if his world hadn't just been spun around and dropped on its head, strode into the stall with one of the saddlebags that he'd dropped when she'd jumped him.

Unable to trust his voice, he focused on saddling them up. No need to rope Lucy or the foal. Lucy, he knew would follow them, and the foal would follow his mom.

Placing his hands around Chelsea's waist to help her up into the saddle was almost his undoing. With her arms raised to the reins and pommel, her jacket slid up and her waist was slender and warm in the circle of his hands.

Her smile was mischievous as he climbed up on his own mount. "What?"

"I just wanted you to know, that kiss wasn't a thanks for how wonderfully you took care of me last night."

"Then what was it?"

She turned Snowflake and with a skilled nudge, sent her down the trail at an easy walk. "That," she called back over her shoulder, the only sound in the still morning other than the clopping of the horses' hooves. "That was just a preview. Like coming attractions at the movies."

Any ability to speak that Doug thought he'd regained was washed away. If that was a preview, he couldn't wait for the main feature. But the ranch was a long way off.

He looked back at Lucy and her foal who'd fallen in behind. "How fast can you walk?"

The horse declined to answer, instead settling into a slow shuffle.

15

So, the lost is found," Mark greeted her cheerfully as Chelsea entered the ranch house kitchen.

"Seems so." It had taken seven hours to walk Lucy back. A long cold ride, but under a broken sky rather than a freezing rain. It was now mid-afternoon and the sky was once again darkening beneath an overcast. At least she'd be cozy and safe for the next storm.

"Tessa's down for her nap, so you can just relax. Where's Doug?"

"He's out at the isolation barn. He wants to keep the three horses and foal away from the herd until he's sure that they're not contagious."

"Good man."

"The best." Chelsea knew she'd never met a better one.

Mark looked at her curiously, and then headed for the door. "I'll just go and check on him."

"Do you know *anything* about horses?" She didn't know where the tease had come from. Women didn't tease men like

Mark Henderson. But Doug had told her how Mark loved to fish, and almost always used an ATV rather than a horse to get there, so she couldn't resist.

He just winked at her and was gone.

Chelsea took a quiet minute to heat some leftover beef vegetable soup before sitting with it at the big kitchen table. It could seat a dozen without crowding. The kitchen was on the border between a generous farm kitchen and a small commercial one. It was cozy but also designed to feed a hungry hoard. She could imagine dinner parties here filled with laughter and good food.

"What would it be like to live here?" she asked the quiet kitchen. "How happy would it be?"

"Quite happy."

Chelsea startled and almost lost her soupspoon to the floor. For a startled second she thought the kitchen had answered her.

Then she spotted Emily Beale sitting quietly in a deep chair by the kitchen fireplace, a book in her lap. She rose smoothly and came to sit just around the corner of the table from Chelsea.

"The first time I came here, I was in absolute terror."

"You, in terror. Like I'm going to believe that."

Emily's smile was always a surprise and it was this time as well. "Seriously. I was engaged to my co-commander of an elite US Army helicopter team—seriously bad from a regulation point of view—and about to meet his parents, one of whom had served twenty years as a Navy SEAL. I'd never gone fishing, never seen a horse up close, and never been to Montana."

Chelsea toyed with her soup. "This place is so amazing though; that must have helped."

"It did. Though not as much as realizing that Mark knew as little about horses as I did." Now Emily's smile turned rather wicked. "Mac and Ama bought the ranch after Mark had gone to West Point."

"So?" Chelsea tried to picture Mark not perfect at something and wasn't coming up with a good image.

"Let's just say that he ended up head over heels in the river and I didn't."

Chelsea held up a hand in salute, but was shocked when Emily actually high-fived it. "Women rule," Chelsea added weakly.

"We do," Emily agreed and offered her a smile of companionship that felt as crazy as everything else that had happened in the last two days.

"I've been very happy here," Emily continued though more as if she was speaking to herself. "It's a good place, as good as any I've ever been."

"I'll miss the ranch when we go."

Emily nodded, but was studying Chelsea carefully.

"What?"

Emily shook her head.

"Nope." Chelsea grabbed onto her bravery. "You don't get to do that."

"Do what?" Emily pretended all innocence.

Chelsea aimed her soupspoon at Emily, "Have that clear a thought and then not share it."

Emily considered for a long moment and then nodded at how that might be a reasonable demand. "Just remember."

"What?"

"You asked."

Chelsea swallowed hard. Why didn't she think she was going to like what came next? She nodded for Emily to go ahead anyway.

"It isn't the ranch that you'll be missing."

Her soupspoon slipped from nerveless fingers and landed in her bowl with a splash.

"Thought so," Emily remarked drily.

"Couldn't you at least have made it a question?"

Emily shook her head. "Why would I, when it isn't one."

"But we haven't even—"

"Doesn't matter. When it's the right one, the particulars don't matter. Trust me, I know."

"The right what? But—" Chelsea managed weakly wondering why she was trying to argue. She'd never met a man like Doug Daniels, a man who simply shone with the love inside him. He had such a passion for the land and the horses.

During the long, cold ride back from the fishing cabin, she and Doug had warmed the time with stories. He'd told her about his experiences overseas, so different from her own tramp abroad. In all of her travels, she'd never found anyone so easy to be with.

And the way he'd knelt before her in the cabin, naked and beautiful and so worried about offending as he treated her abraded legs with stinky liniment.

The way he'd held her last night. There couldn't be another man anywhere who wouldn't have taken advantage of the situation. But not Doug with his soldier's honor.

"I—"

But Emily was no longer there to explain things to. In the big kitchen was only the warm crackling of the fire, Chelsea, and a bowl of soup.

16

D*oug was slumped on* his couch. The grumbling in his stomach complained about missing dinner up at the main house; too frustrated to whip up something in his own kitchen. He hadn't been able to go because of what else he'd find there. What he was wanting so badly.

The knock on his front door had him racing to answer it. "Is Lucy…o…kay?" The only knock he'd been expecting had been Logan's if Lucy had a relapse. His nervous system was not ready for the vivid redhead standing on his front porch.

"Hi!" Her smile was big and again mischievous.

He had the feeling that he was suddenly in deep trouble.

"Do I get invited in? If not, I'm taking Emily's special home-made pizza back with me. She said that it's one of your favorites."

That's when he focused on the large covered tray Chelsea was carrying. Emily was an amazing cook, had won the hearts of Mac, himself, and every one of the ranch hands with a beef stew on her first visit to the ranch. But it was her from-scratch pizza that blew Doug away.

"Uh—" He looked back up at Chelsea. "I'd like to invite you in, but I don't think that's the best idea. Because if I do—" If he did, he couldn't be accountable for keeping his hands off her a second time. Last night he'd liked the brave and competent woman, and lusted after the redheaded knockout. On the long ride back, he'd also come to admire her deeply. She'd made some hard choices on her path, who hadn't. But hers had always come straight from the heart.

"—If you do invite me in," Chelsea picked up for him as she eased him slowly backward with the leading edge of a tray of pizza, "we just might enjoy ourselves beyond all imagining."

"Something like that," he managed.

"Good. I'm counting on it." She kicked off her boots, and carried the tray through his living room and into the kitchen as if she'd always lived here. "You were raised in a barn. Close the door; it's cold out there."

Helpless to argue, he did as she suggested and followed her into the kitchen.

"I'm sorry," she set the tray on top of the cold stove.

"Sorry for what?"

"The pizza and the tiny ranch house tour are going to have to come later. I can't wait any longer." She shed her gloves and jacket and dropped them to the floor. Then she walked straight into his arms.

17

They had cold pizza while sitting among their clothes on the kitchen floor. Doug reheated some after they'd made prolonged use of the living room sofa; long enough to have to restock the fire. They finished the last of the meal on their way upstairs when she went hunting for the bedroom; a search that was gloriously rewarded.

"Did we miss anywhere?" Chelsea lay sprawled over him, sore in so many wonderful ways.

She'd never done anything like this. Never had so much fun having sex either. Doug's blend of powerful yet gentle, of roughly needy and deeply giving had enthralled and sated her like no one before.

"Uh, big bathroom, second bedroom, home office."

"Oh." They'd probably kill each other if they tried for all of them tonight.

"Back porch lit by June moonlight," he mumbled on. "There's a set of waterfalls with a hot spring about a three-hour hike above the fishing cabin that shouldn't scare off a woman who

had hiked in the Himalayas. The open prairie on a warm May afternoon where you'd outshine the sun. I'll show you—"

She put her fingers over his mouth to stop him and he kissed the fingertips.

"I like your imagination," she propped herself up on his chest and looked down into his dark eyes. "So the sex is good."

"Incredible," he agreed.

"You love what you do?"

"I do," he agreed just as equably.

"And you've spent two days and two nights fantasizing about having me beside you forever."

"Yep."

She waited for it. Perhaps it was unfair. Giving a man his favorite food then making love to him multiple times; his defenses were pretty much gone.

But there was no shock of recognition at what she'd just said. No startled disclaimer that he wasn't dumb enough to extrapolate two days into a lifetime.

"Whoa there!" It was supposed to have been a tease.

"As the lady once said," he grinned up at her. "Hello! Not a horse."

"Hold on."

"The way I figure it," she could feel his chest rippling against hers as he spoke, "it's actually been two days and three nights. I think we're closer to sunrise than sunset. So, we've already made it twenty-five percent longer than what you said."

"Douglas," she warned him.

"Just Doug. Nobody calls me Douglas, not even Mom."

"Douglas!"

"Yes, Chelsea?"

"Does it make any sense?"

"Nope. Not a bit," and his voice remained merry.

"Aren't you even a little surprised?"

"Nope."

"Why not?" Chelsea's own thoughts were in such turmoil, they might as well be a wheeling herd of horses.

"Got over it in the barn while taking care of Lucy."

"A *horse* told you that we'd be spending our lives together? Even from horseboy, I'm not buying that one." *Spending our lives together* and still no flinch on his part. She checked in with herself. Even stranger, there wasn't a flinch on her part either.

"No, from Mark."

"Mark?" was all she managed.

"Yep! I was out making sure Lucy and the other horses were all settled in, when he came out to the barn."

"What did he say?" Chelsea was pretty sure she didn't want to know. She went to roll off Doug's chest, but he trapped her in place with a hand resting lightly on her hip. Just enough to tell her she was retreating, not enough that she couldn't get away. *Fine!* She could take it if he could, and rolled back into place.

"He said that you were one of the nicest young women he'd ever met and I'd never find any better. That part I agreed with readily enough," Doug nodded emphatically as if marking such an outrageous statement as simple truth. "And if I was too stupid to see that you were already in love with me, he'd be glad to pound some sense into me."

She let his "love" statement go by for the moment.

"Do you think they set us up?" She wasn't sure if she'd be angry or not, but wanted to know.

"My question too. Mark said no. Emily's not much sneakier than he is, so I'm guessing the answer there is also no. I suspect that we did this to ourselves."

"We…what?" But it was lame and she knew it. Emily had said the same thing, or why else was Chelsea here in bed with Doug?

This time when she pushed away, he let her go.

Chelsea wrapped a blanket around her shoulders and moved to look out the window. The yard rolled away into the darkness. Faint lights marked the barns, a lone porch light up at the main ranch house.

Could she be happy here? Working horses, sharing this gorgeous land with visitors? In a heartbeat.

With this man?

Doug slipped up behind her and wrapped his arms across her shoulders.

How was she supposed to know something like that so quickly?

Even if she already did?

Emily had said she recognized that he was the right one for her. As if she knew what love looked like. Well, if any woman did, it would be Emily Beale.

Chelsea leaned back against Doug—and the rightness was there. It ran so deep that she couldn't imagine being anywhere else.

"So I was thinking," he whispered in her ear.

She hummed with pleasure, couldn't help herself.

"How about we just try each other on for size? You and me."

"And the horses."

She could more feel his laugh than hear it.

"And the horses. We'll agree to make no decisions at all until the snow melts."

"But there isn't any snow," she waved a hand toward the window.

He didn't speak, instead he pointed. In the faint lights, she could see the first flakes spinning down out of the sky.

"A white Christmas," she managed on a tight breath.

He wrapped his arms around her a little more tightly.

Doug was right, they needed time to decide if what was between them was real or not.

But she knew. Her wandering days were done.

A white Christmas together.

Chelsea turned in Doug's arms and kissed him. She knew right down to her heart that this was only the first of so many to come.

Fire Light Cabin Bright

Being on an Interagency Hotshot Crew is brutally tough. These teams, typically twenty in a pack, ride in a small truck called "a box." It is tightly cramped with gear and firefighters, and it is also the very core of their operation. The box delivers these teams as close as they can to a wildfire, then they load up with heavy gear and tools, and walk to the fire—often miles over horrendous terrain. Once there, the battle against the fire is a hand-waged war that allows few breaks and even less sleep.

For five to seven months every summer, they battle fire with only very rare days off. Their pay is unimpressive, but their drive is powerful and heartfelt. These are people who love the wilderness and the firefight and give it all they have—it is also far too dangerous to give it less.

There are some women now qualifying in these positions and proving that they have what it takes.

These are strong, powerful women with a bloody-minded level of perseverance when it comes to fighting fire. For people who like the obscure connections that I sometimes build into my stories, Tori is inspired to fight fire by a brief fling with the smokejumper Akbar the Great in Firehawks book #1 Pure Heat.

And when I was trying to think of who Tori would be running up against in this romance, I wanted someone as opposite from her as I could find. So I chose a writer…and wished him luck.

Thankfully, he had some ideas of what to do with her.

M.L. BUCHMAN
NPR and B&N Top 5 Romance Author of the Year

"Will blow the readers' minds and leave them awestruck."
Romantic Times, Pure Heat
4-1/2 stars

Fire Light Cabin Bright

a Firehawks romance story

1

Just as some days are hotter than others, some fires are hotter than others. And the Checker Mill Fire was a scorcher.

Tori Ellison checked her watch but couldn't see it. Even shining her helmet headlamp on it didn't really help. Her eyes were lack-of-sleep sore and they stung from smoke and salty sweat. She couldn't taste anything but that salt and the char that it collected as it dribbled down her face; the peanut and dark chocolate flavor of her energy bar hadn't lasted more than a few minutes before being overwhelmed.

She'd volunteered to scout what lay over the next ridge while the rest of her Hotshot fire crew crashed out for an hour. She was supposed to go ten minutes out and ten back, but couldn't seem to focus on the watch to tell how long she'd been gone. Three minutes? Fifteen? She no longer knew.

It was zero-dark-thirty, like the military guys said, which was all that really mattered.

Where was here? She wasn't so sure of that either.

She was always doing dumb, impulsive things like this. In college one of her nicknames had been the Energizer Bunny because she'd never had the sense to stop until she dropped. The Bunny part had been shed after she'd punched a particularly obnoxious frat boy hard enough to shatter his nose.

When Tori hit the fire line, her new firefighter nickname was Ginger within three days.

It wasn't that her hair was red—she was a bob-cut blond. The crew chief, Candace Cantrell, had grown up with a Labrador named Ginger who also never knew when to stop.

A low-hanging Douglas fir branch slapped in her face because she was too weary to step around it. At least it was green and smelled of life and fresh pine. She felt bolstered by its presence; it was standing in the cool forest night, trusting her and her team to save it from the encroaching wildfire.

Tori trudged by it and promised to do her best—trudged because trotting was long past her abilities at the moment. They'd come off the Bell Creek Fire along Washington's Skagit River less than forty-eight hours ago and now had been on the Checker Mill for the last thirty-six straight. The Cascade Mountains were rough and she normally liked the challenge… when she was conscious.

She crested the low ridge in a thick stand of trees. It would take a lot of cutting to clear a fire line here if they had to. Too tired to even dodge the branches, she raised her arms in front of her face and ploughed through the heart of the stand.

The trees gave way the moment before her feet snarled in thick vines and she face-planted on the ground.

Her radio crackled, "Ginger, check in."

"Yo, Candace." The ground was soft. Well-tilled soil cool against her cheek. She didn't waste extra effort trying to stand up. It felt so good to lie down, for even a moment.

"Report."

"Hard-ass," Tori teased her. Since it was something the Hotshot team's leader was proud of, it was a safe call. "Trees

much thicker at the ridge. Eight to eighteen-inch diameter Doug fir. Can't see much else."

"Where are you now?"

"Lying on the ground," she looked around to try and be more specific, and was confronted by something large and green. Big enough to completely block her view. "In a zucchini patch."

"A residence?"

That would be bad news. Needing to defend a residence, or worse a neighborhood, could drastically change a fire attack plan.

"Ginger?"

"Hang on. Hang on. Sheesh!" Tori forced her arms beneath her and levered herself upward. They shook with the effort. She'd really tapped herself out this time; right out to the limits.

Once upright she twisted her head side to side to swing the headlamp around.

"I'm in a vegetable garden," she reported.

"You're in *my* vegetable garden, 'Ginger'," a deep male voice sounded from the dark.

She twisted the lamp around and found a mountain man standing about ten feet down a row of tomatoes. Except he wasn't hairy, messy, or clad in rotting animal skins. He wore gym shorts and a frown. She couldn't see his eyes because he had his arm raised to protect them from the glare of her lamp, but from the nose down was very fine. Not a six-pack ab guy, but no extra bits either.

"Who are you? And why were you eavesdropping on my private conversation?"

"My name's Colin James. And if you're in my garden on the radio, how private can your conversation be?"

Tori hit the transmit key, "I'm in Colin's secret garden. And he's just as much of a know-it-all as the one in the book."

"If Dickon shows up, he's mine," Candace replied. "I always had a crush on Dickon."

Tori heard a soft *Hey!* in the background, probably Candace's husband Luke, a top member of the IHC crew.

"What are you doing in my squash?" the man asked from behind his raised arm.

"Well, that's no way to address a lady, Colin."

2

Her voice was the only thing that distinguished her as one. Colin had been lying out in the hammock watching the stars—it was too beautiful and warm a night to stay in his cabin—when he'd heard a hard grunt and rustle from his vegetable garden.

It had sounded human rather than ursine—he didn't worry about bears here…much. When he'd looked, there was a light shining low under his plants. Not stopping for shoes or a flashlight, he raced out to scare away the poacher. He'd put a lot of time and care into his garden and no midnight skulker was going to rob him.

He'd been stopped in his tracks by what he found. Between the zucchini and the pumpkins lay a fully clad firefighter, and one that was making no effort to get up.

By the reflected light off the nearby leaves, Colin had seen a hardhat that might have once been yellow under all the soot. The firefighter wore similarly colored jacket and pants, heavy boots, and a small pack. One hand clutched a nasty-looking axe and the other a handheld radio.

And then the firefighter had spoken and turned out to be a she. Named Ginger.

"If you're lying in my vegetables, I'll address you any way I choose. And get that light out of my eyes."

"Oh, sorry," she turned the lamp toward the ground.

He had a brief glimpse of an oval face and a hint of blond hair before she flicked it off and they were plunged into darkness. He blinked hard, but his night vision was shot and wouldn't be back for several minutes.

He couldn't see, but he could hear that she hadn't moved.

"Are you planning to just lie there all night among my veggies?"

She giggled. "You have a very comfortable garden." A firefighter who giggled.

"What are you doing here anyway?"

"Uh," Ginger paused. "I'm here because…" she sounded as if she was trying to figure that out for herself, "…oh, yeah. I'm here because there's a forest fire in the next valley over. I'm the scout."

Suddenly a dozen things he hadn't paid any real attention to earlier in the day made sense. He'd kept smelling wood smoke, but no one in their right mind would have their fireplace going on such a hot day. Besides, he was pretty sure that he had no neighbors for a long way in any direction.

Also there had been clouds to the north, but he hadn't really paid attention. The novel was finally going well and he hadn't been outside all day. Yet another reason he'd retired to the hammock with the sunset. Now though, he remembered that the clouds had been an odd color for a lightning storm, too dark.

There'd also been the sound of helicopters, but they were often used in logging operations. Maybe not so much today.

"How close?" he swallowed hard.

"A mile or so. I seem to have lost track."

"What kind of a firefighter are you?"

"An exhausted one."

"Here," he reached out into the dark. "Let me help you up."

Somehow they found each other's hands. But when he braced a foot forward to pull her up, he stepped barefoot right on a planting stake. He tumbled forward onto her with an exclamation on his part and a curse on hers.

"That's your idea of being helpful?" she grumbled from where she lay beneath him in the dark.

"Ginger, this is Candace," the radio squawked loudly in his ear. "Are we looking at an individual or a community? I don't show anything on the map."

He tried to roll off her, but the big zucchini bush stopped him. When he shifted the other direction, he partly rolled onto her fireaxe.

"Hey Candace. I can confirm an individual. Clumsy, but cute."

"I'm not—" Well, maybe he was being a klutz. But he hadn't exactly been prepared for a female firefighter lying on the dirt in his garden.

"Ginger!" The woman on the radio was sounding irritated.

"Hang on." Then Ginger reached up to assist him in getting off her, and clipped him fairly solidly on the jaw with a leather-gloved fist.

He tumbled into the vines.

"Oh crap. I'm sorry." She giggled again even as she groped around in the darkness, grabbed his arm for support, which pulled him back atop her with surprising strength. "Well, isn't that interesting."

This time when he tried to pull himself free, she pulled him down and kissed him. Hard.

3

What was she doing? Tori was deep in the kiss before any part of her brain woke up enough to be rational. The mostly unclad Colin was lying full upon her and, after a brief hesitation, was proving he was an exceptional kisser.

After enjoying the situation for several more moments, she managed a "Whoa." Then she pushed against his shoulders to shift him up and far enough away for her head to stop having ideas about where to go next with this mostly naked man. He tasted deliciously of male and toothpaste—a welcome relief from her own salt sweat and char—but it was dumb as could be for her to randomly kiss a total stranger.

Colin didn't resist as she pushed him back. Kept going until he was kneeling between her legs.

"Um," she had nothing to add to that. And she was almost tired enough to drag him back down on her.

"Ginger!"

"Spoilsport," she told the radio without keying the transmit key.

"She's persistent," Colin observed from nearby in the darkness. Her eyes had recovered enough to make out his outline against the stars.

"You have no idea. She needs to know…" something.

"I live alone here. Solo cabin. Is the fire coming my way?"

"Candace," somehow Tori had held onto the radio during the kiss. "It's a solo cabin of a man who tastes like mountain spring water."

"You kissed him?"

"Either I did or he did. I'm a little fuzzy on the details."

"Uh-huh," Candace wasn't buying it. "I'll send Luke up with a couple saws. Make sure the site is prepped for best defense. The fire has slowed and will be good until dawn. We'll get air attack to lay down a perimeter as soon as the helos are back aloft with the sunrise. Take an hour break."

"Roger that."

Now the question was, what to do with an hour?

4

On the gas camping stove, Colin had heated up the leftover chili he'd been planning to have for lunch tomorrow. The woman across the table was wolfing it down while it was still scalding hot as if she hadn't eaten in a week.

"Don't they feed you?"

"Only between fires. No time during a burn. You cook this?" She mumbled around a mouthful, halfway through the bowl.

"Yes, my chef is off this week."

"It's good," she drank back a glass of water in a single gulp. "Really good." She slowed down and began looking around the candlelit cabin. "I see the butler is off this week too."

He looked around and grimaced. "I'm not generally this messy."

He was ten miles up a dead-end road and an hour-long hike on a steep trail past that. He hadn't exactly prepared for visitors. And it wasn't that bad. His sheets were still spread on the couch, his clothes piled on the chair, and the floor hadn't been swept in a while. But the dishes were clean and the food all stowed.

His desk was a train wreck, but that was always the case when he was in the middle of writing a novel.

At least he'd taken a wash in the stream recently. Colin rubbed at his chin. Okay, should have shaved somewhere in the last few days, but how was he supposed to have known that he was going to have his first-ever visitor in five summers.

"You're not exactly all spic-and-span yourself," he told her.

She'd staggered into his cabin, dumping hardhat, jacket, and axe across the threshold. The cotton shirt she wore underneath was both sweat- and soot-stained. But it clung to her in amazing ways. The easy strength she'd revealed in the vegetable garden was evident in her athlete's shoulders. Her curves were feminine and sleek; as unlike his ex-wife as could be.

Mirella had been voluptuous…and needy as hell. The latter had made him feel the powerful protector at first, but what had started out as charming had become a cloying emptiness in the woman that could never be assuaged. He'd been on the verge of running and damn the expenses, when she'd decided to fill that emptiness with another man. He was still smarting from the whole mess—despite his lucky escape—and was not looking for another woman.

But looking *at* the woman before him was proving to be a pleasure.

"You're staring."

He was. "I am," he shrugged an apology. "You offer a lot to look at, Ginger." Her fitness, her curves, the face that would have looked merely nice on any lesser woman. Ginger's face was alive with emotion; smile or sarcasm, her feelings showed easily past the deep exhaustion.

"That's not my name."

"But on the radio…" he trailed off at her self-deprecating smile.

"Nickname I earned for being as dumb as a dog. Tori Ellison," she held out a hand and he shook it, "I never know when to quit."

"Easy answer, never."

5

Tori looked up at Colin sharply. It *was* the easy answer, but no one else ever understood that.

Hotshot crews were trained to keep going no matter what, right until the hallucinations of exhaustion set in, and she was still twenty-four hours from that state. But for everyone else, it was always a challenge to keep going. To push harder.

Instead, Tori always saw it as never having "quit" as an option. It made all the difference in the world, but she'd never been able to explain that to anyone satisfactorily.

"What makes you say that?" she asked carefully as she continued to eat the magnificent chili.

Colin looked around his cabin as if he'd stored the answer to her question somewhere in the room.

It was a sweet setup. A generous one-room; a mountain cabin without being primitive. Big windows that told of a magnificent southern view hidden by the darkness. They sat at a small table for two that might be more workbench than dining table. A hand pump at the sink spoke volumes. But there were also shelves of

books, a cozy wood stove, and, perched at a desk cluttered with paper and books, sat a small laptop computer—the only sign of electricity in the whole cabin. She spotted the large battery and would bet that there was a solar panel somewhere outside that fed it. The laptop, she decided, was the focus of the room. The rest was disorganized, not because he was a slob—for the kitchen was immaculate—but because he didn't care.

He too had turned to the desk as if the answer was there somewhere but he couldn't see it. No. He saw it clearly, but wasn't sure about sharing it.

"Writer's cabin," she guessed.

He nodded, then froze like a animal wondering if it was too late to escape the fire.

"Published."

A very careful nod.

So, not a comfortable topic. Which meant he was either a total failure or a major success. If the former, the kitchen wouldn't be so neat…or the desk so messy; a failure would fail in multiple ways. So, a success that he didn't want to reveal, that had him living in a remote cabin with a vegetable garden.

She returned to her study of him rather than his cabin. Not a burden. Tori had thought he was good-looking by the light of her head lamp, but his arm had hidden his best feature. Warm brown eyes lively with a sharp brain behind them. A writer's brain. But they were also warm with emotion, and each time they drifted down her body, more and more heat was revealed there.

"I dated a writer once," she said without thinking first.

"I hate him already," Colin offered the comment amiably.

"He was okay. But he didn't understand about perseverance." Tori had learned enough while dating Andy to know that writing was all about perseverance. She could see that Colin was surprised she knew that reality.

"You're frustrating me at the moment," he remarked and it sounded like a topic change, so she let it be.

"Good. Only one thing this girl likes more than frustrating a handsome, successful man."

"What's that?"

And suddenly Tori was the one who wanted the topic change. She knew that she was far too tired if she'd let that slip out. She'd gotten into firefighting courtesy of a brief fling with a smokejumper. He'd been fun enough, but their brief foray into the wilderness had been life changing.

Tori had always like the outdoors. She'd earned dual degrees in botany and ecology before that trip. To hang with a group of firefighters deep in the wilderness had been an option she'd never thought of until she met the smokie in a bar. He'd offered the briefest glimpse of a life in that uncontemplated world of wildland firefighting.

She'd even found it easy to fall in with the typical firefighter talk once she became one. But there was still a woman with a dream who'd been born on that trip.

The smokie's bosses had been along on the trip, a pair of heli-aviation pilots. A man and woman and their little daughter. Neither spoke much, but their unity—their perfect togetherness—had been such a daunting vision, that it had set the bar impossibly high. She wanted what they had.

Candace and Luke were another couple that felt that way—the only other example she'd ever met.

So, she trained and became a hotshot. On the teams she laughed and teased, and occasionally played the "I fight wildfires for a living" card to pick up a handsome man in a bar. But there was a part of her that dreamed of finding that "right man" someday.

That was the thing that this girl wanted more than frustrating a handsome, successful man.

Not a chance she'd be admitting that out loud though.

6

Colin watched her sleep.

He'd offered the couch, but she didn't want to mess it up with her soot-stained clothes. Instead, she landed in his back-porch hammock and was out in seconds. He parked himself in an Adirondack chair on the back porch and again took in the night.

The stars that he'd been watching to the east, were blocked to the west—the direction Tori had arrived from—by dark clouds. They weren't black, as clouds usually were at night, but glowed red along the bottoms as if they still caught the last of the long-past sunset.

Fire. They glowed red with fire. The hints of wood smoke from this morning were more constant, though still swirled aside by the gentle night breezes. Close, but not too close. Staying far away, he hoped.

He should go inside. Pack his notes and laptop in a bag so that he could grab it and go if he had to. But he couldn't break the easy comfort of sitting and watching Tori sleep.

The charge on his body guaranteed that any chance of sleep for himself lay a long way off. Pretty, motivated, tenacious, and smart were only a few of the adjectives he cataloged on her behalf. She'd synthesized what he was all about with very few clues, and then had the decency to read that he didn't want to talk about it.

Mirella, despite being the one who'd cheated on him, had wanted a big piece of who he was when she left. She wanted rights to any books he'd written while they were together and any number of other things that his attorney had refused to give up. By the time the acrimonious battle was complete, Colin had paid her nothing and she had convinced him that the only reason she'd ever been with him had been avarice. He liked to think that hadn't been the case. But however it had started, it had nothing to do with love.

He wondered what Tori was like when she wasn't drugged with exhaustion. Still beautiful. Still thoughtful. Still tenacious. He was surprised that he'd very much like to discover more despite swearing off women.

Colin knew too little to make conjectures, but he had enjoyed every waking moment they'd had together, at the table and even lying in the garden's dirt.

That kiss. That brief, spectacular kiss. That had been one thing about Mirella, the sex with her was always fantastic. She might have needed something he couldn't supply to send her seeking another man's bed, but the woman had been built hot and made to last.

He'd kissed Tori for approximately three seconds, and it washed any lingering, lonely-night fantasies of Mirella right out of his mind. If kissing Tori was that good, what would the rest of it be like?

"Been alone in the woods too long," he told the night quietly.

Colin came to the mountain cabin to write. To get away from people and the city and the distractions. He'd been seriously considering wintering over this year despite the harsh winters

that sometimes swept the heights of the Cascades. Mirella had come to the cabin once, and departed rapidly. He'd guess that if he chose to winter over, Tori would be right there with him. And loving it.

He watched over her until a small light came bobbing toward him through the darkness. A tall man wearing a headlamp came up to the porch, flashed his light on Tori's sleeping face and then a quick scan around—without blinding Colin—before dousing the light.

"Name's Luke," the man stepped forward and offered a hand. His shake was strong, firefighter strong.

"Colin."

"She actually looks sweet when she's asleep," Luke commented.

"How about when she's awake?"

"Still sweet," Luke chuckled. "Telling you, something's gotta be wrong with the woman to be so consistently pleasant and cheery, but I haven't found it yet. She's a born firefighter. Thanks for watching over her, not that this one needs it."

"What does she need?"

7

Tori would have to pay Luke back for the "sweet" wisecrack.

"Don't think it's my place to be giving away any of the lady's secrets," Luke was telling Colin. "Why? You got an interest?"

Tori lay very still and awaited the answer.

"Might."

Colin *might* have an interest? All she'd done was punched him, kissed him…spectacularly, eaten his chili, and passed out in his hammock for an hour. She was about to rouse herself, despite how comfortable she was feeling, and give these two a quick whack with an axe handle just for being so male, when Luke finally replied.

"If you want to find a better person than Victoria Ellison, you're too late; I already married her. My Candace." Then Luke slapped her on the calf. "Rise and shine, Ginger. We've got some trees to trim. You're first up swamping."

Tori made a groan for Luke's benefit. When cutting line, one person was the sawyer, and the other hauled everything they cut

as far from the fire line as possible. They'd switch off after every tank of fuel, but going from nice soft hammock to swamping was a rude awakening.

But Luke's compliment was high praise indeed; he was crazy about Candace and deservedly so. She didn't know that Luke thought that highly of her as well.

Luke tramped off toward the trees.

Tori waited a moment by Colin, wishing she could see him better.

"Thanks for taking me in," she didn't know what else to say.

"You're welcome any time," he sounded surprised at this own words.

Whether she was unwilling to risk another supercharged, mega-turbo kiss, or the hour's sleep had been sufficient for her common sense to return, she merely shook his hand and turned for the trees.

He *might* have an interest?

It was stupid. It was based on nothing at all.

The only problem she could think of was that she *might* be having an interest as well.

8

Colin brewed coffee, pulled on boots and work clothes, and headed up the slope to join them. He did pack his grab bag and leave it inside the door just in case.

The coffee was taken, appreciated, and drunk while still too hot.

Tori, who was running the chain saw by the time he arrived, didn't even shut off the saw when she knocked her coffee back like a drug, then returned the mug with the briefest of nods. He almost didn't recognize her in the soft pre-dawn light. For one thing, she was back in her full helmet and gear. But also, she was in Ginger-mode. She was moving full tilt and nothing was going to break her focus. He knew that feeling and did his best not to feel rejected by her lack of acknowledgement.

She had cleats on her boots and a heavy belt that wrapped around the fir. She scaled up the tree to the lowest dead branches, then nipped them off with the saw. Moving the belt higher, the next dead branches dropped to the ground. In moments she was fifty feet in the air and a thick pile of dead branches had accumulated around the base of the tree.

Luke was at the prior tree, gathering up the dead branches and dragging them in the direction of the cabin. Douglas firs grew tall, and the lower branches often died off, yet still hung on for years.

Colin grabbed a bundle of branches and followed Luke. Luke had found the cliff edge below the cabin and dumped the branches over which then tumbled to the bottom. Even if they somehow caught fire there, all they'd do was scorch some rock. Colin pitched his load over and they walked back together.

"I don't get what we're doing."

"Ladder fuels," Luke replied. "Fire wants to burn and climb up a tree. Get rid of undergrowth and it has less to burn, stays cooler on the ground. Cut away the deadwood and it has nothing to climb. The real problem happens when it reaches the crown. Hard to fight a crown fire from the ground."

With two of them swamping, they made quick work of what had already been cut.

Luke fired up a second saw and began clearing the undergrowth. Colin couldn't keep up with both of them, but whenever Tori or Luke ran out of fuel, they'd help him catch up as part of their refueling. Mid-morning he knew trouble was coming when a tanker plane roared by low overhead and dumped a broad swath of retardant on the trees. For a quarter mile, the big jet plane sent down a shower of the dark red liquid in an impossibly dense downpour.

Tori arrived beside him as the tanker finished the run and turned back for its next load. "Retardant coats the wood and keeps the oxygen from reaching it. No oxygen means no fire."

"Then what have we been doing here?" Colin waved at the trees, at the whole area they'd been parking-out.

"Layers of defense, like chapters in a book. Chapter One, we have a fireline cut about a half mile back. We're hoping to narrow the blaze, maybe even knock it out of the crown because it's running high and hot at the moment. Chapter Two, hopefully most of it dies when it hits the retardant line. Chapter Three,

if we can really slow it down here, it won't do much more than mow the grass in your meadow before we can extinguish it. End of story."

He'd been right about smart and kind. She'd thought to switch her words into his metaphor to make sure he understood it easily, rather than assuming he could cross to her side of the fence.

"And if all three chapters fail? What's the fourth?" He'd miss his cabin. He'd rebuild, but there were a lot of good memories here; he could hear the stories that had been written in this idyllic spot.

She pointed up at the sky.

A small helicopter painted black with red flames came pounding up the hill. They watched it together as it flew over his cabin, a huge, bright-orange bucket on a cable dangled far below. The pilot didn't even slow down, just released the load of water dead-center on his roof. It soaked down the shingles and poured off the eaves in waves.

"That's the epilogue, just in case the fire didn't get the message or tries to throw a few hot embers your way."

Colin could see it clearly. All of the different pieces and how they fit together as neatly as any story.

But he couldn't stop looking at the quietly competent woman he'd been working beside all morning.

"Just in case I don't get a chance to say it later, I meant what I said. You're welcome anytime."

9

T ori didn't know what she was doing. It was her first break in weeks. The fire season was running hot and heavy, but Candace had finally declared that enough was enough and shuttered the Leavenworth Hotshots for five days. Thirty days without a break, they were all so punchy that safety was becoming an issue.

Tori had thought about hanging out in town like usual. But she didn't want the noise and the bars. She wanted the quiet that a smokejumper had introduced her to an age ago.

By the time she parked her battered Toyota pickup beside the shiny Jeep Wrangler, Tori at least knew her destination. As if she hadn't looked up the access road on a topo map the moment she'd gotten off the Checker Mill Fire.

She spent most of the hour's hike up his trail telling herself she was being an idiot. A kiss, one bowl of chili, and one fire killed right at the very edge of a vegetable patch. That's all there was between them.

But each day on the fires since, she'd been watching Candace and Luke. And each time she thought about the second kiss, the

one after the fire—the more she knew that she at least had to answer the question that she and Colin had written between them.

The climb to his cabin followed a fast-running stream and then stretched out over a long green meadow. The fire had been killed in the woods. The last lines of trees stood green as well except for some char on the bark. For once, the fire's story had gone exactly as she'd predicted it.

His kiss had been a place of peace that had felt so right, so perfect. Part of it was the land, most of it was the man. The last of it was that he was the sort of man who had chosen this gorgeous stretch of mountainside for himself.

But how would he react to her arrival? Was he just being so thankful to be rescued from the fire that he'd have invited Medusa to come visit?

It had taken her a week after the Checker Mill Fire to make the mental connection. Tori had pulled one of Colin Steele's thrillers off her own bookshelf to discover that the man pictured on the back had introduced himself to her as Colin James.

And that had almost kept her away.

She didn't want to arrive as some sycophant, fan-girl no matter how much she enjoyed his novels. Yet here she was anyway, despite telling herself to stay away. Tori almost turned from his front porch and headed back down the trail, which was beyond stupid. She closed her eyes, trudged up the steps, and struck out at the door.

There.

Now she'd knocked and there was no backing away without looking even beyond stupider than she felt. Stupiderist? Even by Ginger standards, this was extreme.

And she kept standing there.

And standing there.

She knocked again, harder.

Still nothing.

Well, she hadn't driven and hiked and nerved herself up to quit so easily. *Perseverance,* she reminded herself and stalked off to the back side of the cabin.

10

C*olin looked up the* moment she came into view. It was like that utterly impossible moment that always occurred between hero and heroine that he could never resist writing. First sight of each other at the same instant.

Even without the fire gear, he'd know her anywhere. There was a confidence, a surety to her stride unlike any other woman he'd ever known, or written. She came around the corner of his cabin as if she'd always been there, always belonged.

He stayed where he was and waited while she crossed the back porch, shed her pack, and came up into the vegetable garden. She wore hiking boots, shorts atop some of the longest legs he'd ever seen, and a light t-shirt luridly aflame, but patterned like a checker board. Across her chest it announced the Checker Mill Fire and the dates. The second date was the last time he'd seen her; twenty days and three hours ago.

His gaze finally made it up to her eyes as she arrived in front him.

"Great t-shirt."

"I brought one for you. You worked it too."

"Is that why you're here?"

"No. Nor is it because I know who you are."

Colin froze. Here it comes. All of the fantasy and hopes had just become meaningless.

"I've read a lot of your books. I thought you should know. I almost stayed away because of that."

"You what?" He hadn't expected that. "Then why are you here?"

She reached out and brushed her fingertips along his cheek. Not hot like fire, but rather cool like his stream, a caress that calmed and anchored him in this moment.

"You feel it too, don't you?" Tori asked softly.

He could only nod.

She closed the final step that separated them. When she slid her arms around his neck and kissed him, it was a scene right out of fiction. He'd never imagined anyone feeling so right in his arms.

Colin knew that Tori Ellison never stopped once she found what she wanted. She'd keep right on fighting fire or whatever came next in her life just as he'd always be writing.

When they lay down together on the garden path, he knew that he needed her as much as the blank page needed words. And their story would have many, many pages.

Where Dreams Are Well Done

I believe that this is truly the end to the Angelo's Hearth series. After the other stories were done, I thought that I had told all of the stories I wanted to tell from this cast of characters.

Three things collided to bring this story about.

One: In the second book in the series, Where Dreams Reside *(which is Angelo's own love story), he hires a new expediter. An expediter is a crucial position in a high-end restaurant. They make sure that every plate going out is perfect, in the right order, and at the right moment. A good expediter can make a restaurant, and a bad one can break it.*

When we first met Luisa, she was introduced as the evil twin of the fair Graziella (also see Blaze Atop Swallow Hill Lookout *above). So here I had a sharp-tongued, utterly driven, and very beautiful woman on the line. She remained just that way*

throughout multiple books. And she is perhaps the toughest woman I've ever written, right up there with Kee Stevenson in The Night Stalkers.

Two: In the last book of that sereis, Where Dreams Are Written, *Angelo suddenly decides that he wants a third restaurant. That means he'll need a new chef to keep the first one going. Well, I found Sam the prep chef who worked in Angelo's kitchen back in book #3,* Maria's Christmas Table. *(Yes, a writer's mind really works this way, connecting odd bits and pieces in curious ways.)*

Three: With the earlier short story, Where Dreams Are Sewn, *I had "ended" the series with Perrin's fashion boutique. And while that was a major setting in the series, the heart of the series lay in Angelo's Hearth Italian Ristorante.*

That's the setting where I felt the series should finally close: where it began, in Angelo's.

M.L. BUCHMAN

B&N and NPR Top 5 Romance Author of the Year

"One of our favorite
authors."
Romantic Times Book Reviews

Where Dreams
Are Well Done

an Angelo's Hearth romance story

1

What is wrong with you? I needed that fish three minutes ago. Did you learn to cook in a cave?"

"You needed it twelve seconds ago," Sam shot back. And decided against telling Luisa Valenti that she wasn't going to get it for another thirty seconds. Besides, the kitchen's *aboyeur* was busy dressing the plate of pappardelle with wild boar ragù that he'd just handed across the line.

"Fine, then where's my trio of sea scallops and squid-ink pasta?" She didn't even stop for a breath. "Though why everyone at a table would order the same dish is beyond sad. Just keep cooking the way you are and maybe we'll never see such dweebs ever again."

"Open your pretty eyes, Luisa," he teased her as his *sous chef* Marlys slid the three matching plates onto the warmer shelf that separated his station from Luisa's.

She rolled those beautiful brown eyes at him, making it clear that she knew he was trying to distract her from the laggard glazed halibut.

A glance down the cook line either way told him that they were running a little rough, but okay. He dropped another two orders of orzo into a pot of boiling water to help out Valerie. He also passed a tray of stuffed and breaded squash blossoms from Tony to Valerie as she turned to the deep fryer, saving her three extra steps she didn't have time for. He dropped the next two pieces of fish into pans for Marlys and accepted the two plates of sea bass ready for saucing.

He'd never have dared talked to Luisa that way while he was still a prep chef. He'd noticed her of course, there were only a dozen staff at Angelo's Tuscan Hearth Ristorante, including the three waitstaff, but his duties had mainly been in the morning before the restaurant opened. Angelo or Manuel would do the shopping at daybreak, then all of the proteins and produce would arrive for him to prep. When Manuel had shifted to Angelo's new restaurant, Sam had been pulled into the lunch line.

Up until this morning he'd thought he was just being trained to fill in where needed. Tonight they'd dropped him instead of Marlys into the Executive Chef slot for dinner service because Angelo couldn't make it. If he had time, he'd be freaking out right now, but he didn't.

Over the last month he'd worked with Marlys the *grillardin* cooking the meats and Valerie at *entremetier*—the hot appetizers, soups, and pasta station, one of the keys to an Italian restaurant. He'd almost died at the *sous chef* position—keeping the *saucier's* eight pans always filled with whatever had to be sautéed to perfection, because Angelo accepted nothing less, which required being part magician, part juggler, and part octopus.

As a prep cook, the menu had been drilled into this head. Yes, it was always changing based on what was freshest in Seattle's Pike Place Market just out the back door, but there was a style, a flavor, a feel to Angelo's cooking that made sense once he understood it.

He'd even done some turns as the Executive Chef for lunch service. The lighter fare becoming second nature with practice.

That was when he first bumped heads with Luisa.

There was no way to miss Luisa's presence in the kitchen. It was the *aboyeur's* job to expedite service and did she ever. Luisa had every order in her head, never having to check a ticket twice. And she was very vocal about not getting everything in the exact order she'd called for it. Table Seven had a simple ragù, a pan-fried swordfish on a bed of angel hair pasta with one of Angelo's signature sauces that had to be made the moment before service, and a grilled lamb and baby asparagus with a Gorganzola cheese drizzle—and Luisa would throw a fit if they weren't all ready in the same five seconds even though they took drastically different amounts of time to cook. Actually, in the same three seconds.

But with Luisa in charge, there was never an undressed plate or a missed order. She was just as amazing as she looked. And as dangerous.

He slid across the missing halibut with a honey-rosemary-chestnut glaze and the accompanying bowl of the wild boar ragù.

"Finally!" she huffed at him.

Being the sole target for her ire was daunting. Everyone on this side of the cook line answered to him, but he answered to the fair Luisa.

He still didn't know how he'd landed in the Executive Chef slot through a dinner service. He'd entered the kitchen and Luisa had simply told him, "Angelo's busy tonight. It's your cook line." He'd taken his first breath about an hour into the meal, but hadn't had time yet to take a second one.

She finished dressing the plates with berry compote traced in an elegant line around the outline of the halibut. With immaculate timing, Graziella breezed in from the front of house, barely breaking stride as she gathered the completed dishes, and whisked back out.

It was a shock every single time to see them together. Two slender, beautiful Italian women with golden skin and lush dark hair that reached the middle of their backs. They could have

been twins. Except Graziella was as gracious and patient as her name, unflappable under even the most dire circumstances. Luisa's heritage must be at least part Roman, as in Roman candle. Incendiary.

"What are you paying attention to, Chef?" she snapped at him.

Luisa hadn't looked up at him, but he'd been watching her and not his line and somehow she knew. A quick glance showed him that his momentary lapse to admire his *aboyeur* had just caused him more trouble.

"Fire three halibut and two sea bass, a lamb, a beef tenderloin, and two scallop."

Marlys grimaced, but hustled to get them all going.

It was too late, the next five tables were going to be all out of sync and he was going to catch hell for it.

2

Luisa kept her head down to hide her smile.

She remembered her first day here, Angelo and Manuel purposely messing with her, testing her. Angelo would finish a fish and then sit on it for thirty seconds just to break her rhythm. When she'd chewed him out over the line, he'd merely smiled and handed it across.

Manuel had mixed up three different orders, just to see if she'd catch it, like she was that dense. She'd ripped him a new one and he'd told Angelo to hire her on the spot.

That had been a year ago and Angelo's Tuscan Hearth Ristorante was now a well-oiled machine, reproducing the chef's magic to the table with class and consistency.

And at last it was finally turnabout, her turn to be testing someone else to Angelo's stratospheric standards. Manuel had moved to run the new restaurant by the Seattle Center and Angelo wanted to focus on a third restaurant he was creating. She was the one who had suggested pulling Sam Walsh out of the prep role.

His prep had always been immaculate.

Then she'd spotted him making a quick lunch for himself and recognized the instinctual skill of his actions.

She'd spent three years in Italy studying restaurants, eking out every penny she could to continue doing so. Every three months she went to a new restaurant and volunteered to shadow and assist the *aboyeur* in exchange for food and a place to sleep. Many times she'd ultimately been offered a permanent job, but there'd been so much to learn, so many regional varieties of food, so many different chefs to study that she'd always refused. She'd worked her way from Rome down to Puglia, into Sicily and back up the west coast to Tuscany, Liguria, and the Piedmont before the money ran out.

She knew what a real chef looked like, even if Angelo and Manuel were too dense to notice what was in their midst.

When Sam had been momentarily called away, she'd snuck a forkful of his lunch and been stunned. No prep cook should be able to cook like that. Especially not a tall handsome one with auburn hair and such agile hands; he looked Irish not Italian for crying out loud. Of course Manuel was Mexican, but so were most of the sous chefs and line cooks in high-end restaurants throughout the United States.

She'd stolen a second forkful, carried it over to Angelo, and fed it to him. His eyes had gone wide, then thoughtful when she pointed at Sam returning to his meal.

Sam spent some time looking for his fork, never spotting that she'd stolen it. It had been awfully cute.

But the fact that she'd been the one to discover him, meant that he had to perform even better than Manuel and Angelo did. If she had to hound him twice as hard to meet that standard, he'd just have to get used to it. And he had.

Until she'd distracted him.

She handed off the next order to one of Graziella's waiters and then did her best not to laugh at the cascading disaster his lapse had caused.

Luisa was used to men watching her. Hadn't thought much of it except when she wanted to take a likely candidate home with her. At least not until everyone started calling her Graziella's twin sister. When "evil twin sister" had slipped out of someone's mouth, she could only sigh in acknowledgment of the sad truth.

Grace and graciousness had never been in Luisa's genetic makeup; her "good twin" was elegantly second-generation Italian and Luisa was third-generation bitch. But that anyone thought she was that beautiful was still a surprise. And that Sam had noticed so much that he'd lost the thread of the meal…

"C'mon, Sam," she called out to distract herself. "It's your first dinner as Exec Chef. Don't drop the ball on me now."

He didn't snarl or glare at her. He didn't even frown. He simply turned to assist the suddenly overwhelmed Marlys trying to simultaneously fire nine dishes in eight pans.

Luisa moved down the line to chat with the *patissier* about a new dessert idea she'd had. She'd give Sam a little time to recover, but not too much.

3

W*hen the last two-plate* order had slid across to Luisa, Sam was ready to collapse.

Someone slapped him hard on the back and shoved a cold beer into his hand. A round of applause sounded down the line.

And across the cook line from him, even the bane of his existence was applauding with those elegantly fine hands of hers. He tipped his bottle to her in silent salute; they both knew he couldn't have done it without her help.

"So," Angelo and Manuel came in through the back door which had been left open to the warm September night, "Let's see how you did." He took up the two plates of the final order and handed one to Manuel.

A test? Tonight had been a test?

For what?

The two chefs tasted, chewed, swallowed, and then tasted each others' dishes.

"He didn't follow your recipe," Manuel pointed at the smoked eggplant and shrimp ravioli.

"No, he didn't," Angelo closed his eyes for a moment. "Walnut, no. Chestnut." He opened one eye to glare at Sam. "How much?"

"A single light grating over the eggplant before I smoked it." He knew it was taking liberties, but it had seemed right. Now he was less sure.

"Seems odd to me," Manuel replied.

Angelo harrumphed in agreement.

Sam was starting to get really worried, he knew chefs who'd been fired for tinkering with the Head Chef's recipes. Then he spotted Luisa's expression. She winked at him. After all of the abuse she'd unloaded on him during the meal, she winked at him.

Greatly encouraged, he winked back, then waited for Angelo and Manuel to get to the point.

"Going to have to change the damned recipe now," Angelo grumbled, but Sam could finally tell that he was pleased.

Angelo waved him out from behind the cook line. When he reached them, Angelo shook his hand. Manuel gave one of his quiet nods that Sam had long since learned was his form of high praise. Graziella had joined them by that point and kissed him on both cheeks before tasting some of Manuel's dinner and sighing happily. She slid an arm around her husband's waist and he held the bowl so that she could take another mouthful.

Sam felt himself wilting a little every time he saw them. Manuel and Graziella were so sweet together. When would he ever find something like they had? Based on results to date, *never* was his best guess.

"Looks like you were right, Luisa," Angelo looked at the *aboyeur.*

"Told you," was her pert reply.

"Told him what?" Sam asked.

But she just kept grinning at him.

"Told you what?" he asked Angelo.

"I think you two are ready to run this restaurant. Interested?"

Sam looked at Luisa and saw the stunned look on her face, perfectly mirroring what he was feeling right at that moment. Well, something had finally put her in her place.

Angelo's Tuscan Hearth was top-rated as was the Piedmont Hearth which Manuel now ran across town. And not just top in the foodie Pacific Northwest, but nationally.

"We…*two?*" Luisa managed a bare whisper.

Her gaze slid to Angelo then back to him, her dark eyes gone wide.

Not to do this for a single night, but every night?

He tipped his head in the slightest question to her. Angelo was right, he couldn't have done it alone. But to run such a restaurant had been his goal since forever.

"You game?" he managed, his own whisper no louder than hers had been.

The astonishment shifted through a hundred stages on her beautiful face through consideration, weighing factors, acceptance, and finally a blinding smile that stunned him right back on his heels. He'd known she could smile…but not like that.

"Hell yeah!" was her verdict.

They traded high fives and a quick hug as the rest of the crew cheered for them.

4

As far as Luisa knew, she and Sam had never actually touched, always separated by the width of the cook line. But his brief hug was warm and sincere. His big hands had wrapped briefly around her waist and she could still feel the impression of their easy strength. He'd smelled of the cook line; flavor and spice.

If she'd been seeing anyone, it might have had less impact on her thoughts. But she hadn't been. Not for several months even before she'd stolen that taste of Sam's lunch. Then Angelo had assigned her to lay out his training because he was too busy with his plans for the next restaurant, and that had preoccupied her thoughts.

So, she'd made sure that "Angelo's" official schedule rotated Sam through every position until he could do each as well as the station chef normally posted there.

And it had worked. Worked beyond her wildest imaginings actually.

"Run the restaurant?" she whispered to herself, but it couldn't be real.

She cleaned and prepped her station for the next day, tossed out the sauces and garnish that wouldn't survive overnight, stowed everything else where it belonged. The other chefs were tending to their own stations. She'd demanded end-to-end ownership, if you needed more pans or towels or a sharper knife in mid-shift and didn't have it, it was your own damn fault. Angelo himself had been the slowest to adopt the change, but now agreed it was the best way.

"Run the restaurant?" She'd been *aboyeur* for a year, the longest she'd ever stayed anywhere, but that was a long way from running a restaurant.

She was staring at her immaculate station when a hand landed against the small of her back—she knew it was Sam's by the feel alone—and swept her from the kitchen and out into the main room. It was dark, the last of the diners were gone. The shadowed room was spotless. The tables already set with the lunch service cloths and tableware.

Only one table remained candlelit, the one closest to the hearth that was the centerpiece of Angelo's. The fireplace was a simple affair of stone and brass that anchored the room and gave it a lush ambience.

The table contained a chef's meal: a cutting board of crackers and several varieties of cheese, and a bottle of wine uncorked to breathe. Two glasses.

"You trying to woo me, Sam Walsh?" she asked as he held out her chair for her.

"No!" He startled as if she'd just whacked him with a wooden spoon. "Trying to be nice; sort of to say thanks for getting me this chance and maybe try to figure out what's next. Is nice too foreign for you, Luisa Valenti?"

He teased her. He'd actually teased her, which was quite a step for Sam Walsh. She considered several acerbic replies as he settled into the chair opposite and began pouring the wine.

"Way too foreign," she sighed. "I don't think that *nice* runs very deep in my bones."

He snorted out a laugh, "Might have noticed. Pity. However, they're such very *nice* bones."

She squinted at him over her wine glass, he'd chosen an Oregon Pinot Noir—her favorite Northwest wine. Luisa was starting to realize that Sam missed very little.

A part of her was offended by his easy agreement that she wasn't nice, even if most of her sadly agreed. But the compliment she hadn't been ready for.

"Now you're teasing me? Because I know that Sam Walsh never flirted with anyone." Which had stumped her at first. It was a standard part of her repertoire when trying to make male chefs behave. She'd had to find different buttons to push with Sam. And he was such a decent guy, she often fell back on simple cajoling. She didn't have a lot of experience with decent guys.

He slumped in his chair and rubbed at his face, "I'm so strung out that I must not know what I'm doing. Flirting with you doesn't sound like me, does it?"

"No, it doesn't. Why is that? Aren't I flirtable?"

5

*S*am *looked at the* Mona Lisa beautiful woman across the table from him; right down to the enigmatic smile. The candlelight played across her dusky skin made it far too easy to imagine how *all* of her skin might look in such light.

He always felt oversized and awkward around her. And now? She was waiting for his explanation of…

He sipped at the wine. He'd noticed early on that when multiple American wines were circulating, this was the one Luisa always chose. He'd become partial to it himself. It had a fruity body and a low acidity that…had him thinking again about the sleek body and high acidity tongue that sat across the table from him.

"Well? Why haven't you flirted with me?" She did her best to sound offended, but she was also too busy looking pleased with their sudden change of circumstances to really pull it off.

And he was just tired enough to actually answer her question.

"I don't because I can remember every single thing about you since you walked in through that door."

"You what?"

"I was coming out of the cooler with a salmon almost as big around as you are. You breezed in looking like you already knew more about the restaurant than the guys running it. Turned out you did. Gorgeous, opinionated, and feisty as hell. Want me to tell you what you were wearing that day?"

She was looking at him with as much surprise as she'd shown at Angelo's announcement that the two of them were taking over the restaurant's operations.

"Forget it. Never mind. Let's talk about the restaurant," he took a large swallow of the wine. Should have kept his dumb mouth shut.

Luisa was continuing to eye him carefully. "No," she said it slowly, "let's talk about this."

Sam refreshed his glass, then set it aside because after everything else tonight, the alcohol was only going to make him even stupider.

"Talk to me, Sam." Luisa's voice sounded soft, uncertain—something he didn't even know she was capable of feeling until Angelo had blind-sided her with the offer.

Sam would rather—nothing came to mind. Climb the highest alligator? Wrestle the fiercest mountain? He was really exhausted.

"Sam," a flat, insistent tone.

"There's that tone," he acknowledged. "The utter surety of it. Woman who knows what she wants. Never thought you'd notice some lame prep cook."

"Of course I noticed you."

"Not really. I was just a lowly minion; not a chance that you'd actually see me. This was only supposed to be a damned temp job anyway."

"I—Wait!—What?"

"I was just back in Seattle for a few months. Spent a couple years in San Francisco at Acquerello. Then a year cooking for Batali in New York. Graziella got me the prep spot here while I figured out what I wanted to do next." Sam tried to think of some

way to derail the story, but couldn't come up with a way now that it was rolling. *In the shitter now, boy!* was all he could think.

"But if you could do that, why did you stay here as a prep cook?"

"Yeah, good question. I eventually learned just how damn good Angelo was and realized this is where I was supposed to be. Watching that man build a sauce is a serious education; way beyond even my coursework at ICI in Calabria."

"You graduated from ICI?"

"Top of class," why was he bragging to her? Impressing Luisa wasn't something a guy was dumb enough to even try; she always knew what she wanted and just went for it.

"*Eventually,*" she drew the word out, "you were impressed with Angelo. But not at first?"

He shouldn't have said that either. He just shook his head and decided that the glass of wine would do more good inside him that it would sitting on the table.

"Because at first…" she was working it out.

Even shutting up wasn't going to do him any good. Luisa was too smart. He'd said too much. The moment he'd opened his mouth, he'd said too much.

"At first…" tasting it like a fine wine, half-lidded eyes, pursed lips. The way he'd always imagined she'd look in that half breath before a kiss.

"You sure it's not too late to talk about the restaurant instead?" But he knew it was.

Then he saw it click. Those stunning dark eyes zeroed in on him.

He shrugged, "Got me."

"You stayed because of me."

He nodded.

"Oh god. I really need a glass of wine."

"In your hand, Luisa."

She looked down at it in surprise, then knocked back a large swallow before returning her attention to him.

"It's not helping," she said as if it was his fault; which was probably true.

"Noticed that myself."

"Why didn't you say anything?"

6

Luisa was trying desperately to make head or tails of what Sam was saying. But it wasn't working.

"I saw the kind of men who waited for you after work. Slick, urban," Sam waved a hand at himself. He wore jeans, leather shoes battered and stained with too many hours in the kitchen, and a casual flannel shirt rolled up at the sleeves that had replaced his chef's jacket.

He was right; he didn't fit with what she'd always reached for. Looking the way she did, it was always easy to take almost any man she wanted off the shelf. She came from desperate poverty and kept picking some Mr. Rich-and-Successful. But they never seemed to fit when she tried them on.

"Did what I could to get over it. Then Angelo hooked me with his food," again one of those easy shrugs of his powerful shoulders.

"And you told me tonight because you're—"

"Too exhausted to think before I speak." Then he glared at his half-empty wine glass. "I thought I was ready. Could handle whatever you…"

He cut himself off and looked up at her. She could see that the wine had nothing to do with what Sam was telling her.

"I wasn't ready for one thing," he continued.

"What was that?" She tried to guess. The hug had been nice, even promising. But if he was one of those guys who'd built a whole fantasy on such a brief contact, then he was in for a rude awakening. Life wasn't that easy.

"I've been cooking since I was six. Never wanted to do anything else. But I wasn't ready to find out that you believed in me. I still amazed that you're the one who saw..." he waved helplessly toward the kitchen.

Luisa knew she was in so much trouble. It wasn't the easy answer at all. Of course with Sam Walsh it couldn't be, could it? And once she'd seen him, and began tasting his food regularly...

"So I have a question," his voice was little more than a low rumble.

Was she ready for this? She wasn't going to just jump into bed with him because he'd...touched her with his answer. Of course, it hurt that he was right, she'd have brushed off a mere prep cook. But the way he'd cooked tonight? That man, she couldn't help but notice.

For once not trusting her voice, she nodded for him to continue.

"Now that I've been dumb enough to drop that fat in the fryer, are you still okay running this restaurant with me, executive chef and *aboyeur*?"

At her nod of assent—how could she not find a way to make such an opportunity work—he sighed with relief.

"Well that's something, anyway," he mumbled softly into his glass.

It was clear that he was speaking to himself, so she did her best to pretend she couldn't guess what else he was thinking.

7

For two months it was enough. A cool September turned into a cold November and they worked their asses off.

Their every waking thought was consumed by the restaurant. At first, Manuel or Angelo had joined Sam on the daily shopping expeditions in the middle of his night. The restaurant closed at ten, they were cleaned up and out the door by eleven, and except for a few fantasies about Luisa, he'd be asleep by one. Up at six in the chill, predawn darkness to get the pick of the market at Pike Place. Asleep from seven to ten, if he got back to sleep, and then into the restaurant.

Soon, he and Manuel had agreed to alternate mornings and do the shopping for both restaurants. But even on their off days, they were likely to bump into one another at the market, just seeing if there was anything particularly special that day.

Luisa showed up one morning, looking gloriously rumpled in sweatpants, a heavy sweater, and desperately clutching a large thermal mug of coffee. She was good, spotting some possibilities for the Daily Fresh menu that he'd missed. It was soon a routine

to shop together and go out for breakfast afterward; watch the late sunrise and the early tourists while someone else cooked them breakfast. He slept less, but in those quiet mornings is when he grew to know his *aboyeur.*

At first they discussed Angelo's. But soon they were discussing travels, different chefs and their restaurants, eventually they even wandered into past lovers. At the restaurant it was all business, but for the few hours between the shopping and the start of lunch service, that was theirs alone.

Luisa worked with Graziella on the restaurant operations. He worked with the other chefs and his prep chef replacement. In the second month he started running dish variations by Angelo, who was often found experimenting with a new dish on a small side stove. The new restaurant was going to be Southern Italian, a broad departure from the Tuscan and Piedmont themes of the first two, and that needed a lot of prep. Soon they were collaborating and testing dishes together; Graziella and Luisa offering their own insightful palates to the process of turning dishes into a menu. Sam had never so enjoyed the simple craft of cooking as those moments with Angelo.

At the end of the second month, Angelo sat the two of them down and laid out a chart. He didn't need to say a thing. Rather than any dip in sales, there'd been a slow and steady increase. Angelo then laid down four new reviews that left no question he and Luisa were doing well as a team.

Angelo had shaken Sam's hand and kissed Luisa on top of the head before leaving them once again alone at the small table deep in the shadows of the closed restaurant. The silence stretched long after Angelo shut off the kitchen lights and left.

"I like that you still keep me on my toes on the cook line," Sam finally said to break the stillness. "Don't stop doing that."

"Deal," Luisa grinned at him. "Just don't stop blowing my mind with your new dishes." It was practically a caress that he felt right down to his heart. He loved that she loved his cooking.

After another overlong pause, it was Luisa who broke the silence.

"Do you think…" she trailed off.

"I don't know," he answered, fairly sure they were discussing the same topic. It had become like a live wire, or perhaps a tug of war across the cook line. The tension had built between them until he wasn't sleeping that much before the shopping trips either. They'd worked out such a deep cook line communication that it was, well, almost sexual. He couldn't think of any other way to describe it. It was sexual in every way…except the complete lack of sex.

But would that blow apart their working relationship? Since that one brief hug, they hadn't touched so much as a fingertip: not in the restaurant, not while shopping, and not over breakfast.

"All I can think…" All he could think was how she'd look at the moment of perfect ecstasy and just how much he'd like to be the one to help her find it.

"Yes?" her voice was practically pleading with him.

"We have to try. Because if we don't…"

"…we'll both go stark, raving mad," she finished for him.

Sam could only nod.

Again the long silence. His turn to break it.

"Not to sound crass, but…Your place or mine?"

Her smile quirked at that, "Mine is closer by about a block."

"All the difference in the world," he rose to his feet, and held out a hand to help her to her feet.

She looked at his hand, and then up into his eyes, studying him before she took it.

The shock was visceral, though not electric. Neither was it cooking fire hot. It was simply such a powerful feeling of rightness that he didn't stop there. With the slightest tug, he kept pulling her in until she lay against his chest, her head on his shoulder. He bent down to bury his face in her hair. Her slender form felt so perfect in his hands that he wrapped her tightly against him and simply held on.

"This had better be worth it, Sam Walsh."

"No guts, no glory, Luisa Valenti."

"I'm thinking…"

"…that your apartment is too far away."

A restaurant floor was no place to bed a woman for the first time. Then he remembered a luxurious sofa in the entryway for waiting patrons.

He swept her up in his arms and carried her there. The soft light of the fire reached just well enough that he was able to see that she looked as incredible under her clothes as even his wildest fantasies had thought.

Sam made love to her while the rain storm rattled the front doors.

8

Luisa felt thoroughly ravaged and pleasantly trashy. They'd made it back to her studio apartment around three a.m. and passed out. The shopping alarm had gone off at five-thirty. They'd been half dressed by the time they remembered it was Monday and the restaurant was closed for the next two days. They hadn't gone back to sleep for a long while. The chill November rains gave them every reason to stay inside. So they did.

Now it was Tuesday mid-afternoon. Delivery pizza and Chinese cartons were scattered in among their discarded clothes. Damp towels from the most erotic shower of her life had been tossed on the floor as well. A shocking amount of protection was now in her garbage can hidden by a discrete Kleenex. And the most amazing chef and lover of her life lay draped across her like a man dead. They hadn't even gotten dressed—other than a quick robe for the food deliveries—in the last thirty-six hours.

Sam mumbled something unintelligible in her ear, rolled onto his side, then scooped her back against his chest and buried his face in her hair. He couldn't seem to get enough of

that. He'd charmed her in so many ways that she could easily get lost in it.

She rubbed a hand along the back of his arm where it curled around her waist and held her tight. Her studio apartment had little going for it, other than a bed, a chair, and a dresser. It hadn't mattered to Sam. She idly wondered what his place was like. Did he have a masterful kitchen or did he care just as little for what lay beyond the restaurant as she did? No, Sam Walsh would have a one bedroom, maybe even a two. It would be messy around the edges, but the kitchen would be immaculate. He'd cook for her there.

Luisa didn't want to be charmed, not really. She wanted amazing sex and a challenging career. She had dreams, restaurant dreams. She knew how a restaurant should run—had learned an immense amount running Angelo's—and could easily imagine being *aboyeur* to a whole chain of them. But she couldn't imagine a chef-lover in that picture. Or at least she never had, which didn't make it any easier now.

Perhaps it didn't matter, her lovers never lasted long. They'd cross some line and she'd throw them out. Or they couldn't handle one of her caustic quips and they'd be gone.

Maybe that had already been dealt with. The blinders were definitely gone after what she and Sam had done to each other in the last day and a half.

"It's morning," she said even though the afternoon light—the first break from rain in days—was streaming in her west-facing window and warming the bed deliciously.

"Uh-huh," he grunted in her hair.

"So?"

"Wha?"

"So, do you still respect me?"

"I," he nuzzled the back of her neck. She could feel against her backside that other parts of him were impossibly waking up as well. "I respect your sexual prowess no end. If I live through the afternoon, I'll upgrade that from respect to worship."

It was hard to argue with that. Sam was a perfect blend of gentle, creative, and sheer stamina. His body was built to order by any woman, but what he could do with those strong chef's hands of his had to be classified as pure glory.

Even as she gave herself to his roving hands and carnal intentions, she couldn't help wondering how she'd be screwing this up.

She knew it would be her.

Sam was far too nice a guy to take care of that for her.

9

"*You what?*" *Sam strangled* painfully on the last word and grabbed his throat to stop himself from grabbing Luisa's right across the cook line.

"I got a job offer," she repeated more calmly than she ever called out an order.

Sam looked up and down the line. Marlys was staring at Luisa while searing a piece of mahi-mahi for table fourteen. He pointed to get her attention back on the fish before it burned.

Valerie, Tony, Vic, even Marko the dishwasher had all ground to a halt.

He looked down at the empty plate in front of him and for the life of him couldn't remember what went on it. He looked back up at Luisa.

"And you tell me now?"

She shrugged as if his world wasn't falling apart. They were practically living together. For three months they had been together every night, usually at his place because he had a kitchen and a decent sofa.

"Where?"

"L.A. at first; maybe Vegas and overseas after that. Wolfgang Puck wants an experienced *aboyeur* to vet and enhance the operation of two of his high-end restaurants at the Bel-Air and the Ritz-Carlton. If that works out well, I'd expand into Spago and Cut."

It was the entire fine dining line with one of the leading restaurateurs in the country.

"It's just talk at this point," she finished prepping a plate and turned to hand it off.

But her timing was off; neither Graziella nor the other waiters were anywhere to be seen. It was his only clue that she was not nearly as calm or cool about this as she was pretending. Luisa never missed her cues, not once in the last three months since they'd become lovers had she eased up on pushing him for perfection in the kitchen. She'd also trained someone for Manuel and was helping Angelo do interviews for his Southern Italian Hearth over in Bellevue at the top of one of the towers.

"Just talk? You said you had an offer."

She shrugged negligently and he knew she was gone.

He wished to god he knew why.

He did his best to focus back on the service, managed to get the mahi-mahi with the right sauce and vegetables, even if it was on the wrong type of plate. He didn't change it and she didn't say a word before dressing it with a truffle oil finish.

All Sam knew was that there was small box in his pocket that had been burning a hole there for two days since he'd picked up the ring. And his plans to give it to her tonight, when they'd have the next two days off to celebrate, had just been burned past recognition.

10

Luisa hadn't been prepared for her own pain when Sam had shut her out. Not that she could blame him; it had been one of her least smooth exits in history. But this wasn't working for her. He was too close, too real, too important. She'd sworn no man would ever get in the way of her dreams. The Old Boys Club of restaurant chefs would never limit her options. She'd prove—

But the scar that she'd just sliced across Sam's heart was so visible; she'd never imagined anything that bad.

He didn't say another word to her. Instead he returned to creating perfect food like an emotionless machine. She hadn't had the heart to offer a single prod or nudge—he didn't give her cause to, except that one plate. His part of the service was perfect. Machine perfect.

As the last plated dish crossed the line, he turned to Marlys, whispered something to her, and was gone out the back door before she could think what to say. Marlys began cleaning up his station without looking at her once.

No one else on the line was talking to her either.

Luisa cleaned up her station amidst the echoing silence and retreated as quickly as she could.

She entered her apartment too weary to turn on the light. She also couldn't bear seeing one of Sam's forgotten jackets over the back of a chair or the silly mobile he'd bought at the Pike Place Market and hung in her window, made entirely of twisted vintage forks and spoons.

Two days and nights later she was still sitting in the dark when an envelope was slid under her door. She'd hurried barefoot to the peephole, but there was nothing to see and she couldn't bear to open the door. She slid down until she sat beside the envelope with her back to the door.

Inside there was a check. It was two-week's severance pay, plus an extra month's pay for bonus. Signed by Angelo.

A post-it had been stuck to the front of the check.

Two words, no signature, but she'd recognize the handwriting anywhere.

"Good luck."

She caught the next flight to Los Angeles and wore sunglasses the whole way to hide her bloodshot eyes.

11

S*am finished the last* serving of the night and began cleaning up his station. He bantered a bit with Marlys; let the line see he was fine—after two months, he'd better be.

Graziella came up to him as he was finishing the cleanup on his station.

"I know," he told her. "I know. I'll put out an ad tomorrow for a new *aboyeur*. I just couldn't face doing it before. But I really want to thank you for covering, Graziella. You're amazing."

"I am amazing. Thank you for noticing."

He managed a smile. Her quick hug was surprising and kind.

She then nodded toward the front of house. "Someone waiting to see you." And she was gone.

Sam double-checked the kitchen. He was the last one, so he flipped off all except the safety night light and pushed out into the restaurant.

The lights were out. The fire was still going and a single candle burned on the table for two close beside it. A lone woman sat at the table facing away from the kitchen.

For a moment he wondered how Graziella had circled around so fast, but then he knew. His stomach clenched so hard that he couldn't breathe and had to hold onto the door frame to remain upright. He considered moving back through the door, but Luisa sat so still. Even the sound of the swinging door behind him didn't cause her to turn, as if she'd shatter at the least movement.

He circled the long way around the fire so that he didn't approach her from behind. Her face was as frozen as the rest of her. She was normally so animated that she looked unnatural in her stillness.

Sam wanted to yell at her; spit out all of the hard hateful words that had rattled around inside him but never found any target. But the candle picked up the tracks of the silent tears that she made no effort to brush aside, if she was even aware of them. At a loss for what else to do, he sat down across from her.

He saw her swallow hard, several times, but he'd be damned if he'd be first to speak. If he was, he couldn't trust what would come out.

She nodded once, twice, as if trying to confirm something to herself, then began in a soft voice. Not quite looking at him, as if she didn't dare.

"My parents threw me out when I was fourteen. Boys and drugs and never going to school and crap like that was what they said. Maybe. But I know they also couldn't afford to feed me. I learned fast what cold and hungry were like. Got pretty desperate. Finally tried to hustle this chef coming out of a crappy restaurant. Offered to trade what I had to give for some food."

Sam wanted to close his eyes. Didn't want to see the hard memories that were crossing Luisa's lovely face, but he couldn't look away.

"Instead he fed me, helped me get a fake ID because I already looked like this, and gave me my first restaurant job. He paid me in food and a place to sleep on his floor. No money for the first six months because he didn't trust me to not buy drugs until I'd been clean a while."

Sam had heard stories like that. Except the offered payment was usually accepted. She'd gotten lucky.

"I don't have the palate to be a chef. But I'm smart. I earned my GED in two years even though I was missing four years of school. And I *saw* how restaurants worked; as clearly as a child's game. I cooked, cleaned, waitressed, did it all. But I was always fascinated by how it all worked. How things flowed."

Sam nodded. He couldn't quite bring himself to tell her just how good she'd been at her job.

"I always dreamed of running a chain. A big group of restaurants, making them function the way…" her voice stumbled and she took a deep gulping breath not far from a sob but continued. "…the way that we functioned. It was almost as good as sex. Better than, until I met you. Those months with you were the best of my life."

"Mine too," Sam managed his first words and she nodded rapidly in response.

"But I didn't understand about boys, men; about a man. About you. I didn't get that what we had wasn't like anything I'd ever had before. I did the job. I worked for Wolfgang in amazing restaurants. And people listened. I *was* good."

"Best I've ever seen," Sam finally admitted. He didn't tend to think ahead, but he found himself trying to second guess this conversation. He wasn't having much luck. There was a thin thread of hope, but it was blended with memory of a pain so intense that it was utterly blinding.

"I got that dream. The dream that a poor, desperate, cokehead girl had held up as a light to find her way out of the tunnel. But I missed the most important part."

"What's that?" Sam held himself very tightly. Even daring to hope hurt like a knife.

"You."

"Me? Just that simply. Me?"

She nodded again, her arms wrapped tightly around her as if she was freezing to death sitting right next to the fire.

"I'm just a dumb chef. You'd better explain it to me. Because last time I checked you're the one who—" He bit off the words. Clamped down on the recriminations that he wanted to spew all over her…because he didn't want to spew them any more.

For better or worse, he knew one truth absolutely.

He loved Luisa Valenti.

God help him.

He took a deep breath and spoke slowly so that he could choose his words carefully.

"We get back together and you're just going to wish you were back with Wolfgang's restaurants. And I don't want to leave Angelo's. He's perhaps the best Italian chef working today and he's given me his Number One restaurant to run. No way to solve that."

"There is. At least I hope there is."

"I'm listening," Sam hoped there was too. He'd never wanted anything in his life as much as he wanted Luisa, but giving up the restaurant dream would only make him bitter. Just as if she gave up hers.

"Oh god, Sam. You're the best man there ever was. I can't believe you're even listening to me. I didn't deserve you."

Sam waited for her to continue, unable to do more.

Impossibly she clenched her arms even tighter until her frame was shaking.

"I had the wrong dream."

"Say what?" That wasn't what he'd been expecting. He'd expected some plea for him to go with her. Or that she'd give up her own dreams, which he'd never allow. That's why he'd written the message he had even if he hadn't been able to bring himself to sign it.

"Angelo's was an amazing experience," he could hear the truth of it in her voice. "I never had so much fun. Angelo, Manuel, Marlys, Graziella, all of them. And that was before I noticed you. Then it just kept getting better. I asked to be busy, I asked to be challenged."

"And Wolfgang's organization does that for you."

"It did, past tense. He offered me consulting at Spago and Cut; flew in himself to do so. I thanked him and then I quit. I gave my two weeks notice three weeks ago. I've been back for a week trying to find the bravery to come and see you."

"You quit?" Sam knew something was wrong here. "Because of me? No! That doesn't work. I won't let you—"

"It wasn't because of you, doofus. It was because of me," she shouted him down just as she so often did on the cook line.

"Because of you?" He still wasn't getting it.

"Because of me," she said more calmly and unwound her arms, finally resting her hands in her lap as if too weary to do more. "I was too young and stupid to understand the most important dream of all, even if you knew it from the very first day."

"I did?"

She smiled at him; that smile she gave right before she was going to unleash mayhem on his cook line just to tease him. That smile that also spilled forth when she'd woken in his arms to find him watching her.

"Yes," she continued. "I forgot to dream about being happy. I was happy with you, so god damn happy that I scared the shit out of myself and ran away. I'm hoping you'll give me a chance to try out that dream again."

"And I'm supposed to trust that?"

She nodded, but the fear was back and she hung her head to study her hands once more. Luisa wasn't afraid of anything.

"How? Please tell me how."

And when she looked up at him, the tears had returned. "Because I learned something new by leaving you that I never would have learned while we were together."

"What was that?"

"How much I love you, Sam Walsh."

And there it was. How could he possibly deny such a statement, especially when it was so clear in his own heart? Had some part of him hoped that she'd be back? Was that why he'd

never filled the *aboyeur* spot no matter how badly they needed it? Apparently so. But he couldn't let her off the hook that easily no matter how much he was planning to.

Sam crossed his arms over his chest and struggled for a disdainful voice when he really wanted to scoop her into his arms and cry for joy.

"So, you think you can just come back, pick up where you left off at Angelo's. As if we'd take you."

Her face fell, so he went for a slightly lighter tone.

"Then you figure you can just slide back into my bed."

"Well," she was too sharp, and caught on from just that tiniest hint, "I did let go of my apartment, so I do kinda need a bed to slide into."

He kept his arms crossed, but backed it up with the smile he was feeling building deep inside him. "You're probably going to want the ring I had in my pocket that night."

She looked at him aghast, "You bought me a ring?"

"Uh-huh."

She covered her mouth in horror, "That same night? Oh god, I'm so sorry. I never thought anyone would ever do something like that for me."

"I did. *Doofus* that I am. And even worse?"

"What's worse than that?"

Sam rose and circled around the table, then knelt and took Luisa's hands.

She looked into his eyes and he knew he was lost. Happily lost.

"Even worse," he confirmed. "I love you so much that I can't imagine life without you."

The smile that broke out on her face was accompanied by a different type of tears. Then she giggled. A bright merry laugh that he'd missed more than anything about her.

"What?" he asked softly.

She leaned down and gave him a kiss seasoned with pure joy.

"I was just thinking, what with you being the best man I've ever met and all…"

"What?"

"I bet you kept the ring."

He had. He'd felt pathetic doing so, but now he knew why he hadn't returned it. Because some part of him had known that something so right could never be denied for long.

Sam lifted her hands to his lips and kissed her—right where he'd be slipping the ring on later tonight.

Wrapping It Up

It has been such a fun year in short fiction for me. I've traveled from the Suez Canal to a Montana Ranch to the heart of Seattle Pike Place Market. I've written about love between spaceships and war between grill chefs. I've researched everything from ice fishing to ice climbing and have learned more about firefighting that I'd ever imagined.

Writing these stories is my single greatest joy outside my family. I hope that you also find them fun!

Wishing you all the best,

M.L. Buchman

About the Author

M. L. Buchman has over 50 novels and 30 short stories in print. His military romantic suspense books have been named Barnes & Noble and NPR "Top 5 of the year" and twice Booklist "Top 10 of the Year," placing two titles on their "Top 101 Romances of the Last 10 Years" list. He has been nominated for the Reviewer's Choice Award for "Top 10 Romantic Suspense of the Year" by RT Book Reviews and was a 2016 RWA RITA finalist. In addition to romance, he also writes thrillers, fantasy, and science fiction.

In among his career as a corporate project manager he has: rebuilt and single-handed a fifty-foot sailboat, both flown and jumped out of airplanes, and designed and built two houses. Somewhere along the way he also bicycled solo around the world.

He is now making his living as a full-time writer on the Oregon Coast with his beloved wife. He is constantly amazed at what you can do with a degree in Geophysics. You may keep up with his writing by subscribing to his newsletter at:
www.mlbuchman.com.

Other works by M. L. Buchman:

If you enjoyed this collection, try:

A SHORT STORY COLLECTION

The IDES *of* MATT 2014

M.L. BUCHMAN

"Buchman's work has catapulted him to the top, and the honor is well deserved ." —*RT Reviews*